T0140421

# Recommender Systems Handbook

Francesco Ricci • Lior Rokach • Bracha Shapira
Editors

# Recommender Systems Handbook

Third Edition

 Springer

*Editors*
Francesco Ricci
Faculty of Computer Science
Free University of Bozen-Bolzano
Bozen-Bolzano, Italy

Lior Rokach
Software and Information Systems
Engineering
Ben-Gurion University of the Negev
Beer-Sheva, Israel

Bracha Shapira
Software and Information Systems
Engineering
Ben-Gurion University of the Negev
Beer-Sheva, Israel

ISBN 978-1-0716-2199-8         ISBN 978-1-0716-2197-4   (eBook)
https://doi.org/10.1007/978-1-0716-2197-4

This Springer imprint is published by the registered company Springer Science+Business Media, LLC, part of Springer Nature.
The registered company address is: 1 New York Plaza, New York, NY 10004, U.S.A.

*Dedicated to*

*our families in appreciation for their patience and support during the preparation of this handbook*

*and to*

*all our students in appreciation of their ideas, patience, and stimulus for better understanding the topics covered in this handbook*

F.R.
L.R.
B.S.

# Preface

Recommender systems are software tools and techniques providing suggestions for items to be of use to a user. The suggestions provided by a recommender system are aimed at supporting their users in various decision-making processes, such as what items to buy, what music to listen, or what news to read. Recommender systems are valuable means for online users to cope with information overload and help them make better choices. They are now one of the most popular applications of artificial intelligence, supporting information discovery on the Web. Several techniques for recommendation generation have been proposed, and during the last two decades, many of them have also been successfully deployed in commercial environments. Nowadays, all the major Internet players adopt recommendation techniques.

Development of recommender systems is a multi-disciplinary effort which involves experts from various fields such as artificial intelligence, human computer interaction, data mining, statistics, decision support systems, marketing, and consumer behavior.

The first two editions of the handbook, which were published 10 and 6 years ago, were extremely well received by the recommender systems community. The positive reception, along with the fast pace of research in recommender systems, motivated us to further update the handbook. This third edition aims at updating the previously presented material and to show new techniques and applications in the field. The Recommender Systems Handbook is now offered in a greatly revised edition; 11 chapters are totally new, and the remaining chapters are updated versions of selected chapters already published in the second edition.

Despite these revisions, the goal of this handbook remains unaltered. It still aims at presenting both fundamental knowledge and more advanced topics by organizing them in a coherent and unified repository of recommender systems major concepts, theories, methods, trends, challenges, and applications. Its informative, factual pages will provide researchers, students, and practitioners in industry with a comprehensive, yet concise and convenient reference source to recommender systems.

The book describes in detail the classical methods as well as extensions and novel approaches that were more recently introduced. It consists of five parts: General

Recommendation Techniques; Special Recommendation Techniques; Value and Impact of Recommender Systems; Human Computer Interaction; and Applications. The first part presents the most popular and fundamental techniques used nowadays for building recommender systems, such as collaborative filtering, semantic-based methods, recommender systems based on implicit feedback, neural networks, and context-aware methods. The second part comprises chapters on more advanced recommendation techniques, such as session-based recommender systems, adversarial machine learning for recommender systems, group recommendation techniques, reciprocal recommenders systems, natural language techniques for recommender systems, and cross-domain approaches to recommender systems. The third part covers a wide perspective on the evaluation of recommender systems with chapters on methods for evaluating recommender systems, their value and impact, the multistakeholder perspective of recommender systems, and the analysis of the fairness, novelty, and diversity in recommender systems. The fourth part contains a few chapters on the human computer dimension of recommender systems, with papers on the role of explanation, the user personality, and how to effectively support individual and group decision with recommender systems. The last part focusses on application in several important areas, such as, food, music, fashion, and multimedia recommendation.

We would like to thank all authors for their valuable contributions. We would like to express gratitude to all the reviewers who generously provided comments on drafts or counsel otherwise. We would like to express our special thanks to Susan Lagerstrom-Fife and staff members of Springer for their kind cooperation throughout the production of this book. Finally, we wish this handbook will contribute to the growth of this subject; we wish the novices a fruitful learning path, and to those more expert, a compelling application of the ideas discussed in this handbook and a fruitful development of this challenging research area.

Bozen-Bolzano, Italy                                                          Francesco Ricci

Beer-Sheva, Israel                                                              Lior Rokach

Beer-Sheva, Israel                                                              Bracha Shapira
February 2022

# Contents

**Recommender Systems: Techniques, Applications, and Challenges** ..... 1
Francesco Ricci, Lior Rokach, and Bracha Shapira

**Part I  General Recommendation Techniques**

**Trust Your Neighbors: A Comprehensive Survey of
Neighborhood-Based Methods for Recommender Systems** ............... 39
Athanasios N. Nikolakopoulos, Xia Ning, Christian Desrosiers,
and George Karypis

**Advances in Collaborative Filtering** ......................................... 91
Yehuda Koren, Steffen Rendle, and Robert Bell

**Item Recommendation from Implicit Feedback** ........................... 143
Steffen Rendle

**Deep Learning for Recommender Systems** .................................. 173
Shuai Zhang, Yi Tay, Lina Yao, Aixin Sun, and Ce Zhang

**Context-Aware Recommender Systems: From Foundations
to Recent Developments** ....................................................... 211
Gediminas Adomavicius, Konstantin Bauman, Alexander Tuzhilin,
and Moshe Unger

**Semantics and Content-Based Recommendations** .......................... 251
Cataldo Musto, Marco de Gemmis, Pasquale Lops, Fedelucio Narducci,
and Giovanni Semeraro

**Part II  Special Recommendation Techniques**

**Session-Based Recommender Systems** ...................................... 301
Dietmar Jannach, Massimo Quadrana, and Paolo Cremonesi

**Adversarial Recommender Systems: Attack, Defense, and Advances** ...    335
Vito Walter Anelli, Yashar Deldjoo, Tommaso Di Noia,
and Felice Antonio Merra

**Group Recommender Systems: Beyond Preference Aggregation** ........    381
Judith Masthoff and Amra Delić

**People-to-People Reciprocal Recommenders** .............................    421
Irena Koprinska and Kalina Yacef

**Natural Language Processing for Recommender Systems** ................    447
Oren Sar Shalom, Haggai Roitman, and Pigi Kouki

**Design and Evaluation of Cross-Domain Recommender Systems** ........    485
Maurizio Ferrari Dacrema, Iván Cantador, Ignacio Fernández-Tobías,
Shlomo Berkovsky, and Paolo Cremonesi

**Part III   Value and Impact of Recommender Systems**

**Value and Impact of Recommender Systems** ............................    519
Dietmar Jannach and Markus Zanker

**Evaluating Recommender Systems** ..........................................    547
Asela Gunawardana, Guy Shani, and Sivan Yogev

**Novelty and Diversity in Recommender Systems** ..........................    603
Pablo Castells, Neil Hurley, and Saúl Vargas

**Multistakeholder Recommender Systems** ...................................    647
Himan Abdollahpouri and Robin Burke

**Fairness in Recommender Systems** .........................................    679
Michael D. Ekstrand, Anubrata Das, Robin Burke, and Fernando Diaz

**Part IV   Human Computer Interaction**

**Beyond Explaining Single Item Recommendations** ........................    711
Nava Tintarev and Judith Masthoff

**Personality and Recommender Systems** ....................................    757
Marko Tkalčič and Li Chen

**Individual and Group Decision Making and Recommender Systems** ....    789
Anthony Jameson, Martijn C. Willemsen, and Alexander Felfernig

**Part V   Recommender Systems Applications**

**Social Recommender Systems** ...............................................    835
Ido Guy

**Food Recommender Systems** ................................................    871
David Elsweiler, Hanna Hauptmann, and Christoph Trattner

**Music Recommendation Systems: Techniques, Use Cases, and
Challenges**................................................................  927
Markus Schedl, Peter Knees, Brian McFee, and Dmitry Bogdanov

**Multimedia Recommender Systems: Algorithms and Challenges**........  973
Yashar Deldjoo, Markus Schedl, Balázs Hidasi, Yinwei Wei,
and Xiangnan He

**Fashion Recommender Systems** ...........................................  1015
Shatha Jaradat, Nima Dokoohaki, Humberto Jesús Corona Pampín,
and Reza Shirvany

**Index**........................................................................  1057

# Recommender Systems: Techniques, Applications, and Challenges

**Francesco Ricci, Lior Rokach, and Bracha Shapira**

## 1 Introduction

Recommender systems (RSs) are software tools and techniques that provide suggestions for items that are most likely of interest to a particular user [21, 56, 58]. The suggestions usually relate to various decision-making processes, such as what items to buy, what music to listen to, or what online news to read.

"Item" is the general term used to denote what the system recommends to users. An RS normally focuses on a specific type of item (e.g., movies or news articles) and accordingly, its design, its graphical user interface, and the core recommendation technique used to generate the recommendations are all customized to provide useful and effective suggestions for that specific type of item.

RSs are primarily directed at individuals who lack sufficient personal experience or competence to evaluate the potentially overwhelming number of items that a website may offer [58]. A prime example is an e-commerce recommender engine that assists users in selecting items to purchase. On the popular website, Amazon.com, the site employs an RS to personalize the online store for each customer [42, 70]. Since recommendations are usually personalized, different users or user groups benefit from diverse, tailored suggestions. However, there are also non-personalized recommendations, which are much simpler to generate. Typical examples include the top 10 editor selections of books or movies. While they may be useful and effective in certain situations, for example, when not enough information

F. Ricci (✉)
Faculty of Computer Science, Free University of Bozen-Bolzano, Bozen-Bolzano, Italy
e-mail: fricci@unibz.it

L. Rokach · B. Shapira
Department of Software and Information Systems Engineering, Ben-Gurion University of the Negev, Beersheba, Israel
e-mail: liorrk@bgu.ac.il; bshapira@bgu.ac.il

© Springer Science+Business Media, LLC, part of Springer Nature 2022
F. Ricci et al. (eds.), *Recommender Systems Handbook*,
https://doi.org/10.1007/978-1-0716-2197-4_1

about the target user's preferences or interests is available, these types of non-personalized recommendations have not been the primary focus of the RS research.

In their simplest form, personalized recommendations are provided as ranked lists of items. In performing this ranking, RSs try to predict the most suitable products or services, based on the user's preferences and constraints. In order to complete such a computational task, RSs collect information from users regarding their preferences, which are either explicitly expressed, for example, as ratings for products, or inferred by interpreting the actions performed by the user while interacting with the system. For instance, an RS may consider the purchase of a particular item as an implicit sign of the user's preference for the item.

The development of RSs initiated from a rather simple observation: individuals often rely on recommendations provided by others in making routine, daily decisions [56, 68]. For example, it is common to turn to one's peers for recommendations when selecting a book to read; employers rely on recommendation letters in their recruitment decisions; and when selecting a movie to watch, individuals tend to read and rely on the movie reviews written by film critics or ask their friends for recommendations.

In seeking to mimic this behavior, the first RSs applied algorithms in order to leverage recommendations produced by a community of users and deliver these recommendations to an "active" or "target" user looking for suggestions. The recommendations were for items that similar users, or those with similar tastes, liked. This approach is termed collaborative filtering [57], and it is based on the rationale that if the active user agreed with certain users in the past, then other recommendations coming from these similar users should be relevant and of interest to the active user.

As e-commerce websites began to develop, there was a pressing need to provide recommendations derived from filtering the entire range of available alternatives. Users found it difficult to choose between the extremely wide variety of items (products and services) that these websites offered.

The explosive growth and variety of information and e-commerce related services available on the Web (selling, product comparison, auctions, etc.) frequently overwhelm users, leading them to make poor decisions. The large availability of choices, instead of being beneficial, may affect users' wellbeing. It came to be understood that while choice is good, more choice is not always better. Indeed, choice, with its implications of freedom, autonomy, and self-determination, may become burdensome and ultimately cause freedom to be regarded/perceived as a kind of misery-inducing tyranny [66].

In recent years, RSs have proven to be valuable means of coping with the information overload problem. An RS addresses this phenomenon by pointing a user toward new, not-yet-experienced items, or toward items that the user may want to reconsume, which are relevant to the user's current task or context. Upon a user's request, which can be articulated, depending on the recommendation approach, by the user's context and need, RSs generate recommendations using various types of knowledge and data about users, the available items, and previous transactions stored in customized databases. The user can then browse the recommendations, choose whether or not to accept them, and may provide (immediately or at a later stage) implicit or explicit feedback. This user action and feedback can be stored in the RS database for use in generating new recommendations in future user-system interactions.

The study of recommender systems is relatively new compared to research on other classical information system tools and techniques (e.g., databases or search engines). Recommender systems emerged as an independent research area in the mid-1990s [10, 32, 56, 68]. In recent years, interest in recommender systems has dramatically increased, as shown in the examples below:

1. Recommender systems play an important role in widely used Internet sites such as Amazon.com, YouTube, Netflix, Spotify, LinkedIn, Facebook, Tripadvisor, Last.fm, and IMDb. Moreover many media companies are now developing and deploying RSs as part of the services they provide to their subscribers. Like Netflix, that provided real data, and awarded a million dollar prize to the team that substantially improved its recommender system's performance, other companies have shared data for RS-related competitions [41, 48].
2. There are conferences and workshops dedicated specifically to the field, namely the Association of Computing Machinery's (ACM) Conference Series on Recommender Systems (RecSys), established in 2007. This conference is the premier annual event on RSs technology research and applications. In addition, sessions dedicated to RSs are frequently included in more traditional conferences in the area of databases, information systems, user modeling, artificial intelligence, machine learning, data science, and adaptive systems. Additional noteworthy conferences within this scope include: ACM's Special Interest Group on Information Retrieval (SIGIR); User Modeling, Adaptation and Personalization (UMAP); Intelligent User Interfaces (IUI); World Wide Web (WWW); ACM's Special Interest Group on Management of Data (SIGMOD); and the Knowledge and Data Discovery (KDD) conference which is a main venue for data science related research.
3. At institutions of higher education around the world, undergraduate and graduate courses are now dedicated entirely to RSs, and tutorials on RSs are very popular at computer science conferences. A few books on different aspects of recommender systems have been published; some introduced the origin of RSs and RS techniques [38, 65], others cover practical aspects of implementing RSs [27, 33], and there is even a textbook dedicated to the subject [4]. Springer is publishing several books on specific recommender system topics in its series: Springer Briefs in Electrical and Computer Engineering.
4. There have been several special issues of academic journals which cover research and developments in the RS field. Among the journals that have dedicated issues to RSs are: ACM Transactions on Intelligent Systems and Technology (2015); IEEE Intelligent Systems (2019); Electronic Markets (2019); Cognitive Computation (2020); and User Modeling and User Adapted Interaction (2020).

In this introductory chapter, we briefly discuss basic RS ideas and concepts. Our main goal is not to present a self-contained comprehensive survey on RSs but rather to provide a structured overview of the chapters in this handbook and to help the reader navigate the rich and detailed content it contains. The interested reader can also consult other available resources on recommender systems [17, 40, 44, 54, 61]. At the end of this chapter, we identify some research challenges that we believe are particularly important and merit further research attention.

The handbook is divided into five sections: (1) general recommendation techniques; (2) special recommendation techniques; (3) value and impact of recommender systems; (4) human-computer interaction; and (5) applications.

The first section presents the fundamental techniques most commonly used today for building RSs, such as collaborative filtering techniques; semantic and content-based methods, context-aware techniques, and deep neural networks based approaches.

The second section surveys special machine learning techniques and approaches aimed at addressing unique RS challenges and domains, such as session-based RSs, group RSs, people-to-people RSs, cross-domain RSs, adversarial ML in RSs, and the use of NLP in RSs.

The third section discusses the specific value of recommender systems for distinguished actors and the role they play for these actors, who are referred to as stakeholders. This section introduces multi-stakeholder RSs and presents methods and metrics to assess the RS's value for different actors from various perspectives. In addition to reviewing classic evaluation metrics, this section discusses more recently considered qualities, such as diversity, novelty, bias and fairness.

The forth section is dedicated to human-computer interaction aspects of recommender systems and discusses decision-making aware design and approaches to assess and enhance RS explainability as a means of improving users' trust and perception of recommendations. This section also discusses user's personality acquisition and utilization aimed at improving users' RS experience.

Finally, the fifth section presents various recommender system applications and unique use cases. This section contains a list of chapters covering a broad spectrum of recommendation techniques used to recommend food, multimedia, as well as fashion and social RS techniques.

## 2 Recommender Systems' Value

We have defined RSs as software tools and techniques that provide users with suggestions for items that a user may wish to utilize. Now we refine this definition to illustrate a range of possible roles that an RS can play. In fact, in the last years it emerged that RSs must be considered as a platform playing different roles for distinguished actors. The end user of an RS is an important actor but not the only one. In Chapter "Multistakeholder Recommender Systems" is first introduced the notion of a recommendation stakeholder, which is any group or individual that can affect, or is affected by, the delivery of recommendations to users. Three important stakeholders are identified: Consumers (aka users), Providers (aka suppliers) and the

System owners. Consumers are the end users who receive item recommendations. Providers are those actors that supply the recommended items (e.g., a book publisher in Amazon). Finally, the organization itself, which has created the platform and the associated RS in order to match consumers with items is an important actor. For instance, a travel recommender system is typically offered by a travel intermediary, such as, Booking.com or AirBnB, or a destination management organization, such as Visitfinland.com. The first type of actors primarily aims at selling more (hotel rooms and other travel services), while the second type is interested in increasing the number of tourists visiting the various regions of a destination, in a fair and sustainable way [18, 60]. While, the user's primary motivations for accessing such a system would be to find a suitable hotel, interesting events or attractions when visiting a destination or planning the travel to a destination. Actually, the providers and the system owner, can use the RS to influence the choices of the end users and finally determining the success of the platform and the profit of the suppliers.

Chapter "Value and Impact of Recommender Systems" focuses on various ways RSs create value for different stakeholders and also discusses the possible risks and the potentially negative impacts of using such systems. The chapter deals with the organizational and business-oriented perspective by reporting how practical systems are evaluated and which effects have been observed in real-world deployments. Here, we want to give some of the basic reasons as to why both item providers and recommender system owners may want to exploit RSs.

- *Increase the number of items sold.* This is probably the most important function for a commercial RS, i.e., to be able to sell an additional set of items compared to those usually sold without any kind of recommendation. This goal is achieved because the recommended items are likely to suit the user's needs and wants. Presumably the user will recognize this after having tried several recommendations. Non-commercial systems have similar goals, even if often there is no cost for the user that is associated with selecting an item. For instance, a public content network such as BBC (UK) or RAI (IT) aims at increasing the number of news items read on its site. In general, we can say that from the service provider's point of view, the primary goal for introducing an RS is to increase the conversion rate, i.e., the number of users that accept the recommendation and consume an item, compared to the number of simple visitors that just browse through the information.
- *Sell more diverse items.* Another major function of an RS is to enable the user to select items that might be hard to find without a precise recommendation, i.e., to not select only the "blockbusters". For instance, in a tourist RS the service provider is interested in promoting all the places of interest in a tourist area, not just the most popular ones. This is a major objective for a tourism RS maintained by a public destination management organisation. Similar goals are posed to public content network managers. But, attaining this goal could be difficult without an RS, since the system provider at the same time cannot afford the risk of advertising places (or media) that are not likely to suit a particular user's taste.

- *Increase the user satisfaction.* A well designed RS can also improve the experience of the user with the site or the application. The user will find the recommendations interesting, relevant and, with a properly designed human-computer interaction, he or she will also enjoy using the system. The combination of effective, accurate recommendations and a usable interface will increase the user's subjective evaluation of the system. This, in turn, will increase system usage and the likelihood that the recommendations will be accepted.
- *Increase user fidelity.* A user should be loyal to a website which, when visited, recognizes the old customer and treats her as a valued visitor. This is a standard feature of an RS since many RSs compute recommendations, thus leveraging the information acquired from the user during previous interactions such as the user's ratings of items or user's visited pages. Consequently, the longer the user interacts with the site, the more refined the user's model becomes: the system's representation of the user's preferences develops and the effectiveness of the RS output to customize and match to the user's preferences is increased.
- *Better understanding of what the user wants.* Another important function of an RS, which can also be leveraged by other applications of the RS owner, is the description of the user's preferences, which are collected either explicitly or predicted by the system. The system owner, but also the item providers, may decide to reuse this knowledge for a number of other goals, such as improving the management of the item's stock or production. For instance, in the travel domain, destination management organizations can decide to advertise a specific region to new customer sectors or advertise a particular type of promotional message derived by analyzing the data collected by the RS (transactions of the users).

We mentioned above some important motivations as to why the system owner introduces an RS and the item providers offer their services/items through the RS platform. But naturally users will use an RS if it will effectively support their tasks or goals. Consequently an RS must balance the needs of these three stakeholder and offer a service that is valuable to all of them. Hence advanced evaluation techniques must be introduced in order to proper assess the value and the impact of an RS across multiple groups of stakeholders, as it is discussed in Chapters "Multistakeholder Recommender Systems" and "Value and Impact of Recommender Systems".

Without fully describing the range of dimensions that drive the adoption of RSs, which is discussed in Chapter "Value and Impact of Recommender Systems", we want now to focus on the end user side and refer to Herlocker et al. [36], a paper that has become a classical reference in this field. The authors define eleven popular user's tasks that an RS can assist in implementing. Some may be considered as the main or core tasks that are normally associated with an RS, such as offering suggestions for items that may be useful to a user. Others might be considered as more "opportunistic" ways to exploit an RS. As a matter of fact, this task differentiation is similar to what happens with other digital "tools", e.g., a search engine. Search engines primary function is to locate documents that are relevant to the user's information need, but it can also be used to check the importance of a webpage (looking at the position of the page in the result list of a query) or to discover the various usages of a word in a collection of documents.

- *Find Some Good Items:* Recommend to a user some items as a ranked list along with predictions of how much the user would like them (e.g., on a scale of one-to-five stars). This is the main recommendation task that many commercial systems address (see, for instance, Chapters "Music Recommendation Systems: Techniques, Use Cases, and Challenges" and "Multimedia Recommender Systems: Algorithms and Challenges"). We note that some systems do not show the predicted rating, especially when users may not be interested in understanding why the item is recommended (e.g., in music).
- *Find all good items:* Recommend all the items that can satisfy the user needs, in such cases it is insufficient to just find some of them. This is especially true when the number of items is relatively small or when the RS is mission-critical, such as in medical or financial applications. In these situations, in addition to the benefit derived from carefully examining all the possibilities, the user may also benefit from a motivated ranking of these items or from additional explanations that the RS generates (see Chapter "Beyond Explaining Single Item Recommendations").
- *Annotation in context:* Given an existing context, e.g., a list of items, emphasize some of them depending on the user's long-term preferences. For example, a TV recommender system might annotate which TV shows displayed in the electronic program guide (EPG) are worth watching (Chapter "Social Recommender Systems", provides interesting examples of this task).
- *Recommend a sequence:* Instead of focusing on the generation of a single recommendation, the idea is to recommend a sequence of items, one by one or as a whole. Typical examples include recommending, a next point of interest to visit, a TV series, a book on RSs after having recommended a book on data mining, or a compilation of musical tracks (see Chapters "Music Recommendation Systems: Techniques, Use Cases, and Challenges", and "Session-Based Recommender Systems", for more examples and discussion).
- *Recommend a bundle:* Suggest a group of items that fits well together. For instance, a travel plan may be composed of various attractions, destinations, and accommodation services that are located in a delimited area. From the point of view of the user, these various alternatives can be considered and selected as a single travel destination [45].
- *Just browsing:* In this task, the user browses the catalog without any imminent intention of purchasing an item. The task of the RS is to help the user to browse the items that are more likely to fall within the scope of the user's interests for that specific browsing session. This is a task that has also been supported by adaptive hypermedia techniques [20].
- *Find credible recommender:* Some users do not trust recommender systems, thus they play with them to see how good they are at making recommendations. Hence, a certain system may also offer specific functions to let the users test its behavior in addition to those just required for obtaining recommendations.

- *Improve the profile:* This relates to the capability of the user to provide (input) information to the recommender system about what he or she likes and dislikes. This is a fundamental task that is strictly necessary to provide personalized recommendations. If the system has no specific knowledge about the active user, then it can only provide the same recommendations that would be delivered to an "average" user (see Chapter "People-to-People Reciprocal Recommenders" for an application domain where preference elicitation is critical).
- *Express self:* Some users may not care about the recommendations at all. Rather, what is important to them is that they be allowed to contribute with their ratings and express their opinions and beliefs. The user satisfaction for that activity can still act as leverage, resulting in the user's continued loyalty to the application (as we mentioned prior, in discussing the service provider's motivations).
- *Help others:* Some users are happy to contribute with information, e.g., their evaluation of items (ratings), because they believe that the community benefits from their contribution. This could be a major motivation for entering information into a recommender system that is not used routinely. For instance, with an automobile RS, a user who has already purchased a new car is aware that the rating entered in the system is more likely to be useful to other users rather than to oneself, the next time a new-car-purchase is contemplated.
- *Influence others:* In Web-based RSs, there are users whose main goal is to explicitly influence other users into purchasing particular products. As a matter of fact, there are also malicious users that may use the system simply to promote or penalize certain items.

As these various points indicate, the role of an RS within an information system can be quite diverse. This diversity calls for the exploitation of a range of different knowledge sources and techniques. In the next two sections, we discuss the data that an RS manages and the core technique used to identify the right recommendations.

# 3   Data and Knowledge Sources

RSs are information processing systems that actively gather various kinds of data in order to build their recommendations. The data used within an RS can relate to three kinds of objects: items, users, and interactions between the users and the items. Some systems collect users' features, such as their areas of interest, age, or gender, while other gather items' characteristics, such as the genre of a movie or the price of an item. All systems, in order to learn users' preferences and generate recommendations, gather users' interactions with items. In practice, the data and knowledge sources available for an RS can be very diverse, and their effective exploitation depends on the recommendation technique and the available computational resources (see Sect. 4).

**Items** Items are the objects that are recommended. Available items data depends on the complexity of acquiring them, for instance item features, which also depends on the availability of meta-data. For example, in a multimedia RS, items such as images, require special image analysis algorithms to extract their features from raw content. Text-based items, such as news, or items that have long descriptions, s.a., reviews, require special algorithms to analyze the text and extract meaningful relevant features from the text. Chapter "Natural Language Processing for Recommender Systems" describes NLP methods that can be used to extract meaningful information from textual descriptions of items (and users). For example, by using such methods, features can be extracted from experts reviews [55], and they may alleviate the cold-start scenario, i.e., when dealing with items that have not gained yet enough interactions. Chapter "Semantics and Content-Based Recommendations" presents methods for the representation of items (and users) adopting semantic-aware representations of textual content. While early models were not able to extract from the textual content a complete description of the item, nor to encode semantic relationships between terms, semantic aware systems enable to give meaning to information expressed in natural language and obtain a deeper comprehension of the information conveyed by textual content, hence leading to better RSs. Chapter "Semantics and Content-Based Recommendations" reviews the main two approaches for semantic representation: endogenous, i.e., relying on the distribution of keywords in documents, and exogenous, which resolves semantic synonymy, polysemy and other language challenges, by utilizing external knowledge, like taxonomies ontologies or other language related information.

**Users** Users of an RS, as we mentioned above, may have very diverse goals and characteristics. In order to personalize the recommendations and the human-computer interaction, RSs exploit a range of information about the users. This information can be structured in various ways. User data is said to constitute the user model [16, 30]. The user model profiles the user, i.e., encodes her preferences and needs. Various user modeling approaches have been used and, in a certain sense, an RS can be viewed as a tool that generates recommendations by building and exploiting user models [13, 14]. The selection of what information to model depends on the recommendation technique, the availability of data and the efforts required to gather and extract the knowledge.

For instance, in collaborative filtering, a user is modeled as a simple list of his ratings for certain items. In context aware RS (see Chapter "Context-Aware Recommender Systems: From Foundations to Recent Developments") the user model incorporates the contextual information to be utilized during the recommendation process in order to recommend items to users under specific contextual situations. For example, by using the temporal context, a travel recommender system would include contextual features describing the weather, and the time of year when the vacation was consumed, so that a vacation recommendation in winter can be very different from the one recommended in summer. In content-based RSs (see Chapter "Semantics and Content-Based Recommendations") users could be modeled by the content features of the items they consumed, and, similarly to the items, such features may derive from of a variety of tabular and unstructured content.

For example, textual user generated reviews about consumed items may be utilized to enrich the user model [22, 55].

**Interactions** RSs record the interactions between their users and items in log-like data that stores important information generated during the interactions. This information is useful for the recommendation generation algorithm that the system is using. For instance, a log may contain a reference to the item purchased by the user, the events that led to the buying event (e.g., browsing, clicking, inclusion in the basket, etc.), the rating that the user assigned to the item, or any other information that the system collects about the time, the location of the interaction. An important part of the information about the interactions is the user feedback to the item. User feedback can be explicitly provided by users or implicitly inferred from the users' behaviour.

Explicit feedback is known to be more reliable than implicit one and provides a level of preference of the user for an item (e.g., ratings 1–5). However, explicit feedback is often not available, or very sparse, since many users would not bother to provide it. Ratings are the most popular form of explicit feedback data that RSs collect [59]. According to [63], ratings can take on a variety of forms:

- Numerical ratings such as the 1–5 stars used in the RS offered by Amazon.com.
- Ordinal ratings, such as "strongly agree, agree, neutral, disagree, strongly disagree" where the user is asked to select the term that best indicates his or her opinion regarding an item (usually via questionnaire).
- Binary ratings that model choices in which the user is simply asked to decide if a certain item is good or bad.

Implicit feedback relates to the inference of the user's preference for an item from the observed interaction of the user with the item, for instance during a recommendation session (see Chapter "Session-Based Recommender Systems"). Implicit feedback is typically available for many more items than those available for explicit feedback. In fact, various types of user-item interactions (clicking, buying, adding to the basket, and many more) provide some implicit indication of the user's preference. Such feedback can be considered as unary ratings, and does not indicate a specific level of preference. The interactions are typically interpreted as positive implicit preferences. However, the absence of an interaction cannot be simply considered as a sort of negative feedback. Usually, there is not enough data to precisely derive the reasons for a missing interaction: the user may dislike the item, or may have not seen it at all, or may have consumed it already in the past. Chapter "Item Recommendation from Implicit Feedback" describes the challenges of building RSs from implicit feedback, and provides methods to address them. First, since implicit feedback refers to items that the users consumed in the past, the data is noisy and gives a weak signal of positive preferences. Most of the items do not have interactions and considering all these absence of interactions as a weak negative signals of preference leads to high training costs. That makes standard training algorithms hard to apply.

An important data characteristic dimension relates to the time span of the data. A user may have a long-term or a short-term profile depending of his or her interaction with the system. If the user is registered and known to the system, the system can indeed learn user's preferences from his past interactions. However, for scenarios where the user is a guest of the system, not registered, or would like to stay anonymous, the system can learn the user's preferences only from the current session. In such session-based scenarios, the RS aims to tailor item suggestions to the short-terms needs and the predicted intents of the user that are learned from the interactions during the session. This is termed session-based recommendation. In addition, it can happen that the user is know to the system but may change her intents from session to session, depending on her specific task. For instance, in one session a user could like to buy a present for an elderly person (parent), and in a different session the same user may buy items for personal use. Chapter "Session-Based Recommender Systems" elaborates on practical application scenarios for session-based recommender systems, defines the problem, outlines the challenges, and finally describes approaches to address these challenges.

In addition, a user model in context-based RSs, as described in Chapter "Context-Aware Recommender Systems: From Foundations to Recent Developments", is dynamic and changes based on contextual features such as temporal, location, mood or any other relevant features. Usually in context-aware RSs the users are known to the system and have user models that should be altered according to the contextual information to address the short-term preferences emerging in a specific context.

We should stress that utilizing the right data is fundamental to the performance of an RS. As described above, a variety of data and knowledge sources can be leveraged in various RSs techniques. The decision about the data to use for a system, and how to use it should be done carefully while considering availability of data, the recommendation algorithm, the required effort, and the available resources.

# 4 Recommendation Techniques

To implement its core function, identifying useful items for the user, an RS must *predict* that an item is worth recommending. To achieve this goal, the system must be able to predict the utility of some items, or at least compare the utility of some items, and then decide which items to recommend based on this comparison. The prediction step may not be explicit in the recommendation algorithm, but we can still apply this unifying model to describe the general role of an RS. Here, our goal is to provide the reader with a unifying perspective rather than an account of all the different recommendation approaches that are illustrated in this handbook.

To illustrate the prediction step of an RS, consider, for instance, a simple and non-personalized recommendation algorithm that recommends only the most popular songs. The rationale for using this approach is that in the absence of more precise information about the user's preferences, a popular song, i.e., one that is liked (high utility) by many users, will also most likely appeal to a generic user, or at least with a higher likelihood than another randomly selected song. Hence, the utility of such popular songs is predicted to be reasonably high for this generic user.

This view of the core recommendation computation as the prediction of the "utility" of an item for a user has been suggested in [3] and updated in [61] by referring to a more generic concept of "evaluation" of the user for an item, rather than utility. Both papers model this degree of utility (evaluation) of the user $u$ for the item $i$ as a (real-valued) function $R(u, i)$, as it is usually done in collaborative filtering by considering users' ratings for items. Then the fundamental task of an RS is to predict the value of $R$ over pairs of users and items, or in other words, to compute $\hat{R}(u, i)$, where we denote with $\hat{R}$ the estimation, computed by the RS, for the true function $R(u, i)$. Consequently, having computed this prediction for the active user $u$ on a set of items, i.e., $\hat{R}(u, i_1), \ldots, \hat{R}(u, i_N)$, the system will recommend the items $i_{j_1}, \ldots, i_{j_K}$ ($K \leq N$) with the largest predicted utility. $K$ is typically a small number, that is, much smaller than the cardinality of the items set or the items on which a user utility prediction can be computed, i.e., RSs "filter" the items that are recommended to users.

As mentioned above, some RSs do not fully estimate the utility before making a recommendation. Instead, they may apply some heuristics to hypothesize that an item may be of use to a user. This is typical, for instance, in knowledge-based systems. These systems use various kinds of knowledge about users, items, and the utility function itself. However, these approaches are much less prevalent in real-world scenarios. Thus, we will not cover them in this handbook edition (the reader is welcome to read the corresponding chapters in the second edition of this handbook).

It is also important to note that sometimes the user utility for an item depends on other variables, which we generically call "contextual". For instance, the utility of an item for a user can be influenced by the time and user location when the recommendation is requested. For instance, users may be more interested in items (e.g., restaurant) closer to their current location. Consequently, the recommendations must be adapted to these specific additional conditions and as a result, it becomes increasingly more difficult to correctly estimate what the right recommendations are.

This handbook presents several different types of recommender systems that vary in terms of the addressed domains and especially regarding the recommendation algorithm, i.e., how the prediction of the utility of a recommendation is made. Other differences relate to how the recommendations are finally assembled and presented to the user in response to user requests. These aspects are discussed as well, later in this introduction.

To provide an initial overview of the different types of RSs, we want to quote a taxonomy provided by [21] that has become a classical way of distinguishing between recommender systems and referring to them. Burke [21] distinguishes between six different classes of recommendation approaches:

**Content-Based** The system learns to recommend items that are similar to the ones that the user liked in the past. The similarity of items is calculated based on the features associated with the compared items. For example, if a user has positively rated a movie that belongs to the comedy genre, then the system can learn to recommend other movies from this genre [43].

Classic content-based recommendation techniques aim at matching the attributes of the user profile against the attributes of the items. In most cases, the items' attributes are simply keywords that are extracted from the items' descriptions. Semantic indexing techniques represent the item and user profiles using concepts instead of keywords. Chapter "Semantics and Content-Based Recommendations" presents a comprehensive survey of semantic indexing techniques to overcome the main problems of classical keyword-based systems. The authors present two main groups of semantic indexing techniques: exogenous and endogenous. Techniques in the former group rely on the integration of external knowledge sources, such as: ontologies, encyclopedic knowledge (such as Wikipedia) and data from the Linked Data cloud, while techniques in the latter group rely on a lightweight semantic representation based on the hypothesis that the meaning of words depends on their usage in large corpora of textual documents. Chapter "Semantics and Content-Based Recommendations" demonstrates how to utilize semantic approaches to realize a new generation of semantic content-based recommender systems by describing their main potentials and limitations.

**Collaborative Filtering** The original and most simple implementation of this approach [32] makes recommendations to the active user based on items that other users with similar tastes liked in the past. The similarity in the taste of two users is calculated based on the similarity in the rating history of the users. This is the reason why [64] refers to collaborative filtering as "people-to-people correlation." Collaborative filtering is considered to be the most popular and widely implemented technique in RS. Chapter "Trust Your Neighbors: A Comprehensive Survey of Neighborhood-Based Methods for Recommender Systems" presents a comprehensive survey of neighborhood-based methods for collaborative filtering. Neighborhood-based methods focus on relationships between items or, alternatively, between users. An item-based approach models the preference of a user to an item based on ratings of similar items by the same user. Neighborhood-based methods have obtained considerable popularity due to their simplicity, efficiency, and ability to produce accurate and personalized recommendations. Chapter "Trust Your Neighbors: A Comprehensive Survey of Neighborhood-Based Methods for Recommender Systems" describes the main benefits of such methods, as well as their principal characteristics. Moreover, the chapter addresses the essential decisions that are required while implementing a neighborhood-based recommender system and gives practical information on how to make such decisions. Perhaps the decision that has the greatest impact on the rating prediction and computational performance of the recommender system is the choice between a user-based and an item-based method. In typical commercial recommender systems where the number of users exceeds the number of available items, item-based approaches should be preferred since they provide more accurate recommendations, while being more computationally efficient and requiring less frequent updates. On the other hand, user-based methods usually provide more original recommendations, which may lead users to a more satisfying experience [26].

Finally, the problems of sparsity and limited coverage, often observed in large commercial recommender systems, are discussed by exploring two research directions: dimensionality reduction and graph-based techniques. Dimensionality reduction provides a compact representation of users and items that captures their most significant features. An advantage of such an approach is that it allows to obtain meaningful relations between pairs of users or items, even though these users have rated different items, or these items were rated by different users. On the other hand, graph-based techniques exploit the transitive relations in the data. These techniques also avoid the problems of sparsity and limited coverage by evaluating the relationship between users or items that are not directly connected. However, unlike dimensionality reduction, graph-based methods also preserve some of the "local" relations in data.

Chapter "Advances in Collaborative Filtering" presents several techniques available for building Collaborative Filtering (CF) RSs. Specifically, the authors discuss latent factor models, such as matrix factorization (e.g., Singular Value Decomposition, SVD). These methods map both items and users to the same latent factor space. The latent space is then used to explain ratings by characterizing both products and users in term of factors automatically inferred from user feedback. The authors elucidate how SVD can handle additional features of the data, including implicit feedback and temporal information. They also describe techniques to address shortcomings of neighborhood techniques by suggesting more rigorous formulations using global optimization techniques. Utilizing such techniques makes it possible to lift the limit on neighborhood size and to address implicit feedback and temporal dynamics. The resulting accuracy is close to that of matrix factorization models, while offering a number of practical advantages.

Particular attention should be given to the scenario in which the users give no explicit feedback. In Chapter "Item Recommendation from Implicit Feedback" the authors provide a comprehensive review of the core challenges in learning item recommendation from implicit feedback. The authors present various methods to address these challenges: sampling methods for general scoring functions based on pointwise, pairwise and softmax losses are described. Moreover, efficient training algorithms that can be applied to dot product models and square losses are discussed. Finally, they present sublinear time approaches for the retrieval task in which the most recommended items are quickly detected from the whole long item catalogue.

As deep learning gained popularity in various domains such as computer vision, more and more researchers started to incorporate deep learning methods in RS [73]. In particular, collaborative filtering tasks that were solved using SVD methods has been transformed into their corresponding deep learning solution. Since 2010, a variety of deep learning designs and methods have been developed for RS. Deep learning has many advantages in the context of RS. First, it can save time for the practitioners because deep learning requires less feature engineering work. Moreover, it can process unstructured raw data such as text, sound, image and video, which are very common in many RSs.

In Chapter "Deep Learning for Recommender Systems" the authors overview some key methods and highlight the impact of deep learning techniques in the recommender systems field. They present a range of challenging tasks in recommendation, such as cold-start problem, explainability, temporal dynamics, and robustness that can be addressed using deep neural networks. In the first part of the chapter, the authors describe the basics of deep learning techniques that are widely used in RS, including multilayered perceptrons, convolutional neural networks (CNN), recurrent neural networks (RNN), deep reinforcement learning, etc. Then the authors explain how users, items and their interactions can be modeled by a deep learning architecture. Finally, they describe various applications of deep learning in RS-related domains, such as e-commerce, online entertainment, news and point-of-interests.

**Community-Based** This type of systems recommends items based on the preferences of the users friends. This technique follows the epigram, "Tell me who your friends are, and I will tell you who you are" [7, 12]. Evidence suggests that people tend to rely more on recommendations from their friends than on recommendations from similar but anonymous individuals [69]. This observation, combined with the growing popularity of open social networks, is generating a rising interest in community-based systems or, as they are usually referred, social recommender systems [31] (see Chapters "Social Recommender Systems" and "People-to-People Reciprocal Recommenders"). This type of RS models and acquires information about the social relations of the users and the preferences of the users friends. The recommendation is based on ratings that were provided by the user's friends. In fact these RSs are following the rise of social-networks and enable a simple and comprehensive acquisition of data related to the social relations of the users.

There are a few more types of RSs that will not be covered explicitly in this edition, but we shortly discuss them here for the sake of completeness:

**Demographic** This type of systems recommends items based on the demographic profile of the user [17]. The assumption is that different recommendations should be generated for different demographic niches. Many websites adopt simple and effective personalization solutions based on demographics. For example, users are dispatched to particular websites based on their language or country. Or, suggestions may be customized according to the age of the user. While these approaches have been quite popular in the marketing literature, there has been relatively little proper RS research on demographic systems.

**Knowledge-Based** Knowledge-based systems recommend items based on specific domain knowledge about how certain item features meet users' needs and preferences and, ultimately, how the item is useful for the user. Notable knowledge-based recommender systems are case-based [19, 24, 28, 39, 62]. In these systems, a similarity function estimates how much the user's needs (problem description) match the recommendations (solutions of the problem). Here, the similarity score can be directly interpreted as the utility of the recommendation for the user. Knowledge-based systems tend to work better than others at the beginning of their deployment but if they are not equipped with learning components, they may be surpassed by other shallow methods that can exploit the logs of the human/computer interaction (as in CF).

**Hybrid Recommender Systems** These RSs are based on the combination of the above mentioned techniques. A hybrid system combining techniques A and B tries to use the advantages of A to fix the disadvantages of B. For instance, CF methods suffer from new-item problems, i.e., they cannot recommend items that have no ratings. This does not limit content-based approaches since the prediction for new items is based on their description (features) that are typically easily available. Given two (or more) basic RSs techniques, several ways have been proposed for combining them to create a new hybrid system (see [21] for the precise descriptions). Moreover, the emergence of deep learning techniques made it much easier to build hybrid RS [53].

As we have already mentioned, the context of the user when he or she is seeking a recommendation can be used to better personalize the output of the system. For example, in a temporal context, vacation recommendations in winter should be very different from those provided in summer [11]. Or a restaurant recommendation for a Saturday evening with one's friends should be different from that suggested for a workday lunch with co-workers [1]. Chapter "Context-Aware Recommender Systems: From Foundations to Recent Developments" reviews the topic of context-aware recommender systems (CARS). It presents the general notion of context and how it can be modeled in RSs. As it discusses the possibilities of combining several context-aware recommendation techniques into a single unified approach, the authors also provide a case study of one such combined approach.

Three popular different algorithmic paradigms for incorporating contextual information into the recommendation process are discussed: reduction-based (pre-filtering), contextual post filtering, and context modeling. In reduction-based (pre-filtering) methods, only the information that matches the current usage context, e.g., the ratings for items evaluated in the same context, are used to compute the recommendations. In contextual post filtering, the recommendation algorithm ignores the context information. The output of the algorithm is filtered/adjusted to include only the recommendations that are relevant in the target context. In the contextual modeling, the more sophisticated of the three approaches, context data is explicitly used in the prediction model.

## 5   Special Recommendation Techniques

It is clear from the previous sections that RS research is evolving in numerous and diverse directions, and special topics are emerging or becoming more important subjects of investigation. In this handbook, we cover some of these topics. Indeed, some have already been mentioned, such as: context-aware recommendations (Chapter "Context-Aware Recommender Systems: From Foundations to Recent

Developments"), which addresses the challenges and reviews methods to leverage contextual information into the recommendation process; and session-based recommendations (Chapter "Session-Based Recommender Systems") that deals with challenges emerging from the scenario of short-term modeling of anonymous users during sessions. Other important topics are covered in this handbook and we will now briefly introduce them.

Chapter "Group Recommender Systems: Beyond Preference Aggregation" deals with situations in which the system should recommend information or items that are relevant to a group of users rather than to an individual. For instance, an RS may select television programs for a group to view or a sequence of songs to listen to, based on preference models of all the group members. Recommending to groups is clearly more complicated than recommending to individuals. Assuming that we know precisely what is good for individual users, the issue is how to combine the individual user models of the group members. In this chapter, the authors present usage scenario for group recommendation and discuss how to acquire information about individual users preferences in a group setting. The chapter includes a discussion of how the system knows who forms the group, i.e., who is present, and how to display and explain group recommendations. Moreover, the chapter includes a description of methods to help users reach a final decision, thus optimizing multiple criteria, while considering fairness (see also Chapter "Fairness in Recommender Systems"), and adopting methods from the multistakeholder RSs that are described in Chapter "Multistakeholder Recommender Systems".

Chapter "Adversarial Recommender Systems: Attack, Defense, and Advances" discusses the vulnerability of RSs to adversarial attacks where an intelligent and adaptive attacker may deliberately manipulate data thereby violating the underlying stationary assumption of ML-based systems [15], that is, that the training data and test data are sampled from a similar (and possibly unknown) distribution. Adversarial attacks risk the security of RSs, and may compromise the integrity of the system, usually with the goal to harvest recommendation outcomes toward an illegitimate benefit, e.g., pushing some targeted items into the top-K list of users for market penetration. RSs should perhaps be designed with security-awareness so that to face adversarial attacks. The chapter is focused on machine-learned attacks based on the adversarial machine learning paradigm, rather than on traditional hand-crafted shilling attacks. The chapter introduces foundation concepts of adversarial machine learning (AML), and reviews the widely adopted attack and defense strategies focusing on adversarial attacks and defense against recommendation models, as well as on possible evaluation methods and metrics for evaluating them.

Chapter "People-to-People Reciprocal Recommenders" describes a special recommendation scenario where the RS recommends users to other users, such as employers to employees, or matching dating options to pairs of users. Thus, the system is required to satisfy the two parties involved in the recommendation (reciprocity) rather than the common one sided recommendation of items to an individual. The authors discuss the distinctive characteristics of reciprocal RSs, review previous works in various domains including social networks, mentor-mentee matching, recommending students to their colleagues in on-line courses setting, job recommendation, and on-line dating. The authors also present a complete case study in on-line dating.

The idea of relating recommendations in different systems or domains by exploiting user data collected in one system or domain to produce recommendations in another, is at the base of the research on cross-domain recommender systems illustrated in Chapter "Design and Evaluation of Cross-Domain Recommender Systems". In fact, recommender system applications are typically restricted to a particular type of item (movies, food, etc.), however, e-commerce sites like Amazon, eBay, and Alibaba regularly collects user feedback for items across multiple domains. Cross domain preferences of users might also be collected from their social media content or in parallel from several systems (one for each domain). It may, therefore, be beneficial to leverage all the available user data provided in various systems and domains. Recommendation quality can be improved, especially by mitigating the cold start and spare-data challenges where knowledge acquired in a "source" domain can be used in the sparse-data target domain. Moreover, cross-domain RSs can enable cross-selling recommendations, where items from multiple domains are suggested together (e.g., a movie and a book on a recommended singer). In this chapter, the authors formally define the cross-domain recommendation problem, and try to provide a unifying perspective by merging ideas and approaches which arise in distinct disciplines. The chapter provides an analytical categorization of prior work, and identifies open issues for future research.

## 6   Recommender Systems Evaluation

Recommender systems research is being conducted with a strong emphasis on practice and commercial applications. One very important issue related to the practical side of RS deployment is the necessity of evaluating the quality and value of the system. Evaluation is required at different stages of the system's life cycle and for various purposes [2, 36] (see Chapter "Evaluating Recommender Systems").

At design time, evaluation is required to verify the selection of the appropriate approach for the specific system goals, of the various system's stakeholders that have been identified. For instance, if a tourism RS is designed to improve the visitors' knowledge of the destination and increase the visits' count in the less known areas, then a technique characterised by high predictive precision is not in order, while a model that can manage well "new items", i.e., items almost never consumed before is pivotal [46]. In the design phase, evaluation should be implemented off-line and the recommendation algorithms, i.e., their computed recommendations, are compared with the stored user interactions. An off-line evaluation consists of running several algorithms on the same datasets of user interactions and comparing their performances along a set of metrics measuring alternative and complementary dimensions, e.g., precision vs. diversity of the recommendations (see Chapter "Novelty and Diversity in Recommender Systems"). This type of evaluation is usually conducted on existing public benchmark data if appropriate data is available, or, otherwise, on collected data. The design of the off-line experiments should follow known experiment design practices [9] in order to ensure reliable results.

Off-line experiments can give some indications on the quality of the chosen recommendation algorithm in fulfilling the system defined task and goals. However, such an evaluation cannot provide any insight about the user satisfaction, acceptance or experience with the system. That is, the algorithm might be very accurate in solving the core recommendation problem, i.e., identifying relevant items, but for some other reason the system may not be accepted by users, for example, because the performance of the system was not as expected. It is also worth noting that the RS may find the "correct" recommendations for a user, i.e., items that are matching the user needs and wants, but the user may not be able to immediately asses it, especially in those domains where the user may not be able to experience the recommended items at once. For instance, while the songs suggested by a music RS may immediately be played and evaluated, how much a book will make us feel is difficult to assess, before actually reading large part of the book itself.

Therefore, a user-centric evaluation is also required. It can be performed online after the system has been launched, or as a focused user study. During online evaluation, real users interact with the system without being aware of the full nature of the experiment running in the background. It is possible to run various versions of the algorithms on different groups of users for comparison and analysis of the system logs in order to enhance system performance. In addition, most of the RS algorithms include parameters, such as weight thresholds, the number of neighbors in collaborative filtering, etc., requiring constant adjustment and calibration.

Focused user studies are conducted when the online evaluation of the system in the real context of usage is not feasible or too risky. In this type of evaluation, a controlled experiment is planned where a small group of users are asked to perform different tasks with various versions of the system. It is then possible to analyze the user's performance and to distribute questionnaires so that users may report on their experience. In such experiments, it is possible to collect both quantitative and qualitative information about the systems.

In recent years there has been an increased interest in user-centric evaluation procedures and metric for recommender systems. Researchers realized that recommender systems' goals extend beyond the accuracy of the algorithms [40] as tools to provide a helpful and enjoyable, personalized experience that leads to user retention and satisfaction. This approach broadened the range of evaluated aspects of an RS to included aspects such as the form of preference elicitation, the presentation of the recommended results (e.g., one top item, top N items, or predicted ratings), and finally, the evaluation of the explanations provided to the users. Explanations may serve a few goals: the most popular is the justification of results, i.e., explaining to the user why the system decided to recommend a specific item. Other goals may include increasing trust in the system, persuading the user to purchase the recommended item, and helping a user with their decision making. When designing the evaluation of the recommendation explanation, it is therefore important to identify the goal of the explanation and adjust a suitable metric for measuring it.

Chapter "Evaluating Recommender Systems" details the three previously mentioned types of experiments that can be conducted in order to evaluate recommender systems, namely, offline, online and user studies. It presents their advantages and disadvantages, and defines guidelines for choosing the methods for evaluating them by considering the properties that are to be evaluated. Unlike existing discussions of RS evaluation in the literature that usually focuses mainly on the accuracy of an algorithm's prediction [36] and related measures, this chapter is unique in its approach to the evaluation discussion since it focuses on property-directed evaluation. It provides a large set of properties (other than accuracy) that are relevant to the system's success. For each of the properties, the appropriate type of experiment and relevant measures are suggested. Among the list of properties are: coverage, cold start, confidence, trust, novelty, risk, and serendipity. The chapter describes the difficulties and pitfalls of each of the properties and guidelines for the selection of the suitable evaluation type and properties for a given recommendation task and system.

As we have already discussed in Sect. 2, RSs are used by multiple stakeholders and for diverse and sometime competing goals. Chapter "Multistakeholder Recommender Systems" also discusses how system evaluation can take into account the perspectives of multiple stakeholders. A multistakeholder perspective on evaluation, in fact, focuses on additional aspects of system performance that may be quite important. For instance, a multi-sided platforms such as eBay, Etsy, Offer Up, or AirBnB, have a key business requirement of attracting and satisfying the needs of providers as well as users. Hence, a system of this type will also need to evaluate its performance from the provider perspective. The importance of such an evaluation should be evident. For instance, providers whose items are not recommended may experience poor engagement from users and lose interest in participating in the platform. Fewer providers on the platform could then lead to a decline in user interest in the platform as they may find it not comprehensive enough. Or, in an e-commerce platform, the profit of each recommended item may be a factor in ordering and presenting recommendation results. This marketing function of recommender systems was apparent from the start in commercial applications, but rarely included as an element of research systems.

Chapter "Value and Impact of Recommender Systems" completes the treatment of the topic of evaluating RSs by focusing on the capability of RSs to influence the users choices and their behavior. This is especially important is systems where the recommendations are central to the user experience of the service. For instance, in a system like YouTube or Netflix a large fraction of the observed interactions stems from recommendations. RSs can even increase this fraction with a self-enforcing cycle; if the content that users come across during browsing has already been pre-filtered and personalized. The use of recommender systems therefore also bears some risks and can potentially even lead to negative and undesired effects, such as limiting the range of items that are actually recommended, hence negating one of the main goals of RSs: information discovery. The chapter discusses how the impact of recommender systems is currently measured, with a particular focus on measures that are used in practice.

A related and very important topic that also emerged in the last years is the fairness of the RS behaviour (Chapter "Fairness in Recommender Systems"). This defines another dimension for the evaluation of RSs, which goes beyond accuracy or user satisfaction. In fact, RSs pose unique challenges for investigating the fairness and non-discrimination concepts that have been developed in other machine learning literature. The multistakeholder nature of recommender applications, the ranked outputs, the centrality of personalization, and the role of user response complicate the problem of identifying precisely what type of fairness may be relevant, and how to precisely measure and evaluate it. In Chapter "Fairness in Recommender Systems" the authors distinguish various ways a recommender system may be unfair and provide a conceptual framework for identifying the fairness that arise in an application and designing a project to assess and mitigate them.

## 7 Recommender Systems and Human-Computer Interaction

As we have illustrated in the previous sections, researchers have been chiefly concerned with designing a range of technical solutions and leveraging various sources of knowledge to achieve better predictions about the "utility" of the various items for the target user, or what is going to be chosen by the target user, in order to "anticipate" these choices with targeted recommendations. The underlying assumption behind this research activity is that presenting these "correct" recommendations, or the best options, would be sufficient. In other words, the recommendations should speak for themselves and the user should definitely accept the recommendations if they are correct. This is clearly an overly simplified account of the recommendation problem and the delivery and acceptance of recommendations is not so straightforward.

In practice, users need recommendations because they do not have enough knowledge to make an autonomous decision. Consequently, it may not be easy for them to evaluate the proposed recommendation. Hence, various researchers have tried to understand the factors that lead to the acceptance of a recommendation by a given user [8, 25, 35, 50, 67, 71].

Swearingen and Sinha [71] were among the first to point out that the effectiveness of an RS is dependent on factors that go beyond the quality of the prediction algorithm. In fact, the RS must also convince users to try (read, buy, listen, watch, etc.) the recommended items. This, of course, depends on the individual characteristics of the selected items, and therefore on the recommendation algorithm. The process also depends, however, on the particular human/computer interaction supported by the system when the items are presented, compared, and explained. Swearingen and Sinha [71] found that from a user's perspective, an effective recommender system must inspire trust in the system and it must have a system logic that is at least somewhat transparent. Additionally, the authors note that it should point users towards new, not-yet-experienced items, and should provide details about recommended items, including pictures and community ratings, and finally, it should present ways to refine recommendations.

Swearingen and Sinha [71] and other similarly oriented researchers do not diminish the importance of the recommendation algorithm, but claim that its effectiveness should not be evaluated only in terms of the accuracy of the prediction, i.e., with standard and popular Machine Learning or Information Retrieval metrics, such as Mean Absolute Error (MAE), precision, or Normalized Discounted Cumulative Gain (NDCG) (see Chapter "Evaluating Recommender Systems"). Other dimensions should be measured that relate to the acceptance of the recommender system and its recommendations. These ideas have been remarkably well presented and discussed also by McNee et al. [50]. In that work, the authors propose user-centric directions for evaluating recommender systems, including: the similarity of recommendation lists, recommendation serendipity, and the importance of user needs and expectations in a recommender. These seminal papers have played an important role in driving the RSs research towards a more user-centric analysis. In this handbook some, already mentioned contributions, deal with these novel evaluation dimensions: Chapters "Fairness in Recommender Systems", "Novelty and Diversity in Recommender Systems", and "Value and Impact of Recommender Systems".

An essential goal of recommender systems is to help users make better choices [23, 34, 37]. Thus, it is important to understand how people make choices and how the human decision making process can be supported. Chapter "Individual and Group Decision Making and Recommender Systems" begins with a compact overview of the psychology of everyday choice and decision making that is based on a large literature of psychological research and formulated so as to be relevant and accessible to recommender systems research community. The authors explain how recommender systems can be viewed as one of many available tools for facilitating choice. Then, the authors provide a high-level overview of strategies for helping people make better choices, indicating how recommender systems fit into the greater picture of choice. The authors show how an understanding of human decision making can inform research and practice concerning these processes. The revised version of this chapter extends the analysis to choices made by groups and their support by recommender systems for groups.

Personality accounts for the most important way in which individuals differ in their enduring emotional, interpersonal, experiential, attitudinal and motivational styles. Several recent studies have shown that personality can be especially useful at tackling the issues of the cold start problem in RSs and in generating diverse recommendations. Hence, the exploitation of the user personality can improve the user-system interaction in several ways. Chapter "Personality and Recommender Systems" discusses first how personality relates to user preferences and how to use personality in recommender systems. The authors present the Five Factor Model (FFM) of personality. This model appears suitable for usage in RSs as it can be easily quantified in terms of features corresponding to the main factors. The acquisition of the personality factors for an observed user can be made either explicitly through questionnaires or implicitly using machine learning approaches on data derived from social media streams or mobile phone call logs. The authors survey personality acquisition methods, strategies for using personality in recommender systems, and available datasets to use in offline recommender systems experiment.

Finally, Chapter "Beyond Explaining Single Item Recommendations" tackles an additional aspect of the user-centric approach to evaluation and highlights an important aspect of RS: explanations of the recommendation results to the users. It examines the reasons that make an evaluation "good" and the effects that explanations might have on RS acceptance. The chapter first analyses the interaction between the recommender system and the explanation in terms of preference elicitation methods and the presentation of results, as well as the recommendation algorithm. Then, explanation styles are described, along with examples of explanation in existing systems. The goals of explanations are listed from which metrics that measure the success of explanations in achieving these goals are described. The chapter concludes with challenges related to explanations. This includes the context in which explanations should be shown, and a major challenge in evaluating the interaction between acceptance of recommendation and explanations, as well as how to assure that explanations are indeed helpful and do not lead users to make poor decisions.

## 8   Recommender Systems Applications

Recommender systems research, aside from its theoretical contribution, is generally aimed at practically improving industrial RSs and involves research about various practical aspects that apply to the implementation of the systems. Indeed, an RS is an example of large-scale machine learning and data mining algorithms in commercial practice [5, 6]. The common interest in the field, both from the research community and the industry has leveraged the availability of data for research on one hand and the evolution of enhanced algorithms on the other hand. Practical related research in RSs examines aspects that are relevant to different stages in the life cycle of an RS, namely, the design of the system, its implementation, evaluation, maintenance and enhancement during system operation. The Netflix Prize announced in 2006, described in Chapter "Advances in Collaborative Filtering", was an important event for the recommender systems research community and industry, and their mutual interaction. It highlighted the importance of the recommendation of items to users and accelerated the development of many new data mining recommendation techniques. Even though the Netflix Prize initiated a lot of research activities, the prize addressed a simplification of the full recommendation problem. It reduced the recommendation problem to the prediction of user's ratings while optimizing the Root Mean Square Error (RMSE).

The first factor to consider while designing an RS is the application's domain, as it has a major effect on the algorithmic approach that should be taken. In [51] the authors provide a taxonomy of RSs and classify existing RS applications to specific application domains. Based on these specific application domains, we define more general classes of domains for the most common recommender systems applications:

- Entertainment—recommendations for movies, music, games, and IPTV.
- Content—personalized newspapers, recommendation for documents, recommendations of webpages, e-learning applications, and e-mail filters.
- E-commerce—recommendations of products to buy such as books, cameras, PCs etc. for consumers.
- Services—recommendations of travel services, recommendation of experts for consultation, recommendation of houses to rent, or matchmaking services.
- Social—recommendation of people in social networks, and recommendations of content social media content such as tweets, Facebook feeds, LinkedIn updates, and others.

As recommender systems became more popular, interest roused in the potential advantages of new and diverse applications, such as recommending insurance policies, or recommending questions for question-answering systems. As the above list cannot cover all the application domains that are now being addressed by RS techniques: it gives only an initial description of the various types of application domains.

The developer of an RS for a certain application domain should understand the specific facets of the domain, its requirements, application challenges and limitations. Only after analyzing these factors one can be able to select an appropriate recommendation algorithm and to design an effective human-computer interaction. In the current version of the handbook, some of the chapters in this section describe applications of recommender systems in specific domains. Each of these chapters describes the requirements of an RS for a specific domain, its precise challenges and the suitable technologies and algorithms for addressing them.

A popular domain for recommendation is music, presented in Chapter "Music Recommendation Systems: Techniques, Use Cases, and Challenges". Unique features of music items that pose various challenges for recommendations should be considered when designing and evaluating RSs for music. Such challenges include, for example, the short time that it takes a user to gain an opinion about a recommended item, as compared to a movie or a book, or the fact that the same item can be recommended many times. In addition, music can be recommended as a single item, a playlist, and abstracted by genre, performer, or band. Music RSs, as opposed to many other domains, rely heavily on content-based recommendation which brings specific challenges to the domain [49].

Another notable example of RSs that emerged with the diffusion of new communication technologies are recommender systems related to the social web, and specifically those that target the social media domain. With the rise of social networks (e.g., Facebook, LinkedIn, Tweeter, Flickr, and others), users are overloaded with information, activities and interactions. Social recommender systems are RSs that aim at assisting the user in identifying relevant content (e.g., tweets, feeds or images), and engage only in relevant activities and interactions (e.g., discussions, or comments). Apart from the RSs that have been developed to be dedicated to social media, recommender systems in other domains can benefit from the new types of data that social media introduces about users to enhance the quality of standard RSs [31]. The term Social RS covers many types of RSs that are relevant to social media platforms. Chapter "Social Recommender Systems" describes two main types: recommendations of social media content and recommendations of people. For recommendations of social content, the chapter reviews various social content media domains, and provides a detailed case study and insights learned from a recommender system operated in the enterprise which suggests mixed social media items. The chapter lists three different types of people recommendations, namely, the recommendation of familiar people (e.g., classmates, family members) that are not connected in the network; recommendations of interesting people (to connect with, or follow), and recommendations of strangers (to date, to hire, or for various other purposes). It explains the complexity of people-recommendation and lists key topics that should be considered and should be further investigated. The list includes: the need for explanation, privacy concerns, social relationships, trust and reputation, as well as the need to define special evaluation measures.

In Chapter "Multimedia Recommender Systems: Algorithms and Challenges" the authors present multimedia recommender systems. These systems utilize multimodal data (e.g., image, text, audio, video) to represent the items. This representation is then used in a content-based filtering RS or a hybrid RS that combines collaborative filtering and content-based filtering. Multimedia recommender systems are commonly used in various application domains. For example, they are frequently used to recommend videos on entertainment platforms. They are also used for recommending photos on social media platforms or recommending paintings to visit in a (virtual) museum. Other domains include recommending news videos and short user-generated videos. Typical examples of audio recommendation include music creation (e.g., recommending sounds such as drum loops to a creator of electronic music), music consumption (e.g., automatic playlist generation), and information (e.g., podcast recommendation). Multimedia recommender systems can be used to extract features for obtaining a proxy multimedia representation of items to recommend non-media items. For example, many fashion and food RSs are leveraging the visual modality to enhance recommendation. In Chapter "Multimedia Recommender Systems: Algorithms and Challenges" the authors survey the state-of-the-art research related to multi-media RS and, in particular focusing on techniques that integrate item or user side information into a hybrid recommender. They also present various methods for automatically extracting features from multimedia data and creating an item representation based on various modalities (image, video, and audio).

Some new application domains for recommendation evolved with the emergence multi-media RS. One example, detailed in Chapter "Fashion Recommender Systems", is presented by fashion RSs. Due to the growth of fashion e-commerce sites, there is an increasing need for personalized clothing recommendations that match users preferences. Fashion RSs can help users in two main scenarios. Firstly, they can help users to find alternative items to a given item (e.g. evening dress) that has the same style or look and feel. The second common scenario in fashion RS is complementary item recommendations, i.e., helping a user find fashion items that complement a given one. For example, a belt and a tie that go with a given suit. Most of the state-of-the-art fashion RSs that are explored in Chapter "Fashion Recommender Systems" are based on deep learning methods due to their ability to extract meaningful features from the items (such as image, video, text, audio). Deep learning methods overcome the limitations of more traditional recommendation methods (e.g. cold start in nearest neighbors methods) and outperform them.

Another detailed example of RSs designed for a specific domain is described in Chapter "Food Recommender Systems", which deals with recommender systems for food. The authors first present the problem of food recommender systems by detailing the numerous possible scenarios such as: (a) cooking RSs that given available time and ingredients, aim to fit the users taste preferences, health needs (e.g., allergies and intolerance), religious (e.g., Kosher food) and other dietary restrictions (e.g., vegetarian); (b) grocery RSs; and (c) restaurant RSs help groups of people who want to visit a restaurant together as a social. The authors review several food RS solutions proposed in the literature including well performing algorithms: traditional algorithms for user-item ranking problem that are adapted to the food domain and context-aware RS algorithms that take into consideration time and space. In addition, they present algorithms that can change behavioral habits into healthier one in attempt to prevent illnesses associated with being overweight, diabetes and hypertension. The authors also present the importance of user interfaces in food RSs as demonstrated in various recent researches. Several successful user-interface for food RS are presented.

# 9  Challenges

The list of newly emerging and challenging RS research topics is not limited to those described in the previous sections of this introduction. Moreover, it is impossible to cover all of them within this short introduction. The reader is referred to the final discussion sections that are included in almost all of the chapters published in this handbook for other important challenges for RS research.

Below, we briefly introduce some of them, which we deem as particularly important.

## 9.1  Preference Acquisition

A number of open issues are related to the critical stage of acquiring reliable information about the user preferences in order to generate a useful user profile.

It is clear that in many real-world applications of RSs, implicit feedback is much more readily available and requires no extra effort on the user's side. For instance, on a web site it is easy to log the users visiting a URL, or clicking on an ad.

The system can treat these actions as a form of positive feedback to the displayed items. That information contains relevant information for predicting future users' actions. For that reason many recent RSs focus only on the use of implicit feedback. This topic is largely covered in Chapter "Item Recommendation from Implicit Feedback", which is fully dedicated to this subject, but also in Chapters "Session-Based Recommender Systems" and "Advances in Collaborative Filtering". In cases only implicit feedback is available, the core computational problem becomes the prediction of the probability that a user will interact with a given item, by performing a target action, such as listening to a music track or browsing a web page. In such a setting, very often, there is no clear negative signal, i.e., an action that signal a lack of interest for an item, hence the available data is either positive or missing. The missing data includes both items that the user explicitly chose to ignore because they were not appealing and items that would have been perfect recommendations but were never presented to the user or discovered by her. Hence, also when managing implicit feedback one has to consider the biases of the available data and the implication for the construction of effective and fair systems. These topics are further discussed in Chapters "Novelty and Diversity in Recommender Systems", "Value and Impact of Recommender Systems", "Multistakeholder Recommender Systems", and "Fairness in Recommender Systems". But more effective solutions for tackling it must be further developed.

Notwithstanding the large availability of implicit feedback, this data cannot completely substitute the usage of explicitly user-made evaluations, at last for some item categories. This is signalled by the fact that many web sites and applications of RSs are still offering to their users the possibility to enter explicitly their opinion on the recommended or consumed items. Moreover, in certain application domains explicit feedback is still central. This is clearly illustrated in Chapters "Social Recommender Systems", "Group Recommender Systems: Beyond preference aggregation", and "People-to-People Reciprocal Recommenders". These are application domains or settings where the RS cannot totally operate as a black box; users are interacting also with themselves, not only with items. In these applications the RS acts as a mediator between users, enabling the users to better understand each other and browse recommendations that are often referring to actions or preferences of other users. This is clearly seen in a group RS where the goal of the system is to support the choices of a group of users, which often requires the reciprocal understanding of the group members opinions and preferences. Hence, collecting, storing, extracting relevant information from explicit feedback of the group members is pivotal. Ample discussion of these topics is provided also in Chapter "Individual and Group Decision Making and Recommender Systems", where the authors clearly expand the role of a RS beyond the narrow functionality of predicting the most relevant items for each user.

Therefore, simplifying the cognitive cost of preference acquisition is of primary importance. For an RS to achieve good recommendation performance, users typically need to provide the system with a certain amount of feedback about their preferences (e.g., in the form of item ratings). Therefore, it is important to measure the costs and benefits of adopting alternative rating approaches and scales, and to find an optimal solution to meet the needs of both the users, the suppliers and the platform owners. In fact, the acquired preference can have a profound impact on the quality and distribution of the recommendations for all the involved stakeholders. This topic is discussed extensively in Chapters "Multistakeholder Recommender Systems", "Fairness in Recommender Systems" and "Novelty and Diversity in Recommender Systems", where the current research challenges receive ample discussion.

An interesting direction of research for easing the preference acquisition process consists of the exploitation of personality, mood and emotions. This is becoming a popular topic, especially because it is clear that more and more techniques will be developed in order to automatically and unobtrusively acquire such information. In Chapter "Personality and Recommender Systems", the authors stress the challenge of acquiring personality information in a nonintrusive fashion. Nowadays, only the longest questionnaires, which consist of around one hundred questions, can provide an accurate evaluation of the user's personality. Hence, non-intrusive approaches are necessary and the research in this area is just starting. Mining user activity for extracting personality information is an option, but also the fast penetration of portable devices and new sensor types, which can potentially life-logging the user's activity, can offer a promising platform that is worth exploring.

Another line of research aimed at tackling the cold start problem and reducing the user preference elicitation effort is cross-domain recommender systems. These techniques (see Chapter "Design and Evaluation of Cross-Domain Recommender Systems") could be used as an alternative path to user preferences' elicitation as they are able to build detailed user profiles without the need to collect explicit user assessment of the target domain items.

## 9.2  Interaction

A major challenge that RS research is now facing, and is clearly addressed in several chapters of this handbook, is that we still need to broaden the scope of research to the system aspects of a recommender system. This means that aside from the algorithms, which are used to predict the user preference and behaviour, and compute the recommendations, the mechanism through which users provide their input and the means by which they receive the systems output, play a significant role and can play an even larger role in determining the success or failure of a recommender system. We still need to better understand the general qualities of alternative solutions to preference elicitation, as we mentioned previously, but also to recommendation presentation, and to develop personalized solutions for these phases of the interaction with the system.

It must be observed that while interacting with a recommender system, users make various types of decisions. The most important one is surely selecting an item from the recommendation list. But, before making the final decision, users often have to decide how to explore the information space and what additional input they could provide to the system. For instance, they could have to select a specific group of recommendations (e.g., the recommendations about items recently added to the catalogue), or to indicate search conditions, such as the category of searched products, that can help to narrow down the typically large set of received recommendations. Moreover, users often do not know or do not reflect on their preferences beforehand, especially when users approach an RS for information discovery. In such cases, the system-supported interaction and visualization contribute to the user construction of their preferences within a specific recommendation session. As it has been illustrated in Chapter "Individual and Group Decision Making and Recommender Systems", there are several challenges with the full support of user decision making in a recommender system. Our understanding of the situational context generated by the system and its effect on item selection processes is still incomplete and we need to better connect RS research to psychology and decision making disciplines. While it is clear that an RS helps to make decisions, there is still the need for further develop research that takes theories from decision psychology and cognitive psychology into account when explaining users' preference construction and decision making process in the context of recommender systems [52].

Considering the user interaction with the recommender system, the topic of explaining the system recommendations still poses a number of interesting and open issues. For instance, it is still not completely clear whether explanations bring more overall benefits than risks. In Chapter "Beyond Explaining Single Item Recommendations" it is shown that explanations are part of a cyclical process: the explanations affect the acceptance of particular recommendations, the users' mental model of the system, and in turn, this affects the ways users interact with the explanations. But, whether the users are influenced in such a way that their choices are improved, and not biased, is not clear. Explanations may even increase the information overload that RSs are supposed to tame, or make the decision problem, in some cases, even more complex. Moreover, while some research has been conducted on explaining recommendations to individual users, explaining recommendations for a group is still an unexplored subject [29, 72]. For instance, one might think that showing how satisfied other group members are, and how this prediction is incorporated in the recommendations selected for a group, could improve users' understanding of the recommendation process and perhaps make it easier to accept items they do not like. However, users need for privacy is likely to conflict with their need for transparency. Moreover, showing the preferences of other users may move the group discussion on the preferences, i.e., how much the system is taking care of the other group member preferences, rather than on the value of the recommended items. We definitely need more research on these topics.

In a discussion about the human-computer interaction in recommender systems, we cannot forget the issue of the assessment of the value of the recommendations, which is not only related to what extent the recommended items are liked by the user. The general relationships between the value of the RS, the supported interaction and the instruments adopted to measure if the RS meets the goal of their users, are discussed in two chapters of this book (Chapters "Value and Impact of

Recommender Systems" and "Evaluating Recommender Systems"). In fact, there is a number of specific factors that may influence the real usefulness of a recommendation during a specific human-computer interaction. For instance, the time value of recommendations, which is partially discussed in the chapter on context-aware RS (Chapter "Context-Aware Recommender Systems: From Foundations to Recent Developments"), refers to the fact that a given set of recommendations may not be applicable forever but there could be a time interval when these items can be recommended. This is clear, for instance, when it comes to news items: people want to be informed about the most recent events and news loose their value even one day after the initial announcement. The time value of a recommendation is clearly dependent on the novelty and diversity of the recommended items. We still need more theoretical, methodological and algorithmic developments on these aspects. For instance, modelling feature-based novelty in probabilistic terms in order to unify discovery and familiarity models would be an interesting line for future work. Aspects such as the time dimension during which items may recover part of their novelty value, or the variability among users regarding their degree of novelty-seeking are examples of issues that require further research and are mentioned in Chapter "Novelty and Diversity in Recommender Systems".

## 9.3   New Recommendation Tasks

The application of recommender systems is still dominated by solutions for recommending relatively simple and inexpensive products, like movies, music, news and books. While there are systems managing more complex item types, such as financial investment or travel, these item categories are considered as atypical cases. Inevitably, complex domains require more elaborated solutions. Complex products are typically configurable or offered in several variants. This feature still poses a challenge to RS, which are instead designed to consider different configurations as different items. Identifying the more suitable configuration requires reasoning on the relations between configurations (classifying and grouping items) and calls for addressing the specificity of the human decision making task generated by the selection of a configuration. In this handbook, Chapter "Fashion Recommender Systems" focuses on such a complex type of products. For instance in the fashion domain, users rarely think about what to wear and buy in isolation, so the RS must enable users to find fashion items that complement a given one. In general, addressing new types and more structured types of items can call for the introduction of new and interesting research lines. For instance, Chapter "People-to-People Reciprocal Recommenders" clearly shows how different the recommendation techniques must be in domains where reciprocal recommendations are needed, as in dating or job finding applications.

As we already indicated in the previous editions of this handbook, RSs that optimise a sequence of recommendations, e.g., a new book recommendation every week, are not frequent and we believe that this is still an open issue. A considerable amount of new research has focused on Session-based RS, as it is clearly illustrated in Chapter "Session-Based Recommender Systems". But, the current research is still motivated by the goal to predict the sequential choice behaviour of users, which is extracted from the available data, rather than to construct sequences of recommendations that have specific properties that the users may appreciate, such as nudging the user towards better and better choices, or more diverse choices. This is an important dimension in many application, for instance in eLearning but even more clearly in the food domain. For instance, while some initial attempts to build a complete meal plan have been tried, still current food recommender systems are not able to effectively generate such sequences of recommendations (see Chapter "Food Recommender Systems").

Moreover, it is important to study the sequential dimension of users' decision making both within a recommendation session and between recommendation sessions. Here, we want to further note the importance of such a topic in group RSs (see Chapter "Group Recommender Systems: Beyond Preference Aggregation"). In these systems, sequential recommendations are a natural setting, since stable groups, such as friends or families, repeatedly choose items of the same type, e.g., when deciding where to go for vacation or what to eat at home. More research is needed on algorithms and user interfaces for producing coherent sequences of recommendations. In particular, one should model the effect on users of several contextual conditions such as the manner in which already-shown-items could influence the user evaluation of the next recommendations, or the social role and relationships of the group members.

Most of the popular RSs are now accessed through mobile devices that follow their owners throughout their daily life, and are always within an arm's reach of their owners. In this scenario, RSs can proactively send notifications to their users about items of potential interest that are relevant because of the contextual situation of the user. The challenge is finding true relevant items for the user situation and not overburdening the user with a stream of irrelevant interruptions. To address this goal, we must better exploit implicit feedback derived from user usage, which means, we should try to build models that can better generalise from the observed behaviour and identify novel and diverse recommendations (see Chapter "Novelty and Diversity in Recommender Systems", and [47]). But, such systems should also learn to better identify contextual situations that require a push of the recommendations. We believe that this depends on the detection of contextual changes that are significant to the user and therefore justify a recommendation. For instance, when it is the ideal time for a pause in writing a paper, i.e., the context is changing from work to leisure, a recommendation of a relevant, or personalized, article of sports news can be delivered. Understanding when context changes, or it could be forced to change, and when a user may be receptive to a recommendation push is a challenging issue for further research. As it is also suggested in Chapter "Context-Aware Recommender Systems: From Foundations to Recent Developments", in order to develop these new and compelling context aware systems, we need to explore novel engineering solutions, including: novel data structures, storage systems, user interface components and service oriented architectures.

Finally, we hope that this handbook, as a useful tool for practitioners and researchers, will contribute to further developing knowledge in this exciting and useful research area and provide a baseline for further exploring the above mentioned issues. Currently the research on RSs has greatly benefited from the combined interest and efforts that industry and academia have invested in this field. We therefore wish the best to both groups as they read this handbook and we hope that it will attract even more researchers to work in this highly interesting and challenging field.

# References

1. G. Adomavicius, K. Bauman, B. Mobasher, F. Ricci, A. Tuzhilin, M. Unger, Workshop on context-aware recommender systems, in *RecSys 2020: Fourteenth ACM Conference on Recommender Systems, Virtual Event, Brazil, September 22–26, 2020*, ed. by R.L.T. Santos, L.B. Marinho, E.M. Daly, L. Chen, K. Falk, N. Koenigstein, E.S. de Moura (ACM, New York, 2020), pp. 635–637
2. G. Adomavicius, A. Tuzhilin, Personalization technologies: a process-oriented perspective. Commun. ACM **48**(10), 83–90 (2005)
3. G. Adomavicius, A. Tuzhilin, Toward the next generation of recommender systems: a survey of the state-of-the-art and possible extensions. IEEE Trans. Knowl. Data Eng. **17**(6), 734–749 (2005)
4. C.C. Aggarwal, *Recommender Systems - The Textbook* (Springer, New York, 2016)
5. X. Amatriain, Mining large streams of user data for personalized recommendations. SIGKDD Explor. Newsl. **14**(2), 37–48 (2013)
6. X. Amatriain, J. Basilico, Past, present, and future of recommender systems: an industry perspective, in *Proceedings of the 10th ACM Conference on Recommender Systems, Boston, MA, USA, September 15–19, 2016*, ed. by S. Sen, W. Geyer, J. Freyne, P. Castells (ACM, New York, 2016), pp. 211–214
7. O. Arazy, N. Kumar, B. Shapira, Improving social recommender systems. IT Prof. **11**(4), 38–44 (2009)
8. H. Asoh, C. Ono, Y. Habu, H. Takasaki, T. Takenaka, Y. Motomura, An acceptance model of recommender systems based on a large-scale internet survey, in *Advances in User Modeling - UMAP 2011 Workshops, Girona, July 11–15, 2011, Revised Selected Papers* (2011), pp. 410–414
9. R.A. Bailey, *Design of Comparative Experiments* (Cambridge University Press, Cambridge, 2008)
10. M. Balabanovic, Y. Shoham, Content-based, collaborative recommendation. Commun. ACM **40**(3), 66–72 (1997)
11. L. Baltrunas, F. Ricci, Experimental evaluation of context-dependent collaborative filtering using item splitting. User Model. User-Adapt. Interact. **24**(1–2), 7–34 (2014)
12. D. Ben-Shimon, A. Tsikinovsky, L. Rokach, A. Meisels, G. Shani, L. Naamani, Recommender system from personal social networks, in *AWIC, Advances in Soft Computing*, vol. 43, ed. by K. Wegrzyn-Wolska, P.S. Szczepaniak (Springer, New York, 2007), pp. 47–55
13. S. Berkovsky, T. Kuflik, F. Ricci, Mediation of user models for enhanced personalization in recommender systems. User Model. User-Adapted Interact. **18**(3), 245–286 (2008)
14. S. Berkovsky, T. Kuflik, F. Ricci, Cross-representation mediation of user models. User Model. User-Adapted Interact. **19**(1–2), 35–63 (2009)

15. B. Biggio, I. Corona, B. Nelson, B.I., Rubinstein, D. Maiorca, G. Fumera, G. Giacinto, F. Roli, Security evaluation of support vector machines in adversarial environments, in *Support Vector Machines Applications* (Springer, New York, 2014), pp. 105–153
16. D. Billsus, M. Pazzani, Learning probabilistic user models, in *UM97 Workshop on Machine Learning for User Modeling* (1997). http://www.dfki.de/bauer/um-ws/
17. J. Bobadilla, F. Ortega, A. Hernando, A. Gutierrez, Recommender systems survey. Knowl. Based Syst. **46**(0), 109–132 (2013)
18. J. Borràs, A. Moreno, A. Valls, Intelligent tourism recommender systems: a survey. Expert Syst. Appl. **41**(16), 7370–7389 (2014)
19. D. Bridge, M. Göker, L. McGinty, B. Smyth, Case-based recommender systems. Knowl. Eng. Rev. **20**(3), 315–320 (2006)
20. P. Brusilovsky, Methods and techniques of adaptive hypermedia. User Model. User-Adapted Interact. **6**(2–3), 87–129 (1996)
21. R. Burke, Hybrid web recommender systems, in *The Adaptive Web* (Springer, Berlin, 2007), pp. 377–408
22. M. Chelliah, S. Sarkar, Product recommendations enhanced with reviews, in *Proceedings of the Eleventh ACM Conference on Recommender Systems* (2017), pp. 398–399
23. L. Chen, M. de Gemmis, A. Felfernig, P. Lops, F. Ricci, G. Semeraro, Human decision making and recommender systems. TiiS **3**(3), 17 (2013)
24. L. Chen, P. Pu, Critiquing-based recommenders: survey and emerging trends. User Model. User-Adapt. Interact. **22**(1–2), 125–150 (2012)
25. D. Cosley, S.K., Lam, I., Albert, J.A., Konstant, J. Riedl, Is seeing believing? How recommender system interfaces affect users' opinions, in *Proceedings of the CHI 2003 Conference on Human factors in Computing Systems, Fort Lauderdale, FL* (2003), pp. 585–592
26. M.D. Ekstrand, F.M. Harper, M.C. Willemsen, J.A. Konstan, User perception of differences in recommender algorithms, in *Eighth ACM Conference on Recommender Systems, RecSys '14, Foster City, Silicon Valley, CA*, 06–10 Oct 2014, pp. 161–168
27. K. Falk, *Practical Recommender Systems* (Manning Publications, Shelter Island, 2019)
28. C. Feely, B. Caulfield, A. Lawlor, B. Smyth, Using case-based reasoning to predict marathon performance and recommend tailored training plans, in *Case-Based Reasoning Research and Development - 28th International Conference, ICCBR 2020, Salamanca, Spain, June 8–12, 2020*, ed. by I. Watson, R.O. Weber. Lecture Notes in Computer Science, vol. 12311 (Springer, New York, 2020), pp. 67–81
29. A. Felfernig, N. Tintarev, T.N.T. Trang, M. Stettinger, Designing explanations for group recommender systems. CoRR abs/2102.12413 (2021)
30. G. Fisher, User modeling in human-computer interaction. User Model. User-Adapted Interact. **11**, 65–86 (2001)
31. J. Golbeck, Generating predictive movie recommendations from trust in social networks, in *Trust Management, 4th International Conference, iTrust 2006 Proceedings, Pisa*, 16–19 May 2006, pp. 93–104
32. D. Goldberg, D. Nichols, B.M. Oki, D. Terry, Using collaborative filtering to weave an information tapestry. Commun. ACM **35**(12), 61–70 (1992)
33. S.K. Gorakala, M. Usuelli, *Building a Recommendation System with R* (Packt Publishing Ltd., Birmingham, 2015)
34. N. Hazrati, M. Elahi, F. Ricci, Simulating the impact of recommender systems on the evolution of collective users' choices, in *HT '20: 31st ACM Conference on Hypertext and Social Media, Virtual Event, 13–15 July 2020*, ed. by U. Gadiraju (ACM, New York, 2020), pp. 207–212
35. J. Herlocker, J. Konstan, J. Riedl, Explaining collaborative filtering recommendations, in *Proceedings of ACM 2000 Conference on Computer Supported Cooperative Work* (2000), pp. 241–250
36. J.L. Herlocker, J.A. Konstan, L.G. Terveen, J.T. Riedl, Evaluating collaborative filtering recommender systems. ACM Trans. Inf. Syst. **22**(1), 5–53 (2004)

37. A. Jameson, Recommender systems seen through the lens of choice architecture, in *Proceedings of the Joint Workshop on Interfaces and Human Decision Making for Recommender Systems, IntRS 2015, Co-located with ACM Conference on Recommender Systems (RecSys 2015), Vienna, 19 Sept 2015, CEUR Workshop Proceedings*, vol. 1438, ed. by J. O'Donovan, A. Felfernig, N. Tintarev, P. Brusilovsky, G. Semeraro, P. Lops (2015), p. 1. CEUR-WS.org

38. D. Jannach, M. Zanker, A. Felfernig, G. Friedrich, *Recommender Systems: An Introduction* (Cambridge University Press, Cambridge, 2010)

39. J.L. Jorro-Aragoneses, M. Caro-Martínez, J.A. Recio-García, B. Díaz-Agudo, G. Jiménez-Díaz, Personalized case-based explanation of matrix factorization recommendations, in *Case-Based Reasoning Research and Development - 27th International Conference, ICCBR 2019, Otzenhausen, Germany, 8–12 Sept 2019, Proceedings*, vol. 11680, ed. by K. Bach, C. Marling. Lecture Notes in Computer Science (Springer, New York, 2019), pp. 140–154

40. J.A. Konstan, J. Riedl, Recommender systems: from algorithms to user experience. User Model. User-Adapted Interact. **22**(1–2), 101–123 (2012)

41. Y. Koren, R.M. Bell, C. Volinsky, Matrix factorization techniques for recommender systems. IEEE Comput. **42**(8), 30–37 (2009)

42. G. Linden, B. Smith, J. York, Amazon.com recommendations: item-to-item collaborative filtering. IEEE Internet Comput. **7**(1), 76–80 (2003)

43. P. Lops, M. de Gemmis, G. Semeraro, Content-based recommender systems: state of the art and trends, in *Recommender Systems Handbook*, ed. by F. Ricci, L. Rokach, B. Shapira, P.B. Kantor (Springer, New York, 2011), pp. 73–105

44. L. Lu, M. Medo, C.H. Yeung, Y.C. Zhang, Z.K. Zhang, T. Zhou, Recommender systems. Phys. Rep. **519**(1), 1–49 (2012)

45. T. Mahmood, F. Ricci, A. Venturini, Learning adaptive recommendation strategies for online travel planning, in *Information and Communication Technologies in Tourism 2009* (Springer, New York, 2009), pp. 149–160

46. D. Massimo, F. Ricci, Clustering users' pois visit trajectories for next-poi recommendation, in *Information and Communication Technologies in Tourism 2019, ENTER 2019, Proceedings of the International Conference in Nicosia, Cyprus, 30 Jan–1 Feb 2019*, ed. by J. Pesonen, J. Neidhardt (Springer, New York, 2019), pp. 3–14

47. D. Massimo, F. Ricci, Enhancing travel experience leveraging on-line and off-line users' behaviour data, in *IUI '20: 25th International Conference on Intelligent User Interfaces, Cagliari, 17–20 March 2020, Companion* (ACM, New York, 2020), pp. 65–66

48. J. McAuley, C. Targett, Q. Shi, A. Van Den Hengel, Image-based recommendations on styles and substitutes, in *Proceedings of the 38th International ACM SIGIR Conference on Research and Development in Information Retrieval* (2015), pp. 43–52

49. B. McFee, T. Bertin-Mahieux, D.P. Ellis, G.R. Lanckriet, The million song dataset challenge, in *Proceedings of the 21st International Conference Companion on World Wide Web, WWW '12 Companion* (ACM, New York, 2012), pp. 909–916

50. S.M. McNee, J. Riedl, J.A. Konstan, Being accurate is not enough: how accuracy metrics have hurt recommender systems, in *CHI '06: CHI '06 Extended Abstracts on Human Factors in Computing Systems* (ACM Press, New York, 2006), pp. 1097–1101

51. M. Montaner, B. López, J.L. de la Rosa, A taxonomy of recommender agents on the internet. Artif. Intell. Rev. **19**(4), 285–330 (2003)

52. M. Otsuka, T. Osogami, A deep choice model, in *Proceedings of the Thirtieth AAAI Conference on Artificial Intelligence, 12–17 Feb 2016, Phoenix, Arizona*, ed. by D. Schuurmans, M.P. Wellman (AAAI Press, Menlo Park, 2016), pp. 850–856

53. T.K. Paradarami, N.D. Bastian, J.L. Wightman, A hybrid recommender system using artificial neural networks. Expert Syst. Appl. **83**, 300–313 (2017)

54. D.H. Park, H.K. Kim, I.Y. Choi, J.K. Kim, A literature review and classification of recommender systems research. Expert Syst. Appl. **39**(11), 10059–10072 (2012)

55. M. Perano, G.L. Casali, Y. Liu, T. Abbate, Professional reviews as service: a mix method approach to assess the value of recommender systems in the entertainment industry. Technol. Forecast. Soc. Change **169**, 120800 (2021)

56. P. Resnick, N. Iacovou, M. Suchak, P. Bergstrom, J. Riedl, Grouplens: an open architecture for collaborative filtering of netnews, in *Proceedings ACM Conference on Computer-Supported Cooperative Work* (1994), pp. 175–186
57. P. Resnick, N. Iacovou, M. Suchak, P. Bergstrom, J. Riedl, Grouplens: an open architecture for collaborative filtering of netnews, in *Proceedings of the 1994 ACM Conference on Computer Supported Cooperative Work* (1994), pp. 175–186
58. P. Resnick, H.R. Varian, Recommender systems. Commun. ACM **40**(3), 56–58 (1997)
59. M. Reusens, W. Lemahieu, B. Baesens, L. Sels, A note on explicit versus implicit information for job recommendation. Decis. Support Syst. **98**, 26–35 (2017)
60. F. Ricci, Travel recommender systems. IEEE Intell. Syst. **17**(6), 55–57 (2002)
61. F. Ricci, Recommender systems: models and techniques, in *Encyclopedia of Social Network Analysis and Mining*, ed. by R. Alhajj, J.G. Rokne, 2nd edn. (Springer, New York, 2018)
62. F. Ricci, D. Cavada, N. Mirzadeh, A. Venturini, Case-based travel recommendations, in *Destination Recommendation Systems: Behavioural Foundations and Applications*, ed. by D.R. Fesenmaier, K. Woeber, H. Werthner (CABI, Wallingford, 2006), pp. 67–93
63. J.B. Schafer, D. Frankowski, J. Herlocker, S. Sen, Collaborative filtering recommender systems, in *The Adaptive Web* (Springer, Berlin, 2007), pp. 291–324
64. J.B. Schafer, J.A. Konstan, J. Riedl, E-commerce recommendation applications. Data Min. Knowl. Disc. **5**(1/2), 115–153 (2001)
65. M. Schrage, *Recommendation Engines* (MIT Press, Cambridge, 2020)
66. B. Schwartz, *The Paradox of Choice* (ECCO, New York, 2004)
67. M. van Setten, S.M. McNee, J.A. Konstan, Beyond personalization: the next stage of recommender systems research, in *IUI*, ed. by R.S. Amant, J. Riedl, A. Jameson (ACM, New York, 2005), p. 8
68. U. Shardanand, P. Maes, Social information filtering: algorithms for automating "word of mouth", in *Proceedings of the Conference on Human Factors in Computing Systems (CHI'95)* (1995), pp. 210–217
69. R.R. Sinha, K. Swearingen, Comparing recommendations made by online systems and friends, in *DELOS Workshop: Personalisation and Recommender Systems in Digital Libraries* (2001)
70. B. Smith, G. Linden, Two decades of recommender systems at amazon.com. IEEE Internet Comput. **21**(3), 12–18 (2017)
71. K. Swearingen, R. Sinha, Beyond algorithms: an HCI perspective on recommender systems, in *Recommender Systems, papers from the 2001 ACM SIGIR Workshop, New Orleans, LA*, ed. by J.L. Herlocker (2001)
72. T.N.T. Tran, M. Atas, A. Felfernig, V.M. Le, R. Samer, M. Stettinger, Towards social choice-based explanations in group recommender systems, in *Proceedings of the 27th ACM Conference on User Modeling, Adaptation and Personalization, UMAP 2019, Larnaca, Cyprus, 9–12 June 2019*, ed. by G.A. Papadopoulos, G. Samaras, S. Weibelzahl, D. Jannach, O.C. Santos (ACM, New York, 2019), pp. 13–21
73. S. Zhang, L. Yao, A. Sun, Y. Tay, Deep learning based recommender system: a survey and new perspectives. ACM Comput. Surv. **52**(1), 1–38 (2019)

# Part I
# General Recommendation Techniques

# Trust Your Neighbors: A Comprehensive Survey of Neighborhood-Based Methods for Recommender Systems

**Athanasios N. Nikolakopoulos, Xia Ning, Christian Desrosiers, and George Karypis**

## 1 Introduction

The appearance and growth of online markets has had a considerable impact on the habits of consumers, providing them access to a greater variety of products and information on these goods. While this freedom of purchase has made online commerce into a multi-billion dollar industry, it also made it more difficult for consumers to select the products that best fit their needs. One of the main solutions proposed for this information overload problem are recommender systems, which provide automated and personalized suggestions of products to consumers.

The recommendation problem can be defined as estimating the response of a user for unseen items, based on historical information stored in the system, and suggesting to this user *novel* and *original* items for which the predicted response is *high*. User-item responses can be numerical values known as ratings (e.g., 1–5 stars), ordinal values (e.g., strongly agree, agree, neutral, disagree, strongly disagree) representing the possible levels of user appreciation, or binary values (e.g.,

A. N. Nikolakopoulos
Amazon, Seattle, WA, USA

X. Ning
Biomedical Informatics Department, Computer Science and Engineering Department, The Ohio State University, Columbus, OH, USA
e-mail: ning.104@osu.edu

C. Desrosiers (✉)
Software Engineering and IT Department, École de Technologie Supérieure, Montreal, QC, Canada
e-mail: christian.desrosiers@etsmtl.ca

G. Karypis
Computer Science & Engineering Department, University of Minnesota, Minneapolis, MN, USA
e-mail: karypis@cs.umn.edu

© Springer Science+Business Media, LLC, part of Springer Nature 2022
F. Ricci et al. (eds.), *Recommender Systems Handbook*,
https://doi.org/10.1007/978-1-0716-2197-4_2

like/dislike or interested/not interested). Moreover, user responses can be obtained explicitly, for instance, through ratings/reviews entered by users in the system, or implicitly, from purchase history or access patterns [45, 87]. For the purpose of simplicity, from this point on, we will call rating any type of user-item response.

Item recommendation approaches can be divided in two broad categories: personalized and non-personalized. Among the personalized approaches are *content-based* and *collaborative filtering* methods, as well as *hybrid* techniques combining these two types of methods. The general principle of content-based (or cognitive) methods [4, 8, 48, 70] is to identify the common characteristics of items that have received a favorable rating from a user, and then recommend to this user unseen items that share these characteristics. Recommender systems based purely on content generally suffer from the problems of *limited content analysis* and *over-specialization* [79]. Limited content analysis occurs when the system has a limited amount of information on its users or the content of its items. For instance, privacy issues might refrain a user from providing personal information, or the precise content of items may be difficult or costly to obtain for some types of items, such as music or images. Another problem is that the content of an item is often insufficient to determine its quality. Over-specialization, on the other hand, is a side effect of the way in which content-based systems recommend unseen items, where the predicted rating of a user for an item is high if this item is similar to the ones liked by this user. For example, in a movie recommendation application, the system may recommend to a user a movie of the same genre or having the same actors as movies already seen by this user. Because of this, the system may fail to recommend items that are different but still interesting to the user. More information on content-based recommendation approaches can be found in chapter "Semantics and Content-based Recommendations" of this book.

Instead of depending on content information, collaborative (or social) filtering approaches use the rating information of other users and items in the system. The key idea is that the rating of a target user for an unseen item is likely to be similar to that of another user, if both users have rated other items in a similar way. Likewise, the target user is likely to rate two items in a similar fashion, if other users have given similar ratings to these two items. Collaborative filtering approaches overcome some of the limitations of content-based ones. For instance, items for which the content is not available or difficult to obtain can still be recommended to users through the feedback of other users. Furthermore, collaborative recommendations are based on the quality of items as evaluated by peers, instead of relying on content that may be a bad indicator of quality. Finally, unlike content-based systems, collaborative filtering ones can recommend items with very different content, as long as other users have already shown interest for these different items.

Collaborative filtering approaches can be grouped in two general classes of *neighborhood* and *model*-based methods. In neighborhood-based (memory-based [10] or heuristic-based [2]) collaborative filtering [19, 20, 32, 45, 51, 55, 73, 75, 79], the user-item ratings stored in the system are directly used to predict ratings for unseen items. This can be done in two ways known as *user-based* or *item-based* recommendation. User-based systems, such as GroupLens [45], evaluate the interest of a target user for an item using the ratings for this item by other users, called *neighbors*, that have similar rating patterns. The neighbors of the target user are typically the users whose ratings are most correlated to the target user's ratings. Item-based approaches [20, 51, 75], on the other hand, predict the rating of a user for an item based on the ratings of the user for similar items. In such approaches, two items are similar if several users of the system have rated these items in a similar fashion.

In contrast to neighborhood-based systems, which use the stored ratings directly in the prediction, model-based approaches use these ratings to learn a predictive model. Salient characteristics of users and items are captured by a set of model parameters, which are learned from training data and later used to predict new ratings. Model-based approaches for the task of recommending items are numerous and include Bayesian Clustering [10], Latent Semantic Analysis [33], Latent Dirichlet Allocation [9], Maximum Entropy [89], Boltzmann Machines [74], Support Vector Machines [28], and Singular Value Decomposition [6, 46, 69, 85, 86]. A survey of state-of-the-art model-based methods can be found in chapter "Advances in Collaborative Filtering" of this book.

Finally, to overcome certain limitations of content-based and collaborative filtering methods, hybrid recommendation approaches combine characteristics of both types of methods. Content-based and collaborative filtering methods can be combined in various ways, for instance, by merging their individual predictions into a single, more robust prediction [8, 71], or by adding content information into a collaborative filtering model [1, 3, 59, 64, 67, 81, 88]. Several studies have shown hybrid recommendation approaches to provide more accurate recommendations than pure content-based or collaborative methods, especially when few ratings are available [2].

## *1.1 Advantages of Neighborhood Approaches*

While recent investigations show state-of-the-art model-based approaches superior to neighborhood ones in the task of predicting ratings [46, 84], there is also an emerging understanding that good prediction accuracy alone does not guarantee users an effective and satisfying experience [31]. Another factor that has been identified as playing an important role in the appreciation of users for the recommender system is *serendipity* [31, 75]. Serendipity extends the concept of novelty by helping a user find an interesting item he or she might not have otherwise discovered. For example, recommending to a user a movie directed by his favorite director constitutes a novel recommendation if the user was not aware of that movie, but is likely not serendipitous since the user would have discovered that movie on his own. A more detailed discussion on novelty and diversity is provided in chapter "Novelty and Diversity in Recommender Systems" of this book.

Model-based approaches excel at characterizing the preferences of a user with latent factors. For example, in a movie recommender system, such methods may determine that a given user is a fan of movies that are both funny and romantic, without having to actually define the notions "funny" and "romantic". This system would be able to recommend to the user a romantic comedy that may not have been known to this user. However, it may be difficult for this system to recommend a movie that does not quite fit this high-level genre, for instance, a funny parody of horror movies. Neighborhood approaches, on the other hand, capture local associations in the data. Consequently, it is possible for a movie recommender system based on this type of approach to recommend the user a movie very different from his usual taste or a movie that is not well known (e.g., repertoire film), if one of his closest neighbors has given it a strong rating. This recommendation may not be a guaranteed success, as would be a romantic comedy, but it may help the user discover a whole new genre or a new favorite actor/director.

The main advantages of neighborhood-based methods are:

- **Simplicity:** Neighborhood-based methods are intuitive and relatively simple to implement. In their simplest form, only one parameter (the number of neighbors used in the prediction) requires tuning.
- **Justifiability:** Such methods also provide a concise and intuitive justification for the computed predictions. For example, in item-based recommendation, the list of neighbor items, as well as the ratings given by the user to these items, can be presented to the user as a justification for the recommendation. This can help the user better understand the recommendation and its relevance, and could serve as basis for an interactive system where users can select the neighbors for which a greater importance should be given in the recommendation [6]. The benefits and challenges of explaining recommendations to users are addressed in chapter "Beyond Explaining Single Item Recommendations" of this book.
- **Efficiency:** One of the strong points of neighborhood-based systems are their efficiency. Unlike most model-based systems, they require no costly training phases, which need to be carried at frequent intervals in large commercial applications. These systems may require pre-computing nearest neighbors in an offline step, which is typically much cheaper than model training, providing near instantaneous recommendations. Moreover, storing these nearest neighbors requires very little memory, making such approaches scalable to applications having millions of users and items.
- **Stability:** Another useful property of recommender systems based on this approach is that they are little affected by the constant addition of users, items and ratings, which are typically observed in large commercial applications. For instance, once item similarities have been computed, an item-based system can readily make recommendations to new users, without having to re-train the system. Moreover, once a few ratings have been entered for a new item, only the similarities between this item and the ones already in the system need to be computed.

While neighborhood-based methods have gained popularity due to these advantages,[1] they are also known to suffer from the problem of limited coverage, which causes some items to be never recommended. Also, traditional methods of this category are known to be more sensitive to the sparseness of ratings and the cold-start problem, where the system has only a few ratings, or no rating at all, for new users and items. Section 5 presents more advanced neighborhood-based techniques that can overcome these problems.

## 1.2   Objectives and Outline

This chapter has two main objectives. It first serves as a general guide on neighborhood-based recommender systems, and presents practical information on how to implement such recommendation approaches. In particular, the main components of neighborhood-based methods will be described, as well as the benefits of the most common choices for each of these components. Secondly, it presents more specialized techniques on the subject that address particular aspects of recommending items, such as data sparsity. Although such techniques are not required to implement a simple neighborhood-based system, having a broader view of the various difficulties and solutions for neighborhood methods may help making appropriate decisions during the implementation process.

The rest of this document is structured as follows. In Sect. 2, we first give a formal definition of the item recommendation task and present the notation used throughout the chapter. In Sect. 3, the principal neighborhood approaches, predicting user ratings for unseen items based on regression or classification, are then introduced, and the main advantages and flaws of these approaches are described. This section also presents two complementary ways of implementing such approaches, either based on user or item similarities, and analyzes the impact of these two implementations on the accuracy, efficiency, stability, justifiability and serendipity of the recommender system. Section 4, on the other hand, focuses on the three main components of neighborhood-based recommendation methods: rating normalization, similarity weight computation, and neighborhood selection. For each of these components, the most common approaches are described, and their respective benefits compared. In Sect. 5, the problems of limited coverage and data sparsity are introduced, and several solutions proposed to overcome these problems are described. In particular, several techniques based on dimensionality reduction and graphs are presented. Finally, the last section of this document summarizes the

---

[1] For further insights of some of the key properties of neighborhood-based methods under a probabilistic lens, see [12]. Therein, the interested reader can find a probabilistic reformulation of basic neighborhood-based methods that elucidates certain aspects of their effectiveness; delineates innate connections with item popularity; while also, allows for comparisons between basic neighborhood-based variants.

principal characteristics and methods of neighorhood-based recommendation, and gives a few more pointers on implementing such methods.

## 2  Problem Definition and Notation

In order to give a formal definition of the item recommendation task, we introduce the following notation. The set of users in the recommender system will be denoted by $\mathcal{U}$, and the set of items by $\mathcal{I}$. Moreover, we denote by $\mathcal{R}$ the set of ratings recorded in the system, and write $\mathcal{S}$ the set of possible values for a rating (e.g., $\mathcal{S} = [1, 5]$ or $\mathcal{S} = \{\text{like, dislike}\}$). Also, we suppose that no more than one rating can be made by any user $u \in \mathcal{U}$ for a particular item $i \in \mathcal{I}$ and write $r_{ui}$ this rating. To identify the subset of users that have rated an item $i$, we use the notation $\mathcal{U}_i$. Likewise, $\mathcal{I}_u$ represents the subset of items that have been rated by a user $u$. Finally, the items that have been rated by two users $u$ and $v$, i.e. $\mathcal{I}_u \cap \mathcal{I}_v$, is an important concept in our presentation, and we use $\mathcal{I}_{uv}$ to denote this concept. In a similar fashion, $\mathcal{U}_{ij}$ is used to denote the set of users that have rated both items $i$ and $j$.

Two of the most important problems associated with recommender systems are the *rating prediction* and *top-N* recommendation problems. The first problem is to predict the rating that a user $u$ will give his or her unrated item $i$. When ratings are available, this task is most often defined as a regression or (multi-class) classification problem where the goal is to learn a function $f : \mathcal{U} \times \mathcal{I} \rightarrow \mathcal{S}$ that predicts the rating $f(u, i)$ of a user $u$ for an unseen item $i$. Accuracy is commonly used to evaluate the performance of the recommendation method. Typically, the ratings $\mathcal{R}$ are divided into a *training* set $\mathcal{R}_{\text{train}}$ used to learn $f$, and a *test* set $\mathcal{R}_{\text{test}}$ used to evaluate the prediction accuracy. Two popular measures of accuracy are the *Mean Absolute Error* (MAE):

$$\text{MAE}(f) = \frac{1}{|\mathcal{R}_{\text{test}}|} \sum_{r_{ui} \in \mathcal{R}_{\text{test}}} |f(u, i) - r_{ui}|, \tag{1}$$

and the *Root Mean Squared Error* (RMSE):

$$\text{RMSE}(f) = \sqrt{\frac{1}{|\mathcal{R}_{\text{test}}|} \sum_{r_{ui} \in \mathcal{R}_{\text{test}}} (f(u, i) - r_{ui})^2}. \tag{2}$$

When ratings are not available, for instance, if only the list of items purchased by each user is known, measuring the rating prediction accuracy is not possible. In such cases, the problem of finding the best item is usually transformed into the task of recommending to an active user $u_a$ a list $L(u_a)$ containing $N$ items likely to interest him or her [20, 75]. The quality of such method can be evaluated by splitting the items of $\mathcal{I}$ into a set $\mathcal{I}_{\text{train}}$, used to learn $L$, and a test set $\mathcal{I}_{\text{test}}$. Let $T(u) \subset \mathcal{I}_u \cap \mathcal{I}_{\text{test}}$

be the subset of test items that a user $u$ found relevant. If the user responses are binary, these can be the items that $u$ has rated positively. Otherwise, if only a list of purchased or accessed items is given for each user $u$, then these items can be used as $T(u)$. The performance of the method is then computed using the measures of *precision* and *recall*:

$$\text{Precision}(L) = \frac{1}{|\mathcal{U}|} \sum_{u \in \mathcal{U}} |L(u) \cap T(u)| / |L(u)| \tag{3}$$

$$\text{Recall}(L) = \frac{1}{|\mathcal{U}|} \sum_{u \in \mathcal{U}} |L(u) \cap T(u)| / |T(u)|. \tag{4}$$

A drawback of this task is that all items of a recommendation list $L(u)$ are considered equally interesting to user $u$. An alternative setting, described in [20], consists in learning a function $L$ that maps each user $u$ to a list $L(u)$ where items are *ordered* by their "interestingness" to $u$. If the test set is built by randomly selecting, for each user $u$, a single item $i_u$ of $\mathcal{I}_u$, the performance of $L$ can be evaluated with the *Average Reciprocal Hit-Rank* (ARHR):

$$\text{ARHR}(L) = \frac{1}{|\mathcal{U}|} \sum_{u \in \mathcal{U}} \frac{1}{\text{rank}(i_u, L(u))}, \tag{5}$$

where $\text{rank}(i_u, L(u))$ is the rank of item $i_u$ in $L(u)$, equal to $\infty$ if $i_u \notin L(u)$. A more extensive description of evaluation measures for recommender systems can be found in chapter "Evaluating Recommender Systems" of this book.

## 3 Neighborhood-Based Recommendation

Recommender systems based on neighborhood automate the common principle that similar users prefer similar items, and similar items are preferred by similar users. To illustrate this, consider the following example based on the ratings of Fig. 1.

*Example 1* User Eric has to decide whether or not to rent the movie "Titanic" that he has not yet seen. He knows that Lucy has very similar tastes when it comes to movies, as both of them hated "The Matrix" and loved "Forrest Gump," so he asks

**Fig. 1** A "toy example" showing the ratings of four users for five movies

| | The Matrix | Titanic | Die Hard | Forrest Gump | Wall-E |
|---|---|---|---|---|---|
| John | 5 | 1 | | 2 | 2 |
| Lucy | 1 | 5 | 2 | 5 | 5 |
| Eric | 2 | ? | 3 | 5 | 4 |
| Diane | 4 | 3 | 5 | 3 | |

her opinion on this movie. On the other hand, Eric finds out he and Diane have different tastes, Diane likes action movies while he does not, and he discards her opinion or considers the opposite in his decision.

## 3.1 User-Based Rating Prediction

User-based neighborhood recommendation methods predict the rating $r_{ui}$ of a user $u$ for an unseen item $i$ using the ratings given to $i$ by users most similar to $u$, called nearest-neighbors. Suppose we have for each user $v \neq u$ a value $w_{uv}$ representing the preference similarity between $u$ and $v$ (how this similarity can be computed will be discussed in Sect. 4.2). The $k$-nearest-neighbors ($k$-NN) of $u$, denoted by $\mathcal{N}(u)$, are the $k$ users $v$ with the highest similarity $w_{uv}$ to $u$. However, only the users who have rated item $i$ can be used in the prediction of $r_{ui}$, and we instead consider the $k$ users most similar to $u$ that *have rated* $i$. We write this set of neighbors as $\mathcal{N}_i(u)$. The rating $r_{ui}$ can be estimated as the average rating given to $i$ by these neighbors:

$$\hat{r}_{ui} = \frac{1}{|\mathcal{N}_i(u)|} \sum_{v \in \mathcal{N}_i(u)} r_{vi}. \tag{6}$$

A problem with (6) is that is does not take into account the fact that the neighbors can have different levels of similarity. Consider once more the example of Fig. 1. If the two nearest-neighbors of Eric are Lucy and Diane, it would be foolish to consider equally their ratings of the movie "Titanic," since Lucy's tastes are much closer to Eric's than Diane's. A common solution to this problem is to weigh the contribution of each neighbor by its similarity to $u$. However, if these weights do not sum to 1, the predicted ratings can be well outside the range of allowed values. Consequently, it is customary to normalize these weights, such that the predicted rating becomes

$$\hat{r}_{ui} = \frac{\sum\limits_{v \in \mathcal{N}_i(u)} w_{uv} r_{vi}}{\sum\limits_{v \in \mathcal{N}_i(u)} |w_{uv}|}. \tag{7}$$

In the denominator of (7), $|w_{uv}|$ is used instead of $w_{uv}$ because negative weights can produce ratings outside the allowed range. Also, $w_{uv}$ can be replaced by $w_{uv}^{\alpha}$, where $\alpha > 0$ is an amplification factor [10]. When $\alpha > 1$, as is it most often employed, an even greater importance is given to the neighbors that are the closest to $u$.

*Example 2* Suppose we want to use (7) to predict Eric's rating of the movie "Titanic" using the ratings of Lucy and Diane for this movie. Moreover, suppose the similarity weights between these neighbors and Eric are respectively 0.75 and 0.15. The predicted rating would be

$$\hat{r} = \frac{0.75 \times 5 + 0.15 \times 3}{0.75 + 0.15} \simeq 4.67,$$

which is closer to Lucy's rating than to Diane's.

Equation (7) also has an important flaw: it does not consider the fact that users may use different rating values to quantify the same level of appreciation for an item. For example, one user may give the highest rating value to only a few outstanding items, while a less difficult one may give this value to most of the items he likes. This problem is usually addressed by converting the neighbors' ratings $r_{vi}$ to normalized ones $h(r_{vi})$ [10, 73], giving the following prediction:

$$\hat{r}_{ui} = h^{-1} \left( \frac{\sum\limits_{v \in \mathcal{N}_i(u)} w_{uv} h(r_{vi})}{\sum\limits_{v \in \mathcal{N}_i(u)} |w_{uv}|} \right). \tag{8}$$

Note that the predicted rating must be converted back to the original scale, hence the $h^{-1}$ in the equation. The most common approaches to normalize ratings will be presented in Sect. 4.1.

## 3.2   User-Based Classification

The prediction approach just described, where the predicted ratings are computed as a weighted average of the neighbors' ratings, essentially solves a *regression* problem. Neighborhood-based *classification*, on the other hand, finds the most likely rating given by a user $u$ to an item $i$, by having the nearest-neighbors of $u$ vote on this value. The vote $v_{ir}$ given by the $k$-NN of $u$ for the rating $r \in \mathcal{S}$ can be obtained as the sum of the similarity weights of neighbors that have given this rating to $i$:

$$v_{ir} = \sum_{v \in \mathcal{N}_i(u)} \delta(r_{vi} = r) w_{uv}, \tag{9}$$

where $\delta(r_{vi} = r)$ is 1 if $r_{vi} = r$, and 0 otherwise. Once this has been computed for every possible rating value, the predicted rating is simply the value $r$ for which $v_{ir}$ is the greatest.

*Example 3*  Suppose once again that the two nearest-neighbors of Eric are Lucy and Diane with respective similarity weights 0.75 and 0.15. In this case, ratings 5 and 3 each have one vote. However, since Lucy's vote has a greater weight than Diane's, the predicted rating will be $\hat{r} = 5$.

A classification method that considers normalized ratings can also be defined. Let $\mathcal{S}'$ be the set of possible normalized values (that may require discretization), the predicted rating is obtained as:

$$\hat{r}_{ui} = h^{-1} \left( \arg\max_{r \in \mathcal{S}'} \sum_{v \in \mathcal{N}_i(u)} \delta(h(r_{vi}) = r) \, w_{uv} \right). \tag{10}$$

## 3.3 Regression vs. Classification

The choice between implementing a neighborhood-based regression or classification method largely depends on the system's rating scale. Thus, if the rating scale is continuous, e.g. ratings in the *Jester* joke recommender system [25] can take any value between $-10$ and 10, then a regression method is more appropriate. On the contrary, if the rating scale has only a few discrete values, e.g. "good" or "bad," or if the values cannot be ordered in an obvious fashion, then a classification method might be preferable. Furthermore, since normalization tends to map ratings to a continuous scale, it may be harder to handle in a classification approach.

Another way to compare these two approaches is by considering the situation where all neighbors have the same similarity weight. As the number of neighbors used in the prediction increases, the rating $r_{ui}$ predicted by the regression approach will tend toward the mean rating of item $i$. Suppose item $i$ has only ratings at either end of the rating range, i.e. it is either loved or hated, then the regression approach will make the safe decision that the item's worth is average. This is also justified from a statistical point of view since the expected rating (estimated in this case) is the one that minimizes the RMSE. On the other hand, the classification approach will predict the rating as the most frequent one given to $i$. This is more risky as the item will be labeled as either "good" or "bad". However, as mentioned before, risk taking may be be desirable if it leads to serendipitous recommendations.

## 3.4 Item-Based Recommendation

While user-based methods rely on the opinion, of like-minded users to predict a rating, item-based approaches [20, 51, 75] look at ratings given to similar items. Let us illustrate this approach with our toy example.

*Example 4* Instead of consulting with his peers, Eric instead determines whether the movie "Titanic" is right for him by considering the movies that he has already seen. He notices that people that have rated this movie have given similar ratings to the movies "Forrest Gump" and "Wall-E". Since Eric liked these two movies he concludes that he will also like the movie "Titanic".

This idea can be formalized as follows. Denote by $\mathcal{N}_u(i)$ the items rated by user $u$ most similar to item $i$. The predicted rating of $u$ for $i$ is obtained as a weighted average of the ratings given by $u$ to the items of $\mathcal{N}_u(i)$:

$$\hat{r}_{ui} = \frac{\sum\limits_{j \in \mathcal{N}_u(i)} w_{ij}\, r_{uj}}{\sum\limits_{j \in \mathcal{N}_u(i)} |w_{ij}|}. \tag{11}$$

*Example 5* Suppose our prediction is again made using two nearest-neighbors, and that the items most similar to "Titanic" are "Forrest Gump" and "Wall-E," with respective similarity weights 0.85 and 0.75. Since ratings of 5 and 4 were given by Eric to these two movies, the predicted rating is computed as

$$\hat{r} = \frac{0.85 \times 5 + 0.75 \times 4}{0.85 + 0.75} \simeq 4.53.$$

Again, the differences in the users' individual rating scales can be considered by normalizing ratings with a $h$:

$$\hat{r}_{ui} = h^{-1}\left( \frac{\sum\limits_{j \in \mathcal{N}_u(i)} w_{ij}\, h(r_{uj})}{\sum\limits_{j \in \mathcal{N}_u(i)} |w_{ij}|} \right). \tag{12}$$

Moreover, we can also define an item-based classification approach. In this case, the items $j$ rated by user $u$ vote for the rating to be given to an unseen item $i$, and these votes are weighted by the similarity between $i$ and $j$. The normalized version of this approach can be expressed as follows:

$$\hat{r}_{ui} = h^{-1}\left( \arg\max_{r \in \mathcal{S}'} \sum\limits_{j \in \mathcal{N}_u(i)} \delta(h(r_{uj}) = r)\, w_{ij} \right). \tag{13}$$

## 3.5   User-Based vs. Item-Based Recommendation

When choosing between the implementation of a user-based and an item-based neighborhood recommender system, five criteria should be considered:

- **Accuracy:** The accuracy of neighborhood recommendation methods depends mostly on the ratio between the number of users and items in the system. As will be presented in Sect. 4.2, the similarity between two users in user-based methods, which determines the neighbors of a user, is normally obtained by comparing the ratings made by these users on the same items. Consider a system that has 10, 000 ratings made by 1000 users on 100 items, and suppose, for the purpose of this analysis, that the ratings are distributed uniformly over the

**Table 1** The average number of neighbors and average number of ratings used in the computation of similarities for user-based and item-based neighborhood methods. A uniform distribution of ratings is assumed with average number of ratings per user $p = |\mathcal{R}|/|\mathcal{U}|$, and average number of ratings per item $q = |\mathcal{R}|/|\mathcal{I}|$

|              | Avg. neighbors                                                              | Avg. ratings             |
| ------------ | --------------------------------------------------------------------------- | ------------------------ |
| User-based   | $(|\mathcal{U}| - 1)\left(1 - \left(\frac{|\mathcal{I}| - p}{|\mathcal{I}|}\right)^p\right)$ | $\frac{p^2}{|\mathcal{I}|}$ |
| Item-based   | $(|\mathcal{I}| - 1)\left(1 - \left(\frac{|\mathcal{U}| - q}{|\mathcal{U}|}\right)^q\right)$ | $\frac{q^2}{|\mathcal{U}|}$ |

**Table 2** The space and time complexity of user-based and item-based neighborhood methods, as a function of the maximum number of ratings per user $p = \max_u |\mathcal{I}_u|$, the maximum number of ratings per item $q = \max_i |\mathcal{U}_i|$, and the maximum number of neighbors used in the rating predictions $k$

|              |                       | Time                       |                    |
| ------------ | --------------------- | -------------------------- | ------------------ |
|              | Space                 | Training                   | Online             |
| User-based   | $O(|\mathcal{U}|^2)$  | $O(|\mathcal{U}|^2 p)$     | $O(|\mathcal{I}|k)$ |
| Item-based   | $O(|\mathcal{I}|^2)$  | $O(|\mathcal{I}|^2 q)$     | $O(|\mathcal{I}|k)$ |

items.[2] Following Table 1, the average number of users available as potential neighbors is roughly 650. However, the average number of common ratings used to compute the similarities is only 1. On the other hand, an item-based method usually computes the similarity between two items by comparing ratings made by the same user on these items. Assuming once more a uniform distribution of ratings, we find an average number of potential neighbors of 99 and an average number of ratings used to compute the similarities of 10.

In general, a small number of high-confidence neighbors is by far preferable to a large number of neighbors for which the similarity weights are not trustable. In cases where the number of users is much greater than the number of items, such as large commercial systems like *Amazon.com*, item-based methods can therefore produce more accurate recommendations [21, 75]. Likewise, systems that have less users than items, e.g., a research paper recommender with thousands of users but hundreds of thousands of articles to recommend, may benefit more from user-based neighborhood methods [31].

- **Efficiency:** As shown in Table 2, the memory and computational efficiency of recommender systems also depends on the ratio between the number of users and items. Thus, when the number of users exceeds the number of items, as is it most often the case, item-based recommendation approaches require much less memory and time to compute the similarity weights (training phase) than user-based ones, making them more scalable. However, the time complexity of the online recommendation phase, which depends only on the number of available

---

[2] The distribution of ratings in real-life data is normally skewed, i.e. most ratings are given to a small proportion of items.

items and the maximum number of neighbors, is the same for user-based and item-based methods.

In practice, computing the similarity weights is much less expensive than the worst-case complexity reported in Table 2, due to the fact that users rate only a few of the available items. Accordingly, only the non-zero similarity weights need to be stored, which is often much less than the number of user pairs. This number can be further reduced by storing for each user only the top $N$ weights, where $N$ is a parameter [75] that is sufficient for satisfactory coverage on user-item pairs. In the same manner, the non-zero weights can be computed efficiently without having to test each pair of users or items, which makes neighborhood methods scalable to very large systems.

- **Stability:** The choice between a user-based and an item-based approach also depends on the frequency and amount of change in the users and items of the system. If the list of available items is fairly static in comparison to the users of the system, an item-based method may be preferable since the item similarity weights could then be computed at infrequent time intervals while still being able to recommend items to new users. On the contrary, in applications where the list of available items is constantly changing, e.g., an online article recommender, user-based methods could prove to be more stable.

- **Justifiability:** An advantage of item-based methods is that they can easily be used to justify a recommendation. Hence, the list of neighbor items used in the prediction, as well as their similarity weights, can be presented to the user as an explanation of the recommendation. By modifying the list of neighbors and/or their weights, it then becomes possible for the user to participate interactively in the recommendation process. User-based methods, however, are less amenable to this process because the active user does not know the other users serving as neighbors in the recommendation.

- **Serendipity:** In item-based methods, the rating predicted for an item is based on the ratings given to similar items. Consequently, recommender systems using this approach will tend to recommend to a user items that are related to those usually appreciated by this user. For instance, in a movie recommendation application, movies having the same genre, actors or director as those highly rated by the user are likely to be recommended. While this may lead to safe recommendations, it does less to help the user discover different types of items that he might like as much.

Because they work with user similarity, on the other hand, user-based approaches are more likely to make serendipitous recommendations. This is particularly true if the recommendation is made with a small number of nearest-neighbors. For example, a user $A$ that has watched only comedies may be very similar to a user $B$ only by the ratings made on such movies. However, if $B$ is fond of a movie in a different genre, this movie may be recommended to $A$ through his similarity with $B$.

# 4   Components of Neighborhood Methods

In the previous section, we have seen that deciding between a regression and a classification rating prediction method, as well as choosing between a user-based or item-based recommendation approach, can have a significant impact on the accuracy, efficiency and overall quality of the recommender system. In addition to these crucial attributes, three very important considerations in the implementation of a neighborhood-based recommender system are (1) the normalization of ratings, (2) the computation of the similarity weights, and (3) the selection of neighbors. This section reviews some of the most common approaches for these three components, describes the main advantages and disadvantages of using each one of them, and gives indications on how to implement them.

## 4.1   Rating Normalization

When it comes to assigning a rating to an item, each user has its own personal scale. Even if an explicit definition of each of the possible ratings is supplied (e.g., 1="strongly disagree," 2="disagree," 3="neutral," etc.), some users might be reluctant to give high/low scores to items they liked/disliked. Two of the most popular rating normalization schemes that have been proposed to convert individual ratings to a more universal scale are *mean-centering* and *Z-score*.

### 4.1.1   Mean-centering

The idea of mean-centering [10, 73] is to determine whether a rating is positive or negative by comparing it to the mean rating. In user-based recommendation, a raw rating $r_{ui}$ is transformation to a mean-centered one $h(r_{ui})$ by subtracting to $r_{ui}$ the average $\bar{r}_u$ of the ratings given by user $u$ to the items in $\mathcal{I}_u$:

$$h(r_{ui}) = r_{ui} - \bar{r}_u.$$

Using this approach the user-based prediction of a rating $r_{ui}$ is obtained as

$$\hat{r}_{ui} = \bar{r}_u + \frac{\sum\limits_{v \in \mathcal{N}_i(u)} w_{uv}\,(r_{vi} - \bar{r}_v)}{\sum\limits_{v \in \mathcal{N}_i(u)} |w_{uv}|}. \tag{14}$$

In the same way, the *item*-mean-centered normalization of $r_{ui}$ is given by

$$h(r_{ui}) = r_{ui} - \bar{r}_i,$$

**Fig. 2** The *user* and *item* mean-centered ratings of Fig. 1

*User* mean-centering:

|       | The Matrix | Titanic | Die Hard | Forrest Gump | Wall-E |
|-------|------------|---------|----------|--------------|--------|
| John  | 2.50       | -1.50   |          | -0.50        | -0.50  |
| Lucy  | -2.60      | 1.40    | -1.60    | 1.40         | 1.40   |
| Eric  | -1.50      |         | -0.50    | 1.50         | 0.50   |
| Diane | 0.25       | -0.75   | 1.25     | -0.75        |        |

*Item* mean-centering:

|       | The Matrix | Titanic | Die Hard | Forrest Gump | Wall-E |
|-------|------------|---------|----------|--------------|--------|
| John  | 2.00       | -2.00   |          | -1.75        | -1.67  |
| Lucy  | -2.00      | 2.00    | -1.33    | 1.25         | 1.33   |
| Eric  | -1.00      |         | -0.33    | 1.25         | 0.33   |
| Diane | 1.00       | 0.00    | 1.67     | -0.75        |        |

where $\bar{r}_i$ corresponds to the mean rating given to item $i$ by user in $\mathcal{U}_i$. This normalization technique is most often used in item-based recommendation, where a rating $r_{ui}$ is predicted as:

$$\hat{r}_{ui} = \bar{r}_i + \frac{\sum\limits_{j \in \mathcal{N}_u(i)} w_{ij} (r_{uj} - \bar{r}_j)}{\sum\limits_{j \in \mathcal{N}_u(i)} |w_{ij}|}. \tag{15}$$

An interesting property of mean-centering is that one can see right-away if the appreciation of a user for an item is positive or negative by looking at the sign of the normalized rating. Moreover, the module of this rating gives the level at which the user likes or dislikes the item.

*Example 6* As shown in Fig. 2, although Diane gave an average rating of 3 to the movies "Titanic" and "Forrest Gump," the user-mean-centered ratings show that her appreciation of these movies is in fact negative. This is because her ratings are high on average, and so, an average rating correspond to a low degree of appreciation. Differences are also visible while comparing the two types of mean-centering. For instance, the item-mean-centered rating of the movie "Titanic" is neutral, instead of negative, due to the fact that much lower ratings were given to that movie. Likewise, Diane's appreciation for "The Matrix" and John's distaste for "Forrest Gump" are more pronounced in the item-mean-centered ratings.

### 4.1.2 Z-score Normalization

Consider, two users $A$ and $B$ that both have an average rating of 3. Moreover, suppose that the ratings of $A$ alternate between 1 and 5, while those of $B$ are always

3. A rating of 5 given to an item by $B$ is more exceptional than the same rating given by $A$, and, thus, reflects a greater appreciation for this item. While mean-centering removes the offsets caused by the different perceptions of an average rating, $Z$-score normalization [30] also considers the spread in the individual rating scales. Once again, this is usually done differently in user-based than in item-based recommendation. In user-based methods, the normalization of a rating $r_{ui}$ divides the *user*-mean-centered rating by the standard deviation $\sigma_u$ of the ratings given by user $u$:

$$h(r_{ui}) = \frac{r_{ui} - \overline{r}_u}{\sigma_u}.$$

A user-based prediction of rating $r_{ui}$ using this normalization approach would therefore be obtained as

$$\hat{r}_{ui} = \overline{r}_u + \sigma_u \frac{\sum\limits_{v \in \mathcal{N}_i(u)} w_{uv} (r_{vi} - \overline{r}_v)/\sigma_v}{\sum\limits_{v \in \mathcal{N}_i(u)} |w_{uv}|}. \tag{16}$$

Likewise, the $z$-score normalization of $r_{ui}$ in item-based methods divides the *item*-mean-centered rating by the standard deviation of ratings given to item $i$:

$$h(r_{ui}) = \frac{r_{ui} - \overline{r}_i}{\sigma_i}.$$

The item-based prediction of rating $r_{ui}$ would then be

$$\hat{r}_{ui} = \overline{r}_i + \sigma_i \frac{\sum\limits_{j \in \mathcal{N}_u(i)} w_{ij} (r_{uj} - \overline{r}_j)/\sigma_j}{\sum\limits_{j \in \mathcal{N}_u(i)} |w_{ij}|}. \tag{17}$$

### 4.1.3 Choosing a Normalization Scheme

In some cases, rating normalization can have undesirable effects. For instance, imagine the case of a user that gave only the highest ratings to the items he has purchased. Mean-centering would consider this user as "easy to please" and any rating below this highest rating (whether it is a positive or negative rating) would be considered as negative. However, it is possible that this user is in fact "hard to please" and carefully selects only items that he will like for sure. Furthermore, normalizing on a few ratings can produce unexpected results. For example, if a user has entered a single rating or a few identical ratings, his rating standard deviation will be 0, leading to undefined prediction values. Nevertheless, if the rating data is not overly sparse, normalizing ratings has been found to consistently improve the predictions [30, 34].

Comparing mean-centering with $Z$-score, as mentioned, the second one has the additional benefit of considering the variance in the ratings of individual users or items. This is particularly useful if the rating scale has a wide range of discrete values or if it is continuous. On the other hand, because the ratings are divided and multiplied by possibly very different standard deviation values, $Z$-score can be more sensitive than mean-centering and, more often, predict ratings that are outside the rating scale. Lastly, while an initial investigation found mean-centering and $Z$-score to give comparable results [30], subsequent analysis showed $Z$-score to have more significant benefits [34].

Finally, if rating normalization is not possible or does not improve the results, another possible approach to remove the problems caused by the rating scale variance is *preference-based filtering*. The particularity of this approach is that it focuses on predicting the relative preferences of users instead of absolute rating values. Since, an item preferred to another one remains so regardless of the rating scale, predicting relative preferences removes the need to normalize the ratings. More information on this approach can be found in [16, 23, 37, 38].

## 4.2 Similarity Weight Computation

The similarity weights play a double role in neighborhood-based recommendation methods: (1) they allow to select trusted neighbors whose ratings are used in the prediction, and (2) they provide the means to give more or less importance to these neighbors in the prediction. The computation of the similarity weights is one of the most critical aspects of building a neighborhood-based recommender system, as it can have a significant impact on both its accuracy and its performance.

### 4.2.1 Correlation-Based Similarity

A measure of the similarity between two objects $a$ and $b$, often used in information retrieval, consists in representing these objects in the form of a vector $\mathbf{x}_a$ and $\mathbf{x}_b$ and computing the *Cosine Vector* (CV) (or *Vector Space*) similarity [4, 8, 48] between these vectors:

$$\cos(\mathbf{x}_a, \mathbf{x}_b) = \frac{\mathbf{x}_a^\top \mathbf{x}_b}{||\mathbf{x}_a|| \cdot ||\mathbf{x}_b||}.$$

In the context of item recommendation, this measure can be employed to compute user similarities by considering a user $u$ as a vector $\mathbf{x}_u \in \Re^{|I|}$, where $\mathbf{x}_{ui} = r_{ui}$ if user $u$ has rated item $i$, and 0 otherwise. The similarity between two users $u$ and $v$ would then be computed as

$$CV(u, v) = \cos(\mathbf{x}_u, \mathbf{x}_v) = \frac{\sum\limits_{i \in \mathcal{I}_{uv}} r_{ui} \, r_{vi}}{\sqrt{\sum\limits_{i \in \mathcal{I}_u} r_{ui}^2 \sum\limits_{j \in \mathcal{I}_v} r_{vj}^2}}, \tag{18}$$

where $\mathcal{I}_{uv}$ once more denotes the items rated by both $u$ and $v$. A problem with this measure is that is does not consider the differences in the mean and variance of the ratings made by users $u$ and $v$.

A popular measure that compares ratings where the effects of mean and variance have been removed is the *Pearson Correlation* (PC) similarity:

$$PC(u, v) = \frac{\sum\limits_{i \in \mathcal{I}_{uv}} (r_{ui} - \bar{r}_u)(r_{vi} - \bar{r}_v)}{\sqrt{\sum\limits_{i \in \mathcal{I}_{uv}} (r_{ui} - \bar{r}_u)^2 \sum\limits_{i \in \mathcal{I}_{uv}} (r_{vi} - \bar{r}_v)^2}}. \tag{19}$$

Note that this is different from computing the CV similarity on the $Z$-score normalized ratings, since the standard deviation of the ratings in evaluated only on the common items $\mathcal{I}_{uv}$, not on the entire set of items rated by $u$ and $v$, i.e. $\mathcal{I}_u$ and $\mathcal{I}_v$. The same idea can be used to obtain similarities between two items $i$ and $j$ [20, 75], this time by comparing the ratings made by users that have rated both these items:

$$PC(i, j) = \frac{\sum\limits_{u \in \mathcal{U}_{ij}} (r_{ui} - \bar{r}_i)(r_{uj} - \bar{r}_j)}{\sqrt{\sum\limits_{u \in \mathcal{U}_{ij}} (r_{ui} - \bar{r}_i)^2 \sum\limits_{u \in \mathcal{U}_{ij}} (r_{uj} - \bar{r}_j)^2}}. \tag{20}$$

While the sign of a similarity weight indicates whether the correlation is direct or inverse, its magnitude (ranging from 0 to 1) represents the strength of the correlation.

*Example 7* The similarities between the pairs of users and items of our toy example, as computed using PC similarity, are shown in Fig. 3. We can see that Lucy's taste in movies is very close to Eric's (similarity of 0.922) but very different from John's (similarity of $-0.938$). This means that Eric's ratings can be trusted to predict Lucy's, and that Lucy should discard John's opinion on movies or consider the opposite. We also find that the people that like "The Matrix" also like "Die Hard" but hate "Wall-E". Note that these relations were discovered without having any knowledge of the genre, director or actors of these movies.

The differences in the rating scales of individual users are often more pronounced than the differences in ratings given to individual items. Therefore, while computing the item similarities, it may be more appropriate to compare ratings that are centered on their *user* mean, instead of their *item* mean. The *Adjusted Cosine* (AC) similarity [75], is a modification of the PC item similarity which compares user-mean-centered ratings:

**Fig. 3** The *user* and *item* PC
similarity for the ratings of
Fig. 1

*User-based* Pearsoncorrelation

|       | John   | Lucy   | Eric   | Diane  |
|-------|--------|--------|--------|--------|
| John  | 1.000  | -0.938 | -0.839 | 0.659  |
| Lucy  | -0.938 | 1.000  | 0.922  | -0.787 |
| Eric  | -0.839 | 0.922  | 1.000  | -0.659 |
| Diane | 0.659  | -0.787 | -0.659 | 1.000  |

*Item-based* Pearsoncorrelation

|            | The Matrix | Titanic | Die Hard | Forrest Gump | Wall-E |
|------------|------------|---------|----------|--------------|--------|
| Matrix     | 1.000      | -0.943  | 0.882    | -0.974       | -0.977 |
| Titanic    | -0.943     | 1.000   | -0.625   | 0.931        | 0.994  |
| DieHard    | 0.882      | -0.625  | 1.000    | -0.804       | -1.000 |
| ForrestGump| -0.974     | 0.931   | -0.804   | 1.000        | 0.930  |
| Wall-E     | -0.977     | 0.994   | -1.000   | 0.930        | 1.000  |

$$AC(i, j) = \frac{\sum\limits_{u \in \mathcal{U}_{ij}} (r_{ui} - \overline{r}_u)(r_{uj} - \overline{r}_u)}{\sqrt{\sum\limits_{u \in \mathcal{U}_{ij}} (r_{ui} - \overline{r}_u)^2 \sum\limits_{u \in \mathcal{U}_{ij}} (r_{uj} - \overline{r}_u)^2}}.$$

In some cases, AC similarity has been found to outperform PC similarity on the prediction of ratings using an item-based method [75].

### 4.2.2 Other Similarity Measures

Several other measures have been proposed to compute similarities between users or items. One of them is the *Mean Squared Difference* (MSD) [79], which evaluate the similarity between two users $u$ and $v$ as the inverse of the average squared difference between the ratings given by $u$ and $v$ on the same items:

$$\text{MSD}(u, v) = \frac{|\mathcal{I}_{uv}|}{\sum\limits_{i \in \mathcal{I}_{uv}} (r_{ui} - r_{vi})^2}. \tag{21}$$

While it could be modified to compute the differences on normalized ratings, the MSD similarity is limited compared to PC similarity because it does not allows to capture negative correlations between user preferences or the appreciation of different items. Having such negative correlations may improve the rating prediction accuracy [29].

Another well-known similarity measure is the *Spearman Rank Correlation* (SRC) [42]. While PC uses the rating values directly, SRC instead considers the ranking of these ratings. Denote by $k_{ui}$ the rating rank of item $i$ in user $u$'s list

**Table 3** The rating prediction accuracy (MAE) obtained on the *MovieLens* dataset using the Mean Squared Difference (MSD), Spearman Rank Correlation and Pearson Correlation (PC) similarity measures. Results are shown for predictions using an increasing number of neighbors $k$

| $k$ | MSD | SRC | PC |
|-----|--------|--------|--------|
| 5 | 0.7898 | 0.7855 | 0.7829 |
| 10 | 0.7718 | 0.7636 | 0.7618 |
| 20 | 0.7634 | 0.7558 | 0.7545 |
| 60 | 0.7602 | 0.7529 | 0.7518 |
| 80 | 0.7605 | 0.7531 | 0.7523 |
| 100 | 0.7610 | 0.7533 | 0.7528 |

of rated items (tied ratings get the average rank of their spot). The SRC similarity between two users $u$ and $v$ is evaluated as:

$$SRC(u, v) = \frac{\sum\limits_{i \in \mathcal{I}_{uv}} (k_{ui} - \overline{k}_u)(k_{vi} - \overline{k}_v)}{\sqrt{\sum\limits_{i \in \mathcal{I}_{uv}} (k_{ui} - \overline{k}_u)^2 \sum\limits_{i \in \mathcal{I}_{uv}} (k_{vi} - \overline{k}_v)^2}}, \qquad (22)$$

where $\overline{k}_u$ is the average rank of items rated by $u$.

The principal advantage of SRC is that it avoids the problem of rating normalization, described in the last section, by using rankings. On the other hand, this measure may not be the best one when the rating range has only a few possible values, since that would create a large number of tied ratings. Moreover, this measure is typically more expensive than PC as ratings need to be sorted in order to compute their rank.

Table 3 shows the user-based prediction accuracy (MAE) obtained with MSD, SRC and PC similarity measures, on the *MovieLens*[3] dataset [29]. Results are given for different values of $k$, which represents the maximum number of neighbors used in the predictions. For this data, we notice that MSD leads to the least accurate predictions, possibly due to the fact that it does not take into account negative correlations. Also, these results show PC to be slightly more accurate than SRC. Finally, although PC has been generally recognized as the best similarity measure, see e.g. [29], subsequent investigation has shown that the performance of such measure depended greatly on the data [34].

### 4.2.3   Considering the Significance of Weights

Because the rating data is frequently sparse in comparison to the number of users and items of a system, it is often the case that similarity weights are computed using

---

[3] http://www.grouplens.org/.

only a few ratings given to common items or made by the same users. For example, if the system has $10,000$ ratings made by $1000$ users on $100$ items (assuming a uniform distribution of ratings), Table 1 shows us that the similarity between two users is computed, on average, by comparing the ratings given by these users to a *single* item. If these few ratings are equal, then the users will be considered as "fully similar" and will likely play an important role in each other's recommendations. However, if the users' preferences are in fact different, this may lead to poor recommendations.

Several strategies have been proposed to take into account the *significance* of a similarity weight. The principle of these strategies is essentially the same: reduce the magnitude of a similarity weight when this weight is computed using only a few ratings. For instance, in *Significance Weighting* [30, 53], a user similarity weight $w_{uv}$ is penalized by a factor proportional to the number of commonly rated item, if this number is less than a given parameter $\gamma > 0$:

$$w'_{uv} = \frac{\min\{|\mathcal{I}_{uv}|, \gamma\}}{\gamma} \times w_{uv}. \tag{23}$$

Likewise, an item similarity $w_{ij}$, obtained from a few ratings, can be adjusted as

$$w'_{ij} = \frac{\min\{|\mathcal{U}_{ij}|, \gamma\}}{\gamma} \times w_{ij}. \tag{24}$$

In [29, 30], it was found that using $\gamma \geq 25$ could significantly improve the accuracy of the predicted ratings, and that a value of 50 for $\gamma$ gave the best results. However, the optimal value for this parameter is data dependent and should be determined using a cross-validation approach.

A characteristic of significance weighting is its use of a threshold $\gamma$ determining when a weight should be adjusted. A more continuous approach, described in [6], is based on the concept of *shrinkage* where a weak or biased estimator can be improved if it is "shrunk" toward a null-value. This approach can be justified using a Bayesian perspective, where the best estimator of a parameter is the posterior mean, corresponding to a linear combination of the prior mean of the parameter (null-value) and an empirical estimator based fully on the data. In this case, the parameters to estimate are the similarity weights and the null value is zero. Thus, a user similarity $w_{uv}$ estimated on a few ratings is shrunk as

$$w'_{uv} = \frac{|\mathcal{I}_{uv}|}{|\mathcal{I}_{uv}| + \beta} \times w_{uv}, \tag{25}$$

where $\beta > 0$ is a parameter whose value should also be selected using cross-validation. In this approach, $w_{uv}$ is shrunk proportionally to $\beta/|\mathcal{I}_{uv}|$, such that almost no adjustment is made when $|\mathcal{I}_{uv}| \gg \beta$. Item similarities can be shrunk in the same way:

$$w'_{ij} = \frac{|\mathcal{U}_{ij}|}{|\mathcal{U}_{ij}| + \beta} \times w_{ij}, \tag{26}$$

As reported in [6], a typical value for $\beta$ is 100.

### 4.2.4 Considering the Variance of Ratings

Ratings made by two users on universally liked/disliked items may not be as informative as those made for items with a greater rating variance. For instance, most people like classic movies such as "The Godfather" so basing the weight computation on such movies would produce artificially high values. Likewise, a user that always rates items in the same way may provide less predictive information than one whose preferences vary from one item to another.

A recommendation approach that addresses this problem is the *Inverse User Frequency* [10]. Based on the information retrieval notion of *Inverse Document Frequency* (IDF), a weight $\lambda_i$ is given to each item $i$, in proportion to the log-ratio of users that have rated $i$:

$$\lambda_i = \log \frac{|\mathcal{U}|}{|\mathcal{U}_i|}.$$

In the *Frequency-Weighted Pearson Correlation* (FWPC), the correlation between the ratings given by two users $u$ and $v$ to an item $i$ is weighted by $\lambda_i$:

$$\text{FWPC}(u, v) = \frac{\sum\limits_{i \in \mathcal{I}_{uv}} \lambda_i (r_{ui} - \bar{r}_u)(r_{vi} - \bar{r}_v)}{\sqrt{\sum\limits_{i \in \mathcal{I}_{uv}} \lambda_i (r_{ui} - \bar{r}_u)^2 \sum\limits_{i \in \mathcal{I}_{uv}} \lambda_i (r_{vi} - \bar{r}_v)^2}}. \tag{27}$$

This approach, which was found to improve the prediction accuracy of a user-based recommendation method [10], could also be adapted to the computation of item similarities. More advanced strategies have also been proposed to consider rating variance. One of these strategies, described in [36], computes the factors $\lambda_i$ by maximizing the average similarity between users.

### 4.2.5 Considering the Target Item

If the goal is to predict ratings with a user-based method, more reliable correlation values can be obtained if the target item is considered in their computation. In [5], the user-based PC similarity is extended by weighting the summation terms corresponding to an item $i$ by the similarity between $i$ and the target item $j$:

$$\text{WPC}_j(u, v) = \frac{\sum\limits_{i \in \mathcal{I}_{uv}} w_{ij}\,(r_{ui} - \bar{r}_u)(r_{vi} - \bar{r}_v)}{\sqrt{\sum\limits_{i \in \mathcal{I}_{uv}} w_{ij}\,(r_{ui} - \bar{r}_u)^2 \sum\limits_{i \in \mathcal{I}_{uv}} w_{ij}\,(r_{vi} - \bar{r}_v)^2}}. \tag{28}$$

The item weights $w_{ij}$ can be computed using PC similarity or obtained by considering the items' content (e.g., the common genres for movies). Other variations of this similarity metric and their impact on the prediction accuracy are described in [5]. Note, however, that this model may require to recompute the similarity weights for each predicted rating, making it less suitable for online recommender systems.

## 4.3  Neighborhood Selection

The number of nearest-neighbors to select and the criteria used for this selection can also have a serious impact on the quality of the recommender system. The selection of the neighbors used in the recommendation of items is normally done in two steps: (1) a global filtering step where only the most likely candidates are kept, and (2) a per prediction step which chooses the best candidates for this prediction.

### 4.3.1  Pre-filtering of Neighbors

In large recommender systems that can have millions of users and items, it is usually not possible to store the (non-zero) similarities between each pair of users or items, due to memory limitations. Moreover, doing so would be extremely wasteful as only the most significant of these values are used in the predictions. The pre-filtering of neighbors is an essential step that makes neighborhood-based approaches practicable by reducing the amount of similarity weights to store, and limiting the number of candidate neighbors to consider in the predictions. There are several ways in which this can be accomplished:

- **Top-$N$ filtering:** For each user or item, only a list of the $N$ nearest-neighbors and their respective similarity weight is kept. To avoid problems with efficiency or accuracy, $N$ should be chosen carefully. Thus, if $N$ is too large, an excessive amount of memory will be required to store the neighborhood lists and predicting ratings will be slow. On the other hand, selecting a too small value for $N$ may reduce the coverage of the recommendation method, which causes some items to be never recommended.
- **Threshold filtering:** Instead of keeping a fixed number of nearest-neighbors, this approach keeps all the neighbors whose similarity weight's magnitude is greater than a given threshold $w_{\min}$. While this is more flexible than the previous filtering technique, as only the most significant neighbors are kept, the right value of $w_{\min}$ may be difficult to determine.

- **Negative filtering:** In general, negative rating correlations are less reliable than positive ones. Intuitively, this is because strong positive correlation between two users is a good indicator of their belonging to a common group (e.g., teenagers, science-fiction fans, etc.). However, although negative correlation may indicate membership to different groups, it does not tell how different are these groups, or whether these groups are compatible for some other categories of items. While certain experimental investigations [30, 31] have found negative correlations to provide no significant improvement in the prediction accuracy, in certain settings they seem to have a positive effect (see e.g., [83]). Whether such correlations can be discarded depends on the data and should be examined on a case-by-case basis.

Note that these three filtering approaches are not exclusive and can be combined to fit the needs of the recommender system. For instance, one could discard all negative similarities *as well as* those that are not in the top-$N$ lists.

### 4.3.2   Neighbors in the Predictions

Once a list of candidate neighbors has been computed for each user or item, the prediction of new ratings is normally made with the $k$-nearest-neighbors, that is, the $k$ neighbors whose similarity weight has the greatest magnitude. The choice of $k$ can also have a significant impact on the accuracy and performance of the system.

As shown in Table 3, the prediction accuracy observed for increasing values of $k$ typically follows a *concave* function. Thus, when the number of neighbors is restricted by using a small $k$ (e.g., $k < 20$), the prediction accuracy is normally low. As $k$ increases, more neighbors contribute to the prediction and the variance introduced by individual neighbors is averaged out. As a result, the prediction accuracy improves. Finally, the accuracy usually drops when too many neighbors are used in the prediction (e.g., $k > 50$), due to the fact that the few strong local relations are "diluted" by the many weak ones. Although a number of neighbors between 20 to 50 is most often described in the literature, see e.g. [29, 31], the optimal value of $k$ should be determined by cross-validation.

On a final note, more serendipitous recommendations may be obtained at the cost of a decrease in accuracy, by basing these recommendations on a few very similar users. For example, the system could find the user most similar to the active one and recommend the new item that has received the highest rated from this user.

## 5   Advanced Techniques

The neighborhood approaches based on rating correlation, such as the ones presented in the previous sections, have three important limitations:

- **Limited Expressiveness:** Traditional neighborhood-based methods determine the neighborhood of users or items using some predefined similarity measure like cosine or PC. Recommendation algorithms that rely on such similarity measures have been shown to enjoy remarkable recommendation accuracy in certain settings. However their performance can vary considerably depending on whether the chosen similarity measures conform with the latent characteristics of the dataset onto which they are applied.
- **Limited coverage:** Because rating correlation measures the similarity between two users by comparing their ratings for the same items, users can be neighbors *only if* they have rated common items. This assumption is very limiting, as users having rated a few or no common items may still have similar preferences. Moreover, since only items rated by neighbors can be recommended, the coverage of such methods can also be limited. This limitation also applies when two items have only a few or no co-ratings.
- **Sensitivity to sparse data:** Another consequence of rating correlation, addressed briefly in Sect. 3.5, is the fact that the accuracy of neighborhood-based recommendation methods suffers from the lack of available ratings. Sparsity is a problem common to most recommender systems due to the fact that users typically rate only a small proportion of the available items [7, 26, 76, 77]. This is aggravated by the fact that users or items newly added to the system may have no ratings at all, a problem known as *cold-start* [78]. When the rating data is sparse, two users or items are unlikely to have common ratings, and consequently, neighborhood-based approaches will predict ratings using a very limited number of neighbors. Moreover, similarity weights may be computed using only a small number of ratings, resulting in biased recommendations (see Sect. 4.2.3 for this problem).

A common solution for latter problems is to fill the missing ratings with default values [10, 20], such as the middle value of the rating range, or the average user or item rating. A more reliable approach is to use content information to fill out the missing ratings [18, 26, 45, 54]. For instance, the missing ratings can be provided by autonomous agents called *filterbots* [26, 45], that act as ordinary users of the system and rate items based on some specific characteristics of their content. The missing ratings can instead be predicted by a content-based approach [54]. Furthermore, content similarity can also be used "instead of" or "in addition to" rating correlation similarity to find the nearest-neighbors employed in the predictions [4, 50, 71, 82]. Finally, data sparsity can also be tackled by acquiring new ratings with active learning techniques. In such techniques, the system interactively queries the user to gain a better understanding of his or her preferences. A more detailed presentation of interactive and session-based techniques is given in chapter "Session-Based Recommender Systems" of this book. These solutions, however, also have their own drawbacks. For instance, giving a default values to missing ratings may induce bias in the recommendations. Also, item content may not be available to compute ratings or similarities.

This section presents two approaches that aim to tackle the aforementioned challenges: *learning-based* and *graph-based* methods.

## 5.1 Learning-Based Methods

In the methods of this family the similarity or affinity between users and items is obtained by defining a parametric model that describes the relation between users, items or both, and then fits the model parameters through an optimization procedure.

Using a learning-based method has significant advantages. First, such methods can capture high-level patterns and trends in the data, are generally more robust to outliers, and are known to generalize better than approaches solely based on local relations. In recommender systems, this translates into greater accuracy and stability in the recommendations [46]. Also, because the relations between users and items are encoded in a limited set of parameters, such methods normally require less memory than other types of approaches. Finally, since the parameters are usually learned offline, the online recommendation process is generally faster.

Learning-based methods that use neighborhood or similarity information can be divided in two categories: factorization methods and adaptive neighborhood learning methods. These categories are presented in the following sections.

### 5.1.1 Factorization Methods

Factorization methods [6, 7, 17, 25, 46, 60, 76, 85, 86] address the problems of limited coverage and sparsity by projecting users and items into a reduced latent space that captures their most salient features. Because users and items are compared in this dense subspace of high-level features, instead of the "rating space," more meaningful relations can be discovered. In particular, a relation between two users can be found, even though these users have rated different items. As a result, such methods are generally less sensitive to sparse data [6, 7, 76].

There are essentially two ways in which factorization can be used to improve recommender systems: (1) factorization of a sparse *similarity* matrix, and (2) factorization of a user-item *rating* matrix.

Factorizing the Similarity Matrix

Neighborhood similarity measures like the correlation similarity are usually very sparse since the average number of ratings per user is much less than the total number of items. A simple solution to densify a sparse similarity matrix is to compute a low-rank approximation of this matrix with a factorization method.

Let $W$ be a symmetric matrix of rank $n$ representing either user or item similarities. To simplify the presentation, we will suppose the latter case. We wish to approximate $W$ with a matrix $\hat{W} = QQ^{\top}$ of lower rank $k < n$, by minimizing the following objective:

$$E(Q) = ||W - QQ^{\top}||_F^2 \tag{29}$$

$$= \sum_{i,j} \left( w_{ij} - \mathbf{q}_i \mathbf{q}_j^{\top} \right)^2,$$

where $||M||_F = \sqrt{\sum_{i,j} m_{ij}^2}$ is the matrix Frobenius norm. Matrix $\hat{W}$ can be seen as a "compressed" and less sparse version of $W$. Finding the factor matrix $Q$ is equivalent to computing the eigenvalue decomposition of $W$:

$$W = VDV^{\top},$$

where $D$ is a diagonal matrix containing the $|\mathcal{I}|$ eigenvalues of $W$, and $V$ is a $|\mathcal{I}| \times |\mathcal{I}|$ orthogonal matrix containing the corresponding eigenvectors. Let $V_k$ be a matrix formed by the $k$ principal (normalized) eigenvectors of $W$, which correspond to the axes of the $k$-dimensional latent subspace. The coordinates $\mathbf{q}_i \in \Re^k$ of an item $i$ in this subspace is given by the $i$-th row of matrix $Q = V_k D_k^{1/2}$. Furthermore, the item similarities computed in this latent subspace are given by matrix

$$\hat{W} = QQ^{\top}$$

$$= V_k D_k V_k^{\top}. \tag{30}$$

This approach was used to recommend jokes in the Eigentaste system [25]. In Eigentaste, a matrix $W$ containing the PC similarities between pairs of items is decomposed to obtain the latent subspace defined by the $k$ principal eigenvectors of $W$. A user $u$, represented by the $u$-th row $\mathbf{r}_u$ of the rating matrix $R$, is projected in the plane defined by $V_k$:

$$\mathbf{r}_u' = \mathbf{r}_u V_k.$$

In an offline step, the users of the system are clustered in this subspace using a recursive subdivision technique. Then, the rating of user $u$ for an item $i$ is evaluated as the mean rating for $i$ made by users in the same cluster as $u$. This strategy is related to the well-known spectral clustering method [80].

Factorizing the Rating Matrix

The problems of cold-start and limited coverage can also be alleviated by factorizing the user-item rating matrix. Once more, we want to approximate the $|\mathcal{U}| \times |\mathcal{I}|$ rating matrix $R$ of rank $n$ by a matrix $\hat{R} = PQ^{\top}$ of rank $k < n$, where $P$ is a $|\mathcal{U}| \times k$ matrix of *users* factors and $Q$ a $|\mathcal{I}| \times k$ matrix of *item* factors. This task can be formulated as finding matrices $P$ and $Q$ which minimize the following function:

$$E(P, Q) = ||R - PQ^\top||_F^2 \tag{31}$$

$$= \sum_{u,i} \left( r_{ui} - \mathbf{p}_u \mathbf{q}_i^\top \right)^2.$$

The optimal solution can be obtained by the Singular Value Decomposition (SVD) of $R$: $P = U_k D_k^{1/2}$ and $Q = V_k D_k^{1/2}$, where $D_k$ is a diagonal matrix containing the $k$ largest singular values of $R$, and $U_k$, $V_k$ respectively contain the left and right singular vectors corresponding to these values.

However, there is significant problem with applying SVD directly to the rating matrix $R$: most values $r_{ui}$ of $R$ are undefined, since there may not be a rating given to $i$ by $u$. Although it is possible to assign a default value to $r_{ui}$, as mentioned above, this would introduce a bias in the data. More importantly, this would make the large matrix $R$ dense and, consequently, render impractical the SVD decomposition of $R$. A common solution to this problem is to learn the model parameters using only the known ratings [6, 46, 84, 86]. For instance, suppose the rating of user $u$ for item $i$ is estimated as

$$\hat{r}_{ui} = b_u + b_i + \mathbf{p}_u \mathbf{q}_i^\top, \tag{32}$$

where $b_u$ and $b_i$ are parameters representing the user and item rating biases. The model parameters can be learned by minimizing the following objective function:

$$E(P, Q, \mathbf{b}) = \sum_{r_{ui} \in \mathcal{R}} (r_{ui} - \hat{r}_{ui})^2 + \lambda \left( ||\mathbf{p}_u||^2 + ||\mathbf{q}_i||^2 + b_u^2 + b_i^2 \right). \tag{33}$$

The second term of the function is as a regularization term added to avoid overfitting. Parameter $\lambda$ controls the level of regularization. A more comprehensive description of this recommendation approach can be found in chapter "Advances in Collaborative Filtering" of this book.

The SVD model of Eq. (32) can be transformed into a similarity-based method by supposing that the profile of a user $u$ is determined implicitly by the items he or she has rated. Thus, the factor vector of $u$ can be defined as a weighted combination of the factor vectors $\mathbf{s}_j$ corresponding to the items $j$ rated by this user:

$$\mathbf{p}_u = |\mathcal{I}_u|^{-\alpha} \sum_{j \in \mathcal{I}_u} c_{uj} \mathbf{s}_j. \tag{34}$$

In this formulation, $\alpha$ is a normalization constant typically set to $\alpha = 1/2$, and $c_{uj}$ is a weight representing the contribution of item $j$ to the profile of $u$. For instance, in the SVD++ model [46] this weight is defined as the bias corrected rating of $u$ for item $j$: $c_{uj} = r_{ui} - b_u - b_j$. Other approaches, such as the FISM [39] and NSVD [69] models, instead use constant weights: $c_{uj} = 1$.

Using the formulation of Eq. (34), a rating $r_{ui}$ is predicted as

$$\hat{r}_{ui} = b_u + b_i + |\mathcal{I}_u|^{-\alpha} \sum_{j \in \mathcal{I}_u} c_{uj} \, \mathbf{s}_j \mathbf{q}_i^\top. \tag{35}$$

Like the standard SVD model, the parameters of this model can be learned by minimizing the objective function of Eq. (33), for instance, using gradient descent optimization.

Note that, instead of having both user and item factors, we now have two different sets of item factors, i.e., $\mathbf{q}_i$ and $\mathbf{s}_j$. These vectors can be interpreted as the factors of an asymmetric item-item similarity matrix $W$, where

$$w_{ij} = \mathbf{s}_i \mathbf{q}_j^\top. \tag{36}$$

As mentioned in [46], this similarity-based factorization approach has several advantages over the traditional SVD model. First, since there are typically more users than items in a recommender system, replacing the user factors by a combination of item factors reduces the number of parameters in the model, which makes the learning process faster and more robust. Also, by using item similarities instead of user factors, the system can handle new users without having to re-train the model. Finally, as in item-similarity neighborhood methods, this model makes it possible to justify a rating to a user by showing this user the items that were most involved in the prediction.

In FISM [39], the prediction of a rating $r_{ui}$ is made without considering the factors of $i$:

$$\hat{r}_{ui} = b_u + b_i + \left( |\mathcal{I}_u| - 1 \right)^{-\alpha} \sum_{j \in \mathcal{I}_u \backslash \{i\}} \mathbf{s}_j \mathbf{q}_i^\top. \tag{37}$$

This modification, which corresponds to ignoring the diagonal entries in the item similarity matrix, avoids the problem of having an item recommending itself and has been shown to give better performance when the number of factors is high.

### 5.1.2 Neighborhood-Learning Methods

Standard neighborhood-based recommendation algorithms determine the neighborhood of users or items directly from the data, using some pre-defined similarity measure like PC. However, subsequent developments in the field of item recommendation have shown the advantage of learning the neighborhood automatically from the data, instead of using a pre-defined similarity measure [43, 46, 56, 72].

Sparse Linear Neighborhood Model

A representative neighborhood-learning recommendation method is the SLIM algorithm, developed by Ning et al. [65]. In SLIM, a new rating is predicted as a sparse aggregation of existing ratings in a user's profile,

$$\hat{r}_{ui} = \mathbf{r}_u \mathbf{w}_i^\top, \tag{38}$$

where $\mathbf{r}_u$ is the $u$-th row of the rating matrix $R$ and $\mathbf{w}_j$ is a sparse row vector containing $|\mathcal{I}|$ aggregation coefficients. Essentially, the non-zero entries in $\mathbf{w}_i$ correspond to the neighbor items of an item $i$.

The neighborhood parameters are learned by minimizing the squared prediction error. Standard regularization and sparsity are enforced by penalizing the $\ell_2$-norm and $\ell_1$-norm of the parameters. The combination of these two types of regularizers in a regression problem is known as elastic net regularization [90]. This learning process can be expressed as the following optimization problem:

$$\underset{W}{\text{minimize}} \quad \frac{1}{2}\|R - RW\|_F^2 + \frac{\beta}{2}\|W\|_F^2 + \lambda\|W\|_1$$
$$\text{subject to} \quad W \geq 0 \tag{39}$$
$$\text{diag}(W) = 0.$$

The constraint $\text{diag}(W) = 0$ is added to the model to avoid trivial solutions (e.g., $W$ corresponding to the identity matrix) and ensure that $r_{ui}$ is not used to compute $\hat{r}_{ui}$ during the recommendation process. Parameters $\beta$ and $\lambda$ control the amount of each type of regularization. Moreover, the non-negativity constraint on $W$ imposes the relations between neighbor items to be positive. Dropping the non-negativity as well as the sparsity constraints has been recently explored in [83], and was shown to work well on several datasets with small number of items with respect to users. Note, however, that without the sparsity constraint the resulting model will be fully dense; a fact that imposes practical limitations on the applicability of such approaches in large item-space regimes (Fig. 4).

Sparse Neighborhood with Side Information

Side information, such as user profile attributes (e.g., age, gender, location) or item descriptions/tags, is becoming increasingly available in e-commerce applications. Properly exploited, this rich source of information can significantly improve the performance of conventional recommender systems [1, 3, 81, 88].

Item side information can be integrated in the SLIM model by supposing that the co-rating profile of two items is correlated to the properties encoded in their side information [67]. To enforce such correlations in the model, an additional requirement is added, where both the user-item rating matrix $R$ and the item side

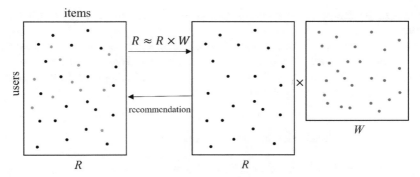

**Fig. 4** A simple illustration of SLIM. The method works by first building an item-to-item model, based on $R$. Intuitively, this item model expresses each item (i.e., each column of the original rating matrix $R$) as a *sparse linear combination* of the rest of the items (i.e., the other columns of $R$). Then, given $W$, new recommendations for a target user $u$ can be readily produced by multiplying the row corresponding to user $u$ (i.e. the $u$-th row of $R$), with the learned item model, $W$

information matrix $F$ should be reproduced by the same sparse linear aggregation. That is, in addition to satisfying $R \sim RW$, the coefficient matrix $W$ should also satisfy $F \sim FW$. This is achieved by solving the following optimization problem:

$$\underset{W}{\text{minimize}} \quad \frac{1}{2}\|R - RW\|_F^2 + \frac{\alpha}{2}\|F - FW\|_F^2 + \frac{\beta}{2}\|W\|_F^2 + \lambda\|W\|_1$$

$$\text{subject to} \quad W \geq 0, \tag{40}$$

$$\text{diag}(W) = 0.$$

The parameter $\alpha$ is used to control the relative importance of the user-item rating information $R$ and the item side information $F$ when they are used to learn $W$.

In some cases, requiring that the aggregation coefficients be the same for both $R$ and $F$ can be too strict. An alternate model relaxes this constraints by imposing these two sets of aggregation coefficients to be similar. Specifically, it uses an aggregation coefficient matrix $Q$ such that $F \sim FQ$ and $W \sim Q$. Matrices $W$ and $Q$ are learned as the minimizers of the following optimization problem:

$$\underset{W,Q}{\text{minimize}} \quad \frac{1}{2}\|R - RW\|_F^2 + \frac{\alpha}{2}\|F - FQ\|_F^2 + \frac{\beta_1}{2}\|W - Q\|_F^2$$

$$+ \frac{\beta_2}{2}\left(\|W\|_F^2 + \|Q\|_F^2\right) + \lambda\left(\|W\|_1 + \|Q\|_1\right) \tag{41}$$

$$\text{subject to} \quad W, Q \geq 0,$$

$$\text{diag}(W) = 0, \ \text{diag}(Q) = 0.$$

Parameter $\beta_1$ controls how much $W$ and $Q$ are allowed to be different from each other.

In [67], item reviews in the form of short texts were used as side information in the models described above. These models were shown to outperform the SLIM method without side information, as well as other approaches that use side information, in the top-$N$ recommendation task.

### Global and Local Sparse Neighborhood Models

A global item model may not always be sufficient to capture the preferences of every user; especially when there exist subsets of users with diverse or even opposing preferences. In cases like these training *local item models* (i.e., item models that are estimated based on user subsets) is expected to be beneficial compared to adopting a single item model for all users in the system. An example of such a case can be seen in Fig. 5.

GLSLIM [13] aims to address the above issue. In a nutshell, GLSLIM computes top-$N$ recommendations that utilize user-subset specific models (local models) and a global model. These models are jointly optimized along with computing the

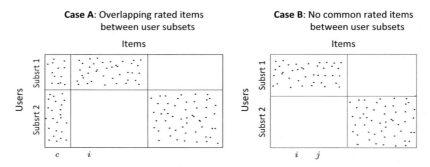

**Fig. 5** GLSLIM Motivating Example. The figure shows the training matrices $R$ of two different datasets. Both contain two user subsets. Let's assume that we are trying to compute recommendation scores for item $i$, and that the recommendations are computed using an item-item similarity-based method. Observe that in Case A there exist a set of items that have been rated solely by users that belong to Subset 1, while also a set of items which have been rated by users in both Subsets. Notice that the similarities of items $c$ and $i$ will be different when estimated based on the feedback of (**a**) Subset 1 alone; (**b**) Subset 2 alone; and, (**c**) the complete set of users. Specifically, their similarity will be zero for the users of Subset 2 (as item $i$ is not rated by the users in that subset), but it will be e.g., $l_{ic} > 0$ for the users of Subset 1, as well as e.g., $g_{ic} > 0$ when estimated globally, with $g_{ic}$ being potentially different than the locally estimated $l_{ic}$. Combining global and local item-item similarity models, in settings like this could help capture potentially diverse user preferences which would otherwise be missed if only a single global model, was computed instead. On the other hand, for datasets like the one pictured in Case B the similarity between e.g., items $i$ and $j$ will be the same, regardless of whether it is estimated globally, or locally for Subset 1, since both items have been rated only by users of Subset 1

user-specific parameters that weigh their contribution in the production of the final recommendations. The underlying model used for the estimation of local and global item similarities is SLIM.

Specifically, GLSLIM estimates a global item-item coefficient matrix $S$ and also $k$ local item-item coefficient matrices $S^{p_u}$, where $k$ is the number of user subsets and $p_u \in \{1, \ldots, k\}$ is the index of the user subset, for which a local matrix $S^{p_u}$ is estimated. The recommendation score of user $u$, who belongs to subset $p_u$, for item $i$ is estimated as:

$$\tilde{r}_{ui} = \sum_{l \in R_u} g_u s_{li} + (1 - g_u) s_{li}^{p_u}. \tag{42}$$

Term $s_{li}$ depicts the global item-item similarity between the $l$-th item rated by $u$ and the target item $i$. Term $s_{li}^{p_u}$ captures the item-item similarity between the $l$-th item rated by $u$ and target item $i$, based on the local model that corresponds to user-subset, $p_u$, to which user $u$ belongs. Finally, the term $g_u \in [0, 1]$ is the personalized weight of user $u$, which controls the involvement of the global and the local components, in the final recommendation score.

The estimation of the global and the local item models, the user assignments to subsets, and the personalized weights is achieved by alternating minimization. Initially, the users are separated into subsets, using a clustering algorithm. Weights $g_u$ are initialized to 0.5 for all users, in order to enforce equal contribution of the global and the local components. Then the coefficient matrices $S$ and $S^{p_u}$, with $p_u \in \{1, \ldots, k\}$, as well as personalized weights $g_u$ are estimated, by repeating the following two step procedure:

**Step 1:** **Estimating local and global models:** The training matrix $R$ is split into $k$ training matrices $R^{p_u}$ of size $|\mathcal{U}| \times |\mathcal{I}|$, with $p_u \in \{1, \ldots, k\}$. Every row $u$ of $R^{p_u}$ coincides with the $u$-th row of $R$, if user $u$ belongs in the $p_u$-th subset; or is left empty, otherwise. In order to estimate the $i$-th column, $\mathbf{s}_i$, of matrix $S$, and the $i$-th columns, $\mathbf{s}_i^{p_u}$, of matrices $S^{p_u}$, $p_u \in \{1, \ldots, k\}$, GLSLIM solves the following optimization problem:

$$\underset{\mathbf{s}_i, \{\mathbf{s}_i^1, \ldots, \mathbf{s}_i^k\}}{\text{minimize}} \ \frac{1}{2} \left\| \mathbf{r}_i - \mathbf{g} \odot R\mathbf{s}_i - \mathbf{g}' \odot \sum_{p_u=1}^k R^{p_u} \mathbf{s}_i^{p_u} \right\|_2^2 + \frac{1}{2}\beta_g \|\mathbf{s}_i\|_2^2 + \lambda_g \|\mathbf{s}_i\|_1 +$$
$$+ \sum_{p_u=1}^k \frac{1}{2}\beta_l \left\| \mathbf{s}_i^{p_u} \right\|_2^2 + \lambda_l \left\| \mathbf{s}_i^{p_u} \right\|_1 \tag{43}$$

subject to $\mathbf{s}_i \geq 0$, $\mathbf{s}_i^{p_u} \geq 0$, $\forall p_u \in \{1, \ldots, k\}$
$[\mathbf{s}_i]_i = 0$, $[\mathbf{s}_i^{p_u}]_i = 0$, $\forall p_u$

where $\mathbf{r}_i$ is the $i$-th column of $R$; and, $\beta_g$, $\beta_l$ are the $l_2$ regularization hyperparameters corresponding to $S$, $S^{p_u}$, $\forall p_u \in \{1, \ldots, k\}$, respectively. Finally, $\lambda_g$, $\lambda_l$ are the $l_1$ regularization hyperparameters controlling the sparsity of $S$, $S^{p_u}$ $\forall p_u \in \{1, \ldots, k\}$, respectively. The constraint $[\mathbf{s}_i]_i = 0$ makes sure that when computing $r_{ui}$, the element $r_{ui}$ is not used. Similarly, the constraints $[\mathbf{s}_i^{p_u}]_i = 0$ $\forall p_u \in \{1, \ldots, k\}$, enforce this property for the local sparse coefficient matrices as well.

**Step 2:** **Updating user subsets:** With the global and local models fixed, GLSLIM proceeds to update the user subsets. While doing that, it also determines the personalized weight $g_u$. Specifically, the computation of the personalized weight $g_u$, relies on minimizing the squared error of Eq. (42) for user $u$ who belongs to subset $p_u$, over all items $i$. Setting the derivative of the squared error to 0, yields:

$$g_u = \frac{\sum_{i=1}^{m} \left(\sum_{l \in R_u} s_{li} - \sum_{l \in R_u} s_{li}^{p_u}\right) \left(r_{ui} - \sum_{l \in R_u} s_{li}^{p_u}\right)}{\sum_{i=1}^{m} \left(\sum_{l \in R_u} s_{li} - \sum_{l \in R_u} s_{li}^{p_u}\right)^2}. \tag{44}$$

GLSLIM tries to assign each user $u$ to every possible subset, while computing the weight $g_u$ that the user would have, if assigned to that subset. Then, for every subset $p_u$ and user $u$, the training error is computed and the user is assigned to the subset for which this error is minimized (or remains to the original subset, if no difference in training error occurs).

Steps 1 and 2, are repeated until the number of users who switch subsets, in Step 2, becomes smaller than 1% of $|\mathcal{U}|$. It is empirically observed that initializing subset assignments with the CLUTO [40] clustering algorithm, results in a significant reduction of the number of iterations till convergence.

Furthermore, a comprehensive set of experiments conducted in [13] explore in detail the qualitative performance of GLSLIM, and suggest that it improves upon the standard SLIM, in several datasets.

## 5.2 Graph-Based Methods

In graph-based approaches, the data is represented in the form of a graph where nodes are users, items or both, and edges encode the interactions or similarities between the users and items. For example, in Fig. 6, the data is modeled as a bipartite graph where the two sets of nodes represent users and items, and an edge connects

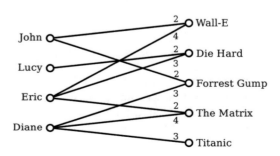

**Fig. 6** A bipartite graph representation of the ratings of Fig. 1 (*only ratings with value in* {2, 3, 4} *are shown*)

user $u$ to item $i$ if there is a rating given to $i$ by $u$ in the system. A weight can also be given to this edge, such as the value of its corresponding rating. In another model, the nodes can represent either users or items, and an edge connects two nodes if the ratings corresponding two these nodes are sufficiently correlated. The weight of this edge can be the corresponding correlation value.

In these models, standard approaches based on correlation predict the rating of a user $u$ for an item $i$ using only the nodes directly connected to $u$ or $i$. Graph-based approaches, on the other hand, allow nodes that are not directly connected to influence each other by propagating information along the edges of the graph. The greater the weight of an edge, the more information is allowed to pass through it. Also, the influence of a node on another should be less if the two nodes are further away in the graph. These two properties, known as *propagation* and *attenuation* [27, 35], are often observed in graph-based similarity measures.

The transitive associations captured by graph-based methods can be used to recommend items in two different ways. In the first approach, the proximity of a user $u$ to an item $i$ in the graph is used directly to evaluate the relevance of $i$ to $u$ [21, 27, 35]. Following this idea, the items recommended to $u$ by the system are those that are the "closest" to $u$ in the graph. On the other hand, the second approach considers the proximity of two users or item nodes in the graph as a measure of similarity, and uses this similarity as the weights $w_{uv}$ or $w_{ij}$ of a neighborhood-based recommendation method [21, 52].

### 5.2.1   Path-Based Similarity

In path-based similarity, the distance between two nodes of the graph is evaluated as a function of the number and of paths connecting the two nodes, as well as the length of these paths.

Let $R$ be once again the $|U| \times |I|$ rating matrix, where $r_{ui}$ is the rating given by user $u$ to an item $i$. The adjacency matrix $A$ of the user-item bipartite graph can be defined from $R$ as

$$A = \begin{pmatrix} 0 & R^{\top} \\ R & 0 \end{pmatrix}.$$

The association between a user $u$ and an item $i$ can be defined as the sum of the weights of all distinctive paths connecting $u$ to $v$ (allowing nodes to appear more than once in the path), whose length is no more than a given maximum length $K$. Note that, since the graph is bipartite, $K$ should be an odd number. In order to attenuate the contribution of longer paths, the weight given to a path of length $k$ is defined as $\alpha^k$, where $\alpha \in [0, 1]$. Using the fact that the number of length $k$ paths between pairs of nodes is given by $A^k$, the user-item association matrix $S_K$ is

$$S_K = \sum_{k=1}^{K} \alpha^k A^k$$

$$= (I - \alpha A)^{-1} (\alpha A - \alpha^K A^K). \tag{45}$$

This method of computing distances between nodes in a graph is known as the *Katz* measure [41]. Note that this measure is closely related to the *Von Neumann Diffusion* kernel [22, 44, 47]

$$K_{\text{VND}} = \sum_{k=0}^{\infty} \alpha^k A^k$$

$$= (I - \alpha A)^{-1} \tag{46}$$

and the *Exponential Diffusion* kernel

$$K_{\text{ED}} = \sum_{k=0}^{\infty} \frac{1}{k!} \alpha^k A^k$$

$$= \exp(\alpha A), \tag{47}$$

where $A^0 = I$.

In recommender systems that have a large number of users and items, computing these association values may require extensive computational resources. In [35], spreading activation techniques are used to overcome these limitations. Essentially, such techniques work by first activating a selected subset of nodes as starting nodes, and then iteratively activating the nodes that can be reached directly from the nodes that are already active, until a convergence criterion is met.

Path-based methods, as well as the other graph-based approaches described in this section, focus on finding relevant associations between users and items, not predicting exact ratings. Therefore, such methods are better suited for item retrieval tasks, where explicit ratings are often unavailable and the goal is to obtain a short list of relevant items (i.e., the top-$N$ recommendation problem).

### 5.2.2   Random Walk Similarity

Transitive associations in graph-based methods can also be defined within a probabilistic framework. In this framework, the similarity or affinity between users or items is evaluated as a probability of reaching these nodes in a random walk. Formally, this can be described with a first-order Markov process defined by a set of $n$ states and a $n \times n$ transition probability matrix $P$ such that the probability of jumping from state $i$ to $j$ at any time-step $t$ is

$$p_{ij} = \Pr\big(s(t+1) = j \mid s(t) = i\big).$$

Denote $\pi(t)$ the vector containing the state probability distribution of step $t$, such that $\pi_i(t) = \Pr(s(t) = i)$, the evolution of the Markov chain is characterized by

$$\pi(t+1) = P^\top \pi(t).$$

Moreover, under the condition that $P$ is row-stochastic, i.e. $\sum_j p_{ij} = 1$ for all $i$, the process converges to a stable distribution vector $\pi(\infty)$ corresponding to the positive eigenvector of $P^\top$ with an eigenvalue of 1. This process is often described in the form of a weighted graph having a node for each state, and where the probability of jumping from a node to an adjacent node is given by the weight of the edge connecting these nodes.

## Itemrank

A recommendation approach, based on the PageRank algorithm for ranking Web pages [11], is ItemRank [27]. This approach ranks the preferences of a user $u$ for unseen items $i$ as the probability of $u$ to visit $i$ in a random walk of a graph in which nodes correspond to the items of the system, and edges connects items that have been rated by common users. The edge weights are given by the $|\mathcal{I}| \times |\mathcal{I}|$ transition probability matrix $P$ for which $p_{ij} = |\mathcal{U}_{ij}|/|\mathcal{U}_i|$ is the estimated conditional probability of a user to rate and item $j$ if it has rated an item $i$.

As in PageRank, the random walk can, at any step $t$, either jump using $P$ to an adjacent node with fixed probability $\alpha$, or "teleport" to any node with probability $(1 - \alpha)$. Let $\mathbf{r}_u$ be the $u$-th row of the rating matrix $R$, the probability distribution of user $u$ to teleport to other nodes is given by vector $\mathbf{d}_u = \mathbf{r}_u/\|\mathbf{r}_u\|$. Following these definitions, the state probability distribution vector of user $u$ at step $t+1$ can be expressed recursively as

$$\pi_u(t+1) = \alpha P^\top \pi_u(t) + (1-\alpha)\mathbf{d}_u. \tag{48}$$

For practical reasons, $\pi_u(\infty)$ is usually obtained with a procedure that first initializes the distribution as uniform, i.e. $\pi_u(0) = \frac{1}{n}\mathbf{1}$, and then iteratively updates $\pi_u$, using (48), until convergence. Once $\pi_u(\infty)$ has been computed, the system recommends to $u$ the item $i$ for which $\pi_{ui}$ is the highest.

## Average First-Passage/Commute Time

Other distance measures based on random walks have been proposed for the recommendation problem. Among these are the *average first-passage time* and the *average commute time* [21, 22]. The average first-passage time $m(j|i)$ [68] is the average number of steps needed by a random walker to reach a node $j$ for the first time, when starting from a node $i \neq j$. Let $P$ be the $n \times n$ transition probability matrix, $m(j|i)$ can be obtained expressed recursively as

$$
m(j \mid i) = \begin{cases} 0 & , \text{ if } i = j \\ 1 + \sum_{k=1}^{n} p_{ik} \, m(j \mid k) , & \text{otherwise} \end{cases}
$$

A problem with the average first-passage time is that it is not symmetric. A related measure that does not have this problem is the average commute time $n(i, j) = m(j \mid i) + m(i \mid j)$ [24], corresponding to the average number of steps required by a random walker starting at node $i \neq j$ to reach node $j$ for the first time and go back to $i$. This measure has several interesting properties. Namely, it is a true distance measure in some Euclidean space [24], and is closely related to the well-known property of resistance in electrical networks and to the pseudo-inverse of the graph Laplacian matrix [21].

In [21], the average commute time is used to compute the distance between the nodes of a bipartite graph representing the interactions of users and items in a recommender system. For each user $u$ there is a directed edge from $u$ to every item $i \in \mathcal{I}_u$, and the weight of this edge is simply $1/|\mathcal{I}_u|$. Likewise, there is a directed edge from each item $i$ to every user $u \in \mathcal{U}_i$, with weight $1/|\mathcal{U}_i|$. Average commute times can be used in two different ways: (1) recommending to $u$ the item $i$ for which $n(u, i)$ is the smallest, or (2) finding the users nearest to $u$, according to the commute time distance, and then suggest to $u$ the item most liked by these users.

### 5.2.3  Combining Random Walks and Neighborhood-Learning Methods

Motivation and Challenges

Neighborhood-learning methods have been shown to achieve high top-$n$ recommendation accuracy while being scalable and easy to interpret. The fact, however, that they typically consider only direct item-to-item relations imposes limitations to their quality and makes them brittle to the presence of sparsity, leading to poor itemspace coverage and substantial decay in performance. A promising direction towards ameliorating such problems involves treating item models as graphs onto which random-walk-based techniques can then be applied. However directly applying random walks on item models can lead to a number of problems that arise from their inherent mathematical properties and the way these properties relate to the underlying top-$n$ recommendation task.

In particular, imagine of a random walker jumping from node to node on an item-to-item graph with transition probabilities proportional to the proximity scores depicted by an item model $W$. If the starting distribution of this walker reflects the items consumed by a particular user $u$ in the past, the probability the walker lands on different nodes after $K$ steps provide an intuitive measure of proximity that can be used to rank the nodes and recommend items to user $u$ accordingly.

Concretely, if we denote the transition probability matrix of the walk $S = \text{diag}(W\mathbf{1})^{-1}W$ where $\mathbf{1}$ is used to denote the vector of ones, personalized recommendations for user $u$ can be produced e.g., by leveraging the $K$-step landing distribution of a walk rooted on the items consumed by $u$;

$$\pi_u^\top = \phi_u^\top S^K, \qquad \phi_u^\top = \frac{\mathbf{r}_u^\top}{\|\mathbf{r}_u^\top\|_1} \qquad (49)$$

or by computing the limiting distribution of a random walk with restarts on $S$, using $\phi_u^\top$ as the restarting distribution. The latter approach is the well-known personalized PageRank model [11] with teleportation vector $\phi_u^\top$ and damping factor $p$, and its stationary distribution can be expressed [49] as

$$\pi_u^\top = \phi_u^\top \sum_{k=0}^{\infty}(1 - p)p^k S^k. \qquad (50)$$

Clearly, both schemes harvest the information captured in the $K$-step landing probabilities $\{\phi_u^\top S^k\}_{k=0,1,\dots}$. But, how do these landing probabilities behave as the number of steps $K$ increases? For how long will they still be significantly influenced by user's preferences $\phi_u^\top$?

Markov chain theory ensures that when $S$ is irreducible and aperiodic the landing probabilities will converge to a *unique* stationary distribution irrespectively of the initialization of the walk. This means that for large enough $K$, the $K$-step landing probabilities will no longer be "personalized," in the sense that they will become independent of the user-specific starting vector $\phi_u^\top$. Furthermore, long before reaching equilibrium, the quality of these vectors in terms of recommendation will start to plummet as more and more probability mass gets concentrated to the central nodes of the graph. Note, that the same issue arises for simple random walks that act directly on the user-item bipartite network, and has lead to methods that typically consider only very short-length random walks, and need to explicitly re-rank the $K$-step landing probabilities, in order to compensate for the inherent bias of the walk towards popular items [14]. However, longer random-walks might be necessary to capture non-trivial multi-hop relations between the items, as well as to ensure better coverage of the itemspace.

The RecWalk Recommendation Framework

RecWalk [61, 62] addresses the aforementioned challenges, and resolves this long- vs short-length walk dilemma through the construction of a *nearly uncoupled random walk* [58, 63] that gives full control over the stochastic dynamics of the walk towards equilibrium; provably, and irrespectively of the dataset or the specific item model onto which it is applied. Intuitively, this allows for prolonged and effective

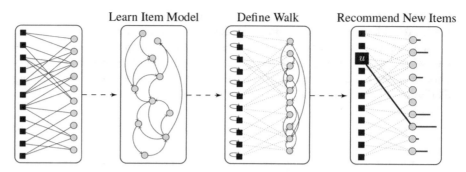

**Fig. 7** RecWalk Illustration. Maroon colored nodes correspond to users; Gold colored nodes correspond to items

exploration of the underlying network while keeping the influence of the user-specific initialization strong.[4]

From a random-walk point of view, the RecWalk model can be described as follows: Consider a random walker jumping from node to node on the user-item bipartite network. Suppose the walker currently occupies a node $c \in \mathcal{U} \cup \mathcal{I}$. In order to determine the next step transition the walker tosses a biased coin that yields heads with probability $\alpha$ and tails with probability $(1 - \alpha)$:

1. If the coin-toss yields *heads*, then:

   a. if $c \in \mathcal{U}$, the walker jumps to one of the items rated by the current user (i.e., the user corresponding to the current node $c$) uniformly at random;
   b. if $c \in \mathcal{I}$, the walker jumps to one of the users that have rated the current item uniformly at random;

2. If the coin-toss yields *tails*, then:

   a. if $c \in \mathcal{U}$, the walker stays put;
   b. if $c \in \mathcal{I}$, the walker jumps to a related item abiding by an *item-to-item transition probability matrix* $M_{\mathcal{I}}$, that is defined in terms of an underlying item model.

The stochastic process that describes this random walk is defined to be a homogeneous discrete time Markov chain with state space $\mathcal{U} \cup \mathcal{I}$; i.e., the transition probabilities from any given node $c$ to the other nodes, are fixed and independent of the nodes visited by the random walker before reaching $c$. An illustration of the RecWalk model is given in Fig. 7.

The transition probability matrix $P$ that governs the behavior of the random walker can be usefully expressed as a weighted sum of two stochastic matrices $H$

---

[4] The mathematical details behind the particular construction choices of RecWalk that enforce such desired mixing properties can be found in [61, 62].

and $M$ as

$$P = \alpha H + (1 - \alpha)M \tag{51}$$

where $0 < \alpha < 1$, is a parameter that controls the involvement of these two components in the final model. Matrix $H$ can be thought of as the transition probability matrix of a simple random walk on the user-item bipartite network. Assuming that the rating matrix $R$ has no zero columns and rows, matrix $H$ can be expressed as

$$H = \mathrm{Diag}(A1)^{-1}A, \qquad \text{where } A = \begin{pmatrix} & R \\ R^\top & \end{pmatrix}. \tag{52}$$

Matrix $M$, is defined as

$$M = \begin{pmatrix} I & \\ & M_{\mathcal{I}} \end{pmatrix} \tag{53}$$

where $I \in \Re^{U \times U}$ the identity matrix and $M_{\mathcal{I}} \in \Re^{I \times I}$ is a transition probability matrix designed to capture relations between the items. In particular, given an item model with non-negative weights $W$ (e.g., the aggregation matrix produced by a SLIM model), matrix $M_{\mathcal{I}}$ is defined using the following stochasticity adjustment strategy:

$$M_{\mathcal{I}} = \frac{1}{\|W\|_\infty}W + \mathrm{Diag}\left(1 - \frac{1}{\|W\|_\infty}W1\right). \tag{54}$$

The first term divides all the elements by the maximum row-sum of $W$ and the second enforces stochasticity by adding residuals to the diagonal, appropriately. The motivation behind this definition is to retain the information captured by the relative differences of the item-to-item relations in $W$. This prevents items that are loosely related to the rest of the itemspace to disproportionately influence the inter-item transitions and introduce noise to the model.[5]

In RecWalk the recommendations are produced by exploiting the information captured in the successive landing probability distributions of a walk initialized in a user-specific way. Two simple recommendation strategies that were considered in [61] are:

RecWalk$^{\text{K-step}}$:   The recommendation score of user $u$ for item $i$ is defined to be the probability the random walker lands on node $i$ after $K$ steps, given that

---

[5] From a purely mathematical point-of-view the above strategy promotes desired spectral properties to $M_{\mathcal{I}}$ that are shown to be intertwined with recommendation performance. For additional details see [62].

the starting node was $u$. In other words, the recommendation score for item $i$ is given by the corresponding elements of

$$\pi_u^\top = \mathbf{e}_u^\top P^K \tag{55}$$

where $\mathbf{e}_u \in \mathfrak{R}^{U+I}$ is a vector that contains the element 1 on the position that corresponds to user $u$ and zeros elsewhere. The computation of the recommendations is performed by $K$ sparse-matrix-vector products with matrix $P$, and it entails $\Theta(K \, \mathrm{nnz}(P))$operations, where $\mathrm{nnz}(P)$ is the number of nonzero elements in $P$.

RecWalk$^{\mathrm{PR}}$:   The recommendation score of user $u$ for item $i$ is defined to be the element that corresponds to item $i$ in the limiting distribution of a random walk with restarts on $P$, with restarting probability $\eta$ and restarting distribution $\mathbf{e}_u$:

$$\pi_u^\top = \lim_{K \to \infty} \mathbf{e}_u^\top \left( \eta P + (1 - \eta) \mathbf{1} \mathbf{e}_u^\top \right)^K. \tag{56}$$

The limiting distribution in (56) can be computed efficiently using e.g., the power method, or any specialized PageRank solver. Note that this variant of RecWalk also comes with theoretical guarantees for item-space coverage for every user in the system, regardless of the base item model $W$ used in the definition of matrix $M_{\mathcal{I}}$ [62].

In [62] it was shown that both approaches manage to boost the quality of several base item models on top of which they were built. Using fsSLIM [66] with small number of neighbors as a base item model, in particular, was shown to achieve state-of-the-art recommendation performance, in several datasets. At the same time RecWalk was found to dramatically increase itemspace coverage of the produced recommendations, in every considered setting. This was true both for RecWalk$^{\mathrm{K-step}}$, as well as for RecWalk$^{\mathrm{PR}}$.

### 5.2.4   User-Adaptive Diffusion Models

**Motivation** Personalization of the recommendation vectors in the graph-based schemes we have seen thus far, comes from the use of a user-specific initialization, or a user-specific restarting distribution. However, the underlying mechanism for propagating user preferences, across the itemspace (i.e., the adopted diffusion function, or the choice of the $K$-step distribution) is *fixed* for every user in the system. From a user modeling point of view this translates to the implicit assumption that every user explores the itemspace in *exactly the same* way—overlooking the reality that different users can have different behavioral patterns. The fundamental premise of PerDif [57] is that the latent item exploration behavior of the users can be captured better by *user-specific* preference propagation mechanisms; thus, leading to improved recommendations.

`PerDif` proposes a simple model of personalized item exploration subject to an underlying item model. At each step the users might either decide to go forth and discover items related to the ones they are currently considering, or return to their base and possibly go down alternative paths. Different users, might explore the itemspace in different ways; and their behavior might change throughout the exploration session. The following stochastic process, formalizes the above idea:

**The PerDIF Item Discovery Process** Consider a random walker carrying a bag of $K$ biased coins. The coins are labeled with consecutive integers from 1 to $K$. Initially, the random walker occupies the nodes of graph according to distribution $\phi$. She then flips the 1st coin: if it turns heads (with probability $\mu_1$), she jumps to a different node in the graph abiding by the probability matrix $P$; if it turns tails (with probability $1 - \mu_1$), she jumps to a node according to the probability distribution $\phi$. She then flips the 2nd coin and she either follows $P$ with probability $\mu_2$ or 'restarts' to $\phi$ with probability $(1 - \mu_2)$. The walk continues until she has used all her $K$ coins. At the $k$-th step the transitions of the random walker are completely determined by the probability the $k$-th coin turning heads ($\mu_k$), the transition matrix $P$, and the restarting distribution $\phi$. Thus, the stochastic process that governs the position of the random walker over time is a time-inhomogeneous Markov chain with state space the nodes of the graph, and transition matrix at time $k$ given by

$$G(\mu_k) = \mu_k P + (1 - \mu_k)\mathbf{1}\phi^\top. \tag{57}$$

The node occupation distribution of the random walker after the last transition can therefore be expressed as

$$\pi^\top = \phi^\top G(\mu_1) G(\mu_2) \cdots G(\mu_K). \tag{58}$$

Given an item transition probability matrix $P$, and a user-specific restarting distribution $\phi_u$, the goal is to find a set of probabilities $\mu_u = (\mu_1, \ldots, \mu_K)$ so that the outcome of the aforementioned item exploration process yields a meaningful distribution over the items that can be used for recommendation. `PerDif` tackles this task as follows:

**Learning the Personalized Probabilities** For each user $u$ we randomly sample one item she has interacted with (henceforth referred to as the 'target' item) alongside $\tau_{neg}$ unseen items, and we fit $\mu_u$ so that the node occupancy distribution after a $K$-step item exploration process rooted on $\phi_u$ (cf (58)) yields high probability to the target item while keeping the probabilities of the negative items low. Concretely, upon defining a vector $\mathbf{h}_u \in \Re^{\tau_{neg}+1}$ which contains the value 1 for the target item and zeros for the negative items, we learn $\mu_u$ by solving

$$\begin{aligned} \underset{\mu_u \in \Re^K}{\text{minimize}} \quad & \left\| \phi_u^\top G(\mu_1) \cdots G(\mu_K) E_u - \mathbf{h}_u^\top \right\|_2^2 \\ \text{subject to} \quad & \mu_i \in (0, 1), \quad \forall i \in [1, \ldots, K] \end{aligned} \tag{59}$$

where $\mu_i = [\mu_u]_i, \forall i$, and $E_u$ is a $(I \times (\tau_{neg} + 1))$ matrix designed to select and rearrange the elements of the vector $\boldsymbol{\phi}_u^\top G(\mu_1) \cdots G(\mu_K)$ according to the sequence of items comprising $\mathbf{h}_u$. Upon obtaining $\boldsymbol{\mu}_u$, personalized recommendations for user $u$ can be computed as

$$\boldsymbol{\pi}_u^\top = \boldsymbol{\phi}_u^\top G(\mu_1) \cdots G(\mu_K). \tag{60}$$

Leveraging the special properties of the stochastic matrix $G$ the above non-linear optimization problem can be solved efficiently. In particular, it can be shown [57] that the optimization problem (59) is equivalent to

$$\underset{\omega_u \in \Delta_{++}^{K+1}}{\text{minimize}} \left\| \boldsymbol{\omega}_u^\top S_u E_u - \mathbf{h}_u^\top \right\|_2^2$$

where $\Delta_{++}^{K+1} = \{x : x^\top \mathbf{1} = 1, x > 0\}$ and

$$S_u = \begin{pmatrix} \boldsymbol{\phi}_u^\top \\ \boldsymbol{\phi}_u^\top P \\ \boldsymbol{\phi}_u^\top P^2 \\ \vdots \\ \boldsymbol{\phi}_u^\top P^K \end{pmatrix}, \qquad \boldsymbol{\omega}_u \equiv \boldsymbol{\omega}_u(\boldsymbol{\mu}_u) = \begin{pmatrix} 1 - \mu_K \\ \mu_K (1 - \mu_{K-1}) \\ \mu_K \mu_{K-1} (1 - \mu_{K-2}) \\ \vdots \\ \mu_K \cdots \mu_2 (1 - \mu_1) \\ \mu_K \cdots \mu_2 \mu_1 \end{pmatrix}.$$

The above result simplifies learning $\boldsymbol{\mu}_u$ significantly. It also lends `PerDif` its name. In particular, the task of finding personalized probabilities for the item exploration process, reduces to that of finding *personalized diffusion coefficients* $\boldsymbol{\omega}_u$ over the space of the first $K$ landing probabilities of a walk rooted on $\boldsymbol{\phi}_u$ (see definition of $S_u$). Afterwards $\boldsymbol{\mu}_u$ can be obtained in linear time from $\boldsymbol{\omega}_u$ upon solving a simple forward recurrence [57]. Taking into account the fact that in recommendation settings $K$ will typically be small and $\boldsymbol{\phi}_u$, $P$ sparse, building 'on-the-fly' $S_u E_u$ row-by-row, and solving the $(K + 1)$-dimensional convex quadratic problem

$$\text{PERDIF}^{\text{FREE}} \ : \ \underset{\omega_u \in \Delta_{++}^{K+1}}{\text{minimize}} \left\| \boldsymbol{\omega}_u^\top S_u E_u - \mathbf{h}_u^\top \right\|_2^2 \tag{61}$$

can be performed very efficiently (typically in a matter of milliseconds even in large scale settings).

Moreover, working on the space of landing probabilities can also facilitate parametrising the diffusion coefficients within a family of known diffusions. This motivates the parameterized variant of `PerDif`

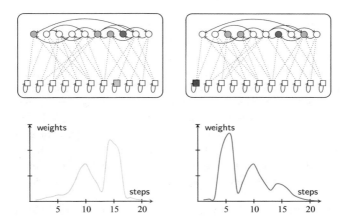

**Fig. 8** Personalized diffusions on the user-item bipartite network

$$\text{PERDIF}^{\text{PAR}} : \quad \underset{\gamma_u \in \Delta_+^L}{\text{minimize}} \, \|\gamma_u^\top D S_u E_u - \mathbf{h}_u^\top\|_2^2 \tag{62}$$

with $\Delta_+^L = \{y : y^\top \mathbf{1} = 1, y \geq 0\}$ and $D \in \mathfrak{R}^{L \times (K+1)}$ defined such that its rows contain preselected diffusion coefficients (e.g., PageRank [11] coefficients for several damping factors, heat kernel [15] coefficients for several *temperature* values etc.), normalized to sum to one. Upon obtaining $\gamma_u$, vector $\omega_u$ can be computed as $\omega_u^\top = \gamma_u^\top D$ (Fig. 8).

While PERDIF$^{\text{FREE}}$ learns $\omega_u$ by weighing the contributions of the landing probabilities directly, PERDIF$^{\text{PAR}}$ constrains $\omega_u$ to comprise a user-specific mixture of predetermined such weights (i.e., the rows of $D$), thus allowing one to endow $\omega_u$ with desired properties, relevant to the specific recommendation task at hand. Furthermore, the use of matrix $D$ can improve the robustness of the personalized diffusions in settings where the recommendation quality of the individual landing distributions comprising $S_u$ is uneven across the $K$ steps considered.

Besides, its merits in terms of recommendation accuracy, personalizing the diffusions within the `PerDif` framework can also provide useful information arising from the analysis of the learned diffusion coefficients, $\omega_u$. In particular, the dual interpretation of the model parameters ($\mu_u$ in the item exploration space; and, $\omega_u$ in the diffusion space) allows utilizing the learned model parameters to identify users for which the model will most likely lead to poor predictions, at training-time—thereby affording preemptive interventions to handle such cases appropriately. This affords a level of transparency that can prove particularly useful in practical settings (for the technical details on how this can be achieved see [57]).

# 6   Conclusion

One of the earliest approaches proposed for the task item recommendation, neighborhood-based recommendation still ranks among the most popular methods for this problem. Although quite simple to describe and implement, this recommendation approach has several important advantages, including its ability to explain a recommendation with the list of the neighbors used, its computational and space efficiency which allows it to scale to large recommender systems, and its marked stability in an online setting where new users and items are constantly added. Another of its strengths is its potential to make serendipitous recommendations that can lead users to the discovery of unexpected, yet very interesting items.

In the implementation of a neighborhood-based approach, one has to make several important decisions. Perhaps the one having the greatest impact on the accuracy and efficiency of the recommender system is choosing between a user-based and an item-based neighborhood method. In typical commercial recommender systems, where the number of users far exceeds the number of available items, item-based approaches are typically preferred since they provide more accurate recommendations, while being more computationally efficient and requiring less frequent updates. On the other hand, user-based methods usually provide more original recommendations, which may lead users to a more satisfying experience. Moreover, the different components of a neighborhood-based method, which include the normalization of ratings, the computation of the similarity weights and the selection of the nearest-neighbors, can also have a significant influence on the quality of the recommender system. For each of these components, several different alternatives are available. Although the merit of each of these has been described in this document and in the literature, it is important to remember that the "best" approach may differ from one recommendation setting to the next. Thus, it is important to evaluate them on data collected from the actual system, and in light of the particular needs of the application.

Modern machine-learning-based techniques can be used to further increase the performance of neighborhood-based approaches, by automatically extracting the most representative neighborhoods based on the available data. Such models achieve state-of-the-art recommendation accuracy, however their adoption imposes additional computational burden that needs to be considered in light of the particular characteristics of the recommendation problem at hand. Finally, when the performance of a neighborhood-based approach suffers from the problems of limited coverage and sparsity, one may explore techniques based on dimensionality reduction or graphs. Dimensionality reduction provides a compact representation of users and items that captures their most significant features. An advantage of such approach is that it allows to obtain meaningful relations between pairs of users or items, even though these users have rated different items, or these items were rated by different users. On the other hand, graph-based techniques exploit the transitive relations in the data. These techniques also avoid the problems of sparsity and limited coverage by evaluating the relationship between users or items that are not "directly connected". However, unlike dimensionality reduction, graph-based methods also preserve some of the "local" relations in the data, which are useful in making serendipitous recommendations.

# References

1. R.P. Adams, G.E. Dahl, I. Murray, Incorporating side information into probabilistic matrix factorization using Gaussian processes, in *Proceedings of the 26th Conference on Uncertainty in Artificial Intelligence*, ed. by P. Grünwald, P. Spirtes (2010), pp. 1–9
2. G. Adomavicius, A. Tuzhilin, Toward the next generation of recommender systems: a survey of the state-of-the-art and possible extensions. IEEE Trans. Knowl. Data Eng. **17**(6), 734–749 (2005)
3. D. Agarwal, B.C. Chen, B. Long, Localized factor models for multi-context recommendation, in *Proceedings of the 17th ACM SIGKDD international conference on Knowledge discovery and data mining, KDD '11* (ACM, New York, NY, 2011), pp. 609–617. http://doi.acm.org/10.1145/2020408.2020504
4. M. Balabanović, Y. Shoham, Fab: content-based, collaborative recommendation. Commun. ACM **40**(3), 66–72 (1997)
5. L. Baltrunas, F. Ricci, Item weighting techniques for collaborative filtering, in *Knowledge Discovery Enhanced with Semantic and Social Information* (Springer, New York, 2009), pp. 109–126
6. R. Bell, Y. Koren, C. Volinsky, Modeling relationships at multiple scales to improve accuracy of large recommender systems, in *KDD '07: Proceedings of the 13th ACM SIGKDD International Conference on Knowledge Discovery and Data Mining* (ACM, New York, NY, 2007), pp. 95–104
7. D. Billsus, M.J. Pazzani, Learning collaborative information filters, in *ICML '98: Proceedings of the 15th International Conference on Machine Learning* (Morgan Kaufmann Publishers Inc., San Francisco, CA, 1998), pp. 46–54
8. D. Billsus, M.J. Pazzani, User modeling for adaptive news access. User Model. User-Adapted Interact. **10**(2–3), 147–180 (2000)
9. D.M. Blei, A.Y. Ng, M.I. Jordan, Latent dirichlet allocation. J. Mach. Learn. Res. **3**, 993–1022 (2003)
10. J.S. Breese, D. Heckerman, C. Kadie, Empirical analysis of predictive algorithms for collaborative filtering, in *Proceedings of the 14th Annual Conference on Uncertainty in Artificial Intelligence* (Morgan Kaufmann, 1998), pp. 43–52
11. S. Brin, L. Page, The anatomy of a large-scale hypertextual Web search engine. Comput. Netw. ISDN Syst. **30**(1–7), 107–117 (1998)
12. R. Cañamares, P. Castells, A probabilistic reformulation of memory-based collaborative filtering: Implications on popularity biases, in *Proceedings of the 40th International ACM SIGIR Conference on Research and Development in Information*. Retrieval (2017), pp. 215–224
13. E. Christakopoulou, G. Karypis, Local item-item models for top-n recommendation, in *Proceedings of the 10th ACM Conference on Recommender Systems, RecSys '16* (Association for Computing Machinery, New York, NY, 2016), p. 6774. https://doi.org/10.1145/2959100.2959185
14. F. Christoffel, B. Paudel, C. Newell, A. Bernstein, Blockbusters and wallflowers: accurate, diverse, and scalable recommendations with random walks, in *Proceedings of the 9th ACM Conference on Recommender Systems, RecSys '15* (Association for Computing Machinery, New York, NY, 2015), p. 163170. https://doi.org/10.1145/2792838.2800180

15. F. Chung, The heat kernel as the pagerank of a graph. Proc. Natl. Acad. Sci. **104**(50), 19735–19740 (2007)
16. W.W. Cohen, R.E. Schapire, Y. Singer, Learning to order things, in *NIPS '97: Proceedings of the 1997 Conference on Advances in Neural Information Processing Systems* (MIT Press, Cambridge, MA, 1998), pp. 451–457
17. P. Cremonesi, Y. Koren, R. Turrin, Performance of recommender algorithms on top-n recommendation tasks, in *Proceedings of the Fourth ACM Conference on Recommender Systems* (2010), pp. 39–46
18. M. Degemmis, P. Lops, G. Semeraro, A content-collaborative recommender that exploits wordnet-based user profiles for neighborhood formation. User Model. User-Adapt. Interact. **17**(3), 217–255 (2007)
19. J. Delgado, N. Ishii, Memory-based weighted majority prediction for recommender systems, in *Proceedings of the ACM SIGIR'99 Workshop on Recommender Systems* (1999)
20. M. Deshpande, G. Karypis, Item-based top-N recommendation algorithms. ACM Trans. Inf. Syst. **22**(1), 143–177 (2004)
21. F. Fouss, J.M. Renders, A. Pirotte, M. Saerens, Random-walk computation of similarities between nodes of a graph with application to collaborative recommendation. IEEE Trans. Knowl. Data Eng. **19**(3), 355–369 (2007)
22. F. Fouss, L. Yen, A. Pirotte, M. Saerens, An experimental investigation of graph kernels on a collaborative recommendation task, in *ICDM '06: Proceedings of the 6th International Conference on Data Mining* (IEEE Computer Society, Washington, DC, 2006), pp. 863–868
23. Y. Freund, R.D. Iyer, R.E. Schapire, Y. Singer, An efficient boosting algorithm for combining preferences, in *ICML '98: Proceedings of the 15th International Conference on Machine Learning* (Morgan Kaufmann Publishers Inc., San Francisco, CA, 1998), pp. 170–178
24. F. Gobel, A. Jagers, Random walks on graphs. Stoch. Process. Appl. **2**, 311–336 (1974)
25. K. Goldberg, T. Roeder, D. Gupta, C. Perkins, Eigentaste: a constant time collaborative filtering algorithm. Inf. Retr. **4**(2), 133–151 (2001)
26. N. Good, J.B. Schafer, J.A. Konstan, A. Borchers, B. Sarwar, J. Herlocker, J. Riedl, Combining collaborative filtering with personal agents for better recommendations, in *AAAI '99/IAAI '99: Proceedings of the 16th National Conference on Artificial Intelligence* (American Association for Artificial Intelligence, Menlo Park, CA, 1999), pp. 439–446
27. M. Gori, A. Pucci, Itemrank: a random-walk based scoring algorithm for recommender engines, in *Proceedings of the 2007 IJCAI Conference* (2007), pp. 2766–2771
28. M. Grcar, B. Fortuna, D. Mladenic, M. Grobelnik, k-NN versus SVM in the collaborative filtering framework. Data Sci. Classif. 251–260 (2006). http://db.cs.ualberta.ca/webkdd05/proc/paper25-mladenic.pdf
29. J. Herlocker, J.A. Konstan, J. Riedl, An empirical analysis of design choices in neighborhood-based collaborative filtering algorithms. Inf. Retr. **5**(4), 287–310 (2002)
30. J.L. Herlocker, J.A. Konstan, A. Borchers, J. Riedl, An algorithmic framework for performing collaborative filtering, in *SIGIR '99: Proceedings of the 22nd Annual International ACM SIGIR Conference on Research and Development in Information Retrieval* (ACM, New York, NY, 1999), pp. 230–237
31. J.L. Herlocker, J.A. Konstan, L.G. Terveen, J.T. Riedl, Evaluating collaborative filtering recommender systems. ACM Trans. Inf. Syst. **22**(1), 5–53 (2004)
32. W. Hill, L. Stead, M. Rosenstein, G. Furnas, Recommending and evaluating choices in a virtual community of use, in *CHI '95: Proceedings of the SIGCHI Conference on Human Factors in Computing Systems* (ACM Press/Addison-Wesley Publishing Co., New York, NY, 1995), pp. 194–201
33. T. Hofmann, Collaborative filtering via Gaussian probabilistic latent semantic analysis, in *SIGIR '03: Proceedings of the 26th Annual International ACM SIGIR Conference on Research and Development in Information Retrieval* (ACM, New York, NY, 2003), pp. 259–266
34. A.E. Howe, R.D. Forbes, Re-considering neighborhood-based collaborative filtering parameters in the context of new data, in *CIKM '08: Proceeding of the 17th ACM Conference on Information and Knowledge Management* (ACM, New York, NY, 2008), pp. 1481–1482

35. Z. Huang, H. Chen, D. Zeng, Applying associative retrieval techniques to alleviate the sparsity problem in collaborative filtering. ACM Trans. Inf. Syst. **22**(1), 116–142 (2004)

36. R. Jin, J.Y. Chai, L. Si, An automatic weighting scheme for collaborative filtering, in *SIGIR '04: Proceedings of the 27th Annual International ACM SIGIR Conference on Research and Development in Information Retrieval* (ACM, New York, NY, 2004), pp. 337–344

37. R. Jin, L. Si, C. Zhai, Preference-based graphic models for collaborative filtering, in *Proceedings of the 19th Annual Conference on Uncertainty in Artificial Intelligence (UAI-03)* (Morgan Kaufmann, San Francisco, CA, 2003), pp. 329–33

38. R. Jin, L. Si, C. Zhai, J. Callan, Collaborative filtering with decoupled models for preferences and ratings, in *CIKM '03: Proceedings of the 12th International Conference on Information and Knowledge Management* (ACM, New York, NY, 2003), pp. 309–316

39. S. Kabbur, X. Ning, G. Karypis, Fism: factored item similarity models for top-n recommender systems, in *Proceedings of the 19th ACM SIGKDD International Conference on Knowledge Discovery and Data Mining, KDD '13* (ACM, New York, NY, 2013), pp. 659–667. http://doi.acm.org/10.1145/2487575.2487589

40. G. Karypis, Cluto-a clustering toolkit. Tech. rep., Minnesota Univ Minneapolis, Dept of Computer Science (2002)

41. L. Katz, A new status index derived from sociometric analysis. Psychometrika **18**(1), 39–43 (1953)

42. M. Kendall, J.D. Gibbons, *Rank Correlation Methods*, 5th edn. (Charles Griffin, London, 1990)

43. N. Koenigstein, Y. Koren, Towards scalable and accurate item-oriented recommendations, in *Proceedings of the 7th ACM Conference on Recommender Systems, RecSys '13* (ACM, New York, NY, 2013), pp. 419–422. http://doi.acm.org/10.1145/2507157.2507208

44. R.I. Kondor, J.D. Lafferty, Diffusion kernels on graphs and other discrete input spaces, in *ICML '02: Proceedings of the Nineteenth International Conference on Machine Learning* (Morgan Kaufmann Publishers Inc., San Francisco, CA, 2002), pp. 315–322

45. J.A. Konstan, B.N. Miller, D. Maltz, J.L. Herlocker, L.R. Gordon, J. Riedl, GroupLens: applying collaborative filtering to usenet news. Commun. ACM **40**(3), 77–87 (1997)

46. Y. Koren, Factorization meets the neighborhood: a multifaceted collaborative filtering model, in *KDD'08: Proceeding of the 14th ACM SIGKDD International Conference on Knowledge Discovery and Data Mining* (ACM, New York, NY, 2008), pp. 426–434

47. J. Kunegis, A. Lommatzsch, C. Bauckhage, Alternative similarity functions for graph kernels, in *Proceedings of the International Conference on Pattern Recognition* (2008)

48. K. Lang, News Weeder: learning to filter netnews, in *Proceedings of the 12th International Conference on Machine Learning* (Morgan Kaufmann publishers Inc., San Mateo, CA, 1995), pp. 331–339

49. A.N. Langville, C.D. Meyer, *Google's PageRank and Beyond: The Science of Search Engine Rankings* (Princeton University Press, Princeton, 2011)

50. J. Li, O.R. Zaiane, Combining usage, content, and structure data to improve Web site recommendation, in *Proceedings of the 5th International Conference on Electronic Commerce and Web Technologies (EC-Web)* (2004)

51. G. Linden, B. Smith, J. York, Amazon.com recommendations: item-to-item collaborative filtering. IEEE Intern. Comput. **7**(1), 76–80 (2003)

52. H. Luo, C. Niu, R. Shen, C. Ullrich, A collaborative filtering framework based on both local user similarity and global user similarity. Mach. Learn. **72**(3), 231–245 (2008)

53. H. Ma, I. King, M.R. Lyu, Effective missing data prediction for collaborative filtering, in *SIGIR '07: Proceedings of the 30th Annual International ACM SIGIR Conference on Research and Development in Information Retrieval* (ACM, New York, NY, 2007), pp. 39–46

54. P. Melville, R.J. Mooney, R. Nagarajan, Content-boosted collaborative filtering for improved recommendations, in *18th National Conference on Artificial intelligence* (American Association for Artificial Intelligence, Menlo Park, CA, 2002), pp. 187–192

55. A. Nakamura, N. Abe, Collaborative filtering using weighted majority prediction algorithms, in *ICML '98: Proceedings of the 15th International Conference on Machine Learning* (Morgan Kaufmann Publishers Inc., San Francisco, CA, 1998), pp. 395–403

56. N. Natarajan, D. Shin, I.S. Dhillon, Which app will you use next?: Collaborative filtering with interactional context, in *Proceedings of the 7th ACM Conference on Recommender Systems, RecSys '13* (ACM, New York, NY, 2013), pp. 201–208. http://doi.acm.org/10.1145/2507157. 2507186

57. A.N. Nikolakopoulos, D. Berberidis, G. Karypis, G.B. Giannakis, Personalized diffusions for top-n recommendation, in *Proceedings of the 13th ACM Conference on Recommender Systems, RecSys '19* (Association for Computing Machinery, New York, NY, 2019), p. 260268 https://doi.org/10.1145/3298689.3346985

58. A.N. Nikolakopoulos, J.D. Garofalakis, Ncdawarerank: a novel ranking method that exploits the decomposable structure of the web, in *Proceedings of the Sixth ACM International Conference on Web Search and Data Mining, WSDM '13* (Association for Computing Machinery, New York, NY, 2013), p. 143152. https://doi.org/10.1145/2433396.2433415

59. A.N. Nikolakopoulos, J.D. Garofalakis, Top-n recommendations in the presence of sparsity: An ncd-based approach, in *Web Intelligence*, vol. 13 (IOS Press, Amsterdam, 2015), pp. 247–265

60. A.N. Nikolakopoulos, V. Kalantzis, E. Gallopoulos, J.D. Garofalakis, Eigenrec: generalizing puresvd for effective and efficient top-n recommendations. Knowl. Inf. Syst. **58**(1), 59–81 (2019)

61. A.N. Nikolakopoulos, G. Karypis, Recwalk: nearly uncoupled random walks for top-n recommendation, in *Proceedings of the Twelfth ACM International Conference on Web Search and Data Mining, WSDM '19* (Association for Computing Machinery, New York, NY, 2019), p. 150158. https://doi.org/10.1145/3289600.3291016

62. A.N. Nikolakopoulos, G. Karypis, Boosting item-based collaborative filtering via nearly uncoupled random walks. ACM Trans. Knowl. Discov. Data **14**(6) (2020). https://doi.org/10.1145/3406241

63. A.N. Nikolakopoulos, A. Korba, J.D. Garofalakis, Random surfing on multipartite graphs, in *2016 IEEE International Conference on Big Data (Big Data)* (2016), pp. 736–745

64. A.N. Nikolakopoulos, M.A. Kouneli, J.D. Garofalakis, Hierarchical itemspace rank: exploiting hierarchy to alleviate sparsity in ranking-based recommendation. Neurocomputing **163**, 126–136 (2015)

65. X. Ning, G. Karypis, Slim: sparse linear methods for top-n recommender systems, in *Proceedings of 11th IEEE International Conference on Data Mining* (2011), pp. 497–506

66. X. Ning, G. Karypis, Slim: sparse linear methods for top-n recommender systems, in *2011 IEEE 11th International Conference on Data Mining (ICDM)*. (IEEE, New York, 2011), pp. 497–506

67. X. Ning, G. Karypis, Sparse linear methods with side information for top-n recommendations, in *Proceedings of the Sixth ACM Conference on Recommender Systems, RecSys '12* (ACM, New York, NY, 2012), pp. 155–162. http://doi.acm.org/10.1145/2365952.2365983

68. J.R. Norris, *Markov Chains*, 1st edn. (Cambridge University Press, Cambridge, 1999)

69. A. Paterek, Improving regularized singular value decomposition for collaborative filtering, in *Proceedings of the KDD Cup and Workshop* (2007)

70. M. Pazzani, D. Billsus, Learning and revising user profiles: The identification of interesting Web sites. Mach. Learn. **27**(3), 313–331 (1997)

71. M.J. Pazzani, A framework for collaborative, content-based and demographic filtering. Artif. Intell. Rev. **13**(5–6), 393–408 (1999)

72. S. Rendle, C. Freudenthaler, Z. Gantner, S.T. Lars, BPR: Bayesian personalized ranking from implicit feedback, in *Proceedings of the Twenty-Fifth Conference on Uncertainty in Artificial Intelligence, UAI '09* (AUAI Press, Arlington, VA, 2009), pp. 452–461

73. P. Resnick, N. Iacovou, M. Suchak, P. Bergstrom, J. Riedl, GroupLens: an open architecture for collaborative filtering of netnews, in *CSCW '94: Proceedings of the 1994 ACM Conf. on Computer Supported Cooperative Work* (ACM, New York, NY, 1994), pp. 175–186

74. R. Salakhutdinov, A. Mnih, G. Hinton, Restricted Boltzmann machines for collaborative filtering, in *ICML '07: Proceedings of the 24th International Conference on Machine Learning* (ACM, New York, NY, 2007), pp. 791–798

75. B. Sarwar, G. Karypis, J. Konstan, J. Reidl, Item-based collaborative filtering recommendation algorithms, in *WWW '01: Proceedings of the 10th International Conference on World Wide Web* (ACM, New York, NY, 2001), pp. 285–295

76. B.M. Sarwar, G. Karypis, J.A. Konstan, J.T. Riedl, Application of dimensionality reduction in recommender systems A case study, in *ACM WebKDD Workshop* (2000)

77. B.M. Sarwar, J.A. Konstan, A. Borchers, J. Herlocker, B. Miller, J. Riedl, Using filtering agents to improve prediction quality in the grouplens research collaborative filtering system, in *CSCW '98: Proceedings of the 1998 ACM Conference on Computer Supported Cooperative Work* (ACM, New York, NY, 1998), pp. 345–354

78. A.I. Schein, A. Popescul, L.H. Ungar, D.M. Pennock, Methods and metrics for cold-start recommendations, in *SIGIR '02: Proceedings of the 25th Annual International ACM SIGIR Conference on Research and Development in Information Retrieval* (ACM, New York, NY, 2002), pp. 253–260

79. U. Shardanand, P. Maes, Social information filtering: algorithms for automating "word of mouth". in *CHI '95: Proceedings of the SIGCHI Conference on Human factors in Computing Systems* (ACM Press/Addison-Wesley Publishing Co., New York, NY, 1995), pp. 210–217

80. J. Shi, J. Malik, Normalized cuts and image segmentation. IEEE Trans. Pattern Anal. Mach. Intell. **22**(8), 888–905 (2000)

81. A.P. Singh, G.J. Gordon, Relational learning via collective matrix factorization, in *Proceeding of the 14th ACM International Conference on Knowledge Discovery and Data Mining* (2008), pp. 650–658. http://doi.acm.org/10.1145/1401890.1401969

82. I.M. Soboroff, C.K. Nicholas, Combining content and collaboration in text filtering, in *Proceedings of the IJCAI'99 Workshop on Machine Learning for Information Filtering* (1999), pp. 86–91

83. H. Steck, Embarrassingly shallow autoencoders for sparse data, in *The World Wide Web Conference* (2019), pp. 3251–3257

84. G. Takács, I. Pilászy, B. Németh, D. Tikk, Major components of the gravity recommendation system. SIGKDD Exploration Newslett. **9**(2), 80–83 (2007)

85. G. Takács, I. Pilászy, B. Németh, D. Tikk, Investigation of various matrix factorization methods for large recommender systems, in *Proceedings of the 2nd KDD Workshop on Large Scale Recommender Systems and the Netflix Prize Competition* (2008)

86. G. Takács, I. Pilászy, B. Németh, D. Tikk, Scalable collaborative filtering approaches for large recommender systems. J. Mach. Learn. Res. (Spec. Top. Mining Learn. Graphs Relat.) **10**, 623–656 (2009)

87. L. Terveen, W. Hill, B. Amento, D. McDonald, J. Creter, PHOAKS: a system for sharing recommendations. Commun. ACM **40**(3), 59–62 (1997)

88. J. Yoo, S. Choi, Weighted nonnegative matrix co-tri-factorization for collaborative prediction, in *Advances in Machine Learning*, ed. by Z.H. Zhou, T. Washio. Lecture Notes in Computer Science, vol. 5828 (Springer Berlin/Heidelberg, 2009), pp. 396–411

89. C.L. Zitnick, T. Kanade, Maximum entropy for collaborative filtering, in *AUAI '04: Proceedings of the 20th Conference on Uncertainty in Artificial Intelligence* (AUAI Press, Arlington, VA, 2004), pp. 636–643

90. H. Zou, T. Hastie, Regularization and variable selection via the elastic net. J. R. Stat. Soc. Ser. B **67**(2), 301–320 (2005)

# Advances in Collaborative Filtering

**Yehuda Koren, Steffen Rendle, and Robert Bell**

## 1 Introduction

Collaborative filtering (CF) methods produce user specific recommendations of items based on patterns of ratings or usage (e.g., purchases) without need for exogenous information about either items or users. While well-established methods work adequately for many purposes, we present several extensions available to practitioners who are for the best possible recommendations.

Recommender systems rely on various types of input. Most convenient is high quality *explicit feedback*, where users directly report on their interest in products. For example, users can provide star ratings for products or can indicate their preferences for TV shows by hitting thumbs-up/down buttons. Because explicit

---

Robert Bell is retired.

This article includes copyrighted materials, which were reproduced with permission of ACM and IEEE. The original articles are:

R. Bell and Y. Koren [3], © 2007 IEEE. Reprinted by permission.

Y. Koren [26], © 2008 ACM, Inc. Reprinted by permission. http://doi.acm.org/10.1145/1401890.1401944

Y. Koren [27], © 2009 ACM, Inc. Reprinted by permission. http://doi.acm.org/10.1145/1557019.1557072

---

Y. Koren (✉)
Google, Israel
e-mail: yehudako@gmail.com

S. Rendle
Google Research, Mountain View, CA, USA
e-mail: srendle@google.com

R. Bell
Berkeley Heights, NJ, USA

© Springer Science+Business Media, LLC, part of Springer Nature 2022
F. Ricci et al. (eds.), *Recommender Systems Handbook*,
https://doi.org/10.1007/978-1-0716-2197-4_3

feedback is not always available, many recommenders infer user preferences from the more abundant *implicit feedback*, which indirectly reflect opinion through observing user behavior [38]. Types of implicit feedback include purchase history, streaming activities, and even lower-fidelity indications like browsing history, search patterns, or long views. For example, a user who purchased many books by the same author probably enjoys reading that author.

In order to establish recommendations, CF systems need to relate two fundamentally different entities: items and users. There are two primary approaches to facilitate such a comparison, which constitute the two main techniques of CF: *the neighborhood approach* and *latent factor models*. Neighborhood methods focus on relationships between items or, alternatively, between users. An item-item approach models the preference of a user to an item based on ratings of similar items by the same user. Latent factor models, such as matrix factorization (aka, SVD), comprise an alternative approach by transforming both items and users to the same latent factor, or *embedding*, space. The latent space tries to explain ratings by characterizing both item and users on factors automatically inferred from user feedback (also known as item- and user-embeddings).

Producing more accurate prediction methods requires deepening their foundations and reducing reliance on arbitrary decisions. In this chapter, we describe a variety of improvements to the primary CF modeling techniques. Yet, the quest for more accurate models goes beyond this. At least as important is the identification of all the signals, or features, available in the data. Conventional techniques address the sparse data of user-item ratings. Accuracy significantly improves by also utilizing other sources of information, such as meta-data associated with the item. Another prime example includes all kinds of temporal effects reflecting the dynamic, time-drifting nature of user-item interactions. No less important is accounting for hidden feedback such as which items users chose to rate (regardless of rating values). Rated items are not selected at random, but rather reveal interesting aspects of user preferences, going beyond the numerical values of the ratings.

Throughout this work, we describe the proposed techniques for the rating prediction task and more specifically for the Netflix Prize competition. This competition began in October 2006 and fueled the progress in the field of collaborative filtering. For the first time, the research community gained access to a large-scale, industrial strength data set of 100 million movie ratings—attracting thousands of scientists, students, engineers and enthusiasts to the field. The nature of the competition has encouraged rapid development, where innovators built on each generation of techniques to improve prediction accuracy. Because all methods were judged by the same rigid yardstick on common data, the evolution of more powerful models has been especially efficient.

First, Sect. 2 introduces the problem and notation. Then, Sect. 3 describes simple baseline predictors which are extended with time modeling. Section 4 surveys matrix factorization techniques, which combine implementation convenience with a relatively high accuracy. This section describes the theory and practical details behind those techniques. In addition, much of the strength of matrix factorization models stems from their natural ability to handle additional features of the data, including implicit feedback and temporal information. This section describes in detail how to enhance matrix factorization models to address such features.

Section 5 turns attention to neighborhood methods. The basic methods in this family are well known, and to a large extent are based on heuristics. We present a more advanced method, which uses the insights of common neighborhood methods, with global optimization techniques typical to factorization models. This method allows lifting the limit on neighborhood size, and also addressing implicit feedback and temporal dynamics. The resulting accuracy is close to that of matrix factorization models, while offering some practical advantages.

Pushing the foundations of the models to their limits reveals surprising links among seemingly unrelated techniques. We elaborate on this in Sect. 6 to show that, at their limits, user-user and item-item neighborhood models may converge to a single model. Furthermore, at that point, both become equivalent to a simple matrix factorization model. The connections reduce the relevance of some previous distinctions such as the traditional broad categorization of matrix factorization as "model based" and neighborhood models as "memory based".

In Sect. 7 we present *factorization machines*, and describe how simple feature engineering can derive equivalent models to the CF approaches we have discussed so far. The expressive power of factorization machines can also be used to solve other recommender tasks such as cold start recommendation by adding meta data to the input features or context aware recommendation by adding contextual data.

Finally, in Sect. 8 we discuss how the patterns identified in this chapter have been applied to other recommendation tasks, such as item recommendation, where the best items for a user should be retrieved, or settings where additional data, such as context or meta data, needs to be taken into account. We also discuss more recent evaluations, showing that more than a decade after their invention, the proposed methods of this chapter still yield state-of-the art results.

# 2 Preliminaries

We start by defining the rating prediction task and refer to Sect. 8.2 for a discussion about item recommendation. We are given ratings for $m$ users (aka customers) and $n$ items (aka products). We reserve special indexing letters to distinguish users from items: for users $u, v$, and for items $i, j, l$. A rating $r_{u,i}$ indicates the preference by user $u$ of item $i$, where high values mean stronger preference. For example, values can be integers ranging from 1 (star) indicating no interest to 5 (stars) indicating a strong interest. We distinguish predicted ratings from known ones, by using the notation $\hat{r}_{u,i}$ for the predicted value of $r_{u,i}$. The goal of rating prediction is to learn a function $\hat{r}_{u,i}$ that can predict ratings for any user-item pair $(u, i)$. The model parameters, $\theta$, of this function are learned from $\mathcal{K} = \{(u, i) \mid r_{u,i} \text{ is known}\}$, the set of $(u, i)$ pairs for which $r_{u,i}$ is known. Usually the vast majority of ratings are unknown. For example, in the Netflix data 99% of the possible ratings are missing because a user typically rates only a small portion of the movies. The model parameters, $\theta$, are learned by fitting the predicted rating to the previously observed ratings, for example using a least squares objective

$$\underset{\theta}{\text{argmin}} \sum_{(u,i)\in\mathcal{K}} (r_{u,i} - \hat{r}_{u,i})^2. \tag{1}$$

However, our goal is to generalize those in a way that allows us to predict future, unknown ratings. Thus, caution should be exercised to avoid overfitting the observed data. We achieve this by regularizing the learned parameters, whose magnitudes are penalized:

$$\underset{\theta}{\text{argmin}} \sum_{(u,i)\in\mathcal{K}} (r_{u,i} - \hat{r}_{u,i})^2 + \lambda||\theta||^2. \tag{2}$$

Regularization of the model parameters $\theta$ is controlled by a constant $\lambda$. As the value grows, regularization becomes heavier. The exact value of $\lambda$ is determined by cross validation. Often it is beneficial to use different regularization values $\lambda_1, \lambda_2, \ldots$ for different parts of the model parameters, for example parameters that model user behavior might have a different regularization strength than parameters that model items.

In addition to the rating, we investigate settings where each rating $r_{u,i}$ is annotated with the time $t_{u,i}$ of the rating event. One can use different time units, based on what is appropriate for the application at hand. For example, when time is measured in days, then $t_{u,i}$ counts the number of days elapsed since some early time point. For each of the recommender algorithms in this chapter, first we will introduce them independent of the time of the rating event and then present extensions to improve the quality by taking temporal effects into account.

Furthermore, each user $u$ is associated with a set of items denoted by $R(u) := \{i \mid (u, i) \in \mathcal{K}\}$, which contains all the items for which ratings by $u$ are available. Likewise, $R(i)$ denotes the set of users who rated item $i$. Sometimes, we also use a set denoted by $N(u)$, which contains all items for which $u$ provided an implicit preference (items that $u$ rented/purchased/watched, etc.).

## 2.1 The Netflix Data

In order to compare the relative accuracy of algorithms described in this chapter, we evaluated all of them on the Netflix data of more than 100 million date-stamped movie ratings performed by anonymous Netflix customers between Nov 11, 1999 and Dec 31, 2005 [6]. Ratings are integers ranging between 1 and 5. The data spans 17,770 movies rated by over 480,000 users. Thus, on average, a movie receives 5600 ratings, while a user rates 208 movies, with substantial variation around each of these averages. To maintain compatibility with results published by others, we adopt some standards that were set by Netflix. First, quality of the results is usually measured by the root mean squared error (RMSE):

$$\sqrt{\sum_{(u,i) \in TestSet} (r_{u,i} - \hat{r}_{u,i})^2 / |TestSet|}$$

a measure that puts more emphasis on large errors compared with the alternative of mean absolute error. Root mean squared error (RMSE), (Consider Chap. "Evaluating Recommender Systems" for a comprehensive survey of alternative evaluation metrics of recommender systems.)

We report results on a test set provided by Netflix (also known as the Quiz set), which contains over 1.4 million recent ratings. Compared with the training data, the test set contains many more ratings by users that do not rate much and are therefore harder to predict. In a way, this represents real requirements for a CF system, which needs to predict new ratings from older ones, and to equally address all users, not just the heavy raters.

The Netflix data is part of the Netflix Prize competition, where the benchmark is Netflix's proprietary system, Cinematch, which achieved a RMSE of 0.9514 on the test set. The grand prize was awarded to a team that managed to drive this RMSE below 0.8563 (10% improvement) after almost three years of extensive efforts. Achievable RMSE values on the test set lie in a quite compressed range, as evident by the difficulty to win the grand prize. Nonetheless, there is evidence that small improvements in RMSE terms can have a significant impact on the quality of the top few presented recommendations [26, 28].

## 2.2 Implicit Feedback as Auxiliary Data

This chapter is centered on explicit user feedback. Nonetheless, when additional sources of implicit feedback are available, they can be exploited for better understanding user behavior. This helps to combat data sparseness and can be particularly helpful for users with few explicit ratings. We describe extensions for some of the models to address implicit feedback.

For a dataset such as the Netflix data, the most natural choice for implicit feedback would probably be movie rental history, which tells us about user preferences without requiring them to explicitly provide their ratings. For other datasets, browsing or purchase history could be used as implicit feedback. However, such data is not available to us for experimentation. Nonetheless, a less obvious kind of implicit data does exist within the Netflix dataset. The dataset does not only tell us the rating values, but also *which* movies users rate, regardless of *how* they rated these movies. In other words, a user implicitly tells us about her preferences by choosing to voice her opinion and vote a (high or low) rating. This creates a binary matrix, where "1" stands for "rated", and "0" for "not rated". While this binary data may not be as informative as other independent sources of implicit feedback, incorporating this kind of implicit data does significantly improves prediction accuracy. The benefit of using the binary data is closely related to the fact that ratings are not missing at random; users deliberately choose which items to rate (see Marlin et al. [36]).

It should be noted, that here we discussed implicit feedback as auxiliary information to learn a rating prediction model. This has to be distinguished from the use of implicit information in item recommendation models. Item recommendation is the task of retrieving the best $k$ items from the set of all items for a given user. Item recommendation models are typically trained from implicit feedback. In this case, the implicit feedback is the main source and used as the label. Such item recommendation models could use the rating data, if present, as auxiliary data. We will focus on rating prediction in the next sections and will return to item recommendation in Sect. 8.2.

# 3    Baseline Predictors

CF models try to capture the interactions between users and items that produce the different rating values. However, much of the observed rating values are due to effects associated with either users or items, independently of their interaction. A principal example is that typical CF data exhibit large user and item biases—i.e., systematic tendencies for some users to give higher ratings than others, and for some items to receive higher ratings than others.

We will encapsulate those effects, which do not involve user-item interaction, within the *baseline predictors* (also known as *biases*). Because these predictors tend to capture much of the observed signal, it is vital to model them accurately. Such modeling enables isolating the part of the signal that truly represents user-item interaction, and subjecting it to more appropriate user preference models.

Denote by $\mu$ the overall average rating. A baseline prediction for an unknown rating $r_{u,i}$ is denoted by $b_{u,i}$ and accounts for the user and item effects:

$$b_{u,i} = \mu + b_u + b_i \tag{3}$$

The parameters $b_u$ and $b_i$ indicate the observed deviations of user $u$ and item $i$, respectively, from the average. For example, suppose that we want a baseline predictor for the rating of the movie Titanic by user Joe. Now, say that the average rating over all movies, $\mu$, is 3.7 stars. Furthermore, Titanic is better than an average movie, so it tends to be rated 0.5 stars above the average. On the other hand, Joe is a critical user, who tends to rate 0.3 stars lower than the average. Thus, the baseline predictor for Titanic's rating by Joe would be 3.9 stars by calculating $3.7 - 0.3 + 0.5$. In order to estimate $b_u$ and $b_i$ one can solve the least squares problem

$$\operatorname*{argmin}_{b_*} \sum_{(u,i)\in\mathcal{K}} (r_{u,i} - \mu - b_u - b_i)^2 + \lambda_1 \left( \sum_u b_u^2 + \sum_i b_i^2 \right).$$

Here, the first term $\sum_{(u,i)\in\mathcal{K}}(r_{u,i} - \mu + b_u + b_i)^2$ strives to find $b_u$'s and $b_i$'s that fit the given ratings. The regularizing term—$\lambda_1(\sum_u b_u^2 + \sum_i b_i^2)$—avoids overfitting by penalizing the magnitudes of the parameters. This least square problem can be solved fairly efficiently by the method of stochastic gradient descent (described in Sect. 4.1).

For the Netflix data the mean rating ($\mu$) is 3.6. As for the learned user biases ($b_u$), their average is 0.044 with standard deviation of 0.41. The average of their absolute values ($|b_u|$) is: 0.32. The learned item biases ($b_i$) average to $-0.26$ with a standard deviation of 0.48. The average of their absolute values ($|b_i|$) is 0.43.

An easier, yet somewhat less accurate way to estimate the parameters is by decoupling the calculation of the $b_i$'s from the calculation of the $b_u$'s. First, for each item $i$ we set

$$b_i = \frac{\sum_{u\in R(i)}(r_{u,i} - \mu)}{\lambda_2 + |R(i)|}.$$

Then, for each user $u$ we set

$$b_u = \frac{\sum_{i\in R(u)}(r_{u,i} - \mu - b_i)}{\lambda_3 + |R(u)|}.$$

Averages are shrunk towards zero by using the regularization parameters, $\lambda_2, \lambda_3$, which are determined by cross validation. Typical values on the Netflix dataset are: $\lambda_2 = 25$, $\lambda_3 = 10$.

## 3.1 Time Changing Baseline Predictors

So far, the baseline predictors ignored any temporal variation. Now, we discuss biases that each vary over time, specifically: (1) user biases $b_u(t)$, and (2) item biases $b_i(t)$. Much of the temporal variability is included within the baseline predictors, through two major temporal effects. The first addresses the fact that an item's popularity may change over time. For example, movies can go in and out of popularity as triggered by external events such as the appearance of an actor in a new movie. This is manifested in our models by treating the item bias $b_i$ as a function of time. The second major temporal effect allows users to change their baseline ratings over time. For example, a user who tended to rate an average movie "4 stars", may now rate such a movie "3 stars". This may reflect several factors including a natural drift in a user's rating scale, the fact that ratings are given in relationship to other ratings that were given recently and also the fact that the identity of the rater within a household can change over time. Hence, it is often sensible to take the parameter $b_u$ as a function of time. This induces a template for a time sensitive baseline predictor for $u$'s rating of $i$ at day $t_{u,i}$:

$$b_{u,i}(t_{u,i}) = \mu + b_u(t_{u,i}) + b_i(t_{u,i}) \tag{4}$$

Here, $b_u(\cdot)$ and $b_i(\cdot)$ are real valued functions that change over time. The exact way to build these functions should reflect a reasonable way to parameterize the involving temporal changes. Our choice in the context of the movie rating dataset demonstrates some typical considerations.

A major distinction is between temporal effects that span extended periods of time and more transient effects. In the movie rating case, we do not expect movie likeability to fluctuate on a daily basis, but rather to change over more extended periods. On the other hand, we observe that user effects can change on a daily basis, reflecting inconsistencies natural to customer behavior, especially when rating items in batches. This requires finer time resolution when modeling user-biases compared with a lower resolution that suffices for capturing item-related time effects.

We start with our choice of time-changing item biases $b_i(t)$. We found it adequate to split the item biases into time-based bins, using a constant item bias for each time period. The decision of how to split the timeline into bins should balance the desire to achieve finer resolution (hence, smaller bins) with the need for enough ratings per bin (hence, larger bins). For the movie rating data, there is a wide variety of bin sizes that yield about the same accuracy. In our implementation, each bin corresponds to roughly ten consecutive weeks of data, leading to 30 bins spanning all days in the dataset. A day $t$ is associated with an integer $\text{Bin}(t)$ (a number between 1 and 30 in our data), such that the movie bias is split into a stationary part and a time changing part

$$b_i(t) = b_i + b_{i,\text{Bin}(t)}. \tag{5}$$

While binning the parameters works well on the items, it is more of a challenge on the users side. On the one hand, we would like a finer resolution for users to detect very short lived temporal effects. On the other hand, we do not expect enough ratings per user to produce reliable estimates for isolated bins. Different functional forms can be considered for parameterizing temporal user behavior, with varying complexity and accuracy.

One simple modeling choice uses a linear function to capture a possible gradual drift of user bias. For each user $u$, we denote the mean date of rating by $t_u$. Now, if $u$ rated a movie on day $t$, then the associated time deviation of this rating is defined as

$$\text{dev}_u(t) = \text{sign}(t - t_u) \cdot |t - t_u|^\beta.$$

Here $|t - t_u|$ measures the number of days between dates $t$ and $t_u$. We set the value of $\beta$ by cross validation; in our implementation $\beta = 0.4$. We introduce a single new parameter for each user called $\alpha_u$ so that we get our first definition of a time-dependent user-bias

$$b_u^{(1)}(t) = b_u + \alpha_u \cdot \mathrm{dev}_u(t) . \tag{6}$$

This simple linear model for approximating a drifting behavior requires learning two parameters per user: $b_u$ and $\alpha_u$.

A more flexible parameterization is offered by splines. Let $u$ be a user associated with $n_u$ ratings. We designate $k_u$ time points—$\{t_1^u, \ldots, t_{k_u}^u\}$—spaced uniformly across the dates of $u$'s ratings as kernels that control the following function:

$$b_u^{(2)}(t) = b_u + \frac{\sum_{l=1}^{k_u} e^{-\sigma|t-t_l^u|} b_{t_l}^u}{\sum_{l=1}^{k_u} e^{-\sigma|t-t_l^u|}} \tag{7}$$

The parameters $b_{t_l}^u$ are associated with the control points (or, kernels), and are automatically learned from the data. This way the user bias is formed as a time-weighted combination of those parameters. The number of control points, $k_u$, balances flexibility and computational efficiency. In our application we set $k_u = \lfloor n_u^{0.25} \rfloor$, letting it grow with the number of available ratings. The constant $\sigma$ determines the smoothness of the spline; we set $\sigma = 0.3$ by cross validation.

So far we have discussed smooth functions for modeling the user bias, which mesh well with *gradual concept drift*. However, in many applications there are *sudden drifts* emerging as "spikes" associated with a single day or session. For example, in the movie rating dataset we have found that multiple ratings a user gives in a single day, tend to concentrate around a single value. Such an effect need not span more than a single day. The effect may reflect the mood of the user that day, the impact of ratings given in a single day on each other, or changes in the actual rater identity in multi-person accounts. To address such short lived effects, we assign a single parameter per user and day, absorbing the day-specific variability. This parameter is denoted by $b_{u,t}$. Notice that in some applications the basic primitive time unit to work with can be shorter or longer than a day.

In the Netflix movie rating data, a user rates on 40 different days on average. Thus, working with $b_{u,t}$ requires, on average, 40 parameters to describe each user bias. It is expected that $b_{u,t}$ is inadequate as a standalone for capturing the user bias, since it misses all sorts of signals that span more than a single day. Thus, it serves as an additive component within the previously described schemes. The time-linear model (6) becomes

$$b_u^{(3)}(t) = b_u + \alpha_u \cdot \mathrm{dev}_u(t) + b_{u,t} . \tag{8}$$

Similarly, the spline-based model becomes

$$b_u^{(4)}(t) = b_u + \frac{\sum_{l=1}^{k_u} e^{-\sigma|t-t_l^u|} b_{t_l}^u}{\sum_{l=1}^{k_u} e^{-\sigma|t-t_l^u|}} + b_{u,t} . \tag{9}$$

A baseline predictor on its own cannot yield personalized recommendations, as it disregards all interactions between users and items. In a sense, it is capturing the portion of the data that is less relevant for personalizing recommendations. Nonetheless, to better assess the relative merits of the various choices of time-dependent user-bias, we compare their accuracy as standalone predictors. In order to learn the involved parameters we minimize the associated regularized squared error by using stochastic gradient descent. For example, in our actual implementation we adopt rule (8) for modeling the drifting user bias, thus arriving at the baseline predictor

$$b_{u,i} = \mu + b_u + \alpha_u \cdot \mathrm{dev}_u(t_{u,i}) + b_{u,t_{u,i}} + b_i + b_{i,\mathrm{Bin}(t_{u,i})} . \tag{10}$$

To learn the involved parameters, $b_u$, $\alpha_u$, $b_{u,t}$, $b_i$ and $b_{i,\mathrm{Bin}(t)}$, one should solve

$$\min \sum_{(u,i)\in\mathcal{K}} (r_{u,i} - \mu - b_u - \alpha_u\mathrm{dev}_u(t_{u,i}) - b_{u,t_{u,i}} - b_i - b_{i,\mathrm{Bin}(t_{u,i})})^2$$

$$+ \lambda_4(b_u^2 + \alpha_u^2 + b_{u,t_{u,i}}^2 + b_i^2 + b_{i,\mathrm{Bin}(t_{u,i})}^2) .$$

Here, the first term strives to construct parameters that fit the given ratings. The regularization term, $\lambda_4(b_u^2 + \dots)$, avoids overfitting by penalizing the magnitudes of the parameters, assuming a neutral 0 prior. Learning is done by a stochastic gradient descent algorithm running 20–30 iterations, with $\lambda_4 = 0.01$.

Table 1 compares the ability of various suggested baseline predictors to explain signal in the data. As usual, the amount of captured signal is measured by the root mean squared error on the test set. As a reminder, test cases come later in time than the training cases for the same user, so predictions often involve extrapolation in terms of time. We code the predictors as follows:

- *static*, no temporal effects: $b_{u,i} = \mu + b_u + b_i$.
- *mov*, accounting only for movie-related temporal effects: $b_{u,i} = \mu + b_u + b_i + b_{i,\mathrm{Bin}(t_{u,i})}$.
- *linear* , linear modeling of user biases: $b_{u,i} = \mu + b_u + \alpha_u \cdot \mathrm{dev}_u(t_{u,i}) + b_i + b_{i,\mathrm{Bin}(t_{u,i})}$.
- *spline*, spline modeling of user biases: $b_{u,i} = \mu + b_u + \dfrac{\sum_{l=1}^{k_u} e^{-\sigma|t_{u,i}-t_l^u|} b_{t_l}^u}{\sum_{l=1}^{k_u} e^{-\sigma|t_{u,i}-t_l^u|}} + b_i + b_{i,\mathrm{Bin}(t_{u,i})}$.
- *linear+*, linear modeling of user biases and single day effect: $b_{u,i} = \mu + b_u + \alpha_u \cdot \mathrm{dev}_u(t_{u,i}) + b_{u,t_{u,i}} + b_i + b_{i,\mathrm{Bin}(t_{u,i})}$.

**Table 1** Comparing baseline predictors capturing main movie and user effects. As temporal modeling becomes more accurate, prediction accuracy improves (lowering RMSE)

| Model | Static | Mov | Linear | Spline | Linear+ | Spline+ |
|-------|--------|-----|--------|--------|---------|---------|
| RMSE | 0.9799 | 0.9771 | 0.9731 | 0.9714 | 0.9605 | 0.9603 |

- *spline+*, spline modeling of user biases and single day effect: $b_{u,i} = \mu + b_u +$
$\dfrac{\sum_{l=1}^{k_u} e^{-\sigma|t_{u,i}-d_l|} b_{t_l}^u}{\sum_{l=1}^{k_u} e^{-\sigma|t_{u,i}-t_l^u|}} + b_{u,t_{u,i}} + b_i + b_{i,\mathrm{Bin}(t_{u,i})}.$

The table shows that while temporal movie effects reside in the data (lowering RMSE from 0.9799 to 0.9771), the drift in user biases is much more influential. The additional flexibility of splines at modeling user effects leads to better accuracy compared to a linear model. However, sudden changes in user biases, which are captured by the per-day parameters, are most significant. Indeed, when including those changes, the difference between linear modeling ("linear+") and spline modeling ("spline+") virtually vanishes.

Beyond the temporal effects described so far, one can use the same methodology to capture more effects. A primary example is capturing periodic effects. For example, some products may be more popular in specific seasons or near certain holidays. Similarly, different types of television or radio shows are popular throughout different segments of the day (known as "dayparting"). Periodic effects can be found also on the user side. As an example, a user may have different attitudes or buying patterns during the weekend compared to the working week. A way to model such periodic effects is to dedicate a parameter for the combinations of time periods with items or users. This way, the item bias of (5), becomes

$$b_i(t) = b_i + b_{i,\mathrm{Bin}(t)} + b_{i,\mathrm{period}(t)} .$$

For example, if we try to capture the change of item bias with the season of the year, then $\mathrm{period}(t) \in \{\text{fall, winter, spring, summer}\}$. Similarly, recurring user effects may be modeled by modifying (8) to be

$$b_u(t) = b_u + \alpha_u \cdot \mathrm{dev}_u(t) + b_{u,t} + b_{u,\mathrm{period}(t)} .$$

However, we have not found periodic effects with a significant predictive power within the movie-rating dataset, thus our reported results do not include those.

Another temporal effect within the scope of basic predictors is related to the changing scale of user ratings. While $b_i(t)$ is a user-independent measure for the merit of item $i$ at time $t$, users tend to respond to such a measure differently. For example, different users employ different rating scales, and a single user can change his rating scale over time. Accordingly, the raw value of the movie bias is not completely user-independent. To address this, we add a time-dependent scaling feature to the baseline predictors, denoted by $c_u(t)$. Thus, the baseline predictor (10) becomes

$$b_{u,i} = \mu + b_u + \alpha_u \cdot \mathrm{dev}_u(t_{u,i}) + b_{u,t_{u,i}} + (b_i + b_{i,\mathrm{Bin}(t_{u,i})}) \cdot c_u(t_{u,i}) . \qquad (11)$$

All discussed ways to implement $b_u(t)$ would be valid for implementing $c_u(t)$ as well. We chose to dedicate a separate parameter per day, resulting in: $c_u(t) = c_u + c_{u,t}$. As usual, $c_u$ is the stable part of $c_u(t)$, whereas $c_{u,t}$ represents day-specific variability. Adding the multiplicative factor $c_u(t)$ to the baseline predictor lowers RMSE to 0.9555. Interestingly, this basic model, which captures just main effects disregarding user-item interactions, can explain almost as much of the data variability as the commercial Netflix Cinematch recommender system, whose published RMSE on the same test set is 0.9514 [6].

### 3.1.1 Predicting Future Days

Our models include day-specific parameters. An apparent question would be how these models can be used for predicting ratings in the future, on new dates for which we cannot train the day-specific parameters? The simple answer is that for those future (untrained) dates, the day-specific parameters should take their default value. In particular for (11), $c_u(t_{u,i})$ is set to $c_u$, and $b_{u,t_{u,i}}$ is set to zero. Yet, one wonders, if we cannot use the day-specific parameters for predicting the future, why are they good at all? After all, prediction is interesting only when it is about the future. To further sharpen the question, we should mention the fact that the Netflix test sets include many ratings on dates for which we have no other rating by the same user and hence day-specific parameters cannot be exploited.

To answer this, notice that our temporal modeling makes no attempt to capture future changes. All it is trying to do is to capture transient temporal effects, which had a significant influence on past user feedback. When such effects are identified they must be tuned down, so that we can model the more enduring signal. This allows our model to better capture the long-term characteristics of the data, while letting dedicated parameters absorb short term fluctuations. For example, if a user gave many higher than usual ratings on a particular single day, our models discount those by accounting for a possible day-specific good mood, which does not reflects the longer term behavior of this user. This way, the day-specific parameters accomplish a kind of data cleaning, which improves prediction of future dates.

## 4    Matrix Factorization Models

Latent factor models approach collaborative filtering with the holistic goal to uncover latent features that explain observed ratings; examples include pLSA [23], neural networks [46], Latent Dirichlet Allocation [7], and models that are induced by factorization of the user-item ratings matrix. Matrix factorization models have gained popularity, thanks to their attractive accuracy and scalability.

In information retrieval, matrix factorization by singular value decomposition is well established for identifying latent semantic factors [16]. However, applying matrix factorization to explicit ratings in the CF domain raises difficulties due to the high portion of missing values. Conventional singular value decomposition is undefined when knowledge about the matrix is incomplete. Moreover, carelessly addressing only the relatively few known entries is highly prone to overfitting. Earlier works relied on imputation [25, 47], which fills in missing ratings and makes the rating matrix dense. However, imputation can be very expensive as it significantly increases the amount of data. In addition, the data may be considerably distorted due to inaccurate imputation. Hence, more recent works [5, 8, 18, 26, 39, 46, 52] suggested modeling directly only the observed ratings, while avoiding overfitting through an adequate regularized model.

In this section we describe several matrix factorization techniques, with increasing complexity and accuracy. We start with the basic model, a direct factorization of the rating matrix. Then, we show how to integrate implicit feedback in order to increase prediction accuracy. Finally we deal with the fact that customer preferences for products may drift over time. Product perception and popularity are constantly changing as new selection emerges. Similarly, customer inclinations are evolving, leading them to ever redefine their taste. This leads to a factor model that addresses temporal dynamics for better tracking user behavior.

## 4.1 Factorizing the Rating Matrix

Matrix factorization models map both users and items to a joint latent factor space of dimensionality $f$, such that user-item interactions are modeled as inner products in that space. The latent space tries to explain ratings by characterizing both products and users on factors automatically inferred from user feedback. For example, when the products are movies, factors might measure obvious dimensions such as comedy vs. drama, amount of action, or orientation to children; less well defined dimensions such as depth of character development or "quirkiness"; or completely uninterpretable dimensions.

Accordingly, each item $i$ is associated with a vector $\mathbf{q}_i \in \mathbb{R}^f$, and each user $u$ is associated with a vector $\mathbf{p}_u \in \mathbb{R}^f$. For a given item $i$, the elements of $\mathbf{q}_i$ measure the extent to which the item possesses those factors, positive or negative. For a given user $u$, the elements of $\mathbf{p}_u$ measure the extent of interest the user has in items that are high on the corresponding factors (again, these may be positive or negative). The resulting dot product,[1] $\mathbf{q}_i^T \mathbf{p}_u$, captures the interaction between user $u$ and item $i$—i.e., the overall interest of the user in characteristics of the item. The final rating is created by also adding in the aforementioned baseline predictors that depend only on the user or item (here, dropping temporal baselines for simplicity). Thus, a rating is predicted by the rule

$$\hat{r}_{u,i} = \mu + b_i + b_u + \mathbf{q}_i^T \mathbf{p}_u \, . \tag{12}$$

---

[1] Recall that the dot product between two vectors $\mathbf{x}, \mathbf{y} \in \mathbb{R}^f$ is defined as: $\mathbf{x}^T \mathbf{y} = \langle \mathbf{x}, \mathbf{y} \rangle = \sum_{k=1}^{f} x_k \cdot y_k$.

In order to learn the model parameters $(b_u, b_i, \mathbf{p}_u$ and $\mathbf{q}_i)$ we minimize the regularized squared error

$$\underset{b_*, \mathbf{q}_*, \mathbf{p}_*}{\mathrm{argmin}} \sum_{(u,i) \in \mathcal{K}} \left[ (r_{u,i} - \mu - b_i - b_u - \mathbf{q}_i^T \mathbf{p}_u)^2 + \lambda_5 (b_i^2 + b_u^2 + \|\mathbf{q}_i\|^2 + \|\mathbf{p}_u\|^2) \right].$$

The constant $\lambda_5$, which controls the extent of regularization, is usually determined by cross validation. Minimization is typically performed by either stochastic gradient descent or alternating least squares.

Alternating least squares techniques rotate between fixing the $\mathbf{p}_u$'s to solve for the $\mathbf{q}_i$'s and fixing the $\mathbf{q}_i$'s to solve for the $\mathbf{p}_u$'s. Notice that when one of these is taken as a constant, the optimization problem is quadratic and can be optimally solved; see [3, 5].

An easy stochastic gradient descent optimization was popularized by Funk [18] and successfully practiced by many others [26, 39, 46, 52]. The algorithm loops through all ratings in the training data. For each given rating $r_{u,i}$, a prediction $(\hat{r}_{u,i})$ is made, and the associated prediction error $e_{u,i} \stackrel{\text{def}}{=} r_{u,i} - \hat{r}_{u,i}$ is computed. For a given training case $r_{u,i}$, we modify the parameters by moving in the opposite direction of the gradient, yielding:

- $b_u \leftarrow b_u + \gamma \cdot (e_{u,i} - \lambda_5 \cdot b_u)$
- $b_i \leftarrow b_i + \gamma \cdot (e_{u,i} - \lambda_5 \cdot b_i)$
- $\mathbf{q}_i \leftarrow \mathbf{q}_i + \gamma \cdot (e_{u,i} \cdot \mathbf{p}_u - \lambda_5 \cdot \mathbf{q}_i)$
- $\mathbf{p}_u \leftarrow \mathbf{p}_u + \gamma \cdot (e_{u,i} \cdot \mathbf{q}_i - \lambda_5 \cdot \mathbf{p}_u)$

When evaluating the method on the Netflix data, we used the following values for the meta parameters: $\gamma = 0.005, \lambda_5 = 0.02$. Henceforth, we dub this method "SVD". Note that this is not a conventional singular value decomposition as discussed in the introduction of Sect. 4.

A general remark is in place. One can expect better accuracy by dedicating separate learning rates ($\gamma$) and regularization ($\lambda$) to each type of learned parameter. Thus, for example, it is advised to employ distinct learning rates to user biases, item biases and the factors themselves. A good, intensive use of such a strategy is described in Takács et al. [53]. When producing exemplary results for this chapter, we did not use such a strategy consistently, and in particular many of the given constants are not fully tuned.

## 4.2 Matrix Factorization with Implicit Feedback

Prediction accuracy is improved by considering also implicit feedback, which provides an additional indication of user preferences. This is especially helpful for those users that provided much more implicit feedback than explicit one. As explained in Sect. 2.2, even in cases where independent implicit feedback is absent, one can capture a significant signal by accounting for which items users rate, regardless of their rating value. This led to several methods [26, 39, 45] that modeled a user factor by the identity of the items they have rated. Here we focus on the SVD++ method [26], which was shown to offer accuracy superior to SVD.

To this end, a second set of item factors is added, relating each item $i$ to a factor vector $y_i \in \mathbb{R}^f$. Those new item factors are used to characterize users based on the set of items that they rated. The exact model is as follows:

$$\hat{r}_{u,i} = \mu + b_i + b_u + \mathbf{q}_i^T \left( \mathbf{p}_u + |R(u)|^{-\frac{1}{2}} \sum_{j \in R(u)} \mathbf{y}_j \right) \tag{13}$$

The set $R(u)$ contains the items rated by user $u$.

Now, a user $u$ is modeled as $p_u + |R(u)|^{-\frac{1}{2}} \sum_{j \in R(u)} \mathbf{y}_j$. We use a free user-factors vector, $\mathbf{p}_u$, much like in (12), which is learnt from the given explicit ratings. This vector is complemented by the sum $|R(u)|^{-\frac{1}{2}} \sum_{j \in R(u)} \mathbf{y}_j$, which represents the perspective of implicit feedback. Since the $\mathbf{y}_j$'s are centered around zero (by the regularization), the sum is normalized by $|R(u)|^{-\frac{1}{2}}$, in order to stabilize its variance across the range of observed values of $|R(u)|$.

Model parameters are determined by minimizing the associated regularized squared error function through stochastic gradient descent. We loop over all known ratings in $\mathcal{K}$, computing:

- $b_u \leftarrow b_u + \gamma \cdot \left( e_{u,i} - \lambda_6 \cdot b_u \right)$
- $b_i \leftarrow b_i + \gamma \cdot \left( e_{u,i} - \lambda_6 \cdot b_i \right)$
- $\mathbf{q}_i \leftarrow \mathbf{q}_i + \gamma \cdot (e_{u,i} \cdot (\mathbf{p}_u + |R(u)|^{-\frac{1}{2}} \sum_{j \in R(u)} \mathbf{y}_j) - \lambda_7 \cdot \mathbf{q}_i)$
- $\mathbf{p}_u \leftarrow \mathbf{p}_u + \gamma \cdot (e_{u,i} \cdot \mathbf{q}_i - \lambda_7 \cdot \mathbf{p}_u)$
- $\forall j \in R(u):$
  $\mathbf{y}_j \leftarrow \mathbf{y}_j + \gamma \cdot (e_{u,i} \cdot |R(u)|^{-\frac{1}{2}} \cdot \mathbf{q}_i - \lambda_7 \cdot \mathbf{y}_j)$

When evaluating the method on the Netflix data, we used the following values for the meta parameters: $\gamma = 0.007$, $\lambda_6 = 0.005$, $\lambda_7 = 0.015$. It is beneficial to decrease step sizes (the $\gamma$'s) by a factor of 0.9 after each iteration. The iterative process runs for around 30 iterations until convergence.

Several types of implicit feedback can be simultaneously introduced into the model by using extra sets of item factors. For example, if a user $u$ has a certain kind of implicit preference to the items in $N^1(u)$ (e.g., she rented them), and a different type of implicit feedback to the items in $N^2(u)$ (e.g., she watched their trailers), we could use the model

$$\hat{r}_{u,i} = \mu + b_i + b_u + \mathbf{q}_i^T \left( \mathbf{p}_u + |N^1(u)|^{-\frac{1}{2}} \sum_{j \in N^1(u)} \mathbf{y}_j^{(1)} + |N^2(u)|^{-\frac{1}{2}} \sum_{j \in N^2(u)} \mathbf{y}_j^{(2)} \right).$$

$$\tag{14}$$

The relative importance of each source of implicit feedback will be automatically learned by the algorithm by its setting of the respective values of model parameters.

## 4.3  Time-Aware Factor Model

The matrix-factorization approach lends itself well to modeling temporal effects, which can significantly improve its accuracy. In Sect. 3.1 we discussed the way time affects baseline predictors, $b_u(t)$ and $b_i(t)$. However, temporal dynamics go beyond this, they also affect user preferences and thereby the interaction between users and items. Users change their preferences over time. For example, a fan of the "psychological thrillers" genre may become a fan of "crime dramas" a year later. Similarly, humans change their perception on certain actors and directors. This type of evolution is modeled by taking the user factors (the vector $\mathbf{p}_u$) as a function of time. Once again, we need to model those changes at the very fine level of a daily basis, while facing the built-in scarcity of user ratings. In fact, these temporal effects are the hardest to capture, because preferences are not as pronounced as main effects (user-biases), but are split over many factors.

We modeled each component of the user preferences $\mathbf{p}_u(t)^T = (p_{u,1}(t), \ldots, p_{u,f}(t))$ in the same way that we treated user biases. Within the movie-rating dataset, we have found modeling after (8) effective, leading to

$$p_{u,k}(t) = p_{u,k} + \alpha_{u,k} \cdot \mathrm{dev}_u(t) + p_{u,k,t} \quad k = 1, \ldots, f \,. \tag{15}$$

Here $p_{u,k}$ captures the stationary portion of the factor, $\alpha_{u,k} \cdot \mathrm{dev}_u(t)$ approximates a possible portion that changes linearly over time, and $p_{u,k,t}$ absorbs the very local, day-specific variability.

Note that we do specify static item characteristics, $\mathbf{q}_i$, because we do not expect significant temporal variation for items, which, unlike humans, are static in nature.

At this point, we can tie all pieces together and extend the SVD++ factor model by incorporating the time changing parameters. The resulting model will be denoted as *timeSVD++*, where the prediction rule is as follows:

$$\hat{r}_{u,i} = \mu + b_i(t_{u,i}) + b_u(t_{u,i}) + \mathbf{q}_i^T \left( \mathbf{p}_u(t_{u,i}) + |R(u)|^{-\frac{1}{2}} \sum_{j \in R(u)} \mathbf{y}_j \right) \tag{16}$$

The exact definitions of the time drifting parameters $b_i(t)$, $b_u(t)$ and $\mathbf{p}_u(t)$ were given in (5), (8) and (15). Learning is performed by minimizing the associated squared error function on the training set using a regularized stochastic gradient descent algorithm. The procedure is analogous to the one involving the original SVD++ algorithm. Time complexity per iteration is still linear with the input size, while wall clock running time is approximately doubled compared to SVD++, due to the extra overhead required for updating the temporal parameters. Importantly, the convergence rate was not affected by the temporal parameterization, and the process converges in around 30 iterations.

## 4.4 Comparison

In Table 2 we compare results of the three algorithms discussed in this section. First is SVD, the plain matrix factorization algorithm. Second, is the SVD++ method, which improves upon SVD by incorporating a kind of implicit feedback. Finally is timeSVD++, which accounts for temporal effects. The three methods are compared over a range of factorization dimensions ($f$). All benefit from a growing number of factor dimensions that enables them to better express complex movie-user interactions. Note that the number of parameters in SVD++ is comparable to their number in SVD. This is because SVD++ adds only item factors, while the complexity of our dataset is dominated by the much larger set of users. On the other hand, timeSVD++ requires a significant increase in the number of parameters, because of its refined representation of each user factor. Addressing implicit feedback by the SVD++ model leads to accuracy gains within the movie rating dataset. Yet, the improvement delivered by timeSVD++ over SVD++ is consistently more significant. We are not aware of any single algorithm in the literature that could deliver such accuracy. Further evidence of the importance of capturing temporal dynamics is the fact that a timeSVD++ model of dimension 10 is already more accurate than an SVD model of dimension 200. Similarly, a timeSVD++ model of dimension 20 is enough to outperform an SVD++ model of dimension 200.

## 4.5 Summary

In its basic form, matrix factorization characterizes both items and users by vectors of factors inferred from patterns of item ratings. High correspondence between item

**Table 2** Comparison of three factor models: prediction accuracy is measured by RMSE (lower is better) for varying factor dimensionality ($f$). For all models, accuracy improves with growing number of dimensions. SVD++ improves accuracy by incorporating implicit feedback into the SVD model. Further accuracy gains are achieved by also addressing the temporal dynamics in the data through the timeSVD++ model

| Model | $f=10$ | $f=20$ | $f=50$ | $f=100$ | $f=200$ |
|---|---|---|---|---|---|
| SVD | 0.9140 | 0.9074 | 0.9046 | 0.9025 | 0.9009 |
| SVD++ | 0.9131 | 0.9032 | 0.8952 | 0.8924 | 0.8911 |
| Timesvd++ | 0.8971 | 0.8891 | 0.8824 | 0.8805 | 0.8799 |

and user factors leads to recommendation of an item to a user. These methods deliver prediction accuracy superior to other published collaborative filtering techniques. At the same time, they offer a memory efficient compact model, which can be trained relatively easy. Those advantages, together with the implementation ease of gradient based matrix factorization model (SVD), made this the method of choice within the Netflix Prize competition.

What makes these techniques even more convenient is their ability to address several crucial aspects of the data. First, is the ability to integrate multiple forms of user feedback. One can better predict user ratings by also observing other related actions by the same user, such as purchase and browsing history. The proposed SVD++ model leverages multiple sorts of user feedback for improving user profiling.

Another important aspect is the temporal dynamics that make users' tastes evolve over time. Each user and product potentially goes through a distinct series of changes in their characteristics. A mere decay of older training instances cannot adequately identify communal patterns of behavior in time changing data. The solution we adopted is to model the temporal dynamics along the whole time period, allowing us to intelligently separate transient factors from lasting ones. The inclusion of temporal dynamics proved very useful in improving the quality of predictions, more than various algorithmic enhancements.

## 5   Neighborhood Models

The most common approach to CF is based on neighborhood models. Chapter "Trust Your Neighbors: A Comprehensive Survey of Neighborhood-Based Methods for Recommender Systems" provides an extensive survey on this approach. Its original form, which was shared by virtually all earlier CF systems, is user-user based; see [20] for a good analysis. User-user methods estimate unknown ratings based on recorded ratings of like-minded users.

Later, an analogous item-item approach [35, 48] became popular. In those methods, a rating is estimated using known ratings made by the same user on similar items. Better scalability and improved accuracy make the item-item approach more favorable in many cases [3, 48, 52]. In addition, item-item methods are more amenable to explaining the reasoning behind predictions. This is because users are familiar with items previously preferred by them, but do not know those allegedly like-minded users. We focus mostly on item-item approaches, but the same techniques can be directly applied within a user-user approach; see also Sect. 5.4.2.

The structure of this section is as follows. First, we describe how to estimate the similarity between two items, which is a basic building block of most neighborhood techniques. Then, we move on to the widely used similarity-based neighborhood method, which constitutes a straightforward application of the similarity weights. We identify certain limitations of this similarity based approach.

While most neighborhood methods are local in their nature—concentrating only on a small subset of related ratings—it appears that an improved accuracy is achieved by employing a more global viewpoint, leading to a method described in Sect. 5.3. This newer approach, which is based on a more rigorous global optimization, facilitates two new possibilities. First, is a neighborhood model with vastly improved space requirements, as described in Sect. 5.4. Second is a treatment of temporal dynamics within the neighborhood model, leading to better prediction accuracy, as described in Sect. 5.5.

## 5.1 Similarity Measures

Central to most item-item approaches is a similarity measure between items. Frequently, it is based on the Pearson correlation coefficient, $\rho_{i,j}$, which measures the tendency of users to rate items $i$ and $j$ similarly. Since many ratings are unknown, some items may share only a handful of common observed raters. The empirical correlation coefficient, $\hat{\rho}_{i,j}$, is based only on the common user support. It is advised to work with residuals from the baseline predictors (the $b_{u,i}$'s; see Sect. 3) to compensate for user- and item-specific deviations. Thus the approximated correlation coefficient is given by

$$\hat{\rho}_{i,j} = \frac{\sum_{u \in U(i,j)} (r_{u,i} - b_{u,i})(r_{u,j} - b_{u,j})}{\sqrt{\sum_{u \in U(i,j)} (r_{u,i} - b_{u,i})^2 \cdot \sum_{u \in U(i,j)} (r_{u,j} - b_{u,j})^2}} . \tag{17}$$

The set $U(i, j) \stackrel{\text{def}}{=} R(i) \cap R(j)$ contains the users who rated both items $i$ and $j$.

Because estimated correlations based on a greater user support are more reliable, an appropriate similarity measure, denoted by $s_{i,j}$, is a shrunk correlation coefficient of the form

$$s_{i,j} \stackrel{\text{def}}{=} \frac{n_{i,j} - 1}{n_{i,j} - 1 + \lambda_8} \hat{\rho}_{i,j} . \tag{18}$$

The variable $n_{i,j} = |U(i, j)|$ denotes the number of users that rated both $i$ and $j$. A typical value for $\lambda_8$ is 100.

Such shrinkage can be motivated from a Bayesian perspective; see Section 2.6 of Gelman et al. [19]. Suppose that the true $\rho_{i,j}$ are independent random variables drawn from a normal distribution,

$$\rho_{i,j} \sim N(0, \tau^2)$$

for known $\tau^2$. The mean of 0 is justified if the $b_{u,i}$ account for both user and item deviations from average. Meanwhile, suppose that

$$\hat{\rho}_{i,j} | \rho_{i,j} \sim N(\rho_{i,j}, \sigma_{i,j}^2)$$

for known $\sigma_{i,j}^2$. We estimate $\rho_{i,j}$ by its posterior mean:

$$E(\rho_{i,j}|\hat{\rho}_{i,j}) = \frac{\tau^2 \hat{\rho}_{i,j}}{\tau^2 + \sigma_{i,j}^2}$$

the empirical estimator $\hat{\rho}_{i,j}$ shrunk a fraction, $\sigma_{i,j}^2/(\tau^2 + \sigma_{i,j}^2)$, of the way toward zero.

Formula (18) follows from approximating the variance of a correlation by $\sigma_{i,j}^2 = 1/(n_{i,j} - 1)$, the value for $\rho_{i,j}$ near 0.

Notice that the literature suggests additional alternatives for a similarity measure [48, 52].

## 5.2 Similarity-Based Interpolation

Here we describe the most popular approach to neighborhood modeling, and apparently also to CF in general. Our goal is to predict $r_{u,i}$—the unobserved rating by user $u$ for item $i$. Using the similarity measure, we identify the $k$ items rated by $u$ that are most similar to $i$. This set of $k$ neighbors is denoted by $S^k(i; u)$. Note that $S^k(i; u)$ excludes the item $i$ that we want to predict. The predicted value of $r_{u,i}$ is taken as a weighted average of the ratings of neighboring items, while adjusting for user and item effects through the baseline predictors

$$\hat{r}_{u,i} = b_{u,i} + \frac{\sum_{j \in S^k(i;u)} s_{i,j}(r_{u,j} - b_{u,j})}{\sum_{j \in S^k(i;u)} s_{i,j}} \tag{19}$$

$$= b_{u,i} + \sum_{j \in S^k(i;u)} \frac{s_{i,j}}{\sum_{l \in S^k(i;u)} s_{i,l}}(r_{u,j} - b_{u,j}).$$

Note the dual use of the similarities for both identification of nearest neighbors and as the interpolation weights in Eq. (19).

Sometimes, instead of relying directly on the similarity weights as interpolation coefficients, one can achieve better results by transforming these weights. For example, we have found at several datasets that squaring the correlation-based similarities is helpful. This leads to a rule like: $\hat{r}_{u,i} = b_{u,i} + \frac{\sum_{j \in S^k(i;u)} s_{i,j}^2(r_{u,j} - b_{u,j})}{\sum_{j \in S^k(i;u)} s_{i,j}^2}$.

Toscher et al. [55] discuss more sophisticated transformations of these weights.

Similarity-based methods became very popular because they are intuitive and relatively simple to implement. They also offer the following two useful properties:

1. *Explainability.* The importance of explaining automated recommendations is widely recognized [21, 54]; see also Chap. "Beyond Explaining Single Item Recommendations". Users expect a system to give a reason for its predictions,

rather than present "black box" recommendations. Explanations not only enrich the user experience, but also encourage users to interact with the system, fix wrong impressions and improve long-term accuracy. The neighborhood framework allows identifying which of the past user actions are most influential on the computed prediction.

2. *New ratings.* Item-item neighborhood models can provide updated recommendations immediately after users enter new ratings. This includes handling new users as soon as they provide feedback to the system, without needing to retrain the model and estimate new parameters. This assumes that relationships between items (the $s_{i,j}$ values) are stable and barely change on a daily basis. Notice that for items new to the system we do have to learn new parameters. Interestingly, this asymmetry between users and items meshes well with common practices: quality systems are expected to provide immediate recommendations to new users, or to instantly refresh recommendations as new ratings are entered by existing users. On the other hand, it is reasonable to require a waiting period before recommending items new to the system.

However, standard neighborhood-based methods raise some concerns:

1. The similarity function ($s_{i,j}$), which directly defines the interpolation weights, is arbitrary. Various CF algorithms use somewhat different similarity measures, trying to quantify the elusive notion of user- or item-similarity. As an example, suppose that a particular item is predicted perfectly by a subset of the neighbors. In that case, we would like the predictive subset to receive all the weight, but that is impossible for bounded similarity scores like the Pearson correlation coefficient.

2. Previous neighborhood-based methods do not account for interactions among neighbors. Each similarity between an item $i$ and a neighbor $j \in S^k(i; u)$ is computed independently of the content of $S^k(i; u)$ and the other similarities: $s_{il}$ for $l \in S^k(i; u) - \{j\}$. For example, suppose that our items are movies, and the neighbors set contains three movies that are highly correlated with each other (e.g., sequels such as "Lord of the Rings 1–3"). An algorithm that ignores the similarity of the three movies when determining their interpolation weights, may end up essentially triple counting the information provided by the group.

3. By definition, the interpolation weights sum to one, which may cause overfitting. Suppose that an item has no useful neighbors rated by a particular user. In that case, it would be best to ignore the neighborhood information, staying with the more robust baseline predictors. Nevertheless, the standard neighborhood formula uses a weighted average of ratings for the uninformative neighbors.

4. Neighborhood methods may not work well if the variability of ratings differs substantially among neighbors.

Some of these issues can be fixed to a certain degree, while others are more difficult to solve within the basic framework. For example, the third item, dealing with the sum-to-one constraint, can be alleviated by using the following prediction rule:

$$\hat{r}_{u,i} = b_{u,i} + \frac{\sum_{j \in S^k(i;u)} s_{i,j}(r_{u,j} - b_{u,j})}{\lambda_9 + \sum_{j \in S^k(i;u)} s_{i,j}} \tag{20}$$

The constant $\lambda_9$ penalizes the neighborhood portion when there is not much neighborhood information, e.g., when $\sum_{j \in S^k(i;u)} s_{i,j} \ll \lambda_9$. Indeed, we have found that setting an appropriate value of $\lambda_9$ leads to accuracy improvements over (19). Nonetheless, the whole framework here is not justified by a formal model. Thus, we strive for better results with a more rigorous approach, as we describe in the following.

## 5.3   A Global Neighborhood Model

In this subsection, we introduce a neighborhood model based on global optimization. The model offers an improved prediction accuracy, with additional advantages that are summarized as follows:

1. No reliance on arbitrary or heuristic item-item similarities. The new model is cast as the solution to a global optimization problem.
2. Inherent overfitting prevention or "risk control": the model reverts to robust baseline predictors, unless a user entered sufficiently many relevant ratings.
3. The model can capture the totality of weak signals encompassed in all of a user's ratings, not needing to concentrate only on the few ratings for most similar items.
4. The model naturally allows integrating different forms of user input, such as explicit and implicit feedback.
5. A highly scalable implementation (Sect. 5.4) allows linear time and space complexity, thus facilitating both item-item and user-user implementations to scale well to very large datasets.
6. Time drifting aspects of the data can be integrated into the model, thereby improving its accuracy; see Sect. 5.5.

### 5.3.1   Building the Model

We gradually construct the various components of the model, through an ongoing refinement of our formulations. Previous models were centered around *user-specific* interpolation weights—$s_{i,j} / \sum_{l \in S^k(i;u)} s_{i,l}$ in (19)—relating item $i$ to the items in a user-specific neighborhood $S^k(i; u)$. In order to facilitate global optimization, we would like to abandon such user-specific weights in favor of global item-item weights independent of a specific user. The weight from $j$ to $i$ is denoted by $w_{i,j}$ and will be learned from the data through optimization. An initial sketch of the model describes each rating $r_{u,i}$ by the equation

$$\hat{r}_{u,i} = b_{u,i} + \sum_{j \in R(u)} (r_{u,j} - b_{u,j})w_{i,j} . \tag{21}$$

This rule starts with the crude, yet robust, baseline predictors $(b_{u,i})$. Then, the estimate is adjusted by summing over *all*[2] ratings by $u$.

Let us consider the interpretation of the weights. Usually the weights in a neighborhood model represent interpolation coefficients relating unknown ratings to existing ones. Here, we adopt a different viewpoint, that enables a more flexible usage of the weights. We no longer treat weights as interpolation coefficients. Instead, we take weights as part of adjustments, or *offsets*, added to the baseline predictors. This way, the weight $w_{i,j}$ is the extent by which we increase our baseline prediction of $r_{u,i}$ based on the observed value of $r_{u,j}$. For two related items $i$ and $j$, we expect $w_{i,j}$ to be high. Thus, whenever a user $u$ rated $j$ higher than expected $(r_{u,j} - b_{u,j}$ is high), we would like to increase our estimate for $u$'s rating of $i$ by adding $(r_{u,j} - b_{u,j})w_{i,j}$ to the baseline prediction. Likewise, our estimate will not deviate much from the baseline by an item $j$ that $u$ rated just as expected $(r_{u,j} - b_{u,j}$ is around zero), or by an item $j$ that is not known to be predictive on $i$ ($w_{i,j}$ is close to zero).

This viewpoint suggests several enhancements to (21). First, we can use the form of binary user input, which was found beneficial for factorization models. Namely, analyzing which items were rated regardless of rating value. To this end, we add another set of weights, and rewrite (21) as

$$\hat{r}_{u,i} = b_{u,i} + \sum_{j \in R(u)} [(r_{u,j} - b_{u,j})w_{i,j} + c_{i,j}]. \qquad (22)$$

Similarly, one could employ here another set of implicit feedback, $N(u)$—e.g., the set of items rented or purchased by the user—leading to the rule

$$\hat{r}_{u,i} = b_{u,i} + \sum_{j \in R(u)} (r_{u,j} - b_{u,j})w_{i,j} + \sum_{j \in N(u)} c_{i,j}. \qquad (23)$$

Much like the $w_{i,j}$'s, the $c_{i,j}$'s are offsets added to the baseline predictor. For two items $i$ and $j$, an implicit preference by $u$ for $j$ leads us to adjust our estimate of $r_{u,i}$ by $c_{i,j}$, which is expected to be high if $j$ is predictive on $i$.

Employing global weights, rather than user-specific interpolation coefficients, emphasizes the influence of missing ratings. In other words, a user's opinion is formed not only by what he rated, but also by what he did not rate. For example, suppose that a movie ratings dataset shows that users that rate "Shrek 3" high also gave high ratings to "Shrek 1–2". This will establish high weights from "Shrek 1–2" to "Shrek 3". Now, if a user did not rate "Shrek 1–2" at all, his predicted rating for "Shrek 3" will be penalized, as some necessary weights cannot be added to the sum.

For the prior model (19) that interpolated $r_{u,i} - b_{u,i}$ from $\{r_{u,j} - b_{u,j} | j \in S^k(i; u)\}$, it was necessary to maintain compatibility between the $b_{u,i}$ values and

---

[2] The item $i$ should be excluded from the summation over $R(u)$. To simplify notation, we omit this detail in the remainder of this section.

the $b_{u,j}$ values. However, here we do not use interpolation, so we can decouple the definitions of $b_{u,i}$ and $b_{u,j}$. Accordingly, a more general prediction rule would be: $\hat{r}_{u,i} = \tilde{b}_{u,i} + \sum_{j \in R(u)} (r_{u,j} - b_{u,j})w_{i,j} + c_{i,j}$. The constant $\tilde{b}_{u,i}$ can represent predictions of $r_{u,i}$ by other methods such as a latent factor model. Here, we suggest the following rule that was found to work well:

$$\hat{r}_{u,i} = \mu + b_u + b_i + \sum_{j \in R(u)} [(r_{u,j} - b_{u,j})w_{i,j} + c_{i,j}] \qquad (24)$$

Importantly, the $b_{u,j}$'s remain constants, which are derived as explained in Sect. 3. However, the $b_u$'s and $b_i$'s become parameters that are optimized much like the $w_{i,j}$'s and $c_{i,j}$'s.

We have found that it is beneficial to normalize sums in the model leading to the form

$$\hat{r}_{u,i} = \mu + b_u + b_i + |R(u)|^{-\alpha} \sum_{j \in R(u)} [(r_{u,j} - b_{u,j})w_{i,j} + c_{i,j}]. \qquad (25)$$

The constant $\alpha$ controls the extent of normalization. A non-normalized rule ($\alpha = 0$), encourages greater deviations from baseline predictions for users that provided many ratings (high $|R(u)|$). On the other hand, a fully normalized rule ($\alpha = 1$), eliminates the effect of number of ratings on deviations from baseline predictions. In many cases it would be a good practice for recommender systems to have greater deviation from baselines for users that rate a lot. This way, we take more risk with well-modeled users that provided much input. For such users we are willing to predict quirkier and less common recommendations. At the same time, we are less certain about the modeling of users that provided only a little input, in which case we would like to stay with safe estimates close to the baseline values. Our experience with the Netflix dataset shows that best results are achieved with $\alpha = 0.5$, as in the prediction rule

$$\hat{r}_{u,i} = \mu + b_u + b_i + |R(u)|^{-\frac{1}{2}} \sum_{j \in R(u)} [(r_{u,j} - b_{u,j})w_{i,j} + c_{i,j}]. \qquad (26)$$

As an optional refinement, complexity of the model can be reduced by pruning parameters corresponding to unlikely item-item relations. Let us denote by $S^k(i)$ the set of $k$ items most similar to $i$, as determined by e.g., a similarity measure $s_{i,j}$ or a natural hierarchy associated with the item set. Additionally, we use $R^k(i; u) \overset{\text{def}}{=} R(u) \cap S^k(i)$.[3] Now, when predicting $r_{u,i}$ according to (26), it is expected that

---

[3] Notational clarification: With other neighborhood models it was beneficial to use $S^k(i; u)$, which denotes the $k$ items most similar to $i$ among those rated by $u$. Hence, if $u$ rated at least $k$ items, we will always have $|S^k(i; u)| = k$, regardless of how similar those items are to $i$. However, $|R^k(i; u)|$ is typically smaller than $k$, as some of those items most similar to $i$ were not rated by $u$.

the most influential weights will be associated with items similar to $i$. Hence, we replace (26) with

$$\hat{r}_{u,i} = \mu + b_u + b_i + |\mathbf{R}^k(i; u)|^{-\frac{1}{2}} \sum_{j \in \mathbf{R}^k(i;u)} [(r_{u,j} - b_{u,j})w_{i,j} + c_{i,j}]. \tag{27}$$

When $k = \infty$, rule (27) coincides with (26). However, for other values of $k$ it offers the potential to significantly reduce the number of variables involved.

### 5.3.2  Parameter Estimation

Prediction rule (27) allows fast online prediction. More computational work is needed at a pre-processing stage where parameters are estimated. A major design goal of the new neighborhood model was facilitating an efficient global optimization procedure, which prior neighborhood models lacked. Thus, model parameters are learned by solving the regularized least squares problem associated with (27):

$$\min_{b_*, w_*, c_*} \sum_{(u,i) \in \mathcal{K}} \left[ \left( r_{u,i} - \mu - b_u - b_i - |\mathbf{R}^k(i; u)|^{-\frac{1}{2}} \right. \right.$$
$$\left. \times \sum_{j \in \mathbf{R}^k(i;u)} \left( (r_{u,j} - b_{u,j})w_{i,j} + c_{i,j} \right) \right)^2$$
$$\left. + \lambda_{10} \left( b_u^2 + b_i^2 + \sum_{j \in \mathbf{R}^k(i;u)} w_{i,j}^2 + c_{i,j}^2 \right) \right] \tag{28}$$

An optimal solution of this convex problem can be obtained by least square solvers, which are part of standard linear algebra packages. However, we have found that the following simple stochastic gradient descent solver works much faster. Let us denote the prediction error, $r_{u,i} - \hat{r}_{u,i}$, by $e_{u,i}$. We loop through all known ratings in $\mathcal{K}$. For a given training case $r_{u,i}$, we modify the parameters by moving in the opposite direction of the gradient, yielding:

- $b_u \leftarrow b_u + \gamma \cdot \left( e_{u,i} - \lambda_{10} \cdot b_u \right)$
- $b_i \leftarrow b_i + \gamma \cdot \left( e_{u,i} - \lambda_{10} \cdot b_i \right)$
- $\forall j \in \mathbf{R}^k(i; u)$:
  $$w_{i,j} \leftarrow w_{i,j} + \gamma \cdot \left( |\mathbf{R}^k(i; u)|^{-\frac{1}{2}} \cdot e_{u,i} \cdot (r_{u,j} - b_{u,j}) - \lambda_{10} \cdot w_{i,j} \right)$$
  $$c_{i,j} \leftarrow c_{i,j} + \gamma \cdot \left( |\mathbf{R}^k(i; u)|^{-\frac{1}{2}} \cdot e_{u,i} - \lambda_{10} \cdot c_{i,j} \right)$$

The meta-parameters $\gamma$ (step size) and $\lambda_{10}$ are determined by cross-validation. We used $\gamma = 0.005$ and $\lambda_{10} = 0.002$ for the Netflix data. Another important parameter is $k$, which controls the neighborhood size. Our experience shows that increasing $k$ always benefits the accuracy of the results on the test set. Hence, the choice of $k$ should reflect a tradeoff between prediction accuracy and computational cost. In Sect. 5.4 we will describe a factored version of the model that eliminates this tradeoff by allowing us to work with the most accurate $k = \infty$ while lowering running time.

A typical number of iterations throughout the training data is 15–20. As for time complexity per iteration, let us analyze the most accurate case where $k = \infty$, which is equivalent to using prediction rule (26). For each user $u$ and item $i \in R(u)$ we need to modify $\{w_{i,j}, c_{i,j} | j \in R(u)\}$. Thus the overall time complexity of the training phase is $O(\sum_u |R(u)|^2)$.

### 5.3.3  Comparison of Accuracy

Experimental results on the Netflix data with the globally optimized neighborhood model, henceforth dubbed GlobalNgbr, are presented in Fig. 1. We studied the model under different values of parameter $k$. The solid black curve with square symbols shows that accuracy monotonically improves with rising $k$ values, as root mean squared error (RMSE) falls from 0.9139 for $k = 250$ to 0.9002 for $k = \infty$. (Notice that since the Netflix data contains 17,770 movies, $k = \infty$ is

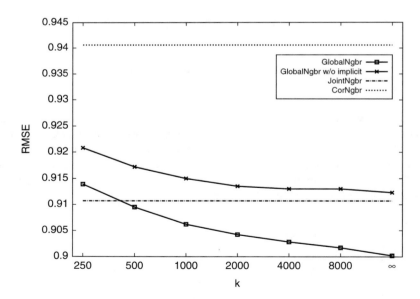

**Fig. 1** Comparison of neighborhood-based models. Accuracy is measured by RMSE on the Netflix test set, so lower values indicate better performance. We measure the accuracy of the globally optimized model (GlobalNgbr) with and without implicit feedback. RMSE is shown as a function of varying values of $k$, which dictates the neighborhood size. The accuracy of two other models is shown as two horizontal lines; for each we picked an optimal neighborhood size

equivalent to $k = 17,769$, where all item-item relations are explored.) We repeated the experiments without using the implicit feedback, that is, dropping the $c_{i,j}$ parameters from our model. The results depicted by the black curve with X's show a significant decline in estimation accuracy, which widens as $k$ grows. This demonstrates the value of incorporating implicit feedback into the model.

For comparison we provide results of a similarity-based neighborhood model, CorNgbr, as decribed in Sect. 5.2), and a more accurate neighborhood model, JointNgbr [3], that learns local interpolations weights. For both these two models, we tried to pick optimal parameters and neighborhood sizes, which were 20 for CorNgbr, and 50 for JointNgbr. The results are depicted by the dotted and dashed lines, respectively. It is clear that the popular CorNgbr method is noticeably less accurate than the other neighborhood models. On the opposite side, GlobalNgbr delivers more accurate results even when compared with JointNgbr, as long as the value of $k$ is at least 500. Notice that the $k$ value (the $x$-axis) is irrelevant to the previous models, as their different notion of neighborhood makes neighborhood sizes incompatible. Yet, we observed that while the performance of GlobalNgbr keeps improving as more neighbors are added, this was not true with the two other models. For CorNgbr and JointNgbr, performance peaks with a relatively small number of neighbors and declines therafter. This may be explained by the fact that in GlobalNgbr, parameters are directly learned from the data through a formal optimization procedure that facilitates using many more parameters effectively.

Finally, let us consider running time. Previous neighborhood models require very light pre-processing, though, JointNgbr [3] requires solving a small system of equations for each provided prediction. The new model does involve pre-processing where parameters are estimated. However, online prediction is immediate by following rule (27). Pre-processing time grows with the value of $k$. Figure 2 shows typical running times per iteration on the Netflix data, as measured on a single processor 3.4 GHz Pentium 4 PC.

## 5.4 A Factorized Neighborhood Model

In the previous subsection we presented a more accurate neighborhood model, which is based on prediction rule (26) with training time complexity $O(\sum_u |R(u)|^2)$ and space complexity $O(m + n^2)$. (Recall that $m$ is the number of users, and $n$ is the number of items.) We could improve time and space complexity by sparsifying the model through pruning unlikely item-item relations. Sparsification was controlled by the parameter $k \leqslant n$, which reduced running time and allowed space complexity of $O(m + nk)$. However, as $k$ gets lower, the accuracy of the model declines as well. In addition, sparsification required relying on an external, less natural, similarity measure, which we would have liked to avoid. Thus, we will now show how to retain the accuracy of the full dense prediction rule (26), while significantly lowering time and space complexity.

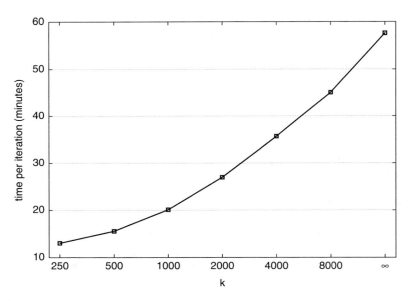

**Fig. 2** Running time per iteration of the globally optimized neighborhood model, as a function of the parameter $k$

### 5.4.1 Factoring Item-Item Relationships

We factor item-item relationships by associating each item $i$ with three vectors: $\mathbf{q}_i, \mathbf{x}_i, \mathbf{y}_i \in \mathbb{R}^f$. This way, we confine $w_{i,j}$ to be $\mathbf{q}_i^T \mathbf{x}_i$. Similarly, we impose $c_{i,j} = \mathbf{q}_i^T \mathbf{y}_j$. Essentially, these vectors strive to map items into an $f$-dimensional latent factor space where they are measured against various aspects that are revealed automatically by learning from the data. By substituting this into (26) we get the following prediction rule:

$$\hat{r}_{u,i} = \mu + b_u + b_i + |\mathrm{R}(u)|^{-\frac{1}{2}} \sum_{j \in \mathrm{R}(u)} [(r_{u,j} - b_{u,j})\mathbf{q}_i^T \mathbf{x}_j + \mathbf{q}_i^T \mathbf{y}_j] \qquad (29)$$

Computational gains become more obvious by using the equivalent rule

$$\hat{r}_{u,i} = \mu + b_u + b_i + \mathbf{q}_i^T \left( |\mathrm{R}(u)|^{-\frac{1}{2}} \sum_{j \in \mathrm{R}(u)} (r_{u,j} - b_{u,j})\mathbf{x}_j + \mathbf{y}_j \right). \qquad (30)$$

Notice that the bulk of the rule ($|\mathrm{R}(u)|^{-\frac{1}{2}} \sum_{j \in \mathrm{R}(u)} [(r_{u,j} - b_{u,j})\mathbf{x}_j + \mathbf{y}_j]$) depends only on $u$ while being independent of $i$. This leads to an efficient way to learn the model parameters. As usual, we minimize the regularized squared error function associated with (30)

$$\operatorname*{argmin}_{\mathbf{q}_*,\mathbf{x}_*,\mathbf{y}_*,b_*} \sum_{(u,i)\in\mathcal{K}} \left[ \left( r_{u,i} - \mu - b_u - b_i - \mathbf{q}_i^T \left( |\mathrm{R}(u)|^{-\frac{1}{2}} \sum_{j\in\mathrm{R}(u)} (r_{u,j} \right. \right. \right.$$

$$\left. \left. \left. - b_{u,j}) \mathbf{x}_j + \mathbf{y}_j \right) \right)^2 \right.$$

$$\left. + \lambda_{11} \left( b_u^2 + b_i^2 + \|\mathbf{q}_i\|^2 + \sum_{j\in\mathrm{R}(u)} \|\mathbf{x}_j\|^2 + \|\mathbf{y}_j\|^2 \right) \right]. \qquad (31)$$

Optimization is done by a stochastic gradient descent scheme, which is described in the following pseudo code:

```
LearnFactorizedNeighborhoodModel(Known ratings: r_{u,i}, rank: f)
% For each item i compute q_i, x_i, y_i ∈ ℝ^f
% which form a neighborhood model
Const #Iterations = 20, γ = 0.002, λ = 0.04
% Gradient descent sweeps:
for count = 1, ..., #Iterations do
    for u = 1, ..., m do
        % Compute the component independent of i:
        p_u ← |R(u)|^{-1/2} Σ_{j∈R(u)} (r_{u,j} − b_{u,j}) x_j + y_j
        sum ← 0
        for all i ∈ R(u) do
            r̂_{u,i} ← μ + b_u + b_i + q_i^T p_u
            e_{u,i} ← r_{u,i} − r̂_{u,i}
            % Accumulate information for gradient steps on x_i, y_i:
            sum ← sum + e_{u,i} · q_i
            % Perform gradient step on q_i, b_u, b_i:
            q_i ← q_i + γ · (e_{u,i} · p_u − λ · q_i)
            b_u ← b_u + γ · (e_{u,i} − λ · b_u)
            b_i ← b_i + γ · (e_{u,i} − λ · b_i)
        for all i ∈ R(u) do
            % Perform gradient step on x_i:
            x_i ← x_i + γ · (|R(u)|^{-1/2} · (r_{u,i} − b_{u,i}) · sum − λ · x_i)
            % Perform gradient step on y_i:
            y_i ← y_i + γ · (|R(u)|^{-1/2} · sum − λ · y_i)
return {q_i, x_i, y_i | i = 1, ..., n}
```

The time complexity of this model is linear with the input size, $O(f \cdot \sum_u(|\mathrm{R}(u)|)) = O(f \cdot |\mathcal{K}|)$, which is significantly better than the non-factorized model that required time $O(\sum_u |\mathrm{R}(u)|^2)$. We measured the performance of the model on the Netflix data; see Table 3. Accuracy is improved as we use more

**Table 3** Performance of the factorized item-item neighborhood model. The models with $\geqslant 200$ factors slightly outperform the non-factorized model, while providing much shorter running time

| #factors | 50 | 100 | 200 | 500 |
|---|---|---|---|---|
| RMSE | 0.9037 | 0.9013 | 0.9000 | 0.8998 |
| Time/iteration | 4.5 min | 8 min | 14 min | 34 min |

factors (increasing $f$). However, going beyond 200 factors could barely improve performance, while slowing running time. Interestingly, we have found that with $f \geqslant 200$ accuracy negligibly exceeds the best non-factorized model (with $k = \infty$). In addition, the improved time complexity translates into a big difference in wall-clock measured running time. For example, the time-per-iteration for the non-factorized model (with $k = \infty$) was close to 58 minutes. On the other hand, a factorized model with $f = 200$ could complete an iteration in 14 minutes without degrading accuracy at all.

The most important benefit of the factorized model is the reduced space complexity, which is $O(m + nf)$—linear in the input size. Previous neighborhood models required storing all pairwise relations between items, leading to a quadratic space complexity of $O(m + n^2)$. For the Netflix dataset which contains 17,770 movies, such quadratic space can still fit within core memory. However, some commercial recommenders process a much higher number of items. For example, large video sharing platforms, music streaming services or online shops offer millions of items. Storing the entire set of all pairwise relations becomes infeasible and instead, a sparse version of the pairwise relations can be considered. To this end, only values relating an item to its top-$k$ most similar neighbors are stored thereby reducing space complexity to $O(m + nk)$. However, a sparsification technique will inevitably degrade accuracy by missing important relations, as demonstrated in the previous section. In addition, identification of the top $k$ most similar items in such a high dimensional space is a non-trivial task that can require considerable computational efforts. All these issues do not exist in our factorized neighborhood model, which offers a linear time and space complexity without trading off accuracy.

The factorized neighborhood model resembles some latent factor models. The important distinction is that here we factorize the item-item relationships, rather than the ratings themselves. The results reported in Table 3 are comparable to those of the widely used SVD model, but not as good as those of SVD++; see Sect. 4. Nonetheless, the factorized neighborhood model retains the practical advantages of traditional neighborhood models discussed earlier—the abilities to explain recommendations and to immediately reflect new ratings.

As a side remark, we would like to mention that the decision to use three separate sets of factors was intended to give us more flexibility. Indeed, on the Netflix data this allowed achieving most accurate results. However, another reasonable choice could be using a smaller set of vectors, e.g., by requiring: $q_i = x_i$ (implying symmetric weights: $w_{i,j} = w_{j,i}$).

### 5.4.2    A User-User Model

A user-user neighborhood model predicts ratings by considering how like-minded users rated the same items. Such models can be implemented by switching the roles of users and items throughout our derivation of the item-item model. Here, we would like to concentrate on a user-user model, which is dual to the item-item model of (26). The major difference is replacing the $w_{i,j}$ weights relating item pairs, with weights relating user pairs:

$$\hat{r}_{u,i} = \mu + b_u + b_i + |R(i)|^{-\frac{1}{2}} \sum_{v \in R(i)} (r_{v,i} - b_{v,i}) w_{u,v} \tag{32}$$

The set $R(i)$ contains all the users who rated item $i$. Notice that here we decided to not account for implicit feedback. This is because adding such feedback was not very beneficial for the user-user model when working with the Netflix data.

User-user models can become useful in various situations. For example, some recommenders may deal with items that are rapidly replaced, thus making item-item relations very volatile. On the other hand, a stable user base enables establishment of long term relationships between users. An example of such a case is a recommender system for web articles or news items, which are rapidly changing by their nature; see, e.g., [15]. In such cases, systems centered around user-user relations are more appealing.

In addition, user-user approaches identify different kinds of relations that item-item approaches may fail to recognize, and thus can be useful on certain occasions. For example, suppose that we want to predict $r_{u,i}$, but none of the items rated by user $u$ is really relevant to $i$. In this case, an item-item approach will face obvious difficulties. However, when employing a user-user perspective, we may find a set of users similar to $u$, who rated $i$. The ratings of $i$ by these users would allow us to improve prediction of $r_{u,i}$.

The major disadvantage of user-user models is computational. Since typically there are many more users than items, pre-computing and storing all user-user relations, or even a reasonably sparsified version thereof, is overly expensive or completely impractical. In addition to the high $O(m^2)$ space complexity, the time complexity for optimizing model (32) is also much higher than its item-item counterpart, being $O(\sum_i |R(i)|^2)$ (notice that $|R(i)|$ is expected to be much higher than $|R(u)|$). These issues have rendered user-user models as a less practical choice.

**A Factorized Model**    All those computational differences disappear by factorizing the user-user model along the same lines as in the item-item model. Now, we associate each user $u$ with two vectors $p_u, z_u \in \mathbb{R}^f$. We assume the user-user relations to be structured as: $w_{u,v} = p_u^T z_v$. Let us substitute this into (32) to get

$$\hat{r}_{u,i} = \mu + b_u + b_i + |R(i)|^{-\frac{1}{2}} \sum_{v \in R(i)} (r_{v,i} - b_{v,i}) \mathbf{p}_u^T \mathbf{z}_v . \tag{33}$$

**Table 4** Performance of the factorized user-user neighborhood model

| #factors | 50 | 100 | 200 | 500 |
|---|---|---|---|---|
| RMSE | 0.9119 | 0.9110 | 0.9101 | 0.9093 |
| Time/iteration | 3 min | 5 min | 8.5 min | 18 min |

Once again, an efficient computation is achieved by including the terms that depends on $i$ but are independent of $u$ in a separate sum, so the prediction rule is presented in the equivalent form

$$\hat{r}_{u,i} = \mu + b_u + b_i + \mathbf{p}_u^T |R(i)|^{-\frac{1}{2}} \sum_{v \in R(i)} (r_{v,i} - b_{v,i}) \mathbf{z}_v . \tag{34}$$

In a parallel fashion to the item-item model, all parameters are learned in linear time $O(f \cdot \sum_i |R(i)|) = O(f \cdot |\mathcal{K}|)$. The space complexity is also linear with the input size being $O(n + mf)$. This significantly lowers the complexity of the user-user model compared to previously known results. In fact, running time measured on the Netflix data shows that now the user-user model is even faster than the item-item model; see Table 4. We should remark that unlike the item-item model, our implementation of the user-user model did not account for implicit feedback, which probably led to its shorter running time. Accuracy of the user-user model is significantly better than that of the widely-used correlation-based item-item model that achieves RMSE=0.9406 as reported in Fig. 1. Furthermore, accuracy is slightly better than the variant of the item-item model, which also did not account for implicit feedback (dotted curve in Fig. 1). This is quite surprising given the common wisdom that item-item methods are more accurate than user-user ones. It appears that a well implemented user-user model can match speed and accuracy of an item-item model. However, our item-item model could significantly benefit by accounting for implicit feedback.

**Fusing Item-Item and User-User Models** Since item-item and user-user models address different aspects of the data, overall accuracy is expected to improve by combining predictions of both models. Such an approach was previously suggested and was shown to improve accuracy; see, e.g. [5, 57]. However, past efforts were based on blending the item-item and user-user predictions during a post-processing stage, after each individual model was trained independently of the other model. A more principled approach optimizes the two models simultaneously, letting them know of each other while parameters are being learned. Thus, throughout the entire training phase each model is aware of the capabilities of the other model and strives to complement it. Our approach, which states the neighborhood models as formal optimization problems, allows doing that naturally. We devise a model that sums the item-item model (29) and the user-user model (33), leading to

$$\hat{r}_{u,i} = \mu + b_u + b_i + |R(u)|^{-\frac{1}{2}} \sum_{j \in R(u)} [(r_{u,j} - b_{u,j})\mathbf{q}_i^T \mathbf{x}_j + \mathbf{q}_i^T \mathbf{y}_j]$$

$$+ |R(i)|^{-\frac{1}{2}} \sum_{v \in R(i)} (r_{v,i} - b_{v,i})\mathbf{p}_u^T \mathbf{z}_v . \tag{35}$$

Model parameters are learned by stochastic gradient descent optimization of the associated squared error function. Our experiments with the Netflix data show that prediction accuracy is indeed better than that of each individual model. For example, with 100 factors the obtained RMSE is 0.8966, while with 200 factors the obtained RMSE is 0.8953.

Here we would like to comment that our approach allows integrating the neighborhood models also with completely different models in a similar way. For example, in [26] we showed an integrated model that combines the item-item model with a latent factor model (SVD++), thereby achieving improved prediction accuracy with RMSE below 0.887. Therefore, other possibilities with potentially better accuracy should be explored before considering the integration of item-item and user-user models.

## 5.5 Temporal Dynamics at Neighborhood Models

One of the advantages of the item-item model based on global optimization (Sect. 5.3), is that it enables us to capture temporal dynamics in a principled manner. As we commented earlier, user preferences are drifting over time, and hence it is important to introduce temporal aspects into CF models.

When adapting rule (26) to address temporal dynamics, two components should be considered separately. First component, $\mu + b_i + b_u$, corresponds to the baseline predictor portion. Typically, this component explains most variability in the observed signal. Second component, $|R(u)|^{-\frac{1}{2}} \sum_{j \in R(u)} (r_{u,j} - b_{u,j})w_{i,j} + c_{i,j}$, captures the more informative signal, which deals with user-item interaction. As for the baseline part, nothing changes from the factor model, and we replace it with $\mu + b_i(t_{u,i}) + b_u(t_{u,i})$, according to (5) and (8). However, capturing temporal dynamics within the interaction part requires a different strategy.

Item-item weights ($w_{i,j}$ and $c_{i,j}$) reflect inherent item characteristics and are not expected to drift over time. The learning process should capture unbiased long term values, without being too affected from drifting aspects. Indeed, the time changing nature of the data can mask much of the longer term item-item relationships if not treated adequately. For instance, a user rating both items $i$ and $j$ high within a short time period, is a good indicator for relating them, thereby pushing higher the value of $w_{i,j}$. On the other hand, if those two ratings are given five years apart, while the user's taste (if not her identity) could considerably change, this provides less evidence of any relation between the items. On top of this, we would argue that those considerations are pretty much user-dependent; some users are more consistent than others and allow relating their longer term actions.

Our goal here is to distill accurate values for the item-item weights, despite the interfering temporal effects. First we need to parameterize the decaying relations between two items rated by user $u$. We adopt exponential decay formed by the function $e^{-\beta_u \cdot \Delta t}$, where $\beta_u > 0$ controls the user specific decay rate and should be learned from the data. We also experimented with other decay forms, like the more computationally-friendly $(1 + \beta_u \Delta t)^{-1}$, which resulted in about the same accuracy, with an improved running time.

This leads to the prediction rule

$$\hat{r}_{u,i} = \mu + b_i(t_{u,i}) + b_u(t_{u,i}) + |R(u)|^{-\frac{1}{2}} \sum_{j \in R(u)} e^{-\beta_u \cdot |t_{u,i} - t_{u,j}|}((r_{u,j} - b_{u,j})w_{i,j} + c_{i,j}).$$

(36)

The involved parameters, $b_i(t_{u,i}) = b_i + b_{i,\text{Bin}(t_{u,i})}$, $b_u(t_{u,i}) = b_u + \alpha_u \cdot \text{dev}_u(t_{u,i}) + b_{u,t_{u,i}}$, $\beta_u$, $w_{i,j}$ and $c_{i,j}$, are learned by minimizing the associated regularized squared error

$$\sum_{(u,i) \in \mathcal{K}} \left[ \left( r_{u,i} - \mu - b_i - b_{i,\text{Bin}(t_{u,i})} - b_u - \alpha_u \text{dev}_u(t_{u,i}) - b_{u,t_{u,i}} - \right.\right.$$

$$|R(u)|^{-\frac{1}{2}} \sum_{j \in R(u)} e^{-\beta_u \cdot |t_{u,i} - t_{u,j}|}((r_{u,j} - b_{u,j})w_{i,j} + c_{i,j}) \Bigg)^2 +$$

$$\lambda_{12} \left( b_i^2 + b_{i,\text{Bin}(t_{u,i})}^2 + b_u^2 + \alpha_u^2 + b_{u,t}^2 + w_{i,j}^2 + c_{i,j}^2 \right) \Bigg].$$

(37)

Minimization is performed by stochastic gradient descent. We run the process for 25 iterations, with $\lambda_{12} = 0.002$, and step size (learning rate) of 0.005. An exception is the update of the exponent $\beta_u$, where we are using a much smaller step size of $10^{-7}$. Training time complexity is the same as the original algorithm, which is: $O(\sum_u |R(u)|^2)$. One can tradeoff complexity with accuracy by sparsifying the set of item-item relations as explained in Sect. 5.3.

As in the factor case, properly considering temporal dynamics improves the accuracy of the neighborhood model within the movie ratings dataset. The RMSE decreases from 0.9002 [26] to 0.8885. To our best knowledge, this is significantly better than previously known results by neighborhood methods. To put this in some perspective, this result is even better than those reported by using hybrid approaches such as applying a neighborhood approach on residuals of other algorithms [3, 39, 55]. A lesson is that addressing temporal dynamics in the data can have a more significant impact on accuracy than designing more complex learning algorithms. Compared to the matrix factorization counterparts the neighborhood methods show a higher RMSE, typically by about 0.01. For example, 0.8998 for the factorized item-item neighborhood model vs 0.8911 for SVD++, and 0.8885 for best neighborhood model with time dynamics vs 0.8799 for timeSVD++.

We would like to highlight an interesting point. Let $u$ be a user whose preferences are quickly drifting ($\beta_u$ is large). Hence, old ratings by $u$ should not be very influential on his status at the current time $t$. One could be tempted to decay the weight of $u$'s older ratings, leading to "instance weighting" through a cost function like

$$\sum_{(u,i)\in\mathcal{K}} e^{-\beta_u \cdot |t - t_{u,i}|}\left[\left(r_{u,i} - \mu - b_i - b_{i,\mathrm{Bin}(t_{u,i})} - b_u - \alpha_u \mathrm{dev}_u(t_{u,i}) - \right.\right.$$

$$\left.\left. b_{u,t_{u,i}} - |R(u)|^{-\frac{1}{2}} \sum_{j \in R(u)} ((r_{u,j} - b_{u,j})w_{i,j} + c_{i,j})\right)^2 + \lambda_{12}(\cdots)\right].$$

Such a function is focused at the *current* state of the user (at time $t$), while de-emphasizing past actions. We would argue against this choice, and opt for equally weighting the prediction error at all past ratings as in (37), thereby modeling *all* past user behavior. Therefore, equal-weighting allows us to exploit the signal at each of the past ratings, a signal that is extracted as item-item weights. Learning those weights would equally benefit from all ratings by a user. In other words, we can deduce that two items are related if users rated them similarly within a short time frame, even if this happened long ago.

## 5.6   Summary

This section follows a less traditional neighborhood based model, which unlike previous neighborhood methods is based on formally optimizing a global cost function. The resulting model is no longer localized, considering relationships between a small set of strong neighbors, but rather considers all possible pairwise relations. This leads to improved prediction accuracy, while maintaining some merits of the neighborhood approach such as explainability of predictions and ability to handle new ratings (or new users) without re-training the model.

The formal optimization framework offers several new possibilities. First, is a factorized version of the neighborhood model, which improves its computational complexity while retaining prediction accuracy. In particular, it is free from the quadratic storage requirements that limited past neighborhood models.

Second addition is the incorporation of temporal dynamics into the model. In order to reveal accurate relations among items, a proposed model learns how influence between two items rated by a user decays over time. Much like in the matrix factorization case, accounting for temporal effects results in a significant improvement in predictive accuracy.

## 6  Between Neighborhood and Factorization

This chapter is structured around two different approaches to CF: factorization and neighborhood. Each approach evolved from different basic principles, which led to distinct prediction rules. We also argued that factorization can lead to somewhat more accurate results, while neighborhood models may have some practical advantages. In this section we will show that despite those differences, the two approaches share much in common.

Let us consider the matrix factorization model of Sect. 4.1 (SVD), based on

$$\hat{r}_{u,i} = \mathbf{q}_i^T \mathbf{p}_u . \tag{38}$$

For simplicity, we ignore here the baseline predictors, but one can easily reintroduce them or just assume that they were subtracted from all ratings at an earlier stage.

We arrange all item-factors within the $n \times f$ matrix $Q = [\mathbf{q}_1 \mathbf{q}_2 \ldots \mathbf{q}_n]^T$. Similarly, we arrange all user-factors within the $m \times f$ matrix $P = [\mathbf{p}_1 \mathbf{p}_2 \ldots \mathbf{p}_m]^T$. We use the $n_u \times f$ matrix $Q[u]$ to denote the restriction of $Q$ to the items rated by $u$, where $n_u = |\mathrm{R}(u)|$. Let the vector $\mathbf{r}_u \in \mathbb{R}^{n_u}$ contain the ratings given by $u$ ordered as in $Q[u]$. Now, by activating (38) on all ratings given by $u$, we can reformulate it in a matrix form

$$\hat{\mathbf{r}}_u = Q[u]\mathbf{p}_u \tag{39}$$

For $Q[u]$ fixed, $\|\mathbf{r}_u - Q[u]\mathbf{p}_u\|_2$ is minimized by

$$\mathbf{p}_u = (Q[u]^T Q[u])^{-1} Q[u]^T \mathbf{r}_u$$

In practice, we will regularize with $\lambda \geqslant 0$ to get

$$\mathbf{p}_u = (Q[u]^T Q[u] + \lambda I)^{-1} Q[u]^T \mathbf{r}_u .$$

By substituting $\mathbf{p}_u$ in (39) we get

$$\hat{\mathbf{r}}_u = Q[u](Q[u]^T Q[u] + \lambda I)^{-1} Q[u]^T \mathbf{r}_u . \tag{40}$$

This expression can be simplified by introducing some new notation. Let us denote the $f \times f$ matrix $(Q[u]^T Q[u] + \lambda I)^{-1}$ as $W^u$, which should be considered as a weighting matrix associated with user $u$. Accordingly, the weighted similarity between items $i$ and $j$ from $u$'s viewpoint is denoted by $s_{i,j}^u = \mathbf{q}_i^T W^u \mathbf{q}_j$. Using this new notation and (40) the predicted preference of $u$ for item $i$ by SVD is rewritten as

$$\hat{r}_{u,i} = \sum_{j \in \mathrm{R}(u)} s_{i,j}^u r_{u,j} . \tag{41}$$

We reduced the matrix factorization model into a linear model that predicts preferences as a linear function of past actions, weighted by item-item similarity. Each past action receives a separate term in forming the prediction $\hat{r}_{u,i}$. This is equivalent to an item-item neighborhood model. Quite surprisingly, we transformed the matrix factorization model into an item-item model, which is characterized by:

- Interpolation is made from *all* past user ratings, not only from those associated with items most similar to the current one.
- The weight relating items $i$ and $j$ is factorized as a product of two vectors, one is related to $i$ and the other to $j$.
- Item-item weights are subject to a user-specific normalization, through the matrix $W^u$.

Those properties support our findings on how to best construct a neighborhood model. First, we showed in Sect. 5.3 that best results for neighborhood models are achieved when the neighborhood size (controlled by constant $k$) is maximal, such that all past user ratings are considered. Second, in Sect. 5.4 we touted the practice of factoring item-item weights. As for the user-specific normalization, we used a simpler normalizer: $n_u^{-0.5}$. It is likely that SVD suggests a more fundamental normalization by the matrix $W^u$, which would work better. However, computing $W^u$ would be expensive in practice. Another difference between our suggested item-item model and the one implied by SVD is that we chose to work with asymmetric weights ($w_{i,j} \neq w_{j,i}$), whereas in the SVD-induced rule: $s_{i,j}^u = s_{j,i}^u$.

In the derivation above we showed how SVD induces an equivalent item-item technique. In a fully analogous manner, it can induce an equivalent user-user technique, by expressing $q_i$ as a function of the ratings and the user factors. This brings us to three essentially equivalent models: SVD, item-item and user-user. Beyond linking SVD with neighborhood models, this also shows that user-user and item-item approaches, at the limit once well designed, are equivalent.

This last relation (between user-user and item-item approaches) can also be approached intuitively. Neighborhood models try to relate users to new items by following chains of user-item adjacencies. Such adjacencies represent preference- or rating-relations between the respective users and items. Both user-user and item-item models act by following exactly the same chains. They only differ in which "shortcuts" are exploited to speed up calculations. For example, recommending itemB to user1 would follow the chain user1–itemA–user2–itemB (user1 rated itemA, which was also rated by user2, who rated itemB). A user-user model follows such a chain with pre-computed user-user similarities. This way, it creates a "shortcut" that bypasses the sub-chain user1–itemB–user2, replacing it with a similarity value between user1 and user2. Analogously, an item-item approach follows exactly the same chain, but creates an alternative "shortcut", replacing the sub-chain itemA–user2–itemB with an itemA–itemB similarity value.

Another lesson here is that the distinction that deems neighborhood models as "memory based", while taking matrix factorization and the likes as "model based" is not always appropriate, at least not when using accurate neighborhood models that are model-based as much as SVD. In fact, the other direction is also true. The better matrix factorization models, such as SVD++, are also following memory-based habits, as they sum over all in-memory ratings when doing the online prediction; see rule (13). Hence, the traditional separation between "memory based" and "model based" techniques is not appropriate for categorizing the techniques surveyed in this chapter.

So far, we concentrated on relations between neighborhood models and matrix factorization models. However, in practice it may be beneficial to break these relations, and to augment factorization models with sufficiently different neighborhood models that are able to complement them. Such a combination can lead to improved prediction accuracy [4, 26]. A key to achieve this is by using the more localized neighborhood models, where the number of considered neighbors is limited. The limited number of neighbors might not be the best way to construct a standalone neighborhood model, but it makes the neighborhood model different enough from the factorization model in order to add a local perspective that the rather global factorization model misses.

## 7   Factorization Machines

The methods so far have been specifically designed for recommender systems. Now, we want to relate them to classical machine learning methods. First we show that linear regression can be used to obtain the baseline rating predictors $b_{u,i}$. Depending on the choice of input features, linear regression covers the different variants described in Sect. 3. Next, we introduce factorization machines and show how to obtain sophisticated rating predictor models such as SVD, SVD++, time-aware SVD++, factorized neighborhood models by changing the input feature vector. Surprisingly, the feature engineering required to obtain even seemingly complex models like time-aware SVD++ is very simple. Moreover, factorization machines do not require to derive learning algorithms for each model variation, but they can be used as libraries where only the input data and hyperparameters needs to be specified. This makes state-of-the art recommender system modeling easily accessible to practitioners.

### 7.1   Baseline Predictors Through Linear Regression

The solutions proposed in this section are based on feature engineering where each instance (here a user-item pair $(u, i)$) is described by a real valued feature vector $\mathbf{x} \in \mathbb{R}^p$ of length $p$ that is applied to a standard solver. We start our discussion with linear regression and show how they lead to the baseline predictors described in Sect. 3.

The linear model $\hat{r} : \mathbb{R}^p \rightarrow \mathbb{R}$ over an input vector $\mathbf{x} \in \mathbb{R}^p$ is

$$\hat{r}(\mathbf{x}) = \mu + \sum_{j=1}^{p} \mathbf{b}_j \, x_j = \mu + \mathbf{b}^T \mathbf{x}. \tag{42}$$

where $\mu \in \mathbb{R}$ is the global bias and $\mathbf{b} \in \mathbb{R}^p$ a linear contribution of each input variable, e.g., $b_j$ contains the effect for the variable $x_j$.

Obviously, the choice of $\mathbf{x}$ specifies what dependencies the linear model learns. For the collaborative filtering applications discussed before, the variable $\hat{r}$ that we want to predict is the rating and the prediction should depend on the user $u$ and item $i$. To apply linear regression, we need to encode the user $u$ and item $i$ in a real valued vector $\mathbf{x}$. Both user $u$ and item $i$ are categorical variables. The typical encoding of a categorical variable by a real-valued vector is *one-hot* encoding, where each possible choice of the variable is assigned one dimension in the feature vector (e.g., each dimension is assigned to a user), meaning the length $p$ of the feature vector $\mathbf{x}$ is equal to the number of possible choices for the categorical variable. When we want to encode that a instance is about a particular user $u$, the dimension of the user $u$ in $\mathbf{x}$ gets activated by assigning it a value of $x_u = 1$, and the remaining dimensions are assigned $x_j = 0$ for $j \neq u$. Likewise if we deal with two categorical variables, (e.g., user and item), the feature vector is a concatenation of the one-hot encoding of the user and the one hot encoding for the item. For sure, this can be extended to more variables and also other types of variables can be mixed.

To simplify notation, when dealing with several input variables, we describe $\mathbf{x}$ as a concatenation of several vectors, $[\mathbf{x}^A, \mathbf{x}^B, \ldots]$ and make the same notational assumption on $\mathbf{b}$ as well, i.e., that $\mathbf{b} = [\mathbf{b}^A, \mathbf{b}^B, \ldots]$. This makes the logic structure in the input feature more obvious and avoids introducing offsets. The equivalent linear regression model is

$$\hat{r}([\mathbf{x}^A, \mathbf{x}^B, \ldots]) = \mu + \sum_{j=1}^{p^A} \mathbf{b}_j^A \, x_j^A + \sum_{j=1}^{p^B} \mathbf{b}_j^B \, x_j^B + \ldots \tag{43}$$

### 7.1.1 Baseline with User and Item Biases

We now discuss several choices of the encoding $\mathbf{x}$ that recover the baseline predictors mentioned earlier. First, let $\mathbf{x}$ encode just the user $u$ and item $i$, i.e., $\mathbf{x} = [\mathbf{x}^U, \mathbf{x}^I]$ where the length of $\mathbf{x}^U$ is equal to the number of users, $m$, and the length of $\mathbf{x}^I$ is equal to the number of items, $n$. For a particular user-item pair $(u, i)$, all entries of $\mathbf{x}$ will be zeros besides the entry of $x_u^U = 1$ and $x_i^I = 1$. When we apply the linear model from Eq. (43), all elements $b_j^U$ from $\mathbf{b}^U$ where $j \neq u$ cancel out because $x_j^U = 0$ and only $b_u^U$ associated with the activated feature $x_u^U = 1$ remains. Likewise for the item part, only the linear effect for the activated item $i$ remains. The linear regression model is equivalent to

$$\hat{r}([\mathbf{x}^U, \mathbf{x}^I]) = \mu + b_u^U + b_i^I, \tag{44}$$

which is equivalent to the baseline predictor in (3).

### 7.1.2 Time Changing Baseline Predictors

Similarly, the time changing baseline predictors from Sect. 3.1 can be recovered by linear regression and a specific encoding. We will discuss the choices presented in Table 1 from a feature engineering perspective.

- The *mov* model is equivalent to a linear regression model where $\mathbf{x}$ encodes three categorical variables, one for the user $\mathbf{x}^U$, the item $\mathbf{x}^I$, and a third categorical variable encoding the time-bin for this item $\mathbf{x}^{I,T}$. Here, $\mathbf{x}^{I,T}$ has a separate choice for each item and time-bin combination. With the setting introduced in Sect. 3.1, $\mathbf{x}^{I,T}$ has a length of about $30 \cdot n$, i.e., for every item about 30 entries, where exactly one item-time bin combination gets activated. Applying the linear model to this feature vector encoding $\mathbf{x} = [\mathbf{x}^U, \mathbf{x}^I, \mathbf{x}^{I,T}]$ will result in an equivalent model to the temporal *mov* baseline.
- The *linear time* model can be achieved by concatenating an additional categorical feature for each user, where the value of the activated dimension is $x_u^{U,T} = \text{dev}_u(t_{u,i})$. Feeding $\mathbf{x} = [\mathbf{x}^U, \mathbf{x}^I, \mathbf{x}^{I,T}, \mathbf{x}^{U,T}]$ into a linear regression results in a model equivalent to the *linear time* model.
- For the *spline* model, instead of adding one dimension per user, we would add $k_u$ many (following Sect. 3.1) and set their feature values to $x_{u,l}^{U,T} = e^{-\sigma|t-t_l^u|} / \sum_{l'=1}^{k_u} e^{-\sigma|t-t_{l'}^u|}$.
- The feature-based representation of the *linear+* or *spline+* model adds a one-hot encoded categorical variable for the cross product between user id and day of the rating.

### 7.1.3 Global Neighborhood Model

The global neighborhood model can be expressed by a linear model as well. It corresponds to a feature vector $\mathbf{x}^{I,R}$ of size $n^2$, where each element encodes the strength of the effect from item $j$ to item $l$. For predicting the rating of a particular user-item pair $(u, i)$, all elements in $x_{i,j}^{I,R}$ are activated where $j \in R(u)$. The value of the activated element is $x_{i,j}^{I,R} = |R(u)|^{-\frac{1}{2}}(r_{u,j} - b_{u,j})$. Similarly, implicit feedback $N(u)$ can be added with another feature vector $\mathbf{x}^{I,N}$ of size $n^2$ where activated elements follow the same pattern but are set to $x_{i,j}^{I,N} = |R(u)|^{-\frac{1}{2}}$. The resulting linear regression on $\mathbf{x} = [\mathbf{x}^U, \mathbf{x}^I, \mathbf{x}^{I,R}, \mathbf{x}^{I,N}]$ is equivalent to the global neighborhood model of Eq. (26).

## 7.2 Factorization Machines for Collaborative Filtering

The previous discussion has shown how the baseline predictors can be achieved by linear regression with feature engineering. Now we present how the factorized models, SVD, SVD++, timeSVD++ and factorized neighborhood models, can be derived in a similar way. We start with introducing factorization machines.

Factorization machines (FM) [40] are a generalization of linear regression that include all pairwise interactions between the input variables. The general form of a FM over a p-dimensional feature vector $\mathbf{x} \in \mathbb{R}^p$ is

$$\hat{r}(\mathbf{x}) := \mu + \sum_{j=1}^{p} b_j \, x_j + \sum_{j=1}^{p} \sum_{l>j} \langle \mathbf{v}_j, \mathbf{v}_l \rangle \, x_j \, x_l \tag{45}$$

Where $V \in \mathbb{R}^{p \times f}$ are model parameters that are learned and $f$ is the embedding dimension. The first part of a FM is equivalent to a linear regression, the second part contains the pairwise interactions $x_j \, x_l$ between all input coordinates. An FM can be evaluated efficiently in linear time with respect to the non-zeros in $\mathbf{x}$, see [40] for details. While most instances of factorization machines use pairwise interactions, they can be extended to model higher order interactions as well using tensor models [40].

Next, we show how state of the art collaborative filtering models can be obtained with FM by feature engineering. Again, to simplify notation, we describe $\mathbf{x}$ as a concatenation of vectors $\mathbf{x} = [\mathbf{x}^A, \mathbf{x}^B, \ldots]$ and the corresponding model parameters use the same notation: $\mathbf{b} = [\mathbf{b}^A, \mathbf{b}^B, \ldots]$ and $V = [V^A, V^B, \ldots]$.

### 7.2.1 Encoding SVD and Neighborhood Models

Most of the models described throughout this chapter can be recovered by a factorization machine by simple feature engineering. First, to recover the SVD model, we reuse the one-hot encoding of user $\mathbf{x}^U$ and item $\mathbf{x}^I$ in the feature vector $\mathbf{x} = [\mathbf{x}^U, \mathbf{x}^I]$ from the previous discussion about baseline predictors using linear regression. That means the vector $\mathbf{x}$ has two non-zeros, one of the non-zeros identifies the user $u$ and the other the item $i$. Applying a factorization machine to this vector $\mathbf{x}$ and canceling all terms where either $x_j^U = 0$ or $x_j^I = 0$ results in:

$$\hat{r}([\mathbf{x}^U, \mathbf{x}^I]) = \mu + b_u^U + b_i^I + \langle \mathbf{v}_u^U, \mathbf{v}_i^I \rangle \tag{46}$$

which is equivalent to the SVD model with biases (see Eq. (12)).

Next, to recover SVD++ we need to include implicit information about the items the user has rated. This set of items (="bag of words"), $R(u)$, can be encoded by a multi-hot vector $\mathbf{x}^N$ of size $n$, where the activated coordinates are normalized by $|R(u)|^{-1/2}$. The FM model applied to this input vector is equivalent to

$$\hat{r}([\mathbf{x}^U, \mathbf{x}^I, \mathbf{x}^N]) = \overbrace{\mu + b_u^U + b_i^I + \langle \mathbf{v}_u^U, \mathbf{v}_i^I \rangle + |R(u)|^{-1/2} \sum_{j \in R(u)} \langle \mathbf{v}_i^I, \mathbf{v}_j^N \rangle}^{=\text{SVD++}} \quad (47)$$

$$+ |R(u)|^{-1/2} \sum_{j \in R(u)} \left( b_j^N + \left\langle \mathbf{v}_j^N, \mathbf{v}_u^U + |R(u)|^{-1/2} \sum_{l \in R(u), l > j} \mathbf{v}_l^N \right\rangle \right). \quad (48)$$

This factorization machine is equivalent to the SVD++ model but contains also some additional terms: bias terms for $b_j^N$ for implicit feedback and also interactions between two implicit feedback items as well as between user and implicit feedback. These additional terms are independent of the target item $i$ and depend only on the user $u$. In this sense, they represent a kind of user bias based on implicit feedback instead of the user id. If these new biases are not useful for predicting the rating, the user bias $b_u^U$ might be able to overwrite any negative effect of these additional terms.

Similar to the SVD++ encoding, we can recover the factorized neighborhood model from Sect. 5.4. Instead of encoding the binary implicit information in the feature vector, the ratings are encoded. For the item-based neighborhood model, choose $\mathbf{x}^R$ of length $n$ and for every item $j$ rated by the user $u$ for whom we want to predict a rating, set $x_j = |R(u)|^{-1/2}(r_{u,j} - b_{u,j})$. A factorization machine on the input vector $[\mathbf{x}^I, \mathbf{x}^R]$ is equivalent to the factorized item-based neighborhood model modulo some extra user bias terms similar to the FM-SVD++ model.

For adding time effects, we can follow any of the feature encodings covered in Sect. 7.1.2 and apply an FM on top of these feature vectors. Here, we describe shortly a very simple time encoding that works surprisingly well. Instead of introducing time varying model biases or factors for a user or item, we simply encode time as a categorical variable $\mathbf{x}^D$ that has one entry for each day in the dataset. The entry $x_t^D$ of the day $t$ of a rating is activated. An FM on $\mathbf{x} = [\mathbf{x}^U, \mathbf{x}^I, \mathbf{x}^D]$ is equivalent to

$$\hat{r}([\mathbf{x}^U, \mathbf{x}^I, \mathbf{x}^T]) = \mu + b_u^U + b_i^I + b_t^D + \langle \mathbf{v}_u^U, \mathbf{v}_i^I \rangle + \langle \mathbf{v}_u^U, \mathbf{v}_t^D \rangle + \langle \mathbf{v}_i^I, \mathbf{v}_t^D \rangle. \quad (49)$$

The effect $\langle \mathbf{v}_u^U, \mathbf{v}_t^D \rangle$ can be seen as a factorized version of the time-varying baseline predictor $b_u(t)$—a similar argument can be made for items. For sure, a more sophisticated (and larger) time model could be build by adding a time encoding such as the cross product between users and day or a spline as discussed in Sect. 7.1.2.

Obviously, adding the vector $\mathbf{x}^N$, which encodes implicit information, results in a time-aware SVD++ model.

### 7.2.2    Attribute-Aware "Hybrid" Models

Most models in this chapter represent items or users by a unique identifier. However, in many applications additional information about the items or users exists. Such information can be a movie's metadata like its genres, actors, director, release year, etc. Such information can be helpful for improving the recommender model. For example, the model could learn that a user likes to watch movies starring a particular actor, or that a user prefers recent movies over older ones. Similar to the effects we have described before, attributes can show up as biases, e.g., an actor is in general popular, or tied to a user, e.g., a user likes an actor less than usual. The resulting models, which blend collaborative filtering (users behavior) with content filtering (user and item attributes), are commonly known as *hybrid recommendation models*.

A factorization machine can add attribute information to any of the models discussed above. The most straightforward use of attributes is to encode them in the feature vector with the rules discussed earlier in this section: (1) a categorical variable such as the *director* can be encoded by a one-hot vector similar to the user or item identifier. (2) A set such as the *genres* of a movie can be encoded by a multi-hot vector with the dimension of the overall number of genres and where the genres of the particular movie are non-zero. The vector can be normalized similarly to the implicit vectors. (3) A numerical variable such as the *release date* can be treated as a categorical variable, or an interpolation method such as the *linear* or *spline* model could be chosen. More sophisticated encoding schemes might help to learn more complex patterns. In general, with factorization machines, experimenting with feature encodings is easy because feature exploration is done in the preprocessing step and does not require to change the learning algorithm. When several attributes are available, they can be encoded separately and concatenated in a vector $\mathbf{x}^A$ that holds all attribute information associated with a rating event.

After the attribute information has been encoded in a real valued feature vector $\mathbf{x}^A$, it can be simply concatenated to any of the previously discussed models. For example, it can be appended to $\mathbf{x}^U$ and $\mathbf{x}^I$ to obtain a matrix factorization model with attributes similar to [1]. Or a complex attribute-aware timeSVD++ can be achieved by concatenating $[\mathbf{x}^U, \mathbf{x}^I, \mathbf{x}^T, \mathbf{x}^N, \mathbf{x}^A]$. Depending on the application, instead of adding attributes to existing models, it might make sense to replace some of the variables by attributes. For example, in a pure cold start setting, where all items are new, it is beneficial not to use $\mathbf{x}^I$ for representing items at all but instead rely on the item's attributes, $\mathbf{x}^A$. This will ensure that during training none of the item effects are captured in $V^I$ but instead item effects are captured in the attribute embeddings $V^A$.

## 7.3    *Learning Factorization Machines*

An advantage of a feature based method such as FMs is that we can design and implement learning algorithms for the general model and then apply this algorithm to solve any feature vector. We describe some common algorithms briefly.

The models presented in Sects. 4 and 5 were optimized by stochastic gradient descent (SGD). Individual learning algorithms were derived for each model. Now, we show an SGD algorithm for FM which can handle any input vector. The algorithm simply iterates over the feature vectors $\mathbf{x}$ of all user-item pairs from the training data $\mathcal{K}$, computes the error $e(\mathbf{x}) = r_{u,i} - \hat{r}(\mathbf{x})$ and applies the following updates:

$$\mu \leftarrow \mu + \gamma \cdot (e(\mathbf{x}) - \lambda\mu) \tag{50}$$

$$\mathbf{b} \leftarrow \mathbf{b} + \gamma \cdot (e(\mathbf{x})\mathbf{x} - \lambda\mathbf{b}) \tag{51}$$

$$\mathbf{v}_j \leftarrow \mathbf{v}_j + \gamma \cdot (e(\mathbf{x})x_j \sum_{l \neq j} x_l \mathbf{v}_l - \lambda\mathbf{v}) = \mathbf{v}_j + \gamma \cdot (e(\mathbf{x})x_j(\mathbf{x}V - x_j\mathbf{v}_j) - \lambda\mathbf{v})$$

$$\tag{52}$$

An efficient implementation of this algorithm takes the sparsity of $\mathbf{x}$ into account and iterates only over the non-zero elements of $\mathbf{x}$. Such a sparse implementation makes the general FM model as computational efficient as the specialized models, e.g., FM-SVD has the same computational complexity as the SVD algorithm sketched in Sect. 4.1.

Similarly, we can derive alternating least squares algorithms [41] or Bayesian methods [17] for the general FM model class. With a sparse implementation, they have the same computational complexity as the SGD algorithm. An added benefit of Bayesian methods is that they can avoid the time-consuming search for regularization parameters by making inference over their values. More details about applying these learning methods to FMs can be found in [41]. Solvers for FMs are available as standalone tools or as part of machine learning frameworks. Practitioners can reuse such existing implementations to design advanced CF solutions.

## 7.4   Comparison

As we have discussed before, FMs recover the advanced CF methods presented in Sects. 4 and 5 by simple feature engineering. Table 5 shows results for a Bayesian FM with different choices of input features. The table discusses four simple features: user id, item id, implicit item information and the day variable. The results were generated with the Bayesian MCMC solver using the software package libFM [41]. As expected, the results of the factorization machine variations (Table 5) show similar behavior to the previous experiments (Table 2). The FM-SVD and FM-SVD++ models show slightly better quality due to Bayesian inference, and the FM-timeSVD++ shows slightly worse quality compared to timeSVD++ from Sect. 4.3 likely because it uses a simpler representation of time.

**Table 5** Bayesian factorization machines, results from [42], all results with $f = 128$, additional results for larger $f$ and more features see [42]

| Input features for FM | Related CF model | RMSE |
|---|---|---|
| User $\mathbf{x}^U$, item $\mathbf{x}^I$ | SVD | 0.8937 |
| User $\mathbf{x}^U$, item $\mathbf{x}^I$, implicit $\mathbf{x}^N$ | SVD++ | 0.8865 |
| User $\mathbf{x}^U$, item $\mathbf{x}^I$, day $\mathbf{x}^D$ | timeSVD | 0.8873 |
| User $\mathbf{x}^U$, item $\mathbf{x}^I$, implicit $\mathbf{x}^N$, day $\mathbf{x}^D$ | timeSVD++ | 0.8809 |

## 7.5   Discussion

This section illustrated how factorization machines can mimic the collaborative models introduced in Sects. 4 and 5 by feature engineering. Interestingly, the features are not opaque but very intuitive. For instance, a simple categorical one-hot encoding of user and item id yields SVD, an additional bag of words encoding of implicit feedback yields SVD++ and a bag of ratings results in factorized KNN. Time modeling corresponds to bucketizing the time variable, or building explicit crosses. Factorization machines make it trivial to combine all these ideas by simply concatenating the input representations. This suggests a powerful framework for extending state-of-the art CF models to new domains and experimenting with new ideas. For example, cold start recommendation can be addressed by adding a feature representation of item meta data to the input vector $\mathbf{x}$, context aware recommendation by adding contextual information. Section 8.3 discusses shortly some other applications that can benefit from the patterns identified in this chapter.

## 8   Beyond Rating Prediction for the Netflix Prize

This chapter used rating prediction for the Netflix prize to motivate and derive new collaborative filtering approaches. For sure, the field of recommender system is much more diverse with many different problem settings. We will discuss how the lessons and patterns described in this chapter generalize to other settings such as solving item recommendation tasks, and building models for cold-start, sequential or context-aware recommendation. We start by covering more recent results for rating prediction before discussing other settings such as item recommendation and context-aware recommendation.

## 8.1   Recent Results for Rating Prediction

A decade after the Netflix prize concluded, matrix factorization and nearest neighbor methods as described in this work, still play an important role in collaborative filtering. Unfortunately, recent publications on CF methods are usually not evaluated on the Netflix prize, so it is not possible to compare their performance against the large body of work published on the Netflix prize. We are not aware of any published method that provides substantially better results on the Netflix prize than the approaches described in this chapter.

Instead of reusing the Netflix prize benchmark, more recent work on rating prediction uses different benchmarks. Movielens 10M with a 90:10 split has become one of the commonly used benchmarks. While new CF methods proposed over the last years initially suggested big improvements, with a proper setup the proposed methods in this chapter still yield the best results. Table 6 shows an overview of the Movielens 10M benchmark. The first group of results are obtained by rating prediction methods that have been proposed between 2013 and 2019. The table reports the numbers that the authors of the corresponding papers achieve with their methods. The second group of results [44] are the collaborative filtering methods discussed in this chapter. This second group of results was obtained with a factorization machine with different choices of features to mimic SVD, SVD++, timeSVD++, etc. as described in Sect. 7. The choices of features and the setup of the method is analogously to the experiments conducted earlier on the Netflix prize (Table 5). As can be seen, even the simple Bayesian SVD model is as good as any of the newly proposed methods. The time-aware and implicit versions suggested in this chapter further improve the results substantially.

**Table 6** Movielens 10M results: First group are newly proposed methods between 2013 and 2019. Second group are models proposed in this chapter obtained by a Bayesian FM. See [44] for details

| Method | RMSE | Comment |
|---|---|---|
| I-AUTOREC [49] | 0.782 | Result from [49] |
| LLORMA [29] | 0.7815 | Result from [29] |
| WEMAREC [9] | 0.7769 | Result from [9] |
| I-CFN++ [51] | 0.7754 | Result from [51] |
| MPMA [10] | 0.7712 | Result from [10] |
| CF-NADE 2 layers [58] | 0.771 | Result from [58] |
| SMA [31] | 0.7682 | Result from [31] |
| GLOMA [11] | 0.7672 | Result from [11] |
| ERMMA [32] | 0.7670 | Result from [32] |
| AdaError [33] | 0.7644 | Result from [33] |
| MRMA [30] | 0.7634 | Result from [30] |
| Bayesian SVD | 0.7633 | FM features: user, item |
| Bayesian timeSVD | 0.7587 | FM features: user, item, day |
| Bayesian SVD++ | 0.7563 | FM features: user, item, R(u) |
| Bayesian timeSVD++ | 0.7523 | FM features: user, item, day, R(u) |
| Bayesian timeSVD++ flipped | 0.7485 | FM features: user, item, day, R(u), R(i) |

## 8.2   *Item Recommendation*

While algorithmic research on recommender systems during the Netflix prize was focused on rating prediction, more recent research shifted the focus to top-k item recommendation. The goal of top-k item recommendation is to retrieve the $k$ most relevant items for a user from the full catalogue of items. Item recommendation models are usually not learned from rating data (e.g., star ratings) but from binary implicit feedback, where the positive labels are the items a user has selected in the past and the negatives are all remaining items. Item recommendation depends on a scoring function that is used to compute a user's preference score for each item, and then the items are ranked by this score. The main difference between item recommendation and rating prediction is how the scoring function is learned. Chapter "Item Recommendation from Implicit Feedback" covers the unique challenges and learning methods for item recommendation in detail.

Besides this differences in learning, item recommendation and rating prediction share the same types of scoring functions, and the functions, $\hat{r}$, as well as the patterns described in this chapter can be apply to item recommendation. We will discuss shortly some adaption of matrix factorization and neighborhood methods for item recommendation. The simplest adaption of the methods proposed in this chapter to item recommendation is to create a dense binary 'rating' matrix where all items that a user has selected in the past get a 'rating' of 1 and the remaining ones a 'rating' of 0. Applying the SVD technique described in Sect. 4.1 to this dense data is equivalent to the PureSVD [13] method proposed for item recommendation. Adding weights to the loss function and downweighting the negative class, is equivalent to the popular iALS [24] method. Applying the neighborhood model from Sect. 5.3 is similar to the recently proposed EASE [50] and also to SLIM [37]. The baseline models in Sect. 3 result in *most-popular* recommenders. It should be noted, that even though the aforementioned methods share the same scoring function and optimization objective, they propose more efficient learning algorithms tailored to the unique computational challenges of item recommendation. The models proposed in this chapter can also be used with other losses. For example, pairwise losses have been proposed for SVD, neighborhood models [43] and for factorization machines in general [2, 40].

Similar to the findings of Sect. 8.1, a recent meta study [14] found that matrix factorization (iALS) and neighborhood methods (e.g., SLIM and EASE) achieve highly competitive results for item recommendation outperforming many newly proposed methods.

To summarize, the functions $\hat{r}_{u,i}$ that were introduced in Sects. 3–7 can be seen as a *scoring function*. The same classes of scoring functions can be used for rating prediction and item recommendation. The difference between an item recommender and a rating prediction model is how the parameters of the scoring function are learned and how the scoring function is applied. Empirically, classes of scoring functions that have shown good performance for rating prediction, also work well for item recommendation. This indicates that patterns learned from one task are helpful for the other task as well.

## 8.3  Generalizing to Other Recommendation Tasks

The previous subsection covered the difference between the prediction task, where either items are retrieved/ranked or ratings are predicted. Another large body of recommender tasks deal with problems where the input is more complex. In this case, we focus on the variables that are used to make a prediction. So far, we have discussed the basic case where the input is a user and an item. Section 3.1 also extended this with time. Now we show how the patterns introduced in this chapter to build high-quality scoring functions can be applied to other recommender tasks where auxiliary information is available to extend the scoring function. Factorization machines are especially suited for these applications because they can encode additional information as part of the input vector.

For example, cold-start recommendation requires to make prediction for items or users with little or no training data. This is typically solved by adding some side information such as meta data of items to the scoring function. With a factorization machine, this meta data can be encoded and appended to the vector $\mathbf{x}$. For sequential recommendation, where the prediction should depend on the previous user actions, Markov chains are a convenient choice and it was shown that a factorized Markov chain [40] recommender can be simply achieved by adding the previous actions to a factorization machine, similar to the pattern that this chapter established for implicit feedback $R(u)$. Similarly, for context aware recommendation where the prediction depends on the current situation, e.g., location, mood, time, the context can be added as additional variables to a factorization machine resulting in a context-aware scoring function.

## 8.4  Neural Networks

In general, the scoring function $\hat{r}$ can be modeled by any function, including neural networks. Neural networks have been used for more than a decade as recommender system models but have gained increased attention in recent years. Restricted Boltzmann Machines [46] have been successfully applied to the Netflix Prize competition achieving a quality similar to the basic SVD model. Furthermore, blending predictions of matrix factorization and Restricted Boltzmann Machines has proved to yield meaningfully better accuracy than each method on its own. Another popular network structure are autoencoders that have been used both for rating prediction [49] (see also Sect. 8.1) and item recommendation [34]. Shallow autoencoders are closely related to learned neighborhood models (see Sect. 8.2) such as EASE [50], SLIM [37] and the global neighbor approach from Sect. 5.3.

Another direction for modeling the scoring function of a recommender system is applying a multi-layer perceptron over a learned input representation [12]. Here, categorical variables such as a user id or movie id are represented by a learned embedding vector, categorical set variables are represented by an aggregation (e.g., average) over learned embedding vectors, and numerical variables are appended either directly or with some transformation such as a logarithm. The concatenated vector is then passed through a multi-layer perceptron. This approach is similar to factorization machines as discussed in Sect. 7, and the embedding lookups of a neural network are analogous to the multiplication of the one-hot-encoding of the FM input vector $\mathbf{x}$ with the embeddings $V$. The main difference is that a FM has a fixed interaction structure over this learned representation, while the interaction structure is learned for the neural network. Similar to factorization machines, additional information such as context or attributes can be easily added to the input representation of the neural network. For the task of item recommendation, it is generally beneficial to separate the representation of the user (query) side from the item side and merge them at the very end of the network with a dot product. This will enable very efficient retrieval. More details on such dot product models can be found in Chap. "Item Recommendation from Implicit Feedback".

The use of sequence data has also been studied extensively in neural networks, most notably in the area of natural language processing, and the network patterns established there have been applied successfully for recommendations, including recurrent neural networks [22] and transformers [56].

# References

1. D. Agarwal, B.-C. Chen, Regression-based latent factor models, in *Proceedings of the 15th ACM SIGKDD International Conference on Knowledge Discovery and Data Mining, KDD'09* (Association for Computing Machinery, New York, 2009)
2. I. Bayer, X. He, B. Kanagal, S. Rendle, A generic coordinate descent framework for learning from implicit feedback, in *Proceedings of the 26th International Conference on World Wide Web, WWW'17* (2017), pp. 1341–1350
3. R.M. Bell, Y. Koren, Scalable collaborative filtering with jointly derived neighborhood interpolation weights, in *Seventh IEEE International Conference on Data Mining (ICDM 2007)* (2007), pp. 43–52
4. R.M. Bell, Y. Koren, Lessons from the netflix prize challenge. SIGKDD Explor. Newsl. **9**(2), 75–79 (2007)
5. R. Bell, Y. Koren, C. Volinsky, Modeling relationships at multiple scales to improve accuracy of large recommender systems, in *Proceedings of the 13th ACM SIGKDD International Conference on Knowledge Discovery and Data Mining, KDD'07* (Association for Computing Machinery, New York, 2007), pp. 95–104
6. J. Bennett, S. Lanning, The netflix prize, in *In KDD Cup and Workshop in Conjunction with KDD* (2007)
7. D.M. Blei, A.Y. Ng, M.I. Jordan, Latent dirichlet allocation. J. Mach. Learn. Res. **3**, 993–1022 (2003)
8. J. Canny, Collaborative filtering with privacy via factor analysis, in *Proceedings of the 25th Annual International ACM SIGIR Conference on Research and Development in Information Retrieval, SIGIR'02* (Association for Computing Machinery, New York, 2002), pp. 238–245

9. C. Chen, D. Li, Y. Zhao, Q. Lv, L. Shang, WEMAREC: accurate and scalable recommendation through weighted and ensemble matrix approximation, in *Proceedings of the 38th International ACM SIGIR Conference on Research and Development in Information Retrieval, SIGIR'15* (ACM, New York, 2015), pp. 303–312

10. C. Chen, D. Li, Q. Lv, J. Yan, S.M. Chu, L. Shang, MPMA: mixture probabilistic matrix approximation for collaborative filtering, in *Proceedings of the Twenty-Fifth International Joint Conference on Artificial Intelligence, IJCAI'16* (AAAI Press, Palo Alto, 2016), pp. 1382–1388

11. C. Chen, D. Li, Q. Lv, J. Yan, L. Shang, S. Chu, GLOMA: embedding global information in local matrix approximation models for collaborative filtering, in *AAAI Conference on Artificial Intelligence* (2017)

12. P. Covington, J. Adams, E. Sargin, Deep neural networks for youtube recommendations, in *Proceedings of the 10th ACM Conference on Recommender Systems, RecSys'16* (Association for Computing Machinery, New York, 2016), pp. 191–198

13. P. Cremonesi, Y. Koren, R. Turrin, Performance of recommender algorithms on top-n recommendation tasks, in *Proceedings of the Fourth ACM Conference on Recommender Systems, RecSys'10* (Association for Computing Machinery, New York, 2010), pp. 39–46

14. M.F. Dacrema, S. Boglio, P. Cremonesi, D. Jannach, A troubling analysis of reproducibility and progress in recommender systems research. ACM Trans. Inf. Syst. **39**(2), Article 20 (2021). https://doi.org/10.1145/3434185

15. A.S. Das, M. Datar, A. Garg, S. Rajaram, Google news personalization: scalable online collaborative filtering, in *Proceedings of the 16th International Conference on World Wide Web, WWW'07* (Association for Computing Machinery, New York, 2007), pp. 271–280

16. S. Deerwester, S.T. Dumais, G.W. Furnas, T.K. Landauer, R. Harshman, Indexing by latent semantic analysis. J. Am. Soc. Inf. Sci. **41**(6), 391–407 (1990)

17. C. Freudenthaler, L. Schmidt-Thieme, S. Rendle, Bayesian factorization machines, in *Proceedings of the NIPS Workshop on Sparse Representation and Low-rank Approximation* (2011)

18. S. Funk, Netflix update: try this at home (2006). http://sifter.org/~simon/journal/20061211.html

19. A. Gelman, J.B. Carlin, H.S. Stern, D.B. Rubin, *Bayesian Data Analysis* (Chapman and Hall, London, 1995)

20. J.L. Herlocker, J.A. Konstan, A. Borchers, J. Riedl, An algorithmic framework for performing collaborative filtering, in *Proceedings of the 22nd Annual International ACM SIGIR Conference on Research and Development in Information Retrieval, SIGIR'99* (Association for Computing Machinery, New York, 1999), pp. 230–237

21. J.L. Herlocker, J.A. Konstan, J. Riedl, Explaining collaborative filtering recommendations, in *Proceedings of the 2000 ACM Conference on Computer Supported Cooperative Work, CSCW'00* (Association for Computing Machinery, New York, 2000), pp. 241–250

22. S. Hochreiter, J. Schmidhuber, Long short-term memory. Neural Comput. **9**(8), 1735–1780 (1997)

23. T. Hofmann, Latent semantic models for collaborative filtering. ACM Trans. Inf. Syst. **22**(1), 89–115 (2004)

24. Y. Hu, Y. Koren, C. Volinsky, Collaborative filtering for implicit feedback datasets, in *Proceedings of the 2008 Eighth IEEE International Conference on Data Mining, ICDM'08* (2008), pp. 263–272

25. D. Kim, B.-J. Yum, Collaborative filtering based on iterative principal component analysis. Expert Syst. Appl. **28**(4), 823–830 (2005)

26. Y. Koren, Factorization meets the neighborhood: a multifaceted collaborative filtering model, in *Proceedings of the 14th ACM SIGKDD International Conference on Knowledge Discovery and Data Mining, KDD'08* (Association for Computing Machinery, New York, 2008), pp. 426–434

27. Y. Koren, Collaborative filtering with temporal dynamics, in *Proceedings of 15th ACM SIGKDD International Conference on Knowledge Discovery and Data Mining* (ACM, New York, 2009), pp. 447–456

28. Y. Koren, Factor in the neighbors: scalable and accurate collaborative filtering. ACM Trans. Knowl. Discov. Data **4**(1), 1–24 (2010)
29. J. Lee, S. Kim, G. Lebanon, Y. Singer, Local low-rank matrix approximation, in *Proceedings of the 30th International Conference on International Conference on Machine Learning - Volume 28, ICM'13*. JMLR.org. (2013), pp. II–82–II–90
30. D. Li, C. Chen, W. Liu, T. Lu, N. Gu, S. Chu, Mixture-rank matrix approximation for collaborative filtering, in *Advances in Neural Information Processing Systems 30*, ed. by I. Guyon, U.V. Luxburg, S. Bengio, H. Wallach, R. Fergus, S. Vishwanathan, R. Garnett (Curran Associates, Red Hook, 2017), pp. 477–485
31. D. Li, C. Chen, Q. Lv, J. Yan, L. Shang, S.M. Chu, Low-rank matrix approximation with stability, in *Proceedings of the 33rd International Conference on Machine Learning - Volume 48, ICML'16*. JMLR.org (2016), pp. 295–303
32. D. Li, C. Chen, Q. Lv, L. Shang, S. Chu, H. Zha, ERMMA: expected risk minimization for matrix approximation-based recommender systems, in *AAAI Conference on Artificial Intelligence* (2017)
33. D. Li, C. Chen, Q. Lv, H. Gu, T. Lu, L. Shang, N. Gu, S.M. Chu, AdaError: an adaptive learning rate method for matrix approximation-based collaborative filtering, in *Proceedings of the 2018 World Wide Web Conference, WWW'18*, Republic and Canton of Geneva, Switzerland. International World Wide Web Conferences Steering Committee (2018), pp. 741–751
34. D. Liang, R.G. Krishnan, M.D. Hoffman, T. Jebara, Variational autoencoders for collaborative filtering, in *Proceedings of the 2018 World Wide Web Conference, WWW'18*, Republic and Canton of Geneva, CHE. International World Wide Web Conferences Steering Committee (2018)
35. G. Linden, B. Smith, J. York, Amazon.com recommendations: item-to-item collaborative filtering. IEEE Internet Comput. **7**(1), 76–80 (2003)
36. B.M. Marlin, R.S. Zemel, S. Roweis, M. Slaney, Collaborative filtering and the missing at random assumption, in *Proceedings of the Twenty-Third Conference on Uncertainty in Artificial Intelligence, UAI'07* (AUAI Press, Arlington, 2007), pp. 267–275
37. X. Ning, G. Karypis, SLIM: Sparse linear methods for top-n recommender systems, in *Proceedings of the 2011 IEEE 11th International Conference on Data Mining, ICDM'11* (IEEE Computer Society, Washington, 2011), pp. 497–506
38. D.W. Oard, J. Kim, Implicit feedback for recommender systems, in *Proceedings of 5th DELOS Workshop on Filtering and Collaborative Filtering* (1998), pp. 31–36
39. A. Paterek, Improving regularized singular value decomposition for collaborative filtering, in *Proceedings of KDD Cup and Workshop* (2007)
40. S. Rendle, Factorization machines, in *Proceedings of the 2010 IEEE International Conference on Data Mining, ICDM'10* (IEEE Computer Society, Washington, 2010), pp. 995–1000
41. S. Rendle, Factorization machines with libFM. ACM Trans. Intell. Syst. Technol. **3**(3), 57:1–57:22 (2012)
42. S. Rendle, Scaling factorization machines to relational data, in *Proceedings of the 39th International Conference on Very Large Data Bases, PVLDB'13*. VLDB Endowment (2013), pp. 337–348
43. S. Rendle, C. Freudenthaler, Z. Gantner, L. Schmidt-Thieme, BPR: Bayesian personalized ranking from implicit feedback, in *Proceedings of the Twenty-Fifth Conference on Uncertainty in Artificial Intelligence, UAI'09* (AUAI Press, Arlington, 2009), pp. 452–461
44. S. Rendle, L. Zhang, Y. Koren, On the difficulty of evaluating baselines: a study on recommender systems. CoRR, abs/1905.01395 (2019)
45. R. Salakhutdinov, A. Mnih, Probabilistic matrix factorization, in *Proceedings of the 20th International Conference on Neural Information Processing Systems, NIPS'07* (Curran Associates, Red Hook, 2007), pp. 1257–1264
46. R. Salakhutdinov, A. Mnih, G. Hinton, Restricted boltzmann machines for collaborative filtering, in *Proceedings of the 24th International Conference on Machine Learning, ICML'07* (Association for Computing Machinery, New York, 2007), pp. 791–798

47. B.M. Sarwar, G. Karypis, J.A. Konstan, J.T. Riedl, Application of dimensionality reduction in recommender system – a case study, in *WEBKDD'2000* (2000)
48. B. M. Sarwar, G. Karypis, J. Konstan, J. Riedl, Item-based collaborative filtering recommendation algorithms, in *Proceedings of the 10th International Conference on World Wide Web, WWW'01* (Association for Computing Machinery, New York, 2001), pp. 285–295
49. S. Sedhain, A.K. Menon, S. Sanner, L. Xie, AutoRec: autoencoders meet collaborative filtering, in *Proceedings of the 24th International Conference on World Wide Web, WWW'15 Companion* (ACM, New York, 2015), pp. 111–112
50. H. Steck, Embarrassingly shallow autoencoders for sparse data, in *The World Wide Web Conference, WWW'19* (Association for Computing Machinery, New York, 2019), pp. 3251–3257
51. F. Strub, J. Mary, R. Gaudel, Hybrid recommender system based on autoencoders. CoRR, abs/1606.07659 (2016)
52. G. Takács, I. Pilászy, B. Németh, D. Tikk, Major components of the gravity recommendation system. SIGKDD Explor. Newsl. **9**(2), 80–83 (2007)
53. G. Takács, I. Pilászy, B. Németh, D. Tikk, Matrix factorization and neighbor based algorithms for the netflix prize problem, in *Proceedings of the 2008 ACM Conference on Recommender Systems, RecSys'08* (Association for Computing Machinery, New York, 2008), pp. 267–274
54. N. Tintarev, J. Masthoff, A survey of explanations in recommender systems, in *Proceedings of the 2007 IEEE 23rd International Conference on Data Engineering Workshop, ICDEW'07* (IEEE Computer Society, Washington, 2007), pp. 801–810
55. A. Töscher, M. Jahrer, R. Legenstein, Improved neighborhood-based algorithms for large-scale recommender systems, in *Proceedings of the 2nd KDD Workshop on Large-Scale Recommender Systems and the Netflix Prize Competition, NETFLIX'08* (Association for Computing Machinery, New York, 2008)
56. A. Vaswani, N. Shazeer, N. Parmar, J. Uszkoreit, L. Jones, A.N. Gomez, Ł. Kaiser, I. Polosukhin, Attention is all you need, in *Advances in Neural Information Processing Systems, 30*, ed. by I. Guyon, U.V. Luxburg, S. Bengio, H. Wallach, R. Fergus, S. Vishwanathan, R. Garnett. (Curran Associates, Red Hook, 2017), pp. 5998–6008
57. J. Wang, A.P. de Vries, M.J.T. Reinders, Unifying user-based and item-based collaborative filtering approaches by similarity fusion, in *Proceedings of the 29th Annual International ACM SIGIR Conference on Research and Development in Information Retrieval, SIGIR'06* (Association for Computing Machinery, New York, 2006), pp. 501–508
58. Y. Zheng, B. Tang, W. Ding, H. Zhou, A neural autoregressive approach to collaborative filtering, in *Proceedings of the 33rd International Conference on Machine Learning - Volume 48, ICML'16*, JMLR.org (2016). pp. 764–773

# Item Recommendation from Implicit Feedback

## Steffen Rendle

## 1  Introduction

Item recommendation, also known as top-n recommendation, is the task to select the best items from a large catalogue of items to a user in a given context. For example, an e-commerce website recommends products to their customers, or a video platform might want to select videos that matches a user's interests. Item recommendation models are usually learned from implicit feedback. Instead of recommending based on explicit preferences such as *a user gave 4 stars to a movie*, the recommendations are based on the past interactions of the user with the system. For example, recommendation can be based on past products purchased by a customer, or videos watched by a user. A good recommender system is often contextual, considering the situation at which a recommendation should be made, e.g., what product is the user currently browsing on, or time information such as weekday vs weekend.

Item recommendation can be formulated as a context-aware ranking problem where the whole set of items should be ordered given a query context. In traditional collaborative filtering, the context would be the user, in more complex cases, the context might be user and time, a user and location, the sequence of previously selected items by a user, etc. To cover these cases, this chapter uses the general concept of a context as a placeholder. A scoring function that expresses the preference between a context and an item is used to order the items. This chapter covers learned scoring functions that are optimized from implicit feedback. The core challenges of learning from implicit feedback are to formulate an optimization objective and to design algorithms that can handle large item catalogues. First,

S. Rendle (✉)
Google Research, Mountain View, CA, USA
e-mail: srendle@google.com

© Springer Science+Business Media, LLC, part of Springer Nature 2022
F. Ricci et al. (eds.), *Recommender Systems Handbook*,
https://doi.org/10.1007/978-1-0716-2197-4_4

implicit feedback contains weak signals about a user's preference: the selected items in the past give a weak indicator of what a user likes. Implicit feedback usually does not contain negative interactions, instead weak negatives are derived from all the remaining items, i.e., the items that a user has not interacted with. Considering all remaining items as weak negative signal leads to the second challenge of high training costs. Formulating the item recommendation problem as a standard machine learning task is very costly because every positive signal from the implicit feedback entails negative signal over all items. That makes off-the-shelf training algorithms hard to apply to item recommendation models. Item recommenders are usually trained either (a) using sampling or (b) specialized algorithms taking into account model and loss structure.

This chapter starts with a detailed discussion of the item recommendation problem, covering the scoring function, the retrieval task, common evaluation metrics, and a summary of the core challenges in learning item recommendation from implicit feedback. The main part of this chapter focuses on the training problem. Sampling methods for general scoring functions based on pointwise, pairwise and softmax losses are described. The sampling distribution and weighting function are important for fast convergence and for aligning the algorithm better with the evaluation metrics. Section 5 discusses more efficient training algorithms that can be applied to dot product models and square losses. Finally, at application time, item recommenders need to be able to retrieve the highest scoring items quickly from the whole item catalogue. Section 6 discusses sublinear time approaches for the retrieval task.

## 2   Problem Definition

The goal of item recommendation is to retrieve a subset of interesting items for a given context $c \in C$ from a set of items $I$. This chapter uses the general concept of a context as a placeholder to cover many common item recommendation scenarios. In the simplest case, the context could be the user, in more complex cases, the context might be user and time, a user and location, the sequence of previously selected items by a user, etc. Also the input representation of a context and an item is flexible, it could be a user id or item id, but also more complex such as image pixels of an item, a textual description, etc. Such attributes are especially important in the cold start problem. The discussion in this chapter is independent of these choices and should be able to accommodate most item recommendation settings.

### 2.1   Recommender Modeling

The preference of a context $c \in C$ to an item $i \in I$ is given by a scoring function

**Table 1** Frequently used notation

| Symbol | Description |
|---|---|
| $I$ | Set of items |
| $C$ | Set of context |
| $S \subseteq C \times I$ | Set of implicit observations |
| $I_c$ | Set of items selected in context $c$, i.e., $I_c := \{i : (c, i) \in S\}$ |
| $C_i$ | Set of context that selected item $i$, i.e., $C_i := \{c : (c, i) \in S\}$ |
| $\hat{y}(i\|c)$ | (Learnable) scoring function over context-item pairs |
| $\theta$ | Model parameters of the scoring function $\hat{y}$ |
| $\phi(c)$ | (Learnable) context representation/ embedding |
| $\psi(i)$ | (Learnable) item representation/ embedding |
| $\langle \cdot, \cdot \rangle$ | Inner/ dot/ scalar product between two vectors or two matrices |
| $\otimes$ | Outer product between two vectors |
| $\delta(b)$ | Indicator function, 1 if $b$ is true, 0 otherwise |

$$\hat{y}(c, i) = \hat{y}(i|c), \quad \hat{y} : C \times I \to \mathbb{R} \tag{1}$$

The choice and design of the scoring function $\hat{y}$ is the problem of recommender system modeling. Most of the discussion here is independent of the particular choice of $\hat{y}$. Chapter "Advances in Collaborative Filtering" focuses on recommender system modeling and contains some examples of possible scoring functions.

The scoring function is parametrized[1] by a set of model parameters $\theta \in \mathbb{R}^p$. Learning the model parameters is a core problem within item recommendation and the main focus of this chapter. The first derivative of $\hat{y}(i|c)$ with respect to a parameter vector $\theta$ is denoted as $\nabla_\theta \hat{y}(i|c)$, and the second derivative as $\nabla_\theta^2 \hat{y}(i|c)$. Table 1 contains an overview of frequently used symbols.

### 2.1.1 Dot Product Models

While most of this chapter does not make any assumption about the scoring function, many recommender models have additional structure which is discussed now. For item recommendation, $\hat{y}$ is often decomposable into a dot product between a context and item representation

$$\hat{y}(i|c) = \langle \phi(c), \psi(i) \rangle \tag{2}$$

where context and items are represented by $d$ dimensional embeddings

$$\phi : C \to \mathbb{R}^d, \quad \psi : I \to \mathbb{R}^d. \tag{3}$$

---

[1] To be precise, $\hat{y}$ should be $\hat{y}_\theta$, but for convenience, the subscript is omitted in this chapter.

Again, $\phi$ and $\psi$ are parametrized by $\theta$. Such models are also known as two tower models, or dual encoders [30].

Unless stated otherwise, the methods described in this chapter will be general and are *not* restricted to dot product models. However, dot product models will be revisited frequently as they have very desirable properties for item recommendation.

### 2.1.2 Examples

Two common examples for scoring functions with a dot product are matrix factorization and two tower deep neural networks. First, the scoring function of matrix factorization is $\hat{y}(i|c) = \langle \mathbf{w}_c, \mathbf{h}_i \rangle$ where each context (or item) is directly represented by a $d$ dimensional embedding vector, $\mathbf{w}_c \in \mathbb{R}^d$ (or $\mathbf{h}_i \in \mathbb{R}^d$). These embedding matrices, $W \in \mathbb{R}^{C \times d}$ and $H \in \mathbb{R}^{I \times d}$ are the model parameters $\theta$. A matrix factorization is a dot product model with $\phi(c) = \mathbf{w}_c$ and $\psi(i) = \mathbf{h}_i$.

Second, a general two tower deep neural network (DNN), $\hat{y}(i|c) = \langle \phi(c), \psi(i) \rangle$, where $\phi(c)$ is a learnable function that extracts a $d$ dimensional representation for each context, and analogously $\psi(i)$ is a learned representation for items. The model parameters of these functions are $\theta$. For example $\psi(i)$ could be a multi layer perceptron that generates an embedding based on the features of an item (e.g., genre, actors, director, release year of a movie)—here the model parameters $\theta$ would be the weight matrices of the hidden layers. Other examples for $\psi(i)$ or $\phi(c)$ would be a convolutional network to extract spatial features from an image and map it to an embedding, or a recurrent model that builds an embedding representation based on sequential data such as previous purchases of a user or textual descriptions of an item.

## 2.2 Applying Item Recommenders

The scoring function $\hat{y}$ can be used to rank items for a given context. Let $\hat{r}(i|c)$ be the rank/position of item $i$ in the sorted list of items $I$ given context $c$ and scoring function $\hat{y}$, i.e., $\hat{r} : I \rightarrow \{1, \dots, |I|\}$. Here $r$ is a bijective mapping—in case of ties, there is some resolution, e.g., random. Thus, the inverse function gives the item ranked at a certain position, e.g, $\hat{r}^{-1}(3|c)$ would be the item ranked at the 3rd position for a context $c$.

When a recommendation model $\hat{y}$ is used for the task of item recommendation, it needs to find the $n$ highest scoring items for a given context $c$. That means it has to compute the items $\hat{r}^{-1}(1|c), \hat{r}^{-1}(2|c), \dots, \hat{r}^{-1}(n|c)$. For example, a recommendation platform might want to show the user the 20 items with the highest score. Section 6 will discuss this problem of computing the top $n$ items in more detail.

For a good user experience there are other considerations besides high scores when selecting items. For example, diversity of results and slate optimization are important factors [11, 28]. A naive implementation just returns the top scoring items. While each item that is shown to a user might be individually a good choice, the combination of items might be suboptimal. For example, showing item $i$ might make item $j$ less attractive if $i$ and $j$ are very close or even interchangeable. Instead, it might be better to choose an item $l$ that complements $i$ better—even if $l$ has a lower score than $j$. Diversification of result sets is an example to avoid some of these effects. Commonly, algorithms for slate optimization and diversification are built on top of an item recommender. The remainder of this chapter will not discuss these important issues further, instead the chapter will focus on the top $n$ item recommendation task.

## 2.3  Evaluation

The quality of an item recommendation algorithm is typically measured by a ranking metric over the positions at which it ranks relevant items. For a given context $c$ and a ground truth set of relevant items $\{i_1, i_2, \ldots\}$, the ranks of these relevant items, $R := \{\hat{r}(i_1|c), \hat{r}(i_2|c), \ldots\}$, are computed and then a metric is applied. Several popular choices are discussed next. Chapter "Evaluating Recommender Systems" contains an exhaustive discussion about evaluation and more metrics.

*Precision at position n* measures the fraction of relevant items among the top $n$ predicted items:

$$\mathrm{Prec}(R)_n = \frac{|\{r \in R : r \leq n\}|}{n}. \tag{4}$$

*Recall at position n* measures the fraction of all relevant items that were recovered in the top $n$:

$$\mathrm{Recall}(R)_n = \frac{|\{r \in R : r \leq n\}|}{|R|}. \tag{5}$$

*Average Precision at n* measures the precision at all ranks that hold a relevant item:

$$\mathrm{AP}(R)_n = \frac{1}{\min(|R|, n)} \sum_{i=1}^{n} \delta(i \in R)\mathrm{Prec}(R)_i. \tag{6}$$

*Normalized discounted cumulative gain (NDCG) at n* places an inverse log reward on all positions that hold a relevant item:

$$\mathrm{NDCG}(R)_n = \frac{1}{\sum_{i=1}^{\min(|R|,n)} \frac{1}{\log_2(i+1)}} \sum_{i=1}^{n} \delta(i \in R) \frac{1}{\log_2(i+1)}. \tag{7}$$

*Area under the ROC curve (AUC)* measures the likelihood that a random relevant item is ranked higher than a random irrelevant item.

$$\text{AUC}(R) = \frac{1}{|R|(|I| - |R|)} \sum_{r \in R} \sum_{r' \in (\{1,...,|I|\} \setminus R)} \delta(r < r') \tag{8}$$

These metrics differ in how much emphasize they place on particular ranks. Most metrics focus strongly on the top ranks and give almost no rewards for items that appear late in the result list. AUC is an exception as it has a slow (linear) decay. A more detailed discussion about these metrics can be found in [14].

The choice of the metric depends on the application. For retrieval tasks, head-heavy metrics such as Precision, Recall, AP or NDCG are better choices than AUC because users will likely investigate only the highest ranked items. If the retrieved results are directly shown, NDCG and AP can be better metrics than Precision and Recall because they emphasize the top ranked results stronger, e.g., distinguish between the first and second position. If the retrieval stage is followed by a reranking step, metrics such as Recall with $n$ equal to the number of candidates that will be reranked can be a better choice than metrics that care about the ranks within the top $n$ such as NDCG or AP.

## 2.4   Learning from Implicit Feedback

This chapter covers implicit feedback between contexts and items. Let $S \subseteq C \times I$ be a set of such observations, where $(c, i) \in S$ is an implicit feedback pair. Examples for $S$ are clicks on links, video watches, purchases of items. Implicit observations provide a (weak) positive signal of a user's preferences. For example, if a user watches a video, it means that this video matches to the user's taste to some degree. Likewise, the items a user has never considered in the past are indicative of their taste as well. These non-selected items contain both the items that a user is not interested in and items that should be recommended to the user in the future. A core problem of learning item recommendation from implicit feedback is to define a training objective for trading off the selected with the non-selected items. Section 3 describes such strategies. Each of these strategies defines the loss over all selected and non-selected items, i.e., not just over $S$ but over the whole set $C \times I$. This leads to a second core challenge of deriving efficient learning algorithms.

This differentiates item recommendation from the learning to rank literature [5] where a smaller set of candidates needs to be reranked. This reranking of a small set of candidates has different problem characteristics than item recommendation because (1) it can be done directly on clickthrough data [12] avoiding the problem to define labels on unobserved data and (2) it is computationally simpler because it does not involve the full catalogue of all items. Such models serve a different purpose than the item recommendation models that are discussed in this chapter, and applying them directly to retrieval over the whole corpus is prone to low retrieval quality due to folding [29]. However, they are very useful as a second stage reranker on top of the retrieval models discussed in this chapter. For example, [6] describes such a two stage architecture where item recommendation is trained from implicit feedback and a second stage reranks the top $n$ retrieved items based on impression data. In this case, the primary goal of item recommendation is to achieve a high recall within the top $n$ results because the final ranking of these candidates is determined by the second stage ranker.

Another class of recommender system problems is based on explicit feedback, most notably *rating prediction*. Here, explicit preferences are provided such as *a user u assigns 5 stars to a movie i, but only 2 stars to movie j.* This is a different problem from item recommendation from implicit feedback. The training objective is simpler because it can be defined directly on the explicit feedback, avoiding the costly optimization over all items. These aspects makes it similar to the learning to rank problem, however it differs from learning to rank in the metrics. A commonality of item recommendation from implicit feedback and recommenders over explicit feedback, is that they share similar design patterns for the scoring function. For example, matrix factorization or item-based collaborative filtering is popular for both problems.

## 2.5 Core Challenges of Item Recommendation

To summarize, the core challenges of item recommendation are:

1. Formulating a loss function over implicit feedback. This requires to define a strategy to trade off selected with unselected items while considering the ranking metric.
2. Learning algorithms that can handle the large number of unselected items efficiently.

# 3 Learning Objectives

This section describes three popular approaches for casting the item recommendation problem into a supervised machine learning problem.

## 3.1 Pointwise Loss

The observations, $S$, can be directly casted into a binary classification problem by treating the observed feedback $S$ as positive instances and all non selected items,

$(C \times I) \setminus S$ as weak negatives. Each training instance has a positive or negative label and an example weight. Any positive pair $(c, i) \in S$, is assigned a positive label, e.g., $y = 1$ or, if available, another positive signal such as the strength of the interaction, e.g., number of times selected, or time spent. Additionally all non-observed items $j$ for the context are included with a negative label, e.g., $y = 0$ (for regression) or $y = -1$ (for classification). A weight $\alpha$ can be assigned to each context-item tuple to indicate the confidence about the training case. Usually, the confidence in the non-observed pairs is lower than for observed one. That balances the large amount of $O(|C \times I|)$ negatives with the smaller amount of $|S|$ positives. Other weighting schemes can be used to downweight popular items to promote more tail items in the result, or reduce the weight of users with many positive interactions so that the experience is not dominated by the behavior of heavy users.

From a regression point of view, a good scoring model predicts scores close to $y$. From a probabilistic point of view, the binary event $p(i = \text{true}|c)$ is modeled through $\hat{y}$, e.g., $p(i = \text{true}|c) = \sigma(\hat{y}(i|c))$. These ideas can be formalized as a pointwise loss

$$L(\boldsymbol{\theta}, S) := \sum_{c \in C} \sum_{i \in I} \alpha(c, i) \, l(\hat{y}(i|c), y(c, i)) + \lambda(\boldsymbol{\theta}) \tag{9}$$

where $\alpha : C \times I \rightarrow \mathbb{R}^+$ is a weight function, $y$ the label function (e.g., $y(c, i) = \delta((c, i) \in S)$), and $\lambda$ is some regularization function and $l$ is a loss function, e.g.,

$$l^{\text{square}}(\hat{y}, y) = (\hat{y} - y)^2 \tag{10a}$$

$$l^{\text{logistic}}(\hat{y}, y) = \ln(1 + \exp(-y\,\hat{y})) \tag{10b}$$

$$l^{\text{hinge}}(\hat{y}, y) = \max(0, 1 - y\,\hat{y}) \tag{10c}$$

This casts the problem of item recommendations to a standard binary classification or regression problem. What makes this problem special is that the number of training examples is huge $O(|C|\,|I|)$. The number of items can be in the millions for large scale problems which makes application of standard learning methods challenging.

Examples for recommender system algorithms that use a pointwise loss are iALS [10], SLIM [16], or iCD [2].

## 3.2 Pairwise Loss

For retrieval, it is not of primary interest if the score $\hat{y}$ is exactly 1 or 0. Instead the relative ordering of the items in a context is of central importance. Given a context $c$, pairwise methods compare all pairs of items $(i, j)$ where $(c, i) \in S$, $(c, j) \notin S$. A successful scoring function, $\hat{y}$, assigns a higher score to the observed item $i$ than the unobserved one: $\hat{y}(i|c) > \hat{y}(j|c)$. For the final retrieval, the items are simply sorted by the score.

A general pairwise loss function can be defined as:

$$L(\boldsymbol{\theta}, S) := \sum_{(c,i) \in S} \sum_{j \in I} \alpha(c, i, j) \, l(\hat{y}(i|c) - \hat{y}(j|c), 1) + \lambda(\boldsymbol{\theta}) \tag{11}$$

where $\alpha$ is a weight function. See Eq. (10) for some example losses $l$.

The number of pairs in this loss is $O(|S|\,|I|) \supseteq O(|C|\,|I|)$. That means it is at least as expensive as pointwise training, and it faces the same computational challenges.

## 3.3 Softmax Loss

From a probabilistic point of view, pointwise training models the binary event $p(i = \text{true}|c)$ through $\hat{y}$, e.g., $p(i = \text{true}|c) = \sigma(\hat{y}(i|c))$. Pairwise training models the conditional probability of the binary comparison $p(i > j|c)$ through $\hat{y}$, e.g., $p(i > j|c) = p(\hat{y}(i|c) > \hat{y}(j|c)) = \sigma(\hat{y}(i|c) - \hat{y}(j|c))$. A third option is to model the conditional multinomial event $p(i|c)$, i.e., $\sum_{j \in I} p(j|c) = 1$ and $\forall j : 0 \leq p(j|c) \leq 1$. The most common function to translate a real valued score, such as $\hat{y}$, to a multinomial distribution is the softmax:

$$p(i|c) = \frac{\exp(\nu\,\hat{y}(i|c))}{\sum_{j \in I} \exp(\nu\,\hat{y}(j|c))} \propto \exp(\nu\,\hat{y}(i|c)). \tag{12}$$

The numerator of the softmax is often referred to as the *partition function*. $\nu \in \mathbb{R}^+$ of the softmax is an (inverse) temperature parameter. Small choices of $\nu$ (=high temperature) make the distribution more uniform. Larger choices (=low temperature) make it more spiky and concentrated around the large values. For $\nu \to \infty$, softmax approaches the *maximum* operator. And for this limit the probability is concentrated at one index

$$p(i|c) = \delta\left(i = \underset{j}{\text{argmax}}\,\hat{y}(j|c)\right) \tag{13}$$

This aligns well with ranking metrics that focus on the top ranked positions. Softmax is widely used in multiclass classification problems, including natural language models where the classes are words, or image classification where the labels are categories.

To keep the notation short, the remainder of this chapter ignores $\nu$ because usually $\hat{y}$ can absorb any scaling effect of $\nu$ by increasing the norm of $\hat{y}$. From the softmax definition of the multinomial probability follows the loss function through the negative log likelihood:

$$L(\boldsymbol{\theta}, S) = -\sum_{(c,i)\in S} \ln p(i|c) + \lambda(\boldsymbol{\theta})$$

$$= -\sum_{(c,i)\in S} \ln \frac{\exp(\hat{y}(i|c))}{\sum_{j\in I} \exp(\hat{y}(j|c))} + \lambda(\boldsymbol{\theta})$$

$$= -\sum_{(c,i)\in S} \left[ \hat{y}(i|c) - \ln \sum_{j\in I} \exp(\hat{y}(j|c)) \right] + \lambda(\boldsymbol{\theta}) \qquad (14)$$

Again, the loss contains a double sum over $S$ and $I$, so the number of elements is $O(|S|\,|I|)$, or $O(|C|\,|I|)$ if the partition function is computed once per context.

## 4  Sampling Based Learning Algorithms

Algorithms based on sampling are the most common approaches for learning recommender systems from implicit feedback. Their basic idea is to iterate over the positive examples $(c, i) \in S$, sample negatives $j$ with respect to a sampling distribution $q(j|c)$ and perform a weighted gradient step. This section discusses algorithms for pointwise, pairwise and softmax losses. The choice of $q$ together with a weighting function $\tilde{\alpha}$ are key for designing the learning algorithms. These two choices are related to $\alpha$ in the loss formulation. Many different instances of these algorithms have been proposed by varying $q$ and $\tilde{\alpha}$ and some popular choices are discussed in this section. Most of the results of this section apply to any scoring function, with the exception of Sects. 4.2.4 and 4.3.1 that are restricted to dot-product models.

### 4.1  Algorithms for Pointwise Loss

The pointwise loss in Eq. (9) is defined over all context-item combinations, $C \times I$. The elements in this set are either a positive observation, if $(c, i) \in S$ or an implicit negative one otherwise. The positive observations are usually informative while the negative provide weaker feedback and are assigned a smaller weight so that they contribute less to the loss. A sampling based learning algorithm iterates over the positive observations and samples $m$ negatives. Algorithm 4.1 sketches this idea.

## Pointwise SGD

1: **repeat**
2:     sample $(c, i) \in S$                                              ▷ positive item
3:     $\theta \leftarrow \theta - \eta \left[ \tilde{\alpha}(c, i) \nabla_\theta l(\hat{y}(c, i), 1) + \nabla_\theta \lambda(\theta) \right]$
4:     **for** $l \in \{1, \ldots, m\}$ **do**                              ▷ $m$ negative items
5:         sample $j$ from $q(j|c)$
6:         $\theta \leftarrow \theta - \eta \left[ \tilde{\alpha}(c, j) \nabla_\theta l(\hat{y}(c, j), 0) + \nabla_\theta \lambda(\theta) \right]$
7:     **end for**
8: **until** converged

The main algorithm iterates over positive items as well as samples of negative items and performs an update step on each of them. The proportion of positive and negative items is controlled by $m$. It can help to sample more than one negative item per positive item because there are overall more negative items than positive ones and often each negative item is less informative than a positive one. Each update step moves the model parameters in the opposite direction of the gradient with respect to the loss. This process is known as stochastic gradient descent (SGD). The size of an update step is controlled by the learning rate $\eta \in \mathbb{R}^+$. Larger values of $\eta$ lead to faster progress but if the value is too large, the algorithm diverges. The iterative updates are repeated until some stopping criterion is met, e.g., a fixed number of repetitions, or the learning progress is less than a predefined threshold. The learning progress can be defined as the difference in a metric over two sufficiently spaced evaluations over the training set or a holdout set. It should be noted that due to the stochastic updates, the progress even on the training loss is noisy and not monotonically decreasing.

The main design choices of the algorithm are the sampling distribution $q$ and the weight $\tilde{\alpha}$. Their choice is related to $\alpha$ from the loss. If $q$ and $\tilde{\alpha}$ are independent of the model parameters $\theta$, then the relationship between the three values is

$$\alpha(c, j) = \begin{cases} \tilde{\alpha}(c, j), & (c, j) \in S \\ m|I_c|q(j|c)\tilde{\alpha}(c, j), & (c, j) \notin S \end{cases} \quad (15)$$

where the sampling probability $q(j|c) = 0$ for $(c, j) \in S$. This shows how a weighted loss can be either achieved by changing $q$ or $\tilde{\alpha}$. For example, if the sampling probability $q$ is uniform

$$\alpha(c, j) = \begin{cases} \tilde{\alpha}(c, j), & (c, j) \in S \\ m\tilde{\alpha}(c, j), & (c, j) \notin S \end{cases} \quad (16)$$

A non-uniform sampling might be beneficial because most of the items are trivially negative, e.g., it is unlikely that a customer likes a random product from a large catalogue. Such samples are trivial for the model to predict and presenting an already correctly classified item during training has a (near) zero gradient, so overall convergence will be slow. Sampling items that have a higher chance to be considered by a user are more informative during learning.

Besides uniform sampling, popularity based sampling is another common approach. Here, the items are sampled proportional to their empirical frequency in the dataset $q(j|c) \propto |C_j|^\beta$ with a squashing exponent $\beta$ to desharpen the distribution. Section 4.2 describes some more sophisticated sampling algorithms.

### 4.1.1   Batching

The SGD algorithm presented so far performs one update step per context-item pair. Each step is computationally cheap. Modern hardware can benefit from larger units of computations, so grouping several training examples can result in more efficient utilization of hardware. For the proposed SGD algorithm, several positive and negative items can be sampled and grouped into a batch of training examples, and then their updates are performed in parallel. This scheme is especially useful if the same set of negatives is shared in the batch—a context independent sampler, e.g., frequency based, naturally fits here. Even more, instead of sampling from a global distribution, it is also common to use the positive items in the batch as the negatives. For example, if the batch contains positive observations $(c_1, i_1), (c_2, i_2), \ldots$, then $\{i_1, i_2, \ldots\}$ are treated as the negatives for this batch, e.g., for $c_1$, the negatives would be $(c_1, i_2), (c_1, i_3), \ldots$. In expectation, this corresponds to choosing the empirical frequency sampler for $q$. To summarize, batching does not improve the computational costs from a complexity perspective but it lowers the wall time on modern hardware.

### 4.1.2   Omitted Details

For simplicity, Algorithm 4.1 was sketched with a constant learning rate, $\eta$, but more complex schedules can be applied as well. Also depending on the form of $\hat{y}$, it is common to update only the subset of model parameters that affects $\hat{y}(i|c)$ for the particular choice of $(c, i)$. For example, in an embedding model, usually only a subset of the embeddings are involved in scoring a pair $(c, i)$, and it is common to apply the gradient step only to this subset of embeddings involved. These details are omitted because they are orthogonal to the problem of item recommendation from implicit feedback.

## 4.2   Algorithms for Pairwise Loss

The pairwise loss (Eq. 11) is a double sum over elements $(c, i) \in S$ and items $j \in I$. This scheme can be directly used for sampling and gradient steps can be performed for each item-pair $(i, j)$. Algorithm 4.2 shows a SGD algorithm for optimizing a pairwise loss.

**Pairwise SGD**

---

1: **repeat**
2:      sample $(c, i) \in S$
3:      sample $j$ from $q(j|c, i)$
4:      $\theta \leftarrow \theta - \eta \left[ \tilde{\alpha}(c, i, j) \nabla_\theta l(\hat{y}(c, i) - \hat{y}(c, j), 1) + \nabla_\theta \lambda(\theta) \right]$
5: **until** converged

---

The design choices of this algorithm are the sampling distribution of negatives, $q(j|c, i)$, and the weight of the gradient $\tilde{\alpha}(c, i, j)$. These two choices are related to the weight $\alpha(c, i, j)$ in the loss (Eq. (11)). Similar to the pointwise loss, if $q(j|c, i)$ and $\tilde{\alpha}(c, i, j)$ are independent of the model parameters then choosing $q$ and $\tilde{\alpha}$ implies a pairwise loss with $\alpha(c, i, j) = q(c, i, j) \tilde{\alpha}(c, i, j)$. Again, fixing two of the quantities will determine the third. For pairwise optimization, several variations with model dependent sampling and weighting have been proposed.

The next subsections discuss several popular variations of the algorithm. The first one samples items uniformly and has no weights. The second scheme adds a weight to consider head-heavy metrics. The third algorithm samples uniformly from items close or above the positive item, the sampling scheme is used to estimate the rank which is used as a weight. Finally, a sampler that oversamples items based on their rank is presented.

### 4.2.1   Uniform Sampling Without Weight

The most simple instance of this algorithm uses a constant weight $\tilde{\alpha}(c, i, j) = 1$ and a uniform sampling probability $q(j|c) \propto 1$, which optimizes an unweighted loss with $\alpha(c, i, j) = 1$. This loss optimizes the area under the ROC curve, AUC, or equivalently, the likelihood that a random relevant item is ranked correctly above a random irrelevant item. This loss and algorithm is often referred to as *Bayesian Personalized Ranking (BPR)* [20].

There are two major issues with this approach:

1. Uniform sampling will sample irrelevant items that can be easily distinguished by the model and thus the gradient of such pairs becomes 0. The progress of the algorithm will be very slow because most of the pairs are not helpful.

2. Metrics that focus on the top items are often preferred for evaluating item recommendation. As discussed in Sect. 2.3, AUC is not a top-heavy metric. For example, for AUC, the benefit for moving a relevant item from position 1000 to 991 is the same as moving on item from position 10 to 1. For a metric such as NDCG, the second change is much more beneficial and gives a much larger improvement in the metric. In most applications, the second change would be considered more important.

Several improvements are discussed next.

### 4.2.2 Uniform Sampling with Weights

A popular approach in the learning to rank community is LambdaRank [5] which weights the gradient by its influence on a ranking metric. Let $M$ be a metric as defined in Sect. 2.3, then the weight of a pair of items $i, j$ given a context $c$ is $\tilde{\alpha}(c, i, j) = |M(\hat{r}(i|c)) - M(\hat{r}(j|c))|$. That means the weight is directly tied to the metric because it measures the influence of flipping the order of the two items. A practical difficulty of this approach is that the rank of the items $i$ and $j$ needs to be computed. For item recommendation with a large number of items this becomes very costly. This is less of a problem for learning to rank problems where only a small set of items are reranked. An adaption of LambdaRank to item recommendation was proposed in [32].

### 4.2.3 Adaptive Sampling with Weights

WARP (Weighted Approximate-Rank Pairwise) [27] is a pairwise algorithm specifically designed for item recommendation. The motivation of the WARP loss is to directly learn a ranking metric. Each rank is associated with a penalty, $\gamma$, where the better the predicted rank of the relevant item, the lower the penalty. This penalty can be chosen to align with the ranking metric [25]. For example, the penalty can be linearly increasing (as in AUC) or be a step function for a metric such as precision@k, or increase by the reciprocal of the position.

WARP uses an online algorithm for optimizing this loss. First, the sampling distribution is non-uniform. The algorithm tries to sample an irrelevant item $j$ such that it is ranked close to or above the relevant item, i.e., $\hat{y}(j|c) + 1 > \hat{y}(i|c)$. This choice is aligned with the hinge loss (Eq. (10c)), where the gradient is zero if the item is correctly ranked above the margin. The authors propose a rejection sampling algorithm that draws $j$ uniformly, and if it does not meet the condition, it discards $j$ and samples another item. This means the sampling distribution is $q(j|i, c) \propto \delta(\hat{y}(j|c) + 1 > \hat{y}(i|c))$.

Sampling an item $j$ that fulfills the requirement $\hat{y}(j|c) + 1 > \hat{y}(i|c)$ becomes increasingly hard, the better the model is and the more items are in the catalogue. The authors propose to stop the sampling algorithm after $|I|$ repetitions. The number of rounds until the algorithm terminates can be used to estimate the rank of the relevant item: $r(i|c) \approx \lfloor (\#\text{rounds} - 1)/|I| \rfloor$. The penalty, $\gamma$, of the estimated rank is used as the weight in the SGD algorithm:

$$\tilde{\alpha}(c, i, j) = \gamma(\lfloor (\#\text{rounds} - 1)/|I| \rfloor) \tag{17}$$

The definition of $q$ avoids the problem of sampling pairs that have no influence on the gradient step. However, while the algorithm avoids steps on non-influential items, the cost for finding a relevant item are high because it needs to score all discarded items. Usually, scoring an item and updating an item have similar computational complexity. This gives WARP at most a constant speedup over a trivial method that does not discard items. However, sampling is only one part of the WARP algorithm and the more important part is that WARP uses the estimated rank in the weight for the gradient. This way, WARP optimizes a loss that is more reflective of ranking metrics than uniformly weighted methods such as BPR.

### 4.2.4 Adaptive Sampling Without Weights

As an improvement of the BPR algorithm, [19] propose to sample items based on their rank, here the sampling distribution is $q(j|c) \propto \exp(-r(j|c)/\gamma)$, while keeping the weight constant $\tilde{\alpha} = 1$. While LambdaRank and WARP introduce a weight on the rank by $\tilde{\alpha}$, here the weight is implicitly introduced through $q$ that depends on the rank. Note that this sampling distribution depends on the context and the current choice of model parameters. As the model learns, the sampling distribution adapts.

The novelty of this work is an algorithm to approximately sample from $q(j|c)$ in constant time. Unlike all the methods that have been discussed in this chapter so far and that were applicable to any scoring function, this work assumes a dot product model $\langle \boldsymbol{\phi}(c), \boldsymbol{\psi}(i) \rangle$. Next, the sampling algorithm is described briefly and more background can be found in [19].

For a given context, it first samples an embedding dimension $f^*$ according to

$$q(f|c) = |\phi_{c,f}|\sigma_f, \quad \sigma_f^2 = \text{Var}(\psi_{.,f}) \tag{18}$$

The motivation for this choice is that for a given context embedding $\boldsymbol{\phi}(c)$ different embedding dimensions $f \in \{1, \ldots, d\}$ are expected to have different contributions on the score and consequently on the rank. The larger an entry $|\phi_{c,f}|$ of the context embedding, the more influential is this embedding dimension. The sign is ignored at this step because without looking at the item embeddings it is meaningless. Each dimension is normalized by the variance of the item embeddings to ensure that dimensions with larger entries are overweighted in the sampling step.

Then an item is sampled according to the rank induced by $\psi(i)_{.,f^*}$. This is done by first sampling a desired rank $r^*$ from $\exp(-r/\gamma)$, this step is independent of the scoring function. Then the $r^*$-largest item from $\psi(i)_{.,f}$ is returned—or if the sign of $\phi_{c,f^*}$ is negative, the $r^*$-smallest (=most negative). This last step can be achieved by storing a sorted list of items for each embedding dimension. This sorted list is independent of the context and can be refreshed occasionally. In total, this is an amortized constant time algorithm for sampling an item approximately from $q(j|c) \propto \exp(-r(j|c)/\gamma)$.

## 4.3 Algorithms for Sampled Softmax

A common strategy for training a model over a softmax loss is to sample $m$ negatives $\{j_1, \ldots, j_m\}$ from a distribution $q(j|c)$ and to compute the partition function on this smaller sample. A key difference of sampled softmax to negative sampling for pointwise and pairwise losses is that for softmax the sample is applied inside a logarithm (see Eq. (14)). That makes it difficult to get unbiased estimates. Sampled softmax is commonly used with a correction to the scores and for a sample $\{j_1, \ldots, j_m\}$ of $m$ negative items, and for an observation, $(c, i) \in S$, a gradient step with respect to the following loss is taken:

$$-\hat{y}(i|c) + \ln \left( \exp(\hat{y}(i|c)) + \sum_{l=1}^{m} \frac{\exp(\hat{y}(j_l|c))}{m\, q(j_l|c)} \right) \qquad (19)$$

The correction $\frac{1}{m\,q(j|c)}$ ensures that if $m \to \infty$, then sampled softmax is unbiased [3]. However, typically a small (finite) set of negatives is sampled and thus sampled softmax is biased. This means no matter how many training epochs the algorithm is run, as long as $m$ is constant, sampled softmax will converge to a different solution than full softmax. It was shown that the only way to avoid this issue is to use $q(j|c) = p(j|c)$, i.e., to use the softmax distribution itself for sampling [4]. In this case, sampled softmax is unbiased for any sample size $m$. For sure, sampling from $p$ is expensive and not a viable option. Nevertheless, it shows that the bias can be reduced by either increasing $m$ and/or using a sampling distribution closer to $p(j|c)$.

In practice, when optimizing a recommender system with sampled softmax, the sample size $m$ is an important hyperparameter because it directly impacts what loss is optimized. Typically, a large sample size, $m$, is chosen to mitigate the bias. Again, the most common sampling distributions are (squashed) popularity (="unigram") sampling or in-batch sampling (=popularity sampling where $m =$ batchsize). Two more sophisticated sampling approaches that have been specifically designed for softmax are discussed in Sects. 4.3.1 and 4.3.2.

Algorithm 4.3 sketches pseudo code for optimizing a model with SGD for sampled softmax.

**Sampled Softmax SGD**

1: **repeat**
2:     sample $(c, i)$ from $S$
3:     sample $\{j_1, \ldots, j_m\}$ from $q(j|c)$
4:     $\boldsymbol{\theta} \leftarrow \boldsymbol{\theta} - \eta \nabla_{\boldsymbol{\theta}} \left[ -\hat{y}(i|c) + \ln\left( \exp(\hat{y}(i|c)) + \sum_{l=1}^{m} \frac{\exp(\hat{y}(j_l|c))}{m\, q(j_l|c)} \right) + \lambda(\boldsymbol{\theta}) \right]$
5: **until** converged

---

### 4.3.1 Kernel Based Sampling

As discussed before, the sampling distribution is very important for sampled softmax optimization because it lowers the number of samples, $m$, required while keeping the bias low. The ideal sampling distribution would be $p(j|c, \boldsymbol{\theta}) \propto \exp(\hat{y}(j|c))$—or for dot product models $p(j|c, \boldsymbol{\theta}) \propto \exp(\langle \boldsymbol{\phi}(c), \boldsymbol{\psi}(j) \rangle)$. This formulation can be seen as a kernel and a mapping function $\pi : \mathbb{R}^d \to \mathbb{R}^D$ from the original embedding space to a larger space can be introduced, such that

$$p(j|c, \boldsymbol{\theta}) \propto \exp(\langle \boldsymbol{\phi}(c), \boldsymbol{\psi}(j) \rangle) \approx \langle \pi(\boldsymbol{\phi}(c)), \pi(\boldsymbol{\psi}(j)) \rangle \propto q(j|c, \boldsymbol{\theta}) \qquad (20)$$

Such a decomposition allows to compute the (approximated) partition function as:

$$\sum_{j \in I} \exp(\hat{y}(j|c)) \approx \sum_{j \in I} \langle \pi(\boldsymbol{\phi}(c)), \pi(\boldsymbol{\psi}(j)) \rangle = \left\langle \pi(\boldsymbol{\phi}(c)), \underbrace{\sum_{j \in I} \pi(\boldsymbol{\psi}(j))}_{\mathbf{z} \in \mathbb{R}^D} \right\rangle$$

$$= \langle \pi(\boldsymbol{\phi}(c)), \mathbf{z} \rangle$$

The key here is that $\mathbf{z}$ is independent of the context and can be precomputed. Then the sampling probability of an element j, $q(j|c)$, can be computed in $O(D)$ time. Based on this observation, a divide and conquer algorithm can be derived to sample j from $q(j|c, \boldsymbol{\theta})$ in $O(D \log_2 |I|)$ time (see [4] for details). Quadratic expansion [4] and random feature maps [17] have been explored for $\pi$.

### 4.3.2 Two-Pass Sampler

Another sampling strategy is a two stage sampler [1], where first a large set of $M > m$ items is sampled and only the $m$ most promising candidates are accepted. The first set is sampled from the squashed empirical frequency distribution, typically 1–10% of the items $I$ are sampled. Then all items on this large set are scored against a batch of context and a smaller set of the $m < M$ highest scoring items is returned. For efficient computation, it was proposed to distribute the computation of the scores of the $M$ items over several machines, preferably machines with GPUs. Finally, sampled softmax is applied on the smaller set. More details can be found in [1].

### 4.3.3   Relation of Sampled Softmax to Pairwise Loss

Finally, a relationship between sampled softmax and a pairwise loss is highlighted. When the number of negative samples in sampled softmax is $m = 1$, then

$$
-\hat{y}(i|c) + \ln \left( \exp(\hat{y}(i|c)) + \frac{\exp(\hat{y}(j|c))}{q(j|c)} \right) = \ln \frac{\exp(\hat{y}(i|c)) + \frac{\exp(\hat{y}(j|c))}{q(j|c)}}{\exp(\hat{y}(i|c))}
$$

$$
= l^{\text{logistic}} \left( \hat{y}(i|c) - \hat{y}(j|c) + \ln q(j|c), 1 \right)
$$

Which is a pairwise logistic loss where the prediction of the negative item is shifted by the correction term $\ln q(j|c)$.

## 5   Efficient Learning Algorithms for Special Cases

This section investigates efficient learning algorithms for dot product models, i.e., $\hat{y}(i|c) = \langle \boldsymbol{\phi}(c), \boldsymbol{\psi}(i) \rangle$, and square losses. First efficient alternating least squares and gradient descent learning algorithms are presented for pointwise losses. Finally, their application to pairwise square losses is discussed briefly.

### 5.1   Pointwise Square Loss

The following restrictions are made: (1) the weights and labels for all unobserved tuples $(c, i) \notin S$ is constant: let their weight be $\alpha_0$ and the label 0. It follows

$$
L(\boldsymbol{\theta}, S) = \sum_{c \in C} \sum_{i \in I} \alpha(c, i)(\hat{y}(i|c) - y(c, i))^2 + \lambda(\boldsymbol{\theta})
$$

$$
= \sum_{(c,i) \in S} [\alpha(c, i)(\hat{y}(i|c) - y(c, i))^2 - \alpha_0 \hat{y}(i|c)^2]
$$

$$
+ \alpha_0 \sum_{c \in C} \sum_{i \in I} \hat{y}(i|c)^2 + \lambda(\boldsymbol{\theta})
$$

This can be further simplified to

$$\tilde{L}(\boldsymbol{\theta}, \tilde{S}) = \sum_{(c,i,\alpha,y)\in\tilde{S}} \alpha(\hat{y}(i|c) - y)^2 + \alpha_0 \sum_{c\in C} \sum_{i\in I} \hat{y}(i|c)^2 + \lambda(\boldsymbol{\theta}) \tag{21}$$

With $\tilde{S} = \{(c, i, \alpha(c, i) - \alpha_0, y\alpha(c, i)/(\alpha(c, i) - \alpha_0)) : (c, i) \in S\}$ and both $\tilde{L}$ and L share the same optimum (see [2] for more details). The remainder of this section uses this simplified formulation.

### 5.1.1  Gramian Trick

The first part of the loss $\tilde{L}$ depends only on the small set of observed positive data and is cheap to compute. The second part appears to be expensive with $|C||I|$ terms and a naive computation is prohibitively costly. However, the *Gramian trick* makes the computation of the second part very efficient [2]:

$$\alpha_0 \sum_{c\in C} \sum_{i\in I} \hat{y}(i|c)^2 = \alpha_0 \sum_{c\in C} \sum_{i\in I} \langle \boldsymbol{\phi}(c), \boldsymbol{\psi}(i) \rangle^2 \tag{22}$$

$$= \alpha_0 \left\langle \sum_{c\in C} \boldsymbol{\phi}(c) \otimes \boldsymbol{\phi}(c), \sum_{i\in I} \boldsymbol{\psi}(i) \otimes \boldsymbol{\psi}(i) \right\rangle \tag{23}$$

$$= \alpha_0 \left\langle G^C, G^I \right\rangle \tag{24}$$

with Gram matrices

$$G^C = \sum_{c\in C} \boldsymbol{\phi}(c) \otimes \boldsymbol{\phi}(c), \quad G^I = \sum_{i\in I} \boldsymbol{\psi}(i) \otimes \boldsymbol{\psi}(i) \tag{25}$$

The advantage of this *Gramian trick* is that the sum over all context-item combination is simplified as a the dot product over two Gram matrices, where each matrix has size $d \times d$. The cost for computing the Gramians[2] is $O(d^2(|I| + |C|))$. The cost for computing the loss, $\tilde{L}$, assuming the Gramians are known is $O(|S|d + d^2)$. The overall costs are dominated by the loss over the positive examples, $|S|$, as long as $d \leq |S|/(|C| + |I|)$. This means the loss over *all* examples (including the implicit negative ones) can be computed without paying the computational costs for the negative ones.

The remainder of this section derives efficient solvers for the loss with the Gramian formulation

---

[2] The analysis here ignores the cost for computing the embeddings $\boldsymbol{\phi}(c)$ and $\boldsymbol{\psi}(i)$. The derived results have a linear complexity in the costs for computing the embeddings.

$$\tilde{L}(\boldsymbol{\theta}, \tilde{S}) = \sum_{(c,i,\alpha,y) \in \tilde{S}} \alpha(\hat{y}(i|c) - y)^2 + \alpha_0 \left\langle G^C, G^I \right\rangle + \lambda(\boldsymbol{\theta}) \qquad (26)$$

The Gramian trick is related to the Kernel softmax (see Sect. 4.3.1) where $\exp(\langle \boldsymbol{\phi}(c), \boldsymbol{\psi}(i) \rangle) \approx \langle \pi(\boldsymbol{\phi}(c)), \pi(\boldsymbol{\psi}(i)) \rangle$. In particular, using the square function instead of the exp, $\langle \boldsymbol{\phi}(c), \boldsymbol{\psi}(i) \rangle^2 = \langle \pi(\boldsymbol{\phi}(c)), \pi(\boldsymbol{\psi}(i)) \rangle$ for $\pi(\mathbf{x}) = \mathbf{x} \otimes \mathbf{x}$.

### 5.1.2   Coordinate Descent/ALS Solver for Multilinear Models

The Gramian trick can be used to derive efficient alternating least squares solvers for the family of multilinear models. This section starts with a recap of multilinear models and then derives a learning algorithm. Besides multilinearity, it is assumed that the regularization $\lambda(\boldsymbol{\theta})$ is a L2 regularization, i.e., $\lambda(\boldsymbol{\theta}) = \lambda ||\boldsymbol{\theta}||^2$.

Multilinear Model

Following [2], the derivation in this subsection makes the assumption that $\boldsymbol{\phi}(c)$ and $\boldsymbol{\psi}(i)$ do not share model parameters. In particular for every scalar coordinate $\tilde{\theta}$ from the vector $\boldsymbol{\theta}$ of model parameters

$$\forall c \in C : \nabla_{\tilde{\theta}} \boldsymbol{\phi}(c) = 0 \quad \text{or} \quad \forall i \in I : \nabla_{\tilde{\theta}} \boldsymbol{\psi}(i) = 0 \qquad (27)$$

To simplify notation, it is further assumed that the model is multilinear in the model parameters. That means for each model parameter $\tilde{\theta}$ the model can be written in a linear form

$$\hat{y}(i|c) = g(i|c) + \tilde{\theta} \, \nabla_{\tilde{\theta}} \hat{y}(i|c), \quad \nabla_{\tilde{\theta}}^2 \hat{y}(i|c) = 0 \qquad (28)$$

Often, models are not just linear in scalars but linear in subvectors $\tilde{\boldsymbol{\theta}}$ of the model parameters $\boldsymbol{\theta}$

$$\hat{y}(i|c) = g(i|c) + \langle \tilde{\boldsymbol{\theta}}, \nabla_{\tilde{\boldsymbol{\theta}}} \hat{y}(i|c) \rangle, \quad \nabla_{\tilde{\boldsymbol{\theta}}}^2 \hat{y}(i|c) = 0 \qquad (29)$$

For example, matrix factorization: $\hat{y}(i|c) = \langle \mathbf{w}_u, \mathbf{h}_i \rangle$ is linear in $\mathbf{w}_u$ or $\mathbf{h}_i$. Some other examples for multilinear models are factorization machines [18], or tensor factorization models such as PARAFAC [8] and Tucker Decomposition [24].

Optimization

With these prerequisites, efficient alternating least squares solvers can be derived. Let $\tilde{\boldsymbol{\theta}}_1, \tilde{\boldsymbol{\theta}}_2, \ldots$ be a partition of the model parameters $\boldsymbol{\theta}$ such that the model is linear in each $\tilde{\boldsymbol{\theta}}$. From this follows that the optimal values for $\tilde{\boldsymbol{\theta}}$ have a closed form solution that can be obtained by a linear regression solver. An efficient computation of the least square solution needs to consider the Gramian trick. This will be discussed next. The solution can be derived by finding the root of $\nabla_{\tilde{\theta}} L(\theta, S)$. A single iteration of Newton's method finds its solution:

$$\tilde{\theta}^* = \tilde{\theta} - (\nabla_{\tilde{\theta}}^2 \tilde{L}(\theta, \tilde{S}))^{-1} \nabla_{\tilde{\theta}} \tilde{L}(\theta, \tilde{S})$$

The computation of the sufficient statistics $\nabla_{\tilde{\theta}}^2 \tilde{L}(\theta, \tilde{S})$ and $\nabla_{\tilde{\theta}} \tilde{L}(\theta, \tilde{S})$ is now discussed for a vector of model parameters $\tilde{\theta}$ from the context side. The equations for model parameters from the item side are analogously.

$$\nabla_{\tilde{\theta}} \tilde{L}(\theta, \tilde{S}) \overset{(*)}{=} \sum_{(c,i,\alpha,y) \in \tilde{S}} \alpha(\hat{y}(i|c) - y)\nabla_{\tilde{\theta}} \hat{y}(i|c) + \alpha_0 \sum_{c \in C} \phi(c) G^I (\nabla_{\tilde{\theta}} \phi(c))^t + \lambda \tilde{\theta}$$

$$\nabla_{\tilde{\theta}}^2 \tilde{L}(\theta, \tilde{S}) \overset{(**)}{=} \sum_{(c,i,\alpha,y) \in \tilde{S}} \alpha \nabla_{\tilde{\theta}} \phi(c) \otimes \nabla_{\tilde{\theta}} \phi(c) + \alpha_0 \sum_{c \in C} (\nabla_{\tilde{\theta}} \phi(c)) G^I (\nabla_{\tilde{\theta}} \phi(c))^t + \lambda I$$

where (*) uses $\nabla_{\tilde{\theta}} \hat{y}(i|c) = \langle \nabla_{\tilde{\theta}} \phi(c), \psi(i) \rangle$ and (**) uses of $\nabla_{\tilde{\theta}}^2 \hat{y}(i|c) = 0$.

From these equations follows the generic ALS algorithm from implicit data. The algorithm iterates over parameters from the context side, computes the first and second derivative as outlined above and performs the Newton update step. The same steps are repeated for parameters from the item side.

## iALS-Pointwise for Multilinear Models

1: **repeat**
2:      $G^I \leftarrow \sum_{i \in I} \psi(i) \otimes \psi(i)$
3:      **for** $\tilde{\theta} \in \{\tilde{\theta}_1, \tilde{\theta}_2, \ldots\}$ **do**      ▷ Iterate over parameters of the context side
4:          $\nabla_{\tilde{\theta}} \leftarrow \lambda \tilde{\theta}$
5:          $\nabla_{\tilde{\theta}}^2 \leftarrow \lambda I$
6:          **for** $c \in C$ where $\nabla_{\tilde{\theta}} \phi(c) \neq 0$ **do**
7:              $\nabla_{\tilde{\theta}} \leftarrow \nabla_{\tilde{\theta}} + \alpha_0 \phi(c) G^I (\nabla_{\tilde{\theta}} \phi(c))^t$
8:              $\nabla_{\tilde{\theta}}^2 \leftarrow \nabla_{\tilde{\theta}}^2 + \alpha_0 (\nabla_{\tilde{\theta}} \phi(c)) G^I (\nabla_{\tilde{\theta}} \phi(c))^t$
9:          **end for**
10:          **for** $(c, i, \alpha, y) \in \tilde{S}$ where $\nabla_{\tilde{\theta}} \phi(c) \neq 0$ **do**
11:              $\nabla_{\tilde{\theta}} \leftarrow \nabla_{\tilde{\theta}} + \alpha(\hat{y}(i|c) - y)\langle \nabla_{\tilde{\theta}} \phi(c), \psi(i) \rangle$
12:              $\nabla_{\tilde{\theta}}^2 \leftarrow \nabla_{\tilde{\theta}}^2 + \alpha \nabla_{\tilde{\theta}} \phi(c) \otimes \nabla_{\tilde{\theta}} \phi(c)$
13:          **end for**
14:          $\tilde{\theta} \leftarrow \tilde{\theta} - (\nabla_{\tilde{\theta}}^2)^{-1} \nabla_{\tilde{\theta}}$
15:      **end for**
16:      Perform a similar pass over parameters from the item side
17: **until** converged

If $\hat{y}$ is a matrix factorization model, and if for $\tilde{\theta}$ a user embedding vector, $\mathbf{w}_u$ or item embedding vector $\mathbf{h}_i$ is chosen, then this is equivalent to the iALS algorithm proposed in [10]. This algorithm has a complexity of $O(|S|\, d^2 + (|C| + |I|)\, d^3)$ per epoch. This is much more efficient than the complexity of $O(|C|\,|I|\, d^2 + (|C| + |I|)\, d^3)$ of a naive ALS implementation. The work of [10] was the first to provide this efficient training algorithm for learning matrix factorization from implicit feedback. Later, [9] derived a variation of this algorithm for PARAFAC tensor factorization.

The derivation in this chapter applies to any multilinear model including matrix factorization, PARAFAC, Tucker Decomposition or factorization machines. The coordinate descent version (i.e., choosing a scalar for $\theta$) of this generalized algorithm was proposed in [2]. The alternating least squares version in this chapter is more general because any vector $\theta$ can be chosen, including a vector of size one (i.e., a scalar) which would reduce to a coordinate descent algorithm. Coordinate descent (CD) algorithms have a lower computation complexity than ALS solvers, e.g., $O(|S|\, d + (|C| + |I|)\, d^2)$ for matrix factorization (see [2] for details). However, on modern hardware, vector operations as used by a ALS can be much more efficient than the scalar operations of CD, and with a careful implementations the higher theoretical complexity is not noticable. Moreover, using vectors instead of scalars reduces synchronization points which is very important in distributed algorithms, so ALS methods can be overall more efficient in wall time than their CD equivalents.

### 5.1.3  SGD Solver for General Models

The previous section discussed multilinear models. Now, following [13], this is generalized to any dot-product model $\hat{y}(i|c) = \langle \phi(c), \psi(i) \rangle$. In particular, $\phi(c)$ and $\psi(i)$ can be any structure, including non-linear DNNs. Such structures are commonly optimized by SGD algorithms that iterate over observed examples $S$.

Now, the loss in Eq. (26) is reformulated as a sum over training examples

$$\tilde{L}(\boldsymbol{\theta}, \tilde{S}) = \sum_{(c,i,\alpha,y)\in\tilde{S}} l(c, i, \alpha, y) \tag{30}$$

so that stochastic gradient descent can be applied. To achieve this form, the Gramian term in the loss is rewritten as a sum over training examples:

$$\langle G^C, G^I \rangle = \frac{1}{2}\left( \sum_{c\in C} \phi(c) G^I \phi(c)^t + \sum_{i\in I} \psi(i) G^C \psi(i)^t \right) \tag{31}$$

$$= \sum_{(c,i,\alpha,y)\in\tilde{S}} \frac{1}{2}\left( \frac{1}{|I_c|}\phi(c) G^I \phi(c)^t + \frac{1}{|C_i|}\psi(i) G^C \psi(i)^t \right) \tag{32}$$

Combining this with the loss on the positive observations, results in the final form of the elementwise loss

$$l(c, i, \alpha, y) = \alpha(\hat{y}(i|c) - y)^2 + \frac{\alpha_0}{2}\left(\frac{1}{|I_c|}\boldsymbol{\phi}(c)G^I\boldsymbol{\phi}(c)^t + \frac{1}{|C_i|}\boldsymbol{\psi}(i)G^C\boldsymbol{\psi}(i)^t\right) \tag{33}$$

Now, gradient descent can be applied to this equation. For the purpose of learning $\boldsymbol{\theta}$, $G^C$ and $G^I$ are treated as constants and replaced by estimates $\hat{G}^C$ and $\hat{G}^I$

$$\nabla_{\boldsymbol{\theta}} l(c, i, \alpha, y) = 2\alpha(\hat{y}(i|c) - y)\nabla_{\boldsymbol{\theta}}\hat{y}(i|c)$$

$$+ \alpha_0\left(\frac{1}{|I_c|}\boldsymbol{\phi}(c)\hat{G}^I(\nabla_{\boldsymbol{\theta}}\boldsymbol{\phi}(c))^t + \frac{1}{|C_i|}\boldsymbol{\psi}(i)\hat{G}^C(\nabla_{\boldsymbol{\theta}}\boldsymbol{\psi}(i))^t\right) \tag{34}$$

The Gramian estimates are updated by gradient descent as well. A good Gramian estimate, $\hat{G}$, is close to the true Gramian, $G$. A reasonable objective to enforce closeness is the Frobenius norm $||\hat{G} - G||_F^2$. To make this objective amendable to SGD, the Gramian loss is reformulated as a sum over positive training examples

$$\underset{\hat{G}^C}{\text{argmin}} \left\|\hat{G}^C - G^C\right\|_F^2 = \underset{\hat{G}^C}{\text{argmin}} \left\|\hat{G}^C - \sum_{c \in C}\boldsymbol{\phi}(c) \otimes \boldsymbol{\phi}(c)\right\|_F^2 \tag{35}$$

$$= \underset{\hat{G}^C}{\text{argmin}} \sum_{c \in C}\left\|\hat{G}^C - |C|\boldsymbol{\phi}(c) \otimes \boldsymbol{\phi}(c)\right\|_F^2 \tag{36}$$

$$= \underset{\hat{G}^C}{\text{argmin}} \sum_{(c,i,\alpha,y) \in \tilde{S}}\frac{1}{|I_c|}\left\|\hat{G}^C - |C|\boldsymbol{\phi}(c) \otimes \boldsymbol{\phi}(c)\right\|_F^2 \tag{37}$$

which results in the gradient of the Gramian estimate:

$$\nabla_{\hat{G}^C}\left[\sum_{(c,i,\alpha,y) \in \tilde{S}}\frac{1}{|I_c|}\left\|\hat{G}^C - |C|\boldsymbol{\phi}(c) \otimes \boldsymbol{\phi}(c)\right\|_F^2\right] \tag{38}$$

$$= \sum_{(c,i,\alpha,y) \in \tilde{S}}\frac{1}{|I_c|}\left(\hat{G}^C - |C|\boldsymbol{\phi}(c) \otimes \boldsymbol{\phi}(c)\right) \tag{39}$$

To summarize, when sampling a training example $(c, i, \alpha, y) \in \tilde{S}$, the update rule for the Gramian estimate is

$$\hat{G}^C \leftarrow \hat{G}^C - \eta\frac{1}{|I_c|}\left(\hat{G}^C - |C|\boldsymbol{\phi}(c) \otimes \boldsymbol{\phi}(c)\right) \tag{40}$$

$$= \hat{G}^C \left(1 - \frac{\eta}{|I_c|}\right) + \frac{\eta}{|I_c|}|C|\phi(c) \otimes \phi(c) \tag{41}$$

See [13] for other estimation algorithms.[3]

Algorithm 5.1.3 shows the final SGD procedure.

---

**SGD-Pointwise with Gramian Trick**

---

1: **repeat**
2:     sample $(c, i) \in S$
3:     $\theta \leftarrow \theta - \eta \nabla_\theta l(c, i, \alpha, y)$                    ▷ See Eq. (34)
4:     sample $(c, i) \in S$
5:     $\hat{G}^C \leftarrow \hat{G}^C - \eta \frac{1}{|I_c|} \left( \hat{G}^C - |C|\phi(c) \otimes \phi(c) \right)$
6:     $\hat{G}^I \leftarrow \hat{G}^I - \eta \frac{1}{|C_i|} \left( \hat{G}^I - |I|\psi(i) \otimes \psi(i) \right)$
7: **until** converged

---

Comparing this to the vanilla SGD Algorithm 4.1, it can be seen that there is no negative sampling step necessary when using the Gramian trick. All negatives are already considered in the update step of the positive item through the Gramian (see line 3). In addition, there is the estimation step of the Gramians.

In some sense, negative sampling can be seen as an estimation of Gramians as well: instead of keeping track of an estimate of the Gramian as proposed here, negative sampling rebuilds the Gramian based on only a small sample of $m$ negatives. This is a much less precise estimate of the Gramian than keeping a long-term estimate because the global Gramians are unlikely to change considerably for each step.

## 5.2   Pairwise Square Loss

Finally, efficient solvers for pairwise square losses are shortly discussed. For keeping the derivation simple, it is assumed that $\alpha(c, i, j) = 1$; a generalization to $\alpha(c, i, j) = \alpha_{c,i}\alpha_j$ is simple and follows the same pattern. The pairwise square loss is defined as

---

[3] [13] uses a different weight on each Gramian element which makes the derivation more natural for SGD algorithms. This chapter uses the same Gramian definition as in Eq. (25) to make all results consistent.

$$L(\boldsymbol{\theta}, S) = \sum_{(c,i) \in S} \sum_{j \in I} [\hat{y}(i|c) - \hat{y}(j|c) - 1]^2. \tag{42}$$

This can be reformulated to:

$$L(\boldsymbol{\theta}, S) = \sum_{(c,i) \in S} \sum_{j \in I} [(\hat{y}(i|c) - 1)^2 - 2(\hat{y}(i|c) - 1)\hat{y}(j|c) + \hat{y}(j|c)^2] \tag{43}$$

$$= \sum_{(c,i) \in S} |I|(\hat{y}(i|c) - 1)^2 + \langle G^C, G^I \rangle - 2 \sum_{(c,i) \in S} (\hat{y}(i|c) - 1)\langle \boldsymbol{\phi}(c), \mathbf{z} \rangle \tag{44}$$

with

$$G^C := \sum_{c \in C} |I_c| \boldsymbol{\phi}(c) \otimes \boldsymbol{\phi}(c), \quad G^I := \sum_{i \in I} \boldsymbol{\psi}(i) \otimes \boldsymbol{\psi}(i), \quad \mathbf{z} := \sum_{i \in I} \boldsymbol{\psi}(i) \tag{45}$$

The first two terms in the loss are identical to the pointwise square loss in Eq. (26): a pointwise loss over the positive observations, and the Gramian term. The last term is an additional correction term for pairwise square loss, and its computation is in $O(|S| d)$. The algorithms (both iALS and SGD) introduced for pointwise loss can be extended to take the new term into account. Instead of just storing the Gramian $G^I$, the vector $\mathbf{z}$ has to be stored as well.

Takács and Tikk [23] first introduced an efficient solver for matrix factorization and pairwise square loss. Their derivation is based on a gradient reformulation as previously invented for pointwise square loss by Hu et al. [10]. The derivation in this chapter is more general and makes pairwise square loss applicable to a wider variety of models (multilinear models and general dot-product models) and optimization schemes (ALS, CD, SGD).

## 6  Retrieval with Item Recommenders

The typical application of an item recommender has to return the highest scoring items given a context $c$. This section will first highlight the limitations of a naive brute-force implementation and then discuss efficient algorithms for dot product models.

## 6.1  Limitations of Brute-Force Retrieval

A naive implementation scores all the items, $I$, with $\hat{y}(i|c)$, sorts the scores and returns the highest ranked ones. This is very costly, with a linear dependency on $|I|$.

This makes this method infeasible for online scoring for large or medium sized item catalogues because applications commonly expect the results in a few milliseconds within a user's request. For moderately sized catalogues, the top items can be precomputed offline and stored for each context that could be queried. However, the size of possible query context can be much larger than the set of training context $|C|$ and evaluating and storing all top lists becomes infeasible. For example, for a sequential recommender, the context might be the $k$ most recent clicks of the user, the potential space of all context is $|I|^k$. It is impossible to precompute the top scoring items for all of these $|I|^k$ context. Offline computation becomes also a problem if the recommender system is trained online, e.g., a user's embedding might be updated online which would require recomputing the top scoring items. To summarize, brute-force algorithms are only applicable if the number of items is small or for a static model if the number of context is small. In the former case, brute-force scoring can be applied online, in the latter, brute-force scoring can be applied offline.

## 6.2   Approximate Nearest Neighbor Search

For dot product models, $\hat{y}(i|c) = \langle \boldsymbol{\phi}(c), \boldsymbol{\psi}(i) \rangle$, the problem of finding the top scoring items is equivalent to the well studied field of *maximum inner product search* or *approximate nearest neighbor search* [26]. The nearest neighbor search problem is to find the closest entries in a database of vectors for a query vector. For item recommendation, the context $\boldsymbol{\phi}(c)$ corresponds to the query and the database of vectors are the items $I$ represented by the embeddings $\boldsymbol{\psi}(i)$. These problems are very well studied with efficient sublinear approximate algorithms that can be applied for the item recommendation problem. Typically, solutions are based on narrowing down the set of possible nearest neighbors that needs to be evaluated, e.g., using partitioning algorithms such as trees [15, 31], and efficient scoring using quantization [26].

These efficient sublinear algorithms make dot product models very attractive for large scale item recommendation. If an application requires fast retrieval, dot product models are the best choice.

## 6.3   Dynamic User Model

When a user interacts with a recommender system, the system should be responsive and change the recommendation based on a user's feedback. For example, after watching a video, the system should be able to make better recommendations taking into account this new information. Depending on the type of scoring function, this might require to update the model itself. For example, in a user-item matrix factorization model, the user embedding $\mathbf{w}_u$ would need to be retrained in real time after every interaction—for user-item matrix factorization a projection, Eq. (30) can be applied. Other scoring functions avoid retraining by representing the user as a function of their past behavior. For example, if the user is represented by the average item embedding in their history, the user embedding function $\phi(c)$ is trivial to compute and can be done online at query time. In both cases, the main model can be retrained occasionally offline to propagate all training data through the model.

# 7   Conclusion

This chapter introduced the item recommendation problem and discussed its unique challenges. Several popular approaches to train recommender systems from implicit feedback data were presented. An advantage of sampling based approaches is that they can be applied to most recommender models. That makes them a popular choice for learning item recommenders. Another advantage is that they can be adapted to retrieval metrics by rank based weighting and sampling schemes. However, if the item catalogue is large, sampling based approaches are slow unless they use good samplers which is an open area of research. For special cases the Gramian trick allows to learn item recommenders much more efficently, typically dropping the runtime dependency on the catalogue size $|I|$. However, this trick is only applicable to square losses which can be less effective than a ranking loss such as weighted pairwise or softmax. Dot product models that represent the context and the items by a $d$-dimensional embedding have useful properties throughout learning and retrieval. For learning, their structure can be exploited for faster sampling, or the Gramian trick, and for retrieval it allows real-time retrieval in sublinear time through maximum inner product search algorithms.

As a conclusion of this chapter, a few rules of thumb for applying recommender systems in practice are suggested. Dot product models are a reasonable default choice due to their attractive retrieval properties. The family of dot product models is broad including popular approaches such as matrix factorization but also more complex techniques such as deep neural networks including transformers, recurrent structures, or in general any model that extracts representations which are combined in the final layer with a dot product. For learning, sampling based algorithms such as sampled softmax with a large number of samples $m$ are easy to implement and can result in good ranking quality. If the item catalogue is very large, algorithms based on the Gramian trick can be a better choice, and implementations for models like matrix factorization are simple. In any case, properly tuning and setting up models can be more important than switching to a more complex approach [7, 21, 22].

**Acknowledgments** I would like to thank Nicolas Mayoraz and Li Zhang for helpful comments and suggestions.

# References

1. Y. Bai, S. Goldman, L. Zhang, TAPAS: Two-pass Approximate Adaptive Sampling for Softmax. arXiv: 1707.03073 (2017)
2. I. Bayer, X. He, B. Kanagal, S. Rendle, A generic coordinate descent framework for learning from implicit feedback, in *Proceedings of the 26th International Conference on World Wide Web WWW'17* (2017), pp. 1341–1350
3. Y. Bengio, J.-S. Senecal, Quick training of probabilistic neural nets by importance sampling, in *Proceedings of the Ninth International Workshop on Artificial Intelligence and Statistics, AISTATS 2003, Key West, January 3–6, 2003* (2003)
4. G. Blanc, S. Rendle, Adaptive sampled softmax with kernel based sampling, in *Proceedings of the 35th International Conference on Machine Learning*, ed. by J. Dy, A. Krause. Proceedings of Machine Learning Research, vol. 80, Stockholmsmässan, Stockholm, 10–15 July 2018 (2018), pp. 590–599
5. C.J.C. Burges, From ranknet to lambdarank to lambdamart: an overview. Technical Report MSR-TR-2010-82 (2010)
6. P. Covington, J. Adams, E. Sargin, Deep neural networks for youtube recommendations, in *Proceedings of the 10th ACM Conference on Recommender Systems, RecSys'16* (Association for Computing Machinery, New York, 2016), pp. 191–198
7. M.F. Dacrema, S. Boglio, P. Cremonesi, D. Jannach, A troubling analysis of reproducibility and progress in recommender systems research. ACM Trans. Inf. Syst. **39**(2), 1–49 (2021). https://doi.org/10.1145/3434185
8. R.A. Harshman, Foundations of the PARAFAC procedure: models and conditions for an "explanatory" multi-modal factor analysis. UCLA Work. Pap. Phonetics **16**(1), 84 (1970)
9. B. Hidasi, D. Tikk, Fast ALS-based tensor factorization for context-aware recommendation from implicit feedback, in *Joint European Conference on Machine Learning and Knowledge Discovery in Databases* (Springer, Berlin, 2012), pp. 67–82
10. Y. Hu, Y. Koren, C. Volinsky, Collaborative filtering for implicit feedback datasets, in *Proceedings of the 2008 Eighth IEEE International Conference on Data Mining, ICDM'08* (2008), pp. 263–272
11. R. Jiang, S. Gowal, Y. Qian, T. Mann, D.J. Rezende, Beyond greedy ranking: slate optimization via list-CVAE, in *International Conference on Learning Representations* (2019)
12. T. Joachims, Optimizing search engines using clickthrough data, in *Proceedings of the Eighth ACM SIGKDD International Conference on Knowledge Discovery and Data Mining, KDD'02* (Association for Computing Machinery, New York, 2002), pp. 133–142
13. W. Krichene, N. Mayoraz, S. Rendle, L. Zhang, X. Yi, L. Hong, E. Chi, J. Anderson, Efficient training on very large corpora via gramian estimation, in *7th International Conference on Learning Representations* (2019)
14. W. Krichene, S. Rendle, On sampled metrics for item recommendation, in *Proceedings of the 26th ACM SIGKDD International Conference on Knowledge Discovery & Data Mining, KDD'20* (Association for Computing Machinery, New York, 2020), pp. 1748–1757
15. M. Muja, D.G. Lowe, Scalable nearest neighbor algorithms for high dimensional data. IEEE Trans. Pattern Anal. Mach. Intell. **36**(11), 2227–2240 (2014)
16. X. Ning, G. Karypis, SLIM: sparse linear methods for top-n recommender systems, in *Proceedings of the 2011 IEEE 11th International Conference on Data Mining, ICDM'11* (IEEE Computer Society, Washington, 2011), pp. 497–506
17. A.S. Rawat, J. Chen, F.X.X. Yu, A.T. Suresh, S. Kumar, Sampled softmax with random fourier features, in *Advances in Neural Information Processing Systems*, ed. by H. Wallach, H. Larochelle, A. Beygelzimer, F. d'Alché-Buc, E. Fox, R. Garnett, vol. 32 (Curran Associates, Red Hook, 2019), pp. 13857–13867
18. S. Rendle, Factorization machines, in *Proceedings of the 2010 IEEE International Conference on Data Mining, ICDM'10* (IEEE Computer Society, Washington, 2010), pp. 995–1000

19. S. Rendle, C. Freudenthaler, Improving pairwise learning for item recommendation from implicit feedback, in *Proceedings of the 7th ACM International Conference on Web Search and Data Mining, WSDM'14* (Association for Computing Machinery, New York, 2014), pp. 273–282
20. S. Rendle, C. Freudenthaler, Z. Gantner, L. Schmidt-Thieme, BPR: Bayesian personalized ranking from implicit feedback, in *Proceedings of the Twenty-Fifth Conference on Uncertainty in Artificial Intelligence, UAI'09* (AUAI Press, Arlington, 2009), pp. 452–461
21. S. Rendle, L. Zhang, Y. Koren, On the difficulty of evaluating baselines: a study on recommender systems. CoRR, abs/1905.01395 (2019)
22. S. Rendle, W. Krichene, L. Zhang, J. Anderson, Neural collaborative filtering vs. matrix factorization revisited, in *Proceedings of the 14th ACM Conference on Recommender Systems, RecSys'20* (2020)
23. G. Takács, D. Tikk, Alternating least squares for personalized ranking, in *Proceedings of the Sixth ACM Conference on Recommender Systems, RecSys'12* (Association for Computing Machinery, New York, 2012), pp. 83–90
24. L.R. Tucker, Some mathematical notes on three-mode factor analysis. Psychometrika **31**, 279–311 (1966)
25. N. Usunier, D. Buffoni, P. Gallinari, Ranking with ordered weighted pairwise classification, in *Proceedings of the 26th Annual International Conference on Machine Learning, ICML'09* (Association for Computing Machinery, New York, 2009), pp. 1057–1064
26. J. Wang, W. Liu, S. Kumar, S.-F. Chang, Learning to hash for indexing big data - a survey. CoRR, abs/1509.05472 (2015)
27. J. Weston, S. Bengio, N. Usunier, Wsabie: scaling up to large vocabulary image annotation, in *Proceedings of the Twenty-Second International Joint Conference on Artificial Intelligence - Volume Three, IJCAI'11* (AAAI Press, Palo Alto, 2011), pp. 2764–2770
28. M. Wilhelm, A. Ramanathan, A. Bonomo, S. Jain, E.H. Chi, J. Gillenwater, Practical diversified recommendations on youtube with determinantal point processes, in *Proceedings of the 27th ACM International Conference on Information and Knowledge Management, CIKM'18* (Association for Computing Machinery, New York, 2018), pp. 2165–2173
29. D. Xin, N. Mayoraz, H. Pham, K. Lakshmanan, J.R. Anderson, Folding: why good models sometimes make spurious recommendations, in *Proceedings of the Eleventh ACM Conference on Recommender Systems, RecSys'17* (Association for Computing Machinery, New York, 2017), pp. 201–209
30. J. Yang, X. Yi, D. Zhiyuan Cheng, L. Hong, Y. Li, S. Xiaoming Wang, T. Xu, E.H. Chi, Mixed negative sampling for learning two-tower neural networks in recommendations, in *Companion Proceedings of the Web Conference 2020, WWW'20* (Association for Computing Machinery, New York, 2020), pp. 441–447
31. P.N. Yianilos, Data structures and algorithms for nearest neighbor search in general metric spaces, in *Proceedings of the Fourth Annual ACM-SIAM Symposium on Discrete Algorithms, SODA'93* (Society for Industrial and Applied Mathematics, Philadelphia, 1993), pp. 311–321
32. F. Yuan, G. Guo, J.M. Jose, L. Chen, H. Yu, W. Zhang, Lambdafm: Learning optimal ranking with factorization machines using lambda surrogates, in *Proceedings of the 25th ACM International on Conference on Information and Knowledge Management, CIKM'16* (Association for Computing Machinery, New York, 2016), pp. 227–236

# Deep Learning for Recommender Systems

**Shuai Zhang, Yi Tay, Lina Yao, Aixin Sun, and Ce Zhang**

## 1 Introduction

Since the reinvigoration of neural networks, deep learning has attracted significant attention. It provides a revolutionary step to actualize artificial intelligence and has fundamentally changed the landscape of a number of fields, including computer vision, natural language processing, robotics, and more. Applications of deep learning have become ubiquitous. For example, deep neural networks can be trained to recognize objects in images, translate text between languages, recognize speech, generate images/texts, just to mention a few. Very often, state-of-the-art performances are achieved with deep neural networks.

With this striking success, a variety of deep learning designs and methods have blossomed in the context of recommender systems. It has attracted a huge interest from academia and industry, evidenced by the exponentially increase of public research works and implementations. A growing number of companies are developing and deploying deep learning based algorithms to enhance their recommender systems, so as to attract more customers, improve user satisfaction and retention rate, and as a result boost their revenue. Clearly, deep learning

S. Zhang (✉) · C. Zhang
ETH Zurich, Zurich, Switzerland
e-mail: shuai.zhang@inf.ethz.ch; ce.zhang@inf.ethz.ch

Y. Tay
Google, Mountain View, CA, USA

L. Yao
UNSW Sydney, Sydney, NSW, Australia
e-mail: lina.yao@unsw.edu.au

A. Sun
NTU, Singapore, Singapore
e-mail: axsun@ntu.edu.sg

© Springer Science+Business Media, LLC, part of Springer Nature 2022
F. Ricci et al. (eds.), *Recommender Systems Handbook*,
https://doi.org/10.1007/978-1-0716-2197-4_5

based recommendation models are now in wide use across platforms and domains. Meanwhile, recommendation with deep learning techniques has been the major focus in the research community and the number of scientific publications on this topic grows exponentially.

There are numerous benefits of using deep learning techniques to build recommender systems. Firstly, deep learning requires less feature engineering effort as it can easily process unstructured data such as text, image, sound, video, etc. Secondly, deep learning opens up an opportunity for a range of challenging recommendation tasks. For example, we can solve the cold-start problem, improve the recommendation explainability/robustness, and handle temporal dynamics effortlessly with deep neural networks. Thirdly, deep neural networks have a high degree of flexibility in the sense that multiple neural building blocks can be composed end-to-end so that building powerful models (e.g., multitask models, etc.) becomes more convenient. Just as importantly, the increasing of computational resources as well as the easy access of deep learning computation libraries make the implementation, iteration, and deployment process more handily. All in all, these benefits have led to the rapid and revolutionary development of recommender systems.

In this chapter, we do not intend to give a thorough review of all related methods, but rather to overview various key approaches and highlight the impact of deep learning techniques on the recommender systems field. This chapter is divided into three parts. We begin with the basics of deep learning techniques that are widely adopted in the recommendation area, namely, multilayered perceptrons, convolutional neural networks, recurrent neural networks, graph neural networks, deep reinforcement learning, etc. In the second part, we review how deep learning methods are applied to specific recommendation problems including interaction modeling, user modeling, content representation learning, cold-start problem, and interpretability/robustness enhancement. Thereafter, we describe how recommendations in various application domains such as e-commerce, online entertainment, news, and point-of-interests, can benefit from deep learning techniques.

## 2 Deep Learning for Recommender Systems: Preliminary

We begin with the basics of recommender systems and widely adopted deep learning techniques.

### 2.1 Basics of Recommender Systems

In a typical recommender system setting, there are a bunch of users and a catalog of items. Items can refer to movies, books, jobs, jokes, news articles, music pieces, houses, products, and even services. There usually exist some historical interactions (e.g., purchases, ratings, likes, watches, etc.) between users and items. If there is no interaction record for some users/items, we call them cold-start users/items. Oftentimes, additional information such as contexts like timestamp, item descriptions, and user profiles are available. With these available data, we can learn a recommendation model that anticipates user's preferences and performs personalised recommendations. A good recommender system should have the ability of accurately capturing users' preferences/intentions, modeling the characteristics of items, and fitting the context better. Popular methods such as matrix factorization [64] (Chapters "Advances in Collaborative Filtering" and "Item Recommendation from Implicit Feedback"), user-based/item-based collaborative filtering [92] (Chapter "Trust Your Neighbors: A Comprehensive Survey of Neighborhood-Based Methods for Recommender Systems"), content-based approaches (Chapter "Semantics and Content-based Recommendations"), context-aware recommender systems (Chapter "Context-Aware Recommender Systems: From Foundations to Recent Developments"), cross-domain recommender systems (Chapter "Design and Evaluation of Cross-Domain Recommender Systems"), and session-based recommender systems (Chapter "Session-Based Recommender Systems") have been extensively investigated in the past decades. We refer users to the corresponding chapters for more details.

## 2.2   Basics of Deep Learning Techniques

Deep learning utilizes a hierarchical level of neural networks to carry out the process of machine learning. Neural network is made up of neurons which are inspired by biological neurons. A typical deep neural network consists of multiple layers and each layer has multiple neurons. Neurons between two successive layers are connected. It accumulates signals from a previous layer and the information transmission between layers is controlled with activation functions [36]. Unlike traditional linear methods, activation functions such as *ReLU*, *Sigmoid*, and *Tanh* in neural networks enable machines to process data in a nonlinear way.

At a high level, deep learning is a representation learning approach. It offers the potential to identify complex patterns and relationships hidden in the data. As such, deep learning has been widely used to learn representations of text, image, audio, video, etc. The representation learning process can operate either in a supervised manner or in an unsupervised manner. Based on the way how neurons are organized in the network, there are various neural network architectures. Major (but not exhaustive) neural networks include: multi-layer perceptrons (MLPs), convolutional neural networks (CNNs), recurrent neural networks (RNNs), autoencoders, generative adversarial networks, and graph neural networks [36, 63].

To learn parameters of deep learning models, an algorithm known as back-propagation is widely adopted [36]. Back-propagation computes the derivatives of the loss function like mean squared error loss, cross-entropy loss, or triplet loss, with respect to the weights and biases in a neural network. The computed gradients are then used by optimizers such as gradient descent, stochastic gradient descent, and

Adam to update the corresponding weights and biases. Same as traditional machine learning techniques, deep learning also suffers from the overfitting problem. To overcome this problem, regularization methods such as dropout, batch normalization, and layer normalization are usually employed. We will briefly introduce some widely used neural network architectures in the following sections.

### 2.2.1  Multi-Layer Perceptrons

A multilayer perceptron is composed of multiple perceptrons. It has an input layer to receive the signal, an output layer to make predictions, and an arbitrary number of hidden layers in between. Theoretically, an MLP is able to learn any mapping functions and has been proven to be a universal approximation algorithm. The predictive capability of MLPs comes from the hierarchical and multi-layered structure. It is an ideal representation learning approach as it can represent features at different scales, orders, and resolutions.

MLPs are usually used to model the correlations or dependencies between inputs and outputs. An example of MLP is shown in Fig. 1. The input layer takes the dataset as input and passes it to the next layer. The layer after the input layer is a hidden layer. It is possible to stack many hidden layers to form a very deep architecture. Deep learning refers to learning with networks that have many hidden layers. The last layer is the output layer which outputs a scalar or a vector depending on the task. For example, a regression problem has a single output neuron and there are more than one neuron in binary or multi-class classification tasks. Generally, the network outputs are converted to probabilities with *sigmoid* or *softmax* function.

### 2.2.2  Convolutional Neural Networks

Convolutional neural networks (CNNs) are suitable for grid topology data. For grid-like data such as images, MLPs might not scale well. For example, a small image of size 32 (width) × 32 (height) × 3 (RGB channels) has 3072 dimensions. The full connectivity will bring in a huge number of parameters and is more likely to cause overfitting. CNNs are precisely designed to circumvent this limitation. CNN can capture the locality and spatial-invariance (e.g., position of an object in an image

**Fig. 1** Multi-layer perceptrons

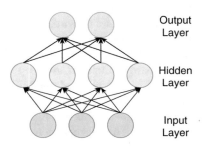

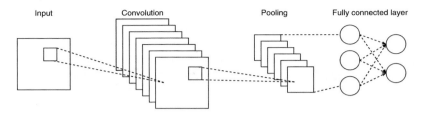

**Fig. 2** Illustration of a convolutional neural network which has a convolutional layer, a pooling layer, and a fully-connected layer

does not change the class of the object) in features with merely a small number of free-parameters. CNNs can also process other types of data structure such as sequences [2].

A typical convolutional neural network consists of an input layer, several convolutional layers and pooling layers, and fully-connected layers (same as MLPs). An example is shown in Fig. 2. Convolutional layers are a major building block of CNNs. A convolution layer defines a window (or filter, kernel), by which the model can examine a subset of input (e.g., a region of an image), and subsequently scan the entire input by looking through this window. The window is small and parameterized so that it needs fewer parameters than fully connected layers. In addition, this window mechanism provides a way to look at the local regions of the input and enables a more efficient learning of local patterns. Oftentimes, we can use multiple windows to get multiple feature mappings. Pooling layer is also an important component in CNNs. Pooling operations are used to compress spatial information rather than extract certain features. It is useful when we care more about whether some features are present or not instead of where they are. Pooling is also operated with windows and various pooling operations including max pooling, average pooling, and minimum pooling are usable. Among them, max pooling is the most commonly used operation as it can return the most notable features.

Recent years have witnessed the emergence of numerous CNN architectures such as LeNet, AlexNet, VGG, Inception, ResNet, ResNeXt, and many more [36]. These networks have got very deep (up to 100 layers) and achieved huge performance boost. We omit the details for brevity.

### 2.2.3 Recurrent Neural Networks

Many practical problems such as time series prediction, sentence classification, machine translation, text generation, and speech recognition, require the ability of modeling sequential and temporal patterns. In these tasks, current outputs depend on previous inputs and outputs. Recurrent neural network is such a model that can deal with sequential dynamics in the data and are becoming increasingly popular.

RNNs allow a model to operate over sequences in the input, output, or both of them [36]. The most important feature of RNNs is the hidden state which remembers

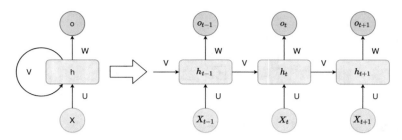

**Fig. 3** Illustration of recurrent neural networks in folded (left) and unfolded (right) forms

what information has been calculated about a sequence. Figure 3 shows the structure of RNNs in both folded and unfolded forms. It has several distinct components: (1) $X$ is the sequential input; (2) $o$ is the output; (3) $h$ is the hidden states; (4) $V$, $W$, $U$ are trainable model parameters. The hidden state $h$ is the memory of the network and is calculated based on the previous hidden state and the current input. A simple calculation is:

$$h_t = f(UX_t + Vh_{t-1}), \tag{1}$$

where $f$ is a nonlinear activation function.

Vanilla RNNs may suffer from problems like gradient vanishing and exploding problems. They are less effective in processing very long sequences. Therefore, there are many variants of RNNs and Long short-term memory networks (LSTMs) [52] and gated recurrent units (GRUs) [22] are the most popular ones. LSTMs are proposed to address the problem of keeping or resetting context. LSTM unit is composed of a cell, an input gate, an output gate, and a forget gate. It utilizes the cell and gate mechanism to memorize or filter long-term dependencies. The sibling architecture GRUs operate in a similar fashion and can achieve the same goal. In some tasks like speech recognition and handwriting recognition, it is important to look beyond the current state, to fix the past. Knowing what is coming next can help understand the context and alleviate the ambiguity faced in the past. This gives rise to bi-directional RNNs [36].

### 2.2.4 Encoder and Decoder Architectures

The encoder-decoder architecture is quite common in the deep learning field. We first introduce auto-encoder (commonly written as autoencoder).

Autoencoder [111] is an unsupervised learning framework which consists of an encoder, a decoder, and a bottleneck layer (see Fig. 4). The output can be viewed as a copy of the input. It aims to reconstruct the input in the output layer by minimizing the reconstruction error measured by the differences between the input and the reconstruction. It is a widely used representation learning approach as it can compress the high-dimensional input into a low-dimensional hidden representation

**Fig. 4** Architecture of
autoencoder

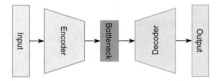

at the bottleneck layer, while maintaining the important characteristics. A variant of autoencoder called variational autoencoder (VAE) can generate new data. VAE encodes the input as a distribution over a latent space and this distribution is regularised to ensure continuity and completeness.

Conceptually, in an typical encoder-decoder architecture, an encoder is used to map the input to a compact hidden representation and a decoder is used to generate a high-dimensional output by up-sampling the compressed representation. It is not exclusive to autoencoder. Many sequence-to-sequence models such as transformer [109] are also structured in this way.

### 2.2.5 Graph Neural Networks

Graph data (e.g., social networks, chemical molecules, knowledge graph, transaction graph, etc.) is commonplace in real world applications. A graph consists of a set of nodes and a set of edges between nodes. Usually, nodes and edges are associated with some feature information for descriptive purposes.

The network architectures discussed earlier (e.g., CNNs or RNNs) cannot handle graph data effectively. As such, graph neural networks (GNNs) [63, 110] come into play. GNNs learn node/graph representations from node (edge) features and graph structures via message propagation and aggregation. The most critical process, message passing, pushes messages from surrounding nodes around a given reference node through its edges. At a single time step, each node is updated by its current presentation and the aggregation of its neighbors' representations. This step is repeated multiple times in deep GNNs, allowing information from multi-hops away to be propagated. A general update rule of GNNs is as follows:

$$h_i^{(\ell+1)} = \sigma(h_i^{(\ell)} W_0^{(\ell)} + \sum_{j \in \mathcal{N}_i} \frac{1}{c_{ij}} h_j^{(\ell)} W_1^{(\ell)}), \qquad (2)$$

where $h_i^{(\ell+1)}$ is the representation of node $i$ at layer $\ell + 1$; $\mathcal{N}_i$ is a set of neighbors of node $i$; $c_{ij}$ is a fixed/trainable norm; $W_*^*$ is a weight matrix.

### 2.2.6 Deep Reinforcement Learning

Deep reinforcement learning [32] is the combination of deep learning and reinforcement learning. Reinforcement learning provides a formalism for behavior

**Fig. 5** The architecture of
deep reinforcement learning

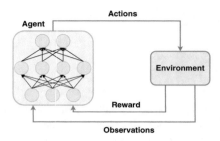

and allows us to solve many types of decision-making problems. Deep learning is powerful in handling unstructured data such as raw sensor or image signals. By integrating deep learning techniques, it allows reinforcement learning to solve more complex problems end-to-end. Deep neural networks enable the agent to get knowledge from raw data and derive efficient representations without handcrafted features and domain heuristics. Thus, deep reinforcement learning opens up the possibility for applications in domains such as robotics, games, healthcare, and many more (Fig. 5).

Reinforcement learning is learning through interaction with an *environment*, The *agent* learns to achieve a goal in a complex *environment*. It employs a trial and error strategy to learn from the consequence of its *actions* and selects its actions on the basis of past experiences (exploitation) and new choices (exploration). The agent receives rewards or penalties for the actions it performs and seeks to learn to select actions that maximize the accumulated reward over time.

### 2.2.7   Adversarial Neural Networks

We are concerned with two techniques: adversarial training and generative adversarial neural networks (GANs) [24].

An adversarial example is an instance with small, intentional feature perturbations that cause a machine learning model to make a mistake. Adversarial examples will cause model performance degradation and, even worse, be dangerous in applications like autonomous driving. Adversarial training is a straightforward way to combat with adversarial examples. It generates adversarial examples during the training process and the model is learned to combat these generated examples such that it cannot be fooled easily after training.

Generative adversarial neural networks (GANs), inspired by the minimax game in game theory, have two neural networks which compete with each other. The two networks in GANs are: generator and discriminator. The generator creates new data instances that resemble the training data, and the discriminator is trained to distinguish the generated fake data from real data. The generator and the discriminator are simultaneously trained with back-propagation. In the end, the generator can generate images that are not present in the training data.

### 2.2.8 Restricted Boltzmann Machines

A restricted Boltzmann machine (RBM) [31] can lean a probability distribution over its input. RBMs have found applications in dimensionality reduction, classification, topic modeling, recommendation, etc. A classical RBM can be considered as a shallow two-layer neural network which in general consists of a visible layer (also known as an input layer) and a hidden layer. The restriction indicates that there is no intra-layer communication. Same as MLPs, activation functions and bias terms can also be used. To optimize the parameters of RBMs, the algorithm contrastive divergence is usually adopted. RBMs are the building blocks of deep belief networks (DBNs) [51] where multiple hidden layers are needed.

# 3 Deep Learning for Recommender Systems: Algorithms

This section focuses on how various challenges (e.g., interaction/user modeling, cold-start problems, robustness, explainability, etc.) in recommender systems can be tackled with deep learning techniques.

## 3.1 Deep Learning for Interaction Modeling

Interaction modeling lives at the heart of recommender systems. Past interactions recorded between users and items such as ratings, purchases, likes, and views are the major source for performing collaborative filtering. As such, a good recommendation model should be able to capture the interactions/crossings between objects/features. Modeling features in their raw form will rarely provide optimal results and interaction modeling becomes necessary.

### 3.1.1 User-Item Interaction Modeling

Interactions between users and items form a user-item interaction matrix. This interaction matrix can be extremely sparse as many interactions are missing. Modeling user item interaction from this partially observed matrix forms the bedrock of collaborative filtering (CF) algorithms. A standard CF approach is to model user item interactions with dot product (e.g., matrix factorization). As an alternative, we can also use deep neural networks to fulfill this goal.

Formally, we adopt the following formula to define the interaction between users and items.

$$s = f(h(g_1(u), g_2(i))), \tag{3}$$

**Table 1** Comparison of various user item interaction modelling approaches

| Models | $h$ | $f$ | $g_*$ |
|---|---|---|---|
| NeuMF [47] | Concatenation: $[U_u; V_i]$ <br> Hadamard product: $U_u \odot V_i$ | MLPs | Embedding look-up |
| He et al. [45] | Outer product: $U_u \otimes V_i$ | CNNs | Embedding look-up |
| AutoRec [93], CDAE [127] | Encoder | Decoder | MLPs |
| DFM [132] | Dot product: $U_u \cdot V_i$ | Identity mapping | MLPs |
| CML [53], LRML [53] | Euclidean distance: <br> $\| U_u - V_i \|_2^2$ | Identity mapping | Embedding look-up |
| HyperML [112] | Hyperbolic distance: <br> $\cosh^{-1}(1 + 2\frac{\|U_u - V_i\|^2}{(1-\|U_u\|^2)(1-\|V_i\|^2)})$ | Identity mapping | Embedding look-up |
| Yao et al. [137] | Searched | MLPs | Embedding look-up |

where $g_*$ is used to get the representation of user/item. For example, if $g_*$ is a sparse embedding look-up function, it will output the corresponding user embedding $U_u$ and item embedding $V_i$; $s$ is the recommendation score; $f$ can be neural networks, identity mapping, etc., and $h$ is the interaction operation such as concatenation, Hadamard product, etc. Table 1 summarizes popular user-item interaction modeling approaches.

Using neural networks to model the interactions between users and items has aroused extensive interest. Recently, there is an increasing popularity to replace dot product with MLPs. For example, NeuMF [47] models the interaction with multilayer perceptron. It takes the concatenation of $U_u$ and $V_i$ or the element-wise multiplication of $U_u$ and $V_i$ as the input of MLPs to capture the interactions. A follow-up work [45] replaces dot product with outer product where the output is treated as an interaction map and CNNs are applied to learn high-order correlations among embedding dimensions. Despite its popularity, a recent study [88] suggests that it is non-trivial to learn the dot product with an MLP, and it might be too costly to be used in production environments.

The autoencoder architecture can also be applied to interaction modeling [93, 127]. These methods take as inputs the columns or rows of the partially observed interaction matrix, then recover the columns/rows in the output. The gradients of unobserved entries are masked out in the training phase. Nonlinear activation functions, dropout, and denoising strategies are usually used to enhance the model expressiveness and robustness. These methods can be regarded as generalizations of latent factor models. Different from conventional autoencoders, the goal of these models is to complete the interaction matrix in a column-wise (row-wise) manner instead of learning compact representations of inputs. The generated scores in

**Fig. 6** Deep learning based
interaction modeling for
recommendation

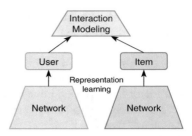

the outputs for missing entries are viewed as the recommendation scores for the corresponding user-item pair.

As shown in Fig. 6, learning user or item representations is also an indispensable step in interaction modeling. Instead of using a simple one-hot identifier to learn user/item embeddings, an alternative way is to enrich the representation information by incorporating all historical interactions of each user/item [132, 148]. For example, we can use the rows (all items that the user rated) or columns (all users who liked the item) as the input and obtain the corresponding user/item representations; the pioneering neural network based recommendation model, RMB-CF [35, 91], aims to learn a user-specific feature vector from the observed interaction matrix (e.g., rating matrix, implicit feedback, etc.) via RBMs. The learned hidden units, in return, can help approximate the distribution of the whole user-item interaction input, thus, making predictions on unobserved data possible.

The idea of word2vec [82] can also be utilized for item representation learning. Representative models are item2vec [4], prod2vec [37], and [38]. In essence, they treat items as words and the sets of items generated from purchase logs/click sequences as sentences and a skip-gram algorithm is utilized to learn the item embeddings. For example, prod2vec learns the product embeddings by minimizing the following objective function:

$$\mathcal{L} = \sum_{s \in S} \sum_{p_i \in s} \sum_{-c \leq j \leq c, j \neq 0} \log P(p_{i+j} | p_i), \tag{4}$$

where $S$ is the entire "sentence" set; $c$ defines the window size; $p_{i+j}$ is the neighboring product of current product $p_i$. The probability $P(p_{i+j} | p_i)$ is obtained with a *softmax* function. Works [4, 38] take a similar form so that the details are omitted for brevity.

Another line of research is modeling interactions between users and items via distance metric learning where a shorter distance between a user and an item indicates stronger fondness. These methods aim to learn the distance $\| U_u - V_i \|_2^2$ between users and items [53]. To enhance the geometrical flexibility, Yi et al. [104] propose adding a trainable relation vector $r$ and the distance function becomes $\| U_u + r - V_i \|_2^2$. Extending this metric learning approach to non-Euclidean space for recommender systems also shows promising performance. An ideal option is the hyperbolic space where hierarchical structures and exponentially expansion properties can be easily captured. Representative methods such as HyperML [112, 145] have shown promising results. We review these methods because most of these methods are constructed and optimized within standard deep learning operators and they can also be integrated into other deep learning architectures.

However, due to the complex nature of interactions in real-world applications, it is difficult for a specific interaction function to have consistently good performance across different application scenarios. As such, Yao et al. [137] propose using automated machine learning (AutoML) to search for the neural interaction functions for different scenarios. The search space includes inner product, Euclidean distance, outer product, concatenation, maximum, minimum, etc. The proposed method can train the search architecture and the recommendation model simultaneously in an end-to-end manner.

### 3.1.2 Feature Interaction Modeling

Apart from the user-item interaction matrix, there are abundant side information/feature[1] available that can be predictive for making recommendations. Nevertheless, using features in their raw form, in general, will not lead to optimal solutions. Two features combined via some operations could be more predictive than the same two features used independently. However, exhaustive manual search for feature interaction is infeasible in real world applications. The capability of learning feature interactions automatically is of great importance to recommendation models.

A popular feature interaction modeling approach is factorization machines which capture the pairwise interactions in linear time complexity. However, modeling higher-order feature interaction with factorization machines can be costly. Recently, there is a growing interest in applying deep neural networks to model feature interaction. For instance, Cheng et al. [19] propose a wide and deep framework to combine MLPs with a linear model (shown in Fig. 7a). In this framework, the deep part captures feature combinations and has good generalization for unseen feature combinations, and the wide part keeps good memorization for feature co-occurrence or correlation. The recommendation score is defined as:

$$s = \sigma(f_{LM}(x_{wide}) + f_{MLP}(x_{deep}))  \tag{5}$$

where $f_{LM}$ and $f_{MLP}$ represent a linear model and multilayered perceptrons respectively. The input feature $x$ is manually separated into $x_{wide}$ and $x_{deep}$.

Obviously, the wide and deep network requires expert knowledge in splitting features into two parts. As such, Guo et al. [40] propose an improved deep and wide architecture, DeepFM (shown in Fig. 7b). In this model, the deep part and the wide have shared input features. The deep part is an MLP utilized to capture the

---

[1] In this section, we mainly refer to categorical features.

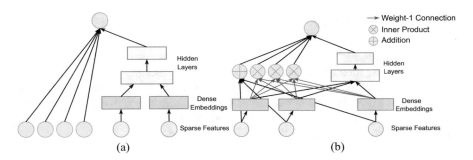

**Fig. 7** Architectures for feature interaction modeling: (**a**) Wide and Deep Model [19]; (**b**) DeepFM [40]

high-level and nonlinear feature interactions and the wide part is an FM. DeepFM can capture both high-level and lower-order feature interactions. A concurrent work NFM [44] also has a similar design as DeepFM. The scoring function of DeepFM is defined as:

$$s = \sigma(f_{FM}(x) + f_{MLP}(x)) \tag{6}$$

where $f_{FM}$ represents factorization machines. Clearly, the two components share the same input.

It is noted that MLPs can only model interactions implicitly which is not necessarily effective for all types of cross features. To circumvent this limitation, Wang et al. [118] propose a deep and cross network (CrossNet) to explicitly apply feature crossings across layers. This model requires no manual feature engineering and will not incur much additional computational cost. The downside of CrossNet is that the interactions come in a bit-wise fashion. As such, Lian et al. [74] design a generic version of CrossNet to enable vector-wise feature interaction. The proposed model, xdeepfm [74], is a combination of a generalized CrossNet and MLPs.

Evidently, not all feature interactions are created equally. Inappropriate feature interactions might be useless and even harmful. To better distinguish the importance of different feature combinations, Xiao et al. [129] propose an attention based deep factorization recommendation algorithm (AFM). This model utilizes a parameterized attention network to get an attention score for each pair of interactions in FM. The output is an attentive summation of all possible interactions. A similar attention based FM model, A3NCF [20], is proposed for the rating prediction task. Essentially, FM model adopts inner product to model the pairwise interaction. Same as the methods used to improve user item interactions, we can replace the interaction function with a few alternatives. For example, HFM [107] uses circular convolution operations or circular correlation operations to replace the inner product in FM. LorentzFM [130] substitutes the inner product with Lorentzian distance and generalizes FM to non-Euclidean space. It is worth noting that AFM, A3NCF, HFM, and LorentzFM lose the efficiency advantage of FM and have quadratic training/inference complexity.

## 3.2    Deep Learning for User Modeling

To provide personalised recommendations to users, a key step is to accurately infer users' demands, intentions, and interests from their profiles and past behaviors. However, the highly dynamic interaction data and extremely entangled/complex user interests hinder the use of traditional machine learning methods. We scratched the surface of user representation learning in the previous section. Here, we will dive deeper into more specific challenges on this topic.

### 3.2.1    Temporal Dynamics Modeling

Many traditional recommendation algorithms ignore the chronological order of user historical interactions. Nonetheless, the user item interactions are essentially sequential. User's short-term interests have huge impact on her decisions. Time context (e.g., holiday, black Friday, etc.) also affects user behaviors. Moreover, items' popularity are dynamic rather than static over time [59]. The temporal dynamics call for sequence-aware recommender systems. Learning preference representation from the sequence of actions becomes the fundamental task in sequence-aware recommendation. The preference learning process is shown in Fig. 8.

To tackle sequential patterns, popular sequence modeling approaches such as RNN come into play. Wu et al. [124] propose recurrent recommender networks (RRN) which uses LSTMs to capture the temporal dependencies for both users and items. In RRN, the state evolution for a user depends on which items she liked previously. An item's state is dependent on the users who liked it in the past. At last, short-term and long-term impacts (e.g., long-term preference of a user, fixed properties of an item) are integrated for final prediction. Using standard LSTMs or GRUs for temporal dynamics modeling in recommender systems might not be the optimal solution since users and items are usually modeled separately. As such, Donkers et al. [27] devise a novel gated GRU cell for personalized next item recommendations. The proposed cell integrates user characteristics into the recommendation model, which is beneficial to the network's predictive power. Thenceforth, customizing the gated cell in recurrent neural networks has gain growing interest. For instance, Bharadhwaj et al. [6] propose a customized gated recurrent unit to capture latent features of users and items. Guo et al. [41] propose

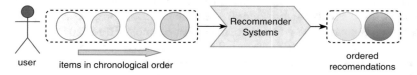

**Fig. 8** Sequence aware recommender systems for temporal dynamics modeling

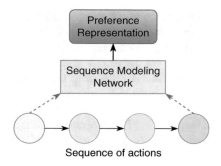

Sequence of actions

PcGRU, an improved GRU, to model the preference drifts and dynamics of user preferences (Fig. 9).

Convolutional neural networks have demonstrated superior performance in a diverse set of sequence modeling tasks such as polyphonic music modeling, character-level language modeling, and word-level language modeling, over canonical recurrent neural networks [2]. CNNs based recommendation models also show promising results in user representation learning from historical sequences. Tang et al. [103] present a convolutional sequence embedding learning approach (Caser) for sequential recommendation. Caser consists of a vertical convolutional network and a horizontal convolutional network to capture both union-level patterns and point-level sequential patterns. You et al. [139] propose a hierarchical temporal convolutional networks (HierTCN) for sequential aware recommendation. HierTCN is composed of a GRU to model the long-term interests across sessions, and a temporal convolutional network [2] to learn dynamic user embeddings. Yan et al. [133] propose a 2-D convolutional network (CosRec) to capture complex item correlations in user's historical actions. Very often, CNNs based sequential learning architectures are more efficient than RNNs based models.

Self-attention makes huge performance boost in the natural language processing field and is gradually becoming a key component in sequence models [109]. This mechanism can also be applied to address the task of sequential recommendation. Atrank [156] takes the lead in using self-attention network to capture temporal dynamics in user behaviors and achieves encouraging results. Kang et al. [61] use multi-head self-attention networks to model user's dynamic interests. The adopted framework is the same as the transformer model [109]. Zhang et al. [146] propose integrating self-attention module into metric learning for sequence dependencies modeling. Huang et al. [58] adopt a self-attention network to model the heterogeneous user behaviors, including diversity of actions and multi-modal property of content. Zhang et al. [150] combine vanilla attention with self-attention to capture feature importance and item transition patterns. The original self-attention model does not consider the time span even with positional encoding. Nonetheless, the absolute time span also matters. Intuitively, if a user has no activities for a relative long time, her last action might have less impact on her current decision. Thus, Xu et al. [131] propose a time kernel to learn functional time representations in a self-attention model. The improved model obtains promising results on sequence-aware recommendation tasks. Another limitation of self-attention is that it can only model left-to-right unidirectional patterns, which might lead to sub-optimal performance. Inspired by the recent success of BERT [26], Sun et al. [99] adopt a BERT-like model which uses a deep bidirectional self-attention module to fuse both left-to-right and right-to-left dependencies in behavior sequences.

There are many other methods for temporal dynamics modeling. For example, Chen et al. [79] propose a hierarchical gating networks to discriminate item importance based on users' preferences via gating mechanism. Chen et al. [16] design a memory-augmented network based sequential recommendation algorithm. It utilizes an external memory matrix to store, access, and manipulate users' historical records in a more explicit fashion. Tang et al. [101] propose a mixture model which combines MLPs, RNNs, CNNs, and self-attention to capture tiny, small, and long range sequential dependencies.

In the sequence-aware recommendation task, users' identifier may be missing in some scenarios. For example, some websites/applications do not require user registration or login. In this case, the system can only record the activities in the current active session. Making recommendations in this scenario is known as session-based recommendations. GRU4Rec [49] is the pioneering work that solves this task with neural networks. Specifically, it adopts GRUs to model the sequential patterns in the sessions and predict the next event. In their extension work, the authors design a new ranking loss that is tailored to RNNs in session-based recommendation settings [48], and propose a parallel RNNs architecture to model both the click sessions and the features of the clicked items [50]. Li et al. [71] present an attention based model to capture users' intentions in the current session. Tuan et al. [108] propose a content-based session recommendation approach with CNNs and a three-dimensional CNN is used to learn representations from item textual descriptions and categorical features.

### 3.2.2 Diverse Interest Modeling

Users might have a diverse set of preferences and be interested in certain aspects of items. For example, a user might like history documentaries and romantic comedies for different reasons. Yet, common modeling approaches usually force all these interests to be encoded into one latent factor. It is critical to enable models to be aware of the diversity in user interests and to have the capability of distinguishing them. To reach the goal, Li et al. [70] present a multi-interest learning approach with dynamic routing. Specifically, to achieve richer user representations, the dynamic routing method disentangles user's interests from her historical interactions. The dynamic routing component acts as a soft clustering approach which groups user's historical behaviors into several clusters, with each cluster corresponding to a particular interest. Similar idea is presented in [8]. The difference is that this work adopts a dynamic routing mechanism from Capsnet [90]. Multiple interests is also investigated in [149] where quaternions are used to represent users. Each component of the quaternion (three imagery components and one real component) represents one aspect of user's interests. In their follow-up work [144], multiple hypercuboids (i.e., concentric hypercuboids, multiple independent hypercuboids) are utilized to represent user interests. The benefits of using hypercuboids is that they can naturally embed the range of preference such as the preferred price range, enabling greater expressiveness. To disentangle the complex user interests, Ma et al. [80] use disentangled variational autoencoder to infer both macro and micro disentanglements for user interests. Macro disentanglement refers to high-level intentions (e.g., buy a book or smartphone) while micro disentanglement reflects low-level factors (e.g., color, size, etc.). Learning disentanglement representations makes the recommendation lists controllable and enriches the interpretability of the learned representations.

## 3.3   Deep Learning for Content Representation Learning

Features (e.g., text, image, sound, video, etc.) associated with items or users can be predictive for recommender systems. Unsurprisingly, these features play a pivotal role in conventional content-based recommender systems. Nonetheless, processing these content and mapping them into latent factors are non-trivial with traditional feature extraction methods. The representation learning capability of deep learning makes it easy to integrate these raw features into modern recommendation models.

### 3.3.1   Textual Feature Extraction

Text data is a rich source of information. Text can be collected from different places such as user reviews, news content, social media, and many more. In recommender systems, text data can be leveraged to better understand items and users, to alleviate the sparsity and even to address the cold-start problem. Processing text data and extracting useful representations can be challenging due to its unstructured nature. Recently, deep learning is becoming the mainstream option for various text processing tasks. Deep learning cannot only process unstructured data easily, but also has the ability to uncover hidden patterns in text data.

Textual description of items (e.g., abstract of an academic paper, plot summary of a movie, news content, tweets) is one of the most used text data in recommender systems.[2] To make use of the textual descriptions, Wang et al. [114] propose a framework, CDL (short for collaborative deep learning), to fuse stacked autoencoder with Bayesian probabilistic matrix factorization (BPMF). An EM-style optimization algorithm is devised to alternatively update the parameters of autoencoder and

---

[2] Since there are a large body of related work on news recommendation, we refer readers to Sect. 4 for more details.

BPMF. Nonetheless, CDL uses bag-of-words representation as the input and ignores the contextual information such as surrounding words and word orders. To overcome the limitations, Kim et al. [62] replaces the autoencoder of CDL with convolutional neural networks. Pre-trained word embeddings (e.g., Glove embeddings) are used for better semantics and expressiveness. Bansal et al. [3] present an end-to-end collaborative filtering model which leverages GRUs to encode the text associated with items into latent vectors for both warm-start and cold-start recommendations. Tan et al. [100] present a quote recommendation method that uses LSTMs to learn the distributed representations for quotes. Lee et al. [67] design a quotes recommender system which combines CNNs and RNNs for quotes process.

Reviews-based recommendation is another typical text-intensive recommendation scenario. On many e-commerce platforms, writing reviews is a strongly encouraged act. The rich semantic information in text reviews cannot be conveyed via the implicit interaction data. For example, users explain the reasons behind their ratings and their additional opinions in reviews. Reviews also provide a wealth of knowledge for prospective customers. In recent years, there has been a widespread interest in using deep learning to exploit reviews for better recommendations. Zheng et al. [155] present a deep cooperative neural network for jointly modeling of users and items reviews. Each user is represented by all reviews she has written and an item is represented by all reviews it received. Specifically, reviews for a single user (item) are concatenated and processed by convolutional networks to get the user(item) representation. To further improve model interpretability, Seo et al. [94] propose adding an attention layer before the convolutional layer. The attention layer contains a local attention window to select informative keywords and a global attention window to filter out irrelevant and noisy words. Yet, the former paradigm of simple concatenation of reviews is unnatural. For example, when deciding if a book should be recommended or not, the user's historical reviews on other books are highly relevant, while her reviews on clothes can be ignored. To this end, Tay et al. [105] present a multi-pointer co-attention network for review based recommendation where a pointer-based learning scheme is employed to extract important and predictive reviews from user and item reviews for subsequent user-item interaction. The pointer-based method is implemented with a gumbel-softmax based pointer mechanism that incorporates discrete vectors within differentiable neural architectures. When reviewing an item, different users may have opinions on different aspects of the item. For example, in restaurant recommendation, some users might care more about tastes while others might pay more attention to locations or ambiences. As such, Guan et al. [39] design an aspect-aware method for review-based recommendation. It extracts the aspects of reviews and uses an attention network to dynamically learn the importance of each aspect.

### 3.3.2 Image Feature Extraction

Images are commonplace on many e-commerce and social platforms. On platforms like Amazon, product images are what users typically scan through most intently.

Images can attract user's attention easily, hence incorporating visual features into recommender systems catches potential preferences of users.

Benefiting from the computer vision research, extracting and processing feature representations from images become easier than ever before. CNNs are the most popular tool for image processing. A number of works attempt to incorporate image features into recommender systems. He et al. [43] incorporate visual features into the conventional Bayesian personalised ranking (BPR) framework and propose VBPR for top-n recommendations. VBPR uses a pretrained-CNN architecture for image pre-processing. Niu et al. [84] propose a neural personalised ranking model for image recommendation on Flickrnote[3] where a hybrid-CNN model is used for visual feature extraction. Lei et al. [69] present a comparative deep learning method for Flickr image recommendation. In this dual network, images are processed with an Alexnet-like CNN architecture. Geng et al. [34] use CNNs to learn visual feature representations for image-based recommendation in social networks on Pinterest. Chu et al. [21] present a visual-aware restaurant recommender system which adopts CNNs to extract visual features from restaurant environment and food images. Additionally, in fashion recommendation domain, incorporation of visual content is also a matter of prime importance. Likewise, CNNs are the top option for feature extraction. For example, McAuley et al. [81], He et al. [42], Yu et al. [140] and Liu et al. [76] all utilize CNNs to infer user's preference/aesthetic on visual styles of fashion products (e.g., clothes. accessories, shoes, etc.). Kang et al. [60] present a system that recommends the best fit products based on 'scene' images (e.g., recommending hats based on a given selfie), where ResNet is employed for image representation learning.

### 3.3.3  Video and Audio Feature Extraction

Deep learning has achieved tremendous success in audio/video analysis such as speech recognition and video surveillance analytics. Compared with the aforementioned features, learning from audio and video features is relatively less common in the context of recommender systems.

Music recommendation is a representative audio-based recommendation scenario. In general, audio information such as rhythm, melody, and timbre is of critical importance to listeners. In the music recommender model designed by Van et al. [85], CNNs are used to extract music features from audio clips. CNNs are also adopted by Huang et al. [56] for music representation learning from acoustic inputs to address the task of music co-listen prediction. CNNs are suited for audio representation learning as they can operate on multiple timescales. Concurrently, Wang et al. [121] use deep belief networks (DBNs) for automatic music feature extraction for music recommendation and they claim that DBNs do well in modeling rhythms and melodies.

---

[3] https://www.flickr.com/.

Video features can also be leveraged to improve recommendations. Usually, videos are converted into a sequence of frames and audio waves. As such, CNNs based models become a desirable option for video analysis. For instance, Xu et al. [17] propose a key frame recommender system to select the key frames from a video for each user. A CNN with five convolutional layers and three fully-connected layers is used for frame representation learning. In the video recommender system proposed by Lee et al. [68], both video frames and audio features are included. The model employs an Inception-v3 network to extract frame features, then it aggregates the frame-level features into video-level ones with average pooling. The audio features are extracted with a modified version of ResNet-50.

In summary, for each type of data, various deep learning techniques can be applied. For text data, sequence modeling approaches are more suitable. While for image, video, and audio data, CNN based architectures are preferable.

## 3.4   Deep Learning for Graph-Structured Data in Recommendation

The interaction data generated from recommender systems can be formulated as a bipartite graph (shown in Fig. 10a). Users and items are two disjoint and independent sets. The interactions between users and items compose the edges of the graph. The user item interaction matrix is the biadjacency matrix of this bipartite graph. Furthermore, connections between users can also formulate a relationship/social network.

The idea of incorporating social networks into recommendation has been circulating for quite a while. Only recently has it been connected to deep learning approaches. In recent years, there is a surge of interest in graph neural networks which has achieved state-of-the-art performances on a number of graph related tasks. Given the interaction logs, it is natural to employ GNNs to learn from the bipartite graph so as to better understand the characteristics of nodes and their relationships. Using GNNs for recommender systems is beneficial for two

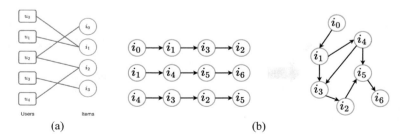

**Fig. 10** Example graphs in recommender systems. (**a**) Interaction graph; (**b**) item session graph

reasons: (1) GNN can better model graph-structural relationships; (2) GNN can help incorporate other graph related data (e.g., social graph, knowledge graph).

Recommendation on the user-item bipartite graph is performed from the link prediction view, where ratings/purchase are represented as links. In doing so, the graph-structural information can be naturally captured. Berg et al. [5] propose a graph convolutional matrix completion model which produces latent factors of user and item nodes via message passing on the bipartite interaction graph. A bilinear decoder with the learnt latent factors is used to reconstruct the links. The same neighbourhood message passage is adopted in Wang et al. [120]. In addition, multi-hop propagation is also taken into consideration in this model. To avoid computation on the entire graph, Ying et al. [138] present a sampling based graph convolution approach for pins recommendation at Pinterest. The proposed model performs a localized convolution by sampling the neighborhood around a node, and dynamically constructing aggregations from the sampled neighborhood. Monti et al. [83] present a multi-graph convolutional neural networks to extract local features from the interaction matrix with row and column similarities encoded by user and item graphs.

GNNs can also be used to capture inter-item relatedness by mining from the items graph. For instance, in session-based recommendation task, the click sequences and transition patterns from sessions can be represented with a session graph (shown in Fig. 10b). As such, it is natural to use graph neural networks to learn click representations [87, 126] in session-aware recommender systems.

Social graph is another type of graph that is useful for recommender systems. When making decisions, users can be easily influenced by their friends. GNNs can be utilized for influence diffusion on social graphs in social recommendations. Fan et al. [28] propose learning user representations by simultaneously aggregating information from her neighbourhood items in the interaction graph and her neighbourhood users in the social network and an attention mechanism is adopted for information aggregation. Influences maybe context-dependent. For example, users might trust different groups of friends for different types of products. To model such dynamic effects, Song et al. [97] propose a dynamic graph attention network to attend the influence of friends with user's short-term actions.

Knowledge graph is also a widely available information source in recommender systems. Knowledge graph is a directed graph of linked data where nodes correspond to entities (e.g., person, movie, etc.) and edges correspond to relations. For example, a triple in a knowledge graph ⟨James Cameron, direct, Titanic⟩ has two entities (James Cameron and Titanic) and one relation (direct). Knowledge graphs are usually associated with items, e.g., movies in this example. As such, this supplementary information can be used to infer connections between items. For example, the main idea behind the model by Wang et al. [117] is to consider items as entities in the knowledge graph, and then the model aggregates neighborhood information for learning entity representations. In their later study [116], they add flexibility to the model by enabling user-specific relation learning. The authors also introduce a label-smoothing regularization to overcome the outfitting issue.

Graph neural networks have exhibited great potential in learning from graph-structured data in recommender systems. However, there are challenges that remain to be solved. A worrisome aspect of these models is the low computational efficiency. Real-world recommendation tasks on the other hand, often come with large scale data, so more efficient GNNs based recommender systems are expected.

## 3.5  Deep Learning for Cold-Start Recommendation

Cold-start problems happen when new users or new items arrive in the system. Because most recommendation models are built on historical interactions, the lack of interaction records for these new items/users makes the cold-start problem challenging. Leveraging the associated side information accounts for the major paradigm in cold-start scenarios.

Bansal et al. [3] present deep recurrent neural networks for cold-start article recommendation. It adopts GRUs to encode text into latent factors, and a multi-task learning framework is proposed to enable both recommendations and item metadata predictions. Instead of incorporating additional content-based objective terms, Volkovs et al. [113] focus on the optimization process. They demonstrate that neural network models can be explicitly trained for cold-start recommendation via dropout. A key observation is that cold-start problem is equivalent to the problem of missing data. As such, they make DNNs generalize to cold-start settings by selecting an appropriate amount of dropout. This approach shows superior performance on both warm start and cold start scenarios. Pan et al. [86] propose a meta-learning approach to address the cold-start problem which trains an embedding generator for new items through gradient-based meta learning. This model learns embeddings by taking as input item content and attributes. Zhang et al. [142] present Star-GCN, a graph neural networks based recommendation model. Star-GCN predicts the embeddings of unseen nodes by the means of masking a part of (or the whole) user/item embeddings and then reconstructing the masked embeddings. Zhang et al. [143] propose a graph based matrix completion algorithm under the inductive setting. It trains GNNs based on one-hop subgraphs, and can generalize to unseen users/items. This approach does not use any side information. However, it cannot address extreme cold-start scenarios because it needs the cold user/item to have a few interactions with neighbors.

## 3.6  Deep Learning for Recommendation: Beyond Accuracy

Beyond accuracy, some other characteristics of recommender systems are also crucial. Here, we review methods that enhance the explainability and robustness of recommender engines using deep learning techniques.

### 3.6.1  Explainable Recommendations

Explainable recommender systems not only generate personalised recommendations but also produce intuitive explanations to the recommendations. In doing so, model transparency is enhanced and the persuasiveness and trustworthiness of recommendations are improved. Moreover, the explanations also provide a way for system designers to diagnose and refine the recommendation algorithms.

It is well known that the internal decision process of deep learning models is hard to control and explain. By circumventing direct explanation on the internal architectures, we can use deep learning methods to generate explicit explanations for the recommendations. For example, we can generate a user-specific sentence for each item in a recommendation list. For example, the model by Li et al. [73] can simultaneously make recommendations and generate readable tips that describe possible reviews a user might give to the recommended items. The model combines a recommendation module, a neural rating regression module, and a text generation module, RNNs. The text generation module is similar to the decoder of sequence-to-sequence models for machine translation, where user reviews are used as the supervision signals of the tips generation module. To generate controllable textual explanations, Li et al. [72] adopt a cue word based GRU [136] to control the generated sentence with a given cue word. For example, for hotel recommendations, given a cue word "room", the generated tips would be "the room is spacious and comfortable". It offers the option to users to select aspects that they care about.

Attention mechanism is also a desirable tool to enhance explanation. It can be employed to highlight critical information of the recommended items. For example, attention mechanism can be used to highlight the prime information in the reviews for each recommendation. Chen et al. [10] propose an attention based approach to pick up valuable and useful reviews from all reviews of a target item. They further improve the accuracy of explanations by considering the dynamics of user preferences. In specific, the model attentively learns the importance of review information according to the user's current state and preference [18]. Attention networks are also used to highlight key phrases/words in the reviews in Seo et al. [94]. Key regions in images can also be highlighted to provide visual explanations. In the area of fashion recommendation, Chen et al. [15] adopt CNNs to extract region representations of each image and use an attention mechanism to select the most impactful regions to the prediction. This model therefore can tell users, for recommended items, which parts that they are likely to be interested in.

The rich linked data in knowledge graphs has also been leveraged to provide tailored explanations for recommendations. Huang et al. [55] present a knowledge graph based approach to enhance explanations in sequential recommendation settings. A memory network is used to capture attribute-level preferences by leveraging external knowledge graphs. This approach provides explanations on which attributes (e.g., genre, album, singer, actor, etc.) are taking effects when making recommendations. Knowledge graph can also used to identify the paths that lead to the recommendations. Huang et al. [57] adopt a self-attention network to explicitly model the knowledge graph aware paths between user-item pairs for path-wise explanations. The attention scores help identify the most influential path for each recommendation. Xian et al. [128] propose a deep reinforcement learning approach for pathfinding in knowledge graphs as the interpretable evidence for recommendations. For example, item $b$ is recommended to user $u$ because of the path: {user $u$ → item $a$ → brand $e$ → item b}. This reasoning path gives explicit explanations for why a certain item is recommended.

### 3.6.2   Robust Recommender Systems

Recent work shows that recommendation models are not robust and are vulnerable to adversarial attacks. Defending against those adversaries lives at the core pertaining to the robustness. This task can be modeled as a minimax game similar to that in GANs where the adversary crafts adversarial examples to degrade recommendation performances while the recommender engine is trained to become robust to such noises.

He et al. [46] propose an adversarial personalized ranking approach to improve Bayesian personalised ranking (BPR) method by performing adversarial training. The adversarial perturbations are added on the embedding vectors of users and items, and are trained to maximize the BPR objective. The recommendation model is trained to minimize the BPR loss plus an additional loss with adversary. This adversarial training method can also be combined with autoencoder [141], which could further improve both the robustness and performance. Adversarial permutations can also be added to the content of users/items. For instance, Tang et al. [102] apply this adversarial training approach to multimedia recommender systems where the adversary adds perturbations on the multimedia content of items to maximize VBPR (visual BPR) loss. The recommendation model is learned by minimizing the VPBR and adversary's loss.

## 3.7   Deep Reinforcement Learning for Recommendation

Recommender systems are usually solved in a supervised learning paradigm. There are two major limitations of this line of work: (1) recommendation is an interactive process; supervised learning will bring the so-called system bias that only observed feedback from the current system is considered. (2) it tends to give myopic recommendations by only recommending catchy items to get immediate response instead of long term user utility. In recent years, DRL has begun to attract attention in the recommender systems community. Using DRL, the recommendation process can be formulated as a dynamic decision-making problem. DRL aims to maximize the overall accumulated reward and considers the long term reward of the recommendations. In order to build a recommender system based on reinforcement

**Fig. 11** Reinforcement learning based recommender systems

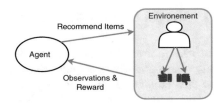

learning, the key elements are defined as: *Agent* is a candidate generator; *State* corresponds to user interests and contexts; *Reward* is defined as user satisfaction; and *action* represents the nomination of items for recommendation. The definition might be different in some cases, we refer readers to corresponding papers (Fig. 11).

To overcome the aforementioned limitations of supervised learning, Chen et al. [12] propose REINFORCE, for recommendations at YouTube. REINFORCE is a policy-gradient algorithm which can scale to production environment with an action spaces up to millions. Chen et al. [11] present a tree-structured policy gradient approach to handle the problem of large discrete action space in recommender systems. In specific, to reduce time complexity, a balanced hierarchical clustering tree is built over items and selecting an item is formulated as finding a path from the root in this tree. Zhao et al. [153] apply DRL to capture distinct contributions of both positive and negative feedback in a sequential interaction setting. During scanning of the recommendation list, users may skip some recommended items. Incorporating such negative feedback helps the system gain better understandings about user's preference. Zhao et al. [152] explore the page-wise recommendation scenario with DRL. The proposed framework DeepPage is able to adaptively optimize a page of items based on user's real-time actions. Zheng et al. [154] present DRN, a reinforcement learning based news recommender method which considers: (1) the dynamic changes of news content and user preferences, (2) incorporating return patterns (to the service) of users, and (3) the increase diversity of recommendations. Xian et al. [128] propose a reinforcement learning approach to find the paths in knowledge graph to enhance the explainability of recommendation. Zou et al. [159] propose PDQ where real customers are replaced with a customer simulator. The simulator is used to simulate the environment and is optimized according to the current recommendation policy. Using a customer simulator reduces the instability of convergence and provides unlimited interactions without involving real users. In many works like [11, 152, 153], the embeddings of users and items are usually pretrained and kept fixed during the training of the reinforcement learning algorithms. However, the pretrained and fixed embeddings cannot reflect the dynamics in user preferences and item characteristics, and might lead to sub-optimal solutions. To solve this problem, Liu et al. [75] propose an end-to-end framework to overcome the limitations. As the embedding component cannot be stably trained with reinforcement learning algorithms, they introduce an additional supervised learning classifier to stabilize the embedding learning process by predicting whether a user offers positive feedback to the recommended item or not.

# 4 Deep Learning for Recommender Systems: Applications

In what follows, we introduce how deep learning techniques are used to build a variety of recommendation applications. We categorize existing publications according to their target domains which are prevalent in today's life. We contextualize closely related domains and review how they address specific recommendation problems in these domains. Table 2 summarizes a representative publication. Note that this is not an exhaustive enumeration.

## 4.1 Deep Learning for E-commerce Recommendation

Recommender systems have become serious business tools in many e-commerce sites such as Amazon,[4] eBay,[5] and Alibaba,[6] and are re-shaping the world of e-commerce. Most large commerce web sites are using recommender systems to help their customers to identify the products they are interested in. Nowadays, deep learning has been playing a primary role in the recommendation process on e-commerce platforms. Here, we give some examples to illustrate how deep learning is used in some e-commerce platforms.[7]

Lake et al. [66] present a deep learning based product embedding approach for product recommendations on Amazon. The model adopts an attention network to select relevant pieces of information from a user's historical interactions. It then learns a joint representation from a specific user-item pair, instead of representing user and items in a common latent space. Both online and offline

**Table 2** Deep learning based recommendation applications and corresponding publications

| Applications | Publications |
|---|---|
| E-commerce | Lake et al. [66], Galron et al. [33], Zhou et al. [156, 157], Wu et al. [125], Li et al. [70], Feng et al. [29, 30], Cen et al. [8], Chen et al. [13, 14] |
| Entertainment | Covington et al. [23], Chen et al. [12], Van et al. [85], Huang et al. [56], Wang et al. [121], Cheng et al. [19], Ying et al. [138], Yang et al. [135] |
| News | Wu et al. [122, 123], An et al. [1], De et al. [98], Wang et al. [115] Hu et al. [54] |
| Point-of-Interests | Yang et al. [134], Wang et al. [119], Chang et al. [9], Liu et al. [77] Zhou et al. [158] |
| Other domains | Shang et al. [95, 96], Biswal et al. [7], Tay et al. [106] |

---

[4] https://www.amazon.com/.

[5] https://www.ebay.com/.

[6] https://www.alibaba.com/.

[7] Note that there is no guarantee that the listed methods are currently in use.

experiments demonstrate promising performance of this method for personalized product recommendation. Galron et al. [33] propose an item embedding network for recommendation on eBay. It has an item embedding network to learn continuous item representations from its features (e.g., attributes, categories, and title tokens), and a prediction network to compute the similarity between seed items (e.g. a recently purchased item) and recommendation candidates. Deep learning based recommender algorithms are also widely employed in Alibaba [8, 13, 14, 29, 30, 70, 156, 157]. Zhou et al. [157] use deep learning to learn user representations from user profiles and her behaviors. A local activation unit is used to attend the related user interests by soft-searching for relevant parts from historical behaviors. In the follow-up work [13, 30, 156], the sequential order of user behaviors is also considered. In specific, sequence modeling approaches RNNs, self-attention mechanism, and transformer are adopted to capture the sequential signals underlying users' behavior data. To enhance the capability of modeling the diversity of user interests, [70] and [8] adopt dynamic routing approaches to learn each user multiple interest representations. Knowledge graph is also leveraged to improve the recommendation performance [29]. Xu et al. [14] explore the use of deep learning in fashion outfit recommendations on Alibaba. In this model, an encoder-decoder model is designed to generate personalized fashion outfits. Wu et al. [125] deploy a deep recurrent neural network based method for recommendation at NetEase (Kaola platform[8]). In this model, deep RNNs are used to model the sequential behaviors of users.

## 4.2 Deep Learning for Online Entertainment Recommendation

Online entertainment platforms are taking over theatrical and home entertainment business. Online entertainment covers a range of domains such as music, video, book, etc. As such, personalized entertainment recommendation is critical to help people narrow the universe of potential items to fit their unique tastes. In recent years, deep learning also plays an important part in online entertainment recommendations.

Covington et al. [23] propose a deep neural network based recommendation method for recommendation on YouTube. It consists of a deep candidate generation module and a separate deep ranking module. Both modules are fully connected deep nonlinear neural networks. The deep candidate generation module is used for candidate selection from the large pool of videos and the ranking module ranks candidates. To improve the long-term user utility and tackle system biases, Chen et al. [12] further propose a deep reinforcement learning based recommendation algorithm which is currently in use on YouTube. Deep learning techniques can also be applied for music recommendations [56, 85, 121]. In specific, these models use CNNs to extract music representations from audio signals. Cheng et al. [19]

---

[8] https://www.kaola.com/.

present a wide and deep learning based model to make app recommendations in the Google play application store. A two-tower neural network with mixed negative sampling is also employed in Google play [135]. Ying et al. [19] design a graph convolutional neural network based recommender that is deployed to perform Pins recommendations at Pinterest.[9]

## 4.3    Deep Learning Based News Recommendation

Reading news online has become a routine in many people's lives. To avoid overwhelming readers, personalization of the recommended articles is important for online news services. Generally, news is text extensive and sometimes also contains images, audio, and video pieces. An encoder that gets the representation for each piece of news is indispensable, and deep learning techniques are suitable tools for news resources.

Wu et al. [122] propose a news recommender model which adopts CNNs to learn news representations from titles. The model applies attention networks to select important words. User representation is learned via an attentive multi-view learning framework from user's search queries, clicked news, and browsed web-pages. They further improve the news recommender by learning both news and users representations with multi-head self-attention [123]. An et al. [1] use an attention-based CNNs method for news representation learning, and incorporate sequential behaviors and short-term preferences of users with GRUs. User's long- and short-term representations are integrated by concatenation, or by using the long-term representation as the initialization of the hidden state of the GRUs. De et al. [98] propose a deep learning architecture for session-based recommendation. This architecture has two modules: a news article representation module and a session-based recommendation module with RNNs. News readers usually have diverse interests. Their interests in different topics or events can be revealed from their historical browsed news. To learn the fine-grained interests, Wang et al. [115] construct representations for each news from multiple semantic views with stacked dilated convolutional encoder, and perform fine-grained matching between candidate news and user's browsed news at different semantic levels. Hu et al. [54] model the diverse interests by disentangling a user's latent preference factors. In specific, they construct a bipartite graph from the user-news interactions and apply graph neural networks to relationships encoding via information propagation. The learned representations are disentangled by a neighborhood routing algorithm to enhance the expressiveness and interpretability.

---

[9] https://www.pinterest.com/.

## 4.4   Deep Learning for Point-of-Interest Recommendation

Point-of-Interest recommendation is useful and acts as a link between internet users and real-world places such as stores, cinemas, restaurants, and tourist attractions. Location-aware apps like Yelp,[10] Foursquare,[11] Google map,[12] and Meituan[13] all provide location based services that are extensively used by millions of users.

Yang et al. [134] adopt fully connected neural networks to model the interaction between users and POIs, and regularize users' and POIs' latent factors with an additional context prediction task. Wang et al. [119] propose to enrich the POIs recommendation by incorporating the visual contents of users' posts (e.g., photos of landmarks and food). In their model, CNNs is used for visual feature extraction which is utilized for user intention identification. Textual information such as reviews/comments on locations can also be utilized. Chang et al. [9] use both multi-head attention mechanism and character-level CNNs to encode user's generated textual content into content embeddings. Those content embeddings are integrated with LSTMs to capture user's overall interests. To model the sequential behaviors, Liu et al. [77] propose a spatial temporal recurrent neural networks to model both time-specific transition and distance specific transition in POIs. Zhou et al. [158] integrate a temporal latent Dirichlet allocation topic model and memory network for personalized POI recommendation. This model integrates the POI-specific geographical influence to enhance recommendations. Zhao et al. [151] present a spatio-temporal gated network (STGN) for POI recommendation by enhancing the long-short term memory network with gating mechanisms. Specifically, a time gate and a distance gate are proposed to control the updates of short-term and long-term preference representations.

## 4.5   Deep Learning Based Recommendation on Other Domains

Besides the domains described above, there exist substantial studies on many other domains. We use healthcare as an example domain here. An important healthcare related application is medication recommendations. Shang et al. [96] present a graph augmented memory networks for medication combination recommendation. It integrates both drug-to-drug interaction knowledge and patient health records to provide safe and personalized recommendation. The problem in medication recommendation is that the records of patients with only single visit (visit hospital only once) are usually discarded. To leverage those data, Shang et al. [95] propose

---

[10] https://www.yelp.com/.

[11] https://foursquare.com/.

[12] https://www.google.com/maps.

[13] http://www.meituan.com/.

a pre-training approach to leverage the single visit health records and fine-tune it for downstream predictive tasks on longitudinal electronic health records from patients with multiple visits. Biswal et al. [7] present doctor2vec to help identify the appropriate doctors for clinical trial based on trial description and patient EHR data. Friends recommendation is also popular in applications such as Twitter and Facebook. For instance, Tay et al. [106] present an attention based GRU model for friends recommendation on Twitter.

## 5 Discussion and Conclusion

Each task in recommender systems has its own challenges and specific factors to consider. This chapter provides a categorised overview and perspective on the published academic literature on deep learning based recommender systems. For example, deep learning is advantageous in interaction modeling by introducing nonlinearity and high-order interactions; user modeling by capturing the temporal dynamics via sequential models and by disentangling user complex interests; representation learning from unstructured content information; interpretability and robustness. We hope that this chapter can shed some light on those who are confronting these tough problems.

Nonetheless, some attractive properties come with the cost of time complexity. Not all models reviewed in the chapter are computationally efficient, and some models have high running time even with the help of GPU acceleration. In addition, deep learning models are black boxes and neurons interact in complex ways to produce the results. It is difficult to interpret the learned model and understand how the outputs are arrived at. Luckily, we can partially bypass this problem by producing human understandable explanations for recommendations. Deep learning based models usually have many hyperparameters to tune. For example, the number of layers, the number of neurons, dropout rate, how to initialize parameters, just to name a few. There are no principles to guide the hyperparameter selection process. Moreover, it usually requires a large amount of data to learn an effective deep learning model. It might lead to sub-optimal solutions if the system does not have enough data available. Expectantly, these limitations can be reduced with the development of deep learning techniques.

Moreover, criticism on the field cannot be overlooked [25, 65, 89, 147]. One major concern is that the evaluation of those newly proposed recommendation models are not rigorous, which is manifested in several aspects. Firstly, the selection of baselines and datasets in most papers are seemingly arbitrary. Authors have free rein over the choices of datasets(e.g., random splits of train/val/test)/baselines. This is understandable to a certain extent when the number of baselines/datasets increases too quickly. Nonetheless, without a fair and comprehensive comparison, the reported improvements do not add up and the advancements do not seem convincing. Secondly, the inappropriate use of evaluation methodology exaggerates the problem. Recent study shows that a popular evaluation methodology based on sampling that are widely adopted to measure the effectiveness of recommendation models does not reflect the true model effectiveness [65, 89]. Having a full grasp of the chosen evaluation methodology is critical in preventing inconsistent results. Thirdly, many traditional methods are not fully tuned for comparison. It is reported that some simple traditional methods can easily outperform neural networks on some tasks with sufficient tuning [78]. Although these issues are not only tied to this field, it is a good practice to bear these pitfalls in mind when evaluating your models. We also encourage that standard benchmarks with consistent evaluation metrics and data splits should be constructed in order to better judge incoming recommendation algorithms. Also, codes and datasets shall be released to enhance reproducibility when appropriate.

This chapter introduces the main deep learning techniques that can be applied in the design of recommender systems. We have reviewed their use in the literature and characterized them from different perspectives. The aim is to give researchers and practitioners an understanding and a thorough summary of the field's breadth and depth. There are broad types of recommender systems; some of them are rarely touched in the moment and some aspects might require more modelling efforts. We hope that deep learning could lead to the advancement of next-generation recommender systems that possess better intelligence and offer better customer experience.

# References

1. M. An, F. Wu, C. Wu, K. Zhang, Z. Liu, X. Xie, Neural news recommendation with long- and short-term user representations, in *Proceedings of the 57th Annual Meeting of the Association for Computational Linguistics, Florence* (2019), pp. 336–345
2. S. Bai, J.Z. Kolter, V. Koltun, An empirical evaluation of generic convolutional and recurrent networks for sequence modeling. Preprint, arXiv:1803.01271 (2018)
3. T. Bansal, D. Belanger, A. McCallum, Ask the GRU: multi-task learning for deep text recommendations, in *Recsys, New York, NY* (2016), pp. 107–114
4. O. Barkan, N. Koenigstein, Item2vec: neural item embedding for collaborative filtering, in *2016 IEEE 26th International Workshop on Machine Learning for Signal Processing (MLSP)* (IEEE, Piscataway, 2016), pp. 1–6
5. R. van den Berg, T.N. Kipf, M. Welling, Graph convolutional matrix completion. CoRR abs/1706.02263 (2017). http://arxiv.org/abs/1706.02263
6. H. Bharadhwaj, H. Park, B.Y. Lim, RecGAN: recurrent generative adversarial networks for recommendation systems, in *Recsys* (2018), pp. 372–376
7. S. Biswal, C. Xiao, L.M. Glass, E. Milkovits, J. Sun, Doctor2vec: dynamic doctor representation learning for clinical trial recruitment, in *Proceedings of AAAI* (2020), pp. 557–564
8. Y. Cen, J. Zhang, X. Zou, C. Zhou, H. Yang, J. Tang, Controllable multi-interest framework for recommendation, in *Proceedings of SIGKDD* (2020)
9. B. Chang, Y. Koh, D. Park, J. Kang, Content-aware successive point-of-interest recommendation, in *Proceedings of the 2020 SIAM International Conference on Data Mining* (SIAM, Philadelphia, 2020), pp. 100–108
10. C. Chen, M. Zhang, Y. Liu, S. Ma, Neural attentional rating regression with review-level explanations, in *Proceedings of WWW, WWW '18, Republic and Canton of Geneva, CHE* (2018), pp. 1583–1592

11. H. Chen, X. Dai, H. Cai, W. Zhang, X. Wang, R. Tang, Y. Zhang, Y. Yu, Large-scale interactive recommendation with tree-structured policy gradient, in *Proceedings of AAAI*, vol. 33 (2019), pp. 3312–3320

12. M. Chen, A. Beutel, P. Covington, S. Jain, F. Belletti, E.H. Chi, Top-k off-policy correction for a reinforce recommender system, in *Proceedings of WSDM, WSDM '19* (Association for Computing Machinery, New York, 2019), p. 456–464

13. Q. Chen, H. Zhao, W. Li, P. Huang, W. Ou, Behavior sequence transformer for e-commerce recommendation in Alibaba, in *Proceedings of the 1st International Workshop on Deep Learning Practice for High-Dimensional Sparse Data, DLP-KDD '19, New York* (2019)

14. W. Chen, P. Huang, J. Xu, X. Guo, C. Guo, F. Sun, C. Li, A. Pfadler, H. Zhao, B. Zhao, POG: personalized outfit generation for fashion recommendation at Alibaba iFashion, in *Proceedings of SIGKDD, KDD '19, New York* (2019), pp. 2662–2670

15. X. Chen, H. Chen, H. Xu, Y. Zhang, Y. Cao, Z. Qin, H. Zha, Personalized fashion recommendation with visual explanations based on multimodal attention network: towards visually explainable recommendation, in *Proceedings of SIGIR, SIGIR'19, New York* (2019), pp. 765–774

16. X. Chen, H. Xu, Y. Zhang, J. Tang, Y. Cao, Z. Qin, H. Zha, Sequential recommendation with user memory networks, in *Proceedings of WSDM* (2018), pp. 108–116

17. X. Chen, Y. Zhang, Q. Ai, H. Xu, J. Yan, Z. Qin, Personalized key frame recommendation, in *Proceedings of SIGIR, New York* (2017), pp. 315–324

18. X. Chen, Y. Zhang, Z. Qin, Dynamic explainable recommendation based on neural attentive models, in *Proceedings of AAAI*, vol. 33 (2019), pp. 53–60

19. H.T. Cheng, L. Koc, J. Harmsen, T. Shaked, T. Chandra, H. Aradhye, G. Anderson, G. Corrado, W. Chai, M. Ispir et al., Wide & deep learning for recommender systems, in *Proceedings of the 1st Workshop on Deep Learning for Recommender Systems* (2016), pp. 7–10

20. Z. Cheng, Y. Ding, X. He, L. Zhu, X. Song, M. Kankanhalli, A3NCF: an adaptive aspect attention model for rating prediction, in *Proceedings of IJCAI* (2018), pp. 3748–3754

21. W.T. Chu, Y.L. Tsai, A hybrid recommendation system considering visual information for predicting favorite restaurants. WWW **20**(6), 1313–1331 (2017)

22. J. Chung, C. Gulcehre, K. Cho, Y. Bengio, Empirical evaluation of gated recurrent neural networks on sequence modeling. Preprint, arXiv:1412.3555 (2014)

23. P. Covington, J. Adams, E. Sargin, Deep neural networks for youtube recommendations, in *Proceedings of Recsys* (2016), pp. 191–198

24. A. Creswell, T. White, V. Dumoulin, K. Arulkumaran, B. Sengupta, A.A. Bharath, Generative adversarial networks: an overview. IEEE Signal Process. Mag. **35**(1), 53–65 (2018)

25. M.F. Dacrema, P. Cremonesi, D. Jannach, Are we really making much progress? A worrying analysis of recent neural recommendation approaches, in *Proceedings of the 13th ACM Conference on Recommender Systems* (2019), pp. 101–109

26. J. Devlin, M.W. Chang, K. Lee, K. Toutanova, Bert: pre-training of deep bidirectional transformers for language understanding, in *Proceedings of NAACL-HLT* (2019)

27. T. Donkers, B. Loepp, J. Ziegler, Sequential user-based recurrent neural network recommendations, in *Proceedings of Recsys* (2017), pp. 152–160

28. W. Fan, Y. Ma, Q. Li, Y. He, E. Zhao, J. Tang, D. Yin, Graph neural networks for social recommendation, in *Proceedings of WWW, New York* (2019), pp. 417–426

29. Y. Feng, B. Hu, F. Lv, Q. Liu, Z. Zhang, W. Ou, ATBRG: adaptive target-behavior relational graph network for effective recommendation, in *Proceedings of SIGIR* (2020)

30. Y. Feng, F. Lv, W. Shen, M. Wang, F. Sun, Y. Zhu, K. Yang, Deep session interest network for click-through rate prediction, in *International Joint Conferences on Artificial Intelligence Organization, IJCAI* (2019), pp. 2301–2307

31. A. Fischer, C. Igel, An introduction to restricted Boltzmann machines, in *Iberoamerican Congress on Pattern Recognition* (Springer, New York, 2012), pp. 14–36

32. V. François-Lavet, P. Henderson, R. Islam, M.G. Bellemare, J. Pineau et al., An introduction to deep reinforcement learning. Found. Trends Mach. Learn. **11**(3–4), 219–354 (2018)

33. D.A. Galron, Y.M. Brovman, J. Chung, M. Wieja, P. Wang, Deep item-based collaborative filtering for sparse implicit feedback (2018)
34. X. Geng, H. Zhang, J. Bian, T.S. Chua, Learning image and user features for recommendation in social networks, in *Proceedings of ICCV* (2015), pp. 4274–4282
35. K. Georgiev, P. Nakov, A non-IID framework for collaborative filtering with restricted Boltzmann machines, in *Proceedings of ICML* (2013), pp. 1148–1156
36. I. Goodfellow, Y. Bengio, A. Courville, *Deep Learning* (MIT Press, 2016). https://www.deeplearningbook.org
37. M. Grbovic, V. Radosavljevic, N. Djuric, N. Bhamidipati, J. Savla, V. Bhagwan, D. Sharp, E-commerce in your inbox: product recommendations at scale, in *Proceedings of the 21th ACM SIGKDD International Conference on Knowledge Discovery and Data Mining* (2015), pp. 1809–1818
38. A. Greenstein-Messica, L. Rokach, M. Friedman, Session-based recommendations using item embedding, in *Proceedings of the 22nd International Conference on Intelligent User Interfaces* (2017), pp. 629–633
39. X. Guan, Z. Cheng, X. He, Y. Zhang, Z. Zhu, Q. Peng, T.S. Chua, Attentive aspect modeling for review-aware recommendation. ACM Trans. Inf. Syst. **37**(3) (2019)
40. H. Guo, R. Tang, Y. Ye, Z. Li, X. He, DeepFM: a factorization-machine based neural network for CTR prediction, in *Proceedings of IJCAI* (2017), pp. 1725–1731
41. X. Guo, C. Shi, C. Liu, Intention modeling from ordered and unordered facets for sequential recommendation, in *Proceedings of WWW* (2020), pp. 1127–1137
42. R. He, J. McAuley, Ups and downs: modeling the visual evolution of fashion trends with one-class collaborative filtering, in *Proceedings of WWW, Republic and Canton of Geneva, CHE* (2016), pp. 507–517
43. R. He, J. McAuley, VBPR: visual Bayesian personalized ranking from implicit Feedback, in *Proceedings of AAAI* (2016), pp. 144–150
44. X. He, T.S. Chua, Neural factorization machines for sparse predictive Analytics, in *Proceedings of SIGIR* (2017), pp. 355–364
45. X. He, X. Du, X. Wang, F. Tian, J. Tang, T.S. Chua, Outer product-based neural collaborative filtering, in *Proceedings of IJCAI* (2018), pp. 2227–2233
46. X. He, Z. He, X. Du, T.S. Chua, Adversarial personalized ranking for recommendation, in *Proceedings of SIGIR, SIGIR '18, New York* (2018), pp. 355–364
47. X. He, L. Liao, H. Zhang, L. Nie, X. Hu, T.S. Chua, Neural collaborative filtering, in *Proceedings of WWW* (2017), pp. 173–182
48. B. Hidasi, A. Karatzoglou, Recurrent neural networks with top-k gains for session-based recommendations, in *Proceedings of CIKM* (2018), pp. 843–852
49. B. Hidasi, A. Karatzoglou, L. Baltrunas, D. Tikk, Session-based recommendations with recurrent neural networks. Preprint, arXiv:1511.06939 (2015)
50. B. Hidasi, M. Quadrana, A. Karatzoglou, D. Tikk, Parallel recurrent neural network architectures for feature-rich session-based recommendations, in *Proceedings of Recsys* (2016), pp. 241–248
51. G.E. Hinton, Deep belief networks. Scholarpedia **4**(5), 5947 (2009)
52. S. Hochreiter, J. Schmidhuber, Long short-term memory. Neural Comput. **9**(8), 1735–1780 (1997)
53. C.K. Hsieh, L. Yang, Y. Cui, T.Y. Lin, S. Belongie, D. Estrin, Collaborative metric learning, in *Proceedings of WWW* (2017), pp. 193–201
54. L. Hu, S. Xu, C. Li, C. Yang, C. Shi, N. Duan, X. Xie, M. Zhou, Graph neural news recommendation with unsupervised preference disentanglement, in *Proceedings of Association for Computational Linguistics* (2020)
55. J. Huang, W.X. Zhao, H. Dou, J.R. Wen, E.Y. Chang, Improving sequential recommendation with knowledge-enhanced memory networks, in *Proceedings of SIGIR, SIGIR '18, New York* (2018), pp. 505–514
56. Q. Huang, A. Jansen, L. Zhang, D.P.W. Ellis, R.A. Saurous, J. Anderson, Large-scale weakly-supervised content embeddings for music recommendation and tagging, in *ICASSP 2020–*

*2020 IEEE International Conference on Acoustics, Speech and Signal Processing (ICASSP)* (2020), pp. 8364–8368

57. X. Huang, Q. Fang, S. Qian, J. Sang, Y. Li, C. Xu, Explainable interaction-driven user modeling over knowledge graph for sequential recommendation, in *Proceedings of the 27th ACM International Conference on Multimedia, MM '19, New York* (2019), pp. 548–556

58. X. Huang, S. Qian, Q. Fang, J. Sang, C. Xu, CSAN: contextual self-attention network for user sequential recommendation, in *Proceedings of the 26th ACM International Conference on Multimedia* (2018), pp. 447–455

59. Y. Ji, A. Sun, J. Zhang, C. Li, A re-visit of the popularity baseline in recommender systems, in *Proceedings of SIGIR* (2020), pp. 1749–1752

60. W.C. Kang, E. Kim, J. Leskovec, C. Rosenberg, J. McAuley, Complete the look: scene-based complementary product recommendation, in *Proceedings of CVPR* (2019)

61. W.C. Kang, J. McAuley, Self-attentive sequential recommendation. in *Proceedings of ICDM* (IEEE, Piscataway, 2018), pp. 197–206

62. D. Kim, C. Park, J. Oh, S. Lee, H. Yu, Convolutional matrix factorization for document context-aware recommendation, in *Proceedings of Recsys, New York* (2016), pp. 233–240

63. T.N. Kipf, M. Welling, Semi-supervised classification with graph convolutional networks. Preprint, arXiv:1609.02907 (2016)

64. Y. Koren, R. Bell, C. Volinsky, Matrix factorization techniques for recommender systems. Computer **42**(8), 30–37 (2009)

65. W. Krichene, S. Rendle, On sampled metrics for item recommendation, in *Proceedings of SIGKDD, KDD '20, New York* (2020), pp. 1748–1757

66. T. Lake, S.A. Williamson, A.T. Hawk, C.C. Johnson, B.P. Wing, Large-scale collaborative filtering with product embeddings. Preprint, arXiv:1901.04321 (2019)

67. H. Lee, Y. Ahn, H. Lee, S. Ha, S.g. Lee, Quote recommendation in dialogue using deep neural network, in *Proceedings of SIGIR, New York* (2016), pp. 957–960

68. J. Lee, S. Abu-El-Haija, B. Varadarajan, A.P. Natsev, Collaborative deep metric learning for video understanding, in *Proceedings of SIGKDD, New York* (2018), pp. 481–490

69. C. Lei, D. Liu, W. Li, Z.J. Zha, H. Li, Comparative deep learning of hybrid representations for image recommendations, in *Proceedings of CVPR* (2016)

70. C. Li, Z. Liu, M. Wu, Y. Xu, H. Zhao, P. Huang, G. Kang, Q. Chen, W. Li, D.L. Lee, Multi-interest network with dynamic routing for recommendation at Tmall, in *Proceedings of CIKM, CIKM '19, New York* (2019), pp. 2615–2623

71. J. Li, P. Ren, Z. Chen, Z. Ren, T. Lian, J. Ma, Neural attentive session-based recommendation, in *Proceedings of CIKM* (2017), pp. 1419–1428

72. L. Li, L. Chen, Y. Zhang, Towards controllable explanation generation for recommender systems via neural template, in *Proceedings of WWW, WWW '20, New York* (2020), pp. 198–202

73. P. Li, Z. Wang, Z. Ren, L. Bing, W. Lam, Neural rating regression with abstractive tips generation for recommendation, in *Proceedings of SIGIR, SIGIR '17, New York* (2017), pp. 345–354

74. J. Lian, X. Zhou, F. Zhang, Z. Chen, X. Xie, G. Sun, xDeepFM: combining explicit and implicit feature interactions for recommender systems, in *Proceedings of SIGKDD* (2018), pp. 1754–1763

75. F. Liu, H. Guo, X. Li, R. Tang, Y. Ye, X. He, End-to-end deep reinforcement learning based recommendation with supervised embedding, in *Proceedings of WSDM, WSDM '20, New York* (2020), pp. 384–392

76. Q. Liu, S. Wu, L. Wang, Deepstyle: learning user preferences for visual recommendation, in *Proceedings of SIGIR, New York* (2017), pp. 841–844

77. Q. Liu, S. Wu, L. Wang, T. Tan, Predicting the next location: a recurrent model with spatial and temporal contexts, in *Thirtieth AAAI Conference on Artificial Intelligence* (2016)

78. M. Ludewig, D. Jannach, Evaluation of session-based recommendation algorithms. User Model. User-Adapted Interact. **28**(4–5), 331–390 (2018)

79. C. Ma, P. Kang, X. Liu, Hierarchical gating networks for sequential recommendation, in *Proceedings of SIGKDD* (2019), pp. 825–833
80. J. Ma, C. Zhou, P. Cui, H. Yang, W. Zhu, Learning disentangled representations for recommendation, in *Proceeding of NeurIPS*, ed. by H. Wallach, H. Larochelle, A. Beygelzimer, F. d'Alché-Buc, E. Fox, R. Garnett (Curran Associates, Red Hook, 2019), pp. 5711–5722
81. J. McAuley, C. Targett, Q. Shi, A. van den Hengel, Image-based recommendations on styles and substitutes, in *Proceeding of SIGIR, New York* (2015), pp. 43–52
82. T. Mikolov, K. Chen, G. Corrado, J. Dean, Efficient estimation of word representations in vector space. Preprint, arXiv:1301.3781 (2013)
83. F. Monti, M. Bronstein, X. Bresson, Geometric matrix completion with recurrent multi-graph neural networks, in *Advances in Neural Information Processing Systems*, ed. by I. Guyon, U.V. Luxburg, S. Bengio, H. Wallach, R. Fergus, S. Vishwanathan, R. Garnett (Curran Associates, Red Hook, 2017), pp. 3697–3707
84. W. Niu, J. Caverlee, H. Lu, Neural personalized ranking for image recommendation, in *Proceedings of WSDM, New York* (2018), pp. 423–431
85. A. van den Oord, S. Dieleman, B. Schrauwen, Deep content-based music recommendation, in *Advances in Neural Information Processing Systems*, ed. by C.J.C. Burges, L. Bottou, M. Welling, Z. Ghahramani, K.Q. Weinberger (2013), pp. 2643–2651
86. F. Pan, S. Li, X. Ao, P. Tang, Q. He, Warm up cold-start advertisements: improving CTR predictions via learning to learn ID embeddings, in *Proceedings of SIGIR, SIGIR'19, New York* (2019), pp. 695–704
87. R. Qiu, Z. Huang, J. Li, H. Yin, Exploiting cross-session information for session-based recommendation with graph neural networks. ACM Trans. Inf. Syst. **38**(3), 1–23 (2020)
88. S. Rendle, W. Krichene, L. Zhang, J. Anderson, Neural collaborative filtering vs. matrix factorization revisited. Preprint, arXiv:2005.09683 (2020)
89. S. Rendle, W. Krichene, L. Zhang, J. Anderson, Neural collaborative filtering vs. matrix factorization revisited, in *Fourteenth ACM Conference on Recommender Systems, RecSys '20, New York* (2020), pp. 240–248
90. S. Sabour, N. Frosst, G.E. Hinton, Dynamic routing between capsules, in *Proceedings of NeurIPS* (2017), pp. 3856–3866
91. R. Salakhutdinov, A. Mnih, G. Hinton, Restricted Boltzmann machines for collaborative filtering, in *Proceedings of ICML* (2007), pp. 791–798
92. B. Sarwar, G. Karypis, J. Konstan, J. Riedl, Item-based collaborative filtering recommendation algorithms, in *Proceedings of WWW* (2001), pp. 285–295
93. S. Sedhain, A.K. Menon, S. Sanner, L. Xie, Autorec: autoencoders meet collaborative filtering, in *Proceedings of WWW, WWW '15 Companion, New York* (2015), pp. 111–112
94. S. Seo, J. Huang, H. Yang, Y. Liu, Interpretable convolutional neural networks with dual local and global attention for review rating prediction, in *Proceedings of Recsys, New York* (2017), pp. 297–305
95. J. Shang, T. Ma, C. Xiao, J. Sun, Pre-training of graph augmented transformers for medication recommendation, in *Proceedings of IJCAI* (2019), pp. 5953–5959
96. J. Shang, C. Xiao, T. Ma, H. Li, J. Sun, Gamenet: graph augmented memory networks for recommending medication combination, in *Proceedings of AAAI*, vol. 33 (2019), pp. 1126–1133
97. W. Song, Z. Xiao, Y. Wang, L. Charlin, M. Zhang, J. Tang, Session-based social recommendation via dynamic graph attention networks, in *Proceedings of WSDM, New York* (2019), pp. 555–563
98. G. de Souza Pereira Moreira, F. Ferreira, A.M. da Cunha, News session-based recommendations using deep neural networks, in *Proceedings of the 3rd Workshop on Deep Learning for Recommender Systems* (2018), pp. 15–23
99. F. Sun, J. Liu, J. Wu, C. Pei, X. Lin, W. Ou, P. Jiang, Bert4rec: sequential recommendation with bidirectional encoder representations from transformer, in *Proceedings of CIKM* (2019), pp. 1441–1450

100. J. Tan, X. Wan, J. Xiao, A neural network approach to quote recommendation in writings, in *Proceedings of CIKM, New York* (2016), pp. 65–74
101. J. Tang, F. Belletti, S. Jain, M. Chen, A. Beutel, C. Xu, H. Chi, E.: Towards neural mixture recommender for long range dependent user sequences, in *Proceedings of WWW* (2019), pp. 1782–1793
102. J. Tang, X. Du, X. He, F. Yuan, Q. Tian, T. Chua, Adversarial training towards robust multimedia recommender system. IEEE Trans. Knowl. Data Eng. **32**(5), 855–867 (2020)
103. J. Tang, K. Wang, Personalized top-n sequential recommendation via convolutional sequence embedding, in *Proceedings of WSDM* (2018), pp. 565–573
104. Y. Tay, L. Anh Tuan, S.C. Hui, Latent relational metric learning via memory-based attention for collaborative ranking, in *Proceedings of WWW* (2018), pp. 729–739
105. Y. Tay, A.T. Luu, S.C. Hui, Multi-pointer co-attention networks for recommendation, in *Proceedings of SIGKDD, New York* (2018), pp. 2309–2318
106. Y. Tay, L.A. Tuan, S.C. Hui, Couplenet: paying attention to couples with coupled attention for relationship recommendation, in *Twelfth International AAAI Conference on Web and Social Media* (2018)
107. Y. Tay, S. Zhang, A.T. Luu, S.C. Hui, L. Yao, T.D.Q. Vinh, Holographic factorization machines for recommendation, in *Proceedings of AAAI*, vol. 33 (2019), pp. 5143–5150
108. T.X. Tuan, T.M. Phuong, 3D convolutional networks for session-based recommendation with content features, in *Proceedings of RecSys, New York* (2017), pp. 138–146
109. A. Vaswani, N. Shazeer, N. Parmar, J. Uszkoreit, L. Jones, A.N. Gomez, Ł. Kaiser, I. Polosukhin, Attention is all you need, in *Proceedings of NeurIPS* (2017), pp. 5998–6008
110. P. Veličković, G. Cucurull, A. Casanova, A. Romero, P. Lio, Y. Bengio, Graph attention networks. Preprint, arXiv:1710.10903 (2017)
111. P. Vincent, H. Larochelle, I. Lajoie, Y. Bengio, P.A. Manzagol, L. Bottou, Stacked denoising autoencoders: learning useful representations in a deep network with a local denoising criterion. J. Mach. Learn. Res. **11**(12), 3371–3408 (2010)
112. L. Vinh Tran, Y. Tay, S. Zhang, G. Cong, X. Li, Hyperml: a boosting metric learning approach in hyperbolic space for recommender systems, in *Proceedings of WSDM* (2020), pp. 609–617
113. M. Volkovs, G. Yu, T. Poutanen, Dropoutnet: addressing cold start in recommender systems, in *Proceedings of NIPS*, ed. by I. Guyon, U.V. Luxburg, S. Bengio, H. Wallach, R. Fergus, S. Vishwanathan, R. Garnett (Curran Associates, Red Hook, 2017), pp. 4957–4966
114. H. Wang, N. Wang, D.Y. Yeung, Collaborative deep learning for recommender systems, in *Proceedings of SIGKDD, New York* (2015), p. 1235–1244
115. H. Wang, F. Wu, Z. Liu, X. Xie, Fine-grained interest matching for neural news recommendation, in *Proceedings of ACL* (2020), pp. 836–845
116. H. Wang, F. Zhang, M. Zhang, J. Leskovec, M. Zhao, W. Li, Z. Wang, Knowledge-aware graph neural networks with label smoothness regularization for recommender systems, in *Proceedings of SIGKDD, New York* (2019), pp. 968–977
117. H. Wang, M. Zhao, X. Xie, W. Li, M. Guo, Knowledge graph convolutional networks for recommender systems, in *Proceedings of WWW, New York* (2019), pp. 3307–3313
118. R. Wang, B. Fu, G. Fu, M. Wang, Deep & cross network for ad click predictions, in *Proceedings of the ADKDD'17, New York* (2017)
119. S. Wang, Y. Wang, J. Tang, K. Shu, S. Ranganath, H. Liu, What your images reveal: exploiting visual contents for point-of-interest recommendation, in *Proceedings of WWW* (2017), pp. 391–400
120. X. Wang, X. He, M. Wang, F. Feng, T.S. Chua, Neural graph collaborative filtering, in *Proceedings of SIGIR, New York* (2019), pp. 165–174.
121. X. Wang, Y. Wang, Improving content-based and hybrid music recommendation using deep learning, in *Proceedings of the 22nd ACM International Conference on Multimedia, New York* (2014), pp. 627–636
122. C. Wu, F. Wu, M. An, T. Qi, J. Huang, Y. Huang, X. Xie, Neural news recommendation with heterogeneous user behavior, in *Proceedings of EMNLP-IJCNLP* (Association for Computational Linguistics, Hong Kong, 2019), pp. 4874–4883

123. C. Wu, F. Wu, S. Ge, T. Qi, Y. Huang, X. Xie, Neural news recommendation with multi-head self-attention, in *Proceedings of EMNLP-IJCNLP* (2019), pp. 6390–6395
124. C.Y. Wu, A. Ahmed, A. Beutel, A.J. Smola, H. Jing, Recurrent recommender networks, in *Proceedings of WSDM* (2017), pp. 495–503
125. S. Wu, W. Ren, C. Yu, G. Chen, D. Zhang, J. Zhu, Personal recommendation using deep recurrent neural networks in NetEase, in *Proceedings of ICDE* (IEEE, Piscataway, 2016), pp. 1218–1229
126. S. Wu, Y. Tang, Y. Zhu, L. Wang, X. Xie, T. Tan, Session-based recommendation with graph neural networks, in *Proceedings of AAAI*, vol. 33 (2019), pp. 346–353
127. Y. Wu, C. DuBois, A.X. Zheng, M. Ester, Collaborative denoising auto-encoders for top-n recommender systems, in *Proceedings of WSDM, New York* (2016), pp. 153–162
128. Y. Xian, Z. Fu, S. Muthukrishnan, G. de Melo, Y. Zhang, Reinforcement knowledge graph reasoning for explainable recommendation, in *Proceedings of SIGIR, SIGIR'19, New York* (2019), pp. 285–294
129. J. Xiao, H. Ye, X. He, H. Zhang, F. Wu, T.S. Chua, Attentional factorization machines: learning the weight of feature interactions via attention networks, in *Proceedings of IJCAI* (2017), pp. 3119–3125
130. C. Xu, M. Wu, Learning feature interactions with Lorentzian factorization machine, in *Proceedings of AAAI* (2019)
131. D. Xu, C. Ruan, E. Korpeoglu, S. Kumar, K. Achan, Self-attention with functional time representation learning, in *Proceedings of NeurIPS* (2019), pp. 15915–15925
132. H.J. Xue, X. Dai, J. Zhang, S. Huang, J. Chen, Deep matrix factorization models for recommender systems, in *Proceedings of IJCAI* (2017), pp. 3203–3209
133. A. Yan, S. Cheng, W.C. Kang, M. Wan, J. McAuley, Cosrec: 2d convolutional neural networks for sequential recommendation, in *Proceedings of CIKM* (2019), pp. 2173–2176
134. C. Yang, L. Bai, C. Zhang, Q. Yuan, J. Han, Bridging collaborative filtering and semi-supervised learning: a neural approach for poi recommendation. in *Proceedings of SIGKDD, KDD '17, New York* (2017), pp. 1245–1254
135. J. Yang, X. Yi, D. Zhiyuan Cheng, L. Hong, Y. Li, S. Xiaoming Wang, T. Xu, E.H. Chi, Mixed negative sampling for learning two-tower neural networks in recommendations, in *Companion Proceedings of the Web Conference 2020* (2020), pp. 441–447
136. L. Yao, Y. Zhang, Y. Feng, D. Zhao, R. Yan, Towards implicit content-introducing for generative short-text conversation systems. in *Proceedings of the 2017 Conference on Empirical Methods in Natural Language Processing* (Association for Computational Linguistics, Copenhagen, 2017), pp. 2190–2199
137. Q. Yao, X. Chen, J.T. Kwok, Y. Li, C.J. Hsieh, Efficient neural interaction function search for collaborative filtering. in *Proceedings of WWW* (2020), pp. 1660–1670
138. R. Ying, R. He, K. Chen, P. Eksombatchai, W.L. Hamilton, J. Leskovec, Graph convolutional neural networks for web-scale recommender systems, in *Proceedings of SIGKDD, New York* (2018), pp. 974–983
139. J. You, Y. Wang, A. Pal, P. Eksombatchai, C. Rosenburg, J. Leskovec, Hierarchical temporal convolutional networks for dynamic recommender systems, in *Proceedings of WWW* (2019), pp. 2236–2246
140. W. Yu, H. Zhang, X. He, X. Chen, L. Xiong, Z. Qin, Aesthetic-based clothing recommendation, in *Proceedings of WWW, Republic and Canton of Geneva, CHE* (2018), pp. 649–658
141. F. Yuan, L. Yao, B. Benatallah, Adversarial collaborative neural network for robust recommendation, in *Proceedings of SIGIR, SIGIR'19, New York* (2019), pp. 1065–1068
142. J. Zhang, X. Shi, S. Zhao, I. King, Star-GCN: stacked and reconstructed graph convolutional networks for recommender systems, in *Proceedings of IJCAI* (AAAI Press, Palo Alto, 2019), pp. 4264–4270
143. M. Zhang, Y. Chen, Inductive matrix completion based on graph neural networks. Preprint, arXiv:1904.12058 (2019)

144. S. Zhang, H. Liu, A. Zhang, Y. Hu, C. Zhang, Y. Li, T. Zhu, S. He, W. Ou, Learning user representations with hypercuboids for recommender systems, in *Proceedings of the 14th ACM International Conference on Web Search and Data Mining* (2020)
145. S. Zhang, Y. Tay, W. Jiang, D.c. Juan, C. Zhang, Switch spaces: learning product spaces with sparse gating. Preprint, arXiv:2102.08688 (2021)
146. S. Zhang, Y. Tay, L. Yao, A. Sun, J. An, Next item recommendation with self-attentive metric learning. in *AAAI Workshop*, vol. 9 (2019)
147. S. Zhang, L. Yao, A. Sun, Y. Tay, Deep learning based recommender system: a survey and new perspectives. ACM Comput. Surv. **52**(1), 1–38 (2019)
148. S. Zhang, L. Yao, A. Sun, S. Wang, G. Long, M. Dong, Neurec: on nonlinear transformation for personalized ranking, in *Proceedings of IJCAI* (2018), pp. 3669–3675
149. S. Zhang, L. Yao, L. Vinh Tran, A. Zhang, Y. Tay, Quaternion collaborative filtering for recommendation, in *Proceedings of IJCAI* (2019), pp. 4313–4319
150. T. Zhang, P. Zhao, Y. Liu, V.S. Sheng, J. Xu, D. Wang, G. Liu, X. Zhou, Feature-level deeper self-attention network for sequential recommendation, in *Proceedings of IJCAI* (2019), pp. 4320–4326
151. P. Zhao, H. Zhu, Y. Liu, J. Xu, Z. Li, F. Zhuang, V.S. Sheng, X. Zhou, Where to go next: a spatio-temporal gated network for next poi recommendation. in *Proceedings of AAAI*, vol. 33 (2019), pp. 5877–5884
152. X. Zhao, L. Xia, L. Zhang, Z. Ding, D. Yin, J. Tang, Deep reinforcement learning for page-wise recommendations, in *Proceedings of Recsys, RecSys '18* (Association for Computing Machinery, New York, 2018), pp. 95–103
153. X. Zhao, L. Zhang, Z. Ding, L. Xia, J. Tang, D. Yin, Recommendations with negative feedback via pairwise deep reinforcement learning, in *Proceedings of SIGKDD, KDD '18, New York* (2018), pp. 1040–1048
154. G. Zheng, F. Zhang, Z. Zheng, Y. Xiang, N.J. Yuan, X. Xie, Z. Li, DRN: a deep reinforcement learning framework for news recommendation. in *Proceedings of WWW* (2018), pp. 167–176
155. L. Zheng, V. Noroozi, P.S. Yu, Joint deep modeling of users and items using reviews for recommendation, in *Proceedings of WSDM, New York* (2017), pp. 425–434
156. C. Zhou, J. Bai, J. Song, X. Liu, Z. Zhao, X. Chen, J. Gao, Atrank: an attention-based user behavior modeling framework for recommendation, in *Proceedings of AAAI* (2018)
157. G. Zhou, X. Zhu, C. Song, Y. Fan, H. Zhu, X. Ma, Y. Yan, J. Jin, H. Li, K. Gai, Deep interest network for click-through rate prediction, in *Proceedings of SIGKDD* (2018), pp. 1059–1068
158. X. Zhou, C. Mascolo, Z. Zhao, Topic-enhanced memory networks for personalised point-of-interest recommendation, in *Proceedings of SIGKDD* (2019), pp. 3018–3028
159. L. Zou, L. Xia, P. Du, Z. Zhang, T. Bai, W. Liu, J.Y. Nie, D. Yin, Pseudo Dyna-Q: a reinforcement learning framework for interactive recommendation, in *Proceedings of WSDM, WSDM '20, New York* (2020), pp. 816–824

# Context-Aware Recommender Systems: From Foundations to Recent Developments

Gediminas Adomavicius, Konstantin Bauman, Alexander Tuzhilin, and Moshe Unger

## 1 Introduction and Motivation

Many existing approaches to recommender systems focus on recommending the most relevant items to individual users and do not take into consideration any contextual information, such as time, place, and the company of other people (e.g., for watching movies or dining out). In other words, traditionally recommender systems deal with applications having only two types of entities, users and items, and do not put them into a context when providing recommendations.

However, in many applications, such as recommending a vacation package, personalized content on a web site, or a movie, it may not be sufficient to consider only users and items—it is also important to incorporate the *contextual information* into the recommendation process in order to recommend items to users under certain *circumstances*. For example, using the temporal context, a travel recommender system would provide a vacation recommendation in the winter that can be very different from the one in the summer. Similarly, in case of personalized content delivery on a website, a user might prefer to read world news when she logs on the website in the morning and the stock market report in the evening, and on weekends to read movie reviews and do shopping, and appropriate recommendations should be provided to her in these different contexts.

G. Adomavicius
Carlson School of Management, University of Minnesota, Minneapolis, MN, USA
e-mail: gedas@umn.edu

K. Bauman (✉)
Fox School of Business, Temple University, Philadelphia, PA, USA
e-mail: kbauman@temple.edu

A. Tuzhilin · M. Unger
Stern School of Business, New York University, New York, NY, USA
e-mail: atuzhili@stern.nyu.edu; munger@stern.nyu.edu

© Springer Science+Business Media, LLC, part of Springer Nature 2022
F. Ricci et al. (eds.), *Recommender Systems Handbook*,
https://doi.org/10.1007/978-1-0716-2197-4_6

These observations are consistent with the findings in behavioral research on consumer decision making in marketing that have established that decision making, rather than being invariant, is contingent on the context of decision making. Therefore, accurate prediction of consumer preferences undoubtedly depends upon the degree to which the recommender system has incorporated the relevant contextual information into a recommendation method.

Over the past 15 years, context-aware recommendation capabilities have been developed by academic researchers and applied in a variety of different application settings, including: movie recommenders [4], restaurant recommenders [97, 134], travel recommenders and tourist guides [22, 42, 90, 104, 110, 141], general music recommenders [71, 84, 103, 108], specialized music recommenders (e.g., for places of interest [33, 130], in-car music [21], or music while reading [39]), mobile information search [48], news recommenders [80], shopping assistants [111], mobile advertising [36], mobile portals [119], mobile app recommenders [73, 130], and many others. In particular, *mobile* recommender systems constitute an important special case of context-aware recommenders, where context is often defined by spatial-temporal information such as location and time, and there exists a large body of literature dedicated specifically to mobile recommender systems (e.g., see [69, 124, 130, 132, 134, 137] for a few representative examples). In this chapter we focus on the issues related to the general area of context-aware recommender systems and, therefore, we do not provide a separate in-depth review of mobile recommender systems. Readers interested in a systematic coverage of mobile recommender systems are referred to [79].

Besides academic research on this topic, companies have also extensively incorporated contextual information into their recommendation engines across different applications. For example, in the movie recommendation domain, Netflix has been using contextual information, such as the time of the day or user's location, for a long time with significant business outcomes. As Reed Hastings, the CEO of Netflix, pointed out in 2012 [63], Netflix can improve the performance of its recommendation algorithms up to 3% when taking into account such contextual information. More recently, it has been observed at Netflix that "contextual signals can be as strong as personal preferences; ...make them central to your system and infrastructure" [19]. Following these observations, Netflix expanded the scope of contextual information being used and the range of ways it is utilized by the company [19]. In particular, Netflix considers the following types of contexts in their recommendation methods: location (country and region within the country), the type of device being used for watching videos, time, cultural/religious/national festivities, attention (if the user is focused on watching the video or it is being played in the background), companion with whom the video is being watched, outside events occurring at the same time (sport events, elections, etc.), weather, seasonality, and user's daily patterns (e.g., commuting to work) [19]. This contextual information is being used in several types of recommendation methods at Netflix, ranging from the Context-Aware Tensor Factorization [72] to Contextual RNNs [30].

In addition to movies, context is also used extensively in music recommendations [33, 39, 59, 84, 85]. For example, when choosing which songs to play for a given user, some streaming platforms infer the mood of the user based on the users' short-term goals and their recent activities. In addition to the inferred mood, Spotify also relies on the following types of contexts in their recommendation algorithms when providing song recommendations, among others: day of the week, time of the day, user's region, type of user's device, and the platform on the device being used for listening [54, 58, 91].

Similarly, LinkedIn uses various types of contextual information, including date, time, location, device/platform and page, to provide career-related recommendations [11]. In fact, context plays a central role in LinkedIn recommendations, the goal for which has been explicitly formulated as "predict probability that a user will respond to an item *in a given context*" [emphasis added] [11].

In this chapter we review the topic of *context-aware recommender systems* (*CARS*). In particular, we discuss the notion of context and how it can be modeled in recommender systems. We also review the major approaches to modeling contextual information in recommender systems and further present recent work on contextual modeling techniques for CARS. Finally, we discuss important directions for possible future research.

## 2   Context in Recommender Systems

### 2.1   What Is Context?

Context is a multifaceted concept that has been studied across different research disciplines, including computer science (primarily in artificial intelligence and ubiquitous computing), cognitive science, linguistics, philosophy, psychology, and organizational sciences. In fact, an entire conference—CONTEXT[1]—is dedicated exclusively to studying this topic and incorporating it into various other branches of science, including medicine, law, and business. Since context is a multidisciplinary concept, each discipline tends to take its own idiosyncratic view that is somewhat different from other disciplines and is more specific than the standard generic dictionary definition of context as "interrelated conditions in which something exists or occurs".[2] Therefore, there exist many definitions of context across various disciplines and even within specific subfields of these disciplines. Bazire and Brézillon [29] present and examine 150 different definitions of context from different fields.

To bring some "order" to this diversity of views on what context is, Dourish [53] introduces taxonomy of contexts, according to which contexts can be classified into the representational and the interactional views. In the *representational* view, context is defined with a predefined set of observable factors, the structure (or

---

[1] https://link.springer.com/conference/context.

[2] https://www.merriam-webster.com/dictionary/context.

schema, using database terminology) of which does not change significantly over time. In other words, the representational view assumes that the contextual factors are identifiable and known a priori and, hence, can be captured and used within the context-aware applications. In contrast, the *interactional* view assumes that the user behavior is induced by an underlying context, but that the context itself is not necessarily observable. Furthermore, Dourish [53] assumes that different types of actions may give rise to and call for different types of relevant contexts, thus assuming a *bidirectional* relationship between activities and underlying contexts: contexts influence activities and different activities giving rise to different contexts.

Turning to recommender systems, one of their goals is estimating user's utility, e.g., expressed as ratings, for the yet-to-be-consumed items, typically by examining prior interactions between users and items. This means that *users* and *items* are fundamental entities considered in recommender systems, and therefore some of their features, i.e., user's personal characteristics and item's content attributes, can have an effect on user's preferences for items. Moreover, aside from the user and item features, a number of other factors that reflect the user's circumstances while consuming the items, may also impact these preferences, such as time, location and weather. These factors, affecting user's preferences besides user and items characteristics, are considered *contextual factors*.

In the remainder of Sect. 2, we further explore the notion of context and focus on what context is in recommender systems. We start with the traditional and popular representational approach to modeling contextual information in Sect. 2.2, explore and describe a broader classification of modeling contextual factors in Sect. 2.3, and discuss the ways to design and obtain contextual factors in Sect. 2.4.

## 2.2  Modeling Contextual Information in RS: Traditional Approach

Rating-based recommender systems typically start with the specification of the initial set of ratings that is either explicitly provided by the users or is implicitly inferred by the system. Once these initial ratings are specified, a recommender system (RS) tries to estimate the rating function

$$R : User \times Item \rightarrow Rating$$

for the (user, item) pairs that have not been rated yet by the users. *Rating* is typically represented by a totally ordered set, and *User* and *Item* are the domains of users and items respectively. Once the function $R$ is estimated for the whole *User* $\times$ *Item* space, a RS can recommend the highest-rated items for each user, possibly also taking into account item novelty, diversity, or other considerations of recommendation quality [115]. We call such canonical systems *two-dimensional* (2D) since they consider only the *User* and *Item* dimensions in the recommendation process.

In context-aware recommender systems, the aforementioned rating estimation process is enhanced by incorporating contextual information (in addition to user and item information) as potential factors that can impact users' ratings for items:

$$R : User \times Item \times Context \rightarrow Rating,$$

where *Context* represents the domain of contextual information used in the recommender system.

It is important to note that *Context* can be modeled in CARS in a variety of different ways, as will be discussed in Sect. 2.3 in more detail. One popular approach—we refer to it as the *traditional (representational) approach to context-aware recommender systems*—assumes that the contextual information, such as time, location, and the company of other people, is explicitly described by a set of pre-defined *contextual factors* (sometimes called contextual dimensions, variables, or attributes), the structure of which does not change over time (i.e., is static). To illustrate this approach, consider the following example.

*Example 1*  Consider the application of recommending movies to users, where users and movies are described by the following factors:

- Movie: the set of all the movies that can be recommended; it is represented by MovieID, but can have additional item content features available (e.g., genre, actors, director, plot keywords).
- User: the people to whom movies are recommended; it is defined by UserID, but can have additional user characteristics available (e.g., demographic and socioeconomic attributes).

Further, the contextual information consists of the following three contextual factors:

- *Location*: the location from which the user watches the movie that is represented by LocationType ("home", "theater", "airplane", and "other").
- *Time*: the time when the movie can be or has been seen; it is represented by Date. Depending on the relevant granularity for a given application, Date can also be aggregated in different ways, such as DayOfWeek (with values Mon, Tue, Wed, Thu, Fri, Sat, Sun) or TimeOfWeek ("weekday" and "weekend").
- *Companion*: represents a person or a group of persons with whom one can see a movie. It is defined by CompanionType with the following values: "alone", "friends", "girlfriend/boyfriend", "parents", "extended family", "co-workers", and "other".

Hence, the star rating assigned to a movie by a person may depend not only on the movie and the user, but also on where and how the movie has been seen, with whom, and at what time, i.e., StarRating $= R$ (UserID, MovieID, LocationType, Date, CompanionType). For example, the type of movie to recommend to college student Jane Doe can differ significantly depending on whether she is planning to

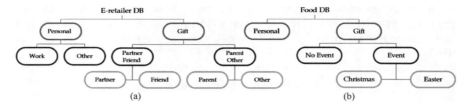

**Fig. 1** Contextual information hierarchical structure: (**a**) e-retailer dataset, (**b**) food dataset [99]

see it on a Saturday night with her boyfriend in a movie theater vs. on a weekday with her parents at home.

In addition, contextual factors can have complex domains of possible values. Although this complexity can take many different forms, one popular defining characteristic is the *hierarchical* structure of contextual information that can be represented as trees, as is done in most of the context-aware recommender and profiling systems, including [4] and [99]. E.g., Example 1 already mentioned that the standard Date values (i.e., calendar dates) can be hierarchically aggregated to DayOfWeek and then further to TimeOfWeek. As another example, Fig. 1 presents a multi-level hierarchy for the PurchasingIntent contextual factor, which allows to specify the purpose of a purchasing transaction in an e-retailer application (Fig. 1a) and a groceries application (Fig. 1b) at different levels of granularity.

In [4, 10], the authors proposed to treat the contextual information as part of a *multidimensional data* (*MD*) *model* within the framework of Online Analytical Processing (OLAP) used for modeling multidimensional databases deployed in data warehousing applications [44, 74]. Mathematically, the OLAP model is defined with an $n$-dimensional *tensor* (see Sect. 4 for subsequent discussion of tensors and their factorization in the context of CARS). In particular, in addition to the classical *User* and *Item* dimensions, additional contextual dimensions, such as *Time, Location*, etc., are also included as part of the tensor. Formally, let $D_1, D_2, \ldots, D_n$ be dimensions, two of these dimensions being *User* and *Item*, and the rest being contextual factors. We define the recommendation space for these dimensions as a Cartesian product $S = D_1 \times D_2 \times \ldots D_n$. Moreover, let *Rating* be a rating domain representing the ordered set of all possible rating values. Then the *rating function* is defined over the space $S$ as $R : D_1 \times \ldots \times D_n \rightarrow Rating$.

Visually, ratings $R(d_1, \ldots, d_n)$ on the recommendation space $S$ can be represented by a multidimensional cube, such as the one shown in Fig. 2. For example, the cube in Fig. 2 stores ratings $R(u, i, t)$ for the recommendation space *User* $\times$ *Item* $\times$ *Time*, where the three tables define the sets of users, items, and times associated with *User, Item*, and *Time* dimensions respectively. For example, rating $R(101, 7, 1) = 6$ in Fig. 2 means that for the user with User ID 101 and the item with Item ID 7, rating 6 was specified during the weekday.

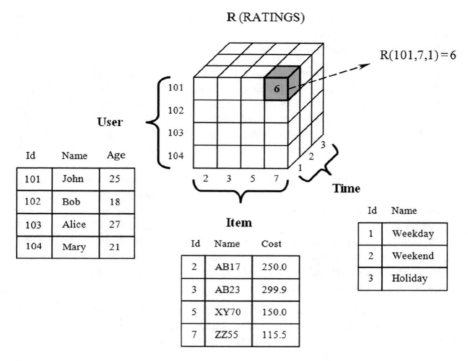

**Fig. 2** Multidimensional (MD) model for the *User × Item × Time* recommendation space

## 2.3 Modeling Contextual Information in RS: Major Approaches

As mentioned earlier, many existing CARS approaches assume the existence of certain contextual factors (sometimes called contextual dimensions, variables, or attributes) that identify the context in which recommendations are provided. As discussed in Sect. 2.2, each contextual factor can be defined by (a) the set of values that the contextual factors take and (b) the structure according to which the values can be meaningfully aggregated, which includes potentially complex hierarchical structures defined using trees or OLAP hierarchies.

A broader classification of major approaches to modeling contextual information, i.e., classification that goes beyond the standard assumption of the explicit availability of predefined contextual factors with stable (static) structure, is based on the following two aspects of contextual factors [9]: (i) what a recommender system may assume (or know) about the structure of contextual factors, and (ii) how the structure of contextual factors changes over time.

The first aspect presumes that a recommender system can have different levels of knowledge about the contextual factors. This may include knowledge of the list of relevant factors, their structure, and lists of their values. Depending on what is known about the contextual factors in a recommender system, we can classify this knowledge into the following categories:

- *Explicit:* The contextual factors relevant to the application, as well as their structure and lists of their values are known *explicitly*. For example, in a restaurant application, the recommender system may use only DayOfWeek, TimeOfDay, Company, and Occasion contextual factors. For each of these factors, the system may know the relevant structure and the complete list of their values, such as using values Morning, Afternoon, Evening, Night for the TimeOfDay variable.
- *Latent:* No information about contextual factors is explicitly available to the recommender system, and it makes recommendations by utilizing only the *latent* knowledge of context in an implicit manner. For example, the recommender system may build a latent predictive model, such as a hierarchical linear or hidden Markov model, to estimate unknown ratings, where context is modeled using latent variables. Alternatively, we can use deep-learning-based embeddings of users and items combined with latent contextual embeddings to provide context-aware recommendations [52, 126].

The second aspect of contextual factors takes into account whether and how their availability and structure change over time. The settings where the contextual factors are stable over time are classified as static, whereas the factors changing over time are classified as dynamic [9]:

- *Static:* The relevant contextual factors and their structure remain the same (stable) over time. For example, in case of recommending a purchase of a certain product, such as a shirt, we can include the contextual factors of Time, PurchasingPurpose, ShoppingCompanion and only these factors during the current modeling and deployment time-frame of the purchasing recommendation application. Furthermore, we assume that the structure of the PurchasingPurpose factor does not change over time: the set of purposes will remain the same throughout the relevant time-frame of the application. The same is applicable to the ShoppingCompanion factor when the same types of shopping companions remain throughout the application.
- *Dynamic:* This is the case when the contextual factors (e.g., the factors themselves, or their structure, or the list of their possible values) change over time. For example, in the online settings, hashtags of the recent user's posts on social-media platforms, such as Twitter, can be considered as contextual factor for recommendations. At each point of time, the system knows only a finite set of values of this contextual factor, but this list changes dynamically over time.

Combining the two aspects of knowledge about the structure of contextual factors results in four major categories of approaches to modeling contextual factors in recommender systems, as indicated in Fig. 3 (top-right), and we describe them below.

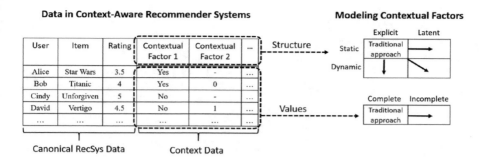

**Fig. 3** Modeling contextual information in recommender systems. Top-right part shows four major modeling approaches: traditional (explicit static), latent static, explicit dynamic, and latent dynamic

**Traditional (i.e., Explicit Static) Approach** As discussed in Sect. 2.2, this approach corresponds to the *representational* view of context [53], which assumes that all the contextual information in a given application can be modeled with a predefined, explicit, finite set of observable factors, where each factor has a well-defined structure and the structure does not change significantly over time (i.e., is static). Vast majority of the first generation of CARS papers has focused on this approach, and it still remains popular because of its simplicity and clarity.

**Latent Static Approach** This approach represents recommendation settings with stable (i.e., static), yet directly unobservable contextual factors. For example, such contextual factors may include implicit context representation extracted from photos for inferring the companion or the mood of the user, or using compressed (latent) mobile sensor data to represent the user current context. The latent approach is mostly used in order to represent context in an efficient and reduced manner from high-dimensional data, where the relationships between the original contextual features are revealed. Because the contextual factor structure is stable, it can be modeled with latent variables, mostly in the form of a vector containing numeric attributes. These latent variables can be learned using machine-learning methods, such as matrix factorization [77], probabilistic latent semantic analysis (PLSA), deep learning (DL), or hierarchical linear models (HLMs). For example, generative models can be used to extract the implicit contextual representation from the interaction between users, items and explicit contexts. As shown in Fig. 4a, the latent context representation $\vec{lc}$ can be extracted from the hidden layer of a mixture attentional constrained auto-encoder [52], and represents the weighted combination of multiple implicit contextual representations. Another way to represent unstructured latent contexts is through hierarchical trees, as suggested by Unger and Tuzhilin [125] and shown in Fig. 4b. A hierarchical latent contextual vector is defined as the path (i.e., a set of contextual situations $s_i$) of the unstructured latent contextual vector from its leaf to the top of the tree. By representing latent contextual information as a set of contextual situations $s_i$ (cluster ids) at different granularity levels extracted from a hierarchical tree [125], we can learn the structure

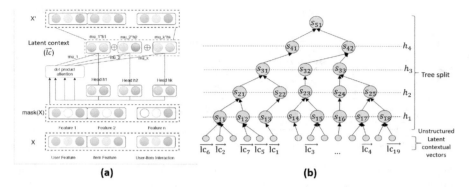

**Fig. 4** Examples of latent context representations: (**a**) Implicit context [52] (**b**) Hierarchical latent context [125]

of latent contextual factors and the semantically meaningful relationships among them. Latent approach to contextual modeling constitutes a popular stream in the current generation of CARS research and, therefore, we will describe it at length below.

**Explicit Dynamic Approach** This approach represents recommendation settings where the explicit structure as well as the list of values of the contextual factor can change over time. For example, in the online settings, the recommender system may consider the website from where a customer arrived at your webpage as an explicit contextual factor PreviousWebsite. Although a customer can arrive to your webpage from numerous different websites, the system may only be aware of a limited set of such websites, each of them serving as a value of the contextual factor PreviousWebsite. Further, this list of previous websites can change dynamically over time, with some new sites being added to the PreviousWebsite variable, while older websites disappearing (and so are the corresponding values of the contextual factor). This makes explicit contextual factor PreviousWebsite dynamic. As yet another example, consider conversational recommender systems that can iteratively collect contextual information from the user. The list of types and values of such contextual information can also change dynamically over time when new contextual factors extracted from recent conversations being constantly added to the list.

One approach to dealing with a dynamically changing contextual information would be to reduce it to the well-studied static case. In particular, the standard static recommendation models can be re-trained once in a while based on the latest dynamically collected information about the context in the domain. Another way to apply static models in dynamic settings is to pre-process the contextual data. For example, values of the PreviousWebsite contextual factor can be generalized to PreviousWebsiteCategory, such as "search," "social-media," etc., which has a finite and stable set of values. Although the reduction of dynamic settings to the well-studied static case may have its benefits, the introduced simplifications could lead to performance limitations. Exploring new models that specifically focus on modeling dynamic context constitutes an important future research direction.

**Latent Dynamic Approach** This approach represents recommendation settings with latent contextual factors that can change over time. The structure of such latent factors may change because of the dynamic nature of the underlying data. For example, in session-based recommendations, information about the latest websites the user visited (including text, images, etc.) could be used to construct low-level latent contextual factors describing characteristics of the user session. Dynamic changes in the list of websites, such as adding new websites, or modifications of their content may lead to corresponding dynamic changes in the structure of the constructed latent contextual factors. In addition, dynamic information collected from sensors, such as IoT devices as well as the availability of wireless networks can be used to construct latent factors and detect user's context [95].

Similarly to the explicit dynamic case, this approach can also be reduced to the static latent one by regular re-training of the standard latent model using the latest information about the data. For example, Hariri et al. [59] use a topic-modeling approach to map sequences of tags produced by user interactions to a sequence of latent topics of tags representations (tags being considered as different contexts) which capture trends in user's current interests. Although the list of such tags changes dynamically, the model described in [59] is static because it is trained periodically on most frequent tags and has to be retrained in order to capture the latest state of the underlying data. Further, some recent work on CARS has made a step towards the actual latent dynamic approach by implementing various models based on reinforcement learning. For example, [140] deals with the highly dynamic nature of news recommendation and proposes an online deep reinforcement learning recommendation framework that considers the context when a news request happens, including time, weekday, and the freshness of the news (the gap between request time and news publish time). In particular, they represent context as a set of latent factors that kept changing over time based on the reinforcement learning policy.

**Complete vs. Incomplete Information** Another important aspect affecting the specifics of context modeling approach in RS is the completeness of knowledge about the actual values of contextual factors for each observed user-item interaction. Therefore, based on the completeness of the observed contextual data, approaches to modeling contextual factors can also be classified into the following categories (Fig. 3 (bottom)):

- *Complete:* The values of all the contextual factors used in the application are known *explicitly* at the time when recommendations are made or when the user provides feedback (i.e., a rating) to the system. For example, in the case of recommending a purchase of a certain product, such as a shirt, the recommender system may use contextual factors Time, PurchasingPurpose, and ShoppingCompanion, and know the actual values of the contextual factors at the recommendation time (e.g., when this purchase is made, with whom, and for whom).

- *Incomplete:* Only a subset of the contextual factors values are known by the system at the time when recommendations are made or when the user provides feedback (i.e., a rating) to the system. In the previous example, the recommender system may know the actual purpose of the purchase and shopping companion only for some users, while explicitly observing the time factor for all of them. Incompleteness of information may be caused by the way it is being collected, such as through user reviews. For example, [61] applies topic modeling to extract users' trip types from their reviews and [27, 28] discover and extract contextual factors discussed in user reviews for an application. In particular, [27] uses dichotomy between specific and generic reviews to identify context related phrases. [28] studies the ways in which contextual information is expressed in user reviews and relies on the obtained insights to develop a Context Parsing method that extracts all relevant contextual factors along with their values from user reviews. However, not all the reviews contain information about the values of all contextual factors, which makes the observed data incomplete.

The majority of the CARS models working with explicit context are designed with an assumption of complete information about contextual values. However, not all those methods could be directly applied to the incomplete-information case and, thus, recommender system may require additional components. For example, it may pre-process the collected data, impute the missing values of the contextual factors (e.g., using a wide variety of data imputation techniques), and subsequently apply the same recommendation methods designed for the complete-information case. For example, missing contextual data could be imputed using the methods similar to the ones used for imputing missing multi-criteria ratings in [83]. Another way to deal with incomplete information is to modify the recommendation model itself, such as applying various relaxation techniques [141]. Some recent CARS methods [124, 125, 133] that use the latent approach apply multiple feature reduction and transformation techniques in order to automatically handle incomplete information in a lower-dimension representation. For example, as shown in Fig. 4a, encoding-decoding techniques can be used to extract contextual latent information for the recommendation process, where the encoder encodes the high-dimensional data with incomplete information to a latent representation and the decoder takes the lower-dimensional data and tries to reconstruct the original high-dimensional data.

**Potential for Future Research** As shown in the top-right portion of Fig. 3, there are four major categories of contextual modeling based on the knowledge about the structure of contextual factors, including traditional (i.e., explicit static), latent static, explicit dynamic, and latent dynamic approaches. The first generation of CARS-related research was mostly focused on the traditional approach of modeling contextual information assuming static and explicitly defined structure. The new wave of research papers on this topic explores the potential of large amounts of data collected through various means, such as user logs or various types of sensors, largely following the latent static approach to modeling contextual factors.

As we discuss in Sect. 4, recent work on the use of reinforcement learning in CARS started exploring the latent dynamic approach. More generally, however, the dynamic approaches (both explicit and latent) to modeling contextual factors have been highly under-explored and, thus, represent a strong potential for novel future research. We will discuss these issues further in Sect. 5.


## 2.4 Designing and Obtaining Contextual Factors

As mentioned earlier, there are several ways to model contextual factors in context-aware recommender systems. The data values for contextual factors can also be obtained in several ways, including:

- *Directly*, i.e., by asking users direct questions or eliciting this information directly from other sources of contextual information. For example, a website may obtain contextual information by asking a user to answer some specific questions before providing a context-aware recommendation. Similarly, a smartphone app may obtain time, location, and motion data from the phone's clock, GPS sensor, and accelerometer, respectively, and weather information can be obtained from a third-party resource by querying it with a specific time and location.
- *By inferring* the context using statistical or data mining methods. In other words, contextual information can be "hidden" in the data in some latent form but can still be *implicitly* used to better estimate the unknown ratings. For example, the household identity of a person flipping the TV channels (husband, wife, son, daughter, etc.) may not be explicitly known to a cable TV company; but it can be modeled as a latent variable using various machine learning methods by observing the TV programs watched and the channels visited. This information can then be used to estimate how much this household would like a particular TV program. As another example of inferring contextual information, consider online reviews, such as provided on Yelp, Amazon, and other popular websites. These reviews contain plenty of contextual information describing specific purchasing or consumption experiences, such a restaurant visits. For example, the user may indicate in a review that she went to the restaurant for dinner with her boyfriend to celebrate his birthday. Some recent work on this topic includes a method for analyzing such online reviews and extracting contextual information from them [27, 28] and a deep learning model that extracts unstructured latent contexts from rich contextual factors such as images and texts for event prediction [96]. Furthermore, [125] proposed a structured representation of latent contextual information that captures contextual situations in a hierarchical manner for context-aware recommendation. Still another approach toward inferring contextual information, albeit for non-RS related problems, was proposed in [75] where temporal contexts were discovered in web-sessions by decomposing these sessions into non-overlapping segments, each segment relating to one specific context. These contexts were subsequently identified using certain optimization and clustering methods.

Regardless of what approach is chosen to model contextual factors in a specific context-aware recommender system, a common challenge for the system designers is that there are many *candidate* contextual factors available for consideration. Thus, the question of which information observable by a recommender system should be considered as relevant, useful context for a given application and used as such in the recommendation process might not be straightforward to answer. In this section, we discuss several broad guidelines that could be helpful to the recommender systems designers in making these decisions.

**Contextual Factors Should Not Be Static Properties of Users and Items** The first guideline represents an observation that, in order to model a certain data attribute as a contextual factor, it should truly be representative of *context* and not be a static characteristic of users or items. For example, in a restaurant recommendation application, the user's dietary restrictions (e.g., vegetarianism or peanut allergy) would be more appropriate to model as user attributes, while the user's intended company for the meal (e.g., with a significant other vs. with small children vs. alone) would be more appropriate to model as contextual factors. A helpful analogy could be made in terms of Entity-Relationship Diagram design considerations in databases: if there is a clear functional dependency between the data attribute and UserID (or ItemID), then it is more natural to model this data attribute as a user (or item) characteristic and not as a contextual factor. Note, however, that dynamic information about the user or item, such as information that user is on vacation this week, would often be considered and modeled as context.

**Contextual Factors Should Be Discoverable Both at System Recommendation Time and at User Feedback Time** Interactions between users and recommender systems can be categorized into the following two broad categories based on their timing and directionality:

- *Recommendation* (system-to-user interaction): the system provides recommendations to the users of the items predicted to be most relevant to them;
- *Feedback* (user-to-system interaction): the users provide feedback to the system about their preferences for the consumed items.

In order to provide genuine context-aware recommendations, the recommender system must be able to use the contextual information (i.e., context in which the user intends to consume, which can be elicited from the user, observed or imputed directly by the system) *at the time of recommendation*. As importantly, however, *at the time of feedback* that the user is providing to the recommender system about a consumed item, the corresponding contextual information (i.e., context in which the item was consumed by the user) also needs to be elicited from the user or observed directly by the system. The latter enables the recommender system to get informative contextual training data for improving its predictive models and for making future context-aware recommendations. In other words, for context-aware recommender systems to be useful, system designers must model context and choose contextual factors in a way that encompasses *both* types of aforementioned interactions.

It is important to note that contextual information that is available (observable, discoverable) during *both* of these interactions may only be a subset of what is available at the recommendation time or at the feedback time separately, depending on the specific application domain. In particular, for the *delayed consumption recommender systems*, where the recommendation interactions and the feedback interactions may occur a substantial time apart, the available contextual information can be significantly different. For example, for a restaurant recommendation application, at the recommendation time it may be more useful to ask for the user's current food-related desires and moods (e.g., "I am in the mood for authentic Italian pizza"). However, at the feedback time, typically more specific contextual details that affected the actual restaurant experience can be collected, e.g., from the user's restaurant review (such as "I ended up getting spaghetti and meatballs", "it was too hot—air conditioning was not working properly", and "the waiter was rude"). Modeling the "common" view of contextual information that *combines* the recommendation-time and the feedback-time perspectives represents an important challenge for context-aware recommender systems designers, especially in the delayed consumption applications. In contrast, this challenge is typically less pronounced for *instant consumption recommender systems*, where the recommendation interactions and feedback interactions occur in close temporal proximity, and where the available contextual information can be treated as essentially identical. For example, for a music streaming platform, the recommendation for the next song can be made based on the user's self-reported current mood, and the user's immediate consumption of the song can be viewed as occurring directly in that same context.

**Contextual Factors Should Be Broadly Relevant** In order to be meaningful for recommendation purposes, a contextual factor should be relevant for a substantial number of users and/or items. This means that the values of a relevant contextual factors should not be constant across different user-item experiences. More importantly, this also means that relevant contextual factors are the ones that affect users' preference judgments (ratings).

Naturally, not all the available contextual factors might be relevant or useful for recommendation purposes. Consider, for example, a book recommender system. Many types of contextual data could potentially be obtained by such a system from book buyers, including: (a) the purpose of buying the book (possible options include for work, for leisure, etc.); (b) planned reading time (weekday, weekend, etc.); (c) planned reading place (at home, at school, on a plane, etc.); (d) the value of the stock market index at the time of the purchase. Clearly some types of contextual information can be more relevant in a given application than some other types. For instance, in this example, the value of a stock market index is likely to be much less relevant as contextual information than the purpose of buying a book.

Because relevance of contextual factors can vary dramatically from application to application (e.g., location as a recommendation context may matter significantly in one recommendation application, but have no impact in another), domain expertise typically plays a major role in identifying a candidate set of contextual factors for a given application. For example, for mobile recommendation applications, the following four general types of contextual information are often considered [9, 55]: *physical context* (e.g., time, position, activity of the user, weather, light conditions, temperature), *social context* (e.g., is the user alone or in the group, presence and role of other people around the user), *interaction media context* (e.g., device characteristics—phone/tablet/laptop/etc., media content type—text/audio/video/etc.), *modal context* (e.g., user's state of mind—mood, experience, current goals).

In addition to using the *manual* approach, e.g., leveraging domain knowledge of the recommender system's designer or a domain expert, there are *computational* approaches to determining the relevance of a given contextual factor. In particular, a wide variety of existing feature selection procedures from machine learning [76, 86, 120] and statistics [43] can be used to identify relevant contextual factors based on existing ratings data. One example of how to decide which contextual factors should be used in a recommendation application (and which should not) is presented in [4], where the authors use a range of variables initially selected by the domain experts as possible candidates for the contextual factors for the application, build multiple recommendation models by incorporating different subsets of contextual factors, and let the predictive performance of these models point to the most advantageous contextual factors. A different approach [94] suggests that, after collecting the data, one could apply various types of statistical tests identifying which of the chosen contextual factors are truly significant in the sense that they indeed affect movie-watching experiences, as manifested by significant deviations in ratings across different values of a contextual factor. Furthermore, [99] demonstrated relevance of the contextual latent factors, such as intent of purchasing a product, by evaluating the predictive performance improvement of the Bayesian network classifier. A similar approach of using latent variables is presented in [14]. [28] tested the relevance of contextual factors extracted from Yelp reviews by comparing the rating prediction performance of models with and without each factor. Those contextual factors that demonstrated significant performance improvement were claimed to be relevant for an application. [125] used a similar approach to show that usage of hierarchical latent contextual representations leads to significantly better recommendations than the baselines for the datasets having high- and medium-dimensional contexts.

Another approach for assessing relevance of contextual information has been proposed by Baltrunas et al. [23], who developed a survey-based instrument that asks the users to judge what their preferences would be in a wide variety of *hypothetical* (i.e., imagined) contextual situations. This allows to collect richer contextual preference information in a short timeframe, evaluate the impact of each contextual factor on user preferences based on the collected data, and include into the resulting context-aware system only those factors that were shown to be important. Even though the collected data includes only hypothetical contextual preferences (i.e., preferences for items that users imagined consuming under certain contextual circumstances), the authors demonstrate that the resulting context-aware recommender system was perceived to be more effective by users as compared to the non-context-aware recommender.

# 3 Paradigms for Incorporating Context in RS

Once the relevant context information has been identified and obtained, the next step is to use context intelligently in order to produce better recommendations. In general, approaches to using contextual information in recommender systems can be broadly categorized into two groups: (1) recommendation via *context-driven querying and search*, and (2) recommendation via *contextual preference elicitation and estimation*. In the first approach, systems typically use contextual information (obtained either directly from the user, e.g., by specifying current mood or interest, or from the environment, e.g., obtaining local time, weather, or current location) to query or search a certain repository of resources (e.g., restaurants) and present the resources that best match a given query (e.g., nearby restaurants that are currently open) to the user. This approach has been used by a wide variety of mobile [3, 42] and travel/tourist recommendations, such as GUIDE [46], INTRIGUE [16], COMPASS [127], and MyMap [51] systems. The second general approach to using contextual information in the recommendation process constitutes techniques that model and learn contextual user preferences, e.g., by observing the interactions of this and other users with the systems or by obtaining preference feedback from the user on various previously recommended items. This approach represents a more recent trend in context-aware recommender systems literature [4, 10, 97, 100, 129, 138]. Some applications may also combine the techniques from both general approaches (i.e., both context-driven querying and search as well as contextual preference elicitation and estimation) into a single system, such as mobile tourist guide UbiquiTO [42] and personalized news recommendations system News@hand [41]. While both general approaches offer a number of research challenges, in the remainder of this chapter we will focus on the second, more recent trend of the contextual preference elicitation and estimation in recommender systems.

To start the discussion of the contextual preference elicitation and estimation techniques for the traditional approach to CARS, note that, in its general form, a canonical 2-dimensional (2D) ($User \times Item$) recommender system can be described as a *function*, which takes partial user preference data as its *input* and produces a list of recommendations for each user as an *output*. Accordingly, Fig. 5 presents a general overview of the canonical 2D recommendation process, which includes three components: data (input), 2D recommender system (function), and recommendation list (output). Note that, as indicated in Fig. 5, after the recommendation function is defined (or constructed) based on the available data, recommendation list

**Fig. 5** General components of the canonical recommendation process

for any given user $u$ is typically generated by using the recommendation function on user $u$ and all candidate items to obtain a predicted rating for each of the items and then by ranking all items according to their predicted rating value. Later in this section, we will discuss how the use of contextual information in each of those three components gives rise to three different paradigms for context-aware recommender systems.

As mentioned in Sect. 2.2, canonical recommender systems are built based on the knowledge of *partial user preferences* presented in the form $< user, item, rating >$. In contrast, context-aware recommender systems are built based on the knowledge of *partial contextual user preferences* and typically deal with data records of the form $< user, item, context, rating >$, that also includes the contextual information in which the item was consumed by this user (e.g., *Context* = Saturday). In addition, context-aware recommender systems may also make use of the structures of context attributes, such as context hierarchies (e.g., Saturday $\rightarrow$ Weekend) mentioned in Sect. 2.2. Based on the presence of this additional contextual data, several important questions arise: How contextual information should be reflected when modeling user preferences? Can we reuse the wealth of knowledge in canonical (non-contextual) recommender systems to generate context-aware recommendations? We will explore these questions in this section in more detail.

In the presence of available contextual information, following the diagrams in Fig. 6, we start with the data having the form $U \times I \times C \times R$, where $C$ is additional contextual dimension and end up with a list of contextual recommendations $i_1, i_2, i_3 \ldots$ for each user. However, unlike the process in Fig. 5, which does not take into account the contextual information, we can apply the information about

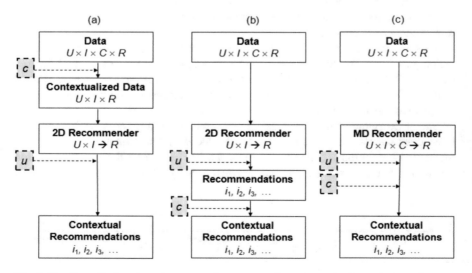

**Fig. 6** Paradigms for incorporating context in recommender systems. (**a**) Contextual pre-filtering. (**b**) Contextual post-filtering. (**c**) Contextual modeling

the current (or expected) context $c$ at various stages of the recommendation process. More specifically, the context-aware recommendation process that is based on contextual user preference elicitation and estimation can take one of the three forms, based on which of the three components the context is used in, as shown in Fig. 6:

- *Contextual pre-filtering* (or contextualization of recommendation input). In this recommendation paradigm (presented in Fig. 6a), contextual information drives data selection or data construction for that specific context. In other words, information about the current context $c$ is used for selecting or constructing the relevant set of data records (i.e., ratings). Then, ratings can be predicted using any canonical 2D recommender system on the selected data.
- *Contextual post-filtering* (or contextualization of recommendation output). In this recommendation paradigm (presented in Fig. 6b), contextual information is initially ignored, and the ratings are predicted using any canonical 2D recommender system on the *entire* data. Then, the resulting set of recommendations is adjusted (*contextualized*) for each user using the contextual information.
- *Contextual modeling* (or contextualization of recommendation function). In this recommendation paradigm (presented in Fig. 6c), contextual information is used directly in the modeling technique as part of rating estimation.

**Contextual Pre-filtering** As shown in Fig. 6a, the contextual pre-filtering approach uses contextual information to select or construct the most relevant 2D (*User* × *Item*) data for generating recommendations. In particular, context $c$ essentially serves as a *query* for selecting (filtering) relevant ratings data. One major advantage of this approach is that it allows deployment of any of the numerous 2D recommendation techniques previously proposed in the literature [6]. For example, following the contextual pre-filtering paradigm, Adomavicius et al. [4] proposed a *reduction-based approach*, which reduces the problem of multidimensional (MD) contextual recommendations to the standard 2D *User* × *Item* recommendation space. An example of a contextual data reduction for a movie recommender system would be: if user $u$ wants to see a movie on Saturday, among the set of existing ratings $D$, *only* the Saturday rating data $D[DayOfWeek = Saturday]$ is used to recommend movies to $u$. Note that this example represents an *exact pre-filter*. In other words, the data filtering query has been constructed using exactly the specified context.

However, the exact context sometimes can be too narrow. Consider, for example, the context of watching a movie with a girlfriend in a movie theater on Saturday or, i.e., $c$ = (Girlfriend, Theater, Saturday). Using this exact context as a data filtering query may be problematic for several reasons. First, certain aspects of the overly specific context may not be significant. For example, user's movie watching preferences with a girlfriend in a theater on Saturday may be exactly the same as on Sunday, but different from Wednesday's. Therefore, it may be more appropriate to use a more general context specification, i.e., Weekend instead of Saturday. And second, exact context may not have enough data for accurate rating prediction, which is known as the "sparsity" problem in recommender systems literature. In other words, the recommender system may not have enough data points about the past movie watching preferences of a given user with a girlfriend in a theater on Saturday.

In these cases, recommender system may utilize the hierarchical structure of contextual factors and apply *generalized pre-filtering* proposed by Adomavicius et al. [4]. This approach suggests to generalize the data filtering query obtained based on a specified context. More formally, let's define $c' = (c'_1, \ldots, c'_k)$ to be a generalization of context $c = (c_1, \ldots, c_k)$ if and only if $c'_i$ is a parent of $c_i$ in the corresponding context hierarchy for every $i = 1, \ldots, k$. Then, $c'$ (instead of $c$) can be used as a data query to obtain contextualized ratings data. Note that there typically exist multiple different possibilities for context generalization, based on the context taxonomy and the desired context granularity. Therefore, choosing the "right" generalized pre-filter becomes an important research problem. For example, Jiang and Tuzhilin [70] examine optimal levels of granularity of customer segments in order to maximize predictive performance of segmentation methods. Applicability of these techniques in the context-aware recommender systems settings constitutes an interesting problem for future research.

So far, we have discussed applying only one pre-filter at a time. However, as it has been well-documented in recommender systems literature, often a combination (a "blend" or an ensemble) of several solutions provides significant performance improvements over the individual approaches [37, 38, 77, 106]. Therefore, it may also be useful to combine several contextual pre-filters into one model at the same time. The rationale for having a number of different pre-filters is based on the fact that, typically there can be multiple different (and potentially relevant) generalizations of the same specific context. Following this idea, Adomavicius et al. [4] use pre-filters based on the number of possible contexts for each rating, and then combine recommendations resulting from each contextual pre-filter. Note that the combination of several pre-filters can be done in multiple ways. For example, for rating estimation in a specific context, (a) one could choose the best-performing pre-filter, or (b) use an aggregate prediction from the entire "ensemble" of pre-filters.

Also note that the contextual pre-filtering approach is related to the problems of building local models in machine learning [13]. Rather than building the global rating estimation model utilizing all the available ratings, this approach builds a *local* rating estimation model that uses only the ratings pertaining to the user-specified criteria in which a recommendation is made (e.g., morning). The advantage of the local model is that it uses *more relevant* data, but at the same time it relies on *fewer* data points. Which of these two trends dominates on a particular segment may depend on the application domain and on the specifics of the available data.

Among other developments, Ahn et al. [12] use a technique similar to the contextual pre-filtering to recommend advertisements to mobile users by taking into account user location, interest, and time, and Lombardi et al. [88] evaluate the effect of contextual information using a pre-filtering approach on the data obtained from an online retailer. Also, Baltrunas and Ricci [18, 25] take a somewhat different approach to contextual pre-filtering in proposing and evaluating the benefits of the *item splitting* technique, where each item is split into several fictitious items based on the different contexts in which these items can be consumed. Similarly to the item splitting idea, Baltrunas and Amatriain [17] introduce the idea of *micro-profiling* (or user splitting), which splits the user profile into several (possibly overlapping) sub-profiles, each representing the given user in a particular context. The predictions are done using these contextual micro-profiles instead of a single user model. The idea of generalized contextual pre-filtering has also been adopted in various studies; for example, Zheng et al. [141] use a similar approach (called context "relaxation") for travel recommendations. Furthermore, Codina et al. [49] provide a different approach to generalize the pre-filtering approach; they leverage semantic similarities between different contextual conditions and compute the recommendations based on the ratings taken not just from the single contextual condition, but from the similar contexts as well.

Although the pre-filtering paradigm has been predominantly used in conjunction with the Explicit Static approach to modeling contextual factors, extending it to more complex context modeling scenarios, such as latent and dynamic cases, constitutes a highly promising future research direction. Ability to pre-filter training data based on any given context—explicit vs. latent, static vs. dynamic—is largely a data querying/matching problem, addressing which would automatically enable the use of any traditional (2D) recommendation algorithms.

**Contextual Post-filtering** As shown in Fig. 6b, the contextual post-filtering approach ignores contextual information when generating the initial ranking of the list of all the candidate items from which top-$N$ recommendations can be made. Then, the contextual post-filtering approach adjusts the obtained recommendation list for each user using the contextual information. The recommendation list adjustments can be made by:

- *Filtering* out recommendations that are irrelevant in a given context, or
- *Re-ranking* the list of recommendations based on a given context.

For example, in a movie recommendation application, if a person wants to see a movie on a weekend, and on weekends she only watches comedies, the system can filter out all non-comedies from the recommended movie list. More generally, the basic idea for contextual post-filtering approaches is to analyze the contextual preference data for a given user in a given context to find specific item usage patterns (e.g., user Jane Doe watches only comedies on weekends) and then use these patterns to adjust the item list, resulting in more "contextual" recommendations, as depicted in Fig. 7.

As with many recommendation techniques, the contextual post-filtering approaches can be classified into heuristic and model-based techniques. *Heuristic* post-filtering approaches focus on finding common item characteristics (attributes) for a given user in a given context, while *model-based* post-filtering approaches build predictive models to calculate the probability with which the user chooses a certain type of item in a given context. These characteristics and probabilities are subsequently used to adjust the recommendations. For example, Panniello et al. [100] provide an experimental comparison of the pre-filtering method versus two post-filtering methods based on several real-world e-commerce datasets. The

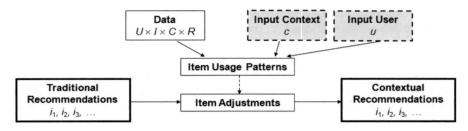

**Fig. 7** Final phase of the contextual post-filtering approach: recommendation list adjustment

empirical results show that the best approach to use (pre- or post-filtering) highly depends on a given application.

Similar to the contextual pre-filtering, a major advantage of the contextual post-filtering approach is that it allows using any of the numerous canonical 2D recommendation techniques previously proposed in the literature [6]. Also, incorporating context generalization techniques into post-filtering techniques and extending them to dynamic and/or latent context settings constitute interesting issues for future research.

**Contextual Modeling** As shown in Fig. 6c, the contextual modeling approach incorporates the contextual information *directly* into the recommendation function as an explicit predictor of a user's rating for an item. While contextual pre-filtering and post-filtering approaches can use canonical 2D recommendation algorithms, the contextual modeling approach gives rise to truly multidimensional recommendation techniques, which essentially represent predictive models (built using decision trees, regressions, probabilistic models, deep learning models, or other techniques) or heuristic calculations that incorporate contextual information in addition to the user and item data. A significant number of novel contextual modeling approaches—based on a variety of heuristics as well as advanced predictive modeling techniques—have been developed over the last five years [102, 129]. In Sect. 4 we present both the classical approaches and more advanced machine learning techniques for contextual modeling, such as tensor factorization and deep learning.

## 4   Contextual Modeling Approaches

Contextual modeling, recognized by its effectiveness in improving performance of recommendations vis-a-vis alternative pre- and post-filtering methods in many cases, has become an increasingly popular paradigm for incorporating context into RS [102, 129]. In the contextual modeling approach, contextual attributes are used in the process of prediction of the ratings by recommendation systems. Contextual modeling challenges include the development of new techniques and mechanisms for: (i) integrating context into canonical recommendation models; (ii) improving rating estimation methods by exploiting context and revealing hidden interactions between users, items, and contexts; and (iii) identifying the contextual factors that should be integrated into the recommendation model. In this section, we will describe different techniques used in CARS research along with their contribution in dealing with cold-start, data sparsity, and scalability problems and the type of contextual attributes used in them.

**Classical Approaches** A number of classical contextual modeling approaches that are based on heuristic as well as model-based techniques can be extended from the two-dimensional (2D) to the multidimensional recommendation settings. For example, [5] proposes a heuristic-based approach that extends the canonical 2D neighborhood-based approach [34, 112] to the multidimensional case, which includes the contextual information, by using an $n$-dimensional distance metric instead of the user-user or item-item similarity metrics traditionally used in such techniques. Another heuristic-based contextual modeling (CM) method is presented in [102], where four variants of the same CM method are considered. Each of these CM methods requires building a contextual profile $prof(u, c)$ for user $u$ in context $c$, and then using the contextual profiles of all the users to find the $N$ nearest neighbors of user $u$ in terms of these profiles in context $c$.

In addition to the heuristic-based contextual modeling techniques, there have been several model-based techniques proposed in the literature. For example, Adomavicius and Tuzhilin [5] present a method of extending a regression-based Hierarchical Bayesian (HB) collaborative filtering model of estimating unknown ratings proposed by Ansari et al. [15] in order to incorporate additional contextual dimensions, such as time and location, into the HB model.

In order to efficiently model user, item, and context interactions, latent factor models suggest to embed users, items, and contexts as tensors (i.e., vectors) in a low-dimensional space of latent (or hidden) features. These models represent both users' preferences, their corresponding items' features and contexts' features in a unified way so that the relevance score of the user-item-context interaction can be measured as an inner product of their vectors in the latent feature space. Such predictive models include context-aware matrix factorization [2] and latent semantic models [122]. For example, matrix-factorization-based approach has been proposed by Baltrunas et al. [24] and Unger et al. [124], who extended the traditional matrix factorization technique to context-aware settings by introducing additional model parameters to model the interaction of the contextual factors with ratings.

In addition to extending the canonical $User \times Item$ model-based collaborative filtering techniques to incorporate the contextual dimensions, there have also been new techniques developed specifically for context-aware recommender systems based on the contextual modeling paradigm. For example, following this paradigm, Oku et al. [97] propose to incorporate additional contextual factors (such as time, companion, and weather) directly into recommendation space and use machine learning techniques to provide recommendations in a restaurant recommender system. In particular, they use the support vector machine (SVM) classification method, which views the set of liked items and the set of disliked items of a user in various contexts as two sets of vectors in an $n$-dimensional space, and constructs a separating hyperplane in this space, which maximizes the separation between the two data sets. The resulting hyperplane represents a classifier for future recommendation decisions (i.e., a given item in a specific context will be recommended if it falls on the "like" side of the hyperplane, and will not be recommended if it falls on the "dislike" side). Furthermore, Oku et al. [97] empirically show that the context-aware SVM significantly outperforms the non-contextual SVM-based recommendation algorithm in terms of predictive accuracy and user's satisfaction with recommendations. Similarly, Yu et al. [138] use contextual modeling approach to provide content recommendations for smart phone users by introducing context as additional model dimensions and using hybrid recommendation technique (synthesizing content-based, Bayesian-classifier, and rule-based methods) to generate recommendations. Also, Hariri et al. [60] employ the Latent Dirichlet Allocation (LDA) model for use in context-aware recommender systems, which allows to model jointly the users, items, and the meta-data associated with contextual information. Finally, another model-based approach is presented in [1] where a Personalized Access Model (PAM) is presented that provides a set of personalized context-based services, including context discovery, contextualization, binding, and matching services, which can be combined to form and deploy context-aware recommender systems.

**Tensor Factorization (TF)** Tensor-based models provide a powerful set of tools for merging various types of contextual information, which increases flexibility, customizability, and quality of recommendation models [56]. Tensor Factorization (TF) extends the canonical two-dimensional Matrix Factorization (MF) problem to the $n$-dimensional problem by incorporating contextual information [72]. The resulting MD rating matrix $\mathcal{R}$ is factored into lower-dimensional representation, where the user, the item and each contextual dimension are represented with lower dimensional feature vectors, as follows:

$$\mathcal{R} \approx \mathcal{G} \times \mathbf{U} \times \mathbf{V} \times \mathbf{C} = \sum_{p=1}^{P} \sum_{q=1}^{Q} \sum_{l=1}^{L} g_{pql} \, \mathbf{u}^{p} \circ \mathbf{v}^{q} \circ \mathbf{c}^{l}$$

The rating matrix is illustrated in Fig. 8a, where $U \in \mathbf{R}^{M \times P}$, $V \in \mathbf{R}^{N \times Q}$, $C \in \mathbf{R}^{K \times L}$ are orthogonal matrices, having similar meaning as the latent feature matrices in the case of SVD. Tensor $\mathcal{G} \in \mathbf{R}^{P \times Q \times L}$ is called the *core tensor* of the Tensor Decomposition (TD). The elements of the core tensor $\mathcal{G}$ represent the latent interactions between different modes, where $P$, $Q$, and $L$ are generally chosen to be smaller than the mode ranks $M$, $N$, and $K$ of the original tensor. Thus, the core tensor $\mathcal{G}$ is often viewed as a compressed representation of the original tensor $\mathcal{R}$.

For example, the Multiverse model [72] is a generic TF-based context-aware model for the rating prediction task that integrates any contextual variables over a finite categorical domain. It demonstrates that high context influence leads

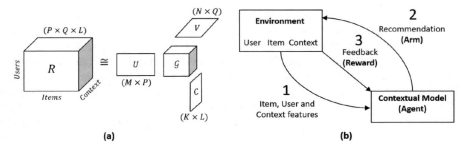

**Fig. 8** Illustration of contextual modeling techniques: (**a**) Tensor factorization, and (**b**) Reinforcement learning

to significantly better rating prediction results using the proposed TF approach. While [72] focuses only on the rating prediction task, the CARTD (Context-Aware Recommendation Tensor Decomposition) model [109, 131] provides a generalized framework for an arbitrary number of contexts and targets an optimal ranking (top-N recommendation) instead of the rating prediction task.

TF can also be used in temporal models that are designed to learn time-evolving patterns in data by taking the time aspect into account [56]. For example, the use of a probabilistic TF model is proposed in several CARS frameworks [117, 136] to allow consideration of all the context simultaneously while modeling the system.

One of the main concerns for the higher-order models is the inevitable growth of computational complexity when the number of contextual dimensions increases. Since context is represented as a low-dimensional tensor, it raises various challenges in terms of context explainability, sparsity, potential for privacy issues, and structure. Improving the robustness of a tensor-based model is challenging due to the sparsity of the observed tensor and the multi-linear nature of TF. For example, Luan et al. [89] show that, if the tensor is big and sparse, then the degree of correlation between entities suffers. To overcome this, they propose a partition-based collaborative TF approach that considers correlations between two points of interest based on location. Specifically, a tensor is partitioned into sub-tensors and then the collaborative TF is applied. [45] proposed a model that combines TF and adversarial learning for context-aware recommendations and showed that their proposed method outperforms standard tensor-based methods and improves the robustness of a recommender model.

Traditional CARS methods model latent contextual information as vectors of high dimensionality and often ignore certain structural properties of latent contextual variables. While trying to reduce dimensionality of the contextual space, it is important to take into account the *structure* of latent contextual variables and the semantically meaningful interrelationships among them [93, 105]. For example, [133] proposed a bias tensor factorization model which uses encoded contextual features that are extracted from a regression tree.

**Deep Learning (DL)** Deep learning techniques have had a major impact on recommendation architectures [26, 139]. In context-aware recommender systems, deep-learning-based approaches are typically used for two purposes: (i) context representation and modeling and (ii) producing predictions by capturing preferences of users over both items and contexts. According to [26], the main reason why DL tends to improve recommendation accuracy is its power of extracting hidden features and jointly combining information from varying sources. Therefore, DL techniques are suitable for handling multi-dimensional problems and constraints that are complex and dynamic in nature.

Since some of the canonical recommender algorithms, such as matrix factorization and factorization machines, can also be expressed as neural architectures [64, 65], several DL-based CARS methods have been proposed to enhance these traditional neural architectures to model contextual representations [50, 62, 126, 135]. For example, [135] proposed a general context-aware algorithm based on factorization machines which combines automatic feature interaction modeling of factorization machine with convolutional neural networks. Figure 9 shows a general framework extending the Neural Collaborative Filtering (NCF) with contextual information [126]. The input of the network is $s_u^{user}$, $s_i^{item}$, and $s_c^{context}$, which are the representations of user profiles, item features, and context information, respectively. The embedding layer consists of low-dimensional representations for each of the inputs. The network is aimed to learn non-linear interactions between users, items, and contexts and can be used for multiple tasks (i.e., rating prediction, top-N recommendation, classification). Here, the scoring function is defined for the rating prediction task as follows:

$$\hat{r}_{uic} = f(U^T \cdot s_u^{user}, V^T \cdot s_i^{item}, C^T \cdot s_c^{context} \mid U, V, C, \theta)$$

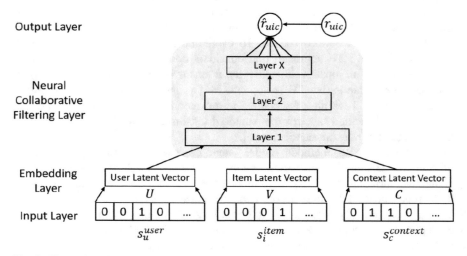

**Fig. 9** Illustration of context-aware recommendation based on deep learning frameworks

where function $f(\cdot)$ represents the multilayer perceptron, and $\theta$ represents the parameters of this network. The network can be trained with weighted square loss (for explicit feedback) or binary cross-entropy loss (for implicit feedback).

Recent state-of-the-art context-aware methods represent the relations between users/items and contexts as a tensor, with which it is difficult to distinguish the impacts of different contextual factors and to model complex, non-linear interactions between contexts and users/items. Therefore, several attention-based recommendation models are used to enhance CARS through adaptively capturing the interactions between contexts and users/items and improve the interpretability of recommendations through identifying the most important contexts [52, 66, 92]. For example, [92] proposed a neural model, named Attentive Interaction Network (AIN), to enhance CARS through adaptively capturing the interactions between contexts and users/items, and [52] proposed an online neural-based recommendation model for ranking recommendations by capturing implicit interactions between users, items, and contexts. The implicit contexts were inferred by a Mixture Attentional Constrained Denoise AutoEncoder (MACDAE) model to combine multiple implicit contextual representations by using generative models. They assumed that different implicit contextual representations contribute differently to the final latent contextual representation and, therefore, their relative contribution can be learned automatically using the attention mechanism.

Context-aware sequential recommendations have been extensively used to monitor the evolution of user tastes over time, which helps to improve the quality of contextual recommendations [107]. Several DL-based techniques can be used to model user activity across time. For example, Recurrent Neural Networks (RNNs) can be used to estimate the next event in a session for a user and can implicitly capture contextual information and combine it with item embeddings [30, 87, 118]. RNNs have positive effects on coverage of recommendations and short-term predictions compared to conventional nearest-neighbor and matrix factorization-based approaches. Such success originates from RNNs' ability to take into account the evolution of users tastes and co-evolution between user, item, and context latent factors.

**Reinforcement Learning (RL)** Most recommendation models consider the recommendation process as static, which makes it difficult to capture users' temporal intentions and to respond to them in a timely manner. In recent years, RL has begun to garner attention [54, 98, 140, 143] in making contextualized recommendations. For example, [54] proposed a contextual bandit algorithm to decide which content to recommend to users, where the reward function takes into consideration contextual information, and [84] proposed a framework for online learning and adaptation to sequential preferences within a listening session in order to tailor the playlist to the listener's current context.

Unlike supervised learning (e.g., classification) that considers prescribed training data, a reinforcement learning (RL) algorithm actively explores its environment to gather information and exploits the learnt knowledge to dynamically make decisions or predictions. Multi-armed bandit is the most thoroughly studied RL problem.

It is inspired by slot machines in a casino: for a bandit (slot) machine with $M$ arms, pulling arm $i$ will result in a random payoff (reward) $r$, sampled from an unknown and arm-specific distribution $p_i$. The objective is to maximize the total reward of the user over a given number of interactions. Figure 8b provides a high-level illustration of contextual modeling recommendation approach using RL. The environment includes information about users, items, and contexts, and a decision-making agent represents the contextual modeling algorithm that produces a list of recommendations. Each time a user requests a recommendation, the information about the user, the contextual factors, and features of candidate items are passed to the agent. The agent will select the best action (i.e., recommending a list of items to a user), fetch user feedback as reward, and observe the new state of the environment. Specifically, the feedback of the user (reward function) is calculated based on the metric being optimized in the environment and the context of the user. Finally, the recommendation list and the user's feedback will be stored in the memory of the agent. Importantly, rather than learn from historical data, the only way the agent can learn which actions are good in different situations is to sequentially try them out and receive a reward from the user.

Previous works have used reinforcement learning to recommend web pages, travel information, books, videos, news, etc. For example, Tripathi et al. [123] combine the potential of reinforcement learning and deep bidirectional recurrent neural networks for automatic personalized video recommendation. They present a context-aware collaborative filtering approach, where the intensity of user's non-verbal emotional response toward the recommended video is captured through interactions and facial expression analysis. In [121], a Q-learning-based travel recommender is proposed, where trips are ranked using a linear function of several content and contextual features including trip duration, price, and country, and the weights are updated using user feedback. Shani et al. use a Markov decision process (MDP) to model the dynamics of user preference for book recommendations [116], where the purchase history is used to define the environment's states, and the generated profit is used as the reward. Liu et al. use MDP to recommend music based on a user's heart rate to help the user maintain it within the normal range [85]. States are defined as different levels of heart rate, and biofeedback is used as rewards. Chi et al. uses MDP to automatically generate playlist [47] and, similarly to [116], states are defined as mood categories of the recent listening history.

## 5   Discussion and Conclusions

Context-awareness is being recognized as an important issue in many recommendation applications, which is evidenced by the increasing number of papers being published on this topic. Looking at the current state of the art in context-aware recommender systems, the main research issues, challenges, and directions can be broadly classified into the following four general categories [8]:

- *Algorithms*, i.e., developing recommendation algorithms that can incorporate contextual information into recommender systems in advantageous ways.
- *Evaluation*, i.e., in-depth performance evaluation of various context-aware recommendation approaches and techniques, their benefits and limitations.
- *Engineering*, i.e., designing general-purpose architectures, frameworks, and approaches to facilitate the development, implementation, deployment, and use of context-aware recommendation capabilities.
- *Fundamentals*, i.e., deeper understanding the notion of context and modeling context in recommender systems.

**Algorithms** Not surprisingly, most of the existing research on context-aware recommender systems can be classified under the "Algorithms" category [8], i.e., researchers have focused primarily on how to take advantage of contextual information in order to improve the quality of recommendations for different recommendation tasks and applications. Compared to "Algorithms", the other three categories have been relatively under-explored, although there have been more work in several other areas in recent years.

**Evaluation** One representative example in the "Evaluation" category is the work by Panniello et al. [102], who performed a comprehensive evaluation and comparison of several contextual pre-filtering, post-filtering, and modeling techniques under variety of conditions, e.g., for different recommendation tasks ("find-all" vs. "top-k"), different recommendation utility metrics (accuracy vs. diversity), the granularity of the processed contextual information, as well as other characteristics. Among many findings, the comparison shows that there is no "universally" best technique under all evaluation dimensions. For example, the contextual modeling and post-filtering approaches demonstrate good accuracy performance in many situations, while the exact pre-filtering approach often exhibits better diversity. However, the contextual modeling approaches tend to achieve a comparatively good balance in the accuracy-diversity trade-off. Another example in the "Evaluation" category is the work by Campos et al. [40], who focused on exploring "time" as one of the most valuable and widely used contextual factors in many recommender systems applications. The authors review common evaluation practices and methodological issues related to the comparative evaluation of time-aware recommender systems. They also demonstrate that the choice of the evaluation conditions impacts the performance ranking of different recommendation strategies and propose a methodological framework for a robust and fair evaluation process. These works represent an important step in the direction of improved, standardized and, thus, more reproducible evaluation procedures for context-aware recommender systems.

Evaluating the CARS performance in real-world settings and in terms of domain-relevant metrics (e.g., based on economic or behavioral outcomes) represents another important direction for CARS research. As an example, [101] study how contextual information affects business outcomes in terms of sales volume, revenue, and customer trust in the provided recommendations using an empirical live study (A/B test) on a sample of customers of a major comics book publisher in Europe. In particular, [101] show that context-aware recommendation techniques outperform canonical (non-contextual) approaches in terms of accuracy, trust, and several economics-based performance metrics across most of their experimental settings. Another interesting example of user studies with context-aware recommender systems is the work by Braunhofer et al. [31]. In this study, the users, after receiving a recommendation from a context-aware system that recommends points-of-interest, were asked to evaluate the system's performance on the following two dimensions: "Does this recommendation fit my preference?" (i.e., "personalization" performance) and "Is this recommendation well-chosen for the situation?" (i.e., "contextualization" performance). In this specific study, the authors show that their proposed technique is able to improve the baseline performance along one of these dimensions—contextualization. In summary, it is important for the CARS community to continue the lines of work described in [31, 101] and to provide strong additional evidence (e.g., via live controlled experiments) of the economic and usability advantages of CARS over the traditional recommendation methods.

One of the major issues that have slowed down the progress of the CARS field in the past was the availability of large-scale publicly available datasets on which novel CARS-based methods could be evaluated. This situation has improved significantly over the last few years when such datasets became publicly available. For example, DePaulMovie [142] and LDOS-CoMoDa [94] datasets contain movie ratings collected along with contextual information; InCarMusic [21], Frappe [20] and STS [32] datasets provide the apps usage logs. Furthermore, some recently published datasets with music listening history, such as the ones from Spotify [35] and Last.fm [128], contain large-scale information about sequence of music listening logs, where previous tracks and other user activity features can be considered as contextual information. Despite this progress, the CARS community would benefit significantly from additional publicly available, large-scale, context-oriented datasets, and this should also be an important priority for the research community.

**Engineering** Much of the work on context-aware recommender systems has been conceptual, where recommendation techniques are developed, tested on some (often limited) data, and shown to perform well in comparison to certain benchmarks. Historically, there has been little work done on the "Engineering" aspect for CARS, i.e., developing novel data structures, efficient storage methods, and new system architectures. One representative example of such system-building work is the study by Hussein et al. [67, 68], where the authors introduce a service-oriented architecture enabling to define and implement a variety of different "building blocks" for context-aware recommender systems, such as recommendation algorithms, context sensors, various filters and converters, in a modular fashion. These building blocks can then be combined and reused in many different ways into systems that can generate contextual recommendations. Another example of such work is Abbar et al. [1], where the authors present a service-oriented approach that implements the Personalized Access Model (PAM) previously proposed by the authors. In addition, there has been a number of research advances in database community in developing frameworks and techniques for building recommender systems functionality (including some context-aware functionality) directly into relational database engines [81, 82, 113, 114], which provides a number of important benefits in terms of storage, query processing, query optimization, and others.

Another important "Engineering" aspect is to develop richer interaction capabilities with CARS that make recommendations more flexible. As compared to canonical recommender systems, context-aware recommenders have two important differences. The first is *increased complexity*, since CARS involve not only users and items in the recommendation process, but also various types of contextual information. Thus, the types of recommendations can be significantly more complex in comparison to the canonical non-contextual cases. For example, in a movie recommendation application, a certain user (e.g., Tom) may seek recommendations for him and his girlfriend of top 3 movies and the best times to see them over the weekend. The second difference is *increased interactivity*, since more information (i.e., context) usually needs to be elicited from the user in the CARS settings. For example, to utilize the available contextual information, a CARS system may need to elicit from the user (Tom) with whom he wants to see a movie (e.g., girlfriend) and when (e.g., over the weekend) before providing any context-specific recommendations. The combination of these two features calls for the development of more flexible recommendation methods that allow the user to express the types of recommendations that are of interest to them rather than consuming standard recommendations that are "hard-wired" into the recommendation engines provided by many current vendors. The second requirement of interactivity also calls for the development of tools allowing users to provide inputs into the recommendation process in an interactive and iterative manner, preferably via some well-defined user interfaces (UI).

Such flexible context-aware recommendations can be supported in several ways. First, Adomavicius et al. [7] developed a *recommendation query language* REQUEST that supports a wide variety of features and allows its users to express in a flexible manner a broad range of recommendations that are tailored to their own individual needs and, therefore, more accurately reflect their interests. In addition, Adomavicius et al. [7] provide a discussion of the expressive power of REQUEST and present a multidimensional recommendation algebra that provides the theoretical basis for this language. Another proposal to provide flexible recommendations is presented in [78], where the FlexRecs system and framework are described. FlexRecs approach supports flexible recommendations over structured data by decoupling the definition of a recommendation process from its execution. In particular, a recommendation can be expressed declaratively as a high-level parameterized workflow containing traditional relational operators and novel recommendation-specific operators that are combined together into a recommendation workflow.

**Fundamentals** "Fundamentals" represents arguably the least developed research direction. Context is a complex notion, and there are many diverse approaches of conceptualizing context in recommender systems, some of them still being debated among the researchers. Although the recommender systems community is gradually converging toward the common definition of context, and this chapter represents an attempt to integrate different approaches into one common framework, additional work is still required for the CARS community to arrive at a more formalized definition of context and, therefore, many important questions still remain to be explored in a principled manner. For example, what are the underlying theoretical underpinnings for context relevance? Are there systematic approaches for identifying relevant contextual factors (i.e., on which to collect data)? If explicit contextual information is not available, when should we model context as a latent factor vs. ignore modeling the context altogether? What are the tradeoffs of different modelling assumptions (e.g., static vs. dynamic context)? The recommender systems community has been moving towards studying some of these questions. For example, Sect. 2 of this chapter covered certain foundational issues. Also, [57] examines in which stages of the recommender system life-cycle incorporation of contextual information into the recommender system is the most useful.

Furthermore, as the CARS field progresses, our understanding of what context is and how to deal with it is evolving. As stated in Sect. 2.3, most of the previous work on CARS has focused on the traditional Explicit-Static approach to modeling contextual factors. More recently, there has been significant work done on how to work with (and take advantage of) latent contextual information in CARS due, primarily, to the popularity of the DL-based methods, as discussed in Sect. 4. In contrast, the topic of *dynamic* context that is important in some applications, such as hashtag-based and news-based recommendations mentioned in Sect. 2.3, is barely explored in CARS. Moving forward, we believe that significantly more research is needed to better understand various aspects of latent and/or dynamic contextual information and how to leverage it in providing better recommendations. Another related and also under-explored research area is how to extract the contextual information from different data sources, such as user reviews, tweets, sensors, IoT devices, etc., that is not explicitly observed and/or recorded as context by the recommender system. This is especially useful when dealing with many different types of contextual factors "buried" in these data sources, most of which are not explicitly known in advance. We expect that much interesting research will be dedicated to these topics within the next several years, keeping the CARS field as vibrant as it has been over the last ten years.

In summary, the field of context-aware recommender systems (CARS) has matured very significantly over the last decade, and many leading companies have incorporated CARS into their recommender systems platforms. However, the field is still developing rapidly, as discussed in this section, and many interesting and practically important research problems still need to be explored further in a principled and comprehensive manner.

# References

1. S. Abbar, M. Bouzeghoub, S. Lopez, Context-aware recommender systems: a service-oriented approach, in *VLDB PersDB Workshop* (2009)

2. M.H. Abdi, G. Okeyo, R.W. Mwangi, Matrix factorization techniques for context-aware collaborative filtering recommender systems: a survey (2018)
3. G.D. Abowd, C.G. Atkeson, J. Hong, S. Long, R. Kooper, M. Pinkerton, Cyberguide: a mobile context-aware tour guide. Wirel. Netw. **3**(5), 421–433 (1997)
4. G. Adomavicius, R. Sankaranarayanan, S. Sen, A. Tuzhilin, Incorporating contextual information in recommender systems using a multidimensional approach. ACM Trans. Inf. Syst. **23**(1), 103–145 (2005)
5. G. Adomavicius, A. Tuzhilin, Incorporating context into recommender systems using multi-dimensional rating estimation methods, in *Proceedings of the 1st International Workshop on Web Personalization, Recommender Systems and Intelligent User Interfaces (WPRSIUI 2005)* (2005)
6. G. Adomavicius, A. Tuzhilin, Toward the next generation of recommender systems: a survey of the state-of-the-art and possible extensions. IEEE Trans. Knowl. Data Eng. **17**(6), 734–749 (2005)
7. G. Adomavicius, A. Tuzhilin, R. Zheng, REQUEST: a query language for customizing recommendations. Inf. Syst. Res. **23**(1), 99–117 (2011)
8. G. Adomavicius, D. Jannach, Preface to the special issue on context-aware recommender systems. User Model. User-Adapt. Interact. **24**(1–2), 1–5 (2014)
9. G. Adomavicius, B. Mobasher, F. Ricci, A. Tuzhilin, Context-aware recommender systems. AI Mag. **32**(3), 67–80 (2011)
10. G. Adomavicius, A. Tuzhilin, Multidimensional recommender systems: a data warehousing approach, in *Electronic Commerce*, ed. by L. Fiege, G. Mühl, U. Wilhelm. Lecture Notes in Computer Science, vol. 2232 (Springer, Berlin, 2001), pp. 180–192
11. D. Agarwal, Scaling machine learning and statistics for web applications, in *Proceedings of the 21th ACM SIGKDD International Conference on Knowledge Discovery and Data Mining, KDD '15* (Association for Computing Machinery, New York, 2015), p. 1621
12. H. Ahn, K. Kim, I. Han, Mobile advertisement recommender system using collaborative filtering: MAR-CF, in *Proceedings of the 2006 Conference of the Korea Society of Management Information Systems* (2006), pp. 709–715
13. E. Alpaydin, *Introduction to Machine Learning* (The MIT Press, London, 2004)
14. S.S. Anand, B. Mobasher, Contextual recommendation. WebMine, LNAI **4737**, 142–160 (2007)
15. A. Ansari, S. Essegaier, R. Kohli, Internet recommendation systems. J. Market. Res. **37**(3), 363–375 (2000)
16. L. Ardissono, A. Goy, G. Petrone, M. Segnan, P. Torasso, Intrigue: personalized recommendation of tourist attractions for desktop and hand held devices. Appl. Artif. Intell. **17**(8), 687–714 (2003)
17. L. Baltrunas, X. Amatriain, Towards time-dependant recommendation based on implicit feedback, in *Workshop on Context-Aware Recommender Systems (CARS 2009), New York* (2009)
18. L. Baltrunas, F. Ricci, Context-dependent items generation in collaborative filtering, in *Workshop on Context-Aware Recommender Systems (CARS 2009), New York* (2009)
19. L. Baltrunas, Keynote: contextualization at netflix, in *Workshop on Context-Aware Recommender Systems at the 13th ACM Conference on Recommender Systems, RecSys '19* (2019)
20. L. Baltrunas, K. Church, A. Karatzoglou, N. Oliver, Frappe: understanding the usage and perception of mobile app recommendations in-the-wild. Preprint, arXiv:1505.03014 (2015)
21. L. Baltrunas, M. Kaminskas, B. Ludwig, O. Moling, F. Ricci, A. Aydin, K.-H. Lüke, R. Schwaiger, Incarmusic: context-aware music recommendations in a car, in *E-Commerce and Web Technologies*, ed. by C. Huemer, T. Setzer. Lecture Notes in Business Information Processing, vol. 85 (Springer, Berlin, 2011), pp. 89–100

22. L. Baltrunas, B. Ludwig, S. Peer, F. Ricci, Context-aware places of interest recommendations for mobile users, in *Design, User Experience, and Usability. Theory, Methods, Tools and Practice*, ed. by A. Marcus. Lecture Notes in Computer Science, vol. 6769 (Springer, Berlin, 2011), pp. 531–540

23. L. Baltrunas, B. Ludwig, S. Peer, F. Ricci, Context relevance assessment and exploitation in mobile recommender systems. Pers. Ubiquit. Comput. **16**(5), 507–526 (2012)

24. L. Baltrunas, B. Ludwig, F. Ricci, Matrix factorization techniques for context aware recommendation, in *Proceedings of the Fifth ACM Conference on Recommender Systems, RecSys '11* (ACM, New York, 2011), pp. 301–304

25. L. Baltrunas, F. Ricci, Experimental evaluation of context-dependent collaborative filtering using item splitting. User Model. User-Adap. Inter. **24**(1–2), 7–34 (2014)

26. Z. Batmaz, A. Yurekli, A. Bilge, C. Kaleli, A review on deep learning for recommender systems: challenges and remedies. Artif. Intell. Rev. **52**(1), 1–37 (2019)

27. K. Bauman, A. Tuzhilin, Discovering contextual information from user reviews for recommendation purposes, in *Proceedings of the ACM RecSys Workshop on New Trends in Content Based Recommender Systems* (2014)

28. K. Bauman, A. Tuzhilin, Know thy context: parsing contextual information from user reviews for recommendation purposes. Inf. Syst. Res. (2021). Forthcoming

29. M. Bazire, P. Brézillon, Understanding context before using it, in *Proceedings of the 5th International Conference on Modeling and Using Context*, ed. by A. Dey et al. (Springer, Berlin, 2005)

30. A. Beutel, P. Covington, S. Jain, C. Xu, J. Li, V. Gatto, Ed.H. Chi, Latent cross: making use of context in recurrent recommender systems, in *Proceedings of the Eleventh ACM International Conference on Web Search and Data Mining* (2018), pp. 46–54

31. M. Braunhofer, M. Elahi, M. Ge, F. Ricci, Context dependent preference acquisition with personality-based active learning in mobile recommender systems, in *Learning and Collaboration Technologies. Technology-Rich Environments for Learning and Collaboration - First International Conference, LCT 2014, Held as Part of HCI International 2014, Heraklion, Crete, Greece, June 22–27, 2014, Proceedings, Part II*, ed. by P. Zaphiris, A. Ioannou. Lecture Notes in Computer Science, vol. 8524 (Springer, Berlin, 2014), pp. 105–116

32. M. Braunhofer, M. Elahi, F. Ricci, STS: a context-aware mobile recommender system for places of interest, in *Posters, Demos, Late-breaking Results and Workshop Proceedings of the 22nd Conference on User Modeling, Adaptation, and Personalization (UMAP2014), Aalborg, Denmark, July 7–11, 2014. CEUR Workshop Proceedings*, vol. 1181, ed. by I. Cantador, M. Chi, R. Farzan, R. Jäschke (2014). CEUR-WS.org

33. M. Braunhofer, M. Kaminskas, F. Ricci, Recommending music for places of interest in a mobile travel guide, in *Proceedings of the Fifth ACM Conference on Recommender Systems, RecSys '11* (ACM, New York, 2011), pp. 253–256

34. J.S. Breese, D. Heckerman, C. Kadie, Empirical analysis of predictive algorithms for collaborative filtering, in *Proceedings of the Fourteenth Conference on Uncertainty in Artificial Intelligence, San Francisco, CA*, vol. 461 (1998), pp. 43–52

35. B. Brost, R. Mehrotra, T. Jehan, The music streaming sessions dataset, in *Proceedings of the 2019 Web Conference* (ACM, New York, 2019)

36. R. Bulander, M. Decker, G. Schiefer, B. Kolmel, Comparison of different approaches for mobile advertising, in *Proceedings of the Second IEEE International Workshop on Mobile Commerce and Services, WMCS '05* (IEEE Computer Society, Washington, DC, 2005), pp. 174–182

37. R. Burke, Hybrid recommender systems: survey and experiments. User Model. User-Adap. Inter. **12**(4), 331–370 (2002)

38. R. Burke, Hybrid web recommender systems, in *The Adaptive Web* (2007), pp. 377–408

39. R. Cai, C. Zhang, C. Wang, L. Zhang, W.-Y. Ma, Musicsense: contextual music recommendation using emotional allocation modeling, in *Proceedings of the 15th International Conference on Multimedia, MULTIMEDIA '07* (ACM, New York, 2007), pp. 553–556

40. P.G. Campos, F. Díez, I. Cantador, Time-aware recommender systems: a comprehensive survey and analysis of existing evaluation protocols. User Model. User-Adapt. Interact. **24**(1–2), 67–119 (2014)
41. I. Cantador, P. Castells, Semantic contextualisation in a news recommender system, in *Workshop on Context-Aware Recommender Systems (CARS 2009), New York* (2009)
42. F. Cena, L. Console, C. Gena, A. Goy, G. Levi, S. Modeo, I. Torre, Integrating heterogeneous adaptation techniques to build a flexible and usable mobile tourist guide. AI Commun. **19**(4), 369–384 (2006)
43. S. Chatterjee, A.S. Hadi, B. Price, *Regression Analysis by Example* (Wiley, New York, 2000)
44. S. Chaudhuri, U. Dayal, An overview of data warehousing and olap technology. ACM Sigmod Rec. **26**(1), 65–74 (1997)
45. H. Chen, J. Li, Adversarial tensor factorization for context-aware recommendation, in *Proceedings of the 13th ACM Conference on Recommender Systems* (2019), pp. 363–367
46. K. Cheverst, N. Davies, K. Mitchell, A. Friday, C. Efstratiou, Developing a context-aware electronic tourist guide: some issues and experiences, in *Proceedings of the SIGCHI conference on Human factors in computing systems* (ACM, New York, 2000), pp. 17–24
47. C. Chi, R.T. Tsai, J. Lai, J.Y. Hsu, A reinforcement learning approach to emotion-based automatic playlist generation, in *2010 International Conference on Technologies and Applications of Artificial Intelligence* (2010), pp. 60–65
48. K. Church, B. Smyth, P. Cotter, K. Bradley, Mobile information access: a study of emerging search behavior on the mobile internet. ACM Trans. Web **1**(1), 4-es (2007)
49. V. Codina, F. Ricci, L. Ceccaroni, Exploiting the semantic similarity of contextual situations for pre-filtering recommendation, in *User Modeling, Adaptation, and Personalization*, ed. by S. Carberry, S. Weibelzahl, A. Micarelli, G. Semeraro. Lecture Notes in Computer Science, vol. 7899 (Springer, Berlin, 2013), pp. 165–177
50. F.S. da Costa, P. Dolog, Collective embedding for neural context-aware recommender systems, in *Proceedings of the 13th ACM Conference on Recommender Systems, RecSys '19* (Association for Computing Machinery, New York, 2019), pp. 201–209
51. B. De Carolis, I. Mazzotta, N. Novielli, V. Silvestri, Using common sense in providing personalized recommendations in the tourism domain, in *Workshop on Context-Aware Recommender Systems (CARS 2009), New York* (2009)
52. X. Ding, J. Tang, T. Liu, C. Xu, Y. Zhang, F. Shi, Q. Jiang, D. Shen, Infer implicit contexts in real-time online-to-offline recommendation, in *Proceedings of the 25th ACM SIGKDD International Conference on Knowledge Discovery & Data Mining, KDD '19* (Association for Computing Machinery, New York, 2019), pp. 2336–2346
53. P. Dourish, What we talk about when we talk about context. Pers. Ubiquit. Comput. **8**(1), 19–30 (2004)
54. P. Dragone, R. Mehrotra, M. Lalmas, Deriving user- and content-specific rewards for contextual bandits, in *The World Wide Web Conference, WWW '19* (Association for Computing Machinery, New York, 2019), pp. 2680–2686
55. B. Fling, *Mobile Design and Development: Practical Concepts and Techniques for Creating Mobile Sites and Web Apps*, 1st edn. (O'Reilly Media, Sebastopol, 2009)
56. E. Frolov, I. Oseledets, Tensor methods and recommender systems. Wiley Interdiscip. Rev. Data Min. Knowl. Discov. **7**(3), e1201 (2017)
57. M. Gorgoglione, U. Panniello, A. Tuzhilin, Recommendation strategies in personalization applications. Inf. Manag. **56**(6), 103143 (2019)
58. C. Hansen, C. Hansen, L. Maystre, R. Mehrotra, B. Brost, F. Tomasi, M. Lalmas, Contextual and sequential user embeddings for large-scale music recommendation, in *Fourteenth ACM Conference on Recommender Systems* (2020), pp. 53–62
59. N. Hariri, B. Mobasher, R. Burke, Context-aware music recommendation based on latent topic sequential patterns, in *Proceedings of the Sixth ACM Conference on Recommender Systems, RecSys '12* (ACM, New York, 2012), pp. 131–138

60. N. Hariri, B. Mobasher, R. Burke,  Query-driven context aware recommendation,  in *Proceedings of the 7th ACM Conference on Recommender Systems, RecSys '13* (ACM, New York, 2013), pp. 9–16
61. N. Hariri, B. Mobasher, R. Burke, Y. Zheng, Context-aware recommendation based on review mining, in *Proceedings of the 9th Workshop on Intelligent Techniques for Web Personalization and Recommender Systems (ITWP 2011)*. Citeseer, (2011), p. 30
62. K. Haruna, M.A. Ismail, S. Suhendroyono, D. Damiasih, A. Pierewan, H. Chiroma, T. Herawan,  Context-aware recommender system: a review of recent developmental process and future research direction. Appl. Sci. **7**(12), 1211 (2017)
63. R. Hastings,  AWS re:Invent 2012, Day 1 Keynote (2012). http://www.youtube.com/watch?v=8FJ5DBLSFe4. YouTube video; see the video at 44:40 min
64. X. He, T.-S. Chua,  Neural factorization machines for sparse predictive analytics,  in *Proceedings of the 40th International ACM SIGIR Conference on Research and Development in Information Retrieval* (2017), pp. 355–364 (2017)
65. X. He, L. Liao, H. Zhang, L. Nie, X. Hu, T.-S. Chua,  Neural collaborative filtering,  in *Proceedings of the 26th International Conference on World Wide Web* (2017), pp. 173–182
66. B. Hu, C. Shi, W.X. Zhao, P.S. Yu,  Leveraging meta-path based context for top- n recommendation with a neural co-attention model, in *Proceedings of the 24th ACM SIGKDD International Conference on Knowledge Discovery & Data Mining, KDD '18* (Association for Computing Machinery, New York, 2018), pp. 1531–1540
67. T. Hussein, T. Linder, W. Gaulke, J. Ziegler,  Context-aware recommendations on rails,  in *Workshop on Context-Aware Recommender Systems (CARS 2009), New York* (2009)
68. T. Hussein, T. Linder, W. Gaulke, J. Ziegler, Hybreed: a software framework for developing context-aware hybrid recommender systems.  User Model. User-Adapt. Interact. **24**(1–2), 121–174 (2014)
69. D. Jannach, K. Hegelich, A case study on the effectiveness of recommendations in the mobile internet, in *Proceedings of the Third ACM Conference on Recommender Systems, RecSys '09* (ACM, New York, 2009), pp. 205–208
70. T. Jiang, A. Tuzhilin, Improving personalization solutions through optimal segmentation of customer bases. IEEE Trans. Knowl. Data Eng. **21**(3), 305–320 (2009)
71. M. Kaminskas, F. Ricci, Contextual music information retrieval and recommendation: State of the art and challenges. Comput. Sci. Rev. **6**(2–3), 89–119 (2012)
72. A. Karatzoglou, X. Amatriain, L. Baltrunas, N. Oliver,  Multiverse recommendation: N-dimensional tensor factorization for context-aware collaborative filtering,  in *Proceedings of the Fourth ACM Conference on Recommender Systems, RecSys '10* (ACM, New York, 2010), pp. 79–86
73. A. Karatzoglou, L. Baltrunas, K. Church, M. Böhmer, Climbing the app wall: enabling mobile app discovery through context-aware recommendations,  in *Proceedings of the 21st ACM International Conference on Information and Knowledge Management, CIKM '12* (ACM, New York, 2012), pp. 2527–2530
74. R. Kimball, M. Ross, *The Data Warehousing Toolkit* (Wiley, New York, 1996)
75. J. Kiseleva, H.T. Lam, M. Pechenizkiy, T. Calders, Discovering temporal hidden contexts in web sessions for user trail prediction, in *Proceedings of the 22Nd International Conference on World Wide Web Companion, WWW '13 Companion, Republic and Canton of Geneva* (2013), pp. 1067–1074. International World Wide Web Conferences Steering Committee
76. D. Koller, M. Sahami,  Toward optimal feature selection,  in *Proceedings of the 13th International Conference on Machine Learning* (Morgan Kaufmann, Burlington, 1996), pp. 284–292
77. Y. Koren, Factorization meets the neighborhood: a multifaceted collaborative filtering model, in *Proceedings of the 14th ACM SIGKDD International Conference on Knowledge Discovery and Data Mining* (ACM, New York, 2008), pp. 426–434
78. G. Koutrika, B. Bercovitz, H. Garcia-Molina,  Flexrecs: expressing and combining flexible recommendations,  in *Proceedings of the 35th SIGMOD international conference on Management of data* (ACM, Providence, 2009), pp. 745–758

79. N. Lathia, *The Anatomy of Mobile Location-Based Recommender Systems* (Springer US, Boston, 2015), pp. 493–510
80. H.J. Lee, S.J. Park, Moners: a news recommender for the mobile web. Expert Syst. Appl. **32**(1), 143–150 (2007)
81. J.J. Levandoski, M.D. Ekstrand, M. Ludwig, A. Eldawy, M.F. Mokbel, J. Riedl, Recbench: benchmarks for evaluating performance of recommender system architectures. Proc. VLDB **4**(11), 911–920 (2011)
82. J.J. Levandoski, A. Eldawy, M.F. Mokbel, M.E. Khalefa, Flexible and extensible preference evaluation in database systems. ACM Trans. Database Syst. **38**(3), 17 (2013)
83. P. Li, A. Tuzhilin, Latent multi-criteria ratings for recommendations, in *Proceedings of the 13th ACM Conference on Recommender Systems, RecSys '19* (Association for Computing Machinery, New York, 2019), pp. 428–431
84. E. Liebman, M. Saar-Tsechansky, P. Stone, The right music at the right time: adaptive personalized playlists based on sequence modeling. MIS Q. **43**(3), 765–786 (2019)
85. H. Liu, J. Hu, M. Rauterberg, Music playlist recommendation based on user heartbeat and music preference, in *2009 International Conference on Computer Technology and Development*, vol. 1 (2009), pp. 545–549
86. H. Liu, H. Motoda, *Feature Selection for Knowledge Discovery and Data Mining* (Springer, New York, 1998)
87. Q. Liu, S. Wu, D. Wang, Z. Li, L. Wang, Context-aware sequential recommendation, in *2016 IEEE 16th International Conference on Data Mining (ICDM)* (IEEE, Piscataway, 2016), pp. 1053–1058
88. S. Lombardi, S.S. Anand, M. Gorgoglione, Context and customer behavior in recommendation, in *Workshop on Context-Aware Recommender Systems (CARS 2009), New York* (2009)
89. W. Luan, G. Liu, C. Jiang, L. Qi, Partition-based collaborative tensor factorization for poi recommendation. IEEE/CAA J. Automat. Sin. **4**(3), 437–446 (2017)
90. T. Mahmood, F. Ricci, A. Venturini, Improving recommendation effectiveness: adapting a dialogue strategy in online travel planning. J. IT Tour. **11**(4), 285–302 (2009)
91. J. McInerney, B. Lacker, S. Hansen, K. Higley, H. Bouchard, A. Gruson, R. Mehrotra, Explore, exploit, and explain: personalizing explainable recommendations with bandits, in *Proceedings of the 12th ACM Conference on Recommender Systems, RecSys '18* (Association for Computing Machinery, New York, 2018), pp. 31–39
92. L. Mei, P. Ren, Z. Chen, L. Nie, J. Ma, J.-Y. Nie, An attentive interaction network for context-aware recommendations, in *Proceedings of the 27th ACM International Conference on Information and Knowledge Management, CIKM '18* (Association for Computing Machinery, New York, 2018), pp. 157–166
93. H.F. Nweke, Y.W. Teh, M.A. Al-Garadi, U.R. Alo, Deep learning algorithms for human activity recognition using mobile and wearable sensor networks: State of the art and research challenges. Expert Syst. Appl. **105**, 233–261 (2018)
94. A. Odic, M. Tkalcic, J.F. Tasic, A. Kosir, Predicting and detecting the relevant contextual information in a movie-recommender system. Interact. Comput. **25**(1), 74–90 (2013)
95. S. Ojagh, M.R. Malek, S. Saeedi, S. Liang, An internet of things (IoT) approach for automatic context detection, in *2018 IEEE 9th Annual Information Technology, Electronics and Mobile Communication Conference (IEMCON)* (IEEE, Piscataway, 2018), pp. 223–226
96. M. Okawa, T. Iwata, T. Kurashima, Y. Tanaka, H. Toda, N. Ueda, Deep mixture point processes: Spatio-temporal event prediction with rich contextual information, in *Proceedings of the 25th ACM SIGKDD International Conference on Knowledge Discovery & Data Mining, KDD '19* (Association for Computing Machinery, New York, 2019), pp. 373–383
97. K. Oku, S. Nakajima, J. Miyazaki, S. Uemura, Context-aware SVM for context-dependent information recommendation, in *Proceedings of the 7th International Conference on Mobile Data Management* (2006), p. 109
98. R.O. Oyeleke, C.-Y. Yu, C.K. Chang, Situ-centric reinforcement learning for recommendation of tasks in activities of daily living in smart homes, in *2018 IEEE 42nd Annual Computer Software and Applications Conference (COMPSAC)*, vol. 2 ( IEEE, Piscataway, 2018), pp. 317–322

99. C. Palmisano, A. Tuzhilin, M. Gorgoglione, Using context to improve predictive modeling of customers in personalization applications. IEEE Trans. Knowl. Data Eng. **20**(11), 1535–1549 (2008)

100. U. Panniello, A. Tuzhilin, M. Gorgoglione, C. Palmisano, A. Pedone, Experimental comparison of pre-vs. post-filtering approaches in context-aware recommender systems, in *Proceedings of the 3rd ACM conference on Recommender Systems* (ACM, New York, 2009), pp. 265–268

101. U. Panniello, M. Gorgoglione, A. Tuzhilin, In carss we trust: How context-aware recommendations affect customers' trust and other business performance measures of recommender systems. Inf. Syst. Res. **27**(1), 182–196 (2016)

102. U. Panniello, A. Tuzhilin, M. Gorgoglione, Comparing context-aware recommender systems in terms of accuracy and diversity. User Model. User-Adapt. Interact. **24**(1–2), 35–65 (2014)

103. H.-S. Park, J.-O. Yoo, S.-B. Cho, A context-aware music recommendation system using fuzzy bayesian networks with utility theory, in *Proceedings of the Third International Conference on Fuzzy Systems and Knowledge Discovery, FSKD'06* (Springer, Berlin, 2006), pp. 970–979

104. M.-H. Park, J.-H. Hong, S.-B. Cho, Location-based recommendation system using bayesian user's preference model in mobile devices, in *Proceedings of the 4th International Conference on Ubiquitous Intelligence and Computing, UIC'07* (Springer, Berlin, 2007), pp. 1130–1139

105. D. Pathak, P. Krahenbuhl, J. Donahue, T. Darrell, A.A. Efros, Context encoders: feature learning by inpainting, in *Proceedings of the IEEE Conference on Computer Vision and Pattern Recognition* (2016), pp. 2536–2544

106. D.M. Pennock, E. Horvitz, Collaborative filtering by personality diagnosis: a hybrid memory-and model-based approach, in *IJCAI'99 Workshop: Machine Learning for Information Filtering* (1999)

107. M. Quadrana, P. Cremonesi, D. Jannach, Sequence-aware recommender systems. ACM Comput. Surv. **51**(4), 1–36 (2018)

108. S. Reddy, J. Mascia, Lifetrak: music in tune with your life, in *Proceedings of the 1st ACM International Workshop on Human-centered Multimedia, HCM '06* (ACM, New York, 2006), pp. 25–34

109. A. Rettinger, H. Wermser, Y. Huang, V. Tresp, Context-aware tensor decomposition for relation prediction in social networks. Soc. Netw. Anal. Min. **2**(4), 373–385 (2012)

110. F. Ricci, Q.N. Nguyen, Mobyrek: a conversational recommender system for on-the-move travelers, in *Destination Recommendation Systems: Behavioural Foundations and Applications* (2006), pp. 281–294

111. S. Sae-Ueng, S. Pinyapong, A. Ogino, T. Kato, Personalized shopping assistance service at ubiquitous shop space, in *Proceedings of the 22nd International Conference on Advanced Information Networking and Applications - Workshops, AINAW '08* (IEEE Computer Society, Washington, DC, 2008), pp. 838–843

112. B. Sarwar, G. Karypis, J. Konstan, J. Reidl, Item-based collaborative filtering recommendation algorithms, in *Proceedings of the 10th International Conference on World Wide Web* (ACM, New York, 2001), pp. 285–295

113. M. Sarwat, J. Avery, M.F. Mokbel, A recdb in action: recommendation made easy in relational databases. Proc. VLDB **6**(12), 1242–1245 (2013)

114. M. Sarwat, J.J. Levandoski, A. Eldawy, M.F. Mokbel, Lars*: an efficient and scalable location-aware recommender system. IEEE Trans. Knowl. Data Eng. **26**(6), 1384–1399 (2014)

115. G. Shani, A. Gunawardana, Evaluating recommendation systems, in *Recommender Systems Handbook* (Springer, Boston, 2011), pp. 257–297

116. G. Shani, D. Heckerman, R.I. Brafman, An MDP-based recommender system. J. Mach. Learn. Res. **6**(Sep), 1265–1295 (2005)

117. P. Sitkrongwong, S. Maneeroj, A. Takasu, Latent probabilistic model for context-aware recommendations, in *2013 IEEE/WIC/ACM International Joint Conferences on Web Intelligence (WI) and Intelligent Agent Technologies (IAT)*, vol. 1 (IEEE, Piscataway, 2013), pp. 95–100

118. E. Smirnova, F. Vasile, Contextual sequence modeling for recommendation with recurrent neural networks, in *Proceedings of the 2nd Workshop on Deep Learning for Recommender Systems, DLRS 2017* (Association for Computing Machinery, New York, 2017), pp. 2–9

119. B. Smyth, P. Cotter, Mp3 - mobile portals, profiles and personalization, in *Web Dynamics* (Springer, Berlin, 2004), pp. 411–433

120. S. Solorio-Fernández, J.A. Carrasco-Ochoa, J.F. Martínez-Trinidad, A review of unsupervised feature selection methods. Artif. Intell. Rev. **53**(2), 907–948 (2020)

121. A. Srivihok, P. Sukonmanee, E-commerce intelligent agent: personalization travel support agent using q learning, in *Proceedings of the 7th International Conference on Electronic Commerce, ICEC '05* (Association for Computing Machinery, New York, 2005), pp. 287–292

122. X. Su, T.M. Khoshgoftaar, A survey of collaborative filtering techniques. Adv. Artif. Intell. **2009** (2009). https://doi.org/10.1155/2009/421425

123. A. Tripathi, T.S. Ashwin, R.M.R. Guddeti, EmoWare: a context-aware framework for personalized video recommendation using affective video sequences. IEEE Access **7**, 51185–51200 (2019)

124. M. Unger, A. Bar, B. Shapira, L. Rokach, Towards latent context-aware recommendation systems. Knowl. Based Syst. **104**(C), 165–178 (2016)

125. M. Unger, A. Tuzhilin, Hierarchical latent context representation for context-aware recommendations. IEEE Trans. Knowl. Data Eng. (2020). https://doi.org/10.1109/TKDE.2020.3022102

126. M. Unger, A. Tuzhilin, A. Livne, Context-aware recommendations based on deep learning frameworks. ACM Trans. Manag. Inf. Syst. **11**(2), 1–15 (2020)

127. M. Van Setten, S. Pokraev, J. Koolwaaij, Context-aware recommendations in the mobile tourist application compass, in *Adaptive Hypermedia*, ed. by W. Nejdl, P. De Bra (Springer, New York, 2004), pp. 235–244

128. G. Vigliensoni, I. Fujinaga, The music listening histories dataset, in *Proceedings of the 18th International Society for Music Information Retrieval Conference, Suzhou* (2017), pp. 96–102

129. N.M. Villegas, C. Sánchez, J. Díaz-Cely, G. Tamura, Characterizing context-aware recommender systems: a systematic literature review. Knowl. Based Syst. **140**, 173–200 (2018)

130. Q. Wang, H. Yin, T. Chen, Z. Huang, H. Wang, Y. Zhao, N.Q.V. Hung, Next point-of-interest recommendation on resource-constrained mobile devices, in *Proceedings of The Web Conference 2020* (2020), pp. 906–916

131. H. Wermser, A. Rettinger, V. Tresp, Modeling and learning context-aware recommendation scenarios using tensor decomposition, in *2011 International Conference on Advances in Social Networks Analysis and Mining* (IEEE, Piscataway, 2011), pp. 137–144

132. W. Woerndl, J. Huebner, R. Bader, D. Gallego-Vico, A model for proactivity in mobile, context-aware recommender systems, in *Proceedings of the Fifth ACM Conference on Recommender Systems, RecSys '11* (ACM, New York, 2011), pp. 273–276

133. W. Wu, J. Zhao, C. Zhang, F. Meng, Z. Zhang, Y. Zhang, Q. Sun, Improving performance of tensor-based context-aware recommenders using bias tensor factorization with context feature auto-encoding. Knowl. Based Syst. **128**, 71–77 (2017)

134. M. Xie, H. Yin, H. Wang, F. Xu, W. Chen, S. Wang, Learning graph-based poi embedding for location-based recommendation, in *Proceedings of the 25th ACM International on Conference on Information and Knowledge Management* (2016), pp. 15–24

135. X. Xin, B. Chen, X. He, D. Wang, Y. Ding, J. Jose, CFM: convolutional factorization machines for context-aware recommendation, in *Proceedings of the 28th International Joint Conference on Artificial Intelligence* (AAAI Press, Palo Alto, 2019), pp. 3926–3932

136. L. Xiong, X. Chen, T.-K. Huang, J. Schneider, J.G. Carbonell, Temporal collaborative filtering with bayesian probabilistic tensor factorization, in *Proceedings of the 2010 SIAM International Conference on Data Mining* (SIAM, Philadelphia, 2010), pp. 211–222

137. H. Yin, Y. Sun, B. Cui, Z. Hu, L. Chen, LCARS: a location-content-aware recommender system, in *Proceedings of the 19th ACM SIGKDD International Conference on Knowledge Discovery and Data Mining* (2013), pp. 221–229

138. Z. Yu, X. Zhou, D. Zhang, C.Y. Chin, X. Wang, J. Men, Supporting context-aware media recommendations for smart phones. IEEE Pervasive Comput. **5**(3), 68–75 (2006)
139. S. Zhang, L. Yao, A. Sun, Y. Tay, Deep learning based recommender system: a survey and new perspectives. ACM Comput. Surv. **52**(1), 1–38 (2019)
140. G. Zheng, F. Zhang, Z. Zheng, Y. Xiang, N.J. Yuan, X. Xie, Z. Li, DRN: a deep reinforcement learning framework for news recommendation, in *Proceedings of the 2018 World Wide Web Conference* (2018), pp. 167–176
141. Y. Zheng, R. Burke, B. Mobasher, Differential context relaxation for context-aware travel recommendation, in *E-Commerce and Web Technologies*, ed. by C. Huemer, P. Lops. Lecture Notes in Business Information Processing, vol. 123 (Springer, Berlin 2012), pp. 88–99
142. Y. Zheng, B. Mobasher, R. Burke, CARSKit: a java-based context-aware recommendation engine, in *2015 IEEE International Conference on Data Mining Workshop (ICDMW)* (IEEE, Piscataway, 2015), pp. 1668–1671
143. F. Zhou, R. Yin, K. Zhang, G. Trajcevski, T. Zhong, J. Wu, Adversarial point-of-interest recommendation, in *The World Wide Web Conference* (2019), pp. 3462–34618

# Semantics and Content-Based Recommendations

Cataldo Musto, Marco de Gemmis, Pasquale Lops, Fedelucio Narducci, and Giovanni Semeraro

## 1 Introduction

Content-based recommender systems (CBRSs) basically rely on descriptive features to build a representation of items and users which is used to generate personalized recommendations. Such recommendations may regard *both* items provided with a textual description (e.g., the plot of a movie) as well as items that are themselves 'textual' (e.g., a news article).

Typically, content-based recommendations are obtained by matching up the attributes of the target user profile, in which preferences and interests are stored, with the attributes of the items. The result is a relevance score that represents the target users' level of interest in those items. In other terms, CBRSs are based on the assumption that user preferences remain stable over time (even when such preferences are constructed during the interaction with the system, as it happens in *conversational approaches*), since they suggest items similar to those a target user already liked in the past.

However, early CBRS models were based on keyword-based approaches exploiting simple term-counting. Accordingly, early models were not able to obtain a complete comprehension of the textual content describing the items nor to encode semantic relationships between terms. In particular, early CBRS show clear limits due to properties of natural language elements, such as:

C. Musto (✉) · M. de Gemmis · P. Lops · G. Semeraro
Università degli Studi di Bari, Bari, Italy
e-mail: cataldo.musto@uniba.it; marco.degemmis@uniba.it; pasquale.lops@uniba.it; giovanni.semeraro@uniba.it

F. Narducci
Politecnico di Bari, Bari, Italy
e-mail: fedelucio.narducci@poliba.it

© Springer Science+Business Media, LLC, part of Springer Nature 2022
F. Ricci et al. (eds.), *Recommender Systems Handbook*,
https://doi.org/10.1007/978-1-0716-2197-4_7

- POLYSEMY, multiple meanings for one word;
- SYNONYMY, multiple words with the same meaning;
- MULTI-WORD EXPRESSIONS, a sequence of two or more words whose properties are not predictable from those of the individual words;
- ENTITY IDENTIFICATION OR NAMED ENTITY RECOGNITION, the difficulty to locate and classify elements mentioned in text into predefined categories;
- ENTITY LINKING OR NAMED ENTITY DISAMBIGUATION, the difficulty of determining the identity (often called the *reference*) of entities mentioned in text.

This is a sharp limitation, since a keyword-based syntactic representation is often not enough to correctly catch the preferences of the users, as well as the informative content conveyed by the items. Of course, a sub-optimal comprehension of the informative content leads to a sub-optimal representation of the items and, in turn, to recommendations which are not accurate. As shown in several literature [86], it is necessary to improve such a representation in order to fully exploit the potential of content-based features and textual data.

With no doubt, we can state that *semantics* represents the theoretical foundation to proceed in this direction, by implementing more advanced models that allow machines to better understand information provided in natural language . In this research line, *semantics-aware recommender systems* represent one of the most innovative lines of research in the area of recommender systems. Indeed, as we stated in our previous contribution to the handbook [37], thanks to these representations it is possible to give meaning to information expressed in natural language and to obtain a deeper comprehension of the information conveyed by textual content.

The goal of this chapter is to integrate and extend the content previously presented in the chapter *"Semantics-aware Content-based Recommender Systems"* [37]. In particular, the goal of this chapter is to focus on *novel* research directions in the area.

The chapter is organized as follows: first, we start the discussion with an historical perspective in the area, then we provide an overview of the main techniques to incorporate semantics into items and user profiles. These approaches can be broadly split in two categories: *exogenous* and *endogenous* approaches. The former relies on the integration of external knowledge sources, such as ontologies, encyclopedic knowledge and data from the Linked Data cloud, while the latter relies on a lightweight semantic representation based on the hypothesis that the meaning of a word depends on its usage in large corpora of textual documents [57, 70].

Both the approaches can be exploited to cope with the issues of keyword-based syntactic representation: as an example, *word sense disambiguation techniques based on linguistic resources*, such as WordNet, can tackle polysemy, synonymy and multi-word expressions. Similarly, techniques to link items to knowledge graphs, such as those based on the exploitation of *ontologies* and *Linked Open Data*, can be helpful to deal with entity identification and entity linking. The same principle holds for *endogenous* representations, since the lightweight semantics learnt based on distributional semantics models can effectively tackle ambiguity issues.

Further details on the methods are provided in Sect. 2. However, this chapter only sketches these techniques since they represent the focus of the previous edition of our work [37]. Indeed, as previously said, a significant part of the chapter is devoted to recent trends in the area of content-based recommendations. Such trends regard: (1) techniques investigating new methods for *representing* content-based features; (2) techniques investigating new *sources* to gather content-based features; (3) new *use cases* for content-based features, such the exploitation of *content* to generate explainable recommendation and to build conversational recommender systems .

## 2   Content-Based Recommender Systems

This section reports an overview of the basic principles for building CBRSs and describes the evolution of the techniques adopted for representing items and user profiles.

### 2.1   The Architecture of a Content-Based Recommender System

The high level architecture of a content-based recommender system is depicted in Fig. 1. The recommendation process is performed in three steps, each of which is handled by a separate component:

- CONTENT ANALYZER—The main responsibility of this component is to represent the content of items (e.g., documents, product descriptions, etc.) coming from information sources in a form suitable for the next processing steps. It extracts features (keywords, n-grams, concepts, . . . ) from item descriptions and produces a structured item representation stored in the repository *Represented Items*. Early CBRSs adopt relatively simple retrieval models, such as the Vector Space Model (VSM), a spatial representation of text documents [166]. In that model, each document is represented by a vector in a multidimensional space, where each dimension corresponds to a term from the overall vocabulary of a given document collection [9, 165]. As we will show in the next sections, this model is often replaced by more recent and more effective *embedding* techniques. This representation is the input to the PROFILE LEARNER and FILTERING COMPONENT. For more details on NLP techniques for recommender systems we also suggest to refer to Chap. "Natural Language Processing for Recommender Systems";
- PROFILE LEARNER—This module collects data representative of the user preferences and builds the user profile, a model that generalizes the observed data. Preferences on items are collected as ratings on discrete scale and stored in a repository (*Feedback*). Usually, the generalization strategy is realized through

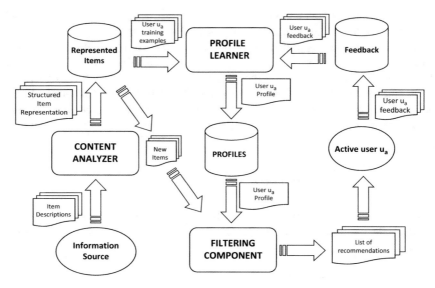

**Fig. 1** High level architecture of a content-based recommender

supervised machine learning algorithms [121], which infer the *user profile* from items and corresponding ratings.

- FILTERING COMPONENT—This module predicts whether a new item is likely to be of interest for the *active user*, the one for whom recommendations are to be calculated. It matches the features in the user profile against those in the item representations and produces a binary or continuous relevance judgment, the latter case resulting in a ranked list of potentially interesting items [77]. Ratings can be gathered on generated recommendations, then the learning process is performed again on the new training set, and the resulting profile is adapted to the updated user interests. The iteration of the feedback-learning cycle over time enables the system to take into account the dynamic nature of user preferences.

## 2.2 Semantics-Aware Content Representation

The Vector Space Model [166] can be useful to develop very simple intelligent information systems, but in order to cope with the above mentioned issues inherently related to natural language and its ambiguity, semantic techniques are crucial to shift from a *keyword-based* to a *concept-based* representation of items and user profiles.

Figure 2 depicts a classification of semantic techniques.

*Endogenous* approaches exploit large corpora of documents to infer the usage of a word, i.e. its *implicit* semantics. The main ideas behind these methods is described in Sect. 2.2.1.

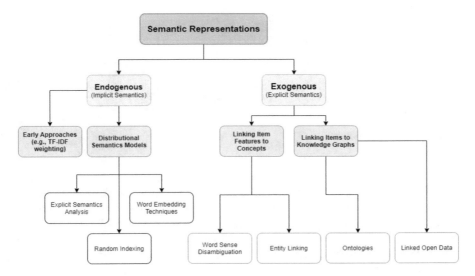

**Fig. 2** Classification of semantic representation techniques

*Exogenous* approaches rely on *external* knowledge sources, such as machine readable dictionaries, taxonomies, thesauri or ontologies, for representing items and user profiles. Section 2.2.2 focuses on the description of these techniques for an *explicit* representation of the semantics.

As previously said, this section provides just a brief overview of these techniques; we suggest to refer to the previous edition of this chapter for a deeper analysis [37]. A complete overview of semantics-aware representation strategies is discussed in [105].

### 2.2.1 Endogenous Semantics

Techniques for *endogenous semantics representation* fall into the general class of Distributional Semantics Models (DSMs), that were originally introduced in computational linguistics and cognitive sciences [169].

These data-driven approaches rely on the so-called *distributional hypothesis*, which states that *"Words that occur in the same contexts tend to have similar meanings"* [70], thus the algorithms that follow these approaches extract information about the meaning of a word by analyzing its usage in large corpora of textual documents.

DSMs are based on Wittgenstein's idea that the meaning of a word is its use in the language [204], and that semantically similar words also share similar contexts of usage [162].

|         | c1 | c2 | c3 | c4 | c5 | c6 | c7 | c8 | c9 |
|---------|----|----|----|----|----|----|----|----|----|
| beer    |    | ✓  | ✓  |    |    | ✓  | ✓  |    |    |
| wine    |    | ✓  | ✓  |    |    | ✓  | ✓  | ✓  |    |
| spoon   | ✓  |    |    | ✓  |    |    |    | ✓  | ✓  |
| glass   | ✓  | ✓  | ✓  |    | ✓  |    |    |    | ✓  |

**Fig. 3** A term-context matrix

These models, also known as *geometrical models*, learn similarities and connections in a totally unsupervised way. Indeed, they represent each term that occurs in a corpus as a vector in a high-dimensional vector space called WordSpace [106].

Given a corpus, usually the WordSpace is built by means of a *term-context matrix*, as the one presented in Fig. 3. Each row represents one term of the vocabulary (obtained by applying an NLP pipeline), while each column is *a context of usage*. Every time a term is used in a particular context, this information is encoded in the matrix. As an example, according to Fig. 3 the term *beer* is used in contexts $c_2$, $c_3$, $c_6$ and $c_7$. We can imagine a context as a fragment of text in which the word occurs. Thus, each term is represented by a *vector* (the corresponding row in the matrix), modeled in a vector space whose dimensions are the columns of the term-context matrix.

Given a WordSpace, a vector space representation of the documents called DocSpace can be also computed. A DocSpace can be obtained by following different strategies: as shown by Sahlgren [164], a document representation may be calculated as the centroid vector of the vector space representation of the words that appear in the document, or as their *weighted sum*. Next, given a WordSpace, the similarity of two terms can be estimated by analyzing the overlap between their usage. According to the example in Fig. 3, we can state that words *beer* and *wine* are very similar since they share a large number of contexts. In practice, the similarity between words is estimated as the proximity between vectors that represent those words in the WordSpace, in accordance with the *similarity-is-proximity* metaphor [164]. Thus, it can be computed in several ways, such as cosine similarity, Manhattan and Euclidean distances, or relative entropy-based measures [122].

We need to further clarify the definition of *context*. Generally speaking, it is *a fragment of text in which a word appears*. In the simplest formulation, the context is the *whole document*. In that case, the *term-context matrix* corresponds to the *term-document matrix* in the classical Vector Space Model [166]. Finer-grained options are possible: the context could be a paragraph, a sentence, a window of surrounding words or even a single word. A survey about different strategies to handle the concept of context is provided in [188].

Among the approaches for *endogenous semantics representation*, it is worth to mention the Explicit Semantic Analysis (ESA). ESA, which became very popular in the early 2010s, builds a semantics-aware representation of words in terms of Wikipedia concepts [61]. In ESA, the representation of the terms is based on the so-called *ESA matrix*, whose rows correspond to the Wikipedia vocabulary (i.e., the set of distinct terms found within Wikipedia articles, after applying basic NLP operations), while columns correspond to Wikipedia pages. Each row represents a term and is called *semantic interpretation vector*. It contains the list of *concepts* (Wikipedia pages) associated to the term, along with the corresponding weights.

The semantics of a document is typically obtained by computing the centroid of the semantic interpretation vectors associated with the individual terms occurring in that document. ESA showed good performance in tasks as text categorization [59], semantic similarity computation [60] and recommendation [143].

One of the main issues with DSMs is represented by the tremendous increase of the size of the term-context matrix as the context gets smaller. *Word Embedding* techniques have been developed to manage that issue. They *project* the original vector space into a smaller *but substantially equivalent* one, thus returning a more compact `WordSpace`. Differently from *pure* DSMs, such as ESA, the new dimensions of the reduced vector space are not human understandable anymore. Popular Word Embedding techniques are Latent Semantic Analysis [95] and Random Indexing [164]. These representations recently got new interest after the introduction of Word2Vec [117] and, as we will show in Sect. 3.1, the use of Word Embedding techniques is today one of the most active research lines in the area of semantics-aware content-based recommender systems.

Word2Vec is a technique that exploits *neural networks* to learn a vector space representation of words. It was first proposed by Tomas Mikolov et al. [117], and it gained a lot of attention in the last years due to the simplicity of the approach and to the effectiveness it obtained in several tasks, including recommendation [129]. In a nutshell, this approach is used to learn (small) word embeddings by exploiting a *two-layer neural network* which is fed with examples gathered from a corpus of textual data to learn the contexts and the linguistic usage of words to generate the embeddings.

A toy example of the neural network exploited by Word2Vec is reported in Fig. 4. Given a corpus of textual data, we define an input layer of size $|V|$, that corresponds to the dimension of the vocabulary of the terms. This means that each term that appears in the corpus is mapped to an element in the input layer. Next, an output layer $|N|$ is created. In this case, $N$ is the size of the embedding we want to obtain at the end of the learning process. The value of $N$ is a parameter of the model and has to be properly tuned. Clearly, the greater the value, the more complex the learning process since a larger number of weights in the network has to be learnt, but the better the resulting representation.[1]

The edges connecting the nodes in the network have different weights. They are initially randomly set and they are updated through the training process. The final

---

[1] The equation *'larger vectors, better representation'* is typically valid. However, when the dimension becomes too large a decrease in the performance can be noted.

**Fig. 4** Structure of the
network

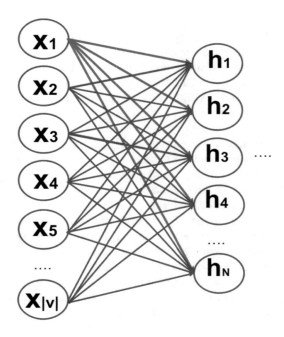

representation of a term is the set of weights that connects its corresponding node
in the input layer to all the nodes in the output layer. Formally, given a term $t_k$ its
representation is given by $[w_{t_k,v_1}, w_{t_k,v_2}, w_{t_k,v_n}]$.

Such a discussion makes immediately emerge the importance of the training
process in Word2Vec, since the network needs to acquire input examples to properly
learn linguistic regularities and to update the weights in the network (and, in turn,
the resulting representation) accordingly.

The training of the network can be carried out by exploiting two different
methodologies, that is to say the Skip-Gram methodology (SG) and the Continuous
Bag of Words (CBOW). The choice of the most suitable technique is a design
choice: according to Mikolov, SG works well when the training set is small and
shows a good accuracy even on rare terms, whereas CBOW is several time faster
than SG and is more accurate for frequent words.

More details about Word2Vec can be found also in [118], while a thorough
discussion about *endogenous methods* can be found in Chapter 3 of the book [105].

### 2.2.2   Exogenous Semantics

Approaches for *exogenous* semantics representations rely on the linguistic, cultural
and background knowledge that is encoded and made available through *external*
knowledge bases.

The main difference between *endogenous* and *exogenous* techniques for semantics-aware representation lies in the nature of the knowledge bases they rely on. In the first case, the semantics is obtained by exploiting *unstructured* data (corpora), and is directly inferred from the available information. In the second, the semantics *comes from the outside*, since it is obtained by mining and exploiting data which are previously encoded in *structured* and external knowledge sources.

Most popular structured knowledge sources today available are:

**WordNet** [54, 119, 120].    It is a lexical database for the English language, made by cognitive scientists, freely available online[2] and extensively used in NLP research [178]. The goal of WordNet is to model *meanings* that can be expressed through the known *word forms*, and to represent the lexical relations that exist among them. The basic building block of WordNet is the SYNSET (SYNonym SET), that encodes a specific *meaning* (concept) though the set of synonym words that can be used to express it.

**BabelNet** [145].    It is a large-scale multilingual encyclopedic dictionary and semantic network.[3] It integrates heterogeneous resources such as WordNet, Wikipedia, Wikidata [192] (described later), Wiktionary and other lexical databases. In a nutshell, the knowledge encoded in BabelNet is represented through a labeled directed graph. *Nodes* are *concepts* extracted from WordNet and Wikipedia (synsets and Wikipages), while *edges* among nodes encode the *semantic relations* coming from WordNet, as well as semantically unspecified relations from hyperlinked text coming from Wikipedia.

**The Linked Open Data (LOD) cloud** [17].    This term was introduced to identify the huge number of datasets released through the Linked Open Data initiative, a project started in the late 2000s that inherits some of the concepts and the ideas of the *Semantic Web*. The LOD project is grounded on two cornerstones: (1) each resource available on the Web should be uniquely referred to through an URI; (2) data have to be encoded and linked by using RDF, acronym of Resource Description Framework. The *nucleus* of the LOD cloud is commonly represented by DBpedia [8], the RDF mapping of Wikipedia that acts as a *hub* for most of the RDF triples made available in the LOD cloud.

**Wikidata** [192].    It is a free, collaborative and multilingual database, built with the goal of turning Wikipedia into a fully structured resource. While DBpedia is almost automatically built by mapping in RDF format the information contained in the Wikipedia infoboxes, Wikidata entries are collaboratively created and maintained by both Wikidata editors and automated bots. Due to its collaborative nature Wikidata is continously updated, while DBpedia is usually updated only twice a year.

As described in Fig. 2, two strategies can be adopted to exploit the data available in the knowledge sources to build a semantics-aware representation of items:

---

[2] http://wordnet.princeton.edu.

[3] http://babelnet.org.

(a) linking item features to concepts or (b) linking items to a knowledge graph.

The goal of the first group of techniques is to associate each feature with its *correct semantics* and to identify more complex concepts expressed in the text. Word Sense Disambiguation (WSD) techniques fall in this category because they tackle the problem of correctly identifying which of the senses of an ambiguous word is invoked in a particular use of the word itself [110]. Several WSD algorithms have been developed; for instance, in [170] the authors exploit WordNet to disambiguate item descriptions used to train a content-based recommender system and to build synset-based user profiles. Entity Linking (EL) methods [155] are also classified in the first group. EL is the task of associating the *mention* of an *entity* in a text to an *entity* of the real world stored in a knowledge base [33]. A systematic review of other techniques and algorithms for EL is provided in [172].

Techniques included in the second group directly link items to nodes in a knowledge graph rather than mapping word forms to word meanings or entities. There is no need to process any textual content because either the item is directly linked to the Linked Open Data cloud or an *ontological* representation of the domain of interest is built, so that items are modeled in terms of *classes* and *relations* that exist in the ontology. More details about exogenous methods can be found in Chapter 4 of the book [105], while a deeper discussion on using LOD and Knowledge Graphs for recommender systems will be provided in Sect. 3.2 of this chapter.

## *2.3   Strengths and Weaknesses of Content-Based Recommendations*

The adoption of the content-based recommendation strategies (especially in their semantics-aware forms) has several advantages when compared to the collaborative one:

- USER INDEPENDENCE—Content-based recommenders exploit solely ratings provided by the active user to build her own profile. Instead, collaborative filtering methods need ratings from other users in order to find the "nearest neighbors" of the active user or to build sophisticated machine learning models. In other terms, they are less prone to *data sparsity* issues and can be more effective when a little amount of data is available.
- TRANSPARENCY—Explanations on how the recommender system works can be provided by explicitly listing content features or descriptions that caused an item to occur in the list of recommendations. Those features are indicators to consult in order to decide whether to trust a recommendation. Conversely, collaborative systems are black boxes since the only explanation for an item recommendation is that unknown users with similar tastes liked that item. More details about recent *explainable* content-based recommendation methods are provided in Sect. 3.4;

- NEW ITEM—Content-based recommenders are capable of recommending items not yet rated by any user. As a consequence, they do not suffer from the first-rater problem, which affects collaborative recommenders which rely solely on users' preferences to make recommendations. Therefore, until the new item is rated by a substantial number of users, the system would not be able to recommend it.

Nonetheless, content-based systems have several shortcomings:

- LIMITED CONTENT ANALYSIS—Content-based techniques have a natural limit in the number and type of features that are associated, whether automatically or manually, with the objects they recommend. No content-based recommendation system can provide suitable suggestions if no descriptive features of the items is available. Of course, recent advances in the area (e.g., pre-trained language models used to train word embeddings and structured features available in knowledge graphs) have partially mitigated this issue, but the *need for content* is still a mandatory requirement for these approaches;
- OVER-SPECIALIZATION—Content-based recommenders have no inherent methods for finding something unexpected. The system suggests items whose scores are high when matched against the user profile, hence the user is going to be recommended items similar to those already rated. This drawback is also called *serendipity* problem, to highlight the tendency of the content-based systems to produce recommendations with a limited degree of novelty. To give an example, when a user has only rated movies directed by Stanley Kubrick, she will be recommended just that kind of movies. A "perfect" content-based technique would rarely find anything *novel*, limiting the range of applications for which it would be useful.
- NEW USER—As other recommendation paradigms, also content-based recommender systems suffer of *cold start*. Indeed, enough ratings have to be collected before a content-based recommender system can really understand user preferences and provide accurate recommendations. Therefore, when few ratings are available, as for a new user, the system will not be able to provide reliable recommendations.

# 3 Recent Developments and New Trends

Recent developments in the area of content-based recommender systems can be roughly split into three research directions: first, as for *endogenous* techniques, the most relevant advances concern methods based on embeddings and distributed representations. Next, as for *exogenous* techniques, recent research focused on methods exploiting the information encoded in knowledge graphs . Finally, it is worth mentioning methods based on multimedia features and user-generated content. In the next sections, relevant work recently presented in these areas will be discussed. Moreover, we will also show how content can be used to improve the user experience by means of natural language explanations and conversational interfaces.

## 3.1 Embeddings and Distributed Representations

As introduced in Sect. 2.2.1, approaches for *endogenous semantics representation* exploit textual content and produce a vector space representation of the items to be recommended as well as of the users.

Nowadays, it is very common to refer to these representation as *embeddings*. This term can be further specialized into *word embeddings* and *sentence embeddings*, depending on whether a representation for each word or for each sentence is built. Due to the effectiveness of these techniques, whose spread has been largely discussed in recent years [69], there are many approaches that exploit embeddings and distributed representations for recommendation tasks. In this section we will provide an overview of relevant work in the area. The section is organized in two parts: First, we introduce early approaches that directly use word and sentence embeddings to feed recommendation models. Next, following the recent trend of deep learning architecture, we show how distributed representations can be used, together with deep neural networks, to generate accurate semantics-aware content-based recommendations.

### 3.1.1 Recommender Systems Based on Word and Sentence Embeddings

The early attempts in the area put their roots in the area of *pure* distributional semantics models (DSMs). In particular, these attempts exploited a DSM to learn a vector space representation of users and items and they *directly* used this representation in a recommendation model.

As an example, McCarey et al. [114] evaluate the effectiveness of Latent Semantic Indexing (LSI) [38] in a content-based recommendation scenario. Similarly, Musto et al. [124, 125] propose an extension of the classical VSM, called *enhanced Vector Space Model* (eVSM), that exploits Random Indexing to build a dense vector space representation of users and items. As shown in the experiments, eVSM overcame other classical content-based filtering techniques, and the findings were confirmed by subsequent experiments where the same approach is evaluated in a context-aware recommendation scenario [127].

By following these attempts, in [111] the authors exploited word embedding techniques to infer the vector-space representations of venues based on venue descriptions and reviews data. This work confirmed the previously presented outcomes, since the experiments showed that the use of content features and word embeddings significantly enhanced the accuracy of venue recommendations.

Next, a significant boost to the research in the area was noted after the introduction of Word2Vec [117]. This technique, based on the principles of *distri-*

*butional semantics models*, exploits a two-layer neural network to learn a vectorial representation of users and items. In this resarch line, Ozsoy et al. [147] proposed the use of Word2Vec to learn word embeddings representing items and user profiles. Their experiments showed that CBRSs based on word embeddings can obtain results comparable to those obtained by other content-based approaches and by algorithms for collaborative filtering based on matrix factorization. Next, in [173] the authors employed Word2Vec to compute the vectors of tags in Tumblr and recommended Tumblr blogs to the users. Next, a context-aware recommender method that extracts contextual information from textual reviews using a word embedding based model is proposed in [181] and a similarity measure inspired by Word2Vec, which is then used to learn the similarity between items, is presented in [116].

A comparative analysis among different word embedding techniques in a content-based recommendation scenario is presented in [130]. In particular, the work compares Latent Semantic Indexing, Random Indexing, and Word2Vec to establish the most effective technique. Results of the experiments in a *movie* and *book* recommendation scenarios show the good performance of the Word2Vec strategy, with the interesting outcome that even a smaller word representation could lead to accurate results. Furthermore, it emerged that the effectiveness of word embedding approaches is directly dependent on the sparsity of the data. This is an expected behavior since content-based approaches can better deal with cold-start situations, and with very sparse datasets they perform better than collaborative filtering or matrix factorization baselines. However, as discussed in [23] and [26], it is necessary to point out that the performance of Word2Vec-based models is strictly dependant on hyperparameter tuning. As shown in both these works, the performance of Word2Vec-based approach are strictly dependant of the optimization of the parameters, so it is fundamental to devote the necessary attention to this step.

Due to the effectiveness shown by Word2Vec, several research exploited Word2Vec to encode sequences of actions or sequences of events (instead of sequences of words) to learn a vector space representation of the items. Even if these approaches do not exploit *content-based features*, we deemed as relevant to briefly discuss them in order to provide a complete overview of the effectiveness and of the flexibility of word embedding approaches. The first work that exploited this analogy is Item2Vec [12], where items are used as words and baskets are used as sentences. Similarly, Grbovic at al. [65] adapted Word2Vec to generate product recommendations (i.e., prod2vec). They treated purchase history of a user as the sentence and each product as the word. This approach is further extended is MetaProd2Vec [190], which is based on Prod2Vec and incorporates side information in both the input and output space of the neural network.

Next to Word2Vec, several work propose recommendation methods based on Doc2Vec [96]. Doc2Vec is a neural approach that shares the same principles of Word2Vec and focuses on the representation of *sentences* and *documents*. As an example, in [90] Doc2Vec is used to learn an embedding representing a news article, based on the text and the title of the news. Next, such a representation is used to feed a hybrid recommendation approach. Similarly, in [179] the authors propose an hybrid approach based on the combination of two techniques inspired by Doc2Vec, that it to say, *user2vec*, which uses item descriptions and usage histories to model users and *context2vec* which uses further metadata on items and users in an attempt to incorporate context into the model. In both the cases, the experiments showed that these approaches outperform all the baselines. Good performance of this technique also emerged in [34], where Doc2Vec is used to represent items in a digital library recommendation scenario. Finally, in [3] the authors exploit Paragraph2Vec to generate a vector space representation of reviews. Next, the resulting feature vectors (the neural embeddings) are used in combination with the rating scores in a hybrid probabilistic matrix factorization algorithm. The proposed methodology is then compared to three other similar approaches on six datasets in order to assess its performance. As shown by the results, the exploitation of reviews embeddings led to an improvement of the performance.

### 3.1.2 Deep Learning Models Based on Word and Sentence Embeddings

Due to the effectiveness shown by recommendation strategies based on pure word and sentence embeddings, several research investigated how to encode distributed representation into more complex recommendation models. One of the first attempts in this research line was due to Kula, who proposed in [94] a hybrid matrix factorisation model representing users and items as linear combinations of their content features' latent factors. As shown by the experiments, the model outperforms both collaborative and content-based models in cold-start or sparse interaction data scenarios.

Next, in parallel with the growth of deep learning architectures and neural models, most of the research effort has been devoted to inject content information into deep architectures. In this area, several work tried to exploit Recurrent Neural Networks (RNNs) [163], which are particularly effective to model sequences of inputs (e.g., audio signals), to encode textual context as well.

As an example, in [180] the authors learn a vector-space representation of textual content based on a Long-Short Term Memory (LSTM) network [79], a specialization of RNNs. The experiments carried out by the authors confirmed again the effectiveness of the model, since the results showed that the use of LSTM networks can further improve the accuracy of the recommendations. This is due to the fact that such architectures are able to learn embeddings that also encode the dependencies between words. These findings have been further investigated in [135], where the authors present a deep content-based recommender system that exploits Bidirectional Recurrent Neural Networks (BRNNs) to learn an effective representation of the items to be recommended based on their textual description. In particular, BRNNs extend RNNs by encoding information about both the preceding and following words, thus leading to a more precise representation. Moreover, the authors further extended such a representation by introducing structured features extracted from the Linked Open Data (LOD) cloud, as the genre of a book, the director of a movie and so on. In the experimental session the effectiveness of the approach is evaluated in a top-N recommendation scenario, and the results showed that the approach obtained very competitive results by overcoming the baselines.

The use of LSTMs to model textual content in CBRS is also investigated in Almahairi et al. [4], who used LSTMs to model textual content (textual *reviews*, in that case) to feed a collaborative recommendation algorithm and by Bansal et al. [10], which applies a bidirectional *Gated Recurrent Unit (GRU)* network to encode item description and an embedding for each tag associated to the item.

Next to the approach based on the exploitation of Recurrent Neural Networks, we can also cite a plenty of research which is based on the use of Convolutional Neural Networks. As an example, in [211], the authors present Deep Cooperative Neural Networks (DeepCoNN), a deep model that consists of two parallel neural networks coupled in the last layers. One of the networks focuses on learning user behaviors exploiting reviews written by the user, while the other one learns item properties from the reviews written for the item. Experimental results demonstrated that DeepCoNN significantly outperforms all the baselines. A similar approach based on the processing of item reviews is also presented in [30] and in [103], where a dual attention mutual learning between ratings and reviews for item recommendation, named DAML, is proposed. In this case, the authors utilize local and mutual attention of the convolutional neural network to jointly learn the features of reviews to enhance the interpretability of the proposed model.

The use of Convolutional Neural Networks is also investigated in [194], where a deep knowledge-aware network (DKN) that incorporates knowledge graph representation for news recommendation is presented, and in [205], where the authors use a CNN network to learn hidden representations of news articles based on their titles.

Generally speaking, all these works gave evidence of the effectiveness of embeddings and distributed representation for recommendation tasks. For the sake of completeness, it is worth mentioning recent attempts in the area of *contextual word representations*. They differ from classic word embeddings since they are able to learn a *context-aware* representation of words which depends on the other words which occur in the sentences. Accordingly, they are able to better handle *ambiguity issues* in content representation. This process is possible because of the use of large pre-trained language models, which can learn highly transferable and task-agnostic properties of a language [51]. The adoption of *contextual word representation* techniques in recommendation tasks is a relatively new direction. As an example, in [71] the authors present a comparative evaluation among several techniques, such as BERT [44], SciBERT [16], ELMo [152], USE [25] and InferSent Sentence Encoders [35]. Experiments show that the sole consideration of semantic information from these encoders does not lead to improved recommendation performance over the traditional BM25 technique [160], while their integration enables the retrieval of a set of relevant papers that may not be retrieved by the BM25 ranking function. Overall, the best results were obtained by USE. Similarly, Stakhiyevich and Huang in [177] proposed an approach for building user profiles to be used in a personalized recommender system, based on Embeddings from Language Models (ELMo) and user reviews. In this case, cosine similarity has been adopted to calculate the similarity between user interests and item categories over their contextual word representations. Finally, Cenikj et al. in [24] investigate, with a preliminary study, the possibility to enhance a graph-based recommender system for Amazon products with two state-of-the-art representation models: BERT and GraphSage. The results obtained by the authors are encouraging in following the idea to merge graph and contextual word representation techniques.

## 3.2   Linked Open Data and Knowledge Graphs

Linked Open Data provide a great potential to effectively feed filtering algorithms with *exogenous* semantic representations of items. One of the first attempts to leverage Linked Open Data to build recommender systems is *dbrec* [149], a music recommender system based on the *Linked Data Semantic Distance* (LDSD) algorithm [150], which computes the semantic distance between artists referenced in DBpedia. Semantic distance can be seen as a way to compute the *relatedness* between two nodes in a knowledge graph (in this case, two artists), and it is obtained as the linear combination of *direct relationship* (i.e., the amount of direct links between the nodes, such as overlapping properties) and *indirect relationship* (i.e., the amount of shared links through other resources). Given a dataset of artists gathered from DBpedia, LDSD is computed off-line and it is used to provide users with music recommendations. In particular, semantically related artists, that is to say, artists with a low LDSD score, are suggested. As shown in the experiments reported in [149], recommendations based on LDSD provide competitive results with respect to a music recommender system based on Last.fm, thus giving one of the earliest evidences of the effectiveness of recommendation strategies based on knowledge graphs.

By following this research line, in [126], DBpedia is used to enrich the playlists extracted from a Facebook profile with new related artists. Each artist in the original playlist is mapped to a DBpedia node, and other similar artists are selected by taking into account shared properties, such as the genre and the musical category of the artist. Another simpler approach to define a CBRS exploiting Linked Open Data is presented in [46]. The ontological information, encoded via specific properties extracted from DBpedia and LinkedMDB [72], is adopted to perform a *semantic expansion* of the item descriptions, in order to catch implicit relations, not detectable just looking at the nodes directly linked to the item. In that work, the authors used Support Vector Machines to learn user profiles, a technique which tends to be fairly robust with respect to overfitting and can scale up to considerable dimensionalities. The evaluation of different combinations of properties revealed that more properties lead to more accurate recommendations, since this seems to mitigate the limited content analysis issue of CBRSs.

Besides the above mentioned approaches to catch implicit relations which allow to increase the number of common features between items, more sophisticated approaches may be exploited, in order to implement more complex reasoning

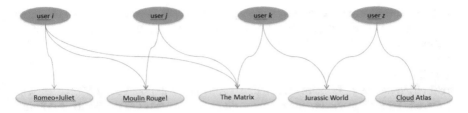

**Fig. 5** Basic bipartite graph representing users, items, and their preferences

over the graphs. In particular, graph-based recommender systems model *users* and *items* as *nodes* in a graph, and *edges* connect users with items according to their preferences. An example of such a data model is reported in Fig. 5.

This basic representation is very similar to that usually adopted for collaborative filtering, and recommendations can be produced by assigning to each item $i \in I$ a *relevance score*.

Given such a formulation, the problem of providing a user with a recommendation can be tackled by exploiting an algorithm that assigns a score to an item node in the graph, such as the *PageRank* algorithm [148]. As an example, given a graph-based data model, the PageRank algorithm can be run and PageRank scores can be sorted in descending order. Next, the $k$ nodes with the highest PageRank scores can be returned by the algorithm as recommendations. However, PageRank has the main problem of being *not personalized*, that is to say, the PageRank score of the item nodes (and, accordingly, the recommendations returned by the algorithm) only depends on the topology as well as on the connections that exist in the graph. A well-known variant of the PageRank, called *PageRank with Priors* [73] can be adopted to tackle this issue, since it allows to get a bias towards some nodes, specifically, the preferences of a specific user. As described in [131, 134], this algorithm can be really effective for recommendation tasks since it can adapt her own behavior on the preferences of the target user. In this scenario, PageRank with Priors is executed for each user and the nodes in the graph are ranked according to their PageRank score, as it happens for classical PageRank algorithm. The list of *item* nodes, not yet *voted* by the target user are provided as recommendations. In this setting, Linked Open Data can be used to enrich graphs by introducing *additional* nodes and edges, in order to come up with more effective representations including new connections resulting from the properties encoded in the LOD cloud. Ideally, we can run this enrichment step again and again, in order to introduce in the graph *non-direct* relationships. However, in [128] it has been shown that the introduction of non-direct relationships leads to an exponential growth of the PageRank running time, without a significant improvement in the precision of the recommendation process. Indeed, a simple and straightforward question may emerge from such a scenario: *Is it necessary to inject all the available properties?* and *Are all the properties equally important to provide users with accurate recommendations?* Hence, similarly to what happens in other settings, e.g., machine learning problems, it is necessary to investigate to what extent each property modeled in the graph improves the accuracy of the recommendation strategy, in order to filter out non-useful connections and select only the most meaningful properties. Hence, a possible strategy to automatically identify the most promising LOD-based features is to exploit *feature-selection* methodologies (e.g., Principal Component Analysis, Support Vector Machines, Chi-Squared Test, Page Rank, Information Gain, Information Gain Ratio, Minimum Redundancy Maximum Relevance) adopted in machine learning, whose goal is to improve the prediction performance of the predictors and to provide faster and more cost-effective predictors. For a more detailed discussion on the impact of feature-selection techniques on accuracy and diversity of recommendations we suggest to refer to [131].

Another way to exploit the information stored in a Knowledge Graph (KG) is to extract *topological features* that can be obtained by mining the bipartite and tripartite graph-based data model, as shown in [133]. Such features, encoding some structural characteristics of the data model can be used next to feed a recommendation framework with these new and interesting features.

In [133], an extensive experimental evaluation has been performed to assess the accuracy of different classification algorithms, namely *Naïve Bayes, Logistic Regression* and *Random Forests,* trained with item representations based on different groups of features, including the topological ones. One of the main outcomes is that bipartite and tripartite features have performance comparable to that of textual features (simple item descriptions) or LOD-based features (extracted from the LOD cloud). Given that the process that computes textual and LOD-based features requires a quite complex NLP pipeline or a mapping of items to `DBpedia`, topological features represent a more lightweight (they are very few) and therefore a more viable alternative for representing items. On the other side, the benefit of injecting the exogenous knowledge coming from the Linked Open Data cloud particularly emerged when data are sparse.

Recently, several work tried to combine the information encoded in knowledge graphs into machine learning and deep learning models. As an example, in [103] the authors proposed KRED, a knowledge-aware recommender systems exploiting an *enhanced* representation based on KGs. In particular, the authors start from a vector-space representation of the item (a news article, in this case) and then enrich the embedding by aggregating information from their *neighborhood* (extracted from the knowledge graph, of course) of the entities mentioned in the article. In this way, extra information coming from the other nodes directly connected to the target one can be encoded in the representation. This task is carried out by an *information distillation layer*, that aggregates the entity embeddings under the guidance of the original item representation, and transforms the item vector into a new one. Next, in [15] the authors present SemAuto, a recommendation model that puts together knowledge graphs and deep learning techniques. SemAuto is based on *autoencoders*, a type of neural network used to unsupervisedly learn a representation for a set of data. An autoencoder is typically split into two parts, a reduction side, that aims to reduce the noise of original representation, and a reconstructing side, that tries to generate from the reduced encoding a representation as close as possible to its original input. In this setting, Bellini et al. introduce the idea of enhancing autoencoders by means of knowledge graphs. In particular, SemAuto is inspired by fully-connected autoencoders and tries to make the model explainable by labeling neurons in hidden layers with the entities available in the knowledge graphs. In other terms, nodes in the hidden layers are replaced by nodes and connections available in a KG. According to authors' claims, this is supposed to improve both the predictive accuracy as well as the explainability of the model. As shown by the results, SemAuto actually outperforms state-of-the-art recommendation algorithms on both accuracy and diversity.

Moreover, a recent trend concerns the application of Graph Neural Networks (or Graph Convolutional Neural Networks [209]) in the area of RSs. In particular, GCNs aim to generalize convolutional neural networks to non-Euclidean domains (such as graphs). In this area, Wang et al. [196] present Knowledge Graph Convolutional Networks (KGCN) for recommender systems. KGCN learn a representation based on: (1) interactions encoded in the user-item matrix; (2) descriptive encoded in a knowledge graph. Generally speaking, KGCN learn a h-order representation of an entity as a mixture of its initial representations and the representation of its neighbors up to $h$ hops away. In order to reduce the computational load of the method, the authors adopt a sampling method to select just a fixed size of neighbors instead of using its full neighbors. In other terms, KGCN is proposed to capture high-order structural proximity among entities in a knowledge graph. So the final representation of an entity is dependent on itself as well as its immediate neighbors, and also takes into account users' personalized and potential interests. As shown in the experiments, KGCN outperform state-of-the-art baselines in movie, book, and music recommendation. A similar intuition is investigated in [197], where the authors present KGAT (Knowledge Graph Attention Network). In this case, *attention mechanisms* are used to model the high order connectivities in KG in an end-to-end fashion. The core of the approach lies in the definition of an attentive embedding propagation layer, which adaptively propagates the embeddings from a node's neighbors to update the node's representation. For further details on recommender systems based on knowledge graphs we suggest to refer to the survey by Guo et al. [68].

Finally, several work investigated *graph embedding* techniques to improve the quality of recommendation algorithms based on knowledge graphs. In particular, graph embedding techniques take a graph as input and produce a vector-space representation of the nodes encoded in the graph. Intuitively, these models merge the flexibility and the effectiveness of vector-space representation with the richness of a graph-based representation. Such an intuition is investigated in [136], where two popular graph embedding techniques, that is to say, *node2vec* [66] and *Laplacian Eigenmaps* [14], are compared in a recommendation task based on a classification framework. In other terms, the vectors returned by the graph embedding techniques are used as positive and negative examples to build a classification model for each user. As shown by the experiments, the information extracted from knowledge graphs significantly improved the effectiveness of the recommendation models. Overall, the approach based on *node2vec* obtained the better results. A similar intuition was proposed by Zhang et al. in [208]. In this case, the authors exploit *TransR*, a popular graph embedding method, together with stacked denoising autoencoders and stacked convolutional autoencoders, to learn a vector-space representation which is then used to feed the final integrated framework, which is termed as Collaborative Knowledge Base Embedding (CKE), that jointly learns the latent representations in collaborative filtering as well as items' semantic representations from the knowledge base. Also in this case, the experiments confirmed the effectiveness of the approach since the proposed method outperformed several widely adopted state-of-the-art recommendation methods.

To conclude, we can clearly state that this overview of recommender systems based on knowledge graphs confirmed that these methodologies can be helpful to obtain a more precise representation of user and items, that leads in turn to a more accurate generation of the recommendations. Moreover, as we will show in the next section, the use of knowledge graphs also allows to build effective *natural language explanations* supporting the recommendations, thus making these information sources as particularly relevant to design and implement semantics-aware content-based recommender systems.

## 3.3 User-Generated Content and Multimedia Features

In the last decade, we observed a number of works that use new types of side information, e.g., user-generated content (UGC) or metadata for recommendation.

With the emergence of the participatory Web (Web 2.0), various types of UGC became available, such as product reviews, user tags and forum discussions. Most of the early works that tried to leverage this information in the recommendation context focused on user-provided *tags*, which were used to enhance user and item profiles, or to explain recommendations to users [18, 88, 191]. In another stream of works, researchers focused on *user reviews* and tried to extract various types of information from them, including semantics or user opinions and sentiments, that can be used in the recommendation process [28]. More details are provided in Sect. 3.3.1.

On the other side, content-based recommender systems dealing with non-textual objects were commonly based on the use of metadata for representing descriptions. However, advances in audio, image and video analysis made it possible to represent multimedia objects by features that were extracted from the objects themselves. Those advanced representations could be effectively adopted to define advanced content-based recommender systems [42]. In Sect. 3.3.2, some pointers to the most relevant work concerning recommender systems leveraging multimedia content are presented and discussed.

### 3.3.1   Content-Based Recommender Systems Leveraging User-Generated Content

Recommender systems focused mostly on using user ratings or item metadata to generate recommendations, but another source of information that has been used to generate more relevant recommendations are *tags*. Tags are keywords that typically describe characteristics of the objects they are applied to, and can be made up of one or more words. Users are free to apply any type and any number of tags to an object, resulting in true bottom-up classification. Social tagging has been applied to many domains such as music, movies, books, and web sites [50, 55, 187, 199], where tags can be used as an additional resource to generate better recommendations. Tags have been effectively integrated into collaborative and content-based algorithms for item recommendations in several different ways. In collaborative filtering based on nearest neighbor algorithms, tags are exploited to aid in the calculation of the user/item similarities, or in some cases even replace the user-item rating matrix entirely, by relying on the less sparse user-tag and item-tag matrices to compute the similarities [91, 142, 187]. Tags have been also integrated into collaborative filtering algorithms based on latent factor models. In [108], the authors propose a tag-augmented version of matrix factorization, which integrates the latent factors of the item tags and ratings to provide a better approximation of the lower-rank user-item matrix. Tags have been also used to link different domains together for generating cross-domain recommendations [55]. Another possible way to include tags is in graph-based algorithms, which include tags as an additional node type and generate recommendations based on the resulting tripartite network, containing users, items and tags. As an example, the FolkRank algorithm runs personalized PageRank to assign a weight to each network node [80].

Tags are also suitable to be included in content-based recommender systems [21, 104]. Being free annotations, tags tend to suffer from syntactic problems, like polysemy and synonymy. Semantic content-based approaches have been proposed to address this problem. In [39], WordNet is adopted to perform Word Sense Disambiguation to content as well as tags, and a Naïve Bayes classifier is exploited to learn a probabilistic model of the user's disambiguated interests. This semantic user profile is then matched against the semantic item representations to locate the most relevant items for the active user.

However, the use of tags may be inadequate, especially when the target user has little historical data, or when the overall data sparsity level is high. A well consolidated approach is to leverage users' reviews for supporting their preferences and improve the recommendation process. User-generated reviews encapsulate rich semantic information such as the possible explanation of the users' preferences and the description of specific item attributes [28], thus they represent a rich source of information about the users' preferences, and can be exploited to build fine-grained user profiles. Therefore, a variety of review-based recommender systems have been proposed in the last years, which incorporate information extracted from user-generated textual reviews into the user modeling and recommending process. This may be beneficial to deal with the problem of data sparsity, cold-start for new users, and even in the case of dense data condition, may help to learn users' latent preference factors by considering the aspect opinions mentioned in the reviews [28]. In the survey by Chen et al. [28], various elements of valuable information that can be extracted from user reviews and that can be utilized by recommender systems have been identified. They range from quite simple information, such as frequently used terms, discussed topics, and overall opinions about reviewed items, to more complex information, such as specific opinions about item features, comparative opinions, reviewers' emotions, and reviews helpfulness.

Among the previous elements, opinions and sentiments expressed by users in their reviews about specific features or *aspects* of the reviewed items represent a promising approach to improve the recommendation process [62]. Aspect-based recommender systems leverage those opinions about aspects to provide improved personalized recommendations. Indeed, when items are evaluated with the same rating value, these systems are able to capture particular strengths and weaknesses of the items and, based on this information, better estimate their relevance for the target user [13, 132].

Aspect-based recommender systems include three main tasks, namely aspect extraction, to identify references to item aspects in user reviews, aspect polarity identification, to identify if the opinion on the aspects is positive, negative or neutral, and aspect-based recommendation, to exploit the extracted aspect opinion information to provide enhanced recommendations. In [78], a thorough investigation of the problem is presented, by separately addressing the three tasks.

Several methods for extracting opinions about items aspects have been proposed in literature [78]. They are classified as:

- vocabulary-based methods, that make use of lists of aspect words, as in [1]
- word frequency-based methods, in which words that have a high appearance frequency are selected as aspects. The methods to identify references to aspects in the reviews range from simple approaches based on the words frequently used in the reviews of a specific domain, to more complex ones based on language models [167], and on the comparison of the use of language when talking about a specific domain with respect to a general topic, in order to identify aspects mentioned in the reviews more often than usual [22]
- syntactic relation-based methods, where syntactic relations between words of a sentence are the basis for identifying aspect opinions, as in the Double Propagation algorithm in [153], which exploits syntactic relations between the words in a review to identify those that correspond to aspects
- topic model-based methods, where topic models are used to extract the main aspects from user reviews. Standard LDA models are modified so different generation distributions can focus on specific parts of the reviews [184, 185].

In the recommendation process aspects may be used in different ways. In [78], aspect-based recommender systems are categorized in approaches:

- enhancing item profiles with aspect opinion information, as in [48], where an item profile is composed of aspects with sentiment and popularity scores, and a case-based recommender matches the user's profile with items whose profiles are highly similar and produce greater sentiment improvements

- modeling latent user preferences on item aspects, as in [112], where a matrix factorization model incorporates hidden topics as a proxy for item aspects. The model aligns latent factors in rating data with latent factors in review texts
- deriving user preference weights from aspect opinions, as in [102], where the weight of an aspect in the user profile is determined by means of two factors, namely how much the user concerns about the aspect, and how much quality the user requires for such aspect. The value of concern is related to the frequency of comments of a user on specific aspects in his/her reviews, while the value of requirement increases when the user frequently rates an aspect lower than other users across different items
- incorporating aspect-level user preferences into recommendation methods, as in [132], where a multi-criteria user- and item-based collaborative filtering algorithm incorporating aspect opinion information is presented and evaluated. Similarity between users or items are computed according to the opinions expressed in the reviews. In [99], the authors propose to use latent multi-criteria ratings generated from user reviews to provide recommendations and capture latent complex heterogeneous user preferences. The latent rating generation process is based on two different models: the one-stage model, which utilizes document hashing to directly compute latent ratings, and the two-stage model, which utilizes a variational autoencoder to map user reviews into latent embeddings and subsequently compresses them into low-dimensional discrete vectors that constitute latent multi-criteria ratings.

### 3.3.2 Content-Based Recommender Systems Leveraging Multimedia Content

Traditional content-based recommender systems are usually fed by text, e.g. the description of a product, the plot of a movie, or the synopsis of a book [37, 104]. Even though audio or visual content are also associated with text to describe the items, they are usually not taken into account to generate recommendations, albeit they might have impact on user preferences.

Similarly to the architecture of classical content-based recommender systems dealing with textual content, a recommender leveraging multimedia content, such as audio or video, analyzes the items in order to represent them in a feature space, even though the pipeline for processing that kind of content is more complex.

One of the most popular tasks in which audio has been used is music recommendation, ranging from the classical task of recommending songs to a user based on her previous interests, to automatic playlist generation or continuation [19, 154].

Research in music recommender systems is becoming more and more interesting, also due to the increasing availability of music streaming services [168]. The challenge to recommend music relies on the fact that music tastes depends on a variety of factors, ranging from personality and emotional state of users [56], to contextual factors, such as weather conditions [20].

Content-based methods have shown to be useful when user feedback information is scarce, as in cold-start scenarios. As a source of content features to recommend music, social tags have been extensively used [92], even though features extracted from audio signals have also been used. The work in [6] presents a hybrid mood-aware music artist recommender system integrating both artists' and users' mood as well as audio features, such as timbre, tempo, loudness, and key confidence attributes to compute artists' pairwise similarities. This is a two-stage recommender system which identifies candidate artists based on the comparison of the mood of the user and the artist, and then re-ranks the list using artist similarity based on audio content of the artists' most popular songs. Deep learning approaches have been increasingly adopted for music recommendations, in particular in content-based systems which learn latent song or artist representations from the audio signal or from textual metadata. In [189], a CNN is adopted to represent each music item using 50-dimensional latent factors vectors, learned from log-compressed Mel spectrograms of music audio. The resulting latent factor representation of items is used together with latent user factors in a standard collaborative filtering fashion, and experiments on the Million Song Dataset show that this seems a viable method for recommending unpopular music.

Research in image recommendation includes approaches exploiting visual content extracted from images to recommend media items, e.g. paintings, or non-media items, e.g. clothes based on the appearance of photos. For example, the work in [2] describes a recommender system which uses as input the observed painting and generates a list of recommended paintings as output. Recommendations are generated using an algorithm resembling the PageRank strategy, which takes into account the past behavior of individual users, the overall behavior of the entire community of users, and intrinsic features of multimedia objects (low-level and semantic similarities). The past behavior of each individual users is given in terms of the browsing history of objects of that specific user, while the behavior of the whole community of users takes into account the browsing history of any user. The features of the multimedia objects have been used to compute their similarity, using low-level visual features, such as color, texture, and shape, and metadata such as painter, genre, and subject. The system has been effectively evaluated for recommending paintings in a virtual museum scenario containing digital reproductions of paintings in the Uffizi Gallery.

Recommending products leveraging visual content is adopted in several domains, such as fashion, food, and tourism [42]. In the fashion domain, McAuley et al. [113] propose a recommender system to match clothes with accessories by exploiting their images. The authors use freely available data collected from the Amazon, containing millions of relationships between a pool of objects, to identify alternative or complementary pair of products. The gist of the approach is to develop a fashion recommender able to model the human sense of relationship between objects by utilizing the visual appearance of products. More complex works model different levels of a user's preference for different parts of items. In [32], the authors learn a part-based user model based on different partitions of the image to obtain a personalized recommendation model. Experiments on a dataset from Amazon.com including images of helmets, t-shirts, and watches yielded improved results over existing textual or visual recommender systems that disregard appealing differences between parts of products. A similar approach is proposed in [67], where fine-grained facial attributes such as gender, race, eyebrow thickness, skin color, fatness, and hair color are extracted from a frontal face photo to recommend the best-fit eyeglasses using a probabilistic model.

In the food domain, multimodal information such as recipe images or ingredients are taken into account in the recommendation process. The results of a study presented in [49] indicate that preferred recipes could be predicted by leveraging low-level image features and recipe meta-data. Similarly in the tourism domain, visual features extracted from images shared by users are used to model their tastes for enhancing Points of Interests (POI) recommendation. In [195], the authors adopt a CNN pre-trained on ImageNet to extract visual features from images shared on Instagram and a probabilistic matrix factorization algorithm to model interactions between visual content, users and locations, with good results in coping with cold start as well.

Finally, visual characteristics of the content could be also adopted to recommend videos, such as movies, and most of the approaches analyze trailers, movie clips, or posters instead of the full movie. In [40], the authors designed a content-based movie recommender system which takes into account stylistic visual features to distinguish for example between comedy movies, usually made with a large variety of bright colors, and horror films using dark hues. To this purpose, stylistic visual features as shot length, color variation, lighting key, and motion vectors have been adopted, and results of the experiments have shown that low-level visual features provide better recommendations than the high-level features, such as genres.

The work presented in [41] combines audio and visual features with movie metadata, such as genre and cast, into a unique representation, called the Movie Genome, in order to deal with the new item problem. The authors proposed a novel recommendation model, called collaborative-filtering-enriched content-based filtering (CFeCBF), which exploits the collaborative knowledge of videos with interactions (warm items) to weight content information for videos without interactions (cold items). More details about multimedia recommender systems can be found in Chap. "Multimedia Recommender Systems: Algorithms and Challenges".

## 3.4  Transparency and Content-Based Explanations

The importance of content-based information has been largely discussed throughout this chapter. Moreover, a recent and interesting trend that further confirms the effectiveness of such features lies in their exploitation to *explain* or *justify* a recommendation.

As already discussed in Chap. "Beyond Explaining Single Item Recommendations", the idea of providing intelligent information systems with *explanation* facilities was studied from the early 1990s [89], and gained again attention in the light of the recent *General Data Protection Regulation (GDPR)*,[4] which emphasized and regulated the *users' right to explanation* [64] when people face machine learning-based (or, generally speaking, artificial intelligence-based) systems.

This a very relevant problem, since intelligent systems are becoming more and more important in our everyday lives. Accordingly, it is fundamental that the *internal mechanisms* that guide these algorithms are as clear as possible. The need for *transparent* algorithms is even more felt for RS since, as shown by Sinha et al. [175], the more the *transparency* of the algorithm, the more the *trust* the user puts in the system. Similarly, Cramer et al. [36] proved the relationship between the transparency of a RS and users' acceptance of the recommendations.

The interest of the community in the topic is also confirmed by the spread of several research discussing the positive impact of explanations for recommender systems [63, 93, 115]. In this section we will focus our attention on *content-based features*, thus we will provide an overview of recent approaches to *explain* or *justify* a recommendation based on content.

According to the taxonomy of explanation strategies in RSs provided by Friedrich et al. [58], approaches to generate explanations and justifications can be split into two categories: *white box* methodologies, which generate an explanation which is directly connected to the underlying explanation method and *black box* methodologies, where the explanation strategy is not aware (and is independent) of the underlying recommendation model which is used to generate the suggestions. In the following, we will refer to the first as *explainable recommendations* strategies, while the latter are referred to as *post hoc explanation* strategies.

### 3.4.1 Generating Explainable Content-Based Recommendations

The idea behind *explainable recommendation strategies* is to encode content-based features in the recommendation model and to exploit them to generate a natural language explanation that supports the recommendation. One of the early attempts in the area is presented in [182], where the authors proposed a model to generate explanations which exploits the overlap between the features of the user profile and those describing the suggestion.

More recently, several methods were devoted to the explanation of the recommendation coming from matrix factorization techniques. As an example, in [207], the authors propose an approach based on the exploitation of users' reviews, since they extract explicit product features and then aligns each latent dimension in order to explain recommendation coming from matrix factorization techniques.

---

[4] http://ec.europa.eu/justice/data-protection/reform/files/regulation_oj_en.pdf.

However, due to the recent spread of deep learning methods [97], most of the recent research effort focuses on designing *explainable* recommendation methods based on deep neural networks. As an example, in [29] the authors propose NARRE (Neural Attentional Regression model with Review-level Explanations). The core of the approach is an attention mechanism that catches the usefulness of reviews. Such information is used to: (1) predict the interest of the user towards the item and (2) predict the usefulness of each review, simultaneously. Therefore, the reviews labeled as highly-useful are exploited to provide the user with review-level explanations. Similarly, the use of reviews is also investigated in [11], where an approach based on deep neural networks that quantifies the relationship between aspects and reviews is proposed. In particular, the authors build a user-aspect bipartite relation as a bipartite graph. Next, by using dense sub-graph extraction and ranking-based technique an *explainable recommendation* is returned. Finally, in [75] the authors exploited a tripartite modeling of user-item-aspect tuples and used graph-based ranking to find the most relevant aspects of a user that match with relevant aspects of places. These aspects can be used to both drive the recommendation and explanation process.

Finally, several work propose the adoption of *attention mechanisms* as a source to *explain* recommendation. As an example, in [30] the authors propose an Attention-driven Factor Model (AFM), which learns and tunes the user attention distribution over features. This is used to drive both the recommendation and the explanation process, since the features with the highest attention can be used as explanation. Similarly, a method to jointly optimize matrix factorization and attention-based GRU network is proposed in [171]. In this case, the matrix component is used to drive the recommendation while the attention-based mechanism is used to generate a suitable explanation. However, it should be pointed out that some literature [83] argued that *attention* and *explanation* refer to two different concepts, thus the effectiveness of attention modules for explanation purposes tends to be overestimated in current research.

### 3.4.2 Generating Post Hoc Content-Based Explanations

Differently from *explainable* recommendation methods, that directly encode content-based features in the model and use these features to explain a suggestions, *post hoc* explanations strategies generate an explanation *after* the recommendation process, thus they are completely *independent* from the underlying recommender system.

One of the first attempts is discussed in [191], where the authors used *tags* to generate explanations. As shown in [151], these strategies allow to maintain a good predictive accuracy whilst yielding content-based explanations. In this research line, it is also possible to mention the work proposed in [206], where the authors extract causal rules from user history to provide personalized, item-level, post hoc explanations. In this case, the causal explanations are extracted through a perturbation model and a causal rule learning model. Recently, several work tried to exploit *topic models* to generate content-based explanations independent

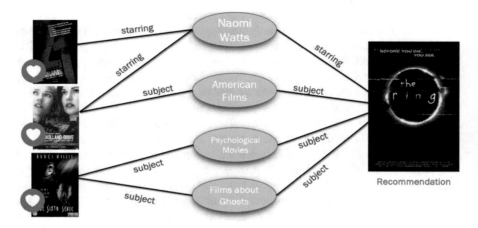

**Fig. 6** Graph-based data model to generate explanations

from the recommendation model: as an example, in [161] the authors collect from Freebase textual data about movies and design a method to map latent factors to the topic emerging from textual content.

Moreover, one of the most promising research direction lies in the extraction of content-based information from open knowledge sources and knowledge graphs. As an example, EXPLOD [130] is a framework which exploits the information available in DBpedia [17] cloud to generate a *natural language explanation*. The methodology is based on a graph in which the items liked by a user are connected to the items recommended through the properties available in DBpedia. Basically, this graph connects the items the user liked and those in her recommendation list through the values of the properties describing those items in the knowledge graph. An example of such a data model is reported in Fig. 6. Given such a data module, the explanation is based on a natural language *template* which is filled in by exploiting the *top-K* overlapping properties.

As an example, by referring to the *toy* example reported in Fig. 6, the output of the framework would have been '*I suggest you The Ring, because you often watch movies starred by Naomi Watts and you have also liked psychological movies, such as The Sixth Sense*'. Similar attempts, based on the exploitation of structured features gathered from knowledge graphs are also presented in [5] and in [107], where the authors specifically designed a strategy for the tourism domain.

As shown by the authors, these explanations significantly overcome other simple strategies to generate content-based explanations [138]. These findings are confirmed in a recent work [174], where several strategies to generate post hoc explanations are compared.

Beyond features gathered from knowledge graphs, other attempts exploited review-based features for explanation purposes. In this research line, Muhammad et al. [123] introduced the concept of *opinionated explanations*, that is to say, explanations mined from user-generated reviews. To this end, the authors identify relevant aspects of the items (e.g., bar, service, parking, etc.), and present the most relevant ones. In this case, the authors did not justify the *reasons* behind the choice of highlighting a specific characteristic. In other terms, this approach lets the users be informed about the main characteristics of the items, but does not *explain* why , *e.g.,* the bar or the service are particularly good.

Differently from the approaches in which aspects in the reviews are manually extracted, in [137], the authors propose a framework to automatically generate review-based justifications of the recommendations. In particular, the authors first run Part-of-Speech (POS) tagging algorithm over the set of the reviews to obtain representative concepts. Next, they implement ranking strategies to identify the most relevant aspects that best describe and characterize the item. Finally, excerpts of users' reviews discussing those aspects with a positive sentiment are dynamically combined to fill in a natural language template in the GENERATION module. This represents the final output of the algorithm, which is provided to the user as *justification* of the recommendation she received.

By referring again to the previous toy example concerning 'The Ring', a likely justification would have been: *I recommend you The Ring because people who liked the movie think that the plot delivers some bone-chilling terror. Moreover, people liked The Ring since the casting is pretty good.*[5] As shown by the authors in their research [141], users tended to prefer review-based explanation to feature-based explanation since the identification of relevant reviews excerpts allowed them to discover new information about the recommended items. Further extensions of this framework have been presented in [139], where the authors adopt text summarization techniques to automatically generate a summary of relevant reviews excerpts, and in [140], where a context-aware extension of the framework, which aims to differentiate the justification based on the different *context of consumption* of the items, is discussed.

## 3.5 Exploiting Content for Conversational Recommender Systems

Conversational Recommender Systems (CRS) are gaining more and more attention in the last few years. This renewed interest probably derives from the massive diffusion of Digital Assistants, such as Amazon Alexa, Siri, Google Assistant, that allow users to execute a wide range of actions through messages in natural language. In fact, the main distinguishing aspect of a CRS, compared to a canonical recommender system, is its capability of interacting with the user during the recommendation process [109] by means of a multi-turn dialog [87]. CRSs make the

---

[5] In this case, we assume that the aspect extraction module would have identified plot and casting as hallmarks of the movie.

interaction more efficient and natural [198], as they are able to help users navigate complex product spaces by iteratively selecting items for recommendation [176].

During the interaction with a recommender system (canonical or CRS), the following four main phases can usually be identified: preference acquisition, recommendation generation, explanation, and user feedback acquisition. Of course, some steps may be optional. For example, explanations are not provided by all recommender systems. The peculiarity of a CRS lies in the fact that the acquisition of user preferences and needs becomes crucial. Indeed, users do not provide the system with all the preferences in one step. Conversely, they usually provide and refine their feedback during the dialog.

In this context, there are two main scenarios:

1. the user starts the dialog by providing characteristics or features that the ideal item should have [146];
2. the system asks questions on constraints of features and the user answers [27, 186]. These two scenarios can be combined during the dialog, of course. For example, the user provides some initial criteria, then receives an initial set of recommendations, and starts to revise the initial preferences since recommendations do not meet the *ideal* item. This is a typical situation in *critiquing-based* approaches, where the user can add new constraints (*tightening* [158]) or *relax* some others [84, 157]. Further details on the recommendation process in an interactive recommender system are analyzed in [74].

From an architectural perspective, a fundamental distinction can be made between systems that implement a *dialog state tracker* and systems that implement an *end-to-end* architecture [47]. The former require a fine grained definition of internal dialog states and precisely defined user intents. As a consequence, they can hardly scale to large domains characterized by dialogs with high variability in terms of language. This kind of systems usually implement a *pipeline-based* architecture made up of a series of modules, each one with its own specific function [47, 101, 202, 210]. Conversely, end-to-end dialog systems do not rely on explicit internal states, thus they do not need state tracking modules. They combine components that are trained on conversational data and that handle all the steps, from understanding the user message to generating a response.

The next section analyzes the impact of content on these two types of approaches and, more generally, on CRSs. As we will see, the role of textual content is different on the ground of the implemented solution.

### 3.5.1   Dialog State-Based CRS

A CRS that explicitly manages the dialog state usually consists of the following modules: *Intent Recognizer, Dialog Manager, Recommender System*, some NLP modules such as an *Entity Recognizer* and a *Sentiment Analyzer*. An example of this modular architecture is implemented in the ConveRSE framework [82] and is shown in Fig. 7.

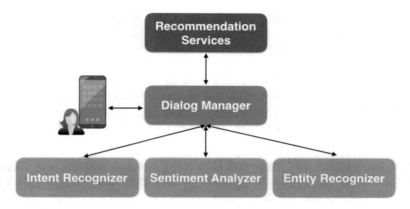

**Fig. 7** An example of modular architecture for a CRS

**Dialog Manager** This is the core component of the architecture since its responsibility is to supervise the whole recommendation process. The *Dialog Manager* (DM) is the component that keeps track of the *dialog state*. It can be viewed as the orchestrator of the system, and strictly depends on the task the dialog agent is dealing with. The DM receives the user message, invokes the components needed for answering to the user request, and returns the message to be shown to the user. When the information for fulfilling the user request is available, the DM returns the message to the client.

According to Williams et al. [203] algorithms for dialog-state-tracking can be based on *hand-crafted rules, generative models,* and *discriminative models.*

Approaches based on *hand-crafted rules* require no training data to be implemented, and this represents a huge advantage in some contexts. Furthermore, designers can incorporate domain knowledge through rule definition. The main weakness of these models is that they consider a single hypothesis for the dialog state. Actually, multiple competing dialog states can occur at any given stage of the conversation, and the choice of the right dialog state represents a key aspect. In fact, the actions the system should take depend on the dialog state. Accordingly, more recent approaches assign a score to each dialog state [193]. However, also the computation of this score requires parameter tuning, which cannot be based on dialog data. This represents an additional limitation for this kind of approaches.

*Generative approaches* assume that the dialog can be modeled through a Bayesian network. These models make independence assumptions (which generally are invalid) in order to reduce the complexity. For example, they can assume that user errors are equally distributed, while some errors are much more frequent than others [200]. Parameters of generative models are often hand-crafted. Usually, the next dialog state depends on the previous state, but some authors introduce terms that accumulate dialog history [201], or terms that express common-sense relationships between user actions [43].

*Discriminative approaches* directly model the distribution over the dialog state, given arbitrary and possibly correlated input features [200]. Discriminative approaches can exploit machine-learning algorithms (e.g., maximum entropy linear classifiers, neural networks, web-style ranking models). These models are able to integrate a large number of features and have the goal of optimizing the predictive accuracy. Most of the approaches encode the dialog history in the feature in order to learn a classifier [200]. Handerson et al. [76] apply a deep neural network as classifier. Some other approaches adopt sequential models such as Markov models [156].

**Intent Recognizer** This component is strictly connected to the Dialog Manager and has the goal of defining the intent of the user from the utterances formulated through a natural language sentence. When the user sends a message, the recommender system must first understand what is the goal of the user and what she wants to express and get by that message. Hence, as first step, the CRS needs to identify the intent of the user.

Four main intents can be identified in a CRS, each one corresponding to a specific step in the recommendation process:

- *preference:* the user is providing a preference on an item or on a feature
- *recommendation:* the user is asking to receive a recommendation
- *critiquing:* the user is providing a feedback on the received recommendation
- *explanation:* the user is asking to receive an explanation on a recommendation.

The Intent Recognizer usually requires a set of utterances for training each intent. From this perspective, intent recognition can be viewed as a text classification task.

**Entity Recognizer** The aim of the *Entity Recognizer* (ER) module is to find relevant entities mentioned in the user sentence, and to link them to the correct concept in a Knowledge Base (KB).

A classical entity linking approach consists of two steps: *spotting* and *disambiguation*.

The *spotting* step analyzes the text in order to discover candidate entities. Specifically, the algorithm detects a sequence of words (*surface form*) that represent an entity and retrieves all the concepts that can be associated to that surface form.
The *disambiguation* step consists in selecting the correct concept for each surface form.

As an example, let us suppose the user writes the following sentence: "I like Michael Jackson and Beat It". "Beat It" is an ambiguous entity since it might be referred to the candidate concept *Beat It by Michael Jackson* or to *Beat It by Sean Kingston*. Hence, this surface form needs to be disambiguated. In order to accomplish this task, a possible strategy the ER might implement is based on the computation of the similarity between the candidate entities (i.e., *Beat It by Michael Jackson* and *Beat It by Sean Kingston* and the other entities in the context (i.e., *Michael Jackson*). In this example, the similarity between *Michael Jackson* and *Beat It by Michael Jackson* will be higher than the one between *Michael Jackson* and *Beat*

*It by Sean Kingston.* Therefore, *Beat It by Michael Jackson* will be chosen. More details on this ER implementation can be found in [144].

**Sentiment Analyzer** The *Sentiment Analyzer* (SA) has the goal of assigning the right sentiment polarity (i.e., positive and negative) to the entities mentioned in the sentences (e.g., singer, song, genre) and identified by the ER module. The difficulty lies in the fact that the same sentence can contain different sentiment tags (even opposite) as well as different entities. For example, given the sentence "I like Rocket Man, but I don't like Gus Dudgeon", the SA identifies a positive sentiment (i.e., like) and a negative one (i.e., don't like). SA should assign the positive sentiment to the entity *Rocket Man* and the negative sentiment to the entity *Gus Dudgeon*.

**Recommendation Services** This component collects the services strictly related to the recommendation process. The *recommendation algorithm* is invoked by the DM when all the information required for generating the recommendations are available.

Henceforth, we will analyze some relevant approaches by highlighting the dialog state they are focused on and the exploited content. As regards the *preference acquisition*, it can be easily imagined that preferences are mostly expressed by the user during the dialog in terms of characteristics that an ideal item should have (e.g., "I would like to have dinner in a restaurant with a beautiful sea view"). Actually, a CRS could also acquire the preferences by asking feedback on individual items [45]. However, preferences expressed in terms of item facets are more interesting from our point of view since content plays a crucial role in that case. This modality of preference acquisition is also known as *slot filling* [87] and it establishes a straightforward relation between CRSs and content-based recommender systems. Indeed, while a preference given on a specific item (e.g., "I like the movie American Beauty") can also be exploited by a collaborative recommendation algorithm, preferences in terms of item facets necessarily need that the recommendation algorithm is able to exploit this information for generating recommendations. From this perspective, it also emerges the similarity between CRSs and constraint-based recommender systems as well as knowledge-based ones. In fact, during the dialog with the user, several CRSs perform *filtering* and *re-ranking* of the items through constraint-based techniques [52]. For example, the knowledge-based recommender environment named CWAdvisor [53] proposes a set of questions to the user, each one associated to set of predefined answers. The goal of this step is to acquire customer properties and constraints, and matching them with product properties. In this way the system is also able to *explain* the recommendation by providing the user with the reasons why a given product suits the customer needs and wishes. In a CRS powered by a content-based recommendation algorithm, the idea is basically the same, with the difference that preferences are acquired through a multi-turn dialog, as can be seen in Adaptive Place Adivsor (APA) [183]. The interaction in APA takes the form of a sequence of questions whose goal is to discard items not interesting for the user. Hence, the goal of the dialog is to acquire attribute-value specifications, such as *cuisine = Chinese*. APA ends the search of the *ideal* item for the user when a small number of items match the constraints and are highly similar. Still on the subject of the adopted *recommendation strategy*, Argal et al. [7] propose a hybrid recommendation algorithm that combines collaborative and content-based paradigms in a CRS. The content-based algorithm is exploited when the system does not register any user activity (cold-start stage), otherwise collaborative recommendations are provided. The content-based algorithm performs a matching of user preferences against product representations stored in Elasticsearch.[6] Recently, Li et al. [98] focused their attention on the *critiquing* step and proposed a latent-linear-critiquing model for CRSs. A user is iteratively provided with item recommendations and attribute descriptions for those items. The user can accept the recommendation or can critique an attribute to generate a new recommendation. The model exploits preferences implicitly revealed from user reviews. User critiques are then transformed in an embeddable term-frequency representation that can be co-embedded with user preferences. Item descriptions are keyphrases from user reviews.

### 3.5.2 End-to-End Systems

End-to-end systems have become popular thanks to the spread of Deep Learning techniques [47]. End-to-end systems are a promising solution because they do not require the development of specific and complex rules for managing the dialog and its states, which makes them easier to port to a new domain.

However, there are still some challenges to face in order to enable them to handle goal-oriented conversations [47]. One of the main obstacles is the lack of datasets for training deep-learning models. Indeed, in this case content plays an *unusual* role compared to the one we are accustomed to observe in a recommender system. Content, in the form of recorded dialogs, is essential for learning at least the two actions of providing preferences and generating recommendations [87]. Recently, three datasets that contain utterances in the movie, book, and music domains have been released.[7] These datasets consist of real dialogs between users and a CRS, collected during an experimental session with the CRS [81]. However, the number of utterances (5,318 movie, 1,862 book, 2,096 music) is probably not enough for learning end-to-end components.

Ritter et al. [159] proposed one of the first approaches which adopts an end-to-end architecture for conversational agents. In that work, the authors defined a data-driven method to generate responses to open-domain linguistic stimuli, based on phrase-based Statistical Machine Translation (SMT). They demonstrated that SMT techniques perform better than information retrieval approaches on the task of response generation. Recently, Jannach and Manzoor [85] analyzed the utterances generated by two novel end-to-end models for CRSs. More specifically, they compared *DeepCRS* [100] and *KBRD* [31]. DeepCRS has a set of sub-components

---

[6] https://www.elastic.co/elasticsearch//.

[7] https://github.com/aiovine/converse-dataset.

for sentence encoding, next-utterance prediction, sentiment classification, and rec-ommendation. KBRD has a dialog system based on a transformer-based sequence-to-sequence module and a knowledge graph which provides knowledge about the domain. A switching network connects the two modules. The comparative analysis between KBRD and DeepCRS showed that one third of the system utterances were not meaningful for each system in the given context, and less than two third of the recommendations were meaningful. As a result, the quality of responses and recommendations of end-to-end CRSs showed ample room for improvement at this stage. Both systems have been trained on the ReDial dataset[8] that is annotated dataset consisting of 10,006 dialogues, 182,150 utterances, 956 users, and 51,699 movie mentions. In each conversation users recommend movies to each other and the number of movies mentioned varies. The dataset has been developed by Amazon Mechanical Turk. In each dialog there are two roles: the recommendation seeker and the recommender. The movie seeker has to explain what kind of movie she likes, and asks for movie recommendation. The recommender tries to understand the seeker tastes, and recommends movies. All exchanges of information and recommendations are made using natural language.

# 4   Discussion and Future Outlook

This chapter covered the recent advances in the area of *semantics-aware content-based recommendations*. As we have shown, most of the current work can be split into three main research lines: (1) techniques investigating new methods for *representing* content-based features; (2) techniques investigating new *sources* to gather content-based features; (3) new *use cases* for content-based features.

In the first group, we have discussed work based on *word embedding techniques*. The use of vector space representations, together with deep architectures, is probably the most popular research direction in the area. In this sense, the adoption of *contextual word representations* (discussed in Sect. 3.1.2), such as BERT, ELMo and so on, is a research direction particularly worthy of attention in the next years. Indeed, these models proved to overcome state-of-the-art techniques on several NLP tasks, thus they are gaining more and more attention in recommendation scenarios as well. In particular, their ability to model the precise meaning of words and sentences can be useful to develop more accurate recommendation strategies.

As for the sources of information, in Sect. 3.2 we introduced approaches based on the exploitation of *knowledge graphs* and techniques based on *user-generated content* and *multimedia features*. In this research line, we expect to see more and more work trying to merge heterogeneous groups of features through deep learning architectures. As an example, we consider as very promising the adoption of *graph embedding techniques*, since they learn a vector space representation that allows to

---

[8] https://redialdata.github.io/website/.

merge word embeddings with the information gathered from knowledge graphs. In Sect. 3.2 we already discussed some work investigating this intuition, but we expect an increase of the research trying to put together collaborative information (e.g., ratings), content-based information (e.g., review data, descriptive text, multimedia features, if any) and structured features available in *knowledge graphs*. In this sense, research investigating the *best* strategies to combine these features is still at an early stage.

Finally, as new use cases we focused our attention on the usage of content to generate *natural language explanations* and to design *conversational recommender systems*. In this case, we emphasize the importance of explanation facilities and more elaborated interaction models in recommender systems research. On the one side, explanations are fundamental to *open* black-box recommendation models, to make users *aware* of the underlying algorithms as well as to further increase their trust. On the other side, more *natural* interaction strategies, based on conversational recommender systems and dialog mechanisms, can be helpful to take the final step towards the adoption of recommendation algorithms in very *sensitive* domains, such as medicine and finance. In this sense, we expect a significant effort in developing personalized conversational agents for both those domains.

## 5    Conclusions

All the literature we discussed throughout the chapter gave evidence of the importance of both content-based features and semantic representations. Indeed, the usage of content can help to design and develop more accurate recommendation models in a broad range of domains and applications, as shown in this chapter, since it is able to effectively tackle typical problems of *pure* collaborative models, such as cold start, new item problems and *opacity* of the recommendation models. We think that content-based information and semantics-aware representation strategies will play an increasingly central role in recommender systems research in the next years.

It is our hope that this chapter may stimulate the research community to adopt and effectively integrate the discussed techniques in several recommendation scenarios in order to foster future innovations in the area of content-based recommender systems.

## References

1. S. Aciar, D. Zhang, S.J. Simoff, J.K. Debenham, Informed recommender: basing recommendations on consumer product reviews. IEEE Intell. Syst. **22**(3), 39–47 (2007)
2. M. Albanese, A. d'Acierno, V. Moscato, F. Persia, A. Picariello, A multimedia recommender system. ACM Trans. Internet Techn. **13**(1), 3:1–3:32 (2013)

3. G. Alexandridis, T. Tagaris, G. Siolas, A. Stafylopatis, From free-text user reviews to product recommendation using paragraph vectors and matrix factorization, in *Companion Proceedings of the 2019 World Wide Web Conference* (2019), pp. 335–343

4. A. Almahairi, K. Kastner, K. Cho, A. Courville, Learning distributed representations from reviews for collaborative filtering, in *Proceedings of the 9th ACM Conference on Recommender Systems* (2015), pp. 147–154

5. M. Alshammari, O. Nasraoui, S. Sanders, Mining semantic knowledge graphs to add explainability to black box recommender systems. IEEE Access **7**, 110563–110579 (2019)

6. I. Andjelkovic, D. Parra, J. O'Donovan, Moodplay: interactive music recommendation based on artists' mood similarity. Int. J. Hum. Comput. Stud. **121**, 142–159 (2019)

7. A. Argal, S. Gupta, A. Modi, P. Pandey, S. Shim, C. Choo, Intelligent travel chatbot for predictive recommendation in echo platform, in *2018 IEEE 8th Annual Computing and Communication Workshop and Conference (CCWC)* (IEEE, Piscataway, 2018), pp. 176–183

8. S. Auer, C. Bizer, G. Kobilarov, J. Lehmann, R. Cyganiak, Z. Ives, *DBpedia: A Nucleus for a Web of Open Data* (Springer, Berlin, 2007)

9. R. Baeza-Yates, B. Ribeiro-Neto, *Modern Information Retrieval*, vol. 463 (ACM Press New York, 1999)

10. T. Bansal, D. Belanger, A. McCallum, Ask the GRU: multi-task learning for deep text recommendations, in *Proceedings of the 10th ACM Conference on Recommender Systems* (2016), pp. 107–114

11. R. Baral, X. Zhu, S. Iyengar, T. Li, ReEL: review-aware explanation of location recommendation, in *Proceedings of the 26th Conference on User Modeling, Adaptation and Personalization* (ACM, New York, 2018), pp. 23–32

12. O. Barkan, N. Koenigstein, Item2vec: neural item embedding for collaborative filtering, in *2016 IEEE 26th International Workshop on Machine Learning for Signal Processing (MLSP)* (IEEE, Piscataway, 2016), pp. 1–6

13. K. Bauman, B. Liu, A. Tuzhilin, Aspect based recommendations: recommending items with the most valuable aspects based on user reviews, in *Proceedings of the 23rd ACM SIGKDD International Conference on Knowledge Discovery and Data Mining, Halifax, NS, Canada, August 13–17, 2017* (ACM, New York, 2017), pp. 717–725

14. M. Belkin, P. Niyogi, Laplacian eigenmaps for dimensionality reduction and data representation. Neural Comput. **15**(6), 1373–1396 (2003)

15. V. Bellini, T. Di Noia, E. Di Sciascio, A. Schiavone, Semantics-aware autoencoder. IEEE Access **7**, 166122–166137 (2019)

16. I. Beltagy, K. Lo, A. Cohan, Scibert: a pretrained language model for scientific text (2019). arXiv:1903.10676

17. C. Bizer, J. Lehmann, G. Kobilarov, S. Auer, C. Becker, R. Cyganiak, S. Hellmann, DBpedia-a crystallization point for the Web of Data. J. Web Semantics **7**(3), 154–165 (2009)

18. T. Bogers, Tag-based recommendation, in *Social Information Access - Systems and Technologies*. Lecture Notes in Computer Science, vol. 10100 (Springer, Berlin, 2018), pp. 441–479

19. G. Bonnin, D. Jannach, Automated generation of music playlists: survey and experiments. ACM Comput. Surv. **47**(2), 26:1–26:35 (2014)

20. M. Braunhofer, M. Kaminskas, F. Ricci, Location-aware music recommendation. Int. J. Multim. Inf. Retr. **2**(1), 31–44 (2013)

21. I. Cantador, A. Bellogín, D. Vallet, Content-based recommendation in social tagging systems, in *Proceedings of the 2010 ACM Conference on Recommender Systems, RecSys 2010, Barcelona, Spain, September 26–30, 2010*, ed. by X. Amatriain, M. Torrens, P. Resnick, M. Zanker (ACM, New York, 2010), pp. 237–240

22. A. Caputo, P. Basile, M. de Gemmis, P. Lops, G. Semeraro, G. Rossiello, SABRE: a sentiment aspect-based retrieval engine, in *Information Filtering and Retrieval*, ed. by C. Lai, A. Giuliani, G. Semeraro. Studies in Computational Intelligence, vol. 668 (Springer, Berlin, 2017), pp. 63–78

23. H. Caselles-Dupré, F. Lesaint, J. Royo-Letelier, Word2vec applied to recommendation: hyperparameters matter, in *Proceedings of the 12th ACM Conference on Recommender Systems* (2018), pp. 352–356

24. G. Cenikj, S. Gievska, Boosting recommender systems with advanced embedding models, in *Companion Proceedings of the Web Conference 2020, WWW'20* (Association for Computing Machinery, New York, 2020), pp. 385–389
25. D. Cer, Y. Yang, S.-Y. Kong, N. Hua, N. Limtiaco, R.S. John, N. Constant, M. Guajardo-Cespedes, S. Yuan, C. Tar, et al., Universal sentence encoder (2018). arXiv:1803.11175
26. B.P. Chamberlain, E. Rossi, D. Shiebler, S. Sedhain, M.M. Bronstein, Tuning word2vec for large scale recommendation systems, In *Fourteenth ACM Conference on Recommender Systems* (2020), pp. 732–737
27. L. Chen, P. Pu, Preference-based organization interfaces: aiding user critiques in recommender systems, in *International Conference on User Modeling* (Springer, Berlin, 2007), pp. 77–86
28. L. Chen, G. Chen, F. Wang, Recommender systems based on user reviews: the state of the art. User Model. User-Adap. Inter. **25**(2), 99–154 (2015)
29. C. Chen, M. Zhang, Y. Liu, S. Ma, Neural attentional rating regression with review-level explanations, in *Proceedings of the 2018 World Wide Web Conference*. International World Wide Web Conferences Steering Committee (2018), pp. 1583–1592
30. J. Chen, F. Zhuang, X. Hong, X. Ao, X. Xie, Q. He, Attention-driven factor model for explainable personalized recommendation, in *The 41st International ACM SIGIR Conference on Research & Development in Information Retrieval* (2018), pp. 909–912
31. Q. Chen, J. Lin, Y. Zhang, M. Ding, Y. Cen, H. Yang, J. Tang, Towards knowledge-based recommender dialog system, in *Proceedings of the 2019 Conference on Empirical Methods in Natural Language Processing and the 9th International Joint Conference on Natural Language Processing, EMNLP-IJCNLP 2019, Hong Kong, China, November 3–7, 2019*, ed. by K. Inui, J. Jiang, V. Ng, X. Wan (Association for Computational Linguistics, Stroudsburg, 2019)
32. H. Chi, C. Chen, W. Cheng, M. Chen, UbiShop: commercial item recommendation using visual part-based object representation. Multim. Tools Appl. **75**(23), 16093–16115 (2016)
33. N. Chinchor, P. Robinson, MUC-7 named entity task definition, in *Proceedings of the 7th Conference on Message Understanding*, vol. 29 (1997), pp. 1–21
34. A. Collins, J. Beel, Document embeddings vs. keyphrases vs. terms: an online evaluation in digital library recommender systems (2019). arXiv:1905.11244
35. A. Conneau, D. Kiela, H. Schwenk, L. Barrault, A. Bordes, Supervised learning of universal sentence representations from natural language inference data (2017). arXiv:1705.02364
36. H. Cramer, V. Evers, S. Ramlal, M. Van Someren, L. Rutledge, N. Stash, L. Aroyo, B. Wielinga, The effects of transparency on trust in and acceptance of a content-based art recommender. User Model. User-Adapt. Interact. **18**(5), 455–496 (2008)
37. M. de Gemmis, P. Lops, C. Musto, F. Narducci, G. Semeraro, Semantics-aware content-based recommender systems, in *Recommender Systems Handbook*, ed. by F. Ricci, L. Rokach, B. Shapira (Springer, Berlin, 2015), pp. 119–159
38. S. Deerwester, S.T. Dumais, G.W. Furnas, T.K. Landauer, R. Harshman, Indexing by latent semantic analysis. J. Am. Soc. Inf. Sci. **41**(6), 391–407 (1990)
39. M. Degemmis, P. Lops, G. Semeraro, P. Basile, Integrating tags in a semantic content-based recommender, in *Proceedings of the 2008 ACM Conference on Recommender Systems, RecSys 2008, Lausanne, Switzerland, October 23–25, 2008*, ed. by P. Pu, D.G. Bridge, B. Mobasher, F. Ricci (ACM, New York, 2008), pp. 163–170
40. Y. Deldjoo, M. Elahi, P. Cremonesi, F. Garzotto, P. Piazzolla, M. Quadrana, Content-based video recommendation system based on stylistic visual features. J. Data Semant. **5**(2), 99–113 (2016)
41. Y. Deldjoo, M.F. Dacrema, M.G. Constantin, H. Eghbal-zadeh, S. Cereda, M. Schedl, B. Ionescu, P. Cremonesi, Movie genome: alleviating new item cold start in movie recommendation. User Model. User Adapt. Interact. **29**(2), 291–343 (2019)
42. Y. Deldjoo, M. Schedl, P. Cremonesi, G. Pasi, Recommender systems leveraging multimedia content. ACM Comput. Surv. **53**(5), 106:1–106:38 (2020)

43. D. DeVault, M. Stone, Managing ambiguities across utterances in dialogue, in *Proceedings of the 11th Workshop on the Semantics and Pragmatics of Dialogue (Decalog 2007)* (2007), pp. 49–56
44. J. Devlin, M.-W. Chang, K. Lee, K. Toutanova, BERT: pre-training of deep bidirectional transformers for language understanding (2018). arXiv:1810.04805
45. L.W. Dietz, S. Myftija, W. Wörndl, Designing a conversational travel recommender system based on data-driven destination characterization, in *ACM RecSys Workshop on Recommenders in Tourism* (2019), pp. 17–21
46. T. Di Noia, R. Mirizzi, V.C. Ostuni, D. Romito, Exploiting the web of data in model-based Recommender Systems, in *Proceedings of the Sixth ACM conference on Recommender Systems* (ACM, New York, 2012), pp. 253–256
47. J. Dodge, A. Gane, X. Zhang, A. Bordes, S. Chopra, A. Miller, A. Szlam, J. Weston, Evaluating prerequisite qualities for learning end-to-end dialog systems (2015). arXiv:1511.06931
48. R. Dong, M. Schaal, M.P. O'Mahony, K. McCarthy, B. Smyth, Opinionated product recommendation, in *Case-Based Reasoning Research and Development - 21st International Conference, ICCBR 2013, Saratoga Springs, NY, USA, July 8–11, 2013. Proceedings*, ed. by S.J. Delany, S. Ontañón. Lecture Notes in Computer Science, vol. 7969 (Springer, Berlin, 2013), pp. 44–58
49. D. Elsweiler, C. Trattner, M. Harvey, Exploiting food choice biases for healthier recipe recommendation, in *Proceedings of the 40th International ACM SIGIR Conference on Research and Development in Information Retrieval, Shinjuku, Tokyo, Japan, August 7–11, 2017*, ed. by N. Kando, T. Sakai, H. Joho, H. Li, A.P. de Vries, R.W. White (ACM, New York, 2017), pp. 575–584
50. M. Enrich, M. Braunhofer, F. Ricci, Cold-start management with cross-domain collaborative filtering and tags, in *E-Commerce and Web Technologies - 14th International Conference, EC-Web 2013, Prague, Czech Republic, August 27–28, 2013. Proceedings*, ed. by C. Huemer, P. Lops. Lecture Notes in Business Information Processing, vol. 152 (Springer, Berlin, 2013), pp. 101–112
51. K. Ethayarajh, How contextual are contextualized word representations? Comparing the geometry of BERT, ELMo, and GPT-2 embeddings, in *Proceedings of the 2019 Conference on Empirical Methods in Natural Language Processing and the 9th International Joint Conference on Natural Language Processing (EMNLP-IJCNLP), Hong Kong, China, Nov. 2019* (Association for Computational Linguistics, Stroudsburg, 2019), pp. 55–65
52. A. Felfernig, G. Friedrich, D. Jannach, M. Zanker, Constraint-based recommender systems, in *Recommender Systems Handbook* (Springer, Berlin, 2015), pp. 161–190
53. A. Felfernig, G. Friedrich, D. Jannach, M. Zanker, An integrated environment for the development of knowledge-based recommender applications. Int. J. Electron. Commerce **11**(2), 11–34 (2006)
54. C. Fellbaum, *WordNet: An Electronic Lexical Database* (MIT Press, 1998)
55. I. Fernández-Tobías, I. Cantador, Exploiting social tags in matrix factorization models for cross-domain collaborative filtering, in *Proceedings of the 1st Workshop on New Trends in Content-based Recommender Systems Co-located with the 8th ACM Conference on Recommender Systems, CBRecSys@RecSys 2014, Foster City, Silicon Valley, California, USA, October 6, 2014.* CEUR Workshop Proceedings, vol. 1245. CEUR-WS.org (2014), pp. 34–41
56. B. Ferwerda, M. Schedl, M. Tkalcic, Personality & emotional states: understanding users' music listening needs, in *Posters, Demos, Late-breaking Results and Workshop Proceedings of the 23rd Conference on User Modeling, Adaptation, and Personalization (UMAP 2015), Dublin, Ireland, June 29 - July 3, 2015*, ed. by A.I. Cristea, J. Masthoff, A. Said, N. Tintarev. CEUR Workshop Proceedings. CEUR-WS.org, vol. 1388 (2015)
57. J.R. Firth, A synopsis of linguistic theory, 1930–1955. *Studies in linguistic analysis* (Oxford, Blackwell, 1957)
58. G. Friedrich, M. Zanker, A taxonomy for generating explanations in recommender systems. AI Mag. **32**(3), 90–98 (2011)

59. E. Gabrilovich, S. Markovitch, Overcoming the brittleness bottleneck using wikipedia: enhancing text categorization with encyclopedic knowledge, in *Proceedings of the Twenty-First National Conference on Artificial Intelligence and the Eighteenth Innovative Applications of Artificial Intelligence Conference* (AAAI Press, Palo Alto, 2006), pp. 1301–1306

60. E. Gabrilovich, S. Markovitch, Computing semantic relatedness using wikipedia-based explicit semantic analysis, in *IJCAI'07: Proceedings of the 20th International Joint Conference on Artifical Intelligence* (2007), pp. 1606–1611

61. E. Gabrilovich, S. Markovitch, Wikipedia-based semantic interpretation for natural language processing. J. Artif. Intell. Res. **34**, 443–498 (2009)

62. G. Ganu, Y. Kakodkar, A. Marian, Improving the quality of predictions using textual information in online user reviews. Inf. Syst. **38**(1), 1–15 (2013)

63. F. Gedikli, D. Jannach, M. Ge, How should I explain? A comparison of different explanation types for recommender systems. Int. J. Hum.-Comput. Stud. **72**(4), 367–382 (2014)

64. B. Goodman, S. Flaxman, European union regulations on algorithmic decision-making and a right to explanation (2016). arXiv:1606.08813

65. M. Grbovic, V. Radosavljevic, N. Djuric, N. Bhamidipati, J. Savla, V. Bhagwan, D. Sharp, E-commerce in your inbox: product recommendations at scale, in *Proceedings of the 21th ACM SIGKDD International Conference on Knowledge Discovery and Data Mining* (2015), pp. 1809–1818

66. A. Grover, J. Leskovec, node2vec: scalable feature learning for networks, in *Proceedings of the 22nd ACM SIGKDD International Conference on Knowledge Discovery and Data Mining* (2016), pp. 855–864

67. X. Gu, L. Shou, P. Peng, K. Chen, S. Wu, G. Chen, iglasses: a novel recommendation system for best-fit glasses, in *Proceedings of the 39th International ACM SIGIR Conference on Research and Development in Information Retrieval, SIGIR 2016, Pisa, Italy, July 17–21, 2016*, ed. by R. Perego, F. Sebastiani, J.A. Aslam, I. Ruthven, J. Zobel (ACM, New York, 2016), pp. 1109–1112

68. Q. Guo, F. Zhuang, C. Qin, H. Zhu, X. Xie, H. Xiong, Q. He, A survey on knowledge graph-based recommender systems. *IEEE Transactions on Knowledge and Data Engineering* (Early Access). (2020). https://doi.org/10.1109/TKDE.2020.3028705

69. L. Gutiérrez, B. Keith, A systematic literature review on word embeddings, in *International Conference on Software Process Improvement* (Springer, Berlin, 2018), pp. 132–141

70. Z.S. Harris, *Mathematical Structures of Language* (Interscience, New York, 1968)

71. H.A.M. Hassan, G. Sansonetti, F. Gasparetti, A. Micarelli, J. Beel, Bert, elmo, use and infersent sentence encoders: the panacea for research-paper recommendation? in *RecSys (Late-Breaking Results)* (2019), pp. 6–10

72. O. Hassanzadeh, M.P. Consens, Linked movie data base, in *Proceedings of the WWW2009 Workshop on Linked Data on the Web, LDOW 2009*, ed. by C. Bizer, T. Heath, T. Berners-Lee, K. Idehen. CEUR Workshop Proceedings, vol. 538, CEUR-WS.org (2009)

73. T.H. Haveliwala, Topic-sensitive pagerank: a context-sensitive ranking algorithm for web search. IEEE Trans. Knowl. Data Eng. **15**(4), 784–796 (2003)

74. C. He, D. Parra, K. Verbert, Interactive recommender systems: a survey of the state of the art and future research challenges and opportunities. Exp. Syst. Appl. **56**, 9–27 (2016)

75. X. He, T. Chen, M.-Y. Kan, X. Chen, Trirank: review-aware explainable recommendation by modeling aspects, in *Proceedings of the 24th ACM International on Conference on Information and Knowledge Management* (ACM, New York, 2015), pp. 1661–1670

76. J. Henderson, O. Lemon, Mixture model pomdps for efficient handling of uncertainty in dialogue management, in *ACL 2008, Proceedings of the 46th Annual Meeting of the Association for Computational Linguistics, June 15–20, 2008, Columbus, Ohio, USA, Short Papers* (The Association for Computer Linguistics, Stroudsburg, 2008), pp. 73–76

77. L. Herlocker, J.A. Konstan, L.G. Terveen, J.T. Riedl, Evaluating Collaborative Filtering Recommender Systems. ACM Trans. Inf. Syst. **22**(1), 5–53 (2004)

78. M. Hernández-Rubio, I. Cantador, A. Bellogín, A comparative analysis of recommender systems based on item aspect opinions extracted from user reviews. User Model. User Adapt. Interact. **29**(2), 381–441 (2019)

79. S. Hochreiter, J. Schmidhuber, Long short-term memory. Neural Comput. **9**(8), 1735–1780 (1997)

80. A. Hotho, R. Jäschke, C. Schmitz, G. Stumme, Information retrieval in folksonomies: search and ranking, in *The Semantic Web: Research and Applications, 3rd European Semantic Web Conference, ESWC 2006, Budva, Montenegro, June 11–14, 2006, Proceedings*, ed. by Y. Sure, J. Domingue. Lecture Notes in Computer Science, vol. 4011 (Springer, Berlin, 2006), pp. 411–426

81. A. Iovine, F. Narducci, M. de Gemmis, A dataset of real dialogues for conversational recommender systems, in *CEUR-WS, Proceedings of the Sixth Italian Conference on Computational Linguistics*, ed. by R. Bernardi, R. Navigli, G. Semeraro , vol. 2481, (2019). CLiC-it 2019

82. A. Iovine, F. Narducci, G. Semeraro, Conversational recommender systems and natural language: a study through the converse framework. Decis. Support Syst. **131**, 113250 (2020)

83. S. Jain, B.C. Wallace, Attention is not explanation (2019). arXiv:1902.10186

84. D. Jannach, Finding preferred query relaxations in content-based recommenders, in *Intelligent Techniques and Tools for Novel System Architectures* (Springer, Berlin, 2008), pp. 81–97

85. D. Jannach, A. Manzoor, End-to-end learning for conversational recommendation: a long way to go? in *Proceedings of the 7th Joint Workshop on Interfaces and Human Decision Making for Recommender Systems Co-located with 14th ACM Conference on Recommender Systems (RecSys 2020), Online Event, September 26, 2020*, ed. by P. Brusilovsky, M. de Gemmis, A. Felfernig, P. Lops, J. O'Donovan, G. Semeraro, M.C. Willemsen. CEUR Workshop Proceedings, vol. 2682, CEUR-WS.org (2020), pp. 72–76

86. D. Jannach, M. Zanker, A. Felfernig, G. Friedrich, *Recommender Systems: An Introduction* (Cambridge University Press, Cambridge, 2010)

87. D. Jannach, A. Manzoor, W. Cai, L. Chen, A survey on conversational recommender systems (2020). arXiv:2004.00646

88. R. Jäschke, L. Marinho, A. Hotho, L. Schmidt-Thieme, G. Stumme, Tag recommendations in folksonomies, in *European Conference on Principles of Data Mining and Knowledge Discovery* (Springer, Berlin, 2007), pp. 506–514

89. H. Johnson, P. Johnson, Explanation facilities and interactive systems, in *Proceedings of the 1st International Conference on Intelligent User Interfaces* (ACM, New York, 1993), pp. 159–166

90. D. Khattar, V. Kumar, M. Gupta, V. Varma, Neural content-collaborative filtering for news recommendation, in *CEUR-WS*, Recent Trends in News Information Retrieval, Proceedings of the Second International Workshop on Recent Trends in News Information Retrieval, Co-Located with 40th European Conference on Information Retrieval (ECIR 2018), ed. by D. Albakour, D. Corney, J. Gonzalo, M. Martinez, B. Poblete, A. Valochas , vol. 2079, (2018), pp. 45–50 NewsIR 2018

91. H. Kim, A. Roczniak, P. Lévy, A. El-Saddik, Social media filtering based on collaborative tagging in semantic space. Multim. Tools Appl. **56**(1), 63–89 (2012)

92. P. Knees, M. Schedl, A survey of music similarity and recommendation from music context data. ACM Trans. Multim. Comput. Commun. Appl. **10**(1), 2:1–2:21 (2013)

93. B.P. Knijnenburg, M.C. Willemsen, Z. Gantner, H. Soncu, C. Newell, Explaining the user experience of recommender systems. User Model. User-Adapt. Interact. **22**(4–5), 441–504 (2012)

94. M. Kula, Metadata embeddings for user and item cold-start recommendations (2015). arXiv:1507.08439

95. T.K. Landauer, P.W. Foltz, D. Laham, An introduction to latent semantic analysis. Discourse Process. **25**(2–3), 259–284 (1998)

96. Q. Le, T. Mikolov, Distributed representations of sentences and documents, in *International Conference on Machine Learning* (2014), pp. 1188–1196

97. Y. LeCun, Y. Bengio, G. Hinton, Deep learning. Nature **521**(7553), 436–444 (2015)

98. H. Li, S. Sanner, K. Luo, G. Wu, A ranking optimization approach to latent linear critiquing for conversational recommender systems, in *Fourteenth ACM Conference on Recommender Systems* (2020), pp. 13–22

99. P. Li, A. Tuzhilin, Learning latent multi-criteria ratings from user reviews for recommendations. *IEEE Transactions on Knowledge and Data Engineering* (Early Access). (2020). https://doi.org/10.1109/TKDE.2020.3030623

100. R. Li, S.E. Kahou, H. Schulz, V. Michalski, L. Charlin, C. Pal, Towards deep conversational recommendations, in *Advances in Neural Information Processing Systems* (2018), pp. 9725–9735

101. B. Liu, I. Lane, An end-to-end trainable neural network model with belief tracking for task-oriented dialog. *Interspeech 2017* (2017), pp. 2506–2510. arXiv: 1708.05956

102. H. Liu, J. He, T. Wang, W. Song, X. Du, Combining user preferences and user opinions for accurate recommendation. Electron. Commer. Res. Appl. **12**(1), 14–23 (2013)

103. D. Liu, J. Lian, S. Wang, Y. Qiao, J.-H. Chen, G. Sun, X. Xie, KRED: knowledge-aware document representation for news recommendations (2019). arXiv:1910.11494

104. P. Lops, M. de Gemmis, G. Semeraro, Content-based recommender systems: state of the art and trends, in *Recommender Systems Handbook*, ed. by F. Ricci, L. Rokach, B. Shapira, P.B. Kantor (Springer, Berlin, 2011), pp. 73–105

105. P. Lops, C. Musto, F. Narducci, G. Semeraro, *Semantics in Adaptive and Personalised Systems - Methods, Tools and Applications* (Springer, Berlin, 2019)

106. W. Lowe, Towards a theory of semantic space, in *Proceedings of the Twenty-Third Annual Conference of the Cognitive Science Society* (Lawrence Erlbaum Associates, 2001), pp. 576–581

107. V. Lully, P. Laublet, M. Stankovic, F. Radulovic, Enhancing explanations in recommender systems with knowledge graphs. Proc. Comput. Sci. **137**, 211–222 (2018)

108. X. Luo, Y. Ouyang, Z. Xiong, Improving neighborhood based collaborative filtering via integrated folksonomy information. Pattern Recognit. Lett. **33**(3), 263–270 (2012)

109. T. Mahmood, F. Ricci, Improving recommender systems with adaptive conversational strategies, in *Proceedings of the 20th ACM Conference on Hypertext and Hypermedia* (ACM, New York, 2009), pp. 73–82

110. C.D. Manning, H. Schütze, *Foundations of Statistical Natural Language Processing*, vol. 999 (MIT Press, Cambridge, 1999)

111. J. Manotumruksa, C. Macdonald, I. Ounis, Modelling user preferences using word embeddings for context-aware venue recommendation (2016). arXiv:1606.07828

112. J.J. McAuley, J. Leskovec, Hidden factors and hidden topics: understanding rating dimensions with review text, in *Seventh ACM Conference on Recommender Systems, RecSys'13, Hong Kong, China, October 12–16, 2013*, ed. by Q. Yang, I. King, Q. Li, P. Pu, G. Karypis (ACM, New York, 2013), pp. 165–172

113. J.J. McAuley, C. Targett, Q. Shi, A. van den Hengel, Image-based recommendations on styles and substitutes, in *Proceedings of the 38th International ACM SIGIR Conference on Research and Development in Information Retrieval, Santiago, Chile, August 9–13, 2015*, ed. by R. Baeza-Yates, M. Lalmas, A. Moffat, B.A. Ribeiro-Neto (ACM, New York, 2015), pp. 43–52

114. F. McCarey, M.Ó. Cinnéide, N. Kushmerick, Recommending library methods: an evaluation of the vector space model (VSM) and latent semantic indexing (LSI), in *International Conference on Software Reuse* (Springer, Berlin, 2006), pp. 217–230

115. D. McSherry, Explanation in recommender systems. Artif. Intell. Rev. **24**(2), 179–197 (2005)

116. M.M. Mehrabani, H. Mohayeji, A. Moeini, A hybrid approach to enhance pure collaborative filtering based on content feature relationship (2020). arXiv:2005.08148

117. T. Mikolov, I. Sutskever, K. Chen, G.S. Corrado, J. Dean, Distributed representations of words and phrases and their compositionality, in *Advances in Neural Information Processing Systems* (2013), pp. 3111–3119

118. T. Mikolov, W.-T. Yih, G. Zweig, Linguistic regularities in continuous space word repre-sentations, in *Proceedings of the 2013 Conference of the North American Chapter of the Association for Computational Linguistics: Human Language Technologies* (2013), pp. 746–751

119. G. Miller, WordNet: An on-line lexical database. Int. J. Lexicography. **3**(4), 235–244 (1990)

120. G. Miller, Wordnet: a lexical database for english. Commun. ACM **38**(11), 39–41 (1995)

121. T.M. Mitchell, *Machine Learning*. McGraw Hill Series in Computer Science (McGraw-Hill, New York, 1997)

122. S. Mohammad, G. Hirst, Distributional measures of semantic distance: a survey (2012). CoRR, abs/1203.1858

123. K.I. Muhammad, A. Lawlor, B. Smyth, A live-user study of opinionated explanations for recommender systems, in *Proceedings of the 21st International Conference on Intelligent User Interfaces* (ACM, New York, 2016), pp. 256–260

124. C. Musto, Enhanced vector space models for content-based recommender systems, in *Proceedings of the 2010 ACM Conference on Recommender Systems, RecSys 2010, Barcelona, Spain, September 26–30, 2010*, ed. by X. Amatriain, M. Torrens, P. Resnick, M. Zanker (ACM, New York, 2010), pp. 361–364

125. C. Musto, G. Semeraro, P. Lops, M. de Gemmis, Random indexing and negative user preferences for enhancing content-based recommender systems, in *E-Commerce and Web Technologies - 12th International Conference, EC-Web 2011, Toulouse, France, August 30 - September 1, 2011. Proceedings*, ed. by C. Huemer, T. Setzer. Lecture Notes in Business Information Processing, vol. 85 (Springer, 2011), pp. 270–281

126. C. Musto, G. Semeraro, P. Lops, M. de Gemmis, F. Narducci, Leveraging social media sources to generate personalized music playlists, in *E-Commerce and Web Technologies - 13th International Conference, EC-Web 2012*. Lecture Notes in Business Information Processing, vol. 123 (Springer, Berlin, 2012), pp. 112–123

127. C. Musto, G. Semeraro, P. Lops, M. de Gemmis, Contextual eVSM: a content-based context-aware recommendation framework based on distributional semantics, in *E-Commerce and Web Technologies - 14th International Conference, EC-Web 2013, Prague, Czech Republic, August 27–28, 2013. Proceedings*, ed. by C. Huemer, P. Lops. Lecture Notes in Business Information Processing (Springer, Berlin, 2013), pp. 125–136

128. C. Musto, P. Basile, P. Lops, M. de Gemmis, G. Semeraro, Linked open data-enabled strategies for top-n recommendations, in *Proceedings of the 1st Workshop on New Trends in Content-Based Recommender Systems Co-located with the 8th ACM Conference on Recommender Systems, CBRecSys@RecSys 2014, Foster City, Silicon Valley, California, USA, October 6, 2014.*, ed. by T. Bogers, M. Koolen, I. Cantador. CEUR Workshop Proceedings, vol. 1245. CEUR-WS.org (2014), pp. 49–56

129. C. Musto, G. Semeraro, M. de Gemmis, P. Lops, Learning word embeddings from wikipedia for content-based recommender systems, in *European Conference on Information Retrieval* (Springer, Berlin, 2016), pp. 729–734

130. C. Musto, G. Semeraro, M. de Gemmis, P. Lops, Learning word embeddings from wikipedia for content-based recommender systems, in *Advances in Information Retrieval - 38th European Conference on IR Research, ECIR 2016, Padua, Italy, March 20–23, 2016. Proceedings*, ed. by N. Ferro, F. Crestani, M. Moens, J. Mothe, F. Silvestri, G.M.D. Nunzio, C. Hauff, G. Silvello. Lecture Notes in Computer Science, vol. 9626 (Springer, Berlin, 2016), pp. 729–734

131. C. Musto, P. Basile, P. Lops, M. de Gemmis, G. Semeraro, Introducing linked open data in graph-based recommender systems. Inf. Process. Manag. **53**(2), 405–435 (2017)

132. C. Musto, M. de Gemmis, G. Semeraro, P. Lops, A multi-criteria recommender system exploiting aspect-based sentiment analysis of users' reviews, in *Proceedings of the Eleventh ACM Conference on Recommender Systems, RecSys 2017, Como, Italy, August 27–31, 2017*, ed. by P. Cremonesi, F. Ricci, S. Berkovsky, A. Tuzhilin (ACM, New York, 2017), pp. 321–325

133. C. Musto, P. Lops, M. de Gemmis, G. Semeraro, Semantics-aware recommender systems exploiting linked open data and graph-based features. Knowl.-Based Syst. **136**, 1–14 (2017)

134. C. Musto, G. Semeraro, M. de Gemmis, P. Lops, Tuning personalized pagerank for semantics-aware recommendations based on linked open data, in *The Semantic Web - 14th International Conference, ESWC 2017*, ed. by E. Blomqvist, D. Maynard, A. Gangemi, R. Hoekstra, P. Hitzler, O. Hartig. Lecture Notes in Computer Science, vol. 10249 (2017), pp. 169–183

135. C. Musto, T. Franza, G. Semeraro, M. de Gemmis, P. Lops, Deep content-based recommender systems exploiting recurrent neural networks and linked open data, in *Adjunct Publication of the 26th Conference on User Modeling, Adaptation and Personalization* (2018), pp. 239–244

136. C. Musto, P. Basile, G. Semeraro, Hybrid semantics-aware recommendations exploiting knowledge graph embeddings, in *International Conference of the Italian Association for Artificial Intelligence* (Springer, Berlin, 2019), pp. 87–100

137. C. Musto, P. Lops, M. de Gemmis, G. Semeraro, Justifying recommendations through aspect-based sentiment analysis of users reviews, in *Proceedings of the 27th ACM Conference on User Modeling, Adaptation and Personalization* (2019), pp. 4–12

138. C. Musto, F. Narducci, P. Lops, M. de Gemmis, G. Semeraro, Linked open data-based explanations for transparent recommender systems. Int. J. Hum.-Comput. Stud. **121**, 93–107 (2019)

139. C. Musto, G. Rossiello, M. de Gemmis, P. Lops, G. Semeraro, Combining text summarization and aspect-based sentiment analysis of users' reviews to justify recommendations, in *Proceedings of the 13th ACM Conference on Recommender Systems* (2019), pp. 383–387

140. C. Musto, G. Spillo, M. de Gemmis, P. Lops, G. Semeraro, Exploiting distributional semantics models for natural language context-aware justifications for recommender systems, in *Proceedings of the 7th Joint Workshop on Interfaces and Human Decision Making for Recommender Systems Co-located with 14th ACM Conference on Recommender Systems (RecSys 2020)* (2020), pp. 65–71

141. C. Musto, M. de Gemmis, P. Lops, G. Semeraro, Generating post hoc review-based natural language justifications for recommender systems. User Model. User-Adapt. Interact. **31**, 629–673 (2021)

142. R.Y. Nakamoto, S. Nakajima, J. Miyazaki, S. Uemura, H. Kato, Y. Inagaki, Reasonable tag-based collaborative filtering for social tagging systems, in *Proceedings of the 2nd ACM Workshop on Information Credibility on the Web, WICOW 2008, Napa Valley, California, USA, October 30, 2008*, ed. by K. Tanaka, T. Matsuyama, E. Lim, A. Jatowt (ACM, New York, 2008), pp. 11–18

143. F. Narducci, C. Musto, G. Semeraro, P. Lops, M. de Gemmis, Exploiting big data for enhanced representations in content-based recommender systems, in *International Conference on Electronic Commerce and Web Technologies* (Springer, Berlin, 2013), pp. 182–193

144. F. Narducci, P. Basile, M. de Gemmis, P. Lops, G. Semeraro, An investigation on the user interaction modes of conversational recommender systems for the music domain. User Model. User-Adapt. Interact. **30**(2), 251–284 (2020)

145. R. Navigli, S.P. Ponzetto, Babelnet: the automatic construction, evaluation and application of a wide-coverage multilingual semantic network. Artif. Intell. **193**, 217–250 (2012)

146. L. Nie, W. Wang, R. Hong, M. Wang, Q. Tian, Multimodal dialog system: generating responses via adaptive decoders, in *Proceedings of the 27th ACM International Conference on Multimedia* (2019), pp. 1098–1106

147. M.G. Ozsoy, From word embeddings to item recommendation (2016). arXiv:1601.01356

148. L. Page, S. Brin, R. Motwani, T. Winograd, *The PageRank Citation Ranking: Bringing Order to the Web.* (Stanford InfoLab, Stanford, 1999)

149. A. Passant, dbrec - music recommendations using dbpedia, in *International Semantic Web Conference, Revised Papers.* Lecture Notes in Computer Science, vol. 6497 (Springer, Berlin, 2010), pp. 209–224

150. A. Passant, Measuring semantic distance on linking data and using it for resources recommendations, in *AAAI Spring Symposium: Linked Data Meets Artificial Intelligence* (AAAI, Menlo Park, 2010), pp. 93–98

151. G. Peake, J. Wang, Explanation mining: post hoc interpretability of latent factor models for recommendation systems, in *Proceedings of the 24th ACM SIGKDD International Conference on Knowledge Discovery & Data Mining* (2018), pp. 2060–2069

152. M.E. Peters, M. Neumann, M. Iyyer, M. Gardner, C. Clark, K. Lee, L. Zettlemoyer, Deep contextualized word representations (2018). arXiv:1802.05365

153. G. Qiu, B. Liu, J. Bu, C. Chen, Opinion word expansion and target extraction through double propagation. Comput. Linguist. **37**(1), 9–27 (2011)

154. M. Quadrana, P. Cremonesi, D. Jannach, Sequence-aware recommender systems. ACM Comput. Surv. **51**(4), 66:1–66:36 (2018)

155. D. Rao, P. McNamee, M. Dredze, Entity linking: finding extracted entities in a knowledge base, in *Multi-Source, Multilingual Information Extraction and Summarization* (Springer, Berlin, 2013), pp. 93–115

156. H. Ren, W. Xu, Y. Yan, Markovian discriminative modeling for cross-domain dialog state tracking, in *2014 IEEE Spoken Language Technology Workshop, SLT 2014, South Lake Tahoe, NV, USA, December 7–10, 2014* (IEEE, Piscataway, 2014), pp. 342–347

157. F. Ricci, Q.N. Nguyen, Acquiring and revising preferences in a critique-based mobile recommender system. IEEE Intell. Syst. **22**(3), 22–29 (2007)

158. F. Ricci, A. Venturini, D. Cavada, N. Mirzadeh, D. Blaas, M. Nones, Product recommendation with interactive query management and twofold similarity, in *International Conference on Case-Based Reasoning* (Springer, Berlin, 2003), pp. 479–493

159. A. Ritter, C. Cherry, W.B. Dolan, Data-driven response generation in social media, in *Proceedings of the Conference on Empirical Methods in Natural Language Processing, EMNLP'11* (Association for Computational Linguistics, Stroudsburg, 2011), pp. 583—593

160. S. Robertson, H. Zaragoza, *The Probabilistic Relevance Framework: BM25 and Beyond* (Now Publishers, Norwell, 2009)

161. M. Rossetti, F. Stella, M. Zanker, Towards explaining latent factors with topic models in collaborative recommender systems, in *Database and Expert Systems Applications (DEXA), 2013 24th International Workshop on* (IEEE, Piscataway, 2013), pp. 162–167

162. H. Rubenstein, J.B. Goodenough, Contextual correlates of synonymy. Commun. ACM **8**(10), 627–633 (1965)

163. D.E. Rumelhart, G.E. Hinton, R.J. Williams, Learning representations by back-propagating errors. Nature **323**(6088), 533–536 (1986)

164. M. Sahlgren, The word-space model: using distributional analysis to represent syntagmatic and paradigmatic relations between words in high-dimensional vector spaces. Ph.D. Thesis, Stockholm University (2006)

165. G. Salton, M. McGill, *Introduction to Modern Information Retrieval* (McGraw-Hill, New York, 1983)

166. G. Salton, A. Wong, C. Yang, A vector space model for automatic indexing. Commun. ACM **18**(11), 613–620 (1975)

167. C. Scaffidi, K. Bierhoff, E. Chang, M. Felker, H. Ng, C. Jin, Red opal: product-feature scoring from reviews, in *Proceedings 8th ACM Conference on Electronic Commerce (EC-2007), San Diego, California, USA, June 11–15, 2007*, ed. by J.K. MacKie-Mason, D.C. Parkes, P. Resnick (ACM, New York, 2007), pp. 182–191

168. M. Schedl, P. Knees, B. McFee, D. Bogdanov, M. Kaminskas, Music recommender systems, in *Recommender Systems Handbook*, ed. by F. Ricci, L. Rokach, B. Shapira (Springer, Berlin, 2015), pp. 453–492

169. H. Schütze, Automatic word sense discrimination. Comput. Linguistics **24**(1), 97–123 (1998)

170. G. Semeraro, M. Degemmis, P. Lops, P. Basile, Combining learning and word sense disambiguation for intelligent user profiling, in *IJCAI 2007, Proceedings of the 20th International Joint Conference on Artificial Intelligence, Hyderabad, India, January 6–12, 2007*, ed. by M.M. Veloso (2007), pp. 2856–2861

171. S. Seo, J. Huang, H. Yang, Y. Liu, Interpretable convolutional neural networks with dual local and global attention for review rating prediction, in *Proceedings of the Eleventh ACM Conference on Recommender Systems* (2017), pp. 297–305

172. W. Shen, J. Wang, J. Han, Entity linking with a knowledge base: issues, techniques, and solutions. IEEE Trans. Knowl. Data Eng. **27**(2), 443–460 (2014)
173. D. Shin, S. Cetintas, K.-C. Lee, Recommending tumblr blogs to follow with inductive matrix completion, in *RecSys Posters* (2014)
174. D. Shmaryahu, I. Negev, G. Shani, B. Shapira, Post-hoc explanations for complex model recommendations using simple methods, in *(RecSys 2020), CEUR-WS*, Interfaces and Human Decision Making for Recommender Systems 2020, Proceedings of the 7th Joint Workshop on Interfaces and Human Decision Making for Recommender Systems, Co-Located with 14th ACM Conference on Recommender Systems, ed. by P. Brusilovsky, M. de Gemmis, A. Felfernig, P. Lops, J. O'Donovan, G. Semeraro, M. C. Willemsen , vol. 2682, (2020), pp. 26–36 IntRS 2020
175. R. Sinha, K. Swearingen, The role of transparency in recommender systems, in *CHI'02 Extended abstracts on Human Factors in Computing Systems* (ACM, New York, 2002), pp. 830–831
176. B. Smyth, L. McGinty, An analysis of feedback strategies in conversational recommenders, in *The Fourteenth Irish Artificial Intelligence and Cognitive Science Conference (AICS 2003)* (Citeseer, 2003)
177. P. Stakhiyevich, Z. Huang, Building user profiles based on user interests and preferences for recommender systems, in *2019 IEEE International Conferences on Ubiquitous Computing Communications (IUCC) and Data Science and Computational Intelligence (DSCI) and Smart Computing, Networking and Services (SmartCNS)* (2019), pp. 450–455
178. M. Stevenson, *Word Sense Disambiguation: The Case for Combinations of Knowledge Sources* (CSLI Publications, Stanford, 2003)
179. S. Stiebellehner, J. Wang, S. Yuan, Learning continuous user representations through hybrid filtering with doc2vec (2017). arXiv:1801.00215
180. A. Suglia, C. Greco, C. Musto, M. de Gemmis, P. Lops, G. Semeraro, A deep architecture for content-based recommendations exploiting recurrent neural networks, in *Proceedings of the 25th Conference on User Modeling, Adaptation and Personalization, UMAP 2017, Bratislava, Slovakia, July 09–12, 2017*, ed. by M. Bieliková, E. Herder, F. Cena, M.C. Desmarais (ACM, New York, 2017), pp. 202–211
181. C. Sundermann, J. Antunes, M. Domingues, S. Rezende, Exploration of word embedding model to improve context-aware recommender systems, in *2018 IEEE/WIC/ACM International Conference on Web Intelligence (WI)* (IEEE, Piscataway, 2018), pp. 383–388
182. P. Symeonidis, A. Nanopoulos, Y. Manolopoulos, MoviExplain: a recommender system with explanations, in *Proceedings of the Third ACM Conference on Recommender Systems* (ACM, New York, 2009), pp. 317–320
183. C.A. Thompson, M.H. Goker, P. Langley, A personalized system for conversational recommendations. J. Artif. Intell. Res. **21**, 393–428 (2004)
184. I. Titov, R.T. McDonald, A joint model of text and aspect ratings for sentiment summarization, in *ACL 2008, Proceedings of the 46th Annual Meeting of the Association for Computational Linguistics, June 15–20, 2008, Columbus, Ohio, USA*, ed. by K.R. McKeown, J.D. Moore, S. Teufel, J. Allan, S. Furui (The Association for Computer Linguistics, Stroudsburg, 2008), pp. 308–316
185. I. Titov, R.T. McDonald, Modeling online reviews with multi-grain topic models, in *Proceedings of the 17th International Conference on World Wide Web, WWW 2008, Beijing, China, April 21–25, 2008*, ed. by J. Huai, R. Chen, H. Hon, Y. Liu, W. Ma, A. Tomkins, X. Zhang (ACM, New York, 2008), pp. 111–120
186. W. Trabelsi, N. Wilson, D. Bridge, F. Ricci, Comparing approaches to preference dominance for conversational recommenders, in *2010 22nd IEEE International Conference on Tools with Artificial Intelligence*, vol. 2 (IEEE, Piscataway, 2010), pp. 113–120.
187. K.H.L. Tso-Sutter, L.B. Marinho, L. Schmidt-Thieme, Tag-aware recommender systems by fusion of collaborative filtering algorithms, in *Proceedings of the 2008 ACM Symposium on Applied Computing (SAC), Fortaleza, Ceara, Brazil, March 16–20, 2008*, ed. by R.L. Wainwright, H. Haddad (ACM, New York, 2008), pp. 1995–1999

188. P.D. Turney, P. Pantel, From frequency to meaning: vector space models of semantics. J. Artif. Intell. Res. **37**(1), 141–188 (2010)
189. A. van den Oord, S. Dieleman, B. Schrauwen, Deep content-based music recommendation, in *Advances in Neural Information Processing Systems*, ed. by C.J.C. Burges, L. Bottou, M. Welling, Z. Ghahramani, K.Q. Weinberger, vol. 26 (Curran Associates, Red Hook, 2013), pp. 2643–2651
190. F. Vasile, E. Smirnova, A. Conneau, Meta-prod2vec: product embeddings using side-information for recommendation, in *Proceedings of the 10th ACM Conference on Recommender Systems* (2016), pp. 225–232
191. J. Vig, S. Sen, J. Riedl, Tagsplanations: explaining recommendations using tags, in *Proceedings of the 14th International Conference on Intelligent User Interfaces* (ACM, New York, 2009), pp. 47–56
192. D. Vrandecic, M. Krötzsch, Wikidata: a free collaborative knowledgebase. Commun. ACM **57**(10), 78–85 (2014)
193. Z. Wang, O. Lemon, A simple and generic belief tracking mechanism for the dialog state tracking challenge: on the believability of observed information, in *Proceedings of the SIGDIAL 2013 Conference, The 14th Annual Meeting of the Special Interest Group on Discourse and Dialogue, 22–24 August 2013, SUPELEC, Metz, France* (The Association for Computer Linguistics, Stroudsburg, 2013), pp. 423–432
194. H. Wang, F. Zhang, X. Xie, M. Guo, DKN: deep knowledge-aware network for news recommendation, in *Proceedings of the 2018 World Wide Web Conference* (2018), pp. 1835–1844
195. S. Wang, Y. Wang, J. Tang, K. Shu, S. Ranganath, H. Liu, What your images reveal: exploiting visual contents for point-of-interest recommendation, in *Proceedings of the 26th International Conference on World Wide Web, WWW 2017, Perth, Australia, April 3–7, 2017*, ed. by R. Barrett, R. Cummings, E. Agichtein, E. Gabrilovich (ACM, New York, 2017), pp. 391–400
196. H. Wang, M. Zhao, X. Xie, W. Li, M. Guo, Knowledge graph convolutional networks for recommender systems, in *The World Wide Web Conference* (2019), pp. 3307–3313
197. X. Wang, X. He, Y. Cao, M. Liu, T.-S. Chua, KGAT: knowledge graph attention network for recommendation, in *Proceedings of the 25th ACM SIGKDD International Conference on Knowledge Discovery & Data Mining* (2019), pp. 950–958
198. P. Wärnestål, User evaluation of a conversational recommender system, in *Proceedings of the 4th IJCAI Workshop on Knowledge and Reasoning in Practical Dialogue Systems* (2005), pp. 32–39
199. R. Wetzker, C. Zimmermann, C. Bauckhage, S. Albayrak, I tag, you tag: translating tags for advanced user models, in *Proceedings of the Third International Conference on Web Search and Web Data Mining, WSDM 2010, New York, NY, USA, February 4–6, 2010*, ed. by B.D. Davison, T. Suel, N. Craswell, B. Liu (ACM, New York, 2010), pp. 71–80
200. J.D. Williams, Challenges and opportunities for state tracking in statistical spoken dialog systems: results from two public deployments. IEEE J. Sel. Top. Signal Process. **6**(8), 959–970 (2012)
201. J.D. Williams, S.J. Young, Partially observable markov decision processes for spoken dialog systems. Comput. Speech Lang. **21**(2), 393–422 (2007)
202. J. Williams, A. Raux, M. Henderson, The dialog state tracking challenge series: a review. Dial. Discourse **7**(3), 4–33 (2016)
203. J.D. Williams, A. Raux, M. Henderson, The dialog state tracking challenge series: a review. Dial. Discourse **7**(3), 4–33 (2016)
204. L. Wittgenstein, *Philosophical Investigations* (Blackwell, Oxford, 1953)
205. C. Wu, F. Wu, M. An, J. Huang, Y. Huang, X. Xie, NPA: neural news recommendation with personalized attention, in *Proceedings of the 25th ACM SIGKDD International Conference on Knowledge Discovery & Data Mining* (2019), pp. 2576–2584
206. S. Xu, Y. Li, S. Liu, Z. Fu, Y. Zhang, Learning post-hoc causal explanations for recommendation (2020). arXiv:2006.16977

207. Y. Zhang, G. Lai, M. Zhang, Y. Zhang, Y. Liu, S. Ma, Explicit factor models for explainable recommendation based on phrase-level sentiment analysis, in *Proceedings of the 37th International ACM SIGIR Conference on Research & Development in Information Retrieval* (2014), pp. 83–92
208. F. Zhang, N.J. Yuan, D. Lian, X. Xie, W.-Y. Ma, Collaborative knowledge base embedding for recommender systems, in *Proceedings of the 22nd ACM SIGKDD International Conference on Knowledge Discovery and Data Mining* (2016), pp. 353–362
209. S. Zhang, H. Tong, J. Xu, R. Maciejewski, Graph convolutional networks: a comprehensive review. Comput. Soc. Netw. **6**(1), 1–23 (2019)
210. T. Zhao, M. Eskenazi, Towards end-to-end learning for dialog state tracking and management using deep reinforcement learning (2016). arXiv:1606.02560 [cs]
211. L. Zheng, V. Noroozi, P.S. Yu, Joint deep modeling of users and items using reviews for recommendation, in *Proceedings of the Tenth ACM International Conference on Web Search and Data Mining* (2017), pp. 425–434

# Part II
# Special Recommendation Techniques

# Session-Based Recommender Systems

Dietmar Jannach, Massimo Quadrana, and Paolo Cremonesi

## 1  Introduction

In session-based recommendation scenarios, the goal is to tailor the suggestions that are made to the users according to their assumed short-term intents during an ongoing usage session. The main inputs in such settings consists of (1) the sequence of interactions with an individual user that were observed by the system since the session started, (2) information about past sessions by other users. Typical application scenarios can, for example, be found in the e-commerce domain, where the shopping goals of a consumer can change from session to session and where these goals can also relate to entirely different product categories in each session. Furthermore, in many cases, the past interests and preferences of an individual user might not even be known to the system, e.g., because it is a first-time user or a user that is not logged-in and thus anonymous. The challenge in such situations therefore consists in making helpful recommendations based on very little amounts of information about the current user. As an extreme case, consider Amazon's *"Customer's who bought ... also bought"* recommendations, where the recommendations are tailored—albeit in a non-personalized way—to the one single item that is currently viewed by the user.

D. Jannach (✉)
University of Klagenfurt, Klagenfurt, Austria
e-mail: dietmar.jannach@aau.at

M. Quadrana
Pandora Media LLC, Oakland, CA, USA
e-mail: mquadrana@pandora.com

P. Cremonesi
Politecnico di Milano, Milan, Italy
e-mail: paolo.cremonesi@polimi.it

© Springer Science+Business Media, LLC, part of Springer Nature 2022                    301
F. Ricci et al. (eds.), *Recommender Systems Handbook*,
https://doi.org/10.1007/978-1-0716-2197-4_8

Such challenging situations do, however, not only arise in the e-commerce domain, but are common in various application areas of recommender systems (RS) such as music, video, or news recommendation. From a historical perspective, early influential technical approaches for session-based recommendation were proposed in the first half of the 2000s, e.g., [81, 87, 98]. However, despite the high practical relevance of the problem, research in this area remained sparse, in particular when compared to the large number of technical works that focus on making non-contextualized recommendations based on long-term preference profiles; cf. Chap. "Advances in Collaborative Filtering". Until about 2015, individual proposals for session-based recommendations were published mostly for the popular domains of e-commerce, music, and news, e.g., [10, 15, 29, 36, 47, 107], or for niche applications like adaptive modeling support for machine learning workflows [49]. In 2015, the data challenge associated with the ACM Conference on Recommender Systems was focusing on specific session-based recommendation problems, and a larger dataset containing anonymous click logs was released. This dataset release and the then-starting boom in deep learning in recommender systems contributed to the largely increased interest in session-based recommendation problems that we observe today.

In this chapter, we review the topic of session-based recommender systems (SBRS) from different perspectives. First in Sect. 2, we characterize the problem setting and its variants in more detail in the context of Sequence-aware Recommender Systems [86]. We then review technical approaches to build SBRS in Sect. 3. Section 4 is afterwards devoted to questions regarding the evaluation and comparison of session-based recommendation approaches. In Sect. 5, finally, we given an outlook on future directions in this area.

# 2   Problem Characterization

Session-based recommendation problems differ from the traditional "matrix-filling" [92] abstraction in two main ways, see also Table 1.

- *Data:* The main input to an SBRS is a time-ordered sequence of logged user interactions.
- *Computational task:* The main task of an SBRS is to make recommendations that are relevant for an ongoing session.

Figure 1 illustrates the problem setting in more detail. The inputs are shown on the left-hand side. First, there are the interactions—for example, purchases or item view actions—that were recorded in the past for an *entire user community*. These can be used to train machine learning models or to apply data mining techniques. SBRS therefore are collaborative filtering (CF) systems [46] or hybrids with a CF component.

Each entry (interaction record) in the log can have additional attributes. The most common attributes are the type of the interaction and which item is affected by the

**Table 1** Comparing session-based recommendation with traditional recommendation scenarios

|       | Matrix-filling recommender system | Session-based recommendation |
|-------|-----------------------------------|------------------------------|
| Data  | User-item rating matrix | Time-ordered sequence of interactions, organized in sessions |
| Task  | Make time-agnostic recommendations for long-term preferences | Make recommendations for ongoing session |

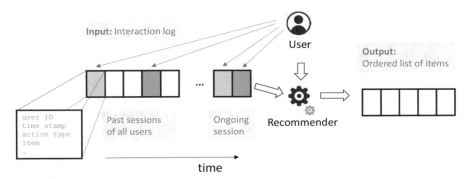

**Fig. 1** High-level overview of sequence-aware recommendation, adapted from [86]

interaction, e.g., a purchase of a certain item. SBRS can in principle also consider interactions that are not item-related, for example, search actions. In Fig. 1, the different shades of grey indicate different types of actions.

The log of past interactions is organized in *usage sessions*, which means that each interaction record has some information attached that describes to which session it belongs. This can, for example, be a session ID, which is explicitly assigned by the application that creates the log. If no session IDs are available, one can use information from browser cookies or similar techniques to identify which records belong together. In some cases, IDs of registered users might assigned to each record, even though the common assumption in SBRS is that users are anonymous.[1]

In cases when there are no explicit session IDs assigned, sessions can be constructed from the log post-hoc with the help of heuristics based, e.g., on cookies. A typical heuristic in some domains is to assume that a new session starts after an extended period of inactivity by the user, e.g., for 30 minutes.

The second input to an SBRS are the interactions that are observed in the *ongoing* usage session. Again, these interactions can have additional attributes, in particular the type of interaction and the affected item. The *computational task* of an SBRS is then to take these currently observed interactions and to make recommendations that are relevant for the ongoing session, using the past interaction data as background

---

[1] Not connecting click data with individual user accounts may in some cases also be beneficial in terms of user privacy and the General Data Protection Regulation in the European Union.

information, e.g., in the form of a machine learning model. The *output* of an SBRS, finally, consists of a rank-ordered list of recommended items as usual.

In [86], the following terminology is used to describe two variants of the underlying recommendation problem:

- *Session-based recommendation:* In this scenario, no user IDs are attached to the interactions or sessions. This means that no personalization of the recommendations according to longer-term preferences is possible. These problems are in the focus of this chapter.
- *Session-aware recommendation:* In this case, the interactions or sessions have user IDs attached, which makes it possible to consider long-term preferences when recommending, as done, e.g., in [51, 83, 85] or [104].

The use of terminology in the literature is however not always consistent. Sometimes, the term *session-based* recommendation is also used for situations where long-term preferences are available, e.g., in [85]. Some authors also use the term *sequential recommendation* in the context of session-based recommendation scenarios. Session-based recommendation problems are sequential both in the sense that the goal is to predict the next interactions in a given session and that the available data is time-ordered. However, not all sequential approaches are necessarily session-based. Instead, they may be based on time-stamped and longer-term user preference information, e.g., in the context of next-POI or next-basket recommendations. *Sequence-aware recommendation* is an umbrella term that covers all discussed problem settings: session-based, session-aware, and sequential recommendation. A formal characterization of sequence-aware problems is given in [86], where the authors also discuss additional computational tasks that can be done based on sequential interaction logs, such as trend detection, repeated recommendation, or the consideration of ordering constraints among the recommendations.

Looking at existing categorizations of recommender systems, SBRS[2] are typically based only on *implicit feedback*, cf. Chap. "Item Recommendation from Implicit Feedback", i.e., they face the usual problem of correctly interpreting the various types of user interactions [52]. SBRS are also *context-aware* systems [3], cf. Chap. "Context-Aware Recommender Systems: From Foundations to Recent Developments". Context adaptation in SBRS is however usually not based on explicit information about the user's contextual situation, e.g., the weather, but on the observed user interactions in the ongoing session, i.e., on the *interactional context*. Finally, there is also some relation to *time-aware* RS [13] where the user actions are time-stamped or time-ordered. Most time-aware RS in the literature are however not session-based, but are based on the user-item rating matrix abstraction. Still, these systems share the problem with SBRS that specific evaluation protocols are required that consider the ordering of the ratings or interactions.

---

[2] These discussions apply both for *session-based* and for *session-aware* recommender systems. For the sake of brevity, we will use the term *session-based* in this chapter, and will explicitly state when not both types of problems are meant.

In sum, the two main challenges of SBRS are the following:

- *Limited preference information:* Since in SBRS no user IDs are available, recommendations have to be made based only on a very small set of observed interactions. This in some ways corresponds to a *user cold-start* problem.
- *Implicit feedback:* There are various types of interactions that can be taken into account and there is some uncertainty attached to the interpretation of the observed user behavior.

## 3 Technical Approaches

Algorithmic approaches of different types were proposed for session-based recommendation problems over the years. Several of these technical approaches were previously also explored for the more general class of sequence-aware recommendation problems. In [86], the authors propose a categorization of existing approaches to sequence-aware recommendation based on the underling family of algorithms that is used to extract or model (sequential) patterns from the logged user interactions. An alternative approach is followed in [116], where the technical approaches for session-based recommendation are categorized into two main classes. We adopt this latter categorization in this section.

- *Model-free approaches*: In this category, we consider all algorithmic approaches based on data-mining techniques that do not involve the computation of a user or session model. Typical examples in this category are methods that are based on the extraction of frequent ordered and unordered patterns from historical session data [36, 47, 119]. Moreover, we include also all methods that perform inference based on nearest neighbors of the current user session [69] in this category.
- *Model-based approaches*: In this category, we consider all machine-learning approaches that explicitly model the user-item interactions within and across sessions. Examples of methods in this class are Markov Models [29, 42], Recurrent Neural Networks [39, 85] and Reinforcement Learning [98, 107].

In what follows, we only consider pure collaborative approaches, which are solely based on the recorded user-item interactions. In principle, all these approaches can be extended or combined with other techniques in order to consider various types of side information, such as item meta-data or contextual information. Examples of model-based techniques that leverage side information are [24] or [40].

## 3.1 Model-Free Approaches

Model-free approaches are often conceptually simpler, easy to implement and sometimes also less resource demanding than model-based ones. In many cases, they do not require the computation of complex mathematical models from historical session data, and they generally rely on well optimized data mining procedures to extract patterns from sessions. Another potential advantage is that they can be easier to analyze and debug than complex models. For example, the recommendations generated by frequent-pattern and rule mining techniques can be straightforwardly explained in the terms of the patterns that were extracted from historical session data. Their simplicity can also make them less prone to over-fitting in situations when the training data is sparse. Nonetheless, despite their simplicity, certain types of model-free algorithms—in particular nearest-neighbor techniques—can lead to competitive or even superior performance when compared to more complex model-based variants according to several offline benchmarks [69, 71, 72].

### 3.1.1 Sequential Pattern Mining

Frequent Pattern Mining (FPM) techniques aim to discover user consumption patterns within large datasets of transactions. These patterns can then be used to predict the most probable actions[3] that the user will take in the session given the sequence of actions they have done so far.

In session-based recommendation scenarios, the most simple approach is to determine only pairwise item co-occurrence frequencies in past sessions (training data) in order to implement buying suggestions of the form "Customers who bought ... also bought". At run time, recommendations are made by looking up which items often appeared in the same session as those in the ongoing session. A simplified version of association rules mining for SBRS, which only considers rules of size two, is provided in [69].

These kinds of recommendations belong to the family of Association Rules [4], which identify items or actions that frequently co-occur in the same session, regardless of the order of their appearance. Based on our definition of the problem as described in Sect. 2, FPM-based algorithms are not session-based in the strict sense, but can be used as building blocks for more advanced approaches. Beyond simple co-occurrence patterns, one can consider Sequential Pattern Mining techniques [5], which take the order of items into account. Finally, Contiguous Sequential Patterns require that the co-occurring items are adjacent in the sequence of actions within a session. A simple yet often effective light-weight form of such *Sequential Rules* of size two was also proposed in [69].

---

[3] For the sake of readability, we use the terms "user action (on an item)" and "user-item interaction" interchangeably.

### 3.1.2  Nearest Neighbors

Nearest-neighbor-based methods, cf. Chap. "Trust Your Neighbors: A Comprehensive Survey of Neighborhood-Based Methods for Recommender Systems", are among the simplest and most effective approaches to session-based recommendation. We describe here two variants, namely *item-based* and *session-based* nearest-neighbor methods.

Item-based methods for session-based recommendation were introduced in [40], in which the similarity between items is computed based on their co-occurrence in a set of training sessions. Every item is associated to a large sparse binary vector representing whether or not the item appeared in each of the training sessions, and the item similarity is simply computed as the cosine similarity between each pair of vectors. The recommendations are computed as the list of most similar items to the *last item* in the user session. The type of recommendations produced by item-based approaches follow closely the "Customers who bought ... also bought" logic.

Session-based methods extend this concept beyond the last item in the session by comparing the *entire* user session with the past sessions in the training data. A comprehensive description of this family of model-free approaches is provided in [69]. In summary, given a session $s$, these methods determine the set $N_s$ of the top-$k$ most similar sessions using a suitable similarity measure $\text{sim}(s_1, s_2)$, such as the Jaccard or cosine similarity over the binary vectors that represent sessions in the item space. The recommendation scores for each item $i \in s$ are computed as:

$$\text{score}_{\text{SKNN}}(i, s) = \sum_{n \in N_s} \text{sim}(s, n) \cdot 1_n(i) \tag{1}$$

where $1_n(i)$ returns 1 if the session $n$ contains $i$ and 0 otherwise. Note that Jaccard and cosine similarity do not take the order of the items in a session into account. In [69] the authors therefore show how variants of this session-based nearest-neighbor method can be built to take into account the order of the user actions. A recent time-aware and sequence-aware variation of this scheme was also discussed in [31].

For both item-based and session-based nearest neighbor methods, the set of neighboring items and sessions can usually be precomputed offline to speed up the processing at prediction time [45].

## 3.2  Model-Based Approaches

Model-based approaches exploit advanced machine learning techniques to represent the complex dynamics of the user actions within and across sessions. Most recent model-based approaches to SBRS fall into the category of *sequence learning* methods, which aim to learn models from sequences of past user actions to predict future ones [74]. Advanced sequence learning methods allow to model complex interactions between user actions in sessions. For example, certain models compute a representation of the user's latent purchase intent in the ongoing session [35, 106]. Other methods explicitly model the temporal dynamics between sessions [112]. In general, model-based algorithms therefore have the potential to identify more complex dependencies and interactions from the logs of historical user sessions.

On the other hand, the greater representational power of these family of techniques often comes at the cost of much larger amounts of training data and computational resources than the model-free counterparts. Moreover, despite the many recent advances in the field of interpretability of machine learning models (see, for example, the work in [122]), most of these models behave as *black-boxes* that are difficult to inspect and to debug when their responses do not conform with our expectations. These issues and the potential unpredictability of the responses require complex systems to constantly monitor their behaviors, which makes their integration into existing production settings non-trivial.

Nevertheless, there exist commercial solutions based, e.g., on Hierarchical Recurrent Neural Networks[4] that simplify the adoption of state-of-the-art session-based recommendation systems in a wide range of real-world applications.

### 3.2.1 Markov Models

Markov Models treat sequential data as a stochastic process over discrete random variables over a finite number of steps. The simplest of these models are first-order Markov Chains, which model the transition probabilities between two *subsequent* actions in a sequence. An example of using first-order Markov Chains for next-item recommendation in the context of SBRS is given in [69]. The model in this work simply relies on the count of the number of times in which an action on an item $m$ happens immediately after an action on item $l$ for every session in the training dataset $S$. In summary, given a session $s = (s_1, s_2, \ldots s_{|s|})$, where $s_{|s|}$ is the last item on which the last action happened, the recommendation score for item $i$ is computed as:

$$
\text{score}_{\text{MC}}(i, s) = \frac{1}{\sum_{p \in S} \sum_{x=1}^{|p|-1} 1_{\text{EQ}}(s_{|s|}, p_x)}
$$
$$
\times \sum_{p \in S} \sum_{x=1}^{|p|-1} 1_{\text{EQ}}(s_{|s|}, p_x) \cdot 1_{\text{EQ}}(i, p_{x+1}) \tag{2}
$$

where the function $1_{\text{EQ}}(a, b)$ indicates whether items $a$ and $b$ are the same. The right-hand side counts how frequently item $i$ immediately follows $s_{|s|}$ in the training

---

[4] https://docs.aws.amazon.com/personalize/latest/dg/native-recipe-hrnn.html (Date visited: 2020/06/01).

data, while the normalization coefficient on the left ensures that scores are valid transition probabilities.

In situations with very large item catalogs, the transitions between pairs of items tend to be extremely sparse, which translates into poor estimates of the above transition probabilities. This issue can be alleviated by using certain heuristics, like skipping, clustering and finite mixture modeling [98], or by factorizing the transition matrix (or tensor) [90]. Markov Chains can be extended beyond first-level interactions, as the user's next action and intent might in fact depend on action previous to the last one. One option are Variable-order Markov Models (VMMs), or context-trees, which allow to consider different sequence lengths in one model [29, 37]. The work in [77] describes the offline and online evaluation of an e-commerce recommender systems based on Bayesian Variable-order Markov Modeling. Suffixes of sessions that were observed in the past are arranged into a tree-shaped structure. Given the current session, the algorithm then computes the probability that a specific item is encountered by aggregating the probabilities from the corresponding nodes in the tree.

Another option are Hidden Markov Models (HMMs), which replace item transition probabilities of plain Markov Chains with the transition probabilities between (consecutive) discrete hidden states, which represent the unobserved hidden state of the user, and a probability distribution from the hidden states to the observed actions. Both probability distributions can be learned from observational data [42].

### 3.2.2 Recurrent Neural Networks

Recurrent Neural Networks (RNNs) are real-valued hidden-state models that are especially suited for processing data of sequential nature, see also Chap. "Deep Learning for Recommender Systems". Similarly to HMMs, they are based on hidden states that are updated for every input in the sequence, and then used to predict the probability of the next item in the sequence. However, RNNs use complex non-linear dynamics, which can largely enhance their learning power. In their simplest variant, the hidden state $h_t$ of the RNN is updated from the current input in the sequence $s_t$ and the previous hidden state as follows:[5]

$$h_t = \sigma(\mathbf{W}^{hs} s_t + \mathbf{W}^{hh} h_{t-1}) \tag{3}$$

where $\mathbf{W}^{hs}$ is the matrix of conventional weights between the input and the hidden layer, $\mathbf{W}^{hh}$ is the matrix of recurrent weights between the hidden layer and itself at adjacent time steps, and $\sigma(\cdot)$ is a non-linear function like the logistic sigmoid function. The updated hidden state is then used to compute the probability of the next item in the sequence simply as follows:

---

[5] Bias terms are omitted to enhance readability.

$$y_t = \text{softmax}(\mathbf{W}^{yh} h_t) \qquad (4)$$

where $\mathbf{W}^{yh}$ is the matrix of conventional weights between the hidden layer and the output layer. There exist several variants of RNNs, like Long-Short Term Memory networks (LSTM) and Gated Recurrent Units (GRU). A comprehensive review can be found in [65].

RNNs are a "natural" option for session-based recommendation given the intrinsic sequential nature of user actions. In fact, various RNN-based models were proposed for SBRS [115]. Many of them are extensions of GRU4Rec [39], the first RNN-based model explicitly devised for session-based recommendation. GRU4Rec models user sessions using Gated Recurrent Units (GRU) [19], which enhance the update Eq. 3 by learning when and how much to update the hidden state at each step.

RNNs are generally trained by minimizing the average cross-entropy loss over all sessions $s \in S$:

$$L(S) = -\frac{1}{|S|} \sum_{s \in S} \sum_{s_t \in s} \log y_t$$

in which all actions $s_t$ in the session are assumed to be relevant to the user. In addition to that, GRU4Rec can be trained using ranking loss functions like BPR [89], TOP1, BPR-max and TOP1-max to directly optimize the ranking of the true next action over a sample of negative ones. Training GRU4Rec using ranking losses leads to better recommendation accuracy, as highlighted in [41]. Training RNNs over large item catalogs and long sessions is computationally intensive. GRU4Rec can be trained efficiently on GPUs thanks to mini-batch negative sampling, which samples the negative items used by the loss function among the items in the current mini-batch of sessions that is being processed. GRU4Rec+[41] adds the opportunity to use an additional fixed set of negative items to further improve the output ranking quality.

Several extensions to GRU4Rec were proposed in the literature. PRNNs [40], for example, train multiple GRU4Rec instances in parallel to learn a multi-modal SBRS. Each instance learns a representation of the session from a different mode, such as item identifiers and features like product images and descriptions. PRNNs then employ different strategies to train and merge each GRU4Rec instance. Another extension is HRNN [73, 85], a *session-aware* model that employs two interconnected GRU4Rec instances to model the user behavior both at session- and user-level. After each session, the user-level GRU4Rec instance uses the hidden representation of each session-level GRU4Rec instance at the end of each session to update the user state. The updated user-level representation is then used to initialize the hidden state of the forthcoming session-level GRU4Rec instance, which from there on behaves as a traditional session-based GRU4Rec model.

Even though RNNs are a natural Neural Network architecture for session-based recommendation, they are not the only option available. Researchers have studied several other architectures like Shallow Networks, Convolutional Neural Networks

(CNNs), Graph Neural Networks (GNNs) and Variational Auto-Encoders (VAE) for session-based and, more generally, for sequence-aware recommendation. See Table 2 for a non-exhaustive list of such models. As can be seen, many neural models were proposed in recent years. For a few of them, reports regarding their practical use are available as well, see e.g., [58].

### 3.2.3  Reinforcement Learning

All the previously described methods learn a representation of the user or session in order to maximize the chances of guessing the immediate next user action(s) correctly. However they rarely, if never, consider the *final goal* of the user session and *long-term effects* of the recommendations when deciding which action(s) to recommend next. By myopically optimizing for the immediate next actions and *rewards*, the recommender can be easily misled to trade the long-term user satisfaction for immediate engagement, with possible detrimental effects like click-baiting [44]. Moreover, the aforementioned families of recommenders operate deterministically, meaning that the recommender will always suggest the same list of items if fed with the same input sequence of user actions.

In practice, recommenders operate in environments with large and dynamic item spaces, in which items are frequently added and removed from the catalog. Think, for example, of the many newly released albums or singles that are added every day to the catalog of popular music streaming services. User preferences tend also to evolve relatively quickly over time, often influenced by external factors like community trends or by the natural evolution of personal interests.

Recommendation algorithms should hence be flexible enough to adapt their beliefs of the user interests and, consequently, their predictions, to the inherent dynamics of the item and user spaces. One strategy often used in practice is to introduce a certain degree of *exploration* into recommendation algorithms in order to gather feedback from users on the newest items, or to refine the knowledge on the current user's interests by exposing them to less *exploitative* items.

Exploration requires recommenders to take "suboptimal" actions w.r.t. the current estimates of the affinity between users and items, which indeed incurs costs in the short-term that will hopefully be offset in the mid- and long-term. Trading short-term engagement with longer-term success lies at the core of the well-known *exploration-exploitation dilemma* [63], and it is one of the guiding principles behind many Reinforcement Learning techniques.

The main goal of Reinforcement Learning (RL) is that of learning an algorithm capable of taking actions in an environment that are optimized towards a long-term goal. At each round, or *trial*, the RL algorithm, or *agent*, observes a *state* based on which it decides which *action* to take next. Actions are taken by following a *policy*, which is often defined as a stochastic function over actions. For each action taken, the agent receives a *reward* from the environment that it is used to refine the policy in a way that the long-term objective can be achieved under certain guarantees.

**Table 2** List of neural network based models for SBRS

| Model | Network type | Task | Short description |
|-------|-------------|------|------------------|
| GRU4Rec [39] | RNN | Session-based | Next-item prediction for anonymous user sessions with Gated Recurrent Units, mini-batch negative sampling and ranking loss functions. |
| GRU4Rec+ [41] | RNN | Session-based | Enhanced GRU4Rec with additional negative sampling and alternative ranking loss functions. |
| PRNN [40] | RNN | Session-based | A multi-modal session-based model based on multiple GRU4Rec models trained in parallel. |
| HRNN [73, 85] | RNN | Session-aware | A hierarchical GRU4Rec model that learns from user's long-term and short-term (session-level) interactions. |
| THRNN [112] | RNN | Session-aware | An enhanced HRNN model with Temporal Point Processes that factors in the temporal distance between consecutive user sessions. |
| SWIWO [43] | Shallow Network | Session-based | A shallow network with a wide-in-wide-out structure. |
| NARM [64] | RNN+Attention | Session-based | A model that combines RNN with attention to jointly learn a global session representation and the assumed main purpose of a session. |
| STAMP [66] | Memory Network | Session-based | Replaces RNN with an attentive memory network that retains the short-term user interests. |
| CASER [105] | CNN | Sequence-based | A model that transforms session sequences into an "image" of item embeddings, then learns sequential patterns through convolutional filters. |
| NextItNet [121] | CNN | Sequence-based | A model that learns long-term dependencies over long user sequences through holed convolutions and residual connections. |
| SVAE [94] | VAE | Sequence-based | A model that encodes sequences of user actions through a recurrent Variational Auto Encoder. |
| SR-GNN [117] | GNN | Session-based | A model that represents session sequences as graphs. Uses a GNN to capture complex item transitions on the session graph. |
| TAGNN [120] | GNN | Session-based | Extends SR-GNN by conditioning the session representation on the target item using an Attention network in combination with a GNN. |

In the context of recommender systems, actions are often thought of as the recommendable items, rewards are the users' feedbacks on the recommended items (such as clicks, listening, add-to-cart, etc.). Long-term goals may be the purchase at the end of a session or the conversion of the listener from trial to premium subscription. The fact that policies are no longer deterministic but *stochastic* inherently allows any RL method to explore the action space more or less effectively. Here, we briefly introduce two RL methods that are frequently used in real-world recommenders: Multi-Arm Bandit (MAB) algorithms and Markov Decision Processes (MDP).

Multi-Arm Bandits (MABs) are widely used for sequential decision making. At each round, or trial, the agent selects one of $K$ *arms* for which it receives a reward, which in the simplest case is binary (e.g., the user clicks/plays the recommended item). The goal of the agent is to maximize the sum of the rewards over time. Since the agent receives rewards only for the arms it selects, it needs to *explore* all arms to identify which are the best ones. At the same time, the selection of suboptimal arms leads to lower rewards, which induces the bandit to *exploit* arms with known rewards. A central element of any Bandit algorithm therefore lies in its arm-selection policy, which has to balance exploration and exploitation. Exploration might decrease the *short-term* recommendation quality as many suboptimal arms can be chosen initially, but in turn can improve the *long-term* quality as more information on the items' rewards is gathered. When using MABs for session-based recommendations, each arm is modeled as an item to be recommended to the user within the session. At each trial, one item is recommended (one arm is chosen) and the algorithm receives as feedback information about whether or not the user interacted with the item. Some examples of arm-selection strategies are $\epsilon$-greedy and Upper Confidence Bounds [8]. An often used variant of Bandit models are Contextual Bandits (CB) [63], which learn personalized arm-selection policies by conditioning the expectations of the rewards on a joint user/item feature vector, called *context*.[6] However, the contextual feature vectors at each round are assumed to be independent from the previous ones, which severely limits the applicability of CB to session-based recommendation scenarios, where the sequence of user actions in the session can be highly important to understand as user's intent or goals.

Markov Decision Processes (MDPs) improve over CB-based approaches by conditioning each action taken by the agent on the sequence of past user actions [17]. An MDP is defined as a four-tuple $(S, I, \mathbf{P}, R)$, where $S$ is a set of states, $I$ is a set of items, $\mathbf{P} : S \times I \times S \to \mathbb{R}$ is a state transition probability and $R : S \times I \to \mathbb{R}$ is the reward function. In an MDP we seek for a policy $\pi(i|s)$ that represents a distribution over the item to recommend $i \in I$ conditioned on the user state $s \in S$ such that the expected cumulative reward obtained by the recommender systems is maximized [17, 98]. There are different families of methods to solve MDP problems,

---

[6] This should not to be confused with the common definition of context in recommender systems, which often refers to the set of contextual variables, such as location, time, etc., that can influence a user's decisions.

such as Q-learning [99] and Policy Gradient [62]. In a straightforward application of MDPs for session based recommendations, the set of states $S$ is defined as the set of all possible sequences of items, and every element in $S$ can be identified with a (possibly infinitely long) user session in terms of interacted items. At each step, the next item to recommend will be sampled from the policy $\pi$, which can be later updated based on the reward received by the user (e.g., the user playing vs. skipping the next recommended song). In SBRS, the MDP can be trained to optimize certain session-based metrics of interest, like the overall time spent listening/watching, or to maximize the chances that the user will perform certain actions throughout the session, such as to purchase one or more products.

The presence of the user state $s$ and of the station-transition function **P** allows to condition the actions taken by the MDP-based recommender on historical user actions. For example, the authors of [98] define the user state as the sequence of the last $k=3$ user interactions, and the state-transition probabilities are inferred from a $k$-order Markov Chain with special correction factors. More recently, thanks to the advancements in Deep Reinforcement Learning [80], the authors of [17] proposed an MDP-based recommender model in which the states and state-transition function are parameterized by an RNN with softmax output. In parallel, another RNN learns a user behavior model by predicting which action the user will most likely take next. The user behavior model is used to correct the biases introduced when learning from logged feedback data through importance weighting (see Sect. 4.5.2 for a further discussion on the topic).

In summary, Reinforcement Learning methods can represent a powerful option in situations with rapidly evolving catalogs of items and when the long-term impact of recommendations is of interest, which makes them suitable for both session-based and session-aware recommendation scenarios. However, the design and implementation of these recommenders is non-trivial, as they, for example, can be sensitive to the definition of the state-transition and reward functions. Other challenges include, for example, the existence of large item spaces or small effect sizes [111].

# 4  Evaluation of Session-Based Recommenders

The main types of evaluation approaches in recommender systems are (1) offline experiments, (2) user studies, and (3) field tests [97]; see also Chap. "Evaluating Recommender Systems".

In academia, "offline" (simulation-based) experiments on datasets containing historical interactions between the users and the system (e.g., explicit ratings, implicit feedback) are predominant. The common offline evaluation methodology when using the matrix completion abstraction is to split the given interaction data into training and test splits, use the training data to learn a model and predict the held-out preferences based on this model. This process is typically repeated a number of times in a cross-validation procedure in order to compute confidence bounds. The relevance of the recommendations generated by an algorithm can then be assessed with the help of error, classification and ranking metrics from machine learning and information retrieval such as RMSE, Precision, Recall, MAP [38]. In addition to these relevance measures, one can analyze a number of other quality factors of the recommendations, e.g., in terms of the diversity of the list, the general popularity of the recommended items, or the algorithm's catalog coverage [95]; see also Chap. "Novelty and Diversity in Recommender Systems".

User (laboratory) studies are an alternative to offline experiments. Such studies are often used to assess the potential impact of recommendations on the actual choices of users, an aspect which cannot be determined based on simulation studies. Finally, field studies (e.g., in the form of A/B tests) are used to analyze the effects of recommenders on their users in real-world environments, see also Chap. "Value and Impact of Recommender Systems". This latter form of evaluation is however comparably rare in academic environments [32].

While in principle all three mentioned evaluation forms (offline studies, user studies, field tests) can be applied for SBRS, slightly different evaluation methodologies are used in offline studies.

## 4.1 Offline Evaluation Protocols

When conducting an offline experiment, different choices have to be made regarding (1) the partitioning of the dataset, (2) the definition of the target interactions, and (3) the metrics that are employed. These choices should be guided by the particularities of the targeted application scenario.

### 4.1.1 Dataset Partitioning

The first step for the evaluation of an SBRS is the generation of the training and test datasets. The former is used to train the recommendation algorithm, whereas the second contains held-out data used exclusively during the evaluation. Additionally, a fraction of the data in the training dataset can be held out for tuning the hyperparameters of the recommendation algorithm (the validation set).

In standard matrix completion setups, the partitioning of the interactions into training and test sets is often done by sampling interactions randomly. Placing 80% of the interactions in the training set and 20% in the test set is a common approach. However, in session-based recommendation scenarios, where the interactions are sequentially ordered and grouped in sessions, an alternative approach is needed. For instance, one should consider the temporal ordering of the interactions and select the most recent interactions for the test set.

Two forms of splitting the data into training and test set are possible: *interaction-level* and *session-level* splitting. Session-level splitting partitions the dataset by holding out entire sessions, i.e., without breaking sessions apart. Interaction-level splitting, in contrast, acts over the individual interactions: some or all of the sessions are split into contiguous subsets that are assigned either to the training or the test set.

When splitting the data at the *interaction-level* two additional options exist. We can assign all interactions before a certain point in time to the training set and the remaining interactions to the test set, as done in [34]. With this procedure, an interaction is therefore assigned to one of the sets independent of the user or the session it might belong to. Or, we can use fixed-size *within-session* splits and place the last $k$ interactions of each session into the test set (leave-$k$-out) and the remaining into the training set, as done in [18]. Similarly, one can apply a time-based splitting criterion at the *session-level*, i.e., we do not split up sessions but consider the timestamp of the first interaction within a session to decide if the session goes in the training or test set [39, 109].

Both interaction-level and session-level partitioning can be applied to either the whole set of users (*hold-out of interactions*) or to a subset of the users (*hold-out of users*). In the latter case one can select a number of *training users* and put all their data into the training set. Interactions of the remaining *test users* are then further split—either at interaction or session-level—into a *user profile* and *test data* [47]. The user profile includes the less recent interactions and is used as input to the recommender while the test data contains the most recent interactions to be predicted.

Mixed approaches are possible and often advised. One can, for example, first adopt a session-level splitting to partition the dataset into training and test sessions. Interactions within each test session are further split with a leave-$k$-out approach into a *session profile* and *test interactions*.

No common standard exists in the community regarding data partitioning procedures. It is however advisable to use a session-level partitioning procedure to avoid that individual user sessions are split up. This is in particular necessary to mimic the behavior of SBRS that are batch-trained on instances of past sessions for efficiency reasons. Moreover, user-level partitioning is preferred to ensure that a recommender is able to provide recommendations to new users (i.e., users with no records of past interactions with the system). Finally, it is advisable to use a partitioning procedure that reflects the task at hand. For instance, within-session splits are useful for pure session-based recommender systems. On the contrary, a mixed approach that combines session-level partitioning and within-session split is preferred for session-aware recommender systems.

The size of the training data can also be chosen based on domain-specific requirements. In domains such as news recommendation or e-commerce, focusing on the most recent interactions in some cases is sufficient or even advisable, e.g., because news can quickly become outdated or because e-commerce shoppers might concentrate on trending or recently added items [45]. In general, the evaluation should be performed with different time splits, in order to evaluate the learning rate of the algorithm (i.e., the minimum amount of training data that provide stable recommendations).

Using a repeated and randomized dataset partitioning for cross-validation is not always possible with SBRS. Sequence or time dependent interactions exist, which make it impossible to randomly assign interactions to the training and test splits. One possible workaround is to partition the dataset into several, potentially overlapping time slices of about the same length ("sliding window") and use the actions at the end of these time slices as test sets [45, 47].

### 4.1.2  Definition of Target Interactions

In traditional evaluation setups, we aim to predict ratings for a set of target items or search for an optimal ranking of these items. In SBRS, we are usually interested in predicting future interactions between a user and the system, where interactions usually have an associated type and item. In addition, the input to the evaluation step is not limited to a user or user-item pair for which a ranked list or rating prediction is sought for, but the input includes a sequence of interactions, e.g., entire sessions from the test set and/or partial sessions from the user or session profile. If such a sequence of interactions is given, the general idea is to hide a subset or all of the interactions of the given session and predict the user's (next) interactions. Two variations of the evaluation scheme can be used. With both variants the first $N$ interactions of the session are used as session profile and "revealed" to the algorithm.

- *Given-N next-item prediction.* With this protocol, the task of the recommender is to predict the immediate next interaction when the first $N$ interactions of the session are revealed to the algorithm [47]. $N$ can be incrementally increased [39, 40, 85]. In same domains, often only the last element of a sequence is hidden (i.e., $N$ = session length$-1$) [36].
- *Given-N next-item prediction with look-ahead.* With this protocol, the task is to predict one of the remaining hidden interactions (which form the look-ahead set) [34, 39]. The order of the hidden interactions is often not considered relevant in terms of the quality of the recommendation. Also in this case, $N$ can be incrementally increased.

Additional variations of these protocols exist. One can, for example, limit the evaluation to interactions of certain types (e.g., purchase actions). Another variant is to hide only the last interaction of a session, which is used in application scenarios to assess the recommendation performance when more information is available ("warm start"). For instance, in the domain of next-track music recommendation, sometimes only the last element of a session is hidden [36].

Generally, one limitation of such "hide-and-predict" evaluation approaches lies in the assumption that there exists exactly one good recommendation. Depending on the domain, this might however be too narrow and it might therefore be reasonable to consider items other than the hidden one as good recommendations as well. This was done, for example, in [58] for the e-commerce domain. In their case, items that were nearly identical to the hidden ground truth item were treated as relevant recommendations as they represented possible shopping alternatives.

### 4.1.3 Evaluation Metrics

The output of an SBRS usually is a ranked list of alternatives, where the items are ordered according to the their likelihood to be the next one that the user will interact with. It is therefore possible to use standard *classification* and *ranking* metrics for performance evaluation; see also Chap. "Evaluating Recommender Systems''. The set of possible metrics include Precision, Recall, Mean Average Rank (MAR), Mean Average Precision (MAP), Mean Reciprocal Rank (MRR), Normalized Discounted Cumulative Gain (NDCG) and the F1 metric. When the application domain requires recommendations to fulfill an explicit or implicit order constraint, ranking metrics (such as MAP, MRR and NDCG) are preferred over classification metrics. Typically, most of the ranking metrics are highly correlated when evaluated with a fixed-size split and their choice does not largely affect the outcomes [21]. Note that differently from many conventional recommender systems, many SBRS recommend also items that the user already knows or has purchased in the past. In these application scenarios, the evaluation protocol has to be adapted in order to consider already known items when computing the metrics.

In many application domains ranking metrics alone cannot fully inform us about the quality of the recommendations. In playlist recommendation, for example, we may want to assess the capability of the recommender to generate good transitions between subsequent songs. Given a current track, there may be many other tracks that are almost equally likely good to be played next, and ranking metrics to not take that aspect into account. This can in turn lead to the effect that more popular tracks are favored by an algorithm [48]. Therefore, alternative metrics derived from natural language processing, like the Average Log-Likelihood (ALL), can be adopted instead [15, 16]. Regarding the ALL, some concerns however exist regarding the true usefulness this measure [10].

Generally, when the goal of a recommender is not limited to predicting the next-best interaction but a sequence or set of interactions that are related to each other, alternative "multi-metric" evaluation approaches are required that can take multiple quality factors into account in parallel. They can consider the order of the recommendations or the internal coherence or diversity of the recommended list as a whole. For instance, in the music domain, one might also be interested that the set of next tracks is coherent not only in itself but with the last played tracks. In many cases, the different quality factors can lead to trade-off situations like "accuracy vs. diversity", which have to be balanced by a recommendation algorithm [54], see also Chap. "Novelty and Diversity in Recommender Systems". Unfortunately, no standards exist yet when it comes to evaluate recommendation lists with respect to, e.g., diversity or coherence. Usually, a selection of meta-data features is used to measure, for example, *intra-list* diversity or the smoothness of the transitions between the objects in the list. However, to what extent such measures reflect the users' quality perceptions sometimes remains open.

Finally, SBRS are often computationally more complex than traditional recommenders. Therefore, the evaluation of SBRS should also discuss the time and space complexity of the methods and report running times for typical training data sets so that the scalability can be assessed.

## 4.2 Evaluation of Reinforcement Learning Approaches

The offline evaluation of recommender systems based on traditional supervised learning algorithms assumes the independence between interactions and recommendations: a user's future interactions with items are not influenced by the recommendations provided to the user, but only by the items in the catalog.

The assumption is not valid anymore when evaluating recommender systems based on reinforcement learning, as the algorithm (agent) is updated based on both the interactions (reward) from the user (environment) at the current time stamp and the recommended items (context, or impressions) at the previous time stamp [114].

Because of the interactive nature of reinforcement learning algorithms, the best way to do an unbiased evaluation is to run the algorithm online on live data, e.g., with A/B testing. However, in practice, this approach is seldom feasible.

Therefore, the offline evaluation of session-based reinforcement learning approaches requires to simulate an online environment in which users dynamically interact with the system whenever a new set of recommendations is presented. However, creating a simulator itself might introduces bias in the results, making the evaluation unreliable [63]. The problem arises because datasets are collected by presenting items to users by using a logging policy (i.e., a recommender system, either personalized or non-personalized,) which is different from the one under test. This leads to the *off-policy policy evaluation problem* in reinforcement learning further discussed in Sect. 4.5.2.

One solution is to apply the *replayer* method, as described in [96, 114]: each user interaction in the historical dataset is replayed to the algorithm under evaluation. If the item recommended by the algorithm under test is identical to the one in the historical dataset, this item is considered as an impression to the user and the reward can be computed. The replayer method is unbiased under the assumption that interactions in the dataset were collected by providing users with random recommendations. However, this assumption is unrealistic in most session-based datasets.

Biases in the results obtained with the replayer method can be removed by using off-policy estimators, such as the Inverse Propensity Score used in [32]. However, this comes at the cost of increased variance in the results.

*RecoGym*, presented recently in [93], is an alternative reinforcement learning simulator that can be used also for session-based recommendations. RecoGym allows to simulate sessions containing both organic interactions (i.e., interactions not affected by recommendations) and *bandit interactions* (inspired by bandit algorithms), thus allowing for a fine control of the trade-off between variance and bias in the evaluation.

## 4.3 Datasets

Datasets suitable for benchmarking SBRS algorithms must satisfy two requirements: (1) interactions must be ordered in time (eventually with an absolute time-stamp) and (2) interactions must be grouped into *usage sessions*. Note that differently from the more traditional top-N recommendation task, datasets for SBRS do not necessarily need to map interactions to users.

In recent years, a number of datasets to benchmark SBRS algorithms have become available to researchers. Table 3 shows the characteristics of a (non-exhaustive) set of public datasets used in existing works.

Very few datasets (e.g., JD, Microsoft Web Data, Delicious and 30Music) associate sessions with users and, as such, can be used to evaluate both session-aware and session-based recommender systems. Most of the datasets do not associate sessions with users. For these datasets, the assumption is that each session corresponds to a unique user. These datasets therefore cannot be used to evaluate session-aware algorithms. Only a few datasets, such as the Diginetica dataset, provide a partial mapping between sessions and users. Most of the datasets provide multiple types of interactions. For instance, the Diginetica dataset contains views, clicks and purchases; the 30Music datasets contains play events and ratings.

Note that in Table 3, we only list datasets where the interactions are already organized into sessions. We omit datasets in which interactions are time-ordered but

**Table 3** Examples of publicly available datasets for session-based recommendation

| Dataset | Domain | Users | Items | Interactions | Sessions | Reference |
|---|---|---|---|---|---|---|
| Diginetica | EC | – | 134M | 4,3M | 574K | [25] |
| ACM RecSys challenge 2015 | EC | – | 38K | 34M | 9.5M | [88] |
| AVITO | EC | – | 4K | 767K | 32K | [9] |
| RetailRocket | EC | – | 417K | 2.8M | 1.4M | [9] |
| JD | EC | 810K | 240K | 11M | 2.7M | [53] |
| Microsoft web data | Web | 40K | 8K | 68K | 37K | [78] |
| Delicious | Web | 8.8K | 3.3K | 60K | 45K | [23] |
| NYC taxi dataset | POI | – | 299 | – | 670K | [82] |
| Art of the mix | Music | – | 218K | – | 29K | [7] |
| 30Music | Music | 40K | 5.6M | 31M | 2.7M | [1, 110] |
| Spotify sequential skip prediction | Music | – | 3.7M | – | 150M | [12, 108] |

EC: E-commerce, Web: Web browsing; POI: Point of Interest

not organized into sessions. For these datasets, some pre-processing logic can be applied to create sessions through heuristics, which, however, makes the comparison of different research works more challenging. Furthermore, the chosen heuristics that determine if two interactions belong to the same usage session have to be chosen with care. In the e-commerce domain, for example, we might consider two interactions with a 30-minute gap between them as belonging to different sessions. This gap might, however, be inappropriate in another application. A user of a video-on-demand service might, for example, watch a number of episodes from the same series back to back. In this case, there might be no interactions by the user for more than 30 minutes while the episode is played. Still, the interactions happening after this episode should be part of the same session.

Finally, in some research works, we observe that time-stamped user-item inter-action data, like the MovieLens or Netflix datasets, are used to evaluate SBRS. These datasets are however not suitable for the task, e.g., because the time stamps do not correspond to the time of watching a movie but to the point in time when the rating was submitted, which can be hours, weeks, or years after the interaction. The prediction would therefore refer to the question which item a user will rate next, which is typically not in the focus of SBRS research.

## 4.4   User-Centric Evaluation of Session-Based Recommendations

Offline evaluations, as discussed so far, have their limitations. Importantly, they cannot inform us about the *quality perception* by users, e.g., if the recommendations were truly considered useful, if they helped to discover something new, reduced the choice difficulty, or increased the users' satisfaction with the system.

User-centric evaluation approaches in the form of controlled randomized experiments are not uncommon in the recommender systems literature in general. Studies like the one in [26] often analyze various quality factors of recommender systems in parallel and, for example, investigate the relationships between quality dimensions such as perceived commendation accuracy, novelty, or diversity. While such user studies usually seek to answer very specific research questions, certain quality dimensions are often common across these dimensions. Pu et al. [84] and Knijnenburg et al. [57] have therefore put forward two alternative general frameworks for user-centric evaluation. These frameworks are not limited to aspects regarding the recommendation algorithms themselves (e.g., regarding accuracy, diversity, and novelty), but also include questions related to the perceived quality of the human-computer interaction and the usefulness of the system as a whole.

In the context of session-based recommendation approaches, the number of user studies and field studies is still very limited. The outcomes of a field study of deploying recommendations of the style "Customers who bought . . . also bought" in the context of a reference item are reported in [60, 61]. The studies revealed, among other aspects, that purchase-based consumption patterns are far more effective than view-based patterns, or that recommender systems of that type can lead to a decrease in sales diversity over time.

Recently, differences between alternative session-based recommendation strategies were explored through user studies in [55] and [70]. In both studies, the application domain was that of music recommendations. The question analyzed in these studies is how playlist continuations, as produced by session-based recommendation algorithms, were perceived by users along different quality dimensions. Moreover, both studies tried to shed light on the question to what extent accuracy results obtained in offline experiments correlate with such quality dimensions.

In [55], 277 study participants were presented with a number of four-item playlists, and for each playlist five alternative continuations were provided. The playlists were previously created and shared by users on different music platforms and originally consisted of at least five tracks. One main task of the participants was to rank the five alternatives in terms of their suitability as a continuation. Four of the alternative continuations were computed with the help of different session-based algorithms, and one was the *true* continuation of the original playlist.

The study led to three main insights. First, the study indicated that it is meaningful to rely on user-created playlists as a "gold standard" for offline evaluations, because the true but hidden continuations were often considered as highly appropriate. Second, it turned out that a hybrid algorithm that considered music metadata was not only favorable in terms of offline accuracy, but was also considered to lead to better continuations according to the users' perceptions. This confirms that offline results were at least to some extent indicative of the users' quality perceptions. Finally, the study also revealed some possible limitations of user-centric evaluations. In particular, it turned out that participants had a tendency to consider tracks as good continuations when they already knew them before the experiment, see also [68] for a discussion of such familiarity effects.

A different study setup for evaluating session-based recommendations was used in [70]. Here, 250 participants were interacting with an online radio service, which was created for the purpose of the study. The participants were first asked to select a start track for the radio, based on which playlists were generated by different algorithms using a between-subjects design. Five different algorithms were used, including three conceptually simple ones, GRU4Rec as a representative of neural techniques, and the recommendations returned from Spotify's API. The participants could then skip or like individual tracks and answered a post-task questionnaire.

The main outcomes can be summarized as follows. First, it was found that conceptually simple methods often led to playlists that the participants liked more than those of more complex approaches. Again, an offline evaluation was also partly indicative of this result. Second, however, it was found that the choice of the optimization goal matters. The tracks played by the simple association rule method, as described above, received many *like* statements, but the radio station itself was considered to be less attractive than others. Third, the recommendations returned from Spotify's API led to very low performance results in an offline experiment, but the radio station was nonetheless highly rated by the participants. The responses in the questionnaires indicate that the capability of Spotify's algorithm to help users discover interesting tracks contribute to the perceived quality of the recommendations.

Overall, user studies can—as the discussed examples show—help address important research questions that cannot be answered through pure offline evaluation approaches. Clearly, user studies can also have their limitations, e.g., regarding the representativeness of the participants of the study. Nonetheless, they are an important instrument that can help us understand if accuracy improvements in offline experiments really lead to better recommendations as perceived by users.

## 4.5   Limitations of Current Research

Current research, as mentioned above, is largely based on offline evaluations, and many papers are published each year that propose increasingly more complex methods for the recommendation task. Estimating the true value that would be achieved when using such methods in real-world environments however remains challenging for at least two reasons. First, methodological issues in applied machine learning can easily lead to progress that is actually very limited [27]. Second, the basis for our evaluations, i.e., the datasets, can be "unnatural" in a sense that what we see in the logs is affected by the circumstances and environment in which the data was collected.

### 4.5.1   Methodological Issues and Limitations of Academic Research

Most published papers on recommendation algorithms are applied machine learning works, where the goal is to demonstrate empirically that a new machine learning model outperforms what is considered the "state-of-the-art" on certain computational measures like Precision or Recall. One problem of this research approach is that there is no exact definition what represents the state-of-the-art (i.e., the baselines in a comparative evaluation). Furthermore, when researchers provide empirical evidence for the superiority of their method, they have some degrees of freedom with respect to the experimental configuration, e.g., regarding the choice of the datasets or the evaluation metric. Finally, another phenomenon observed in the literature is that researchers often invest much effort in fine-tuning their own new model, but do not do the same for the baselines. As a result, it can turn out that the claimed improvements "don't add up" [6], and that newer methods are actually not better than long-known techniques if these are properly tuned. Such problems, which are not new [6], can be found in the recommender systems literature [27, 91], in related areas such as information retrieval [118], and in more distant fields such as time-series forecasting [75].

In the area of session-based recommendation, related phenomena were observed. In [45] and [69], it was for example shown that nearest neighbor techniques are often competitive or even outperform a widely-used RNN-based technique as well as other sequential recommendation approaches. In subsequent works [71, 72], twelve algorithms—six neural and six non-neural ones—were compared under identical conditions on several datasets. Again, it turned out that in the majority of the examined cases the neural methods were outperformed by conceptually simple methods. Interestingly, many of the more recent papers do not consider such simpler baselines, even though they are not only competitive in terms of performance, but also have other advantages such as the possibility to incrementally update the database.

Like argued in [27, 28] it is therefore important to consider baselines from different algorithm families, and not just methods of a certain type, e.g., deep learning based approaches, and to optimize these baselines in a systematic way. Furthermore, it was observed in [27, 28] that reproducibility of published research in recommender systems is not very high, which is also a factor that hampers progress. Therefore, to ensure progress also in session-based recommendation, it is important that researchers share the data and the software artifacts that they used in their empirical evaluations, including all pieces of code used in the evaluation pipeline, from pre-processing, over data splitting, model learning, and evaluation. A corresponding open-source framework for the evaluation of session-based recommender systems called SESSION-REC[7] was made available in [71].

Finally, it is worth pointing out that the offline evaluation practices used in academia represent only a subset of the dimensions and domains on which machine learning algorithms are evaluated in practice. While offline evaluation can be helpful when benchmarking machine learning algorithms under well defined experimental conditions, it can only provide a partial view of their characteristics. In particular, it is not always clear if the obtained results generalize beyond the controlled environment that was used in the evaluation. Session-based recommenders are no exception to this rule. Besides a potential offline-online mismatch discussed in Sect. 4.4, the different families of session-based recommenders described above exploit certain types of *inductive biases* in the data, like any machine learning algorithm. Examples of inductive biases are the sequentiality of user actions which is exploited by RNN-based models, or the notion of similarity between user sessions in nearest-neighbor techniques. The effectiveness of each method is ultimately strongly based on the presence of such inductive biases in the scenario (dataset) under study [79]. It is therefore advisable to test different families of algorithms in order to truly understand what method works best for the task at hand.

### 4.5.2 Potential Biases in Datasets

More and more datasets are nowadays available for conducting offline experiments, as discussed in Sect. 4. However, often little is known about the environment and

---

[7] https://github.com/rn5l/session-rec.

circumstances in which the data was collected. The specific circumstances can, however, have a significant impact on what we observe in the logs. For example, when data is collected on an e-commerce site, at least some interactions may be the result of an advertisement campaign or the effect of a temporary discount [113]. Some clicks or purchase actions may also be triggered by an already existing recommendation system. Likewise, in the music domain, the listening logs that are obtained from Last.fm [1], might be the result of what was played by an automated radio station.

An analysis of an e-commerce log dataset in [51] revealed that about every hundredth view action in the logs originated from a click on a recommendation. More importantly, however, such clicks on recommendations very often led to purchases afterwards, and these purchases were often related to items that were currently discounted or generally trending on the store.

As a result, when models are trained on such datasets, they will eventually reflect these biases, and we might for example end up in re-constructing—at least to a certain extent—the logic of a recommender system or effects of a marketing campaign that took place during the data collection period. Therefore, the trained models might be not as effective as expected when they are deployed in practice.

In recent years, different approaches were put forward to deal with such problems. Possible countermeasures include the use of evaluation metrics that take such biases into account or mechanisms that aim to "de-bias" the data [14, 101, 113]. To end up with more realistic offline evaluations, i.e., evaluations that are more predictive of the online success than current evaluation procedures, different proposals were also made in the context of *off-policy learning* in reinforcement learning and bandit-based recommendation approaches that aim at answering to the counterfactual question "How would the system have performed under a different recommender?" The main idea is that of learning a model from the logged feedback and using it to correct the data biases when training the new model. One of the mostly used techniques is Inverse Propensity Scoring (IPS), which assigns a probability to every user action $(u, i)$ under the logging policy $\pi_0$, i.e., the unknown policy that generated the data, and re-weights the reward of each action taken by the policy $\pi$ of the recommender system with weight $w(i|u) = \frac{\pi(i|u)}{\pi_0(i|u)}$. It can be proven that, under reasonable assumptions, the weighting function leads to an unbiased estimator over the biased data [11].

Researchers have extended IPS in several ways, by e.g., computing better estimates of the weights in conditions of high variance through normalization and capping [102], or by applying it to complex recommendation scenarios like slate recommendation [103]. More recently off-policy correction was also employed to correct biases in sequences of user actions [17, 67].

## 5  Future Directions

Despite the increased research interest in SBRS in recent years, a number of questions and challenges remain open, in particular ones that go beyond algorithmic questions.

### 5.1  Algorithmic Improvements: Considering More Types of Data

In terms of algorithmic approaches, we identify the following main areas where more research is required: (1) the integration of long-term preferences, (2) the inclusion of item meta-data and context, and, (3) the consideration of temporal aspects.

The majority of today's algorithmic proposals focus on situations, where no long-term preference information about the current user is available. While it is often much more important to quickly adapt to the user's *current* interests [47], considering the user's past interests can be beneficial as well. The number of corresponding proposals to build such *session-aware* recommender systems is however still comparably small, and one of the main challenges is to decide how strongly past sessions should be considered in general or in a given session. Some recent model-based and neural approaches are, for example, purely based on collaborative information and the observed user behavior such as purchases or item views [83, 85]; see [59] for a performance comparison. Other approaches in addition try to leverage additional information in the recommendation process. This can both include information about the items [47] and other types of knowledge, e.g., about the social network of the users [50].

The use of *side information*, see also Chap. "Semantics and Content-Based Recommendations", in general appears to be a promising research direction, independent of the question if the long-term preferences are considered or not. In some ways, while numerous model-based approaches were proposed in recent years, the progress that is achieved with pure collaborative models can appear limited [71]. A few methods exist that incorporate side information to build a hybrid system, see e.g., [24, 40] for neural approaches or [107] for an earlier approach based on Markov models. In particular deep learning based techniques seem promising here due to their ability to learn complex relationships and interactions in heterogeneous data. A number of opportunities exist in this context, in particular when new datasets become available that contain detailed item information, e.g., in the news domain [24], or when additional information about the users' behavior can be used, e.g., in terms of their navigation or search activities, see [20].

Explicit knowledge about the user's *context*, e.g., the current time or weather, is another type of information that has not been considered much so far in session-based recommendation. Ample evidence exists that considering the user's current context can be essential to create meaningful recommendations; cf. Chap. "Context-Aware Recommender Systems: From Foundations to Recent Developments". To our knowledge, [100] is the only method that exploits contextual information for general sequence-based recommendation based on RNNs. Context information can be relevant both in *session-based* and *session-aware* scenarios. In session-based recommendation, where there is little known about the user at the beginning of the session, context information like the time of the day or what is currently trending in a certain geographical area can be helpful starting points to deal with the extreme cold-start situation. In session-aware scenarios, context information can be helpful to better guess the current user's intent and to correspondingly base the recommendations on past sessions of the user that match this assumed intent. In the domain of music recommendation for example, users might sometimes be open to explore something new, while on other days they might prefer to listen to their favorite tracks from the past [56]. Here, context information therefore might help to decide on which types of items the recommender system should focus on.

To what extent such past interests are relevant in a given situation can also depend on other *temporal aspects* [51, 86]. In some domains, we might, for example, observe an interest drift over time, and users might not be interested in things anymore they liked in the past. In other domains, like for the recommendation of consumables, we can often observe repeat purchases. In this context, recommender systems can act as reminders, and a crucial question in such a setting relates to the best point in time to remind the user. Finally, in certain domains it might be helpful to consider seasonal aspects in particular when long-term preferences are available which are not relevant in the current season and which therefore should not be considered.

## 5.2  *Improving the Evaluation Methodology*

Given the mentioned limitations of today's research practice, a change in terms of how we evaluate session-based recommender systems is needed. Such extensions of our research practice should include (1) better reproducibility of research results and progress, (2) a stronger focus on the user perception and impact of session-based recommendations, and (3) the consideration of alternative offline procedures that consider the perspective of multiple stakeholders and potential biases in the data.

Regarding *reproducibility and progress* in recommender systems research, some recent works point out in the area of traditional "top-n" reproducibility is actually not high, with less than 40% of papers published at top-level conferences being reproducible "*with reasonable effort*" [27]. Moreover, almost all of these reproducible works were not to be better in terms of prediction accuracy than long-known and much simpler methods. A similar phenomenon can be observed in the literature on session-based recommendation, where many papers report progress over the state-of-the-art, but these improvements do not add up [71]. Future research that is based on offline experiments therefore must follow more rigid rules regarding reproducibility and the choice and optimization of the baselines.

In terms of the evaluation approach, more research is required regarding the *user perception* of session-based recommender systems and the *impact* that different algorithms have on user behavior. User studies as the ones described in Sect. 4 help us learn about user perceptions and intentions and they can help us contrast offline experimental results with these observations. Ultimately, however, such studies to some extent remain proxies to assess how recommender systems would impact users "in the wild". There is no doubt that recommender systems in general can generate substantial business value for providers [33, 44], but only limited academic research so far exists that explores the effects of session-based recommendations on user behavior, e.g., in terms of click-through-rates as in [30].

Clearly, it is typically not possible for academic researchers to test their proposals in the real world through A/B tests. The works reported in [22] and [30], as well as other previous studies, indicate that there is a gap between offline and online performance. Often, the algorithms that work well according to computational metrics in an offline setting, do not lead to the best results in terms of the business value in reality. We are therefore in need of improved offline evaluation strategies and metrics, in particular ones that (1) are better suited to predict the online success of different algorithms [76], (2) consider the value perspective of multiple stakeholders [2], and (3) look at longitudinal effects of recommender [123].

# References

1. 30Music Listening and Playlists Dataset (2015). http://recsys.deib.polimi.it/datasets/. Accessed 15 May 2020
2. H. Abdollahpouri, G. Adomavicius, R. Burke, I. Guy, D. Jannach, T. Kamishima, J. Krasnodebski, L. Pizzato, Multistakeholder recommendation: Survey and research directions. User Model. User-Adapt. Interact. **30**, 127–158 (2020)
3. G. Adomavicius, B. Mobasher, F. Ricci, A. Tuzhilin, Context-aware recommender systems. AI Mag. **32**(3), 67–80 (2011)
4. R. Agrawal, T. Imieliński, A. Swami, Mining association rules between sets of items in large databases, in *Proceedings of the 1993 ACM SIGMOD International Conference on Management of Data* (1993), pp. 207–216
5. R. Agrawal, R. Srikant, Mining sequential patterns, in *Proceedings International Connference on Data Engineering, ICDE'95* (1995), pp. 3–14
6. T.G. Armstrong, A. Moffat, W. Webber, J. Zobel, Improvements that don't add up: ad-hoc retrieval results since 1998, in *Proceedings of the 18th ACM Conference on Information and Knowledge Management, CIKM'09* (2009), pp. 601–610
7. Art of the Mix (2004). http://www.ee.columbia.edu/~dpwe/research/musicsim/. Accessed 15 May (2020)
8. P. Auer, N. Cesa-Bianchi, P. Fischer, Finite-time analysis of the multiarmed bandit problem. Mach. Learn. **47**(2–3), 235–256 (2002)
9. Avito Context Ad Clicks (2015). https://www.kaggle.com/c/avito-context-ad-clicks/. Accessed 15 May 2020
10. G. Bonnin, D. Jannach, Automated generation of music playlists: survey and experiments. ACM Comput. Surv. **47**(2), 26:1–26:35 (2014)

11. L. Bottou, J. Peters, J. Quiñonero-Candela, D.X. Charles, D.M. Chickering, E. Portugaly, D. Ray, P. Simard, E. Snelson, Counterfactual reasoning and learning systems: the example of computational advertising. J. Mach. Learn. Res. **14**(1), 3207–3260 (2013)

12. B. Brost, R. Mehrotra, T. Jehan, The music streaming sessions dataset, in *Proceedings of the TheWebConf* (2019), pp. 2594–2600

13. P.G. Campos, F. Díez, I. Cantador, Time-aware recommender systems: a comprehensive survey and analysis of existing evaluation protocols. User Model. User-Adapt. Interact. **24**(1–2), 67–119 (2014)

14. D. Carraro, D. Bridge, Debiased offline evaluation of recommender systems: a weighted-sampling approach (extended abstract), in *Proceedings of the ACM RecSys 2019 Workshop on Reinforcement and Robust Estimators for Recommendation (REVEAL '19)* (2019)

15. S. Chen, J.L. Moore, D. Turnbull, T. Joachims, Playlist prediction via metric embedding, in *Proceedings of the 18th ACM SIGKDD International Conference on Knowledge Discovery and Data Mining, KDD'12* (2012), pp. 714–722

16. S. Chen, J. Xu, T. Joachims, Multi-space probabilistic sequence modeling, in *Proceedings ACM SIGKDD International Conference on Knowledge Discovery, KDD'13* (2013), pp. 865–873

17. M. Chen, A. Beutel, P. Covington, S. Jain, F. Belletti, E.H. Chi, Top-k off-policy correction for a reinforce recommender system, in *Proceedings of the Twelfth ACM International Conference on Web Search and Data Mining, WSDM'19* (2019), pp. 456–464

18. C. Cheng, H. Yang, M.R. Lyu, I. King, Where you like to go next: successive point-of-interest recommendation, in *Proceedings International Joint Conference on Artificial Intelligence, IJCAI'13* (2013), pp. 2605–2611

19. K. Cho, B. van Merriënboer, Ç. Gülçehre, D. Bahdanau, F. Bougares, H. Schwenk, Y. Bengio, Learning phrase representations using RNN encoder–decoder for statistical machine translation, in *Proceeedindgs of Empirical Methods in Natural Language Processing, EMNLP'14* (2014), pp. 1724–1734

20. CIKM, The CIKM Cup 2016 (2016). https://competitions.codalab.org/competitions/11161. Accessed March 2020

21. P. Cremonesi, Y. Koren, R. Turrin, Performance of recommender algorithms on top-n recommendation tasks, in *Proceedings ACM Conference on Recommender Systems, RecSys'10* (2010), pp. 39–46

22. P. Cremonesi, F. Garzotto, R. Turrin, Investigating the persuasion potential of recommender systems from a quality perspective: an empirical study. Trans. Interact. Intell. Syst. **2**(2), 1–41 (2012)

23. Delicious (2008). http://www.dai-labor.de/en/competence_centers/irml/datasets/. Accessed 15 May 2020

24. G. de Souza Pereira Moreira, D. Jannach, A.M. da Cunha, Contextual hybrid session-based news recommendation with recurrent neural networks. IEEE Access **7**, 169185–169203 (2019)

25. Diginetica CIKM Cup 2016 Dataset (2016). https://competitions.codalab.org/competitions/11161. Accessed 15 June 2020

26. M.D. Ekstrand, F.M. Harper, M.C. Willemsen, J.A. Konstan, User perception of differences in recommender algorithms, in *Proceedings of the 2014 ACM Conference on Recommender Systems, RecSys'14* (2014), pp. 161–168

27. M. Ferrari Dacrema, P. Cremonesi, D. Jannach, Are we really making much progress? A worrying analysis of recent neural recommendation approaches, in *Proceedings of the 13th ACM Conference on Recommender Systems, RecSys'19* (2019), pp. 101–109

28. M. Ferrari Dacrema, S. Boglio, P. Cremonesi, D. Jannach, A troubling analysis of reproducibility and progress in recommender systems research. ACM Trans. Inf. Syst. **39**, 1–49 (2021)

29. F. Garcin, C. Dimitrakakis, B. Faltings, Personalized news recommendation with context trees, in *Proceedings of the ACM Confererence on Recommender Systems, RecSys'13* (2013), pp. 105–112

30. F. Garcin, B. Faltings, O. Donatsch, A. Alazzawi, C. Bruttin, A. Huber, Offline and online evaluation of news recommender systems at swissinfo.ch, in *Proceedings of the ACM Confererence on Recommender Systems, RecSys'14* (2014), pp. 169–176
31. D. Garg, P. Gupta, P. Malhotra, L. Vig, G. Shroff, Sequence and time aware neighborhood for session-based recommendations: stan, in *Proceedings of the 42nd International ACM SIGIR Conference on Research and Development in Information Retrieval, SIGIR'19* (2019), pp. 1069–1072
32. A. Gilotte, C. Calauzènes, T. Nedelec, A. Abraham, S. Dollé, Offline a/b testing for recommender systems, in *Proceedings of the ACM International Conference on Web Search and Data Mining, WSDM'18* (2018), pp. 198–206
33. C.A. Gomez-Uribe, N. Hunt, The Netflix recommender system: algorithms, business value, and innovation. Trans. Manag. Inf. Syst. **6**(4), 13:1–13:19 (2015)
34. M. Grbovic, V. Radosavljevic, N. Djuric, N. Bhamidipati, J. Savla, V. Bhagwan, D. Sharp, E-commerce in your inbox: product recommendations at scale, in *Proceedings ACM SIGKDD International Conference on Knowledge Discovery and Data Mining, KDD'15* (2015), pp. 1809–1818
35. X. Guo, C. Shi, C. Liu, Intention modeling from ordered and unordered facets for sequential recommendation, in *Proceedings of The Web Conference 2020, WWW'20, New York, NY, USA,* (2020), pp. 1127–1137
36. N. Hariri, B. Mobasher, R. Burke, Context-aware music recommendation based on latent topic sequential patterns, in *Proceedings of the Sixth ACM Conference on Recommender Systems, RecSys'12* (2012), pp. 131–131
37. Q. He, D. Jiang, Z. Liao, S.C.H. Hoi, K. Chang, E.-P. Lim, H. Li, Web query recommendation via sequential query prediction, in *Proceedings International Conference on Data Engineering, ICDE'09* (2009), pp. 1443–1454
38. J.L. Herlocker, J.A. Konstan, L.G. Terveen, J.T. Riedl, Evaluating collaborative filtering recommender systems. ACM Trans. Inf. Syst. **22**(1), 5–53 (2004)
39. B. Hidasi, A. Karatzoglou, L. Baltrunas, D. Tikk, Session-based recommendations with recurrent neural networks, in *Proceedings International Conference on Learning Representatinos, ICLR'16* (2016)
40. B. Hidasi, M. Quadrana, A. Karatzoglou, D. Tikk, Parallel recurrent neural network architectures for feature-rich session-based recommendations, in *Proceedings ACM Conference on Recommender Systems, RecSys'16* (2016), pp. 241–248
41. B. Hidasi, A. Karatzoglou, Recurrent neural networks with top-k gains for session-based recommendations, in *Proceedings of the 27th ACM International Conference on Information and Knowledge Management, CIKM'18* (2018), pp. 843–852
42. M. Hosseinzadeh Aghdam, N. Hariri, B. Mobasher, R. Burke, Adapting recommendations to contextual changes using hierarchical hidden markov models, in *Proceedings of the 9th ACM Conference on Recommender Systems, RecSys'15* (2015), pp. 241–244
43. L. Hu, L. Cao, S. Wang, G. Xu, J. Cao, Z. Gu, Diversifying personalized recommendation with user-session context, in *Proceedings of the Twenty-Sixth International Joint Conference on Artificial Intelligence, IJCAI-17* (2017), pp. 1858–1864
44. D. Jannach, M. Jugovac, Measuring the business value of recommender systems. ACM Trans. Manag. Inf. Syst. **10**(4), 1–23 (2019)
45. D. Jannach, M. Ludewig, When recurrent neural networks meet the neighborhood for session-based recommendation, in *Proceedings of the 11th ACM Conference on Recommender Systems, RecSys'17* (2017), pp. 306–310
46. D. Jannach, M. Zanker, Collaborative filtering: matrix completion and session-based recommendation tasks, in *Collaborative Recommendations: Algorithms, Practical Challenges and Applications*, ed. by S. Berkovsky, I. Cantador, D. Tikk (World Scientific, Singapore, 2019), pp. 1–38
47. D. Jannach, L. Lerche, M. Jugovac, Adaptation and evaluation of recommendations for short-term shopping goals, in *Proceedings of the ACM Conference on Recommender Systems, RecSys'15* (2015), pp. 211–218

48. D. Jannach, L. Lerche, I. Kamehkhosh, M. Jugovac, What recommenders recommend: an analysis of recommendation biases and possible countermeasures. User Model. User-Adapt. Interact. **25**(5), 427–491 (2015)

49. D. Jannach, M. Jugovac, L. Lerche, Supporting the design of machine learning workflows with a recommendation system. ACM Trans. Interact. Intell. Syst. **6**(1), 1–35 (2016)

50. D. Jannach, I. Kamehkhosh, L. Lerche, Leveraging multi-dimensional user models for personalized next-track music recommendation, in *Proceedings of the ACM Symposium on Applied Computing, ACM SAC2017* (2017)

51. D. Jannach, M. Ludewig, L. Lerche, Session-based item recommendation in e-commerce: on short-term intents, reminders, trends, and discounts. User-Model. User-Adapt. Interact. **27**(3–5), 351–392 (2017)

52. D. Jannach, L. Lerche, M. Zanker, Recommending based on implicit feedback, in *Social Information Access*, ed. by P. Brusilovsky, D. He (Springer, Berlin, 2018)

53. JD Dataset (2019). https://github.com/alicogintel/SDM. Accessed 15 May 2020

54. M. Jugovac, D. Jannach, L. Lerche, Efficient optimization of multiple recommendation quality factors according to individual user tendencies. Exp. Syst. Appl. **81**, 321–331 (2017)

55. I. Kamehkhosh, D. Jannach, User perception of next-track music recommendations, in *Proceedings of the 2017 Conference on User Modeling Adaptation and Personalization, UMAP'17* (2017), pp. 113–121

56. K. Kapoor, V. Kumar, L. Terveen, J.A. Konstan, P. Schrater, I like to Explore Sometimes: adapting to dynamic user novelty preferences, in *Proceedings of the 9th ACM Conference on Recommender Systems* (2015), pp. 19–26

57. B.P. Knijnenburg, M.C. Willemsen, Z. Gantner, H. Soncu, C. Newell, Explaining the user experience of recommender systems. User Model. User-Adapt. Interact. **22**, 441–504 (2012)

58. P. Kouki, I. Fountalis, N. Vasiloglou, X. Cui, E. Liberty, K. Al Jadda, From the lab to production: a case study of session-based recommendations in the home-improvement domain, in *Fourteenth ACM Conference on Recommender Systems, RecSys'20* (2020), pp. 140–149

59. S. Latifi, N. Mauro, D. Jannach, Session-aware recommendation: a surprising quest for the state-of-the-art. Inf. Sci. **573**, 291–315 (2021)

60. D. Lee, K. Hosanagar, Impact of recommender systems on sales volume and diversity, in *Proceedings of the International Conference on Information Systems, ICIS 2014* (2014)

61. D. Lee, K. Hosanagar, How do recommender systems affect sales diversity? A cross-category investigation via randomized field experiment. Inf. Syst. Res. **30**(1), 239–259 (2019)

62. S. Levine, V. Koltun, Guided policy search, in *International Conference on Machine Learning* (2013), pp. 1–9

63. L. Li, W. Chu, J. Langford, X. Wang, Unbiased offline evaluation of contextual-bandit-based news article recommendation algorithms, in *Proceedings of the Fourth ACM International Conference on Web Search and Data Mining, WSDM'11* (2011), pp. 297–306

64. J. Li, P. Ren, Z. Chen, Z. Ren, T. Lian, J. Ma, Neural attentive session-based recommendation, in *Proceedings of the 2017 ACM on Conference on Information and Knowledge Management, CIKM'17* (2017), pp. 1419–1428

65. Z.C. Lipton, J. Berkowitz, C. Elkan, A critical review of recurrent neural networks for sequence learning (2015). CoRR, 1506.00019

66. Q. Liu, Y. Zeng, R. Mokhosi, H. Zhang, STAMP: short-term attention/memory priority model for session-based recommendation, in *Proceedings ACM SIGKDD International Conference on Knowledge Discovery and Data Mining, KDD'18* (2018), pp. 1831–1839

67. Y. Liu, A. Swaminathan, A. Agarwal, E. Brunskill, Off-policy policy gradient with state distribution correction (2019). arXiv:1904.08473

68. B. Loepp, T. Donkers, T. Kleemann, J. Ziegler, Impact of item consumption on assessment of recommendations in user studies, in *Proceedings of the 12th ACM Conference on Recommender Systems, RecSys'18* (2018), pp. 49–53

69. M. Ludewig, D. Jannach, Evaluation of session-based recommendation algorithms. User-Model. User-Adapt. Interact. **28**(4–5), 331–390 (2018)

70. M. Ludewig, D. Jannach, User-centric evaluation of session-based recommendations for an automated radio station, in *Proceedings of the 13th ACM Conference on Recommender Systems, RecSys'19* (2019)
71. M. Ludewig, N. Mauro, S. Latifi, D. Jannach, Performance comparison of neural and non-neural approaches to session-based recommendation, in *Proceedings of the 13th ACM Conference on Recommender Systems, RecSys'19* (2019), pp. 462–466
72. M. Ludewig, N. Mauro, S. Latifi, D. Jannach, Empirical analysis of session-based recommendation algorithms. User Model. User-Adapt. Interact. **31**, (2021)
73. Y. Ma, B.M. Narayanaswamy, H. Lin, H. Ding, Temporal-contextual recommendation in real-time, in *Proceedings of the 26th ACM SIGKDD International Conference on Knowledge Discovery & Data Mining, KDD'20* (2020), pp. 2291–2299
74. N.R. Mabroukeh, C.I. Ezeife, A taxonomy of sequential pattern mining algorithms. ACM Comput. Surv. **43**(1), 1–41 (2010)
75. S. Makridakis, E. Spiliotis, V. Assimakopoulos, Statistical and machine learning forecasting methods: concerns and ways forward. PloS One **13**(3), e0194889 (2018)
76. A. Maksai, F. Garcin, B. Faltings, Predicting online performance of news recommender systems through richer evaluation metrics, in *Proceedings of the ACM Confererence on Recommender Systems, RecSys'15* (2015), pp. 179–186
77. S. Martin, B. Faltings, V. Schickel, Context-tree recommendation vs matrix-factorization: algorithm selection and live users evaluation, in *Proceedings of the AAAI Conference on Artificial Intelligence*, vol. 33 (2019), pp. 9534–9540
78. Microsoft Anonymous Web Data (1998). http://kdd.ics.uci.edu/databases/msweb/msweb.html. Accessed 15 May 2020
79. T.M. Mitchell, *The Need for Biases in Learning Generalizations*. Department of Computer Science, Laboratory for Computer Science Research (1980)
80. V. Mnih, K. Kavukcuoglu, D. Silver, A. Graves, I. Antonoglou, D. Wierstra, M. Riedmiller, Playing atari with deep reinforcement learning (2013) . arXiv:1312.5602
81. B. Mobasher, H. Dai, T. Luo, M. Nakagawa, Using sequential and non-sequential patterns in predictive web usage mining tasks, in *Proceedings of IEEE International Conference on Data Mining, ICDM'02* (2002), pp. 669–672
82. NYC Taxi Trips (2013). http://www.andresmh.com/nyctaxitrips/. Accessed 15 May 2020
83. T.M. Phuong, T.C. Thanh, N.X. Bach, Neural session-aware recommendation. IEEE Access **7**, 86884–86896 (2019)
84. P. Pu, L. Chen, R. Hu, A user-centric evaluation framework for recommender systems, in *Proceedings of the 5th ACM Conference on Recommender Systems, RecSys'11* (2011), pp. 157–164
85. M. Quadrana, A. Karatzoglou, B. Hidasi, P. Cremonesi, Personalizing session-based recommendations with hierarchical recurrent neural networks, in *Proceedings of the ACM Conference on Recommender Systmes, RecSys'17* (2017)
86. M. Quadrana, P. Cremonesi, D. Jannach, Sequence-aware recommender systems. ACM Comput. Surv. **54**, 1–36 (2018)
87. R. Ragno, C.J.C. Burges, C. Herley, Inferring similarity between music objects with application to playlist generation, in *Proceedings of the 7th ACM SIGMM International Workshop on Multimedia Information Retrieval, MIR'05* (2005), pp. 73–80
88. RecSys Challenge 2015 (2015). http://2015.recsyschallenge.com/challenge.html. Accessed 15 May 2020
89. S. Rendle, C. Freudenthaler, Z. Gantner, L. Schmidt-Thieme, BPR: Bayesian personalized ranking from implicit feedback, in *Proceedings of the Conference on Uncertainty in Artificial Intelligence, UAI'09* (2009), pp. 452–461
90. S. Rendle, C. Freudenthaler, L. Schmidt-Thieme, Factorizing personalized markov chains for next-basket recommendation, in *Proceedings of the World Wide Web Conference, WWW'10* (2010), pp. 811–820
91. S. Rendle, L. Zhang, Y. Koren, On the difficulty of evaluating baselines: a study on recommender systems (2019). CoRR, abs/1905.01395

92. P. Resnick, N. Iacovou, M. Suchak, P. Bergstrom, J. Riedl, Grouplens: an open architecture for collaborative filtering of netnews, in *Proceedings of the 1994 ACM Conference on Computer Supported Cooperative Work, CSCW'94* (1994), pp. 175–186

93. D. Rohde, S. Bonner, T. Dunlop, F. Vasile, A. Karatzoglou, Recogym: a reinforcement learning environment for the problem of product recommendation in online advertising (2018). arXiv:1808.00720

94. N. Sachdeva, G. Manco, E. Ritacco, V. Pudi, Sequential variational autoencoders for collaborative filtering, in *Proceedings of the Twelfth ACM International Conference on Web Search and Data Mining, WSDM'19* (2019), pp. 600–608

95. A. Said, D. Tikk, K. Stumpf, Y. Shi, M.A. Larson, P. Cremonesi, Recommender systems evaluation: a 3d benchmark, in *RUE Workshop at ACM RecSys 2012* (2012), pp. 21–23

96. J. Sanz-Cruzado, P. Castells, E. López, A simple multi-armed nearest-neighbor bandit for interactive recommendation, in *Proceedings of the 13th ACM Conference on Recommender Systems, RecSys'19* (2019), pp. 358–362

97. G. Shani, A. Gunawardana, Evaluating recommendation systems, in *Recommender Systems Handbook* (Springer, Berlin, 2011), pp. 257–297

98. G. Shani, D. Heckerman, R.I. Brafman, An MDP-based recommender system. J. Mach. Learn. Res. **6**, 1265–1295 (2005)

99. D. Silver, A. Huang, C.J. Maddison, A. Guez, L. Sifre, G. Van Den Driessche, J. Schrittwieser, I. Antonoglou, V. Panneershelvam, M. Lanctot, et al., Mastering the game of go with deep neural networks and tree search. Nature **529**(7587), 484–489 (2016)

100. E. Smirnova, F. Vasile, Contextual sequence modeling for recommendation with recurrent neural networks (2017). CoRR, abs/1706.07684

101. H. Steck, Training and testing of recommender systems on data missing not at random, in *Proceedings of the 16th ACM SIGKDD International Conference on Knowledge Discovery and Data Mining, KDD'10* (2010), pp. 713–722

102. A. Swaminathan, T. Joachims, The self-normalized estimator for counterfactual learning, in *Advances in Neural Information Processing Systems, NIPS'15* (2015), pp. 3231–3239

103. A. Swaminathan, A. Krishnamurthy, A. Agarwal, M. Dudik, J. Langford, D. Jose, I. Zitouni, Off-policy evaluation for slate recommendation, in *Advances in Neural Information Processing Systems* (2017), pp. 3632–3642

104. P. Symeonidis, L. Kirjackaja, M. Zanker, Session-aware news recommendations using random walks on time-evolving heterogeneous information networks. User Model. User-Adapt. Interact. **30**, 727–755 (2020)

105. J. Tang, K. Wang, Personalized top-n sequential recommendation via convolutional sequence embedding, in *Proceedings of the Eleventh ACM International Conference on Web Search and Data Mining, WSDM'18* (2018), pp. 565–573

106. M.M. Tanjim, C. Su, E. Benjamin, D. Hu, L. Hong, J. McAuley, Attentive sequential models of latent intent for next item recommendation, in *Proceedings of The Web Conference 2020, WWW'20* (2020), pp. 2528–2534

107. M. Tavakol, U. Brefeld, Factored MDPs for detecting topics of user sessions, in *Proceedings of the 8th ACM Conference on Recommender Systems, RecSys'14* (2014), pp. 33–40

108. The Music Streaming Sessions Dataset (2019). https://www.aicrowd.com/challenges/spotify-sequential-skip-prediction-challenge-old. Accessed 10 June 2020

109. R. Turrin, A. Condorelli, R. Pagano, M. Quadrana, P. Cremonesi, Large scale music recommendation, in *Proceedings of the LSRS Workshop at ACM RecSys 2015* (2015)

110. R. Turrin, M. Quadrana, A. Condorelli, R. Pagano, P. Cremonesi, 30music listening and playlists dataset, in *ACM RecSys 2015 Posters* (2015)

111. F. Vasile, D. Rohde, O. Jeunen, A. Benhalloum, A gentle introduction to recommendation as counterfactual policy learning, in *Proceedings of the 28th ACM Conference on User Modeling, Adaptation and Personalization, UMAP'20* (2020), pp. 392–393

112. B. Vassøy, M. Ruocco, E. de Souza da Silva, E. Aune, Time is of the essence: a joint hierarchical RNN and point process model for time and item predictions, in *Proceedings of the Twelfth ACM International Conference on Web Search and Data Mining, WSDM'19* (2019), pp. 591–599

113. M. Wan, J. Ni, R. Misra, J. McAuley, Addressing marketing bias in product recommendations, in *Proceedings of the 13th International Conference on Web Search and Data Mining, WSDM'20* (2020), pp. 618–626

114. Q. Wang, C. Zeng, W. Zhou, T. Li, S.S. Iyengar, L. Shwartz, G.Y. Grabarnik, Online interactive collaborative filtering using multi-armed bandit with dependent arms. IEEE Trans. Knowl. Data Eng. **31**(8), 1569–1580 (2018)

115. M. Wang, P. Ren, L. Mei, Z. Chen, J. Ma, M. de Rijke, A collaborative session-based recommendation approach with parallel memory modules, in *Proceedings of the 42nd International ACM SIGIR Conference on Research and Development in Information Retrieval, SIGIR'19* (2019), pp. 345–354

116. S. Wang, L. Cao, Y. Wang, A survey on session-based recommender systems (2019). CoRR, abs/1902.04864

117. S. Wu, Y. Tang, Y. Zhu, L. Wang, X. Xie, T. Tan, Session-based recommendation with graph neural networks, in *Proceedings of the Thirty-Third AAAI Conference on Artificial Intelligence, AAAI* (2019), pp. 346–353

118. W. Yang, K. Lu, P. Yang, J. Lin, Critically examining the neural hype: weak baselines and the additivity of effectiveness gains from neural ranking models, in *Proceedings of the 42nd International ACM SIGIR Conference on Research and Development in Information Retrieval, SIGIR'19* (2019), pp. 1129–1132

119. G.-E. Yap, X.-L. Li, P.S. Yu, Effective next-items recommendation via personalized sequential pattern mining, in *Proceedings International Conference on Database Systems for Advanced Applications, DASFAA'12* (2012), pp. 48–64

120. F. Yu, Y. Zhu, Q. Liu, S. Wu, L. Wang, T. Tan, TAGNN: target attentive graph neural networks for session-based recommendation, in *Proceedings of the 43rd International ACM SIGIR Conference on Research and Development in Information Retrieval, SIGIR'20* (2020)

121. F. Yuan, A. Karatzoglou, I. Arapakis, J.M. Jose, X. He, A simple convolutional generative network for next item recommendation, in *Proceedings of the 12th ACM International Conference on Web Search and Data Mining, WSDM'19* (2019), pp. 582–590

122. M. Zaheer, A. Ahmed, Y. Wang, D. Silva, M. Najork, Y. Wu, S. Sanan, S. Chatterjee, Uncovering hidden structure in sequence data via threading recurrent models, in *Proceedings ACM International Conference on Web Search and Data Mining, WSDM'19* (2019), pp. 186–194

123. J. Zhang, G. Adomavicius, A. Gupta, W. Ketter, Consumption and performance: understanding longitudinal dynamics of recommender systems via an agent-based simulation framework. Inf. Syst. Res. **31**, 76–101 (2020)

# Adversarial Recommender Systems: Attack, Defense, and Advances

Vito Walter Anelli, Yashar Deldjoo, Tommaso Di Noia, and Felice Antonio Merra

## 1 Introduction

Machine learning (ML) models are increasingly deployed in online systems due to their ability to generalize on new data. In recommender systems (RS), latent factor models (LFM), such as matrix factorization (MF), are among the most prominent ML techniques that have been utilized in various recommendation settings. From an algorithmic point-of-view, these models try to find latent factors of users and items whose interactions can explain an unknown preference. Nearly three decades of research on RS has resulted in various recommendation models that aim to exploit the user interaction data and side-information [39, 86] to improve recommendation performance accuracy and beyond-accuracy aspect [58]. Notwithstanding their great success, inherent to machine-learned algorithms, lies the fundamental assumption of *data stationarity*, that is, both the training data and test data are sampled from a similar (and possibly unknown) distribution [17]. In an adversarial setting, however, an intelligent and adaptive attacker may deliberately manipulate data—thereby violating the stationarity assumption—aiming to undermine RS operation and compromise the system's integrity. As RS are adopted in many life-affecting decision-making scenarios, this vulnerability raises several issues whether ML-based techniques can be safely adopted in different recommendation scenarios or if they must (and can) be redesigned to exhibit desired security behavior to be trustworthy in the face of aggressive provocation from determined attackers.

Authors have equally contributed to the chapter. They are listed in alphabetical order.

V. W. Anelli · Y. Deldjoo · T. Di Noia (✉) · F. A. Merra
Polytechnic University of Bari, Bari, Italy
e-mail: vitowalter.anelli@poliba.it; yashar.deldjoo@poliba.it; tommaso.dinoia@poliba.it; felice.merra@poliba.it

**Table 1** Different categories of attacks and example research in each case

| Attack type | Example research |
| --- | --- |
| *Hand-engineered shilling attacks* | |
| • Attack by leveraging interaction data | [61, 76] |
| • Attack by exploiting semantic data | [4] |
| • Studying the impact of data characteristics on shilling attacks | [38] |
| • Detection and defense of shilling attack | [2, 15, 20, 21, 117, 118] |
| *Machine-learned data poisoning optimization* | |
| • Factorization-based models | [31, 34, 45, 56, 57, 64, 67, 92] |
| • Reinforcement Learning models | [23, 88, 115] |
| • Other recommendation models | [32, 44, 108] |
| • Defense | [67] |
| *Adversarial machine-learned attacks* | |
| • Adversarial perturbations on model parameters | [42, 55, 110] |
| • Adversarial perturbation on content data | [5, 7, 75, 93] |
| • Defense, robustification, and evaluation | [10, 55, 93] |

Security of RS has been studied in two different contexts in the history of RS research in the last three decades, the one related to hand-crafted shilling attacks since 2000 [19, 38, 89], the other one to machine-learned adversarial attacks starting from 2016 [34, 55, 115]. A third and recently emerging area is also ML-based approaches for shilling attacks [44, 64], which are, to some degree, similar to adversarial attacks from the viewpoint of using ML-learned techniques to alter the recommendation performance with minimal data variations. Shilling attacks utilize the similarity patterns between users' rating profiles and *manually* injected fake rating patterns identical to those already in the system such that it can push the desired item into the recommendation list of users (push attack) or recommend non-relevant items to create a mistrust on a system (nuke attack). In contrast, attacks based on adversarial machine learning (AML) focus on learning additive perturbations, that once injected into the data they can manipulate data stationary assumption and alter the recommendation results toward an engineered and often malicious outcome [37, 41]. Table 1 summarizes attack types with some corresponding references.

Despite the similarities between ML classification and recommendation learning tasks, there are considerable differences/challenges in adversarial attacks on RS compared with ML and the degree to which the subject has been studied in the respective communities:

• *Poisoning vs. adversarial attack.* As we said before, in the beginning, the main focus of the RS research community has been on *hand-engineered* fake user profiles (a.k.a. shilling attacks) against rating-based CF [38]. Given a URM with $n$ real users and $m$ items, the goal of a shilling attack is to augment a fraction of malicious users $\lfloor \alpha n \rfloor$ ($\lfloor . \rfloor$ is the floor operation) to the URM ($\alpha \ll 1$) in which each malicious use profile can contain ratings to a maximum number of $C$ items.

The ultimate goal is to harvest recommendation outcomes toward an illegitimate benefit, e.g., pushing some targeted items into the top-$K$ list of users for market penetration. Shilling attacks against RS had established literature, and their development face two main milestones: the first one—since the early 2000s—where the literature was focused on building hand-crafted fake profiles whose rating assignment follow different strategy according to random, popular, love-hate, bandwagon attacks among others [52]; the second research direction started in 2016 when the first ML-optimized attack was proposed by Li et al., [64] on factorization-based RS. This work reviews a novel type of data poisoning attack that applies the adversarial learning paradigm for generating poisoning input data. Nonetheless, given their significant impact against modern recommendation models, the research works focusing on *machine-learned adversarial attacks* against RS have recently received considerable attention from the research community.

- *CF vs. classification models.* Attacks against classification tasks focus on enforcing the wrong prediction of individual instances in the data. In RS, however, the mainstream attacks rely on CF principles, i.e., mining similarities in opinions of like-minded users to compute recommendations. This interdependence between users and items can, on the one hand, *improve robustness* of CF, since predictions depend on a group of instances, not on an individual one and, on the other hands, may cause *cascade effects*, where attacks on a single user may impact other neighbor users [34].

- *Granularity and application type.* Adversarial examples created, e.g., for image classification tasks, are empowered based on a continuous real-valued representation of image data (i.e., pixel values), but in RS, the raw values are user/item IDs and ratings that are discrete. Perturbing these discrete entities is infeasible since it may lead to changing the input semantics, e.g., loosely speaking applying $ID + \delta$ can result in a new user $ID$. Therefore, existing adversarial attacks in the field of ML are not transferable to the RS problems trivially. Furthermore, in the context of Computer Vision (CV)—attacks against images—the perturbations often need to be "human-imperceptible" or "inconspicuous" (i.e., may be visible but not suspicious) [104]. How we can capture these nuances for designing attacks in RS remains an open challenge.

In the following, we present the outline of the current book chapter: in Sect. 2, we present foundation concepts to adversarial machine learning (AML) by presenting the common goal of empirical risk minimization (ERM) in supervised learning ML task, contrast it with the adversarial perceptive and present the countermeasure strategies. Afterward, we review the widely adopted attack and defense strategies built on top of the ERM problem. Then, in Sect. 3, we present state of the art in adversarial attack and defense against recommendation models. We also present machine-learned data poisoning attacks. In Sect. 4, we present evaluation protocols of adversarial attacks and defenses in recommendation tasks. Finally, in Sect. 5, we conclude the book chapter and summarize new arising challenges.

Please also note that the previous edition of this book contained a chapter named "Robust collaborative recommendation" [19] that is at some level relevant to the current book chapter since both address security issues in RS. However, in [19] the focus is on hand-crafted shilling attacks, whereas in the current book chapter, we focus primarily on modern machine-learned attacks based on the AML paradigm.

## 2 Foundations

Adversarial attacks were first investigated in 2013 in [90] by Szegedy et al. They discovered that, given an image, when added some carefully selected perturbations that are barely perceptible to the human eye, a well-trained deep neural network (DNN) could misclassify the adversarial image with high confidence. For instance, the attacker may perturb pixels of a pandas image not to be perceived by a human observer and obtain gibbon as the classification result with high confidence. These outcomes were strikingly stunning since it was expected that state-of-the-art DNNs generalize well on unknown data and do not alter the class of a given test image that is marginally perturbed using a cheap analytical approach. Szegedy et al. coined the term 'adversarial examples' and presented an optimization-based system using box-constrained L-BFGS to learn such perturbations. At this time, it was believed, as suggested by Szegedy et al., that *non-linearity* of neural networks is the main reason for their adversarial vulnerability.

A year later, Goodfellow et al. [50] proposed a counterintuitive hypothesis, informing *linearity* of neural networks—instead of their non-linearity—as the main reason for adversarial behavior. This claim, which is commonly known as the 'linearity hypothesis' in the literature, was supported by the fact that the design of neural networks intentionally encourages linear behavior, especially when using activation functions like Relu MaxOut. In other words, although these functions make the models theoretically non-linear, they are trained to function in the linear region of the activation function to counter phenomenons like the vanishing of gradients. To support this hypothesis, the authors demonstrated that the FGSM attack that worked based on the linearity assumption of neural networks was sufficient to fool deep neural networks, supporting their argument that neural networks behave like a linear classifier.

As attack strategies introduced in this chapter work primarily on classification tasks, we discuss the foundation concepts for a classification problem to keep this chapter self-contained. In a classical *supervised learning* setting, we are given a paired training dataset $\mathcal{D} = \{(\mathbf{x}_i, y_i)\}_{i=1}^{n}$ where $\mathbf{x}_i \in \mathcal{X} \subseteq R^d$ is a feature vector in the *input space* $\mathcal{X}$ and $y_i \in \mathcal{Y}$ is the corresponding label in some *output space* $\mathcal{Y}$. For instance, in binary classification $\mathcal{Y}$ is binary and used as $\mathcal{Y} = \{-1, +1\}$. Each pair in $\mathcal{D}$ is assumed to be generated i.i.d.[1] from an unknown distribution $P$,

---

[1] Independent and identically distributed (i.i.d.).

i.e., $(\mathbf{x}, y) \sim P$. We also assume that we are given a suitable loss function $\mathscr{L}(., .)$, for instance the cross-entropy loss for a neural network. The goal is to find a good candidate function $f(\mathbf{x}; \theta)$ that minimizes the following empirical risk

$$\min_{\theta} \; \mathbb{E}_{(\mathbf{x},y) \sim P} \; \mathscr{L}(f(\mathbf{x}; \theta), y) \tag{1}$$

where $\mathbb{E}_{(\mathbf{x},y) \sim P}$ is commonly termed *expected risk* of the classifier, $\theta$ is the model parameter and $y$ is the class label for the input sample $\mathbf{x}$. As $P$ it is often unknown, we use $\mathscr{D}$ in order to learn the suitable candidate function $f(\mathbf{x}, \theta)$. The training objective function can be formulated as the following optimization problem,

$$\min_{\theta} \sum_{(\mathbf{x}_i, y_i) \in \mathscr{D}} \mathscr{L}(f(\mathbf{x}_i; \theta), y_i) \tag{2}$$

where $f(\mathbf{x}_i; \theta)$ and $y_i$ are the predicted and class label for the sample $i$.

## 2.1 Adversarial Perspective

Empirical risk minimization (ERM) has been found a powerful means to tune classifiers with small population risk. Regretfully, ERM is not capable of producing models that are robust against adversarially crafted samples. In an adversarial setting, we can define a general form of the objective for adversarial attacks based on Eq. 2, which aims to force a trained model to make a wrong prediction under a minimal perturbation budget. The problem of adversarial attacks can be formally stated as follows.

**Definition 1** Given a learned classifier $f(\mathbf{x}; \theta)$ and an instance from the dataset $(\mathbf{x}, y) \in \mathscr{D}$, the attacker takes the sample $\mathbf{x}$ and produces an adversarial version $\mathbf{x}_{adv} = \mathbf{x} + \delta$ such that $f(\mathbf{x}_{adv}; \theta) \neq f(\mathbf{x}; \theta)$. The attacker aims to do this within a minimal perturbation budget, i.e., $\|\delta\|_p \leq \epsilon$ where $\|.\|_p$ is the $p$-norm. The attacker's objective can then be formally stated as follows,

$$\arg\max_{\delta} \; \mathscr{L}(f(\mathbf{x} + \delta; \theta), y), \quad s.t., \; \|\delta\|_p \leq \epsilon, \tag{3}$$

where $\epsilon$ is the perturbation budget, typically chosen as a small value.

∎

## 2.2 Taxonomy of Adversarial Attacks

There have been several proposals to categorize attacks against ML algorithms. The most prominent categorization is based on the following dimensions: *timing*, *knowledge*, and *goal*. We use these dimensions as the basis to distill the main characteristics of adversarial attacks and introduce a taxonomy of attack systems that highlights the various aspects of this area.

### 2.2.1 Attack's Timing

The first crucial aspect in modeling attacks is *when* they occur in the learning pipeline of the ML system. This consideration gives rise to a dichotomy, which is central to attacks on ML models: *attacks on models*—or, more precisely, decisions made by the learned model—and attacks on *algorithms*, occurring before model training by modifying part of the training data used for training [98]. These two categories are respectively known as *evasive* and *data poisoning* attacks:

- **Evasive attack:** The attacks aim to avoid detection—or evade the decisions made by the learned model, thus the name evasive—by directly manipulating malicious test samples. Therefore, these attacks occur after the ML model is trained or in the inference (test) phase. The model is fixed, and the attacker cannot alter the model parameter or structure.
- **Poisoning attack:** These attacks happen before the ML model is trained. The attacker can add false data points (or *poisons*) into the model training data, causing the trained model to produce an erroneous prediction. Poisoning attacks have been explored in the literature for a variety of tasks [98], such as (1) attacks on binary classification for tasks such as label flipping or against kernelized SVM, (2) attacks on unsupervised learning such as clustering and anomaly detection and, (3) attacks on matrix completion task in RS.

### 2.2.2 Attacker's Knowledge

When modeling attacks, the second important consideration is what–or how much–information the adversary has about the learning model, the algorithm, or the training data they aim to attack. This distinction leads to the following classification: *white-box*, *black-box*, and *gray-box* attacks.

1. **Perfect-knowledge (PK) adversary**: A perfect-knowledge (or *white-box*) adversary assumes that the attacker has precise information about the learned model (the actual classifier), including, e.g., the features, the learning algorithm, hyperparameters, among others. The adversary can subsequently adjust the attack strategy to account for the defense. In the field of cybersecurity, it has been shown that assuming attacker having no knowledge—or security by obscurity—

is ineffective [35]; on the opposite hand, if a defender can be robust to PK attacks, it will surely be robust to more knowledge-limited attacks; thus these threat models offer natural reasons for consideration. Therefore, a PK attack is the most potent possible threat model.

2. **Limited-knowledge (LK) adversary**: A limited-knowledge (or *gray-box*) adversary has some, albeit incomplete, level of knowledge about the system under attack. For instance, the attacker may know the classifier (or its type) or the training data used, but not simultaneously. However, it is assumed that the adversary can collect and build a surrogate dataset $D' = \{\mathbf{x}'_i, y'_i\}_{i=1}^{n}$ from the same distribution $p$ from which $\mathscr{D}$ was drawn. This dataset can be used to train a classifier that should be similar to the actual defender in place [46].

3. **Zero-knowledge (ZK) adversary**: A zero-knowledge (or *black-box*) adversary has no information about the learned model or the algorithm used by the learner before developing the attack.

For clarification, we present a simplifying description of the above threat models for adversarial attacks, given by Biggio et al. in [16]. Let $f$ be a learned unsecured model and $d$ a detector component, for instance a classifier for anomaly detection, implemented to secure $f$. Possible scenarios for the previous classification are depicted below:

1. A PK attacker gets full knowledge of the victim model $f$ and its security granted by the detector $d$, knows the model parameters of $d$, and uses this information to craft adversarial samples to evade $d$ and corrupt $f$.

2. A LK attacker is aware $f$ is being secured with a detection component $d$, knows the training scheme of both models, but his knowledge is *limited* by a denied access to the detector and the victim model (or the exact training data). In this scenario, the adversary may not be able to craft effective samples that thoroughly address his malicious goal.

3. A ZK adversary is assumed to generate adversarial examples to attack the unsecured model $f$ being not aware that a detector $d$ is in place. $d$ will successfully protect $f$ if it can detect and reject all the adversarial samples.

### 2.2.3 Attacker's Goals

While the attacker's objective to execute an attack on ML systems may encompass a broad spectrum of possible goals, we can distill attacks into two major classes in terms of the attacker's goals: **targeted attacks** and **untargeted (reliability) attacks**. In both attack types, the attacker's main attempt is to maximize mistakes in the learning algorithm's decisions with respect to the ground truth.

**Definition 2 (Targeted Adversarial Attack)** Given a trained classifier $f(\mathbf{x}; \theta)$ and a test instance from the dataset $\mathbf{x}_0 \in \mathscr{D}$ where $f(\mathbf{x}_0; \theta) = y_0$, the goal of the attacker is to apply a change in the label for $\mathbf{x}_0$ to a specific target label $y_T \neq y_0$, known as *misclassification label*. We can formulate the problem as

$$\min_{\delta} \quad \|\delta\|$$

$$\text{s.t.:} \quad f(\mathbf{x}_0 + \delta; \theta) = y_T \tag{4}$$

Note that for images, a second constraint such as $\mathbf{x}_0 + \delta \in [0, 1]^n$ is used, to impose a value-clipping constraint, where its goal is to bound the adversarial samples into a predefined range so that the perturbation remains human-imperceptible. Alternatively, the above problem can be expressed in an unconstrained optimization problem formulation

$$\min_{\delta: \|\delta\| \leq \epsilon} \quad \mathcal{L}(f(\mathbf{x}_0 + \delta; \theta), \ y_T) \tag{5}$$

One can note that in this case, the attacker aims to *minimize* the loss between adversarial prediction and the misclassification label $y_T$.

∎

**Definition 3 (Untargeted Attack)** The goal of the attacker in untargeted attack is to induce any misclassfication, such that

$$\min_{\delta} \quad \|\delta\|$$

$$\text{s.t.:} \quad f(\mathbf{x}_0 + \delta; \theta) \neq y_0 \tag{6}$$

where $\mathbf{x}_0 \in \mathscr{D}$ is a test instance and $y_0$ is the true class label, such that $f(\mathbf{x}_0; \theta) = y_0$. Similarly, here also we can formulate the untargeted adversarial attack as an unconstrained optimization problem where the goal of the attacker is to *maximize* the loss between the adversarial term and $y_0$

$$\max_{\delta: \|\delta\| \leq \epsilon} \quad \mathcal{L}(f(\mathbf{x}_0 + \delta; \theta), \ y_0) \tag{7}$$

∎

## 2.3 Adversarial Robustness: A Unified View of Adversarial Attacks and Defenses

In this section, we aim to study the adversarial defenses of ML systems. We start by presenting an optimization view of *adversarial defense*—or more precisely *adversarial robustness*—through the lens of robust optimization and use a min-max problem formulation to capture the notion of security against adversarial attacks in a principled manner. This formulation allows us to be precise about the type of security guarantee we would like to achieve, i.e., the broad class of attacks we want to resist. The empirical risk minimization (ERM) formulation introduced in

Eq. 1 does not yield models that are robust against adversarial examples. To reliably train models that are robust to adversarial attacks, it is necessary to redefine the ERM paradigm appropriately. In adversarial supervised learning, we assume that the adversary can modify training data distribution in a particular manner. For example, the adversary can adjust the input feature vectors to cause prediction errors.

**Definition 4 (Adversarial Empirical Risk Minimization (AERM))** Given a trained classifier $f(\mathbf{x}; \theta)$ and a test instance sampled from the distribution $(\mathbf{x}, y) \sim P$ that the classifier has been trained on, the attack model $x_{adv} = A_f(\mathbf{x}; \theta)$ takes $\mathbf{x} \in \mathbb{R}^d$ and projects it into data in the adversarial target set $Z \subset \mathscr{X} \times \mathscr{Y}$ aimed at increasing the prediction loss. The resulting AERM problem can be formulated as

$$\min_{\theta} \ p(\theta), \ \text{where} \ p(\theta) = \mathbb{E}_{(\mathbf{x}, y) \sim P} \left[ \max_{\mathbf{x}_{adv}} \ \mathscr{L}(A_f(\mathbf{x}; \theta), y) \mid (\mathbf{x}, y) \in Z \right] \quad (8)$$

where $\theta \in \mathbb{R}^d$ is the model parameter associated with $f$. The most commonly studied attack model so far is the *additive* adversarial perturbation in the form $\mathbf{x}_{adv} = \mathbf{x} + \delta$, where $\delta \in S \subseteq \mathbb{R}^d$ is the adversarial perturbation taken from the allowed adversarial set $S$. In addition, since $P$ is unknown, therefore we use the training dataset $\mathscr{D} = \{(\mathbf{x}_i, y_i)\}_{i=1}^n$ as a proxy, thus

$$\min_{\theta} \ p(\theta), \ \text{where} \ \underbrace{p(\theta)}_{\text{adversarial loss}} = \sum_{(\mathbf{x}_i, y_i) \in \mathscr{D}} \underbrace{[ \max_{\delta \in S} \ \mathscr{L}(f(\mathbf{x}_i + \delta; \theta), \ y_i) ]}_{\text{worst-case prediction loss } \mathscr{L}} \quad (9)$$

commonly represented in the following **min-max formulation** to capture the notion of security against adversarial attacks in a principled approach

$$\underbrace{\min_{\theta} \sum_{(\mathbf{x}_i, y_i) \in \mathscr{D}} \underbrace{[ \max_{\delta : \|\delta\| \leq \epsilon} \ \mathscr{L}(f(\mathbf{x}_i + \delta; \theta), \ y_i) ]}_{\text{worst-case loss } \mathscr{L} = \textbf{optimal attack}}}_{\textbf{robust classification} \text{ against adversarial attack}} \quad (10)$$

where $\delta : \|\delta\| \leq \epsilon$ represents the adversarial perturbation bounded by $\epsilon$, or the perturbation budget.

∎

The AERM problem defined above can be viewed as a composition of an **inner maximization attack** and **outer minimization defensive** problem. The inner maximization is the definition of adversarial attack presented in Sect. 2.1, which aims to find perturbation $\delta$ that maximizes the prediction loss of the model $f$. The outer minimization aims to find model parameters that minimize the adversarial loss produced by the inner attack problem or, more precisely, robust optimization using adversarial training.

## 2.4 Definition of Adversarial Attack and Countermeasure Strategies

In this section, we aim to formally introduce and define the most prominent attack and defense strategies used in ML systems commonly used in image classification task. The subsequent sections will discuss the same type of attack and defense strategies used against RS methods, with adaptation techniques to make them usable for recommendation tasks.

### 2.4.1 Attack Models

Various adversarial attack methods that aim to find a non-random perturbation $\delta$ to produce an adversarial example $x_{adv} = x + \delta$ that can cause an erroneous prediction (e.g., misclassification) are formally presented in this section. These attack methods all aim to solve the inner-maximization problem in Eq. 10 or more precisely the targeted and untargeted attack problems given by Eqs. 5 and 7.

**Definition 5 (Limited-Memory Broyden–Fletcher–Goldfarb–Shanno (L-BFGS))** The L-BFGS attack is the crucial work studied by Szegedy et al. [90], the one that captured researchers' attention the first time on the vulnerability of DNNs for image recognition tasks. Its main goal is to find a minimally distorted adversarial example based on a $l_2$ distance using an intuitive box-constrained optimization problem.

$$\min_{\delta} \ \|\delta\|_2^2$$
$$\text{s.t.: } f(\mathbf{x} + \delta) = y_T \text{ and } \mathbf{x} + \delta \in [0, M]^n \tag{11}$$

where $n$ is the number of features, $M$ is the maximum value among the pixels in the image, and $y_T$ is the misclassification label. From an implementation point of view, the above problem has two constraints and is hard to solve. The following equivalent with one constraint is solved instead.

$$\min_{\delta} \ c \cdot \|\delta\|_2^2 + \mathscr{L}(f(\mathbf{x}; \theta), y_T)$$
$$\text{subject to } \mathbf{x} + \delta \in [0, M]^n \tag{12}$$

the constraint $\mathbf{x} + \delta \in [0, M]^n$ is addressed by utilizing a box-constrained optimizer and a line-search that finds the best parameter $c$.

∎

**Definition 6 (Fast Gradient Sign Method (FGSM))** The FGSM attack model [50] was originally designed to exploit the 'linearity' of DNNs in the higher dimensional space. The goal of Goodfellow et al. [50] was to solve Eq. 6

(untargeted attack) by adding arbitrary perturbation to the original clean input with the $\ell_\infty$-bound constraint (i.e., $||\delta||_\infty \leq \epsilon$) such that the training loss of the target model increases, thus reducing classification confidence and improving the likelihood of inter-class confusion. While there is no guarantee that increasing the training loss by a certain amount will yield misclassification, this is nevertheless a sensible direction to exercise since the prediction error of a wrongly classified sample is, by definition, larger than the correctly classified one. The key idea in *untargeted FGSM* is to use a first-order approximation of the loss function and utilize the sign of the gradient function to construct adversarial samples for the adversary's target classifier $f$, obtaining.

$$\mathbf{x}_{adv} = \mathbf{x} + \epsilon \cdot \text{sign}(\nabla_x \mathscr{L}(f(\mathbf{x}; \theta), \ y)) \tag{13}$$

where $\epsilon$ (perturbation level) represents the attack strength and $\nabla_x$ is the gradient of the loss function w.r.t. input sample $\mathbf{x}$, $y$ is the correct label and sign(.) is the sign operator. The corresponding approach for *targeted FGSM* [60] is

$$\mathbf{x}_{adv} = \mathbf{x} - \epsilon \cdot \text{sign}(\nabla_x \mathscr{L}(f(\mathbf{x}; \theta), \ y_T)) \tag{14}$$

where $y_T$ is the target misclassification class label for sample $\mathbf{x}$.

∎

Several variants of the FGSM has been proposed in the literature [28, 104]. For instance, the fast gradient value (FGV) method [82], which instead of using the sign of the gradient vector in FGSM, uses the actual value of the gradient vector to modify the adversarial change, or basic iterative method (BIM) [60] (a.k.a. iterative FGSM) that applies FGSM attack multiple times *iteratively* using a small step size and within a total acceptable input perturbation level, according to

$$\mathbf{x}_{adv}^{t+1} = FGSM(\mathbf{x}_{adv}^t) \tag{15}$$

Projected gradient descent (PGD)[68] is similar to BIM attack. It starts from a random position in the clean image neighborhood.

**Definition 7 (Carlini and Wagner (CW-$\ell_0$, CW-$\ell_2$, CW-$\ell_\infty$))** Carlini and Wagner is a powerful attack model for finding adversarial perturbation under three various distance metrics ($\ell_0$, $\ell_2$, $\ell_\infty$). Its key insight is similar to L-BFGS as it transforms the constrained optimization problem into an empirically chosen loss function to form an unconstrained optimization problem as

$$\min_\delta \left( ||\delta||_p^p + c \cdot h(\mathbf{x} + \delta, \ y_T) \right) \tag{16}$$

where $h(\cdot)$ is the candidate loss function.                                           ∎

The C&W attack has been used with several norm-type constraints on perturbation $l_0, l_2, l_\infty$ among which the $l_2$-bound constraint has been reported to be most effective [25, 26, 29]. The CW-$\ell_2$ problem formulation for a targeted attack aiming is given by

$$\min_{\delta} \left( \|\mathbf{x}_{adv} - \mathbf{x}\|_2^2 - c \cdot h(\mathbf{x}_{adv}, y_T) \right)$$

$$h(\mathbf{x}_{adv}) = \max \left( \max_{i \neq t} Z\{\mathbf{x}_{adv_i}\} - Z\{\mathbf{x}_{adv_t}\}, -K \right) \qquad (17)$$

$$\mathbf{x}_{adv} = tanh(arctanh(\mathbf{x}) + \delta) + 1))$$

where $Z(x)$ denotes the logit corresponding to $i$-th class. By increasing the classification confidence $K$, the adversarial sample will be misclassified with higher confidence.

### 2.4.2   Adversarial Countermeasure Strategies

From a broad perspective, the defensive mechanism against adversarial attacks can be classified into one of the following approaches:

- **Increasing robustness of learning model:** These methods aim to formulate models that can correctly classify both adversarial and clean samples. At their heart, many of these methods attempt to create models less sensitive to irrelevant data variations, e.g., by regularizing models to mitigate the attack surface and bound responsiveness to samples that lie off the data manifold.
- **Detection:** While a robust classifier correctly labels adversarial perturbed samples, robustness may be alternatively achieved by *detection of adversarial examples*. A sample that is detected as an adversarial example is rejected.

Note that a recurring hypothesis about the cause for adversarial samples is that these examples lie off the data manifold and are sampled from a different distribution [84]. The learning model has no exposure to such off-manifold regions during training time, and hence its behavior can be controlled arbitrarily. While robust classification aims to map the data back to on-manifold data (e.g., natural image manifold) and recover its actual label, the detection methods treat the problem as an *anomaly detection problem*, only requiring to determine whether the input is an on-manifold data or reject it otherwise.

### Increasing Robustness of Learning Model

For what concerns the robustness of the learning model, we discuss two categories of algorithms for classification task:

1. **Robust optimization:** This is a theoretically grounded framework, as it aims to integrate robustness into learning. As such, robust optimization provides the means to guarantee or certify robustness in a principled manner. We discuss

adversarial training as one of the most common approaches for robust optimization against adversarial attacks here.

2. **Distillation** this is a heuristic approach for making gradient-based attacks more difficult to execute by effectively rescaling the output function to ensure that gradients become unstable [42].

**Robust Optimization and Adversarial Training**

In Sect. 2.3, we presented a unified framework for adversarial attack and defense by showing a zero-sum game between the learning model, in which we can formulate adversarial robustness as a *robust optimization* problem that seeks to find a solution for the worst-case input perturbation with respect to the set of allowed perturbation (i.e., $\delta \in S$)

$$\min_{\theta} \sum_{(\mathbf{x}_i, y_i) \in \mathscr{D}} \max_{\delta \in S} \mathscr{L}(f(\mathbf{x}_i + \delta; \theta), \ y_i) \tag{18}$$

where $\mathscr{L}$ is the loss function, $f$ is the learning model characterized by the parameter $\theta$, and $(\mathbf{x}_i, y_i) \in \mathscr{D}$ is a training sample. Typically $\|\delta\|_p \leq \epsilon$ known as the adversarial budget is used to represent the set $S$ where $\|\cdot\|_p$ is the $l_p$ norm.

Based on the above intuition, adversarial regularization, also known as *adversarial training* (AT) was proposed by Goodfellow et al. [50] for formulating a robust classifier. AT is a data augmentation training process, where one tries at each update to approximately solve the inner attack problem in Eq. 18, generating adversarial examples and injecting them into training data. Assuming worst-case loss function to be

$$\mathscr{L}_{wc}(f(\mathbf{x} + \delta; \theta), y) = \max_{\delta: \|\delta\| \leq \epsilon} \mathscr{L}(f(\mathbf{x} + \delta; \theta), y)$$

AT utilizes $\mathscr{L}_{wc}$ as a regularization component to explicitly trade-off between robustness (on adversarial samples) and accuracy (on non-adversarial data).

$$\min_{\theta} [\mathscr{L}(f(\mathbf{x}; \theta), y) + \lambda \mathscr{L}_{wc}(f(\mathbf{x}; \theta), y)] \tag{19}$$

where $0 < \lambda < 1$ is the regularization coefficient, controlling the amount of trade-off. The robustness achieved by AT strongly depends on the strength of the adversarial examples used. For instance, training on fast non-iterative attacks such as FGSM may yield robustness against non-iterative attacks, and not against PGD attacks [60, 85]. Madry et al. in [68] showed that training on multi-step PGD adversaries achieves state-of-the-art robustness levels against $l_\infty$ attacks.

Adversarial Training of BPR-MF

BPR is the state-of-the-art method for personalized ranking implicit feedbacks. The main idea behind BPR is to maximize the distance between positively and negatively rated items. Given the training dataset $D$ composed of positive and negative items for each user, and the triple $(u, i, j)$ (user $u$, a positive item $i$ and negative item $j$), the BPR objective function is defined as

$$\mathcal{L}_{BPR}(\mathcal{D}|\Theta) = \arg\max_{\Theta} \sum_{(u,i,j)\in\mathcal{D}} \ln \sigma(\hat{x}_{ui}(\Theta) - \hat{x}_{uj}(\Theta)) - \lambda \|\Theta\|^2 \quad (20)$$

where $\sigma$ is the logistic function, and $\hat{x}_{ui}$ is the predicted score for user $u$ on item $i$ and $\hat{x}_{uj}$ is the predicted score for user $u$ on item $j$; $\lambda \|\Theta\|^2$ is a regularization method to prevent over-fitting.[2] Adversarial training of BPR-MF similar to Eq. 19 can be formulated as

$$\mathcal{L}_{APR} = \min_{\theta} \sum_{(u,i,j)\in D} [\underbrace{\mathcal{L}_{BPR}(\mathcal{D}|\Theta) + \lambda \underbrace{\max_{\delta:\|\delta\|\leq\epsilon} \mathcal{L}_{BPR}(\mathcal{D}|\Theta + \delta)}_{\text{optimal attack model against BPR}}]}_{\text{optimal robustness preserving defensive}} \quad (21)$$

■

# 3   A Classification of Adversarial Attacks on Defenses of RS

The majority of the attack strategies that would be introduced in this chapter were initially conceived in the computer vision (CV) domain for the image classification task. Adversarial examples created for images, however, are empowered given the continuous real-valued nature of image data, i.e., pixel values, but the input data of recommender models are mostly discrete features (i.e., user ID, item ID, and other categorical variables) [55]. Perturbing these discrete features by applying noise is meaningless since it may change their semantics. Therefore, existing adversarial attacks in the field of CV do not apply to the RS problems trivially. Instead, they are applied at a deeper level—e.g., at the level of their intrinsic model parameters rather than the extrinsic inputs [55].

In the current section, we adopt a pragmatic approach to classify the research articles conducting adversarial attacks against RS and discuss each category's attack

---

[2] As it can be noted, BPR can be viewed as a classifier on the triple $(u, i, j)$, where the goal of the learner is to classify the difference $\hat{x}_{ui} - \hat{x}_{uj}$ as correct label +1 for a positive triple sample and 0 for a negative instance.

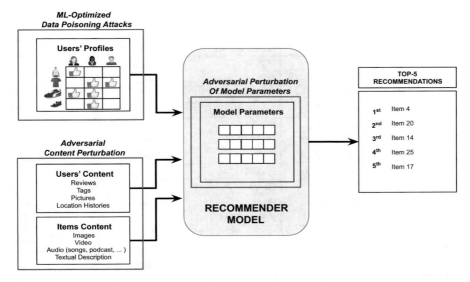

**Fig. 1** A notional view of the possible injection adversarial perturbations on (**a**) users profiles, (**b**) their content data, and (**c**) the learned model parameters. Depending on where adversarial perturbations have been applied, different adversarial attack types, as mentioned in the upper part of Table 2 have been proposed in the literature

and defensive strategies. Our classification entails categorization of attacks based on their *level of granularity* according to:

- Adversarial perturbation of model parameters;
- Adversarial perturbation of content;
- ML-optimized data poisoning attacks.

Figure 1 shows a schematic representation of the three possible points of adversarial perturbations. While the perturbation of model parameters belongs to evasion attacks (decision-time attacks), the URM's perturbation and the content data are considered poisoning attacks (data poisoning attacks) since they are added before the model's training. Moreover, to provide an overview, Table 2 introduces the adversarial attacks, which have been used over the last few years in RS research. It highlights the reviewed research articles according to the following dimensions:

- **Approach.** This column lists the authors' approach name to the proposed adversarial attack/defensive mechanism and provides the reference to the main paper.
- **Attack Model.** This column represents the main *attack* methods applied on various recommendation models. They include *FGSM* and *CW* (cf. Sect. 2.4.1), and generative attacks based on generative adversarial networks (GANs).
- **RS Model.** Given the machine-learned (optimization)-based approach for most of the considered papers, this column represents the core recommendation

**Table 2** Classification of approaches that use adversarial learning for attacking and defending RS models or ML-optimized poisoning attack

| Category | Approach | Attack model | RS model | Defense |
|---|---|---|---|---|
| AML Model parameter perturbation | APR [55] | FGSM | LFM | Adv.Train. |
| | FGACAE [110] | FGSM | AE | Adv.Train. |
| | ACAE [109] | FGSM | AE | Adv.Train. |
| | FNCF [42] | FGSM | NN | Distillation |
| | AdvIR [78] | FGSM | LFM | Adv.Train. |
| | AMASR [95] | FGSM | NN | Adv.Train. |
| | ATMBPR [99] | FGSM | LFM | Adv.Train. |
| | SACRA [65] | FGSM | LFM | Adv.Train. |
| | RAP [13] | Generative | LFM | Adv.Train. |
| AML Content feature perturbation | AMR [93] | FGSM | LFM | Adv.Train |
| | TAaMR [75] | FGSM, BIM, PGD | LFM | × |
| | VAR [5] | FGSM, PGD, C&W | LFM | Adv.Train. |
| ML-optimized Poisoning Attack | S-attack [45] | Hit Ratio maximization | LFM | SVM Class. |
| | LOKI [115] | Reinforcement Learning | Black-box | × |
| | PoisonRec [88] | Reinforcement Learning | Black-box | × |

prediction function according to LFM and non-linear models such as the ones based on *auto-encoder* (AE) and *neural network* (NN).

- **Defense Approach.** This column characterizes the countermeasure defensive approach against adversarial attacks. We have found that *adversarial training (a.k.a. adversarial regularization)* as the most adopted defensive approach irrespective of the attack model, while *distillation* being adopted only by a single paper [42].

## 3.1 Adversarial Perturbation of Model Embedding Parameters

While the attacks presented in Sect. 2.4.1 commonly evaluate the effectiveness of perturbation generated on loss function used for image classification task such as the cross-entropy of a neural network, in the case of RS, these strategies have been re-adapted in order to learn perturbation with respect to the loss function specific to RS tasks. For instance, using the matrix factorization (MF) model trained with BPR (known as MF-BPR)—the state-of-the-art ranking-based criterion for item recommendation—several works have investigated the robustness of embedding parameters added to user embeddings $p_u = p_u + \delta$, item embeddings $q_i = q_i + \delta$ or both in which $p_u, q_i \in \mathbb{R}^F$ are $F$-dimensional embedding factors.

**Attacks**

One of the first adversarial attack strategies against the recommender system's parameters was described by He et al. [55]. The authors studied the robustness of

BPR-MF [80] to adversarial perturbation on the users and items representation in the latent space. Their attack, built on the **FGSM** by Goodfellow et al. [50], was the first proposal to address adversarial perturbation in recommender systems formally. The proposed FGSM-based strategy approximates $\mathscr{L}$ by linearizing it around an initial zero-matrix perturbation $\Delta_0$ and applying the max-norm constraint. The adversarial perturbation $\Delta^{adv}$ is defined as:

$$\Delta^{adv} = \epsilon \frac{\Pi}{\|\Pi\|} \quad \text{where} \quad \Pi = \frac{\partial \mathscr{L}(\Theta + \Delta_0)}{\partial \Delta_0} \tag{22}$$

where $\| \cdot \|$ is the $l_2-$norm. After the calculation of $\Delta^{adv}$, He et al. added this perturbation to the current model parameters $\Theta^{adv} = \Theta + \Delta^{adv}$ and generated the recommendation lists with this perturbed model parameter. He et al. in [55] demonstrated that perturbation obtained from the FGSM with $\epsilon = 0.5$ can impair the accuracy of item recommendation by an amount equal to $-26.3\%$. Inspired by the effectiveness of this attack, several works have performed a similar perturbation against different recommender approaches such as visual-based recommender [91], tensor-factorization machines [30], deep-learning models [109, 110].

Similar to the advances in CV, the single-step FGSM attack has been extended with an iterative strategy in the recommender domain. The authors of [6] proposed an FGSM-based *iterative* strategy to create more effective $\epsilon$-clipped perturbations. The initial model parameters are defined as

$$\Theta_0^{adv} = \Theta + \Delta_0 \tag{23}$$

Starting from this initial state of model parameters, the authors introduce an element-wise clipping function $Clip_{\Theta,\epsilon}$ to limit the perturbation of each original embedding value inside the $[-\epsilon, +\epsilon]$ interval, a step size $\alpha$ which is the maximum perturbation budget of each iteration. The first iteration ($k = 1$) is then defined by:

$$\Theta_1^{adv} = Clip_{\Theta,\epsilon} \left\{ \Theta_0^{adv} + \alpha \frac{\Pi}{\|\Pi\|} \right\} \quad \text{where} \quad \Pi = \frac{\partial \mathscr{L}(\Theta + \Delta_0)}{\partial \Delta_0} \tag{24}$$

and they generalized the $k$-th iteration of the $K$-iterations multi-step attack as:

$$\Theta_k^{adv} = Clip_{\Theta,\epsilon} \left\{ \Theta_{k-1}^{adv} + \alpha \frac{\Pi}{\|\Pi\|} \right\} \quad \text{where} \quad \Pi = \frac{\partial \mathscr{L}(\Theta + \Delta_{k-1}^{adv})}{\partial \Delta_{k-1}^{adv}} \tag{25}$$

where $k \in [1, 2, \ldots, K]$, $\Delta_k^{adv}$ is the adversarial perturbation at the $k$-th iteration, and $\Theta_k^{adv}$ is the sum of the original model parameters $\Theta$ with the perturbation at the $k$-th iteration. Inspired by the advances of AML in iterative attacks, they considered two different versions of multi-step optimized adversarial perturbation: the Basic Iterative Method (BIM) [59] and the Projected Gradient Descent (PGD) [68] approaches. The authors have demonstrated the iterative adversarial strategies to be considerably more effective than the single-step FGSM method such that a state-of-the-art model-based recommender model (i.e., BPR-MF) can be impaired/weakened so much so that their performance become worse than a random recommender (a more than 90% of degradation of accuracy measures). It is possible to verify the drastic performance reduction generated by these iterative strategies against others state-of-the-art model-based recommenders similarly to the application of the single-step strategy in several models [30, 91, 109, 110].

The C&W attack has also been adapted to recommender systems by Du et al. [42], showing how it may contaminate the model performance in the testing phase. Similar to the previous strategies, the authors slightly changed the C&W approach to adapt to a recommender model (i.e., NCF in their experiments [54]). The C&W optimization problem is formulated as:

$$\min_{\Delta^{adv}} \quad ||\Delta^{adv}||$$

$$\text{s. t.} \quad f(\Theta + \Delta^{adv}) > 0.5$$

where $f(\cdot)$ is the prediction function to mark an item relevant to a user. The authors demonstrated that the attacks got a Success Rate close to 100% in inverting the predicted importance ($f(\cdot)$) of each user-item pairs.

It is clear from the above-summarized research that model-based recommenders are highly vulnerable to limited adversarial perturbations applied to model parameters. An attacker may access the model and completely misuse a recommender's utility by slightly perturbing their latent factors. Furthermore, while these settings may be complicated to be present in a real scenario, previous attacks have also demonstrated another important aspect of model-based recommenders: the instability of the training. In fact, authors [55] have claimed that the weakness of these perturbations needs particular study and attention by researchers and practitioners. The loss of a considerable part of accuracy within such small perturbations might be generated in a real scenario with few real (benevolent) users that, with their actions, are causing a model update that will get a great negative change in performance.

**Defenses**

The two defense mechanisms previously described have been proposed as a defensive solution also in the recommendation settings. Respectively, the Adversarial Personalized Ranking (APR) by He et al. [55], is the first method that modified the classical loss function of a recommender model (i.e., BPR-MF) by integrating the adversarial training procedure, and the stage-wise hints training by Du et al. [42], which has been inspired by the defensive distillation of deep neural networks.

To protect against FGSM attack, He et al. [55] proposed an adversarial training strategy to make the model robust to parameters perturbations. Similar to the adversarial training procedure proposed in the computer vision field [50], the authors proposed an adversarial training procedure by adding an adversarial regularization term.

The application of adversarial training has been demonstrated to make the model more robust to FGSM attacks. For instance, He et al. [55] demonstrated a reduction of attack effectiveness on BPR-MF by more than 90%. This training procedure has been applied to several recommender models, e.g., BPR-MF (AMR) [55], VBPR (AMR [91]), CDAE (FG-ACAE [109, 110]), PITF [79] (ATF [30]), etc. Furthermore, the generalization ability of the adversarially trained model that gets it less influenced by parameter perturbations may also improve accuracy performance (e.g., nDCG and HR). The adversarial training procedure by He et al. [55] has been further modified to solve different issues. Tran et al. [95] proposed a *flexible* $\epsilon$-bounded perturbations by multiplying $\epsilon$ with the standard deviation of the perturbed parameter ($\Theta$). Xu et al. [107] designed a *directional adversarial training* procedure to perturb the parameters towards $k$-nearest neighbors in the embedding space such that the directions of perturbations are bounded inside a small collaborative-aware region.

The adversarial training procedure has been applied effectively in defending the recommender systems considering only FGSM perturbations on model parameters. However, the *iterative adversarial perturbations* proposed by Anelli et al. [6] have demonstrated that the adversarial training procedure proposed by He et al. [55] fails with iterative perturbations with the same $\epsilon$-bounded budget perturbations (e.g., accuracy performance degraded by more than 50% when the model is adversarially trained).

Another defense strategy proposed by Du et al. [42] is a form of **defensive distillation** to make a deep recommender model (i.e., NCF) more robust to C&W attacks. The stage-wise hints training procedure transfers (distills) the knowledge learned from a teacher model into a student (architectural smaller) model. The hints, or modules, are the set of parameters transferred from the teacher to the student model in the stage-wise procedure. For instance, the items and users' latent vectors can be treated as two different modules. Furthermore, the authors have integrated the student model with a *noisy layer* for increasing the robustness of parameters against the perturbations. The defense procedure has been demonstrated to reduce the success rate of the C&W attacks compared to the baseline version of the recommender.

It is important to mention that several defense strategies, as well as hundreds of adversarial attack strategies, have been designed and implemented in different domains (e.g., computer vision, speech recognition, and test processing) [104], and only a few of them have been already adapted and evaluated in recommendation tasks.

## 3.2   Adversarial Perturbations on Content Data

The second category of adversarial attacks against recommender systems is associated with the perturbation of content data related to users and items. Content-based and Hybrid recommender systems are the two categories of recommender that are influenced by the quality of content data. For instance, images are gaining significant success in the class of fashion, and food recommendation since the high-level features extracted from convolutional neural networks have been demonstrated to influence users' preferences. For instance, McAuley et al. have proposed different visual-based recommender models [18, 53, 70] to demonstrate how much visual features have a significant impact on users' behavior and are effective in improving the recommendation performance in cold settings. Furthermore, deep features extracted from music/video audio signal [63, 96], text features extracted from tweets [112], reviews [113], or news articles [22] demonstrate an increasing dependence of recommender models from the quality, and 'benevolence' of users and items (side) data. An immediate question would be how these models can be affected when the input data is adversarially perturbed? Are the security measures proposed to protect the feature extractor (e.g., AE, CNN, and RNN) capable of making the recommender robust to such attacks? The answer is open for further investigations.

**Attacks**

In this section, we will present an example of adversarial perturbation of product images used as the input of a state-of-the-art visual-based recommender system, named Visual Bayesian Personalized Ranking (VBPR) [53], which is one of the main recommenders that has been tested under adversarial settings. Tang et al. [91] have demonstrated that adding a human-imperceptible untargeted adversarial perturbation on a single product image it would likely be ranked much lower than before. Di Noia et al. [75] have investigated the application of targeted adversarial attacks on product images (i.e., FGSM and PGD) such that the CNN classifier would have classified the images of a low-recommended category of products with the category of more popular products. The authors demonstrated that an imperceptible alteration of the picture would increase more than three times the probability of a product being recommended. Liu et al. [66] proposed recommendation-aware attack strategies to evaluate minimal perturbation directly considering the output recommendations.

The pivotal research on the impact of adversarial attacks on image data in the case of recommender systems is justifiable by the fact that the computer vision field where the adversarial machine learning starts to get the interest from the research community. A few works focus on attacks on data other than images such as text, audio signal, and video. For instance, Carlini et al. [27] presented an adversarial attack strategy to craft targeted adversarial audio samples such that the transcription is entirely different while preserving the original soundtrack.

We may imagine an adversary modifying the audio signal of tracks in a music recommender system such that the text is associated with an explicit (illegal) content or makes a love song (with an original romantic text) classified as a heavy metal track. Another example might be related to the injection of adversarial reviews in a review based recommender system. For instance, Gao et al. [47] proposed a black-box attack strategy to generate a spam movie review, which is classified as a positive message by the RNN-based classifier. This algorithm might drastically change the recommendations of a review-based recommender since spam review will alter the actual feedback from users. Following the same analysis, video adversarial samples [102] for video recommender, adversarial attack on graph-structured data [36] for social-based recommender, and adversarial input sequences for RNN [77] for sequence-aware recommendations are some intuitive examples that any hybrid/content.based recommender systems may be affected by adversarial attacks on their input data.

**Defenses**
This section defines the main principles to follow to protect a hybrid/content-based recommender system from the adversarial input samples. We define three main strategies: implement state-of-the-art (1) robustification procedure and (2) detection techniques of adversarial samples commonly used in the research field of used ML component to extract the content features, and (3) propose novel recommender models robust to adversarial samples.

*Increasing the Robustness of the Feature Extractor* The first category of defense mechanisms consist of selecting from the feature extractor research field (e.g., computer vision in the case of image classifier [1] and natural language processing for text classification [116]) the state-of-the-art approaches to make models more robust to adversarial samples. The main idea of robustification is to make the model able to be sensitive as little as possible to adversarial and real inputs data. Common strategies are:

- Model Robustification, e.g., Adversarial Training, and Defensive Distillation (see Sect. 3.1)
- Others, e.g., Deep Contractive Network [51], Data Compression [43], and Data Randomization [106]

It is essential to clarify to the practitioner that each defense strategy has to be evaluated based on the definition of the adversary threat model [24].

However, the application of robustification techniques of deep neural networks does not guarantee the recommender model's robustness. For instance, Anelli et al.[7] studied the efficacy of common defense strategies in the computer vision field, e.g., Adversarial Training and Free Adversarial Training, to the image feature extractor component used in state-of-the-art visual-based recommender systems (e.g., VBPR). After defining the adversary threat model, the authors demonstrated that the attacks still remain effective in manipulating the performance towards the adversary's malicious goal, leaving us an open challenge related to the study of defense mechanisms on the DNNs that effectively protect a recommender.

*Detecting Adversarial Samples* The second category of defensive mechanisms is composed of detection techniques. They aim to reject input data when classified as adversarial samples. The main detection techniques are:

- Classical ML Detector: PCA, Softmax, and Reconstruction of Adversarial Images;
- Adversarial Deep Detector: train a DNN to classify original/adversarial samples;
- Distributional Detection: filter out adversarial samples by comparing the distribution of the original images to the adversarial ones. The main methods are based on Kernel Density and Bayesian Uncertainty Estimation.

It is worth noticing that detection techniques have to be chosen based on the adversary threat model. For instance, Carlini and Wagner [26] verified the vulnerability of ten detection techniques under three adversary threat models. Furthermore, within this experimental setting, the authors have been able to demonstrate that the ten tested detection methods might be utterly useless in a scenario with a strong adversary (white-box attacks).

*Increasing Robustness of the Recommender Models* The last strategy to evaluate is identifying robustification methods to protect the model from the perturbability of content data. For instance, Tang et al. [91] modeled the adversarial perturbation directly applied to items' images with the perturbation of the features used in the visual-based recommender model. In this regard, the authors have used the adversarial training strategy presented in Sect. 2.4.2 by adversarially regularizing with respect to the image features. To the best of our knowledge, this is the first attempt to robustify a recommender model to limit the impact of input perturbation.

## 3.3 Machine-Learned Data Poisoning Attacks

Up to this section, the analysis fostered a broad class of attacks on machine learning models. An essential characteristic of those attacks is that they occur during the learning phase when the model is updated. However, another broad class of attacks targets the learning algorithms by manipulating the data used for training these models. The practitioners generally refer to these kinds of attacks as Poisoning attacks. For the sake of clarity, it is beneficial to consider that there are several categories of poisoning attacks. These categories are mainly related to the kinds of modifications on training data the adversary can perform. Moreover, some poisoning attack models will typically either impose a constraint on the number of modifications or a modification penalty; they may also constrain what the attacker can modify about the data (e.g., feature vectors and labels, only feature vectors or only labels), and what kinds of modifications are admissible (e.g., insertion only, or arbitrary modification).

The first kind of attack is named **Label modification attack**: this attack lets the attacker alter some supervised learning labels (within a specific overall budget).

Then, there is the **Poison insertion attack**, in which the attacker can attach some new poisoned feature vectors to the original data, with (or without) the corresponding target label. Another common kind of attack is the **Data modification attack**, where the attacker can modify some existing feature vectors or the labels (or both) in the original data before the model training. Finally, in **Boiling frog attacks**, the attacker exploits the frequent retrainings to poison the model at each retraining imperceptibly. In the literature, several methods have been proposed to perform Poisoning Attacks. Most of them consist of empirical techniques developed to conduct the attacks. Nevertheless, since these algorithms do not provide for any optimization (or adversarial learning), they remain partially outside the section's scope. Here, we drive the reader's attention to all those techniques that not only propose a novel Poisoning Attack but also propose *a specific optimization procedure to maximize the adversary's goal automatically*. To this extent, the underlying data model is crucial to define a proper optimization criterion. For this reason, we first introduce the methods that involve factorization models since this recommendation model is widely adopted for recommender systems, and the first attempts to realize the Poisoning Attack Optimization have been realized by exploiting this model.

### 3.3.1 Data Poisoning Optimization on Factorization-Based Recommenders

This section presents a systematic review of techniques for computing near-optimal data poisoning attacks for factorization-based recommendation models. In general, all these works share a factorization-based recommendation model, consisting of a very sparse URM matrix and the factorized vectors representing the users and the items. Despite the commonalities related to the underlying primary model and some mathematical choices for the gradient computation, the approaches we present in this section differ in other aspects. For a matrix consisting of $m$ users and $n$ items, the attacker is capable of adding a small fraction $\alpha \cdot m$ of malicious users to the training data matrix, and each malicious user is allowed to rate at most $B$ items with each preference bounded in the recommender specific range $[-\Lambda, \Lambda]$.

The first relevant work was proposed by Li et al. [64] in 2016. Their intuition was to exploit the techniques for optimizing malicious data-driven attacks proposed in previous literature regarding machine learning algorithms' security. Since poisoning attack model optimization can be modeled as a constrained optimization problem, some previous works in the ML field focus on how to approximately compute implicit gradients of the solution of an optimization problem based on *first-order KKT conditions*. Li et al. exploit this mathematical background and define a new unified optimization framework for recommender systems for computing optimal attack strategies. In this pioneering work, the authors exploit the *Projected Gradient Ascent (PGA)* method for solving the problem of maximizing the attacker utility. They characterize several attacker utilities, including **availability attacks**, where the goal is to increase the prediction error, and **integrity attacks**, where item-specific objectives are considered. They assume an attacker with knowledge of both the learner's learning algorithms and parameters (that follow the *Kerckhoffs' principle* to ensure reliable vulnerability analysis in the worst case). Finally, they provide the optimization techniques, based on *first-order Karush-Kuhn-Tucker (KKT) conditions*, for two popular minimization algorithms: alternating minimization and nuclear norm minimization.

Data poisoning optimization was further investigated for factorization-based social recommendation models. In Hu et al. [57], the authors have considered two different types of attack actions for the attacker: **(1)** injecting fake ratings and **(2)** generating fake relationships with normal users. In the loss function, the model considers an additional factor that exploits the cosine similarity between the user's friends and the target items. In the targeted poisoning attack, the attacker aims to promote an item to target users by injecting fake ratings and edges. We assume the recommender system recommends a list of items to the target user, consisting of $N$ unrated items with the largest predicted rating scores. If the item is in the list, the attacker has successfully attacked the target user. Therefore, to attack the target user, the attacker's goal is to maximize the hit probability of finding the target item in the recommendation list. Since it is hard to model the hit probability directly, analogously to prior works [44], they adopt the *Wilcoxon-Mann-Whitney loss* [12] to approximate it. However, the final goal is to model the attack as an optimization problem. Let the attacker control a set of malicious nodes. Hence, the attacker aims to find the optimal rating score vector and edge weight vector for all malicious nodes to minimize the total attack loss. The authors formulate the targeted poisoning attack as a *bi-level* optimization problem that considers the attack loss for each target user.

For the sake of completeness, several works in this section employ an approximation of the adversarial gradient. However, Tang et al. [92] have recently proposed a method that exploits the same adversarial gradient.

Furthermore, to avoid simple malicious node detection methods based on the number of ratings, the authors assume that each malicious node can rate *at most* a certain number of items. While for bi-level optimization problems with continuous variables, the literature proposes approximate solutions by exploiting gradient descent methods based on the lower-level problem *KKT conditions*, such methods cannot be directly applied to this problem since it involves discrete variables. Therefore, they propose a framework to approximately solve the problem. Indeed, the authors avoid optimizing the fake data (i.e., the rating scores and edge weights) of all malicious nodes simultaneously. Instead, they propose to optimize the fake data sequentially. Given the original rating scores and edge weights, as well as the fake rating scores and edge weights that were added so far, they find the rating scores and edge weights for the next malicious node to minimize the attacker's loss. Then, they alternatively generate fake rating scores with fixed fake edge weights for each malicious node and then generate fake edge weights with fixed fake rating scores until convergence.

Another critical and complex recommendation scenario is Cross-Domain recommendation. Even more here, where the recommendation takes place by leveraging information in two different domains, personal data security, and privacy are crucial investigation fields. In this sense, Chen et al.[31] have proposed a data poisoning attack framework to conduct several kinds of attacks. Again here, according to the *Kerckhoffs' principle*, they have considered the worst-case scenario in which the attackers have full knowledge of the attacked learning algorithm, namely *perfect knowledge attacks*. In detail, the attacker can inject a *certain number* of malicious users in the source domain and observe the effects on the learning algorithm and the model in the target domain. They rely on a particular factorization model, the *Collective Matrix Factorization*, that is a popular approach for dealing with pairwise relational data. Specifically, the two factorized models (in the source and the target domain) are computed simultaneously by exploiting a joint loss function. Hence, they formulate the optimal attack problem as a *bi-level* optimization problem. The outer optimization maximizes the attackers' utilities, while the inner optimization maximizes the poisoned data models' recommendation utility. In particular, the authors focus on three kinds of data poisoning attacks:

- **Availability attack**: Attackers try to maximize the estimated error in the target domain. The utility function is here formulated as the total amount of perturbation of attacked items scores on unobserved elements.
- **Integrity attack:** Attackers' goal is to increase (or decrease) the popularity of a subset of items.
- **Hybrid attack:** A mixture of availability attack and integrity attack.

The trade-off factors between the availability and the integrity attack in the utility function provide remarkable freedom. Indeed, the factors can be negative whether the attackers want to increase their items' popularity while perturbing less other recommendations to avoid detection.

They then transform the bi-level optimization problem to a single-level constrained optimization problem by exploiting the KKT conditions.

The adoption of learning-to-rank optimization criteria in Recommender Systems most certainly represents the major advance in the last decade. In 2020, Fang et al. [45] have proposed a relevant work that tackles the data poisoning optimization for the top-N recommendation. Here, the attacker's goal is to promote a particular item to as many normal users as possible and maximize the hit ratio, which is defined as the fraction of normal users whose top-N recommendation lists include the target item. The assumption is that the attacker is able to inject *some* fake users into the recommender system. Moreover, as in the other works, the attacker has *full knowledge* of the target recommender system (e.g., all the rating data, the recommendation algorithm). This work introduces several novelties and ideas.

- On the one hand, they propose to use a loss function to approximate the number of users to whom the target item will be recommended. They adopt a standard solution to solve the discrete rating problem: relaxing the integer rating scores to continuous variables and convert them back to integer rating scores after solving the reformulated optimization problem.

- On the other hand, they exploit the influence function approach inspired by the interpretable machine learning literature that shows that the top-N recommendations are mainly affected by a subset $S$ of influential users.

For this reason, they refer to the proposed attack as **S-attack**. Finally, given $S$, they propose a gradient-based optimization algorithm to determine the fake users' rating scores. In detail, they optimize the rating scores for fake users *sequentially* instead of optimizing for all the fake users simultaneously. In particular, they optimize the rating scores of a single fake user and add the fake user to the recommender system. Even here, the *hit ratio maximization problem* (*HRM* problem) is challenging, even if only one fake user is considered. To address the problem, they use a differentiable loss function to approximate the hit ratio (Wilcoxon-Mann-Whitney loss). However, the most important novelty is that, instead of using all regular users, they only use a selected subset of influential users to solve the HRM problem. Finally, they develop a gradient-based method to solve the HRM problem to determine the fake user's rating scores. Indeed, it has been observed [62, 101] that different training samples have different contributions to the solution quality of an optimization problem. The authors also propose a "standard" technique to detect fake users. Specifically, they extract six features: *RDMA* [33], *WDMA* [73], *WDA* [73], *TMF* [73], *FMTD* [73], and *MeanVar* [73] for each user from its ratings. Then, they build a training dataset consisting of an equal number of fake users generated by the attack and randomly sample regular users for each attack. Afterward, the training dataset is used to train a classifier for fake users' detection (here, a *Support Vector Machine*-based classifier).

The work presented in [56] is a novel **defensive strategy** that leverages *trim learning* to make matrix factorization resistant to data poisoning. *Trim learning* is a learning method that is robust against the data poisoning attack. Indeed, *trim learning* exploits the statistical difference between normal users and fake users as well as the differences between normal and fake items to learn a model while excluding the malicious information. Hence, the authors have integrated this learning method in the learning algorithm for matrix factorization, named *trim matrix factorization algorithm*. Finally, they have tested the defensive strategy's efficacy against some popular kinds of attacks: *Random*, *Average*, *Bandwagon*, and *Obfuscated Attacks*.

Differently from the previous works, Liu et al. [67] develop a **robustness analyzer** for factorization machines. Given a trained Factorization Machine (FM), and a perturbation space, the goal is to provide a robustness certificate which states whether the FM's prediction is robust or non-robust against data poisoning attacks. Once an instance is certifiably robust, its label does not change irrespectively to the attack models exploited in the given perturbation space. They decided to focus their work on factorization machines (FMs) thanks to their ability to handle discrete and categorical features. A common solution to deal with these features is to convert them to binary features via one-hot encoding or multi-hot encoding. Since the number of possible values is large, the resulting discrete feature vector can be high-dimensional and sparse. FMs incorporate the interactions between features building an accurate recommendation model. In detail, the FM prediction formula considers the combination of multiple individual features (i.e., it creates a plethora of derived features). The problem arises when an attacker *slightly modifies the input data*, and very similar instances receive very different predictions. In this paper, the authors model the recommendation problem as a binary classification. Therefore, attacked instances might be classified into completely different classes. These possible unreliable results can significantly limit the applicability of Factorization-based recommenders. In this respect, the authors sensed the importance of investigating adversarial perturbations' effects on the factorization machines. To certificate the robustness of an FM, they derive the *upper/lower bounds* that can be reached by all the possible perturbations (with a certain budget). If the bound is *certifiably robust*, no perturbation can change the prediction of the instance.

### 3.3.2  Data Poisoning Optimization and Reinforcement Learning

In contrast to the previous section, the attacks based on Reinforcement Learning (RL) show several similarities. Indeed, irrespective of the specific learning algorithm, the description of the method here requires to describe several aspects:

- the **attacker's knowledge** and **capability** (that are common also to the other attack families);
- the **state space**;
- the **action space**;
- the **reward utility**.

Beyond these commonalities, it is worth noting that Reinforcement Learning is a broad research field, and the specific techniques that are adopted are typically different depending on the scenario, the available information, and the capabilities of the attacker to act in the environment. Here we focus on two attack frameworks, **LOKI** [115] and **PoisonRec** [88].

The reinforcement learning-based framework **LOKI** learns an attack agent to generate adversarial user behavior sequences for data poisoning attacks. Differently from the other attack methods that we have previously analyzed, reinforcement learning algorithms *leverage the feedback from the recommendation systems*, instead of the whole architecture and parameters, *to learn the agent's policy*. The first limitation of this approach is that the attacker hardly controls the target recommendation system to be retrained to get the feedback and update the attack strategy. Moreover, recommendation services generally limit feedback frequency. Unfortunately, a reinforcement learning-based framework requires a large amount of feedback to train a policy function efficiently. For these reasons, the authors decided to circumvent the exploitation of real feedback from the target recommendation system to train a policy. In detail, they build *a local recommender simulator* to mimic the target model and make the reinforcement framework get reward feedback from the recommender simulator instead of the target recommendation system. The recommender simulator is built as an ensemble of *several representative* *recommendation models* to be agnostic to the real recommender system. This choice is based on the assumption that if two recommenders can produce similar recommendation results on a given dataset, then the adversarial samples generated for one of the recommenders could be used to attack the other. However, even though they rely on a local recommender, the number of retrainings needed for a reinforcement learning framework makes the retraining impracticable. Therefore, the authors develop an *outcome estimator* based on the influence function. The outcome estimator estimates the influence of the injected adversarial samples on the attack outcomes.

Let us summarize the main aspects of the scenario that the authors analyze, and that makes the current scenario differ from the previous ones:

1. First, they suppose a specific recommendation scenario, the *next-item recommendation task*; In this setting, given the existing sequences, the next-item recommendation's goal is to produce an ordered item list, which predicts the next items that the user will choose;
2. The attacker's goal is to promote a set of target items to as many target users as possible;
3. The attackers can inject fake users into the recommendation system. These users can visit or rate items, but these actions are limited to make the profile unnoticeable;
4. The attacker can access the full activity history of all the users in the recommendation system. The critical difference with a classic attack is that this knowledge is not known a priori;
5. As in previous works, even here, the number of fake users is limited;
6. Finally, the attacker does not know when the recommendation model is retrained and can only receive a limited amount of feedback from the black box recommendation model.

In the setting depicted by the authors, the data poisoning problem can be formulated as the creation of new sequential patterns that involve the target items in the training set of the target recommendation system. Consequently, the adversarial samples' generation is a *multi-step decision process*, in which the generator ought to select specific actions for the controlled users to *maximize attack outcome*. From the perspective of reinforcement learning, the *goal is to learn a policy function to generate sequential adversarial user behavior samples*, which can maximize the target users' averaged display rate. Hence, the reinforcement learning-based framework **LOKI** consists of a *recommender simulator*, an *outcome estimator*, and an *adversarial sample generator*. The adversarial attack against a local recommender simulator is essentially interpreted as a *multi-step decision problem*. This means that the attack process should be modeled as a *Markov Decision Process* (MDP). Typically, an MDP is defined by a **set of states**, a **set of actions**, the **transition probabilities**, and the **rewards**. In the considered scenario, the **Action space** consists of the possible actions that the agent can act. Since considering every action strategy for each item is particularly expensive and inefficient, the authors propose to divide the item set into groups and use the set of all the groups as action space. Indeed, they show that *adversarial samples do not necessarily need to match the same sample pattern*.

The **state** is defined as the sequence of actions that precede the current step. Finally, since the RL framework should learn a policy that increases the estimated prediction scores of target items, the authors propose a **reward function** that considers the *predicted weighted average influence of the target samples*. Additionally, **actions** are represented by means of an *embedding layer*.

In contrast with the two-fold motivation behind LOKI, the authors of **PoisonRec** focus on the myriad of different recommender algorithms currently available. In their opinion, indeed, it would be *"very difficult and also time-consuming to design effective attacks for each of them directly"*. To learn effective attack strategies on different complex black-box recommender systems, they resort to *Model-free reinforcement learning*, which requires just a little knowledge of the recommender system. From this perspective, they pose only *light constraints* to their recommendation scenario:

- In fact, the authors suppose a limited knowledge about the recommender system that could even be a black-box;
- They assume to know nothing about the logs, the system components, and the recommendation algorithm.
- Beyond this, the attackers can only retrieve basic item information like item title, item description, and an item's sales volume.

**PoisonRec** repeatedly injects fake user actions into a recommender system while improving the attack strategy by exploiting the rewards. In detail, as observed with LOKI, the authors of PoisonRec model the sequential attack behavior trajectory as an MDP. However, the choice of the **reward function** is not trivial at all. They take advantage of a widely-adopted metric, the number of *Page View* (PV), that measures items' exposure within a certain period in the online recommender system. Indeed, the attacker's goal will be *maximizing the Page Views* for certain target items. Although **PoisonRec** can not obtain PV for other users when attacking a real recommender system due to the black-box setting, PV is a usually available statistical indicator. Indeed, the attackers know how many users have viewed their products and other platform-specific indicators. In **PoisonRec**, the *attack trajectory* is defined as an MDP, with the additional constraint that *all the attackers will share the same policy network*. In this MDP, the **state space** consists of the current attacker and all the selected sequential actions performed so far. The **action space** is the union between the items in the catalog and the target items. Finally, as mentioned earlier, the **reward function** corresponds to the target items' PVs. In the policy network, the authors exploit two neural networks, an *LSTM* network and a *deep neural network* (DNN).

### 3.3.3   Data Poisoning Optimization Poisoning with Other Recommendation Families

Although the factorization-based and the reinforcement learning-based data poisoning methods have aroused much research interest in the last years, even other recommendation families deserve to be in the spotlight.

For example, **graph-based recommender systems** are becoming increasingly popular in the last decade. The reasons for their success are manifold.

On the one hand, the graphs let designers develop a *complex recommender system that considers heterogeneous classes*. Indeed, it is possible to build a *multi-partite graph* in which each partition contains *ontologically* different entities.

Second, we have witnessed a flourishing of advanced techniques to explore, summarize, and embed graphs. These techniques are now a swiss knife in the hands of the recommender systems practitioner.

In 2017, Yang et al. [108] proposed injecting fake co-visitations into the system. They supposed a bounded number of fake co-visitations that an attacker could inject, focusing on the items (and the number of fake co-visitations) to inject. By exploiting a strategy similar to the previous approaches, they modeled the attack as a constrained linear optimization problem to perform attacks with maximal threats.

Later, Fang et al. [44] have considered a graph-based recommender system in which the recommendation is realized by exploiting the stationary probabilities generated by random walks. As in the other works, they suppose the presence of an attacker who wants to *promote/demote a set of items*. Even here, they need to define a differentiable loss that might drive the learning. The choice falls on one of the most popular recommendation metrics, the *Hit Ratio*, which could be used to indicate the attack's efficacy. However, the *Hit Ratio* itself is not differentiable. Thus, the authors decide to rely on a proxy function, the *Wilcoxon-Mann-Whitney loss*, known to optimize the ranking performance. Once the model and the optimization function is defined, the authors focus on solving the optimization problem. The resulting optimization procedures *updates the weights of the edges* (the probabilities) *iteratively*. At the end of the learning process, the higher weights will represent the near-optimal path to the items that will compose the filler item set.

The authors also develop a method for detecting fake users inspired by *Sybil detection* in social networks. In detail, they propose a behavior-based method. First, they extract a set of features from a user's rating scores. Then, they train a classifier to distinguish between normal and fake users. In this respect, they have considered the features proposed in the previous literature: *Rating Deviation from Mean Agreement* (RDMA) [33], *Weighted Degree of Agreement* (WDA) [73], *Weighted Deviation from Mean Agreement* (WDMA) [73], *Mean-Variance* (MeanVar) [73], *Filler Mean Target Difference* (FMTD) [73].

*Neighborhood-based recommendation algorithms* is the main topic of [32]. The rationale behind this work is to push a set of items by taking advantage of the neighborhood mechanism. While there are several poisoning attacks to k-NN methods, to date, this represents the first attempt to *learn an optimal set of fake users* automatically for k-NNs through a data poisoning attack optimization problem. Indeed, Chen et al. [32] first define the data poisoning attack as an optimization problem, and then approximate the optimization problem to generate fake users. The authors introduce realistic constraints to limit the number of fake users and the maximum number of filler items. The authors then propose three different loss functions to approximate the *Hit Ratio*: (**1**) based on similarities, (**2**) based on user ratings, (**3**) a combination of both.

# 4 Evaluation

In this section, we discuss and analyze the methodologies to evaluate adversarial recommender systems. First, we present the experimental settings commonly adopted in the literature. Then, we introduce the evaluation metrics for evaluating the three-fold adversary's goals.

## 4.1 The Experimental Setting

Beyond the classical recommendation evaluation protocols , we need to consider a few other evaluation dimensions in Adversarial Learning . Due to the specific experimental scenarios, it is crucial to measure the recommendation's performance both in a *clean* and in an *attacked* setting. For this reason, a first evaluation takes place before running the attack or the defense. Thereby, we can evaluate the recommender's overall performance to assess the *a posteriori* impact of the system under evaluation. Since adversarial learning can support the either attack and defensive strategies, the evaluation settings can be categorized as:

- **Evaluation of Attack Strategies.** It concerns the comparison of the efficacy of adversarial attack methods **A** against a recommendation system **R**. It is common to measure it through the variation of performance before and after the execution of **A**. In this evaluation, it is necessary to check that the attacks have the same adversary threat models, i.e., the knowledge and the capabilities. For instance, comparing a *Zero-Knowledge* with a *Perfect-Knowledge* adversary is meaningless since typically, the more knowledge an attacker has, the more powerful attacks are.
- **Evaluation of a Defense Strategy.**

  - *Evaluation of the Robustification.* It compares the effectiveness of a defense strategy **D** against an adversarial attack method **A** on a recommender system **R**. Here, the evaluation goal is to evaluate the impact of **A** against both **R** and **R** + **D** (i.e., the defended recommender system). It typically entails evaluating the variation of the performance in the clean setting of **R** and **R** + **D**. The variation gives an idea of the influence of the defense strategy in the clean setting.

- *Evaluation of Detection Techniques.* To evaluate the detection of malicious input data (e.g., images, audio tracks, fake profiles), it is common to adopt accuracy metrics like *precision* and *recall*. Moreover, even where the method correctly removes the malicious data, it is crucial to investigate the effects of not detecting the malicious sample (if the techniques were not applied).

## 4.2 Evaluation Metrics

The comparison of attack algorithms has its roots in the adversary's goal. As previously introduced in Sect. 3, the adversarial attack strategies may differ profoundly depending on the domain, the recommendation task, and the aspects the attacker decides to target. One adversary may want to reduce the overall recommendation performance or influence the item recommendation task to push/nuke a product or a set of products. This kind of attack can be evaluated by exploiting protocols that consider usual recommendation metrics. These protocols are not sufficient to cover all the cases. For instance, an adversary may attack the content data (e.g., images, music tracks, reviews) to influence the automatically recognized classes that categorize the item. It is needed to resort to additional metrics to measure the human-perceptibility or other aspects of the adversarial variation in such cases. Even though it is impossible to exhaust all the possible evaluation scenarios, the following sections are a broad overview of the multiple scenarios we can deal with in an adversarial learning setting.

### 4.2.1 Impact on Overall Recommendation Performance

The final goal of these attack strategies is to alter the system's recommendation performance. For the sake of simplicity (and even because this is the generally adopted perspective), here we consider the situation in which the attacker aims to reduce the recommendation effectiveness. In the literature, the worsening of recommendation performance is evaluated with the percentage variation of two classical rank-wise accuracy measures for the item recommendation task: *Hit Ratio* ($HR@k$) and *normalized Discounted Cumulative Gain* ($nDCG@k$). Let $HR_i@k$ be the hit ratio measured in the clean setting, and let $HR_f@k@k$ be the same metric evaluated after the application of an adversarial attack. The percentage variation of the hit ratio ($d_{HR@k}$) is defined as follows

$$d_{HR@k} = \frac{HR_f@k - HR_i@k}{HR_i@k} \times 100 \tag{26}$$

Analogously, the percentage variation of the normalized Discounted Cumulative Gain ($d_{nDCG@k}$) is defined as:

$$d_{nDCG@k} = \frac{nDCG_{f@k} - nDCG_i@k}{nDCG_i@k} \times 100 \qquad (27)$$

For both metrics, the more negative values there are, the more impactful the attack is. In fact, these two metrics serve to study both adversarial attacks on model parameters [30, 55] and content data [91].

Since the adversary may also aim to alter other recommendation aspects, it is possible to apply the percentage variation measures to any accuracy and beyond accuracy metrics. Anelli et al. [6] evaluated the impact of iterative adversarial perturbations by analyzing a broad spectrum of metrics. There, the analysis covers item coverage [48], Shannon Entropy [94], and expected free discovery [97], as well as precision and recall accuracy measures.

### 4.2.2   Impact on the Recommendability of Item Categories

The second type of evaluation has a much specific focus. It considers the performance variation when the adversary targets to push or nuke an item or a set of items. It is worth noticing that this evaluation is different from evaluating the effect of shilling attacks. The item or segment of items is selected in those attacks without considering any relation to the items' content. On the other hand, the evaluation of these attacks considers that content data can be perturbed. The injected perturbation tries to fool the classifier used to extract the high-level features used in a hybrid/content-based recommendation system. For instance, in a real scenario, the attacker may adversarially perturb source product images such that the image feature extractor component will misclassify them. The new classification may make the item fall into a different category that is more popular or less popular compared to the source products. While under a general machine learning perspective, it would be essential to evaluate the attacker's capability to move the item into another class, in a recommendation scenario, the core aspect is the recommendation performance variation of the classes involved in the attack. Analogously, a similar evaluation can occur in a music recommender system when the adversary perturbs soundtracks, where the deep neural classifiers will misclassify them with different song genres (e.g., heavy metal instead of pop). Following this line, a plethora of metrics could be proposed to assess the category-specific recommendation performance. Here, for the sake of space, the focus remains on the two most adopted metrics in adversarial literature on recommendation systems: *Hit Ratio* and *nDCG*.

The **Category Hit Ratio** (CHR@k) [75] is a variant of $HR$ that evaluates the fraction of adversarially perturbed items in the top-$K$ recommendations.

**Definition 8 (Category Hit Ratio)** Let $\mathscr{C}$ be the set of classes for a classifier, $\mathscr{I}_c = \{i \in \mathscr{I} \mid x_i$ is classified as $c \in \mathscr{C}\}$ be the set of items whose content information $x_i$

is classified as $c$. The categorical hit ($chit$) is defined as:

$$chit(u, k) = \begin{cases} 1, & \text{if } k\text{-th item} \in \mathscr{I}_c \\ 0, & \text{if } k\text{-th item} \notin \mathscr{I}_c \end{cases} \tag{28}$$

Analogously, the $CHR_u@K$ is defined as:

$$CHR_u@K = \frac{1}{K}\sum_{k=1}^{K} chit(u, k) \tag{29}$$

*Categorical hit* ($chit(u, k)$) is a 0/1-valued function that returns 1 when the item in the $k$-th position of the top-$K$ recommendation list of the user $u$ is in the set of attacked items not-interacted by $u$. Since *Category Hit Ratio* does not consider the ranking (and relevance) of recommended items, **Category normalized Discounted Cumulative Gain** has been proposed [7], that assigns a gain factor to each considered ranking position. By considering a relevance threshold $\tau$, each item $i \in \mathscr{I}_c$ has an ideal relevance value of:

$$idealrel(i) = 2^{(s_{max}-\tau+1)} - 1 \tag{30}$$

where $s_{max}$ is the maximum possible score for the items in the recommended list. By considering a recommendation list provided to the user $u$, the relevance $rel(\cdot)$ of a suggested item $i$ is defined as:

$$rel(k) = \begin{cases} 2^{(s_{ui}-\tau+1)} - 1, & \text{if } k\text{-th item} \in \mathscr{I}_c \\ 0, & \text{if } k\text{-th item} \notin \mathscr{I}_c \end{cases} \tag{31}$$

where $k$ is the position of the item $i$ in the recommendation list. In Information Retrieval, the *Discounted Cumulative Gain* ($DCG$) is a metric of ranking quality that measures the usefulness of a document based on its relevance and its position in the result list. Analogously, Category Discounted Cumulative Gain ($CDCG$) is:

$$CDCG_u@K = \sum_{k=1}^{K} \frac{rel(k)}{\log_2(1+k)} \tag{32}$$

Since recommendation results may vary in length depending on the user, it is not possible to compare performance among different users, so the cumulative gain at each position should be normalized across users. In this respect, an *Ideal Category Discounted Cumulative Gain* ($ICDCG@K$) is defined as follows:

$$ICDCG@K = \sum_{k=1}^{min(K,|\mathscr{I}_c|)} \frac{rel(k)}{\log_2(1+k)} \tag{33}$$

$ICDCG@N$ indicates the score obtained by an ideal recommendation list that contains only relevant items. We can finally introduce the *normalized Category Discounted Cumulative Gain* (*nCDCG*) defined as:

$$nCDCG_u@K = \frac{1}{ICDCG@K} \sum_{k=1}^{K} \frac{rel(k)}{\log_2(1+k)} \qquad (34)$$

$nCDCG_u@K$ is ranged in the [0, 1] interval, where values close to 1 mean that the attacked items are recommended in higher positions (e.g., the attack is adequate).

### 4.2.3 Qualitative Evaluation of Perturbed Content

In this section, we present qualitative measures for assessing the perceptibility of adversarially perturbed content data in multimedia recommenders. For instance, accuracy-oriented evaluation metrics are insufficient to evaluate the scenario of attacking a recommendation system by perturbing content data. Indeed, an adversary might prepare a powerful attack that effectively alters the recommendation performance. However, the degradation of the perturbed content would be too evident. For instance, the final user would become aware of the attack because the image of a recommended product is altered or the audio track of a song is noisy. For this reason, qualitative perceptual metrics can evaluate the user perceptibility of the content perturbations. Moreover, these metrics can help to design an effective detector for maliciously manipulated content. In the following, we introduce a set of commonly adopted metrics with three distinct content categories: images, audio, and text.

**Images** For evaluating the quality of image perturbations, it is crucial to assess if users can detect a manipulated image. In an offline evaluation setting, the following metrics are often adopted:

- *Distance Metrics*. [111] The norm distances are widely adopted in the computer vision community to assess the perceptibility of perturbations. Based on the norm value $p$, these metrics are usually identified as $l_p$-norm distances.
- *Peak Signal-To-Noise Ratio* (PSNR) [103] is a more easily interpretable, logarithmic version of the Mean Squared Error (MSE).
- *Structural Similarity Index* (SSIM) [100] is a metric based on the assumption that humans are sensitive to the image's *structure*.
- *Learned Perceptual Image Patch Similarity* (*LPIPS*) [114] produces a perceptual distance value between two similar images by leveraging on (1) knowledge extracted from convolutional layers inside state-of-the-art CNNs and (2) collected human visual judgments about pairs of similar images.

**Audio** The metrics measure if the adversarial perturbation has exceeded the hearing thresholds. Unfortunately, the signal-to-noise ratio (SNR) is insufficient to determine the amount of perceptible noise since it does not represent a subjective metric. The main offline evaluation metrics are:

- *Distortion Metric.* [74] It quantifies the distortion introduced by adversarial perturbation by exploiting a $l_\infty$ distance metric.
- *Psychoacoustic hearing thresholds* [83] calculates the differences between the original and the modified signal spectrum to the threshold of human perception.
- *Perceptual Evaluation of Speech Quality (PESQ)* [81] integrates the perceptual analysis measurement system with the perceptual speech quality measure.

**Text**  For text-based content data (e.g., reviews given by users, product descriptions, tweets, or social posts), the evaluation is usually conducted with specific variants of $l_p$-norm distance measures. Indeed, small changes in a text (e.g., words, characters) are more readily perceptible. These evaluation metrics [116] have to measure the minimum perturbation amount to fool the system while remaining human-unperceivable (the lexical, syntactic, and semantic correctness must be preserved even with small distance-based measures). Among them, we cite: *Norm-based measurement, Grammar and syntax related measurement, Semantic-preserving measurement, Edit-based measurement, Jaccard similarity coefficient.*

## 5   Conclusion and the Road Ahead

Combined with the growing abundance of large-high-quality datasets, substantial technical breakthroughs over the last few decades have made machine learning (ML) a vital tool across a broad range of tasks, including computer vision and natural language processing and recommender systems (RS). However, success has been accompanied by a significant new arising challenge: *"many ML applications are adversarial in nature"* [98].

Modern RS utilize latent factor models (LFMs) such as matrix factorization (MF) as the core predictor and optimize it with pairwise ranking objectives, such as the Bayesian personalized ranking (BPR). This combination of LFM+BPR optimization is the most prominent adopted strategy to date to drive item recommendation tasks in different settings and domains. However, despite their great success, it has been shown that this combination leads to recommender models that are not *robust* in the face of adversarial attacks, i.e., subtle but non-random perturbation of model parameters that are found via (solving) an attack strategy, leading to the resultant model to produce erroneous predictions. Therefore, to enhance the robustness of recommender models, a new optimization framework has been designed, named adversarial personalized ranking (APR) [55], where it has been shown that it can not only perform more robust against adversarial samples but also produce a better generalization performance on clean samples.

Throughout this book chapter, we have reviewed the current state of the art in adversarial attacks against recommendation models. Adversarial machine learning for recommender systems (AML-RecSys) [3] combines best practices in ML and security to improve data security in RS tasks. We provided a taxonomy of adversarial RS that classifies AML-RecSys according to a novel classification point, level of granularity, according to (i) adversarial perturbation of model parameters, (ii) adversarial perturbation of content and, (iii) ML-optimized data poisoning attacks in which (iii) is an additional contribution we offer in this work to give a complete picture of AML for the security of RS. This novel classification axis provides a pragmatic approach to implement adversarial attacks against a specific given recommendation model.

Despite the many new and challenging results reached in the field we have seen in this chapter, there are still open issues and research directions that need further investigation [41]. Among them, we can indeed mention the following open challenges.

*Find New Attack/Defense Models in the RS Domain to Better Exploit the Theoretical and Practical Results Obtained in the ML/CV Field* First, we need to understand what is for RS the equivalent notion of *human-imperceptible or inconspicuous* we find in CV. As a second aspect, images are represented as continuous-valued data, while in RS, we deal with discrete data representing user profiles.

*Assess the Effects of Adversarial Attacks on Item/User Side Information* Most of the proposed strategies mainly deal with the collaborative data available in the user-item matrix. Nevertheless, modern recommendation models exploit a wealth of side-information beyond the user-item matrix, such as multimedia content presented in Chapter "Multimedia Recommender Systems: Algorithms and Challenges", social-connections, semantic data, among others [86]. Investigating the impact of adversarial attacks against these heterogeneous data remains an interesting open challenge.

*Definition of the Attack Threat Model* The research in the RS community misses a common evaluation approach for attacking/defending scenarios such as the one introduced by Carlini at el. [24]. For instance, it is essential to define a standard threat model to establish the attacker's knowledge and capabilities to make the attack (or defense) reproducible and comparable with novel proposals.

*Model Stealing* Another relevant area to the security of RS concerns the idea of model stealing in which the attacker can query a classifier, e.g., a web-based RS, and use the classifications labels (i.e., recommendation results) for reverse-engineering the model. While in adversarial attacks and data poisoning attacks, which we presented in this book chapter, the adversary typically has some knowledge about the structure of the model and tries to learn the model parameters, in model-stealing attacks, these assumptions are relaxed. The goal of these systems is to build a functionally equivalent classifier without knowing about the model type, structure, and parameters [87].

*Effects on Beyond Accuracy Metrics* Most of the research on adversarial ML and RS focuses on accuracy metrics, e.g., HR and nDCG. The impact on beyond accuracy metrics could be, in principle, the main objective of a new breed of attack strategies aiming to compromise the diversity/novelty of results.

*Scalability and Stability of Learning* We identify the need to further explore the stability learning problems in the discrete item sampling strategy to train the generator of a GAN architecture. This has already been identified as a big problem when GAN-based recommenders are applied in real scenarios with huge catalogs. A point of study may be that of novel GAN models proposed in computer vision (e.g., WGAN [11], LSGAN [69], and BEGAN [14]).

*Attacks on User Privacy and Fairness of RS and Protection Against It* Another investigation area is related to protecting user privacy, in the face of growing attention to user ownership of data after major data impeaches such as Cambridge Analytica [49] and GDPR proposed by the European Union. Recently attempts have been made to build machine-learned recommendation models that offer a privacy-by-design architecture, such as federated learning [8, 9], or the ones based on differential privacy. Additionally, in the light of adversarial setting, more attention has been made to build a privacy-preserving framework that can protect users from adversaries that aim to infer, or reconstruct, their historical interactions and social connections, consider for example [71, 72]. Fair and unbiased recommendations [40] are also similar related concerns of users—and item providers—in the RecSys community, see Chapter "Fairness in Recommender Systems". To address these concerns, recent attempts are emerging that try to use AML to provide an unbiased ranking list [119] or build fairness-aware systems for specific domains such as news [105].

# References

1. N. Akhtar, A. Mian, Threat of adversarial attacks on deep learning in computer vision: a survey. IEEE Access **6**, 14410–14430 (2018)
2. M. Aktukmak, Y. Yilmaz, I. Uysal, Quick and accurate attack detection in recommender systems through user attributes, in *RecSys* (ACM, New York, 2019), pp. 348–352
3. V.W. Anelli, Y. Deldjoo, T. Di Noia, F.A. Merra, Adversarial learning for recommendation: Applications for security and generative tasks - concept to code, in *RecSys 2020: Fourteenth ACM Conference on Recommender Systems, Virtual Event, Brazil, September 22–26, 2020* (ACM, New York, 2020), pp. 738–741
4. V.W. Anelli, Y. Deldjoo, T. Di Noia, E.D. Sciascio, F.A. Merra, Sasha: Semantic-aware shilling attacks on recommender systems exploiting knowledge graphs, in *The Semantic Web - 17th International Conference, ESWC 2020, Heraklion, Crete, Greece, May 31–June 4, 2020, Proceedings* (2020), pp. 307–323
5. V.W. Anelli, T. Di Noia, D. Malitesta, F.A. Merra, Assessing perceptual and recommendation mutation of adversarially-poisoned visual recommenders (short paper), in *DP@AI*IACEUR Workshop Proceedings*, vol. 2776, CEUR-WS.org (2020), pp. 49–56
6. V.W. Anelli, A. Bellogín, Y. Deldjoo, T. Di Noia, F.A. Merra, Msap: Multi-step adversarial perturbations on recommender systems embeddings, in *The International FLAIRS Conference Proceedings (FLAIRS 2021)*, vol. 34 (2021)

7. V.W. Anelli, Y. Deldjoo, T. Di Noia, D. Malitesta, F.A. Merra, A study of defensive methods to protect visual recommendation against adversarial manipulation of images, in *SIGIR 2021* (ACM, New York, 2021)

8. V.W. Anelli, Y. Deldjoo, T. Di Noia, A. Ferrara, F. Narducci, Federank: User controlled feedback with federated recommender systems, in *Advances in Information Retrieval - 43rd European Conference on IR Research, ECIR 2021, Virtual Event, March 28—April 1, 2021, Proceedings, Part I*. Lecture Notes in Computer Science, vol. 12656 (Springer, Berlin, 2021), pp. 32–47

9. V.W. Anelli, Y. Deldjoo, T. Di Noia, A. Ferrara, F. Narducci, How to put users in control of their data in federated top-n recommendation with learning to rank, in ed. by C.-C. Hung, J. Hong, A. Bechini, E. Song, *SAC '21: The 36th ACM/SIGAPP Symposium on Applied Computing, Virtual Event, Republic of Korea, March 22–26, 2021* (ACM, New York, 2021), pp. 1359–1362

10. V.W. Anelli, Y. Deldjoo, T. Di Noia, F.A. Merra, Understanding the effects of adversarial personalized ranking optimization method on recommendation quality, in *AdvML 2021: 3rd Workshop on Adversarial Learning Methods for Machine Learning and Data Mining, Virtual Event, August 14–18, 202q* (2021)

11. M. Arjovsky, S. Chintala, L. Bottou, Wasserstein GAN. CoRR, abs/1701.07875 (2017)

12. L. Backstrom, J. Leskovec, Supervised random walks: Predicting and recommending links in social networks, in ed. by I. King, W. Nejdl, H. Li, *Proceedings of the Forth International Conference on Web Search and Web Data Mining, WSDM 2011, Hong Kong, China, February 9–12, 2011* (ACM, New York, 2011), pp. 635–644

13. G. Beigi, A. Mosallanezhad, R. Guo, H. Alvari, A. Nou, H. Liu, Privacy-aware recommendation with private-attribute protection using adversarial learning, in *WSDM '20: The Thirteenth ACM International Conference on Web Search and Data Mining, Houston, TX, USA, February 3–7, 2020* (2020), pp. 34–42

14. D. Berthelot, T. Schumm, L. Metz, BEGAN: boundary equilibrium generative adversarial networks. CoRR abs/1703.10717 (2017)

15. R. Bhaumik, C. Williams, B. Mobasher, R. Burke, Securing collaborative filtering against malicious attacks through anomaly detection, in *Proceedings of the 4th Workshop on Intelligent Techniques for Web Personalization (ITWP'06), Boston*, vol. 6 (2006), p. 10

16. B. Biggio, I. Corona, D. Maiorca, B. Nelson, N. Srndic, P. Laskov, G. Giacinto, F. Roli, Evasion attacks against machine learning at test time, in ed. by H. Blockeel, K. Kersting, S. Nijssen, F. Zelezný, *Machine Learning and Knowledge Discovery in Databases - European Conference, ECML PKDD 2013, Prague, Czech Republic, September 23–27, 2013, Proceedings, Part III*. Lecture Notes in Computer Science, vol. 8190 (Springer, Berlin, 2013), pp. 387–402

17. B. Biggio, I. Corona, B. Nelson, B.I.P. Rubinstein, D. Maiorca, G. Fumera, G. Giacinto, F. Roli, Security evaluation of support vector machines in adversarial environments. CoRR abs/1401.7727 (2014)

18. J. Bourdeau, J. Hendler, R. Nkambou, I. Horrocks, B.Y. Zhao (eds.), *Proceedings of the 25th International Conference on World Wide Web, WWW 2016, Montreal, Canada, April 11–15, 2016* (ACM, New York, 2016)

19. R. Burke, M.P. O'Mahony, N.J. Hurley, Robust collaborative recommendation, in ed. by Ricci et al., *Recommender Systems Handbook* (Springer, Berlin, 2015), pp. 961–995

20. Y. Cai, D. Zhu, Trustworthy and profit: a new value-based neighbor selection method in recommender systems under shilling attacks. Decision Support Syst. **124**, 113112 (2019)

21. J. Cao, Z. Wu, B. Mao, Y. Zhang, Shilling attack detection utilizing semi-supervised learning method for collaborative recommender system. World Wide Web **16**(5–6), 729–748 (2013)

22. S. Cao, N. Yang, Z. Liu, Online news recommender based on stacked auto-encoder, in ed. by G. Zhu, S. Yao, X. Cui, S. Xu, *16th IEEE/ACIS International Conference on Computer and Information Science, ICIS 2017, Wuhan, China, May 24–26, 2017* (IEEE Computer Society, Washington DC, 2017), pp. 721–726

23. Y. Cao, X. Chen, L. Yao, X. Wang, W.E. Zhang, Adversarial attacks and detection on reinforcement learning-based interactive recommender systems, in J. Huang, Y. Chang, X. Cheng, J. Kamps, V. Murdock, J.-R. Wen, Y. Liu, *Proceedings of the 43rd International ACM SIGIR Conference on Research and Development in Information Retrieval, SIGIR 2020, Virtual Event, China, July 25–30, 2020* (ACM, New Yrok, 2020), pp. 1669–1672

24. N. Carlini, A. Athalye, N. Papernot, W. Brendel, J. Rauber, D. Tsipras, I.J. Goodfellow, A. Madry, A. Kurakin, On evaluating adversarial robustness. CoRR abs/1902.06705 (2019)

25. N. Carlini, D.A. Wagner, Defensive distillation is not robust to adversarial examples. CoRR abs/1607.04311 (2016)

26. N. Carlini, D.A. Wagner, Adversarial examples are not easily detected: Bypassing ten detection methods, in *Proceedings of the 10th ACM Workshop on Artificial Intelligence and Security, AISec@CCS 2017, Dallas, TX, USA, November 3, 2017* (2017), pp. 3–14

27. N. Carlini, D.A. Wagner, Audio adversarial examples: Targeted attacks on speech-to-text, in *2018 IEEE Security and Privacy Workshops, SP Workshops 2018, San Francisco, CA, USA, May 24, 2018* (2018), pp. 1–7

28. A. Chakraborty, M. Alam, V. Dey, A. Chattopadhyay, D. Mukhopadhyay, Adversarial attacks and defences: a survey. CoRR, abs/1810.00069 (2018)

29. P.-Y. Chen, H. Zhang, Y. Sharma, J. Yi, C.-J. Hsieh, ZOO: Zeroth order optimization based black-box attacks to deep neural networks without training substitute models, in *Proceedings of the 10th ACM Workshop on Artificial Intelligence and Security, AISec@CCS 2017, Dallas, TX, USA, November 3, 2017* (2017), pp. 15–26

30. H. Chen, J. Li, Adversarial tensor factorization for context-aware recommendation, in *RecSys* (ACM, New York, 2019), pp 363–367

31. H. Chen, J. Li, Data poisoning attacks on cross-domain recommendation, in ed. by W. Zhu, D. Tao, X. Cheng, P. Cui, E.A. Rundensteiner, D. Carmel, Q. He, J.X. Yu, *Proceedings of the 28th ACM International Conference on Information and Knowledge Management, CIKM 2019, Beijing, China, November 3–7, 2019* (ACM, New York, 2019), pp. 2177–2180

32. L. Chen, Y. Xu, F. Xie, M. Huang, Z. Zheng, Data poisoning attacks on neighborhood-based recommender systems. CoRR abs/1912.04109 (2019)

33. P.-A. Chirita, W. Nejdl, C. Zamfir, Preventing shilling attacks in online recommender systems, in ed. by A. Bonifati, D. Lee, *Seventh ACM International Workshop on Web Information and Data Management (WIDM 2005), Bremen, Germany, November 4, 2005* (ACM, New York, 2005), pp. 67–74

34. K. Christakopoulou, A. Banerjee, Adversarial attacks on an oblivious recommender, in *Proceedings of the 13th ACM Conference on Recommender Systems, RecSys 2019, Copenhagen, Denmark, September 16–20, 2019*, (2019), pp. 322–330

35. C. Clavier, Secret external encodings do not prevent transient fault analysis, in ed. by P. Paillier, I. Verbauwhede, *Cryptographic Hardware and Embedded Systems:CHES 2007, 9th International Workshop, Vienna, Austria, September 10–13, 2007, Proceedings.* Lecture Notes in Computer Science, vol. 4727 (Springer, Berlin, 2007), pp. 181–194

36. H. Dai, H. Li, T. Tian, X. Huang, L. Wang, J. Zhu, L. Song, Adversarial attack on graph structured data, in ed. by J.G. Dy, A. Krause, *Proceedings of the 35th International Conference on Machine Learning, ICML 2018, Stockholmsmässan, Stockholm, Sweden, July 10–15, 2018. Proceedings of Machine Learning Research* PMLR, vol. 80 (2018), pp. 1123–1132

37. Y. Deldjoo, T. Di Noia, F.A. Merra, Adversarial machine learning in recommender systems (AML-RecSys), in *WSDM '20: The Thirteenth ACM International Conference on Web Search and Data Mining, Houston, TX, USA, February 3–7, 2020* (ACM, 2020), pp. 869–872

38. Y. Deldjoo, T. Di Noia, E.D. Sciascio, F.A. Merra, How dataset characteristics affect the robustness of collaborative recommendation models, in *Proceedings of the 43rd International ACM SIGIR conference on research and development in Information Retrieval, SIGIR 2020, Virtual Event, China, July 25–30, 2020* (ACM, New York, 2020), pp. 951–960

39. Y. Deldjoo, M. Schedl, P. Cremonesi, G. Pasi, Recommender systems leveraging multimedia content. ACM Comput. Surv. **53**(5), 106:1–106:38 (2020)

40. Y. Deldjoo, V.W. Anelli, H. Zamani, A. Bellogín, T. Di Noia, A flexible framework for evaluating user and item fairness in recommender systems. User Model. User-Adapted Int. **31**, 457–511 (2021)

41. Y. Deldjoo, T. Di Noia, F.A. Merra, A survey on adversarial recommender systems: from attack/defense strategies to generative adversarial networks. *ACM Computing Surveys* **54**, 1–38 (2021)

42. Y. Du, M. Fang, J. Yi, C. Xu, J. Cheng, D. Tao, Enhancing the robustness of neural collaborative filtering systems under malicious attacks. IEEE Trans. Multimedia **21**(3), 555–565 (2019)

43. G.K. Dziugaite, Z. Ghahramani, D.M. Roy, A study of the effect of JPG compression on adversarial images. CoRR abs/1608.00853 (2016)

44. M. Fang, G. Yang, N.Z. Gong, J. Liu, Poisoning attacks to graph-based recommender systems, in *ACSAC* (ACM, 2018), pp. 381–392

45. M. Fang, N.Z. Gong, J. Liu, Influence function based data poisoning attacks to top-n recommender systems, in ed. by Y. Huang, I. King, T.-Y. Liu, M. van Steen, *WWW '20: The Web Conference 2020, Taipei, Taiwan, April 20–24, 2020* (ACM / IW3C2, New York/Geneva, 2020), pp. 3019–3025

46. C. Frederickson, M. Moore, G. Dawson, R. Polikar, Attack strength vs. detectability dilemma in adversarial machine learning, in *2018 International Joint Conference on Neural Networks, IJCNN 2018, Rio de Janeiro, Brazil, July 8–13, 2018* (IEEE, Piscataway, 2018), pp. 1–8

47. J. Gao, J. Lanchantin, M.L. Soffa, Y. Qi, Black-box generation of adversarial text sequences to evade deep learning classifiers, in *2018 IEEE Security and Privacy Workshops, SP Workshops 2018, San Francisco, CA, USA, May 24, 2018* (2018), pp. 50–56

48. M. Ge, C. Delgado-Battenfeld, D. Jannach, Beyond accuracy: Evaluating recommender systems by coverage and serendipity, in *Proceedings of the 2010 ACM Conference on Recommender Systems, RecSys 2010, Barcelona, Spain, September 26–30, 2010* (2010), pp. 257–260

49. F. González, Y. Yu, A. Figueroa, C. López, C.R. Aragon, Global reactions to the cambridge analytica scandal: A cross-language social media study, in *WWW* (2019)

50. I.J. Goodfellow, J. Shlens, C. Szegedy, Explaining and harnessing adversarial examples, in *3rd International Conference on Learning Representations, ICLR 2015, San Diego, CA, USA, May 7–9, 2015, Conference Track Proceedings* (2015)

51. S. Gu, L. Rigazio, Towards deep neural network architectures robust to adversarial examples, in *ICLR (Workshop)* (2015)

52. I. Gunes, C. Kaleli, A. Bilge, H. Polat, Shilling attacks against recommender systems: a comprehensive survey. Artif. Intell. Rev. **42**(4), 767–799 (2014)

53. R. He, J.J. McAuley, VBPR: Visual bayesian personalized ranking from implicit feedback, in ed. by D. Schuurmans, M.P. Wellman, *Proceedings of the Thirtieth AAAI Conference on Artificial Intelligence, February 12–17, 2016, Phoenix, Arizona, USA* (AAAI Press, Palo Alto, 2016), pp. 144–150

54. X. He, L. Liao, H. Zhang, L. Nie, X. Hu, T.-S. Chua, Neural collaborative filtering, in *WWW* (ACM, New York, 2017), pp. 173–182

55. X. He, Z. He, X. Du, T.-S. Chua, Adversarial personalized ranking for recommendation, in *SIGIR* (ACM, New York, 2018), pp. 355–364

56. S. Hidano, S. Kiyomoto, Recommender systems robust to data poisoning using trim learning, in ed. by S. Furnell, P. Mori, E.R. Weippl, O. Camp, *Proceedings of the 6th International Conference on Information Systems Security and Privacy, ICISSP 2020, Valletta, Malta, February 25–27, 2020*, SCITEPRESS (2020), pp. 721–724

57. R. Hu, Y. Guo, M. Pan, Y. Gong, Targeted poisoning attacks on social recommender systems, in *2019 IEEE Global Communications Conference, GLOBECOM 2019, Waikoloa, HI, USA, December 9–13, 2019* (IEEE, Piscataway, 2019), pp. 1–6

58. Y. Koren, R. Bell, Advances in collaborative filtering, in *Recommender Systems Handbook* (Springer, Berlin, 2015), pp. 77–118

59. A. Kurakin, I.J. Goodfellow, S. Bengio, Adversarial examples in the physical world, in *5th International Conference on Learning Representations, ICLR 2017, Toulon, France, April 24–26, 2017, Workshop Track Proceedings* (2017)

60. A. Kurakin, I.J. Goodfellow, S. Bengio, Adversarial machine learning at scale, in *5th International Conference on Learning Representations, ICLR 2017, Toulon, France, April 24–26, 2017, Conference Track Proceedings* (2017)

61. S.K. Lam, J. Riedl, Shilling recommender systems for fun and profit, in *Proceedings of the 13th International Conference on World Wide Web, WWW 2004, New York, NY, USA, May 17–20, 2004* (2004), pp. 393–402

62. À. Lapedriza, H. Pirsiavash, Z. Bylinskii, A. Torralba, Are all training examples equally valuable? CoRR abs/1311.6510 (2013)

63. J. Lee, S. Abu-El-Haija, B. Varadarajan, A. Natsev, Collaborative deep metric learning for video understanding, in ed. by Y. Guo, F. Farooq, *Proceedings of the 24th ACM SIGKDD International Conference on Knowledge Discovery & Data Mining, KDD 2018, London, UK, August 19–23, 2018* (ACM, New York, 2018), pp. 481–490

64. B. Li, Y. Wang, A. Singh, Y. Vorobeychik, Data poisoning attacks on factorization-based collaborative filtering, in *Advances in Neural Information Processing Systems 29: Annual Conference on Neural Information Processing Systems 2016, December 5–10, 2016, Barcelona, Spain* (2016), pp. 1885–1893

65. R. Li, X. Wu, W. Wang, Adversarial learning to compare: Self-attentive prospective customer recommendation in location based social networks, in *WSDM '20: The Thirteenth ACM International Conference on Web Search and Data Mining, Houston, TX, USA, February 3–7, 2020* (2020), pp. 349–357

66. Z. Liu, M.A. Larson, Adversarial item promotion: Vulnerabilities at the core of top-n recommenders that use images to address cold start. CoRR abs/2006.01888 (2020)

67. Y. Liu, X, Xia, L. Chen, X. He, C. Yang, Z. Zheng, Certifiable robustness to discrete adversarial perturbations for factorization machines, in ed. by J. Huang, Y. Chang, X. Cheng, J. Kamps, V. Murdock, J.-R. Wen, Y. Liu, *Proceedings of the 43rd International ACM SIGIR Conference on Research and Development in Information Retrieval, SIGIR 2020, Virtual Event, China, July 25–30, 2020* (ACM, New York, 2020), pp. 419–428

68. A. Madry, A. Makelov, L. Schmidt, D. Tsipras, A. Vladu, Towards deep learning models resistant to adversarial attacks, in *6th International Conference on Learning Representations, ICLR 2018, Vancouver, BC, Canada, April 30 - May 3, 2018, Conference Track Proceedings* (2018)

69. X. Mao, Q. Li, H. Xie, R.Y.K. Lau, Z. Wang, S.P. Smolley, Least squares generative adversarial networks, in *IEEE International Conference on Computer Vision, ICCV 2017, Venice, Italy, October 22–29, 2017* (2017), pp. 2813–2821

70. J.J. McAuley, C. Targett, Q. Shi, A. van den Hengel, Image-based recommendations on styles and substitutes, in *Proceedings of the 38th International ACM SIGIR Conference on Research and Development in Information Retrieval, Santiago, Chile, August 9–13, 2015* (2015), pp. 43–52

71. X. Meng, S. Wang, K. Shu, J. Li, B. Chen, H. Liu, Y. Zhang, Personalized privacy-preserving social recommendation, in *AAAI* (2018)

72. X. Meng, S. Wang, K. Shu, J. Li, B. Chen, H. Liu, Y. Zhang, Towards privacy preserving social recommendation under personalized privacy settings, World Wide Web **22**, 2853–2881 (2019)

73. B. Mobasher, R.D. Burke, R. Bhaumik, C. Williams, Toward trustworthy recommender systems: An analysis of attack models and algorithm robustness. ACM Trans. Int. Techn. **7**(4), 23 (2007)

74. P. Neekhara, S. Hussain, P. Pandey, S. Dubnov, J.J. McAuley, F. Koushanfar, Universal adversarial perturbations for speech recognition systems, in ed. by G. Kubin, Z. Kacic, *Interspeech 2019, 20th Annual Conference of the International Speech Communication Association, Graz, Austria, 15–19 September 2019*, ISCA (2019), pp. 481–485

75. T. Di Noia, D. Malitesta, F.A. Merra, Taamr: Targeted adversarial attack against multimedia recommender systems, in *50th Annual IEEE/IFIP International Conference on Dependable Systems and Networks Workshops, DSN Workshops 2020, Valencia, Spain, June 29–July 2, 2020* (IEEE, 2020), pp. 1–8

76. M.P. O'Mahony, N.J. Hurley, G.C. M. Silvestre, Recommender systems: Attack types and strategies, in *Proceedings, The Twentieth National Conference on Artificial Intelligence and the Seventeenth Innovative Applications of Artificial Intelligence Conference, July 9–13, 2005, Pittsburgh, Pennsylvania, USA* (2005), pp. 334–339

77. N. Papernot, P.D. McDaniel, A. Swami, R.E. Harang, Crafting adversarial input sequences for recurrent neural networks, in ed. by J. Brand, M.C. Valenti, A. Akinpelu, B.T. Doshi, B.L. Gorsic, *2016 IEEE Military Communications Conference, MILCOM 2016, Baltimore, MD, USA, November 1–3, 2016* (IEEE, Piscataway, 2016), pp. 49–54

78. D.H. Park, Y. Chang, Adversarial sampling and training for semi-supervised information retrieval, in *The World Wide Web Conference, WWW 2019, San Francisco, CA, USA, May 13–17, 2019* (2019), pp. 1443–1453

79. S. Rendle, L. Schmidt-Thieme, Pairwise interaction tensor factorization for personalized tag recommendation, in *WSDM* (ACM, New York, 2010), pp. 81–90

80. S. Rendle, C. Freudenthaler, Z. Gantner, L. Schmidt-Thieme, BPR: Bayesian personalized ranking from implicit feedback, in *UAI 2009, Proceedings of the Twenty-Fifth Conference on Uncertainty in Artificial Intelligence, Montreal, QC, Canada, June 18–21, 2009* (2009), pp. 452–461

81. A.W. Rix, J.G. Beerends, M.P. Hollier, A.P. Hekstra, Perceptual evaluation of speech quality (PESQ)-a new method for speech quality assessment of telephone networks and codecs, in *IEEE International Conference on Acoustics, Speech, and Signal Processing, ICASSP 2001, 7–11 May, 2001, Salt Palace Convention Center, Salt Lake City, Utah, USA, Proceedings* (IEEE, Piscataway, 2001), pp. 749–752

82. A. Rozsa, E.M. Rudd, T.E. Boult, Adversarial diversity and hard positive generation, in *2016 IEEE Conference on Computer Vision and Pattern Recognition Workshops, CVPR Workshops 2016, Las Vegas, NV, USA, June 26–July 1, 2016* (IEEE Computer Society, Washington DC, 2016), pp. 410–417

83. L. Schönherr, K. Kohls, S. Zeiler, T. Holz, D. Kolossa, Adversarial attacks against automatic speech recognition systems via psychoacoustic hiding, in *26th Annual Network and Distributed System Security Symposium, NDSS 2019, San Diego, California, USA, February 24–27, 2019* (The Internet Society, Reston, 2019)

84. A.C. Serban, E. Poll, Adversarial examples: a complete characterisation of the phenomenon. CoRR abs/1810.01185 (2018)

85. A. Shafahi, M. Najibi, A. Ghiasi, Z. Xu, J.P. Dickerson, C. Studer, L.S. Davis, G. Taylor, T. Goldstein, Adversarial training for free! in *Advances in Neural Information Processing Systems 32: Annual Conference on Neural Information Processing Systems 2019, NeurIPS 2019, 8–14 December 2019, Vancouver, BC, Canada* (2019), pp. 3353–3364

86. Y. Shi, M. Larson, A. Hanjalic, Collaborative filtering beyond the user-item matrix: a survey of the state of the art and future challenges. ACM Comput. Surv. **47**(1), 3:1–3:45, (2014)

87. Y. Shi, Y. Sagduyu, A. Grushin, How to steal a machine learning classifier with deep learning, in *2017 IEEE International Symposium on Technologies for Homeland Security (HST)* (IEEE, Piscataway, 2017), pp. 1–5

88. J. Song, Z. Li, Z. Hu, Y. Wu, Z. Li, J. Li, J. Gao, Poisonrec: An adaptive data poisoning framework for attacking black-box recommender systems, in *36th IEEE International Conference on Data Engineering, ICDE 2020, Dallas, TX, USA, April 20–24, 2020* (IEEE, Psicataway, 2020), pp. 157–168

89. A.P. Sundar, F. Li, X. Zou, T. Gao, E.D. Russomanno, Understanding shilling attacks and their detection traits: a comprehensive survey. IEEE Access **8**, 171703–171715 (2020)

90. C. Szegedy, W. Zaremba, I. Sutskever, J. Bruna, D. Erhan, I.J. Goodfellow, R. Fergus, Intriguing properties of neural networks, in *ICLR* (2014)

91. J. Tang, X. Du, X. He, F. Yuan, Q. Tian, T. Chua,  Adversarial training towards robust multimedia recommender system, in *IEEE Transactions on Knowledge and Data Engineering* **325**, 1–1 (2019)
92. J. Tang, H. Wen, K. Wang,  Revisiting adversarially learned injection attacks against recommender systems,  in *Fourteenth ACM Conference on Recommender Systems* (2020), pp. 318–327
93. J. Tang, X. Du, X. He, F. Yuan, Q. Tian, T.-S. Chua,  Adversarial training towards robust multimedia recommender system. IEEE Trans. Knowl. Data Eng. **32**(5), 855–867 (2020)
94. N. Tintarev, J. Masthoff, Explaining recommendations: Design and evaluation, in ed. by Ricci et al., *Recommender Systems Handbook* (Springer, Berlin, 2015), pp. 353–382
95. T. Tran, R. Sweeney, K. Lee,  Adversarial mahalanobis distance-based attentive song recommender for automatic playlist continuation,  in *Proceedings of the 42nd International ACM SIGIR Conference on Research and Development in Information Retrieval, SIGIR 2019, Paris, France, July 21–25, 2019* (2019), pp. 245–254
96. A. van den Oord, S. Dieleman, B. Schrauwen, Deep content-based music recommendation, in ed. by C.J.C. Burges, L. Bottou, Z. Ghahramani, K.Q. Weinberger, *Advances in Neural Information Processing Systems 26: 27th Annual Conference on Neural Information Processing Systems 2013. Proceedings of a meeting held December 5–8, 2013, Lake Tahoe, Nevada, United States* (2013), pp. 2643–2651
97. S. Vargas, P. Castells, Rank and relevance in novelty and diversity metrics for recommender systems, in ed. by B. Mobasher, R.D. Burke, D. Jannach, G. Adomavicius,*Proceedings of the 2011 ACM Conference on Recommender Systems, RecSys 2011, Chicago, IL, USA, October 23–27, 2011* (ACM, New York, 2011), pp. 109–116
98. Y. Vorobeychik, M. Kantarcioglu, *Adversarial Machine Learning.* Synthesis Lectures on Artificial Intelligence and Machine Learning (Morgan & Claypool Publishers, San Rafael, 2018)
99. J. Wang, P. Han,  Adversarial training-based mean bayesian personalized ranking for recommender system. IEEE Access **8**, 7958–7968 (2020)
100. Z. Wang, A.C. Bovik, H.R. Sheikh, E.P. Simoncelli, Image quality assessment: from error visibility to structural similarity. IEEE Trans. Image Process. **13**(4), 600–612 (2004)
101. T. Wang, J. Huan, B. Li,  Data dropout: Optimizing training data for convolutional neural networks, in ed. by L.H. Tsoukalas, É. Grégoire, M. Alamaniotis, *IEEE 30th International Conference on Tools with Artificial Intelligence, ICTAI 2018, 5–7 November 2018, Volos, Greece* (IEEE, Piscataway, 2018), pp. 39–46
102. Z. Wei, J. Chen, X. Wei, L. Jiang, T.-S. Chua, F. Zhou, Y.-G. Jiang,  Heuristic black-box adversarial attacks on video recognition models,  in *The Thirty-Fourth AAAI Conference on Artificial Intelligence, AAAI 2020, The Thirty-Second Innovative Applications of Artificial Intelligence Conference, IAAI 2020, The Tenth AAAI Symposium on Educational Advances in Artificial Intelligence, EAAI 2020, New York, NY, USA, February 7–12, 2020* (AAAI Press, Palo Alto, 2020), pp. 12338–12345
103. S. Winkler, P. Mohandas, The evolution of video quality measurement: From PSNR to hybrid metrics. IEEE Trans Broadcasting **54**(3), 660–668 (2008)
104. R.R. Wiyatno, A. Xu, O. Dia, A. de Berker,  Adversarial examples in modern machine learning: a review. CoRR abs/1911.05268 (2019)
105. C. Wu, F. Wu, X. Wang, Y. Huang, X. Xie, Fairness-aware news recommendation with decomposed adversarial learning, in *Proceedings of the AAAI Conference on Artificial Intelligence*, **35**(5), 4462–4469 (2021)
106. C. Xie, J. Wang, Z. Zhang, Y. Zhou, L. Xie, A.L. Yuille, Adversarial examples for semantic segmentation and object detection,  in *ICCV* (IEEE Computer Society, Washington, DC, 2017), pp. 1378–1387
107. Y. Xu, L. Chen, F. Xie, W. Hu, J. Zhu, C. Chen, Z. Zheng, Directional adversarial training for recommender systems, in *ECAI 2020* (2020)
108. G. Yang, N.Z. Gong, Y. Cai, Fake co-visitation injection attacks to recommender systems, in *NDSS* (2017)

109. F. Yuan, L. Yao, B. Benatallah, Adversarial collaborative auto-encoder for top-n recommendation, in *International Joint Conference on Neural Networks, IJCNN 2019 Budapest, Hungary, July 14–19, 2019* (2019), pp. 1–8

110. F. Yuan, L. Yao, B. Benatallah, Adversarial collaborative neural network for robust recommendation, in *Proceedings of the 42nd International ACM SIGIR Conference on Research and Development in Information Retrieval, SIGIR 2019, Paris, France, July 21–25, 2019* (2019), pp. 1065–1068

111. X. Yuan, P. He, Q. Zhu, X. Li, Adversarial examples: attacks and defenses for deep learning. IEEE Trans. Neural Netw. Learning Syst. **30**(9), 2805–2824 (2019)

112. Q. Zhang, J. Wang, H. Huang, X. Huang, Y. Gong, Hashtag recommendation for multimodal microblog using co-attention network, in ed. by C. Sierra, *Proceedings of the Twenty-Sixth International Joint Conference on Artificial Intelligence, IJCAI 2017, Melbourne, Australia, August 19–25, 2017*, ijcai.org (2017), pp. 3420–3426

113. L. Zheng, V. Noroozi, P.S. Yu, Joint deep modeling of users and items using reviews for recommendation, in *Proceedings of the Tenth ACM International Conference on Web Search and Data Mining, WSDM 2017, Cambridge, United Kingdom, February 6–10, 2017* (2017), pp. 425–434

114. R. Zhang, P. Isola, A.A. Efros, E. Shechtman, O. Wang, The unreasonable effectiveness of deep features as a perceptual metric, in *CVPR 2018* (2018)

115. H. Zhang, Y. Li, B. Ding, J. Gao, Practical data poisoning attack against next-item recommendation, in *WWW '20: The Web Conference 2020, Taipei, Taiwan, April 20–24, 2020* (2020), pp. 2458–2464

116. W.E. Zhang, Q.Z. Sheng, A. Alhazmi, C. Li, Adversarial attacks on deep-learning models in natural language processing: a survey. ACM Trans. Intell. Syst. Technol. **11**(3), 1–41 (2020)

117. W. Zhou, J. Wen, Q. Xiong, M. Gao, J. Zeng, SVM-TIA a shilling attack detection method based on SVM and target item analysis in recommender systems. Neurocomputing **210**, 197–205 (2016)

118. W. Zhou, J. Wen, Q. Qu, J. Zeng, T. Cheng, Shilling attack detection for recommender systems based on credibility of group users and rating time series. PloS one **13**(5), e0196533 (2018)

119. Z. Zhu, J. Wang, J. Caverlee, Measuring and mitigating item under-recommendation bias in personalized ranking systems, in *SIGIR* (2020)

# Group Recommender Systems: Beyond Preference Aggregation

Judith Masthoff and Amra Delić

## 1 Introduction

Most work on recommender systems to date focuses on recommending items to individual users. For instance, they may select a book for a particular user to read based on a model of that user's preferences in the past. Here, preferences are considered to be either implicit (e.g., clicks, purchase, viewing time, etc.), or explicit (e.g., likes, ratings, rankings, etc.) [1]. The challenge recommender system designers traditionally faced is how to decide what would be optimal for an individual user. A lot of progress has been made on this, as evidenced by other chapters in this handbook (e.g., [2–4]).

In this chapter, we go one-step further. There are many situations when it would be good if we could recommend to a group of users rather than to an individual. For instance, a recommender system may select television programmes for a group to view or a sequence of songs to listen to, based on individual preferences of all group members. Recommending to groups is more complicated than recommending to individuals. Assuming that we know perfectly what is good for individual users, the issue arises how to define and find what is also good for a group composed by the same individuals. The usual approach to solve this issue usually revolves around combining individual user preferences into a group preference model, or recommendations tailored for individuals into group recommendations. Methods or algorithms that combine individual user preferences or recommendations are called preference aggregation strategies. In this chapter, we will discuss how group

J. Masthoff (✉)
Utrecht University, Utrecht, Netherlands
e-mail: j.f.m.masthoff@uu.nl

A. Delić
University of Sarajevo, Sarajevo, Bosnia and Herzegovina
e-mail: adelic@etf.unsa.ba

© Springer Science+Business Media, LLC, part of Springer Nature 2022
F. Ricci et al. (eds.), *Recommender Systems Handbook*,
https://doi.org/10.1007/978-1-0716-2197-4_10

recommendation works, what its problems are, and what advances have been made. Interestingly, we will show that group recommendation techniques have uses similarly to what has been found for individual recommendations. So, even if you are developing recommender systems aimed at individual users you may still want to read on (perhaps reading Sect. 8 first will convince you).

In recent years, there has been increased attention to so called "multistakeholder" recommendation [5] also in detail discussed in Chapter "Multistakeholder Recommender Systems", and it is worth noting its relationship to group recommendations. Multistakeholder recommendation aims to include the perspectives and utilities of multiple stakeholders, i.e., "any group or individual that can affect, or is affected by, the delivery of recommendations to users" [5]. In that sense, group recommendation is a special case where each individual group member is a stakeholder, and the recommendations are designed to satisfy distinct stakeholders in the group.

Moreover, there is increasing attention in machine learning, and hence also in recommender systems, on fairness, which is mainly concerned with ensuring all subgroups of users (e.g., different genders) are treated equally, and are given the same opportunities with the provided recommendations (e.g., job recommendations). This topic is especially interesting in the context of group recommendation (and other multistakeholder recommendation) where fairness-related criteria become relevant on multiple sides, perhaps also from multiple perspectives [6]. In group recommenders fairness is usually considered as a characteristic of an aggregation strategy which ensures that none of the group members is impaired by the group recommendations [7, 8].

There are other issues to consider when building a group recommender system which are outside the scope of this chapter. In particular:

- *How to acquire information about individual users' preferences.* The usual recommender techniques can be used (such as explicit ratings and implicit feedback collection, see other handbook chapters). There is a complication in that it is difficult to infer an individual's preferences when a group uses the system, but inferences can be made during individual use combined with a probabilistic model when using it in company. An additional complication is that an individual's ratings may depend on the group they are in. For instance, a teenager may be very happy to watch a programme with his younger siblings, but may not want to see it when with his friends.
- *How will the system know who is present?* Different solutions exist, such as users explicitly logging in, probabilistic mechanisms using the time of day to predict who is present, the use of tokens and tags, etc. [9]. More sophisticated approaches have been used in recent years. For example, the GAIN system divides the group into a known subgroup (users which it knows are there) and an unknown subgroup (users that cannot be recognized but should be there statistically) [10]. A group recommender in a public display system recognizes the gender, emotions and group structures of people present (which are alone and which with others) [11].

- *How to present and explain group recommendations?* As seen in this hand-book's chapter on explanations, there are already many considerations when presenting and explaining *individual* recommendations. The case of group recommendations is even more difficult. More discussion on explaining group recommendations is provided in [12] and under Challenges in our final section.
- *How to help users to settle on a final decision?* In some group recommenders, users are given group recommendations, and based on these recommendations negotiate what to do. In other group recommenders this is not an issue (see Sect. 2.3 on the difference between passive and active groups). An overview of how users' decisions can be aided is provided in [12].

The next section highlights usage scenarios of group recommenders, and pro-vides a classification of group recommenders inspired by differences between the scenarios. Section 3 discusses strategies for aggregating models of individual users to allow for group recommendation, what strategies have been used in existing systems, and what was learned from experiments in this area. Section 4 deals with the issue of order when we want to recommend a sequence of items. Section 5 provides an introduction into the modeling of affective state, including how an individual's affective state can be influenced by the affective states of other group members. Section 6 explores how such a model of affective state can be used to build more sophisticated aggregation strategies. Section 7 discusses other group attributes (such as personality of users) that can be used in aggregation strategies. Section 8 shows how group modeling and group recommendation techniques can be used when recommending to an individual user. Section 9 concludes this chapter and discusses future challenges.

## 2    Usage Scenarios and Classification of Group Recommenders

There are many circumstances in which recommendation to a group is needed rather than to an individual. Below, we present two scenarios that inspired our own work in this area, discuss the scenarios underlying related work, and provide a classification of group recommenders inspired by differences between the scenarios.

### 2.1    Usage Scenario 1: Interactive Television

Interactive television offers the possibility of personalized viewing experiences. For instance, instead of everybody watching the same news program, it could be personalized to the viewer. For Judith, this could mean adding more stories about the Netherlands (where she comes from), China (a country that fascinates her after having spent some holidays there) and football, but removing stories about cricket (a sport she hardly understands) and local crime. Similarly, music programs could be adapted to show music clips that a user actually likes.

There are two main differences between traditional recommendation as it applies to say PC-based software and the interactive TV scenarios sketched above. Firstly, in contrast to the use of PCs, television viewing is largely a family or social activity. So, instead of adapting the news to an individual viewer, the television would have to adapt it to the group of people sitting in front of it at that time. Secondly, traditional work on recommendation was often concerned recommending one particular thing to the user, so for instance, which movie the user should watch. In the scenarios sketched above, the television needs to adapt a sequence of items (news items, music clips) to the viewer. The combination of recommending to a group and recommending a sequence is very interesting, as it may allow you to keep all individuals in the group satisfied by compensating for items a particular user dislikes with other items in the sequence which they do like.

## 2.2   Usage Scenario 2: Ambient Intelligence

Ambient intelligence deals with designing physical environments that are sensitive and responsive to the presence of people. For instance, consider the case of a bookstore where sensors detect the presence of customers identified by some portable device (e.g., a Bluetooth-enabled mobile phone, or a fidelity card equipped with an active RFID tag). In this scenario, there are various sensors distributed among the shelves and sections of the bookstore which are able to detect the presence of individual customers. The bookstore can associate the identification of customers with their profiling information, such as preferences, buying patterns and so on.

With this infrastructure in place, the bookstore can provide customers with a responsive environment that would adapt to maximize their well-being with a view to increasing sales. For instance, the device playing the background music should take into account the preferences of the group of customers within hearing distance. Similarly, LCD displays scattered in the store show recommended books based on the customers nearby, the lights on the shop's display window (showing new titles) can be rearranged to reflect the preferences and interests of the group of customers watching it, and so on. Clearly, group adaptation is needed, as most physical environments will be used by multiple people at the same time.

## 2.3   Usage Scenarios Underlying Related Work

In this section we discuss the scenarios underlying some of the best known group recommender systems:

- MUSICFX [13] chooses a radio station for background music in a fitness center, to suit a group of people working out at a given time. This is similar to the Ambient Intelligence scenario discussed above.
- POLYLENS [14] is an extension of MOVIELENS. MOVIELENS recommends movies based on an individual's taste as inferred from ratings and social filtering. POLYLENS allows users to create groups and ask for group recommendations.
- INTRIGUE [15] recommends places to visit for tourist groups taking into account characteristics of subgroups within that group (such as children and the disabled).
- The TRAVEL DECISION FORUM [16] helps a group to agree on the desired attributes of a planned joint holiday. Users indicate their preferences on a set of features (like sport and room facilities). For each feature, the system aggregates the individual preferences, and users interact with embodied conversational agents representing other group members to reach an accepted group preference.
- The COLLABORATIVE ADVISORY TRAVEL SYSTEM (CATS) [17] also recommends a joint holiday. Users consider holiday packages, and critique their features (e.g., 'like the one shown but with a swimming pool'). Based on these critiques, which are combined into a group preference model, the system recommends other holidays. Users also select holidays they like for other members to see, and these are annotated with how well they match the preferences of each group member.
- YU'S TV RECOMMENDER [18] recommends a television programme for a group to watch. It bases its recommendation on the individuals' preferences for programme features (such as genre, actors, keywords).
- The GROUP ADAPTIVE INFORMATION AND NEWS system (GAIN) [10] adapts the display of news and advertisements to the group of people near it.
- The REMINISCENCE THERAPY ENHANCED MATERIAL PROFILING IN ALZHEIMERS AND OTHER DEMENTIAS system (REMPAD) [19] recommends multimedia material to be used by a facilitator in a group reminiscence therapy session, based on the suitability of material for individual participants as inferred from their date of birth, locations lived in, and interest vectors.
- HAPPYMOVIE [20] recommends movies to groups, by enriching the group preference model with the individuals' personality (assertiveness and cooperativeness) and the relationship strengths (they call this social trust) between individuals.
- INTELLIREQ [21] supports groups in deciding which software requirements to implement. Users can view and discuss recommendations for group decisions based on already defined user preferences.
- WHERE2EAT [22] is a mobile group recommender for restaurants. It implements *"interactive multi-party critiquing"*, which allows group members to generate proposals and counter-proposals until an agreement is reached.
- CHOICLA [23] is a group decision support environment that allows users to configure decision tasks and decision-making process in a domain-independent setting.

- HOOTLE [24] is a hotel group recommender that supports negotiation about the features of a desired hotel. After aggregating individual preferences into a group model, content-based filtering is used to generate group recommendations.
- STSGROUP (South Tyrol Suggests for Group) [25–27] is a chat-based, context-aware mobile application for recommending Points of Interest (POIs). An innovative feature of this system is that it combines users' long-term preferences (based on users' previous individual interactions with the system), with the dynamic preferences elicited during the group decision-making process.
- TOURREC [28] recommends routes to individuals and groups. A route is generated by finding POIs which best fit the user profile and the contextual information, and are located between the user-defined start and end point. Then, the system connects POIs in recommendations tailored for individuals. To generate a group recommendation, various social choice strategies were implemented to aggregate users' individual travel preferences. The system also allows the group to split up during the trip, so that every member is still able to visit their own favorite POIs.

## 2.4   A Classification of Group Recommenders

The scenarios provided above differ on several dimensions, which provide a way to classify group recommender systems:

- *Individual preferences are known versus developed over time.* In most scenarios, the group recommender starts with individual preferences. In contrast, in CATS, individual preferences develop over time, using a critiquing style approach. Others have also adopted this critiquing approach (e.g., [22]). In INTELLIREQ, individual preferences can be influenced by the group discussion and the group recommendation based on preferences defined so far. In STSGROUP individual preferences are known, but group-related preferences are developed over time, and the profile of an individual user is also updated based on the group interaction.
- *Recommended items are experienced by the group versus presented as options.* In the Interactive TV scenario, the group experiences the news items. In the Ambient Intelligence, GAIN, and MUSICFX scenarios, they experience the music and advertisements. In contrast, in the other scenarios, they are presented with a list of recommendations, e.g., POLYLENS.
- *The group is passive versus active.* In most scenarios, the group does not interact with the way individual preferences are aggregated. However, in the TRAVEL DECISION FORUM and CATS the group negotiates the group model. In INTELLIREQ, the group does not influence the aggregation, but may influence the ratings provided.
- *Negotiation versus the single-shot recommendations.* Here we distinguish whether the system allows the group to interact with the suggested

recommendations, thus, allowing them to go into several rounds of negotiation [29]. In MUSICFX, recommendations are delivered to a group without an option to interact with the system (e.g., skip a recommended item). In a similar fashion, POLYLENS delivers recommendations in a "single-shot". In contrast, TRAVEL DECISION FORUM, CATS, INTELLIREQ, STSGROUP allow users to discuss proposals made by the system.

- *Recommending a single item versus a sequence.* In the scenarios of MUSICFX, POLYLENS, and YU'S TV RECOMMENDER it is sufficient to recommend individual items: people normally only see one movie per evening, radio stations can play forever, and YU'S TV RECOMMENDER chooses one TV program only. In contrast, in our Interactive TV scenario, a sequence of items is recommended, for example making up a complete news broadcast. Similarly, in INTRIGUE and TOURREC, it is quite likely that a tourist group would visit multiple attractions during their trip, so would be interested in a sequence of attractions to visit. Also, in the Ambient Intelligence scenario it is likely that a user will hear multiple songs, or see multiple items on in-store displays. In GAIN, the display shows multiple items simultaneously; additionally, the display is updated every 7 min, so people are likely to see a sequence as well.

DeCampos et al.'s classification of group recommenders also distinguishes between passive and active groups [30]. In addition, it uses two other dimensions:

- *How individual preferences are estimated.* They distinguish between content-based and collaborative filtering. Of the systems mentioned above, POLYLENS and HAPPYMOVIE use collaborative filtering; the others use content-based filtering (e.g., REMPAD).
- *Whether profiles or recommendations are aggregated.* In the first case, profiles, containing group members' individual preferences usually expressed numerically for each item, are aggregated into a group model. In the second case, recommendations are produced for individuals and then aggregated into a group recommendation. The two approaches are slightly different, since preferences are usually numerical values, while recommendations are items. However, recommendations can be generated (1) as ranked lists, and then individual lists are aggregated into the group list, or (2) as numerical estimations assigned to individual recommendations which then enable their aggregation in a similar fashion as aggregating profiles. They mention INTRIGUE and POLYLENS as aggregating recommendations, while the others tend to aggregate profiles. Aggregating profiles can happen in multiple ways. In this chapter, we will look at the aggregation of preference ratings. It is also possible to aggregate content: for example, GroupReM aggregates individuals' tag cloud profiles to produce a group tag cloud profile [31]. It is also possible to use a combination of aggregating profiles and aggregating recommendations: [32] proposes a hybrid switching approach that uses aggregated recommendations when user data is sparse and aggregated profiles otherwise. Following their example, [33] also uses a combination.

These two dimensions are related to how the group recommender is implemented rather than being inherent to the usage scenario. In this chapter, we focus on aggregating profiles, but the same aggregation strategies apply when aggregating recommendations. The material presented in this chapter is independent of how the individual preferences are obtained.

## 3   Aggregation Strategies

The main problem group recommendation needs to solve is how to adapt to the group as a whole based on information about individual users' likes and dislikes. For instance, suppose the group contains three people: Peter, Jane and Mary. Suppose a system is aware that these three individuals are present and knows their interest in each of a set of items (e.g., music clips or advertisements). Table 1 gives example ratings on a scale of 1 (really hate) to 10 (really like). Which items should the system recommend, given time for four items?

### 3.1   Overview of Aggregation Strategies

Many strategies exist for aggregating individual ratings into a group rating (e.g., used in elections and when selecting a party leader). For example, the Least Misery Strategy uses the minimum of ratings to avoid misery for group members (Table 2). These strategies can then be used to rank items according to their relevance for the group at hand, hence to estimate what should be recommended to the group.

Eleven aggregation strategies inspired by Social Choice Theory are summarized in Table 3 (see [9] for more details). In [34], aggregation strategies are classified into (1) *majority-based strategies* that use the most popular items (e.g., Plurality Voting), (2) *consensus-based strategies* that consider the preferences of all group members (e.g., Average, Average without Misery, Fairness), and (3) *borderline strategies* that only consider a subset (e.g., Dictatorship, Least Misery, Most Pleasure).

**Table 1** Example of individual ratings for ten items (A to J)

|       | A  | B | C | D | E  | F | G | H | I  | J |
|-------|----|---|---|---|----|---|---|---|----|---|
| Peter | 10 | 4 | 3 | 6 | 10 | 9 | 6 | 8 | 10 | 8 |
| Jane  | 1  | 9 | 8 | 9 | 7  | 9 | 6 | 9 | 3  | 8 |
| Mary  | 10 | 5 | 2 | 7 | 9  | 8 | 5 | 6 | 7  | 6 |

**Table 2** Example of the least misery strategy

|              | A  | B | C | D | E  | F | G | H | I  | J |
|--------------|----|---|---|---|----|---|---|---|----|---|
| Peter        | 10 | 4 | 3 | 6 | 10 | 9 | 6 | 8 | 10 | 8 |
| Jane         | 1  | 9 | 8 | 9 | 7  | 9 | 6 | 9 | 3  | 8 |
| Mary         | 10 | 5 | 2 | 7 | 9  | 8 | 5 | 6 | 7  | 6 |
| Group rating | 1  | 4 | 2 | 6 | 7  | 8 | 5 | 6 | 3  | 6 |

**Table 3** Overview of aggregation strategies

| Strategy | How it works | Example |
|---|---|---|
| Plurality/majority voting | Uses 'first past the post': repetitively, the item with the most votes is chosen. | A is chosen first, as it has the highest rating for the majority of the group, followed by E (which has the highest rating for the majority when excluding A) |
| Average | Averages individual ratings | B's group rating is 6, namely (4+9+5)/3 |
| Multiplicative | Multiplies individual ratings | B's group rating is 180, namely 4*9*5 |
| Borda count | Counts points from items' rankings in the individuals' preference lists, with bottom item getting 0 points, next one up getting one point, etc. | A's group rating is 17, namely 0 (last for Jane) + 9 (first for Mary) + 8 (shared top 3 for Peter) |
| Copeland rule | Counts how often an item beats other items (using majority vote[a]) minus how often it looses | F's group rating is 5, as F beats 7 items (B,C,D,G,H,I,J) and looses from 2 (A,E) |
| Approval voting | Counts the individuals with ratings for the item above approval threshold (e.g. 6) | B's group rating is 1 and F's is 3 |
| Least Misery | Takes the minimum of individual ratings | B's group rating is 4, namely the smallest of 4,9,5 |
| Most pleasure | Takes the maximum of individual ratings | B's group rating is 9, namely the largest of 4,9,5 |
| Average without Misery | Averages individual ratings, after excluding items with individual ratings below a certain threshold (say 4). | J's group rating is 7.3 (the average of 8,8,6), while A is excluded because Jane hates it |
| Fairness | Items are ranked as if individuals are choosing them in turn. | Item E may be chosen first (highest for Peter), followed by F (highest for Jane) and A (highest for Mary) |
| Most respected person (or Dictatorship) | Uses the rating of the most respected individual. | If Jane is the most respected person, then A's group rating is 1. If Mary is most respected, then it is 10 |

[a] If the majority of group members have a higher rating for an item X than for an item Y, then item X beats item Y

## 3.2 Aggregation Strategies Used in Related Work

Most of the related work uses one of the aggregation strategies in Table 3 (sometimes with a small variation), and they differ in the one used:

- INTRIGUE uses a weighted form of the Average Strategy. It bases its group recommendations on the preferences of subgroups, such as children or disabled.

It takes the average, with weights depending on the number of people in the subgroup and the subgroup's relevance (children and disabled were given a higher relevance).

- POLYLENS uses the Least Misery Strategy, assuming groups of people going to watch a movie together tend to be small and that a small group tends to be as happy as its least happy member.
- MUSICFX uses a variant of the Average Without Misery Strategy. Users rate all radio stations, from +2 (really love this music) to −2 (really hate this music). These ratings are converted to positive numbers (by adding 2) and then squared to widen the gap between popular and less popular stations. An Average Without Misery Strategy is used to generate a group list: the average of ratings is taken but only for those items with individual ratings all above a threshold. To avoid starvation and always picking the same station, a weighted random selection is made from the top stations of the list.
- YU'S TV RECOMMENDER uses a variant of the Average Strategy. It bases its group recommendation on individuals' ratings of program features: −1 (dislikes the feature), +1 (likes the feature) and 0 (neutral). The feature vector for the group minimizes its distance compared to individual members' feature vectors (see [18] for detail). This is similar to taking the average rating per feature.
- The TRAVEL DECISION FORUM has implemented multiple strategies, including the Average Strategy and the Median Strategy. The Median Strategy (not in Table 3) uses the middle value of the ratings. So, in our example, this results in group ratings of 10 for A, and 9 for F. The Median Strategy was chosen because it is non-manipulable: users cannot steer the outcome to their advantage by deliberately giving extreme ratings that do not truly reflect their opinions. In contrast, for example, with the Least Misery strategy devious users can avoid getting items they dislike slightly, by giving extremely negative ratings. The issue of manipulability is most relevant when users provide explicit ratings, used for group recommendation only, and are aware of others' ratings, all of which is the case in the TRAVEL DECISION FORUM. It is less relevant when ratings are inferred from user behavior, also used for individual recommendations, and users are unaware of the ratings of others (or even of the aggregation strategy used).
- In CATS, users indicate through critiquing which features a holiday needs to have. For certain features, users indicate whether they are required (e.g., ice skating required). For others, they indicate quantities (e.g., at least 3 ski lifts required). The group model contains the requirements of all users, and the item which fulfills most requirements is recommended. Users can also completely discard holidays, so, the strategy has a Without Misery aspect.
- GAIN uses a variant of the Average Strategy, with different weights for users that the system knows are near the system and for unrecognized users who should be there statistically.
- The REMPAD system, having such a specific and especially sensitive task, has to reduce any negative effects that the suggested media material could cause for individuals in the therapy group, and to this end uses the Least Misery Strategy.

- HAPPYMOVIE uses the Average Strategy, but prior to the aggregation, the system updates individuals' ratings as a result of influence in the group which is estimated according to group members' personality (assertiveness and cooperativeness) and their inter-personal trust levels.
- INTELLIREQ allows participants to select their preferred configuration of software requirements, defined upon multiple dimensions, and applies Plurality Voting.
- WHERE2EAT does not aggregate individual preferences of group members. Instead, it allows group members to adjust (critique) the features of a restaurant proposed by a fellow group member in the previous step, and to make a counter proposal. The counter proposal is automatically explained to the fellow group member in terms of differences to her original proposal.
- CHOICLA supports a configurable, domain-independent group decision-making process, where the initiator defines the decision task as well as some basic settings of the decision process. The settings also include the selection of the aggregation strategy, and some of the available options are Majority Vote, Average Vote, Least Misery, Most Pleasure.
- HOOTLE aggregates group members' individual preferences on the set of desirable attributes of a hotel. Hence, each group member gives an importance score for her most important attributes, or she can put a veto on an attribute. However, group members can express their preferences only on a number of attributes where the number depends on how restrictive the attribute is for the output of the recommender system. To aggregate individual preferences the system uses a variant of the Borda Count, so that it accounts not only for the individuals' importance scores, but also for vetoed options and restrictions [35].
- STSGROUP is a content-based recommender system which considers users' individual long-term preferences, as well as session-based preferences which the group members expressed as reactions on each-others' proposals (i.e., best choice, like, dislike). At the beginning of the discussion process, group members' individual utility vectors are aggregated to a group utility vector with the Average Strategy. However, when the interaction begins, the weighted average strategy is used, and the weights are calculated based on the proportion of the user's actions (POI proposals, POI evaluations and POI comments) over the total number of actions acquired in the group [29].
- TOURREC introduces two extensions to the standard aggregation strategies that allow groups to separate during their POI sequence route. In the first one, *Split*, after aggregating individual "profits" to visit a POI, and generating group recommendations, individual recommendations are also generated, which the system uses to check whether there are individuals for whom it would pay off to exchange a certain POI on the group route with their individually more preferred POI, but making sure that the distance to and from that POI is reasonable. In the second one, *Connect Segments (CS)*, individual recommendations are aggregated in a way that each group member has two of their favourite POIs in the group route.

It should be noted that YU'S TV RECOMMENDER, the TRAVEL DECISION FORUM and INTELLIREQ aggregate preferences for each feature without using the idea of fairness: loosing out on one feature is not compensated by getting your way on another. In addition to the strategies in Table 3, more complex strategies and recommendation algorithms have been used:[1]

*Heuristics-Based Methods*  In this group of strategies, we combine methods that make certain assumptions about group decision-making process, individual and group preferences, and accordingly define the aggregation strategy. The *graph-based ranking* algorithm [36] uses (1) a graph with users and items as nodes, with positive links between users and items rated above the user's average item rating, and negative links for items rated below (with weights of how much above/below), (2) a user neighborhood graph linking users with similar rating patterns, and (3) an item neighborhood graph linking items that have been rated similarly. Group recommendations are based on two random walks over the graphs, with the assumption that highly visited items over positive links would tend to be liked by the group, and items highly visited by a random walk over negative links would tend to be disliked by the group. The *Spearman footrule rank* aggregation [37] defines that the aggregated list for the group is a list with minimum distance to the individual lists. The Spearman footrule distance between two lists is the summation of absolute differences between the ranks of the items in the lists. The *Nash equilibrium* [38] models group members as players in a non-cooperative game, and players' actions as item recommendations (choosing from their top 3 items). Group satisfaction is achieved by finding the Nash equilibrium in the game. Finally, the *purity* and *completeness* strategies [39] are statistics-based strategies. The purity is a statistical dispersion strategy that aims to satisfy as many group members' preferences as possible (considering the deviation in preferences). The completeness models group recommendation as a negotiation, favoring high scores whilst penalizing large differences between members.

*Influence-Based Methods*  In this group of strategies, we combine approaches which define preference-based influence and use it in the aggregation method. The *Pre-GROD* and *GROD* [40] first calculate a pairwise similarity matrix, considered as a "trust" matrix, where elements represent how much importance the group members assign to the opinion of others in the group. As it is assumed that members update their opinions according to the influence in the group, the approach updates their individual ratings according to the similarity matrix until a "stability point" (consensus) is reached. Group scores for individual items are the average of the updated opinions. When a group consensus cannot be reached, the GROD approach updates the similarity matrix by adding elements that ensure consensus reaching. The *Preference network (PrefNet)* and *Non-linear preferences (Non-lin)* methods [41] modify importance and preferences of group members (respectively) prior to applying the standard aggregation strategies, e.g., average. PrefNet represents user

---

[1] These strategies are too complicated to fully explain here, see the original papers for details.

preferences in a network structure based on group members' pairwise similarities, and weights group members according to their centrality in that network - giving more importance to those who share a greater deal of preferences with others in the group. Non-lin transforms individual preferences to a non-linear scale, thus allowing stronger positive influence of highly ranked items and stronger negative influence of items ranked at the bottom of individual lists. Finally, *MAGReS* (Multi-Agent Group Recommender System) [42] implements user agents that follow the *Monotonic Concession Protocol* (MPC). Agents suggest items according to the utility value which that item has for their corresponding user, and assesses items proposed by other agents in the group. An agreement is reached when an agent suggests an item which is at least as good, for any other agent, as their own current suggestion. If an agreement is not reached an agent makes a concession. The procedure is repeated until an agreement is found or until no agent can concede anymore.

*Matrix Factorization and Deep Learning Methods* The *After Factorization (AF)*, *Before Factorization (BF)* and *Weighted Before Factorization (WBF)* methods [43] all employ Matrix Factorization (MF), with the only difference whether the individual preferences are aggregated (with standard aggregation strategies) prior or after the MF. The interesting approach here is WBF which aggregates users' individual ratings, but by assigning more weights to items that have been rated by the majority of the group, and have similar ratings by the users in the group. The *AGREE* approach [44] uses a neural network to learn the aggregation strategy and item-dependant weights of each group member. The network extends the structure of the well known Neural Collaborative Filtering (NCF) approach [45] by learning not only the user-item interactions, but as well, group-item interactions. *MoSAN* [46] is another neural approach that models member user-user interactions with the help of attention networks, for ad-hoc groups which did not have previous interactions. For each group, a set of sub-attention networks is created (one for each group member). The network aims to capture decisions of each group member, given the decisions of others in that group, hence the influence. Finally, the *GroupIM* approach [47] trains a neural network to compute probabilities that a group would interact with an item by minimizing (1) the group recommendation loss between history of group-item interactions and the predicted item-probabilities, (2) the contextually weighted user-item loss, and (3) maximizing the mutual information between the group and the member user.

*Fairness-Maximization Methods* The *SPGreedy* and *EFGreedy* strategies [8] use a greedy algorithm to add items to a package recommendation that maximize fairness proportionality or envy-freeness. Fairness proportionality ensures that each group member gets at least $m$ items that she likes, compared to items not in the package, whole envy-freeness looks into the other group members and makes sure that for each member there is a sufficient number of items that she likes more than other group members. The *Greedy-LM* and *Greedy-Var* strategies [7], in a similar fashion, use a greedy algorithm and scalarization method to find a Pareto Efficient solution to a multi-objective optimization problem which maximizes the social welfare of a group and fairness. The social welfare is defined as the average of group members' individual utilities, while fairness is defined with the least misery approach (Greedy-LM), or with the variance of the individual utilities (Greedy-Var).

## 3.3 Which Strategy Performs the Best

Though some exploratory evaluation of MUSICFX, POLYLENS and CATS has taken place, for none of these systems it has been investigated how effective their strategy really is, and what the effect would be of using a different strategy. The experiments presented in this section shed some light on this question. Moreover, systems like WHERE2EAT, CHOICLA, STSGROUP, and HOOTLE each adopt a quite unique approach of delivering recommendations, and hence do not evaluate the performance of different aggregation strategies, but they do evaluate in their own context the usability of the system and performance of the proposed approach. In contrast, some evaluation of YU'S TV RECOMMENDER has taken place [18]. They found that their aggregation worked well when the group was quite homogeneous, but that results were disliked when the group was quite heterogeneous. This is as we would expect, given the Average Strategy will make individuals quite happy if they are quite similar, but will cause misery when tastes differ widely.

In [9], a series of experiments were conducted to investigate which strategy from Table 3 is the best in terms of (perceived) group satisfaction. The Experiment 1 (see Fig. 1) investigated how people would solve the group recommendation problem, using the User as Wizard evaluation method [48]. Participants were given individual ratings identical to those in Table 1. These ratings were chosen to be able to distinguish between strategies. Participants were asked which items the group should watch, if there was time for one, two,..., seven items. Participants' decisions were compared and rationale with those of the aggregation strategies. It was found that participants cared about fairness, and about preventing misery and starvation ("this one is for Mary, as she has had nothing she liked so far"). Participants' behavior reflected that of several of the strategies (e.g. the Average, Least Misery, and Average Without Misery were used), while other strategies (e.g. Borda count, Copeland rule) were clearly not used.[2]

In Experiment 2 (see Fig. 2), participants were given item sequences chosen by the aggregation strategies as well as the individual ratings in Table 1. They rated how satisfied they thought the group members would be with those sequences, and explained their ratings. We found that the Multiplicative Strategy performed the best, in the sense that it was the only strategy for which *all* participants thought its sequence would keep all members of the group satisfied. Borda count, Average, Average without Misery and Most Pleasure also performed quite well. Several

---

[2] This does not necessarily mean that these strategies are bad, as complexity can also play a role. In fact, in Experiment 2 Borda count was amongst the best performing strategies.

**Fig. 1** Experiment 1: which sequence of items do people select if given the system's task

**Fig. 2** Experiment 2: What do people like?

strategies (such as Copeland rule, Plurality voting, Least misery) could be discarded as they clearly were judged to result in misery for group members.

The participants' judgments were also compared with predictions of simple satisfaction modeling functions. Amongst other, it was found that more accurate predictions[3] resulted from using:

- quadratic ratings,[4] which e.g., makes the difference between a rating of 9 and 10 bigger than that between a rating of 5 and 6.
- normalization,[5] which takes into account that people rate in different ways, e.g., some always use the extremes, while others only use the middle of the scale.

---

[3] In terms of satisfaction functions predicting the same relative satisfaction scores for group members as predicted by participants, see [9] for details.

[4] A rating r was transformed into $(r-\text{scale\_midpoint})^2$ if $r \geq$ scale_midpoint, and $-(r-\text{scale\_midpoint})^2$ if r<scale_midpoint.

[5] A rating r was transformed by a user u into r × (TotalRatingsAverage ÷ TotalRatings(u)), where TotalRatingsAverage is the sum for all items of the average ratings by all users, and TotalRatings(u) is the sum for all items of u's rating.

In [49], a further study using simulated users based on models of affective state (see next Section) was conducted. It was found that the Multiplicative Strategy performed the best.

There are other studies investigating the effect of different aggregation strategies from Table 3, and the more advanced methods. Table 4 provides an overview of evaluations of aggregation strategies, but before looking into the results of these studies, we will talk a bit more about evaluating group recommender systems.

## 3.4 Evaluating Group Recommender Systems

To evaluate and measure the quality of group recommender systems user studies as well as off-line experiments have been employed [50]. However, how to properly evaluate a group recommender system is still a major research topic.

### 3.4.1 User Studies

A user study is carried out when the criteria used to measure the system performance are related to system usability and user experiences (e.g., perceived user's satisfaction or recommendation quality), such as those in WHERE2EAT [22], CHOICLA [23], STSGROUP [25], or TOURREC [28]. This type of study can be conducted by directly interviewing participants or through crowd sourcing sites such as Amazon Mechanical Turk [51]. This approach, however, cannot be the sole method for evaluating the efficacy of a system as the approach does not scale. Nevertheless, user studies are crucial in determining the success of a proposed approach, to evaluate whether or not the system is even accepted by the users [1].

### 3.4.2 Off-Line Evaluations

As when evaluating single-user recommendations, off-line evaluations are also used in group recommenders research. However, these approaches are hindered since there is a lack of publicly available data sets that capture the preferences of users in actual group settings.

As one of the solutions for this problem, researchers have used so called synthetic groups. Synthetic groups are artificial groups sampled from standard data sets such as MovieLens [33, 37]. The main challenge of this solution is how to properly evaluate effectiveness of group recommendations once they are delivered to such groups. One approach is to compare the group recommendations with the joint group assessment of the recommended items, however this is indeed problematic when in essence the actual group choice is not known. Therefore, it is assumed that the "true" group preferences can be derived from individual preferences with the Average Strategy, which is exactly the problem that group recommenders aim

**Table 4** Evaluation of aggregation strategies

| Who | Domain | Evaluation methodology | Groups | Strategies | Results |
|---|---|---|---|---|---|
| Masthoff [9] | TV | Experiment 1 above user as wizard | Size: 3 Friends Heterogeneous | All from Table 3. | USED: average, average without misery, least misery. NOT USED: borda, approval, plurality, copeland |
| Masthoff [9] | TV | Experiment 2 above user as wizard | Size: 3 friends heterogeneous | All from Table 3. | BEST: multiplicative. OK: borda, average, average without misery, most pleasure. WORST: copeland, plurality, least misery |
| Masthoff and Gatt [49] | TV | Simulated users metric: satisfaction functions. | Size: 3 friends heterogeneous | All from Table 3. | BEST: multiplicative. WORST: borda, plurality, most pleasure |
| Senot et al. [34] | TV | Historic TV use, including individual and group data | Size: 2–5 Family groups | Plurality voting, least misery, most pleasure, dictatorship, average | BEST: average for most groups, Dictatorship in 20% of groups. WORST: most pleasure, least misery |
| Bourke et al. [53] | TV movies | User study. Metric: average of individual satisfaction | Size: 3,5,10 Types: experts, high similarity, social relationships | Multiplicative, borda, approval voting, least misery, most pleasure, respect | BEST: multiplicative and respect. WORST: most pleasure. Least misery better in larger groups than most pleasure |
| De Pessemier et al. [33] | Movies | Synthetic data metric: average of individual satisfaction | Size: 2,5 | Average, average without misery, dictatorship, least misery, most pleasure | BEST: average and average without misery. Dictatorship low accuracy when aggregating recommendations |
| Berkovsky and Freyne [32] | Recipes | User study | Size: 2,3,4 family types: homogeneous, heterogeneous | Average, weighted average (based on activity, roles, etc.) | BEST: Weighted average with weights based on activity |

to solve, and we do not know which aggregation strategy truly represents the group opinion. Unsurprisingly, those studies tend to find that the Average Strategy performs well (as do strategies that resemble it). Another approach is to compare group recommendations with the individual preferences. It assumes that group members' individual satisfaction with the group recommendation depends only on their individual preferences, and the match/mismatch of these individual preferences with the group recommendations, therefore, that preferences of individuals are independent from the group, which is actually not the case in most scenarios. In fact, the opinions or judgments of users in a group are likely to be influenced by the other group members (i.e., emotional contagion and conformity effects [49]). As a result, specific approaches come into play, as the two experiments that we previously elaborated.

There are notable exceptions to using synthetic groups, where researchers collected data about individual preferences as well as about what those individuals select when in groups with others. For instance, in [34], data with a history of TV use by individuals and groups was collected. In [44], two data sets were used, (1) Mafengwo is collected from a website of the same name, where users can record their traveled venues, create or join a group travel, and at the moment it is not publicly available; (2) CAMRa2011 contains movie ratings of individuals and household and is publicly available. In [46], another data set containing information about real groups was used, i.e., the Plancast data, collected from the event-based social network, contains information about events and participants of those events, hence each event is considered a group and event participants as group members. Finally, [52] provides a detailed description of a study collecting individual and group preferences about travel destinations, and simulates real face-to-face group interactions where the goal is to select a destination to visit jointly. In addition the data contains individual and pairwise characteristics such as personality and social relationships. This provides a more accurate view on what actually happens in groups. The only drawback of that approach is that what happens in real groups does not necessarily lead to optimal group satisfaction. For example, in [34], when a Dictatorship Strategy is used (as seems to have happened in 20% of their groups), this may have left others in the group unsatisfied, and it is possible that the group as a whole would have been more satisfied if a different approach had been used (though sometimes due to for example participant personality, individuals may well be satisfied when Dictatorship is used). This raises the question whether group recommenders should mimic what happens in real groups or should try to do better.

In the context of interactive group recommenders, Nguyen and Ricci [27] proposed a group discussion simulation model where the impact of (1) alternative combinations of long-term and session-based preferences, and (2) different individual group members' behavioral types (conflict resolution types) on the recommendation performance in different group scenarios was studied.

Having this in mind, Table 4 also contains information about the used data sets and the domain. Moreover, most studies compare group sizes and often also compare between homogeneous groups (where users' preferences are similar) and more heterogeneous groups. Studies typically find that aggregation strategies perform better for more homogeneous and smaller groups.

More advanced methods, that we previously shortly introduced, were as well evaluated, and usually by comparing them to the standard aggregation strategies from Table 3, if applicable. When such a comparison was not applicable, for instance for the neural network approaches, then the comparisons were done within the group of methods (e.g., AGREE was compared against NCF combined with standard aggregation strategies, Greedy-LM and Greedy-Var were compared against SPGreedy and EFGreedy, among others).

## 4   Impact of Sequence Order

As mentioned in Sect. 2, we are also interested in recommending a *sequence* of items. In Sect. 3, we have tackled the issue on what items to select if there is time for a certain number of items. For example, for a personalized news program on TV, a recommender may select seven news items to be shown to the group. To select the items, it can use an aggregation strategy (such as the Multiplicative Strategy) to combine individual preferences, and then select the seven items with the highest group ratings.

In this section, we are interested in the *order* of items in the sequence. For example, once seven news items have been selected, the question arises in what order to show them in the news program. Many options exist: for instance, the news program could show the items in descending order of group rating, starting with the highest rated item and ending with the lowest rated one. Or, it could mix up the items, showing them in a random order. An example of this approach, for package recommendations, was implemented in [8], where the authors use the greedy algorithm that in each iteration adds an item with a maximum possible utility for the group, where utility could be the highest average rating, the highest fairness score, etc.

However, the problem is actually far more complicated than that. Firstly, in responsive environments, the group membership changes continuously, so deciding on the next seven items to show based on the current members seems not a sensible strategy, as in the worse case, none of these members may be present anymore when the seventh item is shown.

Secondly, overall satisfaction with a sequence may depend more on the order of the items than one would expect. For example, for optimal satisfaction, we may need to ensure that our news program has:

- *A good narrative flow.* It may be best to show topically related items together. For example, if we have two news items about Michael Jackson (say about his funeral and about a tribute tour) then it seems best if these items are presented together. Similarly, it would make sense to present all sports' items together.

- *Mood consistency.* It may be best to show items with similar moods together. For example, viewers may not like seeing a sad item (such as a soldier's death) in the middle of two happy items (such as a decrease in unemployment).
- *A strong ending.* It may be best to end with a well-liked item, as viewers may remember the end of the sequence most.

Similar ordering issues arise in other recommendation domains, e.g., a music programme may want to consider rhythm when sequencing items. The recommender may need additional information (such as items' mood, topics, rhythm) to optimize ordering. It is beyond the topic of this chapter to discuss how this can be done (and is very domain specific). We just want to highlight that the items already shown may well influence what the best next item is. For instance, suppose the top four songs in a music recommender were all Blues. It may well be that another Blues song ranked sixth may be a better next selection than a Classical Opera song ranked fifth. Similarly, the group may prefer something from a different music genre after a sequence of songs from one genre, even if the song ranked the best next is of the same genre.

In Experiment 3 (see Fig. 3 and more detail in [9]), it was investigated, in the news domain, how a previous item may influence the impact of the next item. Participants rated a set of news items. They were then shown one news item[6] and rated how interested they were in it and how it made them feel, and re-rated the original items to see if their ratings would have changed. Amongst others, we found

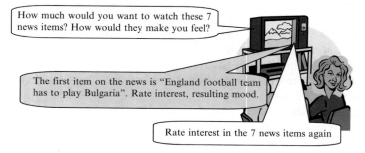

**Fig. 3** Experiment 3: Investigating the effect of mood and topic

---

[6] In a between-subject design, two different topics were used evoking different moods.

that mood (resulting from the previous item) and topical relatedness can influence ratings for subsequent news items.

This means that aggregating individual profiles into a group profile should be done repeatedly, every time a decision needs to be made about the next item to display. So, instead of first selecting say seven items to show and then deciding on the order, only one item is selected, and then it needs to be decided which item from *all* remaining ones is the best to show next, given that the first item may have an impact on the ratings of the remaining ones.

## 5   Modeling Affective State

When recommending to a group of people, you cannot give everybody what they like all the time. However, you do not want anybody to get too dissatisfied. For instance, in a shop it would be bad if a customer were to leave and never come back, because they really cannot stand the background music. Many shops currently opt to play music that nobody really hates, but most people not love either. This may prevent loosing customers, but would not result in increasing sales. An ideal shop would adapt the music to the customers in hearing range in such a way that they get songs they really like most of the time (increasing the likelihood of sales and returns to the shop). To achieve this, it is unavoidable that customers will occasionally get songs they hate, but this should happen at a moment when they can cope with it (e.g., when being in a good mood because they loved previous songs). Therefore, it is important to monitor continuously how satisfied each group member is. Of course, it would put an unacceptable burden on the customers if they had to rate their satisfaction (on music, advertisements, etc.) all the time. Similarly, measuring this satisfaction via sensors (such as heart rate monitors or facial expression classifiers) is not yet an option, as they tend to be too intrusive, inaccurate or expensive. So, it was proposed to model group members' satisfaction; predicting it based on what we know about their likes and dislikes.

Modeling individual's satisfaction is a topic on its own, even without the group dimension considered. For instance, in [49] various satisfaction functions to model an individual's satisfaction with a sequence of items were evaluated. The satisfaction function that performed the best defines the satisfaction of a user with new item after having seen a sequence of items, as a weighted sum of the satisfaction with the previously seen items, and the estimated impact that the new item will have on the user's satisfaction. The function implements two crucial effects (1) the effect of satisfaction decay over time, and (2) the influence of the user's satisfaction after experiencing previous items on the impact (perception) of a new item.

Motivated by these findings, works [54] and [55] implemented a sequence recommendation approach for a group of people by accounting for the decaying effect (i.e., items that were experienced more recently have a greater impact on the user's overall satisfaction with the recommended sequence). In [54], a sequence of artworks on a certain path within a museum is recommended. Here, in addition to modeling satisfaction and the decay effect, the physical order of the artworks has to be considered. To this end, the museum (artworks and paths) has been modelled as a Directed Acyclic Graph, where the goal is to find a maximum satisfaction path bounded by the time available for the museum visit. User satisfaction depends on the previously seen artworks on the path, as well as on the next artwork added to that path. The authors showed that sequence recommendations with a decay effect performed evidently better than a sequence of TOP-k recommendations. In [55], two methods were proposed, i.e., "Balancing without Decay" and "Balancing with Decay". The first step for both methods is to generate candidate items, that is, those with highest aggregated ratings for a group. In the second step, satisfaction of each user in a group is modelled as a cumulative satisfaction without (with) decay, for previously experienced items and for the next item that is to be added to the sequence. To select the next item for the sequence, pairwise differences between group members' cumulative satisfactions is summed, and the item with the lowest sum of differences is added next. Evaluated with a user study, the authors showed that the "Balancing with Decay" method outperformed "Balancing without Decay".

## 5.1  Effects of the Group on an Individual's Satisfaction

In the previously described approaches the satisfaction of other users in the group was not taken into account, which may well influence a user's satisfaction. As argued in [49] based on social psychology, two main processes can take place.

**Emotional Contagion**  Firstly, the satisfaction of other users can lead to so-called emotional contagion: other users being satisfied may increase a user's satisfaction (e.g., if somebody smiles at you, you may automatically smile back and feel better as a result). The opposite may also happen: other users being dissatisfied may decrease a user's satisfaction (e.g., if you are watching a film with a group of friends then the fact that your friends are clearly not enjoying it may negatively impact your own satisfaction). Emotional contagion may depend on your personality (some people are more easily contaged than others), and your relationship with the other person. Anthropologists and social psychologists have found substantial evidence for the existence of four basic types of relationships, see Fig. 4. In Experiment 5 (see Fig. 5), participants were given a description of a hypothetical person they were watching TV with (using the relationship types in Fig. 4) and asked how their own emotion would be impacted (on a scale from 'decrease a lot' to 'increase a lot') by that person's strong positive or negative emotions (see detail in [49]). Results confirmed that emotional contagion indeed depends on the relationship you have: you are more likely to be contaged by somebody you love (such as your best friend) or respect (such as your mother or boss) than by somebody you are on equal footing with or are in competition with.

**Conformity**  Secondly, the opinion of other users may influence your own expressed opinion, based on the so-called process of conformity. Figure 6 shows the

**Fig. 4** Types of relationship

**Fig. 5** Experiment 5: Impact of relationship type on emotional contagion

famous conformity experiment by Asch [56]. Participants were given a very easy task to do, such as decide which of the four lines has the same orientation as the line in Card A. They thought they were surrounded by other participants, but in fact the others where part of the experiment team. The others all answered the question before them, picking the same wrong answer. It was shown that most participants then pick that same wrong answer as well.

Two types of conformity exist: (1) normative influence, in which you want to be part of the group and express an opinion like the rest of the group even though inside you still belief differently, and (2) informational influence, in which your own opinion changes because you believe the group must be right. Informational influence would change your own satisfaction, while normative influence can change the satisfaction of others through emotional contagion because of the (insincere) emotions you are portraying.

More complicated satisfaction functions are presented in [49] to model emotional contagion and both types of conformity. These functions also serve as a basis for work in [57].

**Fig. 6** Conformity experiment by Asch

## 6 Using Satisfaction Inside Aggregation Strategies

Once you have an accurate model of the individual users' satisfaction, which predicts how satisfied each group member is after a sequence of items, it would be nice to use this model to improve the group aggregation strategies. For instance, the aggregation strategy could set out to please the member of the group who is least satisfied with the sequence of items chosen so far. This can be done in many different ways, and we have only started to explore this issue. For example:

- *Strongly Support Grumpiest strategy.* This strategy picks the item which is *most liked* by the least satisfied member. If multiple of these items exist, it uses one of the standard aggregation strategies, for instance the Multiplicative Strategy, to distinguish between them.
- *Weakly Support Grumpiest strategy.* This strategy selects the items that are *quite liked* by the least satisfied member, for instance items with a rating of 8 or above. It uses one of the standard aggregation strategies, such as the Multiplicative Strategy, to choose between these items.
- *Weighted strategy.* This strategy assign weights to users depending on their satisfaction, and then use a weighted form of a standard aggregation strategy. For instance, Table 5 shows the effect of assigning double the weight to Jane when using the Average Strategy. Note that weights are impossible to apply to a strategy such as the Least Misery Strategy.

**Table 5** Results of average strategy with equal weights and with twice the weight for Jane

|                        | A   | B   | C   | D   | E   | F   | G   | H   | I   | J   |
|------------------------|-----|-----|-----|-----|-----|-----|-----|-----|-----|-----|
| Peter                  | 10  | 4   | 3   | 6   | 10  | 9   | 6   | 8   | 10  | 8   |
| Jane                   | 1   | 9   | 8   | 9   | 7   | 9   | 6   | 9   | 3   | 8   |
| Mary                   | 10  | 5   | 2   | 7   | 9   | 8   | 5   | 6   | 7   | 6   |
| Average (equal weights)| 7   | 6   | 4.3 | 7.3 | 8.7 | 8.7 | 5.7 | 7.7 | 6.7 | 7.3 |
| Average (Jane twice)   | 5.5 | 6.8 | 5.3 | 8.3 | 8.3 | 8.8 | 5.8 | 8   | 5.8 | 7.5 |

In [58], this was discussed in more detail, an agent-based architecture was proposed for applying these ideas to the Ambient Intelligence scenario, and an implemented prototype was described. Preliminary work in [59], also uses a strategy which balances user satisfaction. Clearly, empirical research is needed to investigate the best way of using affective state inside an aggregation strategy.

# 7 Incorporating Group Attributes

Above, we discussed how an individual's satisfaction can be influenced by others in the group due to emotional contagion and normative behavior. Individuals' personality (e.g., propensity to emotional contagion) and social relationships between individuals played a role in this. For instance, in [60, 61], the authors showed that group members' individual satisfaction with group decisions is not only related to the match/mismatch between their own individual preferences and the group choice, but also to their personality types, and strength of their social relationships to others in the group. Moreover, in [62], it was demonstrated that the group score on *Conscientiousnes* (i.e., the extent to which one is precise, careful and reliable, or rather sloppy, careless, and undependable), and diversity of group members' preferences are significant predictors of the decision-reaching approach the groups adopted in their travel-related discussions (more details about various personality models most often used in recommender systems can be found in Chapter "Personality and Recommender Systems"). Individuals' personality and social relationships were incorporated into the models of satisfaction [49], which were then used in aggregation strategies [58]. Instead of using group attributes indirectly, via satisfaction models, it is also possible to incorporate them more directly into aggregation strategies.

Firstly, attributes can be used of individual group members, typically giving more weight to certain group members than others in the preference aggregation step:

- *Demographics and Roles*. As mentioned above, INTRIGUE [15] distinguishes different user types (children, adults with and without disability), and uses higher weights for more vulnerable user types. The recipe group recommender in [32] distinguishes between user roles (applicant, partner, child) and varies the weights based on their presumed level of engagement with the system (lowest for child, highest for applicant). An analysis of groups whose behavior corresponded to a Dictatorship strategy in [34] showed many cases in which teenagers acted as dictators in the company of children and where adults acted as dictators in the company of teenagers or children. So, different roles may influence what happens in groups, though [34] does not present the group composition when Dictatorship is not used.

- *Personality: Assertiveness and Cooperativeness.* HAPPYMOVIE [20] uses how *assertive* (extent to which person attempts to satisfy own concerns) and how *cooperative* (extent to which person attempts to satisfy others concerns) group members are, and gives a higher weight to assertive members and a lower weight to cooperative members. Moreover, findings from a simulation study [63] indicated that (1) groups of similar preferences whose members have non-cooperative conflict resolution styles achieved a lower utility compared to groups whose members have cooperative styles; (2) groups of diverse preferences and non-cooperative members often select recommendations that ensure equal utility losses for everybody, which may result in a greater loss on average, while the opposite happened for cooperative groups; 3) groups of mixed conflict resolution styles manage to make the group decision with the smallest average individual's utility loss, even though the difference in their utility is the largest.
- *Expertise.* As reported in [64], according to Social Psychology expertise may provide influence, so in normal group processes experts may have more influence on the group's decision.[7] Gatrell et al. [65] apply higher weights in the generation of recommendations to people with more expertise.[8] They infer expertise from activity, namely the number of movies rated, but only considering a pre-selected set of movies (i.e., 100 popular items). The recipe group recommender in [32] also uses higher weights for family members who have engaged more. In a similar fashion, in [66] group members' expertise and their corresponding weights were evaluated from the group member's activity in the system, i.e., the more item-ratings a group member provided the greater the weight would be. It is noteworthy that none of these approaches observed the group behavior or the actual activity of a group member in the group.
- *Personal impact.*[9] Liu et al. [67] incorporate the concept of *personal impact* into their group recommender algorithm, to model that different members will have different impacts on group decisions. They consider decisions made in the past to decide on personal impact. Similarly, in [68], personal impact is based on the match/mismatch between users' individual choice and the group choice in which

---

[7] Additionally, it seems plausible (but requires investigation) that users would be more dissatisfied with disliked selected items when their expertise is higher than that of other group members.

[8] This may work well when using an Additive or Multiplicative strategy, but does not really work for the Least Misery and Most Pleasure strategies used in [65], and hence unfortunately some of the formulas in [65] which incorporate expertise lack validity.

[9] Personal impact is not completely distinct from Role: somebody's role may influence their personal impact. However, it is still possible for people with the same official roles to have a different cognitive centrality.

the user has participated in the past. For example, if a user was a member of six different groups and her preferred option was selected as the group choice in three out of six cases which will be her weight in the impact-based model. Herr et al. [64] advocate using the social psychology concept of *cognitive centrality*: the degree to which a group member's cognitive information is shared within the group. They propose that the degree of centrality can be used to infer a person's importance and that more important members should be given higher weights. In fact, the underlying idea is that group members who have more in common with others in the group are given more weights, thus as well, items that are liked by these people should be items that are liked by others in the group, and items that are disliked by these members should be items that are disliked by others as well. This approach, applied to group members' shared preferences, was evaluated in [41], indicating that this kind of weighting approach can provide valuable information to aggregation strategies. This approach was even employed for individual recommendations [69], where a group formation method based on the pairwise similarity of users' ratings was proposed.

Secondly, people may behave differently in different groups, for instance depending on their closeness to the group, hence they might want to satisfy others in close relationships, while in not so close groups, they simply might want to make sure no body is too dissatisfied. To this end, attributes can be used of the group as a whole, typically using a different aggregation strategy based on the type of group:

• *Relationship strength.* The authors of [65, 70] advocate using different aggregation strategies depending on the group's relationship strength. They propose to use a Maximum Pleasure strategy for groups with a strong relationship (such as couples and close friends' groups), a Least Misery strategy for groups with a weak relationship (such as first-acquaintance groups), and an Average strategy for groups with an intermediate relationship (such as acquaintance groups).

• *Relationship type.* Wang et al. [57] distinguish between *positionally homogeneous*[10] and *positionally heterogeneous* groups. In positionally homogeneous groups, such as friend and tourist groups, the position of members is equal. In heterogeneous groups, such as family groups, the position of members is unequal. They also distinguish between *tightly-coupled* (strong relationship: members are close and intercommunication is important) and *loosely-coupled* (weak relationship: members are relatively estranged, and intercommunication is less frequent and less important) groups. Based on these two dimensions, Wang et al. define four different group types: tightly-coupled homogeneous (e.g., a friends' group), loosely coupled homogeneous (e.g., a tourist group), tightly-coupled heterogeneous (e.g., a family group), and loosely coupled heterogeneous (e.g., a staff group including managers).

---

[10] We have added the word positional to the terms homogeneous and heterogeneous used in [57] to avoid confusion with the earlier use of these words to indicate how diverse group preferences are.

Thirdly, attributes of people pairs in the group can be used, typically to adjust the ratings of an individual in light of the rating of the other person in the pair:

- *Relationship strengths.* HAPPYMOVIE [20] uses relationship strengths (which they call *social trust*) into its aggregation strategy, adapting individuals' ratings based on the ratings of others depending on the relationship strength between individuals. In [71] social relationships within the group are also considered as an indicator of influence. However, in comparison to the HAPPYMOVIE model, the authors considered three social aspects: (1) Social trust as the affective relationship between pairs of group members, (2) Social similarity between pairs of group members, and (3) Social centrality of each group member in the social environment.
- *Personal impact.* The concept of personal impact from [67] mentioned above can also be used in this way. In [51], it was demonstrated that positive and negative "shifts" from group member's initial preferences can happen, depending whether the relationship to another person in the group is a friendly or conflicting one. Ioannidis et al. [72] argue that people will be influenced by some people in their group more than others. Their group recommender uses a cascading process, where the group can see the votes that have already been cast and by whom, and can comment on alternatives. They and Ye et al. [73] learn social influence values for use in the group aggregation strategy. Rossi et al. [74] also assume that some members are more dominant than the others when making group decisions. Hence they propose a model that exploits group members' interactions in online social networks to determine the impact of each member, and to weight their preferences in the aggregation strategies accordingly.

A study presented in [75] aims to consider these factors simultaneously, i.e., the combination of the effects of group members' personality, expertise, pairwise social relationships (intimacy) and preference similarity in a group preference model. A clear improvement in comparison to the standard aggregation strategies was demonstrated.

# 8 Applying Group Recommendation to Individual Users

So, what if you are developing an application that recommends to a single user? Group recommendation techniques can be useful in three ways: (1) to aggregate multiple criteria, (2) to solve the so-called cold-start problem, (3) to take into account opinions of others.

## 8.1 Multiple Criteria

Sometimes it is difficult to give recommendations because the problem is multidimensional: multiple criteria play a role. For instance, in a news recommender

**Table 6** Ratings on criteria
for 10 news items

|          | A  | B | C | D | E  | F | G | H | I  | J |
|----------|----|---|---|---|----|---|---|---|----|---|
| Topic    | 10 | 4 | 3 | 6 | 10 | 9 | 6 | 8 | 10 | 8 |
| Location | 1  | 9 | 8 | 9 | 7  | 9 | 6 | 9 | 3  | 8 |
| Recency  | 10 | 5 | 2 | 7 | 9  | 8 | 5 | 6 | 7  | 6 |

**Table 7** Average strategy ignoring unimportant criterion location

|         | A  | B   | C   | D   | E   | F   | G   | H | I   | J |
|---------|----|-----|-----|-----|-----|-----|-----|---|-----|---|
| Topic   | 10 | 4   | 3   | 6   | 10  | 9   | 6   | 8 | 10  | 8 |
| Recency | 10 | 5   | 2   | 7   | 9   | 8   | 5   | 6 | 7   | 6 |
| Group   | 10 | 4.5 | 2.5 | 6.5 | 9.5 | 8.5 | 5.5 | 7 | 8.5 | 7 |

system, a user may have a preference for location (being more interested in stories close to home, or related to their favorite holiday place). The user may also prefer more recent news, and have topical preferences (e.g., preferring news about politics to news about sport). The recommender system may end up with a situation such as in Table 6, where different news story rate differently on the criteria. Which news stories should it now recommend?

Table 6 resembles the one we had for group recommendation above (Table 1), except that now instead of multiple users we have multiple criteria to satisfy. It is possible to apply our group recommendation techniques to this problem. However, there is an important difference between adapting to a group of people and adapting to a group of criteria. When adapting to a group of people, it seems sensible and morally correct to treat everybody equally. Of course, there may be some exceptions, for instance when the group contains adults as well as children, or when it is somebody's birthday. But in general, equality seems a good choice, and this was used in the group adaptation strategies discussed above. In contrast, when adapting to a group of criteria, there is no particular reason for assuming all criteria are as important. It is even quite likely that not all criteria are equally important to a particular person. Indeed, in an experiment we found that users treat criteria in different ways, giving more importance to some criteria (e.g., recency is seen as more important than location) [76]. So, how can we adapt the group recommendation strategies to deal with this? There are several ways in which this can be done:

- Apply the strategy to the most respected criteria only. The ratings of unimportant criteria are ignored completely. For instance, assume criterion Location is regarded unimportant, then its ratings are ignored. Table 7 shows the result of the Average Strategy when ignoring Location.
- Apply the strategy to all criteria but use weights. The ratings of unimportant criteria are given less weight. For instance, in the Average Strategy, the weight of a criterion is multiplied with its ratings to produce new ratings. For instance, suppose criteria Topic and Recency were three times as important as criterion Location. Table 8 shows the result of the Average Strategy using these weights.

**Table 8** Average strategy with weights 3 for topic and recency and 1 for location

|          | A   | B   | C   | D   | E   | F   | G   | H   | I   | J   |
|----------|-----|-----|-----|-----|-----|-----|-----|-----|-----|-----|
| Topic    | 30  | 12  | 9   | 18  | 30  | 27  | 18  | 24  | 30  | 24  |
| Location | 1   | 9   | 8   | 9   | 7   | 9   | 6   | 9   | 3   | 8   |
| Recency  | 30  | 15  | 6   | 21  | 27  | 24  | 15  | 18  | 21  | 18  |
| Group    | 8.7 | 5.1 | 3.3 | 6.8 | 9.1 | 8.6 | 5.6 | 7.3 | 7.7 | 7.1 |

**Table 9** Unequal average without misery strategy with location unimportant and threshold 6

|          | A   | B   | C   | D   | E   | F   | G   | H   | I   | J   |
|----------|-----|-----|-----|-----|-----|-----|-----|-----|-----|-----|
| Topic    | 10  | 4   | 3   | 6   | 10  | 9   | 6   | 8   | 10  | 8   |
| Location | 1   | 9   | 8   | 9   | 7   | 9   | 6   | 9   | 3   | 8   |
| Recency  | 10  | 5   | 2   | 7   | 9   | 8   | 5   | 6   | 7   | 6   |
| Group    | 7   |     |     | 7.3 | 8.6 | 8.6 |     | 7.6 | 6.6 | 7.3 |

In case of the Multiplicative Strategy, multiplying the ratings with weights does not have any effect. In that strategy, it is better to use the weights as exponents, so replace the ratings by the ratings to the power of the weight. Note that in both strategies, a weight of 0 results in ignoring the ratings completely, as above.

- Adapt a strategy to behave differently to important versus unimportant criteria: Unequal Average Without Misery. Misery is avoided for important criteria but not for unimportant ones. Assume criterion Location is again regarded as unimportant. Table 9 shows the results of the Unequal Average Without Misery strategy with threshold 6.

We have some evidence that people's behavior reflects the outcomes of these strategies [76], however, more research is clearly needed in this area to see which strategy is the best. Also, more research is needed to establish when to regard a criterion as "unimportant".

## 8.2 Cold-Start Problem

A major concern for recommender systems, as discussed in several other chapters in this book, is the so-called cold-start problem: to adapt to a user, the system needs to know what the user liked in the past. The group recommendation techniques presented in this chapter provide an alternative solution. When a user is new to the system, we simply provide recommendations to that new user that would keep the whole group of existing users happy (i.e., all the existing users in the system). We assume that our user will resemble one of our existing users, though we do not know which one, and that by recommending something that would keep everybody happy, the new user will be happy as well.[11]

Gradually, the system will learn about the new user's tastes, for instance, by them rating our recommended items or, more implicitly, by them spending time on the items or not. We provide recommendations to the new user that would keep the group of existing users happy including the new user (or more precisely, the person we now assume the new user to be). The weight attached to the new user will be low initially, as we do not know much about them yet, and will gradually increase. We also start to attach less weight to existing users whose taste now evidently differs from our new user.

A small-scale study using the MovieLens data set was conducted, in order to explore the effectiveness of this approach. Five movies and twelve users who had rated them were randomly selected: ten users as already known to the recommender, and two as new users. Using the Multiplicative Strategy on the group of known users, movies were ranked for the new users. Results were encouraging: the movie ranked highest was in fact the most preferred movie for the new users, and also the rest of the ranking was fine given the new users' profiles. Applying weights led to a further improvement of the ranking, and weights started to reflect the similarity of the new users with known users. More detail on the study and on applying group adaptation to solve the cold-start problem is given in [77]. A follow on study in [78] confirmed the usefulness of this method. The use of aggregate ratings to solve the cold-start problem is also discussed in [79].

## 8.3   Virtual Group Members

Finally, group adaptation can also be used when adapting to an individual by adding virtual members to the group. For instance, parents may want to influence what television their children watch. They may not mind their children watching certain entertainment programmes, but may prefer them watching educational programmes. When the child is alone, a profile representing the parent's opinions (about how suitable items are for their child) can be added to the group as a virtual group member, and the TV could try to satisfy both, establishing a balance between the opinions of the parent and child. Similarly, a virtual group member with a profile produced by a teacher could be added to a group of learners.

---

[11] This initially offers the user non-personalized recommendations, however not necessarily by purely using popularity (e.g. Average without Misery can be used and fairness principles can be applied towards the other group members when recommending a sequence).

# 9    Conclusions and Challenges

Group recommendation is a relatively new research area. This chapter is intended as an introduction in the area, and the various techniques to deliver group recommendations.

## 9.1    Main Issues Raised

The main issues raised in this chapter are:

- Adapting to groups is needed in many scenarios such as interactive TV, ambient intelligence, recommending to tourist groups, etc. Inspired by the differences between scenarios, group recommenders can be classified using multiple dimensions.
- Many strategies exist for aggregating individual preferences (see Table 3), and some perform better than others. Users seem to care about avoiding misery and fairness.
- Existing group recommenders differ on the classification dimensions and in the aggregation strategies used. See Table 10 for an overview.
- When recommending a sequence of items, aggregation of individual profiles has to occur at each step in the sequence, as earlier items may impact the ratings of later items.
- It is possible to construct satisfaction functions to predict how satisfied an individual will be at any time during a sequence. However, group interaction effects (such as emotional contagion and conformity) can make this complicated.
- It is possible to evaluate in experiments how good aggregation strategies and satisfaction functions are, though this is not an easy problem.
- Group aggregation strategies are not only important when recommending to groups of people, but can also be applied when recommending to individuals, e.g. to prevent the cold-start problem and deal with multiple criteria.

## 9.2    Caveat: Group Modeling

The term "group modeling" is also used for work that is quite different from that presented in this chapter. A lot of work has been on modeling common knowledge between group members (e.g. [80, 81], modeling how a group interacts (e.g. [82, 83]) and group formation based on individual models (e.g. [82, 84]).

**Table 10** Group recommender systems

| System | Usage scenario | Classification | | | | Strategy used |
|---|---|---|---|---|---|---|
| | | Preferences known | Direct Experience | Group Active | Recommends Sequence | |
| MUSICFX [13] | Chooses radio station in fitness center based on people working out | Yes | Yes | No | No | Average Without Misery |
| POLYLENS [14] | Proposes movies for a group to view | Yes | No | No | No | Least Misery |
| INTRIGUE [15] | Proposes tourist attractions to visit for a group based on characteristics of subgroups (such as children and the disabled) | Yes | No | No | Yes | Average |
| TRAVEL DECISION FORUM [16] | Proposes a group model of desired attributes of a planned joint vacation and helps a group of users to agree on these | Yes | No | Yes | No | Median |
| YU'S TV REC. [18] | Proposes a TV program for a group to watch based on individuals' ratings for multiple features | Yes | No | No | No | Average |
| CATS [17] | Helps users choose a joint holiday, based on individuals' critiques | No | No | Yes | No | Counts requirements met Uses Without Misery |
| MASTHOFF'S [9, 49] | Chooses a sequence of music video clips for a group to watch | Yes | Yes | No | Yes | Multiplicative etc. |
| GAIN [10] | Displays information and advertisements adapted to the group present | Yes | Yes | No | Yes | Average |
| REMPAD [19] | Proposes multimedia material for a group reminiscence therapy session | Yes | No | No | No | Least Misery |
| HAPPYMOVIE [20] | Recommends movies to groups | Yes | No | No | No | Average |
| INTELLIREQ [21] | Supports groups in deciding which requirements to implement | No | No | Yes | Yes | Plurality voting |

## 9.3   Challenges

Compared to work on individual recommendations, group recommendation is still quite a novel area. The work presented in this chapter is only a starting point. There are many challenging directions for further research, including:

- *Recommending item sequences to a group.* The work in [9, 49] and the preliminary work in [59] seem to be the only work to date on recommending balanced *sequences* that address the issue of fairness. Even though sequences are important for the usage scenario of INTRIGUE, their work has not investigated making sequences balanced nor has it looked at sequence order. Clearly, a lot more research is needed on recommending and ordering sequences, in particular on how already shown items should influence the ratings of other items. Some of this research will have to be recommender domain specific.
- *Modeling of affective state.* There is a lot more work needed to produce validated satisfaction functions. The work presented in this chapter and [49] is only the starting point. In particular, large scale evaluations are required, as are investigations on the affect of group size.
- *Incorporating satisfaction within an aggregation strategy* As noted in Sect. 6, there are many ways in which satisfaction can be used inside an aggregation strategy. We presented some initial ideas in this area, but extensive empirical research is required to investigate this further.
- *Explaining group recommendations: Transparency and Privacy* One might think that accurate predictions of individual satisfaction can also be used to improve the recommender's transparency: showing how satisfied other group members are could improve users' understanding of the recommendation process and perhaps make it easier to accept items they do not like. However, as also confirmed in a recent study by Najafian et al. [85], users' need for privacy is likely to conflict with their need for transparency. An important task of a group recommender system is to avoid embarrassment. Users often like to conform to the group to avoid being disliked (we discussed normative conformity as part of Sect. 5.1 on how others in the group can influence an individual's affective state). In [49], it was investigated how different group aggregation strategies may affect privacy. More work is needed on explanations of group recommendations, in particular on how to balance privacy with transparency and scrutability. Chapter "Beyond Explaining Single Item Recommendations" provides more detail on the different roles of explanations in recommender systems [86].
- *User interface design.* An individual's satisfaction with a group recommendation may be increased by good user interface design. For example, when showing an item, users could be shown what the next item will be (e.g., in a TV programme through a subtitle). This may inform users who do not like the current item that they will like the next one better. There is also a need for additional research on interfaces for supporting group decision making (for initial research see [87]).
- *Group aggregation strategies for cold-start problems.* In Sect. 8.2, we have sketched how group aggregation can be used to help solve the cold-start problem.

However, the study in this area was very limited, and a lot more work is required to validate and optimize this approach.

- *Dealing with uncertainty.* In this chapter, we have assumed that we have accurate profiles of individuals' preferences. For example, in Table 1, the recommender knows that Peter's rating of item B is 4. However, in reality we will often have probabilistic data. For example, we may know with 80% certainty that Peter's rating is 4. Adaptations of the aggregation strategies may be needed to deal with this. DeCampos et al try to deal with uncertainty by using Bayesian networks [30]. However, they have so far focused on the Average and Plurality Voting strategies, not yet tackling the avoidance of misery and fairness issues.

- *Dealing with group attributes.* In Sect. 7, we have discussed work on incorporating group attributes in group recommender systems. Additionally, as mentioned in [88], users may well have different preferences in the context of a particular group then when they are alone. Clearly more research is needed in this area.

- *Empirical Studies.* More empirical evaluations are vital to bring this field forwards. It is a challenge to design well-controlled, large scale empirical studies in a real-world setting, particularly when dealing with group recommendations and affective state. It is likely that different aggregation strategies may be effective for different kinds of groups and for different application domains (see [87] for initial work on group recommender application domains). Almost all research so far has either been on a small scale, in a contrived setting, using synthetic groups (with the problem of using an Average metric, see Sect. 3.4) or lacks control.

**Acknowledgments** Judith Masthoff's research has been partly supported by Nuffield Foundation Grant No. NAL/00258/G.

# References

1. F. Ricci, L. Rokach, B. Shapira, Recommender systems: Introduction and challenges, in ed. by F. Ricci, L. Rokach, B. Shapira, *Recommender Systems Handbook*, 2nd edn. (Springer, Berlin, 2015), pp. 1–34
2. M.F. Dacrema, I. Cantador, I. Fernandez-Tobias, S. Berkovsky, P. Cremonesi, Design and evaluation of cross-domain recommender systems, in *Recommender Systems Handbook*, ed. by F. Ricci, L. Rokach, B. Shapira (Springer, New York, 2022)
3. C. Musto, M. de Gemmis, P. Lops, F. Narducci, G. Semeraro, Semantics and content-based recommendations, in *Recommender Systems Handbook*, ed. by F. Ricci, L. Rokach, B. Shapira (Springer, New York, 2022)
4. A.N. Nikolakopoulos, X. Ning, C. Desrosiers, G. Karypis, Trust your neighbors: a comprehensive survey of neighborhood-based methods for recommender systems, in *Recommender Systems Handbook*, ed. by F. Ricci, L. Rokach, B. Shapira (Springer, New York, 2022)
5. H. Abdollahpouri, G. Adomavicius, R. Burke, I. Guy, D. Jannach, T. Kamishima, J. Krasnodebski, L. Pizzato, Multistakeholder recommendation: Survey and research directions in *User Modeling and User-Adapted Interaction* (Springer, Berlin, 2020), pp. 127–158
6. R. Burke, Multisided fairness for recommendation. Preprint arXiv:1707.000931 (2017)

7. L. Xiao, Z. Min, Z. Yongfeng, G. Zhaoquan, L. Yiqun, M. Shaoping, Fairness-aware group recommendation with pareto-efficiency, in *Proceedings of the 11th ACM Conference on Recommender Systems* (ACM, New York, 2017), pp. 107–115

8. D. Serbos, S. Qi, N. Mamoulis, E. Pitoura, P. Tsaparas, Fairness in package-to-group recommendations, in *Proceedings of the 26th International Conference on World Wide Web* (ACM, New York, 2017), pp. 371–379

9. J. Masthoff, Group modeling: selecting a sequence of television items to suit a group of viewers. User Model. User-Adap. Interac. **14**, 37–85 (2004)

10. B. De Carolis, Adapting news and advertisements to groups, in *Pervasive Advertising* (Springer, Berlin, 2011), pp. 227–246

11. E. Kurdyukova, S. Hammer, E. André, Personalization of content on public displays driven by the recognition of group context, in *Ambient Intelligence* (Springer, Berlin, 2012), pp. 272–287

12. A. Jameson, B. Smyth, Recommendation to groups, in ed. by P. Brusilovsky, A. Kobsa, W. Njedl, *The Adaptive Web Methods and Strategies of Web Personalization* (Springer, Berlin, 2007), pp. 596–627

13. J. McCarthy, T. Anagnost, MusicFX: An Arbiter of Group Preferences for Computer Supported Collaborative Workouts, in *Proceedings of the 1998 ACM conference on Computer supported cooperative work, CSCW*, Seattle, WA (1998), pp. 363–372

14. M. O' Conner, D. Cosley, J.A. Konstan, J. Riedl, PolyLens: A Recommender System for Groups of Users. ECSCW, Bonn (2001), pp. 199–218. As Accessed on http://www.cs.umn. edu/Research/GroupLens/poly-camera-final.pdf

15. L. Ardissono, A. Goy, G. Petrone, M. Segnan, P. Torasso, Tailoring the recommendation of tourist information to heterogeneous user groups, in ed. by S. Reich, M. Tzagarakis, P. De Bra, *Hypermedia: Openness, Structural Awareness, and Adaptivity, International Workshops OHS-7, SC-3, and AH-3*. Lecture Notes in Computer Science, vol. 2266 (Springer, Berlin, 2002), pp. 280–295

16. A. Jameson, More than the sum of its members: Challenges for group recommender systems, in *International Working Conference on Advanced Visual Interfaces, Gallipoli* (2004)

17. K. McCarthy, L. McGinty, B. Smyth, M. Salamo, The needs of the many: A case-based group recommender system, in *European Conference on Case-Based Reasoning* (Springer, Berlin, 2006), pp. 196–210

18. Z. Yu, X. Zhou, Y. Hao, J. Gu, TV program recommendation for multiple viewers based on user profile merging. User Model. User-Adap. Int. **16**, 63–82 (2006)

19. A. Bermingham, J. O'Rourke, C. Gurrin, R. Collins, K. Irving, A.F. Smeaton, Automatically recommending multimedia content for use in group reminiscence therapy, in *Proceedings of the 1st ACM International Workshop on Multimedia Indexing and Information Retrieval for Healthcare* (ACM, New York, 2013), pp. 49–58

20. L. Quijano-Sanchez, J.A. Recio-Garcia, B. Diaz-Agudo, G. Jimenez-Diaz, Social factors in group recommender systems. ACM Trans. Intell. Syst. Technol. **4**(1), 8 (2013)

21. A. Felfernig, C. Zehentner, G. Ninaus, H. Grabner, W. Maalej, D. Pagano, F. Reinfrank, Group decision support for requirements negotiation, in *Advances in User Modeling* (Springer, Berlin, 2012), pp. 105–116

22. F. Guzzi, F. Ricci, R. Burke, Interactive multi-party critiquing for group recommendation, in *Proceedings of the Fifth ACM Conference on Recommender Systems* (ACM, New York, 2011), pp. 265–268

23. M. Stettinger, Choicla: Towards domain-independent decision support for groups of users, in *Proceedings of the 8th ACM Conference on Recommender Systems* (ACM, New York, 2014), pp. 425–428

24. J.O.Á. Márquez, J. Ziegler, Negotiation and reconciliation of preferences in a group recommender system. J. Inf. Proc. **26**, 186–200 (2018). Information Processing Society of Japan

25. T.N. Nguyen, F. Ricci, Combining long-term and discussion-generated preferences in group recommendations, in *Proceedings of the 25th Conference on User Modeling, Adaptation and Personalization* (ACM, New York, 2017), pp. 377–378

26. T.N. Nguyen, F. Ricci, A chat-based group recommender system for tourism, in *Information and Communication Technologies in Tourism* (Springer, Berlin, 2017), pp. 17–30
27. T.N. Nguyen, F. Ricci, Situation-dependent combination of long-term and session-based preferences in group recommendations: An experimental analysis, in *Proceedings of the 33rd Annual ACM Symposium on Applied Computing* (ACM, New York, 2018), pp. 1366–1373
28. D. Herzog, W. Wörndl, User-centered evaluation of strategies for recommending sequences of points of interest to groups, in *Proceedings of the 13th ACM Conference on Recommender Systems* (ACM, New York, 2019), pp. 96–100
29. T.N. Nguyen, Supporting Group Discussions with Recommendation Techniques (2019). Free University of Bozen-Bolzano
30. L.M. de Campos, J.M. Fernandez-Luna, J.F. Huete, M.A. Rueda-Morales, Managing uncertainty in group recommending processes. User Model. User-Adap. Interac. **19**, 207–242 (2009)
31. M.S. Pera, Y.K. Ng, A group recommender for movies based on content similarity and popularity. Inf. Process. Manag. **49**(3), 673–687 (2013)
32. S. Berkovsky, J. Freyne, Group-based recipe recommendations: Analysis of data aggregation strategies, in *Proceedings of the Fourth ACM Conference on Recommender Systems* (ACM, New York, 2010), pp. 111–118
33. T. De Pessemier, S. Dooms, L. Martens, Comparison of group recommendation algorithms. Multimed. Tools Appl. **72**, 2497–2541 (2014)
34. C. Senot, D. Kostadinov, M. Bouzid, J. Picault, A. Aghasaryan, C. Bernier, Analysis of strategies for building group profiles, in *Proceedings of User Modeling, Adaptation, and Personalization* (Springer, Berlin, 2010), pp. 40–51
35. J.O.Á. Márquez, J. Ziegler, Preference elicitation and negotiation in a group recommender system, in *Proceedings of IFIP Conference on Human-Computer Interaction* (Springer, Berlin, 2015), pp. 20–37
36. H.N. Kim, M. Bloess, A. El Saddik, Folkommender: a group recommender system based on a graph-based ranking algorithm. Multimedia Syst. **19**, 1–17 (2013)
37. L. Baltrunas, T. Makcinskas, F. Ricci, Group recommendations with rank aggregation and collaborative filtering, in *Proceedings of the Fourth ACM Conference on Recommender Systems* (ACM, New York, 2010), pp. 119–126
38. L.A. Carvalho, H.T. Macedo, Users' satisfaction in recommendation systems for groups: An approach based on noncooperative games, in *Proceedings of the 22nd International Conference on World Wide Web Companion* (2013), pp. 951–958
39. M. Salamó, K. McCarthy, B. Smyth, Generating recommendations for consensus negotiation in group personalization services. Personal Ubiquitous Comput. **16**(5), 597–610 (2012)
40. J. Castro, J. Lu, G. Zhang, Y. Dong, L. Martínez, Opinion dynamics-based group recommender systems. IEEE Trans. Syst. Man Cybe. Syst. **48**, 2394–2406 (2017)
41. A. Delic, F. Ricci, J. Neidhardt, Preference networks and non-linear preferences in group recommendations, in *Proceedings of IEEE/WIC/ACM International Conference on Web Intelligence* (ACM, New York, 2019), pp. 403–407
42. S. Schiaffino, D. Godoy, J.A.D. Pace, Y. Demazeau, A MAS-based approach for POI group recommendation in LBSN, in *Proceedings of the International Conference on Practical Applications of Agents and Multi-Agent Systems* (Springer, Berlin, 2020), pp. 238–250
43. F. Ortega, A. Hernando, J. Bobadilla, J.H. Kang, Recommending items to group of users using matrix factorization based collaborative filtering. J. Inf. Sci. **345**, 313–324 (2016)
44. D. Cao, X. He, L. Miao, Y. An, C. Yang, R. Hong, Attentive group recommendation, in *Proceedings of The 41st International ACM SIGIR Conference on Research & Development in Information Retrieval* (ACM, New York, 2018), pp. 645–654
45. X. He, L. Liao, H. Zhang, L. Nie, X. Hu, T.-S. Chua, Neural collaborative filtering, in *Proceedings of the 26th International Conference on World Wide Web* (2017), pp.173–182
46. L. Vinh Tran, T.-A. Nguyen Pham, Y. Tay, Y. Liu, G. Cong, X. Li, Interact and decide: Medley of sub-attention networks for effective group recommendation, in *The Proceedings of the 42nd International ACM SIGIR Conference on Research and Development in Information Retrieval* (ACM, New York, 2019), pp. 255–264

47. A. Sankar, Y. Wu, Y. Wu, W. Zhang, H. Yang, H. Sundaram, GroupIM: A mutual information maximization framework for neural group recommendation, in *Proceedings of the 43rd International ACM SIGIR Conference on Research and Development in Information Retrieval* (ACM, New York, 2020), pp. 1279–1288

48. J. Masthoff, The user as wizard: A method for early involvement in the design and evaluation of adaptive systems, in *Fifth Workshop on User-Centred Design and Evaluation of Adaptive Systems* (2006)

49. J. Masthoff, A. Gatt, In pursuit of satisfaction and the prevention of embarrassment: affective state in group recommender systems. User Model. User-Adapt. Interac. **16**, 281–319 (2006)

50. C. Trattner, A. Said, L. Boratto, A. Felfernig, Evaluating group recommender systems, in ed. by A. Felfernig, L. Boratto, M. Stettinger, M. Tkalcic, *Group Recommender Systems: An Introduction*, 1st edn. (Springer, Berlin, 2018), pp. 59–71

51. F. Barile, J. Masthoff, S. Rossi, A detailed analysis of the impact of tie strength and conflicts on social influence, in *Adjunct Publication of the 25th Conference on User Modeling, Adaptation and Personalization* (ACM, New York, 2017), pp. 227–230

52. A. Delic, J. Neidhardt, T.N. Nguyen, F. Ricci, An observational user study for group recommender systems in the tourism domain. J. Inf. Technol. Tourism **19**, 87–116 (2018). Springer

53. S. Bourke, K. McCarthy, B. Smyth, Using social ties in group recommendation, in *Proceedings of the 22nd Irish Conference on Artificial Intelligence and Cognitive Science*. University of Ulster-Magee. Intelligent Systems Research Centre (2011)

54. S. Rossi, F. Barile, C. Galdi, L. Russo, Artworks sequences recommendations for groups in museums, in *Proceedings of 12th International Conference on Signal-Image Technology & Internet-Based Systems* (IEEE, Piscataway, 2016), pp. 455–462

55. A. Piliponyte, F. Ricci, J. Koschwitz, Sequential music recommendations for groups by balancing user satisfaction, in *Proceedings of UMAP Workshops* (2013). CEUR

56. S.E. Asch, Studies of independence and conformity: a minority of one against a unanimous majority. Pschol. Monogr. **70**, 1–70 (1956)

57. Z. Wang, X. Zhou, Z. Yu, H. Wang, H. Ni, Quantitative evaluation of group user experience in smart spaces. Cybern. Syst. An Int. J. **41**(2), 105–122 (2010)

58. J. Masthoff, W.W. Vasconcelos, C. Aitken, F.S. Correa da Silva, Agent-based group modelling for ambient intelligence, in *AISB Symposium on Affective Smart Environments, Newcastle* (2007)

59. A. Piliponyte, F. Ricci, J. Koschwitz, Sequential music recommendations for groups by balancing user satisfaction, in *Proceedings of the Workshop on Group Recommender Systems: Concepts, Technology, Evaluation at UMAP13* (2013), pp. 6–11

60. A. Delic, J. Neidhardt, L. Rook, H. Werthner, M. Zanker, Researching individual satisfaction with group decisions in tourism: Experimental evidence, in *Information and Communication Technologies in Tourism* (Springer, Berlin, 2017), pp. 73–85

61. A. Delic, J. Masthoff, J. Neidhardt, H. Werthner, *Proceedings of the 26th Conference on User Modeling, Adaptation and Personalization* (ACM, New York, 2018), pp. 121–129

62. A. Delic, J. Neidhardt, H. Werthner, Group decision making and group recommendations, in *Proceedings of IEEE 20th Conference on Business Informatics (CBI)* (IEEE, Piscataway, 2018), pp. 79–88

63. T.N. Nguyen, F. Ricci, A. Delic, D. Bridge, Conflict resolution in group decision making: insights from a simulation study. User Model. User-Adap. Interac. **29**, 895–941 (2019).

64. S. Herr, A. Rösch, C. Beckmann, T. Gross, Informing the design of group recommender systems, in *CHI12 Extended Abstracts on Human Factors in Computing Systems* (ACM, New York, 2012), pp. 2507–2512

65. M. Gartrell, X. Xing, Q. Lv, A. Beach, R. Han, S. Mishra, K. Seada, Enhancing group recommendation by incorporating social relationship interactions, in *Proceedings of the 16th ACM International Conference on Supporting Group Work* (ACM, New York, 2010), pp. 97–106

66. I. Ali, S.-W. Kim, Group recommendations: approaches and evaluation, in *Proceedings of the 9th International Conference on Ubiquitous Information Management and Communication* (ACM, New York, 2015), pp. 1–6
67. X. Liu, Y. Tian, M. Ye, W.C. Lee, Exploring personal impact for group recommendation, in *Proceedings of the 21st ACM International Conference on Information and Knowledge Management* (ACM, New York, 2012), pp. 674–683
68. E. Quintarelli, E. Rabosio, L. Tanca, Recommending new items to ephemeral groups using contextual user influence, in *Proceedings of the 10th ACM Conference on Recommender Systems* (ACM, New York, 2016), pp. 285–292
69. H. Mahyar, K.E. Ghalebi, S.M. Morshedi, S. Khalili, R. Grosu, A. Movaghar, Centrality-based group formation in group recommender systems, in *Proceedings of the 26th International Conference on World Wide Web Companion* (2017), pp. 1187–1196. International World Wide Web Conferences Steering Committee.
70. J.S. Zhang, M. Gartrell, R. Han, Q. Lv, S. Mishra, GEVR: An event venue recommendation system for groups of mobile users, in *Proceedings of the ACM on Interactive, Mobile, Wearable and Ubiquitous Technologies* (ACM, New York, 2019), pp. 1–25
71. I. Alina Christensen, S. Schiaffino, Social influence in group recommender systems. J. Online Inf. Rev. **38**, 524–542 (2014). Emerald Group Publishing Limited.
72. S. Ioannidis, S. Muthukrishnan, J. Yan, A consensus-focused group recommender system. Preprint arXiv:1312.7076 (2013)
73. M. Ye, X. Liu, W.C. Lee, Exploring social influence for recommendation: A generative model approach, in *Proceedings of the 35th International ACM SIGIR Conference on Research and Development in Information Retrieval* (ACM, New York, 2012), pp. 671–680
74. S. Rossi, F. Barile, A. Caso, et al., Dominance weighted social choice functions for group recommendations. ADCAIJ: Adv. Distrib. Comput. Artif. Intell. J. **4**, 65–79 (2015). Ediciones Universidad de Salamanca (España)
75. J. Guo, Y. Zhu, A. Li, Q. Wang, W. Han, A social influence approach for group user modeling in group recommendation systems. IEEE Intell. Syst. **31**, 40–48 (2016). IEEE
76. J. Masthoff, Selecting news to suit a group of criteria: An exploration, in *4th Workshop on Personalization in Future TV - Methods, Technologies, Applications for Personalized TV, Eindhoven* (2004)
77. J. Masthoff, Modeling the multiple people that are me, in ed. by P. Brusilovsky, A.Corbett, F. de Rosis, *Proceedings of the 2003 User Modeling Conference, Johnstown, PA* (Springer, Berlin, 2003), pp. 258–262
78. W.L. de Mello Neto, A. Nowé, Insights on social recommender system, in *Proceedings of the Workshop on Recommendation Utility Evaluation: Beyond RMSE, at ACM RecSyS12* (2012), pp. 33–38
79. A. Umyarov, A. Tuzhilin, Using external aggregate ratings for improving individual recommendations. ACM Trans. Web **5**, 1–45 (2011)
80. J. Introne, R. Alterman, Using shared representations to improve coordination and intent inference. User model. User-Adap. Interac. **16**, 249–280 (2006)
81. S. Suebnukarn, P. Haddawy, Modeling individual and collaborative problem-solving in medical problem-based learning. User Model. User-Adap. Interac. **16**, 211–248 (2006)
82. T. Read, B. Barros, E. Bárcena, J. Pancorbo, Coalescing individual and collaborative learning to model user linguistic competences. User Model. User-Adap. Interac. **16**, 349–376 (2006)
83. A. Harrer, B.M. McLaren, E. Walker, L. Bollen, J. Sewall, Creating cognitive tutors for collaborative learning: steps toward realization. User Model User-Adap. Interac. **16**, 175–209 (2006)
84. E. Alfonseca, R.M. Carro, E. Martín, A. Ortigosa, P. Paredes, The impact of learning styles on student grouping for collaborative learning: a case study. User Model. User-Adap. Interac. **16**, 377–401 (2006)
85. S. Najafian, O. Inel, N. Tintarev, Someone really wanted that song but it was not me! Evaluating which information to disclose in explanations for group recommendations, in *Proceedings of the 25th International Conference on Intelligent User Interfaces Companion* (2020), pp. 85–86

86. N. Tintarev, J. Masthoff, Beyond explaining single item recommendations, in *Recommender Systems Handbook*, ed. by F. Ricci, L. Rokach, B. Shapira (Springer, New York, 2022)
87. M. Stettinger, G. Ninaus, M. Jeran, F. Reinfrank, S. Reiterer, WE-DECIDE: A decision support environment for groups of users, in *Recent Trends in Applied Artificial Intelligence* (Springer, Berlin, 2013), pp. 382–391
88. J. Gorla, N. Lathia, S. Robertson, J. Wang, Probabilistic group recommendation via information matching, in *Proceedings of the 22nd International Conference on World Wide Web* (2013), pp. 495–504

# People-to-People Reciprocal Recommenders

Irena Koprinska and Kalina Yacef

## 1 Introduction

Recommending people to people is the core task of many social websites and platforms. Examples include finding friends, professional contacts and communities to follow on social networks; matching people in online dating websites, job applicants with employers, mentors with mentees, and students in online learning courses. While social networks such as Facebook and LinkedIn aim at connecting people by creating $n$-to-$n$ relationships, other applications such as online dating websites aim at matching people to create 1-to-1 relationships.

Most people-to-people recommendations, and especially the 1-to-1 recommendations, involve creating relationships that are reciprocal, i.e. where both parties can express their likes and dislikes and a good match requires satisfying the preferences of both parties. For instance, in the process of hiring someone for a job, both the candidate and the company offering the job need to assess each other, deciding whether the candidate is fit for the position and vice-versa. In online dating, reciprocity is fundamental—users will build a successful relationship only if both parties are interested in each other. Matching students in education may also require reciprocity in order to maximise learning benefits.

The key role of reciprocity for recommending people to people has only recently being recognised. In this paper we discuss the distinctive nature of reciprocal recommenders, review the previous work and present a case study in online dating.

I. Koprinska (✉) · K. Yacef
School of Computer Science, The University of Sydney, Sydney, NSW, Australia
e-mail: irena.koprinska@sydney.edu.au; kalina.yacef@sydney.edu.au

© Springer Science+Business Media, LLC, part of Springer Nature 2022                421
F. Ricci et al. (eds.), *Recommender Systems Handbook*,
https://doi.org/10.1007/978-1-0716-2197-4_11

## 2 Reciprocal vs. Traditional Recommenders

Reciprocal recommenders must satisfy the preferences and needs of the two parties involved in the recommendation. In contrast, the traditional items-to-people recommenders are one sided and must satisfy only the preference of the person for whom the recommendation is generated. Table 1 summarizes the differences between the two types of recommenders; a comprehensive comparison can be found in [1].

The user behaviour is highly dependent on whether the domain is reciprocal or not. The success of a traditional book recommender is dependent only on the person receiving the recommendation. On the other hand, in a reciprocal domain such as online dating, the user receiving the recommendation knows that the success depends on both parties and this influences his/her behaviour. In addition, users in reciprocal domains may choose to act proactively by taking the initiative to connect with other users or to remain reactive and wait for contact.

Another difference is that for traditional recommenders, users generally have no reason to provide detailed information about themselves (user profile) and their preferences. In contrast, for reciprocal recommenders, there is a clear need and benefit for providing rich user profiles. These profiles might be inaccurate (e.g. due to a lack of self-awareness or desire to have a more attractive profile) and reciprocal recommenders need to account for that.

In traditional recommenders, satisfied and loyal users are likely to repeatedly use the site, allowing it to build rich user model by exploiting the explicitly and implicitly stated user preferences. In contrast, in reciprocal domains people may leave the site permanently after a successful recommendation. For example, a person who successfully finds a lifelong spouse on a dating website or who finds a long term job on a job website may not need to use these sites after that. This creates a paradox for this service provides who want their service to be the best for their users and therefore achieve what they are set to do. But at the same time, if they do provide the best recommendations, users may not use their services for long, possibly affecting revenue. On the other hand, happy users will refer the services

**Table 1** Main differences between reciprocal and traditional recommenders

| Traditional recommenders | Reciprocal recommenders |
| --- | --- |
| Success is determined solely by the user seeking the recommendation | Success is determined by both users—the subject and object of the recommendation |
| Users have no reason to provide detailed explicit user profiles. | Users are expected to provide detailed self-profiles and preferences |
| Satisfied users are likely to return for more recommendations. Better recommendations mean more engagement with the system in the future | Users may leave the system after a successful recommendation. Better recommendations might mean less engagement with the system in the future |
| The same item can be recommended to all users | Popular users should not be recommended to too many users |

to new users, and are likely to use the service again if there is a future need for it. This is a clear multi-objective optimization problem. However, it is important to highlight that both objectives, i.e. (1) good successful recommendations for users and (2) short term revenue goals, should not be optimized in equal weights, since an optimization for short-term revenue is likely to hurt the service in the long term, while optimizing for the goodness of users may actually benefit the whole service. The key to the multi-objective optimization here is to keep short-term revenue high without decreasing user satisfaction.

Finally, in reciprocal domains it is important that users are not recommended to others in a way that may cause them to be overloaded with recommendations. The popularity bias, which may be an issue for traditional recommender systems, becomes an exacerbated problem for reciprocal recommender systems. For instance, if a highly qualified person is recommended to every single job position that he/she fits, this person is likely to be burdened by the amount of contacts and leave the website. A similar situation can occur for popular users in a dating website. These users are important as they represent the best for each service, therefore they should only be recommended to other users when the recommender is absolutely sure that these users will reciprocate the contact. We note that this is a distinguishing feature of reciprocal recommenders compared to non-reciprocal people-to-people recommenders such as social networks where users can manage $n$-to-$n$ connections.

## 3  Previous Work

In this section we review previous work, focusing on reciprocal recommenders but also mention some key non-reciprocal people-to-people recommenders which are relevant and illustrate important characteristics of reciprocal domains. A survey paper on reciprocal recommenders has recently become available [2]. We note, however, that it covers both reciprocal and non-reciprocal people-to-people recommenders.

### 3.1  Social Networks

In the broad area of social matching [3, 4], see also Chapter "Social Recommender Systems", recommending people to other people has a clear link with reciprocal recommenders because the quality of a match is determined by both parties involved in the match. However, most of the existing work on social matching tailors recommendations only to the needs of one of the parties. Just a few papers mention the need for reciprocity and even fewer attempt to act on it.

Chen at al. [5] presented algorithms for recommending people on the enterprise social network Beehive, which allows users to connect to co-workers, post new information or comment on shared information. Two recommendation approaches were studied: content-based and collaborative filtering, and four algorithms were compared. The content-based approach assumes that if two people post content on similar topics, they are likely to be pleased to get connected. It is based on similarity of textual content posted by the user on Beehive and additional information such as job description and location. The collaborative filtering is a typical friend-of-friend approach and uses only linking information from the social network. It is based on the intuition that if many of $A$'s connections are connected to $B$, then $A$ may like to connect to $B$ too. The results show that all approaches increased the number of connections, compared to a control group that received no recommendations. The content-based approach was more successful in recommending contacts that were unknown to each other, while the collaborative filtering approach was more successful in finding known contacts. It is important to note that the befriending in Beehive is non-reciprocal, i.e. any user can connect with another user without the consent of the second user. However, there are still important reciprocal social considerations as noted by the authors, e.g. before adding a contact, one has to consider how the other person would perceive this action and whether they will reciprocate the connection, and also how the new contact will be perceived by the other people using the social network service.

Guy at al. [6] studied people recommendations on another enterprise social network site. They proposed an algorithm for recommending "strangers"—people who the user does not know but may be interested in, based on mutual interests. The similarity is computed based on shared activities such as bookmarking the same page, commenting on the same blog, reading the same file, using the same tag, being tagged with the same tag and being member of the same community. The results showed that the recommender was able to achieve its goal; the users commented on its usefulness for connecting them with experienced people to learn from, becoming aware of others with similar expertise, projects and roles in other places or departments. Similarly to [5], this is not a reciprocal recommender since adding connections did not require the consent of both party, however considering reciprocity when making and accepting recommendations is important due to the social dynamics in the enterprise.

Fazel-Zarandi et al. [7] studied different social drivers to predict collaborators for scientific research. These drivers were grounded in social science theories and included: level of expertise (based on topic and publications), friend-of-friend (based on previous collaborations, homophily (similar gender, affiliation, tenure status and co-citation), social exchange (dyalic relationship based on demand and supply of resources), and follow the crowd (popularity). A prototype recommender system was developed and evaluated using data from grant applications and publication histories. Homophily was found to be the most important factor for predicting collaboration, followed by expertise. This work is relevant as many of the social drivers may be considered reciprocal as aspects such as homophily, friend-of-friend and even level of expertise can be reciprocal, e.g. the mutually beneficial relationship between students and mentors in scientific collaborations.

## 3.2   Mentor-Mentee Matching

The i-Help system [8] helped students find people who could assist them with university courses, e.g. with first year computer science problems. It matched helpers with helpees in a non-reciprocal manner by considering the attributes of the helpers and the preferences of the helpees. For the helpers, it stored or inferred attributes such as knowledge of the topic, interests, cognitive style, eagerness to help, helpfulness, availability, and current load. The information was collected from several sources including self-evaluation and peer feedback in previous help sessions. An initial ranked list of potential helpers was produced. It was then refined by considering the preferences of the helpee, e.g. the importance of criteria such as helpfulness and urgency, the preferred and banned helpers. A final list of five potential helpers was compiled and the first of them to reply became the helper.

The PHelpS system [9] is an earlier prototype of i-Help. It was used in a workplace to train staff in how to use a new data management system. The candidate helpers were filtered based on their knowledge of the task, availability and load using a constraint solver. The list was presented to the helpee who chose the helper. Both i-Help and PHelpS relied on rich user models encoding the expertise and preferences of helpers and helpees.

Li [10] considered the task of matching mentors with mentees on the online platform Codementor which provides a 1-to-1 help for software developers. Users (mentees) post requests for help with programming issues and are matched with a sutable mentor who can help. Reciprocity is taken into account by considering the requirements of both sides. The recommendation problem is divided into two tasks—mentor willingness prediction and mentee acceptance prediction, which are solved simultaneously, generating a ranked list of mentors. Four groups of predictive feature are proposed: (1) availability of mentors in terms of time and availability of mentees in terms of budget, (2) capability of mentors to deal with the request based on the mentor's expertise, ratings by past mentees and past mentoring sessions, (3) activity of mentors—how frequently the mentor has expressed willingness to tackle request and how frequently the mentor was accepted by mentees and (4) proximity of the mentor to the current request—high values are assigned to mentors who have tackled similar requests in the past and are similar to other mentors who have tackled similar requests. The features were extensively evaluated using various supervised machine learning methods and shown to be effective for both tasks, with activity and proximity found to be most effective. The proposed recommendation approach was also evaluated online on Codementor for two weeks, showing accurate mentor prediction, faster selection and higher satisfaction by mentees.

## 3.3   Online Learning Courses

Labarthe et al. [11] studied the effect of a reciprocal peer recommender to foster students' persistence and success in a Massive Online Open Course (MOOCs). This work is based on the finding that social isolation is a contributing factor for poor learning experience and attrition in MOOCs, and that this is further exacerbated by the difficulty of finding the right people to interact with in a newly-formed MOOC community. The authors hypothesized that helping students to connect with other students in the MOOC would alleviate this problem. Their peer recommender provided each student with an individual list of potentially interesting contacts, created in real-time based on their own profile and activities. They evaluated the effect of the reciprocal recommender in four dimensions of learners' persistence: attendance, completion, success and participation. Results of their controlled study (n = 8376) showed improvement across all these four factors: students who received peer recommendations were much more likely to persist and engage in the MOOC than if those who did not [12]. The same team then investigated further what recommendation strategy had the most impact [13]. They conducted another controlled study (n = 2025) comparing three recommendation strategies: one using a socio-demographic-background similarity (gender, geo-location, education level and prior MOOC experience), another one based on the current stage of progress in the MOOC (to facilitate students contacting each other for specific questions about the content of the course), and a third one recommending random people to each other. Their findings showed that the socio-demographic-based peer recommender had a higher success than the other two.

Prabhakar et al. [14] proposed a reciprocal recommender algorithm for a MOOC using a similarity matrix based on the learners' preferences. They compute a reciprocal score for each pair of users that is the harmonic mean of the distance scores between them, where the distance score represents how the preferences of the first user matches the attributes of the other user. They found that the reciprocal algorithm tends to recommend to a given user people who also have that user in their recommendation list, which is as expected given that the reciprocity was promoted. Further work is needed to evaluate the recommender in real settings, as it is yet to be deployed with real users.

## 3.4   Job Recommendation

Malinowski et al. [15] investigated the problem of matching people and jobs and argued that the matching should be reciprocal, considering the preferences of both the job seeker and the recruiter. They built two recommender systems. The first one recommended job seekers (i.e. their resumes/profiles) to job descriptions of a particular recruiter. To create training data, a recruiter manually labelled a set of resumes as either fit or not fit for a list of jobs, based on demographic, educational, job experience, language, technology skills and other attributes. The second system recommended jobs to job seekers. To create training data, the job candidates were asked to rank a set of job descriptions indicating how well the jobs fitted their preferences. In both cases the authors used the expectation maximisation algorithm to build the prediction model. The two recommender systems were evaluated separately, showing promising prediction accuracy results. Several methods for combining the two recommendations were proposed but were not implemented and evaluated.

Hong at al. [16] developed and deployed iHR—an online recruitment system, linking job seekers with recruiters, and providing reciprocal recommendations, in addition to content-based and colaborative filtering. For job seeks, iHR collects information from three sources: (1) user profile and job preferences provided by the job seeker, (2) information automatically extracted from the uploaded resume and (3) behavioural information from the user's activity on the site, e.g. keywords used for searching, preferred profiles viewed and recommendations clicked. The reciprocal recommender takes into consideration three properties: *reciprocity, availability* and *diversity*. *Reciprocity* uses a relevance score which matches the preferences and user profiles of both parties; *availability* limits the number of recommendations during a given time period for both parties so that they are not overwhelmed; *diversity* is calculated based on the similarity of the job seekers—the goal is to recommend candidates with different personal strengths. The reciprocal recommendation task is formulated as an optimisation problem by considering the three properties.

The reciprocal recommender is evaluated on a large dataset for 3 years (almost 200,000 job seekers and 47,000 recruiters) and compared with other state-of-the-art methods, showing that it was the most accurate method and, hence, demonstrating the benefits of reciprocal recommendations. To assess the user experience, a user study based on surveys was also conducted comparing the content-based, collaborative filtering and reciprocal approaches, in terms of relevance, interpretability, diversity and ordering of the results. The reciprocal recommender received the most positive feedback. Hence, it outperformed the other methods in both accuracy and user experience. iHR is one of the first systems that incorporates reciprocal recommenders for recruitment. Importantly, it was also deployed in practice for the Xiamen Talent Service Center in China.

## 3.5   Online Dating

The reported research on building recommender systems for online dating is growing, with most of the papers published in the last 10 years. Since reciprocity is fundamental for online dating, the majority of work on reciprocal recommenders is in this area.

One of the first studies of recommender systems for online dating is the work of Brozovsky and Petricek [17] who evaluated the performance of two collaborative filtering approaches (item-to-item and user-to-user). They used data from a Czech online dating site containing about 12 million ratings and 195,00 users. The results showed that both algorithms outperformed the baselines based on random and mean predictions. A user study was also conducted confirming that the users preferred the collaborative filtering approaches. The authors mentioned the need for reciprocity, but did not explore it.

Another early work is [18], which formulated the matchmaking task as an information retrieval problem, where user profiles were ranked with respect to a given ideal partner profile (i.e. explicit user preferences). Using historical data, a training set of matches (pairs of users represented with their profile attributes) was created and labelled as relevant and non-relevant. A match was considered relevant if users exchanged contact information, and irrelevant if one of the users inspected the other user's profile but did not send a message or if he/she sent a message but the other user did not reply. A machine learning classifier (ensemble of boosted regression trees) was built and used to predict the relevance of new matches; given a new user, the potential candidates were ranked based on their predicted score. The approach was evaluated using data from an online dating website. The authors described the reciprocal aspect of their work as two-sided relevance and stressed its usefulness for ranking candidates in matchmaking problems. Also relevant is the work of McFee and Lanckriet [19] who proposed a method for learning optimal distance metrics and evaluated it an online dating data. However, reciprocity was not discussed in the paper; its main focus is the new general algorithm for learning distance metrics rather than the online dating application.

In [20], our research group proposed the content-based system RECON which is one of the first reciprocal recommenders for online dating. RECON uses both user profiles and user interactions for matching people; to produce recommendations for a given user, it extracts his/her implicit preferences (i.e. the preferences that are inferred from the interactions with the other users) in terms of attribute values and then matches them with the profiles of the other users. Two one-sided compatibility scores are calculated and then combined using the harmonic mean to produce a *reciprocal compatibility score* of the two users. Using data from an Australian dating website, we showed that reciprocity improved both the success rate and recall of the recommender (see further in the article how the success rate was computed).

In [21], we proposed CCR—a recommender system for online dating that combines content-based and collaborative filtering approaches and utilises both user profiles and user interactions (see the Case Study section for more details).

In another early work, Kim et al. [22] created a people recommender system for social networking websites where users can reply positively or negatively to messages from other users. The system was evaluated on data from an online dating website. The authors distinguish between recommender systems for one-way and two-way interaction. They propose an approach for a two-way interaction that considers both the interest of the sender and the interest of the recipient of message, and makes recommendations by combining them with a weighted harmonic mean. The method uses both user profiles and information about previous user interactions. For a given user, it finds the best matching values for every attribute and then combines them in a rule that can be used to generate recommendations. The proposed method was more accurate than a baseline, where users simply browse the site to search for people to connect to.

Cai et al. [23] developed a neighbour-based collaborative filtering approach, called SocialCollab, which considers the preferences of both sides. It is based on similarity of users in terms of attractiveness and taste. Two users are similar in *attractiveness* if they are liked by a common group of users, and similar in *taste* if they like a common group of users. To generate a recommendation for a user $A$, the SocialCollab algorithm considers all potential candidates $R$. For each candidate in $R$ it first finds two groups of similar users, in terms of attractiveness and taste; the candidate is added to the recommendation list for $A$ if there is at least one similar user in both groups that reciprocally liked $A$. The recommendations are ranked according to the number of similar users. SocialCollab was evaluated using data from a comercial online dating websiye for 2 weeks and shown to outperform standard collaborative filtering, confirming the importance of reciprocity in people-to-people recommenders. Cai et al. [23] improved on these results by using gradient descent to learn the relative contribution of similar users in the ranking of the recommendations given by SocialCollab. In the same domain, the work of [24] have reported improvements over Cai et al. by using a recommendation method based on tensor decomposition.

Building upon ideas from RECON and SocialCollab, Xia et al. [25] proposed content-based and collaborative filtering algorithms. Their content-based algorithm is very similar to RECON and the four collaborative filtering algorithms are based on the notion of attractiveness and taste similarity to SocialCollab with some extensions. An evaluation using a large dataset of 200,000 users from an online dating website in China showed that the collaborative filtering methods were more accurate than the content-based method and RECON. The reciprocal collaborative filtering methods proposed by Zhao at al. [26] are also based on attractiveness and taste as in SocialCollab and share similar ideas. The evaluation was done on a dataset of 47,00 users from a popular online dating website, containing user interactions for a period of 6 months.

Alsaleh et al. [27] used clustering to group the male users based on their attributes and the female users based on their preferences. The recomendations were generated by matching the male clusters with the female clusters based on user interactions, and recommending cluster members using compatibility scores. In their subsequent paper [28], a tensor space model was developed for finding latent relationships between users based on user attributes and interactions. The results showed that the proposed model was more accurate than SocialCollab [23] and other recommendation methods and baselines.

Kutty at al. [29] developed a novel reciprocal recommender for online dating based on graph theory and machine learning. They studied in depth the properties of online dating networks and proposed a novel representation, called attributed bipartite graph, to model the user profiles and interactions. In this representation, the nodes correspond to the users, the node attributes include both user profile attributes (e.g. age, gender, education, etc.) and user preferences, and the links correspond to the user interactions. To generate the recommendations, three algorithms are combined: (1) node-based similarity algorithm to predict similarity between bipartite nodes, (2) k-means similar node checking to help with the cold-star problem and (3) a reciprocal node compatibility algorithm that matches the users using a support vector machine prediction algorithm. An extensive evaluation using a large online dating website shows the advantage of the proposed reciprocal approach compared to collaborative filtering and other state-of-the-art recommendation methods.

Alanazi and Bain [30, 31] investigated reciprocal recommenders that consider temporal aspects of user behavior. The motivation is that user behaviour and preferences are not static but change over time. Using Hidden Markov Models (HMM), in their early work [30] the recommendation problem was represented is a graph, where the nodes are the users and the edges are the links between them; given the current state, the goal is to predict the next state, and in particular the new links (matches). In their subsequent work [31], Alanazi and Bain proposed to combine the content-based HMM part with a collaborative filtering part to improve speed and scalability to large datasets. In the proposed hybrid recommender CFHMM-HRT, the collaborative filtering part generates an initial list of recommendations, which is then validated by the trained HMM recommender to output a smaller list, that is ranked to produce the final list of recommendations. CFHMM-HRT was evaluated using a large dataset from an online dating website, showing that it was time efficient and generated considerably more accurate recommendations compared to its content-based and collaborative filtering counterparts.

Vitale at al. [32] formulated the reciprocal recommendation problem as a sequential learning problem. The learning consists of a sequence of rounds; at each round, a user from one of the parties becomes active and based on past feedback, the learning algorithm recommends a user from the other party. The goal is to uncover as many reciprocal matches as possible, and to do this quickly. The paper introduces assumptions for effective learning and an algorithm called SMILE, designed under these assumptions. SMILE is analysed theoretically in terms of computational complexity and limitations on the number of matches. An experimental evaluation is also conducted using synthetic datasets and the real dataset used in [17].

Kleinerman et al. [33] proposed a novel reciprocal recommender method called Reciprocal Weighted Score (RWS), which finds the optimal balance between the interests of both users, in order to create a successful interaction. It calculates two scores: (1) CF—the interest of user A in the recommended user B using collaborative filtering and (2) PR—the likelihood that B will reply positively to A, using a machine learning method. These two scores are combined in a weighed sum, where the weights are calculated separately for each user by formulating the task as an optimization problem. The results show the importance of individual weighting for successful interactions—the weights of the two components vary considerably between users, with most of the women having a higher weight for the CF score and most of the men having a higher weight for the PR score. The machine learning task is formulated as a binary classification problem: (1) positive reply and (2) negative or no reply, with AdaBoost employed as a classifier. The feature vector consists of 54 features describing the sender, receiver and their activities on the platform, selected using domain knowledge and machine learning feature selection methods. The evaluation was done online on a operational online dating platform (Doovdevan), and the results were compared with the reciprocal collaborative filtering method of Xia et al. [25] which captures the mutual interest of the two users (referred to as RCF). The results showed that RWS was more effective than RCF—it generated a higher number of successful recommendations. An extension of this work was presented in [34]; it includes an offline evaluation of RWS, and a reciprocal explanation module which was combined with RWS and extensively evaluated (see also Sect. 6).

Reciprocal recommenders based on latent factors have also been recently proposed. Neve and Palomares [35] developed LFRR—a collaborative filtering algorithm which learns two latent factor models—one for the male preferences and another one for the female preferences towards the opposite group. Each latent model uses matrix factorization and gradient descent to minimize the error on the known ratings, and then makes predictions about the likelihood of a successful interaction between two users by calculating the dot product of the user feature vectors. Similarly to RWS [33], RCF [25] and RECON [20], LFRR computes two unidirectional preference scores from the latent models, which are then combined in a single reciprocal preference score. Four aggregation functions were investigated: arithmetic, geometric and harmonic mean, and cross-ratio uninorm; an additional study exploring these aggregation functions is [36]. The evaluation was conducted offline using data from the Japanese online dating platform Pairs, comparing LFRR with RCF. The results showed that LFRR performed similarly to RCF in terms of accuracy but was faster at generating the recommendations, and also that the type of aggregation function significantly affects the results. In [37] Ramanathan et al. report on the development and deployment of a latent factor reciprocal system similar to LFRR, but which learns from both positive and negative preferences, not only positive as LFRR. It was compared with LFRR in an offline evaluation showing an improvement, and then deployed on a popular online dating platform in Japan with more than 5 million users.

Physical appearance and attractiveness are important factors for successful relationships. Although user photos are widely used in online dating platforms, they have rarely been used in recommender systems, perhaps because photos are more complex to analyse (different angles, quality, etc.) and attractiveness is subjective. Recently, Neve at al. [38] proposed ImRec—a reciprocal recommender that utilizes user photos to train a machine learning classifier (Siamese convolutional neural network), followed by a Random Forest algorithm, to predicts user A's preference for user B, based on A's previous positive and negative interactions. Two uni-directional scores are calculated and subsequently combined in in a single reciprocal score. An offline evaluation on data from the Pairs platform (also used in their previous work [35, 36]) showed that InRec outperformed RECON which doesn't image data, and also LFRR but only when very little data is available about the user (less than 5 positive and negative interactions). This shows the effectiveness if ImRec for addressing the cold-start problem for new users and suggests that ImRec and LFRR can be useful in a switching hybrid system based on the number of interactions. The authors emphasize the importance of providing effective recommendations on the first day—it was found that the Pairs users often decide if to commit to the platform based on their experience during the first 24 hours.

## 4   A Case Study in Online Dating

Online dating services, e.g. Match.com, eHarmony, RSVP, Zoosk, OkCupid, PlentyOfFish, Meetic, Badoo, Tinder and Bumble, are used by millions of people and their popularity is growing. Their revenue is also steadily increasing, e.g. according to [39] the online dating revenue in US was $602 million dollars in 2020 and is projected to reach $755 billion by 2024. In 2020 44.2 million Americans used online dating and this is expected to increase to 53.3 million by 2024. A recent study [40] showed that meeting online has become the most popular way heterosexual couples meet in US, overtaking traditional ways such as meeting through friends and family.

To find dating partners using an online dating website or app, users usually provide information about themselves (*user profile*) and their preferred partner (*user preferences*); an example using predefined attributes is shown in Table 2. The *explicit* user preferences are the preferences stated by the user as shown in Table 2. The *implicit* user preferences are inferred from the interactions of the user with other users and may be quite different to the explicit user preferences (e.g. when a user contacts exclusively short people who smoke in spite of stating in their preferences that they are looking for tall people who do not smoke).

We worked with a major Australian dating site where the user interactions include four steps:

1. Creating a user profile and specifying the explicit user preferences—New user Bob creates an account on the website and provides information about himself (user profile) and his preferred dating partner (explicit user preferences) using a set of predefined attributes such as the ones shown in Table 2. He can also add some textual information to expand on his tastes and personality.
2. Browsing the user profiles of other users for interesting matches—Bob finds Alice and decides to contact her.
3. Mediated interaction—Bob chooses a message from a predefined list, e.g. *I'd like to get to know you, would you be interested?* We call these messages *Expressions of Interest (EOI)*. Alice can reply with a predefined message that is either positive (e.g. *I'd like to know more about you.*) or negative (e.g. *I don't think we are a good match.*) or may not reply at all. When an EOI receives a positive reply, we say that the interest is *reciprocated*.

   We define an interaction between users A and B as *successful* if $A$ has sent an EOI to $B$ and $B$ has responded positively. Similarly, an interaction between A and B is *unsuccessful* if $A$ has sent an EOI to $B$ and $B$ has responded negatively.

**Table 2** User profile and explicit user preferences

| Bob | My details (Who I am?) | My ideal partner details (Who I am looking for?) |
|---|---|---|
| Age | 44 years old | 35–46 years old |
| Location | Sydney | within 20 km |
| Height | 175 cm | at most 175 cm |
| Body type | Athletic | Slim, average, athletic |
| Smoking | Trying to quit | Trying to quit, Don't smoke |
| Relationship status | Divorced | Single, divorced, widowed, separated |
| Have children | Have children who don't live at home | Have children who don't live at home, Have children living at home, Have no children |
| | *How many*: 2 | |
| | *Age range*: 18–23 years old | |
| Personality | Social | Social, average |
| Eye colour | Blue | – |
| Hair color | Brown | – |
| Nationality | Australian | – |

4. Unmediated interaction—Typically after a successful interaction, Bob or Alice buys tokens from the website to send each other unmediated messages. This is the only way to exchange contact details and develop further their relationship.

While the relationship, once taken offline, may or may not become successful for Bob and Alice, reaching the fourth step is crucial and necessary to make it possible for them to find out. It is also the extent to which the dating website can go.

A major hurdle for progressing through these steps is that users must find, fairly quickly, relevant users among the hundreds of thousands available. Failure to do so can result in a loss of interest ("there is no-one who I like"), or a feeling of rejection when they contact people who don't reciprocate ("noone wants to talk to me"). Therefore, an efficient reciprocal recommender algorithm is essential for a good customer experience.

## 4.1 CCR: Content-Collaborative Reciprocal Recommender for Online Dating

CCR is our Content-Collaborative Reciprocal recommender [21]. It uses information from the user profile and user interactions to recommend potential matches for a given user. The content-based part computes similarities between users based on their profiles. The collaborative filtering part uses the interactions of the set of similar users, i.e. who they like/dislike and are liked/disliked by, to produce

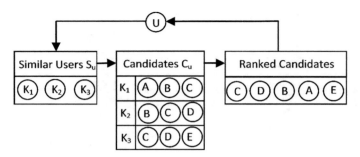

**Fig. 1** The CCR recommender

the recommendation. The recommender is *reciprocal* as it considers the likes and dislikes of both sides and aims to match users so that the pairing has a high chance of success.

The main assumption of CCR, reflected in steps 1 and 2 below, is that a pair of users who have similar profiles will *reciprocally like* (meaning "like and be liked by") the same type of people (in terms of user profiles), i.e. if $U$ has a similar profile to $K_1$ and $K_1$ reciprocally likes $A$, $B$ and $C$, then $U$ will reciprocally like $A$, $B$ and $C$. We tested this hypothesis in [21] using correlation analysis and a large dataset of more than 7000 users and 167,00 EOI, and found that indeed similar people are reciprocally liked by the same type of people.

To generate a recommendation list for user $U$, CCR uses the following steps, also shown in Fig. 1:

1. **Generating similar users based on user profiles**

    This step produces a set of $K$ users who have the most similar profile to $U$, i.e. that have the lowest possible distance to U. We use a modified version of the K-Nearest Neighbor algorithm, with seven attributes (*age, height, body type, education level, smoker, have children and marital status*) and a distance measure specifically developed for these attributes. For example, in Fig. 1 the set of similar users $S_u$ for user $U$ consists of $K_1$, $K_2$ and $K_3$.

2. **Generating recommendation candidates based on user interactions**

    This step produces a set $C_u$ of candidate users for recommending to $U$. For every user $K_i$ in $S_u$, we compute the list of all users with whom $K_i$ had reciprocal interest with and add it to the set of candidates $C_u$. For example in Fig. 1, $K_1$ and $A$ liked each other, so did $K_1$ and $B$, $K_1$ and $C$ and so on, resulting in a recommendation candidate set for $U$ of $\{A, B, C, D, E\}$ with a frequency of 1, 2, 3, 2, 1 respectively.

3. **Ranking the candidates**

    This step uses a ranking method to order the candidates based on their desirability, and provide meaningful recommendations for $U$. Figure 1 shows a ranking method based on frequency—$C$ is ranked the highest because it is the most frequent candidate in $C_u$.

**Ranking Method Support** We developed and compared a number of ranking methods, including two that utilise the explicit and implicit user preferences, and found that the Support ranking method was the best in spite of its simplicity [41]. The Support ranking method is based on the interactions between the group of similar users $S_u$ and the group of candidates. Users are added to the candidates pool if they have responded positively to at least one $S_u$ user or have received a positive reply from at least one $S_u$ user. However, some candidates might have received an EOI from more than one $S_u$ user, responding to some positively and to others negatively. Thus, some candidates have more successful interactions with $S_u$ than others. The Support ranking method computes the support of $S_u$ for each candidate. The support score for $X$ is the number of positive interactions minus the number of negative interactions that $X$ had with $S_u$. The higher the score for $X$, the more reciprocally liked is $X$ by $S_u$. The candidates are then sorted in descending order based on their support score.

### 4.1.1 Evaluation

**Data** To evaluate the performance of CCR, we used a real dataset from our Australian website partner from active users who were Sydney residents and interacting with people of different genders. For each run of the experiment, the dataset is partitioned into two distinct sets, training and test, containing approximately 2/3 and 1/3 of the users, respectively. Each training/test partition contains an even distribution of males and females. Each set was also evenly assigned users who were more popular than their cohort in terms of EOI sent and/or received. The data characteristics are shown in Table 3.

EOIs and their responses from users in the training set to users in the test set and vice versa were removed to ensure fair evaluation. Users in either the test or training set who no longer meet the minimum number of EOI required were removed. This resulted in the removal of less than 1% of the users before the split into training and test sets. Information about the interactions of the users from the test set is never included when ranking the candidates for this user to ensure clear separation between the two sets.

**Table 3** Data characteristics

| | |
|---|---|
| Total users | 216,662 |
| Male users | 119,102 (54.97%) |
| Female users | 97,560 (45.03%) |
| EOIs | 167,810 |
| Successful EOIs | 24,079 (25.59%) |
| Users sent/received at least 1 EOI | 7322 |
| Male users sent/received at least 1 EOI | 3965 |
| Female users sent/received at least 1 EOI | 3357 |

**Selected Attributes and Distance Measure** The original dataset consists of 39 user profile attributes. After a preliminary analysis of the distribution of these attributes to assess their importance and suitability, we manually selected 7 attributes: 2 numeric (*age and height*) and 5 nominal: *body type* (*values: slim, average, overweight*), *education level* (*secondary, technical, university*), *smoker* (*yes, no*), *have children* (*yes, no*) and *marital status* (*single, previously married*). Some attribute values were merged together, e.g. the values *overweight* and *largish* of *body type* were merged into *overweight*.

To measure the similarity between the profiles of users *A* and *B*, we used a distance measure that considers the differences between all attributes, but weights higher the age difference. For nominal attributes, we used a binary representation and Hamming distance and for numerical attributes we used a function of absolute differences; for more details, see [21].

**Performance Measures** For a user *U* we define the following sets:

- Successful EOI sent by *U*, *successful_sent*: The set of users whom *U* sent an EOI where the response was positive.
- Successful EOI received by *U*, *successful_recv*: The set of users who sent an EOI to *U* where the response was positive.
- Unsuccessful EOI sent by *U*, *unsuccessful_sent*: The set of users whom *U* sent an EOI where the response was negative.
- Unsuccessful EOI received by *U*, *unsuccessful_recv*: The set of users who sent an EOI to *U* where the response was negative.
- All successful and unsuccessful EOI for *U*: *successful* = *successful_sent* + *successful_recv*, *U*: *unsuccessful* = *unsuccessful_sent* + *unsuccessful_recv*

For each user *U* from the test set, a list of *N* ordered recommendations *N_rec* is generated. The successful and unsuccessful EOI for *U* in this list are: *successful@N* = *successful* ∩ *N_rec* and *unsuccessful@N* = *unsuccessful* ∩ *N_rec*.

Then, the *success rate* at *N* (i.e. given the *N* recommendations) is defined as:

$$success\,Rate@N[\%] = \frac{\#successful@N}{\#successful@N + \#unsuccessful@N} \tag{1}$$

Hence, given a set of *N* ordered recommendations, the success rate at *N* is the number of correct recommendations over the number of interacted recommendations (correct or incorrect).

For comparison we use the following baseline: the success rate of the recommender using a random set of *K* users in $S_u$ as opposed to *K* nearest neighbors in step 1 of the CCR algorithm (see Fig. 1). The random set of *K* users is used to generate candidates that are then ranked, i.e. there is no change in steps 2 and 3.

Each experiment has been run ten times; the reported success rate is the average over the ten runs.

**Results** We evaluated the performance of CCR for different number of recommendations *N* (from 10 to 500) and different number of minimum number of EOI sent

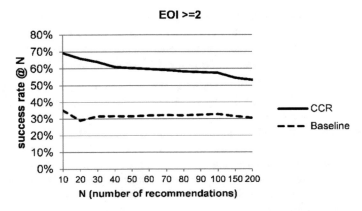

**Fig. 2** CCR success rate results for *minEOI_sent* = 2

by a user *minEOI_sent* (from 1 to 20) and compared it to the baseline success rate of the recommender using a random set of $K$ users in $S_u$ as opposed to the $K$ nearest neighbours. As an example, Fig. 2 shows the success rate for *minE-OI_sent* = 2. We found that CCR significantly outperforms the baseline for all cases. For example, for $N = 10$ and *minEOI_sent* = 2, the success rate of CCR is 69.26% and the baseline success rate is 35.19%.

As the number of recommendations $N$ increases from 10 to 500, the success rate decreases by 10–20%. This means that the best recommendations are at the top of the list and adding more recommendations only dilutes the success rate. Hence, our ranking criterion is useful and effective. In practice, the success rate for a smaller $N$, e.g. $N = 10$–30 is very important as this is the typical $N$ presented to the user. Unsuccessful recommendations, especially recommendations leading to rejection can be very discouraging.

Our results also show that as the number of *minEOI_sent* increases from 1 to 20, the success rate trends are very similar. However, for users who sent more EOIs, the success rate is slightly lower (e.g. 60.16% for *minEOI_sent* = 10 and 58.54% *minEOI_sent* = 20, for $N = 10$). This can be explained by the fact that the highly active users may be less selective.

In all experiments we used $K = 100$ and $C = 250$. With these parameters it took approximately 100 milliseconds to generate the recommendation list for a user which confirms the efficiency of our algorithm for generation of similar users and candidate recommendations.

## 4.2 Explicit and Implicit User Preferences

User preferences are essential for building a successful recommender. We investigate which type of user preferences (explicit or implicit) is more predictive of success in user interaction. More details can be found in [41]. A related investigation is [42] where we defined a histogram-based model of implicit preferences learned from successful interactions only (as opposed to both successful and unsuccessful as in CCR), and evaluated them in the context of a content-based recommender.

**Explicit User Preferences** We define the explicit preferences of a user $U$ as the vector of attribute values specified by $U$. The attributes and their possible values are predefined by the website.

In our study we used all attributes except *location*, i.e. 19 attributes - 2 numeric (*age* and *height*) and 17 nominal (*marital status, have children, education level, occupation industry, occupation level, body type, eye color, hair color, smoker, drink, diet, ethnic background, religion, want children, politics, personality and have pets*). For simplicity we only considered people from Sydney and interactions between people from different genders.

**Implicit User Preferences** We learn the implicit user preferences from the user interaction data by applying a Bayesian classification method; an overview of data mining methods for recommender systems is provided in [43].

The implicit user preferences of a user $U$ are represented by a binary classifier which captures $U$'s likes and dislikes. It is trained on $U$'s previous successful and unsuccessful interactions. The training data consists of all users $U+$ with whom $U$ had successful interactions and all users $U-$ with who $U$ had unsuccessful interactions during a given time period. Each user from $U+$ and $U-$ is one training example; it is represented as a vector of user profile attribute values and labeled as either Success (successful interaction with $U$) or Failure (unsuccessful interaction with $U$). We used the same 19 user profile attributes as the explicit user preferences listed in the previous section. Given a new instance, user $U_{new}$, the classifier predicts how successful the interaction between $U$ and $U_{new}$ will be (class *Success* or *Failure*).

As a classifier we employed NBTree [44] which is a hybrid classifier combining decision tree and Naïve Bayes. As in decision trees, each node of a NBTree corresponds to a test for the value of a single attribute. Unlike decision trees, the leaves of a NBTree are Naïve Bayes classifiers instead of class labels. NBTree was shown to be more accurate than both decision trees and Naïve Bayes, while preserving the interpretability of the two classifiers, i.e. providing an easy to understand output which can be presented to the user [44].

### 4.2.1   Are Explicit Preferences Good Predictors of User Interactions?

**Data**  To evaluate the predictive power of the explicit preferences we consider users who have sent or received at least 1 EOI during a one-month period (March 2010). We further restrict this subset to users who reside in Sydney to simplify the dataset. These two requirements are satisfied by 8012 users (called *target users*) who had 115,868 interactions, of which 46,607 (40%) were successful and 69,621 (60%)

**Table 4** Matching $U$'s explicit preferences with $U_{int}$'s profile for numeric attributes

| $U$'s preference for *height* | 155–175 | 155–175 | 155–175 |
|---|---|---|---|
| $U$'s value in profile for *height* | 160 | 180 | Unspecified |
| Matching outcome ($U$, $U_{int}$) | Match | Non-match | Match |

**Table 5** Matching $U$'s explicit preferences with $U_{int}$'s profile for nominal attributes

| $U$'s preference for *body type* | Slim, average | Slim, average | Slim, average | Slim, average |
|---|---|---|---|---|
| $U_{int}$'s value in profile for *body type* | Slim | Average | Overweight | unspecified |
| Matching outcome ($U$, $U_{int}$) | Match | Match | Non-match | Match |

were unsuccessful. Each target user $U$ has a set of interacted users $U_{int}$, consisting of the users $U$ had interacted with.

**Method** We compare the explicit preferences of each target user $U$ with the profile of the users in $U_{int}$ by calculating the number of matching and non-matching attributes.

While users can specify only a single value for a given attribute in their profile, e.g. *height = 170* or *body = athletic*, more than one value can be specified in their preferences—a set of values for a nominal attribute, e.g. *body = slim* or *athletic*, and a range of values for a numeric attribute, e.g. height = 155–175. Taking this into account, the matching between the preferences of $U$ and the profile of $U_{int}$ for a given attribute is done as follows. For a numeric attribute, $U_{int}$ matches $U$'s preferences if $U_{int}$'s value falls within $U$'s range or $U_{int}$ has not specified a value (see the examples in Table 4). For a nominal attribute, $U_{int}$ matches $U$'s preferences if $U_{int}$'s value has been included in the set of values specified by $U$ or $U_{int}$ has not specified a value (see the examples in Table 5). An attribute is not considered if $U$ has not specified a value for it. The preferences of $U_{int}$ match the profile of $U$ if all attributes match; otherwise, they do not match.

**Results** As shown in Table 6, 59.40% of all interactions occur between users with non-matching preferences and profiles. A further examination of the successful and unsuccessful interactions reveals that:

- In 61.86% of all successful interactions $U$'s explicit preferences did not match $U_{int}$'s profile.
- In 42.25% of all unsuccessful interactions $U$'s explicit preferences matched the $U_{int}$'s profile.

Suppose that we use the matching of the user profiles and preferences to try to predict if an interaction between two users will be successful or not (if the profile and preferences match -> successful interaction; if the profile and preferences do not match -> unsuccessful interaction). The accuracy will be 49.43% (17,775+39,998 /115,868). This is lower than the baseline accuracy of always predicting the majority class (ZeroR baseline) which is 59.78%. A closer examination of the

**Table 6** Explicit preferences—results

|  | $U$'s preferences and $U_{int}$'s profile match | $U$'s preferences and $U_{int}$'s profile do not match | Total |
|---|---|---|---|
| Successful interactions | 17,775 (38.14%) | 28,832 (61.86%) (false positives) | 46,607 (all successful interactions) |
| Unsuccessful interactions | 29,263 (42.25%) (false negatives) | 39,998 (57.75%) | 69,261 (all unsuccessful interactions) |
| Total | 47,038 (40.60%) | 68,830 (59.40%) | 115,858 (all interactions) |

misclassifications shows that the proportion of false positives is higher than the proportion of false negatives, although the absolute numbers are very similar.

In summary, the results show that the explicit preferences are not a good predictor of the success of interaction between users. This is consistent with [18].

### 4.2.2   Are Implicit Preferences Good Predictors of User Interactions?

**Data**  To evaluate the predictive power of the implicit preferences we consider users who have at least 3 successful and 3 unsuccessful interactions during a one-month period (February 2010). This dataset was chosen so that we could test on the March dataset used in the study of the implicit preferences above. Here too, we restrict this subset to users who reside in Sydney. These two requirements are satisfied by 3881 users, called *target users*. The training data consists of the interactions of the target users during February; 113,170 interactions in total, 30,215 positive and 72,995 negative. The test data consists of the interactions of the target users during Match; 95,777 interactions in total, 34,958 positive (37%, slightly less than the 40% in the study above) and 60,819 negative (63%, slightly more than the 60% in the study above). Each target user $U$ has a set of *interacted users $U_{int}$*, consisting of the users $U$ had interacted with.

**Method**  For each target user $U$ we create a classifier by training on $U$'s successful and unsuccessful interactions from February as described in Sect. 3.2. We then test the classifier on $U$'s March interactions. This separation ensures that we are not training and testing on the same interactions.

**Results**  Table 7 summarizes the classification performance of the NBTree classifier on the test data. It obtained an accuracy of 82.29%, considerably higher than the ZeroR baseline of 63.50% and the accuracy of the explicit preferences classifier. In comparison to the explicit preferences, the false positives drop from 61.86% to 30.14%, an important improvement in this domain since recommendations leading to rejection are very discouraging; the false negatives drop from 42.25 to 9.97%.

**Table 7** Classification performance of NBTree on test set

| <– classified as | Successful interactions | Unsuccessful interactions | Total |
|---|---|---|---|
| Successful interactions | 24,060 (68.83%) | 10,538 (30.14%) (false positives) | 34,958 (all successful interactions) |
| Unsuccessful interactions | 6064 (9.97%) (false negatives) | 54,755 (90.03%) | 60,819 (all unsuccessful interactions) |

In summary, the results show that the implicit preferences are a very good predictor of the success of user interactions, and considerably more accurate than the explicit preferences.

# 5 Conclusions

People-to-people reciprocal recommenders are an important class of recommenders which have emerged fairly recently. In this paper we discussed their characteristics (a more comprehensive analysis is available in [1]) and present an overview of recent work in this area. To illustrate different aspects of this type of recommenders and how to take account of the reciprocity and build an effective reciprocal recommender, we presented a case study in online dating, using a large dataset from a major Australian online dating website.

In particular, we presented a case study using CCR, a reciprocal recommender system for an online dating we have developed, that combines content-based and collaborative filtering, and utilises data from both user profiles and user interactions. It is based on our finding that people with similar profiles are reciprocally liked by people with similar profiles. CCR achieved success rate of 64.24–69.26% for different number of EOI, significantly outperforming the baseline success rate of 23.44–35.19%. An important advantage of CCR is that it addresses the cold start problem of new users joining the website by being able to provide recommendations immediately, based on the profile of the new user, which is very important for engaging the new users.

We also studied the differences between the implicit and explicit user preferences. We found that the explicit user preferences, stated by the user, are not a good predictor of the success of user interactions, achieving an accuracy of 49.43%. In contrast, the implicit user preferences, that are learned from successful and unsuccessful previous user interactions, using a probabilistic classifier, were a very good predictor of the success of user interactions, achieving an accuracy of 89.29%.

# 6   Challenges and Future Directions

There are many research questions that arise from designing reciprocal recommenders, some of which are the same as for standard recommenders, and others
are inherent to the reciprocity aspect.

**Popular Users**   In reciprocal recommenders, some user profiles need to be handled
with care, e.g. popular users should not be recommended too often, as they are likely
to be overwhelmed and unresponsive. This problem does not normally occur in non-
reciprocal domains or even people-to-people recommenders that are not reciprocal,
e.g. Twitter.

**Bait-Profiles**   Another issue is that some users, especially popular users, may hide
bait-profiles, created by criminals to lure people into trusting them in romance
scams. The detection of scamming in the online dating industry is a high priority
and requires the recommender systems to ensure they do not favour bait-profiles
over authentic user profiles [45]. Although this issue is very important in online
dating, where people are particularly vulnerable when seeking relationships, it can
also be an issue in other people-to-people recommendations.

**Explicit and Implicit Preferences**   The predictive power of explicit and implicit
user preferences needs further investigation. Not all explicit user preferences are
equally important; if the user can specify the importance of the attributes in the
explicit preferences, this information can be used to improve the prediction of
successful and unsuccessful interactions. A comparison of the explicit and implicit
user preferences would also be beneficial, e.g. (1) to find if there are some latent
factors that are difficult to capture, and also (2) to make users aware when their
stated explicit preferences are very different than their implicit preferences, and
adjust the explicit preferences accordingly. It is also worth investigating if our
findings about the explicit and implicit preferences carry over to other people-to-
people reciprocal domains.

**Data Sources and Information Fusion**   In order to increase the efficiency and
relevance of reciprocal recommenders, a number of other data sources should also
be explored—for instance, the use of temporal information (e.g. how quickly users
respond to EOIs), or the use of photos and free text to refine the quality of the
implicit user profiles. Analysing the importance of each data source and developing
appropriate aggregation and weighting methods [46] is another avenue for future
work.

**Novelty and Diversity**   Although providing unexpected recommendations has
been identified as a useful property of traditional recommender systems (Chapter
"Novelty and Diversity in Recommender Systems"), it is not clear how much
novelty and serendipity is needed in reciprocal recommenders. In contrast with
traditional domains, in reciprocal domains, users provide more information about
themselves in their user profiles and explicit user preferences. Recommending
surprising matches that do not satisfy these preferences may be seen as unacceptable

by some users, and reduce their trust [47] in the system. Some other users, however, may welcome suggestions of people different to the ones they think they like. One way to safely allow novelty and serendipity is to explicitly inform the users when the recommendations deviate from their explicit preferences.

**User Personality** Considering user personality (Chapter "Personality and Recommender Systems") in reciprocal recommender systems is another interesting direction for future work. Some online dating website assess personality by asking users to complete long and intrusive questionnaires, and then match users based on personality type. It will be useful to acquire user personality implicitly in a non-obtrusive way [48], e.g. from free text comments in the user profile; book, movie and sport preferences; writing style, text sentiment, punctuation and grammar; user activity level and interactions.

**Context-Based Recommendations** User preferences depend on the context and also change over time. For example, when on holidays, travelling abroad, the user may want to connect with other travellers to explore the area together, and these other people may have different profiles than the ones the user typically connects with. Context-aware recommender approaches (Chapter "Context-Aware Recommender Systems: From Foundations to Recent Developments") used in non-reciprocal domains can be extended and applied for reciprocal tasks.

**Multi-Stakeholder Recommendations** In people-to-people reciprocal recommenders there are often conflicting incentives. For example, in online dating the best experience for the user is to find a lifelong spouse quickly and leave the dating platform permanently, while for the owners of the online dating platform it might be better to have longer staying or repeated users to maximize revenue. The emerging sub-area of multi-stakeholder recommenders [49] is concerned with integrating the perspectives of multiple stakeholders, not only the users of the system.

**Providing Explanation** Investigating methods for effective explanations in reciprocal recommenders, and how to balance privacy and transparancy (Chapter "Beyond Explaining Single Item Recommendations"), is another direction for future work. First, in reciprocal recommenders it is more challenging to provide explanations without raising privacy concerns. For examples, it is not appropriate to tell Bob that Alice was recommended to him because she liked another user Daniel who is similar to Bob. Second, methods for providing *reciprocal explanations* (why both parties are expected to benefit from the match) were recently introduced by Kleinerman et al. [34, 50], who also conducted an extensive evaluation in both simulated and operational online dating environment. The results showed that the best explanation method depends on the user—in environments where the cost (e.g. emotional cost—fear of rejection) is high, reciprocal explanations are superior to the traditional one-sided as they make the user more confident in the positive outcome. However, when the fear of rejection was removed in the simulation, the one-sided explanation was better. These results are interesting and open avenues for future research.

**Acknowledgments** This work was supported by the Smart Services Cooperative Research Centre. We also thank Joshua Akehurst, Luiz Pizzato and Judy Kay for their contributions to this work.

# References

1. L. Pizzato, T. Rej, J. Akehurst, I. Koprinska, K. Yacef, and J. Kay, Recommending people to people: the nature of reciprocal recommenders with a case study in online dating. User Model. User-Adap. Interact. **23**, 447–488 (2013)
2. I. Palomares, C. Porcel, L. Pizzato, I. Guy, E. Herrera-Viedma, Reciprocal recommender systems: analysis of state-of-art literature, chalelnges and opportunities on social recommendation (2020). https://arxiv.org/abs/2007.16120
3. L. Terveen, D.W. McDonald, Social matching: a framework and research agenda. ACM Trans. Comput.-Human Interact. **12**,401–434 (2005)
4. I. Guy, People recommendation on social media, in ed. by P. Brusilovsky, D. He, *Social Information Access*. Lecture Notes in Computer Science, vol 10100 (Springer, Berlin, 2018)
5. J. Chen, W. Geyer, C. Dugan, M.G.I. Muller, I. Guy, Make new friends, but keep the old: recommending people on social networking sites, in *Proceedings of the International Conference on Computer-Human Interaction (CHI)* (2009)
6. I. Guy, S. Ur, I. Ronen, A. Perer, M. Jacovi, Do you want to know?: Recommending strangers in the enterprise, in *Proceedings of the ACM Conference on Computer Supported Cooperative Work (CSCW)* (2011)
7. M. Fazel-Zarandi, H.J. Devlin, Y. Huang, N. Contractor, Expert recommendation based on social drivers, social network analysis, and semantic data representation, in *Proceedings of the Second International Workshop on Information Heterogeneity and Fusion in Recommender Systems (HetRec)* (ACM, New York, 2011), pp. 41–48
8. S. Bull, J.E. Greer, G.I. McCalla, L. Kettel, J. Bowes, User modelling in I-help: what, why, when and how, in *Proceedings of the International Conference on User Modeling* (2001), pp. 117–126
9. J. Greer, G. McCalla, J. Collins, V. Kumar, P. Meagher, J. Vassileva, Supporting peer help and collaboration in distributed workplace environments. Int. J. Artif. Intell. Edu. **9**, 159–177 (1998)
10. C.-T. Li, Mentor-spotting: Recommending expert mentors to mentees for live trouble-shooting in codementor, in *Knowledge and Information Systems*, vol. 61 (Springer, Berlin, 2020), pp. 799–820
11. H. Labarthe, F. Bouchet, R. Bachelet, K. Yacef, Does a peer recommender foster students' engagement in MOOCs?, in *Proceedings of the International Conference in Educational Data Mining (EDM)* (2016), pp. 418–423
12. H. Labarthe, R. Bachelet, F. Bouchet, K. Yacef, Increasing MOOC completion rates through social interactions: A recommendation system, in *Proceedings of EMOOCS Fourth European MOOCS Stakeholders Summit, Graz, Austria*, (2016), pp. 471–480
13. F. Bouchet, H. Labarthe, K. Yacef, R. Bachelet, Comparing peer recommendation strategies in a MOOC, in *Proceedings of the 25th Conference on User Modeling, Adaptation and Personalization (UMAP)*, (2017), pp. 418–423
14. S. Prabhakar, G. Spanakis, O. Zaiane, Reciprocal recommender system for learners in massive open online courses (MOOCs), in *Proceedings of Advances in Web-Based Learning (ICWL)* (2017), pp. 157–167
15. J. Malinowski, T. Keim, O. Wendt, T. Weitzel, Matching people and jobs: A bilateral recommendation approach, in *Proceedings of the 39th Annual Hawaii International Conference on System Sciences* (2006)
16. W. Hong, L. Lei, T. Li, W. Pan, iHR: An online recruiting system for xiamen talent service center, in *Proceedings of the 19th ACM SIGKDD International Conference on Knowledge Discovery and Data Mining (KDD)* (2013), pp.1177–1185

17. L. Brozovsky, V. Petricek, Recommender system for online dating service, in *Proceedings of Znalosti Conference*, Ostrava (2007)
18. F. Diaz, D. Metzler, S. Amer-Yahia, Relevance and ranking in online dating systems, in *Proceedings of the 33rd international ACM Conference on Research and Development in Information Retrieval (SIGIR)* (2010)
19. B. McFee, G.R.G. Lanckriet, Metric learning to rank, in *Proceedings of the International Conference on Machine Learning (ICML)* (2010)
20. L. Pizzato, T. Rej, T. Chung, I. Koprinska, J. Kay, RECON: A reciprocal recommender for online dating, in *Proceedings of the ACM Conference on Recommender Systems (RecSys)*, Barcelona (2010)
21. J. Akehurst, I. Koprinska, K. Yacef, L. Pizzato, J. Kay, T. Rej, CCR: A content-collaborative reciprocal recommender for online dating, in *Proceedings of the 22nd International Joint Conference on Artificial Intelligence (IJCAI)*, Barcelona (2011)
22. Y.S. Kim, A. Mahidadia, P. Compton, X. Cai, M. Bain, A. Krzywicki, W. Wobcke, People recommendation based on aggregated bidirectional intentions in social network site, in *Proceedings of the Pacific Rim Knowledge Acquisition Workshop (PKAW)* (Springer, Berlin, 2010), pp. 247–260
23. X. Cai, M. Bain, A. Krzywicki, W. Wobcke, Y.S. Kim, P. Compton, A. Mahidadia, Collaborative filtering for people to people recommendation in social networks, in *Proceedings of the Australasian Joint Conference on Artificial Intelligence (AI)* (Springer, Berlin, 2010), pp. 476–485
24. S. Kutty, L. Chen, R. Nayak, A people-to-people recommendation system using tensor space models, in *Proceedings of the 27th Annual ACM Symposium on Applied Computing (SAC)* (2012)
25. P. Xia, B. Liu, Y. Sun, C. Chen, Reciprocal recommendation system for online dating, in *Proceedings of the IEEE/ACM International Conference on Advances in Social Networks Analysis and Mining (ASONAM)* (2015), pp. 234–241
26. P. Xia, B. Liu, Y. Sun, C. Chen, User recommendations in reciprocal and bipartite social networks-an online dating case study. IEEE Intell. Syst. **29**, 27–35 (2014)
27. S. Alsaleh, R. Nayak, Y. Xu, L. Chen, Improving matching process in social network using implicit and explicit user information, in *Proceedings of the 13th Asia-Pacific Conference on Web Technologies and Applications (APWeb)* (2011), pp. 313–320
28. S. Kutty, L. Chen, R. Nayak, A people-to-people recommendation system using tensor space model, in *Proceedings of the 27th Annual ACM Symposium on Applied Computing* (2012), pp. 187–192
29. S. Kutty, R. Nayak, L. Chen, A people-to-people matching system using graph mining techniques. World Wide Web **17**, 311–349 (2014)
30. A. Alanazi, M. Bain, A people-to-people content-based reciprocal recommender using hidden Markov models, in *Proceedings of the 7th ACM Conference on Recommender Systems, (RecSys)* (2013)
31. A. Alanazi, M. Bain, A scalable people-to-people hybrid reciprocal recommender using hidden Markov models, in *Proceedings of the Second Workshop on Machine Learning Methods for Recommender Systems, in conjunction with SIAM International Conference on Data Mining* (2016)
32. F. Vitale, N. Parotsidis, C. Gentile, Online reciprocal recommendation with theoretical performance guarantee, in *Proceedings of at the 32nd Conference on Neural Information Processing Systems (NeurIPS)* (2018)
33. A. Kleinerman, A. Rosenfeld, F. Ricci, S. Kraus, Optimally balancing receiver and recommended users' importance in reciprocal recommender systems, in *Proceedings of the 12th ACM Conference on Recommender Systems (RecSys)* (2018), pp. 131–139
34. A. Kleinerman, A. Rosenfeld, F. Ricci, S. Kraus, Supporting users in finding successful matches in reciprocal recommender systems, in *User Modeling and User-Adapted Interaction* (Springer, Berlin, 2020)

35. J. Neve, I. Palomares, Latent factor models and aggregation operations for collaborative filtering in reciprocal recommender systems, in *Proceedings of the 13th ACM Conference on Recommender Systems (RecSys)* (2019), pp. 219–227

36. J. Neve, I. Palomares, Aggregation strategies in user-to-user reciprocal recommender systems, in *Proceedings of the International Conference on Systems, Man and Cybernetics (SMC)* (2019), pp. 4031–4036

37. R. Ramanathan, N.K. Shinada, S.K. Palanianppan, Building a reciprocal recommenderation system at scale from scratch: Learning from one of Japan's prominent dating applications, in *Proceedings of the 14th ACM Conference on Recommender Systems (RecSys)* (2020), pp. 566–567

38. J. Neve, R. McConville, ImRec: Learning reciprocal preferences using images, in *Proceedings of the 14th ACM Conference on Recommender Systems (RecSys)* (2020), pp. 170–179

39. Statista (Online dating in the United States - Statistics and Facts), 2019, https://www.statista.com/topics/2158/online-dating/

40. M.C. Rosenfeld, R.J. Thomas, S. Hausen, Disintermediating your friends: How online dating in the united states displaces other ways of meeting. Proc. Nat. Acad. Sci. **116**(36), 17753–17758 (2019)

41. J. Akehurst, I. Koprinska, K. Yacef, L. Pizzato, J. Kay, T. Rej, Explicit and implicit user preferences in online dating, in ed. by L. Cao, J. Huang, J. Bailey, Y. Koh, J. Luo, *New Frontiers in Applied Data Mining*. Springer Lecture Notes in Computer Science, vol. 7104 (Springer, Berlin, 2012), pp. 15–27

42. L. Pizzato, T. Chung, T. Rej, I. Koprinska, K. Yacef, J. Kay, Learning user preferences in online dating, in *Proceedings of the Preference Learning (PL-10) Workshop, European Conference on Machine Learning and Principles and Practice of Knowledge Discovery in Databases (ECML PKDD)* (2010)

43. X. Amatriain, A. Jaimes, N. Oliver, J.M. Pujol, Data mining methods for recommender systems, in ed. by F. Ricci, L. Rokach, B. Shapira, *Recommender Systems Handbook*, 2nd edn. (Springer, Berlin, 2015), pp 39–72

44. R. Kohavi, Scaling up the accuracy of Naive-Bayes classifiers: A decision-tree hybrid, in *Proceedings of the International Conference on Knowledge Discovery in Databases (KDD)* (1996), pp. 202–207

45. L. Pizzato, J. Akehurst, C. Silvestrini, K. Yacef, I. Koprinska, J. Kay, The effect of suspicious profiles on people recommenders, in *Proceedings of the 20th Conference on User Modeling, Adaptation, and Personalization (UMAP)*. Lecture Notes in Computer Science, vol. 7379 (Springer, Berlin, 2012), pp. 225–236

46. G. Beliakov, T. Calvo, S. James, Aggregation of preferences in recommender systems, in ed. by F. Ricci, L. Rokach, B. Shapira, *Recommender Systems Handbook*, 2nd edn. (Springer, Berlin, 2015), pp 705–734

47. P. Victor, M. De Cock, C. Cornelis, Trust and recommendations, in ed. by F. Ricci, L. Rokach, B. Shapira, *Recommender Systems Handbook*, 2nd edn. (Springer, Berlin, 2015), pp 645–676

48. S. Berkovsky, R. Taib, I. Koprinska, E. Wang, Y. Zeng, J. Li, S. Kleitman, Detecting personality traits using eye-tracking data, in *Proceedings of the 2019 CHI Conference on Human Factors in Computing Systems (CHI)* (2019), pp. 1–12

49. H. Abdollahpouri, G. Adomavicius, R. Burke, I. Guy, D. Jannach, T. Kamishima, J. Krasnodebski, L. Pizzato, Multistakeholder recommendation: survey and research directions. User Model. User-Adap. Interac. **30**, 127–158 (2020)

50. A. Kleinerman, A. Rosenfeld, S. Kraus, Providing explanations foor recomemndations in reciprocal environments, in *Proceedings of the 12th ACM Conference on Recommender Systems (RecSys)* (2018), pp. 22–30

# Natural Language Processing for Recommender Systems

Oren Sar Shalom, Haggai Roitman, and Pigi Kouki

## 1 Introduction

Recommender systems process all types of signals at their disposal in order to suggest the most relevant items to users. These signals can take diverse forms, from categorical to numerical values, from tabular data to unstructured data. Arguably, the most significant form of content information is textual data, which may hold detailed and invaluable information. Examples include user-generated reviews, which elaborate on experiences of users with items, and textual description on items which can detail an assortment of relevant properties. While the information residing within the textual signals is tremendous, we need to employ advanced techniques in order to extract meaningful insights from it. This is where Natural Language Processing (NLP) comes in as a useful tool to this goal. NLP [18] is a scientific field focusing on automatically processing and analyzing textual data, thus it can be of great help for recommender systems. In this chapter, we break down the various types of inputs that a recommender system can take and identify cases where NLP capabilities can potentially assist each input type. Then, for each such case, we present methods to incorporate the textual input into the recommender system, analyze it and indicate its relative advantages and limitations.

Processing efforts of recommender systems can be coarsely divided into an offline modeling phase and an online recommendation phase. The former aims to model users' preferences and items' traits while the latter considers the current

O. S. Shalom (✉)
Facebook, Tel Aviv-Yafo, Israel

H. Roitman
Buyer Experience Research Department, eBay Research, Netanya, Israel

P. Kouki
Data Science Department, Relational AI, Berkeley, CA, USA

© Springer Science+Business Media, LLC, part of Springer Nature 2022
F. Ricci et al. (eds.), *Recommender Systems Handbook*,
https://doi.org/10.1007/978-1-0716-2197-4_12

state and ephemeral needs of the user. The output contains the recommendations themselves, possibly alongside with some additional means to persuade the user. We now list the possible inputs and outputs while focusing on text:

1. Offline input: collaborative filtering algorithms collect past usage patterns in order to model users and items. This data may include *textual reviews*, which detail the multifaceted experiences of users with items.

   Additionally, content-based filtering utilizes side-information on items, where such information may include textual description of the items. This topic is covered in Chap. 7 and we refer the interested reader to that chapter.
2. Online input: *conversational recommenders* allow the users to detail their current needs in free-text.
3. Output: *explanations* can significantly increase the effectiveness of recommenders, and text generation is one of the prominent techniques to achieve this goal.

This chapter is organized as follows. In Sect. 2 we cover the most prominent approaches for review-based recommenders. Section 3 highlights conversational recommenders. In Sect. 4 we detail on methods to generate explanations.

> This chapter discusses only scenarios pertaining to *applications* of NLP for recommender systems. We acknowledge there are *techniques* that were originally designed for NLP but can be useful for recommenders as well, however they are out of scope for this chapter. For instance, Transformers [72] were designated to process tokens and afterwards were adapted to recommenders to process user sessions.
>
> Some recommenders allow users to write textual feedback on the *recommendations* rather than on items. That is, the users may specify the relevancy level of the suggested recommendations. However, this is a rare scenario (a notable example is [24]) and, therefore, is out of scope for this chapter.

## 2   User Generated Reviews

User-item experiences can be complex and multifaceted, where the user may have a different opinion on various aspects of the item. As an example from the movies domain, a user may not like the special effects because they are either too bombastic or too moderate; likewise, the user may have justified opinions on the acting, directing, plot, etc. To better model users and items, recommenders should capture the entire users' impressions of items. Namely, it is important to infer not only to what extent a user likes an item, but also *why* and *which* factors led that impression. User ratings, the most widely used type of explicit feedback, inevitably lose valuable data because they have to summarize that rich experience into a single scalar.

Compared to that, textual reviews are feasibly the most elaborate type of feedback and they allow to fully describe the multifaceted experience of users with items. Concretely, a textual review may unveil any of the following.

1. User preferences (e.g., "I prefer drama movies")
2. Traits of the discussed item ("starred by Tom Hanks")
3. The matching between them ("amazing soundtrack")
4. The context of the interaction ("I watched this movie *with my friends*")

The unstructured form of textual reviews allows them to detail the multifaceted experiences of users with items. But at the same time, it is non-trivial to integrate this input source alongside with traditional collaborative filtering data.

Broadly speaking, as discussed in Chap. 15 there are two major tasks for recommender systems: ranking and rating prediction. In this chapter we review the most prominent approaches to incorporate textual reviews for these two tasks, and explain how the special characteristics of reviews are exploited to improve performance. Furthermore, since each review holds relatively rich information, then few interactions are required to model users and items, which helps to mitigate the cold start problem. Unless otherwise stated, the input to any of the described algorithms is a list of $(u, i, r_{ui}, t_{ui})$ quartets, each of which indicates an event where user $u$ interacted with item $i$, assigned it an explicit rating $r_{ui}$ and complemented it a textual review $t_{ui}$.

## 2.1 Affinity to Sentiment Analysis

Sentiment analysis is an NLP task aiming to automatically quantify the emotions or opinions expressed within textual data [1]. In its most basic form, it converges to a binary classification problem, where the predicted sentiment is either positive or negative. Occasionally, polarity precision is not sufficient and more fine-grained prediction is required. That is, the sentiment score may span multiple ordinal categories e.g., on a 1–5 Likert scale.

Fine-grained sentiment analysis resembles the rating prediction task of review-based recommenders, and we would like to stress the differences between the two tasks. While both tasks have access to the textual data during training time, they differ at inference time. Sentiment analysis predictions are made with respect to a given text; while a recommender predicts the rating a user would give to an item *before* they have experienced with it. That basically means that the textual review written by the user for the item is *not available* at prediction time. Hence, the rating prediction task is much harder as predictions are made solely based on past behavior, without an accompanying text.

## 2.2 Traditional Methods

In recent years, with the advent of deep learning, state-of-the-art methods for most NLP tasks apply deep learning (DL) techniques, and review-based recommenders are no exception. However, for completeness, and, to better understand the intuition behind DL approaches, we briefly survey some of the most influential traditional approaches.

Several early works employ topic modeling to represent users and items. McAuley et al. [47] define as a document the set of all reviews written on a certain item and apply LDA [4] on the corpus induced by the entire catalog. The topic distribution of each item is considered as its latent features, while the users' latent features are estimated by optimizing rating prediction with gradient descent. A drawback of this approach stems from the fact that the item vectors are unmindful to the ratings. This makes them sub-optimal and, as a result, the user vectors are sub-optimal as well. This drawback was remedied in a later work [2], which considers the ratings and the reviews simultaneously.

The Ratings Meet Reviews (RMR) algorithm [43] is a probabilistic generative model that combines a topic model with a rating model. It applies topic modeling on item review text in order to find the distribution of each item over the latent topics. A rating $r_{ui}$ given by user $u$ to item $i$ is assumed to be generated by a Gaussian mixture model: user $u$ is associated with a Gaussian per latent topic and the topic distribution of item $i$ serves as the mixture weights. However, this process neglects the expressed sentiments. For example, two items with the exact same topic distributions, but opposite sentiments will have the same predicted ratings for any user.

Diao et al. [20] proposed a probabilistic model based on collaborative filtering and topic modeling. It uncovers the relevant aspects in the given domain, finds the interest distribution of users and content distribution of item and finally infers per-aspect sentiment as expressed in the reviews. A limitation of this algorithm is its inability to incorporate explicit ratings, which omits a valuable signal.

Overall, the aforementioned approaches exhibit two shortcomings:

1. Overlook context. These approaches take a bag-of-words approach, and process individual tokens or n-grams while ignoring their order. In contrast, deep learning approaches are able to process each token with respect to its *context* tokens.
2. Operate solely based on lexical similarity. This is a pitfall because semantically similar reviews may have a low lexical overlap. On the other hand, deep learning approaches are based on distributed representations, which capture the *semantics* of the words.

## 2.3   Deep Learning: Preliminaries

Per the fundamental limitations of the traditional methods, it comes as no surprise that state-of-the-art review-based recommender are based on deep learning approaches. This subsection presents few definitions and notations required to understand these approaches. These include the definition of Text Processor and some remarks on machine learning techniques.

### 2.3.1  Text Processor

All notable work in recent years on recommender systems that leverage text (e.g., reviews) rely on a deep learning *Text Processor* unit. Given an input text, this unit represents it by a dense vector which captures the most meaningful aspects with respect to the task at hand.

There are various plausible architectures for the Text Processor, e.g., based on LSTM [25] or Transformers [72]. Most commonly used is the relatively simple Convolutional Neural Network (CNN) Text Processor [32]. This architecture first projects the texts to a latent space by using word embeddings. Then, it applies multiple filters of variable lengths over all sliding windows, each responding to different semantic meanings. Next, a max-pooling operation is performed, to obtain a fixed size vector and to make the algorithm position-invariant. Finally, a fully-connected layer transforms the output to the desired latent space. Jacovi et al. [30] and Kim [32] provide a detailed analysis for the user of convolutional neural networks using text in classification tasks.

We reiterate that different architectures (e.g., LSTM-based) can be used interchangeably.

### 2.3.2  Machine Learning Annotation

Fully Connected Layer: to ease the notation throughout this chapter, we denote a fully connected layer with output dimensionality of $k$ by $FC^k(\cdot)$. This layer refers to the classic neural network component, in which all $k'$ input neurons connect to all $k$ neurons in the output layer. The input neurons are represented by a matrix $X$ and the layer computes $\sigma(XW + b)$, where $W \in \mathbb{R}^{k' \times k}$ and $b \in \mathbb{R}^k$ are model parameters and $\sigma$ is a nonlinear activation function, defaulted to ReLU [52] in this chapter.

Optimization: all described algorithms use conventional techniques for improving the speed, stability and generalization of the models. This includes dropout layers [69], regularization and adaptive learning rate optimization algorithms like Adam [33]. Additionally, other techniques which were not applied in the original papers such as batch normalization [29], could be utilized to further improve performance. Since these optimization techniques are not the essence of the models, we omit them from the descriptions of the algorithms. Also, modern models iterate through the training data in mini-batches. To facilitate the notation, we simply describe the process for a single instance.

## 2.4  Review-Based Recommenders for Rating Prediction

We now proceed to the descriptions of landmarks review-based recommenders. We iteratively explain each of these methods, explain how they work, analyze their strengths but also indicate possible weaknesses (which motivates continuous improvements).

### 2.4.1 *DeepCoNN*

The seminal algorithm *DeepCoNN* [84] was the first to apply deep learning capabilities in order to incorporate textual reviews for recommenders, and it has inspired a great amount of follow-up work. The basic intuition behind this algorithm is that generated textual reviews encapsulate essential information about both users and items. Therefore, two instances of text processors $\Gamma_U$ and $\Gamma_I$ are learned to extract meaningful information on the users and the items, respectively. Specifically, to represent user $u$, all reviews written by that user are concatenated to form text $t_u$, which holds all necessary information to model $u$. Then, the text is fed to $\Gamma_U$, which identifies the user preferences as expressed in their textual reviews and generates $p_u$, the distributed (vectorial) representation of that user. The representation $q_i$ of item $i$ is done similarly by concatenating all reviews written on that item and applying the item text processor:

$$p_u = \Gamma_U(t_u) \qquad p_i = \Gamma_I(t_i)$$

Given the user and item vectors $p_u$ and $q_i$, most CF algorithms predict their associated rating using a dot product (plus possibly the global mean and the corresponding user and item bias terms). *DeepCoNN* takes a different approach to combine these vectors. It concatenates them as $z = p_u \parallel q_i$ and predicts the rating using a Factorization Machine (FM) [61]. FM can model all the second-order interactions in a given vector in linear time, which may add some expressiveness power to simple inner product. Formally, the predicted rating is given by:

$$\hat{r} = FM(z) = \mu + \sum_{j=1}^{|z|} w_j z_j + \sum_{j=1}^{|z|-1} \sum_{k=j+1}^{|z|} \langle v_j, v_k \rangle \, z_j z_k \qquad (1)$$

where $\mu$ is the global computed average rating, $w$ models the strength of each component in $z$ and $v_i$ is a learned vector of some predefined size, corresponding to each component $z_i$, which models the second order interactions in $z$. Training is done end-to-end, where the objective is to minimize the absolute prediction error. Evaluation on several benchmark datasets asserted the viability of user-generated reviews to improve rating prediction of future interactions, as it outperformed all previous methods to incorporate reviews.

### 2.4.2 *TransNets*

If at test time, the target textual review $t_{ui}$ written by user $u$ for item $i$ was available to the model, it could have improved accuracy (see Sect. 2.1). Although textual reviews are not available at test time, they are still available at training time and *TransNets* [8] enhances *DeepCoNN* by harnessing them.

To this end, another subnetwork is trained, dubbed as the Target Network, which aims to reconstruct $t_{ui}$ from $p_u$ and $q_i$. This empowers the recommender since it allows to model users and items such that their representations can reconstruct the target reviews, simulating a situation where the target review is available at test time.

Since the semantics of the target review is of key importance rather than the actual raw text, at training time this subnetwork *transforms* the target review using another instance of a Text Processor and its representation $z_T$ is to be reconstructed. First, the representation of the target network $z_T$ has to capture the meaning of the text reviews. This is achieved by generating representations with predictive capabilities of the ratings. Formally:

$$z_T = \Gamma_T(t_{ui})$$
$$r_T = FM_T(z_T)$$

In view of the fact that $Z_T$ should guide $p_u$ and $q_i$ and not vice versa, training is carefully done. Per each training instance, the parameters of the model are divided into 3 groups, and they are updated sequentially:

1. The parameters of the target network, which comprises $\Gamma_T$ and $FM_T$, are updated by minimizing the rating prediction error: $| r_T - r |$.
2. The parameters of the target network remain fixed and $\Gamma_U$ and $\Gamma_I$ are learned by minimizing the loss: $(z - z_T)^2$.
3. The parameters of $FM$ are updated by minimizing $| \hat{r} - r |$.

We emphasize that these sequential sub-steps occur every instance, rather than training to convergence each subnetwork.

The Target Network functions only to improve the training process and does not participate at inference time, as the target review is not available at this point. Hence, predictions are computed exactly as in *DeepCoNN*. *TransNets* adds another optimization technique: it passes $z$ through stacked fully-connected layers, to allow modeling of more complex interactions between the user and the item.

### 2.4.3 Extended TransNets

Thus far the discussed models ignore the *identities* of the users and the items and representations are based exclusively on the review texts. However, the identities may supply usable information. After all, pure collaborative filtering methods rely solely on this signal. For this reason, the Extended TransNets (*TransNet-Ext*) model introduces embedding matrices for users and items, $\Omega_U$ and $\Omega_I$, respectively. The latent representation of a user-item pair is supplemented with the concatenation of the user and item embeddings:

$$\omega_u = \Omega_U(u) \qquad \omega_i = \Omega_I(i)$$
$$z_{ext} = \omega_u \parallel \omega_i \parallel z$$

Now $z_{ext}$ is used to predict the rating as in Eq. 1. The embedding matrices do not stand on their own, but rather they are complementary to the textual representations. That is, vector $p_u$ represents the preferences of user $u$ as captured by the text processor. However, this vector does not perfectly reflect the true preferences of the user, whether because of limitation of the text processor or simply because the textual reviews do not contain all required information to accurately analyze user $u$. The role of vector $\omega_u$ is to supplement the preferences induced by the texts, very much like the deviation matrix presented in [26]. A similar analogy applies also for the item embedding matrix $\Omega_I$.

## 2.5 State-of-the-Art of Review-Based Recommendations

### 2.5.1 Motivation

A common drawback to the aforementioned work and more recent work CF [9, 11, 12, 64, 73] is the required effort to process a single review during training. The feed-forward operation of a review given by user $u$ to item $i$ includes all reviews written by $u$ and the reviews written on $i$. Let $w$ be the average number of words per review and $l_I$ and $l_U$ be the average number of reviews associated with items and users, respectively. Hence the running time complexity to process a single review is $O(w \cdot (l_I + l_U))$.

If a typical user writes tens of reviews, an item could appear in thousands of reviews and a review comprises more than one hundred tokens, then processing a *single* review requires hundreds of thousands operations. This scalability issue might pose an insurmountable problem for real-life recommenders, as training on large datasets is infeasible and therefore a solution to reduce training time is required.

As we detail in this section, the Matching Distribution by Reviews (*MDR*) [65] takes a different perspective on reviews, which both eliminates the running time issue and improves accuracy.

### 2.5.2 Intuition

Previous algorithms assume that reviews describe item characteristics and user preferences and therefore apply a dedicated text processor for each of these two entity types. Comparing to that, *MDR* claims that reviews concentrate on the *matching* between them, i.e., explain the extent at which the user liked various aspects in the item. This observation is very beneficial since in some cases it is possible to infer the matching between the user and the item regarding a specific trait, even though the individual user preferences or item traits are not disclosed in the review. Consider for example the following review snippet from the movie domain: "I loved the soundtrack". It is clear the soundtrack of the movie fits the preferences of the user, although no information is given on the soundtrack itself, like type or duration.

In each domain there could be an enormous amount of factors relevant to generating recommendations. In our running example from the movie domain, such factors include genre, shooting location, plot complexity, animation style, etc. A single review is unlikely to supply adequate information on all relevant factors. For example, by reading a review, it may be apparent to what extent a user likes the genre but less clear whether they enjoyed the animation style. *MDR* also models the inevitable uncertainty in processing individual reviews. This is done by predicting a *distribution* rather than just a point estimate, which does not allow to encode uncertainty.

As motivated above, both textual reviews and explicit ratings refer to the matching level between the user and the item, but they differ in the level of thoroughness. A textual review is an elaborate form of feedback, as it reasons and details the factors that made the user form their judgment on the interacted item. Therefore, it is only natural to view the reviews as augmented *labels* to train the collaborative filtering algorithm, in addition to explicit ratings.

### 2.5.3 Algorithm Overview

The algorithm consists of two phases: the first, learns how to model the textual reviews and, the second, utilizes the modeled reviews as augmented labels. These phases are learned sequentially, i.e., after the parameters of the first phase are learned, they are held fixed and then the second phase's parameters are learned.

This scheme alleviates the running time issues raised in Sect. 2.5.1, as the whole algorithm has linear time complexity. The first phase is linear in the number of tokens and the second phase is linear in the number of reviews.

In this chapter we detail the steps of the algorithm and explain the intuition behind them. We do not, however, give the theoretical justifications for them. The curious reader can see the proofs in the original paper [65].

### 2.5.4 Phase I: Distribution of Matching Vectors

As motivated above, a textual review can bring to light the matching between a user and an item across the latent features. The purpose of this phase is to automatically infer this information. To avoid clutter, subscripts $u$ and $i$ are omitted when their existence is clear from the context. To this end, *MDR* passes each review $t$ through a text processor and obtains a matching vector $m$. Each component $f$ in the matching vector reflects the matching between the user and the item with regards to the $f$-th latent feature. Therefore the sum of the components in the matching vector predicts the overall satisfaction of the user with the item, which is also reflected by the explicit rating given by the user to the item. That is $\hat{r} = \sum_f m^f$, where $m^f$ stands for the $f$-th component in vector $m$.

The problem with this approach is it assumes all latent features can be adequately estimated by any review. However, a specific review $t$ may be ambiguous or simply does not cover all latent features. To account for this, *MDR* finds the *distribution* of the matching vectors. The distribution of choice is multivariate normal distribution $\mathcal{N}(\mu(t), \Sigma(t))$ with a diagonal covariance matrix, which can be parameterized by two vectors: mean $\mu(t)$ and variance $\Sigma(t)$. When the model has high confidence in the predicted matching value of a specfic factor, then the associated variance of this factor would be low, and vice versa. Given text $t$, the distribution of matching vectors is simply inferred by computing $e$, the output of a text processor on the review. Then two different fully connected layers are applied, to yield the parameters of the mean and variance $\mu(t)$ and $\Sigma(t)$.

$$e = \Gamma(t) \tag{2}$$

$$\mu(t) = FC(e) \qquad\qquad \Sigma(t) = FC(e)$$

In stochastic gradient descent, a single training example represents the whole distribution of the training data. By the same token, a single sampled vector $z \sim \mathcal{N}(\mu(t), \Sigma(t))$ can represent the whole distribution of matching vectors. Sampling is a non-differentiable operation, thereby it would eliminate all gradients and hamper the process of back-propagation. Therefore, the reparameterization trick [34] is exerted. This means that the sampling operation is done on a newly introduced input layer, that samples a vector $\epsilon \sim \mathcal{N}(0, I)$. Then, the appropriate vector $z$ is mapped by scaling and shifting $\mathcal{N}(0, I)$ to align with $\mathcal{N}(\mu(t), \Sigma(t))$. Finally, the matching vector $m$ is obtained by feeding $z$ to a fully connected layer.

$$\epsilon \sim \mathcal{N}(0, I) \qquad\qquad z = \mu(t) + \Sigma^{\frac{1}{2}}(t) \odot \epsilon$$

$$m = FC(z) \tag{3}$$

The predicted rating is the sum of the elements in the matching vector where conventionally, also bias terms and global mean are added: $\hat{r}_{ui} = \sum_f m_{ui}^f + b_i + b_u + \mu$. The loss function asks to minimize mean squared error.

### 2.5.5   Phase II: Collaborative Filtering with Augmented Labels

This phase applies a rating prediction algorithm, while benefiting from augmented labels that were derived in the previous phase.

Latent factor models represent users and items by some vectors of fixed size. Each component $f$ in this space refers to a latent feature and the components in a user vector measure to what extent the user *prefers* these latent features.

Comparably, components in an item vector measure the extent to which the item *holds* these latent features.

While any CF algorithm could be used in this phase, the authors of *MDR* chose the simple *SVD* [58] as the underlying engine. This algorithm asks to minimize the squared error: $(r_{ui} - \hat{r}_{ui})^2$, where the predicted rating is computed as: $\hat{r}_{ui} = p_u \cdot q_i + b_u + b_i + \mu$. The inner product between user vector $p_u$ and item vector $q_i$ indicates the affinity between them. Inner product equals to the sum over the components in the element-wise product: $p_u \cdot q_i = \sum_f (p_u \odot q_i)^f$. As such, the element-wise product indicates the matching between the user and the item over the latent features. We dub the vector $p_u \odot q_i$ as the *collaborative matching vector*.

Using phase I of *MDR*, each review $t_{ui}$ is represented by a distribution of matching vector. The mean of this distribution, dubbed as the *textual matching vector* represents the matching between the user and the item over the latent features and therefore can be utilized as an augmented label for the collaborative filtering algorithm. This is done by minimizing the distance between the collaborative matching vectors and the textual matching vectors. All in all, the loss function is defined as:

$$\mathcal{L}_{ui} = (\hat{r}_{ui} - r_{ui})^2 + \alpha \cdot \| p_u \odot q_i - m_{ui} \|_2^2 \qquad (4)$$

Where hyperparameter $\alpha$ controls the relative importance of terms in the loss function. This loss function contains two labels: explicit rating and textual matching vector. While the former gives a coarse direction to the optimization process, the latter provides a direction per each of the $f$ factors, which can significantly improve performance. Indeed evaluation on several benchmark dataset proves the superiority of this approach over all previous baselines.

## 2.6  Empirical Evaluation

This section compares the performance of the strongest review-based models on several benchmark datasets. The first dataset is `Yelp17`, which contains restaurant reviews, introduced in the Yelp Challenge.[1] Each of the other three datasets is a different domain taken from the latest release of Amazon reviews[2] [48], which contains product reviews from Amazon website. These datasets vary in size and sparsity, where complete statistics can be found at [65].

Table 1 summarizes the empirical results. It shows the viability of incorporating textual reviews, as all review-based algorithms improve SVD, which does not incorporate reviews. Furthermore, *MDR* outperforms all other baselines, in addition to improved running time complexity.

---

[1] https://www.yelp.com/dataset-challenge.

[2] http://jmcauley.ucsd.edu/data/amazon.

**Table 1**  MSE comparison with baselines. Best results are indicated in bold

|              | SVD    | *DeepCoNN* | TransNetTransNet | TransNetTransNet-Ext | TransNetMDR | Improvement |
|--------------|--------|------------|------------------|----------------------|-------------|-------------|
| Yelp         | 1.8661 | 1.7045     | 1.6387           | 1.5913               | **1.4257**  | 10.4%       |
| A-Electronics| 1.8898 | 2.0774     | 1.8380           | 1.7781               | **1.5329**  | 13.8%       |
| A-Clothes    | 1.5212 | 1.7044     | 1.4487           | 1.4780               | **1.2837**  | 13.1%       |
| A-Movies     | 1.4324 | 1.5276     | 1.3599           | 1.2691               | **1.1782**  | 7.2%        |

## 2.7  Review-Based Recommenders for Ranking

In this subsection we explain how to incorporate user generated reviews for the ranking problem. Naturally, to achieve optimal performance on this problem, the training procedure should be optimized directly for *ranking*. However, all hitherto mentioned work optimize *rating prediction* while neglecting the ranking problem.

As a side note, we would like to stress that any existing method for the rating prediction problem can be adjusted for ranking. This can be simply done by adding a ranking loss (e.g., the pairwise BPR loss [62]) as an additional term of the loss function, or by more advanced techniques like [26, 67]. However, this direction has not been significantly investigated in the context of review-based recommenders and is therefore omitted from this chapter.

Recently Chuang et al. [15] suggested the Text-aware Preference Ranking (TPR). This method simultaneously optimizes two main objectives: user-item ranking and item-word ranking. The former aims to rank items in the history of the user higher than missing items. The latter seeks to rank words that appear in an item's review higher than other words. However, this method takes a bag-of-words approach, which may not capture the semantics of the entire review, and is not clear how more elaborate methods can be integrated instead. Furthermore, TPR cannot incorporate explicit ratings. Since usually textual reviews are accompanied with explicit ratings, leaving this signal out might lead to sub-optimal performance.

We now turn to the description of the main algorithm of this subsection. Zhang et al. [81] presented a Joint Representation Learning (JRL) framework to incorporate heterogeneous input signals. In the context of this chapter, the considered input sources are reviews (denoted as $V_1$) and ratings ($V_2$).

The algorithm is learned end-to-end and comprises the following three steps:

1. Create input-source specific representations for users and items.
2. Integrate representations to have a single representation.
3. Apply ranking optimization on the integrated representations.

We now elaborate on each of the aforementioned steps.

### 2.7.1 Input Source Modeling

In each input source $k$ (either reviews or ratings), any user $u$ and item $i$ is represented by a dedicated vector $p_u^k$ and $q_i^k$, respectively. These vectors aim to optimize loss function $\mathcal{L}_k$.

#### Modeling of Textual Reviews

Each review $t_{ui}$ is represented in an unsupervised manner by vector $\boldsymbol{t}_{ui}$ (denoted in bold).

The authors of JRL adopted the PV-DBOW model [41]. This is a generative model, where review's representation should maximize the likelihood of its comprising words. To this end, each word $w$ in the vocabulary is represented by a learned vector $\boldsymbol{w}$. Then, the log probability of having word $w$ in a review is approximated by the negative sampling (NEG) procedure: $\log P(w|t_{ui}) = \log \sigma(\boldsymbol{w}^T \boldsymbol{t}_{ui}) + t \cdot \mathbb{E}_{w_N \sim P_V}[\log \sigma(-\boldsymbol{w}_N^T \boldsymbol{t}_{ui})]$, where $t$ is the number of negative samples and $P_V$ is the noise distribution. In this paper $t$ was set to 5 and $P_V$ is the unigram distribution raised to the 3/4rd power. Hence, the objective for a given review is:

$$\mathcal{L}_1(u, i) = \sum_{w \in t_{ui}} f_{w,t_{ui}} \log P(w|t_{ui}) \tag{5}$$

where $f_{w,t_{ui}}$ counts the number of occurrences of $w$ in the review.

Then user $u$ is represented by an average of their reviews: $p_u^1 = \dfrac{1}{|R_u^1|} \sum_{i \in R_u^1} \boldsymbol{t}_{ui}$, where $R_u^1$ is the set of items reviewed by user $u$. In a similar fashion item vectors are computed.

#### Modeling of Numerical Ratings

Modeling users and items according their explicit ratings is rather straightforward. The predicted rating given by user $u$ to item $i$ is calculated by a two-layer fully connected network: $\hat{r}_{ui} = FC^1(FC^t(p_u^2 \odot q_i^2))$, with $\odot$ denoting the element-wise product, ELU [17] as the activation function and $t$ is a hyperparameter that determines the dimensionality of the hidden layer. The objective in this input source is to minimize the squared prediction error:

$$\mathcal{L}_2(u, i) = (\hat{r}_{ui} - r_{ui})^2 \tag{6}$$

### 2.7.2   Integrated Representation

The integrated representation of each user and item is obtained by a simple concatenation of the representations across the input sources: $p_u = p_u^1 \parallel p_u^2$ and $q_i = q_i^1 \parallel q_i^2$. We address to a limitation of this approach in Sect. 2.7.4.

### 2.7.3   Ranking Optimization

Until this point, the embeddings are not optimized for ranking, but to model textual data ($\mathcal{L}_1$) and to predict ratings ($\mathcal{L}_2$). Therefore, JRL incorporates the widely adopted BPR loss [62] as the pairwise learning-to-rank method. For each observed user-item interaction $(u, i^+) \in R$, a random item $i^-$ such that $(u, i^-) \notin R$ is sampled as a negative sample. The objective is to distinguish between positive and negative pairs:

$$\mathcal{L}_{RANK}(u, i) = \log \sigma (p_u \cdot q_{i+} - p_u \cdot q_{i-}) \tag{7}$$

For more details on the BPR loss, we refer the reader to Chap. 3. Personalized recommendations are generated by ordering all items according to their predicted scores. Combining the input-source dependant loss functions in Eqs. 5 and 6 together with the ranking objective presented in Eq. 7, gives us the objective function of JRL:

$$\mathcal{L} = \sum_{(u,i) \in R} \mathcal{L}_{RANK}(u, i) + \mathcal{L}_1(u, i) + \mathcal{L}_2(u, i) \tag{8}$$

### 2.7.4   In-Depth Analysis

Decoupling Input Sources

Albeit not mentioned in the original paper, while the algorithm assumes the existence of several modalities, they do not need to be aligned. For instance, a user may leave explicit ratings for some items, to write textual reviews on other items and the final user representation will consider all of these interactions. This is in contrast to all previously described algorithms in this chapter, that require each interaction to comprise both explicit ratings and textual reviews.

Fusion of Representations

The personalized score is computed as: $s = p_u \cdot q_i = (p_u^1 \parallel p_u^2) \cdot (q_i^1 \parallel q_i^2) = p_u^1 \cdot q_i^1 + p_u^2 \cdot q_i^2$. Hence, it is the sum of the scores given by two pairs of representations, and at prediction time there is no interaction between these two representations. Perhaps the model could be improved if the integrated representation was done by feeding the input-source representations to fully connected layers, which allow to capture non-linear correlations between the representations.

### Generalized Modeling of Textual Reviews

Modeling of textual reviews is done by PV-DBOW, which does not obtain state-of-the-art results on document representation. Fortunately, JRL is agnostic to the choice of text modeling, and any other unsupervised could be seamlessly used. For instance, reviews can be modeled using AutoEncoders [42] or contextual embedding models (e.g., GPT-3 [6]).

### 2.7.5  Empirical Evaluation

We now compare relevant baselines to asses the vitality of JRL. All experiments were conducted on various domains of the Amazon review dataset presented in Sect. 2.6. For each dataset, 70% of the interactions of each user were randomly selected for training and the remaining were kept for test. All users are presented with $K = 10$ recommendations and a recommended item is marked as "correct" if it resides within the test items of that user. We report two prevalent top-K evaluation measures:

- HT: Hit-ratio, which is the percentage of users with at least a single correct recommendations.
- NDCG: considers also the position of the correct recommendations.

Additional details on these evaluation measures can be found in Chap. 15. The first baseline is BPR [62], a seminal ranking algorithm that does not leverage reviews. Another baseline, which was presented in Sect. 2.4.1, is *DeepCoNN* [84]. This baseline does incorporate reviews, but is optimized for the rating prediction task.

Table 2 shows the performance of each algorithm. First, it gives another evidence for the contribution of review to recommender systems, as *DeepCoNN*

**Table 2** Ranking comparison. Best results are indicated in bold

|  | BPR | | DeepCoNN | | JRL | |
|---|---|---|---|---|---|---|
| Measure | HT | NDCG | HT | NDCG | HT | NDCG |
| A-Movies | 4.421% | 1.267% | 10.522% | 3.800% | **13.245%** | **4.334%** |
| A-CDs | 8.554% | 2.009% | 13.857% | 4.218% | **16.774%** | **5.378%** |
| A-Clothes | 1.767% | 0.601% | 3.286% | 1.310% | **4.634%** | **1.735%** |
| A-Cellular | 5.273% | 1.998% | 9.913% | 3.636% | **10.940%** | **4.364%** |
| A-Beauty | 8.241% | 2.753% | 9.807% | 3.359% | **12.776%** | **4.396%** |

consistently outperforms BPR across all datasets. Most notably is the benefit of having a designated algorithm for ranking, as JRL gains a substantial improvement comparing to other algorithms.

## 2.8 Discussion and Future Outlook

This section showed the tremendous value of user-generated reviews in modeling users and items for both rating prediction and ranking. Over the past years, the recommender systems community has unceasingly improved the means to incorporate textual reviews, which results in increasing performance and reduction in running time complexity. Currently, reviews are an integral part of the modeling process of state-of-the-art production systems and we anticipate a plethora of work in this direction, towards more accurate recommenders.

We should note that most existing review-based algorithms assume the availability of explicit ratings. At first, it might seem like a hard assumption that limits applicability. However, platforms that collect textual user reviews usually also collect explicit numeric ratings, and hence we are not restricted by the additional requirement of having explicit ratings.

So far, existing work in this field relies on datasets such that *all* interactions are associated with reviews. However, this leads to a *selection bias* because only a subset of the users tend to write reviews, while the rest of the population is ignored. These users may differ in their tastes and usage patterns from the general population and therefore the system may suffer from poor performance on the general population. Furthermore, a user may be more likely to devote time for writing a review if the experience with the item was extremely positive or negative. This is another form of a potential selection bias, where extreme cases are overemphasized. Designers of real-life recommenders should consider these aspects and integrate also data that represents the general population.

Looking to the future, with the great predictive power of reviews, comes a potential threat by malevolent stakeholders. As textual reviews add invaluable information on top of explicit ratings, an attack that generates reviews might be considerably striking. Hence, there is an inherent future need for designing systems that can effectively detect adversarial reviews.

Another line of research could be to utilize reviews in order to break down the items and to identify fine-grained recommendable units. Bauman et al. [3] for example suggested a bag-of-words approach to identify aspects of consumption. Then for example, beyond recommending a restaurant the system could suggest specific dishes. This is an interesting problem that has not been widely studied yet, with a variety of potential applications (e.g., suggest a hotel as well as a room with a specific view or to recommend a specific chapter in a handbook).

# 3   Conversational Preference Elicitation

The main purpose of the preference elicitation reference elicitation process is to allow end-users to reveal their preferences by interacting with the recommender system. Such interaction may be implemented through a conversational interface using natural language.

While the preference elicitation task may be viewed as a kind of *intent detection* task, there are two fundamental differences between the two. First, intent detection may be performed also in ad-hoc systems that have no interaction with the user, e.g., predicting query intent [5]. Second, in the context of conversational systems, intent detection is usually treated as a classification task, which is bounded by a predefined set of intents (classes) [44]. This in comparison to the preference elicitation task which may require to reason over a large set of possible user preferences.

Overall, we describe three main types of preference elicitation approaches that utilize NLP methods: *critiquing*, *facet-based* and *question-based*. We start by shortly discussing conversational recommender systems as the environment in which such preference elicitation process is implemented.

## 3.1   Conversational Recommender Systems

Conversation recommender systems (CRS) have become quite common nowadays and are implemented by natural language interfaces (e.g., chat-bots) or speech (e.g., intelligent assistants). Existing conversational systems can be roughly classified as *chit-chat*, *informational* or *task-oriented*. In a sense, a conversation recommender system is both informational and task-oriented. For example, a CRS may be utilized for recommending news to a user, hence satisfying some information need. On the other hand, a CRS may be utilized to fulfill some user goal such as reserving a restaurant, visiting a location or purchasing some item.

Utilizing a CRS is specially effective whenever user preferences are either ill-defined or unclear (e.g., in a cold-start scenario). In a conversational setting, users may express their preferences towards items using a natural language interface. At the same time, the recommender system can further utilize the same natural language interface to interact with its end-users to clarify their needs and provide better personalization.

Different from "traditional" recommender systems, which are usually passive in the sense that they operate in one-shot interaction paradigm, a CRS allows an *interactive*, *mixed-initiative*, dialogue with its end-users. This in turn, allows for varying the recommendation until enough evidence on user's preferences is gathered. Alternatively, a CRS may adapt its previous recommendations to the change in user's preferences as reflected in the ongoing conversation.

A typical CRS has two main components: *Preference-Elicitation* and *Recommendation*. The role of the preference elicitation component is to gather enough information about user preferences for maximizing the predicted user utility. The recommendation component is responsible to provide recommendations based on preference information that was actively curated during the conversation with the user. This component may be also tightly integrated with the preference elicitation component, driving the latter's decisions on which user preferences should be clarified (e.g., present options for item filtering, ask for feedback, etc).

In the rest of this section, we mainly focus on the preference elicitation process using natural language understanding and the representation of derived user preferences for recommendation during a conversational recommendation setting. For broader details on the general scope of conversational recommender systems, the interested reader is referred to recent surveys [31, 59].

## 3.2 Critiquing

In the critiquing setting, the user is allowed to define constraints on items immediately after those were recommended. For example, in a restaurant reservation setting, the user may refine recommended restaurants by their location, price or rating. Using a critique-based conversational recommendation strategy, users incrementally define their preferences towards recommendable items. In a sense, critiquing allows users to modify their learned "static" preferences and adjust the recommender system to their current "dynamic" tastes [77].

A CRS usually presents to the user several options for refinement based on the properties of the pool of recommendable items (e.g., size, color, price, etc). A simple approach to implement a critiquing strategy is to show user options for selection or fill a manual form. User input is usually assumed to define her critiques to be treated as negative feedback on recommended items.

While it is more intuitive for users to interact with CRS in a natural language, such communication form has been less common in earlier critiquing-based recommender systems [10]. Natural language-based interaction is more complex to model, as the set of specified user critiques may be still subjective, ambiguous or unbounded. Hence, most previous works have been focused on system-suggested (predefined and bounded) critiques or user-initiated critiques [10].

Critiques may be automatically derived from past user interactions [63]. Such experience-based methods are implemented by analyzing the properties of items that were successfully recommended in previous similar conversational sessions. Yet, most critique methods strongly rely on the availability of a item metadata (e.g., catalog) or assume a fixed set of critiques.

An alternative and more flexible way to automatically obtain potential critiques is to curate them from textual sources associated with recommendable items. A popular approach is to extract *keyphrases* from item reviews [46, 77]. Keyphrases may be extracted using statistical language-modeling techniques (e.g., identifying terms that are salient in the text), using random-walk methods, etc. For a review on state-of-the-art keyphrase extraction methods, the reader is referred to [49].

We next describe two recent deep-learning methods that exploit keyphrases extracted from item reviews for implementing a critiquing-based conversational preference elicitation process. Using deep-learning techniques allow to better represent both user preferences and feedback (user critiques) in the same latent embedded space, and as a result, improve recommendation accuracy.

### 3.2.1    CE-NCF

As a first method, we describe the *Critiquable and Explainable Neural Collaborative Filtering* (CE-NCF) method [77]. This method has two versions, *deterministic* and *variational*. Here we only describe the deterministic version.

A key assumption which is now made is that, both the observable user $i$ and item $j$ (binary) ratings $r_{ij}$ and (binary) explanation (keyphrases) vector $s_{ij}$ are generated from the same latent representation $z_{ij}$ which is jointly encoded from the latent user $u_i$ and item $v_j$ representations. Initial user and item embeddings can be obtained by any basic method (e.g., [77] used a randomized SVD method that was fined-tuned during end-to-end training). The above assumption is formulated into a deep-learning framework by first encoding each user-item pair into an initial latent representation $\hat{z}_{ij} = f_e(u_i, v_j)$.

Critiquing augments the latent representation which modifies item ratings to better suite user's current preferences. This is achieved in several steps, as follows. First, a prediction function $\hat{s}_{ij} = f_s(\hat{z}_{ij})$ is applied to map the latent representation into explanation predictions for each recommended item to a particular user. The user then takes a critique action which indicates explanations she disagrees with, "zeroing out" the corresponding keyphrases in $\hat{s}_{ij}$. Next, the inverse function $\tilde{z}_{ij} = f_s^{-1}(\tilde{s}_{ij})$ is applied to project back the critiqued explanation to the latent representation. Finally, the model updates both item ratings $\tilde{r}_{ij} = f_r(\tilde{z}_{ij})$ and explanations $\tilde{s}_{ij} = f_s(\tilde{z}_{ij})$. In addition, to more flexibly control the effect of user critiques, whenever a user makes a critiquing action, the latent representation is updated according to the following linear combination: $\tilde{z}_{ij} = \rho\hat{z}_{ij} + (1 - \rho)\tilde{z}_{ij}$, where $\rho \in [0, 1]$ is a hyperparameter.

To train the model end-to-end (together with item recommendation), the following objective is minimized:

$$\min \mathcal{L} = \min \sum_{ij} \mathcal{L}_0(r_{ij}, f_r \circ f_e(u_i, v_j))$$

$$+\lambda_1 \sum_{ij} \mathcal{L}_1(s_{ij}, f_s \circ f_e(u_i, v_j))$$

$$+\lambda_2 \sum_{ij} \mathcal{L}_2(f_e(u_i, v_j), \tilde{f}_s^{-1} \circ f_s \circ f_e(u_i, v_j))$$

$$+\lambda_3 \|\theta\|_2^2,$$

where the three loss functions $\mathcal{L}_0, \mathcal{L}_1$ and $\mathcal{L}_2$ are further taken as Mean Squared Error (MSE) and $\lambda_1$, $\lambda_2$ and $\lambda_3$ are corresponding hyperparameters.

### 3.2.2  Latent Linear Conversational Critiquing

In a sense, each single critiquing step can be viewed as a series of functional transformations that produce a modified prediction $\tilde{r}_i = f_m(r_i, \tilde{s}_i)$ of item preferences for user $i$ given critiqued keyphrases $s_i$ [46]. Having multiple critiquing steps, a user is iteratively provided with the item recommendations $\tilde{r}_i^t$ and, based on those, the user makes a critique action $c_i$ which updates the representation of critiqued keyphrases $\tilde{s}_i^t$ through a cumulative critiquing function: $\tilde{s}_i^t = \psi(s_i, \tilde{s}_i^{t-1}, c_i)$. Alternatively, the user may accept the item recommendation and the current conversation ends.

To derive $\tilde{r}_i$, both user preferences $r_i$ and the critiqued keyphrases are co-embedded to derive a uniform representation $\hat{z}_i$, as follows. First, user ratings $r_i$ are represented by a projected linear embedding $z_i = r_i V$. Next, given user's critique action $c_i$, the cumulative critiquing representation is derived at step $t$ as: $\tilde{s}_i^t = \tilde{s}_i^{t-1} - \max(s_i, 1) \odot c_i^t$ (with $\tilde{s}_i^0 = 0$) and its latent representation is obtained: $\tilde{Z}_i^t = diag(\tilde{s}_i^t) W^T + B$. Each row $\tilde{z}_i^k$ of the matrix $\tilde{Z}_i$ captures the latent representation of the $k$th critiqued keyphrase, and each row of $B$ has an identical bias term $b$. The unified representation $\hat{z}_i$ is then obtained by merging $z_i$ with $\tilde{Z}_i$ as follows:

$$\hat{z}_i = \phi_\lambda(z_i, \tilde{Z}_i) = \lambda_0 z_i + \lambda_1 \tilde{z}_i^1 + \ldots + \lambda_{|K|} \tilde{z}_i^{|K|}.$$

Given $\hat{z}_i$, the updated item ratings are simply given by: $\hat{r}_{ij} = \langle \hat{z}_i, w_j \rangle$ (where $\langle \cdot, \cdot \rangle$ denotes inner product and $w_j$ is the latent representation of item $j$).

$\lambda$ weights can be manually set (e.g., having uniform weight assuming that all critiques have the same importance). Yet a better alternative is to learn $\lambda$ using Linear-Programming (LP) optimization [46].

## 3.3  Facets-Based Preference Elicitation

In a facet-based preference elicitation process, the conversation context defined by user utterances is used to predict which item facets represent the user's preferences. At each step of the conversation, the CRS may decide to either recommend items to the user based on such derived facets or try to clarify user preferences (and hence better learn the item facets that are most relevant).

The facet-based approach allows a more incremental preference elicitation towards targeted item recommendation. Yet, the facet-based approach still requires some memorization effort by users, as users may not be familiar in advanced with all item facets [59] (e.g., what is the meaning of some unit).

In the context of preference elicitation, there are two main challenges that need to be addressed. The first is *mapping user utterances into facets* and the second is *generating clarification questions based on such facets* to be presented to the user whenever more feedback is required.

Mapping user utterances to facets requires to "identify" mentions in the utterance text related to facet types and/or values. As an example, an utterance such as , "*I need a restaurant **near my home***" identifies a location facet with a range of possible values anchored by the user's own location. As another example, a utterance such as "*I would prefer **drama** to **comedy***" defines two facets with a preference order between the user's preferred movie genres.

Facet extraction from query-related text (e.g., search results) has been previously studied in the context of interactive information retrieval [37, 38]. A common facet curation approach is to first apply textual clustering (e.g., K-Means or Hierarchical) [55, 79]. Each cluster then represents a facet and a clustering labeling technique may be applied to extract a short textual description of the facet [7]. Alternative facet extraction techniques include semantic class extraction [57, 66] (using distributional similarity or pattern mining), topic models [74, 76] mention detection [22], named entity recognition [40] and entity linking [36] .

We next shortly describe a more recent approach that is better tailored for CRS. Using deep-learning methods, a give user utterance may be mapped into a set of potential facet type and values [71], as follows. Let $(f, v)$ denote a specific facet-value pair, e.g., *(color, red)*, *(size, small)*, etc. Given a user utterance at time step t, $e_t$, an n-gram vector $z_t$ is first extracted from $e_t$'s text. Next, the sequence of n-grams up to the current time is encoded using an LSTM network into a vector $h_t = LSTM(z_1, z_2, \ldots, z_t)$. To predict facet values probabilities of a given facet $f_i$, a softmax activation layer is then applied.

The set of facet-values probabilities are then used for representing the dialogue state $s_t$. This state is used for belief tracking to decide whether to recommend items based on existing facets or further clarify user preferences by asking questions about a specific facet [71].

The decision whether to recommend an item or clarify user preference towards any of the facets may be implemented using a reinforcement learning approach [71]. To this end, at each step $t$ of the conversation, the CRS agent has $l + 1$ actions it can take. The first $l$ actions capture the decision of the agent on whether to clarify the user's preference regarding the values of one the $l$ possible facets. For example, a question such as "*What **color** do you prefer?*" can be asked given that the color facet has the highest belief in the model.

Alternatively, the CRS agent may decide to provide a recommendation based on existing facet-based belief $s_t$. In [71], such recommendation is implemented using Factorization Machines [61] considering $s_t$ as an additional feature-set. The decision policy of the CRS agent is further learned using the REINFORCE [75] algorithm.

## 3.4   Question-Based Preference Elicitation

In a question-based preference elicitation setting, the CRS aims to unveil user preferences through a series of one or more clarification questions. In a sense, such an approach is similar to the facet-based approach, with two additional options. First, the clarification questions can be topical rather than just asking about existing item facets (properties) [13]. Second, compared to the facet-based approach which asks questions about single facets at a time in an absolute manner, using a relative approach may further allow to identify more clear user preference patterns [14].

We next discuss several recent question-based preference elicitation methods.

### 3.4.1   "System Ask, User Respond" (SAUR)

In the "System Ask, User Respond" (SAUR) setting [82], each conversation is assumed to be initiated by some user information need (e.g., "*Can you find me a **mobile phone** on Amazon?*") and then the CRS agent may either provide a recommendation or start with a series of clarification questions about aspects that are relevant to the user information need (e.g., "*What **operating system** do you prefer?*", "*Do you have requirements on **storage capacity**?*", etc.). Each conversation used for training the model is assumed to end with a successful item recommendation (e.g., judged by analyzing the user's feedback).

Formally, given a conversation session, initiated with user query related to product category $c$ and succeeded with $k$ questions that have been asked so far by the agent ($p_l$) and answered by the user ($q_l$): $Q = p_1, q_1, p_2, q_2, \ldots, p_k, q_k$, the task is to predict the next question to ask $p_{k+1}$. Here, each question is assumed to be about a single product aspect (e.g., operating system) curated from item reviews and each answer has a value specified by the user to that aspect.

While the SAUR model addresses both item recommendation and question generation, we next focus only on the latter sub-task and the interested reader is referred to [82] for the full details.

A given conversation context $Q$ is represented at step $t$ by first concatenating all the text until that step and encoding it with a Gated Recurrent Unit [16] (GRU). Let $c_k$ denote the encoding at current step $k$. Each item $v_j$ is further represented by a textual summary (denoted $s_j^r$) obtained by concatenating its own description with review text associated with it. The representation is then obtained using a GRU with an attention mechanism that allows to make the representation of that item sensitive to a specific conversation context $Q$. Two memory units are further utilized. The first $m_j^1 = GRU(s_j^r, c_k)$ memorizes the relevance of item $v_j$ to $Q$ and used for item recommendation. The second $m_j^2 = GRU(s_j^r, m_j^1)$ serves as question memory for item $v_j$.

The question sub-task now aims to train a model which will maximize the likelihood of the next aspect in the conversation $p_{k+1}$. Such a likelihood is estimated by first concatenating its own representation with $c_k$ and the average memory $\bar{m}^2$

over all items and then applying a two-layer feed forward neural network followed by a softmax layer. Since applying the softmax may be computationally costly due to the large number of aspects to consider, an alternative is to apply a sigmoid with negative sampling proportional to aspect popularity [82].

### 3.4.2 Topic-Based Questions

Most conversational recommender systems ask questions related to properties of end-recommended items. Yet, such an approach may not scale well when the item pool is too large and constantly updated [13] (e.g., videos on YouTube).

An alternative, which might scale better, is to ask users question on topics so user feedback can be propagated among items sharing the same topic [13]. The key idea in such an approach is to present to users a top-N list with topics which they can provide feedback on. User feedback is further captured via user clicks.

Formally, for a given user event history (e.g., video watching) $E = e_1, e_2, \ldots, e_T$, the goal is to estimate the topical distribution of user's interest at time $T + 1$, i.e.: $p(q|E)$. For a large enough data, such distribution can be easily trained using a sequential learning model (e.g., GRU or LSTM). Hence, at questioning time, for a given user event sequence $E'$, questions to be asked can be sampled from the topics with highest likelihood.

Using user's feedback on the displayed top-N topic list, a more personalized recommendation can be made by estimating the likelihood of the next response (items being recommended) $p(r|E', topic = q)$. Such a likelihood can be estimated, for example, by restricting the items being considered for recommendation to those that belong to topic $q$.

### 3.4.3 Asking Absolute vs. Relative Questions

The two methods we discussed so far present questions to users in an absolute fashion one by one in a sequential order [82] or as a list of topics to pick [13] as the conversation progress. Yet, such questioning strategy does not fully capture the relative user preferences. An alternative, therefore, is to present clarification questions to users in a relative form [14]. The key idea here is to pick items (or aspects to ask) which would allow to reveal as much information as possible about user's preferences. For example, a relative question may be given about two items that are relatively far from each other in the latent space.

We next discuss the questioning framework of [14]. This framework has two versions, *Absolute* (recommendation) and *Pairwise*. For our discussion purposes we only focus on the Absolute model and refer the reader to [14] for the second one.

As a first assumption, a latent factor recommendation setting is utilized to implement the underlying Absolute model. Specifically, a simplified version of the *Matchbox Recommender* model [70] is implemented. In a nutshell, in this model each user $i$ is modeled by a bias variable $\alpha_i \sim \mathcal{N}(0, \sigma_2^2)$ and a trait vector $u_i \sim \mathcal{N}(0, \sigma_1^2 \mathbf{I})$. In a similar manner, each item $j$ has a bias variable $\beta_j \sim \mathcal{N}(0, \sigma_2^2)$ and a trait vector $v_j \sim \mathcal{N}(0, \sigma_1^2 \mathbf{I})$. Then, the unobserved affinity between user $i$ and item $j$ is given by $y_{ij} = \alpha_i + \beta_j + u_i^T v_j$. Observations are made $\hat{y}_{ij} \sim \mathcal{N}(y_{ij}, \epsilon_{ij})$, where $\epsilon_{ij}$ models the affinity variance, accounting for noise in user preferences. User (dis)like observation is then given by $\hat{r}_{ij} = \mathbf{1}[\hat{y}_{ij} > 1]$. The hyper-parameters $\sigma_1, \sigma_2$ model the variance in traits and biases. The model variables are learned by maximizing the log-posterior over the item and user variables with fixed hyper-parameters, given the training observations [14]. To "bootstrap" the model, item embeddings are learned offline from logged observations, while user parameters are initialized to the mean user values assuming questions are to be asked in a new user setting [14].

Having defined the underlying recommendation model, we next describe two main options for asking questions about user preferences towards items, either *Absolute* or *Relative*. The key idea is to ask a few questions so both user's preferences and question quality can be learned. Such task is implemented in [14] using ideas that are borrowed from *active learning* (query for labels that provide the highest amount of new information) and *bandit learning* (balance between model exploitation and exploration). In the context of conversational recommenders, such a balance may help focus questions on the most relevant part of the latent space, while still considering that highly preferred items may lie in as of yet unexplored areas of the space [14]. At a given time, the model confidence on user preference towards items $j$ is captured by the current variances of the posterior of the noisy affinities $y_j^{cold}$. As the system asks about an item $j^*$ and observe the user's feedback, the variance of the inferred noisy affinity of this item and of the nearby items in the learned embedding is reduced [14].

The general preference elicitation algorithm is implemented as follows. For a new user $i$, the noisy affinities $y_{ij}$ are inferred. Then, while more questions are allowed to be asked, an item $j^*$ is picked for absolute (relative) question. Several question (item pick) selection strategies are explored in [14], showing preferable results to an approach based on *Thompson Sampling*: $j^* = \arg\max_j \hat{y}_{ij}$. The user feedback $(i, j^*, 0/1)$ is then incorporated into the model to update the noisy affinities $y_i$ according to [70].

To extend to relative preference elicitation, after an item $A$ is selected, we first assume the user *did not like* this item and incorporate a virtual feedback $(i, A, 0)$ into the model. Then, a second item $B$ is selected based on the inferred updated posterior model when it is used as the prior model. Here, the intuition behind such relative question is that the two items the user is asked to give a relative preference on should be relatively far apart in the latent space. This allows the system to both learn user preferences more efficiently while the user is not forced to choose among very similar items [14], hence introducing diversity.

## 3.5   Discussion and Future Outlook

Recent advances in NLP now allow to better understand user preferences through interactive dialogue with the recommendation system. Traditional preference elicitation methods have been implemented with a predefined set of preferences or UI-aided tools (e.g., forms). Using textual sources such as item descriptions or user reviews, now allows to extract and process potential preference-related data (e.g., facets) more easily and integrate such data within existing recommendation methods. Moreover, using the mixed initiative nature of conversation, clarification questions about user preferences can be utilized to allow the recommendation system better personalization to its users.

Preference elicitation using NLP methods can be improved in two main ways. First, most existing works utilize user reviews as a source for preference-related metadata. Other sources may be further utilized and should be explored, including, for example, text curated from user discussions and forums, social media or news.

Second, existing preference elicitation NLP-based methods still utilize very simplistic language models learned by traditional deep learning methods such as CNNs and RNNs. New pre-trained language models such as BERT [19] and GPT-3 [6] should be further explored in the context of this task, allowing to improve textual representation and automatic generation of facets and clarification questions.

## 4   Generating Textual Explanations

As recommendations have become central to shaping decisions, users increasingly demand convincing explanations to help them understand why particular recommendations are made. To this end, the research community has lately focused on studying both how to generate explanations from recommender systems as well as the effect of explanations on influencing user behavior [23, 54]. Chapter 19 provides a detailed overview of explanations in recommender systems. This section complements Chap. 19 by focusing on describing *how* explanations can be generated from text. In more detail, in what follows, we will describe the most representative methods of generating explanations by extracting snippets from text that accompanies an item that is recommended by the system. The input is the user-item matrix as well as the reviews available for the items that are in abundance in most recommender systems. The explanation text can be presented as a description of the item or a comprehensive review (or a set of reviews). The input for most of the algorithms described is the user-item rating matrix as well as the reviews that are written about items. The structure of the Section is as follows: we first discuss deep learning methods that find and show to the user the most useful review (in some cases reviews can be personalized) and then we describe works that extract explicit product features and user opinions from review text.

## 4.1   Review-Level Explanations

User-generated reviews can be viewed, at least in part, as explanations of the ratings given by users. As discussed previously in Sect. 2, recommendation accuracy can be greatly improved when using review text that is available for the items. At the same time, reviews can help towards an additional goal: explaining why an item is recommended to a user. Here, we will discuss literature that has demonstrated the effective application of reviews in the area of explanations ranging from finding and showing to the user the most useful review to more complex methods where the models can automatically generate personalized reviews to show to each user.

Early works that proposed to combine latent dimensions in rating data with review text had a two-fold contribution: first, the approaches could use the text reviews with ratings to provide accurate recommendations and, second, the approaches could use the most useful reviews to show to the users as an explanation for each recommendation (review level explanations). Most of these representative works are the following: McAuley et al. [47] combine latent rating dimensions with latent topics in reviews learned by LDA topic models. The proposed method is able to generate interpretable topics, that can be used to suggest informative reviews. Along the same lines, Diao et al. [20] also propose to combine ratings with review text to identify aspects (relative to topics). The key difference with [47] is that in this work, the authors focus on both learning the aspects as well as learning the sentiments of these aspects (negative vs. positive). From the deep learning community, the neural attentional regression model with review-level explanations model (NARRE) [9] uses the text of the reviews to both predict item ratings and learn the usefulness of the reviews. In this approach, only the highly-useful reviews are provided as explanations to the users. The explanations provided in all the above studies are not personalized, i.e., when the recommender system comes up with a specific item recommendation, all users will be shown the exact same explanation which is the most useful reviews. In what follows we describe works that generate personalized explanations in the form of reviews.

### 4.1.1   Multi-Task Learning for Recommendation and Personalized Explanation

Recently, the research community has shown increased interest in personalized review generation that can serve as a form of explanation to the users. Lu et al. [45] jointly learn recommendations and explanations by introducing a multi-task learning framework. Instead of finding the most useful review, the proposed model generates new reviews that are personalized to the taste of each user. More specifically, the review text is used for learning the user preferences and item properties, and implicit and explicit feedback (e.g., clicks or ratings), is used to learn the level of interest of a user in the attributes of an item. The approach employs a matrix factorization model (see Chap. 3 for more details) that generates the user-item ratings combined with a sequence-to-sequence model that generates a personalized review (that serves as an explanation) for each user-item recommendation.

In order to generate an unbiased personalized review (i.e., explanation), Lu et al. [45] use an adversarial approach over a sequence-to-sequence model, where a review generator and discriminator network are trained simultaneously. For the generator network, which eventually produces the explanations, the set $d_i$ of all reviews written by user $i$ is mapped onto a textual feature vector $\widetilde{U}$ and subsequently fed into a bi-directional Gated Recurrent Unit (GRU) [16] which concatenates the last-step hidden forward and backward activations into a vector-based representation $h_T$ for the review. Overall, given the vector $\widetilde{U}$, the model generates reviews that attempt to maximize the probability of a $T'$-lengthed review $y_{i,1}, \ldots, y_{i,T'}$ given the user's textual vector:

$$p(y_{i,1}, \ldots, y_{i,T'} | \widetilde{U}) = \prod_{t=1}^{T'} p(y_{i,t} | \widetilde{U}_i, y_{i,1}, \ldots, y_{i,t-1})$$

Once the reviews are generated, they are next fed into the discriminator that attempts to discern real from artificially-generated (from the generator) reviews. In this adversarial setting, the discriminator is implemented as a convolutional neural net, typical in text classification tasks. More specifically, the review words are mapped into vectors which are fed into a convolutional, max-pooling and fully-connected projection layers (in this order) while the final output is adjusted using a sigmoid function. Overall, the discriminator network attempts, through sampling of reviews and using policy gradient descent [75], to maximize the function:

$$max_\phi E_{Y \sim p_{data}}[log D_\phi(Y)] + E_{Y' \sim G_\theta}[log(1 - D_\phi(Y'))]$$

where $Y \sim p_{data}$ are the ground-truth (i.e., real) sampled reviews and $Y' \sim G_\theta$ are the generated ones. In the process, the learned textual features and the matrix-factorization-based textual features are regularized in order to allow the sequence-to-sequence model to leverage the user preferences identified from the collaborative filtering step. Finally, personalized recommendation explanations are generated by employing a review decoder that combines the vectors $\widetilde{U}_i$ (i.e., the reviews of user $i$) and $\widetilde{V}_j$ (i.e., the reviews of item $j$) to generate reviews for a given user-item pair.

Another popular way of providing useful explanations to recommendations is by generating (oftentimes personalized) synthetic reviews. Ouyang et al. [56] combine three generative models that provide natural language explanations while leveraging the available helpfulness votes of existing reviews. The approach also takes into account the set of item attributes that may be of interest to the user while generating the reviews. At a high level, the goal is to maximize the likelihood of an explanation $e = (y_1, \ldots, y_l)$ of length $l$ for a given set of input attributes for item $i$ $a_i = (a_i, \ldots, a_{|a|})$:

$$p(e|a) = \prod_{t=1}^{l} p(y_t | y_{i,1}, \ldots, y_{i,t-1}, a)$$

To this end, Ouyang et al. [56], propose three textual models that produce text as both character and word sequences using the user id, item id, item rating and review helpfulness score as input. More specifically, the approach starts in the encoding phase by identifying all word tokens with their positions in the review corpus that will be used both during the encoding (training step) and decoding (generating step). A GCN (generative concatenative) model is employed at the beginning of the process. The goal of the GCN is to learn the relations of attributes and text by considering them as one (concatenated) input. More specifically, the model accepts input of the form $X'_t = [x^r_t : x^a_t]$ where $x^r_t$ is the encoded review text and $x^a_t$ are the one-hot encoded attributes at time step $t$. After the first GCN model, a context and attention model are utilized that aim to transform the attributes to fixed-length [21] embeddings. The goal of these two models is to learn how attributes and text aligns and to provide good initialization weights for the decoder. At the end of the encoding phase, encoded attributes are reshaped to the proper decoder shape: $A = tanh(H[x^a_i, \ldots, x^a_{|a|}] + b_a)$ and are provided as initialization weights for the decoder.

The decoding phase utilizes Recurrent Neural Networks (RNNs) with long short-term memory (LSTM) [28]. Given, at a time $t$, inputs $x_t$, the LSTM cell state $C_{t-1}$, the previous output $H_t$, $W$ the weights, $b$ the bias and $\hat{C}$ the candidate state, the computation of the decoding step proceeds as follows with $\odot$ denoting an element-wise product operation:

$$\hat{C}_t = tanh(W^c_x x_t + W^c_h H_{t-1} + b_c)$$

$$f_t = \sigma(W^f_x x_t + W^f_h H_{t-1} + b_f)$$

$$i_t = \sigma(W^i_x x_t + W^i_h H_{t-1} + b_i)$$

$$C_t = f_t \odot C_{t-1} + i_t \odot C'_t$$

$$o_t = \sigma(W^o_x x_t + W^o_h H_{t-1} + b_o)$$

$$H_t = o_t \odot tanh(C_t)$$

As a final step, for text generation, the decoder output $H_t$ is provided as input to a softmax function that essentially attempts to maximize the probability $p(y_t|y_{<t}, a)$ by greedily inferring the characters and words. As a result, the final explanation review text is produced by inferring the index $Y_t$ of the generated character/word through the following process:

$$p(y_t|y_{i,1}, \ldots, y_{i,t-1}, a) = softmax(W H_t + b)$$

$$Y_t = argmax p(y_t|y_{i,1}, \ldots, y_{i,t-1}, a)$$

### 4.1.2    Providing Explanations for Recommendations in Reciprocal Environments

Reciprocal environments involve applications such as job searching or online dating, where a recommendation should be mutually beneficial to two or more users. Kleinerman et al. [35] study how to generate useful reciprocal explanations for such user-matching (i.e., reciprocal) recommendations on top of two-sided collaborative filtering approaches [53, 78]. In general, the reciprocal setting assumes that, each user $x$, has, for each predefined attribute $a$ in the system, provided their personal values for these attributes $A_x = \{v_a\}$, together with the (implicit or explicit) user's preference $p_{x,a}$ for each of the attributes. In Kleinerman et al. [35], preferences are identified implicitly based on messages sent among users. More specifically, $p_{x,a}$ is the number of messages sent by $x$ to users that have $v_a$ as a value for attribute $a$.

Kleinerman et al. [35] approach the problem of reciprocal explanations for two users $x$ and $y$ by generating two one-sided explanations $e_{x,y}$ and $e_{y,x}$. To this end, they propose two single-sided explanation generation approaches named *Transparent* and *Correlation-based*.

In the *Transparent* approach, in order for the system to explain recommending user $y$ to user $x$, it returns the top-$k$ attributes of $y$ that are considered the most important based on $x$'s preferences $p_{x,a}$.

In the *Correlation-based* approach, the system measures the correlation between a given attribute value $v_a$ in user $y$'s profile and the likelihood of user $x$ indicating that specific attribute as a preference (specifically, sending a message in [35]). For each user $x$, all users $I = i$ that $x$ has interacted with are first identified and two binary metrics are computed: $M_x(i)$ which captures whether $x$ has indicated a preference (sent a message) to $i$, and $S_{x,v_a}(i)$ which captures whether user $i$ has $v_a$ in their profile. Then, given the two binary vectors $M$ and $S$, we compute the Pearson's correlation metric between $M$ and $S$ for each attribute value $v_a$. The top-$k$ attributes with the highest correlation are considered to be the single-sided explanations $e_{x,y}$.

## 4.2    Feature-Level Explanations

In what follows, we describe works that extract explicit product features and user opinions from review text. To infer user opinions, sentiment analysis is used that in some cases also leverages the sentiments of the social connections of the users.

### 4.2.1    Explicit Factor Models for Explainable Recommendation Based on Phrase-Level Sentiment Analysis

Zhang et al. [83] present an approach called Explicit Factor Model. The approach starts by identifying product aspects (features) that may be of interest to the users through sentiment analysis on the reviews and uses learned and latent features to generate both recommendations and aspect-level explanations.

Initially, the approach identifies the feature descriptions from the reviews $\mathcal{F}$, the opinion descriptions ($O$) and the feature sentiments, i.e., ($\mathcal{F}, O$) pairs. The set $\mathcal{L} = (\mathcal{F}, O, \mathcal{S})$ is then used to generate feature, review-sentiment ($\mathcal{F}, \mathcal{S}'$) pairs. Next, a user-feature attention matrix $X$ with $\mathcal{F} = \{F_1, F_2, \ldots, F_p\}$ columns (i.e., features), and $\mathcal{U} = \{u_1, \ldots, u_m\}$ rows (i.e., users) is generated with each element being either 0 if user $u_i$ did not mention feature $F_j$, or the value of a sigmoid function $\sigma(t_{ij})$ otherwise. Similarly, an item-feature quality matrix $Y$ is generated with $\mathcal{P} = \{p_1, \ldots, p_n\}$ rows and $\mathcal{F} = \{F_1, F_2, \ldots, F_p\}$ columns. Again, each element is either 0 if item $p_i$ does review feature $F_j$, or the value of a sigmoid function $\sigma(t_{ij})$. The final factorization model is augmented with additional $r'$ latent factors $H_1 \in \mathbb{R}_+^{m \times r'}$, $H2 \in \mathbb{R}_+^{n \times r'}$ attempting to capture hidden factors affecting the user decision $P = [U_1 \ H_1]$ and $Q = [U_2 \ H_2]$. It also uses the user-item ratings $A$ to capture both explicit and implicit features both for recommendations and explanations:

$$\min_{U_1, U_2, V, H1, H2} \{\|PQ^T - A\|_F^2 + \|Y_1 V^T - X\|_F^2 + \|U_2 V^T - Y\|_F^2$$

$$+ (\|U_1\|_F^2 + \|U_2\|_F^2) + (\|H_1\|_F^2 + \|H_2\|_F^2) + \|V\|_F^2\}$$

$$s.t. \ U_1 \in \mathbb{R}_+^{m \times r}, U_2 \in \mathbb{R}_+^{n \times r}, V \in \mathbb{R}_+^{p \times r}, H_1 \in \mathbb{R}_+^{m \times r'}, H_2 \in \mathbb{R}_+^{n \times r'}$$

Given the solution of the factorization, a recommendation of item $j$ to user $i$ can be computed as: $R_{ij} = \alpha \frac{\sum_{c \in C_i} \hat{X}_{ic} \cdot \hat{Y}_{jc}}{kN} + (1 - \alpha)\hat{A}_{ij}$, where $\hat{X} = U_1 V^T$, $\hat{Y} = U_2 V^T$, $A = U_1 U_2^T + H_1 H2^T$, $N$ is the maximum scale of ratings (e.g. 5), $\alpha$ is a scaling factor, and $C_i$ are the columns of $\hat{X}$ with the $k$ largest values. Explanations are generated by identifying the best and worst-performing features $F_{c_{best}}$ and $F_{c_{worst}}$ of a product $p_j$ for each user $u_i$: $c_{best} = argmax_{c \in C_i} \hat{Y}_{jc}$ and $c_{worst} = argmin_{c \in C_i} \hat{Y}_{jc}$.

### 4.2.2 Social Collaborative Viewpoint Regression with Explainable Recommendations

In addition to concepts, topics, reviews and their sentiments, Ren et al. [60], also utilize the sentiments of users' social connections in order to create a notion of viewpoints, i.e., tuples of concepts, topics, review sentiments and social-connection sentiments. Ren et al. [60] consider the typical setting of users $\mathcal{U} = \{u_1, \ldots, u_U\}$, items $\mathcal{I} = \{i_1, \ldots, i_I\}$, pairs $Q = \{(u, i)\}$ of user-ratings $r_{u,i}$ and reviews $\mathcal{D} = \{d_1, \ldots, d_{|Q|}\}$ with each review being a set of words $d = \{w_1, \ldots, w_{|d|}\}$ and additionally consider a set of trusted social relations $\mathcal{T}_{u_i, u_j}$ where user $u_i$ trusts user $u_j$.

Topics $z$ are defined as a probability distribution over words, while concepts $e$ is a feature in close proximity to a topic. Sentiments $l$ over words $w$ are considered to depend on topics. A viewpoint is a finite mixture over $v = < e, z, l >$ tuples. The user's $u$ rating values $\theta^u$ are then defined as an $R \times V$ matrix with each element $\theta^u_{i,v_j}$ corresponding to the probability of rating $r$, for a given pair of user-viewpoint $(u, v)$. For discovering the concepts, Ren et al., employ a word2vec [50] model while they use a recursive deep model for sentiment analysis [68].

In order to utilize the social connections, a latent factor model is proposed where, in addition to viewpoints, topics, concepts and sentiments, the trusted relations of a user $u$ are modeled by considering the viewpoint distribution of $u$'s social relations $\{\theta_v^{u1}, \ldots, \theta_v^{uF_u}\}$ and a base distribution $\theta^0_{u,v}$. Inference is performed using a Gibbs EM sampler. The E-step approximates the distribution $p(\mathcal{V}, \mathcal{Z}, \mathcal{L}|\mathcal{W}, \mathcal{E}, \mathcal{R}, \mathcal{T}, \mathcal{F})$, while the M-step maximizes each user's $u$ viewpoint distribution $\theta_u$. Once the Gibbs EM sampling is completed, for each user $u$ we compute a recommendation based on the maximum value of predicted rating probabilities: $P(r_{u_i} = r|u, i) = \sum_{v \in \mathcal{V}} \theta^u_{r,v} \cdot \pi_{i,v}$, with $P(v|i) = \pi_{v,i}$ being the viewpoint distribution of item i. Explanations are finally derived by presenting to user $u$ the topics of the recommended items.

### 4.2.3 Review-Aware Explainable Recommendation by Modeling Aspects

Aspects (or features) of recommended items can be very useful in providing good explanations. He et al. [27] utilize textual reviews to identify product aspects which are modeled as a tripartite graph of users, items and aspects. The problem of recommendation and explanation is hence viewed as node ranking over this tripartite graph. He et al. [27], identify the aspects of the items [80] and extract feature, opinion, sentiment tuples $(F, O, S)$. Next, a tripartite graph $G = (U \cup P \cup A, E_{UP} \cup E_{UA} \cup E_{PA})$ is defined, where $U$, $P$ and $A$ are the vertices corresponding to users, items and aspects, while $E_{UP}$, $E_{UA}$ and $E_{PA}$ are the user-to-items, user-to-aspects and items-to-aspects edges respectively. A user $u_i$, rating an item $p_j$ with a review containing aspect $a_k$ is a triangle with edges $e_{ij}, e_{jk}, e_{ik}$ in the graph. The edges are weighted, capturing the strength of relationship among the nodes. Matrices $R$, $Y$ and $X$ represent the weights for $E_{UP}$, $E_{UA}$ and $E_{PA}$ respectively. The goal is to identify a ranking function $f$ that will produce the predicted preference of user $u$ for $p$.

The approach of He et al. [27] (named TriRank) learns $f$ with two useful properties encoded as regularizers: (a) smoothness, i.e., nearby nodes should have similar scores, and, (b) fitting, i.e., the learned function should not cause significant deviation from the observed data. The overall function that the approach aims to optimize through alternating least squares (ALS) is:

$$Q(f) = \alpha \sum_{i,j} r_{ij} \left( \frac{f(u_i)}{\sqrt{d_i^u}} - \frac{f(p_j)}{\sqrt{d_j^p}} \right)^2 + \beta \sum_{j,k} x_{jk} \left( \frac{f(p_j)}{\sqrt{d_j^p}} - \frac{f(a_k)}{\sqrt{d_k^a}} \right)^2$$

$$+ \gamma \sum_{i,k} y_{ik} \left( \frac{f(u_i)}{\sqrt{d_i^u}} - \frac{f(a_k)}{\sqrt{d_k^a}} \right)^2 + \eta_U \sum_i (f(u_i) - u_i^0)^2$$

$$+ \eta_P \sum_j (f(p_j) - p_j^0)^2 + \eta_A \sum_k (f(a_k) - a_k^0)^2$$

The parameters $\alpha$, $\beta$, $\gamma$ capture the desired weight of the smoothness constraint on the user-item, item-aspect and user-aspect edges, while $\eta_U$, $\eta_P$ and $\eta_A$ capture the desired weight of the fitting constraint on users, items and aspects. Explanations are provided as recommended items, i.e., the aspects of a recommended item are presented to the user together with how well the item captures a specific aspect.

## 4.3  Discussion and Future Outlook

All methods described above generate explanations from the review text. As a result, the question here is how one can deal with items that have a very small number of reviews. One possible suggestion is to use additional features that are available for the items (e.g., description, title).

Most of the studies described above are using offline metrics (such as perplexity) that focus on evaluating the goodness of a language model is in evaluating the quality of the explanations. However, a real recommender system cannot rely exclusively on the results of the offline evaluation metrics since there is a large number of other factors that play a significant role in the successful application of explanations. For example, recent studies [39, 51] have shown that users with different personality traits have different preferences in explanations. As a result, more user studies are needed to better understand how the explanations provided by the aforementioned methods are perceived by real end-users of a recommendation engine.

Another very promising direction is for the explanations to account for privacy. Consider, for example, an explanation of the form "We recommend bar Crudo because it is one of the most popular gay bars in the area". It is obvious that such an explanation involves several potential issues regarding the privacy of the user. Document understanding techniques (e.g., GPT-3 [6]) can be deployed to protect users' privacy that may be compromised through explanations.

Another future work direction is to design and implement interactive systems that can foster exploration from the users. The majority of work described above provided natural language explanations in a static context where the user could not interact with the explainable recommender system and give feedback. In addition, most of the work did not allow for further exploration of the provided explanations by the interested user. Scrutability and control are two essential characteristics that an explainable recommender system should offer, thus it is worth for the natural language community to study this approach.

A final promising direction is to study whether explanations should participate in the process of ranking the recommendations. All methods described above first rank the recommendations and then generate the explanations for the top ranked items, i.e., the process of generating explanations does not affect the ranking process. A promising future direction would be to incorporate explanations in the prediction process of the recommender system in order to improve the accuracy and transparency of recommendations.

# References

1. R.K. Bakshi, N. Kaur, R. Kaur, G. Kaur, Opinion mining and sentiment analysis, in *3rd International Conference on Computing for Sustainable Global Development*, INDIACom'16, pp. 452–455 (2016)
2. Y. Bao, H. Fang, J. Zhang, Topicmf: Simultaneously exploiting ratings and reviews for recommendation, in *Proceedings of the 28th AAAI Conference on Artificial Intelligence*, AAAI'14, pp. 2–8 (2014)
3. K. Bauman, B. Liu, A. Tuzhilin, Aspect based recommendations: Recommending items with the most valuable aspects based on user reviews, in *Proceedings of the 23rd ACM SIGKDD International Conference on Knowledge Discovery and Data Mining*, KDD'17, pp. 717–725 (2017)
4. D.M. Blei, A.Y. Ng, M.I. Jordan, Latent dirichlet allocation. J. Mach. Learn. Res. **3**, 993–1022 (2003)
5. D.J. Brenes, D. Gayo-Avello, K. Pérez-González, Survey and evaluation of query intent detection methods, in *Proceedings of the WSDM 2009 Workshop on Web Search Click Data*, WSCD'09, pp. 1–7 (2009)
6. T.B. Brown, B. Mann, N. Ryder, M. Subbiah, J. Kaplan, P. Dhariwal, A. Neelakantan, P. Shyam, G. Sastry, A. Askell, et al., Language models are few-shot learners. Preprint (2020). arXiv:2005.14165
7. D. Carmel, H. Roitman, N. Zwerdling, Enhancing cluster labeling using wikipedia, in *Proceedings of the 32nd International ACM SIGIR Conference on Research and Development in Information Retrieval*, SIGIR '09, pp. 139–146 (2009)
8. R. Catherine, W. Cohen, Transnets: Learning to transform for recommendation, in *Proceedings of the 11th ACM Conference on Recommender Systems*, RecSys'17, pp. 288–296 (2017)
9. C. Chen, M. Zhang, Y. Liu, S. Ma, Neural attentional rating regression with review-level explanations, in *Proceedings of the 2018 World Wide Web Conference*, WWW'18, pp. 1583–1592 (2018)
10. L. Chen, P. Pu, Critiquing-based recommenders: survey and emerging trends. User Model. User Adap. Inter. **22**(1-2), 125–150 (2012)
11. Z. Cheng, Y. Ding, X. He, L. Zhu, X. Song, M.S. Kankanhalli, Aˆ 3ncf: An adaptive aspect attention model for rating prediction, in *Proceedings of the 27th International Joint Conference on Artificial Intelligence*, IJCAI'18, pp. 3748–3754 (2018)
12. Z. Cheng, Y. Ding, L. Zhu, M. Kankanhalli, Aspect-aware latent factor model: Rating prediction with ratings and reviews, in *Proceedings of the 2018 World Wide Web Conference*, WWW'18, pp. 639–648 (2018)
13. K. Christakopoulou, A. Beutel, R. Li, S. Jain, E.H. Chi, Q&R: A two-stage approach toward interactive recommendation, in *Proceedings of the 24th ACM SIGKDD International Conference on Knowledge Discovery & Data Mining*, KDD'18, pp. 139–148 (2018)

14. K. Christakopoulou, F. Radlinski, K. Hofmann, Towards conversational recommender systems, in *Proceedings of the 22nd ACM SIGKDD International Conference on Knowledge Discovery & Data Mining*, KDD'16, pp. 815–824 (2016)

15. Y.-N. Chuang, C.-M. Chen, C.-J. Wang, M.-F. Tsai, Y. Fang, E.-P. Lim, TPR: Text-aware preference ranking for recommender systems, in *Proceedings of the 29th ACM International Conference on Information & Knowledge Management*, CIKM'20, pp. 215–224 (2020)

16. J. Chung, C. Gulcehre, K. Cho, Y. Bengio, Empirical evaluation of gated recurrent neural networks on sequence modeling, in *NIPS 2014 Workshop on Deep Learning* (2014)

17. D.-A. Clevert, T. Unterthiner, S. Hochreiter, Fast and accurate deep network learning by exponential linear units (ELUs), in *4th International Conference on Learning Representations (Poster)*, ICLR'16 (2016)

18. R. Collobert, J. Weston, L. Bottou, M. Karlen, K. Kavukcuoglu, P.P. Kuksa, Natural language processing (almost) from scratch. J. Mach. Learn. Res. **12**, 2493–2537 (2011)

19. J. Devlin, M.-W. Chang, K. Lee, K. Toutanova, BERT: pre-training of deep bidirectional transformers for language understanding, in *Proceedings of the 2019 Conference of the North American Chapter of the Association for Computational Linguistics*, NAACL'19, pp. 4171–4186 (2019)

20. Q. Diao, M. Qiu, C.-Y. Wu, A.J. Smola, J. Jiang, C. Wang, Jointly modeling aspects, ratings and sentiments for movie recommendation (jmars), in *Proceedings of the 20th ACM SIGKDD International Conference on Knowledge Discovery and Data Mining*, KDD'14, pp. 193–202 (2014)

21. L. Dong, S. Huang, F. Wei, M. Lapata, M. Zhou, K. Xu, Learning to generate product reviews from attributes, in *Proceedings of the 15th Conference of the European Chapter of the Association for Computational Linguistics*, EACL'17, pp. 623–632 (2017)

22. P. Ferragina, U. Scaiella, Tagme: on-the-fly annotation of short text fragments (by wikipedia entities), in *Proceedings of the 19th ACM International Conference on Information and Knowledge Management*, CIKM'10, pp. 1625–1628 (2010)

23. G. Friedrich, M. Zanker, A taxonomy for generating explanations in recommender systems. AI Magazine **32**(3), 90–98 (2011)

24. S. Frumerman, G. Shani, B. Shapira, O.S. Shalom, Are all rejected recommendations equally bad? towards analysing rejected recommendations, in *Proceedings of the 27th ACM Conference on User Modeling, Adaptation and Personalization*, UMAP'19, pp. 157–165 (2019)

25. F.A. Gers, J. Schmidhuber, F. Cummins, Learning to forget: Continual prediction with LSTM. Neural Computation **12**(10), 2451–2471 (1999)

26. G. Hadash, O.S. Shalom, R. Osadchy, Rank and rate: multi-task learning for recommender systems, in *Proceedings of the 12th ACM Conference on Recommender Systems*, RecSys'18, pp. 451–454 (2018)

27. X. He, T. Chen, M.-Y. Kan, X. Chen, Trirank: Review-aware explainable recommendation by modeling aspects, in *Proceedings of the 24th ACM International on Conference on Information and Knowledge Management*, CIKM'15, pp. 1661–1670 (2015)

28. S. Hochreiter, J. Schmidhuber, Long short-term memory. Neural Computation **9**(8), 1735–1780 (1997)

29. S. Ioffe, C. Szegedy, Batch normalization: Accelerating deep network training by reducing internal covariate shift, in *Proceedings of the 32nd International Conference on Machine Learning*, volume 37 of *ICML'15*, pp. 448–456 (2015)

30. A. Jacovi, O.S. Shalom, Y. Goldberg, Understanding convolutional neural networks for text classification, in *Proceedings of the 2018 EMNLP Workshop BlackboxNLP: Analyzing and Interpreting Neural Networks for NLP*, BlackboxNLP'18, pp. 56–65 (2018)

31. D. Jannach, A. Manzoor, W. Cai, L. Chen, A survey on conversational recommender systems. Preprint (2020). arXiv:2004.00646

32. Y. Kim, Convolutional neural networks for sentence classification, in *Proceedings of the 2014 Conference on Empirical Methods in Natural Language Processing*, EMNLP'14, pp. 1746–1751 (2014)

33. D.P. Kingma, J. Ba, Adam: A method for stochastic optimization, in *3rd International Conference on Learning Representations*, ICLR'15 (2015)

34. D.P. Kingma, M. Welling, Auto-encoding variational Bayes, in *2nd International Conference on Learning Representations*, ICLR'14 (2014)

35. A. Kleinerman, A. Rosenfeld, S. Kraus, Providing explanations for recommendations in reciprocal environments, in *Proceedings of the 12th ACM Conference on Recommender Systems*, RecSys'18, pp. 22–30 (2018)

36. N. Kolitsas, O.-E. Ganea, T. Hofmann, End-to-end neural entity linking, in *Proceedings of the 22nd Conference on Computational Natural Language Learning*, CoNLL'18, pp. 519–529 (2018)

37. W. Kong, J. Allan, Extracting query facets from search results, in *Proceedings of the 36th International ACM SIGIR Conference on Research and Development in Information Retrieval*, SIGIR'13, pp. 93–102 (2013)

38. W. Kong, J. Allan, Precision-oriented query facet extraction, in *Proceedings of the 25th ACM International on Conference on Information and Knowledge Management*, CIKM'16, pp. 1433–1442 (2016)

39. P. Kouki, J. Schaffer, J. Pujara, J. O'Donovan, L. Getoor, Personalized explanations for hybrid recommender systems, in *Proceedings of the 24th International Conference on Intelligent User Interfaces*, IUI'19, pp. 379–390 (2019)

40. G. Lample, M. Ballesteros, S. Subramanian, K. Kawakami, C. Dyer, Neural architectures for named entity recognition, in *Proceedings of the 2016 Conference of the North American Chapter of the Association for Computational Linguistics*, NAACL'16, pp. 260–270 (2016)

41. Q. Le, T. Mikolov, Distributed representations of sentences and documents, in *The 31st International Conference on Machine Learning*, ICML'14, pp. 1188–1196 (2014)

42. J. Li, T. Luong, D. Jurafsky, A hierarchical neural autoencoder for paragraphs and documents, in *Proceedings of the 53rd Annual Meeting of the Association for Computational Linguistics*, ACL'15, pp. 1106–1115 (2015)

43. G. Ling, M.R. Lyu, I. King, Ratings meet reviews, a combined approach to recommend, in *Proceedings of the 8th ACM Conference on Recommender systems*, RecSys'14, pp. 105–112 (2014)

44. B. Liu, I. Lane, Attention-based recurrent neural network models for joint intent detection and slot filling, in *17th Annual Conference of the International Speech Communication Association*, Interspeech'16, ed. by N. Morgan, pp. 685–689 (2016)

45. Y. Lu, R. Dong, B. Smyth, Why i like it: Multi-task learning for recommendation and explanation, in *Proceedings of the 12th ACM Conference on Recommender Systems*, RecSys'18, pp. 4–12 (2018)

46. K. Luo, S. Sanner, G. Wu, H. Li, H. Yang, Latent linear critiquing for conversational recommender systems, in *Proceedings of The Web Conference 2020*, WWW'20, pp. 2535–2541 (2020)

47. J. McAuley, J. Leskovec, Hidden factors and hidden topics: understanding rating dimensions with review text, in *Proceedings of the 7th ACM conference on Recommender systems*, RecSys'13, pp. 165–172 (2013)

48. J. McAuley, R. Pandey, J. Leskovec, Inferring networks of substitutable and complementary products, in *Proceedings of the 21th ACM SIGKDD International Conference on Knowledge Discovery and Data Mining*, KDD'15, pp. 785–794 (2015)

49. Z.A. Merrouni, B. Frikh, B. Ouhbi, Automatic keyphrase extraction: a survey and trends. J. Intell. Inf. Syst. **54**, 1–34 (2019)

50. T. Mikolov, K. Chen, G.S. Corrado, J. Dean, Efficient estimation of word representations in vector space, in *1st International Conference on Learning Representations*, ICLR'13 (2013)

51. M. Millecamp, K. Verbert, S. Naveed, J. Ziegler, To explain or not to explain: the effects of personal characteristics when explaining feature-based recommendations in different domains, in *Proceedings of the 6th Joint Workshop on Interfaces and Human Decision Making for Recommender Systems*, IntRS'19, pp. 10–18 (2019)

52. V. Nair, G.E. Hinton, Rectified linear units improve restricted boltzmann machines, in *Proceedings of the 27th International Conference on International Conference on Machine Learning*, ICML'10, pp. 807–814 (2010)

53. J. Neve, I. Palomares, Latent factor models and aggregation operators for collaborative filtering in reciprocal recommender systems, in *Proceedings of the 13th ACM Conference on Recommender Systems*, RecSys'19, pp. 219–227 (2019)

54. I. Nunes, D. Jannach, A systematic review and taxonomy of explanations in decision support and recommender systems. User Model. User Adap. Inter. **27**(3–5), 393–444 (2017)

55. S. Osiński, J. Stefanowski, D. Weiss, Lingo: Search results clustering algorithm based on singular value decomposition. Intell. Inf. Syst., 359–368 (2004)

56. S. Ouyang, A. Lawlor, F. Costa, P. Dolog, Improving explainable recommendations with synthetic reviews, in *RecSys Workshop on Deep Learning for Recommender Systems*, DSLR'18 (2018)

57. M. Paşca, E. Alfonseca, Web-derived resources for web information retrieval: From conceptual hierarchies to attribute hierarchies, in *Proceedings of the 32nd International ACM SIGIR Conference on Research and Development in Information Retrieval*, SIGIR'09, pp. 596–603 (2009)

58. A. Paterek, Improving regularized singular value decomposition for collaborative filtering, in *Proceedings of KDD Cup*, pp. 5–8 (2007)

59. F. Radlinski, N. Craswell, A theoretical framework for conversational search, in *Proceedings of the 2017 Conference on Human Information Interaction and Retrieval*, CHIIR'17, pp. 117–126 (2017)

60. Z. Ren, S. Liang, P. Li, S. Wang, M. de Rijke, Social collaborative viewpoint regression with explainable recommendations, in *Proceedings of the 10th ACM International Conference on Web Search and Data Mining*, WSDM'17, pp. 485–494 (2017)

61. S. Rendle, Factorization machines, in *Proceedings of the 2010 IEEE International Conference on Data Mining*, ICDM'10, pp. 995–1000 (2010)

62. S. Rendle, C. Freudenthaler, Z. Gantner, L. Schmidt-Thieme, BPR: bayesian personalized ranking from implicit feedback, in *Proceedings of the 25th Conference on Uncertainty in Artificial Intelligence*, UAI'09, pp. 452–461 (2009)

63. Y. Salem, J. Hong, History-aware critiquing-based conversational recommendation, in *Proceedings of the 22nd International Conference on World Wide Web*, WWW'13 Companion, pp. 63–64 (2013)

64. O.S. Shalom, G. Uziel, A. Karatzoglou, A. Kantor, A word is worth a thousand ratings: Augmenting ratings using reviews for collaborative filtering, in *Proceedings of the 2018 ACM SIGIR International Conference on Theory of Information Retrieval*, SIGIR'18, pp. 11–18 (2018)

65. O.S. Shalom, G. Uziel, A. Kantor, A generative model for review-based recommendations, in *Proceedings of the 13th ACM Conference on Recommender Systems*, pp. 353–357 (2019)

66. S. Shi, H. Zhang, X. Yuan, J.-R. Wen, Corpus-based semantic class mining: distributional vs. pattern-based approaches, in *Proceedings of the 23rd International Conference on Computational Linguistics*, COLING'10, pp. 993–1001 (2010)

67. Y. Shi, M. Larson, A. Hanjalic, Unifying rating-oriented and ranking-oriented collaborative filtering for improved recommendation. Information Sciences **229**, 29–39 (2013)

68. R. Socher, A. Perelygin, J. Wu, J. Chuang, C.D. Manning, A. Ng, C. Potts, Recursive deep models for semantic compositionality over a sentiment treebank, in *Proceedings of the 2013 Conference on Empirical Methods in Natural Language Processing*, EMNLP'13, pp. 1631–1642 (2013)

69. N. Srivastava, G. Hinton, A. Krizhevsky, I. Sutskever, R. Salakhutdinov, Dropout: a simple way to prevent neural networks from overfitting. J. Mach. Learn. Res. **15**(1), 1929–1958 (2014)

70. D.H. Stern, R. Herbrich, T. Graepel, Matchbox: large scale online bayesian recommendations, in *Proceedings of the 18th International Conference on World Wide Web*, WWW'09, pp. 111–120 (2009)

71. Y. Sun, Y. Zhang, Conversational recommender system, in *Proceedings of the 41st International ACM SIGIR Conference on Research & Development in Information Retrieval*, SIGIR'18, pp. 235–244 (2018)
72. A. Vaswani, N. Shazeer, N. Parmar, J. Uszkoreit, L. Jones, A.N. Gomez, Ł. Kaiser, I. Polosukhin, Attention is all you need, in *Advances in Neural Information Processing Systems*, NIPS'17, pp. 5998–6008 (2017)
73. Q. Wang, H. Yin, H. Wang, Q. Viet Hung Nguyen, Z. Huang, L. Cui, Enhancing collaborative filtering with generative augmentation, in *Proceedings of the 25th ACM SIGKDD International Conference on Knowledge Discovery & Data Mining*, KDD'19, pp. 548–556 (2019)
74. X. Wang, D. Chakrabarti, K. Punera, Mining broad latent query aspects from search sessions, in *Proceedings of the 15th ACM SIGKDD International Conference on Knowledge Discovery and Data Mining*, KDD'09, pp. 867–876 (2009)
75. R.J. Williams, Simple statistical gradient-following algorithms for connectionist reinforcement learning. Machine Learning **8**(3–4), 229–256 (1992)
76. F. Wu, J. Madhavan, A. Halevy, Identifying aspects for web-search queries. J. Artif. Intell. Res. **40**, 677–700 (2011)
77. G. Wu, K. Luo, S. Sanner, H. Soh, Deep language-based critiquing for recommender systems, in *Proceedings of the 13th ACM Conference on Recommender Systems*, RecSys'19, pp. 137–145 (2019)
78. P. Xia, B. Liu, Y. Sun, C. Chen, Reciprocal recommendation system for online dating, in *Proceedings of the 2015 IEEE/ACM International Conference on Advances in Social Networks Analysis and Mining 2015*, ASONAM'15, pp. 234–241 (2015)
79. O. Zamir, O. Etzioni, Grouper: A dynamic clustering interface to web search results. Computer Networks **31**(11-16), 1361–1374 (1999)
80. L. Zhang, B. Liu, S.H. Lim, E. O'Brien-Strain, Extracting and ranking product features in opinion documents, in *Proceedings of the 23rd International Conference on Computational Linguistics*, COLING'10, pp. 1462–1470 (2010)
81. Y. Zhang, Q. Ai, X. Chen, W.B. Croft, Joint representation learning for top-n recommendation with heterogeneous information sources, in *Proceedings of the 26th ACM International Conference on Information and Knowledge Management*, CIKM'17, pp. 1449–1458 (2017)
82. Y. Zhang, X. Chen, Q. Ai, L. Yang, W.B. Croft, Towards conversational search and recommendation: System ask, user respond, in *Proceedings of the 27th ACM International Conference on Information and Knowledge Management*, CIKM'18, pp. 177–186 (2018)
83. Y. Zhang, G. Lai, M. Zhang, Y. Zhang, Y. Liu, S. Ma, Explicit factor models for explainable recommendation based on phrase-level sentiment analysis, in *Proceedings of the 37th International ACM SIGIR Conference on Research & Development in Information Retrieval*, SIGIR'14, pp. 83–92 (2014)
84. L. Zheng, V. Noroozi, P.S. Yu, Joint deep modeling of users and items using reviews for recommendation, in *Proceedings of the 10th ACM International Conference on Web Search and Data Mining*, WSDM'17, pp. 425–434 (2017)

# Design and Evaluation of Cross-Domain Recommender Systems

Maurizio Ferrari Dacrema, Iván Cantador, Ignacio Fernández-Tobías, Shlomo Berkovsky, and Paolo Cremonesi

## 1 Introduction

Nowadays, the majority of recommender systems offer recommendations for items belonging to a single domain. For instance, Sky recommends movies and TV series, Zalando recommends clothing, and Spotify recommends songs and playlists. These domain-specific systems have been successfully deployed by numerous websites, and the single-domain recommendation functionality is not perceived as a limitation, but rather pitched as a focus on a certain market segment.

Nonetheless, large e-commerce sites like Amazon and Alibaba often store user feedback for items across multiple domains, and social media users often express their tastes and interests for a variety of topics. It may, therefore, be beneficial to leverage all the available user data provided in various systems and domains, in order to generate more encompassing user models and better recommendations. Instead of treating each domain (e.g., movies, books and music) independently, knowledge acquired in a *source* domain could be transferred to and exploited in another *target* domain. The research challenge of transferring knowledge and the business potential of delivering recommendations spanning multiple domains, have triggered an increasing interest in *cross-domain recommendations*.

M. F. Dacrema · P. Cremonesi (✉)
Politecnico di Milano, Milan, Italy
e-mail: maurizio.ferrari@polimi.it; paolo.cremonesi@polimi.it

I. Cantador · I. Fernández-Tobías
Universidad Autónoma de Madrid, Madrid, Spain
e-mail: ivan.cantador@uam.es; ignacio.fernandezt@uam.es

S. Berkovsky
Macquarie University, Sydney, Australia
e-mail: shlomo.berkovsky@mq.edu.au

© Springer Science+Business Media, LLC, part of Springer Nature 2022          485
F. Ricci et al. (eds.), *Recommender Systems Handbook*,
https://doi.org/10.1007/978-1-0716-2197-4_13

Consider two motivating use cases for cross-domain recommendations. The first refers to the well known cold-start problem, which makes it difficult to generate recommendations due to the lack of sufficient information about users or items. In a cross-domain setting, a recommender may draw on information acquired from other domains to alleviate such a problem, e.g., user's favorite movie genres may be derived from her favorite book genres. The second refers to the generation of personalized cross-selling or bundle recommendations for items from multiple domains, e.g., a movie accompanied by a music album similar to the movie soundtrack. This recommendation may be informed by the user's movie tastes, extracted from rating correlations within a joined movie-music rating matrix.

These use cases are underpinned by an intuitive assumption that there are correspondences between user and item profiles in the source and target domains. This assumption has been validated in several market basket analysis marketing, behavioral, and data mining studies, which uncover dependencies between various domains [69, 78]. Cross-domain recommender systems leverage these dependencies by considering, for example, overlaps between the user or item sets, correlations between user preferences, and similarities of item attributes. Then, they apply a variety of techniques for enriching the knowledge in the target domain and improving the quality of recommendations generated therein.

A vast literature exists on cross-domain recommender systems, with hundreds of papers on the topic being published since 2005. To this end, the goal of the chapter is not to provide a comprehensive review of that body of literature (see Khan et al. [39] for a recent review). Instead, the chapter provides an overview of the scenarios, where cross-domain recommendations are beneficial and categorizes the methodologies available to materialize such recommendations.

The chapter is structured as follows. In Sect. 2 we formulate the cross-domain recommendation problem, describing its main tasks and goals. In Sect. 3 we categorize cross-domain recommendation techniques, and in Sects. 4.1–4.5 we review these techniques. In Sect. 5 we overview cross-domain recommendation evaluation. In Sect. 7 we discuss practical considerations around cross-domain recommenders. Finally, in Sect. 6 we discuss open research directions.

# 2 Formulation of the Cross-Domain Recommendation Problem

The cross-domain recommendation problem has been addressed from various perspectives in different research areas. Aiming to unify these, we provide a generic formulation of the cross-domain recommendation problem, focusing on the existing domain notions (Sect. 2.1), as well as cross-domain recommendation tasks (Sect. 2.2) and goals (Sect. 2.3), and finally discuss the possible scenarios of data overlap between domains (Sect. 2.4).

## 2.1   Definition of Domain

In prior literature, researchers have considered several notions of domain. For instance, some have treated items like *movies* and *books* as belonging to different domains, while others have considered *action movies* and *comedy movies* as different domains. Here we distinguish between several domain notions according to the attributes and types of recommended items. Specifically, *domain* may be defined at three levels (see Fig. 1):

- *Attribute level.* Items are considered as belonging to distinct domains if they differ in the value of certain attributes, e.g., movies might belong to distinct domains if they have different genres, like action and comedy movies. This definition is rather vague and is mainly used as a means to increase the diversity of recommendations (e.g., recommending thrillers to users, who only watch comedies).
- *Item level.* Items are not of the same type, differing in most, if not all, of their attributes. For instance, movies and books belong to different domains, even though they have some common attributes (title, release year).
- *System level.* Items belong to distinct systems, which are considered as different domains. For instance, movies rated in the MovieLens recommender and movies watched in the Netflix streaming service.

In many real world scenarios, the distinction between the above levels is not clear. For instance, we may have one domain that contains cinema movies and another domain that contains TV series. These domains could be considered either at the attribute or system level. Table 1 summarizes the notions of domains, listing

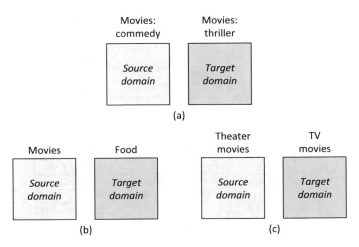

**Fig. 1** Notions of domain according to attributes and types of recommended items. (**a**) *Attribute level*: same type of items (movies) with different values of certain attributes (genre: comedy vs. thriller). (**b**) *Item level*: different types of items (books vs. movies). (**c**) *System level*: same type of items (movies) on different systems (theater vs. TV)

**Table 1** Summary of domain notions, domains, and datasets/systems used in the cross-domain recommendation literature

| Domain notion | Domains | Datasets/Systems | References |
|---|---|---|---|
| *Attribute* | Book categories | *BookCrossing* | Cao et al. [11] |
| | Movie genres | *EachMovie* | Berkovsky et al. [5] |
| | | *MovieLens* | Lee et al. [42] |
| | | | Cao et al. [11] |
| | Education, sport | *HVideo* | Ma et al. [55] |
| *Item* | Books, movies | *LibraryThing, MovieLens* | Zhang et al. [80] |
| | | | Shi et al. [70] |
| | | | Enrich et al. [18] |
| | | *Imhonet* | Sahebi and Brusilovsky [67] |
| | Movies, music | *Facebook* | Shapira et al. [69] |
| | Books, movies, music, TV shows | *Facebook* | Tiroshi and Kuflik [75] |
| | | | Tiroshi et al. [74] |
| | Music, tourism | – | Fernández-Tobías et al. [21] |
| | | | Kaminskas et al. [37] |
| | Books, movies, music | *Amazon* | Hu et al. [33] |
| | | | Loni et al. [52] |
| | | | Zhao et al. [84] |
| | | *Movielens, Douban* | Zhu et al. [86] |
| | Clothing, sport, home | *Amazon* | Liu et al. [50] |
| | Home, office, music | *Amazon* | Wang et al. [77] |
| | Various types of items | – | Winoto and Tang [78] |
| | | | Fu et al. [28] |
| | | | Li et al. [47] |
| | | *Amazon* | Hu et al. [34] |
| | | | Yuan et al. [79] |
| *System* | Movies | *Netflix* | Cremonesi et al. [14] |
| | | *MovieLens, Moviepilot, Netflix* | Pan et al. [64] |
| | Music | *Delicious, Last.fm* | Loizou [51] |
| | | *Blogger, Last.fm* | Stewart et al. [71] |
| | Various domains | *Delicious, Flickr, StumbleUpon, Twitter* | Abel et al. [1] |
| | | | Abel et al. [2] |
| | | *Yahoo! services* | Low et al. [53] |

example papers, along with the type of domain and, when available, the datasets or systems used for experimental evaluation. It can be seen that the focus of past research works has been distributed across the three definitions of domain.

## 2.2 Cross-Domain Recommendation Tasks

Cross-domain recommendation research generally aims to exploit knowledge from a source domain $\mathcal{D}_S$ to perform or improve recommendations in a target domain $\mathcal{D}_T$. Analyzing the literature, we observe that the addressed tasks are diverse, and an agreed-upon definition of the cross-domain recommendation problem has not been formulated yet. Hence, some researchers have proposed models aimed at providing joint recommendations for items belonging to multiple domains, whereas others have developed methods to alleviate the cold-start and sparsity problems in the target domain using information from the source domains.

Aiming to provide a unified formulation of the cross-domain recommendation problem, we define the tasks we identify as providing recommendations across domains. Without loss of generality, we consider two domains $\mathcal{D}_S$ (source) and $\mathcal{D}_T$ (target). The definitions are extensible to multiple source domains. Let $\mathcal{U}_S$ and $\mathcal{U}_T$ denote the sets of users, and let $\mathcal{I}_S$ and $\mathcal{I}_T$ – the sets of items in $\mathcal{D}_S$ and $\mathcal{D}_T$, respectively. The users of a domain are those who interacted with the items in that domain (e.g., ratings, reviews, purchases). Note that not all the items in a domain necessarily need to have interactions with the domain users, as some may have content attributes that establish their membership in the domain.

Sorted in an increasing order of complexity, we distinguish between three recommendation tasks (see Fig. 2):

- *Multi-domain recommendation*: recommend items in both the source and target domains, i.e., items in $\mathcal{I}_S \cup \mathcal{I}_T$ to users in $\mathcal{U}_S \cup \mathcal{U}_T$.
- *Linked-domain recommendation*: recommend items in the target domain to users from the source domain, i.e., items in $\mathcal{I}_T$ to users in $\mathcal{U}_S$, or vice versa, i.e., items in $\mathcal{I}_S$ to users in $\mathcal{U}_T$.
- *Cross-domain recommendation*: recommend items in the target domain to users in the target domain, i.e., items in $\mathcal{I}_T$ to users in $\mathcal{U}_T$.

Multi-domain approaches have focused on the provision of cross-system recommendations, by jointly considering user preferences for items in multiple systems. To generate such recommendations, a significant overlap between user preferences

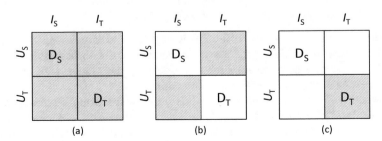

**Fig. 2** Cross-domain recommendation tasks. Colored areas represent the target users and items. (**a**) Multi-domain. (**b**) Linked-domain. (**c**) Cross-domain

across the domains is needed. This is becoming increasingly feasible, since users maintain profiles in various social media systems and there exist interconnecting mechanisms for cross-system interoperability [10] and user identification [9]. The benefits of multi-domain recommendations are also evident in e-commerce, where cross-selling recommendations were shown to boost customer satisfaction/loyalty and businesses profitability [17, 40]. For such purposes, approaches generally aim to aggregate knowledge from the source and target domains.

Linked-domain approaches have been mainly explored to improve recommendations in a target domain where there is a scarcity of user preferences, either at the user/item level (cold-start) or at the community level (data sparsity). To deal with these situations, a common solution is to enrich the available knowledge in the target domain with knowledge imported from the source domain. To generate this type of recommendations, data relations (or overlaps) between the domains are needed, and approaches aim to establish explicit or implicit knowledge-based links between the domains.

Finally, cross-domain approaches have been proposed to provide recommendations in the target domain, where no information about the users is available. In this case, there is no assumption of data relations or overlaps between the domains, and the approaches aim to establish knowledge-based links between domains or to transfer knowledge from the source domain to the target domain.

## 2.3 Cross-Domain Recommendation Goals

From the research and practical perspectives, it is important to match the recommendation algorithms to the task in hand. For this reason, we initially present a taxonomy of cross-domain recommendation goals. The taxonomy is described in a solution-agnostic way: each problem is defined based solely on its goals—disregarding how they are achieved, which will be discussed in Sect. 3.

At the first level of the taxonomy, we consider the three recommendation tasks presented in Sect. 2.2, namely *multi-domain*, *linked-domain*, and *cross-domain* tasks, which are the columns of Table 2. At the second level, we distinguish between the specific goals addressed by cross-domain recommenders, which are the rows of Table 2. We distinguish between the following goals:

- *Addressing the cold-start problem.* This is related to situations, in which the recommender is unable to generate recommendations due to an initial lack of user preferences. One possible solution is to bootstrap the system with preferences from a data source outside the target domain.
- *Addressing the new user problem.* When a user starts using the recommender, this has no knowledge of the user's tastes and preferences, and cannot produce personalized recommendations. This may be solved by exploiting the user's preferences collected in a different domain.

**Table 2** Summary of cross-domain recommendation approaches based on goals and tasks

| Goal | Multi-domain task | Linked-domain task | Cross-domain task |
|---|---|---|---|
| *Cold start* | Wang et al. [77] | | Shapira et al. [69] |
| | | | Hu et al. [34] |
| | | | Zhao et al. [84] |
| *New user* | | Winoto et al. [78] | Berkovsky et al. [4, 5] |
| | | Cremonesi et al. [14] | Berkovsky et al. [6] |
| | | Low et al. [53] | Nakatsuji et al. [58] |
| | | Hu et al. [33] | Cremonesi et al. [14] |
| | | Sahebi et al. [67] | Tiroshi et al. [75] |
| | | | Braunhofer et al. [7] |
| | | | Man et al. [56] |
| | | | Fu et al. [28] |
| *New item* | | | Kaminskas et al. [37] |
| *Accuracy* | Cao et al. [11] | Shi et al. [70] | Pan et al. [64] |
| | Zhang et al. [80] | Pan et al. [59] | Stewart et al. [71] |
| | Li et al. [46] | | Pan et al. [63] |
| | Tang et al. [73] | | Tiroshi et al. [74] |
| | Zhang et al. [82] | | Loni et al. [52] |
| | Liu et al. [49] | | |
| | Taneja et al. [72] | | |
| | Zhu et al. [86] | | |
| | Ma et al. [55] | | |
| | Yuan et al. [79] | | |
| | Li et al. [47] | | |
| | Liu et al. [50] | | |
| *Diversity* | | Winoto et al. [78] | |
| *User model* | | Abel et al. [1] | |
| | | Abel et al. [2] | |

- *Addressing the new item problem.* When a new item is added to a catalog, no prior ratings for the item are available, so it cannot be recommended by a collaborative recommender. This problem is particularly evident when cross-selling new products from different domains.
- *Improving accuracy.* In many domains, the average number of ratings per user and item is low, which may negatively affect the quality of the recommendations. Data collected outside the target domain can increase the rating density, and upgrade the recommendation quality.
- *Improving diversity.* Having similar, redundant items in a recommendation list may degrade user experience. The diversity of recommendations can be improved by considering multiple domains, better covering the range of user preferences.
- *Enhancing user models.* The main goal of cross-domain user modeling applications is to enhance user models. Achieving this goal may have personalization-

oriented benefits, such as discovering new user preferences for the target domain [71] or enhancing similarities between users and items [1, 6].

Table 2 maps the key cross-domain recommendation papers across these tasks and goals. It is evident that cross-domain tasks are mainly used to address the cold-start problem and reduce data sparsity, while linked-domain tasks are used to improve accuracy and diversity.

## 2.4 Cross-Domain Recommendation Scenarios

In order to leverage cross-domain recommendations, source and target domains must be linked. We refer to two domains being

- *collaborative-linked* if there are users or items with interactions in both domains
- *content-linked* if they share users or items with similar attributes
- *context-linked* if interactions share contextual attributes.

The type of overlap between domains may limit the choice of algorithms that can be used for cross-domain recommendations. For instance, if two domains have items or users that share attributes, but no users or items share interactions, collaborative algorithms cannot be used for cross-domain recommendations [15].

We denote by $\mathcal{U}_{ST}$ the set of linked users. In the content-based linkage, two cross-domain users are linked if they share attributes, e.g., linked in a social network, have the same age, or tagged an item with the same tag. In the collaborative case, users are linked if they have interactions in both the source and target domains. Similarly, we denote by $\mathcal{I}_{ST}$ the set of linked items. In the content-based linkage, items are linked if they share attributes, e.g., movies of the same genre. In the collaborative case, items are linked if they have interactions in both the source and target domains. Finally, we denote by $\mathcal{C}_{ST}$ the set of interactions linked by sharing contextual attributes. For example, the same tag might have been used by users from different domains or users might have rated items within the same context [60].

Extending the patterns described in [14], three basic scenarios of data overlap between two domains $S$ and $T$ can be identified:

- *User overlap*: there are linked users, i.e., $\mathcal{U}_{ST} \neq \varnothing$.
- *Item overlap*: there are linked items, i.e., $\mathcal{I}_{ST} \neq \varnothing$.
- *Context overlap*: there are linked interactions, i.e., $\mathcal{C}_{ST} \neq \varnothing$.

Note that not all the combinations of these scenarios are possible: if there is *no overlap* between users, items, and context, i.e., $\mathcal{U}_{ST} = \varnothing$, $\mathcal{I}_{ST} = \varnothing$, and $\mathcal{C}_{ST} = \varnothing$, cross-domain recommendations are not possible, as shown in [15].

The links between items, users and contexts are known a-priori and constitute an explicit input for cross-domain recommendations. However, *deep learning* techniques allows for a fourth scenario, where the domains are linked through semantic concepts extracted from unstructured information, e.g., user reviews for items, rather than by the means of structured attributes. These semantic concepts allow learning a-posteriori implicit relations between domains during the training of the model. We refer to this fourth scenario as *semantic overlap*.

# 3 Categorization of Cross-Domain Recommendation Techniques

Cross-domain recommendations have been addressed from various angles in different research areas. This has entailed the development of an array of approaches, which in many cases are difficult to directly compare due to the diversity in the user preferences they use, the cross-domain scenario they deal with, and the algorithms and data they harness. In this section, we overview the different categorizations proposed in the literature [8, 14, 22, 36, 43]. However, the published reviews and categorizations often do not reflect the complexity of the space. Thus, in the next section we will propose a unifying view for the existing cross-domain recommendation techniques.

Chung et al. presented in their seminal work [12] a framework that provides integrated recommendations for items that may be of different types and belong to different domains. The framework accounts for three levels of integration: *single item type recommendations* that consist of items of the same type, *cross item type recommendations* that consist of items of different types that belong to the same domain, and *cross domain recommendations* that consist of items that belong to different domains. The authors stated that integrated recommendations can be generated by following at least three approaches:

- *General filtering.* Instantiates a recommendation model for multiple item types that may belong to different domains.
- *Community filtering.* Utilizes ratings shared among several communities or systems that may deal with different item types and domains.
- *Market basket analysis.* Applies data mining to extrapolate hidden relations between items of different types/domains and build a model for item filtering.

Loizou identified three main trends in cross-domain recommendation research [51]. The first focuses on compiling unified user profiles for cross-domain recommendations. This is considered as integration of domain-specific user models into a single, unified multiple-domain user model, which is subsequently used for generating recommendations. The second involves profiling user preferences through monitoring their interactions in individual domains, which can be implemented by agents that learn single-domain user preferences and gather them across the domains. The third deals with combining (or mediating) information from several single-domain recommender systems. A number of strategies for mediating single-domain collaborative systems were considered: exchange of ratings, user neighborhoods, user similarities, and recommendations.

Based on these trends, Cremonesi et al. surveyed and categorized cross-domain collaborative systems [14]. They enhanced earlier categorizations by considering a more specific grouping of approaches: (i) Extracting association rules from rating behavior in a source domain, and using the extracted rules to recommend items in a target domain, as proposed by Lee et al. [42]; (ii) Learning inter-domain rating-based similarity and correlation matrices, as proposed by Cao et al. [11] and Zhang et al. [80]; (iii) Combining estimations of rating probability distributions in source domains to generate recommendations in a target domain, as proposed by Zhuang et al. [87]; (iv) Transferring knowledge between domains to address the rating sparsity in a target domain, as proposed by Li et al. [44, 45] and Pan et al. [62, 63].

For the last group, Li surveyed transfer learning techniques for cross-domain collaborative filtering [43]. They proposed a categorization based on the type of domain and distinguished between (i) *system domains* associated with different recommenders and representing a scenario, where the data in a target recommender are sparse, while the data in related recommenders are abundant; (ii) *data domains* associated with multiple sources of heterogeneous data and representing a scenario where user data in source domains can be obtained easier than in a target domain; and (iii) *temporal domains* associated with distinct data periods and representing a scenario where temporal user preference dynamics can be captured. Reflecting these categories, three recommendation strategies differing in the cross-domain knowledge transfer can be considered:

- *Rating pattern sharing.* Factorizes single-domain rating matrices utilizing user and item groups, encodes group-level rating patterns, and transfers knowledge between domains through the encoded patterns [44–46].
- *Rating latent feature sharing.* Factorizes single-domain rating matrices using latent features, shares latent feature spaces across domains, and transfers knowledge between domains through the latent feature matrices [62–64].
- *Domain correlation.* Factorizes single-domain rating matrices using latent features, explores correlations between latent features in individual domains, and transfers knowledge between domains through such correlations [11, 70, 80].

Pan and Yang identified in a survey of transfer learning three key questions [61]: (i) *what* to transfer—which knowledge should be transferred between domains; (ii) *how* to transfer—which algorithms should be exploited to transfer the knowledge; and (iii) *when* to transfer—in which situations the knowledge transfer is beneficial. Focusing on the first two, Pan et al. proposed a two-dimensional categorization of transfer learning-based approaches for cross-domain collaborative filtering [62, 63]. The first dimension takes the type of transferred knowledge into account, e.g., latent rating features, encoded rating patterns, and rating-based correlations. The second considers the algorithm, and distinguishes between adaptive and collective approaches, assuming, respectively, the existence of rating data in the source domain only, and in both the source and target domains.

More recently, Fernández-Tobías et al. stepped beyond collaborative recommendations, considering approaches that establish cross-domain relationships not necessarily based on ratings [22]. They identified three directions to the cross-domain recommendation problem. The first is through the integration of single-domain user preferences into a unified cross-domain user model, which aggregates user profiles from multiple domains and mediates user models across domains. The second direction transfers knowledge from a source domain to a target domain, and includes approaches that exploit recommendations generated for a source domain in a target domain, as well as approaches based on transfer learning [43]. The third direction is around establishing explicit relations between domains, which may be based either on content-based relations between items or rating-based relations between users/items. The authors then proposed a two-dimensional categorization of cross-domain recommendation approaches: (i) relation between domains: content-based relations (item attributes, tags, semantic properties, and feature correlations) vs. rating-based relations (rating patterns, rating latent factors, and rating correlations); and (ii) recommendation task: adaptive models (exploit source domain knowledge to generate recommendations in a target domain) vs. collective models (harness data from several domains to improve recommendations in a target domain).

## 4 Cross-Domain Recommendations Techniques

It is evident that the existing categorizations of cross-domain recommendation are diverse. In this section we reconcile these categorizations in a way that captures and unifies their core ideas. For this, we classify cross-domain recommendation techniques into five categories, generally reflecting their evolution:

- *Merging user preferences.* User preferences from both domains are aggregated in such a way that single-domain recommender systems can be used.
- *Linking domains.* Graph-based approaches, in which nodes from different domains are linked by the means of shared attributes or interactions.
- *Transfer learning.* Knowledge obtained while training a model on the source domain is transferred to make recommendations in the target domain.
- *Co-training of shared features (multi-task learning).* Multiple tasks are solved at the same time in the source and target domains, to learn shared latent features.
- *Deep learning.* Neural network models are used to share/transfer latent features across domains through unstructured semantic features extracted during training.

This classification is not always clear-cut, as some technique capture aspects from different categories. We overview these five categories in the following sections.

### 4.1 Merging User Preferences

Merging user preferences from different source domains is among the most widely used strategies for cross-system personalization, and the most natural way to address the cross-domain recommendation problem (see Fig. 3). This family of techniques requires a partial user overlap between the two domains. Table 3 summarizes the aggregation-based methods presented in this section.

Prior research has shown that rich profiles can be generated for users when multiple sources of personal preferences are combined, revealing tastes and interests not captured in isolated domains [2]. It has been also shown that enriching sparse user preferences in a target domain by adding data from the source domains can improve the generated recommendations under the cold-start and sparsity conditions [67, 69]. This, however, requires considerable user data in multiple domains and methods for merging the data, potentially represented in different ways.

The most favorable scenario for aggregation-based methods implies that different systems share user preferences of the same type and representation [3]. This scenario was addressed by Berkovsky et al. with a mediation strategy for cross-domain collaborative filtering [5]. The authors considered a domain-distributed setting, where a global rating matrix is split, so that single-domain recommenders store locally rating matrices having the same structure. In this setting, a target domain

**Fig. 3** *Merging user preferences.* Data sources from different domains are merged, and traditional single-domain recommender system is used on the merged data

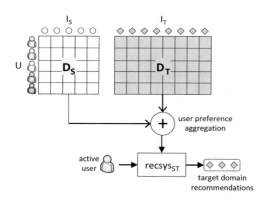

**Table 3** Cross-domain user modeling and recommendation approaches based on merging single-domain user preferences

| Cross-domain approach | Overlap | References |
|---|---|---|
| *Aggregating user ratings into a single multi-domain rating matrix* | $UI$ | Berkovsky et al. [5] |
| | $U$ | Winoto and Tang [78] |
| | $U$ | Sahebi and Brusilovsky [67] |
| | $U$ | Shapira et al. [69] |
| *Using a common representation for users* | $U$ | Abel et al. [1] |
| | $U$ | Abel et al. [2] |
| *Mapping user preferences onto domain-independent features* | $U$ | Loni et al. [52] |

($U$) user overlap, ($UI$) user and item overlap

recommender imports the rating matrices from the source domains, reconstructs the unified rating matrix, and applies collaborative filtering. Although this approach can be seen as a centralized recommender with user data split across multiple domains, smaller rating matrices can be efficiently maintained by local systems and shared with the target domain only when requested.

Berkovsky et al. showed an improvement in the accuracy of the target domain recommendations when aggregating ratings from several domains [5] . This was also observed by Winoto and Tang [78], where the authors collected ratings in several domains and conducted a study that revealed that even when there exists significant overlap between domains, recommendation accuracy in the target domain improved if only ratings from close domains were used. In addition, Winoto and Tang stated that cross-domain recommendations might benefit serendipity and diversity.

Apart from serendipity and diversity, other benefits of cross-domain recommendations have been identified. Sahebi and Brusilovsky examined the impact of the size of user profiles in the source and target domains on the quality of collaborative filtering, and showed that aggregating ratings from several domains improves the accuracy of cold-start recommendations [67]. Similarly, Shapira et al. showed substantial accuracy improvements yielded by aggregation-based methods, when the available user preferences were sparse [69].

Related to these, Abel et al. studied aggregation of tag clouds from multiple systems [1]. They evaluated a number of methods for semantic enrichment of tag overlaps between domains, via tag similarities and association rules deduced from the tagging data across systems. Analyzing commonalities and differences among tag-based profiles, Abel et al. also mapped tags to WordNet categories and DBpedia concepts [2]. They used the mapped tags to build category-based user profiles, which revealed significantly more user information than system-specific profiles.

The final type of cross-domain recommendations based on user preference aggregation is formed by approaches that map user preferences from multiple domains to domain-independent features, and use these feature-based profiles for building models that predict user preferences in the target domain. Loni et al. developed an approach that encoded rating matrices from multiple domains as real-valued feature vectors [52]. With these vectors, an algorithm based on factorization machines [66] found patterns between features from the source and target domains, and produced preference estimations associated with the input vectors.

Overall, this family of techniques constitutes one simple yet effective baseline that should be included in any evaluation protocol, unless the overall size of the merged domains makes the problem intractable.

## 4.2  Linking Domains

Instead of aggregating user preferences directly, several works focused on directed weighted graphs linking user preferences from multiple domains. Such inter-domain correspondences may be established directly using common knowledge, e.g., item

**Fig. 4** *Linking domains.* A
graph is used to link items or
users from different domains

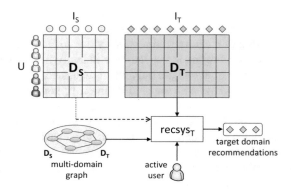

**Table 4** Cross-domain recommendation approaches using graph methods

| Cross-domain approach | Overlap | References |
|---|---|---|
| *Linking user preferences via a multi-domain graph* | U | Nakatsuji et al. [58] |
| | U | Cremonesi et al. [14] |
| | U | Tiroshi et al. [74] |
| | U | Farseev et al. [20] |
| *Building semantic network linking domain concepts* | I | Fernández-Tobías et al. [21] |
| | I | Kaminskas et al. [37] |

(*I*) item overlap, (*U*) user overlap

attributes, semantic networks, association rules, and inter-domain preference-based
similarities or correlations (see Fig. 4). These offer valuable sources of information
for cross-domain reasoning. A recommender could identify potentially relevant
items in the target domain by selecting those that are related to others in the
source domains, and for which the user has expressed a preference. Besides,
inter-domain similarities and correlations can be exploited to adapt or combine
knowledge transferred from different domains. Table 4 summarizes graph-based
methods discussed in this section.

Nakatsuji et al. presented an approach that built domain-specific user graphs,
where nodes were associated with users and edges reflected rating-based user
similarity [58]. Domain graphs were connected via users, who either rated items
in several domains or shared social connections, to create a cross-domain user
graph. Over this graph, a random walk algorithm retrieved items liked by users
associated with the extracted nodes. Cremonesi et al. built a graph, where nodes
were associated with items and edges reflected rating-based item similarity [14].
The inter-domain connections were the edges between pairs of items in different
domains. The authors enhanced inter-domain edges by discovering new edges
and strengthening the existing ones. Tiroshi et al. collected a dataset containing
user preferences in multiple domains extracted from social network profiles [74].
The data was merged into a bipartite user-item graph, and statistical and graph-
based features of users and items were extracted. These features were exploited

by an algorithm that addressed recommendations as a binary classification task. Farseev et al. [20] proposed a cross-network model, which combined the individual behavior, user communities, multiple social media sources, and heterogeneous data, such as text and images. The relationships between users were modeled as a multi-layered graph.

In a realistic setting, items can be heterogeneous, with no common attributes between domains [76]. To address this, more complex, likely indirect relations between items in different domains, may be established. Hence, when suitable knowledge repositories are available, concepts from several domains can be connected by the means of semantic properties, forming networks that explicitly link domains of interest. Along these lines, Fernández-Tobías et al. [21] and Kaminskas et al. [37] developed knowledge-based frameworks of semantic networks linking concepts across domains. These networks were weighted graphs, where nodes with no incoming edges represented concepts from the source domain, and nodes with no outgoing nodes represented concepts from the target domain. The framework facilitated an algorithm that propagated the node weights, in order to identify concepts most related to the source concepts. Implemented on top of DBpedia, the framework was deployed to recommend music suited to places of interest, related through concepts from several domains and spatio-temporal data.

## 4.3  Transfer Learning

We now survey cross-domain recommendation approaches that transfer knowledge between domains, enhancing the information available in the target domain. The knowledge transfer can be done explicitly via common item attributes, implicitly via shared latent features, or by the means of rating patterns transferred between the domains (see Fig. 5). Table 5 summarizes the methods presented in this section.

**Fig. 5** *Transfer learning.* A model is learned in the source domain and used in the target domain

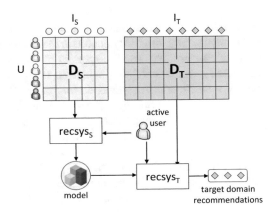

**Table 5** Cross-domain recommendation approaches based on transfer learning. Other transfer learning approaches are described in Table 7

| Cross-domain approach | Overlap | References |
| --- | --- | --- |
| *Aggregating neighbourhoods to generate recommendations* | $U$ | Berkovsky et al. [5] |
| | $UI$ | Tiroshi and Kuflik [75] |
| | $U$ | Shapira et al. [69] |
| *Exploiting user neighborhoods to enhance target user models* | $I$ | Stewart et al. [71] |
| *Combining probabilistic user models* | $U$ | Low et al. [53] |
| *Combining heterogeneous user preferences* | $UI$ | Pan et al. [64] |
| *Extracting association rules from user rating behavior* | $U$ | Lee et al. [42] |
| *Using latent features from source domains to regularize latent features in target domain* | $UI$ | Pan et al. [63] |
| *Using contextual bandits* | $U$ | Liu et al. [49] |
| *Modeling users, items and domains with tensors* | $C$ | Taneja et al. [72] |
| *Using a mapping function between the latent factors of the source and target domains* | $UI$ | Man et al. [56] |

($U$) user overlap, ($I$) item overlap, (C) context overlap, ($UI$) user and item overlap

The key idea behind transfer learning is that importing any user modeling data from source recommenders may benefit a target recommender. For example, in a collaborative system, cross-domain mediation may import the list of nearest neighbors. This example is underpinned by two assumptions: (i) there is overlap of users between domains, and (ii) user similarity spans across domains, i.e., if two users are similar in a source domain, they may be similar also in the target domain [5]. Aggregating the lists of nearest neighbors relies on their data in the target domain only, which may be sparse and result in noisy recommendations. Thus, one could import and aggregate also the degree of their similarity in the source domain.

Weighted k-NN aggregation was further enhanced by Shapira et al. [69]. They used multi-domain Facebook data to produce the set of candidate nearest neighbors, and compute their similarity degree in the source domain. This allowed overcoming the new user problem and the sparsity of ratings in the target domain. The authors compared several weighting schemes, the performance of which was consistent across metrics and recommendation tasks. Tiroshi and Kuflik also harnessed multi-domain Facebook data [75]. They applied random walks to identify source domain-specific neighbor sets, allowing to generate recommendations in the target domain. These cross-domain mediation scenarios assume an overlap in the sets of users. A similar scenario refers to a setting, where items overlap between the source and target domains that paves the way for further mediation. One of them, involving only two systems in the music domain, was studied by Stewart et al. [71]. The authors leveraged the tags assigned on Last.fm to recommend tags on Blogger.

Moving from collaborative to latent factor methods, we highlight two works compatible with the user modeling data mediation pattern. Low et al. developed a probabilistic model combining user information across multiple domains and facilitating personalization in domains with no prior user data [53]. The model was underpinned by a global user profile based on a latent vector and a set of domain-specific factors that eliminated the need for common items or features. Pan et al. dealt with transferring uncertain ratings, i.e., rating range or distribution derived from behavioral logs, using latent features of users and items [64]. The uncertainty was transferred from the source to the target domain and harnessed there as constraints for a matrix factorization model. Taneja et al. [72] proposed a tensor factorization algorithm, which transferred the knowledge in the two domains via the genre of the heterogeneous items, clustered according to their features.

Lee et al. exploited rating patterns for cross-domain recommendation [42]. There, global nearest neighbors were identified by adding the similarity scores from each domain. Then, patterns of items commonly rated together by a set of neighbors were discovered using association rule mining. Finally, rating predictions were computed by the standard user-based collaborative filtering, but enhanced with the rules containing the target items.

Pan et al. addressed the sparsity problem by exploiting user and item information from auxiliary domains, where user feedback might be represented differently [63]. In particular, they studied the case of binary like/dislike preferences in the source domain and 1–5 star ratings in the target domain. They performed singular value decomposition (SVD) in each auxiliary domain, in order to separately compute user and item latent factors, which were shared with the target domain. Specifically, transferred factors were integrated into a factorization of the rating matrix in the target domain and added as regularization terms, so that the characteristics of the target domain could be captured. Man et al. [56] instead addressed the problem of users and items with insufficient interactions in the target domain. The proposed method learned the embeddings with traditional matrix factorization and then trained a linear and non-linear mapping function to compute the embeddings in the target domain given the embeddings in the source domain.

Liu et al. [49] proposed a transferable contextual bandit for both homogeneous and heterogeneous domains. The use of transfer learning aimed to improve the exploitation steered by the policy of the contextual bandit as well as accelerate its exploration in the target domain. The proposed model leveraged both the source and target domain simultaneously via common users and items, and then exploited a translation matrix to match the different feature spaces.

Works by Li et al. [45], Moreno et al. [57], Gao et al. [30], Zang et al. [81] and He et al. [31] transferred rating patterns across multiple domains with simultaneous co-clustering of users and items. Clustering was achieved using a tri-factorization of the source rating matrix [16]. Then, knowledge was transferred through a *codebook*, a compact cluster-level matrix computed in the source domain. Missing ratings in the target domain were predicted using the same codebook. However, Cremonesi and Quadrana disproved the effectiveness of codebook-based transfer methods [15], showing that the codebook could not transfer knowledge when source and target domains did not overlap. Therefore, it is still an open question whether codebook-based approaches cab reliably perform knowledge transfer.

## 4.4 Co-Training of Shared Latent Features

Latent factors shared between domains can be exploited by cross-domain recommenders, as illustrated in Fig. 6. Instead of transferring knowledge, shared models can be learned simultaneously in both the source and target domains. Table 6 summarizes the co-training methods presented in this section.

Pan et al. proposed to learn latent features simultaneously in multiple domains [62]. Both user and item factors were assumed to generate the observed domain ratings and their corresponding random variables were shared between

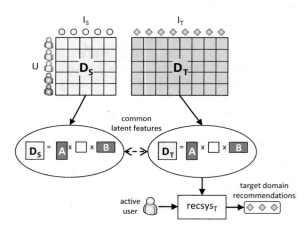

**Fig. 6** *Sharing latent features.* Latent features models are learnt simultaneously in the source and target domains, requiring identical user and/or item features across the domains

**Table 6** Cross-domain recommendation approaches based on co-training. Other co-training approaches are described in Table 7

| Cross-domain approach | Overlap | References |
|---|---|---|
| *Using latent factors to jointly factorize the rating matrices in the source and target domains* | $UI$ | Pan et al. [62] |
| *Extending matrix factorization with latent factors associated to social tags* | $C$ | Enrich et al. [18] |
| | $C$ | Fernández-Tobías and Cantador [23] |
| *Sharing latent features via a user-item-domain tensor factorization* | $U$ | Hu et al. [33] |
| *Constraining matrix factorization with inter-domain similarities* | $U$ | Cao et al. [11] |
| | $U$ | Zhang et al. [80] |
| | $C$ | Shi et al. [70] |

$(U)$ user overlap, $(C)$ context overlap, $(UI)$ user and item overlap probabilistic factorization models of each rating matrix. The factorization method was further extended by incorporating factors capturing domain-specific information, resulting in a tri-factorization scheme. A limitation of the approach was that the users and items from the source and target domains had to be identical. Zhang et al. adapted the probabilistic matrix factorization method to include a probability distribution of user latent factors that encoded inter-domain correlations [80]. One strength of this approach was that user latent factors shared across domains were not needed, allowing more flexibility in the heterogeneity of domains.

Rather than learning implicit correlations in the data, Shi et al. relied on shared social tags to compute cross-domain user-to-user and item-to-item similarities [70]. In similar to previous approaches, rating matrices from the source and target domains were jointly factorized, but user and item latent factors from each domain were restricted, so that their product was consistent with the tag-based similarities. Another way of exploiting co-training is to learn similarities from multiple domains simultaneously. For instance, Cao et al. developed an approach, which implicitly learned inter-domain similarities from the data as model parameters in a non-parametric Bayesian framework [11]. Since user feedback was used to estimate the similarities, user overlap between the domains was required.

Enrich et al. [18] and Fernández-Tobías and Cantador [23] studied the influence of social tags on rating predictions, as a knowledge transfer approach for cross-domain recommendations. The authors presented a number of models based on the SVD++ algorithm [41], aiming to incorporate the effect of tag assignment into rating estimation. The underlying hypothesis was that information about item annotation in a source domain could be exploited to improve predictions in a target domain, as long as a set of common tags between the domains existed. In the proposed models, tag factors were added to the latent item vectors and combined with user latent features to estimate ratings. In all the, models knowledge transfer was performed through the shared tag factors in a collective way, since these were computed jointly for the source and the target domains.

Hu et al. presented a more complex approach that considered also the domain factors [33]. The authors proposed a tensor factorization algorithm that exploited the triadic user-item-domain data. Specifically, they studied the use case, in which the same set of users consumed and rated different types of items. In this case, rating matrices from several domains were simultaneously decomposed into shared user, item, and domain latent factors, and genetic algorithm was then deployed to estimate the optimal weights of the domains.

## 4.5 Deep Learning

Deep learning can be deployed to design sophisticated cross-domain architectures, where collaborative, content-based, and other features are shared across the domains. One of the key advantages of deep learning over traditional machine

**Table 7** Cross-domain recommendation approaches based on deep learning

| Cross-domain approach | Overlap | References |
|---|---|---|
| *Transfer-Learning* | $UI$ | Zhu et al. [85] |
| | $U$ | Yuan et al. [79] |
| | $U$ | Hu et al. [34] |
| | $U$ | Ma et al. [55] |
| | $US$ | Fu et al. [28] |
| | $US$ | Zhao et al. [84] |
| | $U$ | Li et al. [47] |
| | $US$ | Liu et al. [50] |
| *Co-Training* | $U$ | Lian et al. [48] |
| | $U$ or $S$ | Zhu et al. [86] |
| | $US$ | Wang et al. [77] |
| | $UI$ | Kang et al. [38] |

$(U)$ user overlap, $(S)$ semantic overlap, $(UI)$ user and item overlap, $(US)$ user and semantic overlap

learning method is the ability to design cross-domain recommender systems between the domains without an explicit overlap of users, items, or attributes. This is possible when the links between domains are created using features extracted from the data. For instance, considering free-text reviews, users or items may be linked if they share reviews expressing a similar opinion. These algorithms cannot be directly mapped into one of the previous categories, although all of them adopt either *transfer learning* or *co-training* to share knowledge between the domains (Table 7).

### 4.5.1 Transfer Learning

Transfer learning is the most frequently used approach when implementing deep learning algorithms for cross-domain recommender systems. In general, a neural network model trained on the source domain is enriched with layers or components aimed to learn a new model in the target domain.

Zhu et al. computed benchmark latent factors by merging the user and item latent factors in both domains according to their sparsity and neighborhood [85]. Then, a deep function was used to map the benchmark factors with the ones computed in each domain. Hu et al. leveraged unstructured text for collaborative recommendations [34]. The model consisted of memory, transfer and prediction modules. The user and target item were first embedded in a low-dimensional space. The memory module modeled the relations between the text semantics, represented via a dictionary of word embeddings, and user preferences. The transfer module learned a non-linear function transferring knowledge from the embeddings of the source items the user interacted with to the target item. The prediction module was implemented as a multilayered perceptron (MLP) that predicted the user-item rating based on a concatenation of their embeddings. An additional layer considered the outputs of all three modules and computed the predicted user-item score.

Zhao et al. developed a review-based model, where aspects were extracted from reviews and used for knowledge transfer [84]. Users and items were represented by *documents* containing all their reviews. The aspects were extracted using text convolution and attention mechanisms, and the aspects were correlated across domains by training on the overlapping users. For an overlapping user and a target domain item, the user's document in the source domain and the item's document in the target domain were used for rating prediction. In addition to reviews, Fu et al. considered interactions and item features [28]. The model embedded reviews in the user profile encoding and MLP transferred the latent factors. The domains were linked by mapping the embeddings of overlapping users and the MLP generated embeddings in the target domain from the source domain embeddings.

Li et al. proposed a model for bidirectional recommendations, able to generate recommendations both in the source and target domains via a dual transfer learning mechanism [47]. For this, a latent orthogonal mapping represented user preferences in multiple domains in a way that preserved the relations between users across latent spaces and allowed learning user similarity in both domains simultaneously. The model represented users, items, and features by a latent vector computed with an autoencoder based on an MLP. The information was transferred across domains with a latent orthogonal matrix, which preserved the users' similarity and allowed for easy bidirectional transfer through a straightforward inverse. Liu et al. [50] incorporated into a similar model user's aesthetic preferences, associated with personality [13]. These were extracted using a pre-trained aesthetic network and assumed to be domain-independent. The embeddings of the user, source items, target items, and aesthetic traits were used as input for a cross-transfer network connecting the domains and facilitating bidirectional transfer.

### 4.5.2   Co-Training

Co-training in deep learning is used when there is a limited amount of labeled data, but a large amount of unlabeled data. In the case of cross-domain recommender systems, neural networks are used to jointly learn models for the source and target domains, aiming at balancing and complementing their available data.

Multiple tasks are solved at the same time in the source and target domains to learn shared latent features. Zhu et al. developed an algorithm building user and item embeddings from heterogeneous sources: reviews, tags, user profiles, and item features [86]. The embeddings were shared between the domains for overlapping users and items. The model consisted of an embedding layer for interaction and features, a sharing layer connecting the domains, and a layer aggregating domain-specific and shared embeddings. Wang et al. proposed an adversarial model that addressed the cold start and data imbalance problems [77]. The models learned latent representations for users, items and user-item pairs, which were transferred to the target domain. The functions used to transfer the embeddings were learned with a min-max adversarial game, in a way that the embeddings generated from the source and target domains were indistinguishable. Reviews and images of products were used to enhance the latent representation of items. Multiple strategies were proposed for various scenarios of user and item overlaps between the two domains.

Yuan et al. developed a pure collaborative model that transferred rating patterns between domains having the same set of users [79]. First, the embeddings of users and items were computed for each domain independently, and then all these were fed into a neural network that generated recommendations for both domains, balancing extraction of transferable rating patterns and predictive accuracy. Ma et al. studied sequential recommendations in multiple domains [55]. Their model contained a dedicated unit for cross-domain transfer that combined domain-specific models using gated recurrent units (GRUs) and trained according to the data timestamps. The representation learned in the target domain was combined with the one learned in another domain, such that the recommendations were computed using information from both the domains.

Lian et al. [48] proposed a hybrid model, where both collaborative and content-based data were represented in the same latent space and used a deep mapping function between the source and target embeddings. Kang et al. [38] addressed the issue of a limited number of overlapping users by proposing first to learn their embeddings in a metric space, and then to learn a mapping function by taking into account both the overlapping users in a supervised way and the non-overlapping items and users in an unsupervised way, therefore, including in the learning process the user's neighborhood.

# 5    Evaluation of Cross-Domain Recommender Systems

A central topic of research in cross-domain recommender systems lies in the evaluation of recommendation algorithms. Although the type of issues covered by cross-domain recommender systems has gradually expanded over the years, classical rating prediction and top-N recommendation problems still attract most of the attention, and the research community has developed a seemingly standardized way of evaluating these problems.

In the vast majority of published papers, algorithms are compared through offline experiments on historical data. Such experiments are typically easier to conduct than online studies and live evaluations, as they require no interactions with real users [27, 68].[1] With offline experiments, a system is evaluated by analyzing past user preferences. Thus, progress can be claimed if a new algorithm is better at predicting test data than previous ones in terms of predictive error measures (MAE,

---

[1] The reader is referred to Chap. 29 for an extensive discussion on the different methods used to evaluate recommender systems.

**Table 8** Metrics for evaluation of cross-domain recommender systems

| Category | References | |
|---|---|---|
| *Error metrics* | Berkovsky et al. [4, 5] | Hu et al. [33] |
| | Berkovsky et al. [6] | Sahebi et al. [67] |
| | Pan et al. [59] | Shapira et al. [69] |
| | Winoto et al. [78] | Loni et al. [52] |
| | Cao et al. [11] | Man et al. [56] |
| | Nakatsuji et al. [58] | Zhu et al. [85] |
| | Pan et al. [63] | Fu et al. [28] |
| | Zhang et al. [80] | Wang et al. [77] |
| | Li et al. [46] | Yuan et al. [79] |
| | Shi et al. [70] | Li et al. [47] |
| | Pan et al. [64] | Zhao et al. [84] |
| *Ranking metrics* | Abel et al. [1] | Man et al. [56] |
| | Tiroshi et al. [75] | Hu et al. [34] |
| | Abel et al. [2] | Kang et al. [38] |
| | Hu et al. [33] | Ma et al. [55] |
| | Shapira et al. [69] | Zhu et al. [86] |
| | Zhang et al. [82] | Liu et al. [50] |
| | Farseev et al. [20] | |
| *Classification metrics* | Stewart et al. [71] | Farseev et al. [20] |
| | Nakatsuji et al. [58] | Taneja et al. [72] |
| | Cremonesi et al. [14] | Ma et al. [55] |
| | Kaminskas et al. [37] | Li et al. [47] |
| | Tiroshi et al. [74] | |

MSE, etc), classification accuracy measures (Precision, Recall, Fallout, F1, etc), or ranking accuracy measures (MAP, NDCG, etc) [32]. Table 8 provides an overview of the evaluation metrics exploited in prior literature.

In this section, we discuss methods and best practices for offline evaluation of cross-domain recommender systems. The key point to bring up in this context is that such systems cannot be evaluated in a problem-independent way. That is, it is impossible to assess whether a cross-domain recommender system is an appropriate solution without taking into account for what it was intended for. The nature of the evaluation must be connected to the purpose, for which the recommendations were originally conceived, as outlined in Sect. 2.3.

Two key points in the evaluation of cross-domain recommender systems differ significantly from the evaluation of single-domain recommender systems: *data partitioning* and *sensitivity analysis* (e.g., studies of the relative density of domain-specific datasets and degree of user/item overlaps between the domains), as discussed in the following sections.

## 5.1   Data Partitioning

In order to evaluate algorithms offline, it is necessary to simulate the process, where the system makes recommendations, and users evaluate them. This requires a pre-recorded dataset of interactions between users and items. The dataset is then partitioned into the training and test sets. The former is used to build and tune the model used by the recommender algorithm, while the latter is actually used to evaluate the quality of the generated recommendations.

In cross-domain applications, there are (at least) two potentially overlapping datasets: the source dataset $\mathcal{D}_S$ and the target dataset $\mathcal{D}_T$. We assume $\mathcal{D}_S$ and $\mathcal{D}_T$ are chosen according to the recommendation task and goal in hand. For instance, when evaluating a cross-selling recommender, (i) $\mathcal{D}_S$ and $\mathcal{D}_T$ are set at the item level, as described in Sect. 2.1, (ii) contain items of a different nature, like movies and books, and (iii) have overlapping users. On the contrary, when evaluating a cross-domain recommender as a tool to increase recommendation diversity, $\mathcal{D}_S$ and $\mathcal{D}_T$ are set at the item attribute level, with items of the same type, but differing in the value of certain attributes, e.g., comedy and drama movies.

The exact way $\mathcal{D}_S$ and $\mathcal{D}_T$ are partitioned into training and test set depends on the cross-domain scenario and goal:

- **Scenario.** In the case of a multi-domain scenario, where recommendations target both the source and destination domains, the test set must contain interactions from $\mathcal{D}_S$ and $\mathcal{D}_T$. On the contrary, for a cross-domain scenario, test interactions should be collected exclusively from the target domain $\mathcal{D}_T$.
- **Goal.** When the main goal a cross-domain recommender system is to address the new user problem, the profiles of the tested users (i.e., their known ratings) should contain only interactions from $\mathcal{D}_S$. On the contrary, when the main goal is to increase accuracy, the profiles of the tested users should contain also interactions from $\mathcal{D}_T$.

## 5.2   Sensitivity Analysis

Performance of cross-domain recommenders is mainly affected by three parameters: data overlap between the source and target domains, density of the target domain data, and size of the target user's profile. Hence, the evaluation of cross-domain recommenders should analyze sensitivity with respect to these three parameters. Table 9 overviews the the sensitivity analyses reported in the literature.

Most works assume *data overlap* between the source and target domains materialized as an overlap of users, but only a few—Cremonesi et al. [14] and Zhao et al. [83]—study the sensitivity by varying the percentage of overlapping users. Fewer works assume to have the same catalog of items across the domains [6, 14]. Some works [2, 7, 37, 71] studied the case of overlapping features,

**Table 9** Variables for sensitivity analysis of cross-domain recommender systems

| Parameter | References | |
|---|---|---|
| *Overlap between domains* | Cremonesi et al. [14] | Abel et al. [2] |
| | Shi et al. [70] | Zhao et al. [84] |
| *Target domain density* | Pan et al. [59] | Cremonesi et al. [14] |
| | Cao et al. [11] | Shapira et al. [69] |
| | Pan et al. [63] | Hu et al. [34] |
| *User profile size* | Berkovsky et al. [4, 5] | Shi et al. [70] |
| | Berkovsky et al. [6] | Sahebi et al. [67] |
| | Li et al. [44, 45] | |

especially social tags. For example, Shi et al. studied the sensitivity of cross-domain recommendations by varying the number of overlapping tags between 5 and 50 [70].

Some works [6, 44, 45, 67, 70] studied the sensitivity of recommendations as a function of the *user profile size*, i.e., the number of ratings provided by the recipient of the recommendations. This is particularly critical for the cold-start and new user problems. Pan et al. [63] and Abel et al. [2] developed tag-based recommenders, and performed their analysis by varying the number of tags in the user profile in the 10 to 40 and 0 to 150 ranges, respectively. Others conducted a similar analysis on rating-based recommenders: Shi et al. varied the profile size from 20 to 100 ratings [70], Berkovsky et al. varied the profile size from 3% to 33% of ratings [6], and Sahebi et al. [67] varied the profile size in the range of 1 to 20 ratings.

Finally, some works [11, 14, 63, 69] studied the quality of recommendations as a function of the *dataset density*. Cao et al. varied the density of the multi-domain dataset, i.e., the union of the source and target datasets, between 0.2% and 1% [11]. Shapira et al. varied the density of the dataset between 1% and 40%, while evaluating cross-domain algorithms at the 1% density [69]. Cremonesi et al. varied the density of the target domain between 0.1% and 0.9% [14].

## 6 Open Research Questions

This section overviews the frontiers of cross-domain recommendations by providing some guidance to researchers looking for exciting future research directions. Open challenges mainly are in the areas of evaluation, privacy, fairness, session-based recommendations, and datasets.

- **Evaluation.** Close to 90% of works on cross-domain recommenders published since 2016 are based on deep learning. However, several indications show that using increasingly deeper learning methods in recommender systems is not as beneficial as one could expect. For example, four recent papers report that, for single-domain top-N recommendations, neural methods are not superior to long-existing non-neural ones [24–26, 54]. The observation that for certain tasks the reported improvements "don't add up" mainly lies in the choice of weak baselines and their poor optimization. These findings, however, are not limited to single-domain approaches, but also apply to cross-domain approaches. For instance, recognizing the progress made by transfer learning in cross-domain recommendations, it is not uncommon to find papers, where transfer-learning algorithms for cross-domain recommendations are compared against weak baselines [15]. The current methodological approach for benchmarking cross-domain recommender systems seems solid at first sight and suitable to determine if one algorithm outperforms another for a specific combination of goal, task, and overlap between the domains. However, cross-domain recommender systems researchers have ample freedom in selecting their experimental conditions: protocol, datasets, baselines, etc. This complicates reproducibility and direct benchmarking of results. Given these observations regarding potential methodological issues in the evaluation of cross-domain recommenders, we encourage researchers to publish reproducibility studies of cross-domain recommenders.

- **Privacy.** Privacy is an important and challenging consideration for cross-domain recommender systems, as they exploits information collected by multiple platforms. Sharing knowledge between domains can violate privacy policies and increase the risk of privacy leaks. For instance, if a social network is used as a source domain, it can be exploited to breach privacy in the target domain. Moreover, in many scenarios, different domains are managed by different companies, and sharing personal user data between companies may be prohibited or should comply with local policy regulations, such as the General Data Protection Regulation (GDPR) in Europe. However, existing researches on privacy-preserving recommendations are focused almost exclusively on single-domain scenarios, the only exception being the work from Gao et al. [29]. To this end, we call for an increased attention to privacy-preserving cross-domain recommendations.

- **Fairness.** Cross-domain recommender systems, as any recommender system, learn patterns from historical data, which conveys biases in terms of imbalances and inequalities. These biases, if not properly detected and controlled, potentially lead to discrimination and unfairness in recommendations [19]. Cross-domain recommender systems open new challenges in controlling biases and preventing unfairness, as imbalances and inequalities in the source domains might exacerbate unfairness of recommendations in the target domain. For instance, demographic characteristics within the source domain may not be representative of the target population. However, existing work on biases and fairness focused exclusively on single-domain scenarios. Research questions in cross-domain recommender systems are focused, among others, on controlling the effects generated by unbalances between domains, and transparently explaining why a recommender system provides a given result based on data collected from auxiliary domains. Hence, being able to detect, measure, characterize, and mitigate biases in cross-domain recommender systems is largely an open challenge.

- **Sequence-aware recommendations.** An under-explored research field is the one combining sequence-aware and cross-domain recommender systems. Cross-domain sequential recommender systems predict the next item that the user is likely to interact with, based on past sessions behavior across multiple domains [35, 65]. Previous studies have investigated how to link interactions from different domains, regardless of their sequential nature, the only exception being the work of Ren et al. [65]. One of the key challenges in cross-domain sequential recommendation is around grasping and transferring sequential pattern of interactions from the source domain to the target domain. User behavior, in terms of temporal connections between the items they interact with, constitutes a new link that can be exploited to connect different domains and obtain better cross-domain sequence representations.

- **Datasets.** Finally, cross-domain recommender system research often lacks appropriate datasets allowing to assess diverse recommendation scenarios and tasks [39]. Rich datasets are necessary for a reliable evaluation of new cross-domain recommendation approaches, but these are quite scarce and hard to reach in practice. Large-scale cross-domain datasets are typically gathered by big industry players, such Amazon, eBay and Yelp, and these datasets rarely become available to the broader research community. This brings to the fore the emergent need for new datasets, openly available to the research community, allowing to clearly classify their association with the relevant cross-domain recommendation tasks, goals, and scenarios.

# 7 Conclusions

This chapter covered a wide spectrum of models and techniques applicable to cross-domain recommendations. Recommender system practitioners may find this list of overviewed papers and the variety of options overwhelming, when materializing a cross-domain recommender. Therefore, we list few practical considerations that drive the choice of cross-domain recommender systems.

The first consideration deals with the pivotal question of *why* to use cross-domain recommendations, which we have already raised in Sect. 2.3. Different goals require for different cross-domain recommendation approaches. Having a clear vision about the goal of the cross-domain recommender system is critical in designing the correct path and setting the expectations. A related consideration is *if* cross-domain recommendations are needed. The design of cross-domain recommenders is challenging not only from the algorithmic point of view, but also because cross-domain recommenders need access to reliable information, which needs to be collected, cleansed, deduplicated, and reconciled with the target domain data. This is a time consuming and potentially expensive task, which could compromise the benefits of cross-domain recommendations.

Last but not the least, special attention needs to be paid to *ethical* and *privacy* considerations. Transferring data and knowledge between systems may not conform with their privacy policies and existing privacy regulations. Moreover, it may allow malicious attackers not only to get access to a large volume of user data, but also to mine the combined knowledge, uncovering new potentially sensitive information. Developers of a cross-domain recommender system should keep the privacy considerations in mind when designing and evaluating their methods.

# References

1. F. Abel, S. Araújo, Q. Gao, G.-J. Houben, Analyzing cross-system user modeling on the social web, in *11th International Conference on Web Engineering*, pp. 28–43 (2011)
2. F. Abel, E. Helder, G.-J. Houben, N. Henze, D. Krause, Cross-system user modeling and personalization on the social web. User Model. User Adap. Inter. **23**(2-3), 169–209 (2013)
3. S. Berkovsky, T. Kuflik, F. Ricci, Entertainment personalization mechanism through cross-domain user modeling, in *1st International Conference on Intelligent Technologies for Interactive Entertainment*, pp. 215–219 (2005)
4. S. Berkovsky, T. Kuflik, F. Ricci, Cross-domain mediation in collaborative filtering, in *11th International Conference on User Modeling*, pp. 355–359 (2007)
5. S. Berkovsky, T. Kuflik, F. Ricci, Distributed collaborative filtering with domain specialization, in *1st ACM Conference on Recommender Systems*, pp. 33–40 (2007)
6. S. Berkovsky, T. Kuflik, F. Ricci, Mediation of user models for enhanced personalization in recommender systems. User Model. User Adap. Inter. **18**(3), 245–286 (2008)
7. M. Braunhofer, M. Kaminskas, F. Ricci, Location-aware music recommendation. Int. J. Multimedia Inf. Retr. **2**(1), 31–44 (2013)
8. I. Cantador, I. Fernández-Tobías, S. Berkovsky, P. Cremonesi, Cross-domain recommender systems. *Recommender Systems Handbook*, 2nd edn. (Springer, 2015), pp. 919–959
9. F. Carmagnola, F. Cena, User identification for cross-system personalisation. Information Sciences **179**(1-2), 16–32 (2009)
10. F. Carmagnola, F. Cena, C. Gena, User model interoperability: A survey. User Model. User Adap. Inter. **21**(3), 285–331 (2011)
11. B. Cao, N.N. Liu, Q. Yang, Transfer learning for collective link prediction in multiple heterogeneous domains, in *27th International Conference on Machine Learning*, pp. 159–166 (2010)
12. R. Chung, D. Sundaram, A. Srinivasan, Integrated personal recommender systems, in *9th International Conference on Electronic Commerce*, pp. 65–74 (2007)
13. P.T. Costa, R.R. McCrae, Revised NEO personality inventory (NEO-PI-R) and NEO five-factor inventory (NEO-FFI) manual. Psychol. Assess. Resour. (1992)
14. P. Cremonesi, A. Tripodi, R. Turrin, Cross-domain recommender systems, in *11th IEEE International Conference on Data Mining Workshops*, pp. 496–503 (2011)
15. P. Cremonesi, M. Quadrana, Cross-domain recommendations without overlapping data: myth or reality? in *8th ACM Conference on Recommender Systems* (2014)
16. C. Ding, T. Li, W. Peng, H. Park, Orthogonal nonnegative matrix tri-factorizations for clustering, in *12th ACM SIGKDD Conference on Knowledge Discovery and Data Mining*, pp. 126–135 (2006)
17. R. Driskill, J. Riedl, Recommender systems for E-commerce: Challenges and opportunities, in *AAAI'99 Workshop on Artificial Intelligence for Electronic Commerce*, pp. 73–76 (1999)
18. M. Enrich, M. Braunhofer, F. Ricci, Cold-start management with cross-domain collaborative filtering and tags, in *14th International Conference on E-Commerce and Web Technologies*, pp. 101–112 (2013)
19. M.D. Ekstrand, R. Burke, F. Diaz, Fairness and discrimination in recommendation and retrieval, in *Proceedings of the 13th ACM Conference on Recommender Systems*, pp. 576–577 (2019)

20. A. Farseev, I. Samborskii, A. Filchenkov, T. Chua, Cross-domain recommendation via clustering on multi-layer graphs, in *40th International ACM SIGIR Conference on Research and Development in Information Retrieval*, pp. 195–204 (2017)
21. I. Fernández-Tobías, I. Cantador, M. Kaminskas, F. Ricci, A generic semantic-based framework for cross-domain recommendation, in *2nd International Workshop on Information Heterogeneity and Fusion in Recommender Systems*, pp. 25–32 (2011)
22. I. Fernández-Tobías, I. Cantador, M. Kaminskas, F. Ricci, Cross-domain recommender systems: A survey of the state of the art, in *2nd Spanish Conference on Information Retrieval*, pp. 187–198 (2012)
23. I. Fernández-Tobías, I. Cantador, Exploiting social tags in matrix factorization models for cross-domain collaborative filtering, in *1st International Workshop on New Trends in Content-based Recommender Systems* (2013)
24. M. Ferrari Dacrema, S. Boglio, P. Cremonesi, D. Jannach, A troubling analysis of reproducibility and progress in recommender systems research. ACM Trans. Inf. Syst. **39**(2) 49 p. (2021)
25. M. Ferrari Dacrema, P. Cremonesi, D. Jannach, Are we really making much progress? A worrying analysis of recent neural recommendation approaches, in *13th ACM Conference on Recommender Systems*, pp. 101–109 (2019)
26. M. Ferrari Dacrema, F. Parroni, P. Cremonesi, D. Jannach, Critically examining the claimed value of convolutions over user-item embedding maps for recommender systems, in *29th ACM International Conference on Information & Knowledge Management*, pp. 355–363 (2020)
27. J. Freyne, S. Berkovsky, G. Smith, Evaluating recommender systems for supportive technologies, in *User Modeling and Adaptation for Daily Routines*, pp. 195–217 (2013)
28. W. Fu, Z. Peng, S. Wang, Y. Xu, J. Li, Deeply fusing reviews and contents for cold start users in cross-domain recommendation systems, in *33rd AAAI Conference on Artificial Intelligence*, pp. 94–101 (2019)
29. C. Gao, C. Huang, Y. Yu, H. Wang, Y. Li, D. Jin, Privacy-preserving cross-domain location recommendation. Proc. ACM Interactive Mobile Wearable Ubiquit. Technol. **3**(1), 1–21 (2019)
30. S. Gao, H. Luo, D. Chen, S. Li, P. Gallinari, J. Guo, Cross-domain recommendation via cluster-level latent factor model, in *17th and 24th European Conference on Machine Learning and Knowledge Discovery in Databases*, pp. 161–176 (2013)
31. M. He, J. Zhang, P. Yang, K. Yao, Robust transfer learning for cross-domain collaborative filtering using multiple rating patterns approximation, in *11th International Conference on Web Search and Data Mining, WSDM*, pp. 225–233 (2018)
32. J.L. Helocker, J.A. Konstan, L.G. Terveen, J. Riedl, Evaluating collaborative filtering recommender systems. ACM Trans. Inf. Syst. **22**(1), 5–53 (2004)
33. L. Hu, J. Cao, G. Xu, L. Cao, Z. Gu, C. Zhu, Personalized recommendation via cross-domain triadic factorization, in *22nd International Conference on World Wide Web*, pp. 595–606 (2013)
34. G. Hu, Y. Zhang, Q. Yang, Transfer meets hybrid: A synthetic approach for cross-domain collaborative filtering with text, in *The World Wide Web Conference 2019*, pp. 2822–2829 (2019)
35. D. Jannach, P. Cremonesi, M. Quadrana, Session-based recommender systems. *Recommender Systems Handbook*, 3nd edn. (Springer, 2021)
36. S. Jialin Pan, Q. Yang, A survey on transfer learning. IEEE Trans. Knowl. Data Eng. **22**(10), 1345–1359 (2010)
37. M. Kaminskas, I. Fernández-Tobías, F. Ricci, I. Cantador, Ontology-based identification of music for places, in *13th International Conference on Information and Communication Technologies in Tourism*, pp. 436–447 (2013)
38. S. Kang, J. Hwang, D. Lee, H. Yu, Semi-supervised learning for cross-domain recommendation to cold-start users, in *28th ACM International Conference on Information and Knowledge Management, CIKM*, pp. 1563–1572 (2019)
39. M. Khan, R. Ibrahim, I. Ghani, Cross domain recommender systems: A systematic literature review. ACM Comput. Surv. 36:1–36:34 (2017)

40. B. Kitts, D. Freed, M. Vrieze, Cross-sell: A fast promotion-tunable customer-item recommendation method based on conditionally independent probabilities, in *6th ACM SIGKDD Conference on Knowledge Discovery and Data Mining*, pp. 437–446 (2000)
41. Y. Koren, Factorization meets the neighborhood: A multifaceted collaborative filtering model, in *14th ACM SIGKDD Conference on Knowledge Discovery and Data Mining*, pp. 426–434 (2008)
42. C.H. Lee, Y.H. Kim, P.K. Rhee, Web personalization expert with combining collaborative filtering and association rule mining technique. Expert Syst. Appl. **21**(3), 131–137 (2001)
43. B. Li, Cross-domain collaborative filtering: A brief survey, in *23rd IEEE International Conference on Tools with Artificial Intelligence*, pp. 1085–1086 (2011)
44. B. Li, Q. Yang, X. Xue, Can movies and books collaborate? Cross-domain collaborative filtering for sparsity reduction, in *21st International Joint Conference on Artificial Intelligence*, pp. 2052–2057 (2009)
45. B. Li, Q. Yang, X. Xue, Transfer learning for collaborative filtering via a rating-matrix generative model, in *26th International Conference on Machine Learning*, pp. 617–624 (2009)
46. B. Li, X. Zhu, R. Li, C. Zhang, X. Xue, X. Wu, Cross-domain collaborative filtering over time, in *22nd International Joint Conference on Artificial Intelligence*, pp. 2293–2298 (2011)
47. P. Li, A. Tuzhilin, DDTCDR: Deep dual transfer cross domain recommendation, in *13th ACM International Conference on Web Search and Data Mining*, pp. 331–339 (2020)
48. J. Lian, F. Zhang, X. Xie, G. Sun, CCCFNet: A content-boosted collaborative filtering neural network for cross domain recommender systems, in *26th International Conference on World Wide Web Companion*, pp. 817–818 (2017)
49. B. Liu, Y. Wei, Y. Zhang, Z. Yan, Q. Yang, Transferable contextual bandit for cross-domain recommendation, in *32nd Conference on Artificial Intelligence (AAAI)*, pp. 3619–3626 (2018)
50. J. Liu, P. Zhao, F. Zhuang, Y. Liu, V. Sheng, J. Xu, X. Zhou, H. Xiong, Exploiting aesthetic preference in deep cross networks for cross-domain recommendation, in *The Web Conference 2020*, pp. 2768–2774 (2020)
51. A. Loizou, How to recommend music to film buffs: enabling the provision of recommendations from multiple domains. Ph.D. thesis, University of Southampton (2009)
52. B. Loni, Y. Shi, M.A. Larson, A. Hanjalic, Cross-domain collaborative filtering with factorization machines, in *36th European Conference on Information Retrieval* (2014)
53. Y. Low, D. Agarwal, A.J. Smola, Multiple domain user personalization, in *17th ACM SIGKDD Conference on Knowledge Discovery and Data Mining*, pp. 123–131 (2011)
54. M. Ludewig, N. Mauro, S. Latifi, D. Jannach, Empirical analysis of session-based recommendation algorithms. User Model. User Adap. Inter., 1–33 (2020)
55. M. Ma, P. Ren, Y. Lin, Z. Chen, J. Ma, M. De Rijke, $\pi$-Net: A parallel information-sharing network for shared-account cross-domain sequential recommendations, in *42nd International ACM SIGIR Conference on Research and Development in Information Retrieval*, pp. 685–694 (2019)
56. T. Man, H. Shen, X. Jin, X. Cheng, Cross-domain recommendation: An embedding and mapping approach, in *26th International Joint Conference on Artificial Intelligence, IJCAI*, pp. 2464–2470 (2017)
57. O. Moreno, B. Shapira, L. Rokach, G. Shani, TALMUD: transfer learning for multiple domains, in *21st ACM Conference on Information and Knowledge Management*, pp. 425–434 (2012)
58. M. Nakatsuji, Y. Fujiwara, A. Tanaka, T. Uchiyama, T. Ishida, Recommendations over domain specific user graphs, in *19th European Conference on Artificial Intelligence*, pp. 607–612 (2010)
59. S.J. Pan, J.T. Kwok, Q. Yang, Transfer learning via dimensionality reduction, in *23rd AAAI Conference on Artificial Intelligence*, pp. 677–682 (2008)
60. R. Pagano, P. Cremonesi, M. Larson, B. Hidasi, D. Tikk, A. Karatzoglou, M. Quadrana, The contextual turn: From context-aware to context-driven recommender systems, in *10th ACM Conference on Recommender Systems*, pp. 249–252 (2016)

61. S.J. Pan, Q. Yang, A survey on transfer learning. IEEE Trans. Knowl. Data Eng. **22**(10), 1345–1359 (2010)
62. W. Pan, N.N. Liu, E.W. Xiang, Q. Yang, Transfer learning to predict missing ratings via heterogeneous user feedbacks, in *22nd International Joint Conference on Artificial Intelligence*, pp. 2318–2323 (2011)
63. W. Pan, E.W. Xiang, N.N. Liu, Q. Yang, Transfer learning in collaborative filtering for sparsity reduction, in *24th AAAI Conference on Artificial Intelligence*, pp. 210–235 (2010)
64. W. Pan, E.W. Xiang, Q. Yang, Transfer learning in collaborative filtering with uncertain ratings, in *26th AAAI Conference on Artificial Intelligence*, pp. 662–668 (2012)
65. Z. Ren, L. Zhao, J. Ma, M. de Rijke, Mixed information flow for cross-domain sequential recommendations. ACM Trans. Knowl. Discov. Data **1**(1), (2020)
66. S. Rendle, Factorization machines with libFM. ACM Trans. Intell. Syst. Tech. **3**(3), 1–22 (2012)
67. S. Sahebi, P. Brusilovsky, Cross-domain collaborative recommendation in a cold-start context: The impact of user profile size on the quality of recommendation, in *21st International Conference on User Modeling, Adaptation, and Personalization*, pp. 289–295 (2013)
68. G. Shani, A. Gunawardana, Evaluating recommendation systems. *Recommender Systems Handbook*, pp. 257–297 (2011)
69. B. Shapira, L. Rokach, S. Freilikhman, Facebook single and cross domain data for recommendation systems. User Model. User Adap. Inter. **23**(2-3), 211–247 (2013)
70. Y. Shi, M. Larson, A. Hanjalic, Tags as bridges between domains: Improving recommendation with Tag-induced cross-domain collaborative filtering, in *19th International Conference on User Modeling, Adaption, and Personalization*, pp. 305–316 (2011)
71. A. Stewart, E. Diaz-Aviles, W. Nejdl, L.B. Marinho, A. Nanopoulos, L. Schmidt-Thieme, Cross-tagging for personalized open social networking, in *20th ACM Conference on Hypertext and Hypermedia*, pp. 271–278 (2009)
72. A. Taneja, A. Arora, Cross domain recommendation using multidimensional tensor factorization. Expert Syst. Appl., 304–316 (2018)
73. J. Tang, J. Yan, L. Ji, M. Zhang, S. Guo, N. Liu, X. Wang, Z. Chen, Collaborative users' brand preference mining across multiple domains from implicit feedbacks, in *25th AAAI Conference on Artificial Intelligence*, pp. 477–482 (2011)
74. A. Tiroshi, S. Berkovsky, M.A. Kaafar, T. Chen, T. Kuflik, Cross social networks interests predictions based on graph features, in *7th ACM Conference on Recommender Systems*, pp. 319–322 (2013)
75. A. Tiroshi, T. Kuflik, Domain ranking for cross domain collaborative filtering, in *20th International Conference on User Modeling, Adaptation, and Personalization*, pp. 328–333 (2012)
76. D. Vallet, S. Berkovsky, S. Ardon, A. Mahanti, M.A. Kafaar, Characterizing and predicting viral-and-popular video content, in *24th ACM International on Conference on Information and Knowledge Management*, pp. 1591–1600 (2015)
77. C. Wang, M. Niepert, H. Li, RecSys-DAN: Discriminative adversarial networks for cross-domain recommender systems. IEEE Trans. Neural Netw. Learn. Syst. **31**(8), 2731–2740 (2020)
78. P. Winoto, T. Tang, If you like the devil wears prada the book, will you also enjoy the devil wears prada the movie? A study of cross-domain recommendations. N. Gener. Comput. **26**, 209–225 (2008)
79. F. Yuan, L. Yao, B. Benatallah, DARec: Deep domain adaptation for cross-domain recommendation via transferring rating patterns, in *28th International Joint Conference on Artificial Intelligence*, pp. 4227–4233 (2019)
80. Y. Zhang, B. Cao, D.-Y. Yeung, Multi-domain collaborative filtering, in *26th Conference on Uncertainty in Artificial Intelligence*, pp. 725–732 (2010)
81. Y. Zang, X. Hu, LKT-FM: A novel rating pattern transfer model for improving non-overlapping cross-domain collaborative filtering, in *Machine Learning and Knowledge Discovery in Databases - European Conference*, pp. 641–656 (2017)

82. X. Zhang, J. Cheng, T. Yuan, B. Niu, H. Lu, TopRec: domain-specific recommendation through community topic mining in social network, in *22nd International Conference on World Wide Web*, pp. 1501–1510 (2013)
83. L. Zhao, S.J. Pan, E.W. Xiang, E. Zhong, X. Lu, Q. Yang, Active transfer learning for cross-system recommendation, in *27th AAAI Conference on Artificial Intelligence*, pp. 1205–1211 (2013)
84. C. Zhao, C. Li, R. Xiao, H. Deng, A. Sun, CATN: Cross-domain recommendation for cold-start users via aspect transfer network, in *43rd International ACM SIGIR conference on research and development in Information Retrieval*, pp. 229–238 (2020)
85. F. Zhu, Y. Wang, C. Chen, G. Liu, M. Orgun, A deep framework for cross-domain and cross-system recommendations, in *27th International Joint Conference on Artificial Intelligence, IJCAI*, pp. 3711–3717 (2018)
86. F. Zhu, C. Chen, Y. Wang, G. Liu, X. Zheng, DTCDR: A framework for dual-target cross-domain recommendation, in *28th ACM International Conference on Information and Knowledge Management*, pp. 1533–1542 (2019)
87. F. Zhuang, P. Luo, H. Xiong, Y. Xiong, Q. He, Z. Shi, Cross-domain learning from multiple sources: A consensus regularization perspective. IEEE Trans. Knowl. Data Eng. **22**(12), 1664–1678 (2010)

# Part III
# Value and Impact of Recommender Systems

# Value and Impact of Recommender Systems

Dietmar Jannach and Markus Zanker

## 1 Introduction

Automated recommendations are nowadays part of many websites and online services, and they are often a central part of the overall user experience, such as on e-commerce and media streaming sites. The content we come across in the online world is therefore often highly individualized. The applied personalization strategy is usually centrally determined by the recommendation service provider who tailors its own content to consumers.[1]

The challenge therefore consists of striking a balance between economic-oriented goals such as conversion rates, increased sales, or customer retention and keeping users happy with personalized and tailored offerings [41, 69]. A commonly mentioned goal of a recommender system for the latter aspect is therefore to help users find items of interest in situations of information overload. This is according to literature typically assessed by its capability to accurately predict the relevance of individual items for individual users. As Abdollahpouri et al. [1] thus noted, in the best case, a recommender system therefore creates value for the consumers and potentially even further stakeholders while being economically sustainable for the service provider.

---

[1] This is termed a *provider-centric* strategy in [3].

---

D. Jannach
University of Klagenfurt, Klagenfurt, Austria
e-mail: dietmar.jannach@aau.at

M. Zanker (✉)
Free University of Bozen-Bolzano, Bolzano, Italy

University of Klagenfurt, Klagenfurt, Austria
e-mail: markus.zanker@unibz.it

© Springer Science+Business Media, LLC, part of Springer Nature 2022      519
F. Ricci et al. (eds.), *Recommender Systems Handbook*,
https://doi.org/10.1007/978-1-0716-2197-4_14

Independent of such specific goals and their achievement, recommender systems influence the users' choices and their behavior as a matter of principle. In particular in cases where the recommendations are central to the user experience of the service, a large fraction of the observed interactions stems from recommendations. This is, for instance, the case for recommendations at YouTube or Netflix [19, 30]. These fractions can even grow due to self-enforcing cycles in case most of the content that users come across during browsing has already been pre-filtered and personalized in one way or another.

The use of recommender systems therefore also bears some risks and can potentially even lead to negative and undesired effects, such as reinforcing extremist or unhealthy behavioral and consumption patterns. For instance, a hypothesized radicalization pipeline leading users to more extremist video content at YouTube [63], or suggestions of a food recommender that would mostly relate to unhealthy dishes [22].

In this chapter, we review the possible values and the impact of recommender systems for different stakeholders. We consider both *organizational and business value* (e.g., in terms of increased sales or customer retention) as well as different types of *consumer value* (e.g., in the form of reduced search efforts or better decisions), and we describe additional ways in which recommenders may positively or negatively impact the behavior of consumers.

Next, in Sect. 2, we first discuss the different purposes a recommender can serve and in which ways such a system can thereby create value for different stakeholders. Furthermore, we sketch potential risks that can derive from the use of recommendation technology. Section 3 then discusses how the impact of recommender systems is currently measured, with a particular focus on measures that are used in practice. Finally, in Sect. 4, we reflect on our predominant research approaches in academia, where we often largely abstract from the aforementioned general goals and therefore are only partially capable of quantifying the potential impact. Based on this discussion, we correspondingly outline potential ways towards more impact-oriented research methodologies.

# 2   Stakeholders and Value Drivers of Recommender Systems

While a recommender system is effective whenever it creates some *utility* or *value* for one of its stakeholders, the focus of the research literature lies mostly on demonstrating the potential value of recommender systems for *consumers*, i.e., for those receiving recommendations. Correspondingly, the extent of value creation is mostly assessed in terms of consumer-oriented measures, such as the assumed capability of a recommender system to help users finding the most relevant items in their particular situation. However, we argue that recommender systems are applications residing in the core of many e-commerce business models since they are primarily concerned about exploitation of information and matchmaking on virtual markets. Therefore, the impact and created value of recommender systems cannot only be seen from a purely consumer or user perspective, but there are effects on all involved stakeholders. For instance, depending on the pursued recommendation strategy not only the satisfaction of the receivers of recommendations, but also the profitability of the platform or the sales distribution among the items in the product catalog will be potentially impacted. Moreover, the strategy will also have long-term effects on the perception of the integrity and benevolence of the recommendation system itself [9], which in turn will moderate its impact. Therefore, after shortly describing the stakeholders of recommender systems, we will introduce the value driver model of e-commerce business models [6] to structure the discussion of the impact of recommender systems in this chapter.

## 2.1  Stakeholders of Recommender Systems

For every recommender system, there are at least two different types of stakeholders [1]. There are (1) the consumers or users who receive the recommendations, and (2) the organization that provides the recommendations as part of their service. The differentiation between these stakeholders is important. While a recommender system should in the best case create value in parallel for all involved stakeholders, there might be competing goals involved in the process. In other words, the best recommendation for consumers, e.g., one that helps users discover novel things, might not be optimal for the provider, for instance, in terms of generated short-term revenues.

In Table 1, we sketch four types of possible stakeholders, see also [39]. Besides *consumers* and *recommendation service providers*, we also identify *suppliers* and *society*. Suppliers are the providers of item offerings that can be recommended to users. Such suppliers could be content creators, manufacturers, service providers, or also retailers who use the platform of the recommendation service provider for their business activities.

**Table 1**  Stakeholders of Recommender Systems

| Stakeholder | Description |
| --- | --- |
| Consumers | Consumers are the end users of recommendation systems, and their behavior is potentially influenced by the system's recommendations. |
| Recommendation service providers | These are organizations that provide the recommendations as part of their services. They invest in recommendation technology and are the ones in control of the system. |
| Suppliers | These are organizations that provide (some of) the products or services that are recommended to consumers. The recommendation of their offerings can, for instance, influence the overall demand on the market. |
| Society | Depending on the size of the population of end users, recommender systems like on news or social media platforms, can have effects on parts of society as a whole. Note that the impact on society may be more than the aggregate impact on individuals, e.g., when the directly affected consumers act as multipliers of opinions in a society. |

For some applications, no external suppliers might be involved at all, or their interests might not be taken into account in the design of the implemented recommendation strategy. In other scenarios, however, considering the supplier interests in an appropriate way may represent a central problem when designing the recommendation service. This is in particular the case when the recommendation service provider acts as a platform between different suppliers and consumers.

In the travel and tourism domain, for instance, websites of hotel booking platforms often have a recommendation component integrated. The interest of the suppliers—e.g., hotel chains or individual property owners—is to be frequently included in these recommendations, assuming that being listed leads to more sales. In case of suppliers being hotel chains, they might even have specific preferences which hotels are recommended, such as one with an overcapacity during a particular period of time. The recommendation service provider's interest might, in contrast, be to push those hotels where the expected profit margin or commission is highest. The consumers' interest, finally, is to find a hotel that best matches their preferences and interests. The recommendation service provider therefore may have to consider all these interests in parallel. Not considering the consumers' interests to a large enough extent may lead to a limited acceptance of the recommendations and to low conversion rates; resulting in reduced revenues/profit in the short-term and a loss in credibility and impact in the medium and long-term. Not considering the supplier's interest might, on the other hand, lead to dissatisfaction on the supplier's side in the long run.

Finally, *society* as a whole can be another stakeholder that may at least indirectly be affected by recommendations. Today, there are various online services with an enormous reach, including social media sites such as Facebook or Twitter, news aggregation sites like Google News, or media platforms like YouTube. The selection of the content presented on such sites may have significant effects on the users' view of the world, e.g., in terms of political questions, potentially leading to phenomena of filter bubbles or the broad dissemination of fake news.

## 2.2   Value Dimensions of Recommender Systems

The value driver model [6] of e-commerce business models was developed to depict the value-creating transactions by networks of business actors on virtual markets. It basically clusters the value driving aspects of e-commerce business models into four groups: efficiency, complementarities, lock-in and novelty. *Efficiency* generally refers to savings in transaction costs or time due to speed gains, scaling effects and the reduction of information asymmetries on virtual markets. *Complementarities* represent synergy effects due to the combination of different sales channels or product catalogs which creates additional opportunities such as cross-selling of items or follow-up sales offers to customers. *Lock-In* effects are creating value due to customer retention deriving from the avoidance of switching costs or positive network externalities. The latter derives from network connections and the joint use of services. For instance, the fact that collaborative filtering systems become more accurate when exploiting more data and from larger groups of users is an example for such a positive network externality. Finally, *novelty* stands for new architectural configurations and opportunities that would have been unfeasible on non-virtual markets. In the following, we structure the reported value contributions of recommender systems according to these aforementioned categories of the value driver model. As can be seen, recommender systems are making contributions with respect to all value creating aspects of e-commerce business models.

## 2.2.1  Efficiency

One pillar for e-commerce success is the reduction in information asymmetries between buyers and sellers due to the up-to-date and abundant availability of information with an enormous reach and at nearly zero cost. Recommender systems may help to counterbalance the resulting potential overload of information by primarily creating transparency by (relatively) unbiased information offerings, lowering consumers' information search costs and facilitating their decision processes as enlisted in Table 2. However, the impact on users' decision processes is always moderated by the concrete domain and the properties of products as Lee and Hosanagar [53] demonstrated in their large-scale e-commerce study. According to their results utilitarian and experience products, for instance, enjoy a higher lift in awareness due to recommendations as compared to hedonic and search product categories.

Due to the networking of consumers the support of decision making processes may not only target single consumers, but can even address the decision making

**Table 2** Efficiency and the value of recommender systems

| |
|---|
| *Inform in a balanced way:* A recommender system can be tuned to ensure that the recommendations are not biased, e.g., towards certain items or content supporting only one particular opinion. |
| *Help users find objects that match their assumed short-term intent and context:* Recommend items that are relevant in the ongoing user session, sometimes even without long-term preference information. |
| *Help users find objects that match their long-term preferences:* The most explored problem in the literature. Assumes the existence of long-term user profiles, e.g., in the form of a user-item interaction matrix. |
| *Improve decision making:* Decision making can be improved, e.g., by reducing choice difficulty and pre-selecting a small set of options; better choices and choice satisfaction can also be supported by explanations. |
| *Establish group consensus:* In a group recommendation scenario, the purpose of the system can be to make suggestions that are agreeable for all group members. |
| *Cost savings:* Recommender systems allow providers to automate the sales-advisory function at large scale and with zero marginal costs. |
| *Real-time transparency:* Recommender systems allow providers to observe in real-time the immediate effects of their automated sales advisory. |

of groups of users such as in [56]. From the provider perspective the efficiency gains of recommendation technology are primarily in cost savings by automating the traditional sales advisory function being naturally inherent to offline distribution channels like retail or personal selling. In this respect, the recent advances in natural language processing also initiated current research efforts towards chat-based recommendation approaches [36, 56, 60]. For reasons of completeness, however, it needs to be mentioned that the provisioning of recommendation services also incurs some operating costs. Therefore, Goshal et al. [29] made a theoretical analysis on the optimal strategies of consumers that would need to make their choice between personalizing and non-personalizing firms under competition. One of their outcomes is that consumers would weight in lower prices of non-personalizing firms against higher fit costs. Therefore one of their main results is that also the prices and profits of a non-personalizing firm are impacted by the competitor's recommendation system.

Furthermore, the real-time transparency of consumer search also—at least hypothetically—enables providers to immediately react to recent trends and eventually influence them by adapting the recommendation strategies. Since assumptions about purposeful biasing of recommendation agents would undermine consumers' trust in these systems, no research in this respect has been performed or at least reported.

### 2.2.2  Complementarities

The ease of integrating diverse and enormously large product catalogs and service offerings in online marketplaces enables the exploitation of synergies in sales. Again, recommender systems are therefore a central cornerstone in many e-commerce business models to exploit these complementarity effects as summarized in Table 3. Like in the efficiency dimension, a large number of consumer-centric value aspects have been described. Again, however, only anecdotal evidences on generated revenue gains of providers have been made public or scrutinized further in recommender systems research.

### 2.2.3  Lock-In

The *stickiness* or ability to lock-in participants is another core characteristic of online business models. Recommender systems again primarily focus on consumers or users and create value by personalizing their interaction experiences. Table 4 selectively lists reported value aspects of recommender systems. Technically this is facilitated by exploiting users' observed behaviors and data points and making the system an indispensable virtual alter ego that would be lost when switching platform or provider. This is analogous to the stickiness of traditional sales agents in retail knowing their clients for years. Clients over time therefore recognize them as a form of *old friends* whose viewpoints and recommendations they trust. A hidden value

**Table 3** Complementarities and the value of recommender systems

| |
|---|
| *Enable item "discoverability":* A recommender system can help users discover items in the catalog they were not aware of. This can lead to increased demand over time, but also to more engagement. |
| *Help consumers explore or understand the item space:* Sometimes users are not aware of the available options. A recommender system can be tuned to show a diverse set of options. |
| *Create additional demand:* One typical goal of recommendations is to point users to additional items in the catalog, e.g., to stimulate cross-sales. This can lead to increased demand both in the short and in the longer term. |
| *Show alternatives:* Relates to the previous aspect. A recommender system can present alternatives in the context of a reference (currently viewed) item. |
| *Show accessories:* Instead of alternatives, a recommender can point users to accessories of given reference items. |
| *Recommend in sequence:* Create a logical continuation of previously observed user interactions, e.g., recommend next place to visit or next music track to listen to. |
| *Actively notify users of new content:* A recommender might proactively notify users, e.g., through push notifications, with a focus on novel content to stimulate consumption and interaction. |
| *Revenue gain:* based on generated additional business. |

**Table 4** Lock-in and the value of recommender systems

| |
|---|
| *Increase user engagement and activity on the site:* User engagement is often used as a proxy to gauge the effectiveness of a recommender system. When consumers interact with a service, e.g., for music streaming, more frequently customer churn is expected to be lower. |
| *Entertainment:* A recommender system can be entertaining or emotionally satisfying, e.g., by supporting discovery of new content in a convenient way. |
| *Remind consumers of already known items:* In some domains, it can be helpful to remind the user of things they already know or have. The recommender might remind the user of the purchase of consumables or show items the user liked in the past. |
| *Provide a valuable add-on service:* A recommender system can serve as a tool to differentiate the provided service from competitors in the market. High-quality personalized recommendations may lead to increased customer retention and increase the switching costs for customers. |
| *Learn more about the customers:* Personalized recommendations require the collection of customer preference profiles and often a thorough understanding of the specific demands of certain consumer groups. Providing a recommendation service might therefore contribute to a better understanding of customers in general. |

from the provider perspective is therefore in particular the corporate knowledge management aspect that becomes evident in the popularity of recommender systems as an application domain for data science efforts.

### 2.2.4   Novelty

Novelty and the innovation potential of e-commerce business models generally refers to the introduction of new products or services or the establishment of novel process models transforming and creating businesses.

**Table 5** Novelty and the value of recommender systems

| |
| --- |
| *Nudge toward desired behavior:* A recommender might stimulate certain desirable behavior at the user's side, e.g., with respect to healthy behavior, through nudging techniques. |
| *Increase (short-term) business success and promotion of content:* A recommender system can be tuned to steer customer demand and to promote items that are favorable in terms of business-related figures such as revenue or profit. |
| *Change user behavior in desired directions:* By pre-selecting the items in a recommendation list, the choice set for the consumers can be reduced and certain items can be promoted. This may effectively change the consumers' behavior in desired directions, e.g., towards more profitable items. |

Note, however, that with the *novelty* dimension we do not refer to recommender systems being novel themselves, but that they potentially create novel opportunities and business scenarios.

While recommending novel or previously unknown items to users has already been mentioned as an example of value creation due to efficient information processing on virtual markets, the purposeful influencing and biasing of consumers' will by persuasive recommendation and nudging strategies [74] can fall into this novelty category as summarized in Table 5. Provisioning of automated recommendations is nowadays commonplace, particularly in e-commerce and on media streaming or (social) media sites. However, there still exists a number of potential application domains where recommendation techniques are not yet widespread. For instance, providing decision support in the domain of health and well-being, the potential for innovating by inducing behavior change through recommendations is only starting to be developed [65].

## 2.3 Risks of Recommender Systems

While recommender systems are designed to create value and have a positive impact on consumers and organizations, there exist also certain risks that come with the use of recommendation technology. In Table 6, we discuss examples for such risks.

Compared to the analysis of the benefits of recommender systems, potential negative effects have so far not been explored to a large extent in the literature. In [13], for instance, the authors investigated the effects of a malfunctioning music recommender system on consumer trust and behavior. Researchers from the field of Marketing [10] looked at the potential negative monetary effects of recommending the wrong items, such as items that consumers were already likely to purchase anyway. Recommending such items limits the opportunities to promote other items under the assumption that only a limited set of items can be recommended. The aspects of privacy and fairness, finally, have obtained increased interest in recent years in the recommender systems research community, see, for instance, [27] or [12].

**Table 6** Potential risks of recommender systems

| Consumer risks | |
|---|---|
| Poor decisions/choice dissatisfaction | Ultimately, the pre-selection of items or the decision bias introduced by a recommender system may lead to bad decisions or to choice dissatisfaction. |
| Bad user experience/decision difficulty | If the set of recommendations is chosen in an unfortunate way and, e.g., only contains very similar items, this might in parallel lead to a poor user experience and an increased decision difficulty for the consumer. |
| Biased information state | In case the presented options emphasizes one particular range of the available options, the consumer might be left with an incomplete and biased information state. |
| Privacy | Recommendation providers may collect all sorts of user interactions in a comprehensive way and try to connect the information about users across services and sites. This may endanger the privacy of consumers. |
| Organizational risks | |
| Loss of consumer trust | When recommendations are—for an extended period of time—not helpful for consumers or appear biased, they might lose the trust not only in the recommendations but in the organization as a whole. |
| Loss of societal trust | In particular recommendations that appear biased or unfair (e.g., appear to be only advantageous for the provider organization) may lead to a bad reputation of an organization and a loss of trust by society as a whole. |
| Missed opportunities/financial loss | Recommender systems can have significant positive impacts on organizations in terms of business-related figures. A poorly designed recommender system might in contrast lead to missed opportunities, due to a low value of recommendations for users. |
| Societal risks | |
| Filter bubbles & echo chambers | Biased recommendations in particular on news sites or social media platforms may lead to filter bubbles and echo chambers, where the presented information mainly reflects pre-existing interests and viewpoints of individuals or user groups. |
| Algorithmic bias and discrimination | In some application scenarios of recommender systems, the underlying algorithms may reflect or even reinforce uneven distributions in the data, potentially leading to discrimination of certain user groups. |

# 3 Measuring the Impact of Recommender Systems

In the previous section, value contribution and impact of recommender systems in terms of the four categories of the value driver model have been presented. This section will now elaborate on approaches to assess and quantify the impact within these categories as well as their corresponding measures and Key Performance Indicators (KPIs). When seeing a recommender system purely as a machine learning task its evaluation is typically focusing on accuracy related measures. However, nowadays recommender systems research is coming to the conclusion that the impact of personalized content and item recommendations need to be assessed from a multi-disciplinary perspective with a plurality of methods and approaches [76]. Next, we will therefore shortly discuss methodological aspects before moving on to the most widespread measures in practice. We also refer to Chap. "Evaluating Recommender Systems" for a complete and in-depth discussion of state-of-the-art evaluation methodology. They obviously focus on quantifying the value created for providers given that the value needs to surpass or at least match the efforts and investments. However, without creating value for consumers, no sustainable value can be created for providers. Thus, when users are consistently satisfied by, for instance, finding items of interest with ease, also the provider's KPIs will indicate a positive impact. Consequently, a longitudinal approach when quantifying the impact will lead to more reliable results.

The academic literature identifies three main methodologies for evaluating recommender systems (see Chap. "Evaluating Recommender Systems"): (1) field studies (A/B tests), (2) user studies, (3) offline experiments. Field studies are run by recommendation service providers to test the effects of different recommendation strategies on their respective KPIs. We will discuss which KPIs are widespread in industry and which insights were obtained from field studies later in this section. User studies are typically executed in the form of controlled experiments, either in the lab or online. Here, the study participants are randomly assigned to interact with different versions of a recommender system or they interact with the system under different conditions. Typical goals of such user studies are to assess the subjective quality perception of a system or of some of its components and in order to understand how these perceptions might influence the future behavioral intentions of users, see [49, 58]. Such studies often focus on user experience aspects of recommender systems. We will discuss the need for more user-centric studies, which also consider the provider perspective later in Sect. 4.

Offline experiments, finally, are purely data-based and do not require the active involvement of users for their execution. Such experiments are by far the most common ones in literature and we will focus on potential limitations of today's offline experimentation practices when it comes to the assessment of the impact and value of recommender systems.

Most often, offline experiments are used to compare different algorithms in terms of their capability of predicting held-out interaction or preference data. They build on the underlying assumption that when an algorithm is able to rank the presumably more relevant items higher in recommendation lists, the generated recommendations will more likely better match users' interests. Thus, algorithms with higher offline prediction accuracy are supposed to also outperform their weaker offline comparison partners in real-world settings.

However, assessing the impact of a recommender system based on offline experimentation is sometimes seen to be too simplistic and building on problematic assumptions. It is, for instance, relatively straightforward to predict based on past data that a lover of Star Wars movies will watch any new sequel being released. Therefore, recommending a new sequel to this user may—even though the prediction is perfectly accurate—not create much value, neither for the consumer already knowing about the sequel and watching it anyway, nor for the provider who could have promoted another content.

In terms of the evaluation measures, algorithmic research is strongly concentrated on prediction accuracy, i.e., the ability of an algorithm to predict held-out data. Evaluating recommendation algorithms in terms of accuracy is generally meaningful, as discussed above. However, it stands to question (1) if the often small increases in accuracy on selected datasets reported in research reports would truly make a difference if these new algorithms would be deployed in practice,[2] and (2) if the results from offline tests are generally predictive for outcomes of deployed systems [30, 41]. Furthermore, while some algorithms might yield good offline accuracy results, the resulting recommendations may have other, undesired properties such as a bias towards popular items generating only limited additional value for providers [43]. Finally, an orthogonal problem in the context of offline studies seems to lie in methodological problems in applied machine learning research, where we often only observe an "illusion of progress" [32]. While the proposed models become more and more complex, they sometimes actually do not outperform existing methods if evaluated independently [7, 23, 24, 61], which potentially leads to a certain stagnation in algorithm research.

In general, assessing the impact of recommender systems with offline experiments alone has its limits. There are, however, a number of reports on successful real-world deployments of recommender systems. Such reports typically summarize the outcomes of field studies on newly introduced recommender systems or on two system versions A/B tested in parallel. In the following, we provide an overview of selected findings from real-world deployments based on the survey presented in [41]. This overview helps us to understand both (1) which measures and business-oriented KPIs are used in practical environments, and (2) which effect sizes are observed.

## 3.1   Value Dimensions and Measurements in Practical Applications

In field studies, relating the single value drivers to separately measurable effects of the impact of recommendation applications is not always possible, since they cannot be isolated and individually considered. For instance, streaming media providers like Netflix or Spotify have business models based on flat-rate subscription

---

[2] One historical fallacy is that while in other application areas of machine learning, e.g., image classification, every small improvement in accuracy may lead to an improved system, it is less than clear that small improvements in prediction accuracy on past data have a positive impact at all with respect to the effectiveness of a recommender system.

fees. In their case, an individual successful recommendation is not only a sign of an efficient presentation of the enormous catalog space and effective decision making leading to an immediate consumption, but also contributes to the lock-in of the particular subscriber. Taken together this will also effect customer attrition and finally revenues, even if it is not directly measurable. In other cases, e.g., in news recommendation, it can even be difficult to assess in the short term if a recommendation was actually a success. While clicks on recommended articles can be easily measured, we often cannot be entirely sure if the consumer actually liked the content or not or if the consumer would have actually read the article even when it was not recommended.

In Table 7 we relate the measures and KPIs to the value dimensions introduced in Sect. 2. However, note that neither of the mentioned measures does exactly capture and quantify the identified value contributions discussed in the previous section. Let us consider, for instance, the Click-Through-Rate (CTR): it measures that an item presumably caught the user's attention and that it therefore serves as a first and preliminary indicator for a successful match of either user's short-term or long-term interests. Such preference matching therefore relates to the efficiency value dimension that facilitates to cope with information overloading by lowering users' search costs. However, while a click-through is a necessary precondition for indicating a successful match of preferences it is not sufficient. For instance, also some follow-up purchase or consumption (such as reading) would need to be observed additionally in order to have a stronger indication of a successful preference match. Furthermore, if we observe changes in CTR over time we could

**Table 7** Value dimensions and measurements

| Value dimensions | Measurement |
|---|---|
| Efficiency | *Click-through-rate (CTR)*: measures how many clicks a recommendation has garnered. Thus, the CTR indicates that recommendations have been noticed and presumably influenced users' choice due to their timeliness and relevancy. |
| Efficiency | *Adoption and conversion* measures and correlates observed user behavior other than clicks with recommendation success, e.g., when users watch a certain fraction of a movie. Conversion rates measure the fraction of users who take a desired action, e.g., submit a resume after receiving a job recommendation. |
| Efficiency, complementarities | *Sales and revenue*: when recommendations lead to purchases, one can measure corresponding business-related figures such as revenue, sales, or profit. |
| Lock-in | *User engagement*: increased user engagement is often considered as an indicator of customer retention. Engagement can, for instance, be measured in terms of the number or length of usage sessions. |
| Novelty | *Effects on consumption distributions*: recommender systems can lead to desired or undesired changes regarding sales distributions, such as, increased sales of long-tail or already popular items. They can also inspire users to consume different items than they would without a recommender. |

interpret them also as a proxy for an increase/decrease in user engagement that is another value aspect associated to the lock-in dimension. Thus, Table 7 must not be misinterpreted by assuming that the enlisted measures holistically capture already all aspects of the discussed value dimensions. However, the dimensions have to be seen as a structuring mechanism for the main types of measures as they are found in the literature [41]. This emphasizes the need for further research to more systematically develop and define measures that would also be sufficient indicators for the specific value contributions of recommendations.

Our discussion will show that recommender systems can have substantial impact on all these measures, but also that all these measurements have their limitations. Relating measures to the value driving dimensions as done in Table 7 above, can also help to recognize completely uncharted aspects, such as comparing, for instance, the cost aspect of virtual sales advisory and automated recommendations with physical in-shop encounters.

## 3.2   Click-Through-Rate

The CTR is a wide-spread measure used, e.g., in the domain of news recommendation. Results from real-world deployments are reported both for larger news aggregation sites like Google News [18], for business-oriented sites such as Forbes.com [48] and for regional ones like swissinfo.ch [28]. Other domains where the CTR were reported, sometimes in combination with other measures, including recommendations of videos on YouTube [19] or the recommendation of similar offers at eBay [47].

These studies typically report the effects of introducing a new recommender system when compared to the existing one. In the news domain, the reported increase of the CTR is often around 35%. In the study at YouTube [19], however, the difference in CTR between two systems is as high as 200%. In another study at eBay, in contrast, only a 3% improvement was observed [11]. The increase in revenue was however higher (6%).

The huge differences in terms of the reported effects may be explained both by the fact that the systems were deployed in different domains and, more importantly, that different baseline algorithms were used for comparison. In [19], a personalized algorithm for YouTube recommendations was compared to a non-personalized one that recommends the most popular videos on the site to everyone. In [11], in contrast, two more sophisticated techniques were compared, leading to smaller effects.

Independently from this problem of interpreting absolute numbers, using the Click-Through-Rate as the only or main instrument for assessing the value of recommender systems can be misleading. Often, the CTR mainly measures if individual recommendations raised attention or interest in an item. The click counts however cannot tell us if the consumers actually liked the item or if the recommendations will increase the probability that the user will return to the site next time. Increases in the CTR can sometimes also be achieved quite easily, e.g., through clickbait headlines on news portals or by recommending items that are generally popular. Moreover, the visual positioning of the recommendations can have a substantial impact on the CTR as well, as reported in [28]. Here, a more prominent positioning of the recommendation widget doubled the CTR in the short term.

Overall, optimizing for the CTR in the short term might in fact have negative business impacts in the long run, when consumers repeatedly feel misled by the recommendations or when the recommender system fails to draw the consumers' attention to less popular items. Also, as reported in [78], there can be a trade-off between optimizing item rankings for clicks and optimizing according to relevance for the consumer.

## 3.3   Adoption and Conversion Measures

Adoption measures assess the commercial success of a recommendation in a way that is more precise than CTR. In most cases reported in the literature, these measures are specific to the application domain. In the video recommendation domain, providers such as YouTube [19] or Netflix [30] measure the number of rec-ommended videos of which consumers watched at least a certain fraction. In other domains, various types of user actions are interpreted as success indicators for a rec-ommendation, e.g., "add-to-wishlist" events, "bid-through" and "purchase-through" rates on eBay, "cite-through" rates for research paper recommendations [8], booking requests on tourism sites [75], or opened communications on a dating platform [71].

The improvements in terms of such measures are again difficult to interpret on an absolute scale due to different measurement methods, baselines, and application domains. A/B tests sometimes indicate increases of a few percent in terms of the "purchase-through" rate. In other cases, where the user action is not directly leading to increased business value (e.g., "add-to-wishlist" events or "click-out" actions on a marketplace [45]), the differences between two algorithms can be substantial, e.g., 89% in the case of "add-to-wishlist" actions on eBay [47] or over 90% higher click-through rate on tv programs [68].

Like for the CTR, the described adoption measures are often only proxies for business value. An observed "add-to-wishlist" event for a recommendation is not yet a transaction, and one has to be careful not to overestimate the business value of such events. In [40], the results of A/B testing various algorithms for the recommendation of games for mobile phones were presented. The test included a variety of measures including different conversion rates (e.g., recommend-to-purchase), click behavior, game downloads, as well as actual purchases. Among other aspects, it turned out that the number of item view events (i.e., CTR of a game's detail page) incited by a recommendation was *not* indicative for the ultimate business value, which was measured in actually purchased games. In fact, even the stronger signal of a download of a demo version was not a strong predictor, which means that some algorithms created increased interest in certain items, but did not lead to a purchase at the end.

## 3.4   Sales and Revenue

Measuring changes in sales, revenue or similar business-related figures is the most direct way of determining the business value of a recommender system. Such measurements are common when the revenue of a provider is not based on a flat-rate subscription model but on individual item sales or fee-based transactions. In such situations, one cannot only measure the overall effects of different recommendation strategies in an A/B test, but also which recommendations actually led to successful transactions.

A few works in the literature exist that report the outcomes of A/B tests in terms of such business-related measures. In [52], for instance, the authors compared two recommendation algorithms—"view-based" and "purchase based" collaborative filtering—with a baseline condition where no recommendations were provided to customers of an online DVD retailer. The study revealed that the "purchase-based" best strategy led to an increase in sales by 35% (*"for those who purchase"* [52]). Interestingly, when the recommendations were solely based on item view events, no increase in sales was observed compared to the baseline condition. These observations emphasize the importance of algorithm choice and at the same time give an indication of the extent of missed sales opportunities when no recommender is used.

A much more modest, but still relevant increase in terms of revenue was reported for the mobile game recommendation field study mentioned above [40]. In this A/B test, the strongest increase was observed when a very simple content-based algorithm was used, leading to an increase of 3.6% in sales. The study however also revealed that different algorithms should be used in different navigational situations in order to maximize the revenue. Thus, even slightly stronger increases can be achieved, e.g., when a switching hybrid strategy is applied.

A study that examined how context-awareness of content-based recommendations impacted business performance measures was conducted by Panniello et al. [57]. They identified that the increased accuracy and diversity of the context-aware recommendations for comic books positively affected trust in the system which in turn affected purchases.

A number of additional studies in the literature focus on other e-commerce domains like e-grocery or online book stores [66]. The overall direct effects of adding a recommender system are often small, e.g., 0.3% in [20] or 1.8% in [51]. However, there can be major indirect and inspirational effects, e.g., up to 26% in some categories [20]. An interesting observation is furthermore reported in [66], where sales decreased significantly after the recommender was removed from the site.

Overall, we see a strong spread in the reported effects, from almost no effect up to 500% increase in Gross Merchandise Value for a study done at eBay [14]. Comparing the absolute values reported in these studies is challenging because the individual observed changes depend on many factors that are specific to the application domain, business model, market situation etc. A typical limitation of reported studies is that it is not always entirely clear if the observed effects would last beyond the A/B testing period, which is often limited to a few weeks. Another aspect related to the revenue impact of recommender systems is price sensitivity and users' willingness-to-pay (WTP). In this context Adomavicius et al. [4] determined that online recommendations have a positive effect on the recipients' WTP that correlates with the predicted ratings. This effect was observed even in the case of perturbed or manipulated rating predictions.

## 3.5   User Engagement

Discovery support, as mentioned above, can be a main functionality of a recommender system, which in turn leads to increased user engagement and customer retention. User engagement can be assessed in a number of ways, e.g., by the number of visits to the site per month, by the number of consumed content on a media site, or by the length of the individual consumption sessions and the number of particular user actions in such sessions.

Various works in the literature report increased user activity on the site, e.g., in the news domain [28, 48], where the visit lengths sometimes more than double when a recommender system is in place. Substantially increased user activity levels were furthermore observed in [21] and [46] for the music domain, where the number of playlist additions was used as an indicator for user engagement. Additional examples with increased user activity levels due to the presence of recommendation systems can be finally found for social networks and platforms [67, 72].

In many applications, it is a reasonable assumption that higher engagement with a service leads to repeated use in the future. However, the choice of the measurement that serves as a proxy for engagement can be crucial. More clicks (interactions) by a user may, for example, not always be a good sign. Like in the evaluation of a search engine, more interactions might indicate that the system was actually not able to present something relevant to the user. A typical example in the recommendation context would be recommendations that are explicitly designed for discovery, as can be found on music streaming sites. Here, the best case is that the user who is presented with the recommendation list, discovers a new artist and then leaves the discovery module to explore this new artist. As a result, fewer interactions with the recommendation list might indicate that the recommendations were actually good.

Given this need for interpreting the observations of user behavior, research needs to more frequently draw a bow from the underpinnings of cognitive and psychological science to the principled observations and experimentation in order to draw reliable conclusions for algorithm and system adjustments as has been postulated in [76].


## 3.6   Effects on Consumption Distributions

Increases in sales and revenue, as previously reported, are often simply the result of consumers buying *more*, e.g., due to cross-selling links provided by a recommender system. Recommender systems may, however, also inspire users to consume *different* things. Such changes in the distribution of consumption or sales of items can be desirable for different reasons. On the one hand, there might be direct business-related effects when using a price and profit aware recommender system. Such a system might, for example, aim to recommend items that are both a good match for the consumer's preferences and at the same time are more profitable than other items that could be similarly good matches [38].

On the other hand, there might be more indirect effects that come with changes in the consumption distribution due to either enabling customers making new discoveries or re-enforcing rich-get-richer effects [26].

Already the seminal paper of Resnick et al. [62] questioned if the peer groups created by collaborative recommender systems will be permeable or fracture the global village into tribes. Twenty years later Hosanagar et al. [35] identified, at least for the domain of music recommendations, that users consumed in general more and thus also more commonalities among users arose due to the impact of recommendations.

In some application domains like e-commerce, a large fraction of the sales comes from a relatively small number of "blockbuster" items, leading to a long tail of less frequently sold items. Recommender systems have the potential to point consumers to such long tail items and to help them discover parts of the catalog that were previously unknown to them, such as observed in [75]. In transaction-based applications, e.g., on e-commerce sites, this can lead to the indirect effect of more sales in specific categories as discussed above [20]. Such indirect and inspirational effects of recommender systems were also found in the music domain [46]. Generally, supporting discovery and driving customers away from blockbusters is often considered a main purpose of providing recommendations on media streaming services, assuming that discovery leads to more engagement and, ultimately, continued subscriptions [30].

Finally, changes in the consumption behavior of users might also be desirable from a societal perspective. The purpose of a recommender system might, for example, be to entice healthier consumer behavior. In this context, the authors of [22] explored the potential of nudging users towards more healthy dishes through recommendations. Another goal may be to stimulate users to enrich their information consumption behavior, e.g., by presenting news content that covers more than one opinion on a controversial topic to avoid filter bubbles and echo chambers. The study of [25], for instance, identified an increase of individuals' exposure to news from their less preferred side as well as a (modest) increase of ideological distance between individuals when analyzing the news consumption behavior of 50K users based on their browsing histories.

Generally, influencing *short-term* consumer behavior through recommendations to some extent might not be too difficult in a number of application domains. This is in particular the case when the system actually filters out certain options, which cannot be chosen by the consumers anymore. Often, however, a main challenge is to achieve positive long-term effects. Only recommending items that are highly profitable but not too relevant for consumers might, for instance, result in increased short-term profits, but lead to a loss of consumer trust in the long run.

## 4   Towards More Value-Oriented and Impactful Research

Our discussion has shown that there are different dimensions that drive the value creation of recommender systems for their stakeholders. Given this richness, the focus in academic research appears to be narrow. First, there is a strong focus on algorithmic proposals mostly evaluated on historical data. This focus on accuracy— although important—touches however only few value creating dimensions, while, in contrast, questions related to the users' experience or long-term effects of sales distributions are underexplored [39, 42].

Generally, there is no doubt that industry has always picked up proposals from the academic world and that they have successfully implemented and further improved novel machine learning models to serve given organizational goals. A prominent example is the use of matrix factorization techniques, which were explored already in the late 1990s, became popular in the context of the Netflix Prize, and were later broadly adopted in industry. However, as a result of the above-described phenomena, the question arises if—despite the many papers that are published every year—major parts of today's academic research on recommender systems actually have a strong impact in practice.

To be more impactful and relevant in practice, our research approach should therefore be refocused on two dimensions.

1. In terms of the *research scope*, it is important to put more emphasis on the user experience of recommender systems than on algorithms [50], acknowledging that most aspects of recommender systems cannot be evaluated without involving users and without considering its context of use [44].
2. Regarding the *research methodology*, the focus on algorithms led to a certain overreliance on offline experimental designs that are common in applied machine learning research, but which are insufficient to assess the impact of a complex information system like a recommender.

Clearly, it would generally be desirable that more academic proposals were evaluated in real environments. There are, however, a number of other research instruments that are available, and there are several ways in which the research efforts of the community could be refocused in order to be more impactful.

## 4.1 Recommender Systems Research with a Purpose

Looking at the research scope, many of the works published today—in particular in the predominant area of algorithms—is not based on theory or explicitly stated research hypotheses. Higher prediction accuracy of a machine learning model is implicitly equated with progress towards better systems, even in cases where better accuracy is only demonstrated by a very specific experimental configuration of datasets, baseline algorithms, and evaluation procedures. In the midst of such a "leaderboard chasing" culture [54], the fundamental and underlying question "*What is a good recommendation (in a given context)?*" is too rarely asked [37].

As the discussions in Sect. 2 indicate, the same set of recommendations can be useful or not, depending on the goals that one tries to achieve, the perspective that is taken (e.g., consumer vs. provider), and the individual user's preferences and context. Therefore, when evaluating a recommender system, these surrounding factors and the goal that one tries to achieve should be made explicit in order to ensure that the chosen evaluation approach is appropriate [34]. In [37], a four-layer conceptual framework is proposed to guide researchers, both in academia and industry, towards a more goal and value oriented approach when designing and evaluating a recommender system. We summarize the main layers of this framework in Table 8, where we also give examples both from the consumer and provider perspective at each level.

Note the importance of making suitable choices at the lowest layer of this framework, i.e., how the system is evaluated. The framework should help to ensure that there is a "metric-task-purpose" fit, i.e., measuring if the system is actually able to fulfill the goals it intended to serve. Today's research far too often seems to focus solely on the two layers at the bottom of the framework. The system's task usually is considered to help the user to "find good items" [34]. Moreover, how the evaluation of algorithms is done is largely standardized, based on offline protocols and metrics such as Precision, Recall, and RMSE. However, such metrics, while generally useful, can only inform us about differences between algorithms in terms of item retrieval and ranking performance, but not about the created value for individual stakeholders. The proposed framework should therefore help researchers to look beyond the common "find good items" task and extend their research scope to the many other and more specific purposes recommender systems can serve, both from the perspective of the consumer and the provider.

**Table 8** A conceptual framework for goal and value oriented research

| Goals and values | At the top-most strategic level, it is important to understand or define the goals of the system and which value it drives for which group of stakeholders. From an organizational (provider) perspective, this goal could for instance focus on recommending complementarities and increasing revenues or on the lock-in value driver dimension and customer retention. For a consumer, e.g., of a media service, the value could for example simply be entertainment. |
|---|---|
| Recommendation purpose | Depending on these strategic considerations, the specific purpose of the recommender system in this context has to be clarified, i.e., how the recommender system can help to achieve the described goals. If, for instance, one goal is customer retention, one purpose of the system may be to increase the "discoverability" of specific items for consumers and to thereby increase their engagement. |
| System task | At the next, more operational level, the question has to be answered how the system or its components, e.g., an algorithm, has to be designed to support the intended purpose. In case the goal is to support discovery, an algorithm may try to intentionally diversify the recommendations. This might involve prioritizing items that do not have the highest consumption probabilities according to past observations. |
| Evaluation method | The three upper layers finally determine how the system should be evaluated. In the given example of discovery support, one could use a combination of (1) computational metrics to objectively assess the level of novelty of the recommendations, and (2) subjective measures collected through a user study regarding, e.g., the subjective perception of the recommendations and the participant's intention to use the system in the future. |

## 4.2 Utilizing a Richer Methodological Repertoire

Extending the research scope in the described ways, e.g., to consider the impact of recommender systems on different stakeholders, requires a more comprehensive methodological approach than we often observe today. First and foremost, recommender systems are much more than retrieval systems. They exert influence on the choices of users, and whether a recommendation leads to follow-up actions of users or not depends on a variety of factors including individual decision heuristics and biases. Therefore, various aspects of a recommender system can only be addressed with research designs with humans in the loop, such as controlled experiments with users or qualitative and observational studies. On the other hand, recommenders are e-commerce information systems with a clear business impact and even implications for society, which is why we need appropriate evaluation methodologies to assess these effects as well.

In the following, we will outline potential ways forward in terms of how we can address and evaluate the sometimes complex interplay between consumers, organizations, and information systems that is embedded into a societal context.

### 4.2.1   Evaluating with Humans in the Loop

When providing consumers with a recommendation service, we might ask ourselves a number of questions regarding its value for the users, e.g.:

- *How are recommendations perceived by users—do users actually find them helpful?*
- *Do users find recommendations diverse and surprising enough—do they help them discover new things?*
- *Would they appreciate more information justifying why certain items are shown and others not—do users trust recommendations to be fair?*
- *Would they be interested in receiving more recommendations—would they use the service in the future?*

None of these questions, can be confidently answered without involving humans when evaluating the system. Remember here that such user-centric questions are also very important from a provider's perspective, assuming in general that added consumer value directly or indirectly leads to increased value for the operator.

Evaluating from a user-centric or human-computer interaction (HCI) perspective has a long tradition in recommender systems research [50]. User centric research is typically guided by explicitly stated research questions, e.g., to what extent users would value a certain type of explanation. One main first challenge in such research efforts usually is to develop an appropriate experimental design, e.g., a randomized controlled experiment, that helps to reliably answer the stated questions. As a result, user-centric research is much less standardized than typical offline experiments that are conducted to compare the prediction accuracy of algorithms.

Nowadays, at least two problem-independent frameworks for user-centric recommender systems evaluation exist [49, 59]. Researchers benefit from such frameworks for their own works in different ways. These frameworks, for example, provide validated questionnaires for various general quality dimensions of a recommender system, and they also demonstrate how to use broadly-used statistical analysis techniques like Structural Equation Modeling. Nonetheless, user-centric research remains difficult and requires often more effort than offline experimentation in recommender systems research in many ways. Research questions must be stated on theoretical considerations, specific experimental designs have to be designed and defended against reviewers, participants must be recruited that are at least representative for some group of users, and finally there is not always consensus in the community regarding which statistical methods are appropriate for the subsequent analysis.

These difficulties may have led to a certain over-reliance on offline experimentation in our field. As a result, we observe that even though the number of research questions that can be reliably answered with offline experiments alone is actually small, such designs dominate the research landscape. While there are efforts to use computational metrics to capture some of the aspects mentioned above, e.g., regarding the diversity of a set of recommendations, many of the proposed metrics were not validated with humans in the loop. It is, for example, less than clear if the many metrics proposed for novelty, diversity, or serendipity actually correspond to the human perception.

As a result, it remains important that the community focuses more on user-oriented aspects of recommender systems. This is particularly important as there exists a number of research works that indicate that offline metrics like Precision, Recall, or RMSE do *not* necessarily correlate with the quality perception of users [8, 17, 28, 55, 64]. The work of Adomovicius et al. [4] goes even beyond these works by demonstrating in controlled behavioral studies that random recommendations or recommendations that artificially increase the predicted rating lead to a significant increase in the participants' willingness-to-pay for the received song recommendations. Thus, even if in field research the ultimate goal of actual conversion or purchase is reached, we still cannot be entirely confident that the actual recommendation was accurately matching the users' tastes. Here, *multi-modal* evaluation approaches may be advisable that compare results from offline experiments with the user's perceptions from a user study. With respect to the type of user involvement in evaluations, note that researchers are not limited to randomized controlled experiments when it comes to involving humans in the loop. Interviews, surveys, focus groups and various other types of *qualitative* research methods are used in various other fields outside of computer science, and should be considered in recommender systems research more often as well.

### 4.2.2   Re-thinking Data-Based Research

Despite their limitations, offline and data-based research will remain relevant in the future and will not be limited to the assessment of computational aspects such as scalability. Typical offline experiments that are done, e.g., regarding the prediction accuracy of algorithms on historical data, can still serve us to identify algorithms or algorithm variants that we may rule out from A/B testing. However, instead of asking if a new machine learning model can outperform another one by a few percent on given datasets in terms of accuracy, the focus should be shifted to different types of questions. For instance, can we analyze in which ways the recommendations generated by one algorithm are different from those of another one? Evidence exists that algorithms with the same performance in terms of accuracy often recommend very different things to users [43]. Given the characteristics of such recommendations, we may then ask questions if these differences are actually desirable from an organizational perspective. For instance, is recommending already popular items of interest from the business perspective or not? A number of research works were published that go into that direction, including a variety of works that aim at understanding aspects like diversity, novelty, or serendipity. Assuming that the used computational metrics correspond to user perceptions, offline experiments can help us to compare algorithms in these respects and to design new approaches that are able to balance potential trade-offs.

A general limitation of existing offline evaluations is that often little is known about the provenance of the data. Researchers in the domain of music recommendations often base their works on listening logs collected from music services like last.fm. In the e-commerce domain, on the other hand, datasets are often used that contain user interaction logs like item views or purchase events. One major issue with such datasets can be, for instance, that a recommender system or a personalized filtering algorithm was already in place. Such additional attention biases exist both in log-based datasets [15] as well as in rating datasets [5], and researchers have begun in recent years to deal with such phenomena that ultimately lead to biased, and thus not informative evaluation results, see, e.g., [16, 73]. In that context, new forms of offline evaluation approaches have emerged in recent years, including "off-policy" evaluation and "counterfactual reasoning" approaches [16, 31] that promise to lead to results that are more reliable predictors of A/B test outcomes.

Another area where offline experiments could be further explored are simulations. Almost all research today is focusing on short-term effects of recommendations, i.e., if a recommender is able to provide helpful item suggestions in the given situation. In contrast simulation-based approaches, such as agent-based modeling, were already successfully applied in other disciplines like managerial science [70]. Using such simulation approaches allows researchers to investigate, for instance, *longitudinal effects* of recommender systems such as potentially unexpected emergent agent behavior [77] or the impact on choice diversity [33]. These simulation approaches may furthermore also be promising choices to analyze the effects of different recommendation strategies in multi-stakeholder environments (see also Chap. "Multistakeholder Recommender Systems") and eventually even make predictions on societal implications.

Finally, we can often observe a tendency towards over-generalization from results of offline experiments, such as one method being able to improve the state of the art even though this is only shown for a very specific experimental setting. While researchers are generally interested in generalizable solutions, it is very often imperative to closely look at the data and consider the specifics of a domain or application to reach reliable insights. Should we, for instance, in an e-commerce setting remind customers of items they have already inspected in previous sessions? Should we recommend them items that are currently on sale or trending in the community? Answering such research questions often requires an analytical, data science approach to recommender systems research. In some cases, such analytical insights can guide the design of domain-specific algorithms considering the characteristics of successful past recommendations [45], or help to understand the business impact of recommender strategies based on econometric models [2].

## 5 Conclusions

This chapter introduced the different value driving dimensions of e-commerce business models and used them to categorize and structure the various types of value that recommender systems create for their stakeholders. In this chapter we reviewed the most important studies demonstrating the strong influence recommender systems exert on the decision-making behavior of users as well as the potential risks that come with them. While the business value of recommenders in general is undoubted, deeper analysis on the specific value aspects and how to make them transparent with adequate measurements is still in its infancy. The review on the most widely used metrics revealed that they are only loosely related to the various value types associated to recommender systems. Thus, this chapter might also stimulate research to identify the blind spots of value and impacts that are not yet quantifiable with metrics as well as incite efforts to develop more prescriptive advice for practical settings. Furthermore, we found that today's academic research on value and impact of recommender systems suffers from a somewhat limited methodological basis. An over-reliance on offline experimentation and accuracy measures raises questions on the impact of current research work in practice. We therefore argue that a paradigmatic shift is necessary in how we evaluate recommender systems and outline potential ways of re-focusing research in the field both in terms of research scope and applied research methodology.

# References

1. H. Abdollahpouri, G. Adomavicius, R. Burke, I. Guy, D. Jannach, T. Kamishima, J. Krasnodeb-ski, L. Pizzato, Multistakeholder recommendation: Survey and research directions. User Model. User-Adapt. Interact. **30**, 127–158 (2020)
2. P. Adamopoulos, A. Tuzhilin, The business value of recommendations: A privacy-preserving econometric analysis, In *Proceedings of the International Conference on Information Systems (ICIS'15)* (2015)
3. G. Adomavicius, A. Tuzhilin, Personalization technologies: a process-oriented perspective. Commun. ACM **48**(10), 83–90 (2005)
4. G. Adomavicius, J.C. Bockstedt, S.P. Curley, J. Zhang, Effects of online recommendations on consumers' willingness to pay. Inf. Syst. Res. **29**(1), 84–102 (2018)
5. G. Adomavicius, J. Bockstedt, S. Curley, J. Zhang, Reducing recommender systems biases: an investigation of rating display designs. MIS Quart. **43**, 1321–1341 (2019)
6. R. Amit, C. Zott, Value drivers of e-commerce business models, in *Creating Value: Winners in the New Business Environment*, ed. by C. Lucier, R.D. Nixon. (Blackwell, Oxford, 2002), pp. 15–47
7. T.G. Armstrong, A. Moffat, W. Webber, J. Zobel, Improvements that don't add up: ad-hoc retrieval results since 1998, in *Proceedings of the 18th ACM Conference on Information and Knowledge Management, CIKM'09* (2009), pp. 601–610
8. J. Beel, S. Langer, A comparison of offline evaluations, online evaluations, and user studies in the context of research-paper recommender systems, in *Proceedings of the 22nd International Conference on Theory and Practice of Digital Libraries (TPDL'15)* (2015), pp. 153–168
9. I. Benbasat, W. Wang, Trust in and adoption of online recommendation agents. J. AIS **6**, 03 (2005)
10. A.V. Bodapati, Recommendation systems with purchase data. J. Market. Res. **45**(1), 77–93 (2008)
11. Y.M. Brovman, M. Jacob, N. Srinivasan, S. Neola, D. Galron, R. Snyder, P. Wang, Optimizing similar item recommendations in a semi-structured marketplace to maximize conversion, in *Proceedings of the 10th ACM Conference on Recommender Systems, RecSys'16* (2016), pp. 199–202

12. R. Burke, N. Sonboli, A. Ordonez-Gauger, Balanced neighborhoods for multi-sided fairness in recommendation, in *Proceedings of the 1st Conference on Fairness, Accountability and Transparency*. Proceedings of Machine Learning Research, vol. 81 (2018), pp. 202–214

13. P.Y. Chau, S.Y. Ho, K.K. Ho, Y. Yao, Examining the effects of malfunctioning personalized services on online users' distrust and behaviors. Decis. Support Syst. **56**, 180–191 (2013)

14. Y. Chen, J.F. Canny, Recommending ephemeral items at web scale, in *Proceedings of the 34th International ACM SIGIR Conference on Research and Development in Information Retrieval, SIGIR'11* (2011), pp. 1013–1022

15. H.-H. Chen, C.-A. Chung, H.-C. Huang, W. Tsui, Common pitfalls in training and evaluating recommender systems. SIGKDD Explor. Newsl. **19**(1), 37–45 (2017)

16. M. Chen, A. Beutel, P. Covington, S. Jain, F. Belletti, E.H. Chi, Top-k off-policy correction for a reinforce recommender system, in *Proceedings of the Twelfth ACM International Conference on Web Search and Data Mining, WSDM'19* (2019), pp. 456–464

17. P. Cremonesi, F. Garzotto, R. Turrin, Investigating the persuasion potential of recommender systems from a quality perspective: an empirical study. Trans. Interact. Intell. Syst. **2**(2), 1–41 (2012)

18. A.S. Das, M. Datar, A. Garg, S. Rajaram, Google news personalization: scalable online collaborative filtering, in *Proceedings of the 16th International Conference on World Wide Web, WWW'07* (2007), pp. 271–280

19. J. Davidson, B. Liebald, J. Liu, P. Nandy, T. Van Vleet, U. Gargi, S. Gupta, Y. He, M. Lambert, B. Livingston, D. Sampath, The YouTube video recommendation system, in *Proceedings of the 4th Conference on Recommender Systems, RecSys'10* (2010), pp. 293–296

20. M.B. Dias, D. Locher, M. Li, W. El-Deredy, P.J. Lisboa, The value of personalised recommender systems to e-business: a case study, in *Proceedings of the 2008 ACM Conference on Recommender Systems, RecSys'08* (2008), pp. 291–294

21. M.A. Domingues, F. Gouyon, A.M. Jorge, J.P. Leal, J. Vinagre, L. Lemos, M. Sordo, Combining usage and content in an online recommendation system for music in the long tail. Int. J. Multimed. Inf. Retrieval **2**(1), 3–13 (2013)

22. D. Elsweiler, C. Trattner, M. Harvey, Exploiting food choice biases for healthier recipe recommendation, in *Proceedings of the 40th International ACM SIGIR Conference on Research and Development in Information Retrieval, SIGIR'17* (2017), pp. 575–584

23. M. Ferrari Dacrema, P. Cremonesi, D. Jannach, Are we really making much progress? A worrying analysis of recent neural recommendation approaches, in *Proceedings of the 13th ACM Conference on Recommender Systems, RecSys'19* (2019), pp. 101–109

24. M. Ferrari Dacrema, S. Boglio, P. Cremonesi, D. Jannach, A troubling analysis of reproducibility and progress in recommender systems research. ACM Trans. Inf. Syst. **39**(2), 1–49 (2021)

25. S. Flaxman, S. Goel, J.M. Rao, Filter bubbles, echo chambers, and online news consumption. Publ. Opin. Quart. **80**(S1), 298–320 (2016)

26. D. Fleder, K. Hosanagar, Blockbuster culture's next rise or fall: the impact of recommender systems on sales diversity. Manag. Sci. **55**(5), 697–712 (2009)

27. A. Friedman, B. Knijnenburg, K. Vanhecke, L. Martens, S. Berkovsky, Privacy aspects of recommender systems, in *Recommender Systems Handbook*, ed. by F. Ricci, L. Rokach, B. Shapira, 2nd edn. (Springer, Berlin, 2015), pp. 649–688

28. F. Garcin, B. Faltings, O. Donatsch, A. Alazzawi, C. Bruttin, A. Huber, Offline and online evaluation of news recommender systems at swissinfo.ch, in *Proceedings of the 8th ACM Conference on Recommender Systems, RecSys'14* (2014), pp. 169–176

29. A. Ghoshal, S. Kumar, V. Mookerjee, Impact of recommender system on competition between personalizing and non-personalizing firms. J. Manag. Inf. Syst. **31**(4), 243–277 (2015)

30. C.A. Gomez-Uribe, N. Hunt, The Netflix recommender system: algorithms, business value, and innovation. Trans. Manag. Inf. Syst. **6**(4), 1–19 (2015)

31. A. Gruson, P. Chandar, C. Charbuillet, J. McInerney, S. Hansen, D. Tardieu, B. Carterette, Offline evaluation to make decisions about playlist recommendation algorithms, in *Proceedings of the Twelfth ACM International Conference on Web Search and Data Mining, WSDM'19* (2019), pp. 420–428

32. D.J. Hand, Classifier technology and the illusion of progress. Stat. Sci. **21**(1), 1–14 (2006)
33. N. Hazrati, M. Elahi, F. Ricci, Simulating the impact of recommender systems on the evolution of collective users' choices, in *Proceedings of the 31st ACM Conference on Hypertext and Social Media* (2020), pp. 207–212
34. J.L. Herlocker, J.A. Konstan, L.G. Terveen, J.T. Riedl, Evaluating collaborative filtering recommender systems. Trans. Inf. Syst. **22**(1), 5–53 (2004)
35. K. Hosanagar, D. Fleder, D. Lee, A. Buja, Will the global village fracture into tribes? Recommender systems and their effects on consumer fragmentation. Manag. Sci. **60**(4), 805–823 (2014)
36. A. Iovine, F. Narducci, G. Semeraro, Conversational recommender systems and natural language: a study through the ConveRSE framework. Decis. Support Syst. **131**, 113250–113260 (2020)
37. D. Jannach, G. Adomavicius, Recommendations with a purpose, in *Proceedings of the 10th ACM Conference on Recommender Systems, RecSys'16* (2016), pp. 7–10
38. D. Jannach, G. Adomavicius, Price and profit awareness in recommender systems, in *Proceedings of the 2017 Workshop on Value-Aware and Multi-Stakeholder Recommendation (VAMS) at RecSys 2017* (2017)
39. D. Jannach, C. Bauer, Escaping the mcnamara fallacy: towards more impactful recommender systems research. AI Mag. **41**(4), 79–95 (2020)
40. D. Jannach, K. Hegelich, A case study on the effectiveness of recommendations in the mobile internet, in *Proceedings of the 10th ACM Conference on Recommender Systems, RecSys'09* (2009), pp. 205–208
41. D. Jannach, M. Jugovac, Measuring the business value of recommender systems. ACM Trans. Manag. Inf. Syst. **10**(4), 1–23 (2019)
42. D. Jannach, M. Zanker, M. Ge, M. Gröning, Recommender systems in computer science and information systems - a landscape of research, in *Proceedings of the International Conference on Electronic Commerce and Web Technologies, EC-WEB'12* (2012), pp. 76–87
43. D. Jannach, L. Lerche, I. Kamehkhosh, M. Jugovac, What recommenders recommend: an analysis of recommendation biases and possible countermeasures. User Model. User-Adapt. Interact. **25**(5), 427–491 (2015)
44. D. Jannach, P. Resnick, A. Tuzhilin, M. Zanker, Recommender systems - beyond matrix completion. Commun. ACM **59**(11), 94–102 (2016)
45. D. Jannach, M. Ludewig, L. Lerche, Session-based item recommendation in e-commerce: on short-term intents, reminders, trends and discounts. User Model. User-Adapt. Interact. **27**(3), 351–392 (2017)
46. I. Kamehkhosh, G. Bonnin, D. Jannach, Effects of recommendations on the playlist creation behavior of users. User Model. User-Adapt. Interact. **30**, 285–322 (2019)
47. J. Katukuri, T. Könik, R. Mukherjee, S. Kolay, Recommending similar items in large-scale online marketplaces, in *IEEE International Conference on Big Data 2014* (2014), pp. 868–876
48. E. Kirshenbaum, G. Forman, M. Dugan, A live comparison of methods for personalized article recommendation at Forbes.com, in *Proceedings of the 2012th European Conference on Machine Learning and Knowledge Discovery in Databases, ECMLPKDD'12* (2012), pp. 51–66
49. B.P. Knijnenburg, M.C. Willemsen, Z. Gantner, H. Soncu, C. Newell, Explaining the user experience of recommender systems. User Model. User-Adapt. Interact. **22**, 441–504 (2012)
50. J. Konstan, J. Riedl, Recommender systems: from algorithms to user experience. User Model. User-Adapt. Interact. **22**(1–2), 101–123 (2012)
51. R. Lawrence, G. Almasi, V. Kotlyar, M. Viveros, S. Duri, Personalization of supermarket product recommendations. Data Min. Knowl. Disc. **5**(1), 11–32 (2001)
52. D. Lee, K. Hosanagar, Impact of recommender systems on sales volume and diversity, in *Proceedings of the 2014 International Conference on Information Systems, ICIS'14* (2014)
53. D. Lee, K. Hosanagar, How do product attributes and reviews moderate the impact of recommender systems through purchase stages? eBus. eComm. eJ. **67**, 1–659 (2018)

54. J. Lin, The neural hype and comparisons against weak baselines. SIGIR Forum **52**(2), 40–51 (2019)
55. A. Maksai, F. Garcin, B. Faltings, Predicting online performance of news recommender systems through richer evaluation metrics. In *Proceedings of the 9th ACM Conference on Recommender Systems, RecSys'15* (2015), pp. 179–186
56. T.N. Nguyen, F. Ricci, A chat-based group recommender system for tourism. Inf. Technol. Tour. **18**(1–4), 5–28 (2018)
57. U. Panniello, M. Gorgoglione, A. Tuzhilin, Research note—in carss we trust: how context-aware recommendations affect customers' trust and other business performance measures of recommender systems. Inf. Syst. Res. **27**, 1–218 (2016)
58. P. Pu, L. Chen, R. Hu, A user-centric evaluation framework for recommender systems, in *Proceedings of the Fifth ACM Conference on Recommender Systems, RecSys'11* (2011), pp. 157–164
59. P. Pu, L. Chen, R. Hu, A user-centric evaluation framework for recommender systems, in *Proceedings of the 5th Conference on Recommender Systems (RecSys'11)* (2011), pp. 157–164
60. M. Qiu, F.-L. Li, S. Wang, X. Gao, Y. Chen, W. Zhao, H. Chen, J. Huang, W. Chu, Alime chat: a sequence to sequence and rerank based chatbot engine, in *Proceedings of the 55th Annual Meeting of the Association for Computational Linguistics, ACL'17* (2017), pp. 498–503
61. S. Rendle, L. Zhang, Y. Koren, On the difficulty of evaluating baselines: a study on recommender systems (2019). arXiv:1905.01395
62. P. Resnick, N. Iacovou, M. Suchak, P. Bergstrom, J. Riedl, Grouplens: an open architecture for collaborative filtering of netnews, in *Proceedings of the 1994 ACM Conference on Computer Supported Cooperative Work* (1994), pp. 175–186
63. M.H. Ribeiro, R. Ottoni, R. West, V.A. Almeida, W. Meira Jr., Auditing radicalization pathways on YouTube, in *Proceedings of the 2020 Conference on Fairness, Accountability, and Transparency* (2020), pp. 131–141
64. M. Rossetti, F. Stella, M. Zanker, Contrasting offline and online results when evaluating recommendation algorithms, in *Proceedings of the 10th ACM Conference on Recommender Systems, RecSys'16* (2016), pp. 31–34
65. H. Schäfer, S. Hors-Fraile, R.P. Karumur, A. Calero Valdez, A. Said, H. Torkamaan, T. Ulmer, C. Trattner, Towards health (aware) recommender systems, in *Proceedings of the 2017 International Conference on Digital Health* (2017), pp. 157–161
66. G. Shani, D. Heckerman, R.I. Brafman, An MDP-based recommender system. J. Mach. Learn. Res. **6**, 1265–1295 (2005)
67. E. Spertus, M. Sahami, O. Buyukkokten, Evaluating similarity measures: a large-scale study in the orkut social network, in *Proceedings of the Eleventh ACM SIGKDD International Conference on Knowledge Discovery in Data Mining, KDD'05* (2005), pp. 678–684
68. P. Symeonidis, A. Janes, D. Chaltsev, P. Giuliani, D. Morandini, A. Unterhuber, L. Coba, M. Zanker, Recommending the video to watch next: an offline and online evaluation at youtv.de, in *Fourteenth ACM Conference on Recommender Systems, RecSys'20.* (Association for Computing Machinery, New York, 2020), pp. 299–308
69. A. Tuzhilin, Personalization: the state of the art and future directions. Bus. Comput. **3**(3), 3–43 (2009)
70. F. Wall, Agent-based modeling in managerial science: an illustrative survey and study. Rev. Manag. Sci. **10**(1), 135–193 (2016)
71. W. Wobcke, A. Krzywicki, Y. Sok, X. Cai, M. Bain, P. Compton, A. Mahidadia, A deployed people-to-people recommender system in online dating. AI Mag. **36**(3), 5–18 (2015)
72. Y. Xu, Z. Li, A. Gupta, A. Bugdayci, A. Bhasin, Modeling professional similarity by mining professional career trajectories, in *Proceedings of the 20th ACM SIGKDD International Conference on Knowledge Discovery and Data Mining, KDD'14* (2014), pp. 1945–1954
73. L. Yang, Y. Cui, Y. Xuan, C. Wang, S. Belongie, D. Estrin, Unbiased offline recommender evaluation for missing-not-at-random implicit feedback, in *Proceedings of the 12th ACM Conference on Recommender Systems, RecSys'18* (2018), pp. 279–287

74. K.-H. Yoo, U. Gretzel, M. Zanker, *Persuasive recommender systems: conceptual background and implications* (Springer, Berlin, 2012)
75. M. Zanker, M. Bricman, S. Gordea, D. Jannach, M. Jessenitschnig, Persuasive online-selling in quality and taste domains, in *Proceedings of the 7th International Conference on E-Commerce and Web Technologies, EC-Web'06* (2006), pp. 51–60
76. M. Zanker, L. Rook, D. Jannach, Measuring the impact of online personalisation: past, present and future. Int. J. Hum.-Comput. Stud. **131**, 160–168 (2019)
77. J. Zhang, G. Adomavicius, A. Gupta, W. Ketter, Consumption and performance: understanding longitudinal dynamics of recommender systems via an agent-based simulation framework. Inf. Syst. Res. **31**, 76–101 (2020)
78. H. Zheng, D. Wang, Q. Zhang, H. Li, T. Yang, Do clicks measure recommendation relevancy? An empirical user study, in *Proceedings of the Fourth ACM Conference on Recommender Systems, RecSys'10* (2010), pp. 249–252

# Evaluating Recommender Systems

Asela Gunawardana, Guy Shani, and Sivan Yogev

## 1 Introduction

Recommender systems can now be found in many modern applications that expose the user to a huge collections of items. Such systems typically provide the user with a list of recommended items they might prefer, or predict how much they might prefer each item. These systems help users to decide on appropriate items, and ease the task of finding preferred items in the collection.

For example, the online streaming provider Netflix[1] displays lists of recommended movies and TV shows to help the user decide which item to watch next. The online retailer Amazon[2] provides average user ratings for displayed books, and a list of other books that are bought by users who buy a specific book. Microsoft provides many free downloads for users, such as bug fixes, products and so forth. When a user downloads some software, the system presents a list of additional items that are downloaded together. All these systems are typically categorized as recommender systems, even though they provide diverse services.

---

[1] www.Netflix.com.

[2] www.amazon.com.

---

A. Gunawardana
Google, Seattle, WA, USA

G. Shani (✉)
Software and Information Systems Engineering, Ben Gurion University, Beersheba, Israel
e-mail: shanigu@bgu.ac.il

S. Yogev
Outbrain Inc., Haifa, Israel
e-mail: syogev@outbrain.com

© Springer Science+Business Media, LLC, part of Springer Nature 2022
F. Ricci et al. (eds.), *Recommender Systems Handbook*,
https://doi.org/10.1007/978-1-0716-2197-4_15

In the past decade, there has been a vast amount of research in the field of recommender systems, mostly focusing on designing new algorithms for recommendations. An application designer who wishes to add a recommender system to her application has a large variety of algorithms at her disposal, and must make a decision about the most appropriate algorithm for her goals. Typically, such decisions are based on experiments, comparing the performance of a number of candidate recommenders. The designer can then select the best performing algorithm, given structural constraints such as the type, timeliness and reliability of availability data, allowable memory and CPU footprints. Furthermore, most researchers who suggest new recommendation algorithms also compare the performance of their new algorithm to a set of existing approaches. Such evaluations are typically performed by applying some evaluation metric that provides a ranking of the candidate algorithms (usually using numeric scores).

Initially most recommenders have been evaluated and ranked on their prediction power—their ability to accurately predict the user's choices. However, it is now widely agreed that accurate predictions are crucial but insufficient to deploy a good recommendation engine. In many applications people use a recommender system for more than an exact anticipation of their tastes. Users may also be interested in discovering new items, in rapidly exploring diverse items, in preserving their privacy, in the fast responses of the system, and many more properties of the interaction with the recommendation engine. We must hence identify the set of properties that may influence the success of a recommender system in the context of a specific application. Then, we can evaluate how the system preforms on these relevant properties.

In this chapter we review the process of evaluating a recommendation system. We discuss three different types of experiments; offline, user studies and online experiments.

Often it is easiest to perform offline experiments using existing data sets and a protocol that models user behavior to estimate recommender performance measures such as prediction accuracy. A more expensive option is a user study, where a small set of users is asked to perform a set of tasks using the system, typically answering questions afterwards about their experience. Finally, we can run large scale experiments on a deployed system, which we call online experiments. Such experiments evaluate the performance of the recommenders on real users which are oblivious to the conducted experiment. We discuss what can and cannot be evaluated for each of these types of experiments.

We can sometimes evaluate how well the recommender achieves its overall goals. For example, we can check an e-commerce website revenue with and without the recommender system and make an estimation of the value of the system to the website. In other cases, it can also be useful to evaluate how recommenders perform in terms of some specific properties, allowing us to focus on improving properties where they fall short. First, one must show that a property is indeed relevant to users and affect their experience. Then, we can design algorithms that improve upon these properties. In improving one property we may reduce the quality of another property, creating a a trade-off between a set of properties. In many cases it is also difficult to say how these trade-offs affect the overall performance of the system, and we have to either run additional experiments to understand this aspect, or use the opinions of domain experts.

This chapter focuses on property-directed evaluation of recommender algorithms. We provide an overview of a large set of properties that can be relevant for system success, explaining how candidate recommenders can be ranked with respect to these properties. For each property we discuss the relevant experiment types—offline, user study, and online experiments—and explain how an evaluation can be conducted in each case. We explain the difficulties and outline the pitfalls in evaluating each property. For all these properties we focus on ranking recommenders on that property, assuming that better handling the property will improve user experience.

We also review a set of previous suggestions for evaluating recommender systems, describing a large set of popular methods and placing them in the context of the properties that they measure. We especially focus on the widely researched accuracy and ranking measurements, describing a large set of evaluation metrics for these properties. For other, less studied properties, we suggest guidelines from which specific measures can be derived. We provide examples of such specific implementations where appropriate.

The rest of the chapter is structured as follows. In Sect. 2 we discuss the different experimental settings in which recommender systems can be evaluated, discussing the appropriate use of offline experiments, user studies, and online trials. We also outline considerations that go into making reliable decisions based on these experiments, including generalization and statistical significance of results. In Sect. 4 we describe a large variety of properties of recommender systems that may impact their performance, as well as metrics for measuring these properties. Finally, we conclude in Sect. 5.

## 2 Experimental Settings

In this section we describe three levels of experiments that can be used in order to compare several recommenders. The discussion below is motivated by evaluation protocols in related areas such as machine learning and information retrieval, highlighting practices relevant to evaluating recommender systems. The reader is referred to publications in these fields for more detailed discussions [21, 85, 102].

We begin with offline experiments, which are typically the easiest to conduct, as they require no interaction with real users. We then describe user studies, where we ask a small group of subjects to use the system in a controlled environment, and then report on their experience. In such experiments we can collect both quantitative and qualitative information about the systems, but care must be taken to consider various biases in the experimental design. Finally, perhaps the most trustworthy experiment is when the system is used by a pool of real users, typically unaware of the experiment. While in such an experiment we are able to collect only certain types of data, this experimental design is closest to reality.

In all experimental scenarios, it is important to follow a few basic guidelines in general experimental studies:

- **Hypothesis:** before running the experiment we must form an hypothesis. It is important to be concise and restrictive about this hypothesis, and design an experiment that tests the hypothesis. For example, an hypothesis can be that algorithm $A$ better predicts user ratings than algorithm $B$. In that case, the experiment should test the prediction accuracy, and not other factors. Other popular hypothesis in recommender system research can be that algorithm $A$ scales better to larger datasets than algorithm $B$, that system $A$ gains more user trust than system $B$, or that recommendation user interface $A$ is preferred by users to interface $B$.
- **Controlling variables:** when comparing a few candidate algorithms on a certain hypothesis, it is important that all variables that are not tested will stay fixed. For example, suppose that in a movie recommendation system, we switch from using algorithm $A$ to algorithm $B$, and notice that the number of movies that users watch increases. In this situation, we cannot tell whether the change is due to the change in algorithm, or whether something else changed at about the same time. If instead, we randomly assign users to algorithms $A$ and $B$, and notice that users assigned to algorithm $A$ watch more movies than those who are assigned to algorithm $B$, we can be confident that this is due to algorithm $A$.
- **Generalization power:** when drawing conclusions from experiments, we may desire that our conclusions generalize beyond the immediate context of the experiments. When choosing an algorithm for a real application, we may want our conclusions to hold on the deployed system, and generalize beyond our experimental data set. Similarly, when developing new algorithms, we want our conclusions to hold beyond the scope of the specific application or data set that we experimented with. To increase the probability of generalization of the results we must typically experiment with several data sets or applications. It is important to understand the properties of the various data sets that are used. Generally speaking, the more diverse the data used, the more we can generalize the results.

## 2.1 Offline Experiments

An offline experiment is performed by using a pre-collected data set of users choosing or rating items. Using this data set we can try to simulate the behavior of users that interact with a recommendation system. In doing so, we assume that the user behavior when the data was collected will be similar enough to the user behavior when the recommender system is deployed, so that we can make reliable decisions based on the simulation. Offline experiments are attractive because they require no interaction with real users, and thus allow us to compare a wide range of candidate algorithms at a low cost. The downside of offline experiments is that they can answer a very narrow set of questions, typically questions about the prediction power of an algorithm. In particular, we must assume that users' behavior when interacting with a system including the recommender system chosen will be modeled well by the users' behavior prior to that system's deployment. Thus we cannot directly measure the recommender's influence on user behavior in this setting.

Therefore, the goal of the offline experiments is to filter out inappropriate approaches, leaving a relatively small set of candidate algorithms to be tested by the more costly user studies or online experiments. A typical example of this process is when the parameters of the algorithms are tuned in an offline experiment, and then the algorithm with the best tuned parameters continues to the next phase.

### 2.1.1   Data Sets for Offline Experiments

As the goal of the offline evaluation is to filter algorithms, the data used for the offline evaluation should match as closely as possible the data the designer expects the recommender system to face when deployed online. Care must be exercised to ensure that there is no bias in the distributions of users, items and ratings selected. For example, in cases where data from an existing system (perhaps a system without a recommender) is available, the experimenter may be tempted to pre-filter the data by excluding items or users with low counts, in order to reduce the costs of experimentation. In doing so, the experimenter should be mindful that this involves a trade-off, since this introduces a systematic bias in the data. If necessary, randomly sampling users and items may be a preferable method for reducing data, although this can also introduce other biases into the experiment (e.g. this could tend to favor algorithms that work better with more sparse data). Sometimes, known biases in the data can be corrected for by techniques such as reweighing data, but correcting biases in the data is often difficult.

Another source of bias may be the data collection itself. For example, users may be more likely to rate items that they have strong opinions on, and some users may provide many more ratings than others. Furthermore, users tend to rate items that they like, and avoid exploring, and hence rating, items that they will not like. For example, a person who doesn't like horror movies will tend not to watch them, would not explore the list of available horror movies for rental, and would not rate them. Thus, the set of items on which explicit ratings are available may be biased by the ratings themselves. This is often known as the *not missing at random* assumption [63] Once again, techniques such as *resampling* or *reweighting* the test data [98, 99] may be used to attempt to correct such biases.

### 2.1.2   Simulating User Behavior

In order to evaluate algorithms offline, it is necessary to simulate the online process where the system makes predictions or recommendations, and the user corrects the predictions or uses the recommendations. This is usually done by recording historical user data, and then hiding some of these interactions in order to simulate the knowledge of how a user will rate an item, or which recommendations a user will act upon. There are a number of ways to choose the ratings/selected items to be hidden. Once again, it is preferable that this choice be done in a manner that simulates the target application as closely as possible. In many cases, though, we are restricted by the computational cost of an evaluation protocol, and must make compromises in order to execute the experiment over large data sets.

Ideally, if we have access to time-stamps for user selections, we can simulate what the systems predictions would have been, had it been running at the time the data set was collected [13]. We can begin with no available prior data for computing predictions, and step through user selections in temporal order, attempting to predict each selection and then making that selection available for use in future predictions. For large data sets, a simpler approach is to randomly sample test users, randomly sample a time just prior to a user action, hide all selections (of all users) after that instant, and then attempt to recommend items to that user. This protocol requires changing the set of given information prior to each recommendation, which can still be computationally quite expensive.

An even cheaper alternative is to sample a set of test users, then sample a single test time, and hide all items after the sampled test time for each test user. This simulates a situation where the recommender system is built as of the test time, and then makes recommendations without taking into account any new data that arrives after the test time. Another alternative is to sample a test time for each test user, and hide the test user's items after that time, without maintaining time consistency across users. This effectively assumes that the sequence in which items are selected is important, not the absolute times when the selections are made. A final alternative is to ignore time. We would first sample a set of test users, then sample the number $n_a$ of items to hide for each user $a$, and finally sample $n_a$ items to hide. This assumes that the temporal aspects of user selections are unimportant. We may be forced to make this assumption if the timestamps of user actions are not known. All three of the latter alternatives partition the data into a single training set and single test set. It is important to select an alternative that is most appropriate for the domain and task of interest, given the constraints, rather than the most convenient one.

A common protocol used in many research papers is to use a fixed number of known items or a fixed number of hidden items per test user (so called "given $n$" or "all but $n$" protocols). This protocol may be useful for diagnosing algorithms and identifying in which cases they work best. However, when we wish to make decisions on the algorithm that we will use in our application, we must ask ourselves whether we are truly interested in presenting recommendations only for users who have rated exactly $n$ items, or are expected to rate exactly $n$ items more. If that is not the case, then results computed using these protocols have biases that make them unreliable in predicting the performance of the algorithms online, and these protocols should be avoided.

Another suggested protocol, known as one-plus-random [17], is to select for each test user, only a subset of items to test over. For example, in a recommendation task, one may select for each item that the user has chosen, a set of items that the user has not chosen, and rank the chosen item together with the unchosen items. In such cases, a good algorithm would rank the chosen item at higher positions. This protocol is used when ranking the complete set of items requires too much time for conducting rapid experiments. As the number of chosen items in the unchosen subset increases, the result approaches the result over the complete item set. It is important to sample the unchosen items from the true distribution of user-item interactions. For example, the sampling should perhaps follow the popularity distribution of items. Otherwise, given that in many cases the vast majority of items have a very low popularity, almost only unpopular items will be chosen. Then, the experimental results may not generalize to the popular items over which most of the user interactions are conducted.

### 2.1.3   More Complex User Modeling

All the protocols that we discuss above make some assumptions concerning the behavior of users, which could be regarded as a user-model for the specific application. While we discuss only very simple user models, it is possible to suggest more complicated models for user behavior [61]. Using advanced user models we can execute simulations of users interactions with the system, thus reducing the need for expensive user studies and online testing. However, care must be made when designing user-models; First, user-modeling is a difficult task, and there is a vast amount of research on the subject (see, e.g. [23]). Second, when the user model is inaccurate, we may optimize a system whose performance in simulation has little correlation with its performance in practice. While it is reasonable to design an algorithm that uses complex user models to provide recommendations, we should be careful in trusting experiments where algorithms are verified using such complex, difficult to verify, user models.

## 2.2   User Studies

Many recommendation approaches rely on the interaction of users with the system (see, e.g., chapters "Social Recommender Systems", "Food Recommender Systems", "Music Recommendation Systems: Techniques, Use Cases, and Challenges", "Beyond Explaining Single Item Recommendations", "Multimedia Recommender Systems: Algorithms and Challenges"). It is very difficult to create a reliable simulation of users interactions with the system, and thus, offline testing are difficult to conduct. In order to properly evaluate such systems, real user interactions with the system must be collected. Even when offline testing is possible, interactions with real users can still provide additional information about the system performance. In these cases we typically conduct user studies .

We provide here a summarized discussion of the principles of user studies for the evaluation of recommender systems.

A user study is conducted by recruiting a set of test subjects, and asking them to perform several tasks requiring an interaction with the recommender system. While the subjects perform the tasks, we observe and record their behavior, collecting any number of quantitative measurements, such as what portion of the task was completed, the accuracy of the task results, or the time taken to perform the task. In many cases we can ask qualitative questions, before, during, and after the task is completed. Such questions can collect data that is not directly observable, such as whether the subject enjoyed the user interface, or whether the user perceived the task as easy to complete.

A typical example of such an experiment is to test the influence of a recommendation algorithm on the browsing behavior of news stories. In this example, the subjects are asked to read a set of stories that are interesting to them, in some cases including related story recommendations and in some cases without recommendations. We can then check whether the recommendations are used, and whether people read different stories with and without recommendations. We can collect data such as how many times a recommendation was clicked, and even, in certain cases, track eye movement to see whether a subject looked at a recommendation. Finally, we can ask qualitative questions such as whether the subject thought the recommendations were relevant [41, 42].

Of course, in many other research areas user studies are a central tool, and thus there is much literature on the proper design of user studies. This section only overviews the basic considerations that should be taken when evaluating a recommender system through a user study, and the interested reader can find much deeper discussions elsewhere (see. e.g. [9]).

### 2.2.1   Advantages and Disadvantages

User studies can perhaps answer the widest set of questions of all three experimental settings that we survey here. Unlike offline experiments this setting allows us to test the behavior of users when interacting with the recommender system, and the influence of the recommendations on user behavior. In the offline case we typically make assumptions such as "given a relevant recommendation the user is likely to use it" which are tested in the user study. Second, this is the only setting that allows us to collect qualitative data that is often crucial for interpreting the quantitative results. Also, we can typically collect in this setting a large set of quantitative measurements because the users can be closely monitored while performing the tasks.

User studies however have some disadvantages. Primarily, user studies are very expensive to conduct [53]; collecting a large set of subjects and asking them to perform a large enough set of tasks is costly in terms of either user time, if the subjects are volunteers, or in terms of compensation if paid subjects are employed. Therefore, we must typically restrict ourselves to a small set of subjects and a relatively small set of tasks, and cannot test all possible scenarios. Furthermore, each scenario has to be repeated several times in order to make reliable conclusions, further limiting the range of distinct tasks that can be tested.

As these experiments are expensive to conduct we should collect as much data about the user interactions, in the lowest possible granularity. This will allow us later to study the results of the experiment in detail, analyzing considerations that were not obvious prior to the trial. This guideline can help us to reduce the need for successive trials to collect overlooked measurements.

Furthermore, in order to avoid failed experiments, such as applications that malfunction under certain user actions, researchers often execute pilot user studies. These are small scale experiments, designed not to collect statistical data, but to test the systems for bugs and malfunctions. In some cases, the results of these pilot studies are then used to improve the recommender. If this is the case, then the results of the pilot become "tainted", and should not be used when computing measurements in the final user study.

Another important consideration is that the test subjects must represent as closely as possible the population of users of the real system. For example, if the system is designed to recommend movies, the results of a user study over avid movie fans may not carry to the entire population. This problem is most persistent when the participants of the study are volunteers, as in this case people who are originally more interested in the application may tend to volunteer more readily.

However, even when the subjects represent properly the true population of users, the results can still be biased because they are aware that they are participating in an experiment. For example, it is well known that paid subjects tend to try and satisfy the person or company conducting the experiment [81]. If the subjects are aware of the hypothesis that is tested they may unconsciously provide evidence that supports it. To accommodate that, it is typically better not to disclose the goal of the experiment prior to collecting data. Another, more subtle effect occurs when the payment to subjects takes the form of a complete or partial subsidy of items they select. This may bias the data in cases where final users of the system are not similarly subsidized, as users' choices and preferences may be different when they pay full price. Unfortunately, avoiding this particular bias is difficult.

### 2.2.2  Between vs. Within Subjects

As typically a user study compares a few candidate approaches, each candidate must be tested over the same tasks. To test all candidates we can either compare the candidates *between subjects*, where each subject is assigned to a candidate method and experiments with it, or *within subjects*, where each subject tests a set of candidates on different tasks [31].

Typically, within subjects experiments are more informative, as the superiority of one method cannot be explained by a biased split of users between candidate methods. It is also possible in this setting to ask comparative questions about the different candidates, such as which candidate the subject preferred. However, in these types of tests users are more conscious of the experiment, and hiding the distinctions between candidates is more difficult.

Between subjects experiments, also known as *A-B testing* (All Between), provide a setting that is closer to the real system, as each user experiments with a single treatment. Such experiments can also test long term effects of using the system, because the user is not required to switch systems. Thus we can test how the user becomes accustomed to the system, and estimate a learning curve of expertise. On the downside, when running between subjects experiments, typically more data is needed to achieve significant results. As such, between subjects experiments may require more users, or more interaction time for each user, and are thus more costly then within subjects experiments.

### 2.2.3  Variable Counter Balance

As we have noted above, it is important to control all variables that are not specifically tested. However, when a subject is presented with the output of several candidates, as in within subject experiments, we must counter balance several variables.

When presenting several results to the subject, the results can be displayed either sequentially, or together. In both cases there are certain biases that we need to correct for [1]. When presenting the results sequentially the previously observed results influence the user opinion of the current results. For example, if the results that were displayed first seem inappropriate, the results displayed afterward may seem better than they actually are. When presenting two sets of results, there can be certain biases due to location. For example, users from many cultures tend to observe results left to right and top to bottom. Thus, the user may observe the results displayed on top as superior.

A common approach to correct for such untested variables is by using the *Latin square* [9] procedure. This procedure randomizes the order or location of the various results each time, thus canceling out biases due to these untested variables.

### 2.2.4  Questionnaires

User studies allow us to use the powerful questionnaire tool (e.g. Pu et al. [75]). Before, during, and after subjects perform their tasks we can ask them questions about their experience. These questions can provide information about properties that are difficult to measure, such as the subject's state of mind, or whether the subject enjoyed the system.

While these questions can provide valuable information, they can also provide misleading information. It is important to ask neutral questions, that do not suggest a "correct" answer. People may also answer untruthfully, for example when they perceive the answer as private, or if they think the true answer may put them in an unflattering position.

Indeed, vast amount of research was conducted in other areas about the art of questionnaire writing, and we refer the readers to that literature (e.g. Pfleeger and Kitchenham [73]) for more details.

## 2.3 Online Evaluation

In many realistic recommendation applications the designer of the system wishes to influence the behavior of users. We are therefore interested in measuring the change in user behavior when interacting with different recommender systems. For example, if users of one system follow the recommendations more often, or if some utility gathered from users of one system exceeds utility gathered from users of the other system, then we can conclude that one system is superior to the other, all else being equal.

The real effect of the recommender system depends on a variety of factors such as the user's intent (e.g. how specific their information needs are), the user's personality (chapter "Personality and Recommender Systems"), such as how much novelty vs. how much risk they are seeking, the user's context, e.g., what items they are already familiar with, how much they trust the system (chapter "Context-Aware Recommender Systems: From Foundations to Recent Developments"), and the interface through which the recommendations are presented.

Thus, the experiment that provides the strongest evidence as to the true value of the system is an online evaluation, where the system is used by real users that perform real tasks. It is most trustworthy to compare a few systems online, obtaining a ranking of alternatives, rather than absolute numbers that are more difficult to interpret.

For this reason, many real world systems employ an online testing system [52], where multiple algorithms can be compared. Typically, such systems redirect a small percentage of the traffic to different alternative recommendation engine, and record the users interactions with the different systems.

There are a few considerations that must be made when running such tests. For example, it is important to sample (redirect) users randomly, so that the comparisons between alternatives are fair. It is also important to single out the different aspects of the recommenders. For example, if we care about algorithmic accuracy, it is important to keep the user interface fixed. On the other hand, if we wish to focus on a better user interface, it is best to keep the underlying algorithm fixed.

In some cases, such experiments are risky. For example, a test system that provides irrelevant recommendations, may discourage the test users from using the real system ever again. Thus, the experiment can have a negative effect on the system, which may be unacceptable in commercial applications.

For these reasons, it is best to run an online evaluation last, after an extensive offline study provides evidence that the candidate approaches are reasonable, and perhaps after a user study that measures the user's attitude towards the system. This gradual process reduces the risk in causing significant user dissatisfaction.

Online evaluations are unique in that they allow direct measurement of overall system goals, such as long-term profit or user retention. As such, they can be used to understand how these overall goals are affected by system properties such as recommendation accuracy and diversity of recommendations, and to understand the trade-offs between these properties. However, since varying such properties independently is difficult, and comparing many algorithms through online trials is expensive, it can be difficult to gain a complete understanding of these relationships.

## 2.4 Offline-Online Correlations

When developing a new recommender system for an application, it is reasonable to begin with offline experiments over many candidate algorithms. Such candidates may include different approaches to recommendations, such as collaborative filtering and content-based methods, different statistical models, such as matrix factorization, neural network, or kNN algorithms, and various parameter settings for a given algorithm, such as different neighborhood sizes for a kNN approach.

Offline experiments are relatively not costly to conduct, requiring mostly computational resources, such as CPU time or memory. As such, these offline experiments can be used to filter out less successful candidates, allowing us to test less candidates in the more costly user studies, or online evaluation.

However, it is important to be cautious about the ranking of algorithms in the offline case vs. the online case. That is, the most successful algorithm offline may not be the most successful algorithm online [3, 29, 62, 80].

It is very reasonable that the online performance would not be identical to the offline performance. First, a recommender system is designed to influence the behavior of users. As such, users that are presented with a recommendation are expected to behave differently than users who do not get recommendations.

Another reason is that the gathered datasets are often incomplete. First, in many cases the datasets do not contain the items that were presented to the user through other methods, such as a list of editor choice news stories in a news online service. If such lists of items are shown to users, they may steer the users to explore the presented items. These items would then appear as more popular than would had happened if they would not be presented to the user. Finally, it may well be that the train-test split and the filtering techniques used in the offline experiment have added patterns that do not manifest in the application. For example, if one removes users with less than 100 interactions from the offline dataset, then it is reasonable that the predicted performance would not carry to the real application, where users may have an order of magnitude less interactions with the system. As such, it is crucial to make sound choices when designing the offline experiment, to avoid such pitfalls.

Given the above reasons and existing evidence to the different ranking of algorithms offline and online, it is desirable to experiment with as many candidates as possible online. However, avoiding offline experiments altogether does not seem possible. It is hard to see how one tunes the algorithmic parameters, at the very least, offline. Modern methods have many parameters that require tuning, and hence require us to often experiment with many parameter configurations. Currently, offline experiments are the only viable option for conducting this required screening.

## 2.5 Drawing Reliable Conclusions

In any type of experiment it is important that we can be confidant that the candidate recommender that we choose will also be a good choice for the yet unseen data the system will be faced with in the future. As we explain above, we should exercise caution in choosing the data in an offline experiments, and the subjects in a user study, to best resemble the online application. Still, there is a possibility that the algorithm that performed best on this test set did so because the experiment was fortuitously suitable for that algorithm. To reduce the possibility of such statistical mishaps, we must perform significance testing on the results.

### 2.5.1 Confidence and *p*-values

The result of a significance test is a significance level or *p*-value – the probability that the obtained results were due to chance. In practice, we choose a significance test (see below) to match our situation in order to evaluate this probability. Each significance test postulates an underlying random mechanism that may have generated the result. This is termed the null hypothesis. The chosen test then gives us a probability that a result that is at least as good as the one we are testing was produced under the null hypothesis. This probability is the *p*-value. If the *p*-value is below a threshold, we are confident that the null hypothesis is not true, and we deem our results significant. Traditionally, people choose $p = 0.05$ as their threshold, which indicates 95% confidence. More stringent significance levels (e.g. 0.01 or even lower) can be used in cases where the cost of making the wrong choice is higher. Notice, however, that the significance test only tells us that the null hypothesis is unlikely to be true. It does not guarantee that the result was not randomly produced by some other mechanism. Thus, to be confident that we are making meaningful decisions, we need to be careful in choosing a test with a strong null hypothesis that is appropriate for our situation. Below, we discuss how to make this choice. For more details, see, e.g., Bickel and Ducksum [5].

### 2.5.2 Paired Results

In order to perform a significance test that algorithm *A* is indeed better than algorithm *B*, we often use the results of several independent experiments comparing $A$ and $B$. Thus, rather than the aggregate results that we typically use to compare systems, confidence testing requires the results of multiple independent sub-experiments. Indeed, the protocol we have suggested for generating our test data (Sect. 2.1.2) allows us to obtain such a set of results. Assuming that test users are drawn independently from some population, the performance measures of the algorithms for each test user give us the independent comparisons we need. However, when recommendations or predictions of multiple items are made to the same user, it is unlikely that the resulting per-item performance metrics are independent. Therefore, it is better to compare algorithms on a per-user case.

Given such paired per-user performance measures for algorithms $A$ and $B$ a simple test of significance is the **sign test** [21, 59]. To use the sign test, we compute a score (e.g. RMSE for system accuracy) for each user under algorithms $A$ and $B$. The sign test makes no assumption on these scores other than that users are independent, and considers the number of times $A$ beats $B$. The null hypothesis is that whether $A$ beats $B$ or vice-versa is determined by a coin-toss. Thus, it uses the number of times $n_A$ that $A$ beats $B$ (e.g. the number of times that alternative $A$ achieved a lower RMSE than alternative $B$) If we are interested in a pure winner, i.e., that $A$ would achieve a strictly better RMSE than $B$, then draws should count against $A$, that is, they should not be counted in $n_A$. If we are interested in the case where $A$ should do no worse than $B$, then draws should be counted in $n_A$.

Let $n$ be the number of users in $J$ for which the predictions were made. The null hypothesis is that whether $A$ beats $B$ or vice-versa is determined by a coin-toss. We can now compute the probability that we will observe at least $n_A$ times that system $A$ got a better score than system $B$ under the null hypothesis that the two systems are equal using:

$$p = (0.5)^n \sum_{i=n_A}^{n} \frac{n!}{i!(n-i)!} \tag{1}$$

when this $p$-value is below some predefined value (typically, 0.05) we can say that the null hypothesis that the two system have an equal performance is rejected.

The sign test is an attractive choice due to its simplicity, and lack of assumptions over the distribution of cases. When $n_A + n_B$ is large, we can take advantage of large sample theory to approximate equation (1) by a normal distribution. However, this is usually unnecessary with powerful modern computers. Some authors (e.g. [85]) use the term **McNemar's test** to refer to the use of a $\chi^2$ approximation to the two-sided sign test.

Note that sometimes, the sign test may indicate that system A outperforms system B with high probability, even though the average performance of system B is higher than that of system A. This happens in cases where system B occasionally outperforms system A overwhelmingly. Thus, the reason for this seemingly inconsistent result is that the test only examines the probability of one system outperforming the other, without regard to the magnitude of the difference.

The sign test can be extended to cases where we want to know the probability that one system outperforms the other by some amount. For example, suppose that system A is much more resource intensive than system B, and is only worth deploying if it outperforms system B by some amount. We can define "success" in the sign test as A outperforming B by this amount, and find the probability of A not truly outperforming B by this amount as our $p$ value in Eq. (1).

A commonly used test that takes the magnitude of the differences into account is the **paired Student's t-test**, which looks at the average difference between the performance scores of algorithms $A$ and $B$, normalized by the standard deviation of the score difference. Using this test requires that the differences in scores for different users is comparable, so that averaging these differences is reasonable. For small numbers of users, the validity of the test also depends on the differences being Normally distributed. Demšar [21] points out that this assumption is hard to verify when the number of samples is small and that the t-test is susceptible to outliers. He recommends the use of **Wilcoxon signed rank test**, which like the $t$-test, uses the magnitude of the differences between algorithms $A$ and $B$, but without making distributional assumptions on the differences. However, using the Wilcoxon signed rank test still requires that differences between the two systems are comparable between users.

Another way to improve the significance of our conclusions is to use a larger test set. In the offline case, this may require using a smaller training set, which may result in an experimental protocol that is not representative of the amount of training data available after deployment. In the case of user studies, this implies an additional expense. In the case of online testing, increasing the amount of data collected for each algorithm requires either the added expense of a longer trial or the comparison of fewer algorithms.

### 2.5.3 Unpaired Results

The tests described above are suitable for cases where observations are paired. That is, each algorithm is run on each test case, as is often done in offline tests. In online tests, however, it is often the case that users are assigned to one algorithm or the other, so that the two algorithms are not evaluated on the same test cases. The **Mann-Whitney test** is an extension of the Wilcoxon test to this scenario. Suppose we have $n_A$ results from algorithm A and $n_B$ results from algorithm B.

The performance measures of the two algorithms are pooled and sorted so that the best result is ranked first and the worst last. The ranks of ties are averaged. For example if the second through fifth place tie, they are all assigned a rank of 3.5. The Mann-Whitney test computes the probability of the null hypothesis that $n_A$ randomly chosen results from the total $n_A + n_B$ have at least as good an average rank as the $n_A$ results that came from algorithm A.

This probability can be computed exactly be enumerating all $\frac{(n_A+n_B)!}{n_A!n_B!}$ choices and counting the choices that have at least the required average rank, or can be approximated by repeatedly resampling $n_A$ of the results. When $n_A$ and $n_B$ are both large enough (typically over 5), the distribution of the average rank of $n_A$ results randomly selected from a pool of $n_A + n_B$ under the null hypothesis is well approximated by a Gaussian with mean $\frac{1}{2}(n_A + n_B + 1)$ and standard deviation $\sqrt{\frac{1}{12}\frac{n_A}{n_B}(n_A + n_B + 1)}$. Thus, in this case we can compute the average rank of the $n_A$ results from system A, subtract $\frac{1}{2}(n_A + n_B + 1)$, divide by $\sqrt{\frac{1}{12}\frac{n_A}{n_B}(n_A + n_B + 1)}$, and evaluate the standard Gaussian CDF at this value to get the $p$ value for the test.

### 2.5.4 Multiple Tests

Another important consideration, mostly in the offline scenario, is the effect of evaluating multiple versions of algorithms. For example, an experimenter might try out several variants of a novel recommender algorithm and compare them to a baseline algorithm until they find one that passes a sign test at the $p = 0.05$ level and therefore infer that their algorithm improves upon the baseline with 95% confidence. However, this is not a valid inference. Suppose the experimenter evaluated 10 different variants all of which are statistically the same as the baseline. If the probability that any one of these trials passes the sign test mistakenly is $p = 0.05$, the probability that at least one of the ten trials passes the sign test mistakenly is $1 - (1 - 0.05)^{10} = 0.40$. This risk is colloquially known as "tuning to the test set" and can be avoided by separating the test set users into two groups–a development (or tuning) set, and an evaluation set. The choice of algorithm is done based on the development test, and the validity of the choice is measured by running a significance test on the evaluation set.

A similar concern exists when ranking a number of algorithms, but is more difficult to circumvent. Suppose the best of $N + 1$ algorithms is chosen on the development test set. To achieve a confidence $1 - p$ that the chosen algorithm is indeed the best, it must outperform the $N$ other algorithms on the evaluation set with significance $1 - (1 - p)^{1/N}$. This is known as the Bonferroni correction, and should be used when pair-wise significance tests are used multiple times. Alternatively, approaches such as **ANOVA** or the **Friedman test for ranking**, which are generalization of the Student's $t$-test and Wilcoxon's rank test. ANOVA makes strong assumptions about the Normality of the different algorithms' performance measures, and about the relationships of their variances. We refer the reader to Demšar [21] for further discussion of these and other tests for ranking multiple algorithms.

A more subtle version of this concern is when a pair of algorithms are compared in a number of ways. For example, two algorithms may be compared using a number of accuracy measures, a number of coverage measures, etc. Even if the two algorithms are identical in all measures, the probability of finding a measure by which one algorithm seems to outperform the other with some significance level increases with the number of measures examined. If the different measures are independent, the Bonferroni correction mentioned above can be used. However, since the measures are often correlated, the Bonferroni correction may be too stringent, and other approaches such as controlling for false discovery rate [4] may be used.

### 2.5.5 Confidence Intervals

Even though we focus here on comparative studies, where one has to choose the most appropriate algorithm out of a set of candidates, it is sometimes desirable to measure the value of some property. For example, an administrator may want to estimate the error in the system predictions, or the net profit that the system is earning. When measuring such quantities it is important to understand the reliability of your estimates. A popular approach for doing this is to compute **confidence intervals**.

For example, one may estimate that the RMSE of a system is expected to be 1.2, and that it will be between 1.1 and 1.35 with probability 0.95. The simplest method for computing confidence intervals is to assume that the quantity of interest is Gaussian distributed, and then estimate its mean and standard deviations from multiple independent observations. When we have many observations, we can dispense with this assumption by computing the distribution of the quantity of interest with a non-parametric method such as a histogram and finding upper and lower bounds such that include the quantity of interest with the desired probability.

## *2.6 Reporting Results*

Both in academia and in industry it is important to communicate the results of the experiments to peers and management. An example of such a report is the experimental section of almost all academic papers in the recommender system research community. In industry, when the algorithms development team needs to explain their finding to management for making decisions concerning which method to implement into the product, similar reports are needed.

### 2.6.1 Reporting the Experimental Settings

When reporting the experiments, it is crucial to be as clear as possible concerning the experimental settings. A popular rule of thumb is that the report should be sufficiently detailed such that a reader would be able to repeat the experiment.

First, one must describe the datasets that are used in the experiments. It is often considered good practice to experiment with well known public datasets, as this allows us to easily compare the results to previous experiments conducted on the same datasets. However, it is also important to experiment with datasets that are gathered from real users of an application of interest, even though the application owner may not agree to make the data public. Such datasets often capture user behavior that is different than, e.g., the movie rating behavior captured in popular public datasets. It is important to describe the nature of the datasets, the items, the typical behavior of users in the application, how the dataset was collected, and so forth.

A popular, yet questionable practice is to filter a public dataset by, e.g., using only a subset of users who rated more than some arbitrary threshold, or using only a subset of the most popular items. It is important to be clear concerning the reasons for such filters, as they may cause the algorithmic performance to be less correlated with the true performance over all users.

Given the manipulations to the public datasets, as well as the use of private unknown datasets, it is important to report the resulting dataset properties. For example, one should report the number of users, the number of items, and the distribution of interactions between the user and the system, such as provided ratings, or purchases.

In addition, certain algorithms are more appropriate to datasets with some properties, such as high sparsity. In such cases, it is important to report these properties which may affect the method performance.

Following the description of the datasets, one should thoroughly describe the train-test splitting procedure. The procedure for splitting the data can significantly impact the results. In any case, the splitting procedure must be explained in detail, and all the parameters must be specified, including the train-test split ratio, the number of users in the train and test sets, and so forth.

Finally, one should describe the testing procedure. For example, in many experiments we iterate over the users in the test set, and for each user produce a list of recommended items, and then compare this list to the items that the user has chosen in the test set. Alternatively, we may iterate over all items in the user profile in the test set, and for each such item predict its score. This procedure should also be described in sufficient details to allow the reader to reproduce the results.

## 2.6.2 Compared Methods

The recommendation system research community has suggested numerous algorithms for computing recommendation lists or for predicting ratings. Obviously, comparing a new algorithm to all existing algorithms in literature is impractical. Typically, when suggesting a new algorithm one should compare it to several current state-of-the-art competitor. If each new algorithm shows some advantage over the previous best algorithm, then we can conclude that research is advancing the capabilities of recommendation systems.

However, it is also often valuable to compare results to simple competitors. For example, so called memory-based methods, are based on identifying a group of similar users by, e.g., the Pearson correlation [12]. Simple item-item methods are based on identifying similar items using, e.g., the Jaccard correlation [89]. For presenting lists of recommended items, a simple method can use the popularity of items to order the list. Due to their simplicity, these methods do not have numerous tunable parameters, as opposed to modern statistical models, leveraging matrix factorization [55, 78] or neural networks [91, 108].

These simple methods can be used as a common baseline for all reports. Due to their simplicity and minimal tuning, one can often rely on their reported performance to provide an absolute comparable baseline between different experiments ran over different settings, such as datasets, train-test splitting procedures, and so forth.

In practice, in many cases such simple methods still produce competitive results [18]. Even when these methods produce results slightly below modern complex models, it is possible that management would still opt for using a simple algorithm, that is easy to implement and debug, rather than a complex model, that may require much more maintenance later on.

### 2.6.3 Reporting Complete Results

In some cases, mostly in industry, the requirements from the algorithm are strict. For example, the application owner may allow space for only 3 recommended items at a specific location in the application. In such cases, the algorithms development team must find the best solution given these requirements, and it is less important to experiment with other settings.

In many cases, however, both in academia, and in industry, when the requirements are no as strict, that one wishes to understand the behavior of the algorithms over a range of values, such as different recommendation system lengths. Such information can be valuable when making decisions concerning the application. For example, if the precision of the recommendation list supplied by an algorithm drops sharply after 4 recommendations, it is likely that the application manager would not agree to present 5 recommendations.

It is often the case that when methods are compared over a range of settings, for some settings algorithm $A$ is better than $B$, while for other settings algorithm $B$ is better than $A$. In such cases, it is important that this would be presented to both peers and management. Presenting only one or several settings, e.g. only the recall with a recommendation list of length 50, may be misleading ,as the ranking of algorithms for that particular value may not hold for other settings [18]. In fact, reporting only one or several specific settings should be justified in the report.

## 2.7 Standardized Evaluation Frameworks

Mostly, in current research on recommender systems, each team develops their own evaluation implementation. This may cause differences between reported results, not due to inherent differences in the algorithms, but due to particular implementation of the evaluation procedures. In the industry, where each company optimizes for a particular domain, this is not truly a limitation. However, in pure research, this may cause us to prefer one algorithm to another given better reported results, due to differences in evaluation protocol implementation only.

Indeed, many off-the-shelf packages for recommender systems such as MyMediaLite [28], LibRec [32], or Surprise[43], offer, on top of the implementation of many recommendation algorithms, the implementation of many evaluation metrics. Moreover, several researchers have suggested and implemented evaluation frameworks [34, 44, 47, 76, 83, 84, 86]. Currently, however, as can be seen from the multitude of suggested frameworks, none has emerged as a standard in the field. Thus, there is no standard implementation for the evaluation procedure, that would ensure that all algorithms are evaluated equally, and differ truly on their performance, not due to differences in evaluation procedures.

A second possible source for standardized comparison are competitions. For example, in the automated planning community, there is a biannually competition.[3] Researchers submit their algorithm to a centralized engine, where each algorithm is run on a set of problems. Results are then published for all algorithms and problems. In the recommendation systems community there are annual challenges [82], that draw many researchers and practitioners to compete over a particular problem. However, each challenge focuses on a problem with different settings, and the participants are expected to optimize their methods for the particular settings. As such, it is unclear which method provides good results in varying conditions, without problem specific optimizations.

Perhaps the emergence of a competition over a wide range of datasets, as done in the automated planning community, can help in standardizing the recommendation systems research community.

## 3 Evaluation in the Industry

The diversity and scale of recommendation systems usage in the industry have been consistently rising in recent years, with widespread usage of machine learning methods (chapter "Value and Impact of Recommender Systems"). These systems typically use evaluation protocols that are less relevant in the research community. We now discuss evaluation protocols from the perspective of recommendation service providers, which have become a major source for computing and displaying recommendations online. It is a complex scenario involving multiple stakeholders with possibly conflicting goals, which should be represented in the evaluation process.

Commercial recommender systems for online advertising gained substantial impact over the last decade. Native ads overtook display ads to become the most

---

[3] https://www.icaps-conference.org/competitions/.

prominent revenue source [6]. Native ads are ads incorporated into a webpage on a certain platform, such as a news site, recommending either additional content from the same platform or promoted content from advertisers seeking reader attention for gaining various business goals. While some platforms use self managed systems for content recommendations, there are several recommendation service providers, such as Outbrain, Taboola, and Verizon Meida, that serve various platforms, thus allowing advertisers wide exposure with low campaign management overhead [60, 95, 106].

In the common commercial scenario, a platform (the news site) requests recommended content from the provider and displays the content to the reader. When the reader clicks one of the presented content, the advertiser which advertises this content pays a fee to the provider and the platform, usually by a predetermined percentage.

The different stakeholders in the aforementioned scenario have different goals. The reader's main goal is to consume interesting and relevant content safely. The advertiser's main goal is to maximize clicks for a given budget. This is typically measured using $CPC$ (cost per click), defined as:

$$CPC = \frac{\text{Total spend}}{\text{Number of clicks}} \qquad (2)$$

The platform's main goal is to convert page views into revenue. This can be measured using $RPM$ (revenue per thousand impressions):

$$RPM = \frac{\text{Revenue}}{\text{Number of page views}} \times 1000 \qquad (3)$$

Another important measure is $CTR$ (click through rate), defined as:

$$CTR = \frac{\text{Number of clicks}}{\text{Number of page views}} \qquad (4)$$

Hence, from the perspective of the provider, $RPM \propto CPC \times CTR$.

The provider, as the centerpiece of this mechanism, aims to provide recommendations that balance the different goals in an effective way. In terms of user safety, this requires mechanisms to protect the users' privacy, and to ensure that promoted links do not contain harmful content such as adult products, violence, fraud or fake news. In terms of providing advertisers with the tools to succeed, this requires mechanisms that allow targeting specific audiences (using user based, geographic and contextual constraints), and control over the fee to be paid for different ads.

We now delve deeper into an industry test case—Outbrain Inc.—an international native advertising company and one of the largest providers of content recommendations worldwide. We review Outbrain's recommendation system and provide a detailed description of the main evaluation methods used in the system. We further discuss interesting evaluation related issues and insights encountered during the development of the system.

Following the setup described for the similar problem of $CTR$ prediction [35], the recommendation process contains three components: Ranker, Online Joiner and Trainer. The Ranker receives recommendation requests, extracts features from the request and the relevant ads, and uses a prediction model to select the ads. It returns the selected ads and reports the features. The Online Joiner joins the reported features and clicks. The Trainer takes as input the joint features and clicks, and adjusts the prediction model according to the newly observed data periodically.

To select a set of ads to fulfill the request, the system needs a prediction of the expected $RPM$ of each of the relevant ads. As, from the perspective of the provider, $RPM \propto CPC \times CTR$, and assuming each ad has a predefined fee (cost), the prediction model goal is to predict the $CTR$ for each combination of request and ad features.

To handle the amounts of required predictions in a timely manner, the prediction model applies *Logistic Regression* using a subset of the available features. In addition to the model features, several hyper-parameters control different aspects of the learning process, including the learning rate and down sampling of negative samples, that has been shown in [66] to have a relatively small effect on the learning process quality while providing meaningful computational savings.

In theory, it would be best to replace the active prediction model with a newer version after each sample is processed. However, operational limitations dictate processing the samples in batches before uploading a new model. In the balancing act between prediction quality and operational considerations, the current batch size for online learning in Outbrain is 5 minutes. As a safety measure, the prediction of each sample used in the learning phase is tested to fall within reasonable boundaries (between 0.01 and 0.99). The total percentage of extreme predictions (outside the boundaries) generated by the new model on recent samples is calculated, and only if the percentage is below given threshold the new model replaces the active model.

Offline experiments (Sect. 2.1) are constantly performed by data scientists in Outbrain to improve the prediction model. For these experiments to simulate the online process, a set of tools allows fetching historical data, and extracting features using the methods used by the Ranker. Given such data, each data scientist can experiment with algorithmic modifications, such as expanding the feature space, applying different computation engines, modifying the size of the model, or changing hyper parameters of the learning process.

In this offline process, there are several measures that can be used to assess models. In Outbrain, the Relative Information Gain ($RIG$, sometimes referred to as Relative log-loss) is considered to be the most accurate in predicting $RPM$ online. For the sample at time $t$, let $w_t$ be the vector of model parameters, let $p_t$ be the prediction, and let $y_t$ be the label. Then the logistic loss of the sample is given by:

$$LogLoss(w_t) = -y_t \, log(p_t) - (1 - y_t) log(1 - p_t) \qquad (5)$$

Let $p$ be the average empirical $CTR$ in the system. The $RIG$ for the sample at time $t$ is given by:

$$RIG(w_t) = 1 - \frac{LogLoss(w_t)}{p\,log(p) - (1-p)log(1-p)} \qquad (6)$$

The key desired property of off-line measures like $RIG$ is its correlation with key online business metrics. The goal is that a statistically significant increase in offline measure will lead to a statistically significant increase in $RPM$. In Outbrain's system, using $RIG$ in offline experiments correlates highly with RPM gains in online settings. The $AUC$ measure has also shown such correlations, yet it requires a higher computational cost compared to $RIG$, thus allowing less offline experiments. Attempts to use $MRR$ (Eq. (13)) and its weighted variants resulted in weaker correlation to online behavior than $RIG$ and $AUC$. One possible explanation for the inferior results when using $MRR$ is the highly non-homogeneous presentation of recommendations—in some cases there is just a single recommendation presented, and those cases are excluded from $MRR$ measurement, and in other cases the number of recommendations vary significantly and it is hard to normalize between examples. A second possible reason is that $MRR$ does not naturally favor absolute precision, but the final ranking of ads depends on the absolute precision of the predicted $CTR$.

When a new model generates significantly improved $RIG$ in the offline experiment, it is uploaded as an online model and goes through a bootstrapping phase of learning using data from the last 30 days. Once bootstrapping is done, the new model enters an A-B testing (Sect. 2.3) on a small portion of the overall traffic. For a model that presents significant $RPM$ lift in the A-B testing, the percentage of traffic it serves is gradually increased, and upon successful results it becomes the main active model.

As with other recommendation systems, the cold start problem (Sect. 4.3) of introducing new ads to the system requires special treatment. In Outbrain's system this problem is handled by dedicating a predefined percentage of the paid content to new ads. As less data is available for new ads, a separate online model is used to predict their $CTR$, and a separate set of offline experiments are performed to improve this model.

In the context of paid content, some advertisers are interested not only in the clicks made by readers, but also in "conversions"—actions taken in the target page such as site registration or product purchase. For this type of advertisers, an additional model is used to perform $CPC$ prediction which allows automatic adjustment of $CPC$ for each request according to the advertiser's business goals. Offline experiments of the $CPC$ prediction model use $AUC$, for two reasons. First, to allow comparing the new model to the legacy system previously used. Second, because the volume of conversions data is smaller in two orders of magnitude, which makes the computational cost of $AUC$ less prohibitive.

Online learning is applied in Outbrain's system not only for recommending paid content, but also for non-paid content, namely other content pages from the same website. The target metric of the non-paid content model is $CTR$, and it contains possible target web pages features instead of ad features.

## 4   Recommender System Properties

In this section we survey a range of properties that are commonly considered when deciding which recommendation approach to select. As different applications have different needs, the designer of the system must decide on the important properties to measure for the concrete application at hand. Some of the properties can be traded-off, the most obvious example perhaps is the decline in accuracy when other properties (e.g. diversity) are improved. It is important to understand and evaluate these trade-offs and their effect on the overall performance. However, the proper way of gaining such understanding without intensive online testing or defering to the opinions of domain experts is still an open question.

Furthermore, the effect of many of these properties on the user experience is unclear, and depends on the application. While we can certainly speculate that users would like diverse recommendations or reported confidence bounds, it is essential to show that this is indeed important in practice. Therefore, when suggesting a method that improves one of this properties, one should also evaluate how changes in this property affects the user experience, either through a user study or through online experimentation.

Such an experiment typically uses a single recommendation method with a tunable parameter that affects the property being considered. For example, we can envision a parameter that controls the diversity of the list of recommendations. Then, subjects should be presented with recommendations based on a variety of values for this parameter, and we should measure the effect of the parameter on the user experience. We should measure here not whether the user noticed the change in the property, but whether the change in property has affected their interaction with the system. As is always the case in user studies, it is preferable that the subjects in a user study and users in an online experiment will not know the goal of the experiment. It is difficult to envision how this procedure could be performed in an offline setting because we need to understand the user response to this parameter.

Once the effects of the specific system properties in affecting the user experience of the application at hand is understood, we can use differences in these properties to select a recommender.

### 4.1   User Preference

As in this chapter we are interested in the selection problem, where we need to choose on out of a set of candidate algorithms, an obvious option is to run a user study (within subjects) and ask the participants to choose one of the systems [39].

This evaluation does not restrict the subjects to specific properties, and it is generally easier for humans to make such judgments than to give scores for the experience. Then, we can select the system that had the largest number of votes.

However, aside from the biases in user studies discussed earlier, there are additional concerns that we must be aware of. First, the above scheme assumes that all users are equal, which may not always be true. For example, an e-commerce website may prefer the opinion of users who buy many items to the opinion of users who only buy a single item. We therefore need to further weight the vote by the importance of the user, when applicable. Assigning the right importance weights in a user study may not be easy.

It may also be the case that users who preferred system $A$, only slightly preferred it, while users who preferred $B$, had a very low opinion on $A$. In this case, even if slightly more users preferred $A$ we may still wish to choose $B$. To measure this we need non-binary answers for the preference question in the user study. Then, the problem of calibrating scores across users arises.

Finally, when we wish to improve a system, it is important to know why people favor one system over the other. Typically, it is easier to understand that when comparing specific properties. Therefore, while user satisfaction is important to measure, breaking satisfaction into smaller components is helpful to understand the system and improve it.

## 4.2  Prediction Accuracy

Prediction accuracy is by far the most discussed property in the recommendation system literature. At the base of the vast majority of recommender systems lie a prediction engine. This engine may predict user opinions over items (e.g. ratings of movies) or the probability of usage (e.g. purchase).

A basic assumption in a recommender system is that a system that provides more accurate predictions will be preferred by the user. Thus, many researchers set out to find algorithms that provide better predictions.

Assuming accurate and consistent user ratings for items, prediction accuracy is typically independent of the user interface, and can thus be measured in an offline experiment. That being said, the interface used for providing user feedback and preferences over items may influence the gathered ratings [71]. This weakens the generality of the conclusions drawn from such offline experiments. Measuring prediction accuracy in a user study, however, typically measures the accuracy given a set of recommendations or item ratings displayed to the user. This is a different concept from the prediction of user behavior without recommendations, and is closer to the true accuracy in the real system.

We discuss here three broad classes of prediction accuracy measures; measuring the accuracy of ratings predictions, measuring the accuracy of usage predictions, and measuring the accuracy of rankings of items.

### 4.2.1  Measuring Ratings Prediction Accuracy

In some applications, such as in Amazon, we wish to predict the rating a user would give to an item (e.g. 1-star through 5-stars). In such cases, we wish to measure the accuracy of the system's predicted ratings.

**Root Mean Squared Error (RMSE)** is perhaps the most popular metric used in evaluating accuracy of predicted ratings. The system generates predicted ratings $\hat{r}_{ui}$ for a test set $\mathcal{T}$ of user-item pairs $(u, i)$ for which the true ratings $r_{ui}$ are known. Typically, $r_{ui}$ are known because they are hidden in an offline experiment, or because they were obtained through a user study or online experiment. The RMSE between the predicted and actual ratings is given by:

$$\text{RMSE} = \sqrt{\frac{1}{|\mathcal{T}|} \sum_{(u,i)\in\mathcal{T}} (\hat{r}_{ui} - r_{ui})^2} \tag{7}$$

**Mean Absolute Error (MAE)** is a popular alternative, given by

$$\text{MAE} = \frac{1}{|\mathcal{T}|} \sum_{(u,i)\in\mathcal{T}} |\hat{r}_{ui} - r_{ui}| \tag{8}$$

Compared to MAE, RMSE disproportionately penalizes large errors, so that, given a test set with four hidden items RMSE would prefer a system that makes an error of 2 on three ratings and 0 on the fourth to one that makes an error of 3 on one rating and 0 on all three others, while MAE would prefer the second system.

**Normalized RMSE (NMRSE)** and **Normalized MAE (NMAE)** are versions of RMSE and MAE that have been normalized by the range of the ratings (i.e. $r_{max} - r_{min}$). Since they are simply scaled versions of RMSE and MAE, the resulting ranking of algorithms is the same as the ranking given by the unnormalized measures.

**Average RMSE** and **Average MAE** adjust for unbalanced test sets. For example, if the test set has an unbalanced distribution of items, the RMSE or MAE obtained from it might be heavily influenced by the error on a few very frequent items. If we need a measure that is representative of the prediction error on any item, it is preferable to compute MAE or RMSE separately for each item and then take the average over all items. Similarly, one can compute a per-user average RMSE or MAE if the test set has an unbalanced user distribution and we wish to understand the prediction error a randomly drawn user might face.

RMSE and MAE depend only on the magnitude of the errors made. In some applications [20, e.g.], the semantics of the ratings may be such that the impact of a prediction error does not depend only on its magnitude. In such domains it may be preferable to use a suitable distortion measure $d(\hat{r}, r)$ than squared difference or absolute difference. For example in an application with a 3-star rating system where 1 means "disliked," 2 means "neutral" and 3 means "liked," and where recommending an item the user dislikes is worse that not recommending an item a user likes, a distortion measure with $d(3, 1) = 5$, $d(2, 1) = 3$, $d(3, 2) = 3$, $d(1, 2) = 1$, $d(2, 3) = 1$, and $d(1, 3) = 2$ may be reasonable.

### 4.2.2 Measuring Usage Prediction

In many applications the recommender system does not predict the user's preferences of items, such as movie ratings, but tries to recommend to users items that they may use "precision-recall". For example, Netflix suggests a list of movies that may also be interesting, given movies you watched before. In this case we are interested not in whether the system properly predicts the ratings of these movies but rather whether the system properly predicts that the user will add these movies to the queue (use the items).

In an offline evaluation of usage prediction, we typically have a data set consisting of items each user has used. We then select a test user, hide some of her selections, and ask the recommender to predict a set of items that the user will use. We then have four possible outcomes for the recommended and hidden items, as shown in Table 1.

In the offline case, since the data isn't typically collected using the recommender system under evaluation, we are forced to assume that unused items would have not be used even if they had they been recommended—i.e. that they are uninteresting or useless to the user. This assumption may be false, such as when the set of unused items contains some interesting items that the user did not select. For example, a user may not have used an item because she was unaware of its existence, but after the recommendation exposed that item the user can decide to select it. In this case the number of false positives is over estimated.

We can count the number of examples that fall into each cell in the table and compute the following quantities:

$$\textbf{Precision} = \frac{\#tp}{\#tp + \#fp} \tag{9}$$

$$\textbf{Recall (True Positive Rate)} = \frac{\#tp}{\#tp + \#fn} \tag{10}$$

$$\textbf{False Positive Rate } (1 - \textbf{Specificity}) = \frac{\#fp}{\#fp + \#tn} \tag{11}$$

Typically we can expect a trade off between these quantities—while allowing longer recommendation lists typically improves recall, it is also likely to reduce

**Table 1** Classification of the possible result of a recommendation of an item to a user

|          | Recommended         | Not recommended      |
|----------|---------------------|----------------------|
| Used     | True-positive (tp)  | False-negative (fn)  |
| Not used | False-positive (fp) | True-negative (tn)   |

the precision. In applications where the number of recommendations that can be presented to the user is preordained, the most useful measure of interest is Precision at N (often written Precision@N).

In other applications where the number of recommendations that are presented to the user is not preordained, it is preferable to evaluate algorithms over a range of recommendation list lengths, rather than using a fixed length. Thus, we can compute curves comparing precision to recall, or true positive rate to false positive rate. Curves of the former type are known simply as precision-recall curves, while those of the latter type are known as a[4] Receiver Operating Characteristic or curves. While both curves measure the proportion of preferred items that are actually recommended, precision-recall curves emphasize the proportion of recommended items that are preferred while ROC curves emphasize the proportion of items that are not preferred that end up being recommended.

We should select whether to use precision-recall or ROC based on the properties of the domain and the goal of the application; suppose, for example, that an online video rental service recommends DVDs to users. The precision measure describes the proportion of their recommendations were actually suitable for the user. Whether the unsuitable recommendations represent a small or large fraction of the unsuitable DVDs that could have been recommended (i.e. the false positive rate) may not be as relevant as what proportion of the relevant items the system recommended to the user, so a precision-recall curve would be suitable for this application. On the other hand, consider a recommender system that is used for selecting items to be marketed to users, for example by mailing an item to the user who returns it at no cost to themselves if they do not purchase it. In this case, where we are interested in realizing as many potential sales as possible while minimizing marketing costs, ROC curves would be more relevant than precision-recall curves.

Given two algorithms, we can compute a pair of such curves, one for each algorithm. If one curve completely dominates the other curve, the decision about the superior algorithm is easy. However, when the curves intersect, the decision is less obvious, and will depend on the application in question. Knowledge of the application will dictate which region of the curve the decision will be based on.

Measures that summarize the precision-recall or ROC curve such as **F-measure** [101]—the harmonic mean of the equally weighted precision and recall

$$F = \frac{2 \cdot precision \cdot recall}{precision + recall} \tag{12}$$

and the **Area Under the ROC Curve (AUC)** [2] are useful for comparing algorithms independently of application, but when selecting an algorithm for use in a particular task, it is preferable to make the choice based on a measure that reflects the specific needs at hand, such as the actual list length dictated by the application.

---

[4] A reference to their origins in signal detection theory.

An obvious criticism of the above measures for offline experiments is that they only measure exact matches with items in the test set. That is, even if the recommendation algorithm recommended items that are very similar to the chosen item, it obtains no score increase. For example, assume in a movie domain that the system has recommended to the user two movies—"Four Weddings and a Funeral", a romantic comedy, and "Die Hard", an action movie starring Bruce Willis. The user did not chose either movies to watch, but rather watched "Red", another action movie starring Bruce Willis. Intuitively, one can claim that "Die Hard" is a better recommendation that "Four Weddings and a Funeral". However, metrics such as precision and recall do not capture this. One can envision non-binary metrics that take into account the closeness of the recommended item to the consumed item, given some similarity metric, such as content similarity [27].

Precision-Recall and ROC for Multiple Users

When evaluating precision-recall or ROC curves for multiple test users, a number of strategies can be employed in aggregating the results, depending on the application at hand.

In applications where a fixed number of recommendations are made to each user (e.g. when a fixed number of headlines are shown to a user visiting a news portal), we can compute the precision and recall (or true positive rate and false positive rate) at each recommendation list length $N$ for each user, and then compute the average precision and recall (or true positive rate and false positive rate) at each $N$ [87]. The resulting curves are particularly valuable because they prescribe a value of $N$ for each achievable precision and recall (or true positive rate and false positive rate), and conversely, can be used to estimate performance at a given $N$. An ROC curve obtained in this manner is termed a **Customer ROC (CROC) curve** [90].

When different numbers of recommendations can be shown to each user (e.g. when presenting the set of all recommended movies to each user), we can compute ROC or precision-recall curves by aggregating the hidden ratings from the test set into a set of reference user-item pairs, using the recommender system to generate a single ranked list of user-item pairs, picking the top recommendations from the list, and scoring them against the reference set. An ROC curve calculated in this way is termed a **Global ROC (GROC) curve** [90]. Picking an operating point on the resulting curve can result in a different number of recommendations being made to each user.

A final class of applications is where the recommendation process is more interactive, and the user is able to obtain more and more recommendations. This is typical of information retrieval tasks, where the user can keep asking the system for more recommended documents. In such applications, we compute a precision-recall curve (or ROC curve) for each user and then average the resulting curves over users. This is the usual manner in which precision-recall curves are computed in the information retrieval community, and in particular in the influential TREC competitions [103]. Such a curve can be used to understand the trade-off between precision and recall (or false positives and false negatives) a typical user would face.

Hit Rate and Mean Reciprocal Rank

In some applications, the amount of items that a user is expected to consume is very low. In applications such as tourism [79], the user is not expected to select more than one item, e.g., an hotel for a trip. As such, measuring the precision—the portion of chosen items within the recommendation list—may not be appropriate.

In many applications it is sufficient that the list would contain one item that the user chooses for the recommender system to be successful. An alternative to precision@$N$ in such applications is the *hit rate* (or hit ration)—the number of recommendation lists of length $N$ that contained at least one good item.

Precision@$N$, which depends both on $N$—the number of recommended items, as well as on the amount of items that a user chooses, can often be very low, and has an unknown upper bound. The hit rate, however, is bounded by 1, and in many applications researchers have achieved a very high hit rate. Thus, when using hit rate, we can know how close are we to an optimal behavior.

On the other hand, it is simple to increase the hit rate by increasing the number of recommended items. It is important not to measure the hit rate over recommendation lists that are too long to be used in a real application, thus reporting unrealistic performance that cannot be achieved in practice.

It is also important to be caution when using hit rate in applications where there are many items that the user chooses. In such cases, it may be very easy to identify one good item. It might be that the application owner would prefer that as many items in the recommendation list are good items, and hit rate does not distinguish between lists with one good item and lists with many good items.

Another metric that can be used in such settings is the mean reciprocal rank—the mean rank of the first good item in the complete list of possible recommendations:

$$MRR = \frac{1}{|U|} \sum_{u \in U} \frac{1}{r_u} \tag{13}$$

where $r_u$ is the rank (position) of the first item in the recommendation list that $u$ has chosen.

MRR is not sensitive to the list length, which makes it more appropriate for cases where the length of the recommendation list is not dictated by the application owner. In applications where the length of the recommendation list is known, and cannot be changed, the hit rate is arguably better correlated with the real online performance.

### 4.2.3  Ranking Measures

In many cases the application presents to the user a list of recommendations, typically as vertical or horizontal list, imposing a certain natural browsing order. For example, Netflix, shows a set of categories, and in each category, a list of movies that the system predicts the user to like. These lists may be long and the user may need to continue to additional "pages" until the entire list is browsed. In these applications, we are not interested in predicting an explicit rating, or selecting a set of recommended items, as in the previous sections, but rather in ordering items according to the user's preferences. This task is typically known as the ranking of items. There are two approaches for measuring the accuracy of such a ranking. We can try to determine the correct order of a set of items for each user and measure how close a system comes to this correct order, or we can attempt to measure the utility of the system's raking to a user. We first describe these approaches for offline tests, and then describe their applicability to user studies and online tests.

Using a Reference Ranking

In order to evaluate a ranking algorithm with respect to a reference ranking (a correct order), it is first necessary to obtain such a reference.

In cases where explicit user ratings of items are available, we can rank the rated items in decreasing order of ratings. However, there are two problems with this approach. First, most users typically have not rated some (usually most) of the items. Second, in many applications, the user ratings are quantized. For example, in the case of Netflix, each user only rates some of the movies, and the ratings are quantized to a binary scale. Thus, while we know that a movie rated 4 stars is preferred over a movie rated 3 stars, we do not know which (if either) of two 4-star movies is actually preferred by the user. We also know nothing about the user's preferences over most of the movies, which they have not rated.

Constructing reference rankings from usage data also runs into this problem. We can assume that items that the user actually used are preferred to those that the user was aware of but did not use. However, we do not know how to rank unused items that the user is not known to have been aware of (e.g. items that were never presented to the user, or items that were presented in a manner that the user may have easily missed them). We also do not know how to rank used items against other used items, and unused items the user was aware of against other such items.

Such cases where a ranking over items is incompletely specified is described technically as a *partial order*.

Let $\binom{I}{2}$ denote the set of all unordered pairs of items in $I$. Let $\succ$ be a partial order over a set of items $I$. In a partial order, for any two items $i_1$, $i_2$, exactly one of the following three conditions holds:

1. One item is a successor of the other, e.g. $i_1$ is a successor of $i_2$, denoted $i_1 \succ i_2$, typically meaning that $i_1$ is preferred to $i_2$. For example, if the user prefers "Star

Wars IV" to "The Matrix", and the list is ranked by preference, then we can write "Star Wars IV" $\succ$ "The Matrix".

2. If the user prefers the items equally, denoted $i_1 = i_2$. For example, a user may be indifferent as to whether he would get a brand A or brand B laptop, as long as they both have the same amount of memory, or may bid the same amount for two items on an auction at eBay.

3. The items may be incomparable. For example, one may not be able to say whether she prefers the latest Coen brothers movie, or the latest U2 disk. Alternatively, as discussed above, we may not have information about the user's preferences on the pair.

A *total order* over a set of items is an order where for each pair of items $i_1, i_2$, either $i_1 \succ i_2$ or $i_2 \succ i1$. In many cases, the reference ranking is given by a partial order, but the system outputs its recommendations as a total order, although perhaps not on all items. Therefore, we now describe ranking accuracy metrics that allow measurement agreement/disagreement between partial and total orders. To do so, we formally define the concepts of agreement, disagreement, and compatibility.

Let $\succ_1$ and $\succ_2$ be two partial orders over a set of items $I$, where $\succ_1$ is the reference order and $\succ_2$ is the system proposed order. We define an agreement relation between the orders $\succ_1$ and $\succ_2$ with respect to a pair of items as follows:

- The orders $\succ_1$ and $\succ_2$ *agree* on items $i_1$ and $i_2$ if $i_1 \succ_1 i_2$ and $i_1 \succ_2 i_2$.
- The orders $\succ_1$ and $\succ_2$ *disagree* on items $i_1$ and $i_2$ if $i_1 \succ_1 i_2$ and $i_2 \succ_2 i_1$.
- The orders $\succ_1$ and $\succ_2$ *are compatible* on items $i_1$ and $i_2$ if $i_1 \succ_1 i_2$ and neither $i_1 \succ_2 i_2$ nor $i_2 \succ_2 i_1$. In other words the items are either tied or incomparable under at least one of the orders.

The **Normalized Distance based Performance Measure** (NDPM) [104] is commonly used in information retrieval. It differentiates between correct orders of pairs, incorrect orders and ties. Formally, let $\delta_{\succ_1,\succ_2}(i_1, i_2)$ be a distance function between a reference ranking $\succ_1$ and a proposed ranking $\succ_2$ defined as follows:

$$\delta_{\succ_1,\succ_2}(i_1, i_2) = \begin{cases} 0 & \text{if } \succ_1 \text{ and } \succ_2 \text{ agree on } i_1 \text{ and } i_2, \\ 1 & \text{if } \succ_1 \text{ and } \succ_2 \text{ are compatible on } i_1 \text{ and } i_2, \\ 2 & \text{if } \succ_1 \text{ and } \succ_2 \text{ disagree on } i_1 \text{ and } i_2. \end{cases} \qquad (14)$$

The total distance over all item pairs in $I$ is:

$$\beta_{\succ_1,\succ_2}(I) = \sum_{(i_1,i_2)\in\binom{I}{2}} \delta_{\succ_1,\succ_2}(i_1, i_2) \qquad (15)$$

where the summation is over all possible item pairs in $I$ (efficient implementations can sum only over item pairs for which the reference ranking asserts an order).

Let $m(\succ_1) = \text{argmax}_\succ \beta_{\succ_1,\succ}(I)$ be a normalization factor which is the maximal distance that any ranking $\succ$ can have from a reference ranking $\succ_1$. In fact, $m(\succ_1)$ is the number of pairs in $I$ for which the reference ranking asserts an ordering, because the worst possible outcome would be to be wrong on all possible pairs. The NDPM score $NDPM(I, \succ_1, \succ_2)$ comparing a proposed ranking of items $\succ_2$ to a reference ranking $\succ_1$ is

$$NDPM(I, \succ_1, \succ_2) = \frac{\beta_{\succ_1,\succ_2}(I)}{m(\succ_1)} \tag{16}$$

Intuitively, the NDPM measure will give a perfect score of 0 to rankings over the set $I$ that completely agree with the reference ranking, and the worst score of 1 is assigned to a ranking that completely disagrees with the reference ranking. If the proposed ranking does not contain a preference between a pair of items that are ranked in the reference ranking, it is penalized by half as much as providing a contradicting preference.

The proposed ranking is not penalized for containing preferences that are not ordered in the reference ranking. This means that for any pair of items that was not ordered in the reference ranking any ordering predicted by the ranking algorithm is acceptable. This is because we typically display a list within the application. As such the ranking algorithm is expected to output a total, not a partial order, and should not be penalized for being forced to order all pairs.

A potential downside of NDPM in some applications is that it does not consider the location of disagreements in the reference ranking. In some cases it is more important to appropriately order items that should appear closer to the head of the ranked list, than items that are positioned near the bottom. For example, when ranking movies by decreasing preference, it may be more important to properly order the movies that the user would enjoy, than to properly order the movies that the user would not enjoy. It is sometimes important to give different weights to errors depending on their position in the list.

To this end, we can use the **Average Precision (AP) correlation** metric [105], which gives more weight to errors over items that appear at earlier positions in the reference ranking. Formally, let $\succ_1$ be the reference ranking and $\succ_2$ be a proposed ranking over a set of items. The AP measure compares the ordering of each item in the proposed ranking $\succ_2$ with respect to its preceding items (successors) in the reference ranking $\succ_1$.

For each $i_1 \in I$, let the set $Z^{i_1}(I, \succ)$ denote all item pairs $(i_1, i_2)$ in $I$ such that $i_2 \succ i_1$. These are all the items that are preferred over $i_1$ (i.e., preceding terms).

$$Z^i(I, \succ) = \{(i_1, i_2) \mid \forall i_1, i_2 \in I \text{ s.t. } i_2 \succ i_1\} \tag{17}$$

We define the indicator function $\delta(i_1, i_2, \succ_1, \succ_2)$ to equal 1 when $\succ_1$ and $\succ_2$ agree on items $i_1$ and $i_2$, and zero otherwise.

Let $A^{i_1}(I, \succ_1, \succ_2)$ be the normalized agreement score between $\succ_2$ and the reference ranking $\succ_1$ for all items $i_2$ such that $i_2 \succ_1 i_1$.

$$A^{i_1}(I, \succ_1, \succ_2) = \frac{1}{|Z^{i_1}(I, \succ_2)| - 1} \sum_{(i_1, i_2) \in Z^{i_1}(I, \succ_2)} \delta(i_1, i_2, \succ_1, \succ_2) \qquad (18)$$

The AP score of a partial order $\succ_2$ over $I$ given partial order $\succ_1$ is defined as

$$AP(I, \succ_1, \succ_2) = \frac{1}{|I| - 1} \sum_{i \in I} A^i(I, \succ_1, \succ_2) \qquad (19)$$

The AP score gives a perfect score of 1 where there is total agreement between the system proposed ranking and the reference ranking for every item pair above location $i$ for all $i \in \{1 \ldots |I|\}$. The worst score of 0 is given to systems were there is no agreement between the two ranked lists.

In some cases, we may completely know the user's true preferences for some set of items. For example, we may elicit the user's true ordering by presenting the user with binary choices. In this case, when a pair of items are tied in the reference ranking it means that the user is actually indifferent between the items. Thus, a perfect system should not rank one item higher than the other. In such cases, rank correlation measures such as **Spearman's** $\rho$ or **Kendall's** $\tau$ [50, 51] can be used. These measures tend to be highly correlated in practice [26]. is given by

$$\tau = \frac{C^+ - C^-}{\sqrt{C^u}\sqrt{C^s}} \qquad (20)$$

where $C^+$ and $C^-$ are the number of pairs that were correctly, and incorrectly ordered by the system, respectively, $C^u$ is the number of item pairs for which the reference ranking asserts any ordering, and $C^s$ is the number of item pairs for which the evaluated system asserts any ordering.

Utility-Based Ranking

While reference ranking scores a ranking on its correlation with some "true" ranking, there are other criteria for deciding on ordering a list of items. One popular alternative is to order items by decreasing utility. In such cases, we not only care about whether items $i_1$ and $i_2$ were ordered incorrectly, but also about the difference in utility between $i_1$ and $i_2$. It is not as bad to incorrectly order a pair of items with similar utilities, as to incorrectly order items with very different utilities.

It is also common to assume that the utility of a list of recommendations is additive, given by the sum of the utilities of the individual recommendations. The utility of each recommendation is the utility of the recommended item discounted by a factor that depends on its position in the list of recommendations. One example of such a utility is the likelihood that a user will observe a recommendation at position $i$ in the list. It is usually assumed that users scan recommendation lists from the beginning to the end, with the utility of recommendations being discounted more heavily towards the end of the list. The discount can also be interpreted as the probability that a user would observe a recommendation in a particular position in the list, with the utility of the recommendation given that it was observed depending only on the item recommend. Under this interpretation, the probability that a particular position in the recommendation list is observed is assumed to depend only on the position and not on the items that are recommended.

In many applications, the user can use only a single, or a very small set of items, or the recommendation engine is not used as the main browsing tool. In such cases, we can expect the users to observe only a few items of the top of the recommendations list. We can model such applications using a very rapid decay of the positional discount down the list. The R-score metric [12] assumes that the value of recommendations decline exponentially down the ranked list to yield the following score for each user $u$:

$$R_u = \sum_j \frac{\max(r_{u,i_j} - d, 0)}{2^{\frac{j-1}{\alpha-1}}} \tag{21}$$

where $i_j$ is the item in the $j$th position, $r_{u,i}$ is user $u$'s rating of item $i$, $d$ is a task dependent neutral ("don't care") rating, and $\alpha$ is a half-life parameter, which controls the exponential decline of the value of positions in the ranked list. In the case of ratings prediction tasks, $r_{ui}$ is the rating given by the user to each item (e.g. 4 stars), and $d$ is the don't care vote (e.g. 3 stars), and the algorithm only gets credit for ranking items with rating above the "don't care" vote higher than $d$ (e.g. 4 or 5 stars). In usage prediction tasks, $r_{u,i}$ is typically 1 if $u$ selects $i$ and 0 otherwise, while $d$ is 0. Using

$$r_{u,i} = -\log(\text{relative-frequency}(i)) \tag{22}$$

if $i$ is used and 0 otherwise can capture the amount of information in the recommendation [92]. The resulting per-user scores are aggregated using:

$$R = 100 \frac{\sum_u R_u}{\sum_u R_u^*} \tag{23}$$

where $R_u^*$ is the score of the best possible ranking for user $u$.

In other applications the user is expected to read a relatively large portion of the list. In certain types of search, such as the search for legal documents [38], users may look for all relevant items, and would be willing to read large portions of the recommendations list. In such cases, we need a much slower decay of the positional discount. Normalized Discounted Cumulative Gain (NDCG) [45] is a measure from information retrieval, where positions are discounted logarithmically. Assuming each user $u$ has a "gain" $g_{u,i}$ from being recommended item $i$, the average Discounted Cumulative Gain (DCG) for a list of $J$ items is defined as

$$DCG = \frac{1}{N} \sum_{u=1}^{N} \sum_{j=1}^{J} \frac{g_{u,i_j}}{\log_b(j+1)} \tag{24}$$

where $i_j$ is the item at position $j$ in the list. The logarithm base is a free parameter, typically between 2 and 10. A logarithm with base 2 is commonly used to ensure all positions are discounted. NDCG is the normalized version of DCG given by

$$NDCG = \frac{DCG}{DCG^*} \tag{25}$$

where $DCG^*$ is the ideal DCG.

We show the two methods here as they were originally presented, but note that the numerator in the two cases contains a utility function that assigns a value for each item. One can replace the original utility functions with a function that is more appropriate to the designed application. A measure closely related to R-score and NDCG is **Average Reciprocal Hit Rank (ARHR)** [22] which is an un-normalized measure that assigns a utility $1/k$ to a successful recommendation at position $k$. Thus, ARHR decays more slowly than R score but faster than NDCG.

Online Evaluation of Ranking

In an online experiment designed to evaluate the ranking of the recommendation list, we can look at the interactions of users with the system. When a recommendation list is presented to a user, the user may select a number of items from the list. We can now assume that the user has scanned the list at least as deep as the last selection. That is, if the user has selected items 1, 3, and 10, we can assume that the user has observed items 1 through 10. We can now make another assumption, that the user has found items 1,3, and 10 to be interesting, and items 2,4,5,6,7,8, and 9 to be uninteresting (see, e.g. Jung et al. [48]). In some cases we can have additional information whether the user has observed more items. For example, if the list is spread across several pages, and only 20 results are presented per page, then, in the example above, if the user moved to the second page we can also assume that she has observed results 11 through 20 and had found them to be irrelevant.

In the scenario above, the results of this interaction is a division of the list into 3 parts—the interesting items (1,3,10 in the example above), the uninteresting items (the rest of the items from 1 through 20), and the unknown items (21 till the end of the list). We can now use an appropriate reference ranking metric to score the original list. This can be done in two different ways. First, the reference list can contain the interesting items at the top, then the unknown items, and the uninteresting items at the bottom. This reference list captures the case where the user may only select a small subset of the interesting items, and therefore the unknown items may contain more interesting items. Second, the reference list can contain the interesting items at the top, followed by the uninteresting items, with the unknown items completely ignored. This is useful when making unreasonable preference assumptions, such as that some unknown items are preferred to the uninteresting items, may have negative consequences. In either case, it should be borne in mind that the semantics of the reference ranking are different from the case of offline evaluations. In offline evaluations, we have a single reference ranking which is assumed to be correct, and we measure how much each recommender deviates from this "correct" ranking. In the online case, the reference ranking is assumed to be the ranking that the user would have preferred given that were presented with the recommender's ranking. In the offline case, we assume that there is one correct ranking, while in the online case we allow for the possibility of multiple correct rankings.

In the case of utility ranking, we can evaluate a list based on the sum of the utilities of the selected items. Lists that place interesting items with high utility close to the beginning of the list, will hence be preferred to lists that place these interesting items down the list, because we expect that in the latter case, the user will often not observe these interesting items at all, generating no utility for the recommender.

## 4.3  Coverage

As the prediction accuracy of a recommender system, especially in collaborative filtering systems, in many cases grows with the amount of data, some algorithms may provide recommendations with high quality, but only for a small portion of the items where they have huge amounts of data. This is often referred to as the *long tail* or *heavy tail* problem, where the vast majority of the items where selected or rated only by a handful of users, yet the total amount of evidence over these unpopular items is much more than the evidence over the few popular items.

The term coverage can refer to several distinct properties of the system that we discuss below.

### 4.3.1  Item Space Coverage

Most commonly, the term coverage refers to the proportion of items that the recommender system can recommend. This is often referred to as **catalog coverage**. The simplest measure of catalog coverage is the percentage of all items that can ever be recommended. This measure can be computed in many cases directly given the algorithm and the input data set.

A more useful measure is the percentage of all items that are recommended to users during an experiment, either offline, online, or a user study. In some cases it may be desirable to weight the items, for example, by their popularity or utility. Then, we may agree not to be able to recommend some items which are very rarely used anyhow, but ignoring high profile items may be less tolerable.

Another measure of catalog coverage is the **sales diversity** [24], which measures how unequally different items are chosen by users when a particular recommender system is used. If each item $i$ accounts for a proportion $p(i)$ of user choices, the **Gini Index** is given by:

$$G = \frac{1}{n-1} \sum_{j=1}^{n} (2j - n - 1) p(i_j) \tag{26}$$

where $i_1, \cdots i_n$ is the list of items ordered according to increasing $p(i)$. The index is 0 when all items are chosen equally often, and 1 when a single item is always chosen. The Gini index of the number of times each item is recommended could also be used. Another measure of distributional inequality is the **Shannon Entropy**:

$$H = - \sum_{i=1}^{n} p(i) \log p(i) \tag{27}$$

The entropy is 0 when a single item is always chosen or recommended, and $\log n$ when $n$ items are chosen or recommended equally often.

Steck [98] further discusses how accuracy methods can be modified to better model the accuracy in the long tail. He suggests a correction for the bias of users towards the more popular items.

### 4.3.2 User Space Coverage

Coverage can also be the proportion of users or user interactions for which the system can recommend items. In many applications the recommender may not provide recommendations for some users due to, e.g. low confidence in the accuracy of predictions for that user. In such cases we may prefer recommenders that can provide recommendations for a wider range of users. Clearly, such recommenders should be evaluated on the trade-off between coverage and accuracy.

Coverage here can be measured by the richness of the user profile required to make a recommendation. For example, in the collaborative filtering case this could be measured as the number of items that a user must rate before receiving recommendations. This measurement can be typically evaluated in offline experiments.

### 4.3.3 Cold Start

Another related set of issues are the well known cold start problems—the coverage and performance of the system on new items and on new users. Cold start can be considered as a sub problem of coverage because it measures the system coverage over a specific set of items and users. In addition to measuring how large the pool of cold start items or users are, it may also be important to measure system accuracy for these users and items.

Focusing on cold start items, we can use a threshold to decide on the set of cold items. For example, we can decide that cold items are only items with no ratings or usage evidence [90], or items that exist in the system for less than a certain amount of time (e.g., a day), or items that have less than a predefined evidence amount (e.g., less than 10 ratings). Perhaps a more generic way is to consider the "coldness" of an item using either the amount of time it exists in the system or the amount of data gathered for it. Then, we can credit the system more for properly predicting colder items, and less for the hot items that are predicted.

It may be possible that a system better recommends cold items at the price of a reduced accuracy for hotter items. This may be desirable due to other considerations such as novelty and serendipity that are discussed later. Still, when computing the system accuracy on cold items it may be wise to evaluate whether there is a trade-off with the entire system accuracy.

## 4.4  Confidence

Confidence in the recommendation can be defined as the system's trust in its recommendations or predictions [36, 100]. As we have noted above, collaborative filtering recommenders tend to improve their accuracy as the amount of data over items grows. Similarly, the confidence in the predicted property typically also grows with the amount of data.

In many cases the user can benefit from observing these confidence scores [36]. When the system reports a low confidence in a recommended item, the user may tend to further research the item before making a decision. For example, if a system recommends a movie with very high confidence, and another movie with the same rating but a lower confidence, the user may add the first movie immediately to the watching queue, but may further read the plot synopsis for the second movie, and perhaps a few movie reviews before deciding to watch it.

Perhaps the most common measurement of confidence is the probability that the predicted value is indeed true, or the interval around the predicted value where a predefined portion, e.g. 95% of the true values lie. For example, a recommender may accurately rate a movie as a 4 star movie with probability 0.85, or have 95% of the actual ratings lie within $-1$ and $+\frac{1}{2}$ of the predicted 4 stars. The most general method of confidence is to provide a complete distribution over possible outcomes [65].

Given two recommenders that perform similarly on other relevant properties, such as prediction accuracy, is can be desirable to choose the one that can provide valid confidence estimates. In this case, given two recommenders with, say, identical accuracy, that report confidence bounds in the same way, we will prefer the recommender that better estimates its confidence bounds.

Standard confidence bounds, such as the ones above, can be directly evaluated in regular offline trials, much the same way as we estimate prediction accuracy. We can design for each specific confidence type a score that measures how close the method confidence estimate is to the true error in prediction. This procedure cannot be applied when the algorithms do not agree on the confidence method, because some confidence methods are weaker and therefore easier to estimate. In such a case a more accurate estimate of a weaker confidence metric does not imply a better recommender.

*Example 1* Recommenders $A$ and $B$ both report confidence intervals over possible movie ratings. We train $A$ and $B$ over a confidence threshold, ranging of 95%. For each trained model, we run $A$ and $B$ on offline data, hiding a part of the user ratings and requesting each algorithm to predict the missing ratings. Each algorithm produces, along with the predicted rating, a confidence interval. We compute $A_+$ and $A_-$, the number of times that the predicted rating of algorithm $A$ was within and outside the confidence interval (respectively), and do the same for $B$. Then we compute the true confidence of each algorithm using $\frac{A_+}{A_-+A_+} = 0.97$ and $\frac{B_+}{A_-+A_+} = 0.94$. The result indicates that $A$ is over conservative, and computes intervals that are too large, while $B$ is liberal and computes intervals that are too small. As we do not require the intervals to be conservative, we prefer $B$ because its estimated intervals are closer to the requested 95% confidence.

Another application of confidence bounds is in filtering recommended items where the confidence in the predicted value is below some threshold. In this scenario we assume that the recommender is allowed not to predict a score for all values. In a top $n$ recommendation scenario, we may allow a system to sometimes suggest less than $n$ items, because it cannot produce a set of $n$ items with sufficient confidence. In this case the precision of the system is not punished when less results are returned, and the shorter list is expected to result only in lower recall. As such, measuring only precision@N in such problems is insufficient, because algorithms have an incentive to provide less recommendations, or even no recommendations, obtaining a meaningless precision of 1.

We can hence design an experiment around this filtering procedure by comparing the accuracy of two recommenders after their results were filtered by removing low confidence items. In such experiments we can compute a curve, estimating the prediction accuracy (typically precision-recall curves) for each portion of filtered items, or for different filtering thresholds. This evaluation procedure does not require both algorithms to agree on the confidence method.

While user studies and online experiments can study the effect of reporting confidence on the user experience [94], it is difficult to see how these types of tests can be used to provide further evidence as to the accuracy of the confidence estimate.

## 4.5　Trust

While confidence is the system trust in its ratings, in trust we refer here to the user's trust in the system recommendation.[5] For example, it may be beneficial for the system to recommend a few items that the user already knows and likes. This way, even though the user gains no value from this recommendation, she observes that the system provides reasonable recommendations, which may increase her trust in the system recommendations for unknown items. Another common way of enhancing trust in the system is to explain the recommendations that the system provides. Trust in the system is also called the credibility of the system.

If we do not restrict ourselves to a single method of gaining trust, such as the one suggested above, the obvious method for evaluating user trust is by asking users whether the system recommendations are reasonable in a user study [7, 16, 36, 74]. In an online test one could associate the number of recommendations that were followed with the trust in the recommender, assuming that higher trust in the recommender would lead to more recommendations being used. Alternatively, we could also assume that trust in the system is correlated with repeated users, as users who trust the system will return to it when performing future tasks. However, such measurements may not separate well other factors of user satisfaction, and may not be accurate. It is unclear how to measure trust in an offline experiment, because trust is built through an interaction between the system and a user.

## 4.6　Novelty

Novel recommendations (chapter "Novelty and Diversity in Recommender Systems") are recommendations for items that the user did not know about [54]. In applications that require novel recommendation, an obvious and easy to implement approach is to filter out items that the user already rated or used. However, in many cases users will not report all the items they have used in the past. Thus, this simple method is insufficient to filter out all items that the user already knows.

While we can obviously measure novelty in a user study, by asking users whether they were already familiar with a recommended item [14, 46], we can also gain some understanding of a system's novelty through offline experiments. For such an experiment we could split the data set on time, i.e. hide all the user purchases that occurred after a specific point in time. In addition, we can hide some purchases that occurred prior to that time, simulating the items that the user has purchased and is hence familiar with, but did not report their purchase to the system. When recommending, the system is rewarded for each item that was recommended and

---

[5] Not to be confused with trust in the social network research, used to measure how much a user believes another user. Some literature on recommender systems uses such trust measurements to filter similar users [64].

purchased after the split time, but would be punished for each item that was recommended but purchased prior to the split time.

To implement the above procedure we must carefully model the hiding process such that it would resemble the true preference discovery process that occurs in the real system. In some cases the set of purchased items is not a uniform sample of the set of all items the user is familiar with, and such bias should be acknowledged and handled if possible. For example, if we believe that the user will report more purchases of unique items, and less purchases of popular or common items, then the hiding process should tend to hide more popular items.

In using this measure of novelty, it is important to control for accuracy, as irrelevant recommendations may be new to the user, but still worthless. One approach would be to consider novelty only among the relevant items [109].

*Example 2* We wish to evaluate the novelty of a set of movie recommenders in an offline test. As we believe that users of our system rate movies after they watch them, we split the user ratings in a sequential manner. For each test user profile we choose a cutoff point randomly along the time-based sequence of movie ratings, hiding all movies after a certain point in the sequence.

Let us assume that user studies on this imaginary system showed that people tend not to report ratings of movies that they did not feel strongly about, but occasionally also do not report a rating of a movie that they liked or disliked strongly. Therefore, we hide a rating of a movie prior to the cutoff point with probability $1 - \frac{|r-3|}{2}$ where $r \in \{1, 2, 3, 4, 5\}$ is the rating of the movie, and 3 is the neutral rating. We would like to avoid predicting these movies with hidden ratings because the user already knows about them.

Then, for each user, each recommender produces a list of 5 recommendations, and we compute precision only over items after the cutoff point. That is, the recommenders get no credit for recommending movies with hidden ratings that occurred prior to the cutoff point. In this experiment the algorithm with the highest precision score is preferred.

Another method for evaluating novel recommendations uses the above assumption that popular items are less likely to be novel. Thus, novelty can be taken into account by using an accuracy metric where the system does not get the same credit for correctly predicting popular items as it does when it correctly predicts non-popular items [93]. Ziegler et al. [110] and Celma and Herrera [14] also give accuracy measures that take popularity into account.

Finally, we can evaluate the amount of new information in a recommendation together with the relevance of the recommended item. For example, when item ratings are available, we can multiply the hidden rating by some information measurement of the recommended item (such as the conditional entropy given the user profile) to produce a novelty score.

## 4.7 Serendipity

Serendipity is a measure of how surprising the successful recommendations are (chapter "Novelty and Diversity in Recommender Systems"). For example, if the user has rated positively many movies where a certain star actor appears, recommending the new movie of that actor may be novel, because the user may not know of it, but is hardly surprising. Of course, random recommendations may be very surprising, and we therefore need to balance serendipity with accuracy.

One can think of serendipity as the amount of relevant information that is new to the user in a recommendation. For example, if following a successful movie recommendation the user learns of a new actor that she likes, this can be considered as serendipitous. In information retrieval, where novelty typically refers to the new information contained in the document (and is thus close to our definition of serendipity), Zhang et al. [109] suggested to manually label pairs of documents as redundant. Then, they compared algorithms on avoiding recommending redundant documents. Such methods are applicable to recommender systems when some metadata over items, such as content information, is available (chapter "Semantics and Content-Based Recommendations").

To avoid human labeling, we could design a distance measurement between items based on content. Then, we can score a successful recommendation by its distance from a set of previously rated items in a collaborative filtering system, or from the user profile in a content-based recommender [107]. Thus, we are rewarding the system for successful recommendations that are far from the user profile.

*Example 3* In a book recommendation application, we would like to recommend books from authors that the reader is less familiar with. We therefore design a distance metric between a book $b$ and a set of books $B$ (the books that the user has previously read); Let $c_{B,w}$ be the number of books by writer $w$ in $B$. Let $c_B = \max_w c_{B,w}$ the maximal number of books from a single writer in $B$. Let $d(b, B) = \frac{1+c_B-c_{B,w(b)}}{1+c_B}$, where $w(b)$ is the writer of book $b$.

We now run an offline experiment to evaluate which of the candidate algorithms generates more serendipitous recommendations. We split each test user profile— set of books that the user has read—into sets of observed books $B_i^o$ and hidden books $B_i^h$. We use $B_i^o$ as the input for each recommender, and request a list of 5 recommendations. For each hidden book $b \in B_i^h$ that appeared in the recommendation list for user $i$, the recommender receives a score of $d(b, B_i^o)$. Thus the recommender is getting more credit for recommending books from writers that the reader has read less often. In this experiment the recommender that received a higher score is selected for the application.

One can also think of serendipity as deviation from the "natural" prediction [70]. That is, given a prediction engine that has a high accuracy, the recommendations that it issues are "obvious". Therefore, we will give higher serendipity scores to successful recommendations that the prediction engine would deem unlikely.

We can evaluate the serendipity of a recommender in a user study by asking the users to mark the recommendations that they find unexpected. Then, we can also see whether the user followed these recommendations, which would make them unexpected and successful and therefore serendipitous. In an online study, we can assume that our distance metric is correct and evaluate only how distance from the user profile affected the probability that a user will follow the recommendation. It is important to check the effect of serendipity over time, because users might at first be intrigued by the unexpected recommendations and try them out. If after following the suggestion they discover that the recommendations are inappropriate, they may stop following them in the future, or stop using the recommendation engine at all.

## 4.8  Diversity

Diversity is generally defined as the opposite of similarity (chapter "Novelty and Diversity in Recommender Systems"). In some cases suggesting a set of similar items may not be as useful for the user, because it may take longer to explore the range of items. Consider for example a recommendation for a vacation [96], where the system should recommend vacation packages. Presenting a list with 5 recommendations, all for the same location, varying only on the choice of hotel, or the selection of attraction, may not be as useful as suggesting 5 different locations. The user can view the various recommended locations and request more details on a subset of the locations that are appropriate to her.

The most explored method for measuring diversity uses item-item similarity, typically based on item content, as in Sect. 4.7. Then, we could measure the diversity of a list based on the sum, average, min, or max distance between item pairs, or measure the value of adding each item to the recommendation list as the new item's diversity from the items already in the list [10, 110]. The item-item similarity measurement used in evaluation can be different from the similarity measurement used by the algorithm that computes the recommendation lists. For example, we can use for evaluation a costly metric that produces more accurate results than fast approximate methods that are more suitable for online computations.

As diversity may come at the expanse of other properties, such as accuracy [107], we can compute curves to evaluate the decrease in accuracy vs. the increase in diversity.

*Example 4* In a book recommendation application, we are interested in presenting the user with a diverse set of recommendations, with minimal impact to accuracy. We use $d(b, B)$ from Example 3 as the distance metric. Given candidate recommenders, each with a tunable parameter that controls the diversity of the recommendations, we train each algorithm over a range of values for the diversity parameters. For each trained model, we now compute a precision score, and a diversity score as follows; we take each recommendation list that an algorithm produces, and compute the distance of each item from the rest of the list, averaging the result to obtain a diversity score. We now plot the precision-diversity curves of the recommenders in a graph, and select the algorithm with the dominating curve.

In recommenders that assist in information search, we can assume that more diverse recommendations will result in shorter search interactions [96]. We could use this in an online experiment measuring interaction sequence length as a proxy for diversification. As is always the case in online testing, shorter sessions may be due to other factors of the system, and to validate this claim it is useful to experiment with different diversity thresholds using the same prediction engine before comparing different recommenders.

## 4.9   Utility

Many e-commerce websites employ a recommender system in order to improve their revenue by, e.g., enhancing cross-sell. In such cases the recommendation engine can be judged by the revenue that it generates for the website [92]. In general, we can define various types of utility functions that the recommender tries to optimize. For such recommenders, measuring the utility, or the expected utility of the recommendations may be more significant than measuring the accuracy of recommendations. It is also possible to view many of the other properties, such as diversity or serendipity, as different types of utility functions, over single items or over lists. In this chapter, however, we define utility as the value that either the system or the user gains from a recommendation.

Utility can be measured cleanly from the perspective of the recommendation engine or the recommender system owner. Care must be taken, though, when measuring the utility that the user receives from the recommendations. First, user utilities or preferences are difficult to capture and model, and considerable research has focused on this problem [11, 33, 77]. Second, it is unclear how to aggregate user utilities across users for computing a score for a recommender. For example, it is tempting to use money as a utility thus selecting a recommender that minimizes user cost. However, under the diminishing returns assumption [97], the same amount of money does not have the same utility for people with different income levels. Therefore, the average cost per purchase, for example, is not a reasonable aggregation across users.

In an application where users rate items, it is also possible to use the ratings as a utility measurement [12]. For example, in movie ratings, where a 5 star movie is considered an excellent movie, we can assume that a recommending a 5 star movie has a higher utility for the user than recommending a movie that the user will rate with 4 stars. As users may interpret ratings differently, user ratings should be normalized before aggregating across users.

While we typically only assign positive utilities to successful recommendations, we can also assign negative utilities to unsuccessful recommendations. For example, if some recommended item offends the user, then we should punish the system for recommending it by assigning a negative utility. We can also add a cost to each recommendation, perhaps based on the position of the recommended item in the list, and subtract it from the utility of the item.

For any utility function, the standard evaluation of the recommender is to compute the expected utility of a recommendation. In the case where the recommender is trying to predict only a single item, such as when we evaluate the system on time-based splits and try to predict only the next item in the sequence, the value of a correct recommendation should simply be the utility of the item. In the task where the recommender predicts $n$ items we can use the sum of the utilities of the correct recommendations in the list. When negative utilities for failed recommendations are used, then the sum is over all recommendations, successful or failed. We can also integrate utilities into ranking measurements, as discussed in Sect. 4.2.3. Finally, we can normalize the resulting score using the maximal possible utility given the optimal recommendation list.

Evaluating utility in user studies and online is easy in the case of recommender utility. If the utility we optimize for is the revenue of the website, measuring the change in revenue between users of various recommenders is simple. When we try to optimize user utilities the online evaluation becomes harder, because users typically find it challenging to assign utilities to outcomes. In many cases, however, users can say whether they prefer one outcome to another. Therefore, we can try to elicit the user preferences [40] in order to rank the candidate methods.

## *4.10   Risk*

In some cases a recommendation may be associated with a potential risk. For example, when recommending stocks for purchase, users may wish to be risk-averse, preferring stocks that have a lower expected growth, but also a lower risk of collapsing. On the other hand, users may be risk-seeking, preferring stocks that have a potentially high, even if less likely, profit. In such cases we may wish to evaluate not only the (expected) value generated from a recommendation, but also to minimize the risk.

The standard way to evaluate risk sensitive systems is by considering not just the expected utility, but also the utility variance. For example, we may use a parameter $q$ and compare two systems on $E[X] + q \cdot Var(X)$. When $q$ is positive, this approach prefers risk-seeking (also called bold [67]) recommenders, and when $q$ is negative, the system prefers risk-averse recommenders.

## 4.11 Robustness

Robustness (chapter "Adversarial Recommender Systems: Attack, Defense, and Advances") is the stability of the recommendation in the presence of fake information [72], typically inserted on purpose in order to influence the recommendations. As more people rely on recommender systems to guide them through the item space, influencing the system to change the rating of an item may be profitable to an interested party. For example, an owner of an hotel may wish to boost the rating for their hotel. This can be done by injecting fake user profiles that rate the hotel positively, or by injecting fake users that rate the competitors negatively.

Such attempts to influence the recommendation are typically called attacks [58, 69]. Coordinated attacks occur when a malicious user intentionally queries the data set or injects fake information in order to learn some private information of some users. In evaluating such systems, it is important to provide a complete description of the attack protocol, as the sensitivity of the system typically varies from one protocol to another.

In general, creating a system that is immune to any type of attack is unrealistic. An attacker with an ability to inject an infinite amount of information can, in most cases, manipulate a recommendation in an arbitrary way. It is therefore more useful to estimate the cost of influencing a recommendation, which is typically measured by the amount of injected information. While it is desirable to theoretically analyze the cost of modifying a rating, it is not always possible. In these cases, we can simulate a set of attacks by introducing fake information into the system data set, empirically measuring average cost of a successful attack [15, 57].

As opposed to other evaluation criteria discussed here, it is hard to envision executing an attack on a real system as an online experiment. It may be fruitful, however, to analyze the real data collected in the online system to identify actual attacks that are executed against the system.

Another type of robustness is the stability of the system under extreme conditions, such as a large number of requests. While less discussed, such robustness is very important to system administrators, who must avoid system malfunction. In many cases system robustness is related to the infrastructure, such as the database software, or to the hardware specifications, and is related to scalability (Sect. 4.14).

## 4.12 Privacy

In a collaborative filtering system, a user willingly discloses his preferences over items to the system in the hope of getting useful recommendations (chapter "Adversarial Recommender Systems: Attack, Defense, and Advances"). However, it is important for most users that their preferences stay private, that is, that no third party can use the recommender system to learn something about the preferences of a specific user.

For example, consider the case where a user who is interested in the wonderful, yet rare art of growing Bahamian orchids has bought a book titled "The Divorce Organizer and Planner". The spouse of that user, looking for a present, upon browsing the book "The Bahamian and Caribbean Species (Cattleyas and Their Relatives)" may get a recommendation of the type "people who bought this book also bought" for the divorce organizer, thus revealing sensitive private information.

It is generally considered inappropriate for a recommender system to disclose private information even for a single user. For this reason analysis of privacy tends to focus on a worst case scenario, illustrating theoretical cases under which users private information may be revealed. Other researchers [25] compare algorithms by evaluating the portion of users whose private information was compromised. The assumption in such studies is that complete privacy is not realistic and that therefore we must compromise on minimizing the privacy breaches.

Another alternative is to define different levels of privacy, such as $k$-identity [25], and compare algorithms sensitivity to privacy breaches under varying levels of privacy.

Privacy may also come at the expense of the accuracy of the recommendations. Therefore, it is important to analyze this trade-off carefully. Perhaps the most informative experiment is when a privacy modification has been added to an algorithm, and the accuracy (or any other trade-off property) can be evaluated with or without the modification [68].

## 4.13   Adaptivity

Real recommender systems may operate in a setting where the item collection changes rapidly, or where trends in interest over items may shift. Perhaps the most obvious example of such systems is the recommendation of news items or related stories in online newspapers [30]. In this scenario stories may be interesting only over a short period of time, afterwards becoming outdated. When an unexpected news event occurs, such as the tsunami disaster, people become interested in articles that may not have been interesting otherwise, such as a relatively old article explaining the tsunami phenomenon. While this problem is similar to the cold-start problem, it is different because it may be that old items that were not regarded as interesting in the past suddenly become interesting.

This type of adaptation can be evaluated offline by analyzing the amount of information needed before an item is recommended. If we model the recommendation process in a sequential manner, we can record, even in an offline test, the amount of evidence that is needed before the algorithm recommends a story. It is likely that an algorithm can be adjusted to recommend items faster once they become interesting, by sacrificing some prediction accuracy. We can compare two algorithms by evaluating a possible trade-off between accuracy and the speed of the shift in trends.

Another type of adaptivity is the rate by which the system adapts to a user's personal preferences [61], or to changes in user profile [56]. For example, when users rate an item, they expect the set of recommendations to change. If the recommendations stay fixed, users may assume that their rating effort is wasted, and may not agree to provide more ratings. As with the shift in trends evaluation, we can again evaluate in an offline experiment the changes in the recommendation list after adding more information to the user profile such as new ratings. We can evaluate an algorithm by measuring the difference between the recommendation lists before and after the new information was added. The Gini index and Shannon entropy measures discussed in Sect. 4.3 can be used to measure the variability of recommendations made to a user as the user profile changes.

## 4.14  Scalability

As recommender systems are designed to help users navigate in large collections of items, one of the goals of the designers of such systems is to scale up to real data sets. As such, it is often the case that algorithms trade other properties, such as accuracy or coverage, for providing rapid results even for huge data sets consisting of millions of items (e.g. [19]).

With the growth of the data set, many algorithms are either slowed down or require additional resources such as computation power or memory. One standard approach in computer science research is to evaluate the computational complexity of an algorithm in terms of time or space requirements (as done, e.g., in [8, 49]). In many cases, however, the complexity of two algorithms is either identical, or could be reduced by changing some parameters, such as the complexity of the model, or the sample size. Therefore, to understand the scalability of the system it is also useful to report the consumption of system resources over large data sets.

Scalability is typically measured by experimenting with growing data sets, showing how the speed and resource consumption behave as the task scales up (see, e.g. [30]). It is important to measure the compromises that scalability dictates. For example, if the accuracy of the algorithm is lower than other candidates that only operate on relatively small data sets, one must show over small data sets the difference in accuracy. Such measurements can provide valuable information both on the potential performance of recommender systems in general for the specific task, and on future directions to explore.

As recommender systems are expected in many cases to provide rapid recommendations online, it is also important to measure how fast does the system provides recommendations [37, 88]. One such measurement is the throughput of the system, i.e., the number of recommendations that the system can provide per second. We could also measure the latency (also called response time)—the required time for making a recommendation online.

## 5 Conclusion

In this chapter we discussed how recommendation algorithms could be evaluated in order to select the best algorithm from a set of candidates. This is an important step in the research attempt to find better algorithms, as well as in application design where a designer chooses an existing algorithm for their application. As such, many evaluation metrics have been used for algorithm selection in the past.

We describe the concerns that need to be addressed when designing offline and online experiments and user studies. We outline a few important measurements that one must take in addition to the score that the metric provides, as well as other considerations that should be taken into account when designing experiments for recommendation algorithms.

We specify a set of properties that are sometimes discussed as important for the recommender system. For each such property we suggest an experiment that can be used to rank recommenders with regards to that property. For less explored properties, we restrict ourselves to generic descriptions that could be applied to various manifestations of that property. Specific procedures that can be practically implemented can then be developed for the specific property manifestation based on our generic guidelines.

## References

1. R. Bailey, *Design of Comparative Experiments*, vol. 25 (Cambridge University Press, Cambridge, 2008)
2. D. Bamber, The area above the ordinal dominance graph and the area below the receiver operating characteristic graph. J. Math. Psychol. **12**, 387–415 (1975)
3. J. Beel, S. Langer, A comparison of offline evaluations, online evaluations, and user studies in the context of research-paper recommender systems, in *International Conference on Theory and Practice of Digital Libraries* (Springer, New York, 2015), pp. 153–168
4. Y. Benjamini, Y. Hochberg, Controlling the false discovery rate: a practical and powerful approach to multiple testing. J. R. Stat. Soc. Ser. B (Methodological) **57**, 289–300 (1995)
5. P.J. Bickel, K.A. Ducksum, *Mathematical Statistics: Ideas and Concepts* (Holden-Day, San Francisco, 1977)
6. M. Boland, Native ads will drive 74% of all ad revenue by 2021. Business Insider 14, 2016
7. P. Bonhard, C. Harries, J. McCarthy, M.A. Sasse, Accounting for taste: using profile similarity to improve recommender systems, in *CHI '06: Proceedings of the SIGCHI Conference on Human Factors in Computing Systems*, New York, NY, 2006 (ACM, New York, 2006), pp. 1057–1066
8. C. Boutilier, R.S. Zemel, Online queries for collaborative filtering, in *In Proceedings of the Ninth International Workshop on Artificial Intelligence and Statistics*, 2002
9. G.E.P. Box, W.G. Hunter, J.S. Hunter, *Statistics for Experimenters* (Wiley, New York, 1978)
10. K. Bradley, B. Smyth, Improving recommendation diversity, in *Twelfth Irish Conference on Artificial Intelligence and Cognitive Science* (2001), pp. 85–94
11. D. Braziunas, C. Boutilier, Local utility elicitation in GAI models. in *Proceedings of the Twenty-first Conference on Uncertainty in Artificial Intelligence*, Edinburgh, 2005, pp. 42–49
12. J.S. Breese, D. Heckerman, C.M. Kadie, Empirical analysis of predictive algorithms for collaborative filtering, in *UAI*, 1998

13. R. Burke, Evaluating the dynamic properties of recommendation algorithms. in *Proceedings of the Fourth ACM Conference on Recommender Systems*, RecSys '10, New York (ACM, New York, 2010), pp. 225–228
14. Ò. Celma, P. Herrera, A new approach to evaluating novel recommendations, in *RecSys '08: Proceedings of the 2008 ACM Conference on Recommender systems*, New York, NY (ACM, New York, 2008), pp. 179–186
15. P.-A. Chirita, W. Nejdl, C. Zamfir, Preventing shilling attacks in online recommender systems, in *WIDM '05: Proceedings of the 7th Annual ACM International Workshop on Web Information and Data Management*, New York, NY (ACM, New York, 2005), pp. 67–74
16. H. Cramer, V. Evers, S. Ramlal, M. Someren, L. Rutledge, N. Stash, L. Aroyo, B. Wielinga, The effects of transparency on trust in and acceptance of a content-based art recommender. User Model. User-Adapted Interact. 18(5), 455–496 (2008)
17. P. Cremonesi, Y. Koren, R. Turrin, Performance of recommender algorithms on top-n recommendation tasks, in *Proceedings of the Fourth ACM Conference on Recommender Systems* (2010), pp. 39–46
18. M.F. Dacrema, P. Cremonesi, D. Jannach, Are we really making much progress? A worrying analysis of recent neural recommendation approaches, in *Proceedings of the 13th ACM Conference on Recommender Systems* (2019), pp. 101–109
19. A.S. Das, M. Datar, A. Garg, S. Rajaram, Google news personalization: scalable online collaborative filtering, in *WWW '07: Proceedings of the 16th International Conference on World Wide Web*, New York, NY (ACM, New York, 2007), pp. 271–280
20. O. Dekel, C.D. Manning, Y. Singer, Log-linear models for label ranking, in *NIPS'03* (2003), pages 1–1
21. J. Demšar, Statistical comparisons of classifiers over multiple data sets. J. Mach. Learn. Res. 7, 1–30 (2006)
22. M. Deshpande, G. Karypis, Item-based top-N recommendation algorithms. ACM Trans. Inf. Syst. 22(1), 143–177 (2004)
23. G. Fischer, User modeling in human-computer interaction. User Model. User-Adapt. Interact. 11(1–2), 65–86 (2001)
24. D.M. Fleder, K. Hosanagar, Recommender systems and their impact on sales diversity, in *EC '07: Proceedings of the 8th ACM Conference on Electronic Commerce*, New York, NY, 2007 (ACM, New York, 2007), pp. 192–199
25. D. Frankowski, D. Cosley, S. Sen, L. Terveen, J. Riedl, You are what you say: privacy risks of public mentions, in *SIGIR '06: Proceedings of the 29th Annual International ACM SIGIR Conference on Research and Development in Information Retrieval*, New York, NY, 2006 (ACM, New York, 2006), pp. 565–572
26. G.A. Fredricks, R.B. Nelsen, On the relationship between spearman's rho and kendall's tau for pairs of continuous random variables. J. Stat. Plan. Infer. 137(7), 2143–2150 (2007)
27. S. Frumerman, G. Shani, B. Shapira, O. Sar Shalom, Are all rejected recommendations equally bad? towards analysing rejected recommendations, in *Proceedings of the 27th ACM Conference on User Modeling, Adaptation and Personalization* (2019), pp. 157–165
28. Z. Gantner, S. Rendle, C. Freudenthaler, L. Schmidt-Thieme, Mymedialite: a free recommender system library. In *Proceedings of the Fifth ACM Conference on Recommender systems* (2011), pp. 305–308
29. F. Garcin, B. Faltings, O. Donatsch, A. Alazzawi, C. Bruttin, A. Huber, Offline and online evaluation of news recommender systems at swissinfo, in *Proceedings of the 8th ACM Conference on Recommender systems* (2014), pp. 169–176
30. T. George, A scalable collaborative filtering framework based on co-clustering. in *Fifth IEEE International Conference on Data Mining* (2005), pp. 625–628
31. A.G. Greenwald, Within-subjects designs: To use or not to use? Psychol. Bull. 83, 216–229 (1976)
32. G. Guo, J. Zhang, Z. Sun, N. Yorke-Smith, Librec: a java library for recommender systems, in *UMAP Workshops*, vol. 4. Citeseer, 2015

33. P. Haddawy, V. Ha, A. Restificar, B. Geisler, J. Miyamoto, Preference elicitation via theory refinement. J. Mach. Learn. Res. **4**, 317–337 (2003)
34. C. Hayes, P. Cunningham, An on-line evaluation framework for recommender systems. Technical report, Trinity College Dublin, Department of Computer Science, 2002
35. X. He, J. Pan, O. Jin, T. Xu, B. Liu, T. Xu, Y. Shi, A. Atallah, R. Herbrich, S. Bowers, et al., Practical lessons from predicting clicks on ads at facebook, in *Proceedings of the Eighth International Workshop on Data Mining for Online Advertising* (2014), pp. 1–9
36. J.L. Herlocker, J.A. Konstan, J.T. Riedl, Explaining collaborative filtering recommendations, in *CSCW '00: Proceedings of the 2000 ACM conference on Computer Supported Cooperative Work*, New York, NY (ACM, New York, 2000), pp. 241–250
37. J.L. Herlocker, J.A. Konstan, J.T. Riedl, An empirical analysis of design choices in neighborhood-based collaborative filtering algorithms. Inf. Retr. **5**(4), 287–310 (2002). ISSN:1386-4564. http://dx.doi.org/10.1023/A:1020443909834
38. J.L. Herlocker, J.A. Konstan, L.G. Terveen, J.T. Riedl, Evaluating collaborative filtering recommender systems. ACM Trans. Inf. Syst. **22**(1), 5–53 (2004). ISSN:1046-8188. http://doi.acm.org/10.1145/963770.963772
39. Y. Hijikata, T. Shimizu, S. Nishida, Discovery-oriented collaborative filtering for improving user satisfaction, in *IUI '09: Proceedings of the 13th International Conference on Intelligent User Interfaces*, New York, NY (ACM, New York, 2009), pp. 67–76
40. R. Hu, P. Pu, A comparative user study on rating vs. personality quiz based preference elicitation methods, in *IUI 09: Proceedings of the 13th International Conference on Intelligent User Interfaces*, New York, NY (ACM, New York, 2009), pp. 367–372
41. R. Hu, P. Pu, A comparative user study on rating vs. personality quiz based preference elicitation methods, n *IUI* (2009), pp. 367–372
42. R. Hu, P. Pu, A study on user perception of personality-based recommender systems, in *UMAP* (2010), pp. 291–302
43. N. Hug, Surprise: a python library for recommender systems. J. Open Source Softw. **5**(52), 2174 (2020)
44. A. Iovine, F. Narducci, G. Semeraro, Conversational recommender systems and natural language: a study through the converse framework. Decis. Support Syst. **131**, 113250 (2020)
45. K. Järvelin, J. Kekäläinen, Cumulated gain-based evaluation of IR techniques. ACM Trans. Inf. Syst. **20**(4), 422–446 (2002). ISSN:1046-8188. http://doi.acm.org/10.1145/582415.582418
46. N. Jones, P. Pu, User technology adoption issues in recommender systems, in *Networking and Electronic Conference*, 2007
47. M. Jugovac, D. Jannach, M. Karimi, StreamingRec: a framework for benchmarking stream-based news recommenders, in *Proceedings of the 12th ACM Conference on Recommender Systems* (2018), pp. 269–273
48. S. Jung, J.L. Herlocker, J. Webster, Click data as implicit relevance feedback in web search. Inf. Process. Manage. **43**(3), 791–807 (2007)
49. G. Karypis, Evaluation of item-based top-n recommendation algorithms, in *CIKM '01: Proceedings of the Tenth International Conference on Information and Knowledge Management*, New York, NY (ACM, New York, 2001), pp. 247–254
50. M.G. Kendall, A new measure of rank correlation. Biometrika **30**(1–2), 81–93 (1938)
51. M.G. Kendall, The treatment of ties in ranking problems. Biometrika **33**(3), 239–251 (1945)
52. R. Kohavi, R. Longbotham, D. Sommerfield, R.M. Henne, Controlled experiments on the web: survey and practical guide. Data Min. Knowl. Discov. **18**(1), 140–181 (2009)
53. R. Kohavi, A. Deng, B. Frasca, T. Walker, Y. Xu, N. Pohlmann, Online controlled experiments at large scale, in *Proceedings of the 19th ACM SIGKDD International Conference on Knowledge Discovery and Data Mining*, KDD '13, New York, NY, 2013 (ACM, New York, 2013), pp. 1168–1176
54. J.A. Konstan, S.M. McNee, C.-N. Ziegler, R. Torres, N. Kapoor, J. Riedl, Lessons on applying automated recommender systems to information-seeking tasks, in *AAAI*, 2006

55. Y. Koren, R. Bell, C. Volinsky, Matrix factorization techniques for recommender systems. Computer **42**(8), 30–37 (2009)
56. I. Koychev, I. Schwab, Adaptation to drifting user's interests, in *In Proceedings of ECML2000 Workshop: Machine Learning in New Information Age* (2000), pp. 39–46
57. S.K. Lam, J. Riedl, Shilling recommender systems for fun and profit, in *WWW '04: Proceedings of the 13th International Conference on World Wide Web*, New York, NY (ACM, New York, 2004), pp. 393–402
58. S.K. Lam, D. Frankowski, J. Riedl, Do you trust your recommendations? an exploration of security and privacy issues in recommender systems, in *In Proceedings of the 2006 International Conference on Emerging Trends in Information and Communication Security (ETRICS)*, 2006
59. E.L. Lehmann, J.P. Romano, *Testing Statistical Hypotheses*, 3rd edn. Springer Texts in Statistics (Springer, New York, 2005)
60. R. Lempel, Personalization is a two-way street, in *Proceedings of the Eleventh ACM Conference on Recommender Systems* (2017), pp. 3–3
61. T. Mahmood, F. Ricci, Learning and adaptivity in interactive recommender systems. in *ICEC '07: Proceedings of the Ninth International Conference on Electronic Commerce*, New York, NY (ACM, New York, 2007), pp. 75–84
62. A. Maksai, F. Garcin, B. Faltings, Predicting online performance of news recommender systems through richer evaluation metrics, in *Proceedings of the 9th ACM Conference on Recommender Systems* (2015), pp. 179–186
63. B.M. Marlin, R.S. Zemel, Collaborative prediction and ranking with non-random missing data, in *Proceedings of the 2009 ACM Conference on Recommender Systems, RecSys 2009*, New York, NY, October 23–25, 2009, pp. 5–12
64. P. Massa, B. Bhattacharjee, Using trust in recommender systems: An experimental analysis. in *Proceedings of iTrust2004 International Conference* (2004), pp. 221–235
65. M.R. McLaughlin, J.L. Herlocker, A collaborative filtering algorithm and evaluation metric that accurately model the user experience, in *SIGIR '04: Proceedings of the 27th Annual International ACM SIGIR Conference on Research and Development in Information Retrieval*, New York, NY (ACM, New York, 2004), pp. 329–336
66. H.B. McMahan, G. Holt, D. Sculley, M. Young, D. Ebner, J. Grady, L. Nie, T. Phillips, E. Davydov, D. Golovin, et al., Ad click prediction: a view from the trenches, in *Proceedings of the 19th ACM SIGKDD International Conference on Knowledge Discovery and Data Mining* (2013), pp. 1222–1230
67. S.M. McNee, J. Riedl, J.A. Konstan, Making recommendations better: an analytic model for human-recommender interaction. in *CHI '06: CHI '06 Extended Abstracts on Human Factors in Computing Systems*, New York, NY, 2006 (ACM, New York, 2006), pp. 1103–1108
68. F. McSherry, I. Mironov, Differentially private recommender systems: building privacy into the netflix prize contenders. in *KDD '09: Proceedings of the 15th ACM SIGKDD International Conference on Knowledge Discovery and Data Mining*, New York, NY (ACM, New york, 2009), pp. 627–636
69. B. Mobasher, R. Burke, R. Bhaumik, C. Williams, Toward trustworthy recommender systems: an analysis of attack models and algorithm robustness. ACM Trans. Internet Technol. **7**(4), 23 (2007)
70. T. Murakami, K. Mori, R. Orihara, Metrics for evaluating the serendipity of recommendation lists. New Front. Artif. Intell. **4914**, 40–46 (2008)
71. T.T. Nguyen, D. Kluver, T.-Y. Wang, P.-M. Hui, M.D. Ekstrand, M.C. Willemsen, J. Riedl, Rating support interfaces to improve user experience and recommender accuracy, in *Proceedings of the 7th ACM Conference on Recommender Systems*, RecSys '13, New York, NY (ACM, New York, 2013), pp. 149–156
72. M. O'Mahony, N. Hurley, N. Kushmerick, G. Silvestre, Collaborative recommendation: a robustness analysis. ACM Trans. Internet Technol. **4**(4), 344–377 (2004)
73. S.L. Pfleeger, B.A. Kitchenham, Principles of survey research. SIGSOFT Softw. Eng. Notes **26**(6), 16–18 (2001)

74. P. Pu, L. Chen, Trust building with explanation interfaces, in *IUI '06: Proceedings of the 11th International Conference on Intelligent User Interfaces*, New York, NY, 2006 (ACM, New York, 2006), pp. 93–100

75. P. Pu, L. Chen, R. Hu, A user-centric evaluation framework for recommender systems, in *Proceedings of the Fifth ACM Conference on Recommender Systems*, RecSys '11, New York, NY (ACM, New York, 2011), pp. 157–164

76. P. Pu, L. Chen, R. Hu, A user-centric evaluation framework for recommender systems, in *Proceedings of the Fifth ACM Conference on Recommender Systems* (2011), pp. 157–164

77. S. Queiroz, Adaptive preference elicitation for top-k recommendation tasks using gai-networks, in *AIAP'07: Proceedings of the 25th Conference on Proceedings of the 25th IASTED International Multi-Conference*, Anaheim, CA, 2007 (ACTA Press, Calgary, 2007), pp. 579–584

78. S. Rendle, C. Freudenthaler, Z. Gantner, L. Schmidt-Thieme, BPR: Bayesian personalized ranking from implicit feedback, in *UAI '09: Proceedings of the Twenty-Fifth Conference on Uncertainty in Artificial Intelligence*, 2009

79. F. Ricci, Recommender systems in tourism, in *Handbook of e-Tourism* (Springer, Cham, 2020), pp. 1–18

80. M. Rossetti, F. Stella, M. Zanker, Contrasting offline and online results when evaluating recommendation algorithms, in *Proceedings of the 10th ACM Conference on Recommender Systems* (2016), pp. 31–34

81. Margaret L Russell, Donna G Moralejo, and Ellen D Burgess. Paying research subjects: participants' perspectives. J. Med. Ethics **26**(2), 126–130 (2000)

82. A. Said, A short history of the recsys challenge. AI Mag. **37**(4), 102–104 (2017)

83. A. Said, A. Bellogín, Comparative recommender system evaluation: benchmarking recommendation frameworks, in *Proceedings of the 8th ACM Conference on Recommender Systems* (2014), pp. 129–136

84. A. Said, A. Bellogín, Rival: a toolkit to foster reproducibility in recommender system evaluation, in *Proceedings of the 8th ACM Conference on Recommender systems* (2014), pp. 371–372

85. S.L. Salzberg, On comparing classifiers: Pitfalls toavoid and a recommended approach. Data Min. Knowl. Discov. **1**(3), 317–328 (1997)

86. M.R. Santana, L.C. Melo, F.H.F. Camargo, B. Brandão, A. Soares, R.M. Oliveira, S. Caetano, Mars-gym: a gym framework to model, train, and evaluate recommender systems for marketplaces (2020). Preprint. arXiv:2010.07035

87. B. Sarwar, G. Karypis, J. Konstan, J. Riedl, Analysis of recommendation algorithms for e-commerce, in *EC '00: Proceedings of the 2nd ACM Conference on Electronic Commerce*, New York, NY (ACM, New York, 2000), pp. 158–167

88. B. Sarwar, G. Karypis, J. Konstan, J. Reidl, Item-based collaborative filtering recommendation algorithms. in *WWW '01: Proceedings of the 10th International Conference on World Wide Web*, New York, NY (ACM, New York, 2001), pp. 285–295

89. B. Sarwar, G. Karypis, J. Konstan, J. Riedl, Item-based collaborative filtering recommendation algorithms. in *Proceedings of the 10th International Conference on World Wide Web* (2001), pp. 285–295

90. A.I. Schein, A. Popescul, L.H. Ungar, D.M. Pennock, Methods and metrics for cold-start recommendations. in *SIGIR '02: Proceedings of the 25th Annual International ACM SIGIR Conference on Research and Development in Information Retrieval*, New York, NY, 2002 (ACM, New York, 2002), pp. 253–260

91. S. Sedhain, A.K. Menon, S. Sanner, L. Xie, Autorec: autoencoders meet collaborative filtering, in *Proceedings of the 24th International Conference on World Wide Web* (2015), pp. 111–112

92. G. Shani, D. Heckerman, R.I. Brafman, An MDP-based recommender system. J. Mach. Learn. Res. **6**, 1265–1295 (2005)

93. G. Shani, D.M. Chickering, C. Meek, Mining recommendations from the web, in *RecSys '08: Proceedings of the 2008 ACM Conference on Recommender Systems* (2008), pp. 35–42

94. G. Shani, L. Rokach, B. Shapira, S. Hadash, M. Tangi, Investigating confidence displays for top-*n* recommendations. JASIST **64**(12), 2548–2563 (2013)
95. N. Silberstein, O. Somekh, Y. Koren, M. Aharon, D. Porat, A. Shahar, T. Wu, Ad close mitigation for improved user experience in native advertisements, in *Proceedings of the 13th International Conference on Web Search and Data Mining* (2020), pp. 546–554
96. B. Smyth, P. McClave, Similarity vs. diversity, in *ICCBR* (2001), pp. 347–361
97. W.J. Spillman, E. Lang, *The Law of Diminishing Returns* (World Book Company, New York, 1924)
98. H. Steck, Item popularity and recommendation accuracy, in *Proceedings of the Fifth ACM Conference on Recommender Systems, RecSys '11*, New York, NY, 2011 (ACM, New york, 2011), pp. 125–132
99. H. Steck, Evaluation of recommendations: rating-prediction and ranking, in *Seventh ACM Conference on Recommender Systems, RecSys '13*, Hong Kong, China, October 12–16, (2013), pp. 213–220
100. K. Swearingen, R. Sinha, Beyond algorithms: An HCI perspective on recommender systems, in *ACM SIGIR 2001 Workshop on Recommender Systems*, 2001
101. C.J. Van Rijsbergen, *Information Retrieval* (Butterworth-Heinemann, Newton, MA, 1979)
102. E.M. Voorhees, The philosophy of information retrieval evaluation, in *CLEF '01: Revised Papers from the Second Workshop of the Cross-Language Evaluation Forum on Evaluation of Cross-Language Information Retrieval Systems* (Springer, London, 2002), pp. 355–370
103. E.M. Voorhees, Overview of trec 2002, in *Proceedings of the 11th Text Retrieval Conference (TREC 2002), NIST Special Publication 500-251* (2002), pp. 1–15
104. Y.Y. Yao, Measuring retrieval effectiveness based on user preference of documents. J. Am. Soc. Inf. Syst. **46**(2), 133–145 (1995)
105. E. Yilmaz, J.A. Aslam, S. Robertson, A new rank correlation coefficient for information retrieval. in *Proceedings of the 31st Annual International ACM SIGIR Conference on Research and Development in Information Retrieval*, SIGIR '08, New York, NY, 2008 (ACM, New York, 2008), pp. 587–594
106. Y. Zeldes, S. Theodorakis, E. Solodnik, A. Rotman, G. Chamiel, D. Friedman, Deep density networks and uncertainty in recommender systems (2017). Preprint. ArXiv:1711.02487
107. M. Zhang, N. Hurley, Avoiding monotony: improving the diversity of recommendation lists, in *RecSys '08: Proceedings of the 2008 ACM Conference on Recommender Systems* (ACM, New York, NY, 2008), pp. 123–130
108. S. Zhang, L. Yao, A. Sun, Y. Tay, Deep learning based recommender system: a survey and new perspectives. ACM Comput. Surv. (CSUR) **52**(1), 1–38 (2019)
109. Y. Zhang, J. Callan, T. Minka, Novelty and redundancy detection in adaptive filtering, in *SIGIR '02: Proceedings of the 25th Annual International ACM SIGIR Conference on Research and Development in Information Retrieval* (ACM, New York, NY, 2002), pp. 81–88
110. C.-N. Ziegler, S.M. McNee, J.A. Konstan, G. Lausen, Improving recommendation lists through topic diversification, in *WWW 05: Proceedings of the 14th International Conference on World Wide Web* (ACM, New York, 2005), pp. 22–32

# Novelty and Diversity in Recommender Systems

Pablo Castells, Neil Hurley, and Saúl Vargas

## 1 Introduction

Accurately predicting users' interests was the main direct or implicit drive of the recommender systems field in roughly the first decade and a half of the field's development. A wider perspective towards recommendation utility, including but beyond prediction accuracy, started to appear in the literature by the beginning of the 2000s [1, 2], taking views that began to realize the importance of novelty and diversity, among other properties, in the added value of recommendation [3, 4]. This realization grew progressively, reaching an upswing of activity by the turn of the past decade [5–9]. Today we might say that novelty and diversity have become an increasingly frequent part of evaluation practice. They are included increasingly often among the reported effectiveness metrics of new recommendation approaches, and are explicitly targeted by algorithmic innovations time and again [10–12]. It seems difficult to conceive of progress in the recommender systems field without considering these dimensions and further developing our understanding of them. Even though dealing with novelty and diversity remains an active area of research

---

This chapter results from work performed at UAM and is not associated with Amazon.

P. Castells (✉)
Universidad Autónoma de Madrid and Amazon, Madrid, Spain
e-mail: pablo.castells@uam.es

N. Hurley
University College Dublin, Dublin, Ireland
e-mail: neil.hurley@ucd.ie

S. Vargas
Infogrid, Tallinn, Estonia
e-mail: hello@saulvargas.es

© Springer Science+Business Media, LLC, part of Springer Nature 2022
F. Ricci et al. (eds.), *Recommender Systems Handbook*,
https://doi.org/10.1007/978-1-0716-2197-4_16

and development, considerable progress has been achieved in these years in terms of the development of enhancement techniques, evaluation metrics, methodologies, and theory.

In this chapter we analyze the different motivations, notions and perspectives under which novelty and diversity can be understood and defined (Sect. 2). We revise the evaluation procedures and metrics which have been developed in this area (Sect. 3), as well as the algorithms and solutions to enhance novelty and/or diversity (Sect. 4). We analyze the relationship with the prolific stream of work on diversity in Information Retrieval, as a confluent area with recommender systems, and discuss a unifying framework that aims to provide a common basis as comprehensive as possible to explain and interrelate different novelty and diversity perspectives (Sect. 5). We discuss further connections to bias and fairness in Sect. 6. We show some empirical results that illustrate the behavior of metrics and algorithms (Sect. 7), and close the chapter with a summary and discussion of the progress and perspectives in this area, and directions for future research (Sect. 8).

## 2  Novelty and Diversity in Recommender Systems

Novelty can be generally understood as the difference between present and past experience, whereas diversity relates to the internal differences within parts of an experience. The difference between the two concepts is subtle and close connections can in fact be established, depending on the point of view one may take, as we shall discuss. The general notions of novelty and diversity can be particularized in different ways. For instance, if a music streaming service recommends us a song we have never heard before, we would say this recommendation brings some novelty. Yet if the song is, say, a very canonical music type by some very well known singer, the involved novelty is considerably less than we would get if the author and style of the music were also original for us. We might also consider that the song is even more novel if, for instance, few of our friends know about it. On the other hand, a music recommendation is diverse if it includes songs of different styles rather than different songs of very similar styles, regardless of whether the songs are original or not for us. Novelty and diversity are thus to some extent complementary dimensions, though we shall seek and discuss in this chapter the relationships between them.

The motivations for enhancing the novelty and diversity of recommendations are manifold, as are the different angles one may take when seeking these qualities. This is also the case in other fields outside information systems, where novelty and diversity are recurrent topics as well, and considerable efforts have been devoted to casting clear definitions, equivalences and distinctions. We therefore start this chapter by overviewing the reasons for and the possible meanings of novelty and diversity in recommender systems, with a brief glance at related perspectives in other disciplines.

## 2.1  Why Novelty and Diversity in Recommendation

Bringing novelty and diversity into play as target properties of the desired outcome means taking a wider perspective on the recommendation problem concerned with final actual recommendation utility, rather than a single quality such as accuracy [3]. Novelty and diversity are not the only dimensions of recommendation utility one should consider aside from relevance (see e.g. Chap. 15 for a comprehensive survey), but they are fundamental ones [13]. The motivations for enhancing novelty and diversity in recommendations are themselves diverse, and can be founded in the system, user and business perspectives.

### 2.1.1  System Perspective

From the system point of view, user actions as implicit evidence of user needs involve a great deal of uncertainty as to what the actual user preferences really are. User clicks and purchases are certainly driven by user interests, but identifying what exactly in an item attracted the user, and generalizing to other items, involves considerable ambiguity. On top of that, system observations are a very limited sample of user activity, whereby recommendation algorithms operate on significantly incomplete knowledge. Furthermore, user interests are complex, highly dynamic, context-dependent, heterogeneous and even contradictory. Predicting the user needs is therefore an inherently difficult task, unavoidably subject to a non-negligible error rate.

Diversity can be a good strategy to cope with this uncertainty and optimize the chances that at least some item pleases the user, by widening the range of possible item types and characteristics at which recommendations aim, rather than bet for a too narrow and risky interpretation of user actions. For instance, a user who has rated the movie "Rango" with the highest value may like it because—in addition to more specific virtues—it is a cartoon, a western, or because it is a comedy. Given the uncertainty about which of the three characteristics may account for the user preference, recommending a movie of each genre generally pays off more than recommending, say three cartoons. Three hits do not necessarily bring three times the gain of one hit—e.g. the user might rent just one recommended movie anyway—, whereas the loss involved in zero hits is considerably worse than achieving a single hit. From this viewpoint we might say that diversity is not necessarily an opposing goal to accuracy, but in fact a strategy to optimize the gain drawn from accuracy and relevance in matching true user needs in an uncertain environment.

### 2.1.2  User Perspective

From the user perspective, novelty and diversity are generally desirable per se, as a direct source of user satisfaction [4, 14–17]. Consumer behaviorists have long studied the natural variety-seeking drive in human behavior [18]. The explanation of this drive is commonly divided into direct and derived motivations. The former refer to the inherent satisfaction obtained from "novelty, unexpectedness, change and complexity" [19], and a genuine "desire for the unfamiliar, for alternation among the familiar, and for information" [20], linking to the existence of an ideal level of stimulation, dependent on the individual. Satiation and decreased satisfaction results from the repeated consumption of a product or product characteristic in a decreasing marginal value pattern [21]. As preferences towards discovered products are developed, consumer behavior converges towards a balance between alternating choices and favoring preferred products [22]. Derived motivations include the existence of multiple needs in people, multiple situations, or changes in people's tastes [18].

Some authors also explain diversity-seeking as a strategy to cope with the uncertainty about one's own future preference when one will actually consume the choices [23], as e.g. when we choose books and music for a trip. Moreover, novel and diverse recommendations enrich the user experience over time, helping expand the user's horizon. It is in fact often the case that we approach a recommender system with the explicit intent of discovering something new, developing new interests, and learning. The potential problems of the lack of diversity which may result from too much personalization has likewise raised issues in recent years; the concern for so-called echo chambers and filter bubbles [24] is familiar nowadays to the general public. Reconciling personalization with a healthy degree of diversity is certainly part of any approach to deal with these problems.

### 2.1.3 Business Perspective

Diversity and novelty also find motivation in the underlying businesses in which recommendation technologies are deployed. Customer satisfaction indirectly benefits the business in the form of increased activity, revenues, and customer loyalty. Beyond this, product diversification is a well-known strategy to mitigate risk and expand businesses [25]. Moreover, selling in the long tail is a strategy to draw profit from market niches by selling less of more and getting higher profit margins on cheaper products [26].

### 2.1.4 The Limits of Novelty and Diversity

All the above general considerations can be of course superseded by particular characteristics of the specific domain, the situation, and the goal of the recommendations, for some of which novelty and diversity are indeed not always needed. For instance, getting a list of similar products to one we are currently inspecting (e.g. a TV set, a holiday rental, etc.) may help us refine our choice among a large set of very similar options. Recommendations can serve as a navigational aid in this type of situation. In other domains, it makes sense to consume the same or very similar items again and again, such as grocery shopping, clothes, etc. The added value of recommendation is probably more limited in such scenarios though, where other kinds of tools may solve our needs (catalog browsers, shopping list assistants, search engines, etc.), and even in these cases we may appreciate some degree of variation in the mix every now and then. We briefly discuss some specific studies on the motivation and effect of recommendation novelty and diversity on actual users later in Sect. 4.7.

## 2.2 Defining Novelty and Diversity

Novelty and diversity are different though related notions, and one finds a rich variety of angles and perspectives on these concepts in the recommender system literature, as well as other fields such as sociology, economics, or ecology. As pointed out at the beginning of this section, novelty generally refers, broadly, to the difference between present and past experience, whereas diversity relates to the internal differences within parts of an experience. Diversity generally applies to a set of items or "pieces", and has to do with how different the items or pieces are with respect to each other. Variants have been defined by considering different pieces and sets of items. In the basic case, diversity is assessed in the set of items recommended to each user separately (and typically averaged over all users afterwards) [4]. But global diversity across sets of sets of items has also been considered, such as the recommendations delivered to all users [6, 27, 28], recommendations by different systems to the same user [29], or recommendations to a user by the same system over time [30].

The novelty of a set of items can be generally defined as a set function (average, minimum, maximum) on the novelty of the items it contains. We may therefore consider novelty as primarily a property of individual items. The novelty of a piece of information generally refers to how different it is with respect to "what has been previously seen" or experienced. This is related to diversity in that when a set is diverse, each item is "novel" with respect to the rest of the set. Moreover, a system that promotes novel results tends to generate global diversity over time in the user experience; and also enhances the global "diversity of sales" from the system perspective. Multiple variants of novelty arise by considering the fact that novelty is relative to a context of experience, as we shall discuss.

Different nuances have been considered in the concept of novelty. A simple definition of novelty can consist of the (binary) absence of an item in the context of reference (prior experience). We may use adjectives such as unknown or unseen for this notion of identity-based novelty [9]. Long tail notions of novelty are elaborations of this concept, as they are defined in terms of the number of users who would specifically know an item [7, 28, 31]. But we may also consider how different or similar an unseen item is with respect to known items, generally—but not necessarily—on a graded scale. Adjectives such as unexpected, surprising and unfamiliar have been used to refer to this variant of novelty [5, 11, 12]. Unfamiliarity and identitary novelty can be related by trivially defining similarity as equality, i.e. two items are "similar" if and only if they are the same item. Finally, the notion of serendipity is used to mean novelty plus a positive emotional response—in other words, an item is serendipitous if it is novel—unknown or unfamiliar—and relevant [32–34].

The present chapter is concerned with the diversity and novelty involved in recommendations, but one might also study the diversity (in tastes, behavior, demographics, etc.) of the end-user population, or the product stock, the sellers, or in general the environment in which recommenders operate. While some works in the field have addressed the diversity in user behavior [35, 36], we will mostly focus on those aspects a recommender system has a direct hold on, namely the properties of its own output.

## 2.3 Diversity in Other Fields

Diversity is a recurrent theme in several fields, such as sociology, psychology, economics, ecology, genetics or telecommunications. One can establish connections and analogies from some—though not all—of them to recommender systems, and some equivalences in certain metrics, as we will discuss.

Diversity is a common keyword in sociology referring to cultural, ethnic or demographic diversity [37]. Analogies to recommender system settings would apply to the user population, which is mainly a given to the system, and therefore not within our main focus here. In economics, diversity is extensively studied in relation to different issues such as the players in a market (diversity vs. oligopolies), the number of different industries in which a firm operates, the variety of products commercialized by a firm, or investment diversity as a means to mitigate the risk involved in the volatility of investment value [25]. Of all such concepts, product and portfolio diversity most closely relate to recommendation, as mentioned in Sect. 2.1.3, as a general risk-mitigating principle and/or business growth strategy.

Behaviorist psychology has also paid extensive attention to the human drive for novelty and diversity [18]. Such studies, especially the ones focusing on consumer behavior, provide formal support to the intuition that recommender system users may prefer to find some degree of variety and surprise in the recommendations they receive, as discussed in Sect. 2.1.2.

An extensive strand or literature is devoted to diversity in ecology as well, where researchers have worked to considerable depth on formalizing the problem, defining and comparing a wide array of diversity metrics, such as the number of species (richness), Gini-Simpson and related indices, or entropy [38, 39]. Such developments connect to aggregate recommendation diversity perspectives that deal with sets of recommendations as a whole, as we shall discuss in Sects. 3.5 and 5.3.3.

Finally, the issue of diversity has also attracted a great deal of attention in the Information Retrieval (IR) field. A solid body of theory, metrics, evaluation methodologies and algorithms has been developed in this scope in the last decades [40–46], including a dedicated search diversity task in four consecutive TREC editions starting in 2009 [47]. Search and recommendation are different problems, but have much in common: both tasks are about ranking a set of items to maximize the satisfaction of a user need, which may or may not have been expressed explicitly. It has in fact been found that the diversity theories and techniques in IR and recommender systems can be connected [48, 49], as we will discuss in Sect. 5.4. Given these connections, and the significant developments on diversity in IR, we find it relevant to include an overview of this work here, as we will do in Sects. 3 (metrics) and 4 (algorithms).

# 3 Novelty and Diversity Evaluation

The definitions discussed in the previous sections can only get a full, precise and practical meaning when one has given a specific definition of the metrics and methodologies by which novelty and diversity are to be measured and evaluated. We review next the approaches and metrics that have been developed to assess novelty and diversity, after which we will turn to the methods and algorithms proposed in the field to enhance them.

## 3.1 Notation

As is common in the literature, we will use the symbols $i$ and $j$ to denote items, $u$ and $v$ for users, $\mathcal{I}$ and $\mathcal{U}$ for the set of all items and users respectively. By $\mathcal{I}_u$ and $\mathcal{U}_i$ we shall denote, respectively, the set of all items $u$ has interacted with, and the set of users who have interacted with $i$. In general we shall take the case where the interaction consists of rating assignment (i.e. at most one time per user-item pair), except where the distinction between single and multiple interaction makes a relevant difference (namely Sect. 5.2.1). We denote ratings assigned by users to items as $r(u, i)$, and use the notation $r(u, i) = \emptyset$ to indicate missing ratings, as in [50]. We shall use $R$ to denote a recommendation to some user, and $R_u$ whenever we wish or need to explicitly indicate the target user $u$ to whom $R$ is delivered—in other words, $R$ will be a shorthand for $R_u$. By default, the definition of a metric will be given on a single recommendation for a specific target user. For notational simplicity, we omit as understood that the metric should be averaged over all users. Certain global metrics (such as aggregate diversity, defined in Sect. 3.5) are the exception to this rule: they directly take in the recommendations to all users in their definition, and they therefore do not require averaging. In some cases where a metric is the average of a certain named function (e.g. IUF for inverse user frequency, SI for self-information) on the items it contains, we will compose the name of the metric by prepending an "M" for "mean" (e.g. MIUF, MSI) in order to distinguish it from the item-level function.

## 3.2  Average Intra-List Distance

Perhaps the most frequently considered diversity metric and the first to be proposed in the area is the so-called average intra-list distance—or intra-list diversity, ILD (e.g. [2, 4, 51]). The intra-list diversity of a set of recommended items is defined as the average pairwise distance of the items in the set:

$$\text{ILD} = \frac{1}{|R|(|R| - 1)} \sum_{i \in R} \sum_{j \in R} d(i, j)$$

The computation of ILD requires defining a distance measure $d(i, j)$, which is thus a configurable element of the metric. Given the profuse work on the development of similarity functions in the recommender systems field, it is common, handy and sensible to define the distance as the complement of well-understood similarity measures, but nothing prevents the consideration of other particular options. The distance between items is generally a function of item features [4], though the distance in terms of interaction patterns by users has also been considered sometimes [52].

The ILD scheme in the context of recommendation was first suggested, as far as we are aware of, by Smyth and McClave [2], and has been used in numerous subsequent works (e.g. [4, 9, 51, 52]). Some authors have defined this dimension by its equivalent complement intra-list similarity ILS [4], which has the same relation to ILD as the distance function has to similarity, e.g. $\text{ILD} = 1 - \text{ILS}$ if $d = 1 - sim$.

## 3.3  Global Long-Tail Novelty

The novelty of an item from a global perspective can be defined as the opposite of popularity: an item is novel if few people are aware it exists, i.e. the item is far in the long tail of the popularity distribution [7, 31]. Zhou et al. [28] modeled popularity as the probability that a random user would know the item. To get a decreasing function of popularity, the negative logarithm provides a nice analogy with the inverse document frequency (IDF) in the vector-space Information Retrieval model, with users in place of documents and items instead of words, which has been referred to as inverse user frequency (IUF) [53]. Based on the observed user-item interaction, this magnitude can be estimated as $\text{IUF} = -\log_2 |\mathcal{U}_i|/|\mathcal{U}|$. Thus the novelty of a recommendation can be assessed as the average IUF of the recommended items:

$$\text{MIUF} = -\frac{1}{|R|} \sum_{i \in R} \log_2 \frac{|\mathcal{U}_i|}{|\mathcal{U}|} \tag{1}$$

The IUF formula also has a resemblance to the self-information measure of Information Theory, only for that to be properly the case, the probability should add to 1 over the set of items, which is not the case here. We discuss that possibility in Sect. 5.2.1.

## 3.4   User-Specific Unexpectedness

Long-tail novelty translates to non-personalized measures for which the novelty of an item is seen as independent of the target user. It makes sense however to consider the specific experience of a user when assessing the novelty carried by an item that is recommended to her, since the degree to which an item is more or less familiar can greatly vary from one user to the next.

Two perspectives can be considered when comparing an item to prior user experience: the item identity (was this particular item seen before?) or the item characteristics (were the attributes of the item experienced before?). In the former view, novelty is a Boolean property of an item which occurs or not in its totality, whereas the latter allows to appreciate different degrees of novelty in an item even if it was never, itself, seen before.

It is not straightforward to define identity-based novelty on an individual user basis. In usual scenarios, if the system observes the user interact with an item, it will avoid recommending her this item again.[1] This is a rather trivial feature and does not need to be evaluated—if anything, just debugged (e.g. for near-duplicate detection). We may therefore take it for granted, except in particular scenarios where users recurrently consume items—where on the other hand a recommender system may have a more limited range for bringing added value. It would be meaningful though to assess the Boolean novelty of an item in probabilistic terms, considering the user activity outside the system, which in a detailed sense is of course impractical. Long tail novelty can be seen as a proxy for this notion: a user-independent estimate of the prior probability that the user—any user—has seen the item before. Finer, user-specific probability estimation approaches could be explored but have not, to the best of our knowledge, been developed in the literature so far.

An attribute-based perspective is an easier-to-compute alternative for a user-specific novelty definition. Taking the items the user has been observed to encounter, the novelty of an item can be defined in terms of how different it is to the previously encountered items, as assessed by some distance function on the item properties. This notion reflects how unfamiliar, unexpected and/or surprising an item may be based on the user's observed experience. The set-wise distance to the profile items can be defined by aggregation of the pairwise distances by an average, minimum, or other suitable function. For instance, as an average:

---

[1] Of course, what "interaction" means and to what extent it will inhibit future recommendations is application-dependent, e.g. an online store may recommend an item the user has inspected but not bought.

$$\text{Unexp} = \frac{1}{|R||\mathcal{I}_u|} \sum_{i \in R} \sum_{j \in \mathcal{I}_u} d(i, j).$$

Some authors have generalized the notion of unexpectedness to the difference of a recommendation with respect to an expected set of items, not necessarily the ones in the target user profile, thus widening the perspective on what "expected" means [5, 12, 32, 54]. For instance, Murakami et al. [32] define the expected set as the items recommended by a "primitive" system which is supposed to produce unsurprising recommendations. The difference to the expected set can be defined in several ways, such as the ratio of unexpected recommended items:

$$\text{Unexp} = |R - EX|/|R| \tag{2}$$

$EX$ being the set of expected items. Other measures between the recommended and expected set include the Jaccard distance, the centroid distance, the difference to an ideal distance, etc. [5].

## 3.5  Inter-Recommendation Diversity Metrics

In the early 2010s Adomavicius and Kwon [6, 27] proposed measuring the so-called aggregate diversity of a recommender system. This perspective is different from all the metrics described above in that it does not apply to a single set of recommended items, but to all the output a recommender system produces over a set of users. It is in fact a quite simple metric which counts the total number of items that the system recommends.

$$\text{Aggdiv} = \left| \bigcup_{u \in \mathcal{U}} R_u \right| \tag{3}$$

A version Aggdiv@$k$ of the metric can be defined by taking $R_u$ as the top $k$ items recommended to $u$. Since it applies to the set of all recommendations, aggregate diversity does not need to be averaged over users, differently from most other metrics mentioned in these pages.

Aggregate diversity is a relevant measure to assess to what extent an item inventory is being exposed to users. The metric, or close variations thereof, have also been referred to as item coverage in other works [1, 29, 54–56] (see also Chap. 15). This concept can be also related to traditional diversity measures such as the Gini coefficient, the Gini-Simpson's index, or entropy [38, 39], which are commonly used to measure statistical dispersion in such fields as ecology (biodiversity in ecosystems), economics (wealth distribution inequality), or sociology (e.g. educational attainment across the population).

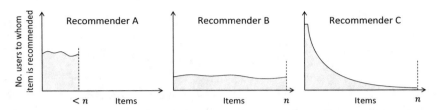

**Fig. 1** By aggregate diversity, recommender B does better than A, but just as well as C. The Gini index, Simpson-Gini and Entropy notice however that B distributes recommendations more evenly over items than C

Mapped to recommendation diversity, such measures take into account not just whether items are recommended to someone, but to how many people and how even or unevenly distributed. To this extent they serve a similar purpose as aggregate diversity as measures of the concentration of recommendations over a few vs. many items. While aggregate diversity counts item exposure to users in a binary way, the Gini index, Simpson-Gini and entropy are more informative as they are sensitive to the amount of users that are recommended each item. Figure 1 illustrates this.

For instance, Fleder and Hosanagar [35] measure sales concentration by the Gini index, which Gunawardana and Shani (Chap. 15) formulate as:

$$\text{Gini} = \frac{1}{|\mathcal{I}| - 1} \sum_{k=1}^{|\mathcal{I}|} (2k - N - 1) p(i_k|s)$$

where $p(i_k|s)$ is the probability of the $k$-th least recommended item being drawn from the recommendation lists generated by a system $s$:

$$p(i|s) = \frac{|\{u \in \mathcal{U} \mid i \in R_u\}|}{\sum_{j \in \mathcal{I}} |\{u \in \mathcal{U} \mid j \in R_u\}|}$$

The Gini index and aggregate diversity have been used in subsequent work such as [57, 58]. Other authors (e.g. [36] or Chap. 15) suggest the Shannon entropy with similar purposes:

$$H = -\sum_{i \in \mathcal{I}} p(i|s) \log_2 p(i|s)$$

Related to this, Zhou et al. [28] observe the diversity of the recommendations across users. They define inter-user diversity (IUD) as the average pairwise Jaccard distance between recommendations to users. In a quite equivalent reformulation of this measure we may define the novelty of an item as the ratio of users to which it is not recommended:[2]

$$\text{IUD} = \frac{1}{|R|} \sum_{i \in R} \frac{|\{v \in \mathcal{U} \mid i \notin R_v\}|}{|\mathcal{U}| - 1} = \frac{1}{|\mathcal{U}| - 1} \sum_{v \in \mathcal{U}} |R - R_v|/|R| \tag{4}$$

Since $|R - R_v|/|R| = 1 - |R \cap R_v|/|R \cup R_v|$, it can be seen that the difference between this definition and the Jaccard-based formulation is basically that the latter has $|R|$ instead of $|R \cup R_v|$ in the denominator, but the above formulation is interesting because it connects to the Gini-Simpson index, as we will show in Sect. 5.3.3.

With a similar metric structure, Bellogín et al. [29] measure the inter-system diversity (ISD), i.e. how different the output of a system is with respect to other systems, in settings where several recommenders are operating. This can be defined as the ratio of systems that do not recommend each item:

$$\text{ISD} = \frac{1}{|R|} \sum_{i \in R} \frac{|\{s \in \mathcal{S} \mid i \notin R^s\}|}{|\mathcal{S}| - 1} = \frac{1}{|\mathcal{S}| - 1} \sum_{s \in \mathcal{S}} |R - R^s|/|R| \tag{5}$$

where $\mathcal{S}$ is the set of recommenders in consideration, and $R^s$ denotes the recommendation to the target user by a system $s \in \mathcal{S}$. This metric thus assesses how different the output of a recommender system is with respect to alternative algorithms. This perspective can be useful, for instance, when an application seeks to distinguish itself from the competition, or when selecting an algorithm to add to an ensemble.

In a different angle, Lathia et al. [30] consider the time dimension in novelty and diversity. Specifically, they study the diversity between successive recommendations by a system to a user, as the ratio of items that were not recommended before:

$$\text{TD} = |R - R'|/|R| \tag{6}$$

The authors distinguish the difference between consecutive recommendations, and the difference between the last recommendation and all prior recommendations. In the former case (which they name "temporal diversity") $R'$ is the recommendation immediately preceding $R$, and in the latter ("temporal novelty") $R'$ is the union of all recommendations to the target user preceding $R$. In both cases, the metric gives a perspective of the ability of a recommender system to evolve with the changes in the environment in which it operates, rather than presenting users the same set of items over and over again.

---

[2] Note that we normalize IUD by $|\mathcal{U}| - 1$ because all items in $R$ are recommended to at least one user (the target of $R$), therefore if we normalized by $|\mathcal{U}|$, the value of the metric for the optimal recommendation would be $(|\mathcal{U}| - 1)/|\mathcal{U}| < 1$. Put another way, $v \in \mathcal{U}$ in the numerator could be as well written as $v \in \mathcal{U} - \{u\}$, which would call for normalizing by $|\mathcal{U} - \{u\}| = |\mathcal{U}| - 1$. The difference is negligible in practice though, and we believe both forms of normalization would be acceptable. The same rationale applies to Eq. 5 next.

Note that IUD, ISD and TD fit as particular cases under the generalized unexpectedness scheme [5] described in the previous section (Eq. 2), where the set $EX$ of expected items would be the items recommended to other users by the same system ($EX = R_v$), to the same user by other systems ($EX = R^s$), or to the same user by the same system in the past ($EX = R'$). One difference is that IUD and ISD take multiple sets $EX$ for each target user (one per user $v$ and one per system $s$ respectively), whereby these metrics involve an additional average over such sets.

In a different perspective, Sanz-Cruzado and Castells [59] research diversity as a global notion across users in the context of social networks, and the effects that recommending people to befriend can have on the evolution of the network structure as a whole. Drawing upon concepts in social network analysis, they consider different diversity angles and how to enhance them. Focusing on weak links as a source of novelty, they analyze the effects that recommending them can have on the diversity of information flowing through the network.

## 3.6 Specific Methodologies

As an alternative to the definition of special-purpose metrics, some authors have evaluated the novelty or diversity of recommendations by accuracy metrics on a diversity-oriented experimental design. For instance, Hurley and Zhang [8] evaluate the diversity of a system by its ability to produce accurate recommendations of difficult items, "difficult" meaning unusual or infrequent for a user's typical observed habits. Specifically, a data splitting procedure is set up by which the test ratings are selected among a ratio of the top most different items rated by each user, "different" being measured as the average distance of the item to all other items in the user profile. The precision of recommendations in such a setting thus reflects the ability of the system to produce good recommendations made up of novel items. A similar idea is to select the test ratings among cold, non-popular long tail items. For instance, Zhou et al. [28] evaluate accuracy on the set of items with less than a given number of ratings. Shani and Gunawardana also discuss this idea in Chap. 15.

## 3.7 Diversity vs. Novelty vs. Serendipity

Even though the distinction between novelty and diversity is not always a fully clean-cut line, We may propose a classification of the metrics described so far as either novelty or diversity measures. ILD can be considered the genuine metric for diversity, the definition of which it applies to the letter. We would also class inter-recommendation metrics (Sect. 3.5) in the diversity type, since they assess how different are recommendations to each other. They do so at a level above an individual recommendation, by (directly or indirectly) comparing sets of recommended items rather than item pairs.

On the other hand, we may consider that long tail and unexpectedness fit in the general definition of novelty: unexpectedness explicitly measures how different each recommended item is with respect to what is expected, where the latter can be related to previous experience. And long tail non-popularity defines the probability that an item is different (is absent) from what a random user may have seen before. The methodologies discussed in the previous section can also be placed in the novelty category, as they assess the ability to properly recommend novel items.

It should also be noted that several authors target the specific concept of serendipity as the conjunction of novelty and relevance [8, 28, 32, 34, 54, 60]. In terms of evaluation metrics, this translates to adding the relevance condition in the computation of the metrics described in Sects. 3.3 and 3.4. In other words, taking the summations over $i \in R \wedge i$ relevant to $u$ in place of just $i \in R$ turns a plain novelty metric (long tail or unexpectedness) into the corresponding serendipity metric.

## 3.8   Information Retrieval Diversity

Differently (at least apparently) from the recommender systems field, diversity in IR has been related to an issue of uncertainty in the user query. Considering that most queries contain some degree of ambiguity or incompleteness as an expression of user needs, diversity is posited as a strategy to cope with this uncertainty by answering as many interpretations of the query as early as possible in the search results ranking. The objective is thus redefined from returning as many relevant results as possible to maximizing the probability that all users (all query interpretations) will get at least some relevant result. This principle is derived from reconsidering the independence assumption on document relevance, whereby returning relevant documents for different query interpretations pays off more than the diminishing returns from additional relevant documents for the same interpretation. For instance a polysemic query such as "table" might be interpreted as furniture or a database concept. If a search engine returns results in only one of the senses, it will satisfy 100% the users who were intending this meaning, and 0% the rest of users. But combining instead a balanced mix of both intents, results will likely meet the needs of most users (getting them well more than half-satisfied), in a typical search where a few relevant results are sufficient to satisfy the user need.

IR diversity metrics have been defined under the assumption that an explicit space of possible query intents (also referred to as query aspects or subtopics) can be represented. In general, the aspects for evaluation should be provided manually, as has been done in the TREC diversity task, where a set of subtopics is provided for each query, along with per-subtopic relevance judgments [47].

Probably the earliest proposed metric was subtopic recall [46], which simply consists of the ratio of query subtopics covered in the search results:

$$S - \text{recall} = \frac{|\{z \in \mathcal{Z} \mid d \in R \wedge d \text{ covers } z\}|}{|\{z \in \mathcal{Z} \mid z \text{ is a subtopic of } q\}|}$$

where $\mathcal{Z}$ is the set of all subtopics. Later on the TREC campaign popularized metrics such as ERR-IA [42] and $\alpha$-nDCG [44], and a fair array of other metrics have been proposed as well. For instance, based on the original definition of ERR in [42], the intent-aware version ERR-IA is:

$$\text{ERR} - \text{IA} = \sum_z p(z|q) \sum_{d_k \in R} p(rel|d_k, z) \prod_{j=1}^{k-1} (1 - p(rel|d_j, z)) \qquad (7)$$

where $p(z|q)$ takes into account that not all aspects need to be equally probable for a query, weighting their contribution to the metric value accordingly. And $p(rel|d, z)$ is the probability that document $d$ is relevant to the aspect $z$ of the query, which can be estimated based on the relevance judgments. E.g. for graded relevance Chapelle [42] proposed $p(rel|d, z) = 2^{g(d,z)-1}/2^{g_{max}}$, where $g(d, z) \in [0, g_{max}]$ is the relevance grade of $d$ for the aspect $z$ of the query. It is also possible to consider simpler mappings, such as a linear map $g(d, z)/g_{max}$, depending on how the relevance grades are defined [9].

Novelty, as understood in recommender systems, has also been addressed in IR, though perhaps not to as much extent as diversity. It is mentioned, for instance, in [61] as the ratio of previously unseen documents in a search result. It is also studied at the level of document sentences, in terms of the non-redundant information that a sentence provides with respect to the rest of the document [62]. Even though the concept is essentially the same, to what extent one may establish connections between the sentence novelty techniques and methodologies, and item novelty in recommendation is not obvious, but might deserve future research.

## 3.9   Proportional Diversity

Closely related to aspect-based diversity in IR, recommendation diversity can be formulated with respect to the different subareas that user interests typically encompass, that we may want to cover in recommendations. An interesting idea in this line is to aim at a coverage of user interest aspects in recommendations that is not necessarily uniform but proportional to a specific desired distribution. A reasonable such distribution can be the proportion of each aspect observed in each user's prior interaction history with items. Vargas et al. [63] and Steck [64] explore different ideas in this direction. The intent-aware scheme [40] in IR, of which the ERR-IA metric [42] described in the previous section is an example, also incorporates the proportionality principle through the $p(z|q)$ term (see Eq. 7), representing the importance of aspect $z$ for the information need $q$—which we can translate to the interests of a user $u$ in a recommendation context. We further discuss later in Sect. 5.4.2 the direct adaptation of IR diversity to recommender systems.

# 4   Novelty and Diversity Enhancement Approaches

Methods to enhance the novelty and diversity of recommendations are reviewed in
this section. It is noteworthy that research in this area has accelerated over the last
number of years. The work can be categorized into methods that re-rank an initial
list to enhance the diversity/novelty of the top items; methods based on clustering;
hybrid or fusion methods; and methods that consider diversity in the context of
learning to rank objectives.

## 4.1   Result Diversification/Re-ranking

One common approach to enhance the diversity of recommendation is the diversification or re-ranking of the results returned by an initial recommender system.
In this approach, a set of candidate recommendations that have been selected on
the basis of relevance, are re-ranked in order to improve the diversity or novelty of
the recommendation, or the aggregate diversity of all recommendations offered by
the system. Generally, work that has taken this approach[4, 51, 65–67] attempts to
optimize the set diversity as expressed by the ILD measure defined in Sect. 3.2.

In the recommendation context, a personalized recommendation is formed for a
given target user $u$, and the relevance of any particular item to the recommendation
depends on $u$. However, for notational simplicity, we will write $f_{rel}(i)$ for the
relevance of item $i$, dropping the dependence on $u$. Given a candidate set $C$, the
problem may be posed to find a set $R \subseteq C$ of some given size $k = |R|$, that
maximizes $div(R)$ i.e.

$$R_{opt} = \underset{R \subseteq C, |R|=k}{\arg \max} \; div(R) \qquad (8)$$

More generally, an objective to jointly optimize for relevance and diversity can
be expressed as:

$$R_{opt}(\lambda) = \underset{R \subseteq C, |R|=k}{\arg \max} \; g(R, \lambda) \qquad (9)$$

where

$$g(R, \lambda) = (1 - \lambda)\frac{1}{|R|} \sum_{i \in R} f_{rel}(i) + \lambda \, div(R)$$

and $\lambda \in [0, 1]$ expresses the trade-off between the average relevance of the items in
the set and the diversity of the set. In Information Retrieval, a greedy construction
approach to solving Eq. 9 is referred to as the maximum marginal relevance (MMR)

**Algorithm 1** Greedy selection to produce a re-ranked list $R$ from an initial set $C$

$R \leftarrow \emptyset$
**while** $|R| < k$ **do**
    $i* \leftarrow \arg\max_{i \in C-R} g(R \cup \{i\}, \lambda)$
    $R \leftarrow R \cup \{i*\}$
**end while**
**return** $R$

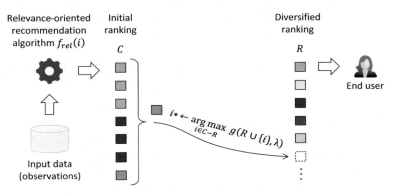

**Fig. 2** Greedy reranking for a tradeoff between relevance and diversity

approach in [41], where relevance is measured with respect to a given query. In the greedy approach, the recommended set $R$ is built in an iterative fashion as follows. Let $R^j$ be the set at iteration $j \in \{1, 2, \ldots, k\}$. The first item in the set is the one that maximizes $f_{rel}(i)$ and the $j$-th item is chosen to maximize $g(R^{j-1} \cup \{i\}, \lambda)$. Algorithm 1 summarizes this approach, illustrated in Fig. 2.

In the context of case-based reasoning, a greedy solution to Eq. 9 is proposed in [2, 68] as a means of selecting a set of cases to solve a given target problem. Using nearest-neighbor user- and item-based collaborative filtering methods to generate the initial candidate set, Ziegler et al. [4] also propose a greedy solution to Eq. 9, as a means of re-ranking the set, terming the method as topic diversification, as they employ a taxonomy-based distance metric. In the context of a publish-subscribe system, Drosou and Pitoura [66] use the formulation in Eq. 9 as a means of selecting a diverse set of relevant items to recommend to a user from a set of items gathered over a particular time window. The method is proposed in the context of image retrieval in [65] and an alternative method to optimize for Eq. 9 is studied in [51], again using an item-based kNN method to generate the candidate set. Also, in [69] a number of different heuristics for solving the maximum diversity problem (Eq. 8) are evaluated and while none out-performs all others in all cases, several succeed in finding very good quality solutions in reasonable time. This work is followed up in [70], where a multiple-pass randomized greedy algorithm is shown to give better performance than the single-pass greedy algorithm.

Rather than maximize as a trade-off between relevance and diversity, [71] takes a more conservative approach of choosing the most diverse subset from a candidate set of items that have equal relevance, thereby maximizing diversity under a constraint of maintaining overall relevance. Similarly, [72] avoids using an explicit weighted trade-off between diversity and relevance and instead presents two algorithms that modify an initial relevance ranking of items to increase diversity.

Though it is difficult to compare directly across the different approaches, as the measures of relevance and pairwise distance differ, researchers have generally found the expected trade-off of increasing diversity and decreasing relevance of the retrieved set as $\lambda$ is decreased towards 0. McGinty and Smyth [2, 68] evaluate the effect of diversifying the recommended set by counting the number of steps it takes a conversational recommender system to reach a given target item. Diversification always performs better than the algorithm that selects items using similarity only. An adaptive method that determines at each step whether or not to diversify gives even better performance. Evaluating on the Book-Crossing dataset, Ziegler et al. [4] found that the accuracy of their system, as measured by precision and recall, dropped with increasing diversification. Zhang and Hurley [51] evaluate on the Movielens dataset; they form test sets of increasing difficulty by splitting each user's profile into training and test sets of items, varying the average similarity of the items in the test set to the items in the training set, and find the diversified algorithm achieves better precision on the more difficult test sets.

**Alternatives to MMR** A number of alternative scoring functions for guiding re-ranking that capture the compromise between relevance and diversity or novelty have been proposed in the literature. For example, [73] computes a weighted sum of a global probability for including a candidate item in $R$ and a local probability dependent on the set of items already in $R$. Definitions of novelty and diversity of news articles based on a distance between concepts in a taxonomy are given in [74] and a replacement heuristic is used to increase the novelty or diversity of the initial ranking by swapping in highly novel/diverse articles. To take account of mutual influence between items, [75] replace the pairwise diversity in the utility function by an estimate of the probability that the pair of items are both liked. Finally, an alternative formulation of the diversity problem encompassing both intra-list dissimilarity and external coverage, is presented in [76] and solved using a greedy algorithm. In this formulation, the items in the set $R$ are selected to cover the candidate, such that there is an item in $R$ within a certain similarity threshold to each item in $C$, under a constraint that all items in $R$ also have a certain minimum pairwise dissimilarity.

**Aggregate Diversity** Targeting aggregate diversity (Eq. 3 in Sect. 3.5), items are re-ranked in [77] using a weighted combination of their relevance and a score based on inverse popularity, or item likeability. Adomavicius and Kwon find that their re-ranking strategies succeed in increasing aggregate diversity at a small cost to accuracy as measured by the precision. A follow-up to this work is presented in [6], in which the aggregate diversity problem is shown to be equivalent to the maximum flow problem on a graph whose nodes are formed from the users and items of the recommendation problem. More recently, Mansoury et al. [56] achieve significant aggregate diversity gains and a good accuracy tradeoff, by a maximum flow approach on the user-item bipartite graph defined by the rating matrix. Other work [78] has investigated how neighborhood filtering strategies and multi-criteria ratings impact on aggregate diversity in nearest-neighbor collaborative filtering algorithms. In this line, Vargas and Castells [58] find out that aggregate diversity is considerably improved by transposing the kNN CF recommendation approach, swapping the role of users and items. The authors show the approach can be generalized to any recommendation algorithm based on a probabilistic reformulation of arbitrary user-item scoring functions which isolates a popularity component.

## 4.2   Using Clustering for Diversification

A method proposed in [79] clusters the items in an active user's profile, in order to group similar items together. Then, rather than recommend a set of items that are similar to the entire user profile, each cluster is treated separately and a set of items most similar to the items in each cluster is retrieved.

A different approach is presented in [80], where the candidate set is again clustered. The goal now is to identify and recommend a set of representative items, one for each cluster, so that the average distance of each item to its representative is minimized.

A nearest-neighbor algorithm is proposed in [81] that uses multi-dimensional clustering to cluster items in an attribute space and select clusters of items as candidates to recommend to the active user. This method is shown to improve aggregate diversity.

A graph-based recommendation approach is described in [82] where the recommendation problem is formulated as a cost flow problem over a graph whose nodes are the users and items of the recommendation. Weights in the graph are computed by a biclustering of the user-item matrix using non-negative matrix factorization. This method can be tuned to increase the diversity of the resulting set, or increase the probability of recommending long-tail items.

## 4.3   Fusion-Based Methods

Since the early days of recommender systems, researchers have been aware that no single recommendation algorithm will work best in all scenarios. Hybrid systems have been studied to offset the strengths of one algorithm against the weaknesses of another (see [83] for example). It may be expected that the combined outputs of multiple recommendation algorithms that have different selection mechanisms, may also exhibit greater diversity than a single algorithm. For example, in [52, 84], recommendation is treated as a multi-objective optimization problem. The outputs of multiple recommendation algorithms that differ in their levels of accuracy, diversity and novelty are ensembled using evolutionary algorithms. As another example, in a music recommendation system called Auralist [33], a basic item-based recommender system is combined with two additional algorithms, in order to promote serendipity (see section below).

## 4.4 Incorporating Diversity in the Ranking Objective

Many recommender algorithms learn a model that optimises an objective representing a loss function that penalises mistakes in the ordering of the items by the model. Several works have examined ways to increase the diversity of the rankings produced by such models. For instance, [85] incorporates a diversity criterion in a regularisation term added to the loss function. In [86], the concept of diversity is integrated into a matrix factorization model, in order to directly recommend item sets that are both relevant and diversified. The loss function of the Bayesian Personalised Ranking (BPR) algorithm [87] is designed to distinguish a set of "relevant items" from a sampled set of "negative" items. Some works e.g. [88, 89] have examined ways to sample the negative items in order to enhance the diversity or novelty of the top $k$ ranked items produced by the model. More recently, Li et al. [10] consider movie and music genre coverage (a diversity dimension) in the expected reward model—the objective function—for a multi-armed bandit recommendation algorithm.

## 4.5 Serendipity: Enabling Surprising Recommendations

A number of algorithms have been proposed in the literature to recommend serendipitous items. For example, in a content-based recommender system, described in [90], a binary classifier is used to distinguish between relevant and irrelevant content. Those items for which the difference in the positive and negative class scores is smallest are determined to be the ones about which the user is most uncertain and therefore the ones that are likely to yield serendipitous recommendations.

Oku and Hattori [91] propose a method for generating serendipitous recommendations that, given a pair of items, uses the pair to generate a recommended set of serendipitous items. Several ways to generate the set are discussed and several ways to rank the items and hence select a top $k$ set are evaluated.

Utility theory is exploited in [5], where the utility of a recommendation is represented as a combination of its utility due to its quality and its utility due to its unexpectedness. A couple of different utility functions are proposed and ways to compute these functions on movie and book recommendation systems are discussed and evaluated.

Other work [92, 93] has investigated the use of graph-based techniques to make serendipitous recommendations in mobile app and music recommendation, respectively.

## 4.6  Other Approaches

A nearest neighbor algorithm called usage-context based collaborative filtering (UCBCF) is presented in [94], which differs from standard item-based CF in the calculation of item-item similarities. Rather than the standard item representation as a vector of user ratings, an item profile is represented as a vector of the $k$ other items with which the item significantly co-occurs in user profiles. UCBCF is shown to obtain greater aggregate diversity than standard kNN and matrix factorization algorithms. A system described in [70] maps items into a utility space and maps a user's preferences to a preferred utility vector. In order to make a diverse recommendation, the utility space is split into $m$ layers in increasing distance from the preferred utility and non-dominated items are chosen from each layer so as to maximize one dimension of the utility vector.

The works discussed so far have considered diversity in terms of the dissimilarity of items in a single recommendation set, or, in the case of aggregate diversity, the coverage of items in a batch of recommendations. Another approach is to consider diversity in the context of the behavior of the system over time. Temporal diversity (Eq. 6 in Sect. 3.5) is investigated by Lathia et al. [30] in a number of standard CF algorithms, and methods for increasing diversity through re-ranking or hybrid fusion are discussed. In a related vein, Mourão et al. [95] explore the "oblivion problem", that is, the possibility that in a dynamic system, items can be forgotten over time in such a way that they recover some degree of the original novelty value they had when they were discovered.

## 4.7  User Studies

It is one thing to develop algorithms to diversify top $N$ lists, but what impact do these algorithms have on user satisfaction? A number of user studies have explored the impact of diversification on users. Topic diversification is evaluated in [4] by carrying out a user survey to assess user satisfaction with a diversified recommendation. In the case of their item-based algorithm, they find that satisfaction peaks around a relevance/diversity determined by $\lambda = 0.6$ in Eq. 9 suggesting that users like a certain degree of diversification in their lists.

While much of the work in diversifying top $N$ lists does not consider the ordering of items in the recommended list, provided an overall relevance is attained, Ge et al.

[54, 96] look at how this ordering affects the user's perception of diversity. In a user study, they experiment with placing diverse items—ones with low similarity to the other items in the list—either in a block or dispersed throughout the list and found that blocking the items in the middle of the list reduces perceived diversity.

The work of Hu and Pu [97] addresses user-interface issues related to augmenting users' perception of diversity. In a user study that tracks eye movements, they find that an organizational interface where items are grouped into categories is better than a list interface in supporting the perception of diversity. In [98], 250 users are surveyed and presented with 5 recommendation approaches, with varying degrees of diversity. They find that users notice differences in diversity and diversity overall improves their satisfaction, but that diverse recommendations may require additional explanations to users who cannot link them back to their preferences.

More recently, in a larger experiment with over 2000 users (after data cleanup) on an e-commerce platform, Chen et al. [34] found serendipity to be a more decisive factor in user satisfaction than novelty or diversity alone. The latter were still important, combined with relevance, as factors of serendipity.

## 4.8 Diversification Approaches in Information Retrieval

Most diversification algorithms proposed in IR follow the same greedy re-ranking scheme as described earlier for recommender systems in Sect. 4.1. The algorithms distinguish from each other in the greedy objective function and the theory behind it. They can be classed into two types based on whether or not the algorithms use an explicit representation of query aspects (as introduced earlier in Sect. 3.8). Explicit approaches draw an approximation of query aspects from different sources, such as query reformulations suggested by a search engine [45], Wikipedia disambiguation entries [99], document classifications [40], or result clustering [100]. Based on this, the objective function of the greedy re-ranking algorithms seeks to maximize the number of covered aspects and minimize the repetition of aspects already covered in previous ranking positions. E.g. xQuAD [45], the most effective algorithm in TREC campaigns, defines its objective function as:

$$f(d_k|S, q) = (1 - \lambda)\, p(q|d_k) + \lambda \sum_z p(z|q)\, p(d_k|q, z) \prod_{j=1}^{k-1} (1 - p(d_j|q, z))$$

where $p(q|d_k)$ stands for the initial search system score, $z$ represents query aspects, $p(z|q)$ weights the contribution on each aspect by its relation to the query, $p(d_k|q, z)$ measures how well document $d_k$ covers aspect $z$, the product after that penalizes the redundancy with previous documents in the ranking covering the same aspect, and $\lambda \in [0, 1]$ sets the balance in the intensity of diversification.

Diversification algorithms that do not explicitly deal with query aspects generally assess diversity in terms of the content of documents. For instance Goldstein and

Carbonell [41] greedily maximize a linear combination of similarity to the query (the baseline search score) and dissimilarity (minimum or average distance) to the documents ranked above the next document. Other non-aspect approaches formulate a similar principle in more formal probabilistic terms [43], or in terms of the trade-off between risk and relevance, in analogy to Modern Portfolio Theory [101] on the optimization of the expected return for a given amount of risk in financial investment. Vargas et al. [48, 49] show that IR diversity principles and techniques make sense in recommender systems and can be adapted to them as well, as we discuss in Sect. 5.4.

# 5  Unified View

As the overview through this chapter shows, a wide variety of metrics and perspectives have been developed around the same concepts under different variants and angles. It is natural to wonder whether it is possible to relate them together under a common ground or theory, establishing equivalences, and identifying fundamental differences. We summarize next a formal foundation for defining, explaining, relating and generalizing many different state of the art metrics, and defining new ones. We also examine the connections between diversity as researched and developed in the Information Retrieval field, and the corresponding work in recommender systems.

## 5.1  General Novelty/Diversity Metric Scheme

As shown in [9] it is possible indeed to formulate a formal scheme that unifies and explains most of the metrics proposed in the literature. The scheme posits a generic recommendation metric $m$ as the expected novelty of the items it contains:

$$m = \frac{1}{|R|} \sum_{i \in R} nov(i|\theta)$$

An item novelty model $nov(i|\theta)$ at the core of the scheme determines the nature of the metric that will result. The scheme further emphasizes the relative nature of novelty by explicitly introducing a context $\theta$. Novelty is relative to a context of experience: (what we know about) what someone has experienced somewhere sometime, where "someone" can be the target user, a set of users, all users, etc.; "sometime" can refer to a specific past time period, an ongoing session, "ever", etc.; "somewhere" can be the interaction history of a user, the current recommendation being browsed, past recommendations, recommendations by other systems, "anywhere", etc.; and "what we know about that" refers to the context of observation, i.e. the available observations to the system. We elaborate next on how such models can be defined, computed, and packed into different metrics.

## 5.2   Item Novelty Models

As discussed in Sect. 3.4, the novelty of an item can be established in terms of whether the item itself or its attributes have been experienced before. The first case, which we may refer to as an issue of simple item discovery, calls for a probabilistic formulation, whereas feature-based novelty, which we shall refer to as an issue of item familiarity, can be more easily defined in terms of a distance model.

### 5.2.1   Item Discovery

In the simple discovery approach, $nov(i|\theta)$ can be expressed in terms of the probability that someone has interacted with the item [9]. This probability can be defined from two slightly different perspectives: the probability that a random user has interacted with the item (to which we shall refer as forced discovery) as in IUF (Eq. 1), or the probability that the item is involved in a random interaction (free discovery). Both can be estimated based on the amount of interactions with the item observed in the system, as a sample of all the interactions the item may have received in the real world. We shall use the notation $p(known|i, \theta)$—the probability that "$i$ is known" by any user given a context $\theta$—for forced discovery, and $p(i|known, \theta)$ for free discovery. Note that these are different distributions, e.g. the latter sums to 1 over the set of all items, whereas the former sums to 1 with $p(\neg known|i, \theta)$. Forced discovery reflects the probability that a random user knows a specific item when asked about it, whereas free discovery is the probability that "the next item" someone discovers is precisely the given item. It is shown in [9] that the metrics induced by either model are quite equivalent in practice, as the two distributions are approximately proportional to each other (exactly proportional if the frequency of user-item pairs is uniform, as is the case e.g. with one-time ratings). In Sect. 7 we shall show some empirical results which confirm this near equivalence in practice.

Now depending on how we instantiate the context $\theta$, we can model different novelty perspectives. For instance, if we take $\theta$ to be the set of available observations of user-item interaction (to be more rigorous, we take $\theta$ to be an unknown user-item interaction distribution of which the observed interactions are a sample), maximum-likelihood estimates of the above distributions yield:

$$p(known|i, \theta) \sim |\{u \in \mathcal{U} \mid \exists t \ (u, i, t) \in \mathcal{O}\}| \ / \ |\mathcal{U}| = |\mathcal{U}_i|/|\mathcal{U}|$$

$$p(i|known, \theta) \sim |\{(u, i, t) \in \mathcal{O}\}| \ / \ |\mathcal{O}| \tag{10}$$

where $\mathcal{U}_i$ denotes the set of all users who have interacted with $i$, and $\mathcal{O} \subset \mathcal{U} \times \mathcal{I} \times \mathcal{T}$ is the set of observed item-user interactions with $i$ (each labeled with a different timestamp $t \in \mathcal{T}$). If the observations consist of ratings, user-item pairs occur only once, and we have:

$$p(known|i, \theta) \sim |\mathcal{U}_i|/|\mathcal{U}| = |\{u \in \mathcal{U} \mid r(u, i) \neq \emptyset\}| / |\mathcal{U}|$$

$$p(i|known, \theta) \sim |\{u \in \mathcal{U} \mid r(u, i) \neq \emptyset\}| / |\mathcal{O}| = |\mathcal{U}_i|/|\mathcal{O}| \tag{11}$$

Both $p(i|known, \theta)$ and $p(known|i, \theta)$ make sense as a measure of how popular an item is in the context at hand. In order to build a recommendation novelty metric based on this, we should take $nov(i|\theta)$ to be a monotonically decreasing function of these probabilities. The inverse probability, dampened by the logarithm function (i.e. $-\log_2 p$) is frequent in the literature [9, 28], but $1 - p$ is also reported as "popularity complement" [9, 52]. The latter has an intuitive interpretation when applied to forced discovery: it represents the probability that an item is not known to a random user. The former also has interesting connections: when applied to forced discovery, it gives the inverse user frequency IUF (see Sect. 3.3). When applied to free discovery, it becomes the self-information (also known as surprisal), an information theory measure that quantifies the amount of information conveyed by the observation of an event.

### 5.2.2 Item Familiarity

The novelty model scheme defined in the previous section considers how different an item is from past experience in terms of strict Boolean identity: an item is new if it is absent from past experience ($known = 0$) and not new otherwise ($known = 1$). There are reasons however to consider relaxed versions of the Boolean view: the knowledge available to the system about what users have seen is partial, and therefore an item might be familiar to a user even if no interaction between them has been observed in the system. Furthermore, even when a user sees an item for the first time, the resulting information gain—the effective novelty—ranges in practice over a gradual rather than binary scale (consider for instance the novelty involved in discovering the movie "Rocky V").

As an alternative to the popularity-based view, we consider a similarity-based model where item novelty is defined by a distance function between the item and a context of experience [9]. If the context can be represented as a set of items, for which we will intentionally reuse the symbol $\theta$, we can formulate this as the distance between the item and the set, which can be defined as an aggregation of the distances to the items in the set, e.g. as the expected value:

$$nov(i|\theta) = \sum_{j \in \theta} p(j|\theta) \, d(i, j)$$

The $p(j|\theta)$ probability enables further model elaborations, or can be simply taken as uniform thus defining a plain distance average.

In the context of distance-based novelty, we find two useful instantiations of the $\theta$ reference set: (a) the set of items a user has interacted with—i.e. the items in his profile—, and (b) the set $R$ of recommended items itself. In the first case, we get a user-relative novelty model, and in the second case, we get the basis for a generalization of intra-list diversity. The notion of expected set in [5] plays a similar role to this idea of $\theta$ context. It is possible to explore other possibilities for $\theta$, such as groups of user profiles, browsed items over an interactive session, items recommended in the past or by alternative systems, etc., which might motivate future work.

## 5.3   Resulting Metrics

As stated at the beginning of this section, having defined a model of the novelty of an item, the novelty or diversity of a recommendation can be defined as the average novelty of the items it contains [9]. Each novelty model, and each context instantiation produce a different metric. In the following we show some practical instantiations that give rise to (hence unify and generalize) metrics described in the literature and covered in Sect. 3. Figure 3 informally illustrates the unified vision we develop in the next subsections.

### 5.3.1   Discovery-Based

A practical instantiation of the item discovery models described in Sect. 5.2.1 consists of taking the novelty context $\theta$ to be the set of user-item interactions

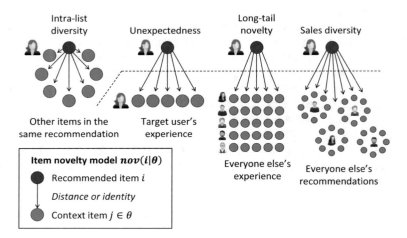

**Fig. 3** Different contexts of reference for item novelty reflect different diversity and novelty notions and result in different metrics

observed by the system. The different discussed variants in the novelty model result in the following practical metric combinations (mean IUF, mean self-information, mean popularity complement):

$$\text{MIUF} = -\frac{1}{|R|} \sum_{i \in R} \log_2 p(known|i, \theta)$$

$$\text{MSI} = -\frac{1}{|R|} \sum_{i \in R} \log_2 p(i|known, \theta) \tag{12}$$

$$\text{MPC} = \frac{1}{|R|} \sum_{i \in R} (1 - p(known|i, \theta))$$

where the probabilities are estimated by Eqs. 10 or 11 depending on the nature of the data. MPC has the advantage of simplicity, a clear interpretation (the ratio of unknown recommended items), and ranges in [0, 1]. MIUF generalizes the metric proposed by Zhou et al. [28] (Eq. 1 in Sect. 3.3), and MSI provides a nice connection to information theory concepts. MPC has the potential shortcoming of a tendency to concentrate its values in a small range near 1, whereas MIUF and MSI deliver less clumped values. We might as well consider the expected popularity complement of free discovery, but that does not have a particularly interesting interpretation or property with respect to the other metrics. In fact, given the discussed near equivalence of free and forced discovery, the three above metrics behave quite similarly to each other, as we will illustrate in Sect. 7 for MSI and MPC.

### 5.3.2   Familiarity-Based

Distance based item novelty models give rise to intra-list diversity and unexpect-edness metrics. As mentioned in Sect. 5.2.2, these metrics simply result from taking, respectively, the recommended items or the target user's profile as the $\theta$ novelty context. The complement of any similarity function between items is potentially suitable to define the distance measure. For instance, with feature-based similarity we may define $d(i, j) = 1 - \cos(i, j)$ for numeric item features, $d(i, j) = 1 - \text{Jaccard}(i, j)$ for Boolean (or binarized) features, and so forth. The distinction between collaborative and content-based similarity deserves attention though, and care should be taken to make a meaningful choice between these two alternatives. Content-based similarity compares items by their intrinsic properties, as described by the available item features. Even though a collaborative similarity measure (which compares items by their common user interaction patterns) might make sense in some particular cases, we would contend that content-based similarity is generally more meaningful to assess the diversity in a way that users can perceive.

### 5.3.3 Further Unification

By explicitly modeling novelty as a relative notion, the proposed framework has a strong unifying potential of further novelty and diversity conceptions. Take for instance the notion of temporal diversity [30] discussed in Sect. 3.5. The metric can be described in the framework in terms of a discovery model where the source of discovery is the past recommendations of the system $\theta \equiv R'$, and novelty is defined as the complement of forced discovery given this context:

$$\frac{1}{|R|} \sum i \in R(1 - p(known|i, R')) = \frac{1}{|R|} \sum_{i \in R} (1 - [i \in R']) = \frac{1}{|R|}|R - R'| = \text{TD}$$

Similarly, for inter-user diversity (Eq. 4 in Sect. 3.5), we take as context the set of recommendations to all users in the system, $\theta \equiv \{R_v | v \in \mathcal{U}\}$. By marginalization over users, and assuming a uniform user prior $p(v) = 1/|\mathcal{U}|$ we have:

$$\frac{1}{|R|} \sum_{i \in R} (1 - p(known \mid i, \{R_v | v \in \mathcal{U}\})) = \frac{1}{|R|} \sum_{i \in R} \sum_{v \in \mathcal{U}} (1 - p(known|i, R_v))p(v)$$

$$= \frac{1}{|R||\mathcal{U}|} \sum_{v \in \mathcal{U}} \sum_{i \in R} (1 - p(known|i, R_v)) = \frac{1}{|R||\mathcal{U}|} \sum_{v \in \mathcal{U}} |R - R_v| = \text{IUD}$$

Inter-system novelty can be obtained in a similar way. So can generalized unexpectedness metrics, in their set difference form (Eq. 2 in Sect. 3.4), by using the expected set as context $\theta$ in place of $R'$, $R_v$ or $R_s$ above.

Biodiversity measures from ecology can also be directly related to some of the recommendation metrics we have discussed. The equivalences hold by equating items to species, and the occurrence of an item in a recommendation as the existence of an individual of the species. In particular, stated in this way, aggregate diversity is the direct equivalent of so called richness, the number of different species that are present in an ecosystem [38, 39].

On the other hand, it can be seen that the Gini-Simpson index (GSI) is exactly equivalent to inter-user diversity. GSI is defined as the probability that two items (individuals) picked at random from the set of recommendations (ecosystem) are different items (species) [38, 39], which can be expressed as a sum over items, or as an average over pairs of recommendations:

$$\text{GSI} = 1 - \sum_{i \in \mathcal{I}} \frac{|\{u \in \mathcal{U} \mid i \in R_u\}|^2}{|\mathcal{U}|(|\mathcal{U}| - 1)k^2} = 1 - \frac{1}{|\mathcal{U}|(|\mathcal{U}| - 1)} \sum_{u \in \mathcal{U}} \sum_{v \in \mathcal{U}} \frac{|R_u \cap R_v|}{|R_u||R_v|}$$

where $k = |R_u|$ assuming they are the same size, or equivalently, considering we are computing GSI@$k$, and we assume item pairs are not sampled from the same recommendation. On the other hand, the average value of IUD over all users is:

**Table 1** Some item novelty and context model instantiations, and the metric they result into

| Metric | Used in | Item novelty model | Context |
|--------|---------|--------------------|---------|
| ILD | [2, 4, 9, 51, 52] | $\sum_{j \in \theta} p(j\|\theta)\, d(i, j)$ | $\theta \equiv R$ |
| Unexp | [5, 9, 32, 54] | | $\theta \equiv$ items in user profile |
| MIUF | [28] | $-\log_2 p(known\|i, \theta)$ | $\theta \equiv$ all observed user-item interaction data |
| MSI | [9] | $-\log_2 p(i\|known, \theta)$ | |
| MPC | [9, 52, 84] | | |
| TD | [30] | | $\theta \equiv$ items recommended in the past |
| IUD/GSI | [28] | $1 - p(known\|i, \theta)$ | $\theta \equiv \{R_u \mid u \in \mathcal{U}\}$ |
| ISD | [29] | | $\theta \equiv \{R^s \mid s \in \mathcal{S}\}$ |

$$\text{IUD} = \frac{1}{|\mathcal{U}|(|\mathcal{U}| - 1)} \sum_{u \in \mathcal{U}} \sum_{v \in \mathcal{U}} \frac{|R_u - R_v|}{|R_u|} = 1 - \frac{1}{|\mathcal{U}|(|\mathcal{U}| - 1)} \sum_{u \in \mathcal{U}} \sum_{v \in \mathcal{U}} \frac{|R_u \cap R_v|}{|R_u|}$$

$$= 1 - k(1 - \text{GSI}) \propto \text{GSI}$$

□

Table 1 summarizes some of the metrics that can be obtained in the unified framework by different instantiations of $\theta$ and item novelty models.

### 5.3.4 Direct Optimization of Novelty Models

The metric scheme described in the previous sections enables the definition of novelty or diversity enhancement re-ranking methods by the greedy optimization of an objective function combining the initial ranking score and the novelty model value, as described earlier in Sect. 4.1:

$$g(i, \lambda) = (1 - \lambda) f_{rel}(i) + \lambda\, nov(i\|\theta)$$

By taking a particular novelty model $nov(i\|\theta)$, one optimizes for the corresponding metric that takes the model at its core. This is an approach to diversity enhancement which is by definition difficult to overcome—in terms of the target metrics—by other re-ranking means.

## 5.4 Connecting Recommendation Diversity and Search Diversity

Recommendation can be formulated as an Information Retrieval task, one where there is no explicit user query. To this extent, and in the aim to find a perspective as comprehensive as possible on the topic at hand in this chapter, it is natural to wonder whether it is possible to establish a connection between the work on diversity in both fields. This question finds affirmative answers in many senses [9, 48, 49]. We summarize here what we find to be the main considerations in this direction: (a) recommendation novelty and diversity can be extended to be sensitive to relevance and rank, (b) IR diversity principles, metrics and algorithms can be adapted to a recommendation setting, and c) personalized search diversity can be formalized as a link between search and recommendation diversity.

### 5.4.1 Rank and Relevance

The novelty and diversity metrics described so far generally lack two aspects: they consider neither the relevance nor the rank position of the items when assessing their contribution to the novelty value of the recommendation. This is in contrast to IR metrics such as ERR-IA and $\alpha$-nDCG which add up the novelty contribution of items only when they are relevant to the query, and apply a rank discount reflecting the assumption that lower ranking positions are less likely to be actually reached by users. Some authors in the recommender systems domain have also proposed to take relevance into account [8, 28, 32, 54, 60], though it is most often not the case, and rank position is generally not taken into account in the reported metrics.

Vargas and Castells [9] show that it is possible to deal with relevance and novelty or diversity together by introducing relevance as an intrinsic feature to the unified metric scheme described in the previous section. This can be done by just replacing "average" by "expected" item novelty at the top level of the scheme, where the novelty of a recommended item should only count when it is actually seen and consumed (chosen, accepted) by the user. The expected novelty is then computed in terms of the probability of choice. If we make the simplifying assumptions that (a) the user chooses an item if and only if she discovers it and likes it, and (b) discovery and relevance are independent, the resulting scheme is:

$$m = C \sum_{i \in R} p(seen|i, R) \; p(rel|i) \; nov(i|\theta) \qquad (13)$$

where $p(rel|i)$ estimates the probability that $i$ is relevant for the user, achieving the desired effect that only relevant novel items count, and $p(seen|i, R)$ estimates the probability that the user will get to see the item $i$ while browsing $R$.

The probability of relevance can be defined based on relevance judgments (test ratings), for instance as $p(rel|i) \sim r(u, i)/r_{max}$, where $r_{max}$ is the maximum possible rating value. Assessing relevance and diversity together has several advantages. It allows for a unified criteria to compare two systems, where separate relevance and novelty metrics may disagree. Furthermore, assessing relevance and novelty together allows distinguishing, for example, between recommendations A and B in Table 2: B can be considered better (relevance-aware MPC = 0.5) since it recommends one useful item (relevant and novel), whereas the items recommended by A (relevance-aware MPC = 0) lack either relevance or novelty. Note that

**Table 2** Toy example recommendations of size two by three systems A, B, C. For each item, the pairs of check and cross marks indicate whether or not the item is relevant (left) and novel (right) to the user (e.g. item 1 of A is relevant and not novel). Below this, the values of MPC are shown with different combinations of rank discount and relevance awareness: plain MPC without relevance or rank discounts, relevance-weighted MPC (without rank discount), and MPC with a Zipfian rank discount (without relevance). The specific expression of the discount function $p(seen|i_k, R)$ and the relevance weight $p(rel|i)$ is shown for each metric variant. The last two rows show the precision of each recommendation, and the harmonic mean of precision and plain MPC

| Rank | A | B | C |
|---|---|---|---|
| 1 | ✓✗ | ✓✓ | ✗✗ |
| 2 | ✗✓ | ✗✗ | ✓✓ |

| Metric | $p(seen \mid i_k, R)$ | $p(rel \mid i)$ | A | B | C |
|---|---|---|---|---|---|
| Plain MPC | 1 | 1 | 1 | 0.5 | 0.5 |
| Relevance-aware MPC | 1 | $r(u,i)/r_{max}$ | 0 | 0.5 | 0.5 |
| Zipfian MPC | $1/k$ | 1 | 0.25 | 0.5 | 0.25 |
| Precision | 1 | $r(u,i)/r_{max}$ | 1 | 0.5 | 0.5 |
| H(Plain MPC, Precision) | – | – | 1 | 0.5 | 0.5 |

an aggregation of separate novelty and relevance metrics would not catch this difference—e.g. the harmonic mean of MPC and precision is 0.5 for both A and B.

On the other hand, the $p(seen|i, R)$ distribution allows the introduction of a browsing model of a user interacting with the ranked recommendations, thus connecting to work on the formalization of utility metrics in IR [42, 102, 103]. The browsing model results in a rank discount which reflects the decreasing probability that the user sees an item as she goes down the ranking. Different models result in discount functions such as logarithmic $p(seen|i_k, R) = 1/\log_2 k$ as in nDCG (see Chap. 15), exponential $p^{k-1}$ as in RBP [103], Zipfian $1/k$ as in ERR [42], and so forth (see [102] for a good compendium and formalization of alternatives). Rank discount allows distinguishing between recommendations B and C in Table 2: B is better (Zipfian MPC $= 0.5$) since it ranks the relevant novel item higher than C (Zipfian MPC $= 0.25$), with higher probability to be seen by the user.

## 5.4.2 IR Diversity in Recommendation

Vargas et al. [48, 49] have shown that the IR diversity principles, metrics and algorithms can be directly applied to recommendation. At a theoretical level, the evidence of user needs implicit in user actions is generally even more ambiguous and incomplete to a recommender system than an explicit query can be to a search system, whereby the rationale of diversifying retrieved results to increase the chances of some relevant result applies here as well.

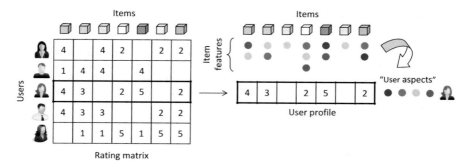

**Fig. 4** User aspects for diversification, as an analogous to query aspects, can be derived from item features and user ratings

At a practical level, it makes as much sense to consider the different aspects of a user's preferences (there are different sides to a person's interests) as it can make for an expressed query. A user interest aspect representation can be drawn from item features in some suitable, meaningful space. This can be done in analogy to the document categories as handled in [40]. Figure 4 illustrates the idea. From this point on, IR diversity metrics such as ERR-IA [42] or subtopic recall [46] can be applied, and aspect-based algorithms such as xQuAD [45] or IA-Select [40] can be adapted, equating users to queries and items to documents. In [104, 105], the aspect representation is explored further, by expressing the trade-off between accuracy and diversity in objectives other than those imported directly from IR, such as the variance minimization objective of modern portfolio theory, as well as proposing an algorithm to learn the aspect probabilities on which these objectives depend.

Non-aspect based diversification methods are applied even more straightforwardly to recommendation, as proved by the equivalence between MMR [41] and methods in the recommender systems field [2, 4] (Sect. 4.1), or the adaptation of the Modern Portfolio Theory from IR [101] to recommendation [82].

### 5.4.3 Personalized Diversity

Recommender and search systems can be seen as extremes in the explicit vs. implicit spectrum of the available evidence of user needs: a recommender system takes no explicit query and relies on observed user choices—implicit evidence of user preferences—as input, whereas basic search systems use just an explicit query. Personalized search represents a middle ground in this spectrum, using both an explicit user query and implicit user feedback and observed actions.

The possibility to consider diversity in the presence of both a query and a user profile has been researched as well [106, 107]. This standpoint has interesting philosophical implications, since personalization can also be seen as a strategy to cope with the uncertainty in a user query. While in a diversification approach the system accepts a situation of uncertainty and adapts its behavior to it, personalization tries to reduce the uncertainty by enhancing the system knowledge about the user need.

Diversity and personalization do not necessarily exclude each other, and can in fact be combined into personalized diversification, as shown in [107]. Vallet and Castells developed and tested a full framework that generalizes IR diversification methods (including xQuAD [45] and IA-Select [40]) into a personalized diversification scheme, by introducing the user as a random variable in the probabilistic formalization of the diversity algorithm [107]. In addition to bridging two theories, the scheme compared favorably empirically to personalization and diversity alone.

## 6   Bias and Fairness

As counterparts of diversity from different perspectives, bias [108–111] and unfairness [112–116] have raised growing concern over the last decade in the recommender systems field, as well as wider disciplines such as Artificial Intelligence or Information Retrieval [117–120]. Fairness in recommendation is covered in Chap. 18, and the reader is referred there for a more comprehensive survey. Bias in recommendation has further angles besides fairness and an in-depth review is beyond the scope of the present chapter We briefly discuss nonetheless here some ways in which novelty, diversity, bias and fairness relate to each other.

### 6.1   Bias in Recommendation

In the context of recommendation, bias can refer to algorithms or to evaluation, and commonly works in opposition to novelty and diversity. A system is said to be biased when it is *systematically* inclined to recommend some items over others, for causes that are—a priori—unrelated to the purpose of recommendation: bringing value to the involved stakeholders (customers, providers, etc.). Seen as a matter of uneven distributions over items this is usually referred to as the *popularity bias*: different researchers have found [109, 121] and explained why [122] many collaborative filtering approaches are structurally biased to concentrate recommendations on majority taste (best sellers, popular trends, etc.), to the detriment of long-tail novelty and sales diversity. The recommendation feedback loop, if not properly handled, adds to this concentration effect [35, 123].

Bias in evaluation means that an experimental procedure is *systematically* favoring some algorithmic approaches over others. Evaluation bias is particularly strong in offline evaluation and is essentially introduced in the sampling of test data for metric evaluation. If test data is sampled from a non-uniform distribution over user-item pairs, algorithms are rewarded for not only predicting user tastes, but also for guessing where the test data is placed—which is irrelevant when assessing the system effectiveness. Ranking-based offline evaluation [108, 110, 111] is commonly exposed to strong item popularity biases, rewarding the behavior of popularity-biased algorithms and encouraging their deployment. From the realization of such biases, new algorithmic approaches [124–126], metrics, and experimental procedures [108, 127–130] have been researched to better cope with biases.

Dealing with popularity biases and caring for novelty are related but different perspectives: the former aims to avoid distortion in evaluating for relevance, while the latter cares about user satisfaction as a direct result of novel experiences, beyond the most popular ones. Interestingly, unexplored equivalences lay between the two perspectives. To see this, let us consider *Inverse propensity scoring* (IPS) [131], a debiasing technique, that is receiving increasing attention in this area [128–130]. IPS corrects for the bias in evaluation by dividing relevance by the "propensity" of relevance to be observed in the computation of offline metrics [129, 130]. For instance, the corrected version of a metric such as precision is:

$$P_{\text{ips}} = \frac{1}{|R|} \sum_{i \in R} \frac{rel_i}{p(obs|i)} \tag{14}$$

where $rel_i = 1$ if $i$ is relevant and zero otherwise, and we make the target user implicit (i.e. the above definition computes precision of a single recommendation, to be averaged then over all users). The term $p(obs|i)$ represents propensity: the probability that the target user is observed interacting with item $i$ as part of the test data for evaluation.

Taking popularity as a simplest, user-independent propensity estimate (as in e.g. [108, 130]) IPS is almost equivalent to relevance-aware MIUF. Combining Eqs. 12 and 13, taking a flat browsing model $p(seen|i, R) = 1$, binary relevance $p(rel|i) = rel_i$, and a typical normalizing constant $C = 1/|R|$, we have:

$$\text{MIUF} = -\frac{1}{|R|} \sum_{i \in R} rel_i \log_2 p(known|i, \theta)$$

$$= \frac{1}{|R|} \sum_{i \in R} rel_i \log_2 \frac{1}{p(known|i, \theta)} \tag{15}$$

If we take a common popularity-based discovery model such as Eqs. 10 or 11 in both the item novelty model and item propensity: $p(known|i, \theta) \equiv |\mathcal{U}_i|/|\mathcal{U}| \equiv p(obs|i)$, we can see the close analogy between Eqs. 14 and 15 above. They are the same except for the logarithm (a monotonic function) in MIUF. In fact, this logarithm could be seen as a form of dampening growth to cope with the well-known high variance problem of IPS and the overdominance of items with smallest $p(obs|i)$, for which modified versions of IPS are often in fact used [127, 128].

We can find sense in this connection between novelty and bias compensation: novelty aims to reward what the user is less likely familiar with, whereas techniques such as IPS aim to reward the recommendation of choices that are less likely to be observed in evaluation. Of course, both conditions are related, since observations of unusual user experiences naturally tend to be scarce in evaluation data.

## 6.2  Fair Recommendation

Biases and poor diversity may have further consequences than a statistical distortion in evaluation or a stale end-user experience. Recommending some items much more frequently than others (low "sales diversity") can be unfair when different providers and creators are involved behind different items (artists on Spotify, vendors on Amazon, owners and operators in the travel and leisure industry, candidates in job recommendation, people in social networks, book authors, research authors, etc.) [112, 114, 115, 132]. Failing to reflect different options, viewpoints or plural opinions (e.g. in recommended news or readings) may work against freedom of choice, critical thinking and healthy societies [24]. In these perspectives, diversity and novelty may not just be a means for business optimization but also an issue of fair opportunity and ethical concern [115, 117, 120, 133].

Fairness can be viewed as a particular case of diversity and as such, some of the views described in the previous sections can be adapted or elaborated upon to measure and enhance different forms of fairness. For instance, fair opportunity can be represented as a particularization and/or elaboration of sales diversity principles discussed in Sect. 3.5. Item vendors, vendor data, news polarity, and other sensitive item features can be represented as item aspects as discussed in Sect. 5.4.2, and handled as diversity objectives [113, 115, 116]. Chapter 18 reviews these and related perspectives in more depth.

## 7  Empirical Metric Comparison

We illustrate the metrics and some of the algorithms described along this chapter with some empirical measurements for a few recommendation algorithms on MovieLens 1M. In the tests we present, ERR-IA and subtopic recall are defined using movie genres as user aspects; ILD and unexpectedness take Jaccard on genres as the distance measure; and aggregate diversity is presented as a ratio over the total number of items, for a better appreciation of differences.

Table 3 shows the metric values for some representative recommendation algorithms: matrix factorization (MF) recommender, based on [134]; a user-based kNN algorithm using the Jaccard similarity, and omitting the normalization by the sum of similarities in the item prediction function; a content-based recommender using movie tags; a most-popular ranking; and random recommendation. We may mainly notice that matrix factorization stands out as the most effective in aggregate diversity and ERR-IA. The latter can be attributed to the fact that this metric takes

**Table 3** Novelty and diversity metrics (at cutoff 10) on a few representative recommendation algorithms in the MovieLens 1M dataset. The highest value of each metric is shown in boldface and table cells are colored in shades of green, darker representing higher values (the values of random recommendation are disregarded in the color and bold font of the rest of rows—though not vice versa—to allow the appreciation of differences excluding the random recommendation)

|         | nDCG   | ILD    | Unexp  | MSI     | MPC    | Aggdiv | IUD    | Entropy | ERR-IA | S-recall |
|---------|--------|--------|--------|---------|--------|--------|--------|---------|--------|----------|
| **MF**     | **0.3161** | 0.6628 | 0.7521 | 9.5908  | 0.8038 | **0.2817** | **0.9584** | **8.5906**  | **0.2033** | 0.5288   |
| **u-kNN**  | 0.2856 | 0.6734 | 0.7785 | 9.0716  | 0.7361 | 0.1589 | 0.8803 | 7.1298  | 0.1800 | **0.5422**   |
| **CB**     | 0.1371 | **0.6825** | 0.7880 | **9.7269**  | **0.8101** | 0.1650 | 0.7762 | 6.2941  | 0.1001 | 0.5378   |
| **PopRec** | 0.1415 | 0.6624 | **0.8451** | 8.5793  | 0.6514 | 0.0183 | 0.4943 | 4.5834  | 0.0773 | 0.5253   |
| **Random** | 0.0043 | **0.7372** | 0.8304 | **13.1067** | **0.9648** | **0.9647** | **0.9971** | **11.7197** | 0.0034 | 0.5055   |

relevance into account, a criteria on which MF achieves the best results of this set (as seen in nDCG).

Content-based recommendation procures the best long tail novelty metrics, confirming a well-known fact [7]. It is not comparably as bad in unexpectedness as one might expect, and this can be attributed to the fact that movie genres (the basis for the unexpectedness distance) and movie tags (the basis for CB similarity) seem not to correlate that much. This, and the good results in terms of ILD can also be related to the suboptimal accuracy of this algorithm as a standalone recommender, which may lend it, albeit to a small degree, some of the flavor of random recommendations. We have checked (outside the reported results) that CB naturally gets the lowest ILD value of all recommenders if the metric uses the same features as the CB algorithm (i.e. movie tags).

Popularity has (almost by definition) the worst results in terms of most novelty and diversity metrics; except in terms of unexpectedness (distance to user profile), which makes sense since this algorithm ignores any data of target users and thus delivers items that are weakly related to the individual profiles. Random recommendation is naturally optimal at most diversity metrics except the one that takes relevance into account (it has low subtopic recall though, because of a bias in MovieLens whereby genre cardinality—therefore subtopic coverage—correlates negatively with popularity). And kNN seems to achieve a good balance of the different metrics. We may also notice that aggregate diversity, IUD and entropy go hand in hand, as one would expect.

In order to give an idea of how related or different the metrics are, Table 4 shows the pairwise Pearson correlation of the metrics on a user basis for the MF recommender. We see that ILD, unexpectedness and subtopic recall tend to go hand in hand, even though they capture different properties as seen previously in the comparison of recommenders in Table 3 (e.g. popularity has very good unexpectedness but very poor ILD). MSI and MPC confirm to be quite equivalent, and IUD (which is equivalent to Gini-Simpson) goes strongly along with these long tail metrics. Note that aggregate diversity and entropy do not have a definition for individual users, and therefore they cannot be included in this table. However, as mentioned before, these measures show strong system-wise correspondence with IUD in Table 3 and, by

**Table 4** Pearson correlation between different metrics (on a user basis) applied to a matrix factorization algorithm on MovieLens 1M. The shades of color (red for negative, green for positive) highlight the magnitude of values

| nDCG | | | | | | | |
|------|------|------|------|------|------|------|-----|
| 0.64 | **ERR-IA** | | | | | | |
| 0.03 | -0.02 | **S-recall** | | | | | |
| 0.03 | -0.09 | 0.71 | **ILD** | | | | |
| 0.07 | -0.06 | 0.62 | 0.85 | **Unexp** | | | |
| 0.02 | 0.09 | -0.21 | -0.21 | -0.20 | **MSI** | | |
| 0.02 | 0.10 | -0.19 | -0.21 | -0.19 | 0.97 | **MPC** | |
| 0.06 | 0.14 | -0.20 | -0.27 | -0.23 | 0.87 | 0.93 | **IUD** |

**Table 5** Novelty and diversity metrics (at cutoff 10) on a few novelty and diversity enhancement algorithms applied to the matrix factorization algorithm on MovieLens 1M. The diversifiers are denoted either by their common name, or by the name of the metric (the item novelty model) they target in their objective function

| | nDCG | ILD | Unexp | MSI | MPC | Aggdiv | ERR-IA | S-recall |
|---|------|------|-------|------|------|--------|--------|----------|
| **MF** | **0.3161** | 0.6628 | 0.7521 | 9.5908 | 0.8038 | 0.2817 | 0.2033 | 0.5288 |
| **+MMR** | 0.2817 | **0.7900** | 0.8089 | 9.6138 | 0.8054 | 0.2744 | 0.1897 | **0.6814** |
| **+Unexp** | 0.2505 | 0.7588 | **0.8467** | 9.6011 | 0.8029 | 0.2483 | 0.1431 | 0.6439 |
| **+MSI** | 0.2309 | 0.6130 | 0.7384 | **10.6995** | **0.8961** | **0.4700** | 0.1583 | 0.4483 |
| **+MPC** | 0.2403 | 0.6233 | 0.7389 | 10.3406 | 0.8818 | 0.3683 | 0.1622 | 0.4696 |
| **+xQuAD** | 0.2726 | 0.6647 | 0.7596 | 9.5784 | 0.8034 | 0.2292 | **0.2063** | 0.6370 |
| **+Random** | 0.0870 | 0.6987 | 0.7698 | 10.2517 | 0.8670 | **0.4836** | 0.0623 | 0.5561 |

transitivity, can be expected to correlate with long-tail metrics as well. The correlation between ERR-IA and nDCG reflects the fact that in addition to aspect diversity ERR-IA takes much into account relevance, which is what nDCG measures.

Finally, and just for the sake of illustration, we see in Table 5 the effect of different novelty/diversity enhancers, applied to the best performing baseline in nDCG, namely matrix factorization. The diversifiers labeled as MMR, Unexp, MSI and MPC are greedy optimizations of the corresponding item novelty model of each metric. xQuAD is an implementation of the algorithm described in [45] using movie genres as aspects, an algorithm which implicitly targets ERR-IA. We arbitrarily set $\lambda = 0.5$ for all algorithms, without a particular motive other than illustrative purposes. We can see that each algorithm maximizes the metric one would expect. The fact that MSI appears to optimize MPC better than MPC itself is because (a) both metrics are almost equivalent, and (b) $\lambda = 0.5$ is not the optimal value for optimization, whereby a small difference seems to tip the scale towards MSI by pure chance. Please note that these results are in no way aiming to approximate optimality or evaluate an approach over another, but rather to exemplify how the different models, metrics and algorithms may work and relate to each other in a simple experiment.

# 8   Conclusion

The consensus is clear in the community on the importance of novelty and diversity as fundamental qualities of recommendations. They are seen today as natural components in the continued improvement and evolution of recommendation technology. Considerable progress has been achieved in the area in defining novelty and diversity from several points of view, devising methodologies and metrics to evaluate them, and developing different methods to enhance them. This chapter aims to provide a wide overview on the work so far, as well as a unifying perspective linking them together as developments from a few basic common root principles.

The area still has wide space for further research. There is room for improving our understanding of the role of novelty and diversity in recommendation, and innovating in theoretical, methodological and algorithmic developments around these dimensions. For instance, modeling feature-based novelty in probabilistic terms in order to unify discovery and familiarity models would be an interesting line for future work. Aspects such as the time dimension, along which items may recover part of their novelty value [95, 135, 136], or the variability among users regarding their degree of novelty-seeking trend [14], are example directions for additional research. Last but not least, continued user studies [4, 15–17, 137] would bring further light on questions such as whether novelty and diversity metrics match the actual user perceptions, the precise extent and conditions in which users appreciate novelty and diversity versus accuracy and other potential dimensions of recommendation effectiveness.

**Acknowledgments** Neil Hurley would like to acknowledge the support of Science Foundation Ireland, under grant no. SFI/12/RC/2289_P2. Pablo Castells was supported by the Spanish Ministry of Science and Innovation, grant no. PID2019-108965GB-I00.

# References

1. J.L. Herlocker, J.A. Konstan, L.G. Terveen, J.T. Riedl, ACM Trans. Inf. Syst. **22**(1), 5–53 (2004)
2. B. Smyth, P. McClave, in *Proceedings of the 4th International Conference on Case-Based Reasoning*, ICCBR 2001 (Springer, London, UK, 2001), pp. 347–361
3. S.M. McNee, J. Riedl, J.A. Konstan, in *CHI 2006 Extended Abstracts on Human Factors in Computing Systems*, CHI EA 2006 (ACM, New York, NY, USA, 2006), pp. 1097–1101
4. C.N. Ziegler, S.M. McNee, J.A. Konstan, G. Lausen, in *Proceedings of the 14th International Conference on World Wide Web*, WWW 2005 (ACM, New York, NY, USA, 2005), pp. 22–32
5. P. Adamopoulos, A. Tuzhilin, Special Issue on Novelty and Diversity in Recommender Systems, ACM Trans. Intell. Syst. Tech. **5**(4) (2014)
6. G. Adomavicius, Y. Kwon, IEEE Trans. Knowl. Data Eng. **24**(5), 896–911 (2012)
7. O. Celma, P. Herrera, in *Proceedings of the 2nd ACM Conference on Recommender Systems*, RecSys 2008 (ACM, New York, NY, USA, 2008), pp. 179–186
8. N. Hurley, M. Zhang, ACM Trans. Internet Tech. **10**(4), 14:1–14:30 (2011)

9. S. Vargas, P. Castells, in *Proceedings of the 5th ACM Conference on Recommender Systems*, RecSys 2011 (ACM, New York, NY, USA, 2011), pp. 109–116
10. C. Li, H. Feng, M.d. Rijke, in *Proceedings of the 14th ACM Conference on Recommender Systems*, RecSys 2020 (ACM, New York, NY, USA, 2020), pp. 33–42
11. P. Li, A. Tuzhilin, ACM Trans. Intell. Syst. Tech. **11**(6) (2020)
12. P. Li, M. Que, Z. Jiang, Y. HU, A. Tuzhilin, in *Proceedings of the 14th ACM Conference on Recommender Systems*, RecSys 2020 (ACM, New York, NY, USA, 2020), pp. 279–288
13. M. Kaminskas, D. Bridge, ACM Trans. Inter. Intell. Syst. **7**(1) (2016)
14. K. Kapoor, V. Kumar, L. Terveen, J.A. Konstan, P. Schrater, in *Proceedings of the 9th ACM Conference on Recommender Systems*, RecSys 2015 (ACM, New York, NY, USA, 2015), pp. 19–26
15. B.P. Knijnenburg, M.C. Willemsen, Z. Gantner, H. Soncu, C. Newell, User Model. User Adap. Inter. **22**(4-5), 441–504 (2012)
16. P. Pu, L. Chen, R. Hu, in *Proceedings of the 5th ACM Conference on Recommender Systems*, RecSys 2011 (ACM, New York, NY, USA, 2011), pp. 157–164
17. M.C. Willemsen, M.P. Graus, B.P. Knijnenburg, User Model. User Adap. Inter. **26**(4), 347–389 (2016)
18. L. McAlister, E.A. Pessemier, J. Consum. Res. **9**(3), 311–322 (1982)
19. S.R. Maddi, in *Theories of Cognitive Consistency: A Sourcebook*, ed. by R.P. Abelson, E. Aronson, W.J. McGuire, T.M. Newcomb, M.J. Rosenberg, P.H. Tannenbaum (Rand McNally, 1968)
20. P.S. Raju, J. Consum. Res. **7**(3), 272–282 (1980)
21. C. Coombs, G.S. Avrunin, Psychological Review **84**(2), 216–230 (1977)
22. P. Brickman, B. D'Amato, J. Pers. Soc. Psychol. **32**(3), 415–420 (1975)
23. B.E. Kahn, J. Intell. Inf. Syst. **2**(3), 139–148 (1995)
24. E. Pariser, *The Filter Bubble: How the New Personalized Web Is Changing What We Read and How We Think* (Penguin Books, 2012)
25. M. Lubatkin, S. Chatterjee, Acad. Manag. J. **37**(1), 109–136 (1994)
26. C. Anderson, *The Long Tail: Why the Future of Business Is Selling Less of More* (Hyperion, 2006)
27. G. Adomavicius, Y. Kwon, INFORMS J. Comput. **26**(2), 351–369 (2014)
28. T. Zhou, Z. Kuscsik, J.G. Liu, M. Medo, J.R. Wakeling, Y.C. Zhang, Proc. Natl. Acad. Sci. **107**(10), 4511–4515 (2010)
29. A. Bellogín, I. Cantador, P. Castells, Information Sciences **221**, 142–169 (2013)
30. N. Lathia, S. Hailes, L. Capra, X. Amatriain, in *Proceedings of the 33rd Annual International ACM SIGIR Conference on Research and Development in Information Retrieval*, SIGIR 2010 (ACM, New York, NY, USA, 2010), pp. 210–217
31. Y.J. Park, A. Tuzhilin, in *Proceedings of the 2th ACM Conference on Recommender Systems*, RecSys 2008 (ACM, New York, NY, USA, 2008), pp. 11–18
32. T. Murakami, K. Mori, R. Orihara, in *New Frontiers in Artificial Intelligence*, ed. by K. Satoh, A. Inokuchi, K. Nagao, T. Kawamura, *Lecture Notes in Computer Science*, vol. 4914 (Springer, Berlin, Heidelberg, 2008), pp. 40–46
33. Y.C. Zhang, D.O. Séaghdha, D. Quercia, T. Jambor, in *Proceedings of the 5th ACM Conference on Web Search and Data Mining*, WSDM 2012 (ACM, New York, NY, USA, 2012), pp. 13–22
34. L. Chen, Y. Yang, N. Wang, K. Yang, Q. Yuan, in *Proceedings of the The World Wide Web Conference*, WWW 2019 (ACM, New York, NY, USA, 2019), pp. 240–250
35. D.M. Fleder, K. Hosanagar, Management Science **55**(5), 697–712 (2009)
36. Z. Szlávik, W. Kowalczyk, M. Schut, in *Proceedings of the 5th AAAI Conference on Weblogs and Social Media*, ICWSM 2011 (The AAAI Press, 2011)
37. D. Levinson, *Ethnic Groups Worldwide: A ready Reference Handbook* (Oryx Press, 1998)
38. G.P. Patil, C. Taillie, J. Am. Stat. Assoc. **77**(379), 548–561 (1982)
39. F. Van Dyke, in *Conservation Biology: Foundations, Concepts, Applications* (Springer Netherlands, Dordrecht, 2008), pp. 83–119

40. R. Agrawal, S. Gollapudi, A. Halverson, S. Ieong, in *Proceedings of the 2nd ACM Conference on Web Search and Data Mining*, WSDM 2009 (ACM, New York, NY, USA, 2009), pp. 5–14
41. J. Carbonell, J. Goldstein, in *Proceedings of the 21st Annual International ACM SIGIR Conference on Research and Development in Information Retrieval*, SIGIR 1998 (ACM, New York, NY, USA, 1998), pp. 335–336
42. O. Chapelle, S. Ji, C. Liao, E. Velipasaoglu, L. Lai, S.L. Wu, Information Retrieval **14**(6), 572–592 (2011)
43. H. Chen, D.R. Karger, in *Proceedings of the 29th Annual International ACM SIGIR Conference on Research and Development in Information Retrieval*, SIGIR 2006 (ACM, New York, NY, USA, 2006), pp. 429–436
44. C.L. Clarke, M. Kolla, G.V. Cormack, O. Vechtomova, A. Ashkan, S. Büttcher, I. MacKinnon, in *Proceedings of the 31st Annual International ACM SIGIR Conference on Research and Development in Information Retrieval*, SIGIR 2008 (ACM, New York, NY, USA, 2008), pp. 659–666
45. R.L. Santos, C. Macdonald, I. Ounis, in *Proceedings of the 19th International Conference on World Wide Web*, WWW 2010 (ACM, New York, NY, USA, 2010), pp. 881–890
46. C.X. Zhai, W.W. Cohen, J. Lafferty, in *Proceedings of the 26th Annual International ACM SIGIR Conference on Research and Development in Information Retrieval*, SIGIR 2003 (ACM, New York, NY, USA, 2003), pp. 10–17
47. C.L. Clarke, N. Craswell, I. Soboroff, in *Proceedings of the 19th Text REtrieval Conference*, TREC 2010 (National Institute of Standards and Technology (NIST), 2010)
48. S. Vargas, P. Castells, D. Vallet, in *Proceedings of the 34th Annual International ACM SIGIR Conference on Research and Development in Information Retrieval*, SIGIR 2011 (ACM, New York, NY, USA, 2011), pp. 1211–1212
49. S. Vargas, P. Castells, D. Vallet, in *Proceedings of the 35th Annual International ACM SIGIR Conference on Research and Development in Information Retrieval*, SIGIR 2012 (ACM, New York, NY, USA, 2012), pp. 75–84
50. G. Adomavicius, A. Tuzhilin, IEEE Trans. Knowl. Data Eng. **17**(6), 734–749 (2005)
51. M. Zhang, N. Hurley, in *Proceedings of the 2nd ACM Conference on Recommender Systems*, RecSys 2008 (ACM, New York, NY, USA, 2008), pp. 123–130
52. A. Veloso, M. Ribeiro, A. Lacerda, E. Moura, I. Hata, N. Ziviani, Special Issue on Novelty and Diversity in Recommender Systems, ACM Trans. Inf. Syst. Tech. **5**(4) (2014)
53. J.S. Breese, D. Heckerman, C. Kadie, in *Proceedings of the 14th Conference on Uncertainty in Artificial Intelligence*, UAI 1998 (Morgan Kaufmann Publishers Inc., San Francisco, CA, USA, 1998), pp. 43–52
54. M. Ge, C. Delgado-Battenfeld, D. Jannach, in *Proceedings of the 4th ACM Conference on Recommender systems*, RecSys 2010 (ACM, New York, NY, USA, 2010), pp. 257–260
55. J.L. Herlocker, J.A. Konstan, A. Borchers, J. Riedl, in *Proceedings of the 22nd Annual International ACM SIGIR Conference on Research and Development in Information Retrieval*, SIGIR 1999 (ACM, New York, NY, USA, 1999), pp. 230–237
56. M. Mansoury, H. Abdollahpouri, M. Pechenizkiy, B. Mobasher, R. Burke, in *Proceedings of the 28th ACM Conference on User Modeling, Adaptation and Personalization*, UMAP 2020 (ACM, New York, NY, USA, 2020), pp. 154–162
57. D. Jannach, L. Lerche, G. Gedikli, G. Bonnin, in *Proceedings of the 21st International Conference on User Modeling, Adaptation and Personalization* (Springer, 2013), pp. 25–37
58. S. Vargas, P. Castells, in *Proceedings of the 8th ACM Conference on Recommender Systems*, RecSys 2014 (ACM, New York, NY, USA, 2014), pp. 145–152
59. J. Sanz-Cruzado, P. Castells, in *Proceedings of the 12th ACM Conference on Recommender Systems*, RecSys 2018 (Vancouver, Canada, 2018), pp. 233–241
60. Y. Zhang, J. Callan, T. Minka, in *Proceedings of the 25th Annual International ACM SIGIR Conference on Research and Development in Information Retrieval*, SIGIR 2002 (ACM, New York, NY, USA, 2002), pp. 81–88
61. R. Baeza-Yates, B. Ribeiro-Neto, *Modern Information Retrieval*, 2nd edn. (Addison-Wesley Publishing Company, USA, 2008)

62. J. Allan, C. Wade, A. Bolivar, in *Proceedings of the 26th Annual International ACM SIGIR Conference on Research and Development in Information Retrieval*, SIGIR 2003 (ACM, New York, NY, USA, 2003), pp. 314–321
63. S. Vargas, L. Baltrunas, A. Karatzoglou, P. Castells, in *Proceedings of the 8th ACM Conference on Recommender Systems*, RecSys 2014 (ACM, New York, NY, USA, 2014), pp. 209–216
64. H. Steck, in *Proceedings of the 12th ACM Conference on Recommender Systems*, RecSys 2018 (ACM, New York, NY, USA, 2018), pp. 154–162
65. T. Deselaers, T. Gass, P. Dreuw, H. Ney, in *Proceedings of the ACM International Conference on Image and Video Retrieval*, CIVR 2009 (ACM, New York, NY, USA, 2009), pp. 39:1–39:8
66. M. Drosou, E. Pitoura, IEEE Data Eng. Bull. **32**(4), 49–56 (2009)
67. M. Drosou, K. Stefanidis, E. Pitoura, in *Proceedings of the 3rd ACM Conference on Distributed Event-Based Systems*, DEBS 2009 (ACM, New York, NY, USA, 2009), pp. 6:1–6:12
68. L. McGinty, B. Smyth, in *Proceedings of the 5th International Conference on Case-based Reasoning*, ICCBR 2003 (Springer, Berlin, Heidelberg, 2003), pp. 276–290
69. M. Drosou, E. Pitoura, Comparing diversity heuristics. Tech. rep., Technical Report 2009-05. Computer Science Department, University of Ioannina (2009)
70. K. Alodhaibi, A. Brodsky, G.A. Mihaila, in *Proceedings of the 43rd Hawaii International Conference on System Sciences*, HICSS 2010 (IEEE Computer Society, Washington, DC, USA, 2010), pp. 1–10
71. D. McSherry, in *Advances in Case-Based Reasoning*, ed. by S. Craw, A. Preece, (Springer, Berlin, Heidelberg, 2002), pp. 219–233
72. C. Yu, L. Lakshmanan, S. Amer-Yahia, in *Proceedings of the 12th International Conference on Extending Database Technology*, EDBT 2009 (ACM, New York, NY, USA, 2009), pp. 368–378
73. Q. Wu, F. Tang, L. Li, L. Barolli, I. You, Y. Luo, H. Li, in *Proceedings of the 26th IEEE Conference on Advanced Information Networking and Applications*, AINA 2012 (IEEE, 2012), pp. 191–198
74. J. Rao, A. Jia, Y. Feng, D. Zhao, in *Web Information Systems Engineering – WISE 2013*, ed. by X. Lin, Y. Manolopoulos, D. Srivastava, G. Huang, *Lecture Notes in Computer Science*, vol. 8180 (Springer, Berlin, Heidelberg, 2013), pp. 209–218
75. A. Bessa, A. Veloso, N. Ziviani, in *String Processing and Information Retrieval*, ed. by O. Kurland, M. Lewenstein, E. Porat, *Lecture Notes in Computer Science*, vol. 8214 (Springer International Publishing, 2013), pp. 17–28
76. M. Drosou, E. Pitoura, Proc. VLDB Endowment **6**(1), 13–24 (2012)
77. G. Adomavicius, Y. Kwon, in *Proceedings of the 1st ACM RecSys Workshop on Novelty and Diversity in Recommender Systems* (2011), DiveRS 2011, pp. 3–10
78. Y. Kwon, J. Intell. Inf. Syst. **18**(3), 119–135 (2012)
79. M. Zhang, N. Hurley, in *Proceedings of the IEEE/WIC/ACM International Joint Conference on Web Intelligence and Intelligent Agent Technology*, WI-IAT 2009 (IEEE Computer Society, Washington, DC, USA, 2009), pp. 508–515
80. R. Boim, T. Milo, S. Novgorodov, in *Proceedings of the 20th ACM Conference on Information and Knowledge Management*, CIKM 2011 (ACM, New York, NY, USA, 2011), pp. 739–744
81. X. Li, T. Murata, in *Proceedings of the 7th International Conference on Computer Science & Education*, ICCSE 2012 (IEEE, 2012), pp. 905–910
82. L. Shi, in *Proceedings of the 7th ACM Conference on Recommender Systems*, RecSys 2013 (ACM, New York, NY, USA, 2013), pp. 57–64
83. J.B. Schafer, J.A. Konstan, J. Riedl, in *Proceedings of the 11th ACM Conference on Information and Knowledge Management*, CIKM 2002 (ACM, New York, NY, USA, 2002), pp. 43–51
84. M.T. Ribeiro, A. Lacerda, A. Veloso, N. Ziviani, in *Proceedings of the 6th ACM Conference on Recommender Systems*, RecSys 2012 (ACM, New York, NY, USA, 2012), pp. 19–26

85. N.J. Hurley, in *Proceedings of the 7th ACM Conference on Recommender Systems*, RecSys 2013 (ACM, New York, NY, USA, 2013), pp. 379–382
86. R. Su, L. Yin, K. Chen, Y. Yu, in *Proceedings of the 7th ACM Conference on Recommender Systems*, RecSys 2013 (ACM, New York, NY, USA, 2013), pp. 415–418
87. S. Rendle, C. Freudenthaler, Z. Gantner, L. Schmidt-Thieme, in *Proceedings of the 25th Conference on Uncertainty in Artificial Intelligence*, UAI 2009 (AUAI Press, Arlington, VA, USA, 2009), pp. 452–461
88. D. Jannach, L. Lerche, I. Kamehkhosh, M. Jugovac, User Model. User Adap. Inter. **25**, 427–491 (2015)
89. J. Wasilewski, N. Hurley, in *Proceedings of the 27th ACM Conference on User Modeling, Adaptation and Personalization*, UMAP 2019 (ACM, New York, NY, USA, 2019), pp. 144–148. https://doi.org/10.1145/3320435.3320468
90. L. Iaquinta, M. de Gemmis, P. Lops, G. Semeraro, M. Filannino, P. Molino, in *Proceedings of the 8th Conference on Hybrid Intelligent Systems*, HIS 2008 (IEEE, 2008), pp. 168–173
91. K. Oku, F. Hattori, in *Proceedings of the 1st ACM RecSys Workshop on Novelty and Diversity in Recommender Systems*, DiveRS 2011 (2011)
92. U. Bhandari, K. Sugiyama, A. Datta, R. Jindal, in *Information Retrieval Technology*, ed. by R.E. Banchs, F. Silvestri, T.Y. Liu, M. Zhang, S. Gao, J. Lang, *Lecture Notes in Computer Science*, vol. 8281 (Springer, Berlin, Heidelberg, 2013), pp. 440–451
93. M. Taramigkou, E. Bothos, K. Christidis, D. Apostolou, G. Mentzas, in *Proceedings of the 7th ACM Conference on Recommender Systems*, RecSys 2013 (ACM, New York, NY, USA, 2013), pp. 335–338
94. K. Niemann, M. Wolpers, in *Proceedings of the 19th ACM SIGKDD Conference on Knowledge Discovery and Data Mining*, KDD 2013 (ACM, New York, NY, USA, 2013), pp. 955–963
95. F. Mourão, C. Fonseca, C. Araújo, W. Meira, in *Proceedings of the 1st ACM RecSys Workshop on Novelty and Diversity in Recommender System*, DiveRS 2011 (2011)
96. M. Ge, D. Jannach, F. Gedikli, M. Hepp, in *Proceedings of the 14th International Conference on Enterprise Information Systems*, ICEIS 2012 (SciTePress, 2012), pp. 201–208
97. R. Hu, P. Pu, in *Proceedings of the 16th International Conference on Intelligent User Interfaces*, IUI 2011 (ACM, New York, NY, USA, 2011), pp. 347–350
98. S. Castagnos, A. Brun, A. Boyer, in *Proceedings of the 3rd International Conference on Advances in Information Mining and Management*, IMMM 2013 (IARIA, Lisbon, Portugal, 2013), pp. 44–50
99. M.J. Welch, J. Cho, C. Olston, in *Proceedings of the 20th International Conference on World Wide Web*, WWW 2011 (ACM, New York, NY, USA, 2011), pp. 237–246
100. J. He, E. Meij, M. de Rijke, J. Assoc. Inf. Sci. Tech. **62**(3), 550–571 (2011)
101. J. Wang, in *Proceedings of the 31st European Conference on Information Retrieval*, ECIR 2009 (Springer, Berlin, Heidelberg, 2009), pp. 4–16
102. B. Carterette, in *Proceedings of the 34th Annual International ACM SIGIR Conference on Research and Development in Information Retrieval*, SIGIR 2011 (ACM, New York, NY, USA, 2011), pp. 903–912
103. A. Moffat, J. Zobel, ACM Trans. Inf. Syst. **27**(1), 2:1–2:27 (2008)
104. J. Wasilewski, N. Hurley, in *Proceedings of the 10th ACM Conference on Recommender Systems*, RecSys 2016 (ACM, New York, NY, USA, 2016), pp. 39–42
105. J. Wasilewski, N. Hurley, in *Adjunct Publication of the 25th Conference on User Modeling, Adaptation and Personalization* (ACM, New York, NY, USA, 2017), pp. 71–76
106. F. Radlinski, S. Dumais, in *Proceedings of the 29th Annual International ACM SIGIR Conference on Research and Development in Information Retrieval*, SIGIR 2006 (ACM, New York, NY, USA, 2006), pp. 691–692
107. D. Vallet, P. Castells, in *Proceedings of the 35th Annual International ACM SIGIR Conference on Research and Development in Information Retrieval*, SIGIR 2012 (ACM, New York, NY, USA, 2012), pp. 841–850

108. H. Steck, in *Proceedings of the 5th ACM Conference on Recommender Systems*, RecSys 2011 (ACM, New York, NY, USA, 2011), pp. 125–132
109. D. Jannach, L. Lerche, I. Kamehkhosh, M. Jugovac, User Model. User Adap. Inter. **25**(5), 427–491 (2015)
110. A. Bellogín, P. Castells, I. Cantador, Information Retrieval **20**(6), 606–634 (2017)
111. R. Cañamares, P. Castells, in *Proceedings of the 41st Annual International ACM SIGIR Conference on Research and Development in Information Retrieval*, SIGIR 2018 (ACM, New York, NY, USA, 2018), pp. 415–424
112. A. Beutel, J. Chen, T. Doshi, H. Qian, L. Wei, Y. Wu, L. Heldt, Z. Zhao, L. Hong, E.H. Chi, C. Goodrow, in *Proceedings of the 25th ACM SIGKDD International Conference on Knowledge Discovery & Data Mining*, KDD 2019 (ACM, New York, NY, USA, 2019), pp. 2212–2220
113. W. Liu, J. Guo, N. Sonboli, R. Burke, S. Zhang, in *Proceedings of the 13th ACM Conference on Recommender Systems*, RecSys 2019 (ACM, New York, NY, USA, 2019), pp. 467–471
114. R. Mehrotra, J. McInerney, H. Bouchard, M. Lalmas, F. Diaz, in *Proceedings of the 27th ACM International Conference on Information and Knowledge Management*, CIKM 2018 (ACM, New York, NY, USA, 2018), pp. 2243–2251
115. N. Sonboli, F. Eskandanian, R. Burke, W. Liu, B. Mobasher, in *Proceedings of the 28th ACM Conference on User Modeling, Adaptation and Personalization*, UMAP 2020 (ACM, New York, NY, USA, 2020), pp. 239–247
116. F. Diaz, B. Mitra, M.D. Ekstrand, A.J. Biega, B. Carterette, in *Proceedings of the 29th ACM International Conference on Information & Knowledge Management*, CIKM 2020 (ACM, New York, NY, USA, 2020), pp. 275–284
117. R. Baeza-Yates, Commun. ACM **61**(6), 54–61 (2018)
118. C. DiCiccio, S. Vasudevan, K. Basu, K. Kenthapadi, D. Agarwal, in *Proceedings of the 26th ACM SIGKDD Conference on Knowledge Discovery and Data Mining*, KDD 2020 (ACM, New York, NY, USA, 2020), pp. 1467–1477
119. R. Epstein, R.E. Robertson, Proc. Natl. Acad. Sci. **112**(33), E4512–E4521 (2015)
120. A. Singh, T. Joachims, in *Proceedings of the 24th ACM SIGKDD International Conference on Knowledge Discovery & Data Mining*, KDD 2018 (ACM, New York, NY, USA, 2018), pp. 2219–2228
121. P. Cremonesi, Y. Koren, R. Turrin, in *Proceedings of the 4th ACM Conference on Recommender Systems*, RecSys 2010 (ACM, New York, NY, USA, 2010), pp. 39–46
122. R. Cañamares, P. Castells, in *Proceedings of the 40th Annual International ACM SIGIR Conference on Research and Development in Information Retrieval*, SIGIR 2017 (ACM, New York, NY, USA, 2017), pp. 215–224
123. A.J.B. Chaney, B.M. Stewart, B.E. Engelhardt, in *Proceedings of the 12th ACM Conference on Recommender Systems*, RecSys 2018 (ACM, New York, NY, USA, 2018), pp. 224–232
124. J.M. Hernández-Lobato, N. Houlsby, Z. Ghahramani, in *Proceedings of the 31st International Conference on Machine Learning*, ICML 2014 (Proc. of Machine Learning Research, Sheffield, UK, 2014), pp. 1512–1520
125. T. Schnabel, A. Swaminathan, A. Singh, N. Chandak, T. Joachims, in *Proceedings of the 33rd International Conference on Machine Learning*, ICML 2016 (Proc. of Machine Learning Research, Sheffield, UK, 2016), pp. 1670–1679
126. D. Liu, P. Cheng, Z. Dong, X. He, W. Pan, Z. Ming, in *Proceedings of the 43rd International ACM SIGIR Conference on Research and Development in Information Retrieval*, SIGIR 2020 (ACM, New York, NY, USA, 2020), pp. 831–840
127. A. Gilotte, C. Calauzènes, T. Nedelec, A. Abraham, S. Dollé, in *Proceedings of the 11th ACM International Conference on Web Search and Data Mining*, WSDM 2018 (ACM, New York, NY, USA, 2018), pp. 198–206
128. A. Gruson, P. Chandar, C. Charbuillet, J. McInerney, S. Hansen, D. Tardieu, B. Carterette, in *Proceedings of the 12th ACM International Conference on Web Search and Data Mining*, WSDM 2019 (ACM, New York, NY, USA, 2019), pp. 420–428

129. A. Swaminathan, A. Krishnamurthy, A. Agarwal, M. Dudík, J. Langford, D. Jose, I. Zitouni, in *Proceedings of the 31st Conference on Neural Information Processing Systems*, NIPS 2017 (Curran Associates, Inc., Red Hook, NY, USA, 2017), pp. 3635–3645

130. L. Yang, Y. Cui, Y. Xuan, C. Wang, S. Belongie, D. Estrin, in *Proceedings of the 12th ACM Conference on Recommender Systems*, RecSys 2018 (ACM, New York, NY, USA, 2018), pp. 279–287

131. P.R. Rosenbaum, D.B. Rubin, Biometrika **70**(1), 41–55 (1983)

132. G.K. Patro, A. Biswas, N. Ganguly, K.P. Gummadi, A. Chakraborty, in *Proceedings of the Web Conference*, WWW 2020 (ACM/IW3C2, 2020), pp. 1194–1204

133. N. Helberger, K. Karppinen, L. D'Acunto, Inf. Commun. Soc. **21**(2), 191–207 (2018)

134. Y. Hu, Y. Koren, C. Volinsky, in *Proceedings of the 8th IEEE International Conference on Data Mining*, ICDM 2008 (IEEE Computer Society, Washington, DC, USA, 2008), pp. 263–272

135. A.P. Jeuland, in *Proceedings of the Educators' Conference*, 43 (American Marketing Association, 1978), pp. 33–37

136. F. Mourão, L. Rocha, C. Araújo, W. Meira, J. Konstan, Information Systems **71**, 137–151 (2017)

137. N. Tintarev, M. Dennis, J. Masthoff, in *User Modeling, Adaptation, and Personalization*, ed. by S. Carberry, S. Weibelzahl, A. Micarelli, G. Semeraro (Springer, Berlin, Heidelberg, 2013), pp. 190–202

# Multistakeholder Recommender Systems

Himan Abdollahpouri and Robin Burke

## 1 Introduction

As described elsewhere in this volume, recommender systems are typically evaluated on their ability to provide items that satisfy the needs and interests of the end user. In some sense, this could be considered a defining characteristic of recommender systems, more generally. At the same time, it is also clear that, in many recommendation domains, the user for whom recommendations are generated is not the only stakeholder in the recommendation outcome. Other users, the providers of products, and even the system's own objectives may need to be considered when these perspectives differ from those of end users.

In many practical settings, such as e-commerce, recommendation is viewed as a form of marketing and, as such, the economic considerations of the retailer will also enter into the recommendation function [47, 61]. A business may wish to highlight products that are more profitable or that are currently on sale, for example. Other businesses may operate as multi-sided platforms, which create value by bringing buyers and sellers together, reducing search and transaction costs [26, 27, 67]. In these cases, keeping both buyers and sellers happy may be essentially to the health of the overall ecosystem. In other applications, system-level objectives such as fairness and balance may be important, and these social-welfare-oriented goals may at times run counter to individual preferences.

While systems incorporating the concerns of multiple stakeholders have been a part of recommender systems research for many years, the identification of multistakeholder recommendation as a particular area of investigation with implications across multiple application areas is relatively new. This chapter describes

H. Abdollahpouri · R. Burke (✉)
University of Colorado, Boulder, CO, USA
e-mail: robin.burke@colorado.edu

© Springer Science+Business Media, LLC, part of Springer Nature 2022
F. Ricci et al. (eds.), *Recommender Systems Handbook*,
https://doi.org/10.1007/978-1-0716-2197-4_17

the key issues in multistakeholder recommendation research and provides an example of multistakeholder evaluation using existing recommendation algorithms. Abdollahpouri et al. [5] offers a survey of different applications in multistakeholder recommendation.

## 2  Recommendation Stakeholders and the Multistakeholder Paradigm

The concept of a stakeholder appears in business management literature as a way to discuss the complexities of corporate governance. According to [32], the term 'stakeholder' appears to have been invented in the early 1960s as a deliberate play on the word 'stockholder' to signify that there are other parties having a 'stake' in the decision-making of the modern, publicly-held corporation in addition to those holding equity positions.

In his classic work, *Strategic Management: A Stakeholder Perspective*, Freeman extends older definitions that emphasize a "stake" as a kind of investment, and instead defines stakeholders as "any groups or individuals that can affect, or are affected by, the firm's objectives" [29, pg. 25]. We adopt this definition for our aims, focusing specifically on recommender systems.

**Definition 1** A *recommendation stakeholder* is any group or individual that can affect, or is affected by, the delivery of recommendations to users.

As recommender systems are elements of an organization's operations, they will necessarily inherit the large and wide-ranging set of stakeholders considered in the management literature. However, only some of these stakeholders will be particularly salient in the generation of recommendations. In this chapter, we will consider three key groups of stakeholders who are particularly close to the recommendation interaction:

**Consumers (aka users)**:    The consumers are the end-users who receive/consume recommendations. They are the individuals whose choice or search problems bring them to the platform, and who expect recommendations to satisfy those needs.

**Providers (aka suppliers)**:    The (item) providers are those entities that supply or otherwise stand behind the recommended objects. Providers can be defined in many different ways, depending on the desired locus of analysis. For example, when recommending movies, the provider might be the movie studio that released it, the director(s) behind it, the actors featured in it, the country of production, or other relevant aspect.

**System**:    The final category is the organization itself, which has created a platform and associated recommender system in order to match consumers with items. The platform may be a retailer, e-commerce site, broker, or other venue where users seek recommendations.

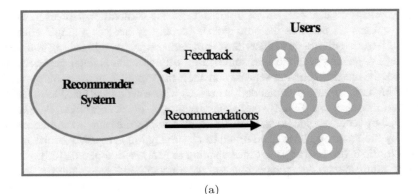

Fig. 1 User-centered schema for recommendation versus multistakeholder schema. (**a**) User-centered view of recommendation. (**b**) Multistakeholder view of recommendation

Of course, none of these stakeholder groups are unitary entities—not even the system, which stands in for various internal groups within an organization who may have different, possibly competing demands on a recommender system. Differentiation among stakeholder groups may be necessary, depending on the application. This taxonomy does, however, highlight an important consequence of a multistakeholder perspective, namely the foregrounding of the multi-sided nature of recommendation.

Figure 1 contrasts the traditional view of recommendations versus the multistakeholder one. As described elsewhere in this volume, for the purposes of evaluation, a recommender system is assumed to generate sets of recommendations for its users and its performance is most often measured with some per-user metric of recommendation quality averaged across all users. In other words the differences among groups of users are often omitted from the evaluation. However, even if we only focus on users as the sole stakeholders, they are not a single entity but rather different individuals with certain characteristics who might belong to different groups (e.g. gender, race etc.), in some cases with associated legal protections. In addition, the multistakeholder schema considers the suppliers of the items as another important party in the recommendation process whose needs and satisfaction should be taken into account. Finally, there may be groups of external *side stakeholders* impacted by recommender system results. For example, the route recommendations given by the Waze app have been found to negatively impact quality of life in areas where traffic is diverted at peak times, leading to resident complaints [28].

The multistakeholder perspective on recommendation can manifest itself in various aspects of recommender systems research and development. We may adopt a multistakeholder view of evaluation: asking the question of how different groups of stakeholders are impacted by the recommendations a system computes. A developer may employ the multistakeholder perspective in designing and optimizing an algo- rithm, incorporating the objectives of multiple parties and balancing among them. Finally, the creation of the recommender system itself may be a multistakeholder endeavor, in which different stakeholders participate in developing a design.

## 2.1  Multistakeholder Evaluation

Recommender systems are generally evaluated from the users' perspective. See Chap. "Evaluating Recommender Systems" for a thorough discussion of this topic. The user-focused methodology is entirely reasonable and logical as, in the end, users are one of the most important stakeholders of any recommender system. We will take it as a given that any recommender system evaluation will include an outcome of this type, the *user summary* evaluation.

A multistakeholder perspective on evaluation, however, brings to light additional aspects of system performance that may be quite important. As mentioned above, multi-sided platforms such as eBay, Etsy, Offer Up, or AirBnB, have a key business requirement of attracting and satisfying the needs of providers as well as users. A system of this type will also need to evaluate its performance from the provider perspective.

Even when only a single group of stakeholders is under consideration, a methodology that relies on a single point estimate of system performance under some metric may miss differences among stakeholder groups. Stakeholder theory recognizes that subgroups within the stakeholder categories may experience a range of different impacts from a firm's decisions [29]. In recommendation, recent work has shown that, depending on the algorithm, male and female users may experience different outcomes in movie recommendations [25]. Therefore, a multistakeholder evaluation is one that includes different stakeholder groups in its assessment, in addition to the user summary evaluation.

**Definition 2** A *multistakeholder evaluation* is one in which the quality of recommendations is assessed across multiple groups of stakeholders, in addition to a point estimate over the user population.

A multistakeholder evaluation may entail the use of different kinds of metrics and different evaluation methodologies than typically encountered in recommender systems research. For example, typical cross-validation methodologies that distribute user profiles between test and training sets may not yield reliable results when assessing outcome for other stakeholders, especially providers. Provider metrics are discussed in greater detail in [1, 5].

## 2.2  Multistakeholder Algorithms

In implementing a given recommender system, a developer may choose to use the metrics associated with multistakeholder evaluation in algorithm design, implementation, and optimization. In general, this will entail balancing the objectives of different stakeholders, as it is unlikely that the optimal solution for one will be the best for all. Some solutions may combine all such objectives into a single optimization, a challenging approach given the methodological complexities noted above; others use a multi-stage approach incorporating different stakeholder concerns throughout a pipeline.

**Definition 3** A *multistakeholder recommendation algorithm* takes into account the preferences of multiple parties when generating recommendations.

Research on multistakeholder algorithms in recommendation predates the invention of this term. For example, Mine et al. [55] discusses the needs of both job seekers and recruiter in providing recommendations on the LinkedIn platform. Other more recent works include [77] and [62], which use a constraint satisfaction approach to distribute recommendation opportunities across providers, and [86], which uses a re-ranking approach to ensure that different groups in the user population are fairly represented in job candidate recommendation. See [5] for a more complete survey.

## 2.3  Multistakeholder Design

Beyond implementing metrics and tuning algorithms, any fielded recommender system will also go through phases of design in which the system's role within a particular platform and its specific requirements are formulated. System designers may choose to engage in a design process such as participatory design [44] in which external stakeholders are incorporated into decision making. Although these techniques are well-established in the HCI community, they have not seen much discussion in machine learning and recommender systems research. The recent workshop on participatory approaches in machine learning indicates the possibility of new momentum around these ideas.[1] Anecdotal information suggests that commercial platforms with multiple stakeholders do engage in stakeholder consultation with item providers in particular [72].

**Definition 4** A *multistakeholder design process* is one in which different recommendation stakeholder groups are identified and consulted in the process of system design.

## 2.4   The Landscape of Multistakeholder Recommendation

The above list of the key recommendation stakeholders provides an outline for understanding and surveying the different types of multistakeholder recommendation. We can conceptualize a recommender system as designed to maximize some combined utility associated with the different stakeholders, and consider how different types of applications yield different stakeholder concerns.

### 2.4.1   Consumer-Side Issues

If we concentrate only on the individuals consuming recommendations, multistakeholder issues arise when there are tradeoffs or disparities between groups of users in the provision of recommendations. For example, [25] explored the performance of recommender system algorithms on users belonging to different demographic groups (gender, age) and observed that some algorithms perform better for certain groups than others. Other researchers have found that users with niche or unusual tastes can be poorly served by particular recommendation algorithms, lapses that are not detectable from point estimates of accuracy measures [2, 31].

  In some settings, such differences in system performance for different users may be considered examples of unfair treatment. In Sect. 5.3.2 we will see that the recommendation algorithms differ in the quality of results for users with different levels of interest in popular items. In particular, we will see that the recommendations for users who have lesser tendency towards popular items are more at odds with those users' expressed range of interests.

---

[1] https://participatoryml.github.io/.

## 2.4.2   Provider-Side Issues

As noted above, multi-sided platforms need to satisfy both the consumers of recommendations and the providers of items that are being recommended. The health of such a platform depends both on a user community and a catalog of items of interest. Providers whose items are not recommended may experience poor engagement from users and lose interest in participating in a given platform. Fewer providers on the platform could then lead to a decline in user interest in the platform as they may find the platform not comprehensive enough. Platforms that facilitate peer-to-peer interactions, such as the online craft marketplace Etsy, may be particularly sensitive to the need to attract and retain sellers.

Depending on the application, providers may have particular audiences in mind as appropriate targets for their items. A well-known example is computational advertising in which advertisers seek particular target markets and ad platforms try to personalize ad presentation to match user interests [85]. In this application, market forces, expressed through auction mechanisms, serve to mediate between supplier interests. In other cases, such as online dating, preferences may be expressed on both sides of the interaction but it is up to the recommender system to perform appropriate allocation.

## 2.4.3   System/Platform Issues

In many real-world contexts, the system may gain some utility when recommending items, and is therefore a stakeholder in its own right. Presumably, an organization creates and operates a recommender system in order to fulfill some business function and generally that is to enhance user experience, lower search costs, increase convenience, etc. This would suggest a consumer stakeholder point of view is sufficient.

However, there are cases in which additional system considerations are relevant, as noted above, and internal stakeholders have an impact on how a recommender system is designed. For example, in an e-commerce platform, the profit of each recommended item may be a factor in ordering and presenting recommendation results. This marketing function of recommender systems was apparent from the start in commercial applications, but rarely included as an element of research systems.

Alternatively, the system may seek to tailor outcomes specifically to achieve particular objectives that are separate from either provider or consumer concerns. For example, an educational site may view the recommendation of learning activities as a curricular decision and seek to have its recommendations fit a model of student growth and development. Its utility may, therefore, be more complex than a simple aggregation of those of the other stakeholders.

### 2.4.4  Side Stakeholders

Complex online ecosystems may involve a number of stakeholders directly impacted by recommendation delivery beyond item suppliers and recommendation consumers. For example, an online food delivery service, such as UberEats, depends on delivery drivers to transport meals from restaurants to diners. Drivers are affected by recommendations delivered to users as the efficiency of routing and the distribution of order load will be a function of which restaurants receive such orders.

### 2.4.5  Tensions Among Stakeholders

Tensions arise when there are tradeoffs across or among stakeholder groups. For exmaple, diverse results may improve provider utility at the cost of making it more difficult for some consumers to find items of interest. Such tradeoffs will typically need to be analyzed and resolved in a domain- and application-dependent manner. An important aspect of multistakeholder design is the question of how to appropriately involve stakeholder groups and their perspectives

Tradeoffs may also exist among groups of similarly positioned stakeholders. For example, providers selling blockbuster products may have interests at odds with providers of niche products. As with other kinds of stakeholder groups, it is difficult to generalize about the specific subgroups, their needs and the associated tensions that will arise in particular applications. Our example below shows examples of disparate mistreatment across both consumer and provider subgroups.

## 3  Related Research

Multistakeholder recommendation brings together research in a number of areas within the recommender systems community and beyond: (1) in economics, the areas of multi-sided platforms and fair division; (2) the growing interest in multiple objectives for recommender systems, including such concerns as fairness, diversity, and novelty; and, (3) the application of personalization to matching problems.

## 3.1  Economic Foundations

The study of the multi-sided business model was crystallized in the work of [67] on what they termed "two-sided markets." Economists now recognize that such contexts are often multi-sided, rather than two-sided, and that "multi-sidedness" is a property of particular business platforms, rather than a market as a whole [26].

Many of today's recommender systems are embedded in multi-sided platforms and hence require a multi-sided approach. The business models of multi-sided platforms are quite diverse, which means it is difficult to generalize about multi-stakeholder recommendation as well. A key element of the success of a multi-sided platform is the ability to attract and retain participants from all sides of the business, and therefore developers of such platforms must model and evaluate the utility of the system for all such stakeholders.

The theory of just division of resources has a long intellectual tradition going back to Aristotle's well-known dictum that "Equals should be treated equally." Economists have invested significant effort into understanding and operationalizing this concept and other related ideas. See [56] for a survey. In recommendation and personalization, we find ourselves on the other side of Aristotle's formulation: all users are assumed unequal and unequal treatment is the goal, but we expect this treatment to be consistent with diverse individual preferences. Some aspects of this problem have been studied under the subject of social choice theory [13]. However, there is not a straightforward adaptation of these classical economic ideas to recommendation applications as the preferences of users may interact only indirectly and in subtle ways. For example, if a music player recommends a hit song to user A, this will not in any way impact its desirability or availability to user B. On the other hand, if a job recommender system recommends an appealing job to user A, it may well have an impact on the utility of the same recommendation to user B who could potentially face an increased competitive environment if she seeks the same position.

## 3.2 Multi-Objective Recommendation

Multistakeholder recommendation is an extension of recent efforts to expand the considerations involved in recommender system design and evaluation beyond simple measurements of accuracy. There is a large body of recent work on incorporating diversity, novelty, long-tail promotion and other considerations as additional objectives for recommendation generation and evaluation. See, for example, [3, 4, 38, 75, 81, 87] and Chap. "Novelty and Diversity in Recommender Systems" in this volume. There is also a growing body of work on combining multiple objectives using constraint optimization techniques, including linear programming. See, for example, [7, 8, 37, 41, 68, 78]. These techniques provide a way to limit the expected loss on one metric (typically accuracy) while optimizing for another, such as diversity. The complexity of these approaches increases exponentially as more constraints are considered, making them a poor fit for the general multistakeholder case. Also, for the most part, multi-objective recommendation research concentrates on maximizing multiple objectives for a single stakeholder, the end user.

Travel recommendation is a classic recommendation application with multistakeholder implications. Recommendations in this domain often must integrate over multiple item providers, such as airlines, hotels, transportation services, restaurants, cultural activities, etc. In addition, items themselves are often capacity-limited (including airline seats, hotel rooms, tickets to events) so it is desirable to spread recommendations across providers. The need for such integration and coordination for IT in travel applications generally is noted in [82]. Specifically in the area of recommendation, Nguyen et al. [58] provides a detailed example of integrating multistakeholder concerns in hotel recommendation incorporating user, system, and provider concerns.

Another area of recommendation that explicitly takes a multi-objective perspective is the area of health and lifestyle recommendation. Multiple objectives arise in this area because users' short-term preferences and their long-term well-being may be in conflict [50, 64]. In such systems, it is important not to recommend items that are too distant from the user's preferences—even if they would maximize health. The goal to be persuasive requires that the user's immediate context and preferences be honored.

Fairness is an example of a system consideration that lies outside the strict optimization of an individual user's personalized results. Therefore, recent research efforts on fairness in recommendation are also relevant to this work [17, 19, 35, 42, 43, 48, 53]. The multistakeholder framework provides a natural "home" for such system-level considerations, which are otherwise difficult to integrate into recommendation. See [5] for a more in-depth discussion.

## 3.3   Personalization for Matching

The concept of multiple stakeholders in recommender systems is suggested in a number of prior research works that combine personalization and matching. The earliest work on two-sided matching problems [69] assumes two sets of individuals, each of which has a preference ordering over possible matches with the other set. This formulation has some similarities to recommendation in two-sided platforms. However, it assumes that all assignments are made at the same time, and that all matchings are exclusive. These conditions are rarely met in recommendation contexts, although extensions to this problem formulation have been developed that relax some of these assumptions in online advertising contexts [15].

Researchers on reciprocal recommendation have looked at bi-lateral considerations to ensure that a recommendation is acceptable to both parties in the interaction. A classical example is on-line dating in which both parties must be interested in order for a match to be successful [63, 83]. Other reciprocal recommendation domains include job seeking [68], peer-to-peer "sharing economy" recommendation (such as AirBnB, Uber and others), on-line advertising [36], and scientific collaboration [51, 80]. See [5] for a detailed discussion.

The field of computational advertising has given considerable attention to balancing personalization with multistakeholder concerns. Auctions, a well-established technique for balancing the concerns of multiple agents in a competitive environment, are widely used both for search and display advertising [54, 85].

However, the real-time nature of these applications and the huge potential user base makes recommender-style personalization computationally intractable in most cases.

## 3.4 Group Recommendation

Although the multistakeholder recommendation distinguishes groups of users in recommendation generation and evaluation, it differs from group recommendation. Group recommendation refers to recommendation scenarios in which groups of individuals are treated as units for the purpose of recommendation generation. Travel in groups in a classic example where a single trip is experienced by multiple individuals and a good recommendation for the group as a whole must take the individuals' preference into account [30]. The challenge in group recommendation is how to deliver a single recommendation that meets, as much as possible, the needs of all members of the group. See [52] and Chap. "Group Recommender Systems: Beyond Preference Aggregation" for additional details.

The challenge in multistakeholder recommendation is quite different. The system is not attempting to craft a single recommendation or recommendation list that satisfies a group of stakeholders. We are concerned more broadly with the system's performance across large subsets of its users and the utility that the group, in general, derives from the system. In addition, group recommendation is only concerns about recommendation recipients / consumers, whereas the multistakeholder perspective incorporates other individuals and groups, including item providers.

## 4 Evaluation

At this point in the development of multistakeholder recommendation research, there is a diversity of methodological approaches and little agreement on basic questions of evaluation. In part, this is a reflection of the diversity of problems that fall under the multistakeholder umbrella. Multistakeholder evaluation and algorithm development do not always use the same methodologies.

A key difficulty is the limited availability of real-world data with multistakeholder characteristics. The reason is these type of data are highly business-critical and often including private information such as margins associated with each supplier and the commissions negotiated by the platform or some demographic information about different stakeholders that can distinguish them from other stakeholders. Close collaboration is required to obtain such sensitive proprietary data. Some researchers have built such collaborations for multistakeholder research, but progress in the field requires replicable experiments that proprietary data does not support. Areas of multistakeholder research that involve public, rather than private, benefit may offer advantages in terms of the availability of data: see, for example, the data sets available from the crowd-funded educational charity site DonorsChoose.org.[2]

## 4.1 Simulation

In the absence of real-world data with associated valuations, researchers have typically turned to simulations. Simulated or inferred supplier data is useful for transforming publicly-available recommendation data sets in standard user, item, rating format into ones that can be used for multistakeholder experimentation. The experiments in [77] provide an example of this methodology: each item in the MovieLens 1M data set was assigned to a random supplier, and the distribution of utilities calculated. To capture different market conditions, the experimenters use two different probability distributions: normal and power-law. There is no accepted standard for producing such simulations and what are reasonable assumptions regarding the distribution of provider utilities or the formulation of system utilities, except in special cases. Other simulation approaches can be seen in [84] and [86].

Researchers have also used objective aspects of data sets to infer proxy attributes for multistakeholder evaluation. In [18], the first organization listed in the production credits for each movie was treated as the provider of that movie—a significant over-simplification of what is a very complex system of revenue distribution in the movie industry. In other work, global metrics such as network centrality [10] have been used to represent system utility for the purposes of multistakeholder evaluation. [20] demonstrated an alternate approach to generate synthetic attribute data based on behavioral characteristics that can be used to evaluate system-level fairness properties.

More sophisticated treatments of profitability and recommendation are to be found in the management and e-commerce literature, some using public data as seen in [6, 23, 59], but these techniques and associated data sets have not yet achieved wide usage in the recommender system community.

## 4.2 Models of Utility

A multistakeholder framework inherently involves the comparison of outcomes across different groups of individuals that receive different kinds of benefits from the system. In economic terms, this entails utility calculation and comparison. As with data, different researchers have made different assumptions about what types of utilities accrue from a recommender system and how they are to be measured.

---

[2] https://data.donorschoose.org/explore-our-impact/.

A standard assumption is that the output of a recommendation algorithm can be treated as an approximation of user utility. Yet, research has confirmed that users prefer diverse recommendation lists [65], a factor in tension with accuracy-based estimates of user utility.

These are, of course, more concentrated on the short-term definitions of utility and things that can be measured from the users feedback in a single run of recommendation generation. More research is therefore required to understand the potential positive and negative long-term effects of multistakeholder recommendation strategies that are not strictly user-focused.

## 4.3   Off-Line Experiment Design

A standard off-line experimental design (discussed in this volume in Chap. "Evaluating Recommender Systems") makes a bit less sense in a multistakeholder context, and this is where the essential asymmetry of the stakeholders comes into play. Suppliers are, in a key sense, passive—they have to wait until users arrive at the system in order to have an opportunity to be recommended. The randomized cross-fold methodology measures what the system can do for each user, given a portion of their profile data, the potential benefit to be realized if the user visits and a recommendation list is generated. Evaluating the supplier side under the same conditions, while a commonly-used methodology, lacks a similar justification.

A more realistic methodology from the suppliers' point of view is a temporal one, that takes into account the history of the system up to a certain time point and examines how supplier utilities are realized in subsequent time periods. See [21] for a comprehensive discussion of time-aware recommender systems evaluation. However, time-aware methods have their own difficulties, forcing the system to cope with cold-start issues possibly outside of the scope of a given project's aims. Therefore, for certain models of utility where the dynamics of suppliers (such as price, profit margin, etc.) may be of less importance, offline evaluation can be still valuable. For example, our evaluation of the popularity bias in recommendation from the perspectives of different stakeholders is done in an offline setting.

## 4.4   User Studies

User studies are another instrument available to researchers that has not been extensively applied to multistakeholder recommender systems. Chapter "Evaluating Recommender Systems" covers user evaluation of recommender systems, with insights that apply to multistakeholder recommendation as well. As usual for such studies, the development of reliable experimental designs is challenging as the participants' decision situation typically remains artificial. Furthermore, as in the study by Azaria et al. [14], familiarity biases might exist—in their study participants were willing to pay more for movies that they already knew—which have been observed for other types of user studies in the recommender systems domain [40]. Ultimately, more field tests—even though they are typically tied to a specific domain and specific business model—are needed that give us more insights into the effects of multistakeholder recommendations in the real world.

## 4.5  Metrics

The building block of multistakeholder evaluation is the measurement of the utility each of the stakeholders gets within a recommendation platform. Common evaluation metrics such as RMSE, precision, NDCG, diversity, etc. are all different ways to evaluate the performance of a recommender system from the user's perspective. (See Chap. "Evaluating Recommender Systems" in this volume for an introduction to recommender system evaluation.) As noted above, these measures are implicitly a form of system utility measure as well: system designers optimize for such measures under the assumptions that (1) higher accuracy metrics correspond to higher user satisfaction and (2) higher user satisfaction contributes to higher system utility through customer retention, trust in the recommendation provided, etc. However, the formulation of multistakeholder recommendation makes it possible to characterize and evaluate system utility explicitly.

Typically, evaluation metrics are averaged over all users to generate a point score indicating the central tendency over all users. However, it is also the case that in a multistakeholder environment additional aspects of the utility distribution may be of interest. For example, in an e-commerce context, providers who receive low utility may leave the eco-system, suggesting that the variance and skew of provider utility may be important as well as the mean. One suggested practice would be to report results for significant user groups within consumer and provider populations, as we will demonstrate in Sect. 5.

### 4.5.1  Provider Metrics

When evaluating the utility of a recommender system for a particular provider, we may take several different stances. One views the recommender as a way to garner user attention. In this case, the relevance of an item to a user may be a secondary consideration. Another perspective views the recommender as a source of potential leads. In this view, recommending an item to uninterested users is of little benefit to the provider. In the first situation, simply counting (with or without a rank-based discount) the number of times a provider's products appears in recommendation lists would be sufficient. In the second situation, the metric should count only those recommendations that were considered "hits", those that appear positively rated in the corresponding test data.

Another provider consideration may be the reach of its recommended products across the user population. A metric could count the number of unique users to whom the provider's items are recommended. Multistakeholder applications may differ in their ability to target specific audiences for their items. In a targeted system, it would make sense to consider reach relative to the target population. For example, in an online dating application where the user can specify desired properties in a match, an evaluation metric might be the fraction out of the target audience receiving the recommendation.

Finally, where the consideration is the accuracy of system's predictions, we can create a provider-specific summary statistic of a measure like RMSE. Ekstrand et al. [24] uses this method to examine differences in error when recommending books by male and female authors. Since the statistic by itself is not that useful for a single provider, a better metric would indicate the provider's position relative to other providers in the overall metric distribution.

### 4.5.2 System Metrics

System utility may, in many cases, be a simple aggregation of the utilities of other parties. However, as noted above, other cases arise where the system has its own targeted utility framework. An important such context is algorithmic fairness: see Chap. "Fairness in Recommender Systems" in this volume. In general, we should not expect that individual consumers or providers will care if the system is fair to others as long as it provides them with good outcomes. Any fairness considerations and related metrics will therefore be ones defined by system considerations.

## 5 Multistakeholder Evaluation Example: Popularity Bias in Recommendation

To illustrate the multistakeholder concept and to demonstrate how a multistakeholder evaluation yields insights not available through other means, the remainder of this chapter analyzes the well-known issue of popularity bias in recommendation from a multistakeholder standpoint. We first briefly introduce the problem of popularity bias and then examine the impact of this algorithmic property on subgroups of users, items, and providers.

The problem of popularity bias in recommendation is well-established: recommender systems often focus on items that are frequently rated, even though there are many other items of value to users [12, 16, 22, 57, 66, 73]. Abdollahpouri [1] discusses this phenomenon and its multistakeholder aspects in detail.

A key reason for popularity bias in recommendation is the substantial skew found in the rating data in many recommendation domains. It is typical, for there to be a small number of popular items which take up the majority of rating interactions while the majority of the items receive little attention from the users. For example, there are relatively few songs and movies that are internationally popular, few people on social media that have millions of followers, a small percentage of research papers that have thousands of citations, and a short list of tourism destinations that attract millions of travelers.

Due to this imbalance property of the rating data, often algorithms inherit this bias and, in many cases, intensify it by over-recommending the popular items, potentially generating a positive feedback loop in which items that already have more attention from users get more attention due to the algorithm's rankings and less well-known items become increasingly ignored [39, 60, 76].

## 5.1  Data

We demonstrate this effect using two publicly available data sets for our experiments: the first one is a sample of the Last.fm (LFM-1b) data set [71] used in [46]. The data set contains user interactions with songs (and the corresponding albums). We used the same methodology in [46] to turn the interaction data into rating data using the frequency of the interactions with each item (more interactions with an item will result in higher rating). In addition, we used albums as the items to reduce the size and sparsity of the item dimension, therefore the recommendation task is to recommend albums to users. We considered the artists associated with each album as the providers. We removed users with less than 20 ratings so only consider users for which we have enough data. The resulting data set contains 274,707 ratings by 2697 users to 6006 albums. Total number of artists is 1998.

The second data set is the MovieLens 1M data set [34].[3] This data set does not have the information about the providers. However, as we mentioned earlier, we considered the director of each movie as the provider of that movie and we extracted that information from the IMDB API. Total number of ratings in the MovieLens 1M data is 1,000,209 given by 6040 users to 3706 movies. Overall, we were able to extract the director information for 3043 movies reducing the ratings to 995,487. The total number of directors is 831.

We randomly split both datasets into 80% training and 20% test set to conduct the experiments.

Figure 2 shows the percentage of users rated different items in MovieLens and Last.fm data sets: the popularity of each item. Items are ranked from the most popular to the least with the most popular item being on the far left. The curve has a long-tail shape [11, 22], indicating that a few popular items are taking up the majority of the ratings while many other items on the far right of the curve have not received much attention. The items are divided into three regions with the items

---

[3] Our experiments showed similar results on MovieLens 20M, and so we continue to use MovieLens 1M for efficiency reasons.

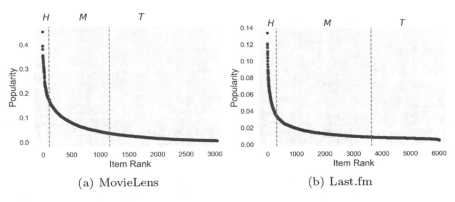

**Fig. 2** Rating distributions for MovieLens and Last.fm datasets. Three item categories based on their popularity are $H$, $M$, and $T$: popular items, relatively popular items, and non-popular items, respectively. (**a**) MovieLens. (**b**) Last.fm

associated with the top 20% of the ratings in the *Head* or *H* category, the middle 60% of the ratings with the *Middle Tail* or *M* category, and the remaining *Tail* (*T*) items.[4] This type of distribution can be found in many other domains such as e-commerce where a few products are best-sellers, online dating where a few profiles attract the majority of the attention, and social networking sites where a few users have millions of followers. A similar type of distribution can be seen in many other domains.

## 5.2 Algorithmic Bias

Though a variety of mechanisms, recommendation algorithms inherit and often amplify the highly-skewed nature of their input. As an example, consider the results of different recommendation algorithms seen in Fig. 3. These four plots contrast item popularity and recommendation popularity for four well-known recommendation algorithms *RankALS* [79], *Biased Matrix Factorization (Biased-MF)* [45], *User-based Collaborative Filtering (User-CF)* [9], and *Item-based Collaborative Filtering (Item-CF)* [70]. All of these algorithms are tuned to achieve their best performance in terms of precision. We used LibRec which is an open source Java library for running these algorithms [33]. Also, in all algorithms the size of the recommended list given to each user is 10.

---

[4] Tail items may also be new items that have yet to find their audience and will eventually become popular. In this way, popularity bias is related to the well-known item cold-start problem in recommendation.

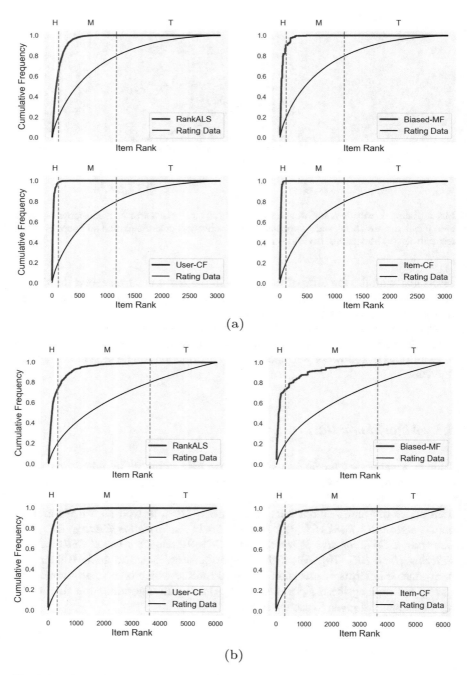

**Fig. 3** Popularity bias amplification (item groups). (**a**) MovieLens. (**b**) Last.fm

The black curve shows the cumulative frequency of ratings for different items at a given rank in the MovieLens 1M data set. The x-axis indicates the rank of each item when sorted from most popular to least popular. The y-axis indicates the cumulative prevalence of that item in the data. As we can see, a few highly ranked items dominate the entire rating history. For instance, only 111 items (less than 3% of the items) take up more than 20% of the ratings.

The blue curves in each figure show popularity bias at work across the four algorithms. In *Item-CF*, no item beyond rank 111 is recommended. The head of the distribution constituting less than 3% of the items are actually taking up 100% of the recommendations. In *User-CF* this number is 99%. The other algorithms are only slightly improved in this regard, where the head items take up more than 64 and 74% of the recommendations in *RankALS* and *Biased-MF*, respectively.

## 5.3 Multistakeholder Impact

It is clear that recommendation algorithms amplify popularity effects found in their input data. Our purpose here is to ask how that effect impacts different stakeholders of a recommender system. To do so, we demonstrate how this bias can impact different groups of items, users, and the suppliers of the items. In particular, we concentrate on the following stakeholder groups:

- Items, differentiated by their popularity: We divide items into $H$, $M$ and $T$ segments.
- Users, segmented by popularity interest: We divide users into equal-sized segments based on their interest in popular items.
- Providers, segmented by the relatively popularity of their associated items: Similar to users, we identify equal-sized groups of providers based on the popularity of the items with which they are associated.

### 5.3.1 Item Groups

In addition to the descriptive plots in Fig. 3, we can quantify the impact on different item groups with a metric *Item Popularity Deviation* (IPD) that looks at groups of items across the popularity distribution. For any item group $c$, $IPD(c) = q(c) - p(c)$ where $q(c)$ is the ratio of recommendations that come from item group $c$ (i.e. $q(c) = \frac{\sum_{u \in U} \sum_{j \in \ell_u} \mathbb{1}(j \in c)}{n \times |U|}$). $p(c)$ is the ratio of ratings that come from item group $c$ (i.e. $p(c) = \frac{\sum_{u \in U} \sum_{j \in \rho_u} \mathbb{1}(j \in c)}{\sum_{u \in U} |\rho_u|}$). The average *IPD* across different groups of items can be measured as:

$$IPD = \frac{\sum_{c \in C} |IPD(c)|}{|C|}, \tag{1}$$

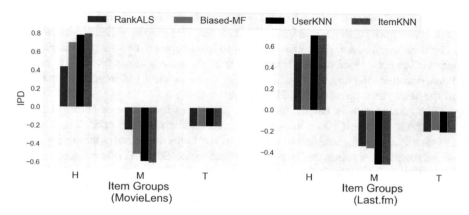

**Fig. 4** Item popularity deviation

In Fig. 4 we can see that all four algorithms have over-recommended $H$ items and under-recommended $M$ and $T$ items on both datasets. With regard to $T$ items, all algorithms have performed equally poorly as they all have under-recommended these items. However, we can see that for $H$ and $M$ items, *RankALS* has a lower deviation than the others followed by *Biased-MF* indicating the superiority of these two over the neighborhood-based models on these item groups.

### 5.3.2 User Groups

We can also take a consumer-/user-centered view of popularity bias Not every user is equally interested in popular items. In cinema, for instance, some might be interested in movies from Yasujiro Ozu, Abbas Kiarostami, or John Cassavetes, and others may enjoy more mainstream directors such as James Cameron or Steven Spielberg.

Figure 5 shows the ratio of rated items for three item categories $H$, $M$, and $T$ in the profiles of different users in the MovieLens 1M and Last.fm data sets. Users are sorted from the highest interest towards popular items to the least and divided into three equal-sized bins $G_{1..3}$ from most popularity-focused to least. The y-axis shows the proportion of each user's profile devoted to different popularity categories. The narrow blue band shows the proportion of each users profile that consists of popular items, and its smooth decrease reflects the way the users are ranked. Note, however, that all groups have rated many items from the middle (green) and tail (red) parts of the distribution, and this makes sense: there are only a small number of really popular movies and even the most blockbuster-focused viewer will eventually run out of them. Figure 6 shows the same information as Fig. 5 but all ratios are calculated as an average value across users in each user group. The differences between different user groups in this figure are more visible.

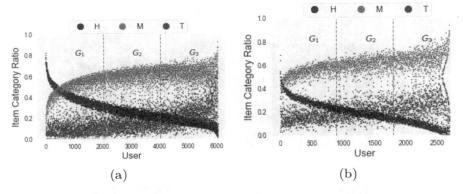

**Fig. 5** Users' Propensity towards item popularity. (**a**) MovieLens. (**b**) Last.fm

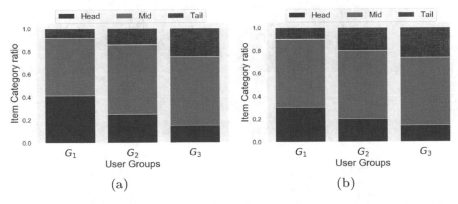

**Fig. 6** User Groups' Propensity towards item popularity. Three users groups based on their interest in popular items are $G_1$, $G_2$, and $G_3$: group with highest interest towards popular items, group with average interest in popular items, and group with the least interest in popular items. (**a**) MovieLens. (**b**) Last.fm

For example, looking at the head ratio (blue color), we can clearly see that users in group $G_1$ has the highest ratio of head items followed by $G_2$ and $G_3$.

Similar to Figs. 6 and 7 shows the ratio of different item groups in the profiles of users in different user groups as an average value. Here, we can clearly see the differences between algorithms in terms of how they have performed for different user groups. The results of *Item-CF* clearly consist of only head items for all user groups.

As we have seen, users have a diversity of interests across popularity groups, which is best represented as a distribution. *User Popularity propensity Deviation (UPD)* measures the quality of recommendation results by looking at the distance between the distribution of item popularity within the user's profile ($P$) and the equivalent distribution across items recommended ($Q$). For this purpose, we employ

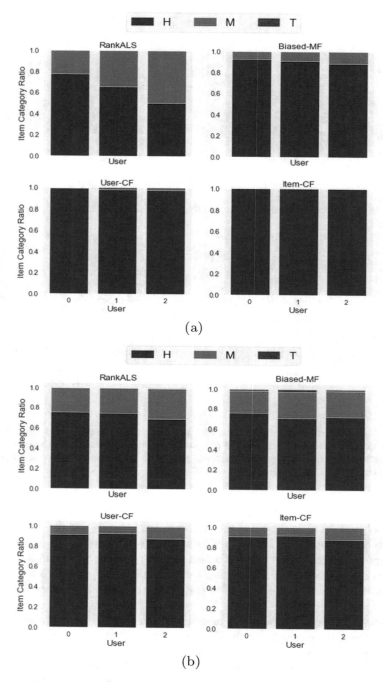

**Fig. 7** Users' centric view of popularity bias (Groups). (**a**) MovieLens. (**b**) Last.fm

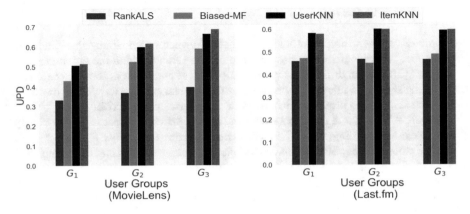

**Fig. 8** User popularity propensity deviation

the Jensen-Shannon divergence ($\mathfrak{J}$), which is a metric of distribution difference. Jensen-Shannon divergence is related to the Kullback-Leibler (KL) divergence but has the benefit of being symmetric.

Given the *KL* as the KL divergence function, the JensenShannon divergence ($\mathfrak{J}$) between two probability distributions $P$ and $Q$ is defined as follows:

$$\mathfrak{J}(P, Q) = \frac{1}{2}KL(P, M) + \frac{1}{2}KL(Q, M), \quad M = \frac{1}{2}(P + Q) \tag{2}$$

With this definition, we can define *UPD* as follows. For any user group $g$, $UPD(g) = \frac{\sum_{u \in g} \mathfrak{J}(P(\rho_u), Q(\ell_u))}{|g|}$. The average *UPD* across different groups of users is:

$$UPD = \frac{\sum_{g \in G} UPD(g)}{|G|} \tag{3}$$

Figure 8 shows the *UPD* results for the different algorithms and data sets. On the Movielens data, we can see that all four algorithms have higher deviation for user group $G_3$ than $G_2$ or $G_1$ meaning the recommendations for niche-focused users are not as a representation of their interests, compared to the ones for other users with more interest in popular content—a pattern we would have not been able to detect without the multistakeholder evaluation. Also, on MovieLens, we can see that the differences between the deviations of different algorithms for this user group are more prominent than for the other groups.

### 5.3.3 Provider Groups

As noted above, multistakeholder analysis in recommendation also includes providers, "those entities that provide or otherwise stand behind the recommended objects" [5]. We can think of many different kinds of contributors standing behind a particular movie. In this study, we consider the director of each movie and the artist for each song as the providers in MovieLens and Last.fm respectively.

Figure 9 is similar to Fig. 3 but shows the rank of different directors by popularity and the corresponding recommendation results from the *Item-CF* algorithm. A similar pattern of popularity bias is evident. The recommendations have amplified the popularity of the popular suppliers (the ones on the extreme left) while suppressing the less popular ones dramatically. Strikingly, movies from just 3 directors (less than 0.4% of the suppliers here) take up 50% of recommendations produced, while the tail of the distribution is seeing essentially zero.

Similar to the item groups, for any provider group, here termed *supplier s*, $SPD(s) = q(s) - p(s)$ where $q(s)$ is the ratio of recommendations that come from items of supplier group $s$ (i.e. $q(s) = \frac{\sum_{u \in U} \sum_{j \in \ell_u} \mathbb{1}(A(j) \in s)}{n \times |U|}$), and $p(s)$ is the ratio of ratings that come from items of supplier group $s$ (i.e. $p(s) = \frac{\sum_{u \in U} \sum_{j \in \rho_u} \mathbb{1}(A(j) \in s)}{\sum_{u \in U} |\rho_u|}$).
The average *SPD* across different groups of suppliers can be calculated as:

$$SPD = \frac{\sum_{i=1}^{|S|} |SPD(S_i)|}{|S|} \qquad (4)$$

Similar to item groups, Fig. 10 all algorithms have performed equally poorly for less popular suppliers ($S_3$). For the other two supplier groups, *RankALS* has a much lower deviation than the others and *ItemKNN* has the highest deviation which is consistent with our analysis before. One interesting observation on MovieLens is that *Biased-MF* and *UserKNN* have very similar *IPD* for *M* items (i.e. items with medium popularity) but *Biased-MF* has significantly lower *SPD* for the suppliers with medium popularity ($S_2$). Again, this is something that only a multistakeholder analysis can show.

## 5.4 Summary

What this example demonstrates is that multistakeholder evaluation can shed light on the differences between different recommendation algorithms that would not be easily detectable without such evaluation. We also saw that different stakeholder groups (e.g. different user groups) were not equally treated using the same algorithm indicating how a single-value metric for all entities in a stakeholder group may hide important information.

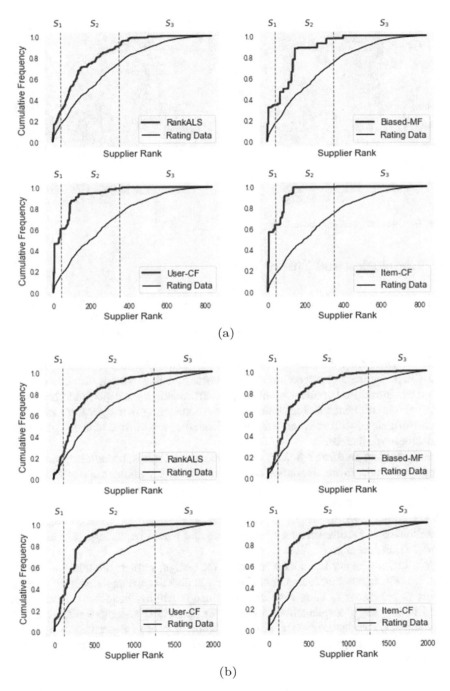

**Fig. 9** Providers' centric view of popularity bias. Three supplier groups based on their popularity are $S_1$, $S_2$, and $S_3$: popular suppliers, relatively popular suppliers, and non-popular suppliers, respectively. (**a**) MovieLens. (**b**) Last.fm

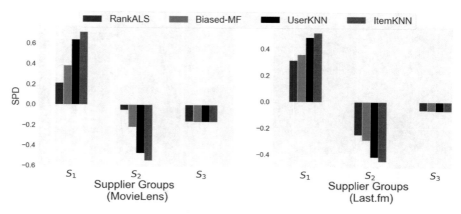

**Fig. 10** Supplier popularity deviation

# 6 Conclusion and Future Work

Recommender systems are embedded in organizational contexts where there may be multiple stakeholders impacted by a system's performance. The research area of multistakeholder recommendation recognizes and integrates the perspective of these stakeholders in recommendation design, generation and evaluation. A closely related area of recommender systems research is fairness-aware recommendation—see Chap. "Fairness in Recommender Systems" in this volume. In this chapter, we have introduced multistakeholder recommendation concepts and focused in particular on multistakeholder evaluation , demonstrating that algorithms may have quite different properties when considered from the perspectives of users, providers, and subgroups thereof.

All areas of multistakeholder recommendation offer substantial opportunities for future work. In-depth studies of particular multistakeholder applications have appeared (including [53]) but are still rare. Additional such studies will be important for developing effective practices around the formulation and operationalization of stakeholder needs. There is very little work detailing the involvement of stakeholders in recommender system design, that would be analogous to emerging work in other areas of AI [49].

While this chapter has concentrated on the design, implementation and evaluation of multistakeholder recommendation , we expect that a complementary research effort in interfaces to such systems will emerge. Multistakeholder concerns will need to be part of explanations delivered to users [74] and other efforts at recommendation transparency. The multistakeholder approach also raises the question of interfaces and system transparency for stakeholders other than recommendation consumers.

**Acknowledgments** The authors were supported in part by the National Science Foundation under Grant Number 191102.

# References

1. H. Abdollahpouri, Popularity bias in recommendation: A multi-stakeholder perspective. Ph.D. Thesis, University of Colorado Boulder (2020). https://arxiv.org/pdf/2008.08551.pdf
2. H. Abdollahpouri, M. Mansoury, Multi-sided exposure bias in recommendation, in *International Workshop on Industrial Recommendation Systems, in Conjunction with ACM KDD* (2020). https://arxiv.org/pdf/2006.15772.pdf
3. H. Abdollahpouri, R. Burke, B. Mobasher, Controlling popularity bias in learning-to-rank recommendation, in *Proceedings of the Eleventh ACM Conference on Recommender Systems* (ACM, New York, 2017), pp. 42–46
4. H. Abdollahpouri, R. Burke, B. Mobasher, Managing popularity bias in recommender systems with personalized re-ranking, in *The Thirty-Second International Flairs Conference* (2019)
5. H. Abdollahpouri, G. Adomavicius, R. Burke, I. Guy, D. Jannach, T. Kamishima, J. Krasnodeb-ski, L. Pizzato, Multistakeholder recommendation: survey and research directions. User Model. User-Adapt. Interact. **30**, 127–158 (2020)
6. P. Adamopoulos, A. Tuzhilin, The business value of recommendations: a privacy-preserving econometric analysis, in *Proceedings of the International Conference on Information Systems, ICIS'15* (2015)
7. D. Agarwal, B.-C. Chen, P. Elango, X. Wang, Click shaping to optimize multiple objectives, in *Proceedings of the 17th ACM SIGKDD International Conference on Knowledge Discovery and Data Mining, KDD'11* (ACM, New York, 2011), pp. 132–140
8. D. Agarwal, B.-C. Chen, P. Elango, X. Wang, Personalized click shaping through lagrangian duality for online recommendation, in *Proceedings of the 35th International ACM SIGIR Conference on Research and Development in Information Retrieval, SIGIR'12* (ACM, New York, 2012), pp. 485–494
9. C.C. Aggarwal, Neighborhood-based collaborative filtering, in *Recommender Systems* (Springer, Berlin, 2016), pp. 29–70
10. L. Akoglu, C. Faloutsos, Valuepick: towards a value-oriented dual-goal recommender system, in *The 10th IEEE International Conference on Data Mining Workshops, Sydney, Australia, ICDM'10* (2010), pp. 1151–1158
11. C. Anderson, *The Long Tail: Why the Future of Business is Selling More for Less* (Hachette Books, Paris, 2006)
12. C. Anderson, *The Long Tail: Why the Future of Business is Selling Less of More* (Hyperion Books, New York, 2008)
13. K.J. Arrow, A. Sen, K. Suzumura, *Handbook of Social Choice and Welfare*, vol. 2 (Elsevier, Amsterdam, 2010)
14. A. Azaria, A. Hassidim, S. Kraus, A. Eshkol, O. Weintraub, I. Netanely, Movie recommender system for profit maximization, in *Proceedings of the 7th ACM Conference on Recommender Systems, RecSys'13* (2013), pp. 121–128
15. M.H. Bateni, Y. Chen, D.F. Ciocan, V. Mirrokni, Fair resource allocation in a volatile marketplace (2016). Available at SSRN 2789380
16. E. Brynjolfsson, Y.J. Hu, M.D. Smith, From niches to riches: anatomy of the long tail. Sloan Manag. Rev. **47**(4), 67–71 (2006)
17. R. Burke, Multisided fairness for recommendation, in *Workshop on Fairness, Accountability and Transparency in Machine Learning (FATML), Halifax, Nova Scotia* (2017), 5pp.
18. R. Burke, H. Abdollahpouri, B. Mobasher, T. Gupta, Towards multi-stakeholder utility evaluation of recommender systems, in *Proceedings of the International Workshop on Surprise, Opposition, and Obstruction in Adaptive and Personalized Systems (SOAP 2016)* (ACM, New York, 2016)

19. R. Burke, N. Sonboli, M. Mansoury, A. Ordoñez-Gauger, Balanced neighborhoods for fairness-aware collaborative recommendation, in *Workshop on Responsible Recommendation (FATRec)* (2017)
20. R. Burke, J. Kontny, N. Sonboli, From amateurs to connoisseurs: modeling the evolution of user expertise through online reviews, in *Workshop on Fairness, Accountability, and Transparency in Recommendation (FATREC) in Conjunction with ACM RecSys* (2018). arXiv:1809.04199
21. P.G. Campos, F. Díez, I. Cantador, Time-aware recommender systems: a comprehensive survey and analysis of existing evaluation protocols. User Model. User-Adapt. Interact. **24**(1), 67–119 (2014)
22. Ò. Celma, P. Cano, From hits to niches? or how popular artists can bias music recommendation and discovery, in *Proceedings of the 2nd KDD Workshop on Large-Scale Recommender Systems and the Netflix Prize Competition* (ACM, New York, 2008), p. 5
23. P.-Y.S. Chen, S. Wu, J. Yoon, The impact of online recommendations and consumer feedback on sales, in *Proceedings of the International Conference on Information Systems ICIS'04* (2004), pp. 711–724
24. M.D. Ekstrand, M. Tian, M.R.I. Kazi, H. Mehrpouyan, D. Kluver, Exploring author gender in book rating and recommendation, in *Proceedings of the 12th ACM Conference on Recommender Systems* (ACM, New York, 2018), pp. 242–250.
25. M.D. Ekstrand, M. Tian, I.M. Azpiazu, J.D. Ekstrand, O. Anuyah, D. McNeill, M.S. Pera, All the cool kids, how do they fit in? Popularity and demographic biases in recommender evaluation and effectiveness, in *Conference on Fairness, Accountability and Transparency* (2018), pp. 172–186
26. D.S. Evans, R. Schmalensee, *Matchmakers: The New Economics of Multisided Platforms* (Harvard Business Review Press, Boston, 2016)
27. D. Evans, R. Schmalensee, M. Noel, H. Chang, D. Garcia-Swartz, Platform economics: essays on multi-sided businesses. Competition Policy International (2011). https://ssrn.com/abstract=1974020
28. E. Fisher, Do algorithms have a right to the city? Waze and algorithmic spatiality. Cultural Stud. **36**(1), 74–95 (2022)
29. R.E. Freeman, *Strategic Management: A Stakeholder Approach* (Cambridge University Press, Cambridge, 2010)
30. I. Garcia, L. Sebastia, E. Onaindia, On the design of individual and group recommender systems for tourism. Exp. Syst. Appl. **38**(6), 7683–7692 (2011)
31. M.A. Ghazanfar, A. Prügel-Bennett, Leveraging clustering approaches to solve the gray-sheep users problem in recommender systems. Exp. Syst. Appl. **41**(7), 3261–3275 (2014)
32. K.E. Goodpaster, Business ethics and stakeholder analysis. Bus. Ethics Quart. **1**, 53–73 (1991)
33. G. Guo, J. Zhang, Z. Sun, N. Yorke-Smith, LibRec: a java library for recommender systems, in *Conference on User Modeling, Adaptation, and Personalization, UMAP'15* (2015)
34. F.M. Harper, J.A. Konstan, The movielens datasets: History and context. ACM Trans. Interact. Intell. Syst. **5**(4), 19 (2015)
35. B. Huang, S. Yao, New fairness metrics for recommendation that embrace differences, in *Workshop on Fairness, Accountability and Transparency in Machine Learning (FATML), Halifax, Nova Scotia* (2017), 5pp.
36. G. Iyer, D. Soberman, J.M. Villas-Boas, The targeting of advertising. Market. Sci. **24**(3), 461–476 (2005)
37. T. Jambor, J. Wang, Optimizing multiple objectives in collaborative filtering, in *Proceedings of the Fourth ACM Conference on Recommender Systems, RecSys'10* (ACM, New York, 2010), pp. 55–62
38. D. Jannach, G. Adomavicius, Recommendations with a purpose, in *Proceedings of the 10th ACM Conference on Recommender Systems* (ACM, New York, 2016), pp. 7–10
39. D. Jannach, L. Lerche, I. Kamehkhosh, M. Jugovac, What recommenders recommend: an analysis of recommendation biases and possible countermeasures. User Model. User-Adapt. Interact. **25**(5), 427–491 (2015)

40. D. Jannach, L. Lerche, M. Jugovac, Item familiarity as a possible confounding factor in user-centric recommender systems evaluation. i-com J. Interact. Media **14**(1), 29–39 (2015). https://doi.org/10.1515/icom-2015-0018
41. Y. Jiang, Y. Liu, Optimization of online promotion: a profit-maximizing model integrating price discount and product recommendation. Int. J. Inf. Technol. Decis. Making **11**(05), 961–982 (2012)
42. T. Kamishima, S. Akaho, Considerations on recommendation independence for a find-good-items task, in *Workshop on Responsible Recommendation (FATRec), Como, Italy* (2017)
43. T. Kamishima, S. Akaho, H. Asoh, J. Sakuma, Correcting popularity bias by enhancing recommendation neutrality, in *Poster Proceedings of the 8th ACM Conference on Recommender Systems, RecSys 2014, Foster City, Silicon Valley, CA, USA, October 6–10* (2014)
44. F. Kensing, J. Blomberg, Participatory design: issues and concerns. Comput. Supp. Coop. Work **7**(3–4), 167–185 (1998)
45. Y. Koren, R. Bell, C. Volinsky, Matrix factorization techniques for recommender systems. Computer **42**(8), 30–37 (2009).
46. D. Kowald, M. Schedl, E. Lex, The unfairness of popularity bias in music recommendation: a reproducibility study, in *European Conference on Information Retrieval* (Springer, Berlin, 2020), pp. 35–42
47. N. Leavitt, Recommendation technology: Will it boost e-commerce? Computer **39**(5), 13–16 (2006)
48. E.L. Lee, J.-K. Lou, W.-M. Chen, Y.-C. Chen, S.-D. Lin, Y.-S. Chiang, K.-T. Chen, Fairness-aware loan recommendation for microfinance services, in *Proceedings of the 2014 International Conference on Social Computing* (2014), pp. 1–4
49. M.K. Lee, D. Kusbit, A. Kahng, J.T. Kim, X. Yuan, A. Chan, D. See, R. Noothigattu, S. Lee, A. Psomas, et al., WeBuildAI: participatory framework for algorithmic governance, in *Proceedings of the ACM on Human-Computer Interaction (CSCW)*, vol. 3 (2019), pp. 1–35
50. Y. Lin, J. Jessurun, B. De Vries, H. Timmermans, Motivate: towards context-aware recommendation mobile system for healthy living, in *2011 5th International Conference on Pervasive Computing Technologies for Healthcare (PervasiveHealth)* (IEEE, Piscataway, 2011), pp. 250–253
51. G.R. Lopes, M.M. Moro, L.K. Wives, J.P.M. De Oliveira, Collaboration recommendation on academic social networks, in *International Conference on Conceptual Modeling* (Springer, Berlin, 2010), pp. 190–199
52. J. Masthoff, Group recommender systems: combining individual models, in *Recommender Systems Handbook* (Springer, Berlin, 2011), pp. 677–702
53. R. Mehrotra, J. McInerney, H. Bouchard, M. Lalmas, F. Diaz, Towards a fair marketplace: counterfactual evaluation of the trade-off between relevance, fairness & satisfaction in recommendation systems, in *Proceedings of the 27th ACM International Conference on Information and Knowledge Management* (ACM, New York, 2018), pp. 2243–2251
54. A. Mehta, A. Saberi, U. Vazirani, V. Vazirani, Adwords and generalized online matching. J. ACM **54**(5), 22 (2007)
55. T. Mine, T. Kakuta, A. Ono, Reciprocal recommendation for job matching with bidirectional feedback, in *2013 Second IIAI International Conference on Advanced Applied Informatics* (2013), pp. 39–44. https://doi.org/10.1109/IIAI-AAI.2013.91
56. H. Moulin, *Fair Division and Collective Welfare* (MIT Press, Cambridge, 2004)
57. T.T. Nguyen, P.-M. Hui, F.M. Harper, L. Terveen, J.A. Konstan, Exploring the filter bubble: the effect of using recommender systems on content diversity, in *Proceedings of the 23rd International Conference on World Wide Web* (ACM, New York, 2014), pp. 677–686
58. P. Nguyen, J. Dines, J. Krasnodebski, A multi-objective learning to re-rank approach to optimize online marketplaces for multiple stakeholders, in *Workshop on Value-Aware and Multistakeholder Recommendation in Conjunction with ACM RecSys* (2017). https://arxiv.org/pdf/1708.00651.pdf
59. G. Oestreicher-Singer, A. Sundararajan, The visible hand? Demand effects of recommendation networks in electronic markets. Manag. Sci. **58**(11), 1963–1981 (2012)

60. Y.-J. Park, A. Tuzhilin, The long tail of recommender systems and how to leverage it, in *Proceedings of the 2008 ACM Conference on Recommender Systems* (2008), pp. 11–18
61. B. Pathak, R. Garfinkel, R.D. Gopal, R. Venkatesan, F. Yin, Empirical analysis of the impact of recommender systems on sales. J. Manag. Inform. Syst. **27**(2), 159–188 (2010)
62. G.K. Patro, A. Chakraborty, N. Ganguly, K. Gummadi, Incremental fairness in two-sided market platforms: on smoothly updating recommendations, in *Proceedings of the AAAI Conference on Artificial Intelligence*, vol. 34 (2020), pp. 181–188
63. L. Pizzato, T. Rej, T. Chung, I. Koprinska, J. Kay, Recon: a reciprocal recommender for online dating, in *Proceedings of the Fourth ACM Conference on Recommender Systems* (ACM, New York, 2010), pp. 207–214
64. V. Ponce, J.-P. Deschamps, L.-P. Giroux, F. Salehi, B. Abdulrazak, Quefaire: context-aware in-person social activity recommendation system for active aging, in *Inclusive Smart Cities and e-Health* (Springer, Berlin, 2015), pp. 64–75
65. P. Pu, L. Chen, R. Hu, A user-centric evaluation framework for recommender systems, in *Proceedings of the Fifth ACM Conference on Recommender Systems* (ACM, New York, 2011), pp. 157–164
66. P. Resnick, R.K. Garrett, T. Kriplean, S.A. Munson, N.J. Stroud, Bursting your (filter) bubble: strategies for promoting diverse exposure, in *Proceedings of the 2013 Conference on Computer Supported Cooperative Work Companion*, (ACM, New York, 2013), pp. 95–100
67. J.-C. Rochet, J. Tirole, Platform competition in two-sided markets. J. Eur. Econ. Assoc. **1**(4), 990–1029 (2003)
68. M. Rodriguez, C. Posse, E. Zhang, Multiple objective optimization in recommender systems, in *Proceedings of the Sixth ACM Conference on Recommender Systems* (ACM, New York, 2012), pp. 11–18
69. A.E. Roth, M. Sotomayor, Two-sided matching, in *Handbook of Game Theory with Economic Applications*, vol. 1 (Elsevier, Amsterdam, 1992), 485–541
70. B. Sarwar, G. Karypis, J. Konstan, J. Riedl, Item-based collaborative filtering recommendation algorithms, in *Proceedings of the 10th International Conference on World Wide Web* (ACM, New York, 2001), pp. 285–295
71. M. Schedl, The lfm-1b dataset for music retrieval and recommendation, in *Proceedings of the 2016 ACM on International Conference on Multimedia Retrieval* (2016), pp. 103–110
72. O. Semerci, A. Gruson, C. Edwards, B. Lacker, C. Gibson, V. Radosavljevic, Homepage personalization at spotify, in *Proceedings of the 13th ACM Conference on Recommender Systems* (ACM, New York, 2019), pp. 527–527
73. G. Shani, A. Gunawardana, Evaluating recommendation systems, in *Recommender Systems Handbook* (Springer, Berlin, 2011), pp. 257–297
74. J. Smith, N. Sonboli, C. Fiesler, R. Burke, Exploring user opinions of fairness in recommender systems, in *Fair & Responsible AI Workshop @ CHI* (2020)
75. B. Smyth, P. McClave, Similarity vs. diversity, in *Case-Based Reasoning Research and Development* (Springer, Berlin, 2001), pp. 347–361
76. H. Steck, Item popularity and recommendation accuracy, in *Proceedings of the Fifth ACM Conference on Recommender Systems* (2011), pp. 125–132
77. Ö. Sürer, R. Burke, E.C. Malthouse, Multistakeholder recommendation with provider constraints, in *Proceedings of the 12th ACM Conference on Recommender Systems* (ACM, New York, 2018), pp. 54–62
78. K.M. Svore, M.N. Volkovs, C.J. Burges, Learning to rank with multiple objective functions, in *Proceedings of the 20th International Conference on World Wide Web, WWW'11* (ACM, New York, 2011), pp. 367–376
79. G. Takács, D. Tikk, Alternating least squares for personalized ranking, in *Proceedings of the Sixth ACM Conference on Recommender Systems* (ACM, New York, 2012), pp. 83–90
80. J. Tang, S. Wu, J. Sun, H. Su, Cross-domain collaboration recommendation, in *Proceedings of the 18th ACM SIGKDD International Conference on Knowledge Discovery and Data Mining* (ACM, New York, 2012), pp. 1285–1293

81. S. Vargas, P. Castells, Rank and relevance in novelty and diversity metrics for recommender systems, in *Proceedings of the Fifth ACM Conference on Recommender Systems* (ACM, New York, 2011), pp. 109–116
82. H. Werthner, A. Alzua-Sorzabal, L. Cantoni, A. Dickinger, U. Gretzel, D. Jannach, J. Neidhardt, B. Pröll, F. Ricci, M. Scaglione, et al., Future research issues in it and tourism. Inf. Technol. Tour. **15**(1), 1–15 (2015)
83. P. Xia, B. Liu, Y. Sun, C. Chen, Reciprocal recommendation system for online dating, in *Proceedings of the 2015 IEEE/ACM International Conference on Advances in Social Networks Analysis and Mining* (ACM, New York, 2015), pp. 234–241
84. S. Yao, B. Huang, Beyond parity: fairness objectives for collaborative filtering, in *Advances in Neural Information Processing Systems*, (2017), pp. 2921–2930
85. S. Yuan, A.Z. Abidin, M. Sloan, J. Wang, Internet advertising: an interplay among advertisers, online publishers, ad exchanges and web users (2012). arXiv:1206.1754
86. M. Zehlike, F. Bonchi, C. Castillo, S. Hajian, M. Megahed, R. Baeza-Yates, Fair: a fair top-k ranking algorithm, in *Proceedings of the 2017 ACM on Conference on Information and Knowledge Management* (ACM, New York, 2017), pp. 1569–1578
87. C.-N. Ziegler, S.M. McNee, J.A. Konstan, G. Lausen, Improving recommendation lists through topic diversification, in *Proceedings of the 14th International Conference on World Wide Web* (ACM, New York, 2005), pp. 22–32

# Fairness in Recommender Systems

**Michael D. Ekstrand, Anubrata Das, Robin Burke, and Fernando Diaz**

## 1 Introduction

Throughout the history of recommender systems, both research and development have examined a range of effects beyond the accuracy and user satisfaction of the system. These effects include the *diversity*, *novelty* and/or *serendipity* of recommendations (see Chapter "Novelty and Diversity in Recommender Systems") and the *coverage* of the system (see Chapter "Evaluating Recommender Systems"), among other concerns.

In recent years, this concern has extended to the *fairness* of the system: are the benefits and resources it provides fairly allocated between different people or organizations it affects? This challenge is connected to the broader set of research

Portions of this chapter are adapted and condensed from: Michael D. Ekstrand, Anubrata Das, Robin Burke and Fernando Diaz (2021), "Fairness in Search and Recommendation", Foundations and Trends® in Information Retrieval, Forthcoming. https://doi.org/10.1561/1500000079.

M. D. Ekstrand (✉)
People and Information Research Team, Boise State University, Boise, ID, USA
e-mail: michaelekstrand@boisestate.edu

A. Das
School of Information, University of Texas at Austin, Austin, TX, USA
e-mail: anubrata@utexas.edu

R. Burke
That Recommender Systems Lab, University of Colorado Boulder, Boulder, CO, USA
e-mail: robin.burke@colorado.edu

F. Diaz
Microsoft Research, Montreal, QC, Canada
e-mail: diazf@acm.org

© Springer Science+Business Media, LLC, part of Springer Nature 2022
F. Ricci et al. (eds.), *Recommender Systems Handbook*,
https://doi.org/10.1007/978-1-0716-2197-4_18

on fairness in sociotechnical systems generally and AI systems more particularly [4, 54], but information access systems, including recommender systems, have their own set of particular challenges and possibilities.

Fair recommendation is not an entirely new concern, but builds on concepts with long precedent in the recommender systems literature. Work on *popularity bias* [15], along with attempts to ensure quality and equity in *long-tail recommendations* [31], can be viewed as a fairness problem to prevent the system from inordinately favoring popular, well-known, and possibly well-funded content creators. Group recommendation (see Chapter "Group Recommender Systems: Beyond Preference Aggregation") is also concerned with ensuring that the various members of a group are treated fairly [48]. The issues we discuss in this chapter go beyond these (important) problems to examine how various biases—particularly *social* biases— can make their way in to the data, algorithms, and outputs of recommender systems.

Our primary goal in this chapter is to guide the reader through the elements needed to measure and possibly improve the fairness of a recommender system. This starts with a clear picture of what, precisely, it means for the system to be unfair with respect to its stakeholders and problem space, and the harms this unfairness may cause. There are many different fairness problems in personalization and recommendation, and each problem has its own particular features and solutions; therefore the details of measuring and providing fairness are as varied as the applications to which they apply. To provide guidance in this space, we discuss the key decisions and resources necessary to carry out a fair recommender systems experiment, grounded in a specific example from our own work, and then survey the research to date in fair recommender systems to provide pointers for further reading.

Like much of recommender systems, fairness in recommendation has significant overlap with fairness in information retrieval and draws heavily from machine learning literature; researchers and practitioners looking to study or improve the fairness of recommendations will do well to pay attention to a broad set of research results.

## 1.1 Example Applications

Throughout this chapter, we will use the problem space of job and candidate search and recommendation to illustrate the challenges and possibilities for producing fair recommendations. Many online platforms attempt to connect job-seekers and employment opportunities in some way. These include dedicated employment-seeking platforms as well as general-purpose professional networking platforms (such as LinkedIn and Xing) for which job-seeking is one (important) component.

Job-seeking is a multi-sided problem—people need good employment, and employers need good candidates—and also has significant fairness requirements that are often subject to regulation in various jurisdictions. Some of the specific fairness concerns for this application include:

- Are recommendations of job opportunities distributed fairly across users?
- Are job/candidate fit scores fair, or do they under- or over-score certain candidates?
- Do users have a fair opportunity to appear in result lists when recruiters are looking for candidates for a job opening?
- Are employers in protected groups (minority-owned businesses, for example) having their jobs fairly promoted to qualified candidates?
- Are there specific fairness concerns that come from regulatory requirements?

Job search is certainly not the only application with fairness concerns, however. The recommendations in music platforms, such as Spotify and Pandora, connect listeners with artists and have significant impact both on the user's listening experience and on the artists' financial and career prospects. A system that does systematically under-promotes particular artists or groups of artists may reduce their visibility and with it harm the financial viability of their artistic endeavors. However, it serves to illustrate many important problems in recommendation fairness in a single setting.

## 1.2   Fundamental Concepts in (Un)Fairness

Algorithmic fairness in general is concerned with going beyond the aggregate accuracy or effectiveness of a system—often, but not always, a machine learning application—to studying the *distribution* of its positive or negative effects on its subjects. Much of this work has been focused on fairness in classification or scoring systems; Mitchell et al. [54] catalog key concepts in fair classification, and Hutchinson and Mitchell [43] situate these concepts in the broader history of fairness in educational testing where many similar ideas were previously developed.

There are many ways to break down the various concepts of algorithmic fairness that have been studied in the existing literature, which we summarize in Table 1. The first is *individual* versus *group* fairness. Individual fairness sets the constraint that *similar individuals should be treated similarly*: given a function that can measure the similarity of two individuals with respect to a task, such as the ability of job applicants to perform the duties of a job, individuals with comparable ability should receive comparable decisions [25]. This decision construct is often probabilistic, to allow for the ability to only select one applicant for any given job; individual fairness in such cases requires that the probability of acceptance be similar for similarly-qualified individuals. This definition depends on the availability of an unbiased comparison of individuals—a significant limitation—and places no requirements on the treatment of dissimilar subjects.

Group fairness considers the system's behavior with respect to the group membership or identities of subjects. Often, this is realized by dividing the observed features into two sets $\langle X_i, A_i \rangle$, where $A_i$ is a set of *sensitive attributes* identifying group membership or identity, often something like race or ethnicity, and $X_i$ is

**Table 1** Summary of concepts in algorithmic fairness and harms

| Individual fairness | Similar individuals have similar experience |
|---|---|
| Group fairness | Different groups have similar experience |
| Sensitive attribute | Attribute identifying group membership |
| *Disparate treatment* | Groups explicitly treated differently |
| *Disparate impact* | Groups receive outcomes at different rates |
| *Disparate mistreatment* | Groups receive erroneous (adverse) effects at different rates |
| Distributional harm | Harm caused by (unfair) distribution of resources or outcomes. |
| Representational harm | Harm caused by inaccurate internal or external representation. |
| Anti-classification | Protected attribute should not play a role in decisions. |
| Anti-subordination | Decision process should actively work to undo past harm. |

the set of other features or covariates. Such work often designates a *protected group*, designated by a particular value of $A_i$, that should be protected from adverse discriminatory decisions. Many of the concepts are often grounded in U.S. legal notions of anti-discrimination [3], but this grounding is not without limitations, such as the resulting technical framings losing connection to the legal concepts that motivated them [69].

The goal of group fairness is to address the impact of group membership on final decision outcomes, but the precise formulations of this goal further divide into several distinct sub-concepts of discriminatory behaviors the system should avoid. *Disparate treatment* is when the sensitive attribute is a direct feature in the decision-making process. *Disparate impact* arises when the outcomes, such as loan approvals, are disparately distributed between groups (e.g. one racial group has a higher loan approval rate than another). It is measured with constructs such as demographic parity (e.g. a resume screening process accepts resumes at the same rate). Both disparate treatment and impact are directly inspired by U.S. anti-discrimination law; disparate impact is not on its own enough to conclusively prove discrimination, but it triggers enhanced scrutiny of a process [3]. *Disparate mistreatment*, on the other hand, looks at the distribution not of decision outcomes but of errors: it arises, for example, when one group has a higher false negative rate than another (for example, when members of one racial group is more likely to be denied loans for which they are qualified). Another disparate mistreatment construct is one sometimes called *equality of opportunity* [38]: the constraint that positive classification be conditionally independent of protected group given the true label—that is, a qualified job candidate from one group is no more likely to be erroneously screened out than a candidate from another.

Orthogonally to group and individual fairness, Crawford [20] divided harms connected to unfairness into two categories: *distributional* harms, where a resource or opportunity is inequitably distributed (as with most of the harms discussed so far), and *representational* harms, where people in the system are represented in a way that is inaccurate and systematically unfair.

The motivations of fairness constructs can also be categorized, and different motivations yield different assessments of their effectiveness and adequacy. U.S. anti-discrimination law is rooted in competing ideas of *anti-classification*, the idea that the influence of a protected attribute should not play a role in decisions, and *anti-subordination*, the idea that current decision-making processes should actively work to reverse the effects of historical patterns of discrimination [3]. These motivations are often present, and sometimes explicitly invoked, in algorithmic fairness literature, although their proper application is not entirely clear [69].

While it is common to talk about "fairness" as the goal of a system designed to improve equity, it is impossible to make a system that is demonstrably and universally "fair". Competing ideals of fairness, the needs of multiple stakeholders, and the fundamentally social and contestable nature of fairness [62] mean that it is not a fully solvable problem or even a well-specified one.

More detailed treatment of algorithmic fairness in general is beyond the scope of this chapter. We refer the reader to Mitchell et al. [54] and Barocas et al. [4] for further (essential) reading.

## 1.3   Multisided Analysis of Recommendation

One significant way in which fair recommendation differs from more general fair machine learning is in the *multisided* nature of recommendation. In many machine learning problems, it is clear who should be treated fairly: the system is making decisions about subjects, such as job or loan applicants. In recommender systems, however, different groups of stakeholders stand to benefit (or lose) from the system's behavior, and assessing the fairness of a recommender system requires identifying the stakeholders who may be treated unfairly. Chapter "Multistakeholder Recommender Systems" of this book provides an introduction to multistakeholder recommendation broadly. For the purposes of this chapter, it is sufficient to note that different stakeholders have different fairness concerns [12]. These include:

- **Consumer fairness**, concerned with treating different users of the system fairly by ensuring that users receive comparable service and no group of users is systematically disadvantaged by the system. For example, a system might be more inaccurate for some groups of users rather than others.
- **Provider fairness**, concerned with treating the creators or providers of items fairly, for example by ensuring that musicians have equitable opportunity for their music to be recommended to users in a streaming music platform.
- **Subject fairness**, concerned with fair treatment of the people or entities that items are about. One example of a domain where this concern may arise is news recommendation, where recommendations that systematically under-represent topics or segments of society can be considered subject-unfair.

These stakeholders' fairness concerns are often considered one at a time, but some work seeks to jointly analyze or provide fairness for multiple stakeholders simultaneously. In this chapter, we organize our literature survey around the stakeholders each work primarily considers.

## 1.4   Unfairness in Recommendation

As noted in Sect. 1.2, it is impossible to make a system fair according to any universal definition. What *can* be done is to measure and avoid specific types of well-defined *un*fairness. Different stakeholder groups can be harmed by the system's unfairness in different ways; aligning with Sect. 1.2, this unfairness can be group or individual, and can result in harms of both distribution and representation.

If a system gives some of its users lower-quality recommendations than others [27, 52], particularly if it systematically gives one socially-salient group of users less useful results than another, then it exhibits unfairness towards consumers that results in a distributional harm (the resource of "relevant information" is not fairly distributed). It may cause representational harm to its consumers if it misrepresents their personal information or interests. It may be unfair towards its consumers if it provides comparably good recommendations, as measured by click-through rate or another suitable metric, but it gives qualitatively different results to different groups, such as being less likely to show women ads for higher-paying jobs [2].

While provider fairness concerns present a relatively straightforward set of harms, there is a lot of nuance in how, precisely, to define the concept, but a system harms producers in an unfair way if it is more likely to recommend content from some creators than others. One well-known example of this is *popularity bias* [15], where the system is more likely to recommend already-popular content so less-popular creators are not as able to make their content visible and obtain recognition and commercial return for their efforts.

But visibility can also be misallocated along other lines, such as systematically under-recommending content by creators of particular gender identities or ethnic groups. This discrepancy can have many causes, including exogenous inequities in opportunity to create content, publicity efforts from publishers and labels, and other systemic societal discrimination; the recommender system reifies and reproduces these inequities as it learns from that data in pursuit of its objective function. Another potential cause is objective functions that do not align with broader social interest—for example, optimizing information distribution in a network to maximize the number of nodes reached may leave significant gaps in access to information [32], a particularly relevant concern for things such as public health awareness campaigns. Misallocation may also arise from feedback loops, where the recommender system learns from user data and amplifies small differences in potential utility into large differences in recommendation exposure.

In Sects. 3–6, we will survey work on each of these (and other) harms. One of the crucial elements of pursuing fairness in research or system development, however, is to be specific about the harms under consideration. Identifying the specific stakeholders in play, and the particular harm or unfairness that we wish to measure or avoid, allows research to make progress and contribute techniques that can be a part of building recommender systems that promote equity instead of reproducing discrimination.

# 2 Studying Fairness

As discussed in Sects. 1.3 and 1.4, there is no one problem of recommender system fairness: there are different stakeholders towards which the system may be unfair, and different ways it may be unfair to them. Our survey of the literature in Sects. 3–7 discusses examples of many different problems and approaches to fair recommendation; in this section, we describe the overall structure of a fair recommendation project and the key decisions that must be made regardless of the specific problem the researcher or developer seeks to address, summarized in Table 2.

As a running example in this section, we use the study of the fairness of collaborative filters with respect to books' authors' gender identities by Ekstrand and Kluver [26], describing the decisions made in that work for each component of the process. We do not claim that this is a perfect example, but it provides a useful context to discuss how these general decision points need to be resolved in the context of a specific study. Also, while we discourage a needless proliferation of new fairness concepts, each individual fair recommendation project will be different due to the intrinsic differences between stakeholders, harms, applications, and data affordances that shape a particular effort.

**Table 2** Key decisions in a fair recommender systems project

| Fairness questions: |
| --- |
| 1. What is the recommendation problem under study? |
| 2. What is the *social goal* of fairness in this setting? |
|     • What stakeholders are considered? |
|     • What harm is to be prevented? |
| 3. How is that goal *operationalized* with specific data, methods, and metrics? |
| 4. What are the findings? |
| 5. How do these findings map back to the social goal? |
| 6. What are the scope and limitations of these methods and results? |

## 2.1  Defining Fairness

The first step in any inquiry is to define precisely the aspects of fairness under study. There are several elements to consider, including:

- The **stakeholders** who should be treated fairly
- The **harms** to quantify and/or reduce
- The **metrics** for measuring and/or optimizing away the harm

A project may target more than one type of unfairness, but each fairness construct needs to be clearly described and account for these dimensions. It is important to be precise about both the social goal of a fairness construct and the way that goal is operationalized into specific methods and metrics. For example:

- To measure and reduce the potential harm that musicians of one ethnicity are more likely to be recommended than musicians of another, we may enforce a constraint that "the probability a song appears in a playlist is independent of musician ethnicity" [46]. We may also measure the exposure [24] or attention [8] providers of different ethnicities receive, and look for absolute disparities (disparate impact) or disproportional disparities based on relevance (disparate mistreatment).
- To measure and reduce the potential representational harm that users are stereotyped by gender in the system's latent feature space, we may measure the accuracy of classifying a user's gender from their embedding and use adversarial learning to minimize this predictability while providing accurate recommendations [6].

The common theme of these operationalizations is that we have taken a concern—reduce a particular harm—and translated it into a specific framework and metric or objective function. This process is inherently reductive, as the resulting metric does not capture everything about the concern. Decisions in the process prioritize some aspects of fairness over others; for example, seeking to make artist groups' exposure proportional to their combined relevance will allow for significant disparities in exposure if they are supported by disparities in relevance judgements (such as user response or ratings), even if the relevance judgements are unfair (due to biases in user preference, presentation biases that affect likelihood to click, or other factors that bias the collection of relevance data).

Such limitations and tradeoffs do not necessarily invalidate a specific fairness construct; indeed, no construct is without them, and there is no well-defined universal concept of fairness. Rather, they are limitations that need to be documented. Clearly describing the concerns, the metrics, their relationship, and the limitations that arise from that choice of metrics will enable readers of a particular research paper to better understand its findings, and apply and extend them in additional situations.

The author gender project focuses on provider fairness as experienced by consumers, and seeks to measure the harm that authors of a particular gender may be under-represented in the recommendation lists the system provides to its users, even if users read may books by authors of that gender. It operationalizes this by measuring the "% Female" for a set $L$ of books (either the books in a user's reading history or the books recommended to them by a collaborative filter):

$$\text{SetBias}(L) = \frac{\text{\# of female-authored books in } L}{\text{\# of known-gender books in } L}$$

This definition captures one particular form of representation, but does not do anything to assess representation along other dimensions (e.g. race or ethnicity), or even capture the full subtlety of gender representation (such as accounting for non-binary gender identities).

Mismatches between a metric and its ultimate goals are common, however, particularly while the field is young and we are still determining how to study fairness and the relative merits of different approaches. We advocate for incrementally improving on the practice of fair recommendation research; perfect alignment and operationalization are likely only achievable in the limit.[1]

## 2.2 Collecting Data

Another major challenge is to find appropriate data for studying the fairness objective under consideration. While data sets for studying recommendation abound, fairness research—particularly group fairness—often requires additional data that is more difficult to come by Olteanu et al. [57] provide a thorough treatment of issues in data collection and augmentation, but here we highlight some specific considerations for fair recommendation research.

Commonly-used data sets are often directly amenable to individual fairness research, because all that is needed is users and items, and possibly item authorship. No further data augmentation is typically necessary.

Group fairness, both for providers and consumers, requires group affiliations or sensitive attributes for the relevant stakeholder entities. Depending on the particular type of group in question, this data may also be sensitive and subject to legal restrictions or ethical concerns for its use and/or distribution. This is particularly true for personal attributes such as gender, sexual orientation, or ethnicity.

Some data sets provide attribute labels directly. The MovieLens 1M data set [39] includes self-described binary gender identities and age brackets for MovieLens

---

[1] In the conference version of the author gender paper, Ekstrand et al. [28] described the goal of promoting equality of opportunity for book authors, but measured fairness through the proportional composition of ranked lists. Exposure-oriented metrics [8, 24] would be a more coherent way of advancing the stated goal. The journal version described here more clearly contextualizes the capabilities and implications of the methods employed, because we're continually advancing our own understanding of these practices as well.

users (subsequent data sets do not, as MovieLens stopped collecting this information from its users). The Last.FM 360K data set collected by Celma [17][2] and the later LastFM 1B data set [61] also include user gender and age.

In other cases, a group can be synthesized from available data. For example, Kamishima et al. [46] looked at fairness with respect to the age of a movie in order to test algorithms intended to be useful for a broader set of fairness problems. This can also be done with genres, keywords, and other information included in a data set. Care is needed, however, because as Selbst et al. [62] argue, we cannot assume techniques for ensuring fairness with respect to one attribute will automatically apply to other attributes even within the same stakeholder class (e.g. a technique that addresses gender discrimination may or may not be effective for racial discrimination and vice versa, depending on the way that discrimination has affected the data generating process), but it is a useful technique for testing out technical possibilities for future repurposing, subject to re-validation on the target problem.

Sometimes, multiple public data sets can be integrated in order to study a fairness problem. For many media domains, there is publicly-available data on content creators such as book authors and film actors; sometimes, this data is in machine-readable form amenable to integration. In the author gender project, Ekstrand and Kluver [26] integrated book ratings from several sources with OpenLibrary, Library of Congress bibliographic records, and the Virtual Internet Authority File to obtain the gender identities of many book authors; the integration tools are publicly available.[3] Because linked author records with gender identities are not available for all books, this project had to limit the final analysis to books where the first author's gender could be uniquely identified; the "% Female" metric is computed over the subset of rated or recommended books for which the author gender identity is known. Further, the available data only records binary gender identities, limiting the study's ability to properly account for authors with non-binary and other gender identities.

Data can also be augmented through crowdsourcing or professional annotation. The TREC Fair Ranking Track [9] uses NIST's professional assessors to obtain information about the authors of scholarly papers from publicly-available sources. This data source is expensive, but depending on the task and source of labor it can result in high-quality annotations. Annotation does require careful attention to defining the legitimate bases for assigning a label; as one example, the Program for Cooperative Cataloging [10] has developed recommended best practices for recording authors' gender identities.

Some authors have used inference techniques to impute demographic labels to users or content producers. Common ways of doing this include statistical gender recognition based on names or computer vision techniques for gender recognition based on portraits or profile pictures. We generally discourage using this source of

---

[2] https://www.upf.edu/web/mtg/lastfm360k.

[3] https://bookdata.piret.info.

data, as it is error-prone, subject to systemic biases [11], reductionistic [35], and fundamentally denies subjects control over their identities.

Another approach sometimes employed is to generate synthetic data, simulating the relevant information needed to assess fairness. This is motivated by the challenges of acquiring data with the appropriate demographic and content characteristics for academic researchers. If it is possible to establish correlates between aspects of user's profiles and demographic characteristics, then demographic labels can be probabilistically associated with user profiles, providing labels for protected and unprotected groups. This approach was demonstrated using job recommendation data by Burke et al. [13].

Finally, recent work has looked at evaluating fairness without linked demographic records, for example by using background data on label distributions to calibrate estimation from proxy attributes [44]. This can work in some cases; Kallus et al. [44] provide methods to identify some of those cases, but the resulting estimates can have very broad error bounds. Hashimoto et al. [40] provide algorithms for fair learning without access to demographic labels, but the minority groups protected may differ from the groups needing protection in a particular application; the implications of this discrepancy will be application- and domain-specific.

No matter the technique used, it is crucial to carefully document the data sources, integration or augmentation decisions, and known limitations of the data. Gebru et al. [33] identify many questions that should be answered for any data set, but are particularly germane to the kinds of data needed for fairness research. Due to the length limitations of many publication venues, this may necessitate appendices or online supplements that document the data sources and strategies more thoroughly than a conference paper allows, but we recommend that the paper itself include sufficient detail to understand and assess the limitations.

## 2.3 Structuring Experiments

The structure of a recommendation fairness study itself tends to follow structures typical for other recommender evaluations (see Chapter "Evaluating Recommender Systems"), particularly studies focusing on non-accuracy analyses such as diversity (Chapter "Novelty and Diversity in Recommender Systems"). The recommender algorithms are trained on available data, produce recommendations for users, and the resulting recommendations are measured for their fairness (and often other properties). This can be done with both offline and online experimental protocols. The only change to existing experimental practice typically required to study fairness is the additional data and metrics.

The author gender study uses a typical offline experimental protocol with user-based sampling. One sample of users is used for tuning algorithm hyperparameters with mean reciprocal rank (MRR) in a leave-one-out protocol, and another sample is used for the fairness study. For the experiment's main results, the authors measure the gender composition of users' reading and rating histories, recommendation lists produced for these users by several collaborative filters, and the relationship between input and output composition (how much the rankings produced by each algorithm reflect the user's historical tendency to read authors of one gender or another).

## 2.4 Reporting Results

The results of a fairness study need to be carefully described and clearly contextualized in light of the study's specific aims and the limitations of the data and metrics. Special care needs to be taken to understand the implications of data and experimental limitations on the results and their generalizability. It is important not to over-claim; *portability* is one of the abstraction traps identified by Selbst et al. [62, §2.2], as results and solutions on one fairness concern do not necessarily apply to another, even if they can both be represented with the same statistical abstraction.

As a practical matter, we recommend clearly separating the social goal and the specific operationalization in both the motivation and discussion of the results in a paper. Being explicit about how the fairness goal is translated into data and measurements, and how specific quantitative results illuminate the social goal, helps reduce risk of the formalism trap [62, §2.3] and provides context for discussing limitations.

The author-gender project is a purely observational, correlational study, and reports findings in light of the capabilities of such methods. It also describes in detail the data set; the coverage and distribution of relevant labels in the data; and limitations of the data, fairness construct and operationalization, and experimental design.

## 3 Consumer Fairness

*Consumer fairness* is the aspect of a recommender system concerned with how recommendations impact consumers and sub-groups of consumers. For example, a marketplace recommending job postings to job seekers may wish to ensure that all its users receive good-quality recommendations and comparable access to job opportunities, particularly marginalized groups such as women and people from minority ethnic backgrounds; such objectives may even be required by law.

## 3.1 Individual Fairness

Since recommender systems' results are personalized to each consumer, users are expected to get different results, we might expect that what counts as "similar" individuals for the purposes of individual fairness in recommendation might be quite fine-grained. For example, a job recommender might be taking into account the user's past history of clicks or views on the site as well as more typical job history or personal data commonly associated with a job application. Recommender systems, especially collaborative ones, are typically built on the assumption that individuals will similar user profiles should indeed receive similar recommendations, so we might expect that individual fairness is a strength of these systems.

However, user behavior is not typically what is understood in the discussion of similar individuals in the context of individual fairness. A person's profile as gathered through interaction with a recommender system might include information that is irrelevant to the match between user and item. To continue with the employment example, a female user might be less likely to click on job descriptions with male-oriented language [41], and this fact may be completely irrelevant to her qualifications. Still, it may be quite difficult to ensure individual fairness in a recommendation setting, as a complete profile of all relevant information that would be needed to compare individuals could be quite challenging to obtain. Almost by definition, if a recommender system, especially a collaborative one, is being deployed, it is because the application is one where user profiles—and similarity—is better obtained through an implicit representation than explicit features.

Therefore, while individual fairness is a worthy objective, we expect that developers trying to ensure consumer-side fairness in recommendation will be primarily interested in framing their fairness concerns relative to group fairness.

## 3.2 Group Fairness

A general approach to group fairness is to consider the benefit or utility that a recommender system delivers to different groups of users. To operationalize this concern, we need to determine how to measure utility and how to compare between groups. In some recommendation domains, it may be important to consider objective qualities of an item in understanding its utility, such as the salary in a job listing. If protected group users receive on the whole lower salary listings, this could be considered unfair regardless of other personalization considerations. Thus, there may be some inherent utility associated with each item to be taken account in the overall utility computation.

Most approaches, however, use the utility construct used for measuring the system's effectiveness, such as an accuracy metric based on ratings, click history, or similar profile data. Some group-fairness evaluations use the predicted rating or click probability estimated by the recommender system itself for a more complete

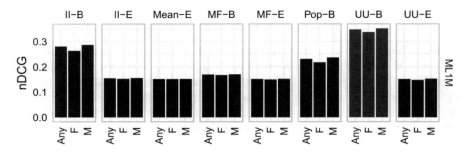

**Fig. 1** Example results from a consumer group-fairness study by Ekstrand et al. [27]. NDCG of algorithms trained on the MovieLens 1M data set is aggregated by user gender

but approximate picture of utility. With a known preference data, utility can be calculated in a number of ways (see also Chapter "Evaluating Recommender Systems"):

- Predicted rating fidelity: How closely do predicted ratings match known ones? The usual approach is to use RMSE or MAE to measure this accuracy.
- Predicted ranking accuracy: How closely do output rankings match known preferences? NDCG is a common utility-based measure for ranking.

Generally, experiments of this kind will look for disparate mistreatment, in the form of greater error or lower ranking quality for the protected group. One example that explores consumer-side fairness is by Ekstrand et al. [27]. They performed an offline top-$N$ evaluation of several collaborative filtering algorithms, computing NDCG and aggregating the results by user demographic (example results shown in Fig. 1). They found significant differences in utility between gender and (in some cases) user age groups, although not always in the same direction. Yao and Huang [73] carried out similar work for prediction accuracy, developing several unfairness metrics over rating prediction errors for protected and unprotected groups. Typically-considered demographic groups are not the only basis for consumer fairness either; Abdollahpouri [1] considers user groups based on their level of interest in popular items and shows that, across recommendation algorithms, users with an interest in less-popular niche items were not receiving recommendations in line with their interests.

## 3.3 Fairness Beyond Accuracy

While much of the work on consumer fairness focuses on the quality of recommendations, some work looks at other aspects of consumer experience that may be discriminatory, such as stereotyping or the specific items users receive. Ali et al. [2] studied the distribution of ads on Facebook to understand potentially discriminatory impact in the visibility of different kinds of ads. They found that even when an advertiser wishes to have fair distribution of their ad, for example to ensure that an ad for a job opening is seen by people of all genders, the combination of relevance optimization and market dynamics results in disparate distribution of ads across racial and gender lines. Nasr and Tschantz [56] describe bidding strategies to attempt mitigate such effects and ensure fair ad distribution even when the platform does not provide it.

Beutel et al. [6] present an approach to learning fair representations in a way that can be applied to consumers, by learning embeddings (such as the user and item embeddings in a recommender system) in an adversarial setting set up to minimize the ability to predict a user's sensitive attribute, such as gender, from their embedding. This has the potential to reduce stereotype effects in resulting recommendations.

## 3.4   More Complex Scenarios

The work described in this section makes assumptions that are often violated in realistic applications. One in particular is the assumption that only a single protected group or sensitive attribute needs to be considered for fairness. But our job recommender, for example, may need to meet constraints having to do with race, gender, religion, and other types of protected categories, as determined by applicable laws and organizational requirements. Legally, the complexities of the interaction of multiple protected categories has been explicated by Crenshaw [21] and others under the framework of *intersectionality*. In the fair machine learning literature, it has been studied under the topic of *rich subgroup fairness* [49]. In recommender systems, there is some research involving sub-group fairness across providers [65]. However, no existing work addresses the compound nature of the disadvantage that Crenshaw highlights as characteristic for individuals who find themselves at the intersection of multiple protected identities.

Another simplification common in consumer-side fairness is assuming that categories are binary (protected vs unprotected) rather than constellations of attributes, including continuous qualities. Some work can naturally deal with multiple attributes—aggregating utility or error by group works fine for multiple groups, and multi-group statistical tests are effective—but optimization approaches typically assume that protected and unprotected groups should have equal results, and generalizing beyond two groups is seldom done.

## 4   Provider Fairness

Providers, the other side of the most common multistakeholder setup of recommender systems, are primarily the suppliers of information or items being recommended. There may be multiple parties who could be considered providers in most scenarios. For example, in news recommendation application, the journalist or publication venue (in a federated system) could be considered the provider. In music recommendation, it could be the individual artist or the record label. In job recommendation, it could be the employer (when candidates are seeking jobs) or the prospective employee (when recruiters are looking for candidates).

The key benefit of a recommender system for providers is that the system provides exposure of their items to recommendation consumers who may be interested in those items. We can think of recommendation opportunities as a resource and the distribution of those opportunities across different groups as the characteristic provider fairness concern. For example, in the candidate recruitment recommendation scenario, when an employer is looking for candidates, different protected groups (e.g., gender and ethnic groups) should be treated fairly. Yet, the challenge remains to provide balanced outcomes between relevant and fairly distributed set of candidates.

Diversity (see Chapter "Novelty and Diversity in Recommender Systems") is related to provider fairness. Some methods of diversity enhancement in recommendation may promote provider fairness, but diversity and fairness respond to different normative concerns. Diversity in recommendation and search results is mainly focused on consumer intent, by presenting results that meet a wide range of users' topical needs. In contrast, provider fairness is motivated by justice concerns to ensure that different providers receive fair opportunity for their content or products to be discovered.

## 4.1 Provider Utility

In contrast to consumer-side fairness, which is concerned with the quality of recommendations delivered to end-users, for provider-side fairness we are concerned with the utility of recommendations to providers. As noted above, a key notion in provider-side fairness is *exposure*: the value of recommendation opportunities given to a particular item or group of items.

Because recommendation lists are often short, a single list contains only a small number of recommendation opportunities; further, due to user attention patterns, not all list positions have equal utility, and only one item can have the benefits of the most valuable ranking in any one ranking [24]. Therefore, provider utility must usually be measured cumulatively (or in some cases amortized) over a sequence of recommendation results delivered to users. This is in contrast to the consumer-side case, where we can often meaningfully evaluate the quality of results for each user by looking only at a single recommendation result.

There are a number of options for computing the exposure/utility associated with a given recommendation opportunity. It may be convenient to assign a fixed utility to all items in a given recommendation list. In this case, the utility accrued to an item is simply the hit rate of its appearance across the recommendation experiment [51]. Utility can also be weighted with a function estimating user attention to an item, which will vary greatly based on its ranking in a recommendation list. This is commonly represented through a logarithmic discount for higher rank items, as captured in the standard NDCG metric, or a geometric discount [8]; for example, the exposure of an item $i$ in a collection $\mathcal{L}$ of rankings under a geometric decay model with patience $0 < \gamma < 1$ can be computed by:

$$\text{Ex}_{\text{geom}}(i|\mathcal{L}) = \sum_{L \in \mathcal{L}} \gamma^{\text{rank}_L(i)}$$

## 4.2 Individual Fairness

The individual fairness principle for producer fairness says that "similar" items (or content producers) should receive similar utility from the recommender system. Defining similarity with respect to the recommendation task, however, requires some subtlety, particularly since each user will like a different set of items (as opposed to nonpersonalized classification settings, where items are similar with respect to a single task such as repaying a loan). One straightforward approach is to define "similar" based on relevance to the user: two items are similar if the same user likes them both. Under this paradigm, we can compare the exposure an item receives to the relevance or utility it provides to users; Biega et al. [8] do this by aggregating each item's attention and relevance over multiple rankings, and measuring the extent to which attention is proportional to relevance (user-centric utility). Diaz et al. [24] use relevance to determine an *ideal exposure*—the exposure items would receive if the system ranked the items perfectly with respect to user utility—and computes the difference between that exposure and the exposure under the actual system:

$$\text{EEL} = \sum_i \left( \text{Ex}_{\text{geom}}(i|\mathcal{L}) - \text{Ex}_{\text{geom}}(i|\mathcal{L}_{\text{ideal}}) \right)^2$$

Another way to think about individual fairness is through latent representation. If individual items in the latent space are clustered together, then the downstream task of recommendation would preserve such characteristics [50]. However, there is open work to explore how to formulate and achieve such representation specifically in the context of recommender systems by taking the trade-offs between personalization and fairness into consideration.

## 4.3 Group Fairness

As with consumer fairness and general ML fairness, provider group fairness seeks to ensure that content providers in different groups are treated fairly. In recommender systems, this needs to be connected to system's personalization goal: recommending items a user does not like to achieve a group-fairness goal is usually not appropriate.

One way to operationalize group-fairness is to measure or enforce statistical independence between the recommendation outcomes and protected attributes [46]. Under this definition, introduced in Sect. 2.1, the recommendations are fair if the probability an item will appear in a particular recommendation list is independent of its sensitive attribute(s).

Another is to measure the extent to which protected groups are equitably represented in recommendation lists. This can be done by looking at the composition of individual recommendation rankings, measuring the divergence between the distribution of groups within a list and a target distribution [14, 22, 70]. Given a recommendation list $L$, these approaches compute a distribution over protected groups $P(A|L)$ and a target $P_\tau(A)$; Sapiezynski et al. [60] call this target a *population estimator*. Given these two distributions, they can be compared using difference in probabilities, odds ratios, KL divergence [22, 70], or cross-entropy [23].

Several approaches have been adopted to deal with the decreasing value of progressively lower ranking positions. Yang and Stoyanovich [70] and Zehlike et al. [74] use simple binomial group distributions (e.g. the fraction of list items produced by the protected group) and average the deviation from the target over successively longer prefixes of the ranking. Sapiezynski et al. [60] integrate rank discounting into $P(A|L)$ itself, so it is the probability of picking an item by a particular group following the browsing model in the rank discount function.

Provider group fairness can also be assessed over multiple rankings through an exposure metric. Aggregating expected exposure by group [8, 24] measures whether each producer *group*—rather than individual producer—receives an appropriate level of exposure.

Lastly, pairwise metrics [7] directly implement the concept of disparate mistreatment over pairs of items, extending BPR loss [58] into a fairness construct. A probabilistic ranking is considered fair if the probability of correctly ranking a relevant item (one the user has clicked) over an irrelevant item is independent of the relevant item's group membership. They define two versions of pairwise fairness: intra-group fairness, where the system is equally correct in its rankings of items within each group, and intra-group fairness, where the system correctly orders a relevant item from one group over an irrelevant item from another.

# 5 Other Stakeholders

While fairness-aware recommendation research to date concentrates largely on consumers and providers, the concept of multistakeholder recommender (see Chapter "Multistakeholder Recommender Systems") includes others raising fairness concerns that a system should address. Indeed, in some settings, regulatory agencies may be the most important stakeholders for deciding the minimum legal standards with respect to fairness and/or non-discrimination that a system must meet. In other cases, the structure of a platform includes stakeholders who are neither consumers nor providers, but are still impacted by specific parameters of transactions on the platform.

For example, consider a food delivery platform such as UberEats[4] [1]. This platform uses recommendations to match consumers with restaurants where they might order food to be delivered, and deliveries are made by Uber's drivers. There may be fairness concerns relative to the consumers or providers here, but there may also be fairness concerns over the set of drivers. These individuals do not participate in the recommendation interaction but the recommendations may impact them; for example, a goal might be to ensure that protected groups among the driver population do not receive fewer orders than others or do not receive a disproportionate number of difficult and/or low-tip jobs.

Another example is *subject fairness*, where the goal is to be fair towards the subjects of the items retrieved. In our news recommendation example, this would arise if we wish to ensure that different topics receive fair coverage in the recommended news articles. This fairness concern may break down along traditional lines of discrimination, such as ethnic groups, or it may arise along geographic lines such as urban vs. rural issues and stories. Karako and Manggala [47] demonstrate an application of a subject fairness constraint in image recommendation. In practice, subject fairness and provider fairness will often be measurable with similar techniques, since both are properties of the items being recommended, but providing one does not necessarily imply the other.

There has been very little published work to date that considers these stakeholders indirectly impacted by recommendation provision, but it is an important direction for future research. The problem of identifying the entities and groups so affected requires significant background data or content analysis.

## 6   Fairness over Time

A recommendation system operates in an iterated, changing environment, continuously making new decisions, gaining fresh users, and losing established users. Analyzing the fairness of decisions at one point in time can overlook temporal patterns of equity. In the context of predictive policing, Ensign et al. [29] theoretically demonstrate how feedback loops can result in fair decision-making.

Hashimoto et al. [40] study feedback loops in production systems. The authors model a population of users iteratively engaging with a system that trains using behavioral data. The model and supporting experiments in the context of predictive typing demonstrate that, over time, machine learning algorithms pay more attention to dominant subgroups of users as they lose under-represented subgroups of users.

---

[4] http://www.ubereats.com/.

The authors propose applying techniques from distributionally robust optimization to achieve more balanced performance, resulting in broad user retention. Zhang et al. [75] extend this work by analyzing the dynamics of fairness in sequential decision-making. Sonboli et al. [64] present an adaptive recommendation approach to multidimensional fairness using probabilistic social choice to control subgroup fairness over time.

Chaney et al. [18] study the homogenization of recommendations in iterative environments. They find that recommendation systems, especially those based on machine learning, increase the consistency in recommendations across different users but also tend to increase the inequity of exposure across items.

## 7   Methods for Mitigation

Unfair biases can enter a recommender system at any step of the process, from data collection to evaluation, and there are approaches to mitigating unfairness at each of them. In Sects. 3 and 4 we discussed different fairness constructs and evaluation metrics for recommender systems; this section turns to methods for mitigating these biases in data and algorithms (user interface and experience dimensions biases, while important, are out of scope for this chapter). There is no one solution for fairness problems; different applications and objectives will require different techniques.

### 7.1   Fairness in Data

There has been limited work to date on data-level fairness interventions for recommender systems. Several techniques have appeared in the broader machine learning literature; they may be applicable directly for reducing bias from aspects of item or user representations, but likely need further adaptation to account for the user- and item-level non-independence of the consumption and preference data used to train recommender systems.

Chen et al. [19] identify different data collection enhancements to cure different kinds of biases resulting from data issues. If there is a group that is under-represented in the training data and therefore receiving lower-quality results, collecting more labeled data from this group can improve the accuracy of their classifications or recommendations. If there is a cluster of users, producers, or items, who experience disparate adverse effects, and that disparity is not explainable by the available features in the data set, then collecting additional features can enhance system fairness.

Other approaches modify the available data in order to reduce unfairness. Feldman et al. [30] propose to perturb the values of insensitive features $X$ so their distributions are independent of a sensitive attribute $A$; if insensitive features are no longer correlated with group membership, then the outcomes of machine learning model trained on them will also be uncorrelated with group membership.

Unfortunately, these techniques have not yet—to our knowledge—been adapted to recommendation settings where much of the learning is done from multiple user interactions with an item. Gebru et al. [33] argue that extensive documentation on motivation for collecting the data, the composition of different components, collection process, and pre-processing can facilitate design decisions to enhance fairness, a lesson which seems as applicable in recommender systems as in other ML applications.

## 7.2   Fairness in Ranking Models

The ranking or scoring model itself can also be modified to improve the fairness of results. This usually takes the form of a multi-objective optimization problem that seeks to simultaneously maximize utility (or relevance) and minimize unfairness under some suitable definition discussed in the preceding sections.

Kamishima et al. [46] does this through *independence*: learning a matrix decomposition that augments the standard regularized reconstruction error loss (see Chapter "Advances in Collaborative Filtering") with an independence term that penalizes correlations between a protected attribute and the system's predicted ratings. They propose multiple non-independence measures, including the difference in means between groups and the mutual information between predicted ratings and sensitive attributes. Independence can also be applied to consumer fairness [45], with the objective that particular item recommendations are independent of users' sensitive attributes. Yao and Huang [73] apply a similar regularization approach to minimize disparate rating prediction errors rather than recommendation errors. Beutel et al. [6] augment pairwise rank loss with a penalty term for disparate rank effectiveness for different groups: this penalty is minimized when the difference in scores between relevant and irrelevant items is uncorrelated with the relevant item's sensitive attribute. It is also possible to directly optimize a learning-to-rank model for equal expected exposure [24].

Constrained optimization approaches such as that of [8] produce rankings that maximize user relevance score, subject to constraints on the cumulative attention received by individual items across different users. This can be done through an integer linear programming formulation.

Fair policy learning applies the multi-objective frame to policy learning for a reinforcement learning agent; the agent is augmented with a constraint that the expected unfairness (as measured by configurable fairness metric) is less than a threshold $\delta$ [63].

## 7.3 Fairness Through Re-ranking

Fairness can also be improved by re-ranking lists that are already optimized for relevance, similar to designs used for enhancing diversity (see Chapter "Novelty and Diversity in Recommender Systems"). There is a basic distinction between approaches that treat the problem as a global optimization task, trying to improve the fairness of an entire collection of recommendation lists, and approaches that are focused on a single list.

An example of the global approach is that of Sürer et al. [68], who suggest a constrained optimization-based method at the post-processing level to enhance fairness for multiple provider groups. This approach is generalizable to multiple constraints, such as (a) the inclusion of items from different provider groups, (b) ensuring a minimum degree of item diversity for each consumer, and (c) avoiding unfairness towards providers from underrepresented populations. The global approach might be suitable in cases where recommendations are all generated at once, such as for a mass personalized email or weekly recommended items list. It could also apply to applications where recommendations are generated in advance and cached.

A more typical approach is to re-rank individual lists as they are generated. Some authors borrow from such approaches as MMR [16] and xQUAD [59] in the information retrieval literature in proposing greedy list extension methods, where the re-ranked list is built by adding items that, at each point, provide an optimal tradeoff between accuracy and fairness. Geyik et al. [34] use a greedy approach to produce rankings of job candidates to achieve a desired (non-discriminatory) distribution of candidates' protected attributes, simultaneously optimizing relevance and fairness. Similarly, Modani et al. [55] re-rank to enhance provider exposure without sacrificing relevance. Zehlike et al. [74] uses a different approach to the re-ranking problem, using the A-star algorithm to achieve distributional fairness in a ranked list at depth K. The objective of the algorithm is to re-arrange ranked items to meet the desired equity distribution of items from different protected groups while maintaining the rank quality.

Liu et al. [51] propose a method that incorporates both provider fairness and user preferences for diverse results. Protected groups are promoted by assigning a higher value to items from such groups in the recommendation objective. The promotion of protected groups is tempered by a factor tied to user profile diversity. This method enhances the trade-off between accuracy and fairness by exposing protected group items to users most likely to be interested in them. Sonboli et al. [65] extend this approach to multiple protected groups defined by different provider features. Each user can be characterized in terms of the opportunity they provide to promote one or another fairness concern across the different groups.

Reranking can also be applied to consumer fairness. Abdollahpouri [1] present a re-ranking approach based on the idea of calibration [67] to improve the fairness for user groups with niche interests.

## 7.4 Fairness Through Engineering

Data and algorithmic interventions are not necessarily the best way to achieve some types of fairness. While opportunities for providers to be exposed is zero-sum for a fixed set of recommendation requests (a recommendation slot given to one provider cannot be given to another), consumer recommendation quality is not: solving disparate recommendation utility by decreasing the quality of some users' recommendations does not necessarily improve the quality of others. Instead, identifying the existence of under-served consumers and reasons their recommendation quality suffers may help in product development, as features which will improve their experience can be prioritized over features that will primarily provide a marginal improvement for users already well-served by the system.

## 8 Conclusion

Fair recommendation is a relatively new but rapidly growing corner of the recommender systems research literature. The work in this space draws from concerns that have long been of interest to recommender systems researchers, such as those motivating long-tail recommendation and the study of popularity bias, but connects it to the emerging field algorithmic fairness and its roots in the broader literature on fairness and discrimination in general.

Fairness and recommendation are both complex, multifaceted problems, and their combination requires particular clarity on the precise effects and harms to be measured and/or mitigated. In particular, the multisided nature and ranked outputs of many recommender systems complicate the problem of assessing their fairness, as we must identify which stakeholders we are concerned with treating fairly and develop a definition of fairness that applies to specific harms in the context of repeated, ranked outputs, among other challenges. The work of fair recommendation often requires data that is not commonly included with recommender system data sets, particularly when seeking to ensure recommendations are fair with respect to sensitive characteristics users or creators, such as their gender or ethnicity.

This work must also be done with great care and compassion to ensure that users and creators are treated with respect and dignity and to avoid various traps that result in overbroad or ungeneralizable claims. We argue that there is nothing particularly new about these requirements, but that thinking about the fairness of recommendations brings to the surface issues that should be considered in all recommendation research and development.

While there has been much useful work in mapping the problem space and addressing certain types of unfairness, there are many open problems in fair recommendation that need attention. Some of the ones we see include:

- *Extending the concepts and methods of fair recommendation research to additional domains, applications, problem framings, and axes of fairness concerns.*

Due to the specific and distinct ways in which social biases and discrimination manifest [62], we cannot assume that findings on one bias translate to another (e.g. findings on race may not apply to ethnicity or geographic location), or that findings on a particular bias in one application will translate to another (e.g. ethnic bias may manifest differently in recommendation vs. NLP classification tasks). Over time, generalizable principles may be discovered and give rise to theories that enable the prediction of particular biases and their manifestations, but at the present time we need to study a wide range of biases and applications to build the knowledge from which such principles may be derived.

- *Deeper study of the development and evolution of biases over time.* Most work— with the exception of fair policy learning and a handful of other studies—focuses on one-shot batch evaluation of recommender algorithms and their fairness. However, recommender behavior is dynamic over time as the system produces recommendations, users respond to them, and it learns from their feedback. This dynamicism means that an initially fair system may become unfair over time if users respond to it in a biased or discriminatory fashion, or that it may move towards a more fair state if users respond well to recommendations that increase overall fairness.

- *Define and study further fairness concerns beyond consumer and provider fairness.* We have identified subject fairness as one additional concern here, but doubt it is the only additional stakeholder whose equity concerns should be considered.

- *Study human desires for and response to fair recommendations.* The first works are beginning to surface in this direction [66], and Harambam et al. [37] explored users' desired features and capabilities for recommendation with concerns that touch on fairness, but at present little is known about what users or content providers expect from a system with respect to its fairness, or how users will respond to fairness-enhancing recommendation interventions.

- *Develop appropriate metrics for recommendation fairness, along with thorough understanding of the requirements and behavior of fairness metrics and best practices for applying them in practical situations.* For example, we believe expected exposure [24] and pairwise fairness [7] are useful frameworks for reasoning about many provider fairness concerns, but there is much work left to do to understand how best to apply and interpret them in offline and online studies.

- *Develop standards and best practices for recommender systems data and model provenance.* Gebru et al. [33] presented the idea of *datasheets* for data sets, thoroughly and carefully documenting data so downstream users can properly assess its applicability, limitations, and the appropriateness of a proposed use. Harper and Konstan [39] provide much of this information for the MovieLens data set, but many other data sets do not have such documentation. Mitchell et al. [53] provide a framework for documenting important properties of trained models, and Yang et al. [71] present a "nutrition label" for (non-personalized) rankings describing their data sources, ranking principles, and other information. These concepts need to be extended to recommender systems and the complex integrated data sets that drive them. Common practices such as pruning may have deep implications for experiment and recommendation outcomes [5], emphasizing the need for careful study of the properties of recommendation data, models, and outputs that should be documented.

- *Engage more deeply with the multidimensional and complex nature of bias.* Most work fair recommendation focuses on single attributes in isolation, often restricting them to binary values. However, the intersection of group memberships often gives rise to particular forms of discrimination and social bias that cannot be explained by any one of the groups alone [21]. Some recent work engages with multiple simultaneous axes of discrimination or fairness [72], but multidimensionality does not fully capture the dynamics invoked by the concept of intersectionality [42]. Further, many categories are complex, unstable, and socially constructed; Hanna et al. [36] present a treatment of some of these complexities in an algorithmic setting, but much work remains to respond to that call and make fairness responsive to these realities.

There is a lot of open space for research in fair recommendation, and this work has the potential for significant improvements to the human and societal impact of the systems this book teaches its readers to build and study.

# References

1. H. Abdollahpouri, Popularity bias in recommendation: a multi-stakeholder perspective. PhD thesis, University of Colorado Boulder, 2020
2. M. Ali, P. Sapiezynski, M. Bogen, A. Korolova, A. Mislove, A. Rieke, Discrimination through optimization: how Facebook's ad delivery can lead to biased outcomes, in *Proceedings of the ACM on Human-Computer Interaction*, vol. 3, no. CSCW (2019), pp. 1–30. https://doi.org/10.1145/3359301
3. S. Barocas, A.D. Selbst, Big data's disparate impact. Calif. Law Rev. **104**(3), 671 (2016). https://doi.org/10.15779/Z38BG31
4. S. Barocas, M. Hardt, A. Narayanan, *Fairness and Machine Learning: Limitations and Opportunities* (2019). https://fairmlbook.org/
5. J. Beel, V. Brunel, Data pruning in recommender systems research: Best-Practice or malpractice? in *ACM RecSys 2019 Late-Breaking Results* (2019)
6. A. Beutel, J. Chen, Z. Zhao, E.H. Chi, Data decisions and theoretical implications when adversarially learning fair representations. Preprint (2017). https://doi.org/1707.00075
7. A. Beutel, E.H. Chi, C. Goodrow, J. Chen, T. Doshi, H. Qian, L. Wei, Y. Wu, L. Heldt, Z. Zhao, L. Hong, Fairness in recommendation ranking through pairwise comparisons, in *Proceedings of the 25th ACM SIGKDD International Conference on Knowledge Discovery & Data Mining* (ACM, New York, 2019). https://doi.org/10.1145/3292500.3330745
8. A.J. Biega, K.P. Gummadi, G. Weikum, Equity of attention: amortizing individual fairness in rankings, in *The 41st International ACM SIGIR Conference on Research & Development in Information Retrieval* (ACM, New York, 2018), pp. 405–414. https://doi.org/10.1145/3209978.3210063
9. A.J. Biega, F. Diaz, M.D. Ekstrand, S. Kohlmeier, Overview of the TREC 2019 fair ranking track, in *Proceedings of the Twenty-Eighth Text REtrieval Conference (TREC 2019)* (2020)

10. A. Billey, M. Haugen, J. Hostage, N. Sack, A.L. Schiff, Report of the PCC ad hoc task group on gender in name authority records. Tech. rep., Program for Cooperative Cataloging (2016). https://www.loc.gov/aba/pcc/documents/Gender_375%20field_RecommendationReport.pdf
11. J. Buolamwini, T. Gebru, Gender shades: intersectional accuracy disparities in commercial gender classification, in *Proceedings of the 1st Conference on Fairness, Accountability, and Transparency, PMLR*, vol. 81 (2018), pp. 77–91
12. R. Burke, Multisided fairness for recommendation. Preprint (2017). https://doi.org/1707.00093
13. R. Burke, J. Kontny, N. Sonboli, Synthetic attribute data for evaluating consumer-side fairness. Preprint (2018). https://doi.org/1809.04199
14. R. Burke, N. Sonboli, A. Ordonez-Gauger, Balanced neighborhoods for multi-sided fairness in recommendation, in *Proceedings of the 1st Conference on Fairness, Accountability and Transparency, PMLR*, vol. 81, ed. by S.A. Friedler, C. Wilson (2018), pp. 202–214
15. R. Cañamares, P. Castells, Should I follow the crowd?: a probabilistic analysis of the effectiveness of popularity in recommender systems, in *The 41st International ACM SIGIR Conference on Research & Development in Information Retrieval* (ACM, New York, 2018), pp. 415–424. https://doi.org/10.1145/3209978.3210014
16. J. Carbonell, J. Goldstein, The use of MMR, diversity-based reranking for reordering documents and producing summaries, in *Proceedings of the 21st Annual International ACM SIGIR Conference on Research and Development in Information Retrieval* (ACM, New York, 1998), pp. 335–336. https://doi.org/10.1145/290941.291025
17. Ò. Celma, *Music Recommendation and Discovery: The Long Tail, Long Fail, and Long Play in the Digital Music Space* (Springer, Berlin, 2010). https://doi.org/10.1007/978-3-642-13287-2
18. A.J.B. Chaney, B.M. Stewart, B.E. Engelhardt, How algorithmic confounding in recommendation systems increases homogeneity and decreases utility, in *Proceedings of the 12th ACM Conference on Recommender Systems* (ACM, New York, 2018), pp. 224–232. https://doi.org/10.1145/3240323.3240370
19. I. Chen, F.D. Johansson, D. Sontag, Why is my classifier discriminatory? in *Advances in Neural Information Processing Systems*, vol. 31, ed. by S. Bengio, H. Wallach, H. Larochelle, K. Grauman, N. Cesa-Bianchi, R. Garnett (2018), pp. 3539–3550
20. K. Crawford, The trouble with bias, in *Neural Information Processing Systems* (2017)
21. K. Crenshaw, Demarginalizing the intersection of race and sex: a black feminist critique of antidiscrimination doctrine, feminist theory and antiracist politics. Univ. Chic. Leg. Forum **1989**, 139–168 (1989)
22. A. Das, M. Lease, A conceptual framework for evaluating fairness in search. Preprint (2019). https://doi.org/1907.09328
23. Y. Deldjoo, V.W. Anelli, H. Zamani, A. Bellogin, T. Di Noia, Recommender systems fairness evaluation via generalized cross entropy, in *Proceedings of the Workshop on Recommendation in Multi-stakeholder Environments at RecSys '19, CEUR-WS*, vol. 2440 (2019)
24. F. Diaz, B. Mitra, M.D. Ekstrand, A.J. Biega, B. Carterette, Evaluating stochastic rankings with expected exposure, in *Proceedings of the 29th ACM International Conference on Information and Knowledge Management* (ACM, New York, 2020). https://doi.org/10.1145/3340531.3411962
25. C. Dwork, M. Hardt, T. Pitassi, O. Reingold, R. Zemel, Fairness through awareness, in *Proceedings of the 3rd Innovations in Theoretical Computer Science Conference* (ACM, New York, 2012), pp. 214–226. https://doi.org/10.1145/2090236.2090255
26. M.D. Ekstrand, D. Kluver, Exploring author gender in book rating and recommendation. User Model. User-Adap. Inter. (2021) https://doi.org/10.1007/s11257-020-09284-2
27. M.D. Ekstrand, M. Tian, I.M. Azpiazu, J.D. Ekstrand, O. Anuyah, D. McNeill, M.S. Pera, All the cool kids, how do they fit in?: Popularity and demographic biases in recommender evaluation and effectiveness, in *Proceedings of the Conference on Fairness, Accountability, and Transparency (PMLR), New York, PMLR*, vol. 81, ed. by S.A. Friedler, C. Wilson (2018), pp. 172–186

28. M.D. Ekstrand, M. Tian, M.R.I. Kazi, H. Mehrpouyan, D. Kluver, Exploring author gender in book rating and recommendation, in *Proceedings of the Twelfth ACM Conference on Recommender Systems* (ACM, New York, 2018). https://doi.org/10.1145/3240323.3240373

29. D. Ensign, S.A. Friedler, S. Neville, C. Scheidegger, S. Venkatasubramanian, Runaway feedback loops in predictive policing, in *Proceedings of the 1st Conference on Fairness, Accountability and Transparency, New York, PMLR*, vol. 81, ed. by S.A. Friedler, C. Wilson (2018), pp. 160–171

30. M. Feldman, S.A. Friedler, J. Moeller, C. Scheidegger, S. Venkatasubramanian, Certifying and removing disparate impact, in *Proceedings of the 21th ACM SIGKDD International Conference on Knowledge Discovery and Data Mining* (ACM, New York, 2015), pp. 259–268. https://doi.org/10.1145/2783258.2783311

31. A. Ferraro, Music cold-start and long-tail recommendation: bias in deep representations, in *Proceedings of the 13th ACM Conference on Recommender Systems* (ACM, New York, 2019), pp. 586–590. https://doi.org/10.1145/3298689.3347052

32. B. Fish, A. Bashardoust, D. Boyd, S. Friedler, C. Scheidegger, S. Venkatasubramanian, Gaps in information access in social networks? in *WWW '19: The World Wide Web Conference* (ACM, New York, 2019), pp. 480–490. https://doi.org/10.1145/3308558.3313680

33. T. Gebru, J. Morgenstern, B. Vecchione, J.W. Vaughan, H. Wallach, H. Daumeé III, K. Crawford, Datasheets for datasets. Preprint (2018). https://doi.org/1803.09010

34. S.C. Geyik, S. Ambler, K. Kenthapadi, Fairness-Aware ranking in search & recommendation systems with application to LinkedIn talent search, in *Proceedings of the 25th ACM SIGKDD International Conference on Knowledge Discovery & Data Mining* (ACM, New York, 2019), pp. 2221–2231. https://doi.org/10.1145/3292500.3330691

35. F. Hamidi, M.K. Scheuerman, S.M. Branham, Gender recognition or gender reductionism?: The social implications of embedded gender recognition systems, in *Proceedings of the 2018 CHI Conference on Human Factors in Computing Systems* (ACM, New York, 2018), p. 8. https://doi.org/10.1145/3173574.3173582

36. A. Hanna, E. Denton, A. Smart, J. Smith-Loud, Towards a critical race methodology in algorithmic fairness, in *Proceedings of the 2020 Conference on Fairness, Accountability, and Transparency* (ACM, New York, 2020), pp. 501–512. https://doi.org/10.1145/3351095.3372826

37. J. Harambam, D. Bountouridis, M. Makhortykh, J. van Hoboken, Designing for the better by taking users into account: a qualitative evaluation of user control mechanisms in (news) recommender systems, in *Proceedings of the 13th ACM Conference on Recommender Systems* (ACM, New York, 2019), pp. 69–77. https://doi.org/10.1145/3298689.3347014

38. M. Hardt, E. Price, N. Srebro, Equality of opportunity in supervised learning, in *Advances in Neural Information Processing Systems* (2016), pp. 3315–3323

39. F.M. Harper, J.A. Konstan, The MovieLens datasets: history and context. ACM Trans. Interact. Intell. Syst. **5**(4), 19:1–19:19 (2015). https://doi.org/10.1145/2827872

40. T. Hashimoto, M. Srivastava, H. Namkoong, P. Liang, Fairness without demographics in repeated loss minimization, in *Proceedings of the 35th International Conference on Machine Learning, Stockholmsmässan, Stockholm Sweden, PMLR*, vol. 80, ed. by J. Dy, A. Krause (2018), pp. 1929–1938

41. T. Hentschel, S. Braun, C.V. Peus, D. Frey, Wording of advertisements influences women's intention to apply for career opportunities. Acad. Manag. Proc. **2014**(1), 15994 (2014). https://doi.org/10.5465/ambpp.2014.15994abstract

42. A.L. Hoffmann, Where fairness fails: data, algorithms, and the limits of antidiscrimination discourse. Inf. Commun. Soc. **22**(7), 900–915 (2019). https://doi.org/10.1080/1369118X.2019.1573912

43. B. Hutchinson, M. Mitchell, 50 years of test (un)fairness: lessons for machine learning, in *FAT 2019: Proceedings of the Conference on Fairness, Accountability, and Transparency* (ACM, New York, 2019), pp. 49–58. https://doi.org/10.1145/3287560.3287600

44. N. Kallus, X. Mao, A. Zhou, Assessing algorithmic fairness with unobserved protected class using data combination. Preprint (2019). https://doi.org/1906.00285

45. T. Kamishima, S. Akaho, Considerations on recommendation independence for a Find-Good-Items task, in *Workshop on Fairness, Accountability and Transparency in Recommender Systems at RecSys 2017* (2017)

46. T. Kamishima, S. Akaho, H. Asoh, J. Sakuma, Recommendation independence, in *Proceedings of the 1st Conference on Fairness, Accountability and Transparency, PMLR*, vol. 81, ed. by S.A. Friedler, C. Wilson (2018), pp. 187–201

47. C. Karako, P. Manggala, Using image fairness representations in Diversity-Based re-ranking for recommendations, in *Adjunct Publication of the 26th Conference on User Modeling, Adaptation and Personalization* (ACM, New York, 2018), pp. 23–28. https://doi.org/10.1145/3213586.3226206

48. M. Kaya, D. Bridge, N. Tintarev, Ensuring fairness in group recommendations by Rank-Sensitive balancing of relevance, in *Fourteenth ACM Conference on Recommender Systems* (ACM, New York, 2020), pp. 101–110, https://doi.org/10.1145/3383313.3412232

49. M. Kearns, S. Neel, A. Roth, Z.S. Wu, An empirical study of rich subgroup fairness for machine learning, in *Proceedings of the Conference on Fairness, Accountability, and Transparency* (ACM, New York, 2019), pp. 100–109. https://doi.org/10.1145/3287560.3287592

50. P. Lahoti, K.P. Gummadi, G. Weikum, iFair: learning individually fair data representations for algorithmic decision making, in *2019 IEEE 35th International Conference on Data Engineering (ICDE)* (2019), pp. 1334–1345. https://doi.org/10.1109/ICDE.2019.00121

51. W. Liu, J. Guo, N. Sonboli, R. Burke, S. Zhang, Personalized fairness-aware re-ranking for microlending, in *Proceedings of the 13th ACM Conference on Recommender Systems* (ACM, New York, 2019). https://doi.org/10.1145/3298689.3347016

52. R. Mehrotra, A. Anderson, F. Diaz, A. Sharma, H. Wallach, E. Yilmaz, Auditing search engines for differential satisfaction across demographics, in *Proceedings of the 26th International Conference on World Wide Web Companion, International World Wide Web Conferences Steering Committee, Republic and Canton of Geneva* (2017), pp. 626–633. https://doi.org/10.1145/3041021.3054197

53. M. Mitchell, S. Wu, A. Zaldivar, P. Barnes, L. Vasserman, B. Hutchinson, E. Spitzer, I.D. Raji, T. Gebru, Model cards for model reporting, in *Proceedings of the Conference on Fairness, Accountability, and Transparency* (ACM, New York, 2019), pp. 220–229. https://doi.org/10.1145/3287560.3287596

54. S. Mitchell, E. Potash, S. Barocas, A. D'Amour, K. Lum, Algorithmic fairness: choices, assumptions, and definitions. Annu. Rev. Stat. Appl. **8** (2020). https://doi.org/10.1146/annurev-statistics-042720-125902

55. N. Modani, D. Jain, U. Soni, G.K. Gupta, P. Agarwal, Fairness aware recommendations on Behance, in *Advances in Knowledge Discovery and Data Mining* (Springer International Publishing, 2017), pp. 144–155. https://doi.org/10.1007/978-3-319-57529-2_12

56. M. Nasr, M.C. Tschantz, Bidding strategies with gender nondiscrimination constraints for online ad auctions, in *Proceedings of the 2020 Conference on Fairness, Accountability, and Transparency* (ACM, New York, 2020), pp. 337–347. https://doi.org/10.1145/3351095.3375783

57. A. Olteanu, C. Castillo, F. Diaz, E. Kıcıman, Social data: biases, methodological pitfalls, and ethical boundaries. Front. Big Data **2**, 13 (2019). https://doi.org/10.3389/fdata.2019.00013

58. S. Rendle, C. Freudenthaler, Z. Gantner, L. Schmidt-Thieme, BPR: Bayesian personalized ranking from implicit feedback, in *Proceedings of the Twenty-Fifth Conference on Uncertainty in Artificial Intelligence* (AUAI Press, Arlington, 2009), pp. 452–461

59. R.L.T. Santos, J. Peng, C. Macdonald, I. Ounis, Explicit search result diversification through sub-queries, in *ECIR 2010: Advances in Information Retrieval*. LNCS, vol. 5993 (Springer, 2010), pp. 87–99. https://doi.org/10.1007/978-3-642-12275-0_11

60. P. Sapiezynski, W. Zeng, E.R. Robertson, A. Mislove, C. Wilson, Quantifying the impact of user attention on fair group representation in ranked lists, in *Companion Proceedings of The 2019 World Wide Web Conference* (ACM, New York, 2019), pp. 553–562. https://doi.org/10.1145/3308560.3317595

61. M. Schedl, The LFM-1b dataset for music retrieval and recommendation, in *Proceedings of the 2016 ACM on International Conference on Multimedia Retrieval* (ACM, New York, 2016), pp. 103–110. https://doi.org/10.1145/2911996.2912004

62. A.D. Selbst, D. Boyd, S.A. Friedler, S. Venkatasubramanian, J. Vertesi, Fairness and abstraction in sociotechnical systems, in *Proceedings of the Conference on Fairness, Accountability, and Transparency - FAT* '19* (ACM, New York, 2019), pp. 59–68. https://doi.org/10.1145/3287560.3287598

63. A. Singh, T. Joachims, Policy learning for fairness in ranking, in *Advances in Neural Information Processing Systems*, vol. 32, ed. by H. Wallach, H. Larochelle, A. Beygelzimer, F. d'Alché Buc, E. Fox, R. Garnett (2019), pp. 5426–5436

64. N. Sonboli, R. Burke, N. Mattei, F. Eskandanian, T. Gao, "and the winner is...": dynamic lotteries for multi-group fairness-aware recommendation. Preprint (2020). https://doi.org/2009.02590

65. N. Sonboli, F. Eskandanian, R. Burke, W. Liu, B. Mobasher, Opportunistic multi-aspect fairness through personalized re-ranking, in *Proceedings of the 28th ACM Conference on User Modeling, Adaptation and Personalization* (ACM, New York, 2020), pp. 239–247. https://doi.org/10.1145/3340631.3394846

66. N. Sonboli, J.J. Smith, F. Cabral Berenfus, R. Burke, C. Fiesler, Fairness and transparency in recommendation: the users' perspective, in *Proceedings of the 29th ACM Conference on User Modeling, Adaptation and Personalization* (2021), pp. 274–279. https://doi.org/10.1145/3450613.3456835

67. H. Steck, Calibrated recommendations, in *Proceedings of the 12th ACM Conference on Recommender Systems* (ACM, 2018), pp. 154–162. https://doi.org/10.1145/3240323.3240372

68. Ö. Sürer, R. Burke, E.C. Malthouse, Multistakeholder recommendation with provider constraints, in *Proceedings of the 12th ACM Conference on Recommender Systems* (ACM, New York, 2018), pp. 54–62. https://doi.org/10.1145/3240323.3240350

69. A. Xiang, I.D. Raji, On the legal compatibility of fairness definitions. Preprint (2019). https://doi.org/1912.00761

70. K. Yang, J. Stoyanovich, Measuring fairness in ranked outputs, in *Proceedings of the 29th International Conference on Scientific and Statistical Database Management* (ACM, New York, 2017), Article 22, pp. 1–6. https://doi.org/10.1145/3085504.3085526

71. K. Yang, J. Stoyanovich, A. Asudeh, B. Howe, H.V. Jagadish, G. Miklau, A nutritional label for rankings, in *Proceedings of the 2018 International Conference on Management of Data - SIGMOD '18* (ACM, New York, 2018), pp. 1773–1776. https://doi.org/10.1145/3183713.3193568

72. K. Yang, J.R. Loftus, J. Stoyanovich, Causal intersectionality for fair ranking. Preprint (2020). http://doi.org/2006.08688

73. S. Yao, B. Huang, Beyond parity: fairness objectives for collaborative filtering, in *Advances in Neural Information Processing Systems*, vol. 30, ed. by I. Guyon, U.V. Luxburg, S. Bengio, H. Wallach, R. Fergus, S. Vishwanathan, R. Garnett (2017), pp. 2925–2934

74. M. Zehlike, F. Bonchi, C. Castillo, S. Hajian, M. Megahed, R. Baeza-Yates, FA*IR: a fair top-k ranking algorithm, in *Proceedings of the 2017 ACM on Conference on Information and Knowledge Management* (ACM, 2017), pp. 1569–1578. https://doi.org/10.1145/3132847.3132938

75. X. Zhang, M. Khaliligarekani, C. Tekin, M. Liu, Group retention when using machine learning in sequential decision making: the interplay between user dynamics and fairness, in *Advances in Neural Information Processing Systems*, vol. 32, ed. by H. Wallach, H. Larochelle, A. Beygelzimer, F. d'Alché Buc, E. Fox, R. Garnett (2019), pp. 15269–15278

# Part IV
# Human Computer Interaction

# Beyond Explaining Single Item Recommendations

Nava Tintarev and Judith Masthoff

## 1 Introduction

Recommender systems such as Amazon, offer users recommendations, or suggestions of items to try or buy. Recommender systems use algorithms which can be categorized as ranking algorithms, which result in the increase in the prominence of some information. Other information (e.g., low rank or low confidence recommendations) are consequently filtered and not shown to people.

However, it is often not clear to a user whether the recommender system's advice is suitable to be followed, e.g., whether it is correct, whether the right information was taken into account, whether crucial information is missing. In other words, there is a large mismatch between the representation of the advice by the system versus the information needs of its users. In this chapter we describe how *explanations*, such as the one depicted in Fig. 1, can help bridge that gap, what we constitute as helpful explanations, as well as the considerations and external factors we must consider when determining the level of helpfulness.

Ultimately, for explanations of recommender systems to be useful, they need to be able to justify the recommendations in a *human understandable* way. This creates a necessity for techniques for automatic generation of satisfactory explanations that are *intelligible* (understandable) for users interacting with the system.

"*Interpretability*" of advice from artificial intelligence systems more broadly has been qualified as the degree to which a human can understand the cause of a decision [79]. However, understanding is rarely an end-goal in itself. Pragmatically, it is

N. Tintarev (✉)
Maastricht University, Maastricht, Netherlands
e-mail: n.tintarev@maastrichtuniversity.nl

J. Masthoff
Utrecht University, Utrecht, Netherlands
e-mail: j.f.m.masthoff@uu.nl

© Springer Science+Business Media, LLC, part of Springer Nature 2022
F. Ricci et al. (eds.), *Recommender Systems Handbook*,
https://doi.org/10.1007/978-1-0716-2197-4_19

**Fig. 1** Explanation in the Pandora system, retrieved November 2020: *"Based on what you've told us so far, we're playing this track because it features solo strings, mystical qualities, minor key tonality, melodic songwriting and intricate melodic phrasing."*

more useful to operationalize the effectiveness of explanations in terms of a specific notion of usefulness or **explanatory goals** such as improved decision support or user trust [123]. One aspect of the intelligibility of an explainable system (often cited for domains such as health) is the ability for users to accurately identify, or correct, an error made by the system. In that case it may be preferable to generate explanations that induce appropriate levels of reliance (in contrast to over- or under-reliance) [138], supporting the user in discarding recommendations when the system is incorrect, but also accepting correct recommendations. The domain affects not only the overall cost of an error, but the cost of a specific type of error (e.g., a false negative might be more harmful than a false positive for a news recommendation if it could be an important government announcement). In a domain such as news, a different goal might be more suitable, such as explanations that facilitate users' epistemic goals (e.g., broadening their knowledge within a topic) [112].

It is sometimes erroneously assumed that explanations need to be completely transparent with regard to the underlying algorithmic mechanisms. However, a transparent explanation is not necessary understandable to an end-user. [11] distinguishes between explanation and justification in the following way: *"a justification explains why a decision is a good one, without explaining exactly how it was made."*. That is, a user-centered explanation may not be fully transparent, but still useful if it fulfills an explanatory goal.

Assessing the effect of explanations on given explanatory goals requires systematic user-centered evaluation. To understand which explanation (e.g., with regard to modality, degree of interactivity, level of detail, and concrete presentational choices) are best for explanations, it is vital to identify which requirements are placed by *individual characteristics*, the *domain*, as well as the *context* in which the explanations are given.

New challenges stemming from recommender systems have revived explanation research, after a decline of studies in expert systems in the 90s. One such development is the increase in data: due to the growth of the web, many systems are being used by thousands of users rather than dozens or just a handful of experts. In addition, new algorithms, in particular in the domain of collaborative filtering, have been adapted and developed. These approaches mitigate domain dependence, and allow for greater generalizability, and are more suitable for large and often sparse datasets.

Research on explanations in recommender systems has been evaluated much more extensively than in previous advisory systems, and in a much wider range of domains varying from e-commerce (e.g., Pandora, Amazon), entertainment (e.g, movies [118]), to services (e.g., financial advice [32]), content (e.g., news [112] and cultural heritage artifacts [27]), as well as social recommendations [42, 43]. In this chapter we supply an overview of existing systems by studying various properties of the existing explanation facilities.

The chapter is structured as follows. Explanations are not strictly decoupled from recommendations themselves, the way preferences are elicited, or the way in which recommendations are presented: these factors influence each other and the explanations that can be generated. Therefore, in the next section we discuss these types of design choices.

This enables us to discuss how these choices interact with different explanation level and styles, including a table of explanations in commercial and academic systems (Sect. 3). We consider three levels: **(1)** individual user, **(2)** contextualization, and **(3)** self-actualization.

We also consider interactive explanations (which may combine explanations styles at all three previous levels). Looking at the different explanation styles we also start to sense that the underlying algorithm of a recommender engine may influence the types of explanations that can be generated.

Next, in Sect. 4, we discuss what defines a good explanation. We list different explanatory criteria, and describe how these have been measured in previous systems. These criteria can also be understood as advantages that explanations may offer to recommender systems, answering the question of *why* to explain. With regard to these criteria, a number of factors influence how effective explanations are, such as personal and situational characteristics, and we dedicate a section to these emerging findings as well in Sect. 5. We conclude with future directions in Sect. 6.

# 2   Presentation of, and Interaction with, Recommendations

Before we can assess what constitutes a useful explanation in Sect. 4, we need to discuss the system context in which explanations are generated. Namely, which explanations can be generated depend on how recommended items are presented, and how much interaction is afforded with recommended items.

Presentation and interaction are often intertwined in an interaction model. For example, one of the design guidelines of Pu et al. states: *"Showing one search result or recommending one item at a time allows for a simple display strategy which can be easily adapted to small display devices; however, it is likely to engage users in longer interaction sessions or only allow them to achieve relatively low decision accuracy."* (Guideline 9, [101]).

## 2.1 Presenting Recommendations

We summarize the ways of presenting recommendations that we have seen for the systems considered in this chapter. While there are a number of possibilities for the *appearance* of the graphical user interface, the actual *structure* of offering recommendations can also vary.

Below we identify categories for structuring the presentation of recommendations, and the kind of explanations that could be generated for them. Each of these may be static or interactive, allowing the user to control factors that influence the resulting recommendations. We discuss this further in relation to interaction with recommendations in Sect. 2.8.

- **Top item.** Perhaps the simplest way to present a recommendation is by offering the user the best item for them. E.g. *"You have been watching a lot of sports, and football in particular. This is the most popular and recent item from the world cup."*
- **Top N-items.** The system may also present several items at once. Items are presented as a list ranked by the predicted relevance to the user. Beyond that, no link between the items is assumed. An implicit assumption is that the user's goal is to find (some but not all) good items, and is satisfied as long as at least one item in the top-N is good enough [47]. *"Given your previous reading behavior it seems you are interested in football and technology. You might like to see the local football results and the gadget of the day."* Note that while this system could be able to explain the rational behind the list as a whole, it could also explain the rational behind each single item.
- **Set or Sequence of items.** Sets and sequences are similar to Top N-items in that the system presents several recommended items at once. Methods for sequences aware-recommendation are also described in chapter "Session-Based Recommender Systems". In sets and sequences, the items are related somehow, and are assumed to be consumed in combination. In the case of a *sequence*, such as a travel itinerary comprised of several points of interest, the order of items is considered important. In contrast, in a *set* such as matching top and trousers the order is not important [135]. *"This itinerary is designed to consider your interests and travel time on foot."; "These trousers fit the style and price range of outfit you already selected."*
- **Similar to top item(s).** Once a user shows a preference for one or more items, the recommender system can offer *similar* items. Unlike a set or package, these recommendations are not expected to be consumed together (and may in fact

aim to replace the original top item). E.g. *"Given you are considering buying this book you might also like... the Hitchhiker's Guide to the Galaxy by Douglas Adams"*.

- **Predicted ratings for all items.** Rather than forcing selections on the user, a system may allow its users to browse all the available options. Recommendations are then presented as predicted ratings on a scale (say from 0 to 5) for each item. A user might query why a certain item, for example local hockey results, is predicted to have a low rating. The recommender system might then generate an explanation like: *"While this is a sports article, it is about hockey, which you do not seem to like!"*.

- **Structured overview.** The recommender system can give a structure which displays trade-offs between items [99, 141]. The advantage of a structured overview is that the user can see how items compare, and what other items are still available if the current recommendation should not meet their requirements. An example of a structured overview can be seen in Fig. 2.

## The most popular product

| | Manufacturer | Price | Processor speed | Battery life | Installed memory | Hard drive capacity | Display size | Weight |
|---|---|---|---|---|---|---|---|---|
| ⊙ | —— | $2'095.00 | 1.67 GHz | 4.5 hour(s) | 512 MB | 80 GB | 38.6 cm | 2.54 kg |

## We also recommend the following products because

### they are cheaper and lighter, but have lower processor speed

| | Manufacturer | Price | Processor speed | Battery life | Installed memory | Hard drive capacity | Display size | Weight |
|---|---|---|---|---|---|---|---|---|
| ○ | —— | $1'499.00 | 1.5 GHz | 5 hour(s) | 512 MB | 80 GB | 33.8 cm | 1.91 kg |
| ○ | —— | $1'739.99 | 1.5 GHz | 4.5 hour(s) | 512 MB | 80 GB | 38.6 cm | 2.49 kg |
| ○ | —— | $1'625.99 | 1.5 GHz | 5 hour(s) | 512 MB | 80 GB | 30.7 cm | 2.09 kg |
| ○ | —— | $1'426.99 | 1.5 GHz | 5 hour(s) | 512 MB | 60 GB | 30.7 cm | 2.09 kg |
| ○ | —— | $1'929.00 | 1.2 GHz | 4 hour(s) | 512 MB | 60 GB | 26.9 cm | 1.41 kg |
| ○ | —— | $1'595.00 | 1 GHz | 5.5 hour(s) | 512 MB | 40 GB | 26.9 cm | 1.41 kg |

### they have higher processor speed and bigger hard drive capacity, but are heavier

| | Manufacturer | Price | Processor speed | Battery life | Installed memory | Hard drive capacity | Display size | Weight |
|---|---|---|---|---|---|---|---|---|
| ○ | —— | $1'220.49 | 1.8 GHz | 5 hour(s) | 1 GB | 100 GB | 38.1 cm | 2.95 kg |
| ○ | —— | $2'148.99 | 2 GHz | 4 hour(s) | 1 GB | 100 GB | 39.1 cm | 2.9 kg |
| ○ | —— | $1'379.00 | 3.3 GHz | 2 hour(s) | 512 MB | 100 GB | 43.2 cm | 4.31 kg |
| ○ | —— | $2'235.00 | 1.8 GHz | 2.5 hour(s) | 1 GB | 100 GB | 43.2 cm | 3.99 kg |
| ○ | —— | $2'319.00 | 1.7 GHz | 4.5 hour(s) | 512 MB | 100 GB | 43.2 cm | 3.13 kg |
| ○ | —— | $2'075.00 | 1.8 GHz | 1.67 hour(s) | 512 MB | 100 GB | 43.2 cm | 4.4 kg |

**Fig. 2** Organizational Structure, this is a form of structured overview which displays trade-offs between items [99]

## 2.2   Preference Elicitation

This section describes different ways in which a user can interact with a recommender system to influence the recommendations that they are given. This
interaction is what distinguishes conversational systems from "single-shot" recommendations. They allow users to elaborate their requirements over the course of an
extended dialog [102] rather than each user interaction being treated independently
of previous history.

While interaction can occur in a recommender system regardless of the explanatory criterion, it may be arguable that this type of interaction is particularly relevant
to the explanation goals of transparency, effectiveness, and scrutability. Justifying
recommendations, or making them transparent, to the user is only half of the
solution, the second half is making the system *scrutable* (see Sect. 4.2) by allowing
the user to make changes. By allowing users to change their preferences the
explanation facility may also increase effectiveness when this leads to more relevant
recommendations.

There are different ways in which a user can give feedback to the system to
let it know how it is doing. Here we expand on the four ways suggested by [74],
supplying examples of current applications. A fifth section on mixed interaction
interfaces is appended to the end of this original list. Note that although there
are more unobtrusive ways to elicit user preferences, e.g. via usage data [92] or
demographics [1], this section focuses on *explicit* feedback from users.

## 2.3   The User Specifies Their Requirements

By allowing a user to directly specify their requirements it is possible to circumvent
the type of faulty assumptions that can be made by a system where the interests of
a user are based on the items they decide to see, or how they interact with them.
An indirect approach depends on the way it collects data, and is sometimes hard to
change in the intended direction. Remember Mr. Iwanyk? This user tried to tame
TiVo's gay fixation by recording war movies and other "guy stuff". As a result
it started to give him documentaries on Joseph Goebbels and Adolf Eichmann. It
stopped thinking he was gay and decided he was a crazy guy reminiscing about the
Third Reich.

One movie recommender system allows users to specify their preferences in plain
English [133], p. 77:

"**User:** I feel like watching a thriller.
**System:** Can you tell me one of your favorite thriller movies.
**User:** Uhm, I'm not sure.
**System:** Okay. Can you tell me one of your favorite actors or actresses?
**User:** I think *Bruce Willis* is good.
**System:** I see. Have you seen *Pulp Fiction*?

**User:** No.
**System:** Pulp Fiction is a thriller starring *Bruce Willis."*

This dialog does not make use of the user's previous interests, nor does it explain *directly*. That is, there is no sentence that claims to be a justification of the recommendation. It does however do so indirectly, by reiterating (and satisfying) the user's *requirements*. The user should then be able to interact with the recommender system, and give their opinion of the recommendation, thus allowing further refinement.

## 2.4   The User Asks for an Alteration

A more direct approach is to allow users to explicitly ask for alterations to recommended items, for instance using a structured overview (see Sect. 2.1). This approach helps the users to find what they want quicker. Users can see how items compare, and see what other items are still available if the current recommendation should not meet their requirements. Have you ever put in a search for a flight, and been told to try other dates, other airports or destinations? This answer does not explain which of your requirements needs to be changed, requiring you to go through a tiring trial-and-error process. If you can see the trade-offs between alternatives from the start, the initial problem can be circumvented.

Some feedback facilities allow users to see how requirement constraints or critiques affect their remaining options. One such system explains the difference between a selected camera and remaining cameras. It describes competing cameras with "Less Memory and Lower Resolution and Cheaper" [73].

Instead of simply explaining to a user that no items fitting the description exist, these systems show what types of items *do* exist. These methods have the advantage of helping users find good enough items, even if some of their initial requirements were too strict.

## 2.5   The User Rates Items

To change the type of recommendations they receive, the user may want to correct predicted ratings, or modify a rating they made in the past. Ratings may be explicitly inputted by the user, or inferred from usage patterns. In a book recommender system a user could see the influence (in percentage) their previous ratings had on a given recommendation [9]. The *influence based explanation* showed which rated titles influenced the recommended book the most (see Table 1). Although this particular system did not allow the user to modify previous ratings, or degree of influence, in the explanation interface, it can be imagined that users could directly change their rating here. Note however, that it would be much harder to modify the degree of

**Table 1** Influence of previous book ratings, on the current book recommendation [9]

| BOOK | YOUR RATING Out of 5 | INFLUENCE Out of 100 |
| --- | --- | --- |
| Of Mice and Men | 4 | 54 |
| 1984 | 4 | 50 |
| Till We Have Faces: A Myth Retold | 5 | 50 |
| Crime and Punishment | 4 | 46 |
| The Gambler | 5 | 11 |

influence, as it is computed: any modification is likely to interfere with the regular functioning of the recommendation algorithm.

## 2.6   The User Gives Their Opinion

A common usability principle is that it is easier for humans to recognize items, than to draw them from memory. Therefore, it is sometimes easier for a user to say what they want or do not want, when they have options in front of them. The options mentioned can be simplified to be mutually exclusive, e.g. either a user likes an item or they do not. It is equally possible to create an explanation facility using a sliding scale.

In previous work in recommender systems a user could for example specify whether they think an item is interesting or not, if they would like to see more similar items, or if they have already seen the item previously [10, 114].

## 2.7   Mixed Interaction Interfaces

Previous work evaluated an interface where users could both specify their requirements from scratch, or make alterations to existing requirements generated by the system, and found that the mixed interface increases efficiency, satisfaction, and effectiveness of decisions compared to only making alterations to existing sets of requirements [19].

Others evaluated a hybrid interface for rating items. In this study, users were able to rate items suggested by the system, as well as search for items to rate themselves. The mixed-initiative model did not outperform either rating model in terms of the accuracy of resulting recommendations. On the contrary, allowing users to search for items to rate increased loyalty (measured in terms of returns to the system, and number of items rated) [75].

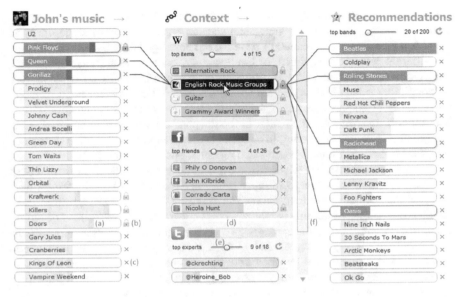

**Fig. 3** Interactive visualization, node-link diagram, used to present data from three recommender components: user profile, algorithm parameters, and resulting recommendations [12]

## 2.8   Interactive Recommendations

In addition to allowing users to make alterations, some interfaces are more interactive and allow users to specify their requirements, make alternations, and rate items [12, 89, 96, 127]. Commonly these support a multi-tier approach where the tiers support different types of preference elicitation.

Figure 3 shows an example of this type of visualization. As seen in the figure, the user can interact with three levels of information. From left to right these are: (1) user profile: favorite artists; (2) algorithm parameters: weighting of top trending items from Wikipedia, Facebook and Twitter; and; (3) resulting recommendations. Changing the user profile or algorithm parameters both change the resulting recommendations. Links between the nodes indicate a relationship or contribution (e.g., between profile and algorithm parameters).

## 3   Explanation Levels and Styles

We identify three *levels of explanation* that a user profile can serve [112]. Each level can contain the information from the previous level and restructure it: individual user (Level 1), contextualization (Level 2), and self-actualization (Level 3) i.e., explanations which aim to fulfill a self-development goal.

By applying a particular algorithm in a recommender systems, certain styles of explanations may be easier to generate since the algorithm can produce the type of information that the explanation style uses. These styles are also more suitable for different levels of explanation. For example, content-based, case-based and knowledge-based explanations are more suitable for individual user explanations (although they contextualize an item with other items). On the other hand collaborative and demographic filtering explanations are more suitable for contextualization with other users. We refer the reader to the chapters on advances in content-based and collaborative-based filtering in chapters "Advances in Collaborative Filtering" and "Semantics and Content-based Recommendations" respectively.

In the following sections we describe explanations that would be supported best by a particular underlying algorithm, or different "explanation styles". We caution however that explanations may follow the "style" of a particular algorithm irrespective of whether or not this is how the recommendations have been retrieved or computed. In other words, the explanation style for a given explanation *may, or may not,* reflect the underlying algorithm by which the recommendations are computed. There often is a divergence between how the recommendations are retrieved and the style of the given explanations. For example, while the recommendations may be based on matrix factorization (i.e., a form of collaborative filtering algorithm), an explanation could be still be formulated in terms of item features (i.e., content-based style). Consequently, this type of explanation would not be consistent with the goal of transparency, but may support other explanatory goals. As noted in the introduction, an explanation which does not focus on transparency, but on other explanatory goals can be seen as a justification.

Transparency is not the only explanatory goal (see Sect. 4 on different explanatory aims and ways to measure them) to consider when deciding upon explanation style. For example, for a given system one might find that collaborative-style explanations are more transparent but that critique-based style explanations are more efficient.

## 3.1 Level 1: Individual User

The first level consists of the raw data that the platform has on the user and the user's history. This data serves as the first layer since it provides the information necessary for more higher level questions or goals that the user might have. On the one hand, the user profile layer simply makes visible the individual user data that the system uses, however, this data can also help the user answer information seeking questions regarding their past consumption behavior (Tables 3).

**Table 2** The *influence based explanation* showed which rated titles influenced the recommended book the most. Although this particular system did not allow the user to modify previous ratings, or degree of influence, in the explanation interface, it can be imagined that users could directly change their rating. Note however, that it would be much harder to modify the degree of influence, as it is computed: any modification is likely to interfere with the regular functioning of the recommendation algorithm [9]

| BOOK | YOUR RATING Out of 5 | INFLUENCE Out of 100 |
|------|----------------------|----------------------|
| Of Mice and Men | 4 | 54 |
| 1984 | 4 | 50 |
| Till We Have Faces: A Myth Retold | 5 | 50 |
| Crime and Punishment | 4 | 46 |
| The Gambler | 5 | 11 |

### 3.1.1 Content-Based Style Explanation

For content-based style explanations the assumed input to the recommender engine are user u's ratings (for a sub-set) of items in I. These ratings are then used to generate a classifier that fits u's rating behavior and use it on i. Recommendations are items that the classifier predicts having the highest ratings.

If we simplify this further, we could say that content-based algorithms consider similarity between items, based on user ratings but considering item properties. In the same spirit, content-based style explanations are based on the items' properties. For example, [115] justifies a movie recommendation according to what they infer is the user's favorite actor (see Table 4). While the underlying approach is in fact a hybrid of collaborative- and content-based approaches, the explanation style suggests that they compute the similarity between movies according to the presence of features in highly rated movies. They elected to present users with several recommendations and explanations (top-N) which may be more suitable if the user would like to make a selection between movies depending on the information given in the explanations (e.g. feeling more like watching a movie with Harrison Ford over one starring Bruce Willis). The interaction model is based on ratings of items.

Other researchers have proposed to personalize explanations based on reviews, enriched by linked open data [82]. The methodology is based on building a graph in which the items liked by a user are connected to the items recommended through the properties available in linked open data. Similarly, [2] propose learning knowledge-based embeddings from hetereogenous knowledge sources, for both recommendation and explanation. In this work, they learned embeddings for product recommendations for different kinds of relationships (e.g., mention, belongs to, produced by, bought together, also bought, also viewed) between items. They did this by linking entities (e.g., user, item, word, brand, category), using a translation function, into a latent space. In contrast to many knowledge-base approaches, the efficiency of the method is computationally comparable to state-of-the-art neural network methods. The generated explanations have not yet been validated with users

**Table 3** Examples of explanations in commercial and academic systems, ordered by explanation style (case-based, collaborative-based, content-based, conversational, demographic-based and knowledge/utility-based.)

| System | Example explanation | Explanation style |
|---|---|---|
| *iSuggest-Usability* [49] | See e.g. Fig. 5 | Case-based |
| *LoveFilm.com* | *"Because you have selected or highly rated: Movie A"* | Case-based |
| *LibraryThing.com* | "Recommended By User X for Book A" | Case-based |
| *Netflix.com* | A list of similar movies the user has rated highly in the past | Case-based |
| *Amazon.com* | *"Customers Who Bought This Item Also Bought ... "* | Collaborative-based |
| *LIBRA* [9] | Keyword style (Tables 7 and 8); Neighbor style (Fig. 11); Influence style (Table 2) | Collaborative-based |
| *MovieLens* [47] | Histogram of neighbors (Fig. 7) | Collaborative-based |
| *Amazon.com* | *"Recommended because you said you owned Book A"* | Content-based |
| *CHIP* [27] | *"Why is 'The Tailor's Workshop recommended to you'? Because it has the following themes in common with artworks that you like: * Everyday Life * Clothes ... "* | Content-based |
| *Moviexplain* [115] | See Table 4 | Content-based |
| MovieLens: "Tagsplanations" [130] | Tags ordered by relevance or preference (see Fig. 4) | Content-based |
| News Dude [10] | *"This story received a [high/low] relevance score, because it contains the words f1, f2, and f3."* | Content-based |
| *OkCupid.com* | Graphs comparing two users according to dimensions such as "more introverted"; comparison of how users have answered different questions | Content-based |
| *Pandora.com* | *"Based on what you've told us so far, we're playing this track because it features a leisurely tempo ... "* | Content-based |
| *Adaptive place Advisor* [117] | Dialog e.g. "Where would you like to eat?" "Oh, maybe a cheap Indian place." | Conversational |
| *ACORN* [134] | Dialog e.g. "What kind of movie do you feel like?" "I feel like watching a thriller." | Conversational |
| INTRIGUE [1] | *"For children it is much eye-catching, it requires low background knowledge, it requires a few seriousness and the visit is quite short. For yourself it is much eye-catching and it has high historical value. For impaired it is much eye-catching and it has high historical value."* | Demographic-based |

(continued)

**Table 3** (continued)

| Qwikshop [73] | "Less Memory and Lower Resolution and Cheaper" | Knowledge/utility-based |
|---|---|---|
| SASY [28] | "...because your profile has: *You are single; *You have a high budget" (Fig. 10) | Knowledge/utility-based |
| Top Case [76] | "Case 574 differs from your query only in price and is the best case no matter what transport, duration, or accommodation you prefer" | Knowledge/utility-based |
| (Internet Provider) [31] | "This solution has been selected for the following reasons: *Webspace is available for this type of connection..." | Knowledge/utility-based |
| "Organizational Structure" [99] | Structured overview: "We also recommend the following products because: *they are cheaper and lighter, but have lower processor speed." (Fig. 2) | Knowledge/utility-based |
| myCameraAdvisor [131] | e.g. "...cameras capable of taking pictures from very far away will be more expensive ..." | Knowledge/utility-based |

**Table 4** Example of an explanation in Moviexplain, using features such as actors, which occur for movies previously rated highly by this user, to justify a recommendation [115]

| Recommended movie title | The reason is the participant | Who appears in |
|---|---|---|
| Indiana Jones and the Last Crusade (1989) | Ford, Harrison | 5 movies you have rated |
| Die Hard 2 (1990) | Willis, Bruce | 2 movies you have rated |

and there is further scope to improve these explanations for beyond accuracy metrics such as diversity.

A more domain independent approach is suggested by [130] who suggest a similarity measure based on user specified keywords, or tags. The explanations used in this study use the relationship between keywords and items (tag relevance), and the relationship between tags and users (tag preference) to make recommendations (see Fig. 4). Tag preference, or how relevant a tag is for a given user, can be seen as a form of content-based explanation, as it is a weighted average of a given user's ratings of movies with that tag. Tag relevance, or how relevant a keyword is for recommending an item, on the other hand is the correlation between (aggregate) users' preference for the tag, and their preference for a movie with which the tag is associated. In this example, showing recommendations as a single top item allows the user to view many of the tags that are related to the item. The interaction model is again based on numerical ratings.

The commercial system Pandora explains its recommendations of songs according to musical properties such as tempo and tonality. These features are inferred from users' ratings of songs. Figure 1 shows an example of this [93]. Here, the user is offered one song at a time (top item) and gives their opinion as "thumbs-up" or "thumbs-down" which also can be considered as numerical ratings.

**Fig. 4** Tagsplanation with
both tag preference and
relevance, but sorted by tag
relevance [130]

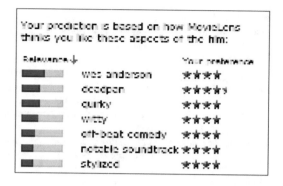

In the classification of [94], content-based style explanations are a type of feature based explanation, since they explain the recommendation in terms of similarity of item features to (features of) previously rated items.

### 3.1.2 Case-Based Reasoning (CBR) Style Explanations

Explanations can also omit mention of detailed item features (e.g. music genre, actor in a movie) and focus primarily on the similar items used to make the recommendation. The items used are thus considered cases for comparison, resulting in case-based style explanations. We note that CBR systems greatly vary with regard to the recommendation algorithm. For example, the FINDME recommender [15] is based on critiquing and the ranking of items in the former *NutKing* system was based on their presence in travel plans of users who expressed similar interests.

While these CBR systems have also used different methods to present their explanations, we recall that this section, and the sections describing the other explanation styles, are focused on the *style* of the explanation rather than the actual underlying algorithm. As such, each of these systems could in theory have had a case-based style explanation.

The "influence based style explanation" of [9] in Table 2 is a type of case-based style explanation. Here, the influence of an item on the recommendation is computed by looking at the difference in the score for the recommended item computed with the influential item, and the score for the recommender item when computed without that influential item. In this case, recommendations were presented as top item, assuming a rating based interaction. Another study computed the similarity between recommended items,[1] and used these similar items as justification for a top item recommendation in the "learn by example" explanations (see Fig. 5) [49]. A recent

---

[1] The author does not specify which similarity metric was used, though it is likely to be a form of rating based similarity measure such as cosine similarity.

**Fig. 5** Learn by example, or case based reasoning [49]

study compared case-based explanations with feature-based explanations. Showing participants previous items (case-based explanations) during the rating process improved accuracy (RMSE) and was considered most useful by participants [86]. In the classification of [94], case-based reasoning style explanations are a type of item style explanations, since they use exemplars of items to justify a recommendation.

### 3.1.3 Knowledge and Utility-Based Style Explanations

Knowledge-based systems reason over a knowledge-base to solve problems through rules in an inference engine. One common category of knowledge-based system are case-based systems which use examples of previous similar situations or cases to predict an outcome or solution. It is therefore arguable that there is a degree of overlap between knowledge-based, content-based (Sect. 3.1.1) and case-based style explanations (Sect. 3.1.2) which can be derived from either type of algorithm depending on the details of the implementation.

For all knowledge and utility-based style explanations the assumed input to the recommender engine are description of user u's needs or interests. The recommender engine then infers a match between the item i and u's needs. One knowledge-based recommender system takes into consideration how camera properties such as memory, resolution and price reflect the available options as well as a user's preferences [73]. Their system may explain a camera recommendation in the following manner: *"Less Memory and Lower Resolution and Cheaper"*. Here, recommendations are presented as a form of structured overview describing the competing options, and the interaction model assumes that users ask for alterations in the recommended items.

Similarly, in the system described in [76] users gradually specify (and modify) their preferences until a top recommendation is reached. This system can generate explanations such as the following for a recommended holiday titled "Case 574":

*"Top Case: Case 574 differs from your query only in price and is the best case no matter what transport, duration, or accommodation you prefer".*
The classification of [94] does not cover this style of explanation.

## 3.2   Level 2: Contextualization

The second level adds to the first by taking a user's specific past behavior and contextualizing it, e.g., within their community or preference space. This level helps the user understand themselves in relation to a context. This has been found to help users for example answer questions about how their consumption habits compare to the overall user base [67, 125].

### 3.2.1   Preference Space

The first kind of contextualization is in relation to the overall preferences in the search space. For example, in our previous work we developed visual explanations that revealed to users those regions of the recommendation space that are unknown to them, i.e. *blind-spots* [125]. Those visualisations do not aim to explain individual items, but instead reveal important properties of the item space as a whole. These summarized (genre-based) explanations visually scale better than explanations for all items in the full search space, and help users make decisions about unseen items.

The chord diagram in Fig. 6 shows media consumption blind-spots (light grey) and the viewing history of a user (dark grey) for individual and combinations of genres. From this diagram we can, among other things, infer that the user: Has not watched any Horror movies; Has not watched any SciFi-Actions; Has watched Drama most. Other possible comparisons include, comparing a user's consumption to those with similar consumption habits or to the platform's publication history.

Another approach has been to explain user preferences by comparing and contrasting parts of the profile [7]. This approach uses a set-based recommendation technique that permits the user model to be explicitly presented to users in natural language. It uses templates like "You (don't) like [first] unless [second]..", resulting explanations such as *"You don't like movies that are tagged as 'adventure', unless they are tagged as 'thriller', such as Twister.".*

### 3.2.2   Demographic-Based Style Explanations

For demographic-based style explanations, the assumed input to the recommender engine is demographic information about user u. From this, the recommendation algorithm identifies users that are demographically similar to u. A prediction for the recommended item i is extrapolated from how the similar users rated this item, and how similar they are to u considering their demographic features.

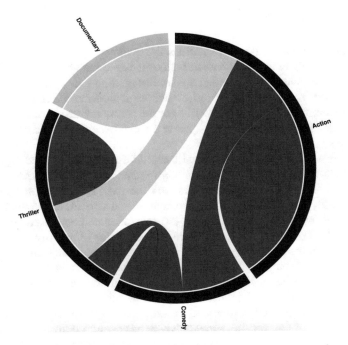

**Fig. 6** Visualization to contextualize a user profile [125]

Surveying a number of systems which use a demographic-based filter e.g. [1, 65, 97], we could only find one which offers an explanation facility: *"For children it is much eye-catching, it requires low background knowledge, it requires a few seriousness and the visit is quite short. For yourself it is much eye-catching and it has high historical value. For impaired it is much eye-catching and it has high historical value."*[1]. In this system recommendations were offered as a structured overview, categorizing places to visit according to their suitability to different types of travelers (e.g. children, impaired). Users can then add these items to their itinerary, but there is no interaction model that modifies subsequent recommendations

To our knowledge, there are no other systems that make use of demographic style explanations. It is possible that this is due to the sensitivity of demographic information; anecdotally we can imagine that many users would not want to be recommended an item based on their gender, age or ethnicity (e.g. *"We recommend you the movie Sex in the City because you are a female aged 20–40."*). The classification of [94] does not cover this style of explanation.

### 3.2.3   Collaborative-Based Style Explanations

For collaborative-based style explanations the assumed input to the recommender engine are user u's ratings of items in I. In user-based collaborative filtering these ratings are used to identify users that are similar in ratings to u. These similar users are often called "neighbors" as nearest-neighbors approaches are commonly used to compute recommendations. A prediction for the recommended item is extrapolated from the neighbors' ratings of i (e.g., using a weighted average over the neighbors predictions).

Commercially, the most well known usage of collaborative-based explanations are the ones used by Amazon.com: *"Customers Who Bought This Item Also Bought … "*. This explanation assumes that the user is viewing an item which they are already interested in, and the explanation and several recommendations are shown just below it. The approach used on Amazon is an item-based collaborative approach, which recommends items based on item (rating) similarity. This approach is different from the approach described above in that it uses the similarity between items (rather than users) to compute a recommendation. The recommendations are presented in the format of similar to top item. In addition, this explanation suggests a preference elicitation model whereby ratings are inferred from purchase behavior rather explicitly requested. (Note that Amazon also supplies recommendations in different aspects of the website, some of which use explicit rating elicitation.)

Herlocker et al. suggested 21 explanation interfaces using text as well as graphics [47]. These interfaces varied with regard to content and style, but a number of these explanations directly referred to the concept of neighbors. Figure 7 for example, shows how neighbors rated a given (recommended) movie, a bar chart with "good", "ok" and "bad" ratings clustered into distinct columns. Again, we see that this explanation is given for a specific way of recommending items, and a particular interaction model: this is a single recommendation (either top item or one item out of a top-N list), and assumes that the users are supplying rating information for items.

**Fig. 7** One out of twenty-one interfaces evaluated for persuasiveness—a histogram summarizing the ratings of similar users (neighbors) for the recommended item grouped by good (5 and 4's), neutral (3s), and bad (2s and 1s), on a scale from 1 to 5 [47]

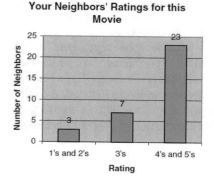

In the classification of [94], collaborative-based style explanations are a type of human explanation style, since they are based on similar users.

### 3.2.4 Interactive Contextualization

Contextualization can benefit from interactive functionalities. In Sect. 2.2 we discussed an interactive node-link, multi-tier interface. Such interfaces also support better explainability capabilities and help contextualize recommendations by visualizing the relationship between different components such as the profile, algorithm parameters, and resulting recommendations.

More advanced visualizations supported by interaction have also been used, where *Algorithm parameters* are represented using Venn diagrams (e.g., [95]), cluster maps (e.g., [128]), and bubble charts (e.g., [54]). While these representations are more visually complex, they can represent more dimensions, and consequently more facets of the recommended items, or even the recommendation methods. For example, each circle in a Venn diagram can depict the items resulting from a different recommendation algorithm [95], making the intersecting items available for interaction. A bubble can represent popularity (size), genre (color), and musical properties (x and y axis) [54]. These kinds of interfaces have been effective in terms of increasing transparency and control among other explanatory criteria (e.g., [53, 57]).

## 3.3   Level 3: Self-Actualization

The third level supports user-centered goals that foster self-actualization. A growing number of studies has tried to connect transparency and explanations with personal or societal values and goals [44, 59, 67, 105, 125]. This level departs from simple information-finding by promoting discovery and exploration. It is additionally *goal-directed* and allows for *user-control* to achieve those goals. One such goal could be a more critical analysis of machine-learning output. In the broader domain of machine-learning, recent work has suggested the use of machine learning "consumer labels", summarizing the quality of systems through dimensions such as accuracy, fairness, generalization, transparency, and robustness [105] (c.f., Fig. 8).

In this self-actualization level, the user has direct *control* over which goal they want to explore, and the recommendations that result from the chosen goal. In the news domain this was applied in relation to epistemic goals, i.e., goals for knowledge development [112]. Four user self-actualization goals were suggested, of which two were studied. The first, *Broaden Horizons*, highlights the need for diversity in news selection [46]. Achieving the goal of Broaden Horizons is made possible by showing users articles, perspectives, and topics that they normally do not read about, and that they may find interesting. The idea is to gradually increase diversity, slowly moving users out of their typical consumption patterns

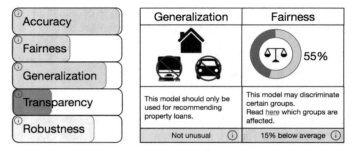

**Fig. 8** Sketch of a machine learning consumer label for a loan prediction application. Left: general overview showing the degree to which certain properties are satisfied (percentages and color-coding), right: details on generalizability and fairness [105]

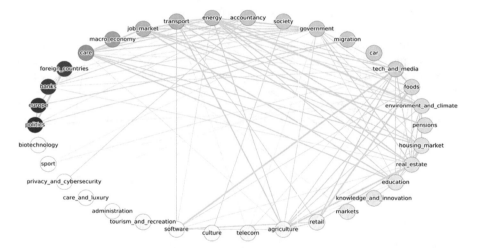

**Fig. 9** User profile. Nodes are news topics, ordered by the user's familiarity with them, and darker color indicates more familiarity. The connections between topics represent their similarity, with thicker lines meaning more similar

[125]. The second, *Discover the Unexplored*, also looks at diversity, but goes a step further to help users discover topics they do not (yet) know about, helping them encounter unknown articles that may spark new interests or provide new contexts, and effectively bringing them to completely new territories. This goal is inspired by the notion of serendipity [104]. A user study suggested that depending on the goal set, users' exhibit different reading intentions when interacting with a visualization such as Fig. 9, when measured in terms of topic familiarity and similarity. However, this study stopped short of studying actual reading behaviors. For users interested learning more about the metrics of diversity and novelty, we also refer to chapter "Novelty and Diversity in Recommender Systems".

## 3.4 Hybrid Style Explanations

There is limited work evaluating combinations of explanation styles. One exception evaluated explanations which considered both Level 1 and Level 2 [112]. These were combinations of different content and collaborative style explanations, for hybrid recommender systems. Such an explanation, in the music domain, was phrased as: "People who listen to your profile items Metallica, Iron Maiden, Rainbow, also listen to Black Sabbath" [62]. The authors found a significant difference between different explanation styles in terms of their persuasiveness.

## 4 Goals and Metrics

Surveying the literature for explanations in recommender systems, we see that recommender systems with explanatory capabilities have been evaluated according to different criteria, and identify seven different goals for explanations of single item recommendations. Here we mention goals that are applicable to single item recommendations, i.e. when a single recommendation is being offered. When recommendations are made for multiple items, such as in a list, additional factors such as diversity (e.g. "this list contains items that are different from each other in order to improve variation") may be relevant.

Table 5 states these goals, some of which are similar to those desired (but not evaluated on) in expert systems, c.f. MYCIN [8]. In Table 6, we summarize examples of previous evaluations of explanations in recommender systems, and the goal by which they have been evaluated.

Previous generations of expert systems were commonly evaluated in terms of user acceptance and the decision support of the system as a whole. User acceptance can be defined in terms of our goals of satisfaction or persuasion. If the evaluation measures acceptance with the system as whole, such as [16] who asked questions such as *"Did you like the program?"*, this reflects user satisfaction. If rather the evaluation measures user acceptance of advice or explanations, as in [140], the criterion can be said to be persuasion.

It is important to identify these goals as distinct, even if they may interact, or require certain trade-offs. Indeed, it would be hard to automatically generate explanations that do well for all of the goals, in reality it is a trade-off.[2] While personalized explanations may lead to greater user satisfaction, they do not necessarily increase effectiveness [39, 121, 122]. Other times, goals that seem to be inherently

---

[2] An exception is claimed in a recent paper which studied all seven explanatory goals introduced in this chapter. The authors found that user-generated (crowd-sourced) explanations were assessed to fulfill multiple criteria to a comparable degree. However, the same paper also found that this assessment (for all criteria) was strongly affected by the overall quality assessment of the recommended item (higher for seen items with a high rating) [6].

**Table 5** Explanatory goals and their definitions

| Aim | Definition |
|---|---|
| Transparency (Tra.) | Explain how the system works |
| Scrutability (Scr.) | Allow users to tell the system it is wrong |
| Trust | Increase users' confidence in the system |
| Effectiveness (Efk.) | Help users make good decisions |
| Persuasiveness (Pers.) | Convince users to try or buy |
| Efficiency (Efc.) | Help users make decisions faster |
| Satisfaction (Sat.) | Increase the ease of use or enjoyment |

related are not necessarily so, for example it has been found that transparency does not necessarily aid trust [27]. For these reasons, while an explanation in Table 6 may have been evaluated for several goals, it may not have achieved them all.

The type of explanation that is given to a user is likely to depend on the goals of the designer of a recommender system. For instance, when building a system that sells books one might decide that user trust is the most important aspect, as it leads to user loyalty and increases sales. For selecting tv-shows, user satisfaction could be more important than effectiveness. That is, in a system focused on pure entertainment it may be more important that a user enjoys using the service, than that they are presented the very best available shows (as long as the shows are "good enough").

In addition, some attributes of explanations may contribute toward achieving multiple goals. For instance, one can measure how *understandable* an explanation is, which can contribute to e.g. user trust, as well as satisfaction.

In this section we describe seven potential aims for explanations (Table 5), and suggest evaluation metrics based on previous evaluations of explanation facilities, or offer suggestions of how existing measures could be adapted to evaluate the explanation facility in a recommender system.

## *4.1 Explain How the System Works: Transparency*

An anecdotal article in the Wall Street Journal titled "*If TiVo Thinks You Are Gay, Here's How to Set It Straight*" describes users' frustration with irrelevant choices made by a video recorder that records programs it assumes its owner will like, based on shows the viewer has recorded in the past.[3] For example, one user, Mr. Iwanyk, suspected that his TiVo thought he was gay since it inexplicably kept recording programs with gay themes. This user clearly deserved an explanation.

An explanation may clarify *how* a recommendation was chosen. In expert systems, such as in the domain of medical decision making, the importance

---

[3] https://www.wsj.com/articles/SB1038261936872356908, retrieved August, 2020.

**Table 6** The goals for which explanations in recommender systems have been evaluated. System names are mentioned if given, otherwise we only note the type of recommended items

|  | Tra. | Scr. | Trust | Efk. | Per. | Efc. | Sat. |
|---|---|---|---|---|---|---|---|
| System (type of items) | | | | | | | |
| (Advice, intrusion detection system) [30] | | | | X | | | |
| (Internet providers) [31] | | X | | | X | | X |
| (Financial advice, Internet providers) [32] | | | | | | X | X |
| (Digital cameras, notebooks computers) [99] | | X | | | | | |
| (Digital cameras, notebooks computers) [100] | | X | X | | | | |
| (Image tags and movies) [106] | X | | | X | | X | |
| (Music) [109] | | X | | | | | |
| (Music) [61] | X | X | | | X | | |
| (Music) [66] | X | | X | | | | |
| (Music) [107] | | | | X | X | | |
| (Movies) [7] | X | X | | | | | |
| (Movies) [6] | X | X | X | X | X | X | X |
| (Movies) [39] | X | | | X | X | X | X |
| (Movies) [121, 122] | | | | X | X | | X |
| (Social network news) [83] | X | X | X | | | | |
| *Adaptive Place Advisor* (restaurants) [117] | | | | X | | X | |
| *ACORN* (movies) [134] | | | | | | | X |
| *CHIP* (cultural heritage artifacts) [26] | X | | X | X | | | |
| *CHIP* (cultural heritage artifacts) [27] | X | | X | | | | X |
| (*expLOD* (Movies) [82] | X | | X | X | X | | |
| *iSuggest-Usability* (music) [49] | X | | | X | | | |
| *LIBRA* (books) [9] | | | | X | | | |
| *MovieLens* (movies) [47] | | | | | X | | X |
| *Moviexplain* (movies) [115] | | | | X | | | X |
| *myCameraAdvisor* [131] | | | X | | | | |
| *Qwikshop* (digital cameras) [73] | | | | X | | X | |
| *SASY* (e.g. holidays) [28] | X | X | | | | | X |
| *Tagsplanations* (movies) [130] | X | | | X | | | |

of transparency has also been recognized [8]. Transparency or the heuristic of "Visibility of System Status" is also an established usability principle [87], and it has been considered in user studies of recommender systems [109].

Vig et al. differentiate between transparency and justification [130]. While transparency should give an honest account of how the recommendations are selected and how the system works, justification can be descriptive and decoupled from the recommendation algorithm. The authors cite several reasons for opting for justification rather than genuine transparency. For example some algorithms that are difficult to explain (e.g. latent semantic analysis where the distinguishing factors are latent and may not have a clear interpretation), protection of trade secrets by system designers, and the desire for greater freedom in designing the explanations.

Cramer et al. studied the effect of transparency on other evaluation goals such as trust, persuasion (acceptance of items) and satisfaction (acceptance) in an art recommender [26, 27]. Transparency itself was evaluated in terms of its effect on actual and perceived understanding of how the system works [27]. Actual understanding was based on the correctness of user answers to interview questions such as "Could you please tell me how the system works...". *Perceived* understanding was extracted from self-reports in questionnaires and interviews, measuring responses to statements such as "I understand what the system bases its recommendations on".

The evaluation of transparency has also been coupled with scrutability (Sect. 4.2) and trust (Sect. 4.3), but we will see in these sections that these goals can be distinct from each other.

## *4.2   Allow Users to Tell the System It Is Wrong: Scrutability*

Explanations may help isolate and correct misguided assumptions or steps. When the system collects and interprets information in the background, as is the case with TiVo, it becomes all the more important to allow the user to modify these assumptions or steps. Explanations can be used in way that helps the users to *correct* reasoning, or make the system *scrutable* [28]. Scrutability is related to the established usability principle of User Control [87]. See Fig. 10 for an example of a scrutable holiday recommender. Here the user can ask why certain assumptions (like a low budget) were made. Selecting this option takes them to a page with a further explanation and an option to modify this in their user model.

While scrutability is very closely tied to the goal of transparency, it deserves to be uniquely identified. Transparency in and of itself does not allow users to modify

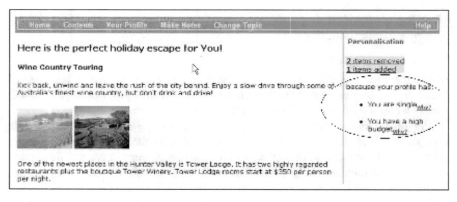

**Fig. 10** Scrutable holiday recommender, [28]. The explanation is in the circled area, and the user profile can be accessed via the "why" links

the reasoning in a system, and some systems may only offer partial transparency together with scrutability. The explanation in Fig. 1 (*"Based on what you've told us so far, we're playing this track because it features a leisurely tempo ... "*) is transparent but not scrutable. Here, the user cannot change the ratings that affected this recommendation. If however, the ratings in Table 8 were changeable, we could argue that the explanation was scrutable. However, it is not (fully) transparent even if they offer some form of justification. There is nothing about the explanations in the table that suggests that the underlying recommendations are based on a Bayesian classifier. In such a case, we can imagine that a user attempts to scrutinize a recommender system, and manages to change their recommendations, but still does not understand exactly what happens within the system. In contrast, [41] made a preliminary attempt to make explanations that are both transparent and scrutable.

Czarkowski found that users were not likely to scrutinize on their own, and that extra effort was needed to make the scrutability tool more visible [28]. In addition, it was easier to get users to perform a given scrutinization task such as changing the personalization (e.g. "Change the personalisation so that only Current Affairs programs are included in your 4:30–5:30 schedule.") Their evaluation included metrics such as task correctness, and if users could express an understanding of what information was used to make recommendations for them. They understood that adaptation in the system was based on their personal attributes stored in their profile, that their profile contained information they volunteered about themselves, and that they could change their profile to control the personalization [28].

## 4.3 Increase Users' Confidence in the System: Trust

A study of users' trust (defined as perceived confidence in a recommender system's *competence*) suggests that users intend to return to recommender systems which they find trustworthy [18]. Trust in the recommender system could also be dependent on the accuracy of the recommendation algorithm [75]. Trust is also sometimes linked with transparency: previous studies indicate that transparency and the possibility of interaction with recommender systems increases user trust [31, 109].

We note however, that there are also cases where transparency and trust were not found to be related [27]. [66] also found that poor explanations could decrease how beneficial they were found by users and led to poor mental models.

Consequently, we do not claim that explanations can fully compensate for poor recommendations, but that they can mitigate their effects on user trust. A user may be more forgiving, and more confident in recommendations, if they understand why a bad recommendation (or one based on low confidence) has been made and can prevent it from occurring again. A user may appreciate when a system is 'frank' and admits that it is not confident about a particular recommendation.

In addition, the interface design of a recommender system may affect its trustworthiness. In a study of factors determining web page credibility, the largest proportion of users' comments (46.1%) referred to the appeal of the overall visual design of a site, including layout, typography, font size and color schemes [38]. Likewise the perceived credibility of a Web article was significantly affected by the presence of a photograph of the author [37]. So, while recommendation accuracy, and the goal of transparency are often linked to the evaluation of trust, design is also a factor that needs to be considered as part of the evaluation.

Questionnaires can be used to determine the degree of trust a user places in a system. An overview of trust questionnaires can be found in [90] which also suggests and validates a five dimensional scale of trust. Note that this validation was done with the aim of using celebrities to endorse products, but was not conducted for a particular domain. Additional validation may be required to adapt this scale to a particular recommendation domain.

A model of trust in recommender systems is proposed in [18, 100], and the questionnaires in these studies consider factors such as intent to return to the system, and intent to save effort. Also [131] query users about trust, but focus on trust related beliefs such as the perceived competence, benevolence and integrity of a virtual adviser. They found that the different trusting beliefs could be improved by explanations with different content. Although questionnaires can be very focused, they suffer from the fact that self-reports may not be consistent with user behavior. In these cases, implicit measures (although less focused) may reveal factors that explicit measures do not.

One such implicit measure could be loyalty, a desirable by-product of trust. One study compared different interfaces for eliciting user preferences in terms of how they affected factors such as loyalty [75]. Loyalty was measured in terms of the number of logins and interactions with the system. Among other things, the study found that allowing users to independently choose which items to rate affected user loyalty. It has also been thought that Amazon's conservative use of recommendations, mainly recommending familiar items, enhances user trust and has led to increased sales [114]. We encourage readers who would like to learn more about trust in recommender systems to read a previous handbook chapter which is dedicated to this topic [129].

## 4.4   Convince Users to Try or Buy: Persuasiveness

Explanations may increase user acceptance of the system or the given recommendations [47]. Both definitions qualify as persuasion, as they are both attempts to influence the user.

[27] evaluated the acceptance of recommended items in terms of how many recommended items were present in a final selection of six favorites. Some authors have measured the related concept of confidence in the recommendation (e.g., "This explanation makes me confident that I will like this artist.") [61]. In another study, of a collaborative filtering- and rating-based recommender system for movies, participants were given different explanation interfaces (e.g. Fig. 7) [47]. This study

directly inquired how likely users were to see a movie (with identifying features such as title omitted) for 21 different explanation interfaces. Persuasion was thus a numerical rating on a 7-point scale.

In addition, it is possible to measure if the evaluation of an item has changed, i.e. if the user rates an item differently after receiving an explanation. Indeed, it has been shown that users can be manipulated to give a rating closer to the system's prediction [25]. It has also been found that confidence information, or how "sure" a system is about the relevance or non-relevance of a recommendation, can influence user ratings. [106] found that participants were more likely to rate an item that was actually non-relevant as relevant if the system said it was very confident. For (truly) relevant items, the participants were also less likely to remain undecided about the relevance of items (a similar pattern was not found for items that were truly irrelevant).

Both studies were in subjective and low investment domains (movies and images), and it is possible that users may be less influenced by incorrect predictions in high(er) cost domains such as cameras.[4] It has also been found that confidence information can influence user ratings. In addition, too much persuasion may backfire once users realize that they have tried or bought items that they do not really want.

Persuasiveness can be measured in a number of ways. For example, it can be measured as the difference between two ratings: the first being a previous rating, and the second a re-rating for the same item but with an explanation interface [25]. Another possibility would be to measure how much users actually try or buy items compared to users in a system without an explanation facility. These metrics can also be understood in terms of the concept of "conversion rate" commonly used in e-Commerce, operationally defined as the percentage of visitors who take a desired action.

## 4.5 Help Users Make Good Decisions: Effectiveness

Rather than simply persuading users to try or buy an item, an explanation may also assist users to make *better* decisions: accepting relevant items and discarding irrelevant ones [106, 121, 122]. Effectiveness is by definition highly dependent on the accuracy of the recommendation algorithm. An effective explanation would help the user evaluate the quality of suggested items according to their own preferences. This would increase the likelihood that the user discards irrelevant options while helping them to recognize useful ones. For example, a book recommender system with effective explanations would help a user to buy books they actually end up liking. Bilgic and Mooney emphasize the importance of measuring the ability of

---

[4] In [120] participants reported that they found incorrect overestimation less useful in high cost domains compared to low cost domains.

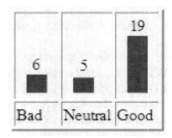

**Fig. 11** The Neighbor Style Explanation - a histogram summarizing the ratings of similar users (neighbors) for the recommended item grouped by good (5 and 4's), neutral (3s), and bad (2s and 1s), on a scale from 1 to 5. The similarity to Fig. 7 in this study was intentional, and was used to highlight the difference between persuasive and effective explanations [9]

**Table 7** The keyword style explanation by [9] for an item (a book). The item is being recommended based on a number of keywords such as "HEART" and "MOTHER". The explanation lists keywords that were used in the description of the item, and that have previously been associated with highly rated items.

| Word | Count | Strength | Explain |
|------|-------|----------|---------|
| HEART | 2 | 96.14 | *Explain* |
| BEAUTIFUL | 1 | 17.07 | *Explain* |
| MOTHER | 3 | 11.55 | *Explain* |
| READ | 14 | 10.63 | *Explain* |
| STORY | 16 | 9.12 | *Explain* |

**Table 8** A more detailed explanation for the strength of a keyword (such as "HEART") which shows after clicking on *"Explain"* in Table 7. The rows represent all the previous items which influence the strength of the keyword [9]

| Title | Author | Rating | Count |
|-------|--------|--------|-------|
| Hunchback of Notre Dame | Victor Hugo, Walter J. Cobb | 10 | 11 |
| Till We Have Faces: A Myth Retold | C.S. Lewis, Fritz Eichenberg | 10 | 10 |
| The Picture of Dorian Gray | Oscar Wilde, Isobel Murray | 8 | 5 |

a system to assist the user in making accurate decisions about recommendations based on explanations such as those in Fig. 11, and Tables 7 and 8 [9]. Effective explanations could also serve the purpose of introducing a new domain, or the range of products, to a novice user, thereby helping them to understand the full range of options [31, 99].

Vig et al. measure perceived effectiveness: *"This explanation helps me determine how well I will like this movie."*[130]. Effectiveness of explanations can also be calculated as the *absence of a difference* between the liking of the recommended item prior to, and after, consumption. For example, in a previous study, users rated a book twice, once after receiving an explanation, and a second time after reading the book [9]. If their opinion on the book did not change much, the system was considered effective. This study explored the effect of the whole recommendation process, explanation inclusive, on effectiveness. The same metric was also used to evaluate whether personalization of explanations (in isolation of a recommender system) increased their effectiveness in the movie domain [120]. In [107] there is a distinction between the likelihood of finding out more about a recommended artist and the actual rating given to the artist after listening to several songs.

While this metric considers the difference between the before and after ratings, it does not discuss the effects of over- contra underestimation.[5] In our work we found that users considered overestimation to be less effective than underestimation, and that this varied between domains. Specifically, overestimation was considered more severely in high investment domains compared to low investment domains. In addition, the strength of the effect on perceived effectiveness varied depending on where on the scale the prediction error occurred [120].

Another way of measuring the effectiveness of explanations has been to test the same system with and without an explanation facility, and evaluate if subjects who receive explanations end up with items more suited to their personal tastes [26]. This approach has been used in work which measured both perceived effectiveness (helpfulness) and performance accuracy (actual effectiveness) [30].

Other work evaluated explanation effectiveness using a metric from marketing [45], with the aim of finding the single *best* possible item (rather than "good enough items" as above) [19]. Participants interacted with the system until they found the item they would buy. They were then given the opportunity to survey the entire catalog and to change their choice of item. Effectiveness was then measured by the fraction of participants who found a better item when comparing with the complete selection of alternatives in the database. So, using this metric, a low fraction represents high effectiveness.

Effectiveness is the criterion that is most closely related to accuracy measures such as precision and recall [26, 115, 117]. In systems where items are easily consumed, these can be translated into recognizing relevant items and discarding irrelevant options respectively [106]. For example, there have been suggestions for an alternative metric of "precision" based on the number of profile concepts matching with user interests, divided by the number of concepts in their profile [26].

## 4.6  Help Users Make Decisions Faster: Efficiency

Explanations may make it *faster* for users to decide which recommended item is best for them. Efficiency is another established usability principle, i.e. how quickly a task can be performed [87]. This criterion is one of the most commonly addressed in the

---

[5] By overestimation we mean that the prediction is higher than the final or actual rating, and underestimation when the prediction is lower than it.

recommender systems literature (see Table 6) given that the task of recommender systems is to find needles in haystacks of information.

Efficiency may be improved by allowing the user to understand the relation between competing options. [73, 76, 99] use so called critiquing, a sub-class of knowledge-based algorithms based on trade-offs between item properties, which lends itself well to the generation of explanations. The rules generated by the algorithm can intuitively be translated to rules such as *"Less Memory and Lower Resolution and Cheaper"* [73]. This explanation can help users to find a cheaper camera more quickly if they are willing to settle for less memory and lower resolution. The efficiency of these explanations is closely tied with the efficiency of the query language, but can also be compared from a user-centered perspective in terms of the number of interactions needed by a user to make a choice.

Efficiency is often used in the evaluation of so-called conversational recommender systems, where users continually interact with a recommender system, refining their preferences (see also Sect. 2.2). In these systems, the explanations can be seen to be implicit in the dialog. Efficiency in these systems can be measured by the total amount of interaction time, and number of interactions needed to find a satisfactory item [117]. Evaluations of explanations based on improvements in efficiency are not limited to conversational systems however. Pu and Chen for example, compared completion time for two explanatory interfaces, and measured completion time as the amount of time it took a participant to locate a desired product in the interface [99].

Other metrics for efficiency also include the number of inspected explanations, and number of activations of repair actions when no satisfactory items are found [31, 103]. Normally, it is not sensible to expose users to all possible recommendations and their explanations, and so users can choose to inspect (or scrutinize) a given recommendation by asking for an explanation. In a more efficient system, the users would need to inspect *fewer* explanations. Repair actions consist of feedback from the user which changes the type of recommendation they receive, as outlined in the sections on scrutability (Sect. 4.2). Examples of user feedback/repair actions can be found in Sect. 2.2.

## 4.7 Make the Use of the System Enjoyable: Satisfaction

Explanations have been found to increase user satisfaction with, or acceptance of, the overall recommender system [31, 47, 109]. [39] studied the effect of various explanatory aims on satisfaction, and found that user-perceived transparency had a significant positive effect on overall satisfaction with the explanation interfaces (but did not find an effect of efficiency or effectiveness on satisfaction). The presence of longer descriptions of individual items has been found to be positively correlated with both the *perceived* usefulness [119], and ease of use of the recommender system [109]. Also, many commercial recommender systems such as those seen in Table 3 are primarily sources of entertainment. In these cases, any extra facility should take notice of the effect on user satisfaction. An explanation for an internet provider, describing the provider in terms of user requirements was evaluated on the criterion of satisfaction [31]. These explanations took into account users' specific constraints, but the explanations did not have a statistically significant effect on satisfaction. However, the authors found other factors improved user satisfaction with the recommendation process, such as product comparisons for which the effect was even strong for participants who already had experience with recommender applications.

When measuring satisfaction, one can directly ask users whether the system is enjoyable to use. Tanaka-Ishii and Frank in their evaluation of a multi-agent system describing a Robocup soccer game ask users whether they prefer the system with or without explanations [116]. Satisfaction can also be measured indirectly by measuring user loyalty [31, 75] (see also Sect. 4.3), and likelihood of using the system for a search task [27].

In measuring explanation satisfaction, it is important to differentiate between satisfaction with the recommendation process,[6] and the recommended products (persuasion) [27, 31].

One (qualitative) way to measure satisfaction with the process would be to conduct usability testing methods such as record a think-aloud protocol for a user conducting a task [69]. In this case, the participants describe their entire experience using the system: what they are looking at, thinking, doing and feeling, as they go about a task such as finding a satisfactory item. Objective notes of everything that users say are taken, without interpretation or influencing the users in any way. Video and voice recordings can also be used to revisit the session and to serve as a memory aid. In such a case, it is possible to identify usability issues and even apply quantitative metrics such as the ratio of positive to negative comments; the number of times the evaluator was frustrated; the number of times the evaluator was delighted; the number of times and where the evaluator worked around a usability problem etc.

It is also arguable that users would be satisfied with a system that offers effective explanations, confounding the two goals. However, a system that aids users in making good decisions, may have other disadvantages that decrease the overall satisfaction (e.g. requiring a large cognitive effort on the part of the user). Fortunately, these two goals can be measured by distinct metrics.

---

[6] Here we mean the entire recommendation process, inclusive of the explanations. We note however that the evaluation of explanations in recommender systems are seldom fully independent of the underlying recommendation process.

## 4.8    Additional Goals

While additional goals have been suggested in the literature, they do not yet form a consistent pattern. The most coherent alternative proposal is suggested in a taxonomy which distinguishes between (i) stakeholder goals, e.g., to increase the user's intention to reuse the system; (ii) user-perceived quality factors, which are system quality factors that can contribute to the achievement of the stakeholder goals (e.g., trust); and (iii) explanation purposes, which are explanation-specific objectives that can contribute to achieving the objectives at the other levels [88]. This taxonomy would benefit from further research to understand how goals at each level interact or overlap.

Another paper measured perceptions of system performance such as perceived accuracy and perceived novelty [61]. In our view these could be seen as user-perceived quality factors, but since they interact strongly with recommendation accuracy we have not considered them in our framework.

A third study measured a goal called engagement. This explanatory goal was defined to users as: *"The explanation helped me discover new information about this movie"* [82]. This can be seen as the stakeholder goal of Education, and is also a criteria of explanations previously identified for case-based reasoning explanations for expert systems (c.f., [111]). These kind of goals for explanations may be more in line with self-actualization (c.f., 'Level 3' in Sect. 3.3) through developing one kind of behavior (e.g., more diverse consumption), or avoiding another (e.g., decreasing bias in recruitment or policing).

The relationship between these goals still needs to be better understood before they can be reclassified.

## 5    Moderating Factors

In the previous section, we discussed how explanations can be useful, by surveying possible explanatory goals. Further, it is prudent to understand the necessary conditions for explanations to be useful. Here, we follow the framework of [60] which differentiates between different user-centered aspects of the recommender systems such as: objective system aspects, user experience, subjective system aspects, interaction, situational, and personal characteristics. Recent studies of explanations in recommender systems have considered both *personal and contextual characteristics* that influence explanation efficacy. We will summarize both types of characteristics below.

**Table 9** Overview of the personal characteristics discussed and their related example measures

| Personal characteristics | Example measures |
| --- | --- |
| Level of experience | Musical sophistication [54, 78, 80] |
| Personality traits | The Big-Five [20, 34], locus of control [77] |
| Demographic characteristics | Age, gender [33, 78] |
| Cognitive skill | Visual working memory [68, 77, 78, 124] |

## 5.1  Personal Characteristics

Beyond recommender systems, the influence of personal characteristics on the performance of users in interactive systems has been researched in depth. These works have investigated a variety of personal characteristics, which we describe below using the classification of [4]: level of experience, personality traits, demographic characteristics, and cognitive ability. An overview of the personal characteristics discussed in this section and their example measures are given in Table 9.

### Experience
Many studies have shown significant effects of the level of experience when interacting with recommender systems. For example, novice users prefer simple and transparent interaction methods [64]. In the domain of music, the *Goldsmiths Musical Sophistication Index* (Gold-MSI)[7] is regarded as an effective way to measure domain expertise of users, and has shown a consistent correlation with interaction with visual elements in recommender interfaces [77, 78]. Related work found that musical sophistication only strengthens the impact of user interface on perceived diversity (moderation effect) when studying the combined effect of controls and visualizations [54].

### Personality Traits
In the Chambers concise dictionary,[8] personality is defined as "a person's nature or disposition; the qualities that give one's character individuality" and it is a key area of research in user modelling and user adaptive systems. Previous work has found that the personality trait *need for cognition* does influence how users interact with explanations [77]: explanations were most beneficial for people with a low need for cognition, while for people with a high need for cognition, the authors observed that explanations could create a lack of confidence. Other work found that people's personalities, modeled as the Big-5 personality traits, also influenced which explanations were more effective [61]: conscientious participants reported being easier to persuade. The participants seemed to be split in terms of neuroticism: calm participants tended to be more receptive of popularity-based explanations while anxious participants tended to be more receptive of item-based collaborative

---

[7] https://www.gold.ac.uk/music-mind-brain/gold-msi/, accessed June 2020.

[8] http://www.chambers.co.uk.

filtering style explanations. Note, chapter "Personality and Recommender Systems" describes research on personality in recommender systems in more depth.

**Demographic Characteristics**

Demographic characteristics categorize users without considering personality, and include categories relating to factors such as age, sex, education level etc. In the domain of music recommendation, previous work has found varying musical preference of different age groups [33]. However, to the best of our knowledge, the effect of age and gender (or other demographic characteristics) on users' interaction with recommendation explanations has not been found in previous research [78].

**Cognitive Capacity**

Cognitive capacities relate to abilities to conduct conscious intellectual activities (such as thinking, reasoning, or remembering) which are needed in acquisition of knowledge, manipulation of information, and reasoning. Such capacities may be particularly relevant to consider for more advanced interfaces or explanation interfaces. Previous work has found *visual working memory (VM)* to be a factor that affects cognitive load when interacting with adaptive interactive systems [21, 68, 124]. Besides, visual working memory and visual literacy have been found to affect users' interaction with visual elements in recommender interfaces [77, 78].

In previous studies, participants with higher visual memory perceived significantly higher diversity, of recommended items, in a more complex visualization compared to participants with low visual working memory [54, 55].

## 5.2    *Situational Characteristics*

Preferences for recommended items in many domains are likely to depend on contextual characteristics such as location and activity. However, most recommender systems do not allow users to adapt recommendations to their current context.

Having additional transparency and control for context such as location, activity, weather, and mood has been found to lead to higher perceived quality and did not increase cognitive load for music recommendations [52]. Moreover, the authors found that users are more likely to modify contexts and their profile during relaxing activities.

Recent work has proposed justifications that differ for varying contextual situations, by automatically learning a lexicon for each contextual setting and by using such a lexicon to diversify the justification. When users selected more than one contextual dimension, selecting which aspects to mention according to context has also been shown to improve explanations for transparency, persuasion, engagement, trust, and effectiveness [81].

Other contextual factors, such as group dynamics, create additional requirements on explanations, such as balancing privacy and transparency [72, 85].

# 6 Future Directions

This section identifies five strands of promising future directions. *Firstly*, a number of new approaches for generating explanations from reviews have proposed in recent years, including deep learning and reinforcement learning. Secondly, it seems likely that future explanations will support the growing trend in social recommendations, such as—(1) recommendations to groups of people (c.f., chapter "Group Recommender Systems: Beyond Preference Aggregation"), (2) recommendations of people (e.g., in domains such as jobs [17, 51], dating [142], and micro-lending [71, 110] (c.f., chapter "Multistakeholder Recommender Systems"); and (3) recommendations of content on social networks based on credibility of people (c.f., chapter "Social Recommender Systems") [56].

*Thirdly*, recommendations like most machine learning is prone toward biases [91] (c.f., chapter "Fairness in Recommender Systems"), and human decision-making is rarely completely rational and suffers from biases, such as confirmation or availability biases [5, 13, 98]. Explanations may increasingly support users dealing with both algorithmic and cognitive bias.

*Fourthly*, recommendations are increasingly made as sets, packages, or sequences (c.f., chapter "Session-based Recommender Systems" on sequence-aware recommendation). Explanations will need then to justify not only the individual items, but also the relationship between them.

*Finally*, research is moving from seeing decision making with AI, as a hybrid collaboration: with the combination of human and machine intelligence, augmenting human intellect and capabilities instead of replacing them and achieving goals that were unreachable by either humans or machines alone [3]. Explanations can be used to balance out the relative strengths and weaknesses of people and recommender systems. For this to be effective, future work should focus on factors affecting over- and under- reliance.

## 6.1 Generating from Reviews

Deep neural networks have been effective for generating explanations from reviews. These promising lines of research would benefit from more thorough evaluation with users. For example, work applying a (capsule) neural network has made a connection between item aspects in reviews, their sentiments, and user evaluations [70]. This approach allows to distinguish between both positive and negative sentiments using an attention mechanism, but has yet to be validated with users. For readers interested in broader uses of deep neural networks for recommender systems, we refer to chapter "Deep Learning for Recommender Systems".

Another approach, which is also model agnostic (applied to both collaborative filtering, and a deep neural network), has been to learn the relationship between "interpretable components" (such as sentences) and explicit feedback (such as ratings), using reinforcement learning [132]. While only evaluated with a small number of participants, this is a promising approach to pursue. Chapter "Semantics and Content-Based Recommendations", while discussing developments in semantics-aware recommendation, also describes the kinds of user generated content that can be used to improve recommender systems. Chapter "Natural Language Processing for Recommender Systems" addresses more specifically recent developments in NLP techniques for recommender systems, while chapter "Multimedia Recommender Systems: Algorithms and Challenges" describes developments in analysis of multimedia content for recommender systems.

## 6.2   Social Recommendations

Explanations can also be helpful in situations where it is difficult to reach an agreement, e.g., when people have diverging preferences in a group of tourists. Previous work on group recommendations found generally there was a better interaction with distributed displays which allowed people to have some interactions on individual devices rather than a shared public display [48]. This highlights the need for privacy. Other recent work also highlights the tension between transparency and privacy for explanations in group decision making [84, 85]. When explaining people-to-people recommendations, it may be important to consider the biases of different stakeholders. E.g., the biases of recruiters as well as of employers when considering job candidates. It is also important to consider whether these relationships are directional, such as in the case of dating—when, if ever, is it suitable to supply an explanation of the type "John likes you so maybe you should give him a try."? There are number of chapters in this handbook for readers who would like to learn more about these subjects. Chapter "People-to-People Reciprocal Recommenders" in this handbook refers to people to people recommendations, chapter "Multistakeholder Recommender Systems" focuses specifically on multistakeholder recommendations and the tensions in the objectives of different stakeholders. Chapter "Fairness in Recommender Systems" lay out the various ways a recommender system may be unfair, also in multistakeholder scenarios, and provide a conceptual framework for identifying the the fairness that arise in an application and designing a project to assess and mitigate them.

The third scenario we consider is explaining content recommendations on social media based on the social network (e.g., "I am recommending this to you because your friend [X] likes it.", which already happens on Facebook). We see here both an opportunity and a challenge: Cialdini's persuasive principle of Liking may cause users to be persuaded by the explanation, but may hinder them from assessing the content critically (i.e., interfering with explanation effectiveness and decision making). We also refer to chapter "Social Recommender Systems" on social recommendations, and recommendations on social media.

There is also a particular challenge here in terms of both determining and conveying the reputability of source, in particular when helping to assess the risk of fake or misrepresenting news. Finally, recommending content in social networks also raises privacy concerns, and a tension with transparency. I.e., to which extent do users want to share their preferences with their network.

## 6.3  Explanations to Deal with Bias

Decision-making at individual, business, and societal levels is influenced by recommender systems. Further, individual cognitive selection strategies and homogeneous networks can amplify bias in customized recommendations, and influence which information users are exposed to [5]. While biases are undesirable in commercial and entertainment domains [29, 63], this is more critical when making recommendations in domains such as jobs and news.

While many kinds of biases may occur, recent research has focused primarily on popularity bias [29, 63], and limited diversity in recommendations [5, 23, 24]. For example, recommender systems have been found in some cases to decrease exposure to more diverse points of view [5, 23, 24], but they also have the potential to increase content and viewpoint diversity [35, 126]. However, exposing users to overly challenging views is more likely to backfire and cause not only rejection of the system but also more extreme views [36, 40]. Therefore it is important for future research to study how to design explanations that are effective for mitigating bias.

So far, design of explanations interfaces has considered the limited diversity as something that interacts with two cognitive biases: confirmation and availability biases. Recent work has evaluated different ways of highlighting parts of the profile the user was unaware of, so called "blind-spots" [67, 112, 125]. Other work has suggested using cross-referencing to assess the credibility of news recommendations [14]. While some work has been done on increasing reflective assessment about online content, this work has yet to be incorporated in recommender systems research [22, 50]. Further work is required to investigate whether an intent to explore unseen content changes behavior in the long term.

## 6.4  Sets and Sequences

As mentioned in Sect. 2.1, sometimes the recommender may show sets (e.g., resulting from package recommendation) or sequences of recommended items (e.g., playlists or itineraries). Research is needed on explanations for such combinations.

When explaining packages, in addition to explaining individual items, one may also need to explain the combinations. For example, for clothing an explanation could be: *"The blue of the top and the green of the bottom go well together, and a striped top goes well with a plain bottom. However, it is a bad idea to combine a formal top with a casual bottom."*

In initial research on explanations for package recommendations in the clothing domain, the impact of different explanation types (including the natural language one exemplified above) was investigated on effectiveness, efficiency, transparency, trust, and satisfaction [136]. They found that including combination aspects in the explanations changed the packages participants selected, so influencing either effectiveness or persuasiveness. They also found that participants varied in their opinions on the different explanation types, pointing towards a need for more research on adapting explanations for packages. This work was preliminary, using indirect experiments and only sets of two items. Far more research is needed on explaining packages, including the impact of domain, and how to explain sets with more items.

When explaining sequences, such as a sequence of activities to do on a trip, one may also need to explain the order of the items (e.g., referencing proximity in tourism), or how the items relate to each other (e.g,. *"your recent recommendations already include several museums"*) [137], in addition to why individual items are included. It becomes even more complicated if the sequence recommendation is made by a group recommender, so recommended to a group to experience together. In such a case, the explanation may also need to address the relationship between the items, and the relationship between group members, as well as privacy concerns [84, 85], and even how group members' affective state may be impacted by the order of items [72].

## 6.5   Over- and Under Reliance

In order for joint decision making between a human and a recommender system to be effective, it is important that people not only accept advice when it is correct but also reject it when it is incorrect.

It is preferable to generate explanations which help users decide whether they should use the decision support. Explanations which enable such *appropriate reliance* can be contrasted with explanations which cause over- reliance (accepting advice even when it is wrong) or under-reliance (rejecting advice even when it is right) [139]. The complexity of decision tasks, the limited cognitive resources of system users, and a tendency to keep decision effort low are related to the phenomenon of *bounded rationality* [108]: Users employ decision heuristics rather than make optimal decisions. Decision making under bounded rationality is a door opener for various types of influences on decision outcomes.

In a classification task, data scientists were found to over-trust and misuse local simplified ("interpretable") machine learning models such as generalized additive models (GAM) and SHapley Additive exPlanations (SHAP) [58]. Off the shelf implementations of these methods are able to indicate feature importance, or a notion of contribution to prediction outcome. Given that many instances of recommendations algorithms can be seen as classification or regression, findings from machine learning may help us understand how they could best be applied in recommender systems. It is not yet known which factors contribute to this kind of over-reliance in the domain of recommender systems. In research on artificial advice givers, the following have been found to influence the level of reliance on advice: the cost of receiving advice (time required for the adviser to give advice), the reliability of the adviser (% of time giving correct advice), and the predictability of the environment [113].

We refer also to chapter "Individual and Group Decision Making and Recommender Systems" on decision making with recommender systems. Further work is required to understand how these findings extend to recommender systems, in order to move us closer to effective decision-making in human-recommender system collaborations.

# References

1. L. Adrissono, A. Goy, G. Petrone, M. Segnan, P. Torasso, Intrigue: personalized recommendation of tourist attractions for desktop and handheld devices. Appl. Artif. Intell. **17**, 687–714 (2003)
2. Q. Ai, V. Azizi, X. Chen, Y. Zhang, Learning heterogeneous knowledge base embeddings for explainable recommendation. Algorithms **11**(9), 137 (2018)
3. Z. Akata, D. Balliet, M. de Rijke, F. Dignum, V. Dignum, G. Eiben, A. Fokkens, D. Grossi, K. Hindriks, H. Hoos, et al., A research agenda for hybrid intelligence: augmenting human intellect with collaborative, adaptive, responsible, and explainable artificial intelligence. Computer **53**(8), 18–28 (2020)
4. N.M. Aykin, T. Aykin, Individual differences in human-computer interaction. Comput. Ind. Eng. **20**(3), 373–379 (1991)
5. E. Bakshy, S. Messing, L.A. Adamic, Exposure to ideologically diverse news and opinion on facebook. Science **348**(6239), 1130–1132 (2015)
6. K. Balog, F. Radlinski, Measuring recommendation explanation quality: the conflicting goals of explanations, in *Proceedings of the 43rd International ACM SIGIR Conference on Research and Development in Information Retrieval, SIGIR '20* (2020)
7. K. Balog, F. Radlinski, S. Arakelyan, Transparent, scrutable and explainable user models for personalized recommendation, in *Proceedings of the 42nd International ACM SIGIR Conference on Research and Development in Information Retrieval (SIGIR '19)* (2019). https://dl.acm.org/citation.cfm?id=3331211
8. S.W. Bennett, A.C. Scott., *The Rule-Based Expert Systems: The MYCIN Experiments of the Stanford Heuristic Programming Project—Specialized Explanations for Dosage Selection*, chap. 19 (Addison-Wesley Publishing Company, Boston, 1985), pp. 363–370
9. M. Bilgic, R.J. Mooney, Explaining recommendations: satisfaction vs. promotion, in *Proceedings of the Wokshop Beyond Personalization, in Conjunction with the International Conference on Intelligent User Interfaces* (2005), pp. 13–18
10. D. Billsus, M.J. Pazzani, A personal news agent that talks, learns, and explains, in *Proceedings of the Third International Conference on Autonomous Agents* (1999), pp. 268–275
11. O. Biran, C. Cotton, Explanation and justification in machine learning: a survey, in *IJCAI-17 Workshop on Explainable AI (XAI)*, vol. 8 (2017)
12. S. Bostandjiev, J. O'Donovan, T. Höllerer, Tasteweights: a visual interactive hybrid recommender system, in *Proceedings of RecSys'12* (ACM, New York, 2012), pp. 35–42
13. D. Bountouridis, M. Makhortykh, E. Sullivan, J. Harambam, N. Tintarev, C. Hauff, Annotating credibility: identifying and mitigating bias in credibility datasets, in *Workshop on Reducing Online Misinformation Exposure (ROME) in Association with SIGIR 2019* (2019)

14. D. Bountouridis, M. Marrero, N. Tintarev, C. Hauff, Explaining credibility in news articles using cross-referencing, in *SIGIR workshop on ExplainAble Recommendation and Search (EARS)* (2018)
15. R.D. Burke, K.J. Hammond, B.C. Young, Knowledge-based navigation of complex information spaces, in *AAAI/IAAI*, vol. 1 (1996), pp. 462–468
16. G. Carenini, V. Mittal, J. Moore, Generating patient-specific interactive natural language explanations, in *Proc. Annu. Symp. Comput. Appl. Med. Care* (1994), pp. 5–9
17. S. Charleer, F. Gutiérrez, K. Verbert, Supporting job mediator and job seeker through an actionable dashboard, in *Proceedings of the 24th International Conference on Intelligent User Interfaces* (2019), pp. 121–131
18. L. Chen, P. Pu, Trust building in recommender agents. in *WPRSIUI in conjunction with Intelligent User Interfaces* (2002), pp. 93–100
19. L. Chen, P. Pu, Hybrid critiquing-based recommender systems, in *Intelligent User Interfaces* (2007), pp. 22–31
20. P.I. Chen, J.Y. Liu, Y.H. Yang, Personal factors in music preference and similarity: User study on the role of personality traits, in *Proc. Int. Symp. Computer Music Multidisciplinary Research (CMMR)* (2015)
21. C. Conati, G. Carenini, E. Hoque, B. Steichen, D. Toker, Evaluating the impact of user characteristics and different layouts on an interactive visualization for decision making, in *Computer Graphics Forum*, vol. 33 (Wiley Online Library, New York, 2014), pp. 371–380
22. P. Connor, E. Sullivan, M. Alfano, N. Tintarev, Motivated numeracy and active reasoning in a western European sample. J. Behav. Publ. Pol. (2020)
23. M.D. Conover, B. Gonçalves, A. Flammini, F. Menczer, Partisan asymmetries in online political activity. EPJ Data Sci. **1**(1), 6 (2012)
24. M.D. Conover, J. Ratkiewicz, M. Francisco, B. Gonçalves, F. Menczer, A. Flammini, Political polarization on twitter, in *Fifth International AAAI Conference on Weblogs and Social Media* (2011)
25. D. Cosley, S.K. Lam, I. Albert, J.A. Konstan, J. Riedl, Is seeing believing?: how recommender system interfaces affect users' opinions, in *CHI, Recommender Systems and Social Computing*, vol. 1 (2003), pp. 585–592
26. H. Cramer, V. Evers, M.V. Someren, S. Ramlal, L. Rutledge, N. Stash, L. Aroyo, B. Wielinga, The effects of transparency on perceived and actual competence of a content-based recommender, in *Semantic Web User Interaction Workshop, CHI* (2008)
27. H.S.M. Cramer, V. Evers, S. Ramlal, M. van Someren, L. Rutledge, N. Stash, L. Aroyo, B.J. Wielinga, The effects of transparency on trust in and acceptance of a content-based art recommender. User Model. User-Adapt. Interact **18**(5), 455–496 (2008)
28. M. Czarkowski, A scrutable adaptive hypertext. Ph.D. thesis, University of Sydney (2006)
29. M.D. Ekstrand, M. Tian, I.M. Azpiazu, J.D. Ekstrand, O. Anuyah, D. McNeill, M.S. Pera, All the cool kids, how do they fit in?: Popularity and demographic biases in recommender evaluation and effectiveness, in *Conference on Fairness, Accountability and Transparency* (2018), pp. 172–186
30. K. Erlich, S. Kirk, J. Patterson, J. Rasmussen, S. Ross, D. Gruen, Taking advice from intelligent systems: The double-edged sword of explanations, in *Intelligent User Interfaces* (2011)
31. A. Felfernig, B. Gula, Consumer behavior in the interaction with knowledge-based recommender applications, in *ECAI 2006 Workshop on Recommender Systems* (2006), pp. 37–41
32. A. Felfernig, E. Teppan, B. Gula, Knowledge-based recommender technologies for marketing and sales. Int. J. Patt. Recogn. Artif. Intell. **21**, 333–355 (2007)
33. B. Ferwerda, M. Tkalcic, M. Schedl, Personality traits and music genre preferences: How music taste varies over age groups, in *1st Workshop on Temporal Reasoning in Recommender Systems (RecTemp) at the 11th ACM Conference on Recommender Systems*, Como, August 31, 2017, vol. 1922 (ACM Digital Library, New York, 2017), pp. 16–20

34. B. Ferwerda, M. Tkalcic, M. Schedl, Personality traits and music genres: What do people prefer to listen to? in *Proceedings of the 25th Conference on User Modeling, Adaptation and Personalization* (ACM, New York, 2017)
35. S. Flaxman, S. Goel, J.M. Rao, Filter bubbles, echo chambers, and online news consumption. Publ. Opin. Quart. **80**(S1), 298–320 (2016)
36. D. Flynn, B. Nyhan, J. Reifler, The nature and origins of misperceptions: Understanding false and unsupported beliefs about politics. Polit. Psychol. **38**, 127–150 (2017)
37. B. Fogg, J. Marshall, T. Kameda, J. Solomon, A. Rangnekar, J. Boyd, B. Brown, Web credibility research: a method for online experiments and early study results, in *CHI 2001* (2001), pp. 295–296
38. B.J. Fogg, C. Soohoo, D.R. Danielson, L. Marable, J. Stanford, E.R. Tauber, How do users evaluate the credibility of web sites?: a study with over 2,500 participants, in *Designing for User Experiences (DUX)*. Focusing on User-to-Product Relationships, vol. 15 (2003), pp. 1–15
39. F. Gedikli, D. Jannach, M. Ge, How should I explain? A comparison of different explanation types of recommender systems. Int. J. Human-Computer Stud. **72**(4), 367–382 (2014)
40. E. Graells-Garrido, M. Lalmas, R. Baeza-Yates, Data portraits and intermediary topics: encouraging exploration of politically diverse profiles, in *Proceedings of the 21st International Conference on Intelligent User Interfaces* (2016), pp. 228–240
41. S. Green, P. Lamere, J. Alexander, F. Maillet, Generating transparent, steerable recommendations from textual descriptions of items, in *Recommender Systems Conference* (2009)
42. I. Guy, I. Ronen, E. Wilcox, Do you know? recommending people to invite into your social network, in *International Conference on Intelligent User Interfaces* (2009), pp. 77–86
43. I. Guy, N. Zwerdling, D. Carmel, I. Ronen, E. Uziel, S. Yogev, S. Ofek-Koifman, Personalized recommendation of social software items based on social relations, in *ACM Conference on Recommender systems* (2009), pp. 53–60
44. J. Harambam, N. Helberger, J. van Hoboken, Democratizing algorithmic news recommenders: how to materialize voice in a technologically saturated media ecosystem. Philos. Trans. R. Soc. A: Math. Phys. Eng. Sci. **376**(2133), 20180088 (2018)
45. G. Häubl, V. Trifts, Consumer decision making in online shopping environments: The effects of interactive decision aids. Market. Sci. **19**, 4–21 (2000)
46. N. Helberger, K. Karppinen, L. D'Acunto, Exposure diversity as a design principle for recommender systems. Inf. Commun. Soc. **21**(2), 191–207 (2018)
47. J.L. Herlocker, J.A. Konstan, J. Riedl, Explaining collaborative filtering recommendations, in *ACM Conference on Computer Supported Cooperative Work* (2000), pp. 241–250
48. D. Herzog, W. Wörndl, A user study on groups interacting with tourist trip recommender systems in public spaces, in *Proceedings of the 27th ACM Conference on User Modeling, Adaptation and Personalization* (ACM, New York, 2019), pp. 130–138
49. M. Hingston, User friendly recommender systems. Master's Thesis, Sydney University (2006)
50. A. Holzer, N. Tintarev, S. Bendahan, S. Greenup, D. Gillet, Digitally scaffolding debate in the classroom. in *Proceedings of the 2018 CHI Conference Extended Abstracts on Human Factors in Computing Systems* (2018)
51. M. Jiang, Y. Fang, H. Xie, J. Chong, M. Meng, User click prediction for personalized job recommendation. World Wide Web **22**(1), 325–345 (2019)
52. Y. Jin, N.N. Htun, N. Tintarev, K. Verbert, Contextplay: evaluating user control for context-aware music recommendation, in *Proceedings of the 27th ACM Conference on User Modeling, Adaptation and Personalization* (2019), pp. 294–302
53. Y. Jin, K. Seipp, E. Duval, K. Verbert, Go with the flow: effects of transparency and user control on targeted advertising using flow charts, in *Proceedings of AVI '16* (ACM, New York, 2016), pp. 68–75
54. Y. Jin, N. Tintarev, N.N. Htun, K. Verbert, Effects of personal characteristics in control-oriented user interfaces for music recommender systems, in *User Modeling and User-Adapted Interaction* (2019), pp. 1–51

55. Y. Jin, N. Tintarev, K. Verbert, Effects of individual traits on diversity-aware music recommender user interfaces, in *Proceedings of the 26th Conference on User Modeling, Adaptation and Personalization* (2018), pp. 291–299
56. B. Kang, J. O'Donovan, T. Höllerer, Modeling topic specific credibility on twitter, in *Proceedings of the 2012 ACM International Conference on Intelligent User Interfaces* (2012), pp. 179–188
57. B. Kang, N. Tintarev, T. Höllerer, J. O'Donovan, What am I not seeing? an interactive approach to social content discovery in microblogs, in *Social Informatics - 8th International Conference, SocInfo 2016*, ed. by E.S. Spiro, Y. Ahn, Bellevue, WA, November 11–14, 2016, Proceedings, Part II. Lecture Notes in Computer Science, vol. 10047(2016), pp. 279–294 . https://doi.org/10.1007/978-3-319-47874-6_20
58. H. Kaur, H. Nori, S. Jenkins, R. Caruana, H. Wallach, J. Wortman Vaughan, Interpreting interpretability: Understanding data scientists' use of interpretability tools for machine learning, in *Proceedings of the 2020 CHI Conference on Human Factors in Computing Systems* (2020), pp. 1–14
59. B.P. Knijnenburg, S. Sivakumar, D. Wilkinson, Recommender systems for self-actualization, in *Proceedings of the 10th ACM Conference on Recommender Systems* (ACM, New York, 2016)
60. B.P. Knijnenburg, M.C. Willemsen, Z. Gantner, H. Soncu, C. Newell, Explaining the user experience of recommender systems. User Model. User-Adapted Interact. **22**(4–5), 441–504 (2012)
61. P. Kouki, J. Schaffer, J. Pujara, J. O'Donovan, L. Getoor, Personalized explanations for hybrid recommender systems, in *Proceedings of the 24th International Conference on Intelligent User Interfaces* (2019), pp. 379–390
62. P. Kouki, J. Schaffer, J. Pujara, J. O'Donovan, L. Getoor, Generating and understanding personalized explanations in hybrid recommender systems. ACM Trans. Interact. Intell. Syst. (TiiS) **10**(4), 1–40 (2020)
63. D. Kowald, M. Schedl, E. Lex, The unfairness of popularity bias in music recommendation: a reproducibility study, in *European Conference on Information Retrieval* (Springer, New York, 2020), pp. 35–42
64. T. Kramer, The effect of measurement task transparency on preference construction and evaluations of personalized recommendations. J. Market. Res. **44**(2), 224–233 (2007)
65. B. Krulwich, The infofinder agent: learning user interests through heuristic phrase extraction. IEEE Intell. Syst. **12**, 22–27 (1997)
66. T. Kulesza, S. Stumpf, M. Burnett, S. Yang, I. Kwan, W.K. Wong, Conference on visual languages and human-centric computing, in *Too Much, Too Little, or Just Right? Ways Explanations Impact End Users' Mental Models* (2013)
67. J. Kumar, N. Tintarev, Using visualizations to encourage blind-spot exploration, in *Recsys Workshop on Interfaces and Decision Making in Recommender Systems* (2018)
68. S. Lallé, C. Conati, G. Carenini, Impact of individual differences on user experience with a visualization interface for public engagement, in Proceedings of the UMAP '17 (ACM, New York, 2017), pp. 247–252
69. C. Lewis, J. Rieman, Task-centered user interface design: a practical introduction. University of Colorado (1994)
70. C. Li, C. Quan, L. Peng, Y. Qi, Y. Deng, L. Wu, A capsule network for recommendation and explaining what you like and dislike. in Proceedings of the 42nd International ACM SIGIR Conference on Research and Development in Information Retrieval (2019), pp. 275–284
71. W. Liu, J. Guo, N. Sonboli, R. Burke, S. Zhang, Personalized fairness-aware re-ranking for microlending. in *Proceedings of the 13th ACM Conference on Recommender Systems* (2019), pp. 467–471
72. J. Masthoff, A. Gatt, In pursuit of satisfaction and the prevention of embarrassment: affective state in group recommender systems. User Model. User-Adapted Interact. **16**(3–4), 281–319 (2006)

73. K. McCarthy, J. Reilly, L. McGinty, B. Smyth, Thinking positively - explanatory feedback for conversational recommender systems. in *Proceedings of the European Conference on Case-Based Reasoning (ECCBR-04) Explanation Workshop* (2004), pp. 115–124

74. L. McGinty, B. Smyth, Comparison-based recommendation. Lecture Notes in Computer Science **2416**, 575–589 (2002)

75. S.M. McNee, S.K. Lam, J.A. Konstan, J. Riedl, Interfaces for eliciting new user preferences in recommender systems. *User Modeling* (2003), pp. 178–187

76. D. McSherry, Explanation in recommender systems. Artif. Intell. Rev. **24**(2), 179–197 (2005)

77. M. Millecamp, N.N. Htun, C. Conati, K. Verbert, To explain or not to explain: the effects of personal characteristics when explaining music recommendations, in *Proceedings of the 2019 Conference on Intelligent User Interface* (ACM, New York, 2019), pp. 1–12

78. M. Millecamp, N.N. Htun, Y. Jin, K. Verbert, Controlling spotify recommendations: effects of personal characteristics on music recommender user interfaces, in *Proceedings of the 26th Conference on User Modeling, Adaptation and Personalization* (ACM, New York, 2018), pp. 101–109

79. T. Miller, Explanation in artificial intelligence: insights from the social sciences. Artif. Intell. **267**, 1–38 (2019)

80. D. Müllensiefen, B. Gingras, J. Musil, L. Stewart, The musicality of non-musicians: an index for assessing musical sophistication in the general population. PloS One **9**(2), e89642 (2014)

81. C. Musto, M. de Gemmis, P. Lops, G. Semeraro, Generating post hoc review-based natural language justifications for recommender systems, in *User Modeling and User-Adapted Interaction* (2020), pp. 1–45

82. C. Musto, F. Narducci, P. Lops, M. De Gemmis, G. Semeraro, Explod: a framework for explaining recommendations based on the linked open data cloud, in *Proceedings of the 10th ACM Conference on Recommender Systems* (2016), pp. 151–154

83. S. Nagulendra, J. Vassileva, Providing awareness, understanding and control of personalized stream filtering in a p2p social network, in *Conference on Collaboration and Technology (CRIWG)* (2013)

84. S. Najafian, A. Delic, M. Tkalčič, N. Tintarev, Factors influencing privacy concern for explanations of group recommendation, in *UMAP* (2021)

85. S. Najafian, O. Inel, N. Tintarev, Someone really wanted that song but it was not me! evaluating which information to disclose in explanations for group recommendations, in *Proceedings of the 25th International Conference on Intelligent User Interfaces Companion* (2020), pp. 85–86

86. T.T. Nguyen, D. Kluver, T.Y. Wang, P.M. Hui, M.D. Ekstrand, M.C. Willemsen, J. Rield, Rating support interfaces to improve user experience and recommender accuracy, in *Recommender Systems Conference* (2013)

87. J. Nielsen, R. Molich, Heuristic evaluation of user interfaces, in *ACM CHI'90* (1990), pp. 249–256

88. I. Nunes, D. Jannach, A systematic review and taxonomy of explanations in decision support and recommender systems. User Modeling and User-Adapted Interaction **27**(3–5), 393–444 (2017)

89. J. O'Donovan, B. Smyth, B. Gretarsson, S. Bostandjiev, T. Höllerer, Peerchooser: visual interactive recommendation, in *Proc. of CHI'08* (ACM, New York, 2008), pp. 1085–1088

90. R. Ohanian, Construction and validation of a scale to measure celebrity endorsers' perceived expertise, trustworthiness, and attractiveness. J. Advert. **19**(3), 39–52 (1990)

91. A. Olteanu, C. Castillo, F. Diaz, E. Kiciman, Social data: biases, methodological pitfalls, and ethical boundaries (2016)

92. D. O'Sullivan, B. Smyth, D.C. Wilson, K. McDonald, A. Smeaton, Improving the quality of the personalized electronic program guide. User Modeling and User-Adapted Interaction **14**, 5–36 (2004)

93. Pandora (2020). http://www.pandora.com
94. A. Papadimitriou, P. Symeonidis, Y. Manolopoulos, A generalized taxonomy of explanation styles for traditional and social recommender systems. Data Mining Knowl. Discov. **24**, 555–583 (2012)
95. D. Parra, P. Brusilovsky, User-controllable personalization: a case study with setfusion. IJHCS'15 **78**, 43–67 (2015)
96. D. Parra, P. Brusilovsky, C. Trattner, See what you want to see: visual user-driven approach for hybrid recommendation, in *Proceedings of the 19th international conference on Intelligent User Interfaces* (ACM, New York, 2014), pp. 235–240
97. M.J. Pazzani, A framework for collaborative, content-based and demographic filtering. Artif. Intell. Rev. **13**, 393–408 (1999)
98. A. Peng, B. Nushi, E. Kıcıman, K. Inkpen, S. Suri, E. Kamar, What you see is what you get? The impact of representation criteria on human bias in hiring, in *Proceedings of the AAAI Conference on Human Computation and Crowdsourcing*, vol. 7 (2019), pp. 125–134
99. P. Pu, L. Chen, Trust building with explanation interfaces. in *IUI'06, Recommendations I* (2006), pp. 93–100
100. P. Pu, L. Chen, Trust-inspiring explanation interfaces for recommender systems. Knowledge-Based Syst. **20**, 542–556 (2007)
101. P. Pu, B. Faltings, L. Chen, J. Zhang, P. Viappiani, Usability guidelines for product recommenders based on example critiquing research, in Recommender Systems Handbook, ed. by F. Ricci, L. Rokach, B. Shapira, P.B. Kantor (Springer US, New York, 2011), pp. 547–576
102. R. Rafter, B. Smyth, Conversational collaborative recommendation - an experimental analysis. Artif. Intell. Rev **24**(3-4), 301–318 (2005)
103. J. Reilly, K. McCarthy, L. McGinty, B. Smyth, Dynamic critiquing, in ECCBR, ed. by P. Funk, P.A. González-Calero. Lecture Notes in Computer Science, vol. 3155 (Springer, New York, 2004), pp. 763–777
104. U. Reviglio, Serendipity as an emerging design principle of the infosphere: challenges and opportunities, in *Ethics and Information Technology* (2019), pp. 1–16
105. C. Seifert, S. Scherzinger, L. Wiese, Towards generating consumer labels for machine learning models, in *2019 IEEE First International Conference on Cognitive Machine Intelligence (CogMI)* (IEEE, New York, 2019), pp. 173–179
106. G. Shani, L. Rokach, B. Shapira, S. Hadash, M. Tangi, Investigating confidence displays for top-n recommendations. J. Am. Soc. Inf. Sci. Technol. **64**, 2548–2563 (2013)
107. A. Sharma, D. Cosley, Do social explanations work? Studying and modeling the effects of social explanations in recommender systems, in *World Wide Web (WWW)* (2013)
108. H.A. Simon, A behavioural model of choice. Quart. J. Econ. **69**(1), 99–118 (1955)
109. R. Sinha, K. Swearingen, The role of transparency in recommender systems, in *Conference on Human Factors in Computing Systems* (2002), pp. 830–831
110. N. Sonboli, F. Eskandanian, R. Burke, W. Liu, B. Mobasher, Opportunistic multi-aspect fairness through personalized re-ranking, in *Proceedings of the 28th ACM Conference on User Modeling, Adaptation and Personalization, UMAP '20* (Association for Computing Machinery, New York, NY, 2020), pp. 239–247. https://doi.org/10.1145/3340631.3394846
111. F. Sørmo, J. Cassens, A. Aamodt, Explanation in case-based reasoning  perspectives and goals. Artif. Intell. Rev. **24**(2), 109–143 (2005)
112. E. Sullivan, D. Bountouridis, J. Harambam, S. Najafian, F. Loecherbach, M. Makhortykh, D. Kelen, D. Wilkinson, D. Graus, N. Tintarev, Reading news with a purpose: explaining user profiles for self-actualization, in *Adjunct Publication of the 27th Conference on User Modeling, Adaptation and Personalization* (2019), pp. 241–245
113. S. Sutherland, C. Harteveld, M. Young, Effects of the advisor and environment on requesting and complying with automated advice. ACM Trans. Interact. Intell. Syst. **6** (2016). Article 27
114. K. Swearingen, R. Sinha, Interaction design for recommender systems, in *Designing Interactive Systems* (2002), pp. 25–28

115. P. Symeonidis, A. Nanopoulos, Y. Manolopoulos, Justified recommendations based on content and rating data, in *WebKDD Workshop on Web Mining and Web Usage Analysis* (2008)
116. K. Tanaka-Ishii, I. Frank, Multi-agent explanation strategies in real-time domains, in *38th Annual Meeting on Association for Computational Linguistics*, (2000), pp. 158–165
117. C.A. Thompson, M.H. Göker, P. Langley, A personalized system for conversational recommendations. J. Artif. Intell. Res. (JAIR) **21**, 393–428 (2004)
118. N. Tintarev, Explaining recommendations, in *User Modeling* (2007), pp. 470–474
119. N. Tintarev, J. Masthoff, Effective explanations of recommendations: user-centered design, in *Recommender Systems* (2007), pp. 153–156
120. N. Tintarev, J. Masthoff, Over- and underestimation in different product domains, in *Workshop on Recommender Systems associated with ECAI* (2008)
121. N. Tintarev, J. Masthoff, Personalizing movie explanations using commercial meta-data, in *Adaptive Hypermedia* (2008)
122. N. Tintarev, J. Masthoff, Evaluating the effectiveness of explanations for recommender systems: methodological issues and empirical studies on the impact of personalization. User Model. User-Adapted Interact. **22**, 399–439 (2012)
123. N. Tintarev, J. Masthoff, Explaining recommendations: design and evaluation, in *Recommender Systems Handbook* (Springer, New York, 2015), pp. 353–382
124. N. Tintarev, J. Masthoff, Effects of individual differences in working memory on plan presentational choices. Front. Psychol. **7**, 1793 (2016)
125. Tintarev, N., Rostami, S., Smyth, B.: Knowing the unknown: visualising consumption blindspots in recommender system, in *ACM Symposium On Applied Computing (SAC)* (2018)
126. N. Tintarev, E. Sullivan, D. Guldin, S. Qiu, D. Odjik, Same, same, but different: algorithmic diversification of viewpoints in news. in *UMAP workshop on Fairness in User Modeling, Adaptation and Personalization, in association with UMAP'18* (2018)
127. K. Verbert, D. Parra, P. Brusilovsky, Agents vs. users: visual recommendation of research talks with multiple dimension of relevance. ACM Trans. Interact. Intell. Syst. (TiiS) **6**(2), 11 (2016)
128. K. Verbert, D. Parra, P. Brusilovsky, E. Duval, Visualising recommendations to support exploration, transparency and controllability, in Proceedings of IUI'13 (ACM, New York, 2013), pp. 351–362
129. P. Victor, M.D. Cock, C. Cornelis, Trust and recommendations, in *Recommender Systems Handbook*, ed. by F. Ricci, L. Rokach, B. Shapira, P.B. Kantor , pp. 547–576 (Springer US, New York, 2011)
130. J. Vig, S. Sen, J. Riedl, Tagsplanations: explaining recommendations using tags, in *Intelligent User Interfaces* (2009)
131. W. Wang, I. Benbasat, Recommendation agents for electronic commerce: effects of explanation facilities on trusting beliefs. J. Manage. Inf. Syst. **23**, 217–246 (2007)
132. X. Wang, Y. Chen, J. Yang, L. Wu, Z. Wu, X. Xie, A reinforcement learning framework for explainable recommendation, in *2018 IEEE International Conference on Data Mining (ICDM)* (IEEE, New York, 2018), pp. 587–596
133. P. Wärnestål, Modeling a dialogue strategy for personalized movie recommendations, in *Beyond Personalization Workshop* (2005), pp. 77–82
134. Wärnestål, P.: User evaluation of a conversational recommender system, in *Proceedings of the 4th Workshop on Knowledge and Reasoning in Practical Dialogue Systems* (2005), pp. 32–39
135. A.T. Wibowo, A. Siddharthan, C. Lin, J. Masthoff, Matrix factorization for package recommendations, in *Proceedings of the RecSys 2017 Workshop on Recommendation in Complex Scenarios (ComplexRec 2017). CEUR-WS (2017)*
136. A.T. Wibowo, A. Siddharthan, J. Masthoff, C. Lin, Understanding how to explain package recommendations in the clothes domain, in *IntRS@ RecSys* (2018), pp. 74–78
137. W. Wörndl, A. Hefele, Generating paths through discovered places-of-interests for city trip planning, in *Information and Communication Technologies in Tourism 2016* (Springer, New York, 2016), pp. 441–453

138. F. Yang, Z. Huang, J. Scholtz, D.L. Arendt, How do visual explanations foster end users' appropriate trust in machine learning? in *Proceedings of the 25th International Conference on Intelligent User Interfaces, IUI '20* (Association for Computing Machinery, New York, NY, 2020), pp. 189–201
139. F. Yang, Z. Huang, J. Scholtz, D.L. Arendt, How do visual explanations foster end users' appropriate trust in machine learning? in *Proceedings of the 25th International Conference on Intelligent User Interfaces, IUI '20* (Association for Computing Machinery, New York, NY, 2020), pp. 189–201
140. L. Ye, P. Johnson, L.R. Ye, P.E. Johnson, The impact of explanation facilities on user acceptance of expert systems advice. MIS Quart. **19**(2), 157–172 (1995)
141. K.P. Yee, K. Swearingen, K. Li, M. Hearst, Faceted metadata for image search and browsing, in *ACM Conference on Computer-Human Interaction* (2003)
142. Y. Zheng, T. Dave, N. Mishra, H. Kumar, Fairness in reciprocal recommendations: a speed-dating study. in *Adjunct Publication of the 26th Conference on User Modeling, Adaptation and Personalization* (2018), pp. 29–34

# Personality and Recommender Systems

**Marko Tkalčič and Li Chen**

## 1 Introduction

As argued in Chapter "Individual and Group Decision Making and Recommender Systems", an important function of recommender systems is to help people make better decisions. It has also been found that an improvement in the rating prediction accuracy (usually measured with metrics such as RMSE, see also Chapter "Evaluating Recommender Systems") does not necessarily mean a better user experience [64]. Furthermore, assessing the recommender systems from a user-centric perspective (e.g., decision confidence and system satisfaction) yields a better picture of the quality of the recommender system under study. Hence, one should better take into consideration user characteristics for optimizing a recommender system. This is why personality, which plays an important role in decision-making [22], has been considered to improve the system.

From its definition in psychology, personality accounts for the individual differences in our enduring emotional, interpersonal, experiential, attitudinal, and motivational styles [48]. Incorporating these differences in the recommender system appears to be a promising choice for delivering personalized recommendations. Furthermore, personality parameters can be quantified as feature vectors, which makes them suitable to use in computer algorithms. Therefore, user personality has been considered in a wide range of aspects of recommender systems. For example, personality has been used to improve user-user similarity calculation in solving the

M. Tkalčič (✉)
Faculty of Mathematics, Natural Sciences and Information Technologies, University of Primorska, Koper, Slovenia
e-mail: marko.tkalcic@famnit.upr.si

L. Chen
Hong Kong Baptist University, Kowloon, Hong Kong
e-mail: lichen@comp.hkbu.edu.hk

© Springer Science+Business Media, LLC, part of Springer Nature 2022          757
F. Ricci et al. (eds.), *Recommender Systems Handbook*,
https://doi.org/10.1007/978-1-0716-2197-4_20

new user problem [44, 97]. It has also been demonstrated that people with different personalities can be more or less inclined to consume novel items, so the degree of diversity in presenting recommended items can be personalized accordingly [102]. Moreover, group modeling based on personality has improved the performance of group recommendations [52, 75, 77]. Indeed, recently the topic of using personality for personalization has gained traction not only in individual research done but also in a dedicated edited volume [94] and tutorials at the RecSys conference [69, 93].

However, the acquisition of personality factors for individual users has been mainly done through extensive questionnaires, which was an obstacle in a day-to-day use of recommender systems. Examples of such questionnaires are the International Personality Item Pool (IPIP) [38] and the NEO Personality Inventory [62]. In recent years, several investigations have been conducted to extract person-ality parameters using machine learning techniques [37, 53, 74]. Valuable sources for detecting the personality of a user without bothering her/him with extensive questionnaires are social media streams (e.g., Facebook [53], blogs [47], Instagram [32], Twitter [74]) and other user-generated data streams (e.g., email [85], drug consumption [35], eye gaze [8]).

In this chapter, we survey the usage of the psychological model of personality to improve recommendation accuracy, diversity, to address the new user problem, to improve cross-domain recommendations, and in group recommendation scenarios. We focus on the tools needed to design such systems, especially on (i) personality acquisition methods and on (ii) strategies for using personality in recommender systems. The chapter is organized as follows. In Sect. 2 we survey various models of personality that were developed and are suitable for recommender systems. In Sect. 3 we present various methods for acquiring personality, which fall in either of the two categories: *implicit* or *explicit*. In Sect. 4 we discuss various strategies that exploit personality and have been used so far in recommender systems. Further, in Sect. 5 we present the challenges that are still ahead in this area. Finally we provide some conclusive thoughts in Sect. 6.

## 2   Personality Model

According to [63], personality accounts for the most important dimensions in which individuals differ in their enduring emotional, interpersonal, experiential, attitudinal, and motivational styles. Translated into the recommender systems terminology, personality can be thought of as a (component of) user profile, which is context-independent (it does not change with time, location or some other contexts—see Chapter "Context-Aware Recommender Systems: From Foundations to Recent Developments" for context in recommender systems) and domain-independent (it does not change through different domains, e.g., books, movies—see also Chapter "Design and Evaluation of Cross-domain Recommender Systems" for personality in cross-domain recommender systems).

**Table 1** Examples of adjectives related to the FFM [63]

| Factor | Adjectives |
| --- | --- |
| Openness (O) | Artistic, curious, imaginative, insightful, original, wide interest |
| Conscientiousness (C) | Efficient, organized, planful, reliable, responsible, thorough |
| Extraversion (E) | Active, assertive, energetic, enthusiastic, outgoing, talkative |
| Agreeableness (A) | Appreciative, forgiving, generous, kind, sympathetic, trusting |
| Neuroticism (N) | Anxious, self-pitying, tense, touchy, unstable, worrying |

Historically, the first reports of studies of individual traits among humans go back to the ancient Greeks with the Hippocrates' Four Humours that eventually led to the personality theory known today as the four temperaments (i.e., Choleric, Sanguine, Melancholic and Phlegmatic) [50].

Today, the Five Factor Model of personality (FFM) [63] is considered one of the most comprehensive and the widely used personality models in recommender systems [15, 27, 43–45, 69, 70, 90, 99, 102]. The FFM is sometimes referred to also as the Big-Five (Big5) model of personality.

## 2.1 The Five Factor Model of Personality

The roots of the FFM lie in the lexical hypothesis, which states that things that are most important in people's lives eventually become part of their language. Studying the usage of language, researchers extracted a set of adjectives that describe permanent traits (see Table 1). With further research, these adjectives were clustered into the five main dimensions: Openness to experience, conscientiousness, extraversion, agreeableness, and neuroticism (the acronym OCEAN is often used) [63].

**Openness to Experience (O)**, often referred to just as Openness, describes the distinction between imaginative, creative people and down-to-earth, conventional people. High O scorers are typically individualistic, non-conforming and are very aware of their feelings. They can easily think in abstraction. People with low O values tend to have common interests. They prefer simple and straightforward thinking over complex, ambiguous and subtle. The sub-factors are imagination, artistic interest, emotionality, adventurousness, intellect, and liberalism.

**Conscientiousness (C)** concerns the way in which we control, regulate and direct our impulses. People with high C values tend to be prudent while those with low values tend to be impulsive. The sub-factors are self-efficacy, orderliness, dutifulness, achievement-striving, self-discipline, and cautiousness.

**Extraversion (E)** tells the degree of engagement with the external world (in case of high values) or the lack of it (low values). The sub-factors of E are friendliness, gregariousness, assertiveness, activity level, excitement-seeking, and cheerfulness. Extrovert people (high score on the E factor) tend to react with enthusiasm and often have positive emotions, while introverted people (low score on the E factor) tend to be quiet, low-key and disengaged in social interactions.

**Agreeableness (A)** reflects individual differences concerning cooperation and social harmony. The sub-factors of A are trust, morality, altruism, cooperation, modesty, and sympathy.

**Neuroticism (N)** refers to the tendency of experiencing negative feelings. People with high N values are emotionally reactive. They tend to respond emotionally to relatively neutral stimuli. They are often in a bad mood, which strongly affects their thinking and decision making (see Chapter "Individual and Group Decision Making and Recommender Systems" for more on decision making). Low N scorers are calm, emotionally stable, and free from persistent bad mood. The sub-factors are anxiety, anger, depression, self-consciousness, immoderation, and vulnerability. The neuroticism factor is sometimes referred to as emotional stability [40].

## 2.2   Other Models of Personality

Other personality models that can be of interest to the recommender system community are the vocational RIASEC (with the main types *realistic, investigative, artistic, social, enterprising*, and *conventional*) model [42], which was used in an e-commerce prototype [10]; and the Bartle model (with the main types *killers, achievers, explorers* and *socializers*), which is suitable for the videogames domain [88].

The Thomas-Kilmann conflict mode personality model has been developed to model group dynamics [89]. The model is composed of the following two dimensions that account for differences in individual behaviour in conflict situations:[1] *assertiveness* and *cooperativeness*. Within this two-dimensional space, subjects are classified into any of these five categories: *competing, collaborating, compromising, avoiding*, or *accommodating*.

Although learning styles per se are not considered as a personality model, they share with personality the quality of being time invariant. In the domain of e-learning, models of learning styles have been used to recommend course materials to students [26]. An example is the Felder and Silverman Learning Style Model [30], which measures four factors: *active/reflective, sensing/intuitive, visual/verbal*, and *sequential/global*.

In addition, some ad-hoc personality models have been proposed in the recommender systems community. For a trendy pictures recommender system, a personality model with two types, the *trend-setters* and the *trend-spotters*, has been proposed, along with a methodology for predicting the personality types from social media networks [84]. Especially in the domain of social networks, there

---

[1] The Thomas-Kilmann conflict mode instrument is available at http://cmpresolutions.co.uk/wp-content/uploads/2011/04/Thomas-Kilman-conflict-instrument-questionaire.pdf.

**Table 2**  Main personality models

| Ref. | Model name | Primary domain | Main types/traits |
|------|-----------|----------------|-------------------|
| [63] | Five Factor Model | General | Openness, conscientiousness, extraversion, agreeableness, and neuroticism |
| [50] | Four temperaments | General | Choleric, Sanguine, Melancholic, and Phlegmatic |
| [42] | RIASEC | Vocational | Realistic, investigative, artistic, social, enterprising, and conventional |
| [88] | Bartle types | Video games | Killers, achievers, explorers, and socializers |
| [30] | Felder and Silverman Learning Style Model | Learning styles | Active/reflective, sensing/intuitive, visual/verbal, and sequential/global |
| [89] | Thomas-Kilmann conflict model | Group/conflict modeling | Assertiveness, cooperativeness |

is a tendency to stress the *influence/susceptibility* aspects of users as the main personality traits (e.g., *leaders/followers*) [5] (Table 2).

## 2.3   *Relationship Between Personality and User Preferences*

A number of studies showed that personality relates strongly with user preferences. Users with different personalities tend to prefer different kinds of content. These relations are domain dependent. Such an information is very valuable when designing a recommender system for a specific domain.

In their study, Rentfrow and Gosling [79] explored how music preferences are related to personality in terms of the FFM model. They categorized music pieces each into one of the four categories: reflective & complex, intense & rebelious, upbeat & conventional, and energetic & rhythmic. The reflective & complex category is related to openness to new experience. Similarly, the intense & rebelious category is also positively related to openness to new experience. However, although this category contains music with negative emotions, it is not related to neuroticism or agreeableness. The upbeat & conventional category was found to be positively related with extraversion, agreeableness, and conscientiousness. Finally, they found that the energetic & rhythmic category is related to extraversion and agreeableness.

The relation between music and personality was also explored by Rawlings et al. [76]. They observed that the extraversion and openness factors are the only ones that explain the variance in the music preferences. Subjects with high openness tend to prefer diverse music styles. Extraversion, on the other hand, was found to be strongly related to preferences for popular music.

Manolios et al. [58] explored the relationship between personal values and user preferences in the music domain. The user preferences were not coded as is usual (e.g., preferences for genres or artists) but as relationships between values, consequences and attributes. Examples of these relationships are the value of conservation is strongly related to the consequence of relaxation, which is strongly related to the attribute emotions; or the value of openness to change is strongly related to the consequence of discovery/stimulation, which is strongly related to the attribute of complexity/originality.

Being different from the above work that focused on studying the effect of personality on user preferences for music genres or styles, Melchiorre and Schedl [65] analyzed its correlations with music audio features. The results show that most of correlations are low to medium, with the strongest effects for the openness trait.

Rentfrow et al. [78] extended the domain to general entertainment, which includes music, books, magazines, films, and TV shows. They categorized the content into the following categories: aesthetic, cerebral, communal, dark, and thrilling. The communal category is positively related to extraversion, agreeableness, and conscientiousness, while being negatively related to extraversion and neuroticism. The aesthetic category is positively related to agreeableness, extraversion and negatively to neuroticism. The dark category is positively related extraversion and negatively to conscientiousness and agreeableness. The cerebral category is related to extraversion, while the thrilling category does not reveal any consistent correlation with personality factors. Cantador et al. [14] also presented the results of an experiment over multiple domains including movies, TV shows, music, and books, based on the myPersonality dataset [53].

In another work [103], the authors surveyed 1706 users on Douban Interest Group,[2] which is a popular Chinese online community where users can join different types of interest groups (e.g., "Sports", "Music", "Health", "Academic") to leave comments or recommend topics to their friends. They found that all the five personality traits as defined in FFM significantly affect users' preferences for group types. For instance, groups about "Health" and "Sports" are more preferred by people who are more self-organized (with high conscientiousness) and extroverted (with high extraversion). Groups related to aesthetics (e.g., "Art" and "Literature") and entertainment (e.g., "Animation", "Music" and "Movie") are more preferred by people who are more creative and aesthetic sensitive (with high openness). Those people also prefer the groups about "Academic" and "Interest". Moreover, people who are more suspicious (with low agreeableness) tend to prefer "Movie" type group, whereas those who are more emotionally unstable (with high neuroticism) are likely to prefer "Costume" type groups.

In an experiment based on a contextual movie recommender system dataset (the CoMoDa dataset [71]), Odic et al. explored the relations between personality factors and the induced emotions in movies in different social contexts [72]. They observed different patterns in experienced emotions for users in different social contexts (i.e., alone vs. not alone) as functions of the extraversion, agreeableness and neuroticism

---

[2] https://www.douban.com/group/explore.

factors. People with different values of the conscientiousness and openness factors did not exhibit different patterns in their induced emotions.

A personal characteristic related to the consumption of multimedia content was identified by Tkalcic et al. [96]. Based on positive psychology research they observed that users differ in their preferences in terms of hedonic quality of the content (pure pleasure, fun) and eudaimonic quality (looking for deeper meaning): making users more or less pleasure or meaning-seekers.

# 3 Personality Acquisition

The acquisition of personality factors is the first major issue in the design of a personality-based recommender system. Generally, the acquisition techniques can be grouped into:

- explicit (or called direct) techniques (questionnaires depending on the model);
- implicit (or called indirect) techniques (e.g., regression/classification based on social media streams).

While explicit techniques provide accurate assessments of the user's personality, they are intrusive and time consuming. Hence, these techniques are useful mainly in laboratory studies or for performing as ground truth data to assess the accuracy of implicit acquisition method.

Implicit techniques, on the other hand, offer an unobtrusive way of acquiring user personality, by normally using machine learning techniques to infer it with features as extracted from users' digital traces. However, the accuracy of these instruments is not high and depends heavily on the quality of the source information (e.g., how often a user tweets).

In this section we survey existing techniques for the acquisition of personality in recommender systems. Table 3 summarizes the methods described in this section.

## 3.1 Explicit Personality Acquisition

A widely used questionnaire for assessing the FFM factors is the International Personality Item Pool (IPIP) set of questionnaires [38]. The IPIP's web page[3] contains questionnaires with 50 and 100 items, depending on the number of questions per factor (10 or 20). The relatively high number of questions makes it an accurate instrument, although it is time consuming for end users to answer. Furthermore, it has been translated in many languages and validated in terms of cross-cultural differences [61].

---

[3] http://ipip.ori.org/.

**Table 3** Personality acquisition methods

| Ref. | Method | Personality model | Source |
|------|--------|-------------------|--------|
| [24, 38–41, 48] | Explicit | FFM | Questionnaires (from 10 questions up) |
| [74] | Implict | FFM | Micro-blogs (Twitter) |
| [4, 53, 83] | Implicit | FFM | Social media (Facebook) |
| [36] | Implicit | FFM | Social media (Weibo) |
| [57] | Implicit | FFM | Role-playing game |
| [25] | Implicit | FFM | Game (Commons Fishing Game) |
| [17] | Implicit | FFM | Mobile phone logs |
| [85] | Implicit | FFM | Emails |
| [45] | Implicit | FFM | Ratings of products in a webstore |
| [23] | Implicit | FFM | Stories |
| [8] | Implicit | FFM | Eye gaze |
| [32] | Implicit | FFM | Images, social media (Instagram) |
| [33] | Implicit | FFM | Privacy preferences |
| [35] | Implicit | FFM | Drugs consumption profile |
| [89] | Explicit | Thomas-Kilmann conflict model | Questionnaire |
| [87] | Explicit | Felder and Silverman Learning Style Model | Questionnaire |

In the questionnaire defined by Hellriegel and Slocum [41], each factor is measured via 5 questions, so there are 25 questions in total regarding the five personality factors. Each factor's value is the average of user's scores on its related five questions. For example, the questions used to assess openness to experience include "imagination", "artistic interests", "liberalism", "adventurousness", and "intellect". Users are required to respond to every question on a 5-point Likert scale (for example, "imagination" is rated from 1 "no-nonsense" to 5 "a dreamer"). John and Srivastava [48] developed a more comprehensive list containing 44 items, called Big Five Inventory (BFI), in which each personality factor is measured by eight or nine questions. For example, the items related to openness to experience are "is original, comes up with new ideas", "is curious about many different things", "is ingenious, a deep thinker", "has an active imagination", etc. (each is rated on a 5-point Likert scale from "strongly disagree" to "strongly agree", under the general question of "*I see Myself as Someone Who* ..."). This questionnaire has been recognized as a well-established measurement of personality traits. The other commonly used public-free instruments include the 100-item Big Five Aspect Scales (BFAS) [24] and the 100 trait-descriptive adjectives [39]). A super-short measure of the FFM is the Ten Item Personality Inventory (TIPI) in which each factor is only assessed by two questions (e.g., openness to experiences is assessed by "open to new experiences, complex" and "conventional, uncreative" on the same Likert scale used in BFI) [40]. This instrument can meet the need for a very short

**Table 4** The ten-items
personality inventory
questionnaire [40]

| FFM factor | Statement: *"I see myself as . . ."* |
|---|---|
| E | Extraverted, enthusiastic. |
| A | Critical, quarrelsome. |
| C | Dependable, self-disciplined. |
| N | Anxious, easily upset. |
| O | Open to new experiences, complex. |
| E | Reserved, quiet. |
| A | Sympathetic, warm. |
| C | Disorganized, careless. |
| N | Calm, emotionally stable. |
| O | Conventional, uncreative. |

measure (e.g., when time is limited), although it may somewhat create diminished psychometric properties. We provide the TIPI questionnaire in Table 4.

A typical example of a commercially controlled instrument is the NEO PI-R (with a 240-items inventory) [19], which cannot only measure the five factors, but also the six facets (i.e., sub-factors) of each factor. For example, extroversion contains six facets: Gregariousness (sociable), Assertiveness (forceful), Activity (energetic), Excitement-seeking (adventurous), Positive emotions (enthusiastic), and Warmth (outgoing). The NEO-FFI instrument, which measures the five factors only (not their related facets), is a 60-item truncated version of NEO PI-R [19].

A quasi-explicit instrument for measuring personality is the approach of using stories. In their work, Dennis et al. [23] developed a set of stereotypical stories, each of which conveys a personality trait from the FFM. Specifically, for each of the five FFM factors they devised a pair of stories, one for a high level of the observed factor and one for the low level of the observed factor. The subject then rates how well each story applies to her/him on a Likert scale from 1 (extremely inaccurate) to 9 (extremely accurate).

Though different instruments have been developed so far, the choice of instrument is highly application-dependent and there is no one-size-fits-all measure. In Sect. 4, we will survey the instruments that have been adopted in recommender systems (e.g., see Table 6).

## 3.2 Implicit Personality Acquisition

Quercia et al. [74] presented the outcomes of a study that shows strong correlations between features extracted from users' micro-blogs and their respective FFM factors. The authors used the myPersonality dataset of 335 users. The dataset contains the users' FFM personality factors and the respective micro-blogs. The authors extracted several features from the micro-blogs and categorized them into the following quantities: listeners, popular, highly-read, and influential. Each of these quantities showed a strong correlation with at least one of the FFM factors.

The authors went a step further into predicting the FFM factors. Using a machine learning approach (the M5-Rules regression and the 10 fold-cross validation scheme), they were able to achieve a predictability in RMSE ranging from 0.69 to 0.88 (on FFM factors ranging from 1 to 5).

Kosinski et al. [53] used the whole myPersonality dataset of over 58,000 users with their respective Facebook activity records to predict the FFM factors of the users. The source dataset was the user-like matrix of Facebook Likes. The authors applied the Singular Value Decomposition (SVD) method to reduce the dimensionality of the matrix and used the Logistic Regression model to predict the FFM factors (along with other user parameters such as gender, age, etc.). Their model was able to predict the traits openness and extraversion with correlations being at least 0.40, while the other traits were predicted with lower accuracy (with correlations no more than 0.30).

An interesting approach was taken by van Lankveld et al. [57] who observed the correlation between FFM factors and the users' behaviour in a videogame. They modified the Neverwinter Nights (a third-person role-playing video game) in order to store 275 game variables for 44 participants. They used variables that recorded conversation behavior, movement behavior and miscellaneous behavior. They found significant correlations between all five personality traits and game variables in all groups.

Chittaranjan et al. [17] used mobile phone usage information for inferring FFM factors. They used call logs (e.g., outgoing calls, incoming calls, average call duration, etc.), SMS logs and application-usage logs as features for predicting the FFM factors. They observed that a number of these features have a significant correlations with the FFM factors. Using the Support Vector Machine classifier, they achieved better results in the prediction of the traits than a random baseline, although the difference was not always significant, which makes the task of inferring personality from call logs a hard one.

Shen et al. [85] attempted to infer the email writer's personality from her/his emails. To preserve privacy, they only extracted high-level aggregated features from email contents, such as bag-of-word features, meta features (e.g., TO/CC/BCC counts, importance of the email, count of words, count of attachments, month of the sent time, etc.), word statistics (e.g., through part-of speech tagging and sentiment analysis), writing styles (in greeting patterns, closing patterns, wish patterns, and smiley words), and speech act scores (for detecting the purpose of work-related emails). These groups of features were then applied to train predictors of the writer's personality, through three different generative models: joint model, sequential model, and survival model. The experiment done on over 100,000 emails showed that the survival model (with label-independence assumption) works best in terms of prediction accuracy and computation efficiency, while joint model performs worst in terms of inferring personality traits such as agreeableness, conscientiousness, and extraversion. The results to some extent infer that the personality traits are relatively distinct and independent from each other. Furthermore, it was found that people with high conscientiousness are inclined to write long emails and use more characters;

people with high agreeableness tend to use more "please" and good wishes in their emails; and people with high neuroticism use more negations.

The set of studies done by Oberlander et al. [47, 68] showed that personality can be inferred also from blog entries. In [47] they used features such as stemmed bigrams, no exclusion of stopwords (i.e., common words) or the boolean presence or absence of features noted (rather than their rate of use) to train the Support Vector Machine classifier. On a large corpus of blogs, they managed to predict the FFM factors (high-low groups for each trait) with an accuracy ranging from 70% (for neuroticism) to 84% (for openness).

With the development of social networking, some researchers have begun to study the correlation between users' personality and their social behavior on the Web (e.g., Facebook, Twitter) [4, 7, 81]. For example, [4] found strong connection between users' personality and their Facebook use through a user survey on 237 students. Participants' personality was self-reported through answering the NEO PI-R questionnaire. The collected personality data were then used to compute correlation with users' Facebook information (such as basic information, personal information, contact information and education, and work information). The results show that extroversion has a positive effect on the number of friends. Moreover, individuals with high neuroticism are more inclined to post their private information (such as photos). The factor openness was found to have positive correlation with users' willingness to use Facebook as a communication tool, and the factor conscientiousness is positively correlated with the number of friends. In [83], a similar experiment was performed. They verified again that extroversion is significantly correlated with the size of a user's social network. Moreover, people tend to choose friends who are with higher agreeableness but similar extroversion and openness.

In [37], the authors developed a method to predict users' personality from their Facebook profile. Among various features, they identified the ones that have a significant correlation with one or more of the FFM personality traits based on studying 167 subjects' public data on Facebook. These features include linguistic features (such as swear words, social processes, affective processes, perceptual processes, etc.), structural features (number of friends, egocentric network density), activities and preferences (e.g., favorite books), and personal information (relationship status, last name length in characters). Particularly, the linguistic analysis of profile text (which is the combination of status updates, About Me, and blurb text) was conducted through Linguistic Inquiry and Word Count (LIWC) program [73], which is a tool to produce statistics on 81 different text features in five psychological categories. They further proposed a regression analysis based approach to predict the personality, in two variations: M5-Rules and Gaussian Processes. The testing shows that the prediction of each personality factor can be within 11% of the actual value. Moreover, M5-Rules acts more effective than Gaussian Processes, with stronger connection to openness, conscientiousness, extroversion, and neuroticism.

Gao et al. [36] proposed a method for inferring the users' personality from their social media contents. To be specific, they obtained 1766 volunteers' personality values and Weibo behavior (which is a popular micro-blog site in China) to train the prediction model. 168 features were extracted from these users' Weibo status, and then classified into categories including status statistics features (e.g., the total number of statuses), sentence-based features (the average number of Chinese characters per sentence), word-based features (the number of emotion words), character-based features (the number of commas, colons, etc.), and LIWC features. They then applied M5-Rules, Pace Regression and Gaussian process to make prediction. The results show that the Pearson correlation between predicted personality and user self-reported personality can achieve 0.4 (i.e., fairly correlated), especially regarding the three traits conscientiousness, extroversion, and openness.

Hu and Pu studied the effect of personality on users' rating behavior in recommender systems [45]. They conducted an online survey and obtained 86 participants' valid ratings on at least 30 items among a set of 871 products (from 44 primary categories) as crawled from gifts.com. The rating behavior was analyzed from four aspects: number of rated items, percentage of positive ratings, category coverage (CatCoverage), and interest diversity (IntDiversity). The CatCoverage is measured as the number of categories of rated items. The IntDiversity reveals the distribution of users' interests in each category, formally defined as the Shannon index according to information theory. They calculated the correlation between users' FFM personality traits and the rating variables through Pearson product-moment. The results identify the significant impact of personality on the way users rate items. Particularly, conscientiousness was found negatively correlated with the number of ratings, category coverage and interest diversity, which indicates that conscientious users are more likely to prefer providing fewer ratings, lower level of category coverage, and lower interest diversity. In addition, agreeableness is positively correlated with the percentage of positive ratings, implying that agreeable people tend to give more positive ratings. All these findings show correlations between personality and rating behavior on the samples collected. However, they did not infer personality from rating behavior in this work.

Dunn et al. [25] proposed, beside an explicit questionnaire, a gamified user interface for the acquisition of personality for recommender systems. Through the Commons Fishing Game (CFG) interface the users were instructed to maximize the amount gathered from a common resource, which was shared amongst a group of players; collectively trying not to deplete this resource. The experiment showed that it is possible to predict extraversion and agreeableness with the described instrument.

More approaches have emerged in the very recent years. Machine learning predictive models have been trained on a wide variety of digital traces. For example, Instagram features, both image-based and comment-based, were used in [32, 34, 86]. A very stable and accurate prediction of personality has been demonstrated in [8], where the authors showed subjects a set of visual stimuli (a subset of pictures from the IAPS dataset [11]) and recorded their eye gaze movements. An interesting approach has been demonstrated by Ferwerda et al. [35], where they showed that personality traits can be predicted even from reports of drug consumption of users.

## 3.3   Datasets for Offline Recommender Systems Experiments

Given that a number of research activities has already been published, there exist some datasets that can be used for developing personality-based recommender systems. The minimal requirements for such a dataset are (a) to include the user-item interaction data (e.g., clicking or ratings) and (b) to include the personality factors associated to the users. In this subsection we survey a number of such datasets, which are summarized in Table 5.

The first dataset containing personality factors was the LDOS-PerAff-1 [98].[4] Based on 52 subjects it contains ratings of images. The user-item matrix has values for all its entries (i.e., sparsity is zero). The dataset contains the corresponding FFM factors for each user. The FFM factors were acquired using the 50-items IPIP questionnaire [38]. Furthermore, all items were selected from the IAPS dataset of images [56] and were annotated with the values of the induced emotions in the valence-arousal-dominance (VAD) space.

The LDOS-CoMoDa (Context Movies Dataset) dataset[5] [54] was released for research on context-aware recommender systems. It contains FFM data of 95 users. The FFM factors were collected using the 50-items IPIP questionnaire [38]. The dataset is also rich in contextual parameters such as time, weather, location, emotions, social state, etc.

A dataset that contains more users is the myPersonality dataset[6] [53]. It contains FFM factors of 38,330 users. The dataset was collected using a Facebook application. It contains the Facebook Likes for each of the users. Furthermore, it contains twitter names for more than 300 users, which opens new possibilities for crawling these users' micro-blogs (as has been done in [74]).

Chittaranjan et al. [17][7] presented a dataset of mobile phone users' logs along with the respective FFM values (as measured using the TIPI questionnaire). The dataset contains information of 177 subjects and their daily phone usage activities (the CDR—call data record) over a period of 17 months on a Nokia N95 smartphone. The phone usage logs contain data related to calls, SMSs and application usage.

Recently, Wang et al. [100] released a dataset[8] containing 11,383 users' feedback on recommendations that they received on a commercial e-commerce application (Mobile Taobao). In addition to obtaining users' self-reported assessments of the recommendation from various aspects (e.g., relevance, novelty, unexpectedness, serendipity, timeliness, satisfaction, and purchase intention), they acquired users'

---

[4] http://markotkalcic.com/resources.html.

[5] https://www.lucami.org/en/research/ldos-comoda-dataset/.

[6] https://sites.google.com/michalkosinski.com/mypersonality (the dataset was stopped sharing in 2018).

[7] The dataset is not publicly available anymore.

[8] https://www.comp.hkbu.edu.hk/~lichen/download/TaoBao_Serendipity_Dataset.html.

**Table 5** Overview of datasets

| Name | Ref. | Domain | Personality model | Number of subjects | Other metadata |
|------|------|--------|-------------------|--------------------|----------------|
| LDOS-PerAff-1 | [98] | Images | FFM | 52 | Item induced emotions in the VAD space |
| LDOS-CoMoDa | [54] | Movies | FFM | 95 | Movie context metadata (location, weather, social state, emotions, etc.) |
| myPersonality | [53] | Social Media (Facebook) | FFM | 38,330 | Twitter names |
| Chittaranjan | [17] | Mobile phone usage | FFM | 117 | Call logs, SMS logs, app logs |
| Taobao Serendipity | [100] | e-Commerce | Curiosity & FFM | 11,383 | User perceptions w.r.t. recommendation relevance, serendipity, satisfaction, purchase intention, etc.; click and purchase behaviors |

curiosity (via Ten-item Curiosity and Exploration Inventory-II (CEI-II) [49]) and FFM personality values (via Ten-Item Personality Inventory (TIPI) [40]).

Furthermore, a number of datasets, not released as datasets per se, exist, as they have been used in the studies reported in this chapter.

## 4 How to Use Personality in Recommender Systems

In this section, we provide an overview of how personality has been used in recommender systems. The most common issues addressed are the cold-start problem and the recommendation diversification. Table 6 summarizes the various strategies described in this section.

**Table 6** Survey of recommender systems using personality

| Author | Recommender system's goal | Personality acquisition method | Approach |
|---|---|---|---|
| Tkalčič et al. [99] | Cold-start problem | IPIP 50 | User-user similarity measure based on personality |
| Hu and Pu [44] | Cold-start problem | TIPI | User-user similarity measure based on personality |
| Elahi et al. [27] & Braunhofer et al. [12] | Cold-start problem | TIPI | Active learning, matrix factorization |
| Tiwari et al. [92] | Cold-start problem | TIPI | User-user similarity based both on personality and demographics |
| Fernández-Tobías et al. [31] | Cold-start problem | TIPI | Injecting personality factors into a matrix factorization algorithm |
| Khwaja et al. [51] | Cold-start problem | TIPI | Recommending well-being activities based on the congruence between the true personality and the projected personality |
| Yusef et al. [104] | Cold-start problem | TIPI | Clustering users using personality-based similarities to reduce sparsity |
| Wu et al. [103] | Diversity | TIPI | Personality-based diversity adjusting approach for recommendations |
| Tintarev et al. [90] | Diversity | NEO IPIP 20 | Personality-based diversity adaptation |
| Cantador et al. [14] | Cross-domain recommendations | NEO IPIP 20 | Similarities between personality-based user stereotypes for genres in different domains |
| Recio-Garcia et al. [77] & Quijano-Sanchez et al. [75] | Group recommendations | Thomas-Kilmann conflict model instrument | Combining assertiveness and cooperativeness into the aggregation function |
| Kompan et al. [52] | Group recommendations | Thomas-Kilmann conflict model instrument and NEO IPIP 20 | Group satisfaction modeling with a personality-based graph model |
| Delic et al. [20] & Delic et al. [21] | Group recommendations | TIPI and Thomas-Kilmann conflict model instrument | An observational study method for understanding the dynamics of group decision-making |
| Roshchina et al. [80] | Recommendations in the traveling domain | TIPI | Text-based personality similarity measure for finding like-minded users |

## 4.1   Addressing the New User Problem

The new user problem occurs when the recommender system does not have enough ratings from a user who has just started to use the system [3]. The problem is present both in content-based recommender systems and in collaborative recommender systems, although it is more difficult to solve within the latter. The system must first have some information about the user, which is usually in the form of ratings. In the case of content-based recommender systems, the lack of ratings implies that, for the observed user, the system does not know the preferences towards the item's features (e.g., the genre). In the case of collaborative filtering, especially in neighborhood methods, the lack of ratings for a new user implies that there are not enough overlapping ratings with other users, which makes it hard to calculate user similarities. So far this problem has been tackled with various techniques such as hybrid methods [3], adaptive learning techniques [28], or simply by recommending popular items [3].

Personality is suitable to address the new user problem. Given the assumption that the user's personality is available (e.g., from another domain), it can be used in collaborative filtering recommender systems.

For instance, personality has been used in a memory-based collaborative filtering recommender system for images [97, 99]. In an offline experiment, the authors acquired explicit FFM factors for each user and calculated the user-user distances (as opposed to similarities) using the weighted distance formula:

$$d(b_i, b_j) = \sqrt{\sum_{l=1}^{5} w_l(b_{il} - b_{jl})^2} \tag{1}$$

where $b_i$ and $b_j$ are the FFM vectors for two arbitrary users ($b_{il}$ and $b_{jl}$ are the individual FFM factors) and $w_l$ are the weights. The weights were computed as the eigenvalues from the principal component analysis on the FFM values of all users. On the given dataset, this approach was statistically equivalent to using standard rating-based user similarity measures.

A similar approach was taken by Hu and Pu [44], but they used a different formula to calculate the user similarities. Concretely, they proposed to use the Pearson correlation coefficient to calculate the user similarities:

$$sim(b_i, b_j) = \frac{\sum_l (b_{il} - \overline{b_i})(b_{jl} - \overline{b_j})}{\sqrt{\sum_l (b_{il} - \overline{b_i})^2}\sqrt{\sum_l (b_{jl} - \overline{b_j})^2}} \tag{2}$$

and linearly combined it with rating-based similarity by controlling the contribution of each similarity measure with the weight $\alpha$. They compared the proposed approach to a rating-based user similarity metric for collaborative filtering recommender systems. On a dataset of 113 users and 646 songs, the personality-based algorithm outperformed the rating-based in terms of mean absolute error, recall and specificity.

Similarly to the approaches proposed in [97] and [44], the authors of [92] used personality and demographics to calculate user-user similarities for recommending movies.

A standard approach to tackle the cold-start problem has been to use the active learning approach [28]. Elahi et al. [27] proposed an active learning strategy that incorporated user personality data. They acquired the personality information using the 10-items IPIP questionnaire through a mobile application. They formulated the rating prediction as a modified matrix factorization approach where the FFM factors are treated as additional users' latent factors:

$$\hat{r}_{ui} = b_i + b_u + q_i^T \cdot (p_u + \sum_l b_l) \tag{3}$$

where $p_u$ is the latent factor of the user $u$, $q_i$ is the latent factor of the item $i$, $b_u$ and $b_i$ are the user's and item's biases, and $b_l$ are the FFM factors. The proposed rating elicitation method outperformed (in terms of Mean Absolute Error) the baseline (the log(popularity)*entropy method) and the random method.

In these examples, personality has been acquired directly with questionnaires. With this approach, the authors have just moved the burden of an initial questionnaire about user ratings to another initial questionnaire (for personality). However, their methods can also be applicable to the condition that the personality is available in advance, for example from other domains or acquired implicitly.

More recently, Fernandez-Tobias et al. [31] compared three approaches to mitigating the new user problem respectively based on (a) personality-based matrix factorization (MF), which improves the recommendation prediction model by directly incorporating user personality into MF; (b) personality-based active learning, which regards personality as the additional preference information for improving the output of recommendation process; (c) personality-based cross-domain recommendation, which exploits personality to enrich the user profile as obtained from auxiliary domains with the aim of compensating for the lack of user preference data in the target domain. They found that all the three personality-based methods achieve performance improvements in real-life datasets, among which the personality-based cross-domain recommendation performs the best.

In [51] the authors presented a recommendation approach for well-being activities, which is a novelty in the domain. According to the authors, the domain suffers from the new user problem. From the algorithmic perspective, the authors matched the user's personality with items (well-being activities) based on previous psychology research. Their assumption is that the true personality of users and the personality exhibited through behaviour are not necessarily the same. If there is an inconsistency between the two personalities, the subjective well-being is low. The true personality of users was collected using a standard questionnaire, whereas the exhibited personality was calculated from behavioural data and previous research on how certain behaviour matches certain personalities. Then they trained an SVM model for predicting the subjective well-being (acquired with a questionnaire as ground-truth) from the discrepancy between the true personalities and the exhibited personalities. In the next step, the authors simulated, for each user, a wide range of activity distributions. For each of these distributions they calculated the subjective well-being that a distribution yields for the active user. The recommended sets of activities were then ranked based on the calculated subjective well-being.

The new user problem was addressed by Yusefi et al. [104] clustering users with similar personalities together and treating all users in a cluster as a single user, hence enlarging the number of ratings per user. When a new user joins the system, s/he is added to one of the clusters that matches best her/his personality. Then a standard collaborative filtering algorithm is used to predict the item ratings. The evaluation of the approach was done on the South Tyrol Suggests dataset [27] with 2534 ratings given by 465 users on 249 items. The authors found out that the system performed best (in terms of MSE) when the number of clusters was between 3 and 7.

Although not addressing directly the new user problem, Roschina et al. [80] designed a collaborative filtering recommender system for hotels where the user similarities were calculated using the users' personalities. In their TWIN (Tell me What I Need) recommender system they used the inverse of the Euclidian distance to calculate user similarities. The novelty of their approach lies in the way they inferred the users' personalities: They trained a machine learning model that used Linguistic Inquiry and Word Count (LIWC) based features extracted from user generated comments.

## 4.2  Personalizing Recommendation Diversity

The impact of personality on users' preferences for recommendation diversity has been investigated first in [15, 90]. Diversity refers to recommending users a diverse set of items,[9] so as to allow them to discover unexpected items more effectively [64]. Related diversity approaches commonly adopt a fixed strategy to adjust the diversity degree within the set of recommendations [2, 46, 105], which however, does not consider that different users might possess different attitudes towards the diversity of items. That limitation motivates the authors of paper [15] to identify whether and how personality might impact users' needs for diversity in recommender systems. For this purpose, they conducted a user survey that involved 181 participants. For each user, they obtained her/his movie selections as well as personality values. Then, two levels of diversity were considered: the diversity in respect to individual attributes (such as the movie's genre, director, actor/actress, etc.); the overall diversity when all attributes are combined. The correlation analysis showed that some personality factors have a significant correlation with users' diversity preferences. For instance, more reactive, excited and nervous persons (high neuroticism) are more inclined to choose diverse directors, and suspicious/antagonistic users (low

---

[9] Here we mainly consider the so called "intra-list diversity" within a set of recommended items (see Chapter "Novelty and Diversity in Recommender Systems").

agreeableness) prefer diverse movie countries. As for the movie's release time, its diversity is preferred by efficient/organized users (high conscientiousness), while for the movie's actor/actress, its diversity is preferred by imaginative/creative users (high openness). At the second level (i.e., overall diversity), no matter how the weights placed on attributes vary, Conscientiousness was shown significantly negatively correlated with it, which means that less conscientious people generally prefer higher level of overall diversity.

Tintarev et al. applied a User-as-Wizard approach to study how people apply diversity to the set of recommendations [90]. Particularly, they emphasized the personality factor "openness to experience" as for its specific role in personalizing the recommendation diversity, because it describes users' imagination, aesthetic sensitivity, attentiveness to inner feelings, preference for variety, and intellectual curiosity (so they assumed that people with higher openness would be more willing to receive novel items). Their experiment was in the form of an online questionnaire with the aid of Amazon's Mechanical Turk (MT) service. 120 users' responses were analyzed. Each of them was required to provide some recommendation to a fictitious friend who is in one of the three conditions: high openness, low openness, and no personality description (baseline). The results did not prove the effect of openness on the overall diversity participants applied, but the authors observed that participants tend to recommend items with high categorical diversity (i.e., across genres) but low thematic diversity (inter-genre) to those who are more open to experience. In other words, users who are low on openness might prefer thematic diversity to categorical variation.

Motivated by the above findings, Wu et al. [103] have attempted to develop an approach to automatically adjusting the degree of recommendation diversity based on the target user's personality. They first validated the relationship between users' five personality traits and their preferences for diversity through a larger-scale user survey (involving 1706 users) on a commercial platform (i.e., Douban's Interest Group where users can join groups with various topics including entertainment, culture, technology, life, and so on). The results showed that the personality traits have significant impact on users' diversity preferences. For instance, more creative (with high openness) and/or more introverted (with low extraversion) person is more inclined to join different types of groups.

The authors have further developed a generalized, dynamic diversity adjusting approach based on user personality [103]. In particular, personality is incorporated into a greedy re-ranking process, by which the system selects the item that can best balance accuracy and personalized diversity at each step, and then produces the final recommendation list to the target user. Concretely, their method is mainly composed of two steps: (1) to predict a user's preference for un-experienced items, and (2) re-rank the items to meet the user's diversity preference. They adopted the greedy re-ranking technique [1], because it cannot only be easily incorporated into the existing recommender algorithms but also explicitly control the level of diversification. Formally, $S$ denotes a candidate item set of size $n$ for a user $u$, which is generated according to her/his predicted preferences for items. $T$ denotes the re-ranked list that user $u$ will finally receive, which includes $N$ items ($N < n$, because the recommendation list $T$ is reproduced from the larger set of candidate items). At each iteration, they add an item that maximizes the objective function $Score_{final}$ with the aim of achieving the trade-off between the user's preference for the item and her/his diversity preference for all items selected so far:

$$Score_{final}(u, i) = \beta * Score_{Pref}(u, i) + (1 - \beta) * Score_{PersonalizedDiv}(u, i)$$
(4)

where $Score_{Pref}(u, i)$ denotes the user $u$'s preference for item $i$, $Score_{PersonalizedDiv}(u, i)$ represents the personalized diversity degree, and the parameter $\beta$ is used to balance the two types of preferences. More details can be found in [103]. Through experiments, they demonstrated that this approach can achieve better performance than related methods (including both non-diversity-oriented and diversity-oriented methods) in terms of both accuracy and diversity metrics.

Another contribution of the above work is that, in addition to standard diversity metrics such as $\alpha\text{-}nDCG$ [18] and $Adaptive\ \alpha\text{-}nDCG$ [29], they proposed a new metric called *Diversity Fitness (DivFit)* in order to more precisely measure the personalization degree of recommendation diversity. It concretely calculates the fitness between the diversity degree $Div_{Rec}(u)$ within the top-N recommendation list and the user's actual diversity preference $Div_{Act}(u)$:

$$DivFit = \frac{1}{k} \sum_{u=1}^{k} |Div_{Act}(u) - Div_{Rec}(u)|$$
(5)

where $k$ is the number of testing users, $Div_{Rec}(u)$ is calculated by means of Shannon Entropy over the types that the recommended items belong to, and $Div_{Act}(u)$ is calculated based on the user $u$'s actual behavior records via Shannon Entropy. A smaller $DivFit$ means that the diversity of the recommendation list has a better fit to the user's actual diversity preference.

## 4.3   Other Applications

As we mentioned in the introduction, personality is domain-independent, i.e., when users are being recommended books or movies, we can use the same personality profile. This can be especially useful in cross-domain recommender systems (see also Chapter "Design and Evaluation of Cross-domain Recommender Systems"). In a study performed by Cantador et al. [14], personality factors were related to domain genres, and similarities between personality-based user stereotypes for genres in different domains were computed. Among the many cross-domain-genres combinations, we can find relations such that people who enjoy humor, mystery and romance books are associated with personality-based stereotypes similar to those for most of the music genres.

Personality can also be useful for group recommendations. As discussed in Chapter "Group Recommender Systems: Beyond Preference Aggregation", recommending items to groups of users is not the same as recommending items to individual users [59]. Beside having to choose among strategies that address users as individuals (e.g., least misery, most pleasure etc.—see Chapter "Group Recommender Systems: Beyond Preference Aggregation" for an extensive overview), the relationships between group members play an important role. Garcia, Sanchez et al. [75, 77] proposed to use the Thomas-Kilmann conflict model [89] to model the relationships between group members in terms of *assertiveness* and *cooperativeness*. They applied the model to three group recommendation approaches (i.e., least misery, minimize penalization and average satisfaction). They collected ground truth data through a user study with 70 students who formed groups, discussed and decided which movies they would watch together in a cinema. The proposed approach showed an increase in prediction accuracy compared to the same techniques without taking into account the conflict personality model.

Similarly, Kompan et al. [52] used the Thomas-Kilmann conflict model and the FFM to model individual users. They modeled the group satisfaction with a graph-based approach where vertices represent users and edges represent user influences based on relationship, personality and actual context. They performed a small-scale user study with users' ratings on movies. The usage of the personality-based group satisfaction model in an average-aggregation strategy-based group recommender system outperformed the same algorithm without the proposed group satisfaction modeling.

An important step in research of group recommender systems has been done by Delic et al. [20], who introduced the influence of the group members on each others' preferences. Their work was based on the aforementioned work of Quijano-Sanchez et al. [75], who used the personal trait of conformity to generate group recommendations. Delic et al. [20] measured the FFM of real users participating in a group decision making process. They found that the choice satisfaction of group members is related to their personality. Furthermore, they found that group members with low assertiveness have a bigger discrepancy between their personal preferences and the final group decision. In their next work, Delic et al. provided further arguments for modeling group dynamics using personality, in particular the Thomas-Killman conflict model, in order to measure how the contagion of preferences works within the group [21, 95]. Conformity has been additionally studied by Nguyen and Ricci [66], who showed, through a simulations study, that different recommendation strategies (in terms of whether long-term preferences should be preferred over short-term preferences) should be applied depending on the conformity level of group participants. They further conducted an experiment [67], which simulated alternative conflict resolution styles as derived from the Thomas–Kilmann conflict model. The results reveal how the conflict resolution style takes roles within groups of members who have similar tastes versus those groups whose members have diverse preferences.

# 5 Open Issues and Challenges

There are quite some open issues and challenges that need to be addressed regarding the usage of personality in recommender systems. In this section we survey these open issues.

## 5.1 Non-intrusive Acquisition of Personality Information

The limitation of traditional explicit acquisition approach is that the required user effort is usually high, especially if we want to obtain their accurate personality profile (e.g., through 100-item Big Five Aspect Scales (BFAS); see Sect. 3.1). Users might be reluctant to follow the time-consuming and tedious procedure to answer all questions, due to their cognitive effort or emotional reason. Thus, the implicit, unobtrusive approach might be more acceptable and effective to build their personality profile. The critical question is then how to accurately derive users' personality traits from the information they have provided. In Sect. 3.2, we discussed various methods, such as the ones based on users' emails or their generated contents and behavior in social networking sites (e.g., Facebook, Twitter). However, the research is still at the beginning stage, and there is large room to improve the existing algorithms' accuracy. One possible solution is to explore other types of information as to their power of reflecting users' personality. For instance, since the significant correlation between users' personality and their rating behavior was proven in [45], the findings might be constructive for some researchers to develop the rating-based personality inference algorithm. The developed method might be further extended to consider the possible impacts of other actions, such as users' browsing, clicking, and selecting behavior in recommender systems. Indeed, it will be interesting to investigate the complementary roles of various resources to fulfill their combinative effect on deriving users' personality. For instance, we may infer users' personality by combining their history data left at different platforms (e.g., the integration of rating behavior, email, and social media content). The different types of information might be heterogenous in nature, so how to effectively fuse them together might be an open issue.

## 5.2 Larger Datasets

The recommender systems oriented datasets containing personality factors of users are still very few (see Sect. 3.3). Furthermore the number of subjects in most datasets is low, ranging from roughly 50 to a little more than 100, with the exception of the myPersonality dataset (38,330 subjects) and the Taobao Serendipity dataset (11,383 subjects). Compared to the huge datasets that the recommender systems community is used to work with (e.g., the MovieLens 25M Dataset[10] with 25 million ratings given by 162,000 users, and the Music Streaming Sessions Dataset with 160 million listening sessions [13]), the lack of bigger datasets is an obvious issue that needs to be addressed.

## 5.3  Cross-Domain Applications

Personality appears to be a natural fit to cross-domain recommender systems (see also Chapter "Design and Evaluation of Cross-domain Recommender Systems"), because personality is domain-independent and can hence be used as a generic user model. Cross-domain applications have been researched in the past and correlations of preferences among different domains have been identified. For example, Winoto et al. [101] observed the relations between the *games*, *TV series* and *movie* domains, while Tiroshi et al. [91] observed the relations between *music*, *movies*, *TV series*, and *books*. The first to explore the potential role of personality in cross-domain applications was Cantador et al. [14], who observed the relations between the FFM factors and preferences in various domains (e.g., movies, TV shows, music, and books). An intuitive continuation of this work is the application of the personalities learned in one domain to another domain to mitigate the cold-start problem.

Another aspect of cross-domain recommendations is cross-application recommendations. In order to be able to transfer the personality profiles between applications, a standardized description of personality should be used. There has been an attempt, the *Personality Markup Language* (PersonalityML), to standardize the description of personality in user models across different domains [6].

## 5.4  Beyond Accuracy

Users are no longer satisfied with seeing items similar to what they preferred before, so showing ones that can be unexpected and surprising to them has increasingly become an important topic in the area of recommender systems [60]. The work on personalized recommendation diversity has revealed the role of personality in enhancing the personalization degree within a set of recommendation in terms of its overall diversity [103]. However, so far little attention has been paid to investigate the relationship between personality and users' perception of a single recommendation. The work lately published in [16] in particular studied the effect of curiosity, because it has been widely regarded as an important antecedent of users' desire for new knowledge or experiences in the field of psychology [9] and found

---

[10] https://grouplens.org/datasets/movielens/.

significantly correlated with the FFM values such as openness, conscientiousness, extraversion, and neuroticism [49]. Through the analysis of curiosity's moderating effect, the authors found that it cannot only strengthen the positive effect of novelty on serendipity, but also that of serendipity on user satisfaction with the recommendation. In other words, it implies that a more curious person will be more likely to perceive a novel recommendation as serendipitous, and be more satisfied with the serendipitous item.

The authors further validated such effect on a larger-scale dataset with over 11,000 users' feedback along with their curiosity and FFM personality values [100]. The results indicated that users with high curiosity, openness, conscientiousness, extraversion, or neuroticism are more likely to perceive the serendipity level of a recommendation higher. Moreover, there exist significant interaction effects between certain item features (e.g., item popularity, item category) and user characteristics on their perceived recommendation serendipity. For instance, users with low curiosity, openness, extraversion, or neuroticism are more sensitive to category difference between the current recommendation and those previously visited by the user.

The above findings thus suggest that personality cannot only be exploited to realize personalized diversity within a recommendation list, but also be likely helpful to achieve personalized serendipity with regard to a single item. For example, for people of different curiosity values, the "surprise" level of a recommendation might be adjusted to meet their propensity towards unexpected discovery. More work can be done in this direction to better optimize user experience with recommender systems.

## 5.5   Privacy Issues

Although all the research done so far on personality in recommender systems touched upon the sensitivity of the data, the issue of privacy has not been addressed properly yet. The fact that, in terms of personality, a user can be tagged as *neurotic* or otherwise with labels that suggest a negative trait making these data very sensitive. Schrammel et al. [82] explored if there were any differences in the degree of disclosure acceptance among users with different personalities, but found no significant differences.

The approaches for implicit personality acquisition, as described in Sect 3.2, raise ethical and privacy concerns that are beyond the scope of this chapter. Two major incidents happened in recent years that raised awareness of the dangers that such automatic processing of user digital traces can bring. In their work on emotional contagions the authors in [55] manipulated several hundreds of thousands of Facebook users without their consent. The second incident was the Cambridge Analytica scandal where the company used methods, similar to the ones described above, for psychological user profiling and political advertising. Therefore, we should increase users' awareness of how their data could be utilized, so they might be able to choose what information they would be willing to disclose online.

However, the study of Ferwerda et al. [33] has shown that personality can be inferred even from information about what a user is willing to disclose and what not. For example, people who score high on extraversion tend to prefer to disclose their preferences for food. People who score high on agreeableness tend not to disclose the places where they have lived. In general, people who score high on openness to experience have a tendency to disclose as little as possible, whereas highly agreeable people have little concerns about disclosing personal information.

# 6   Conclusion

In this chapter we presented the usage of personality in recommender systems. Personality, as defined in psychology, accounts for the most important dimensions in which individuals differ in their enduring emotional, interpersonal, experiential, attitudinal, and motivational styles. It can be acquired using either questionnaires or by inferring implicitly from other sources (e.g., social media streams). The most common model of personality is the Five Factor Model (FFM), which is composed of the five factors: openness, conscientiousness, extraversion, agreeableness, and neuroticism. This model is suitable for recommender systems since it can be quantified with feature vectors that describe the degree each factor is expressed in a user. Furthermore, the FFM (and personality in general) is domain independent. We presented several methods for the acquisition of personality factors, with a special focus on implicit methods. We showcased a number of ways recommender systems have been shown to improve using personality models, especially in terms of the cold-start problem and diversity personalization. Finally, we provided a list of open issues and challenges that need to be addressed in order to improve the adoption of personality in recommender systems.

**Acknowledgments** Part of the work presented in this chapter has received funding from the European Union FP7 programme through the PHENICX project (grant agreement no. 601166), China National Natural Science Foundation (no. 61272365), and Hong Kong Research Grants Council (no. ECS/HKBU211912).

# References

1. G. Adomavicius, Y. Kwon, Toward more diverse recommendations: item re-ranking methods for recommender systems, in *Workshop on Information Technologies and Systems (WITS 2009)*. Citeseer (2009), pp. 417–440
2. G. Adomavicius, Y. Kwon, Improving aggregate recommendation diversity using ranking-based techniques. IEEE Trans. Knowl. Data Eng. **24**(5), 896–911 (2012). https://doi.org/10.1109/TKDE.2011.15
3. G. Adomavicius, A. Tuzhilin, Toward the next generation of recommender systems: a survey of the state-of-the-art and possible extensions. IEEE Trans. Knowl. Data Eng. **17**(6), 734–749 (2005). https://doi.org/10.1109/TKDE.2005.99

4. Y. Amichai-Hamburger, G. Vinitzky, Social network use and personality. Comput. Human Behav. **26**(6), 1289–1295 (2010)
5. S. Aral, D. Walker, Identifying influential and susceptible members of social networks. Science (New York, N.Y.) **337**(6092), 337–341 (2012). https://doi.org/10.1126/science. 1215842
6. M.A.S.N. Nunes, J. Santos Bezerra, A. Adicinéia, PersonalityML: a markup language to standardize the user personality in recommender systems. Rev. Gestão Inovação e Tecnol. **2**(3), 255–273 (2012). https://doi.org/10.7198/S2237-0722201200030006
7. D. Azucar, D. Marengo, M. Settanni, Predicting the big 5 personality traits from digital footprints on social media: a meta-analysis. Pers. Individ. Dif. **124**, 150–159 (2018). https://doi.org/10.1016/j.paid.2017.12.018. http://www.sciencedirect.com/science/article/pii/ S0191886917307328
8. S. Berkovsky, R. Taib, I. Koprinska, E. Wang, Y. Zeng, J. Li, S. Kleitman, Detecting personality traits using eye-tracking data, in *Proceedings of the 2019 CHI Conference on Human Factors in Computing Systems - CHI '19* (2019), pp. 1–12. https://doi.org/10.1145/ 3290605.3300451. http://dl.acm.org/citation.cfm?doid=3290605.3300451
9. D.E. Berlyne, *Conflict, Arousal and Curiosity* (McGraw-Hill, New York, 1960)
10. C. Bologna, A.C.D. Rosa, A.D. Vivo, M. Gaeta, G. Sansonetti, V. Viserta, Personality-based recommendation in E-commerce, in *EMPIRE 2013: Emotions and Personality in Personalized Services* (2013)
11. M.M. Bradley, P.J. Lang, The International Affective Picture System (IAPS) in the study of emotion and attention, in *Handbook of Emotion Elicitation and Assessment, Series in Affective Science*, ed. by J.A. Coan, J.J. Allen, Chap. 2 (Oxford University Press, 2007), pp. 29–46. http://books.google.com/books?hl=en&lr=&id=ChiiBDGyewoC&oi=fnd&pg=PA29&dq =The+international+affective+picture+system+(IAPS)+in+the+study+of+emotion+and+att ention&ots=pJyOP0Y8rD&sig=VJXcIRILIEtevfO38sLZ3rHCNT8%5Cnhttp://books.google .com/books?hl=en&lr=&id=Ch
12. M. Braunhofer, M. Elahi, M. Ge, F. Ricci, Context dependent preference acquisition with personality-based active learning in mobile recommender systems, in *Learning and Collaboration Technologies. Technology-Rich Environments for Learning and Collaboration* (2014), pp. 105–116. https://doi.org/10.1007/978-3-319-07485-6_11
13. B. Brost, R. Mehrotra, T. Jehan, The music streaming sessions dataset, in *The World Wide Web Conference, WWW '19* (Association for Computing Machinery, New York, 2019), pp. 2594–2600. https://doi.org/10.1145/3308558.3313641
14. I. Cantador, I. Fernández-tobías, A. Bellogín, Relating personality types with user preferences in multiple entertainment domains, in *EMPIRE 1st Workshop on "Emotions and Personality in Personalized Services", Rome*, 10 June 2013
15. L. Chen, W. Wu, L. He, How personality influences users' needs for recommendation diversity? in *CHI '13 Extended Abstracts on Human Factors in Computing Systems on - CHI EA '13* (2013), p. 829. https://doi.org/10.1145/2468356.2468505
16. L. Chen, Y. Yang, N. Wang, K. Yang, Q. Yuan, How serendipity improves user satisfaction with recommendations? A large-scale user evaluation, in *The World Wide Web Conference, WWW '19* (Association for Computing Machinery, New York, 2019), pp. 240–250. https:// doi.org/10.1145/3308558.3313469
17. G. Chittaranjan, J. Blom, D. Gatica-Perez, Mining large-scale smartphone data for personality studies. Pers. Ubiquitous Comput. **17**(3), 433–450 (2011). https://doi.org/10.1007/s00779-011-0490-1
18. C.L. Clarke, M. Kolla, G.V. Cormack, O. Vechtomova, A. Ashkan, S. Büttcher, I. MacKinnon, Novelty and diversity in information retrieval evaluation, in *Proceedings of the 31st Annual International ACM SIGIR Conference on Research and Development in Information Retrieval (SIGIR 2008)* (ACM, New York, 2008), pp. 659–666
19. P.T. Costa, R.R. Mccrae, NEO PI-R professional manual, Odessa, FL (1992)

20. A. Delic, J. Neidhardt, T.N. Nguyen, F. Ricci, An observational user study for group recommender systems in the tourism domain. Inf. Technol. Tour. (2018). https://doi.org/10.1007/s40558-018-0106-y. http://link.springer.com/10.1007/s40558-018-0106-y

21. A. Delić, T.N. Nguyen, M. Tkalčič, Group decision-making and designing group recommender systems, in *Handbook of e-Tourism* (Springer International Publishing, Cham, 2020), pp. 1–23. https://doi.org/10.1007/978-3-030-05324-6_57-1. http://link.springer.com/10.1007/978-3-030-05324-6_57-1

22. M. Deniz, An investigation of decision making styles and the five-factor personality traits with respect to attachment styles. Educ. Sci. Theory Pract. **11**(1), 105–114 (2011)

23. M. Dennis, J. Masthoff, C. Mellish, The quest for validated personality trait stories, in *Proceedings of the 2012 ACM International Conference on Intelligent User Interfaces - IUI '12* (ACM Press, New York, 2012). https://doi.org/10.1145/2166966.2167016

24. C.G. DeYoung, L.C. Quilty, J.B. Peterson, Between facets and domains: 10 aspects of the Big Five. J. Pers. Soc. Psychol. **93**(5), 880–896 (2007). https://doi.org/10.1037/0022-3514.93.5.880

25. G. Dunn, J. Wiersema, J. Ham, L. Aroyo, Evaluating interface variants on personality acquisition for recommender systems, in *User Modeling, Adaptation, and Personalization* (2009), pp. 259–270. https://doi.org/10.1007/978-3-642-02247-0_25

26. M.M. El-Bishouty, T.W. Chang, S. Graf, N.S. Chen, Smart e-course recommender based on learning styles. J. Comput. Educ. **1**(1), 99–111 (2014). https://doi.org/10.1007/s40692-014-0003-0

27. M. Elahi, M. Braunhofer, F. Ricci, M. Tkalcic, Personality-based active learning for collaborative filtering recommender systems, in *AI\*IA 2013: Advances in Artificial Intelligence* (2013), pp. 360–371. https://doi.org/10.1007/978-3-319-03524-6_31

28. M. Elahi, V. Repsys, F. Ricci, Rating elicitation strategies for collaborative filtering, in *E-Commerce and Web Technologies* (2011), pp. 160–171

29. F. Eskandanian, B. Mobasher, R. Burke, A clustering approach for personalizing diversity in collaborative recommender systems, in *Proceedings of the 25th Conference on User Modeling, Adaptation and Personalization (UMAP 2017)* (ACM, New York, 2017), pp. 280–284

30. R. Felder, L. Silverman, Learning and teaching styles in engineering education. Eng. Educ. **78**(June), 674–681 (1988)

31. I. Fernández-Tobías, M. Braunhofer, M. Elahi, F. Ricci, I. Cantador, Alleviating the new user problem in collaborative filtering by exploiting personality information. User Model. User-Adapt. Interact. **26**(2–3), 221–255 (2016). https://doi.org/10.1007/s11257-016-9172-z.

32. B. Ferwerda, M. Schedl, M. Tkalcic, Predicting personality traits with instagram pictures, in *Proceedings of the 3rd Workshop on Emotions and Personality in Personalized Systems 2015 - EMPIRE '15*, ed. by M. Tkalčič, B. De Carolis, M. de Gemmis, A. Odić, A. Košir (ACM Press, New York, 2015), pp. 7–10. https://doi.org/10.1145/2809643.2809644. http://dl.acm.org/citation.cfm?doid=2809643.2809644

33. B. Ferwerda, M. Schedl, M. Tkalcic, Personality traits and the relationship with (non-) disclosure behavior on Facebook, in *Proceedings of the 25th International Conference Companion on World Wide Web - WWW '16 Companion* (ACM Press, New York, 2016), pp. 565–568. https://doi.org/10.1145/2872518.2890085

34. B. Ferwerda, M. Tkalcic, Predicting users' personality from instagram pictures, in *Proceedings of the 26th Conference on User Modeling, Adaptation and Personalization - UMAP '18* (ACM Press, New York, 2018), pp. 157–161. https://doi.org/10.1145/3209219.3209248. http://dl.acm.org/citation.cfm?doid=3209219.3209248

35. B. Ferwerda, M. Tkalčič, Exploring the prediction of personality traits from drug consumption profiles, in *Adjunct Publication of the 28th ACM Conference on User Modeling, Adaptation and Personalization, UMAP '20 Adjunct* (Association for Computing Machinery, New York, 2020), pp. 2–5. https://doi.org/10.1145/3386392.3397589

36. R. Gao, B. Hao, S. Bai, L. Li, A. Li, T. Zhu, Improving user profile with personality traits predicted from social media content, in *Proceedings of the 7th ACM Conference on Recommender Systems, RecSys '13* (ACM, New York, 2013), pp. 355–358. https://doi.org/10.1145/2507157.2507219
37. J. Golbeck, C. Robles, K. Turner, Predicting personality with social media, in *Proceedings of the 2011 Annual Conference Extended Abstracts on Human Factors in Computing Systems - CHI EA '11* (2011), p. 253. https://doi.org/10.1145/1979742.1979614
38. L. Goldberg, J. Johnson, H. Eber, R. Hogan, M. Ashton, C. Cloninger, H. Gough, The international personality item pool and the future of public-domain personality measures. J. Res. Personal. **40**(1), 84–96 (2006). https://doi.org/10.1016/j.jrp.2005.08.007
39. L.R. Goldberg, The development of markers for the big-five factor structure. Psychol. Assess. **4**(1), 26–42 (1992)
40. S.D. Gosling, P.J. Rentfrow, W.B. Swann, A very brief measure of the Big-Five personality domains. J. Res. Personal. **37**(6), 504–528 (2003). https://doi.org/10.1016/S0092-6566(03)00046-1. http://linkinghub.elsevier.com/retrieve/pii/S0092656603000461
41. D. Hellriegel, J. Slocum, *Organizational Behavior* (Cengage Learning, New York, 2010)
42. J.L. Holland, *Making Vocational Choices: A Theory of Vocational Personalities and Work Environments* (Psychological Assessment Resources, Washington, DC, 1997)
43. R. Hu, P. Pu, A study on user perception of personality-based recommender systems. User Model. Adapt. Personal. **6075**, 291–302 (2010). https://doi.org/10.1007/978-3-642-13470-8_27
44. R. Hu, P. Pu, Using personality information in collaborative filtering for new users, in *Recommender Systems and the Social Web* (2010), p. 17
45. R. Hu, P. Pu, Exploring relations between personality and user rating behaviors, in *EMPIRE 1st Workshop on "Emotions and Personality in Personalized Services", Rome* 10 June 2013
46. N. Hurley, M. Zhang, Novelty and diversity in top-n recommendation – analysis and evaluation. ACM Trans. Internet Technol. **10**(4), 14:1–14:30 (2011). https://doi.org/10.1145/1944339.1944341
47. F. Iacobelli, A.J. Gill, S. Nowson, J. Oberlander, Large scale personality classification of bloggers, in *Affective Computing and Intelligent Interaction*, ed. by S. D'Mello, A. Graesser, B. Schuller, J.C. Martin. Lecture Notes in Computer Science, vol. 6975 (Springer, Berlin, 2011), pp. 568–577. https://doi.org/10.1007/978-3-642-24571-8
48. O.P. John, S. Srivastava, The Big Five trait taxonomy: history, measurement, and theoretical perspectives, in *Handbook of Personality: Theory and Research*, vol. 2, 2nd edn. ed. by L.A. Pervin, O.P. John (Guilford Press, New York, 1999), pp. 102–138
49. T.B. Kashdan, M.W. Gallagher, P.J. Silvia, B.P. Winterstein, W.E. Breen, D. Terhar, M.F. Steger, The curiosity and exploration inventory-II: development, factor structure, and psychometrics. J. Res. Personal. **43**(6), 987–998 (2009). https://doi.org/10.1016/j.jrp.2009.04.011
50. D. Keirsey, *Please Understand Me 2?* (Prometheus Nemesis, Del Mar, 1998), pp. 1–350
51. M. Khwaja, M. Ferrer, J.O. Iglesias, A. Aldo Faisal, A. Matic, Aligning daily activities with personality: towards a recommender system for improving wellbeing, in *RecSys 2019 - 13th ACM Conference on Recommender Systems (Section 3)* (2019), pp. 368–372. https://doi.org/10.1145/3298689.3347020
52. M. Kompan, M. Bieliková, Social structure and personality enhanced group recommendation, in *UMAP 2014 Extended Proceedings* (2014)
53. M. Kosinski, D. Stillwell, T. Graepel, Private traits and attributes are predictable from digital records of human behavior. Proc. Natl. Acad. Sci. 2–5 (2013). https://doi.org/10.1073/pnas.1218772110
54. A. Košir, A. Odić, M. Kunaver, M. Tkalčič, J.F. Tasič, Database for contextual personalization. Elektrotehniški vestnik **78**(5), 270–274 (2011)

55. A.D.I. Kramer, J.E. Guillory, J.T. Hancock, Experimental evidence of massive-scale emotional contagion through social networks. Proc. Natl. Acad. Sci. USA **111**(29), 8788–90 (2014). https://doi.org/10.1073/pnas.1320040111. http://www.ncbi.nlm.nih.gov/pubmed/24994898 http://www.ncbi.nlm.nih.gov/pubmed/24889601

56. P.J. Lang, M.M. Bradley, B.N. Cuthbert, International affective picture system (IAPS): affective ratings of pictures and instruction manual. Technical Report A-8. Tech. rep., University of Florida, 2005

57. G. van Lankveld, P. Spronck, J. van den Herik, A. Arntz, Games as personality profiling tools, in *2011 IEEE Conference on Computational Intelligence and Games (CIG'11)* (2011), pp. 197–202. https://doi.org/10.1109/CIG.2011.6032007

58. S. Manolios, A. Hanjalic, C.C.S. Liem, The influence of personal values on music taste, in *Proceedings of the 13th ACM Conference on Recommender Systems* (ACM, New York, 2019), pp. 501–505. https://doi.org/10.1145/3298689.3347021. https://dl.acm.org/doi/10.1145/3298689.3347021

59. J. Masthoff, A. Gatt, In pursuit of satisfaction and the prevention of embarrassment: affective state in group recommender systems. User Model. User-Adapt. Interact. J. Personal. Res. **16**(3–4), 281–319 (2006). https://doi.org/10.1007/s11257-006-9008-3

60. C. Matt, T. Hess, A. Benlian, C. Weiß, Escaping from the filter bubble? The effects of novelty and serendipity on users' evaluations of online recommendations (2014). https://EconPapers.repec.org/RePEc:dar:wpaper:66193

61. R. McCrae, I. Allik, *The Five-Factor Model of Personality Across Cultures* (Springer, Berlin, 2002)

62. R.R. McCrae, P.T. Costa, A contemplated revision of the NEO Five-Factor Inventory. Pers. Individ. Dif. **36**(3), 587–596 (2004). https://doi.org/10.1016/S0191-8869(03)00118-1

63. R.R. McCrae, O.P. John, An introduction to the five-factor model and its applications. J. Personal. **60**(2), 175–215 (1992)

64. S.M. McNee, J. Riedl, J.A. Konstan, Being accurate is not enough: How accuracy metrics have hurt recommender systems, in *CHI '06 Extended Abstracts on Human Factors in Computing Systems, CHI EA '06* (ACM, New York, 2006), pp. 1097–1101. https://doi.org/10.1145/1125451.1125659

65. A.B. Melchiorre, M. Schedl, Personality correlates of music audio preferences for modelling music listeners, in *Proceedings of the 28th ACM Conference on User Modeling, Adaptation and Personalization, UMAP '20* (Association for Computing Machinery, New York, 2020), pp. 313–317. https://doi.org/10.1145/3340631.3394874

66. T.N. Nguyen, F. Ricci, Situation-dependent combination of long-term and session-based preferences in group recommendations: an experimental analysis, in *Proceedings of Sac* (2018), pp. 1366–1373. https://doi.org/10.1145/3167132.3167279

67. T.N. Nguyen, F. Ricci, A. Delic, D. Bridge, Conflict resolution in group decision making: insights from a simulation study. User Model. User-Adapt. Interact. **29**(5), 895–941 (2019)

68. S. Nowson, J. Oberlander, Identifying more bloggers: towards large scale personality classification of personal weblogs, in *International Conference on Weblogs and Social Media* (2007)

69. M.A.S. Nunes, R. Hu, Personality-based recommender systems, in *Proceedings of the Sixth ACM Conference on Recommender Systems - RecSys '12* (ACM Press, New York, 2012), p. 5. https://doi.org/10.1145/2365952.2365957

70. M.A.S.N. Nunes, *Recommender Systems Based on Personality Traits: Could Human Psychological Aspects Influence the Computer Decision-Making Process?* (VDM Verlag, Berlin, 2009)

71. A. Odić, M. Tkalčič, J.F. Tasic, A. Košir, Predicting and detecting the relevant contextual information in a movie-recommender system. Interact. Comput. **25**(1), 74–90 (2013). https://doi.org/10.1093/iwc/iws003

72. A. Odić, M. Tkalčič, J.F. Tasič, A. Košir, Personality and social context : impact on emotion induction from movies, in *UMAP 2013 Extended Proceedings* (2013)

73. J.W. Pennebaker, M.E. Francis, R.J. Booth, *Linguistic Inquiry and Word Count: Liwc 2001* (Lawrence Erlbaum Associates, Mahwah, 2001), p. 71

74. D. Quercia, M. Kosinski, D. Stillwell, J. Crowcroft, Our Twitter Profiles, our selves: predicting personality with twitter, in *2011 IEEE Third Int'l Conference on Privacy, Security, Risk and Trust and 2011 IEEE Third Int'l Conference on Social Computing* (IEEE, Piscataway, 2011), pp. 180–185 https://doi.org/10.1109/PASSAT/SocialCom.2011.26

75. L. Quijano-Sanchez, J.A. Recio-Garcia, B. Diaz-Agudo, Personality and social trust in group recommendations, in *2010 22nd IEEE International Conference on Tools with Artificial Intelligence (c)* (2010), pp. 121–126. https://doi.org/10.1109/ICTAI.2010.92

76. D. Rawlings, V. Ciancarelli, Music preference and the five-factor model of the NEO personality inventory. Psychol. Music **25**(2), 120–132 (1997). https://doi.org/10.1177/0305735697252003

77. J.A. Recio-Garcia, G. Jimenez-Diaz, A.A. Sanchez-Ruiz, B. Diaz-Agudo, Personality aware recommendations to groups, in *Proceedings of the Third ACM Conference on Recommender Systems - RecSys '09* (ACM Press, New York, 2009), p. 325. https://doi.org/10.1145/1639714.1639779

78. P.J. Rentfrow, L.R. Goldberg, R. Zilca, Listening, watching, and reading: the structure and correlates of entertainment preferences. J. Personal. **79**(2), 223–58 (2011). https://doi.org/10.1111/j.1467-6494.2010.00662.x

79. P.J. Rentfrow, S.D. Gosling, The do re mi's of everyday life: the structure and personality correlates of music preferences. J. Personal. Soc. Psychol. **84**(6), 1236–1256 (2003). https://doi.org/10.1037/0022-3514.84.6.1236

80. A. Roshchina, J. Cardiff, P. Rosso, TWIN: personality-based intelligent recommender system. J. Intell. Fuzzy Syst. **28**, 2059–2071 (2015). https://doi.org/10.3233/IFS-141484

81. C. Ross, E.S. Orr, M. Sisic, J.M. Arseneault, M.G. Simmering, R.R. Orr, Personality and motivations associated with facebook use. Comput. Hum. Behav. **25**(2), 578–586 (2009)

82. J. Schrammel, C. Köffel, M. Tscheligi, Personality traits, usage patterns and information disclosure in online communities, in *Proceedings of the 23rd British HCI …* (2009), pp. 169–174

83. M. Selfhout, W. Burk, S. Branje, J. Denissen, M. van Aken, W. Meeus, Emerging late adolescent friendship networks and Big Five personality traits: a social network approach. J. Personal. **78**(2), 509–538 (2010). https://doi.org/10.1111/j.1467-6494.2010.00625.x

84. X. Sha, D. Quercia, P. Michiardi, M. Dell'Amico, Spotting trends, in *Proceedings of the Sixth ACM Conference on Recommender Systems - RecSys '12* (ACM Press, New York, 2012), p. 51. https://doi.org/10.1145/2365952.2365967

85. J. Shen, O. Brdiczka, J. Liu, Understanding email writers: personality prediction from email messages, *User Modeling, Adaptation, and Personalization* (2013), pp. 318–330. https://doi.org/10.1007/978-3-642-38844-6_29

86. M. Skowron, M. Tkalčič, B. Ferwerda, M. Schedl, Fusing social media cues, in *Proceedings of the 25th International Conference Companion on World Wide Web - WWW '16 Companion* (ACM Press, New York, 2016), pp. 107–108. https://doi.org/10.1145/2872518.2889368. http://dl.acm.org/citation.cfm?doid=2872518.2889368

87. B.A. Soloman, R.M. Felder, Index of learning styles questionnaire (2014). http://www.engr.ncsu.edu/learningstyles/ilsweb.html

88. B. Stewart, Personality and play styles: a unified model (2011)

89. K.W. Thomas, Conflict and conflict management: reflections and update. J. Organ. Behav. **13**(3), 265–274 (1992). https://doi.org/10.1002/job.4030130307

90. N. Tintarev, M. Dennis, J. Masthoff, Adapting recommendation diversity to openness to experience: a study of human behaviour, in *User Modeling, Adaptation, and Personalization*. Lecture Notes in Computer Science, vol. 7899 (I) (2013), pp. 190–202. https://doi.org/10.1007/978-3-642-38844-6_16

91. A. Tiroshi, T. Kuflik, Domain ranking for cross domain collaborative filtering, in *User Modeling, Adaptation, and Personalization* (2012), pp. 328–333. https://doi.org/10.1007/978-3-642-31454-4_30

92. V. Tiwari, A. Ashpilaya, P. Vedita, U. Daripa, P.P. Paltani, Exploring demographics and personality traits in recommendation system to address cold start problem, pp. 361–369 (2020). https://doi.org/10.1007/978-981-15-0936-0_37. http://link.springer.com/10.1007/978-981-15-0936-0_37

93. M. Tkalčič, Emotions and personality in recommender systems, in *Proceedings of the 12th ACM Conference on Recommender Systems - RecSys '18*, vol. 38 (ACM Press, New York, 2018), pp. 535–536. https://doi.org/10.1145/3240323.3241619. http://link.springer.com/10.1007/978-1-4614-7163-9_110161-1 http://dl.acm.org/citation.cfm?doid=3240323.3241619

94. M. Tkalcic, B.D. Carolis, M.D. Gemmis, A. Odi, A. Košir, *Emotions and Personality in Personalized Services*. Human–Computer Interaction Series (Springer International Publishing, Cham, 2016). https://doi.org/10.1007/978-3-319-31413-6. http://link.springer.com/10.1007/978-3-319-31413-6

95. M. Tkalčič, A. Delić, A. Felfernig, Personality, emotions, and group dynamics, in *Group Recommender Systems an Introduction*, ed. by A. Felfernig, L. Boratto, M. Stettinger, M. Tkalčič (2018), pp. 157–167. https://doi.org/10.1007/978-3-319-75067-5_9. http://link.springer.com/10.1007/978-3-319-75067-5_9

96. M. Tkalcic, B. Ferwerda, M. Tkalčič, B. Ferwerda, M. Tkalcic, B. Ferwerda, M. Tkalčič, B. Ferwerda, Eudaimonic modeling of Moviegoers, in *UMAP '18: 26th Conference on User Modeling, Adaptation and Personalization* (ACM Press, New York, 2018), pp. 163–167. https://doi.org/10.1145/3209219.3209249. http://dl.acm.org/citation.cfm?doid=3209219.3209249

97. M. Tkalcic, M. Kunaver, A. Košir, J. Tasic, Addressing the new user problem with a personality based user similarity measure, in *Joint Proceedings of the Workshop on Decision Making and Recommendation Acceptance Issues in Recommender Systems (DEMRA 2011) and the 2nd Workshop on User Models for Motivational Systems: The Affective and the Rational Routes to Persuasion (UMMS 2011)* (2011)

98. M. Tkalčič, A. Košir, J. Tasič, The LDOS-PerAff-1 corpus of facial-expression video clips with affective, personality and user-interaction metadata. J. Multimodal User Interfaces **7**(1–2), 143–155 (2013). https://doi.org/10.1007/s12193-012-0107-7

99. M. Tkalčič, M. Kunaver, J. Tasič, A. Košir, Personality based user similarity measure for a collaborative recommender system, in *5th Workshop on Emotion in Human-Computer Interaction-Real World Challenges* (2009), p. 30

100. N. Wang, L. Chen, Y. Yang, The impacts of item features and user characteristics on user' perceived serendipity of recommendations, in *Proceedings of the 28th ACM Conference on User Modeling, Adaptation and Personalization, UMAP '20* (Association for Computing Machinery, New York, 2020), pp. 266–274. https://doi.org/10.1145/3340631.3394863.

101. P. Winoto, T. Tang, If you like the devil wears prada the book, will you also enjoy the devil wears prada the movie? A study of cross-domain recommendations. New Gener. Comput. **26**(3), 209–225 (2008). https://doi.org/10.1007/s00354-008-0041-0

102. W. Wu, L. Chen, L. He, Using personality to adjust diversity in recommender systems, in *Proceedings of the 24th ACM Conference on Hypertext and Social Media - HT '13 (May)* (2013), pp. 225–229. https://doi.org/10.1145/2481492.2481521

103. W. Wu, L. Chen, Y. Zhao, Personalizing recommendation diversity based on user personality. User Model. User-Adapt. Interact. **28**(3), 237–276 (2018). https://doi.org/10.1007/s11257-018-9205-x.

104. Z. Yusefi, H. Marjan, K. Afsaneh, Improving sparsity and new user problems in collaborative filtering by clustering the personality factors. Electron. Commer. Res. **18**(4), 813–836 (2018). https://doi.org/10.1007/s10660-018-9287-x.

105. C.N. Ziegler, S.M. McNee, J.A. Konstan, G. Lausen, Improving recommendation lists through topic diversification, in *Proceedings of the 14th International Conference on World Wide Web, WWW '05* (ACM, New York, 2005), pp. 22–32. https://doi.org/10.1145/1060745.1060754

# Individual and Group Decision Making and Recommender Systems

Anthony Jameson, Martijn C. Willemsen, and Alexander Felfernig

## 1 Introduction and Preview

### 1.1 Choice Support as a Central Goal

What is the function of recommender systems? Here are three possible answers:

- *Goal 1:* Given a large set of items, a recommender system should identify a much smaller subset of items that the current user is predicted to like and/or choose.
- *Goal 2:* A recommender system should help its users make *better choices*[1] from large sets of items.
- *Goal 3:* In addition to Goal 2, a recommender system should satisfy other interests of its users (e.g., understanding a space of items, understanding their own tastes, or having a satisfying emotional experience; see Chapters "Value and Impact of Recommender Systems" and "Novelty and Diversity in Recommender Systems" and [57, 66, 96]) as well as interests of other stakeholders such

---

[1] There is no crisp distinction in English between "choices" and "decisions", though the latter term tends to be used more in connection with relatively deliberate choice processes. This chapter uses both terms, depending on which one fits better in a given context.

---

A. Jameson (✉)
Chusable AG, Gamprin, Liechtenstein
e-mail: anthonyjameson@chusable.com

M. C. Willemsen
Eindhoven University of Technology and JADS, Eindhoven, The Netherlands
e-mail: M.C.Willemsen@tue.nl

A. Felfernig
Graz University of Technology, Graz, Austria
e-mail: afelfern@ist.tugraz.at

© Springer Science+Business Media, LLC, part of Springer Nature 2022
F. Ricci et al. (eds.), *Recommender Systems Handbook*,
https://doi.org/10.1007/978-1-0716-2197-4_21

as providers of the items being recommended (e.g., increasing sales or user engagement, adding a differentiating feature to an application, or harvesting data about users; see Chapter "Multistakeholder Recommender Systems" and [1, 4]).

An outsider who sampled publications on recommender systems might think that Goal 1 is the primary goal, since a great deal of research focuses on algorithms for achieving it. But Goal 1 is not an end in itself, and Goal 2 comes closer to describing how recommender systems offer benefit to their users. The authors of this chapter acknowledge the validity of the broader Goal 3, but we focus here on Goal 2, emphasizing the following points:

- There are many ways in which recommender systems can support human choice besides achieving Goal 1 (see Sect. 1.2).
- Exploiting these opportunities requires a comprehensive, systematic understanding of research concerning how people make choices  and how they can be supported while doing so.
- An understanding of this research and its implications for recommender systems is also helpful in connection with the broader Goal 3, since the other possible functions of recommender systems are at least partly related to the function of supporting choice. In particular, efforts to persuade or "nudge" a person to do something that is not necessarily in her[2] own personal interest (e.g., conserve energy or buy a product that is especially profitable for the vendor) can largely be understood as intentionally biased variants of efforts to help someone choose in her own best interest.[3]

## 1.2   Different Degrees of Human Involvement

The ways in which readers will be able to apply this chapter to their own work on recommender systems will depend in part on the nature of the usage scenarios that they deal with. Table 1 describes on a high level four typical ways in which people can make use of recommender systems while making a choice.

Scenarios 3 and 4 have been attracting increasing attention, especially with the current ubiquity of computing devices and the many forms of interaction and communication that they support (see, e.g., the other chapters in the fourth major section of this handbook; the article by Jugovac and Jannach [59]; and the multiyear ACM RecSys workshop series on *Interfaces and Human Decision Making for Recommender Systems*, [15]).

---

[2] In this chapter, generically used personal pronouns alternate systematically between the masculine and feminine forms on an example-by-example basis.

[3] A discussion of the relationship between techniques for *choice support* and for *persuasion* is offered on a general level by Jameson et al. [53, sect. 1.2.2] and in the context of recommender systems by Starke et al. [99].

**Table 1** Four high-level scenarios for recommender systems with increasing amounts of human involvement in decision making

| Scenario | Description | Example |
|---|---|---|
| 1. No involvement | The recommender system autonomously computes recommendations and executes one or more of them | Music is selected automatically for playing in the background |
| 2. Quick selection | The recommender system presents recommendations to the chooser, who then adopts one or more of them after a quick examination | The recommender system recommends several news stories; the chooser clicks on the first one that looks reasonably interesting |
| 3. Deliberate comparison | Given a set of recommendations, the chooser adopts one or more of them after comparing the options more or less thoroughly | The recommender system presents a number of recommended products in a product comparison table; the chooser compares them while inspecting the explanations and the other information provided |
| 4. Iterative requesting of recommendations | Each presentation of recommendations is just one step in an iterative decision making process that may involve the generation of new recommendations on the basis of the chooser's feedback | In the previous scenario, not being satisfied with any of the recommended products, the chooser requests new recommendations and/or looks for further options somewhere else |

The relevance of this chapter is more obvious with Scenarios 3 and 4 than with Scenarios 1 and 2, since Scenarios 3 and 4 involve cognition and actions on the part of choosers that may need to be effectively supported. But even with Scenarios 1 and 2, it is worthwhile to understand the human cognitive activity that is being replaced by the recommender system and what might be lost in the process. For example, the question of what constitutes a good choice (Sects. 1.4 and 4.3) is relevant to the question of what sorts of recommendations ought to be generated.

## 1.3 Preview of This Chapter

As in the previous edition of this handbook [55], the analysis is organized around two previously published models that are intended to make the vast amount of relevant research accessible to those who work on recommender systems and other interactive computing technology:[4]

- The ASPECT model captures the wide variety of ways in which people make everyday choices in terms of six *choice patterns*. Considering these patterns one by one yields new ideas about how recommender systems can support particular aspects of human choice.
- The ARCADES model comprises seven *choice support strategies* which together capture the qualitatively different ways in which it is possible to help people make better choices. One of these strategies is already widely used in recommender systems, but consideration of the other six as well can enable a recommender system to provide more comprehensive choice support.

Whereas the chapter in the previous edition considered only choices made by individual persons, this chapter devotes equal attention to choices made by groups of people and recommender systems designed to support groups (*group recommender systems* for short; see Chapter "Group Recommender Systems: Beyond Preference Aggregation"). The first half is essentially a thoroughly revised and updated version of the first half of the previous chapter. The second half has largely the same structure as the first half, but it focuses on the additional phenomena and considerations that arise in connection with groups.

## *1.4 What Constitutes a Good Choice for an Individual?*

If our goal is to help people make better choices, we should have some idea of when people feel that they have chosen well. A number of researchers have investigated this question (see, e.g., [5, 10, 47, 111]). Although the specific answers and terminology vary, the criteria listed in Table 2 are widely accepted among the relevant researchers (for a more detailed discussion, see [53, sect. 3.6]). Here are some comments on these criteria:

1. *Good outcomes:* This criterion is not as straightforward as it may seem, because what counts as a "good outcome" is in turn surprisingly complex. In this chapter, a good outcome is viewed as one that the chooser is (or would be) satisfied with in retrospect, after having acquired the most relevant knowledge and experience. The emphasis in the recommender systems field on maximizing the accuracy of recommendations can be seen as an attempt to increase the likelihood of satisfactory outcomes.
2. *Limited costs:* Note that a recommender system whose use yields only barely acceptable outcomes can still be considered worth using if it drastically reduces the time and effort required to find such an outcome.

---

[4] Much more detail on these models will be found in the book by Jameson et al. [53], which is freely available via https://chusable.com/foundations.html.

**Table 2** Four widely accepted criteria for what choosers consider to be a "good choice"

| Criterion | Explanation |
|---|---|
| 1. Good outcomes | Choosers want their decisions to yield good outcomes |
| 2. Limited costs | Choosers do not want to invest time and effort in the choice process itself that is out of proportion to the resulting benefits |
| 3. Avoidance of negative affect | Choosers tend to prefer to avoid experiencing unpleasant thoughts and feelings while making a choice |
| 4. Justifiability | Choosers often want to be able to justify the decision that they have made to other persons or to themselves |

3. *Avoidance of negative affect:* Some ways of thinking about a decision can involve distressing thoughts, as when a car buyer considers whether to save money by not purchasing an optional safety feature, noting that doing so will increase the likelihood that a member of his family will be injured (cf. [5, sect. 4]). One benefit of outsourcing parts of the decision process to a recommender system (or to a human advisor) is that the chooser himself does not need to think about such matters.

4. *Justifiability:* Being able to justify a decision to others is often a necessary condition for being able to execute the chosen option (e.g., when an employee requests permission to purchase a particular piece of office equipment that she has selected). And it can help preserve the reputation of the chooser, especially if the outcome of the decision is unexpectedly poor. The explanations (Chapter "Beyond Explaining Single Item Recommendations") provided by many recommender systems can be seen in part as helping the chooser to satisfy this criterion (Sect. 3.3).

In sum, recommender systems are well positioned to help choosers fulfill all of the four main quality criteria for choices. This fact may help to explain their popularity relative to some other forms of choice support that rate poorly with respect to one or more of these criteria (e.g., decision support systems that call for effortful and often frustrating contemplation of trade-offs, which violates two of the four criteria; cf. [111]).

# 2   Choice Patterns and Recommendation to Individuals

The question "How do people make choices?" is surprisingly hard to answer, even if you are familiar with the vast and impressive scientific literature on this topic in psychology, economics, and other fields (see [53, sect. 3.1] for a list of the relevant research areas). Recommender systems researchers and practitioners are in a good position to understand why, since the same difficulty would arise with the question "How can a computer program make recommendations?" In both cases, the top-level answer is: "There are a number of different approaches, and they can be combined in various ways."

With regard to computational recommendation, the various different paradigms (content-based, collaborative, etc.) and the ways of combining them have been described by, among others, Burke [16, 17] and Felfernig et al. [29]; see the section on recommendation techniques in Chapter "Recommender Systems: Techniques, Applications, and Challenges" for a compact discussion. The ASPECT model [53, chap. 3] aims to do something similar for human choice: It distinguishes six human *choice patterns*, which are summarized in Table 3.[5] Each of these patterns is sometimes found in its pure form, but they are often blended together in various ways (see Sect. 2.7).

Distilling out these six choice patterns makes it possible to think in detail about how to support choice when it occurs in accordance with each pattern. As can be seen in Table 3, each pattern involves different concepts and comprises a set of typical processing steps that are mostly different from the steps found in the other patterns. With regard to each pattern, we can ask: What can a recommender system do to help people to execute these steps more successfully?

This approach differs radically from a common way of thinking about human decision making in both the popular scientific literature and that of the recommender systems field: in terms of a list of (often not fully understood) "biases" or forms of "irrationality" which need to be counteracted—or exploited. This latter approach is in some ways like that of a basketball coach who focuses his training on counteracting a dozen typical mistakes that basketball players make (e.g., looking at the ball while dribbling)—as if the players required no help in improving the many other skills that are involved in basketball playing. The ASPECT choice patterns aim to describe all of the things that people do while making choices, most of which are not associated with any particular "bias", so that people can be helped to do all of these things better than they might otherwise.

While introducing the choice patterns, we will refer to the following situation: An English-speaking tourist who is about to visit France would like to buy a French-English dictionary for his or her smartphone from an app store that offers a number of relevant dictionaries.

## 2.1  Attribute-Based Choice

If our example tourist applies the attribute-based pattern, she will view each dictionary as an object that can be described in terms of evaluation-relevant attributes (e.g., number of entries, usability, and price), some of which are more important than others. Each object has a *level* with respect to each attribute, such as a particular number of entries. The chooser can assign a *value* to an object's level

---

[5] ASPECT is an acronym formed from the first letters of the six patterns.

**Table 3** Overview of the six choice patterns that make up the ASPECT model ($C$ = the chooser)

| Attribute-based choice | Socially based choice |
|---|---|
| *Conditions of applicability* | *Conditions of applicability* |
| – The options can be viewed meaningfully as items that can be described in terms of attributes and levels<br>– The (relative) desirability of an item can be estimated in terms of evaluations of its levels of various attributes | – There is some information available about what relevant other people do, expect, or recommend in this or similar situations |
| *Typical procedure* | *Typical procedure* |
| – (Optional:) $C$ reflects in advance about the situation-specific (relative) importance of attributes and/or values of attribute levels<br>– $C$ reduces the total set of options to a smaller *consideration set* on the basis of attribute information<br>– $C$ chooses from among a manageable set of options | – $C$ considers *examples* of the choices or evaluations of other persons<br>– or $C$ considers the *expectations* of relevant people<br>– or $C$ considers explicit advice concerning the options |
| **Consequence-based choice** | **Policy-based choice** |
| *Conditions of applicability* | *Conditions of applicability* |
| – The choices are among actions that will have consequences | – $C$ encounters choices like this one on a regular basis |
| *Typical procedure* | *Typical procedure* |
| – $C$ recognizes that a choice about a possible action can (or must) be made<br>– $C$ assesses the situation<br>– $C$ decides when and where to make the choice<br>– $C$ identifies one or more possible actions (options)<br>– $C$ anticipates (some of) the consequences of executing the options<br>– $C$ evaluates (some of) the anticipated consequences<br>– $C$ chooses an option that rates (relatively) well in terms of its consequences | – [Earlier:] $C$ arrives at a policy for dealing with this type of choice<br>– [Now:] $C$ recognizes which policy is applicable to the current choice situation and applies it to identify the preferred option<br>– $C$ determines whether actually to execute the option implied by the policy |
| **Experience-based choice** | **Trial-and-error-based choice** |
| *Conditions of applicability* | *Conditions of applicability* |
| – $C$ has made similar choices in the past | – The choice will be made repeatedly; or $C$ will have a chance to switch from one option to another even after having started to execute the first option |

(continued)

**Table 3** (continued)

| Experience-based choice | Trial-and-error-based choice |
|---|---|
| *Typical procedure* | *Typical procedure* |
| – *C* applies recognition-primed decision making<br>– or *C* acts on the basis of a habit<br>– or *C* chooses a previously reinforced response<br>– or *C* applies the affect heuristic | – *C* selects an option *O* to try out, either using one of the other choice patterns or (maybe implicitly) by applying an *exploration strategy*<br>– *C* executes the selected option *O*<br>– *C* notices some of the consequences of executing *O*<br>– *C* learns something from these consequences<br>– (If *C* is not yet satisfied:) *C* returns to the selection step, taking into account what has been learned |

of an attribute. Roughly speaking, the chooser will tend to select a dictionary that seems attractive in terms of the values of the levels of (some of) its attributes, with the more important attributes influencing the choice relatively strongly. But there are many specific ways in which people apply the attribute-based pattern, ranging from thoroughly considering each object's levels on many of its attributes to considering only a small sample of the attribute information and selecting an object that looks (relatively) good in terms of the sample. Useful entry points to the literature on attribute-based choice include works by Payne et al. [85], Hsee [50], Pfeiffer [86, chap. 2], Bhatia [11], and Gigerenzer and Todd [41].

Recommendation algorithms of some types likewise make use of information concerning options' attributes and their evaluations by choosers; such algorithms are now often discussed under the heading of *multi-criteria recommender systems* (see [3] for a thorough discussion). These algorithms do not need to imitate human attribute-based processing in order to generate useful sets of recommended options; but a recent algorithmic contribution by Lin et al. [70] explicitly addresses the general observation that humans often apply *noncompensatory* attribute-based strategies, such as focusing mainly on one important attribute and considering other attributes only to narrow down the set of options to be considered. Their new noncompensatory algorithms yield promising results, which suggest that the ways in which humans apply choice patterns may deserve more attention from algorithm designers.

A recommender system can also support attribute-based choice by recommending evaluation criteria instead of options themselves: which attributes should be considered most important and what levels of them should be considered most desirable. For example, a recommender system might in effect tell the chooser "For a person in your situation, a French-English dictionary ought to have at least 30,000 entries". This strategy of recommending evaluation criteria is sometimes found in constraint-based [31] and other knowledge-based recommender systems (see the section on recommendation techniques in Chapter "Recommender Systems: Techniques, Applications, and Challenges").

An understanding of human attribute-based choice can also inform the design of user interfaces for presenting recommendations to choosers in the high-level Scenarios 3 and 4 of Table 1. With some types of option set, it is natural to present the options in a way that facilitates attribute-based choice among them, such as a product comparison table in which each row describes a product and each column shows levels with respect to a particular attribute. Providing functionality for sorting, filtering, and other operations can enable people to apply familiar attribute-based strategies (e.g., eliminating options that are unsatisfactory in terms of some attribute, one attribute at a time, see [107]) even when dealing with a relatively large number of recommended options [86, chap. 5].

## 2.2  Consequence-Based Choice

A different way of thinking about an option is to consider the concrete consequences of choosing it. So instead of contemplating the number of entries offered by a dictionary, our tourist might consider how successfully he is likely to be able to use the dictionary to order meals in restaurants during his vacation. Consequence-based choice involves a number of different issues than attribute-based choice: Among other things, the chooser needs to deal with uncertainty about what consequences will occur if he chooses a particular option and the fact that they may occur in the distant future. And there is often a considerable variety of possible consequences, ranging from objectively describable events to the chooser's affective responses. Entry points to the literature on consequence-based choice include the textbook by Newell et al. [81, chaps. 9–11] and works that focus on the most prominent descriptive model, *prospect theory* [61, 109].

Since long before the development of recommender systems, people making important decisions have been supported by experts in decision analysis (see, e.g., [20]), who apply systematic methods to help them do things like identify relevant causal relationships, estimate the probability of particular consequences of actions, and assign evaluations to these consequences. Computational support for these activities is provided by various forms of *decision support system* (see, e.g., [36]).

A recommender system can in principle employ knowledge-based techniques to support consequence-based choice, basing its recommendations on predicted consequences of options; but the amount of knowledge that needs to be encoded and reasoned about poses a challenge (cf. the comments on knowledge-based recommendation in Chapter "Recommender Systems: Techniques, Applications, and Challenges"). With other types of recommendation algorithm, the consequence-based pattern can be seen as one that human choosers may apply when evaluating the recommended options, as in the dictionary example given above and in Scenarios 3 and 4 of Table 1. One straightforward way of supporting such evaluation is systematically to provide information about consequences that may not be obvious to choosers and about likelihoods that they may have difficulty estimating.

## 2.3  Socially Based Choice

The two preceding patterns can involve some quite elaborate reasoning about the merits of the available options. The remaining four ASPECT patterns describe how people use qualitatively different approaches to arrive at choices in ways that are typically quicker and less effortful.

People often allow their choices to be influenced by the examples, expectations, or advice of others (cf. the three *subpatterns* of the socially based pattern shown as its "Typical Procedure" in Table 3):

1. *Considering social examples:* If many other people have tried a given dictionary and have rated it positively, their ratings can be seen as a summary of a great deal of relevant experience that it would be impractical for the current chooser to acquire herself.
2. *Considering social expectations:* Other people can also have expectations (e.g., as to what is considered cool or politically incorrect) that may influence the chooser.
3. *Considering advice:* Other people (or an artificial agent) may also offer explicit advice.

Entry points to the literature on socially based choice include influential books by Fishbein and Ajzen [33], Cialdini [19, chaps. 4 and 6], and Thaler and Sunstein [104, chap. 3].

A recommendation domain that offers especially clear examples of all three forms of social influence is the domain of fashion recommendation (Chapter "Fashion Recommender Systems").

Some forms of neighborhood-based recommendation (Chapter "Trust Your Neighbors: A Comprehensive Survey of Neighborhood-Based Methods for Recommender Systems") and collaborative filtering (Chapter "Advances in Collaborative Filtering") can be seen as a way of automating the first subpattern of the socially based pattern; but a closer look at this pattern [53, chap. 8] brings to light additional ways in which recommender systems can support this and the other two subpatterns.

1. Whereas recommendation algorithms often (directly or indirectly) consider examples from people who are similar to the current chooser in some respects, the class of similar people is not always the most relevant class: A chooser may want to make choices that are characteristic of a group of people to which he does not (yet) belong (for example, people who are more advanced in a particular domain or who enjoy higher prestige). Some trust-based recommender systems [108] take into account the social relationships between the chooser and the other persons whose opinions and choices are being considered.

2. Social examples are sometimes less relevant than social expectations, which are normally not taken into account by recommendation algorithms (an exception being some work on group recommender systems, discussed in Sect. 4.5.2). For example, for a user who wants to become a well-regarded member of an on-line community, recommendations about how to behave are often better based on the (explicit or implicit) expectations that govern behavior in that community than on the actual typical behavior of members, which may largely fail to conform to these expectations [53, sect. 11.4].

3. To support the advice-taking subpattern, a recommender system can help the chooser to find persons who can provide good advice, as is done in many *people finding* systems (see, e.g., Chapter "Social Recommender Systems"). This subpattern is also relevant to recommender systems on a different level: When the chooser is aware of the fact that she is being offered recommendations, it is natural for her to consider (mostly quickly and intuitively) some of the same questions that she would consider when taking advice from a human advice giver, some of which concern the advice giver's credibility. So designers of recommender systems can benefit from what is known about how people make such assessments (see [113] for a thorough discussion in the context of recommender systems and more generally [12, 60, 102]).

## 2.4 Experience-Based Choice

*Experience-based choice* occurs when the chooser's past experience with the choice situation and/or with particular options directly suggests some particular option. In Table 3, four subpatterns of this pattern are distinguished in the "Typical Procedure", which can be illustrated with four different reasons for the chooser to select the Oxford dictionary on this occasion (see [53, chap. 7], for much more detail):

*Recognition-primed decision making:* The chooser remembers that, the last time he bought a dictionary of this sort, he chose the Oxford dictionary, and it worked out OK (see, e.g., [64]).

*Habit-based choice:* He has become accustomed to purchasing Oxford dictionaries, so the sight of an Oxford dictionary offered alongside others triggers the response of choosing the Oxford dictionary (see, e.g., [110]).

*Choice based on instrumental conditioning:* In the past, buying Oxford dictionaries has led to consequences that were in some way rewarding, even though the chooser may not have been aware of them (see, e.g., [69, chaps. 4–6], [42, chap. 5]).

*Affect-based choice:* Looking at the description of the Oxford dictionary evokes a positive feeling at the time of choosing, even though the chooser may not know why (see, e.g., [97]).

Other books that discuss various aspects of experience-based choice include those of Betsch and Haberstroh [9], Plessner et al. [89], and Gigerenzer [40].

On a high level, content-based (Chapter "Semantics and Content-Based Recommendations") and case-based [98] recommender systems can be seen as taking over (part of) the process of experience-based choice by analyzing the chooser's relevant previous experiences to determine which of the currently available options they suggest.

The first and fourth subpatterns offer ways for a recommender system to support experience-based choice while keeping the chooser more in the loop: by helping the chooser to remember and take into account relevant aspects of her previous experience, such as the specific actions that the chooser has performed in the past and the feelings that she had while performing them. The term *recomindation* [88] has been coined to refer to this approach.

## 2.5   Policy-Based Choice

Sometimes, the choice process can be seen as comprising two phases, which may be separated considerably in time: In the first phase, the chooser arrives at a *policy* for making a particular type of choice (e.g., "When buying a dictionary for your smartphone, always choose the Oxford dictionary if one is available"). Later, when faced with a specific choice to make, the chooser applies the policy [53, chap. 9].

Policy-based choice has been discussed mainly in the literature on organizational decision making, where policies play a more obvious role than they do with individual choice (cf. Sect. 5.5). Some relevant research on individual choices has been conducted in connection with the concepts of *choice bracketing* [93] and *self-control* [91].

A relatively neglected way of supporting a policy-based choice is to recommend a policy to the chooser, such as a particular diet (cf. the discussion in Chapter "Food Recommender Systems" of systems that recommend healthy food) or an exercise regime. This type of recommendation can be especially valuable in that it is often difficult for a chooser to evaluate a possible policy, partly because of the difficulty of anticipating what consequences its application will have in the long run. Providing a striking example, Camerer et al. [18] found that taxi drivers who are in a position to choose how many hours to drive each day often apply a simple policy ("Drive each day until you have earned a fixed target amount of money") that in practice tends to *minimize* rather than maximize their hourly earnings.

An easier and more frequently found type of automated support involves helping the chooser to *apply* a particular policy by (a) enabling the chooser to communicate the policy somehow and (b) automatically helping to execute the policy whenever a relevant case arises. An example would be an application for personalized news reading that allows the reader to assign priorities to particular types of news item so as to influence the news stories that are presented to him.

## 2.6    Trial-and-Error-Based Choice

Especially if none of the other patterns leads readily to a choice, a chooser will sometimes simply (maybe randomly) choose an option and see how well it works out [53, chap. 10]. For example, our dictionary chooser might download a free dictionary and quickly look up a few words. If the experience with the free dictionary is in some way unsatisfactory (e.g., the dictionary entries do not seem informative enough), the chooser is likely to consider which of the other dictionaries she should try out next.

It is useful to view the trial-and-error-based pattern as being applied even in some cases where the chooser does not fully execute the chosen option. For example, our chooser might "try out" a dictionary in the weaker sense of studying its description in the app store and reading its reviews. The choice process and the appropriate forms of support are in many ways similar to those associated with more thorough trials.

Trial-and-error-based choice has been studied from various perspectives in the research literature, mostly not associated with the term "trial and error", as in the works by Rakow and Newell [92], Cohen et al. [21], Pirolli [87], Lindblom [71], and Zwick et al. [115].

One way in which a recommender system can support trial-and-error-based choice is by helping the chooser to decide, at the beginning of each cycle, which option(s) to try out next (cf. Scenario 4 in Table 1)—a type of decision which, upon close inspection, turns out to involve a surprising variety of considerations. An entire class of recommender systems has evolved to provide support for this critical step: A *critiquing-based* recommender system (see, e.g., [80] and the discussion of conversational recommender systems in Chapter "Semantics and Content-Based Recommendations") knows about the properties of a large set of potential options, and the technology enables it to analyze the feedback ("critique") provided by the chooser and find options that are better with respect to the attributes mentioned in the critique (while not being substantially worse in other respects).

The section in Chapter "Trust Your Neighbors: A Comprehensive Survey of Neighborhood-Based Methods for Recommender Systems" on *user-adaptive diffusion models* discusses how a neighborhood-based recommender system can adapt to and support different exploration strategies exhibited by different users.

The trial-and-error pattern is especially applicable in cases where neither the chooser nor any agent supporting the chooser have a clear idea of the chooser's evaluation criteria: These evaluation criteria can be discovered while the user is engaging in trial and error, and they can even change over time as the chooser acquires new experiences (cf. the section on reinforcement learning in Chapter "Session-Based Recommender Systems" and [55, sect. 18.7]), as is typical, for example, in the domain of music recommendation (Chapter "Music Recommendation Systems: Techniques, Use Cases, and Challenges"). Recent work in the movie domain has also shown that a carefully designed interaction dialog can help the user explore new movies [103] or genres [68], especially when supported with conversational interaction (as in [58]; see also Chapter "Semantics and Content-Based Recommendations").

Aside from these already successful applications of recommendation technology in support of the trial-and-error pattern, there are other types of support that could be considered:

1. A recommender system could explicitly recommend an *exploration strategy*: a strategy for choosing the next option to try out (e.g., "In this situation, it seems best to try out the highest-rated dictionaries first, even though they are the most expensive ones"), leaving it to the chooser to apply the exploration strategy.
2. Recommender systems can also support the second main part of the trial-and-error-based pattern, which involves learning from the experience acquired in trying out an option. Among other things, a recommender system can suggest what aspects of the outcome of a trial to attend to—something that is often not obvious. For example, a dictionary chooser might be advised to pay attention to how long it takes him to look up a word, given that this factor will be more important in everyday use of the dictionary than it is while he is trying it out in an artificial situation.

## 2.7 Combinations of Choice Patterns

The six choice patterns are often used in combination, just as different recommendation techniques are often combined to create hybrid recommenders (see [16, 17] and the discussion of recommendation techniques in Chapter "Recommender Systems: Techniques, Applications, and Challenges"). Explicit discussions of forms of combination are rather rare in the psychological research literature (cf. [53, sect. 3.3.7]). Many studies, however, indirectly yield ideas about forms of combination, as does everyday experience. Most people, for example, can remember choice situations in which our experience-based "gut feeling" conflicted with the result of a careful consequence-based analysis, indicating that the two patterns had been applied in parallel and perhaps largely independently of each other. Another common form of combination is a "cascade" in which one pattern (e.g., a simple attribute-based strategy) is used to generate a manageable number of options and then a different choice pattern is used to choose among these options.

Recommender systems can in principle recommend particular (combinations of) choice patterns as being suitable for a given choice situation; see Sect. 3.2 below.

# 3 Choice Support Strategies and Recommendation to Individuals

While discussing the six ASPECT choice patterns, we have considered mainly how their application can be supported by the technology that is most characteristic of recommender systems: algorithms for generating choices and evaluations on behalf of the chooser. But there are several other general approaches to supporting choice, each of which can sometimes be applied fruitfully within recommender systems to provide complementary forms of support.

The ARCADES model is a high-level synthesis of approaches to choice support that have been discussed, studied, and applied both with and without computing technology. Table 4 gives an overview of the seven strategies.[6]

## 3.1 Evaluate on Behalf of the Chooser

The strategy that is typical of recommender systems is called within the ARCADES model *Evaluate on Behalf of the Chooser*. It is being applied whenever there is a prediction of how the chooser would evaluate (e.g., explicitly rate) something if given a chance to evaluate it; or which option(s) the chooser would choose if given a chance to make a choice. Research on recommendation technology has given rise to

**Table 4** Brief explanations of the seven choice support strategies distinguished in the ARCADES model ($C$ = the chooser)

| Name of strategy | Basic idea |
|---|---|
| Access information and experience | Help $C$ to gain access to information and experience that is relevant to the current choice |
| Represent the choice situation | Influence the way in which $C$ perceives the choice situation in such a way that $C$'s processing is facilitated |
| Combine and compute | Process available information computationally in a way that facilitates one or more processing steps of $C$ |
| Advise about processing | Encourage $C$, implicitly or explicitly, to apply one or more choice patterns in a particular way |
| Design the domain | Change the basic reality about which $C$ is choosing so as to make the choice problem easier |
| Evaluate on behalf of the chooser | Take over from $C$ some step in the processing that involves evaluation or choice among alternatives |
| Support communication | Help $C$ communicate more effectively with advisors or other choosers |

---

[6] A detailed introduction to the first six strategies is given by Jameson et al. [53, chap 4]; the seventh strategy is being introduced in this chapter for the first time.

a great variety of algorithms that can predict people's evaluations and choices with useful accuracy—notwithstanding the variety of ways in which people go about making choices, which are in general at best loosely reflected in algorithms for generating recommendations.[7]

The most common way for recommender systems to support choice with this strategy is described in the first three scenarios in Table 1. But as has already been illustrated in our discussion of choice patterns, there are also less commonly used ways of applying this strategy, by evaluating on behalf of the chooser some decision-relevant items (e.g., policies, see Sect. 2.5) that are of a different nature than the options being considered.

## 3.2   Advise About Processing

The other ARCADES strategy that involves a form of recommendation is *Advise About Processing*. The advice being given here concerns not particular options on the domain level but rather ways of applying a particular choice pattern (or combination of patterns). A recommender system can give procedural advice of this sort, in effect telling the user, for example, "In this case, it seems best for you to consider mainly your own past experience and to ignore what you think your friends would do".[8] For a recommender system to provide advice of this sort in a personalized way would make sense in a situation where the chooser could in principle apply any of two or more procedures and the recommender system is able to predict that one is more suitable than the other for the current chooser and/or situation. Research in recent years on ways of giving users more control over the processing of a recommender system (see, e.g., [26, 45, 65]) may give rise to increasing opportunities to apply the *Advise About Processing* strategy in this way.

## 3.3   Access Information and Experience

We now turn to the four ARCADES strategies that are not specifically connected with recommendation technology, though they can all be applied within recommender systems.

One of the most obvious ways of helping people to choose is to provide relevant information about options and/or give them a clearer idea of what sorts

---

[7] Applying the strategy *Evaluate on Behalf of the Chooser* does not always require recommendation technology; for example, a recommendation like "You are advised to close all other applications before proceeding with the installation procedure" is simply formulated once by the designer of the installation procedure.

[8] This type of advice is often given implicitly in the sense that the system provides support for one procedure but not for others.

of experiences they are likely to have if they choose a particular option. Except in Scenario 1 of Table 1, most recommender systems apply this strategy when they are presenting options for evaluation by the chooser. Thinking back to the ASPECT choice patterns reminds us that the types of information, media, and experience that can be provided are by no means restricted to objective information about properties of the available options. For example, to support the consequence-based pattern, a system can give a preview of what it will feel like to watch a particular film; and to support the socially based pattern, it can inform the user about social examples and expectations.

An especially complex and important type of information that a recommender system can provide is an *explanation* of its recommendations (Chapter "Beyond Explaining Single Item Recommendations"). The previous edition of the present chapter [55, sect. 18.4] offers a detailed analysis of this topic from a choice support perspective. Here, we only summarize the main points.

Explanations can take any of several different forms, and they can support choice in three main ways:

1. by providing one or more *arguments* in favor of a recommended option, which a chooser can make use of in whole or in part when thinking about the option herself;
2. by helping the chooser to assess the *reliability of the advice* being given by the recommender system (see, e.g., [8]);
3. by helping the chooser to see how the selection of a given option could be *justified* to other people (fulfilling the fourth criterion for good choices listed in Table 2).

With regard to the second point: If the goal is choice support, it is not in general appropriate to *maximize* the chooser's trust in the recommender system's reliability. Instead, the chooser should be helped to assess the reliability *accurately* and in particular to recognize as such any recommendations that should *not* be adopted. To this end, any purported "explanation" of a recommendation should reflect with reasonable accuracy the way in which the recommender system actually arrived at the recommendation. Otherwise, the account provided may create in the chooser an inaccurate mental model of the recommendation process that makes it more difficult to assess the recommender system's reliability realistically.

## 3.4 Represent the Choice Situation

This high-level strategy takes into account the fact that the particular way in which information about a choice situation is organized (e.g., the way in which items are displayed on a computer screen) can make particular types of processing easier or more difficult. For example, it is often easier to compare options with each other, as opposed to evaluating each option individually, if the options are displayed simultaneously and the information about them is organized so as to facilitate comparison. Shifting from *joint* to *separate evaluation* or vice-versa can have major consequences for processing (see, e.g., [50]).

Since a recommender system almost inevitably contributes to the way in which the choice situation is represented to the chooser, recommender system designers should think about the consequences of particular forms of representation for the chooser's processing. One section in the previous edition of this chapter [55, sect. 18.8] discussed several well-known phenomena that illustrate how the exact way in which a given set of options is presented can influence which option is chosen.

Recently, a strikingly original way of applying the *Represent the Choice Situation* strategy has been introduced by Hilgard et al. [49]: Instead of leaving it to human designers to invent representations that support human choice processes, they develop methods by which an algorithm can learn to generate a helpful representation.

## 3.5  Combine and Compute

Even aside from the recommendation algorithms that it applies, a recommender system can perform various types of computation on the basis of available information whose results can support the chooser's processing in Scenarios 2, 3, and 4 (Table 1). Simple examples include functionality for allowing the user to sort or filter recommended options according to particular attributes (Sect. 2.1). More sophisticated computation is involved, for example, when a set of options is automatically divided into clusters according to inter-option similarity so as to provide the user with a helpful overview.

## 3.6  Design the Domain

The basic idea with this strategy is to design the underlying reality that the chooser is dealing with in a way that makes it easier for him to make good choices—or for a recommender system to generate good recommendations. The difference from the strategy *Represent the Choice Situation* is that you are crafting the options and other aspects of the choice situation themselves, not just the way in which they are presented to the chooser.

Suppose, for example, that you are designing a recommender system that helps members of a particular online social network to choose appropriate privacy settings. Using the strategy *Represent the Choice Situation*, you would try to display the options to the user in a helpful way (e.g., grouping related options together). But if the privacy settings are inherently hard to deal with (for example, if there are a large number of settings that interact in complex ways), even the best representation may confront users with a challenging choice problem, and even the best recommendation algorithm can have a hard time determining which combinations of settings are likely to be best for the chooser (see, e.g., [38]). Applying the strategy *Design the Domain*, you would reconceptualize the set of privacy options themselves—and maybe also the underlying privacy management principles—so as to make the choice problem inherently easier for the chooser and/or for the recommender system [53, sect. 12.2.3].

## 3.7 Support Communication

This strategy covers all attempts to improve the effectiveness and appropriateness of communication among human choosers or between a human chooser and an advisor (which may be a human or a computational system such as a recommender system). The importance of this strategy is especially obvious in the case of group decision making (Sect. 6.7), and in fact this strategy was missing in the previously published six-strategy "ARCADE" model, which considered only individual decision making.

But even in connection with individual decision making, this strategy is relevant in connection with what is often called *preference elicitation* (see, e.g., the discussions of *conversational preference elicitation* in Chapter "Natural Language Processing for Recommender Systems" and of the more general topic of *preference acquisition* in Chapter "Recommender Systems: Techniques, Applications, and Challenges"): An advisor or system tries to elicit from a human chooser information that will help to predict how that person would evaluate particular options. The previous edition of this chapter [55, sect. 18.5] includes a detailed discussion of the various meanings of the term *preferences* (which within the recommender systems field is broadly used but seldom assigned a clear meaning) and the psychological aspects of the most common method of explicit preference elicitation: the elicitation of *ratings*. The last part of that section can be seen as discussing choice support in accordance with the *Support Communication* strategy.

We mention here one illustrative recent development in preference elicitation for recommender systems: Whereas the explicit rating of an (attribute of an) item is the most common and straightforward way for a chooser to communicate explicitly how she evaluates a particular item, sometimes a less direct form of communication has proven useful: Both Graus and Willemsen [43] and Loepp et al. [72] have investigated *choice-based* elicitation for preference modeling for movie recommendation. By having users choose between multiple items, the recommender system is able to go from a "cold start" to build up a preference model interactively by means of a series of choices rather than ratings. More recently, Kalloori et al. [62] investigated the conditions under which elicitation of pairwise judgments is likely to have advantages relative to the elicitation of ratings.

# 4   New Considerations With Regard to Groups

This chapter now shifts its focus from choices of individuals to those made by groups. As can be seen from the rapidly growing number of group recommender systems (see, [30], Chapter "Group Recommender Systems: Beyond Preference Aggregation", and the brief discussion in Chapter "Social Recommender Systems"), there are many scenarios in which there is a need to support the choice of a group of persons.[9] With the increasing ubiquity and interconnectivity of computing devices, the technical obstacles to supporting group choice even in casual everyday situations have been disappearing. But the challenge remains of understanding the processes involved when a group successfully makes a choice as well as the ways in which these processes can be supported.

In the remainder of this chapter, after first taking a broader look at the relevant literature and major differences between groups and individual choosers, we will apply the ASPECT and ARCADES models in the same way as in the previous two sections, this time focusing on issues that are largely specific to groups.

## 4.1   *Relevant Areas of Research and Practice*

The research- and practice-based literature that is relevant to the decisions of groups is found in a number of areas, including the following:

1. Research on group decision making

    Researchers in psychology and other fields have studied what occurs when groups of people make decisions, applying methods ranging from carefully constructed laboratory studies to analyses of historically important decisions. Recent entry points to this literature include the works of Stasser and Abele [100], Forsyth [35, chap. 12], and Brown and Pehrson [14, chaps. 3 and 5].

2. Other research on group dynamics

    The broader field of group dynamics studies many phenomena that arise in groups which are not specific to group decision making but which should still be taken into account in efforts to support it (e.g., the various forms of social influence covered by the socially based choice pattern, discussed in Sects. 2.3 and 5.3). Accessible entry points include chapters in the textbooks by Forsyth [35, chaps. 7 and 17] and Brown and Pehrson [14, chap. 4].

---

[9] With some decisions made by groups, there exist *stakeholders* whose interests need to be taken into account even though they are not among the group members who are participating directly in the group decision making process (see, e.g., [46, chap. 1]). For example, the parents of a family may make decisions about a family outing without fully including the children in the decision making process. Doing justice to the interests of absent stakeholders raises issues that cannot be addressed within this chapter (cf. the discussion of "virtual group members" in Chapter "Group Recommender Systems: Beyond Preference Aggregation").

3. Experience of group facilitators

When important decisions are made by a group within an organization, sometimes a *facilitator* will explicitly guide the discussion. Experienced facilitators have learned a great deal about difficulties that arise in such meetings and about strategies and tactics for minimizing the difficulties and optimizing the procedures. Entry points include the books by Hartnett [46] and Freshley [37].

4. Studies of the use of group decision support systems

Some decision making meetings make use of more or less elaborate software for supporting group decisions. Many *group decision support systems* include functionality that is worth considering for group recommender systems as well, though the concrete application scenarios of group recommender systems tend to involve shorter time frames and less elaborately structured interaction. A good recent entry point to work in this area is the handbook edited by Kilgour and Eden [63]; earlier discussions can be found in the book by French et al. [36].

5. Research on negotiation

Negotiation can be seen as a type of group decision making process where there are acknowledged differences in interests or viewpoints among the participants which require special attention if the group is to arrive at some sort of agreement. Although group recommender systems are only occasionally presented as supporting negotiation (see, e.g., [7, 76]), the relevant decision making processes often do involve some amount of negotiation when differences of this sort come to light. The type of negotiation most relevant to group recommender systems is sometimes called *integrative negotiation* [67, chap. 3]: All parties are assumed to want to arrive at a decision that will be satisfactory to all of them (in contrast to *distributive negotiation*, such as haggling over a price, where one party's gain is in general another party's loss). Accessible entry points to this literature include the textbook by Lewicki et al. [67] and the popular classic book by Fisher et al. [34].

Of these five areas, the one that has attracted the most attention among those who design group recommender systems is the second one: group dynamics research that does not refer specifically to group decision making (cf. Sects. 4.5.2 and 5.3 below).

The first area—group decision making—has attracted less attention overall in connection with group recommender systems, though some researchers have addressed selected concepts such as that of *groupthink* (discussed below in Sect. 5.3). But there has been a recent trend toward increased attention: Delić and colleagues have not only argued for the importance of understanding the details of group decision making but also conducted observational studies of their own with the explicit goal of informing research on group recommender systems [22–24]. Two benefits of this type of empirical research are that it (a) helps to determine which of the numerous concepts and results that can be found in the literature on group decision making are most relevant in typical application scenarios for group recommender systems and (b) may also help to discover previously unknown phenomena that arise in such scenarios.

With regard to the other three areas listed above, we can find fewer cases where the relevant knowledge has been taken into account in the design of group recommender systems.

## 4.2    The Importance of Differences Among Group Members

Many of the new phenomena found in group vs. individual decision making can be seen as resulting from various types of *differences* that are found among members of a group. These include differences with respect to (a) their initial conception of the decision problem, (b) the underlying interests that will influence how they evaluate possible solutions, (c) the relevant knowledge and beliefs that they possess, (d) their ideas about how the decision ought to be made, and (e) their styles of communication. These differences lead to complications in all of the four high-level scenarios introduced in Table 1 that illustrate different degrees of involvement of human choosers:

1. *No involvement:* Even if a recommender system can predict with great accuracy which options each individual group member would like, it is in general not straightforward to predict which option would be most satisfactory to the group as a whole—or even what that concept means (cf. Sect. 4.3).
2. *Quick selection:* Even if a set of recommendations to the group includes obviously satisfactory options for them, acceptance of one or more options by the group requires either (a) some sort of communication among the members or (b) some prior agreement that enables the final choice to be made without such communication [54, 20.5].
3. *Deliberate comparison:* Though each group member could in principle engage in a thorough comparison of options by himself, by doing so he would not be taking into account the interests and knowledge of other group members. Hence it is often worthwhile for members to communicate about the options.
4. *Iterative requesting of recommendations:* This scenario tends to require even more elaborate communication. As was the case with individual choosers, the function of recommendations is to move the decision process forward effectively, not necessarily to propose the options that are most likely to be satisfactory as final choices (Sect. 2.6).

## 4.3    What Constitutes a Good Choice for a Group?

The four criteria listed in Table 2 for judging whether a choice is good involve some additional considerations when groups are involved:

1. *Good outcomes:* Determining what counts as a desirable outcome for a group is complicated by the fact that members can have different interests and beliefs,

which in turn can lead to different assessments of the desirability of any outcome (see the section on *aggregation strategies* in Chapter "Group Recommender Systems: Beyond Preference Aggregation").

2. *Limited costs:* The time devoted to group decisions is typically greater than that for individual decisions, partly because of the need for communication and agreement among group members. So even minor deficiencies in decision procedures can result in significant costs—as participants in traditional decision making meetings within organizations can confirm from everyday experience (cf. [35, sect. 12-1c]).

3. *Avoidance of negative affect:* Groups provide additional ways in which unpleasant thoughts and feelings can arise, partly because of conflicts among interests, beliefs, and communication styles of different group members (cf. Sects. 5.4 and 6.7). In addition to being undesirable in the short term, these affective responses can damage the relationships among the group members, limiting their ability to collaborate again in the future (see Sect. 5.2).

4. *Justifiability:* In a group setting, a justification for a decision can be important not only for ensuring the cooperation of other parties but also for ensuring the commitment of any group members who may be less satisfied with the decision than others. Even more than in the case of individual decision making, a justification will often refer to the appropriateness and fairness of the *procedures* by which the decision was arrived at (cf. Sect. 5.5).

## *4.4 Principles from Research and Practice*

To encourage a focus on design implications for group recommender systems, Table 5 summarizes some of the most relevant results from the areas listed in Sect. 4.1 in terms of seven "principles" for supporting group decision making. While introducing these principles, we will refer occasionally to the following scenario: A group of friends is planning to go on a camping trip together. For this purpose, they will jointly select and purchase a tent on-line.

1. *Support knowledge exchange:* One potential advantage of groups over individual decision makers is that the group members often bring to the table complementary experience and expertise [100, p. 592 ff.], [35, sect. 12.2b], [14, chap. 3]. In our tent choice example, some group members may know little about camping while others have experience with different types of tents and camping conditions. Ideally, all significant relevant knowledge and beliefs would be communicated, so that all members are well-informed and they understand each other's points of view. Unfortunately, research has repeatedly shown that group members typically do not exchange knowledge effectively enough to approach this ideal. Hence designers may want to consider how to encourage and support the exchange of relevant knowledge. Specific ideas will be discussed

**Table 5** Seven widely accepted principles for the support of group decision making

| Principle | Explanation |
|---|---|
| 1. Support knowledge exchange | Support the exchange of relevant knowledge among group members |
| 2. Direct attention to underlying interests | Help group members understand the interests underlying each other's evaluations of specific options |
| 3. Use shared representations | Employ shared representations of the group's view of the decision problem |
| 4. Aim for win–win solutions | Help the group discover (possibly novel) options that satisfy all relevant interests |
| 5. Achieve agreement on decision procedures | Ensure that group members agree that the decision procedures to be applied are appropriate |
| 6. Minimize interpersonal problems | Aim to prevent interaction among group members from becoming frustrating or otherwise unsatisfying |
| 7. Take into account possible nonparticipation | Take into account the possibility that some members may prefer not to go along with the final decision |

in connection with the support strategy *Access Information and Experience* (Sect. 6.3).

2. *Direct attention to underlying interests:* Whereas group members often communicate about specific options (or *positions*) that they like or dislike, it has often been found worthwhile to encourage communication about their more general underlying *interests* [67, chap. 3, pp. 67 ff.], [34, chap. 3]. For example, if one member of the tent-purchasing group, Bob, initially insists that the tent should have two doors while the other members believe that one door is adequate, the group can benefit by understanding the reasons for Bob's interest in two doors.

3. *Use shared representations:* In attempts to apply the first two principles, it has often been found that maintaining some sort of shared representation of the group's view of the decision problem is helpful [25, pp. 20 ff.], [2, pp. 6 ff.], [46, chaps. 7 and 9]. In addition to reminding group members of the facts and interests that have been communicated, the shared representation encourages through its very existence a collaborative attempt to reach agreement.

4. *Aim for win-win solutions:* When conflicts of interest are recognized, it is tempting to try to figure out which of the known options constitutes a "fair" solution in the sense that all group members are roughly equally (dis)satisfied. A basic premise of integrative negotiation is that one should first try to identify some additional option which in the ideal case would make everyone happy [67, chap. 3, pp. 70 ff.], [34, chap. 4]. To return to the tent-door example: If it turns out that Bob wants to be able to leave the tent at any time during the night and believes that it will be easier to do so if there are two doors, someone may suggest a new option: The group buys a one-door tent and agrees that Bob will always

be allowed to sleep next to the door. As this example illustrates, applying this principle often requires expanding the initial conception of the choice problem (Sect. 6.6).

5. *Achieve agreement on decision procedures:* There are in principle a great many procedures that could be applied in any given group decision making process, such as: different forms of communication, different procedures for generating options, and different rules for arriving at a final decision if consensus is not achieved spontaneously. When group members believe in advance that a given set of procedures is fair and appropriate, a group member may accept the final solution even if it is quite different from the one that she would have preferred herself. Hence achieving agreement on decision procedures is one way of dealing with the conflicts that often exist among group members' interests [46, chap. 4], [25, pp. 7–8], [34, chap 5].

6. *Minimize interpersonal problems:* When members of a group communicate, there are various ways in which the interaction can become frustrating or otherwise dissatisfying to at least some members [13], [67, pp. 157 ff.], [46, chap. 2]. These undesirable phenomena can often be avoided through appropriate support for the interaction among group members, comparable to that provided by group decision facilitators, skilled negotiators, and group decision support systems (see the section below on the ARCADES strategy *Support Communication*, Sect. 6.7).

7. *Take into account possible nonparticipation:* In many settings, a group member has the option of removing himself from the decision process [34, chap. 6], [67, chap. 4]. An employee who is having difficulty negotiating a higher salary with her current employer may instead look for another job—or at least let her employer know that she is considering doing so. A member of our example camping group might in effect say "If you would all like to camp in the mountains, feel free to go without me; I don't like camping in cold temperatures, and I have another good vacation option". The relevant concept in the negotiation literature is that of the "best alternative to a negotiated agreement", or *BATNA*. Many ideas have been proposed about how to communicate about BATNAs and how to take them into account during decision making. Some implications for group recommender systems are discussed in connection with the strategy *Design the Domain* (Sect. 6.6).

## 4.5 High-Level Approaches for Group Recommender Systems

In research on group recommender systems that has taken into account some of the knowledge from the areas listed in Sect. 4.1, two broad approaches have emerged.

### 4.5.1    Approach 1: Support Interaction

In this approach, forms of user interaction are designed and recommendation algorithms are adapted with a view to encouraging interaction among group members in accordance with principles like those listed in Table 5. The next two major sections of this chapter will include many examples of this approach, which is applicable mainly in Scenarios 3 and 4 of Table 1.

### 4.5.2    Approach 2: Predict the Results of Interaction

In the pure form of this approach, a recommendation algorithm aims not to *support* particular types of interaction but rather to *predict* their ultimate effects. This approach is most clearly attractive in Scenarios 1 and 2 of Table 1, where at most minimal actual interaction among group members is expected. The goal is essentially to base the recommendation of options on predictions of the choices that the group *would* arrive at if the members did interact, taking into account general knowledge and/or collected data about typical phenomena of group dynamics or group decision making.[10]

One variant of this approach has been explained by Tkalčič et al. [105, Section 9.2]. Referring to Quijano-Sanchez et al. [90] and other works, these authors show how a computational method for predicting the choice of a group can take into account not only the (estimated) evaluations of options by individual group members but also:

- variables such as the size of the group, the diversity of evaluations within the group, and even aspects of the personality of the group members (e.g., their assertiveness; cf. Chapter "Personality and Recommender Systems"); and
- predictions of the impact that these variables would have on other variables if interaction among group members were to occur (e.g., a prediction that each group member's final evaluation of an option will have become more similar to the evaluations of other members).

This approach is in some respects more challenging than Approach 1. Here are some considerations that need to be taken into account:

- Some of the processes that have been (or could be) modeled in this way (e.g., the tendency of some types of group member to have an exaggerated influence on the decision outcome; cf. Sect. 5.3) are considered to be potentially undesirable ones that a skilled facilitator or negotiator would look out for and possibly counteract. If such phenomena are modeled in the recommendation algorithm,

---

[10] Approach 2 can in principle be applied in Scenarios 3 and 4 in support of Approach 1: A group recommender system might be able to support interaction more effectively if it is able to predict what the results of the interaction would be without its support.

an opportunity may be missed to recommend options that are more suitable than the ones that the group members would have arrived at on their own.

- Even when an interaction phenomenon is desirable, successfully taking it into account in the generation of recommendations may not yield the same benefits as actually allowing it to happen. For example, when the expert in a five-person group manages to convince the four nonexperts of the merits of the option favored by the expert, this influence of the expert may well be desirable; but if a recommender system simply recommends the expert's preferred option, some benefits of this interaction will be lost: The nonexperts may not understand as fully why the option in question is the best one, and they will not have been given the opportunity to acknowledge the validity of the expert's arguments and to demonstrate that they fully support his conclusions, thereby creating a feeling of consensus and commitment to the chosen option.
- It is difficult, in most settings, to predict the results of interaction within a specific group quantitatively. Whereas empirical studies in psychology and related fields have demonstrated a great many statistically reliable influences of particular factors on other variables (see, e.g., [35, chaps. 7 and 17], [14, chap. 4]), it is rare that researchers claim to be able to predict particular values of those variables with reasonable accuracy in specific cases.

Moreover, the relevant computational formulas included in recommendation algorithms (see, e.g., [112, sect. 4.1], [114, p. 15]) are necessarily simplified relative to the complexity and variability of the phenomena being predicted.[11]

In cases where a predictive model of some decision-relevant results of interaction has been learned from data collected from groups in a particular application scenario, the model is sometimes found to predict group choices better than baseline methods in that particular scenario (see, e.g., [39, 114]). But unless the target scenarios for real-world applications are very similar to the scenarios in which the model has been learned, it is hard to be confident that the model will be successful in the target scenarios, which are likely to differ in terms of important factors that have been shown to moderate the effects being predicted.

# 5   Choice Patterns and Recommendation to Groups

Though the six ASPECT choice patterns  were originally formulated to account for choices made by individuals, they can help us analyze the complexities of group choices on two levels: (a) The choice made by a group is ultimately the result of more specific choices made by individual group members (e.g., whether to suggest

---

[11] Some computational methods have been presented with reference to example computations but to our knowledge not yet deployed in recommender systems (see, e.g., [78]).

a particular new option), who may apply any ASPECT choice patterns.[12] (b) It sometimes makes sense to characterize an entire group as applying a particular choice pattern, for example when the group discusses and evaluates options in terms of their attributes.

## 5.1 Attribute-Based Choice

Viewing options in terms of their attributes (cf. Sect. 2.1) has a special benefit for groups: It is often possible to represent the differing interests of the group members conveniently in terms of (a) the different importances that they assign to particular attributes and (b) the different levels of each attribute that they consider desirable (e.g., "How many doors should our tent have?"). Group recommender systems that have employed such representations are described by, among others, Ninaus et al. [84, see especially Figure 1] and Samer et al. [95].

A *shared representation* (cf. Principle 3 in Table 5) of this sort can help group members to adapt their evaluation criteria even before they (or a recommendation algorithm) start to search for options that fulfill their criteria (cf. [2]'s characterization of a shared representation as a *transitional object* and [51, 74]). In our tent example, agreeing that no more than one door is needed should make it easier for a recommendation engine to generate satisfactory options that fulfill the agreed-upon criteria. Further discussion of shared representations, including their limitations, will be found in connection with the ARCADES strategy *Represent the Choice Situation* (Sect. 6.4).

## 5.2 Consequence-Based Choice

As is the case with individual decision making (Sect. 2.2), consequence-based choice is an especially knowledge-intensive pattern that is sometimes supported in groups via techniques of decision analysis and decision support (see, e.g., [94], and [2, especially Figure 1]) that are more elaborate than those commonly found in group recommender systems. A potential advantage of group decision making is that each member may be able to contribute consequence-related knowledge that is not available to other members (e.g., about a little-known type of risk); see Sect. 6.3.

A type of consequence that is unique to groups concerns situations in which what happens during the *execution* of a chosen option depends on the behavior and

---

[12] With regard to a choice made by an individual within a group, the contributions of the other group members can largely be viewed as part of the *context* for the choice; hence ideas from the area of context-aware recommendation (Chapter "Context-Aware Recommender Systems: From Foundations to Recent Developments") have some relevance.

interactions of group members during the execution. For example, Masthoff and colleagues (see, e.g., [78, sect. 7] and the section in Chapter "Group Recommender Systems: Beyond Preference Aggregation" about the modeling of affective states) have addressed the fact that whether one group member enjoys a movie selected by the group can depend on whether other members are visibly enjoying it. Predictive methods such as those discussed in connection with the approach "Predict the Results of Interaction" (Sect. 4.5.2) can also be applied to the prediction of consequences of this type. But it is important to distinguish between (a) predicting social influence phenomena that occur during execution of an option, which are part of the consequences of choosing that option; and (b) predicting social influence phenomena that occur during the group decision making process itself (see Sect. 5.3), which is what was discussed in connection with the approach "Predict the Results of Interaction".

Though these phenomena are distinct, the behavior of group members during the execution of an option can be strongly influenced by what happened during the preceding decision making process. For example, if our camping group has used a voting procedure to select a tent that one member strongly objected to, it is possible that the minority member will behave less cooperatively during the actual camping trip when it comes to sharing the space within the tent. Several of the principles in Table 5 aim in part to ensure that all group members will be motivated to execute the selected option in a cooperative way (see also [27] for a broader discussion of "negotiating as if implementation mattered").

## 5.3   Socially Based Choice

Although the socially based pattern can be applied by the entire group (e.g., when a group imitates a decision made by another group), it is at least as interesting to consider how social influence operates *within* a group and affects the decision-relevant actions of individual members. After all, if individual group members adapt their expressed evaluations, their communication of knowledge, and other contributions to the decision making process on the basis of the (expected) behavior and evaluations of other group members, the group decision that ultimately results may depend strongly on these adaptations. These are some of the phenomena that are taken into account within the group recommendation approach "Predict the Consequences of Interaction" (Sect. 4.5.2).

The topic of *conformity* illustrates the complexity of the relevant phenomena. A wide variety of factors have been identified that can influence the extent to which a person will adapt her beliefs, choices, or actions to the examples and expectations of other group members. These factors include (see, e.g., [35, chap. 7], [14, chap. 3]): the extent to which members are aware of each other's identities, the extent to which members like each other, the cohesiveness of the group, the relative size of the subgroup that holds a majority position, and the directness of the interaction among group members.

One general method for avoiding undesired conformity (and other forms of undesired social influence) that is often used in group decision support systems is to ensure the anonymity of contributions in various contexts, such as brainstorming and critiquing of ideas. It has been argued (see, e.g., [2, pp. 14–15]) that in some cases anonymity reduces pressure to conform, partly by leaving it unclear who the nonconforming group member is. But anonymity is not always helpful; for example, sometimes it is useful to know the identity of the source of a piece of advice or information, so as to be able to take that person's credibility into account (cf. Sect. 3.3).

A popular theory that specifically concerns group decision making is Janis's [56] theory of *groupthink*, which describes a situation in which the participants in the group decision making process conform excessively to each other's beliefs, expectations and advice. Though the theory has received mixed support from research, many analysts find it helpful when considering particular contexts. Ideas about how to prevent groupthink from degrading a group's decision success are discussed concisely by Forsyth [35, sect. 12.3f] and Brown and Pehrson [14, chap. 5]. These ideas overlap in part with ideas discussed later in this chapter in connection with the strategies *Access Information and Experience* (Sect. 6.3) and *Support Communication* (Sect. 6.7).

## 5.4   Experience-Based Choice

Of the four subpatterns of experienced-based choice discussed briefly in Sect. 2.4, only the first one, recognition-primed decision making, is straightforwardly applicable to an entire group: If the members of the group remember how they successfully handled a similar choice in the past, they may be inclined to choose essentially the same option on the current occasion. The use of information systems to maintain a *corporate memory* for this purpose has long been commonplace in groups within organizations—understandably, since membership in groups can change over time.[13]

The other three subpatterns of the experience-based pattern, which typically lead to quick choices made with little or no deliberation or awareness, are important mainly in connection with the small choices that individual group members make while participating in a group decision making process. Here are two examples of unfortunate choices of this sort:

- *A habit-based choice:* In accordance with a habit acquired in everyday interaction, a group member says something that unintentionally evokes negative affect

---

[13] Even this simple form of group choice can be made more complex by differences among the interests of group members: If the previously chosen solution was more desirable for some group members than for others, the group may choose a different solution on the current occasion in order to balance the group members' satisfaction over time.

in another group member (e.g., "How could you possibly propose this option?"), even though alternative formulations are available (e.g., "I'd be interested in hearing your reasons for proposing this option").

- *An affect-based choice:* A group member comes to feel that his interests and proposals are not being taken seriously enough. In accordance with the negative affect that she is experiencing, she participates in a generally less constructive way.

Some of the methods applied by skilled negotiators [13], [67, chap. 7] and group facilitators (see, e.g., [46, chap. 6], [37]) can be seen as efforts to replace suboptimal experience-based choice by individual members with behaviors that are better suited to an effective group decision making process.

In many cases where phenomena of this sort are discussed, the problem can be seen as being caused by suboptimal communication among group members. Accordingly, some remedial measures fall under the strategy *Support Communication*; see the examples discussed in the section on that strategy below (Sect. 6.7).

But bad experience-based choices can also be triggered by other events in a group decision making process, such as the proposal of an option that seems unfair to some group members. For this reason, any sort of improvement to the procedures used in group decision making may have at least a side effect of lessening the likelihood of unfortunate experience-based choices.

## 5.5    Policy-Based Choice

The policy-based choice pattern figures more prominently in group decision making than it does in individual decision making, in terms of both attention in published research and visibility in actual decision making processes. In particular, with decisions that are repeatedly made within organizations (see, e.g., March [75, chap. 2]), it is natural for an organization to develop policies that (a) streamline the decision making process; (b) codify the extensive experience that the organization may have with this type of choice; and (c) take into account organizational constraints that may not be clearly known to individual participants.

A type of policy that is especially relevant to group recommender systems concerns procedures for organizing the decision making process. Some procedures are important for ensuring effective and unbiased communication (e.g., who can contribute ideas, at what points in time, and in what manner?), while others aim to ensure that the actual choices made are fair and appropriate (e.g., should the final choice of an option be made via a majority vote, or should the group aim to achieve unanimity?).

In connection with traditional group decision making meetings, it is widely acknowledged that it is desirable for group members to agree at an early stage on the most important policies of these sorts [35, sect. 12-1a], [46, chap. 2]. A group recommender system can deal with such procedures in a variety of ways, including the following, in ascending order of sophistication:

1. Simply build particular procedures into the recommender system without any explanation or justification, trusting that the group members will understand the procedures adequately as they encounter them.
2. Explain and justify the procedures in advance, so that group members feel committed to them from the start.
3. Offer alternative procedures in advance (e.g., several possible voting rules for making the final decision) and give group members a chance to choose the policies that they want to apply.
4. When offering alternative procedures, use recommendation technology to help the group members choose the most appropriate one in each case (e.g., on the basis of data about what final decision rules have led to the most satisfying decisions under various circumstances in the past).

Given that the first approach listed is the most common one for group recommender systems, there appears to be room for improvement if the other approaches are considered as well. The approaches 3 and 4 fall under the ARCADES strategy *Advise About Processing*; some further comments are made in the section on that strategy below (Sect. 6.2).

## 5.6   Trial-and-Error-Based Choice

In the case of groups that communicate in accordance with the highly interactive Scenario 4 in Table 1, the trial-and-error-based choice pattern can take on a form not found with individual choosers (Sect. 2.6): Whenever one or more new options are proposed for the group's consideration, the following types of feedback can come from the group members themselves: (a) relevant knowledge that they possess, either about the specific option in question or about more general aspects of the objects of the domain; and (b) responses that provide insight into their underlying interests. Given the widely acknowledged importance of these two types of information (cf. Principles 1 and 2 in Table 5), it can be worthwhile for a group recommender system to recommend options that are likely to elicit this type of feedback and thereby enable the trial-and-error cycle to converge more quickly.

In a study of how to stimulate knowledge sharing in decision making groups, Atas et al. [6] found that recommending items that are especially *inconsistent* with the group members' initial evaluations yielded a much larger number of comments from group members.

The recommender system that these researchers studied did not follow the critiquing-based paradigm discussed in connection with trial and error for individual choosers (Sect. 2.6); but some group recommender systems have applied adapted variants of this paradigm, including CATS [79] and WHERE2EAT [44]. These systems integrate critiquing in novel ways into the natural process of discussion among group members. For example, in WHERE2EAT, when an individual group member critiques a particular option, the recommender system generates similar options based on that critique, which the group member can in turn offer to the other members as counterproposals that they can in turn critique, and so on.

The more recent chat-based group recommender system for tourists STSGROUP [82, 83] supports group discussions by occasionally offering recommendations based on evaluations expressed by group members, though it does not elicit explicit critiques.

The variety of forms of interaction exhibited by these group recommender systems and their acceptance in user studies suggest that this type of trial and error is a natural and viable paradigm for group recommender systems. Further exploration along these lines can continue to consider typical questions associated with the trial-and-error-based choice pattern, such as: (a) What kinds of option proposals are well suited to yield useful feedback from group members? and (b) How can this feedback be used (by the recommender system and/or by other group members) to generate further proposals that are likely to help the group converge on a mutually satisfactory solution?

## 5.7   Combinations of Choice Patterns

The application of more than one choice pattern, alternately or in parallel, can occur even more readily within a group of choosers than with an individual chooser (Sect. 2.7), simply because different group members can be applying different choice patterns at any given moment. One group member may be thinking primarily about the company's relevant policies while another member is trying to anticipate the concrete consequences of particular options.

Designers of group recommender systems should avoid assuming that all group members will find it natural and appropriate to think about a decision in terms of one particular choice pattern presupposed by the recommender system. Though it may be impractical on the algorithmic level to do justice to all of the choice patterns that might come into play, at least the supplementary communication that occurs—such as free-form verbal communication—should be allowed to be expressive enough to accommodate all potentially applicable choice patterns (cf. Sect. 6.4 below).

## 6   Choice Support Strategies and Recommendation to Groups

The ARCADES strategies  can be applied in an even greater variety of ways with groups than with individuals, in support of the choices of individual group members and those of the group as a whole.

## 6.1   Evaluate on Behalf of the Chooser

The general challenges associated with applying this core support strategy for recommender systems were discussed in connection with the two broad approaches to group recommendation (Sect. 4.5).

## 6.2   Advise About Processing

As was noted in connection with policy-based choice for groups (Sect. 5.5), a group recommender system could in principle follow the example of group facilitators by occasionally recommending a procedure that seems appropriate for a particular part of the decision making process, a minimal precondition being that the recommender system should be able to support more than one procedure for the situation in question. Some group recommender systems fulfill this precondition by making alternative procedures available via a configuration screen (see, e.g., [51, 101]).

Actually recommending a particular procedure to group members in a context-sensitive way requires a further step that would have to be based on some sort of knowledge or data about which procedures work well in which situations. A great deal of relevant information can be found in the research literature summarized in Sect. 4.1, and studies like those of Delić et al. [22–24] conducted in especially relevant settings can provide more targeted guidance.

## 6.3   Access Information and Experience

As was noted in connection with Principle 1 in Table 5, one generally acknowledged benefit of group versus individual decision making is that members of a group can bring to bear complementary decision-relevant knowledge that can be "accessed" during decision making. Research since the mid-1980s has repeatedly demonstrated that group members frequently do not share relevant information to an extent which would be helpful for making a satisfactory final decision (see, e.g., [100, pp. 592 ff.] for a recent review). In particular, a tendency is often observed for information that multiple group members already possess to be discussed and for information possessed by only one group member to be held back.

Various factors have been found to influence the extent of useful knowledge sharing, including: the exact way in which the relevant knowledge is distributed across members; how much each member knows about the relevant expertise of other members and the particular types of knowledge that they possess; the concepts that the group members use to organize relevant information; the extent of the members' motivation to participate in a collaborative way and of their conviction that sharing information is an important way to do so; the extent to which group members are held accountable for the quality of the resulting decision; and even the personality traits and affective states of group members.

Because of the large number of relevant factors, there is no simple guideline that will ensure successful information sharing, but it should help if designers look for ways to encourage and support knowledge sharing that work in their particular application scenarios. As was noted in connection with the trial-and-error-based pattern (Sect. 5.6), some group recommender systems include ways of stimulating the exchange of knowledge and opinions that appear to fit well into interactive group recommendation scenarios.

A different way of applying recommendation technology to the knowledge sharing problem would be for a group recommender system to actively encourage specific group members to contribute particular types of knowledge at relevant times. As with individual choice (Sect. 2.3), techniques for *people recommendation* (Chapter "Social Recommender Systems") are relevant here; but in the case of groups the goal may be to identify a knowledgeable member of the group itself, not just anyone who might be consulted.

Another relevant type of technique discussed in Chapter "Social Recommender Systems" concerns the recommendation of content that is typical of social media, such as multimedia items (see also Chapter "Multimedia Recommender Systems: Algorithms and Challenges"), discussion contributions, and answers to questions. If there has been extensive communication within a group, a group recommender system could at appropriate times remind the group of relevant previous contributions by group members.

A further general way of applying the strategy *Access Information and Experience*, as in connection with individual choice (Sect. 3.3), is to supply explanations of recommendations (Chapter "Beyond Explaining Single Item Recommendations"). Groups introduce additional requirements for explanations (cf. [32] and [54, sect. 20.4]). In particular, one thing that group members often want to know about a recommended option is how well it addresses the interests of the various group members (cf. Sect. 4.3 above). A group member may be interested not only in how well his own interests are satisfied but also, for example, whether another group member is getting a "better deal" or perhaps some third member is being treated unfairly. Jameson and Smyth [54, sect. 20.4] provide concrete examples of how even early group recommender systems used ingenious means to convey this type of information, ranging from simply displaying the predicted or actual ratings of solutions by different group members to elaborately visualizing the ways in which a given option satisfies or fails to satisfy specific interests of particular group members.

## 6.4 Represent the Choice Situation

The discussion of shared representations within the attribute-based pattern (Sect. 5.1) illustrated why the strategy *Represent the Choice Situation* is even more important with groups than with individuals. A different type of representation, used in some group decision support systems for the consequence-based pattern, is a graphical representation of relevant causal relationships (see, e.g., [2, pp. 6 ff.]).

A general challenge here is due to the fact that different group members may want to apply different (sets of) choice patterns at any time (Sect. 5.7). One ambitious approach would be to provide a shared representation that could effectively represent and organize ideas from multiple choice patterns. A relatively simple variant of this approach is found in group recommender systems (e.g., STSGROUP, [82, 83]) that enable members to comment on proposed options verbally using any concepts that they like. One way of adding more structure to this type of representation would be to provide for tagging of the comments in terms that reflect the underlying choice pattern (e.g., "possible negative consequence", "relevant previous choice made by this group", "relevant company policy").

## 6.5 Combine and Compute

A type of computation that is especially characteristic of group decision making is the use of what are sometimes called *aggregation strategies*: Given a set of evaluations of an option by individual group members, a function computes a single index intended to reflect the evaluation of the group as a whole (e.g., the average, minimum, or maximum of the individual evaluations or some more complex function). There has been extensive discussion in the group recommender systems literature of the situations in which particular aggregation strategies are likely to be appropriate (see, e.g., Chapter "Group Recommender Systems: Beyond Preference Aggregation", [28, sect. 2.2], and [54, sect. 20.3]).

Computations based on such strategies often form a part of recommendation algorithms, in which case they help realize the strategy *Evaluate on Behalf of the Chooser*. But in Scenarios 3 and 4 in Table 1, aggregation strategies can be applied in a more interactive way: After a number of options have been considered and rated by all group members, a list of these options sorted in terms of the average evaluation by individual members (or some other aggregation function) can give members a rough idea of which options deserve the most attention and which should be considered for elimination (cf. the discussion of *preference gradient voting* in [46, chap. 5] and the much more elaborate procedure described in [94, chap. 7]).

## 6.6 Design the Domain

Two ways of redefining a choice problem in order to make it easier to make a good choice are especially relevant to groups:

1. *Expanding the option space:* As was noted in connection with Principle 4 in Table 5, a common recommendation in the literature on integrative negotiation is to expand the space of options that can be considered so as to increase the likelihood that an option can be found that satisfies the interests of all parties reasonably well. A classic example is expanding the scope of a salary negotiation to include other benefits such as flexible working time. Our example about agreeing on the number of doors for a tent illustrates this tactic: The space of options was expanded from the space of "tents" to the cross-product of "tents" and "agreements about how to use tents". Masthoff and colleagues ([77] and more briefly Chapter "Group Recommender Systems: Beyond Preference Aggregation") have discussed another type of application of this principle: arranging for members of a group to choose a *sequence* of items to consume (e.g., TV shows) as opposed to choosing each item separately (cf. the discussion of *sequential recommendation* in Chapter "Music Recommendation Systems: Techniques, Use Cases, and Challenges"). In the larger option space of item sequences, it can be easier to find a solution that is considered equitable and satisfactory to all group members.

2. *Allowing changes to the composition of the group:* As was noted in Principle 7 in Table 5, one or more group members may prefer not to participate in a given solution on the grounds that she has a more attractive group-independent alternative (BATNA).

Herzog and Wörndl [48] recently explored a way of taking this idea into account with a tourist group recommender system that plans walking tours through a city: The recommender system sometimes suggests that one group member should "split away" from the group during part of a city walk to visit a location of special interest to him. The authors' user study showed that some people appreciated this practice while others "completely rejected the idea" (p. 99). This result suggests that recommending a solution for only part of a group is a method that should be applied with care (e.g., only with prior consent of all group members).

Where the method is considered to be acceptable, a relevant recommendation paradigm is the *double-sided recommendation*[14] of Lombardi and Vernero [73], in which not only an option is recommended but also a group of persons who are expected to like the option.

---

[14] This concept differs from *people-to-people reciprocal recommendation* (Chapter "People-to-People Reciprocal Recommenders"), in which two or more persons are recommended to each other.

## 6.7   Support Communication

The relevance of the strategy *Support Communication* to group decision making is especially obvious, because of the importance of communication among group members in Scenarios 3 and 4 of Table 1. The relevant research and practice have yielded many examples of ways in which the use of an appropriate form of communication can lead to a more satisfactory group decision process, including controlling the anonymity of contributions (already discussed in Sect. 5.3) and the following measures:

1. When explicit evaluations of options (e.g., ratings, free-text comments) are elicited from group members, communication methods can be chosen that encourage each group member to provide a sincere evaluation, as opposed to a distorted one that is intended to achieve exaggerated influence on the group's decision, as when you rate your number-two option as "terrible" so as to increase the likelihood that your number-one option will be chosen by the group (see [51, sect. 3] for an early discussion and [67, chap. 17, sect. 5] for a more general discussion in the context of negotiation). Tran et al. [106] have recently explored interface design solutions that yield promising results in terms of encouraging sincere evaluations.
2. During brainstorming sessions, there is a well-known rule that new proposals should not be criticized immediately, so that members are not discouraged from generating a large number of options (see, e.g., [35, sect. 10-4]).
3. A method for taking turns while making contributions can be introduced which is perceived as fair and appropriate and which therefore reduces the likelihood of participants becoming frustrated because of not having a chance to contribute [46, chap. 6].
4. Instead of simply allowing unrestricted textual communication, the user interface can make available to group members communication elements that tend to encourage contributions that are helpful rather than disruptive. These can range from carefully designed rating scales, emoticons, and bits of prefabricated text to artificial agents that make use of natural language processing technology (Chapter "Natural Language Processing for Recommender Systems") and/or animated characters (see, e.g., [52]).

## 7   Recapitulation and Concluding Remarks

We hope that this chapter has convinced readers of the following general points:

- There is a greater amount of relevant knowledge about individual and group decision making than the concepts and results that so far have been taken into account in recommender systems research and practice.

- Though the relevant research in psychology and other fields is vast and multi-faceted, it is possible to acquire a grasp of it with the help of comprehensive yet compact and accessible models such as the ASPECT and ARCADES models.
- This knowledge and understanding makes it possible to discover numerous new ways of making recommender systems more effective for the support of human choices, ranging from small improvements of specific aspects of interaction to the application of recommendation technology in novel ways.

Attentive readers may have noticed that many of the issues presented as being typical of group decision making also arise in connection with individual decision making, though in more subtle ways. This fact should not be surprising, given that even a single decision maker can have conflicting interests and apply different choice patterns. Looking for cases like these would be an interesting way to review the material in this chapter; some ideas of this sort are offered near the end of Chapter "Group Recommender Systems: Beyond Preference Aggregation".

When putting the information provided in this chapter to use in their own research or practice, readers will need to consider which of the many issues discussed here are most likely to be relevant to their particular problems, and they will want to consult the cited literature concerning these issues. Some ideas will prove in practice to be more helpful than others in solving the problem being addressed, and readers will need to exploit their own experience and creativity to make effective use of them.

Though there are faster ways of making some types of progress with recommender systems, the field can hardly continue indefinitely with only a partial understanding of the individual and group decision making that it aims to support.

# References

1. H. Abdollahpouri, G. Adomavicius, R. Burke, I. Guy, D. Jannach, T. Kamishima, J. Krasnodebski, L. Pizzato, Multistakeholder recommendation: survey and research directions. User Model. User-Adapt. Interact. **30**(1), 127–158 (2020)
2. F. Ackermann, C. Eden, Group support systems: concepts to practice, in *Handbook of Group Decision and Negotiation*, ed. by D.M. Kilgour, C. Eden (Springer Nature Switzerland AG, Cham, 2020)
3. G. Adomavicius, Y. Kwon, Multi-criteria recommender systems, in *Recommender Systems Handbook*, ed. by F. Ricci, L. Rokach, B. Shapira, 2nd edn. (Springer, Berlin, 2015)
4. X. Amatriain, J. Basilico, Past, present, and future of recommender systems: an industry perspective, in *Proceedings of RecSys 2016* (2016)
5. Q. André, Z. Carmon, K. Wertenbroch, A. Crum, D. Frank, W. Goldstein, J. Huber, L. v. Boven, B. Weber, H. Yang. Consumer choice and autonomy in the age of artificial intelligence and big data. Customer Needs Solut. **5**, 28–37 (2018)
6. M. Atas, A. Felfernig, M. Stettinger, T.N. Tran, Beyond item recommendation: using recommendations to stimulate knowledge sharing in group decisions, in *International Conference on Social Informatics* (2017)
7. P. Bekkerman, S. Kraus, F. Ricci, Applying cooperative negotiation methodology to group recommendation problem, in *Proceedings of the Workshop on Recommender Systems at the 17th European Conference on Artificial Intelligence* (2006)

8. S. Berkovsky, R. Taib, D. Conway, How to recommend? User trust factors in movie recommender systems, in *Proceedings of IUI 2017* (2017)
9. T. Betsch, S. Haberstroh (eds.), *The Routines of Decision Making* (Erlbaum, Mahwah, 2005)
10. J.R. Bettman, M.F. Luce, J.W. Payne, Constructive consumer choice processes. J. Consum. Res. **25**, 187–217 (1998)
11. S. Bhatia, Associations and the accumulation of preference. Psychol. Rev. **120**(3), 522–543 (2013)
12. S. Bonaccio, R.S. Dalal, Advice taking and decision-making: an integrative literature review, and implications for the organizational sciences. Organ. Behav. Hum. Decis. Processes **101**, 127–151 (2006)
13. A.W. Brooks, Emotion and the art of negotiation. Harv. Bus. Rev. **93**(12), 57–64 (2015)
14. R. Brown, S. Pehrson, *Group Processes: Dynamics Within and Between Groups*, 3rd edn. (Wiley Blackwell, Hoboken, 2020)
15. P. Brusilovsky, M. d. Gemmis, A. Felfernig, P. Lops, J. O'Donovan, G. Semeraro, M.C. Willemsen, Interfaces and human decision making for recommender systems, in *Proceedings of RecSys 2020* (2020), pp. 613–618
16. R. Burke, Hybrid recommender systems: survey and experiments. User Model. User-Adapt. Interact. **12**(4), 331–370 (2002)
17. R. Burke, Hybrid web recommender systems, in *The Adaptive Web: Methods and Strategies of Web Personalization*, ed. by P. Brusilovsky, A. Kobsa, W. Nejdl (Springer, Berlin, 2007), pp. 377–408
18. C.F. Camerer, L. Babcock, G. Loewenstein, R.H. Thaler, Labor supply of New York City cab drivers: one day at a time, in *Choices, Values, and Frames*, ed. by D. Kahneman, A. Tversky (Cambridge University Press, Cambridge, 2000)
19. R.B. Cialdini, *Influence: The Psychology of Persuasion* (HarperCollins, New York, 2007)
20. R.T. Clemen, *Making Hard Decisions: An Introduction to Decision Analysis* (Duxbury, Pacific Grove, 1996)
21. J.D. Cohen, S.M. McClure, A.J. Yu, Should I stay or should I go? How the human brain manages the trade-off between exploitation and exploration. Philos. Trans. R. Soc. **362**, 933–942 (2007)
22. A. Delić, J. Masthoff, J. Neidhardt, H. Werthner, How to use social relationships in group recommenders: empirical evidence, in *Proceedings of UMAP 2018* (2018)
23. A. Delić, J. Neidhardt, T.N. Nguyen, F. Ricci, An observational user study for group recommender systems in the tourism domain. Inf. Technol. Tour. **19**, 87–116 (2018)
24. A. Delić, J. Neidhardt, H. Werthner, Group decision making and group recommendations, in *Proceedings of the IEEE 20th Conference on Business Informatics* (2018)
25. C. Eden, Behavioral considerations in group support, in *Handbook of Group Decision and Negotiation*, ed. by D.M. Kilgour, C. Eden (Springer Nature Switzerland AG, Cham, 2020)
26. M.D. Ekstrand, M.C. Willemsen, Behaviorism is not enough, in *Proceedings of RecSys 2016* (2016)
27. D. Ertel, *Getting Past Yes: Negotiating as if Implementation Mattered*. Harvard Business Review (2004), pp. 29–39
28. A. Felfernig, M. Atas, D. Helic, T.N. Tran, M. Stettinger, R. Samer, Algorithms for group recommendation, in *Group Recommender Systems: An Introduction*, ed. by A. Felfernig, L. Boratto, M. Stettinger, M. Tkalčič (Springer International Publishing AG, Cham, 2018)
29. A. Felfernig, M. Atas, M. Stettinger, Decision tasks and basic algorithms, in *Group Recommender Systems: An Introduction*, ed. by A. Felfernig, L. Boratto, M. Stettinger, M. Tkalčič (Springer International Publishing AG, Cham, 2018)
30. A. Felfernig, L. Boratto, M. Stettinger, M. Tkalčič (eds.), *Group Recommender Systems: An Introduction* (Springer International Publishing AG, Cham, 2018)
31. A. Felfernig, G. Friedrich, D. Jannach, M. Zanker, Constraint-based recommender systems, in *Recommender Systems Handbook*, ed. by F. Ricci, L. Rokach, B. Shapira, 2nd edn. (Springer, Berlin, 2015)

32. A. Felfernig, N. Tintarev, T.N. Tran, M. Stettinger, Explanations for groups, in *Group Recommender Systems: An Introduction*, ed. by A. Felfernig, L. Boratto, M. Stettinger, M. Tkalčič (Springer International Publishing AG, Cham, 2018)
33. M. Fishbein, I. Ajzen, *Predicting and Changing Behavior: The Reasoned Action Approach* (Taylor & Francis, New York, 2010)
34. R. Fisher, W.L. Ury, B. Patton, *Getting to YES: Negotiating Agreement Without Giving In*, 3rd edn. (Penguin, New York, 2011)
35. D.R. Forsyth, *Group Dynamics*, 7th edn. (Cengage, Boston, 2019)
36. S. French, J. Maule, N. Papamichail, *Decision Behaviour, Analysis, and Support* (Cambridge University Press, Cambridge, 2009)
37. C. Freshley, *The Wisdom of Group Decisions: 100 Principles and Practical Tips for Collaboration* (Good Group Decisions, Brunswick, 2010)
38. A. Friedman, B. Knijnenburg, K. Vanhecke, L. Martens, S. Berkovsky, Privacy aspects of recommender systems, in *Recommender Systems Handbook*, ed. by F. Ricci, L. Rokach, B. Shapira, 2nd edn. (Springer, Berlin, 2015)
39. M. Gartrell, X. Xing, Q. Lv, A. Beach, R. Han, S. Mishra, K. Seada, Enhancing group recommendation by incorporating social relationship interactions, in *Proceedings of GROUP 2010* (2010)
40. G. Gigerenzer, *Gut Feelings: The Intelligence of the Unconscious* (Penguin, London, 2007)
41. G. Gigerenzer, P.M. Todd (eds.), *Simple Heuristics That Make Us Smart* (Oxford, New York, 1999)
42. M.A. Gluck, E. Mercado, C. Myers, *Learning and Memory: From Brain to Behavior*, 4th edn. (Worth, New York, 2019)
43. M.P. Graus, M.C. Willemsen, Improving the user experience during cold start through choice-based preference elicitation, in *Proceedings of RecSys 2015* (2015)
44. F. Guzzi, F. Ricci, R. Burke, Interactive multi-party critiquing for group recommendation, in *Proceedings of RecSys 2011* (2011)
45. J. Harambam, D. Bountouridis, M. Makhortykh, J.V. Hoboken, Designing for the better by taking users into account: a qualitative evaluation of user control mechanisms in (news) recommender systems, in *Proceedings of RecSys 2019* (2019)
46. T. Hartnett, *Consensus-Oriented Decision-Making The CODM Model for Facilitating Groups to Widespread Agreement* (New Society Publishers, Gabriola Island, 2010)
47. R. Hastie, Problems for judgment and decision making. Annu. Rev. Psychol. **52**, 653–683 (2001)
48. D. Herzog, W. Wörndl, User-centered evaluation of strategies for recommending sequences of points of interest to groups, in *Proceedings of RecSys 2019* (2019)
49. S. Hilgard, N. Rosenfeld, J. Cao, M. Banaji, D.C. Parkes, Learning representations by humans, for humans, in *Workshop on Human-Centric Machine Learning at NeurIPS* (2019)
50. C.K. Hsee, Attribute evaluability: its implications for joint-separate evaluation reversals and beyond, in *Choices, Values, and Frames*, ed. by D. Kahneman, A. Tversky (Cambridge University Press, Cambridge, 2000)
51. A. Jameson, More than the sum of its members: challenges for group recommender systems, in *Proceedings of the International Working Conference on Advanced Visual Interfaces, Gallipoli* (2004), pp. 48–54
52. A. Jameson, S. Baldes, T. Kleinbauer, Two methods for enhancing mutual awareness in a group recommender system, in *Proceedings of the International Working Conference on Advanced Visual Interfaces, Gallipoli* (2004), pp. 447–449
53. A. Jameson, B. Berendt, S. Gabrielli, C. Gena, F. Cena, F. Vernero, K. Reinecke, Choice architecture for human-computer interaction. Found. Trends in Hum. Comput. Interact. **7**(1–2), 1–235 (2014)
54. A. Jameson, B. Smyth, Recommendation to groups, in *The Adaptive Web: Methods and Strategies of Web Personalization*, ed. by P. Brusilovsky, A. Kobsa, W. Nejdl (Springer, Berlin, 2007), pp. 596–627

55. A. Jameson, M. Willemsen, A. Felfernig, M. de Gemmis, P. Lops, G. Semeraro, L. Chen, Human decision making and recommender systems, in *Recommender Systems Handbook*, ed. by F. Ricci, L. Rokach, B. Shapira, 2nd edn. (Springer, Berlin, 2015)
56. I. Janis, *Victims of Groupthink: A Psychological Study of Foreign-Policy Decisions and Fiascoes* (Houghton Mifflin, Boston, 1972)
57. D. Jannach, G. Adomavicius, Recommendations with a purpose, in *Proceedings of RecSys 2016* (2016)
58. Y. Jin, W. Cai, L. Chen, N.N. Htun, K. Verbert, MusicBot: evaluating critiquing-based music recommenders with conversational interaction, in *Proceedings of CIKM 2019* (2019)
59. M. Jugovac, D. Jannach, Interacting with recommenders—overview and research directions. ACM Trans. Interact. Intell. Syst. **7**(3), Article 10 (2017)
60. H. Jungermann, K. Fischer, Using expertise and experience for giving and taking advice, in *The Routines of Decision Making*, ed. by T. Betsch, S. Haberstroh (Erlbaum, Mahwah, 2005)
61. D. Kahneman, A. Tversky, Prospect theory: an analysis of decision under risk. Econometrica **47**(2), 263–295 (1979)
62. S. Kalloori, F. Ricci, R. Gennari, Eliciting pairwise preferences in recommender systems, in *Proceedings of RecSys 2018* (2018)
63. D.M. Kilgour, C. Eden (eds.), *Handbook of Group Decision and Negotiation* (Springer Nature Switzerland AG, Cham, 2020)
64. G. Klein, *Sources of Power: How People Make Decisions* (MIT Press, Cambridge, 1998)
65. B.P. Knijnenburg, N.J. Reijmer, M.C. Willemsen, Each to his own: how different users call for different interaction methods in recommender systems, in *Proceedings of RecSys 2011* (2011)
66. B.P. Knijnenburg, S. Sivakumar, D. Wilkinson, Recommender systems for self-actualization, in *Proceedings of RecSys 2016* (2016)
67. R.J. Lewicki, B. Barry, D.M. Saunders, *Essentials of Negotiation*, 6th edn. (McGraw-Hill Higher Education, New York, 2016)
68. Y. Liang, M.C. Willemsen, Personalized recommendations for music genre exploration, in *Proceedings of UMAP 2019* (2019)
69. D.A. Lieberman, *Human Learning and Memory* (Cambridge University Press, Cambridge, 2012)
70. C. Lin, X. Shen, S. Chen, M. Zhu, Y. Xiao, Non-compensatory psychological models for recommender systems, in *Proceedings of AAAI 2019* (2019)
71. C.E. Lindblom, Still muddling, not yet through. Public Adm. Rev. **39**(6), 517–526 (1979)
72. B. Loepp, T. Hussein, J. Ziegler, Choice-based preference elicitation for collaborative filtering recommender systems, in *Proceedings of CHI 2014* (2014)
73. I. Lombardi, F. Vernero, What and who with: a social approach to double-sided recommendation. Int. J. Hum. Comput. Stud. **101**, 62–75 (2017)
74. N. Mahyar, W. Liu, S. Xiao, J. Browne, M. Yang, S. Dow, ConsensUs: visualizing points of disagreement for multi-criteria collaborative decision making, in *Companion Proceedings of CSCW 2017* (2017)
75. J. March, *A Primer on Decision Making: How Decisions Happen* (The Free Press, New York, 1994)
76. J.O. Márquez, J. Ziegler, Preference elicitation and negotiation in a group recommender system, in *Proceedings of INTERACT 2015* (2015)
77. J. Masthoff, Group modeling: selecting a sequence of television items to suit a group of viewers, in *Personalized Digital Television*, vol. 6 (Springer, Dordrecht, 2004)
78. J. Masthoff, A. Gatt, In pursuit of satisfaction and the prevention of embarrassment: affective state in group recommender systems. User Model. User-Adapt. Interact. **16**(3–4), 281–319 (2006)
79. K. McCarthy, M. Salamó, L. Coyle, L. McGinty, B. Smyth, P. Nixon, CATS: a synchronous approach to collaborative group recommendation, in *Proceedings of FLAIRS 2006* (2006)

80. L. McGinty, J. Reilly, On the evolution of critiquing recommenders, in *Recommender Systems Handbook*, ed. by F. Ricci, L. Rokach, B. Shapira, P.B. Kantor (Springer, Berlin, 2011), pp. 419–453

81. B.R. Newell, D.A. Lagnado, D.R. Shanks, *Straight Choices: The Psychology of Decision Making*, 2nd edn. (Psychology Press, Hove, 2015)

82. T.N. Nguyen, F. Ricci, Dynamic elicitation of user preferences in a chat-based group recommender system, in *Proceedings of SAC 2017* (2017)

83. T.N. Nguyen, F. Ricci, A chat-based group recommender system for tourism. Inf. Technol. Tour. **18**, 5–28 (2018)

84. G. Ninaus, A. Felfernig, M. Stettinger, S. Reiterer, G. Leitner, L. Weninger, W. Schanil, IntelliReq: intelligent techniques for software requirements engineering, in *Prestigious Applications of Intelligent Systems Conference* (2014)

85. J.W. Payne, J.R. Bettman, E.J. Johnson, *The Adaptive Decision Maker* (Cambridge University Press, Cambridge, 1993)

86. J. Pfeiffer, *Interactive Decision Aids in E-Commerce* (Springer, Berlin, 2012)

87. P. Pirolli, *Information Foraging Theory: Adaptive Interaction with Information* (Oxford University Press, New York, 2007)

88. C. Plate, N. Basselin, A. Kröner, M. Schneider, S. Baldes, V. Dimitrova, A. Jameson, Recomindation: new functions for augmented memories, in *Proceedings of AH 2006* (2006), pp. 141–150

89. H. Plessner, C. Betsch, T. Betsch (eds.), *Intuition in Judgement and Decision Making* (Erlbaum, New York, 2008)

90. L. Quijano-Sanchez, J.A. Recio-Garcia, B. Diaz-Agudo, G. Jimenez-Diaz, Social factors in group recommender systems. Trans. Intell. Syst. Technol. **4**(1), Article 8 (2013)

91. H. Rachlin, *The Science of Self-Control* (Harvard, Cambridge, 2000)

92. T. Rakow, B.R. Newell, Degrees of uncertainty: an overview and framework for future research on experience-based choice. J. Behav. Decis. Mak. **23**, 1–14 (2010)

93. D. Read, G. Loewenstein, M. Rabin, Choice bracketing. J. Risk Uncertain. **19**, 171–197 (1999)

94. T. Saaty, K. Peniwati, *Group Decision Making: Drawing Out and Reconciling Differences* (RWS Publications, Pittsburgh, 2008)

95. R. Samer, M. Stettinger, A. Felfernig, Group recommender user interfaces for improving requirements prioritization, in *Proceedings of UMAP 2020* (2020)

96. J. Schaffer, J. O'Donovan, T. Höllerer, Easy to please: separating user experience from choice satisfaction, in *Proceedings of UMAP 2018* (2018)

97. P. Slovic, M. Finucane, E. Peters, D.G. MacGregor, The affect heuristic, in *Heuristics and Biases: The Psychology of Intuitive Judgment*, ed. by T. Gilovich, D. Griffin, D. Kahneman (Cambridge University Press, Cambridge, 2002)

98. B. Smyth, Case-based recommendation, in *The Adaptive Web: Methods and Strategies of Web Personalization*, ed. by P. Brusilovsky, A. Kobsa, W. Nejdl (Springer, Berlin, 2007), pp. 342–376

99. A.D. Starke, M.C. Willemsen, C. Snijders, With a little help from my peers: depicting social norms in a recommender interface to promote energy conservation, in *Proceedings of IUI 2020* (2020)

100. G. Stasser, S. Abele, Collective choice, collaboration, and communication. Annu. Rev. Psychol. **71**, 589–612 (2020)

101. M. Stettinger, Choicla: towards domain-independent decision support for groups of users, in *Proceedings of RecSys 2014* (2014)

102. S.C. Sutherland, C. Harteveld, M.E. Young, Effects of the advisor and environment on requesting and complying with automated advice. ACM Trans. Interact. Intell. Syst. **6**(4), Article 27 (2016)

103. T.T. Taijala, M.C. Willemsen, J.A. Konstan, MovieExplorer: building an interactive exploration tool from ratings and latent taste spaces, in *Proceedings of the 33rd Symposium on Applied Computing* (2018)

104. R.H. Thaler, C.R. Sunstein,  *Nudge: Improving Decisions About Health, Wealth, and Happiness* (Yale University Press, New Haven, 2008)
105. M. Tkalčič, A. Delić, A. Felfernig,  Personality, emotions, and group dynamics,  in *Group Recommender Systems: An Introduction*, ed. by A. Felfernig, L. Boratto, M. Stettinger, M. Tkalčič  (Springer International Publishing AG, Cham, 2018)
106. T.N. Tran, A. Felfernig, V.M. Le, M. Atas, M. Stettinger, R. Samer,  User interfaces for counteracting decision manipulation in group recommender systems, in *Adjunct Proceedings of UMAP 2019* (2019)
107. A. Tversky, Elimination by aspects: a theory of choice. Psychol. Rev. **79**, 281–299 (1972)
108. P. Victor, M.D. Cock, C. Cornelis,  Trust and recommendations,  in *Recommender Systems Handbook*, ed. by F. Ricci, L. Rokach, B. Shapira, P.B. Kantor (Springer, Berlin, 2011), pp. 645–675
109. P. Wakker, *Prospect Theory for Risk and Ambiguity* (Cambridge University Press, Cambridge, 2010)
110. W. Wood, D.T. Neal, A new look at habits and the habit-goal interface. Psychol. Rev. **114**(4), 843–863 (2007)
111. J.F. Yates, E.S. Veinott, A.L. Patalano, Hard decisions, bad decisions: on decision quality and decision aiding,  in *Emerging Perspectives on Judgment and Decision Research*, ed. by S.L. Schneider, J. Shanteau (Cambridge University Press, Cambridge, 2003)
112. M. Ye, X. Liu, W.-C. Lee, Exploring social influence for recommendation: a generative model approach, in *Proceedings of SIGIR 2012* (2012)
113. K.-H. Yoo, U. Gretzel, M. Zanker,  Source factors in recommender system credibility evaluation,  in  *Recommender Systems Handbook*, ed. by F. Ricci, L. Rokach, B. Shapira, 2nd edn. (Springer, Berlin, 2015)
114. J. Zhang, M. Gartrell, R.Y. Han, Q. Lv, S. Mishra, GEVR: an event venue recommendation system for groups of mobile users,  in *Proceedings of the ACM on Interactive, Mobile, Wearable and Ubiquitous Technologies* (2019)
115. R. Zwick, A. Rapoport, A.K. Lo, A.V. Muthukrishnan,  Consumer sequential search: Not enough or too much? Mark. Sci. **22**(4), 503–519 (2003)

# Part V
# Recommender Systems Applications

# Social Recommender Systems

**Ido Guy**

## 1 Introduction

In the "social web" or "Web 2.0" [79], people play a central role by creating content, annotating it with tags, votes (or 'likes'), or comments, joining communities, and connecting with friends and followers. *Social media* websites are proliferating and attract millions of users who author content, post messages, share photos with their friends, and engage in many other types of activities. This rapid growth intensifies the phenomenon of *social overload*, where users of social media are exposed to a huge amount of information and participate in vast amounts of interactions. Social overload makes it harder on the one hand for social media users to choose which sites to engage in and for how long and on the other hand makes it more challenging for social media websites to attract users and retain them.

*Social Recommender Systems (SRS)* are recommender systems that target the social media domain. They aim at coping with the social overload challenge by presenting the most relevant and attractive data to the user, typically by applying personalization techniques. The "marriage" between recommender systems (RS) and social media has many potential benefits for both sides. On the one hand, social media introduces many new types of data and meta-data, such as tags and explicit online relationships, which can be used in a unique manner by RS to enhance their effectiveness. On the other hand, recommender systems are crucial for social media websites to enhance the adoption and engagement by their users and thus play an important role in the overall success of social media. It should be noted that traditional RS, such as user-based collaborative filtering, are social in their nature since they mimic the natural process where we seek advice or suggestions from

I. Guy (✉)
Ben-Gurion University of the Negev, Beer-Sheva, Israel
e-mail: idoguy@acm.org

© Springer Science+Business Media, LLC, part of Springer Nature 2022
F. Ricci et al. (eds.), *Recommender Systems Handbook*,
https://doi.org/10.1007/978-1-0716-2197-4_22

other people [88]. Yet, in this chapter we focus on those recommender systems that are aimed for the social media domain, which we term *social recommender systems* [43].

This chapter focuses on two key areas of SRS, social media content recommendation and people recommendation. We dedicate a section to each of these areas, reviewing the different sub-domains, their unique characteristics, the applied methods, case studies in the enterprise, and open challenges. SRS consist of more areas, such as recommendation of tags and groups (communities), however, these are left beyond the scope of this chapter. The remainder of the chapter is organized as follows: the next two sections discuss in detail content and people recommendation. The following section discusses key aspects characterizing SRS as raised throughout its preceding two sections. The chapter concludes by reviewing emerging SRS domains and open challenges.

## 2    Content Recommendation

Social media introduced many new types of content that can be created and shared by any user in a way that has never been possible before. Users became the center of every social media website and in many cases were the ones creating the actual content of the site: textual content as in Wikipedia and WordPress; photos as in Flickr and Facebook; and video as in YouTube. Users also have a key role in providing feedback and annotating existing content on social media websites. Comments allow users to add their own opinion; votes and ratings allow them to 'like' (or dislike) favourite posts; and tags allow them to annotate the content with keywords that reflect their own viewpoint. These new types of feedback forms allow RS to implicitly infer user preferences and content popularity by analyzing the crowd's feedback.

In the social media era, articulated relationships have become available through social network sites (SNSs) [11] and changed the world of content recommendation. While in the past such relationships could only be partially extracted by surveys and interviews, and later by mining communication patterns from phone logs or email that are highly sensitive privacy-wise, the availability of relationships in social networks allows tapping into one's network of familiar people (Facebook, LinkedIn) or people of interest (Twitter) in a simpler way without infringing privacy. The use of the friend list instead of or alongside the list of similar people as in traditional CF has been broadly proven to be productive for enhancing content recommendations. Sinha and Swearingen [97] were among the first to compare friend-based recommendation with traditional methods and showed their effectiveness for movie and book recommendation. Golbeck [36] showed that friends can be a trusted source for movie recommendation. Groh and Ehmig [38] compared collaborative filtering with friend-based "social filtering" and showed the advantage of the latter for club recommendation within a German SNS. Costa and Ortale [20] use signed social relations (trust and distrust) to produce personalized recommendations using a model-based approach. Yang et al. [110] provide a survey of collaborative filtering based SRS, classifying them into matrix factorization-based approaches and neighbourhood-based approaches. Eirinaki et al. [26] survey large-scale recommender systems that take advantage of the characteristics of the underlying social network. They focus on the variety and volatility of social bonds and tackle the problems of size and speed of change in social graphs. Overall, recommendation based on friends enhance recommendations' accuracy; allow the user to better judge the recommendations since s/he is familiar with the respective people; spare the need for explicit feedback from the user in order to calculate similarity; and help cope with the cold-start problem for new users.

In recent years, studies that harness deep learning approaches to content SRS have emerged. Sun et al. [98] suggested an attentive recurrent network-based approach for temporal social recommendation. They model users' complex dynamic and general static preferences over time by fusing social influence among users with two attention networks. Taneja and Arora [101] prioritize contextual dimensions for user modeling using neural networks and tensor factorization. Tahmasebi et al. [100] suggest a hybrid social recommender system utilizing a deep autoencoder network. The proposed approach uses collaborative and content-based filtering, as well as social influence. Fan et al. [27] propose a Bi-LSTM with attention mechanism to extract "deep" social sequences, which consider information from not only direct neighbors but also distant neighbors. Their approach demonstrates good performance over the Ciao and Epinions datasets.

The remainder of this section reviews key domains of social media content recommendation, such as blogs, microblogs, news, and multimedia. We then briefly discuss group recommendation, which is especially relevant for recommendation of social media content. Following, a case study of social media recommendation within the enterprise is presented in detail. The section concludes with a summary of key points.

## 2.1 Key Domains

**Blogs** Blogs are one of the classic social media applications and a natural ground for recommendation techniques. They typically consist of inherent hierarchy that SRS need to take into account. At the top of this hierarchy is the blog itself, which may be owned by an individual user or a community, and is often focused on a topic or domain. The blog includes different blog posts (or blog entries) that include one article by a single author. The author (and sometimes other users) can usually annotate the post with appropriate tags, which also serve for dissemination to relevant populations. The post's readers can add comments and can often also vote for (or 'like') the post [50]; other authors can use a trackback to reference the post from their own post. In one of the early studies of blog recommendation, Arguello et al. [5] explored personalized recommendation of whole blogs (as opposed to blog posts) using the TrecBlog06 dataset [72]. Given a query that represented the user's topical interests, two document models were explored: the first included a single large document that was based on concatenation of all the blog's posts and the second was based on smaller documents, each representing a single post, while aggregation was made at ranking time. Evaluation indicated that both models performed equally well and that hybridization of both further improved the results.

**Multimedia** Multimedia recommendation is challenging due to the lower amounts of textual data and the extremely large size of the content. One of the most popular social media websites, YouTube, includes an advanced recommender system that drives a large portion of the user traffic and helps direct users to more relevant videos. Davidson et al. [23] stated that the goals of the YouTube recommendations are to be recent and fresh, diverse, and relevant to the user's recent actions. They also stated that users should understand why a video was recommended to them, thus incorporating explanations in the YouTube RS. As described in their paper, YouTube recommendations are based on the user's personal activity on the site and are expanded by a variant of collaborative filtering (CF) over the co-visitation graph. Ranking is done based on a variety of signals for relevance and diversity. A later study examined the impact of recommendation on excessive use of online video streaming services and found it to be high [56].

**Community Question and Answering** Social or community question-and-answering (SQA or CQA) websites, such as StackOverflow, Quora, and Yahoo Answers, allow users to ask various types of questions and receive (and vote for) answers from the crowd. Questions may a wide variety of domains and seeks for information, conversation, or both [7, 45, 51]. As such, they also serve as a fertile ground for different types of recommender systems for both question askers and answerers. The challenge here is twofold: on the one hand, recommend to askers similar previously-asked questions to avoid redundant burden on answerers and spread of similar information in many question pages; on the other hand, recommend answerers with questions they may want to answer and increase overall answer engagement on the website. As one example, Szpektor et al. [99] experimented with recommendation of questions to potential answerers on the Yahoo Answers website. They discovered that topic relevance was not a good enough basis for recommendation. Diversity and freshness also played a key role: on the one hand, a novel and somewhat different question was more likely to arouse answerer's attention and on the other hand it was extremely important for answerers to receive questions that are very fresh, typically only a few minutes old.

**Jobs** LinkedIn is one of the most successful SNSs and as the world's largest professional network it has many unique recommendation challenges, such as of companies and of professional groups. Another specifically interesting example is the recommendation of job opportunities. Such recommendation can have a tremendous influence on people's lives as it can ultimately lead to a career change. Recommendation needs to take into account many aspects, such as location alternatives, candidate's experience, and timing. Wang et al. [106] shed some light on the job recommendation task at LinkedIn and particularly focus on the timing of recommendation. Their statistical model considered the tenure between two successive decisions to estimate the likelihood of a user's decision to make a job transition at a given point. Evaluation used the real-world job application data and demonstrated the effectiveness of their model and the importance of considering the time factor as part of the recommendation process. Olsson et al. [78] expand the scope and define "professional social matching" as the matching of individuals or groups for vocational collaboration and co-creation of value. This covers organizational activities, including recruitment, headhunting, community building, team formation, and individually-driven activities like mentoring, seeking advisory relationships, and general networking.

**News** Social news aggregators such as Digg, Google Reader, Reddit, and Slashdot, allow users to post and rate news articles and surface the most interesting and trending stories. News recommendation is especially challenging due to the need for freshness. Old stories or stories to which the user has already been exposed will be considered bad recommendations, even when relevant to the user's tastes and preferences. The pace of news appearance is very high, while different users have different news consumption rates, which personalization techniques need to take into account. Digg used to be a popular social news aggregation service, allowing its users to submit links to news stories, vote, and discuss them. Aside from promoting the most popular stories to users (by votes), Lerman [68] described the personalized recommender system implemented for Digg that was based on friends and "diggers like me". Recommendations for another popular news website, Google Reader, were described by Liu et al. [71]. They combined CF techniques with "individual filtering" techniques. Evaluation, based on a live trial, indicated that the hybrid approach performed best and improved 38% over a popularity-based baseline. Pure CF was only able to improve 31% on top of the baseline. An increase in return rate was observed due to the hybrid recommendations, however, interestingly, there was no effect on the overall number of stories read on the homepage.

**Microblogs** Microblogging, most famously brought into attention by Twitter, allows user to broadcast short messages. Those messages are typically propagated across a network of followers and "followees", built by the user's ability to follow any another user. On twitter, each message is limited to 140 characters and is called a 'tweet'. The high pace of messages (over half a billion tweets per day), their real-time nature, their concise content, and the lack of metadata and structure, make the challenge of filtering and personalizing the Twitter firehose of unique nature. In one of the earlier studies, Chen et al. [17] explored content recommendation through URLs shared in tweets. They compared 12 algorithms that differed in the following aspects: (1) candidate selection was either based on popular tweets or on tweets from followees and followees-of-followees (FoF); (2) topic relevance was based on cosine similarity between the user and the URL. The user's representation was based on self-tweets or on followees' tweets; (3) social voting was based on the number of user's followees who also follow the author and on author's frequency of tweeting. Results, based on a field study with 44 subjects, indicated that social voting worked better than topic relevance; FoF candidate selection outperformed popularity; and using self tweets for user modeling performed better than using the followees' tweets. The introduction of the 'retweet' feature, which allows user to share another user's tweet with their own audience of followers, provided researchers with direct feedback about the level of interest in an individual tweet. Many studies followed that attempted to use this information to predict "good" tweets. For example, Chen et al. [18] suggested a model for personalized tweet recommendation using "collaborative ranking". The model was based on both explicit and latent features and considered a wide variety of topic-level, social relations, and global factors. Evaluation was based on re-tweet prediction and showed the superiority of the collaborative ranking method over various baselines, such as Latent Dirichlet Allocation and Support Vector Machine. It also indicated that all the three factors are important to consider. In more recent work, Alawad et al. [3] studied the recommendation of "novel" tweets that were not posted or shared by anybody in the user's network. To this end, they created the user's egocentric network up to depth two and applied the transitivity property of the friend-of-a-friend relationship to yield recommendations. Piao and Breslin [84] proposed a learning-to-rank approach for recommending tweets that a user might re-tweet based on a deep neural network with Long Short-Term Memory (LSTM). The rank score was based on both the similarity between the embeddings of a user and the tweet and the similarity between the embeddings of the user and the tweet's author. The 2020 RecSys challenge focused on microblogging, with Twitter releasing around 160 Million public tweets obtained by sub-sampling over a period of two weeks. The specific task was to determine the probability that a user is going to engage with the content of another user via reply, retweet, or 'like' [95].

## 2.2   Group Recommender Systems: Beyond Preference Aggregation

Groups and communities play a central role in social media and often times form the entry gate for participation [89]. This makes group recommendation techniques highly relevant for the SRS domain. Due to this relevance, we briefly review the broad area of group recommendation in this sub-section; in the following section, as part of the enterprise case study, we describe in more detail an example of SRS aimed for communities. Group recommendation targets a group of individuals rather than a single one (Chapter "Group Recommender Systems: Beyond Preference Aggregation"). Example scenarios for group recommendation include friends planning together their "perfect" vacation; a family selecting a movie or a television show to watch together; a group of colleagues choosing a restaurant for an evening outside (or looking for a recipe for a joint meal); or the classic (and less relevant in the era of personal music players) gym problem [73]: selection of a playlist based on the current group of trainees in a fitness center.

Group recommendation poses new challenges compared to individual recommendations. Two of the prominent challenges are the specification of preferences by members and the recommendation generation. Jameson et al. [61] suggested a collaborative interface for members to specify their preferences in a group recommender system for travel, which allowed collaborative editing of the members' preferences. Such an interface holds various benefits: it allows members to persuade others to specify a similar preference to their own, perhaps by giving them information they had previously lacked; it enables to explain and justify a member's preference (e.g., "I can't go hiking due to an injury"); it allows taking into account attitudes and anticipated behavior of other members; and it encourages assimilation to facilitate the reaching of agreement.

The most studied challenge of group recommendation is the generation of recommendations themselves. The two main techniques are profile aggregation and recommendation aggregation. Profile aggregation produces a single profile representative of the group by aggregating the preferences of the different group members. Recommendation aggregation generates a recommendation list for each of the group members and aggregates the list into one single list for the group, typically by using rank aggregation techniques. Berkovsky et al. [10] experimented with these two approaches for recipe recommendation to groups and found that the profile aggregation method was superior over the recommendation aggregation method.

There are various approaches for aggregating member preferences into a single community profile, each with its own pros and cons. Among the prominent approaches are: (1) least misery, which seeks to maximize the minimum ranking of any group member. Obviously, this approach can lead to a recommendation that does not maximize the average rating or the maximum benefit; (2) fairness, which aims at the most equal rating balance across group members. This can lead to a recommendation that gets a low rating by all members of the group; (3) and fusion, which aggregates individual rankings (e.g., by Borda count). Baltrunas et al. [6] compared several techniques for group recommendation using the MovieLens dataset. They examined both profile aggregation and rank aggregation techniques and found the optimal one given a set of parameters, such as the group's size and the similarity among group members. Review of additional studies on group recommendation can be found in a recently published surveys [22, 28].

## 2.3   Case Study: Social Media Recommendation in the Enterprise

In this section, we review a body of research that explored recommendation of mixed social media items within the enterprise, and included three main studies. The first study [53] focused on recommendation based on social relationships . As previously mentioned, social media enables the exposure of different types of social relationships in a way that has never been possible before. The study explored a rich set of indicators for social relationships based on social media data and compared

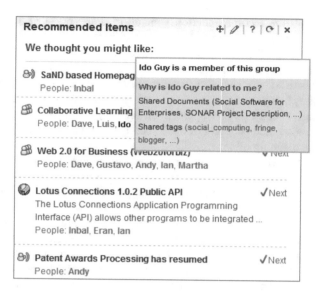

**Fig. 1** Widget for social media item recommendation based on related people

two types of networks as basis for recommendation: familiarity and similarity. The familiarity network was built based on explicit and implicit signals from enterprise social media, such as articulated connection within an enterprise SNS, tagging one another, or co-authorship of the same wiki page. The similarity network was based on common activity in enterprise social media, such as membership in the same communities, usage of the same tags, or commenting on the same blog posts. An "overall" network was also examined, combining the two types of relationships.

The recommendation widget, depicted in Fig. 1, presents the recommendations with explanations, which displays the people who served as the "implicit recommenders" and how they were related to both the user and the recommended item. One of the key research questions of the study was whether explanations influence the instant interest in the recommended items. This was examined by comparing recommendations with and without explanations.

The evaluation was primarily based on a user survey with 290 participants. Figure 2 shows the portion of items rated "interesting" for each of the three network types: familiarity, similarity, and overall. Recommendations from familiar people were found significantly more accurate than recommendations from similar people. The overall network did not improve accuracy on top of the familiarity network. That said, recommendations from similar people were found more diverse and less expected, indicating that the similarity network contributes on other dimensions than accuracy to the recommendation quality [75].

Figure 3 displays the effect of explanations. While explanations have been previously shown to have positive effect on recommendation in the long term, by providing transparency and building trust with the user [57], it was found

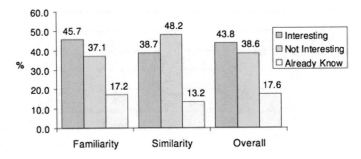

**Fig. 2** Rating results across the three network types

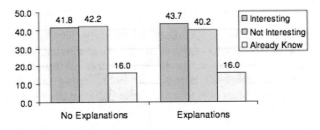

**Fig. 3** Rating results with and without explanations

that recommendations with explanations in this case also increase their instant effectiveness: when the people who serve as implicit recommenders were shown, interest rate in the recommendations grew. This was particularly true for familiar people, following the intuition that seeing a familiar person who is related to a recommended item may increase the likelihood of the user's interest in that item (e.g., "if John has bookmarked the page, there must have been something interesting in it").

After establishing understanding of people-based recommendation, a second study explored the use of tags for the recommendation task and compared tag-based with people-based recommendation [54]. The people-based recommendations were calculated based on a combined network of familiarity and similarity, with a triple-boost given to the familiarity network based on the results of the previous study.

Based on the results of a preliminary study, the tags used for recommendation included those used by the user and those applied to the user by others via an enterprise people tagging application [87]. Experimentation was made with a pure people-based recommender (PBR), a pure tag-based recommender (TBR), two hybrid people-tag recommenders (or-PTBR and and-PTBR), and a popularity baseline (POPBR).

The main comparison results are shown in Fig. 4. In general, all personalization techniques outperformed the popularity-based recommender. In terms of accuracy (interest rate), tag-based recommenders significantly outperformed people-based recommenders. Yet, people-based recommenders showed other benefits, such as

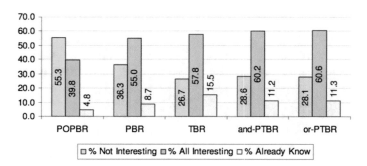

**Fig. 4** Rating results for five different recommenders

increased diversity across item types (tags substantially favored bookmarks), less expected results reflected in lower rates of already-known items, and more effective explanations. Specifically regarding explanations, the effect found for people-based explanations in increasing interest rates was not found for tag-based recommenders. Apparently, seeing the related tags to a recommended item does not have the effect (or extra value) that viewing the related people has. Hybrid recommendations, combining people-based and tag-based approaches, were shown to take the good of both worlds and also achieved the best accuracy with a ratio of around 70% interesting items for the top 16 recommendations.

The third study in the series explored recommendation for online communities rather than for individuals [89]. As mentioned before, online communities have become central to social media experience and much of the social media content is created in the context of a community. In that work, recommendations were generated using group recommendation techniques, but were targeted to the community owners (moderators) only, so that they can share the content with the rest of the members as appropriate. Recommendations were generated using two main techniques. The first considered the members of the communities or a subset of them, and applied profile aggregation using the fusion approach (with advanced scoring) to generate a community profile that included both topics and people. These topics and people in turn served as the basis for recommendation: their most related content items were recommended. In particular, three subsets of the members were examined: all members, all owners, and active members. The second technique was content-based (CB): it considered the title, description, and tags of the community to generate recommendations. Hybrid approaches were also considered, by combining the topics and people from the member-based recommenders with the topics extracted by the content-based recommender into one community profile.

Evaluation was conducted using a large user survey of enterprise community owners and results are summarized in Fig. 5. Hybrid recommenders were generally found to perform better than the pure recommenders. For large communities (100 members or more), it was found that the hybrid profile that considered both active members and community's content performed significantly better than all

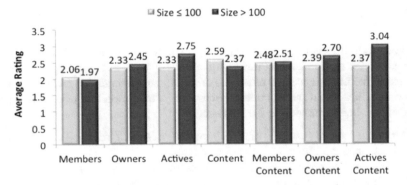

**Fig. 5** Average rating for small vs. large communities across seven community profiles

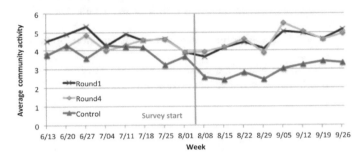

**Fig. 6** Community activity before and after a survey where owners could share social media content with their community

other profiles. The pure active member-based profile was second best for large communities. For small communities (less than 100 members), the pure content profile was the best, followed by the hybrid profile considering all members and the content. These results indicate that for small communities, the content is a strong basis for recommendation and all members are a good representative group for profile aggregation. But for large communities, the content is less effective on its own and the group of all members becomes too disparate, while the group of only active members serves as the best basis for profile aggregation.

In a follow-up study, the impact of such recommendations was inspected over four rounds of recommendation to enterprise community owners [47]. Owners could share recommended content, such as blogs, bookmarks, and forum posts with their community members. As can be seen in Fig. 6, such recommendations, in both the first and fourth round, showed significant effect on overall community activity over a period of the following 8 weeks, compared to a control group whose owners received no recommendations.

## 2.4  Summary

We reviewed different domains for recommendation of social media content and a case study for recommending mixed social media items in the enterprise. We also discussed the importance and relevance of group recommendation techniques when recommending social media content. Below are a few important points we wanted to re-iterate before moving to the next section:

- Articulated social networks play an important role in CF for social media content and enhance traditional CF in various manners.
- Tag-based recommendations are highly effective for producing accurate recommendations and typically outperform regular user-based CF.
- As in Traditional RS, hybrid approaches (e.g., tags + networks, short + long term interests, collaborative + individual filtering) usually enhance recommendation effectiveness.
- A large user-base is desirable and can lead to a strong evaluation on live systems (e.g., A/B testing).
- Accuracy alone is not enough: serendipity, diversity, freshness, and other qualities also play a key role in the success of recommendations.

## 3  People Recommendation

Social recommender systems span beyond content recommendation. As mentioned in the introduction, social overload originates from both information and interaction overload. Since people are the key element that makes the web "social", recommendation of people is a central pillar within the social recommender system domain. Terveen and Mcdonald [102] coined the term "social matching" for recommender systems that recommend people to people. In their work, they explained why a people recommender is a unique RS, which is different than recommendation of other artifacts, and thus deserves its own special attention (see also Chapter "People-to-People Reciprocal Recommenders"). Among other aspects, trust, reputation, privacy, and personal attraction have greater importance when it comes to people recommendation.

Social media sites and in particular SNSs define different types of explicit (or "articulated") relationships among their users. The main dimensions of the relationship types are:

- Symmetric vs. asymmetric. In some sites, such as Facebook and LinkedIn, a relationship between two users is reciprocated. In such a case, one user typically sends an invitation to connect to another user, who needs to accept the invitation. Once the other user accepts, the two are reciprocally connected on the site [81]. On the other hand, asymmetric relationships, such as on Twitter or Pinterest, allow one user to "subscribe to" or "follow" another user. The other user does not necessarily need to follow the first user back and thus many asymmetric relationships are formed.

- Confirmed vs. non-confirmed. Some of the sites require the other side's agreement for connecting or following, while others do not. Typically, symmetric networks require such confirmation and as long as it has not been received, no connection exists. Asymmetric networks do not usually require a confirmation and any user can choose to follow any other user, however there are exceptions to these norms.
- Ad-hoc vs. permanent. Some of the sites encourage connection for an ad-hoc purpose, such as for people to meet at an event or partner for a joint task, while others encourage a long-term relationship that is meant to last over months and years.
- The site's domain. The domain of the SNS has an important influence on the formed network. For example, Facebook is typically used for maintaining social relationships with friends and acquaintances, while LinkedIn is a professional network meant for maintaining business relationships with colleagues and partners. The goals and characteristics of a connection in each of these sites are therefore different, as they would be in SNSs for other domains, such as travel, art, cooking, question and answering, etc.

The different characteristics of people relationships in the different sites require different recommendation techniques. For example, a recommender for people to connect with on Facebook may seek to recommend familiar people, while a recommender for people to follow on Twitter may recommend people the user is interested in, even if they are not familiar. Recommending "celebrities" or popular people is probably a better strategy for a follower-followee network than for a friendship network. A good summary can be found in [41].

In the remainder of this section, we review three key types of people recommendation: recommending people to connect with, recommending people to follow, and recommending strangers to get to know. We describe the unique challenges and characteristics of each of these recommendation types and demonstrate how existing approaches handle them. Before summarizing the key aspects, we briefly discuss two closely related research areas to people recommendation: link prediction and expertise location.

## 3.1 Recommending People to Connect With

The first study that focused on people recommendation in an SNS introduced the "do you know?" (DYK) widget [49]. The widget recommended people to connect to within an enterprise SNS. The action the widget was targeting was clicking a 'connect' button that would trigger an invitation to connect within the SNS, which the other side would need to confirm for the connection to become public. Recommendations were made based on a variety of familiarity signals: org-

848                                                                                    I. Guy

**Fig. 7** The "Do You know?"
(DYK) widget

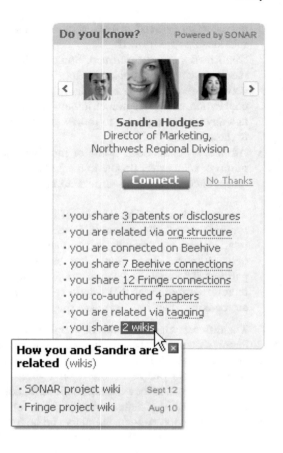

chart relationships (peers, manager-employee, etc.), paper and patent co-authorship, project co-membership, blog commenting, person tagging, mutual connections, connection on another SNS, wiki co-editing, and file sharing. Figure 7 illustrates the widget, which included detailed explanations for each recommendation. The explanations indicated the counts per each of the signals mentioned above and further hovering over an evidence line allowed seeing the specific details (e.g., the wiki pages co-edited) and getting to the actual page of the evidence pieces.

The evaluation of the widget was based on a field study of its use within the Fringe enterprise SNS. Fringe had the "friending" feature before, but did not have a people recommender. The inspected effect on the site was dramatic. Both the number of invitations sent and the number of users who send invitations significantly increased, as can be seen in Figs. 8 and 9. One of the users of the site explained: *"I must say I am a lazy social networker, but Fringe was the first application motivating me to go ahead and send out some invitations to others to connect."* Explanations increased user trust in the system and made them feel more comfortable sending invitations, as one user described: *"If I see more direct connections I'm more likely*

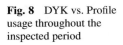

**Fig. 8** DYK vs. Profile usage throughout the inspected period

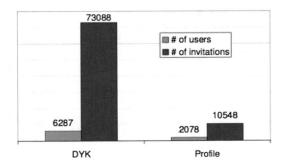

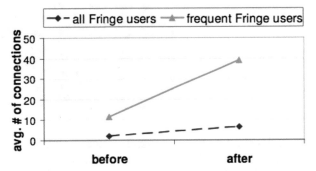

**Fig. 9** Average number of invitations per user before and after the inspected period

*to add them [. . . ] I know they are not recommended by accident."* Overall, there was a substantial increase in the number of connections per user on Fringe. However, a sharp decay of the widget usage was found over time, as excitement of the feature dropped and potential connections were exhausted.

In a follow-up study [16], conducted within a different enterprise SNS, nick-named Beehive, the aggregation algorithm used by the DYK widget (termed 'SONAR') was compared with three other algorithms for people recommendation: (1) Content Matching (CM)—based on cosine similarity of the content created by both users: profile entries, status messages, photos' text, shared lists, job title, location, description, and tags. Word vectors were created by a simple TF-IDF procedure. Latent semantic analysis (LSA) was not shown to produce better results and was not applied since it does not yield intuitive explanations; (2) Content plus Link (CplusL)—combined CM with social links. A social link was defined as a sequence of 3 or 4 users, where for each pair of users in the sequence u1 and u2, either u1 connects to u2, u2 connects to u1, or u1 commented on u2's content; (3) Friend of Friends (FoF)—based on the number of mutual friends, as done in many of the popular SNSs. The FoF algorithm was able to produce recommendations for only 57.2% of the users (compared to 87.7% for SONAR). Figure 10 shows the recommendation widget.

**Fig. 10** People
recommender widget showing
a person recommended using
the CplusL algorithm

Evaluation was based on a user survey and a controlled field study. Figure 11 shows the main survey results. CM and CplusL produced mostly unknown people, while SONAR and FoF produced mostly known individuals. As could be expected, a higher portion of the recommended people who were familiar to the user were rated as good recommendations and resulted in a "connect" action. Yet, the unknown recommended individuals may help discover new potential friends. The overall superiority of algorithms that involve social links over content was clear: only 30.5% of the CM recommendations resulted in a connect action, compared to 40% for CplusL, 47.7% for FoF, and 59.7% for SONAR.

A later study examined the recommendation impact on the network structure [21]. Since recommendations play such a key role in building the network during its early stages, they also substantially influence the structure of the generated network, its characteristics, and measurements. For example, Fig. 12 shows the average degree of recommended connections for each of the four algorithms. FoF is the most biased towards high-degree connections, while CM does not have such bias: it often recommends users with few connections or even none at all. The high-degrees of FoF recommendations lead to a network with fewer nodes and higher average degree compared to the network created by CM recommendations. Another aspect of the effect of recommendations on the network is betweenness centrality, which measures the importance of nodes in the graph [12]: CM and SONAR generate the highest delta in betweenness compared to CplusL and FoF. Regarding demographic characteristics, CM is most biased towards the same country, but least biased towards the same organizational unit, while SONAR substantially increases cross-country and intra-unit connections. The network effects of people recommendations

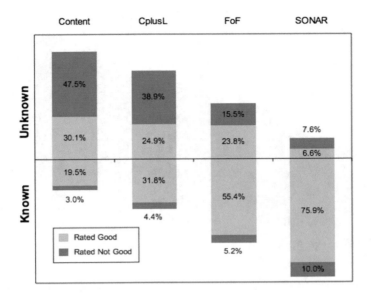

**Fig. 11** Survey results for the four algorithms

**Fig. 12** Degree of
recommended connections
across the four algorithms

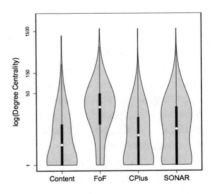

are an important global aspect of a people recommender and need to be considered
when designing a new people recommender system.

Another related study by Freyne et al. focused on recommendation as a means
to increase new users' engagement within an enterprise SNS [32]. That study used
aggregated data external to the SNS in question to recommend both people and
content to new users. Even brand new employees could still get recommendations
based on their initial data, such as their org-chart information (indicating their
peers), location, or organizational unit. The results indicated that combined recommendations have a significant effect in increasing users' visits to the site as well
as their viewing activity and actual contributions to the site (the latter is depicted in
Fig. 13). Interestingly, people recommendations were most effective when focusing
on recommending the most active users, even if they had less familiarity signals

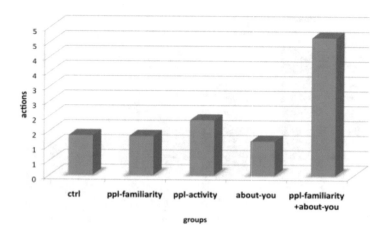

**Fig. 13** Actions over four months

with the user. Yet, as discussed, such recommendations can have a long-term effect on the network structure and lead to a less balanced degree distribution.

Friend recommendation has also become popular on mobile devices, where location often plays a role and makes the recommendations more transient or ad-hoc. Quercia et al. [86] discussed "friendsensing", sensing friends based on Bluetooth information on mobile devices. Friends were recommended based on co-location, while two basic approaches were attempted, taking into account the duration of co-location and its frequency, respectively. A weighted graph was built accordingly and recommendations were generated using that graph based on link analysis (shortest path, page rank, k-markov chain, and HITs). Simulation-based evaluation indicated both basic approaches perform similarly well and way beyond a random baseline.

Gurini et al. [40] proposed a matrix factorization model with temporal dynamics to provide people recommendations. Each dimension in a three-dimensional matrix factorization model represented an "attitude": sentiment, volume, and objectivity. Recommender's accuracy and diversity was shown to increase with attitudes and temporal features.

## 3.2 Recommending Strangers

The focus of the work discussed thus far has been on recommending familiar people one can connect to. As already implied, there could also be value in recommending people the user does not know. StrangerRS [43] attempted to recommend people who are unknown yet interesting within the organization. Such recommendations can be useful in many potential manners, such as, for getting help or advice, reach

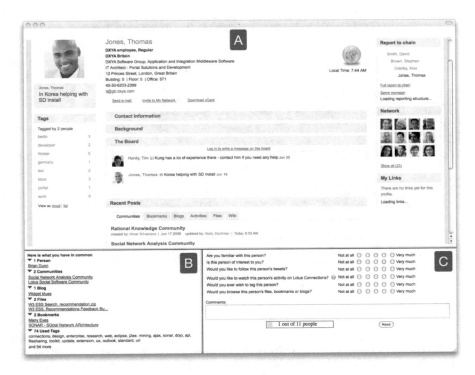

**Fig. 14** User interface of the stranger recommender system

new opportunities, discover new routes for career development, learn about new assets that can be leveraged, connect with subject-matter experts and influencers, cultivate one's organizational social capital, and grow own reputation and influence within the organization. As mentioned before, recommendation of people to connect to within an SNS is mostly effective for the network-building phase. Afterwards one's recommendations become staler, as the network becomes more stable and connection to others becomes less frequent. This is where stranger recommendation can become more relevant and complement the recommendation of familiar individuals, by suggesting interesting people the user does not know, but may want to start getting acquainted with.

Figure 14 shows the user interface of StrangerRS. Since it aimed at recommending strangers, more information about each person was presented, in the form of their full profile page (part A). Evidence for why this person may be interesting was also presented (part B). It included similarity points with that individual, such as common tags, common communities, common files, and others. The action suggested by the recommender was not a connection within the SNS, since it is likely to be too soon to connect to a stranger, but rather it was suggested to view the person's profile, read their blog, or follow them (part C).

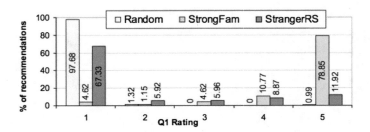

**Fig. 15**  Rating of "strangerness" for StrangerRS and two baselines: random and strong familiarity

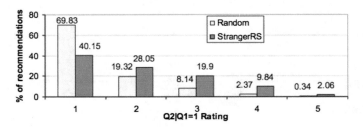

**Fig. 16**  Rating of interest in strangers for RandomRS vs. the random baseline

A successful recommendation by StrangerRS was considered a recommendation of a stranger who might be interesting to the user. These two, almost contradicting, goals were not easy to satisfy and led to a much lower accuracy level than usual familiar people recommendation. Yet, supposedly, the value of a successful recommendation in this case is much higher, since this is no longer just about facilitating connection to a known person, but rather about exposing the user to a new interesting person s/he was not even aware of. The method used for producing the recommendations was based on network composition: the extracted familiarity network was subtracted from the extracted similarity network to produce the recommendations. Jaccard index was the main measure used for similarity between two individuals. Results, depicted in Figs. 15 and 16, indicated that two thirds of the recommended individuals were indeed strangers, yet strangers who were significantly more interesting than a random stranger. Out of 9 recommendations presented to each user, 67% included at least one stranger rated 3 or above in terms of the user's interest, on a 5-point Likert scale.

Stranger recommendation is also a common feature of online dating websites. Pizzato et al. [85] introduced RECON, a reciprocal recommender for online dating. Similarly to the original social matching framework, they specified a few special characteristics of reciprocal recommendations, where people are both the subject and object of recommendations. These included the fact that success is dependent on both sides; the need for both sides to provide their profiles so that matching can occur; and the typical requirement that one individual will not be recommended to too many others. Their evaluation, conducted on a major Australian dating site, was based on four weeks of training and two weeks of testing, where success was determined based on previous user interaction. Generally they found that accounting for reciprocity features improves recommendation accuracy and helps address the cold-start problem. Zheng et al. [112] studied speed dating data, where the user's expectations are well defined, and showed that addressing fairness by trading off utility and performance yields a better recommender system.

## 3.3 Recommending People to Follow

Two studies were the earliest to explore recommendation of people to follow. Hannon et al. [55] used a CB-CF hybridization to recommend "followees" on Twitter. They examined several ways to generate user profiles, based on the user's own tweets, the user's followers, the user's followees, the user's followers' tweets, and the user's followees' tweets. The open source search engine Lucene was used to index users by their profile, after applying TF-IDF to boost distinctive terms or users within the profile. They applied an offline evaluation using a dataset with 20,000 Twitter users. 19,000 were used as a training set and the remaining 1000 were the test users. The different methods were compared based on their ability to predict the user's followees. A slight advantage was observed to profiles that were based on followers and followers' tweets. Hybrid profiles further improved the precision. A small-scale live trial was also conducted where users indicated whom they were likely to follow. On average, hybrid approached reached about 7 out of 30 accurate recommendations.

A second study was performed by Brzozowski and Romero [13], who experimented with the WaterCooler enterprise SNS. During a 24-day live trial period, they observed patterns of 110 users who followed 774 new individuals. The strongest pattern found was of the form $A \leftarrow X \rightarrow B$, meaning that sharing an audience (follower) with another person is a strong reason to follow that person. Most-replied was found as a strong global signal. Similarity and most-read were found as weaker signals for followee recommendation.

Gupta et al. [39] revealed some details about the followee recommender systems in use by Twitter. From an architectural perspective, they noted the decision to process the entire Twitter follower-followee graph in memory using a single server, which contributed to the performance of the feature. They developed an open-source in-memory graph processing engine to traverse the Twitter graph and generate recommendations. The algorithm used was a combination of a random walk and SALSA [67], comparing two approaches: the first gives each user the same influence regardless of the number of users they follow or are followed by and the second gives equal influence to each follower-followee edge. Sanz-Cruzado and Castells [92] study diversity in people recommender systems, linking to prior diversity notions in recommender systems. They focus on the notion of weak ties to enhance structural diversity and demonstrate, using Twitter data, how state-of-the-art recommendation methods compare in such diversity and its tradeoff with accuracy. They show that diverse recommendations result in a corresponding diversity enhancement in the flow of information through the network, with potential alleviation of filter bubbles.


## 3.4 Related Research Areas

Link prediction in social networks is a fertile research area that is closely related to people recommendation and has often been offered to enhance it. The seminal work by Liben-Nowell and Kleinberg [70] formalized it as a task to predict new interactions within a social network based on the existing set of interactions. Experimentation with paper co-authorship networks showed, using an unsupervised learning approach, that the network topology can be effectively used to predict future collaboration. Moving to the social media domain, Leskovec et al. [69] developed models to determine the sign of links (positive or negative) in SNSs where interactions can be positive or negative (Epinions, Slashdot, Wikipedia). Fire et al. [29] experimented with five social media sites, including Facebook, YouTube, and Flickr, and proposed a set of graph-topology features for identifying missing links. This technique was shown to outperform common-friends and Jaccard's coefficient measures, implying it can be useful for recommending new connections. Scellato et al. [93] focused on location-based social networks and suggested a supervised learning framework to predict new links among users and places. In another study of mobile networks, Wang et al. [105] showed that combining network-based features with human mobility features (e.g., user movement across locations) can significantly improve link prediction performance using supervised learning.

It is also worth mentioning the research area of expertise location [64, 74] in the context of people recommendation. Expertise location deals with the problem of finding an expert in a given domain or technical area. It thus falls within the broad search domain, since it is triggered by a user query. Similarly to the difference between content and people recommendation, in expertise location the results are people rather than documents as in content search. For similar reasons to those already discussed in this section, the case of searching for people bears some unique characteristics compared to other content search scenarios and therefore forms its own area of research. Social media serves as a particularly good source for expertise mining and analysis[42] and for providing explanation in the form of expertise evidence [111]. Despite its pertinence to the search field, expertise location is sometimes mixed with people recommendation and in many cases is termed "expert recommendation". It should be noted that a people recommender should be considered as such only when it does not involve a user query and is initiated by the system rather than by the user.

## 3.5 Summary

We have seen that people recommendation is a complex field of study. The fact that it deals with recommending people to themselves bring many interesting aspects to the table. For example, explanations may serve in this case to make users feel more comfortable accepting a recommendation and sending an invitation to connect or start following (in most cases, knowing that the user who has been followed will get a notification about it, even if their approval is not required). We reviewed three types of people recommendations: recommendation of familiar people, for example to connect with on an SNS; recommendation of interesting people, for example to follow in a social media site; and recommendation of strangers, for dating or for getting to know within a community or organization. A social website may transfer between these types of recommendations according to the user's phase within the site. For example, it may be desirable to recommend familiar and interesting people for users in their early stages, so they can build their network of friends or followees. In a later stage, when users start to exhaust their connections, stranger recommendations can help users get to know new individuals and increase their social capital.

## 4 Discussion

In this section, we summarize key SRS-related topics that were brought up throughout the previous two sections on people and content recommendation and suggest directions for future work.

**Explanations** The public nature of social media data enables to provide more transparency into recommendations by showing how they were formed. In some of the enterprise examples we reviewed for both content and people recommendations, explanations were found to have a key role in increasing the instant acceptance rate of recommendations [49, 53]. Studies in the Web domain have shown a similar effect [66, 103]. Beyond that, explanations in RS have been shown to have longer-term effects of building trust relationships with the user [57] (Chapter "Beyond Explaining Single Item Recommendations").

There are also a few challenges with regards to explanations. First, as we have seen, explanations do not always increase accuracy. For example, in our mixed content recommendation study, tag-based explanations did not increase recommendations' ratings. Second, not every recommendation method can provide intuitive explanations; there is usually a trade-off between the method's complexity and the clarity of explanations it can provide. For instance, recommendations that are based on clustering techniques are usually harder to explain. Third, explanations pose challenges in terms of privacy. For example, the YouTube explanations [23] explicitly show videos previously watched by the user, which directly expose information that might be sensitive if watched by another person. Fourth, explanations require extra real-estate on the user interface, which might be particularly challenging on mobile devices; therefore their cost-to-value ratio should be carefully considered when designing the recommender system.

**Privacy** As mentioned several times throughout this chapter, one of the key benefits of social media data is that large portions of it are public and thus can be used for analysis without infringing user privacy, as is the case, for example, with email or file system data. It should be noted, however, that in some countries, public social media information is still considered personal information (PI), when linked to an identity of a real person. This means that analysis and inference from such data may still require explicit user consent. Indeed, aggregation of public data, even if it was previously accessible, may reveal sensitive information the user did not intend to expose. In addition, as just mentioned, explanations aimed for a specific user might reveal very sensitive data, such as browsing or viewing history, when exposed to another person who may watch the screen alongside. Finally, there is much social media data that is still access-restricted. Recommender systems should pay special attention not to infringe the privacy model of the data, to avoid the exposure of sensitive information [25].

**Tags** The work we reviewed indicated that tags, a mechanism introduced by social media to annotate content, such as web pages, photos, or people, can be particularly effective as a basis for recommendation. Tags' ability to concisely summarize user perspective over large content pieces make them a highly valuable resource for producing recommendations [94]. Aside from recommendations, tags have been shown to be useful for other purposes, such as enhancing search or generating "tag clouds" that summarize the common topics of a group of items to the user [63]. Unfortunately, despite their value, tag usage is on the decrease in recent years, with sites such as Delicious becoming less popular and other sites giving less prominence to tags. Tag recommendation techniques [62, 96], which are another type of SRS not discussed in this chapter, should be used to promote tag usage and close the loop: tag recommendations help generate more tags, while these tags, in turn, used to produce other recommendations.

**Social Relationships** One of the most important contributions of social media to recommender systems is the introduction of the explicit (articulated) network . Social network sites , such as Facebook, LinkedIn, and Twitter, allow people to explicitly articulate their connections. As mentioned, there are two main types of connections, one expresses familiarity and the other expresses interest. Both of these articulated networks are very useful for content recommendation, and were shown to enhance traditional CF techniques. They also have other benefits: (1) sparing the need for explicit feedback in the form of ratings to determine the network of similarity, (2) help coping with the new-user cold start problem, in case the network can be used across social media websites, and (3) helping users judge the recommendations, since they originate from people they know or are interested in (also making explanations more effective). On the other hand, as we have seen, recommendations of people to connect with or to follow are essential for enhancing the formation of such explicit relationships. This is a classic demonstration of the mutual relationship between recommender systems and social media discussed in the introduction: on the one hand social media introduces a new type of data that enhances RS; on the other hand RS are essential for generating this type of data.

**Trust and Reputation** The topic of trust has a tremendous importance in the RS domain. Obviously, the best recommendations come from a trusted person. But on the other hand, trust is very challenging to compute as it represents a very abstract and subjective quality between two individuals. Reputation represents a more general concept about a person's perception by others [59]. One way to define it is the aggregation of trust in this person across the entire set of users. Social media and the "wisdom of the crowd" enable to estimate trust and reputation in ways that have not been possible before. Online social relationships and content feedback forms (comments, 'likes', etc.) introduce more signals that can be used to calculate trust and reputation. That said, many of the studies still use rough estimations that are based on controversial assumptions, for example, that a friend on an SNS is a trustworthy individual. Evaluation of trust and reputation is also particularly challenging, as even in the real world people have hard time figuring out who they trust or who has a good reputation. Assuming a network of trust is given, there are growing amounts of research that explore how to use it to enhance CF. The early work of Golbeck [37] suggested to adapt the CF formula in a way that would boost similar users whom the user trusts. More advanced approaches incorporate trust in matrix factorization techniques [60]. Finally, Wu et al. [108] extend collaborative topic regression to model social trust ensemble that reflects true friends in SRS.

**Evaluation** As reviewed throughout this chapter, evaluation of SRS typically uses the common methods in the broader RS domain (Chapter "Evaluating Recommender Systems"). These include offline evaluation, user studies (especially common for SRS), and live field studies or A/B testing. Evaluation measures include RMSE, NDCG, precision, and other commonly used metrics from the RS domain. Looking forward, since social media is characterized by the "wisdom of the crowd", it will only be natural to see more crowdsourcing techniques used for evaluation of SRS. These have become common in many domains in the recent years, including information retrieval (e.g., [4, 14, 65]), however they are not as common yet in RS evaluation. Evaluation that goes beyond accuracy to include serendipity ("surprise"), diversity, novelty, coverage, and other factors is also due in the SRS area [34]. Finally, evaluation over time, which also examines the broader effect of the recommendation on the surrounding ecosystem of users, as demonstrated in [21], is a highly desirable direction. Rather than focusing mostly on recommendation effectiveness, their broader and longer-term influence on the environment should also be considered. As another example to such research, Said and Bellogin [91] started to explore the effect of recipe recommendation within the Allrecipes.com SNS on users' health. This kind of research requires new tools and creative thinking to be brought into the existing set of evaluation methods.

**Recommending Content to Produce** We extensively discussed content recommendation in Sect. 2. Our examples focused on content the user consumes: video, news, questions, social media items, etc. As explained in the introduction, one of the key characteristics of social media is that users are not just the consumers, but also the producers of content. There is a body of research that attempts to recommend users content they may want to produce. Question recommendation in CQA sites, which has already been mentioned in Sect. 2, has a role in encouraging users to produce content in the form of answers. Other works attempted to encourage users to create more profile entries [35], inspire users to write blogs [24], and prompt them to edit articles on Wikipedia [19]. Recommending content to generate is a particularly challenging task since the entry barrier is higher as many social media users are lurkers (only consume content). It is rooted in the area of persuasive technologies and theories such as self determination [90] and behavioral models [30]. Clearly, recommending content to produce has a central role in the symbiosis between recommender systems and social media.

# 5 Emerging Domains and Open Challenges

We conclude this chapter by pointing out potential emerging domains for SRS and a few open challenges on top of the topics discussed throughout this chapter and summarized in the previous section.

## 5.1 Emerging Domains

We enumerate four domains, which we think can serve as a fertile ground for SRS research in the years to come.

**Mobile and Wearables** Recommendations for mobile devices, such as PDAs, have been suggested since the beginning of the millennium. As smartphones and tablets with advanced technologies, such as high-resolution cameras, GPS, and touch screens started to prevail, recommendation technologies adapted themselves, for example, by taking into account the user's location. The combination of mobile and social (sometimes referred to as SoLoMo—social, mobile, and location) holds new opportunities for SRS, which will combine the advanced capabilities of mobile devices with social interaction across these devices. Looking further into the future, wearable devices, such as glasses and watches, are likely to have access to even more personal information that on the one hand will provide more data for SRS to work with, and on the other hand will require more advanced recommendation techniques, so these devices can work appropriately with minimum input from the user.

**Smart TVs** RS have been quite popular in the TV domain for many years. The Netflix prize advanced this domain even further [9]. However, as TVs continue to evolve into "smart TVs", they enable many more social elements, such as sharing and interaction between watchers, which make the new TVs a social medium on its own. This provides a highly interesting opportunity for SRS to make this new generation of televisions even smarter.

**Automotive** The automotive domain is also evolving in recent years. Self-driving cars is arguably the most exciting challenge on the table, but new car models allow more collaboration between cars and their drivers. Being such an advanced instrument, the car itself plays a special role and can sometimes be treated similarly to a person, given all the information gathered through its sensors. As more collaboration is expected to characterize the new generation of smart cars, SRS can play a key role in sparing extra work from drivers and providing cars with more necessary information. We start to see this in social navigation technologies, such as Waze, but this is likely only the beginning.

**Healthcare** The healthcare domain has always been slow to adopt "social", among other things due to the special privacy concerns it entails. On the other hand, it is not hard to imagine how much this domain can benefit from more sharing and collaboration, both among patients and among doctors. In recent years, we start to see a movement towards more openness to medical data sharing. For instance, Yang et al. [109] present a case study for healthcare based on a framework for a social recommender system using both network structure analysis and social context mining. First, microblog users are recommended using exponential graph models. Then, a recommendation list is created by analysing the micro-blog network structure. Finally, sentiment similarities are used to filter the recommendation list and find users who have similar attitudes to the same topic. The recommendation results of diabetes accounts over the Weibo network demonstrate high performance. As it seems that "social healthcare" is taking off, the SRS community should consider how recommendations should be used in this domain, with all the complexities involved and the critical implications of a successful versus wrong recommendation.

## 5.2   Open Challenges

We finally highlight three more challenges for researchers in the SRS area to consider.

**Social Streams** Social streams, such as Twitter or the Facebook newsfeed, syndicate user activity within a social media site or a set of sites. Millions of users who share and interact in social media create a firehose of data in real-time that poses new types of challenges in terms of filtering and personalization. There are different types of streams in terms of the data they contain (homogenous as in

Twitter or heterogeneous as in Facebook), the source of data (a single site or a group of sites), its access-control (public or friends-only), and subscription model (following or "friending"). As demonstrated in the Twitter-related work reviewed in this chapter, the stream's data is different than "traditional" social media content: it represents an activity rather than an artifact or an entity; it is more intensive as one entity (e.g., a wiki page) may have a large amount of activities (e.g., edits); it may be very noisy (e.g., multiple wiki edits might not be of interest); its freshness is key: items that are few days old might already be irrelevant; and it is sparse in content and metadata (e.g., Twitter messages are limited to 140 characters). Due to all these unique characteristics, recommending social stream items becomes a challenge on its own within the SRS domain, and as social information continues to grow, handling this task is becoming both more challenging and more important [2, 31, 44, 48, 52, 80]. On the other hand, the stream data can also be used to model users' interests. Its fresh and concise nature can help build a user model that is up-to-date, identify changes in users' tastes and preferences in real-time, and detect global trends that may influence the recommendation strategy [33, 83].

**Beyond Accuracy and Evaluation over Time** Many of the studies we reviewed focused on measuring the effectiveness of recommendation by their accuracy. As social recommendation proliferate, it is more important than ever to consider the bigger picture when evaluating the value of recommendation. Typical beyond-accuracy measures should be considered, including serendipity, diversity, novelty, and coverage [34, 75]. In addition, the effectiveness of recommendation should be compared against the case where no recommendation would have been provided [8]. Recommendations that can make the user discover and take action regarding an item s/he would not have noticed otherwise, are obviously more valuable. In many of the works we reviewed, evaluation was based on a one-time user survey. Longer term evaluation is required as the results may substantially change over time. Techniques that learn and adapt over time based on user behavior are going to be essential. Additionally, evaluation that examines the broader effect of the recommendation on the surrounding ecosystem of users, as demonstrated in [21, 31, 47, 91] is a highly desirable direction for SRS evaluation. This requires new tools and creative thinking to be brought into the existing evaluation methods.

**Cross-Domain Analysis** As we discussed, migrating data from one social media service to another may go a long way enhancing recommendations and help deal with the cold start problem for new users. Indeed, using another site's network, tags, and other types of information have been performed by various previous systems as mentioned in this chapter. Yet, social media sites differ in many aspects. It is not certain that one's travel network can serve as a reliable source of recommendation for recipes. Similarly, the tags used in a news site context are not necessarily valuable for video recommendation. More research is due to explore the common and different among social media systems and when information can effectively port from one application to another to be used for recommendation. Cross-domain recommendations in RS have always been harder to explore since they require richer datasets and involve more complex use cases and research questions (Chapter "Design and Evaluation of Cross-Domain Recommender Systems"). As social media continues to evolve, it will be more important to explore and better understand these complexities.

**Extraction from User-Generated Content** User-generated content (UGC) is abundant across social media sites, in various forms, such as blog posts, forum threads, community question-answering, reviews, and others. Due to its sheer volume, it is often hard for social media users to find the information they seek for within UGC. Common methods allow searching UGC or sorting it by different criteria such as number of votes or date. Recent work also suggested the extraction of specific information types from UGC to provide specific types of information needs. The extraction of tips—short and practical pieces of advice—has been proposed for CQA content ("how to" questions, in particular) [107], products [58] and, prominently, in the travel domain [46, 113]. Other types of extracted information include personal experiences [82], fun trivial facts [104], locations [15], and product descriptions [76, 77]. Such extractions practically yield recommendations of new content types, such as tips, experiences, descriptions, or trivia facts, to social media users, allowing them to enjoy different aspects unlocked in UGC, which they have no realistic chance to discover by merely traversing or searching. Further use of state-of-the-art NLP methods, such as summarization and translation, can help transform free-form UGC, which is one of the most prominent characteristics of social media, into a source of multiple information types that address different needs of social media users.

**Multistakeholder Recommendation** As most recommender systems apply personalization techniques, they tend to focus on the preferences and needs of a single user. Multistakeholder recommendation has emerged d as a unifying framework for describing and understanding recommendation settings where the end user is not the sole focus [1] (Chapter "Multistakeholder Recommender Systems"). System objectives, such as fairness [112], balance, and profitability receive attention, as well as concerns from other stakeholders, such as the providers or sellers of items being recommended. This extension spans beyond people and group recommendation [41], and requires new optimization targets and evaluation metrics. While multistakeholder issues have surfaced regularly in the history of recommender systems research and have been a constant constraint in fielded applications, the recognition of common threads and research questions has been a more recent occurrence.

# References

1. H. Abdollahpouri, G. Adomavicius, R. Burke, I. Guy, D. Jannach, T. Kamishima, J. Krasnodebski, L. Pizzato, Multistakeholder recommendation: survey and research directions. User Model. User-Adapt. Interact. **30**(1), 127–158 (2020)
2. D. Agarwal, B.-C. Chen, Q. He, Z. Hua, G. Lebanon, Y. Ma, P. Shivaswamy, H.-P. Tseng, J. Yang, L. Zhang. Personalizing LinkedIn feed, in *Proceedings of the 21th ACM SIGKDD International Conference on Knowledge Discovery and Data Mining, KDD '15* (Association for Computing Machinery, New York, 2015), pp. 1651–1660
3. N.A. Alawad, A. Anagnostopoulos, S. Leonardi, I. Mele, F. Silvestri, Network-aware recommendations of novel tweets, in *Proceedings of the 39th International ACM SIGIR Conference on Research and Development in Information Retrieval, SIGIR '16* (Association for Computing Machinery, New York, 2016), pp. 913–916
4. O. Alonso, S. Mizzaro, Can we get rid of TREC assessors? Using Mechanical Turk for relevance assessment, in *Proceedings of the SIGIR 2009 Workshop on the Future of IR Evaluation*, vol. 15 (2009), p. 16
5. J. Arguello, J.L. Elsas, J. Callan, J.G. Carbonell, Document representation and query expansion models for blog recommendation, in *Proceedings of the Second AAAI Conference on Weblogs and Social Media - ICWSM '08* (2008)
6. L. Baltrunas, T. Makcinskas, F. Ricci, Group recommendations with rank aggregation and collaborative filtering, in *Proceedings of the Fourth ACM Conference on Recommender Systems, RecSys '10* (ACM, New York, 2010), pp. 119–126
7. A. Barua, S.W. Thomas, A.E. Hassan, What are developers talking about? An analysis of topics and trends in stack overflow. Empir. Softw. Eng. **19**(3), 619–654 (2014)
8. T. Belluf, L. Xavier, R. Giglio, Case study on the business value impact of personalized recommendations on a large online retailer, in *Proceedings of the Sixth ACM Conference on Recommender Systems, RecSys '12* (ACM, New York, 2012), pp. 277–280
9. J. Bennett, S. Lanning, The Netflix prize, in *Proceedings of KDD Cup and Workshop*, vol. 2007 (2007), p. 35
10. S. Berkovsky, J. Freyne, Group-based recipe recommendations: analysis of data aggregation strategies, in *Proceedings of the Fourth ACM Conference on Recommender Systems, RecSys '10* (ACM, New York, 2010), pp. 111–118
11. D.M. Boyd, N.B. Ellison, Social network sites: definition, history, and scholarship. J. Comput. Mediat. Commun. **13**(1), 210–230 (2007)
12. U. Brandes, A faster algorithm for betweenness centrality. J. Math. Sociol. **25**(2), 163–177 (2001)
13. M.J. Brzozowski, D.M. Romero, Who should i follow? Recommending people in directed social networks, in *ICWSM* (2011)
14. M. Buhrmester, T. Kwang, S.D. Gosling, Amazon's Mechanical Turk a new source of inexpensive, yet high-quality, data? Perspect. Psychol. Sci. **6**(1), 3–5 (2011)
15. J. Capdevila, M. Arias, A. Arratia, Geosrs: a hybrid social recommender system for geolocated data. Inf. Syst. **57**, 111–128 (2016)
16. J. Chen, W. Geyer, C. Dugan, M. Muller, I. Guy, Make new friends, but keep the old: recommending people on social networking sites, in *Proceedings of the SIGCHI Conference on Human Factors in Computing Systems, CHI '09* (ACM, New York, 2009), pp. 201–210
17. J. Chen, R. Nairn, L. Nelson, M. Bernstein, E. Chi, Short and tweet: experiments on recommending content from information streams, in *Proceedings of the SIGCHI Conference on Human Factors in Computing Systems, CHI '10* (ACM, New York, 2010), pp. 1185–1194
18. K. Chen, T. Chen, G. Zheng, O. Jin, E. Yao, Y. Yu, Collaborative personalized tweet recommendation, in *Proceedings of the 35th International ACM SIGIR Conference on Research and Development in Information Retrieval, SIGIR '12* (ACM, New York, 2012), pp. 661–670

19. D. Cosley, D. Frankowski, L. Terveen, J. Riedl, SuggestBot: using intelligent task routing to help people find work in wikipedia, in *Proceedings of the 12th International Conference on Intelligent User Interfaces, IUI '07* (ACM, New York, 2007), pp. 32–41

20. G. Costa, R. Ortale, Model-based collaborative personalized recommendation on signed social rating networks. ACM Trans. Internet Technol. **16**(3), 1–21 (2016)

21. E.M. Daly, W. Geyer, D.R. Millen, The network effects of recommending social connections, in *Proceedings of the Fourth ACM Conference on Recommender Systems, RecSys '10* (ACM, New York, 2010), pp. 301–304

22. S. Dara, C.R. Chowdary, C. Kumar, A survey on group recommender systems. J. Intell. Inf. Syst. **54**(2), 271–295 (2020)

23. J. Davidson, B. Livingston, D. Sampath, B. Liebald, J. Liu, P. Nandy, T. Van Vleet, U. Gargi, S. Gupta, Y. He, et al., The YouTube video recommendation system, in *Proceedings of the Fourth ACM conference on Recommender Systems - RecSys '10* (2010), pp. 293–296

24. C. Dugan, W. Geyer, D.R. Millen, Lessons learned from blog muse: audience-based inspiration for bloggers, in *Proceedings of the SIGCHI Conference on Human Factors in Computing Systems, CHI '10* (ACM, New York, 2010), pp. 1965–1974

25. C. Dwyer, Privacy in the age of Google and Facebook. IEEE Technol. Soc. Mag. **30**(3), 58–63 (2011)

26. M. Eirinaki, J. Gao, I. Varlamis, K. Tserpes, Recommender systems for large-scale social networks: a review of challenges and solutions. Future Gener. Comput. Syst. **78**, 413–418 (2018)

27. W. Fan, Y. Ma, D. Yin, J. Wang, J. Tang, Q. Li, Deep social collaborative filtering, in *Proceedings of the 13th ACM Conference on Recommender Systems, RecSys '19* (Association for Computing Machinery, New York, 2019), pp. 305–313

28. A. Felfernig, L. Boratto, M. Stettinger, M. Tkalčič, *Group Recommender Systems: An Introduction* (Springer, Berlin, 2018)

29. M. Fire, L. Tenenboim, O. Lesser, R. Puzis, L. Rokach, Y. Elovici, Link prediction in social networks using computationally efficient topological features, in *2011 IEEE Third International Conference on Privacy, Security, Risk And Trust (Passat), and 2011 IEEE Third International Conference on Social Computing (SocialCom)* (IEEE, Piscataway, 2011), pp. 73–80

30. B. Fogg, A behavior model for persuasive design, in *Proceedings of the 4th International Conference on Persuasive Technology, Persuasive '09* (ACM, New York, 2009), pp. 40:1–40:7

31. J. Freyne, S. Berkovsky, E.M. Daly, W. Geyer, Social networking feeds: recommending items of interest, in *Proceedings of the Fourth ACM Conference on Recommender Systems, RecSys '10* (ACM, New York, 2010), pp. 277–280

32. J. Freyne, M. Jacovi, I. Guy, W. Geyer, Increasing engagement through early recommender intervention, in *Proceedings of the Third ACM Conference on Recommender Systems, RecSys '09* (ACM, New York, 2009), pp. 85–92

33. S. Garcia Esparza, M.P. O'Mahony, B. Smyth, On the real-time web as a source of recommendation knowledge, in *Proceedings of the Fourth ACM Conference on Recommender Systems, RecSys '10* (ACM, New York, 2010), pp. 305–308

34. M. Ge, C. Delgado-Battenfeld, D. Jannach, Beyond accuracy: evaluating recommender systems by coverage and serendipity, in *Proceedings of the Fourth ACM Conference on Recommender Systems, RecSys '10* (ACM, New York, 2010), pp. 257–260

35. W. Geyer, C. Dugan, D.R. Millen, M. Muller, J. Freyne, Recommending topics for self-descriptions in online user profiles, in *Proceedings of the 2008 ACM Conference on Recommender Systems, RecSys '08* (ACM, New York, 2008), pp. 59–66

36. J. Golbeck, Generating predictive movie recommendations from trust in social networks, in *Proceedings of the 4th International Conference on Trust Management, iTrust'06* (Springer, Berlin, 2006), pp. 93–104

37. J.A. Golbeck, Computing and applying trust in web-based social networks. PhD thesis, College Park, 2005, AAI3178583

38. G. Groh, C. Ehmig, Recommendations in taste related domains, in *Proceedings of the 2007 International ACM Conference on Conference on Supporting Group Work - GROUP '07* (2007), pp. 127–136
39. P. Gupta, A. Goel, J. Lin, A. Sharma, D. Wang, R. Zadeh, WTF: the who to follow service at twitter, in *Proceedings of the 22Nd International Conference on World Wide Web, WWW '13, Republic and Canton of Geneva*. International World Wide Web Conferences Steering Committee (2013), pp. 505–514
40. D.F. Gurini, F. Gasparetti, A. Micarelli, G. Sansonetti, Temporal people-to-people recommendation on social networks with sentiment-based matrix factorization. Future Gener. Comput. Syst. **78**, 430–439 (2018)
41. I. Guy, People recommendation on social media, in *Social Information Access* (Springer, Berlin, 2018), pp. 570–623
42. I. Guy, U. Avraham, D. Carmel, S. Ur, M. Jacovi, I. Ronen, Mining expertise and interests from social media, in *Proceedings of the 22Nd International Conference on World Wide Web, WWW '13, Republic and Canton of Geneva*. International World Wide Web Conferences Steering Committee (2013), pp. 515–526
43. I. Guy, D. Carmel, Social recommender systems, in *Proceedings of the 20th International Conference Companion on World Wide Web - WWW '11* (2011), pp. 283–284
44. I. Guy, R. Levin, T. Daniel, E. Bolshinsky, Islands in the stream: a study of item recommendation within an enterprise social stream, in *Proceedings of the 38th International ACM SIGIR Conference on Research and Development in Information Retrieval, SIGIR '15* (Association for Computing Machinery, New York, 2015), pp. 665–674
45. I. Guy, V. Makarenkov, N. Hazon, L. Rokach, B. Shapira, Identifying informational vs. conversational questions on community question answering archives, in *Proceedings of the Eleventh ACM International Conference on Web Search and Data Mining, WSDM '18* (Association for Computing Machinery, New York, 2018), pp. 216–224
46. I. Guy, A. Mejer, A. Nus, F. Raiber, Extracting and ranking travel tips from user-generated reviews, in *Proceedings of the 26th International Conference on World Wide Web, WWW '17, Republic and Canton of Geneva, CHE*. International World Wide Web Conferences Steering Committee (2017), pp. 987–996
47. I. Guy, I. Ronen, E. Kravi, M. Barnea, Increasing activity in enterprise online communities using content recommendation. ACM Trans. Comput. Hum. Interact. **23**(4), 1–28 (2016)
48. I. Guy, I. Ronen, A. Raviv, Personalized activity streams: sifting through the river of news, in *Proceedings of the Fifth ACM Conference on Recommender Systems, RecSys '11* (ACM, New York, 2011), pp. 181–188
49. I. Guy, I. Ronen, E. Wilcox, Do You Know?: recommending people to invite into your social network, in *Proceedings of the 14th International Conference on Intelligent User Interfaces, IUI '09* (ACM, New York, 2009), pp. 77–86
50. I. Guy, I. Ronen, N. Zwerdling, I. Zuyev-Grabovitch, M. Jacovi, What is your organization "like"? A study of liking activity in the enterprise, in *Proceedings of the 2016 CHI Conference on Human Factors in Computing Systems, CHI '16* (Association for Computing Machinery, New York, 2016), pp. 3025–3037
51. I. Guy, B. Shapira, From royals to vegans: characterizing question trolling on a community question answering website, in *The 41st International ACM SIGIR Conference on Research & Development in Information Retrieval, SIGIR '18* (Association for Computing Machinery, New York, 2018), pp. 835–844
52. I. Guy, T. Steier, M. Barnea, I. Ronen, T. Daniel, Swimming against the streamz: search and analytics over the enterprise activity stream, in *Proceedings of the 21st ACM International Conference on Information and Knowledge Management, CIKM '12* (Association for Computing Machinery, New York, 2012), pp. 1587–1591
53. I. Guy, N. Zwerdling, D. Carmel, I. Ronen, E. Uziel, S. Yogev, S. Ofek-Koifman, Personalized recommendation of social software items based on social relations, in *Proceedings of the Third ACM Conference on Recommender Systems, RecSys '09* (ACM, New York, 2009), pp. 53–60

54. I. Guy, N. Zwerdling, I. Ronen, D. Carmel, E. Uziel, Social media recommendation based on people and tags, in *Proceedings of the 33rd International ACM SIGIR Conference on Research and Development in Information Retrieval, SIGIR '10* (ACM, New York, 2010), pp. 194–201

55. J. Hannon, M. Bennett, B. Smyth, Recommending Twitter users to follow using content and collaborative filtering approaches, in *Proceedings of the Fourth ACM Conference on Recommender Systems, RecSys '10* (ACM, New York, 2010), pp. 199–206

56. M.R. Hasan, A.J. Kumar, Y. Liu, Excessive use of online video streaming services: impact of recommender system use, psychological factors, and motives. Comput. Hum. Behav. **80**, 220–228 (2018)

57. J.L. Herlocker, J.A. Konstan, J. Riedl, Explaining collaborative filtering recommendations, in *Proceedings of the 2000 ACM Conference on Computer Supported Cooperative Work, CSCW '00* (ACM, New York, 2000), pp. 241–250

58. S. Hirsch, S. Novgorodov, I. Guy, A. Nus, Generating tips from product reviews, in *Proceedings of the Fourteenth ACM International Conference on Web Search and Data Mining, WSDM '21* (Association for Computing Machinery, New York, 2021)

59. M. Jacovi, I. Guy, S. Kremer-Davidson, S. Porat, N. Aizenbud-Reshef, The perception of others: Inferring reputation from social media in the enterprise, in *Proceedings of the 17th ACM Conference on Computer Supported Cooperative Work & Social Computing, CSCW '14* (ACM, New York, 2014), pp. 756–766

60. M. Jamali, M. Ester, A matrix factorization technique with trust propagation for recommendation in social networks, in *Proceedings of the Fourth ACM Conference on Recommender Systems, RecSys '10* (ACM, New York, 2010), pp. 135–142

61. A. Jameson, S. Baldes, T. Kleinbauer, Two methods for enhancing mutual awareness in a group recommender system, in *Proceedings of the Working Conference on Advanced Visual Interfaces, AVI '04* (ACM, New York, 2004), pp. 447–449

62. R. Jäschke, L. Marinho, A. Hotho, L. Schmidt-Thieme, G. Stumme, Tag recommendations in folksonomies, in *Knowledge Discovery in Databases: PKDD 2007* (Springer, Berlin, 2007), pp. 506–514

63. O. Kaser, D. Lemire, Tag-cloud drawing: algorithms for cloud visualization. Preprint, arXiv cs/0703109 (2007)

64. H. Kautz, B. Selman, M. Shah, Referral web: combining social networks and collaborative filtering. Commun. ACM **40**(3), 63–65 (1997)

65. A. Kittur, E.H. Chi, B. Suh, Crowdsourcing user studies with Mechanical Turk, in *Proceedings of the SIGCHI Conference on Human Factors in Computing Systems, CHI '08* (ACM, New York, 2008), pp. 453–456

66. J. Kunkel, T. Donkers, L. Michael, C.-M. Barbu, J. Ziegler, Let me explain: impact of personal and impersonal explanations on trust in recommender systems, in *Proceedings of the 2019 CHI Conference on Human Factors in Computing Systems, CHI '19* (Association for Computing Machinery, New York, 2019), pp. 1–12

67. R. Lempel, S. Moran, SALSA: the stochastic approach for link-structure analysis. ACM Trans. Inf. Syst. **19**(2), 131–160 (2001)

68. K. Lerman, Social networks and social information filtering on digg, in *Proceedings of the first AAAI Conference on Weblogs and Social Media - ICWSM '07* (2007)

69. J. Leskovec, D. Huttenlocher, J. Kleinberg, Predicting positive and negative links in online social networks, in *Proceedings of the 19th International Conference on World Wide Web, WWW '10* (ACM, New York, 2010), pp. 641–650

70. D. Liben-Nowell, J. Kleinberg, The link-prediction problem for social networks. J. Am. Soc. Inf. Sci. Technol. **58**(7), 1019–1031 (2007)

71. J. Liu, P. Dolan, E.R. Pedersen, Personalized news recommendation based on click behavior, in *Proceedings of the 15th International Conference on Intelligent User Interfaces - IUI '10* (2010), pp. 31–40

72. C. Macdonald, I. Ounis, The TREC blogs06 collection: creating and analysing a blog test collection. Department of Computer Science, University of Glasgow Tech Report TR-2006-224, vol. 1 (2006), pp. 3–1

73. J.F. McCarthy, T.D. Anagnost, MusicFX: an arbiter of group preferences for computer supported collaborative workouts, in *Proceedings of the 1998 ACM Conference on Computer Supported Cooperative Work, CSCW '98* (ACM, New York, 1998), pp. 363–372

74. D.W. McDonald, M.S. Ackerman, Just talk to me: a field study of expertise location, in *Proceedings of the 1998 ACM Conference on Computer Supported Cooperative Work, CSCW '98* (ACM, New York, 1998), pp. 315–324

75. S.M. McNee, J. Riedl, J.A. Konstan, Being accurate is not enough: how accuracy metrics have hurt recommender systems, in *CHI '06 Extended Abstracts on Human Factors in Computing Systems, CHI EA '06* (ACM, New York, 2006), pp. 1097–1101

76. S. Novgorodov, I. Guy, G. Elad, K. Radinsky, Generating product descriptions from user reviews, in *The World Wide Web Conference, WWW '19* (Association for Computing Machinery, New York, 2019), pp. 1354–1364

77. S. Novgorodov, I. Guy, G. Elad, K. Radinsky, Descriptions from the customers: comparative analysis of review-based product description generation methods. ACM Trans. Internet Technol. **20**(4), 1–31 (2020)

78. T. Olsson, J. Huhtamäki, H. Kärkkäinen, Directions for professional social matching systems. Commun. ACM **63**(2), 60–69 (2020)

79. T. O'Reilly, *What Is Web 2.0* (O'Reilly Media, Sebastopol, 2009)

80. T. Paek, M. Gamon, S. Counts, D.M. Chickering, A. Dhesi, Predicting the importance of newsfeed posts and social network friends, in *AAAI*, vol. 10 (2010), pp. 1419–1424

81. I. Palomares, C. Porcel, L. Pizzato, I. Guy, E. Herrera-Viedma, Reciprocal recommender systems: analysis of state-of-art literature, challenges and opportunities on social recommendation. Preprint, arXiv:2007.16120 (2020)

82. K.C. Park, Y. Jeong, S.-H. Myaeng, Detecting experiences from weblogs, in *Proceedings of the 48th Annual Meeting of the Association for Computational Linguistics* (2010), pp. 1464–1472

83. O. Phelan, K. McCarthy, B. Smyth, Using Twitter to recommend real-time topical news, in *Proceedings of the Third ACM Conference on Recommender Systems, RecSys '09* (ACM, New York, 2009), pp. 385–388

84. G. Piao, J.G. Breslin, Learning to rank tweets with author-based long short-term memory networks, in *Web Engineering*, ed. by T. Mikkonen, R. Klamma, J. Hernández (Springer, Berlin, 2018), pp. 288–295

85. L. Pizzato, T. Rej, T. Chung, I. Koprinska, J. Kay, RECON: a reciprocal recommender for online dating, in *Proceedings of the Fourth ACM Conference on Recommender Systems, RecSys '10* (ACM, New York, 2010), pp. 207–214

86. D. Quercia, L. Capra, FriendSensing: recommending friends using mobile phones, in *Proceedings of the Third ACM conference on Recommender systems - RecSys '09* (2009), pp. 273–276

87. D.R. Raban, A. Danan, I. Ronen, I. Guy, Impression management through people tagging in the enterprise: implications for social media sampling and design. J. Inf. Sci. **43**(3), 295–315 (2017)

88. P. Resnick, H.R. Varian, Recommender systems. Commun. ACM **40**(3), 56–58 (1997)

89. I. Ronen, I. Guy, E. Kravi, M. Barnea, Recommending social media content to community owners, in *Proceedings of the 37th International ACM SIGIR Conference on Research and Development in Information Retrieval, SIGIR '14* (ACM, New York, 2014), pp. 243–252

90. R.M. Ryan, E.L. Deci, Self-determination theory and the facilitation of intrinsic motivation, social development, and well-being. Am. Psychol. **55**(1), 68 (2000)

91. A. Said, A. Bellogín, You are what you eat! tracking health through recipe interactions, in *6th RecSys Workshop on Recommender Systems and the Social Web, RSWeb '14* (2014), p. 4

92. J. Sanz-Cruzado, P. Castells, Enhancing structural diversity in social networks by recommending weak ties, in *Proceedings of the 12th ACM Conference on Recommender Systems, RecSys '18* (Association for Computing Machinery, New York, 2018), pp. 233–241

93. S. Scellato, A. Noulas, C. Mascolo, Exploiting place features in link prediction on location-based social networks, in *Proceedings of the 17th ACM SIGKDD International Conference on Knowledge Discovery and Data Mining, KDD '11* (ACM, New York, 2011), pp. 1046–1054

94. S. Sen, J. Vig, J. Riedl, Tagommenders: connecting users to items through tags, in *Proceedings of the 18th International Conference on World Wide Web, WWW '09* (ACM, New York, 2009), pp. 671–680

95. S. Sidana, A combination of classification based methods for recommending tweets, in *Proceedings of the Recommender Systems Challenge 2020* (Association for Computing Machinery, New York, 2020), pp. 1–5

96. B. Sigurbjörnsson, R. van Zwol, Flickr tag recommendation based on collective knowledge, in *Proceedings of the 17th International Conference on World Wide Web, WWW '08* (ACM, New York, 2008), pp. 327–336

97. R.R. Sinha, K. Swearingen, Comparing recommendations made by online systems and friends, in *DELOS Workshop: Personalisation and Recommender Systems in Digital Libraries*, vol. 106 (2001)

98. P. Sun, L. Wu, M. Wang, Attentive recurrent social recommendation, in *The 41st International ACM SIGIR Conference on Research & Development in Information Retrieval, SIGIR '18* (Association for Computing Machinery, New York, 2018), pp. 185–194

99. I. Szpektor, Y. Maarek, D. Pelleg, When relevance is not enough: promoting diversity and freshness in personalized question recommendation, in *Proceedings of the 22nd International Conference on World Wide Web, WWW '13, Republic and Canton of Geneva*. International World Wide Web Conferences Steering Committee (2013), pp. 1249–1260

100. H. Tahmasebi, R. Ravanmehr, R. Mohamadrezaei, Social movie recommender system based on deep autoencoder network using twitter data. Neural Comput. Appl. **33**, 1607–1623 (2021)

101. A. Taneja, A. Arora, Modeling user preferences using neural networks and tensor factorization model. Int. J. Inf. Manag. **45**, 132–148 (2019)

102. L. Terveen, D.W. McDonald, Social matching: a framework and research agenda. ACM Trans. Comput. Hum. Interact. **12**(3), 401–434 (2005)

103. C.-H. Tsai, P. Brusilovsky, Explaining recommendations in an interactive hybrid social recommender, in *Proceedings of the 24th International Conference on Intelligent User Interfaces, IUI '19* (Association for Computing Machinery, New York, 2019), pp. 391–396

104. D. Tsurel, D. Pelleg, I. Guy, D. Shahaf, Fun facts: automatic trivia fact extraction from wikipedia, in *Proceedings of the Tenth ACM International Conference on Web Search and Data Mining, WSDM '17* (Association for Computing Machinery, New York, 2017), pp. 345–354

105. D. Wang, D. Pedreschi, C. Song, F. Giannotti, A.-L. Barabasi, Human mobility, social ties, and link prediction, in *Proceedings of the 17th ACM SIGKDD International Conference on Knowledge Discovery and Data Mining, KDD '11* (ACM, New York, 2011), pp. 1100–1108

106. J. Wang, Y. Zhang, C. Posse, A. Bhasin, Is it time for a career switch? in *Proceedings of the 22Nd International Conference on World Wide Web, WWW '13, Republic and Canton of Geneva*. International World Wide Web Conferences Steering Committee (2013), pp. 1377–1388

107. I. Weber, A. Ukkonen, A. Gionis, Answers, not links: extracting tips from yahoo! answers to address how-to web queries, in *Proceedings of the Fifth ACM International Conference on Web Search and Data Mining, WSDM '12* (Association for Computing Machinery, New York, 2012), pp. 613–622

108. H. Wu, K. Yue, Y. Pei, B. Li, Y. Zhao, F. Dong, Collaborative topic regression with social trust ensemble for recommendation in social media systems. Knowl. Based Syst. **97**, 111–122 (2016)

109. D. Yang, C. Huang, M. Wang, A social recommender system by combining social network and sentiment similarity: a case study of healthcare. J. Inf. Sci. **43**(5), 635–648 (2017)

110. X. Yang, Y. Guo, Y. Liu, H. Steck, A survey of collaborative filtering based social recommender systems. Comput. Comm. **41**, 1–10 (2014)
111. A. Yogev, I. Guy, I. Ronen, N. Zwerdling, M. Barnea, Social media-based expertise evidence, in *ECSCW 2015: Proceedings of the 14th European Conference on Computer Supported Cooperative Work, 19–23 September 2015, Oslo* (Springer, Berlin, 2015), pp. 63–82
112. Y. Zheng, T. Dave, N. Mishra, H. Kumar, Fairness in reciprocal recommendations: a speed-dating study, in *Adjunct Publication of the 26th Conference on User Modeling, Adaptation and Personalization, UMAP '18* (Association for Computing Machinery, New York, 2018), pp. 29–34
113. D. Zhu, T. Lappas, J. Zhang, Unsupervised tip-mining from customer reviews. Decis. Support Syst. **107**, 116–124 (2018)

# Food Recommender Systems

David Elsweiler, Hanna Hauptmann, and Christoph Trattner

## 1 Introduction to the Food Recommendation Problem

The importance of food to human life cannot be overstated. Food provides sustenance, but more than that it helps shape our identity [10]. Even in the recommender
systems literature, two food recommendation papers—from different groups of
authors—quote the idiom "You are what you eat" in the title [75, 153]. Food forms
the basis of many of our social interactions. Friends tend to have similar eating
habits [30] and our perception of others changes based on what we know about
their diet [169]. Food also has major cultural and religious significance. Different
cultures are associated with differing foods (think haggis, sauerkraut or frog's legs
and we are certain you can associate these meals with particular locations) and food
forms the basis of celebrations and ceremonies regardless of where you are in the
world [11, 210].

The importance of food can also be observed in many of the major challenges
we face in modern society. Health problems ranging from obesity and diabetes
to hypertension, heart disease and cancer, have all been associated with food
consumption. Worrying increases in the prevalence of diet-related diseases suggest

D. Elsweiler
University of Regensburg, Regensburg, Germany
e-mail: david.elsweiler@ur.de

H. Hauptmann
Utrecht University, Utrecht, The Netherlands
e-mail: h.j.hauptmann@uu.nl

C. Trattner (✉)
University of Bergen, Bergen, Norway
e-mail: christoph.trattner@uib.no

© Springer Science+Business Media, LLC, part of Springer Nature 2022
F. Ricci et al. (eds.), *Recommender Systems Handbook*,
https://doi.org/10.1007/978-1-0716-2197-4_23

that people have difficulty finding and maintaining balanced diets. Similarly, the fact that food production accounts for over a quarter greenhouse gas emissions [33], as well as deforestation and loss of bio-diversity [170] have led to the suggestion that changes to individuals' diet on a global scale could be a vital part of the solution to climate change [51, 142, 198]. These facets combine with the fact that, in many places in the world, we have never had such variety of food and food options, yet in other contexts paucity of choice has been documented [40]. Both an abundance and lack of choice have an influence on the food recommendation problem.

All of the points highlighted thus far combine to form the background to this chapter on food recommendation. For all of these reasons and more food recommendation is an important problem. But it is for the very same reasons that the problem is so challenging and worthy of scientific attention. The food recommender literature is still in its infancy and as such no formal theory specifically relating to food recommendation yet exists. There are, however, relevant theories from diverse fields, such as nutritional science (e.g., [36, 61, 140]), psychology (e.g., [164]) cultural science (e.g., [11]), and behavioural economics (e.g. [80, 175]). Food recommender research has in some cases been motivated by these theories and in some cases the results align with theoretical contributions from other fields. Where appropriate we refer to such links. Moreover, practitioners wishing to develop working food recommenders can profit from the results of past research. We summarise lessons that can be learned for these readers.

The chapter is structured as follows. In Sect. 2 we unpack the problem of food recommender systems by detailing the numerous possible facets, which could and should be addressed by researchers. In Sects. 3 and 4 we examine solutions proposed in the literature first, in terms of algorithms and second, via interfaces. In Sect. 5 we summarise evaluation methods, while 6 reviews the resources available for researchers in the field. In Sect. 7 we offer advice to practitioners based on the literature and our experience and present future aspects of research in the area.

## 2  Problem Description

In this section, we systematically define the various forms the food recommendation problem can take. We break the problem down into differing components by defining various user profiles, diverse food items that can be recommended, and different tasks, for which food recommenders can be used. A schematic illustration of the problem is featured in Fig. 1. As presented, the food recommendation problem is context-dependent and must account for different types of inputs, dietary constraints, and interfaces to generate an output, typically in the form of a ranked list. This multi-faceted structure and many other issues make the problem so complex as also acknowledged by the literature [140].

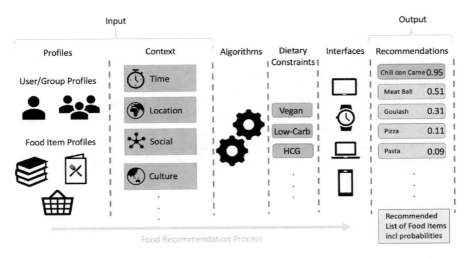

**Fig. 1** Schematic illustration of the food recommendation problem/process from input to output

## 2.1 User Roles and Groups

Food recommendations can be relevant to both individuals [55, 75] or groups of users [43]. Viewing these simply as two individual problems is an oversimplification, however, as these cases can be further teased apart. Individuals can, for instance, make personal decisions, such as what to eat themselves, at any given moment. Individuals can also, though, make decisions on behalf of a group of people, such as a family member deciding what the family will eat for dinner. At a group level, we can make a similar distinction. We may have a group that makes a shared decision, such as deciding which restaurant to visit together. Similarly, in a group setting within a restaurant, individual group members make decisions about what they are going to eat. We know from the literature, however, that what people choose will be influenced by the others present [121].

## 2.2 Item Types

At the lowest level, food items can be of two basic types. We distinguish between a *basic foodstuff*, such as an onion or carrot, and a *food product*, such as a chocolate bar. Both of these basic items can be referred to as *grocery items* in a shopping context, e.g., [5]. Basic items can be grouped into larger compounds to be recommended. The most prominent grouping in the literature is a *recipe*, which contain parts and/or multiples of both food items and food products, which would, in this context, be subsumed under the term of *ingredients*, e.g., [55]. One could argue that a food product is also the result of a recipe mixing different food items.

For practical purposes, however, we define a food product differently since a precise separation into its ingredients is often not possible and typically impractical. When examining the food actually consumed, multiple food items, products, and dishes resulting from recipes can be integrated into one *meal* that is consumed together in one temporal context [50]. The most complex meal form is a *menu*, which contains multiple predefined dishes and possibly appropriate drinks and snacks (food products), e.g., [126]. When addressing a sequence of temporal occasions, a recommender would need to provide a *meal plan*, e.g., [44]. In a set of dishes, as given in a meal or menu, the single ingredients should likely harmonize or complement each other [22]. In a meal plan, on the other hand, the sequence of recommended items should be diversified.

## 2.3   Types of Food Choices

People typically make over 200 food choices every day [195]. Just as a few examples, people have to decide what to eat, when to eat, and where to eat. People also have to decide how to attain the food they wish to eat, to either cook it themselves or to buy it in a pre-prepared state. Then they need to consider where to buy their food. In the supermarket, at the farmers market, in a convenience store, at a restaurant, or should they have it delivered. A lot of these decisions are made out of habit without conscious consideration of the consequences. It is very difficult to formulate a recommender system that can address all of these situations. The recommendation of food is a collection of diverse and distinct sub-domains that each address a different user need. Despite these sub-domains revolving around the user's interaction with food, the needs and goals of the user change in each scenario.

Different scenarios mean changing the means of assessing the suitability of food items both in terms of the type of item being recommended and the constraints considered in the recommendation process. Standing in a supermarket, for example, we may be more interested in the cost of a product compared to similar products. When we are going out with friends and searching for a restaurant, depending on our financial resources, the price may be less relevant than our current location and the mix of preferences in our group. These two simple examples hint at the complex network of constraints that surround a food decision. These incorporate aspects from the users' personality, psychological state, and physical state, through location and socio-demographic factors, to health and environmental priorities. This complexity is reflected in models of food choice in the food and behavioural sciences (see e.g. [61, 140]). In Sects. 3.2, 3.3, and 3.4 we examine such factors in detail, showing how they should influence recommendations and how they can be contradictory.

To illustrate the complexity of food recommendation in the following subsections, we take a detailed look into different food-related scenarios (cooking, grocery, restaurant, and health recommender systems). We examine their associated user profiles (single or group), food item profiles, choice contexts, and recommender tasks.

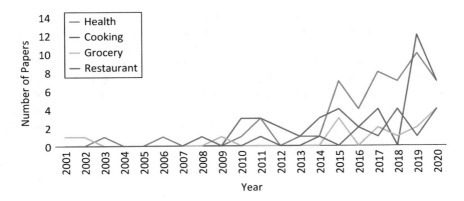

**Fig. 2** Number of papers focusing on one of the four major food recommender subdomains over the past 20 years

The first three scenarios are based on a targeted action in a fixed location context. In the cooking recommender scenario, the user is at home and intending to prepare a meal. In the grocery recommender scenario, the user is visiting a store intending to buy items. In the restaurant recommender scenario, the users are looking for a restaurant that is reachable from their location and provides menu items suiting their food-related goals. The final scenario, health, is independent of a specific location or action. A health recommender in the context of food, adds the dimension of health to the recommendation process in each of the previously listed example scenarios. An overview of the number of works addressing each scenario is given in Fig. 2.

## 2.4 Cooking Recommender

One could imagine different categories of cooking recommender users. The first is a provider (e.g., a parent or carer) cooking a meal for someone other than themselves. This could be an individual or a whole family. Their goal, in this situation, would be to satisfy the eaters' user profile, such as taste preferences and health needs. For example, in a family setting, the preferences of children are known to play an important role [134]. Individuals could also utilise a recommender for their own user profiles. One could imagine many different motivations for using such a system. For example, to diversify or extend one's diet, to learn new cooking skills, or to improve one's habits with respect to health, fitness, or sustainability. Different motivations will need to be serviced with different recommendations, and if systems fail to address these, then it is unlikely the recommender will be used. In a cooking context, the primary item type to be recommended will likely be recipes. The exception being the recommendation of substitute ingredients to make a recipe suitable, e.g., [38, 62]. From these recipe and ingredient profiles, ratings, flavour, and time requirements, are highly relevant. Allergies, intolerances, religious, and other dietary restrictions (e.g., vegetarian) are less likely to play a role since they would already have been addressed when buying the ingredients currently available in the kitchen. Important contextual constraints in a cooking scenario include those set by one's own kitchen, meaning the available food items and kitchen equipment [74]. Additionally, the current cooking skills of the user and the available time play an important role [74]. A typical cooking recommender needs to rank recipes considering these user profiles and contexts. Examples of recipe and substitute recommender systems that target the cooking context are shown in Table 1.

## 2.5    Grocery Recommender

User types in a grocery shopping context are very similar to those in the cooking context. Users are either shopping and making decisions for a whole household or for their own needs. Other relevant user profile criteria that should influence recommendations include: individual or group-specific preferences or nutritional needs [196]. In contrast to the cooking scenario, the grocery recommender focuses on single food items or food products. Such items could, of course, be derived from being ingredients in a recommended recipe. Even though price and quality are the most relevant item properties, the best price and highest quality will not always lead to a buying decision. Some criteria are hard constraints, such as individuals having allergies [162]. Others may be softer, depending on user priorities, such as price, ecological footprint or region. One could imagine systems recommending items from multiple stores (e.g., Google Shopping). In other cases, the physical location or a store preference would represent a hard contextual constraint for recommendations [32]. In this context, one typical use case is to replace food products with healthier but similar alternatives [70] fitting the overall shopping list. Examples of grocery recommender systems are summarized in Table 2.

## 2.6    Restaurant Recommender

A restaurant recommender scenario typically addresses groups of people who want to visit a restaurant together as a social event [138]. Positive user experience depends on user preferences and other aspects, such as location [205], cost [138], and cuisine [207]. It is possible, however, that different group members weigh these priorities differently. While some might prefer restaurants within walking distance, others might require affordable prices or even vegan menu items. As in the grocery context, some constraints are hard, e.g., allergies [99] and others soft, such as service quality [147]. The recommender can either focus on a whole restaurant or base the decision on the selection of available menu items for subsequent dish

**Table 1** Work on recommending recipes, menus, and ingredients to cook

| Ref | Year | User type | Goals | Item type | Dataset | Interface | Evaluation |
|---|---|---|---|---|---|---|---|
| [17] | 2010 | Group of users | Recipes for family meals | Recipe | Wellbeing diet book | Web | Offline evaluation |
| [56] | 2010 | Single user | Individual preference | Recipe | Wellbeing diet book | None | Online survey |
| [55] | 2010 | Single user | Individual preference | Recipe | Wellbeing diet book | None | Online survey |
| [58] | 2011 | Single user | Individual preference | Recipe | Wellbeing diet book | None | Offline evaluation |
| [53] | 2011 | Single user | Individual preference | Recipe | Allrecipes | None | Offline evaluation |
| [189] | 2011 | Health, single user | Similar recipes to previous intake | Recipe | Smulweb | None | Online survey |
| [174] | 2012 | Single user | Individual preference | Recipe | Allrecipes | None | Offline evaluation |
| [94] | 2012 | Single user | Create menu from available ingredients | Menu | Food | None | Offline evaluation |
| [75] | 2013 | Single user | Individual preference including health attitude | Recipe | Quizine | Web | Offline evaluation and online survey |
| [43] | 2014 | Group of users | Conversational group meal planning | Recipe | Wellbeing diet book | Mobile app | None |
| [108] | 2014 | Single user | Individual preference | Recipe | Food | None | Offline evaluation |
| [187] | 2014 | Single user | Individual preference | Recipe | AjinomotoPark | None | Online survey |
| [62] | 2015 | Single user | Ingredient substitution | Ingredient | WikiTaaable | None | None |
| [64] | 2015 | Single user | Individual preference | Recipe | Wellbeing diet book | Mobile app | Offline evaluation and user study |
| [135] | 2015 | Single user | Allergy safe substitutes | Ingredient | Cookpad | None | Offline evaluation |
| [200] | 2015 | Single user | Visual preferences | Recipe | Yummly | None | Offline evaluation and user study |
| [2] | 2016 | Single user | Contextual substitutes | Food item | MyFitnessPal | None | Offline evaluation and online survey |

(continued)

**Table 1** (continued)

| Ref | Year | User type | Goals | Item type | Dataset | Interface | Evaluation |
|---|---|---|---|---|---|---|---|
| [148] | 2016 | Single user | Gender-aware preferences | Recipe | Kochbar | None | Offline evaluation |
| [6] | 2017 | Single user | Contextual substitutes | Food item | INCA 2 | None | Offline evaluation |
| [29] | 2017 | Single user | Contextual (city size) recommendations | Recipe | Kochbar | None | Offline evaluation |
| [117] | 2017 | Single user | Individual preference | Recipe | Wellbeing diet book, food52, Allrecipes | Mobile app | Offline evaluation and user study |
| [19] | 2017 | Single user | Menu of recipes to fit template, preference, and health | Menu | BigOven | None | Offline evaluation |
| [18] | 2019 | Single user | Fitting flavour preferences | Recipe | Allrecipes, MyFitnessPal, yummly, surveyed data | None | Offline evaluation |
| [38] | 2019 | Single user | Ingredients for completing partial recipes | Ingredient | What's cooking | None | Offline evaluation |
| [59] | 2019 | Single user | Information needs during recipe cooking | Recipe and ingredient | Manual simulation | Conversational | User study |
| [76] | 2019 | Single user | Compare and substitute | Recipe and food item | FoodKG | Conversational | Use case |
| [89] | 2019 | Single user | Fitting kitchen context | Recipe | None | None | None |
| [111] | 2019 | Single user | Predict food consumption items | Food item | MyFitnessPal | None | Offline evaluation |
| [113] | 2019 | Single user | Generate personalized recipes | Recipe | Food | None | Offline evaluation and online survey |
| [127] | 2019 | Single user | Fitting flavour preferences | Recipe | Allrecipes, MyFitnessPal, yummly, surveyed data | None | Offline evaluation |

| | | | | | | | |
|---|---|---|---|---|---|---|---|
| [136] | 2019 | Single user | Missing food consumption items | Food item | Intake24 | Web | Offline evaluation |
| [137] | 2019 | Single user | Missing food consumption items | Food item | Intake24 | Web | Online survey and user study |
| [151] | 2019 | Single user | Individual preference | Recipe | Allrecipes | None | Offline evaluation |
| [181] | 2019 | Single user | Individual preference | Recipe | Allrecipes | None | Offline evaluation |
| [16] | 2020 | Single user | Recipes fitting cooking constraints (time, ability, preferences) | Recipe | Manual simulation | Conversational | User study |
| [115] | 2020 | Single user | Food pairing, recipe completion | Food items | CulinaryDB, Flavornet | None | Offline evaluation |
| [122] | 2020 | Single user | Visual preferences | Recipe | Allrecipes, MeishiChina | None | Offline evaluation and case studies |
| [126] | 2020 | Single user | Holistic profile | Menu | UniBa dataset from GialloZafferano | Web | Online survey |
| [139] | 2020 | Single user | Preference, fillingness and health of recipes | Recipe | Spoonacular | Web, conversational | User study |
| [160] | 2020 | Single user | Refrigerator based recipes | Recipe | Manual collection | None | None |
| [210] | 2020 | Single user | Cultural image bias | Recipe | AllRecepies, Xiachufang, Kochbar | None | Online survey |

**Table 2** Work on recommending food products to buy during grocery shopping

| Ref | Year | User type | Goals | Item type | Dataset | Interface | Evaluation |
|-----|------|-----------|-------|-----------|---------|-----------|------------|
| [100] | 2001 | Single user | Revenue boost | Food product | Safeway stores | Personal digital assistant | Field study |
| [114] | 2002 | Single user | Healthier choices | Food product | USDA, manual collection from bills | Printed shopping list + explanations | User study |
| [105] | 2009 | Single user and families | Next purchase | Food product | LeShop. TaFeng, Belgium retailer | None | Offline evaluation |
| [5] | 2015 | Single user and families | Healthy, allergy safe | Food product | Grocery store data lucky's market | Mobile application, augmented reality | User study and online survey |
| [192] | 2015 | Single user | Fruits/vegetables according to functional diet | Food product | Own dataset creation (Mango) | Mobile application, augmented reality | Offline evaluation |
| [196] | 2015 | Single user and families | Healthier choices | Food product | USDA, local supermarket, manually collected | Web | Field study |
| [32] | 2017 | Single user | Cheap offers | Food product | Local supermarket, manually collected | Notifications | Case studies and offline evaluation |
| [193] | 2017 | Single user | Fruits/vegetables according to functional diet | Food product | Own dataset creation (FruitVeg-81) | Mobile application, augmented reality | Offline evaluation and user study |

|  |  |  |  |  |  |  | Use case |
|---|---|---|---|---|---|---|---|
| [69] | 2018 | Single user | Similar products, healthy alternatives, preferences | Food product | None | Mobile application, augmented reality |  |
| [70] | 2019 | Single user | Similar products, healthy alternatives, preferences | Food product | Open food facts | Web, mobile application, augmented reality, hololense | User study |
| [118] | 2019 | Single user | Substitutes with better reviews | Food product | Amazon fine food reviews | None | None |
| [20] | 2020 | Single user and families | Next purchase | Food product | Instacart | None | Offline evaluation |
| [60] | 2020 | Single user | Healthier product choice | Food product | Holoselecta | Mixed reality | User study |
| [102] | 2020 | Single user and families | Diverse but repetitive purchases | Food product | Fresh food | None | Offline evaluation |
| [168] | 2020 | Single user | Completing partial grocery lists with recipes | Recipe, grocery list | AllRecepies | Mobile app | User study |

**Table 3** Work on recommending restaurants and their items

| Ref | Year | User type | Goals | Item type | Dataset | Interface | Evaluation |
|---|---|---|---|---|---|---|---|
| [138] | 2008 | Group of users | Group preferences | Restaurant | Manual collection | Mobile application, top-1-result, map view | Online survey |
| [190] | 2011 | Single user | Timing of restaurant recommendations | Restaurant | Manual collection | Widget, notification, mobile application, mapview, ranked list | Online survey |
| [186] | 2014 | Single user | Prefered menu of dishes in restaurant | Menu | Yelp | None | Offline evaluation |
| [205] | 2016 | Single user | Individual preferences | Restaurant | Baidu map | Mobile application, top-1,5,10-result, map view | Case studies |
| [207] | 2016 | Single user | Individual preferences | Restaurant | Dianping | None | Offline evaluation |
| [147] | 2017 | Single user | Individual preferences | Food trucks | Survey data | None | Offline evaluation |
| [106] | 2018 | Single user | Individual preferences | Menu | Manual collection | None | Offline evaluation |
| [141] | 2018 | Single user | Touristic food experiences | Food items | Survey data | None | Online survey |
| [163] | 2018 | User cohort | Cultural preferences | Restaurant | Twitter, foursquare, google places | Web, map view, ranked list | Online survey |
| [208] | 2018 | User cohort | Customer type preferences | Restaurant | Tripadvisor | None | Offline evaluation |
| [171] | 2019 | Single user | Individual preferences | Restaurant | Dianping | None | Offline evaluation |
| [109] | 2020 | Single user | Avoid majority bias | Restaurant | Yelp | None | Offline evaluation |
| [112] | 2020 | Single user | Aspect-based sentiments | Restaurant | Yelp | Mobile app | Offline evaluation |
| [130] | 2020 | Single user | Individual preference | Food delivery | FOODIE | Mobile app | None |
| [194] | 2020 | Groups of users | Popular menus fitting group and budget | Menu | Manual collection | Mobile app | Offline evaluation and case studies |

choices [106]. Within a chosen restaurant context, the ranking of available menu items could further consider the suitability of multiple items being combined into one meal [194]. Restaurant recommendations are very context-dependent, with location, social context, and temporal, both seasonal and daytime, context changing preferences drastically. Recommender tasks vary depending on the targeted items and supported contexts. Examples of restaurant and restaurant item recommender systems are summarized in Table 3.

## 2.7  Health Recommender

Health can be considered in any of the previously presented scenarios. Users may either want to cook healthier alternatives of a recipe, buy healthier products, or choose a healthy restaurant. The health aspect requires a multitude of additional user profile information, such as age, gender, health status, or history of behaviour. In addition to diverse user profiles, health recommendations are typically targeted at specific user groups, such the elderly [146], children [131], hospital patients [90], individuals with diabetes [8] or simply users aiming for well-being [201]. Health recommender systems must balance user taste preferences with other criteria, such as the nutritional needs of the user [86]. In addition to all abovementioned items, meal planners are especially tightly connected to health since healthy people are often unwilling to invest time and effort into meal planning (see Tables 4 and 5). The temporal context is prominent in the health recommender scenario due to the balancing act of different nutritional requirements over time [54]. While all recommender tasks of the previous scenarios are relevant for healthy recommendations as well, personalization is a crucial aspect of these recommenders' success. One recent advance in this area is the research in personalized nutrition according to genome and microbiome information [204]. Examples of health recommender systems in each of these categories are shown in Table 4 and in Table 5.

## 3  Algorithms

In this section we examine algorithmic solutions to the food recommendation problem. First, in Sect. 3.1 we summarise research on the core problem, which focuses on recommending food items that appeal to users and which has typically been formalised as a prediction or ranking task. Predicting food choices is highly context dependent. In Sect. 3.2, we examine the evidence for which context variables can influence food choices and as such, should be accounted for by food recommenders. Finally, in Sect. 3.4, we look at formalisations of the food recommendation problem that go beyond predicting appreciation to incorporate the complexities outlined in Sect. 2. We conclude each section by reflecting on how the contributions made in the literature relate.

**Table 4** Work on recommending healthy food for special target groups such as patients in hospitals, diabetes patients, children, and elderly. Table is sorted first by category and then by year

| Ref | Year | User type | Goals | Item type | Dataset | Interface | Evaluation |
|---|---|---|---|---|---|---|---|
| [199] | 2014 | Children, single user | Reduce weight of obese children in clinic | Meal Plan | TipTopf (Book) | Mobile App | Preliminary Field Study |
| [131] | 2017 | Children, single user | Healthy nutrients and amount for toddlers | Recipe | FoodFacts, USDA, allrecipes, myrecipes, foodnetwork, bigoven, yummly, saymmm, jaimie oliver, kidspot kitchen, epicurious, cooks, eatingwell | Web | Offline Evaluation |
| [129] | 2019 | Children, single user | Provide children with menus based on previous intake of nutrients | Menu | National food safety information, manual collection | Mobile app, smart plate | User study |
| [101] | 2010 | Diabetes, single user | Food groups and calories for next meal | Food Groups | Ontologies | Desktop | Expert Evaluation |
| [8] | 2018 | Diabetes, single user | Context aware healthy diet and activity | Menu | Ontologies | Mobile App | Case Studies and Expert Evaluation |
| [81] | 2020 | Diabetes, single user | Preference and consistent carbohydrate intake | Recipe | FoodKG | Conversational | None |
| [90] | 2003 | Disease, single user | Expert adjusted hospital menus | Menu | Cbord | Desktop | Case Studies |
| [40] | 2013 | Disease, single user | Menus-items based on eating behaviour in care facilities | Menu-items | None | None | None |

| | | | | | | | |
|---|---|---|---|---|---|---|---|
| [83] | 2018 | Disease, single user | Recipes based on hospital patient illnesses | Recipe | Ontologies, genuinekitchen, hospital data | Mobile app | None |
| [165] | 2019 | Disease, single user | Hypertensive patients | Meal plan | Weighlessmauritius | Mobile app, web | User study |
| [1] | 2006 | Elderly, single user | Avoid malnutrition in older adults | Meal plan | Unknown | Desktop | User study |
| [47] | 2016 | Elderly, single user | Avoid malnutrition in older adults | Meal plan | Ontologies | Desktop | None |
| [144] | 2017 | Elderly, single user | Avoid malnutrition in older adults | Meal plan, grocery list | Portuguese food composition database, manual collection | Mobile app | User study |
| [145] | 2017 | Elderly, single user | Avoid malnutrition in older adults | Meal plan, grocery list | Portuguese food composition database, manual collection | Mobile app | None |
| [146] | 2018 | Elderly, single user | Avoid malnutrition in older adults | Grocery and food delivery | Unknown | Mobile app | None |
| [25] | 2020 | Elderly, single user | Age-centered design for older diabetes patients | Recipe | Manual collection | Mobile app | User study |
| [152] | 2020 | Elderly, single user | Avoid malnutrition in older adults | Meal plan | DGE | Mobile, web, conversational | Interviews |

**Table 5** Work on recommending food for health and wellbeing

| Ref | Year | User type | Goals | Item type | Dataset | Interface | Evaluation |
|---|---|---|---|---|---|---|---|
| [14] | 2011 | Families | Sharing healthy recipes among family members | Recipe | Manual collection | Web | User study |
| [57] | 2011 | Single user | Recommend prefered recipe fitting health rules of a manual meal plan | Recipe | Wellbeing diet book | | User study |
| [188] | 2011 | Single user | Health conditions | Recipe | Cookpad | Mobile app | Offline evaluation and online survey |
| [23] | 2015 | Single user | Genotype and phenotype personalization | Food groups | Manual collection | Email | User study |
| [45] | 2015 | Single user | Healthier recipe alternatives and meal plans | Recipe and meal plan | None | None | None |
| [44] | 2015 | Single user | Meeting daily nutritional requirements with prefered meals | Meal plan | Quizine | Web | Offline evaluation |
| [65] | 2015 | Single user | Caloric requirements and preference fitting recipes | Recipe | Unknown | Mobile app | None |
| [73] | 2015 | Single user | Meeting daily nutritional requirements with prefered meals | Meal plan | Unknown | Web | None |
| [128] | 2015 | Single user | Blood value and activity sensor based meal recommendation | Meal plan | Unknown | TV, mobile app | Preliminary field study |
| [204] | 2015 | Single user | Blood value and microbiome personalization of recipe recommendations | Recipe | Manual collection | None | User study |

| | Year | User | Goal | Output | Data source | Platform | Evaluation |
|---|---|---|---|---|---|---|---|
| [3] | 2016 | Single user | Nutrient optimization based on health profiles | Food items | Health calabrian food database | Web | None |
| [49] | 2016 | Single user | Meal plans for athletes | Meal plan | Unknown | None | Expert evaluation |
| [42] | 2016 | Single user | Genotype and phenotype personalization | Food groups | Manual collection | Email | User study |
| [24] | 2017 | Single user | Genotype and phenotype personalization | Food groups | Manual collection | Email | User study |
| [46] | 2017 | Single user | Healthier recipe alternatives | Recipe | Allrecipes | None | User study |
| [180] | 2017 | Single user | Healthier recipe alternatives and meal plans | Recipe and meal plan | Allrecipes | None | Offline evaluation |
| [201] | 2017 | Single user | Visual preferences and macro-nutrients | Recipe | Yummly | Web | User study |
| [206] | 2017 | Single user | Healthy food group balance | Food groups | Manual collection | Web, mobile app | None |
| [7] | 2018 | Single user | Matching diet and activivty recommendations | Menu | Manual collection | None | None |
| [22] | 2018 | Single user | Harmonized and nutrient optimized menus | Menu | TudoGostoso, allrecipes, tabela de alimentos | None | Offline evaluation |
| [68] | 2018 | Single user | Healthy recipes matching nutrient optimized pseudo-recipe | Recipe | Allrecipes, yummly, USDA | None | Offline evaluation |
| [103] | 2018 | Single user | Personal nutrient requirements | Recipe | KochWiki, BLS | Mobile app | User study |
| [28] | 2019 | Single user | Healthy recipes matching nutrient optimized pseudo-recipe | Recipe | Allrecipes, yummly, USDA | None | Offline evaluation |

(continued)

**Table 5** (continued)

| Ref | Year | User type | Goals | Item type | Dataset | Interface | Evaluation |
|---|---|---|---|---|---|---|---|
| [48] | 2019 | Single user | Healthy food group balance | Food groups | Manual collection | Web, mobile app | User study, interviews, and expert evaluation |
| [63] | 2019 | Single user | Macro-nutrient distribution of calories | Recipe and meal plan | FooDb, philippine food composition table, my food data, USDA manual recipes | Mobile app | User study and expert evaluation |
| [76] | 2019 | Single user | Health and allergy aware recipe recommendr based on available ingredients | Recipe | FoodKG | None | None |
| [85] | 2019 | Single user | Healthy market items and fitting recipes | Food items, recipes | Weibo, unknown | Mobile app | None |
| [91] | 2019 | Single user | Multi-criteria recipe recommender | Recipe | Geniuskitchen | None | Online survey |
| [157] | 2019 | Single user | Personal nutrient requirements and difficulty | Recipe | Alber heijn, nevo | Mobile app | User study |
| [177] | 2019 | Single user | Preference and nutrient requirements for health conditions | Meal plan | Wander, manual collection | Mobile app | Offline evaluation and case studies |
| [9] | 2020 | Single user | Personal recipe generation | Recipe | Food.com | None | Offline evaluation |
| [26] | 2020 | Single user | Preference ratings and calorie content | Recipe | foodRecSys-V1 | None | Offline evaluation |
| [87] | 2020 | Single user | Ethical aspects of personalization | All | None | None | Case studies |
| [150] | 2020 | Single user | Personal food experience | All | NutritionIX, USDA | None | None |

## 3.1  Food Recommendation as a User-Item Ranking Problem

Regardless of the type of food item being recommended, the variant of the food recommendation problem that has received the most research attention is the most basic formulation where only user food preferences are accounted for. The system suggests food items (i.e. recipes, meals, products etc.) that are estimated to appeal to or be appreciated by users. Typically, researchers have formulated this as a prediction task. The aim being to "learn" the preferences of system users and make predictions for how a given user would rate a given recipe. Systems can then be compared with respect to, for example, the extent to which the predicted ratings deviate on average from the actual ratings users provided (e.g., [55, 75]). A second popular approach—and this is becoming the standard option in the literature—is to formulate the recommendation task as a ranking problem (e.g., [180, 181]). Here the aim is to provide a ranking of food items for users where the items ranked highest are predicted to be appropriate for or most likely to be accepted by the user.

Most of the key algorithmic approaches from the general recommender systems literature (i.e. content-based in Chap. "Semantics and Content-based Recommendations", collaborative filtering in Chap. "Advances in Collaborative Filtering", knowledge-based and hybrid approaches) have been evaluated with data relating to food items of different types. There is little clear evidence to suggest that one approach performs better in any particular situation.

For ease of narrative, we focus the examples in this section on recipe recommendation as the majority of the of the literature has focused on this kind of food item. Recipe recommender systems provide recommendations for the user(s) to cook based on their profile(s) and potentially additional constraints, such as context information. Nevertheless the reader should bear in mind that similar core algorithmic approaches have been tested on other food and food-related items, such as restaurants, e.g., [138, 207, 208], cooked meals from menus [88, 92], meal plans [44, 94, 180] and products in supermarkets [100].

When content-based methods are used, items (typically recipes) are represented based on the contained ingredients and the similarity is estimated between items, as well as between items and user profiles modelled in different ways. The representations employed have varied with vector based representations [55, 75], topic model–based representations [95], dependency-tree representations [84], and multi-modal representations that account for different aspects of recipe content [123] all being utilised.

Diverse collaborative filtering-based methods have also been applied, ranging from nearest-neighbour approaches [55, 75] to singular value decomposition [75] and matrix factorization [64, 65] to latent dirichlet allocation and weighted matrix factorization [180]. All of these approaches allow the interaction between user and food items to be exploited in the recommendation process. Finally, hybrid methods have tried to combine the ideas behind the different content and collaborative filtering approaches. Two good examples of hybrid approaches are Freyne and Berkovsky's [55] combination of a user-based collaborative filtering method with a content-base method. The same authors also tested a second hybrid technique where three different recommendation strategies were combined in a single model, with the exact strategy followed being determined by the ratio of the number of items rated by a user and number of items overall. Such hybrid approaches can be very effective. For example, in the experiments reported on by Harvey and colleagues [75], the best performance was achieved using an SVD approach augmented with content-based biases (see also [184]). A further hybrid approach is to exploit knowledge about the user and her goals, preferences and mood, as well as knowledge about the content and properties of food-items. Musto and colleagues evaluated what they refer to as a knowledge-based recommender with mixed results [126]. Many of the factors did not correlate with preferences, but there were hints that knowledge about gender and mood can be useful (we discuss these and other context factors in greater detail below). Other Knowledge-based recommenders have been proposed in the literature (e.g., [72, 82]), but lack of data resources have hindered their utility and also evaluation. Recent developments may change this trend. Haussmann and colleagues [76] recently presented a unified diet related knowledge graph incorporating aspects of foods and ingredients, nutritional knowledge and medical conditions, as well as how these relate. Incorporating such knowledge into the recommendation process could allow better personalised or context sensitive recommendations, for example, to account for available ingredients or cooking equipment.

A detailed overview of the algorithmic approaches that have been tested in a food recommendation context in [179]. Nevertheless, it is difficult to draw direct conclusions when comparing the contributions summarised above as the experiments performed were conducted using different, often small, proprietary datasets. The experiments were, moreover, often setup differently (i.e. to minimise predicted rating errors vs rank problem), with different algorithmic implementations being evaluated. One of the few trends one can find in the results is, for example, that both Harvey et al. [75] and Freyne and Berkovsky [55] report ingredient based CB methods to outperform CF approaches.

Trattner and Elsweiler [181] went some way to resolving the issue of comparability of results by testing several recommender algorithms using standardised implementations and a large, naturalistic dataset sourced from the online recipe portal allrecipes.com. This analysis showed a different outcome to Harvey et al. and Freyne and Berkovsky in that collaborative filtering methods clearly outperformed content-based approaches in their experiments. The results, while contradictory, are not incompatible. Further analyses by Trattner and Elsweiler where the user sample size was varied show that CF methods only start to outperform content approaches when a sample of 637 users was tested (roughly 50% of the data set). In both of the previous studies much smaller samples were employed. While unsurprising this result highlights the dangers of over interpreting studies with small, homogeneous samples.

A second point regarding providing food recommendations purely-based on user taste profiles is that there has been surprisingly little work exploiting different content-based aspects such as how the food looks and how the food tastes. This is despite the fact that we know that human food choice to be highly visual [35, 110] and that flavour preferences vary [107, 202]. Some recent studies have shown that visual information encoded within recipes accessed via online recipes can be used to predict user preferences. For example, Yang and colleagues learned users' visual food preferences and were able to improve predictions with their models [201]. Elsweiler and colleagues showed that low-level image features, such as brightness, sharpness and contrast in photos, can be used to predict user preferences from recipe pairs [46]. Zhang et al. [209] tested the same features along with deep neural networking approaches on the images on three large recipe portals. The results showed that both approaches were able to offer predictive power, but the DNN approaches worked best. There is considerable scope to investigate how visual information can be best combined with other information in hybrid-approaches. Similarly, early work on the relationships between the flavour components in recipes from online portals suggest strong patterns can be observed in tastes, particularly in different geographical locations [4]. Accounting for different aspects of content is particularly important given the low performance found on food datasets.

This is a third striking observation relating to algorithms for food recommendation: Standard recommendation algorithms perform significantly less well when used for recommending food items than on other problems, such as movies or online purchases in an e-commerce context. As a point of comparison, for example, Rendle and colleagues' [143] experiments when evaluating BPR (Bayesian Personalized Ranking from Implicit Feedback) against other modern and benchmark algorithms on movie and online purchase data achieved similar performance on both data sets. The AUC values achieved were consistently above .85 with several algorithms on both collections with the best performance found being 0.89 on the e-commerce data set and 0.93 ). Using the same algorithms on two different recipe datasets, however, attained much poorer performance scores ($AUC_{max}$=0.71). There are many possible reasons for this. One such explanation is that it has less to do with the kind of item being recommended and more to do with the property of the dataset, e.g., how dense the ratings are for users. Trattner and Elsweiler also experimented with this aspect by taking samples with different groups of users and items [181]. This did not, however, influence the maximum performance achieved. Yet another potential explanation is that food taste profiles are less stable than, for example, preferences for movies. As we discuss in detail in the following subsection, food habits are extremely context-dependent—what people eat depends on who they are with, where they are, financial and time constraints, as well as, as we shall see, many further factors.

## 3.2   Context-Dependent Food Recommendations

Context-dependent food recommenders alter the recommendations the provide to account for aspects of context. As has been demonstrated in evaluations of other kinds of recommended items, such as music [15], movies [154] and hotels [104], the appropriateness of recommended food-items is highly context-dependent. Using naturalistic data collected via online food-portals, e.g., [29], as well as laboratory studies, e.g., [75] researchers have gleaned insight into which contextual variables influence the acceptability of food-item recommendations.

Harvey and colleagues [75] studied the reasons why recommendations were considered suitable or unsuitable by participants in a longitudinal, naturalistic study. Many of the reasons provided related not to taste or to the content of the suggestion, but to the relationship between the recipe and the current context. For example, a recipe may have been appealing, but the participant, at that moment, sometimes lacked the time or cooking equipment necessary to prepare the meal, making the recommendation unsuitable.

Time is an important contextual factor for food recommendation. There is strong evidence suggesting that the food items users prefer varies seasonally. Numerous investigations have identified temporal trends in food choices, including the nutritional content, style of food and ingredients contained in recipes. These temporal patterns have been discovered in analyses of food related Tweets [67], in interaction data from online recipe portals [97, 191] and for searches for food using web search engines [197].

Food serves more that sustenance and relates to identity, health and well-being, social relationships and ritual [11]. As such, food choice is culturally embedded and culture becomes an important context variable for food recommender systems [209, 210]. In recommender systems and related fields culture has typically been operationalised using location. Several analyses have taste preferences to vary within (e.g., [185, 212]) and between (e.g., [191, 209]) countries. The size of the city has also been shown to be a location-related factor [29], as has the availability of food in specific locations [39]. Examining ingredients contained within recipes, as well as the flavour components making up ingredients Ahn and colleagues revealed that ingredients that are often paired in recipes vary across geographic regions [4]. Whereas Western cuisines show often use ingredient pairs that share many flavour compounds, East Asian cuisines tend to avoid compound sharing ingredients. Sajadmanesh analysed the content of crawled recipes from 200 different cuisines, identifying strong geographical and cultural similarities on recipes, health indicators, and culinary preferences [155]. These analyses focused on ingredient, flavour and nutritional content of recipes, whereas Zhang and colleagues examined the visual aesthetics of online recipes and how this varies across countries [209]. Again they identified strong regional variation. They found that food images perceived as attractive vary between users of German, American and Chinese food portals, but the visual ideals of German and American users seem to be closer than to those of the users of the Chinese portals. The empirical findings of these food recommender related studies align well with the theories underpinning food choices in other domains, which underline the complex, multi-facted, socially influenced, and personally variable (see e.g. [61, 140]) nature of the problem.

## 3.3    Preferences Vary Across User Groups

The research efforts summarised in Sect. 3.1 have shown that recommendations can be improved by personalising items to user taste profiles. Section 3.2 used further research to argue that personalisation efforts can be improved by accounting for context-factors. Here, we take this one step further, citing evidence suggesting that food recommender systems should behave differently for different groups of users to reflect varying food preferences groups. Gender is a good example of this, with evidence suggesting that male and female users prefer different dishes, make use of different spices and own and utilise different kitchen utensils [148]. Harvey and colleagues grouped users based on their attitude to healthy eating [75]. In their study, a small group of users who identified themselves as being health conscious behaved in a manner which reflected this. These participants rated meals with higher fat and calorific content negatively whereas they rated lower-fat, lower-calorie dishes more favourably. In the remaining participant sample no such relationship was found. Other research indicates that similar groupings will occur in different contexts. For example, hardened meat eaters should be supported differently to those open to transitioning to vegetarian or vegan diets [13]. Even within the latter group users may be grouped by their receptivity to meat-replacement products [159].

## 3.4    Variations on the Food Recommendation Problem

As emphasised in the introduction, what makes food recommendation such an interesting and challenging domain is the fact that the problem itself varies as the user or users have different needs, goals and priorities in different situations.

One way of thinking about more complicated food recommendation situations is to treat these as multi-objective optimisation problems. For example, as people often eat socially, group recommendation becomes important. In group recommendation situations recommendations are optimised to suit multiple taste profiles with the preferences of different users being traded-off or balanced against each other [17, 43]. Similarly, in a health context, the recommendations are not only derived such that they cater for user taste preferences, but also for some additional property that accounts for the healthiness of the food. This is important both for users who wish to prevent illness [45, 65, 180] and those who are ill and wish to manage systems or recover, e.g., diabetes patients [8, 25, 101, 166]. Food choices are increasingly made considering the environmental impact of the dish [13, 79]. Again, as with health-aware systems, recommendations must trade-off the user's food preferences with some measure of environmental impact to satisfy this goal. This could be measured, for instance, in terms of food miles [79], carbon dioxide emissions [13, 41] or some other combined metric of environmental impact, e.g., [34].

As the literature is most pronounced in the area of health, we will focus on this domain. However, with few exceptions the technical solutions that have been employed to derive healthy in food recommendations could equally be applied for sustainability.

**Health-Aware Food Recommendations**

One means of providing health-aware food recommendations is to alter the recommendation algorithms to account for some health property. Ge and colleagues [65] achieve this by means of a calorie balance function of the difference between the calories the user still needs (calculated based on foods eaten that day and an estimation function) and the calories of the recipe. The smaller the difference is, the healthier the recipe is estimated to be. This assumes that calorie balance is one indicator for health. A similar approach was suggested by Elsweiler et al. who proposed optimising recommendations based on a weighted linear combination of recipes predicted to score highly with respect to user taste, and low distance from an estimated nutritional requirement [45].

Trattner and Elsweiler experimented with post-filtering methods where recommendation rankings were altered. In such an approach each item (recipe) for a particular user is re-weighted according to a scoring function relating to one of two health metrics (The WHO or inverse FSA metrics discussed above) for the recipe [180]. Their results demonstrated far superior approaches to the linear combination suggested previously by Elsweiler et al.

Ueta and colleagues [188] presented a system, which aimed at recommending recipes with particular health-related goals in mind. The starting point is a user provided query that provides context for recommendations, e.g., 'I want to cure my acne' or 'I want to recover from my fatigue'. To achieve this a co-occurrence matrix was established between 45 common nutrients and nouns, such as 'cold', 'acne', 'bone' etc. By creating a nutritional profile for the user query, recommendations were provided from a large pool of recipes sourced from the recipe portal Cookpad.com.

Moving away from simply recommending different items by re-scoring existing recipes, Chen and colleagues algorithmically generate "healthy" pseudo recipes. A pseudo-recipe consists of a list of ingredients and respective quantities, with the nutritional values of the pseudo-recipe matching the predefined targets as best is possible [28]. To generate a pseudo-recipe, the authors propose an embedding-based ingredient predictor, which represents all ingredients a latent space and predicts the supplemented ingredients based on the distances of ingredient representations. A second component computes the quantities of the supplemented ingredients. The framework was tested on two large online recipe portals, allrecipes.com and yummly, with the experiments showing that the approach is able to improve the average healthiness of the recommended recipes without requiring any pre-computed nutritional information for the recipes.

Rather than generating completely new recipes, another means of making recipes more healthy is to substitute one or more ingredients within recipes to improve the "healthiness" of individual recipes. Scholars have investigated different methods of generating plausible substitutions. Achananuparp and Weber, for example, used food diary data to test several approaches inspired by the distributional hypothesis in linguistics, that is, foods or ingredients that are consumed in similar contexts are more likely to be similar dietarily and can therefore be treated as substitutes [2]. A crowd-sourced evaluation demonstrated the feasibility of such an approach. A different approach to the same problem was taken by Gaillard and colleagues, who extended a generic case-based reasoning system to handle both ingredient substitution and ingredient amounts using a formal concept analysis and mixed linear optimization [62]. Teng et al. used the user comments associated with recipes to generate an ingredient substitution network [174]. Comments in the form of, for example, "I replaced the egg with soya flour to make the cake vegan" were first parsed patterns matching the form of "replace a with b", "substitute b for a" etc, were isolated and matched against lists of ingredients. This allowed a directed ingredient substitute network to be built representing users' knowledge about which ingredients could be substituted. Rather than examining the feasibility of substitutions, the utility of the network was shown in a recommendation task where the system should predict, from a given pair of similar recipes, which one has higher average rating.

How such substitution approaches work in health contexts, however, have yet to be evaluated. Initial steps in this direction were taken by Kusmierczyk and colleagues whose findings illustrate that to some extent it is possible to recommend a user substitute ingredients based on the their previously uploaded recipes and accounting for context information [98]. A further initial effort was published by who performed clustering analysis of foods with diabetic patients in mind. Employing Self-Organizing Maps and K-mean clustering on nutritional components of food items, in order to provide appropriate food item substitutions for diabetic patients.

A different algorithmic approach with health in mind is not to recommend meals as independent items, but to group them to create dietary plans [44, 57]. This fits with nutritional advice suggesting that individual meals themselves are not unhealthy, but rather should combine to create a balanced diet [54].

Lee et al. [101] proposed a system incorporating an ontology, personalisation and fuzzy logic as means to utilise uncertain data and knowledge to create meal plans for diabetic patients. Domain experts were used to evaluate the output of the system and while the details on the evaluation are minimal, the authors claim that the evidence shows that the proposed approach can work effectively and that the menus can be provided as a reference for patients.

In a research project associated with malnutrition in the elderly, Aberg [1] proposed a menu-planning tool which accounted for several sources of information and constraints, many of which were discussed in Sect. 3.2. Aberg accounted for user taste preferences and dietary restrictions (e.g., allergies); the nutritional make up of recipes; how long a meal would take to prepare, as well as how difficult it is to prepare; the cost and availability of the ingredients and the variety of meals in terms of used ingredients and meal category. To account for all these requirements, Aberg employed a design combining diverse technologies including collaborative filtering, content-based and constraint-based recommendation. The constraint-based component constructed the optimal meal plans. The paper presents a prototype system, which recommends meal-plans to users over particular time periods. Although the authors describe an ongoing user-base, to our knowledge no formal evaluation of the system was published.

Elsweiler and colleagues [44] evaluated a meal planning algorithm for a more general user group whose goal is to nourish themselves for well being. Rather than recommending individual food items, a sequence individual items, optimal under certain criteria are recommended. Starting from a personalised recommended recipe ranking generated from a recommender (as described in Sect. 3.1), the algorithm combines two main meals (for dinner and lunch) with a breakfast, plus an allowance for drinks and snacks such that the user's daily nutritional needs are met. Nutritional needs were calculated firstly by estimating the daily calories that should be consumed by the user and then breaking this down to determine where the calories should be sourced (e.g., from proteins, carbohydrates, fats etc.). A simulated study was devised to test the approach systematically. Plans were created based on recommendations for given user taste profiles mined from a naturalistic dataset, such that they met the needs of diverse personas. Personas were defined user profiles that included details which influence nutritional required nutritional intake, such as height, weight, gender, age, nutritional goal (lose/gain/maintain weight) and activity level (from sedentary to highly active) [44]. While the meal-plan generating component is far simpler than that proposed by Aberg, the evaluation presents analyses on the properties of the generated plans. For example, in addition to testing the feasibility of creating plans that meet theoretical user nutritional needs, the authors explore how plans relate to user taste preferences, as well as diversity

**Recommendations for Behavioural Change**
Many of the technical and empirical contributions described in this section have the goal of behavioural change of some kind. That is, systems aim to alter the eating habits of the user for his or her own benefit. For example, to prevent illnesses associated with being overweight or obese, healthier recipes, meal plans or food items are recommended. In the literature, technical or empirical contributions have been described in conjunction with theories or frameworks from other fields, such as behavioural sciences. In the remainder of this section we describe contributions from the recommender systems literature that reference or were inspired by such concepts.

The behavioural sciences have provided several theoretical frameworks to discuss behavioural change. One such concept, bounded rationality, allows decisions to be influenced by accounting for the cognitive limitations of the decision-maker. Two competing ideas associated with bounded rationality are those of 'nudges' and 'boosts'. 'Nudges' are interventions that steer people in a particular direction while preserving their freedom of choice (e.g., [71, 161]). Boosts, on the other hand, are educational in nature, where interventions provide the user with the knowledge or tools to make better decisions for themselves (see [80] for a detailed comparison of these approaches). In the food recommender literature 'nudging' has been explicitly referenced by Elsweiler, Trattner and Harvey [46] who employed machine learning techniques with low-level image properties of recipes photographs to predict, given pair of similar recipes, which would be preferred by the user. In their experiments they show that when two randomly chosen similar recipes are chosen, users chose the recipe with the highest fat content most often. However, when pairs were selected such that the recipe with the lowest fat content was predicted to be the most visibly attractive, this trend was reversed. As has been demonstrated in diverse contexts from politics to energy consumption [175], this kind of 'nudge' can be a powerful means to influence individual user choices. One major limitation of 'nudges', however, is that no learning takes place and when the nudge is no longer applied 'normal behaviour' returns. There are also major ethical discussions regarding the freedom of choice for users. By educating users via interventions 'boosting' is advantageous in both these respects. To our knowledge, the food recommendation literature has not explicitly referenced 'boosting'. However, explanations for recommendations, which have been suggested by several scholars (see Sect. 4) would correspond to this kind of intervention. For example, an explanation in the form of:

> This recipe was recommended because it contains lentils, which you like. Lentils are an excellent source of B vitamins, iron, magnesium, potassium and zinc.

would correspond to a 'boost' as described by Hertwig and Grüne-Yanoff or an 'educative nudge' as defined by Sunstein [172].

Most studies have been limited to decision-making in individual moments. Few studies have monitored users over a longer time period to establish long-term behavioural change, which would be more difficult and require ingrained personal and social practices to be adapted [157]. Starke argues that whereas it is easy to achieve a smaller changes in behaviour, the example he provides is eating two cookies a day instead of four, moving away from one's ingrained behaviour is challenging. To account for the difficulty of changes, Schäfer and Willemsen [157] proposed the use of the psychometric Rasch model to conceptualise nutrient intake as a single-dimensional construct. They argue that a user's willingness or ability to make changes can be observed via current behaviours. This means that easier changes are associated with high probability of success, while changes associated with higher costs or difficulties are less likely. This can be used to tailor the user's goals and make change more manageable. Schäfer and Willemsen investigated the idea of tailoring the goals of a nutrition assistance system based on the user's abilities according to a Rasch scale. Evaluating two versions of a mobile system that tracked the user's diet and personalised recipe recommendations. The control version targets optimal nutrition and focuses on the six nutrients most necessary of change. The experimental version tailors the advice to the next six achievable nutrients according to a Rasch scale. The results of a two-week study indicate that the tailored advice led to higher success for the focused nutrients and was perceived by users to be more diverse and personalised and therefore more effective.

# 4    Interfaces

The evidence suggests that when and how food recommendations are provided influences how users interact with these. For example, Trattner and colleague's study of how online recipes are interacted with over time emphasises that despite the similar functionality and look and feel of the food portals Kochbar.de and Allrecipes.com, the presentation of recommendations has a strong influence on whether or not they will be bookmarked by users [184]. In this section we wish to provide the reader with an overview of the various interface options that have been proposed for food recommender systems. We summarise different interface components that one can find in the literature, providing examples as necessary. We further give one example of a popular commercial application for each of the scenarios defined in Sect. 2.

## 4.1    Presenting and Accessing Recommendations

Trattner and colleagues [184] demonstrate that since Kochbar only promotes newly uploaded recipes, there is a sharp drop-off in the frequency of bookmarks after a short amount of time has passed. This is not the case on Allrecipes.com where recipes have a longer active lifespan.

The importance of recommendation presentation is also emphasised by Chen and Keung Tsoi's [27] results when comparing three common layout designs: list, grid and pie. Whereas with list and grid format interactions with items tended to be focused on the top-3 recommendations, interactions were more evenly distributed in pie layout. The non-linear formats seemed to be preferred by users and additional effects, such as in increasing users' confidence in their decisions.

Current online food portals typically display recommendations as a simple list (e.g., kochbar.de) or in a grid format (e.g., allrecipes.com, cookpad.com). There are many such portals with similar functionality. Despite being three of the most commonly referred to sites in the recommender systems literature, none of these services provide personalised recommendations to users. Rather, they make general recommendations and combine these with faceted search interface.

Research prototypes described in the literature make use of the same display approaches. One can find recommendations presented in a list format, e.g., [73, 188] and as a grid, e.g., [43]. To our knowledge no system has been published where food item recommendations are presented using the pie layout.

Different pieces of information are typically shown for recommended items. Geleijnse [66], for example, present a graphical representation of the healthfulness of a recipe, which may influence whether this meal is cooked or not.

Elahi and colleagues present an interface, which allows users to clarify their needs using tags [43]. Not only do these tags help users specify their preferences,

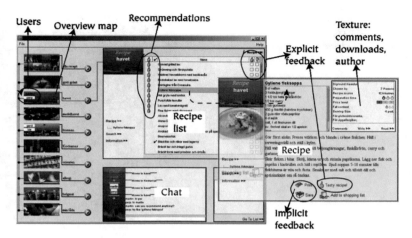

**Fig. 3** Socially navigating recipe collections and recommendations with KALAS [173]

but the authors' experiments show that this extra information can improve the accuracy of recommendations.

Svensson et al. [173] proposed a system named Kalas that allowed recipes in a large database to be navigated socially (see Fig. 3). Different kinds of social navigation were offered in order to study their respective effects on user behavior. Recipes were grouped into sub-collections with specific themes (e.g., vegetarian or spicy food). Traversing recipes could be influenced by other users logged into the system as the real-time presence of others and their navigation trails is displayed. The recommender functionality—also achieved via a socially related collaborative filtering algorithm—affects which recipes appear on the screen. Moreover, recipes can be commented on by users, and the details of past interactions with recipes is shown. Users can also chat about recipes in a chat function. Kalas is one of the few systems to be evaluated in the literature. 302 participants used the system for 200 days, with 18% of cooked recipes coming from recommendations.

One point of note in the KALAS evaluation was that half of the system's users did not understand how recommendations were generated. Explaining recommendations has become an important topic in recommender systems to increase transparency, trust, persuasiveness and satisfaction. We view explanations as a particularly important aspect in food recommender systems, particularly if the goal is to encourage a positive behavioural change. This is underlined by the use of food examples in papers, such as [203]. To our knowledge, however, only a few publications in the food recommender systems make note of explanations.

Elahi et al. [43] present recommendations along with tags as a means to explain recommendations. Moreover, users can explain the ratings they apply using tags. Both Leipold et al. [103] and Schäfer et al. [157] provide textual explanations for the provided recommendations, which emphasise the macro-nutritional benefits of the dish with respect to what is missing in the user's previous intake and at the same

**Fig. 4** Explanations via macro-nutritional content translated from Schäfer et al. [157]

(a)                                                        (b)

**Fig. 5** Explaining restaurant recommendations with geographical context (Vico et al. [190])

time present in the recommended recipe. The example presented in their paper (see Fig. 4) describes the vitamin B2 content.

In a restaurant recommendation context location becomes an important factor with respect to the selection of recommendations. Systems, such as that presented by Vico et al. [190] (see Fig. 5), often use maps as a means to justify or explain recommendations [138, 163, 205].

Research from HCI could provide clues as to how recommendations may be presented in the future including the provision of explanations. Henze and colleagues [78] prototyped different means of augmenting diverse food items with information. In the surveys and focus groups performed to evaluate the concept participants were overwhelmingly positive about the idea and were creative in providing potential use cases.

## 4.2   Eliciting User Needs and Preferences

This section describes different interface variants for establishing the information required to generate suitable recommendations. This includes establishing user taste-preferences, but also additional criteria, such as the nutritional requirements of users.

Typically recommender systems learn user preferences based on past interactions with food items, e.g., ratings [55], bookmarks [180] or tags [43] applied to recipes. Yang et al. [200] raise different problems with this approach in recipe and restaurant recommendation contexts citing the high cognitive and time load on users, as well as a data sparsity issue. Similar issues have been highlighted for lightweight food diary systems where users take photos of the food they eat (e.g., [37, 201]). Yang and colleagues proposed a system called PlateClick to address these issues and explore the advantages of visual strategies for quickly eliciting user food preferences. The aim was to create an engaging experience for users by repeatedly presenting them with algorithmically generated pairs of visually-similar recipes to choose between. They could either indicate their preference or assert that neither recipe appealed. The system was evaluated by means of a field study of 227 users with the results showing the visual comparison method to significantly out-perform baselines. This approach has subsequently been applied in other experiments and systems including [46, 201].

Increased popularity of conversational assistants, such as Siri and Amazon Echo have led to interest in conversational recommendations [31] including for food [16, 59, 156]. Barko-Sherif and colleagues explored the potential for conversational preference elicitation in a food recommender context via a Wizard of Oz Study [16]. Using a between groups design, spoken and text-input chat interfaces were compared where the user interacted with an assistant to explain and refine the criteria of their needs and preferences. This work demonstrates that such interfaces may hold utility. Samagaio presented a RASA-based chatbot that is able to recognise and classify user intentions in a conversation designed to elicit food preferences for recommendation [156].

Other hardware developments with scope to influence food recommender systems relate to wearable and pervasive technology. One popular idea is to make use of wearable cameras to identify consumed food [132, 133]. Further wearable sensors have also been proposed to detect food intake [52]. Smart trays [211] have been proposed for this purpose too. Another use of wearable technology is to determine user calorie burn as an input to food recommendation algorithm [65].

## 4.3   Commercial Applications

In contrast to most research applications, commercial products concerning food recommendations often present rich and interactive interfaces while suffering from a lack of algorithmic recommender solutions. Figure 6 presents a popular commercial

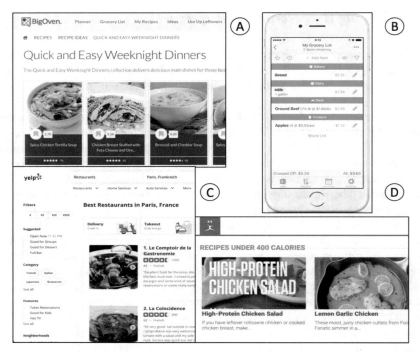

**Fig. 6** Commercial applications for cooking (A, BigOven), grocery shopping (B, AnyList), restaurant visits (C, Yelp), and health (D, MyFitnessPal)

system for each scenario. The applications were selected by taking the first results on searches for the term cooking/grocery/restaurant/healthy food recommendation application. The most popular cooking application, BigOven,[1] provides a large recipe database that can be searched by themes, contexts, or ingredients. It further offers interfaces for meal planning and grocery shopping based on selected recipes. The most popular grocery application, AnyList,[2] offers an interface for organizing grocery items, including their location contextualization, cost aggregation, recipe disaggregation, and a meal planner. The most popular restaurant application, Yelp,[3] offers a ranked list and a map view of recommended restaurants as well as filters for cuisine preference, pricing, distance, and temporal context. In the case of healthy food, the second result, MyFitnessPal,[4] was chosen due to the first choice providing only a glossary. MyFitnessPal offers collections of healthy recipes, including their

---

[1] https://www.bigoven.com/.

[2] https://www.anylist.com/.

[3] https://www.yelp.com/.

[4] https://www.myfitnesspal.com/.

nutritional information, as well as personalized health feedback on current and previous intake when adding these recipes to a personal profile.

## 5 Evaluation

Researchers have different means of evaluating food recommender systems depending on the type of problem they are tackling, the stage of development, and the availability of target group users. An overview of their usage over the past 20 years is shown in Fig. 7. The most frequently used method is the offline evaluation of new algorithms based on compiled dataset (e.g., [181]). Another frequently applied method is to use surveys to collect ratings and feedback for a new recommender system (e.g., [126]). A third common type of evaluation is to perform a user study, which can be varied by type and scales, depending research goals and available resources. These methods are described in detail below. Other methods found in the literature are rarer. These include field studies where the system is used in the wild [196], expert evaluations [49], interviews [48], and case studies [205].

**Offline Evaluations** are used across all types of food recommender systems. However, it relies on the existence of a large ground truth or benchmark dataset. These are most easily accessible via recipe databases, restaurant databases, or online grocery stores. The most frequently targeted variable is, in all cases, the rating of items. The dataset is typically split either by a chronological context or by using k-fold cross-validation. The offline evaluation has also been used in the context of substitutes and health. For substitutes, the ground truth datasets can be derived from large consumption databases or from recipe databases. In the health area, recipe databases can be a good source for offline tests if they contain nutritional information. The benefit of this method is the relatively low effort and cost. One of the biggest limitations of this approach is that it cannot capture the actual rate at which users would follow the given recommendation since there is a discrepancy

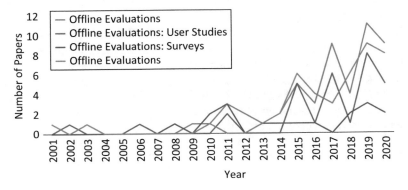

**Fig. 7** Number of papers conducting one of the three major evaluation types over the past 20 years

between online rating behavior and actual preference and behavior. One way to get closer to the actual acceptance rate of the new recommendations with relatively low cost is to conduct online surveys.

**Online Evaluations** come in different forms such as surveys, user studies, field tests, expert evaluations, and interviews. The two main ones used in food recommender systems are surveys and user studies.

*Surveys* have been conducted via Mechanical Turk [126] or by inviting mailing lists to a survey tool [201]. The goal of most surveys being to attain user feedback with respect to a developed recommender system's output compared with that of a baseline or a second variant of the system [181]. Surveys are often used for recipes or menu recommenders where the user has a clear preference in mind (see Table 1). In a health context surveys can be used to correlate the user's reaction to healthy food with other variables such as demographic information or health attitude and intention [5]. Of course, any online survey can also be conducted in person if the required sample size allows for it. The primary advantage of online surveys is that it allows high number of participants to be reached. On the other hand, the quality of feedback can vary largely, which is why safeguards need to be implemented [116] to filter out random responses or contradicting responses.

*User Studies* offer the richest source of insights but are also time and cost-intensive with many pitfalls in terms of the study design. It is unsurprising, therefore, that most user studies conducted on food recommender systems are connected to the area of health where it is difficult to source offline data. Characteristically, user studies conducted in the food context provide the participants with a system that they can use either in predefined sessions [117] or over a longer period [23]. In addition to measuring the user's behavior and interaction with the system, feedback on the system and characteristics of the user are often retrieved via questionnaires or interviews before, during, or after the study. User studies can further target different types of insights. The most prominent approach is to compare two systems with each other in order to validate one system's usefulness or effectiveness, and to gain insights into the user's behavior for system improvements. In a health context, the duration is of high importance. While limited interaction sessions can show the usefulness of a system, the effectiveness is often only measurable after several weeks or longer of using the system [24].

The variables that are measured during user studies typically focus on the user's choices and actions in the system, e.g., preferring one recommendation over another or choosing one item from a list, their perception of the system, e.g., a system usability score, and their food-related behavior when using the system, e.g., dietary diary [157] or grocery bills [114]. One limitation of studying the behavior within such a setting is that many users behave differently when they are part of a study (known as the Hawthorne effect [120]).

# 6 Implementation Resources

In this section we describe implementation resources including datasets and frameworks typically used in the context of food recommender systems research and development. This section also introduces resources to compute and measure, for example, health, flavour and sustainability.

## 6.1 *Recipe Datasets*

To date most food recommender systems research has been performed in the context of recipes [179]. That being said, almost no publicly available datasets exist that make it easy for the research community to perform standardized benchmarks. Frequently resources are sourced online, for example, via recipe portals, such as Allrecipes.[5] Other commonly used Web resources in the European context are Chefkoch.[6] To obtain these data sources, researchers typically have to implement Web crawlers and be careful not to breach the Terms of Services that prohibit sharing of the gathered data or crawling.

Other data sources that have often been used in the literature in the context of food recipe recommendations are Cookpad[7] and Yummly.[8] These services support the research community with an affordable licence model to access their recipe collections.

Non Web-based resources that have been employed includes CSIRO'S Wellbeing Diet Book,[9] which was used by [43, 55]. The dataset is, however, not publicly available.

The Koch-Wiki dataset used in [103] represents a further resource. Unlike the other sources mentioned so far, the Koch Wiki shares data trough a Creative Commons licence, which makes it easy to not only to study the data itself but also to re-use the recipes in ones own research. One limitation is that the data source is in German and does not include any kind of interaction data to perform offline experiments.

Finally, the one million recipe dataset collected by MIT[10] researchers and made publicly available. The dataset comprises over one million recipes including food images and some meta-data and is a web crawl of the most prominent food recipe websites including Allrecipes and Foods. The dataset has to date mostly been used

---

[5] http://www.allrecipes.com.

[6] http://www.chefkoch.de.

[7] http://www.cookpad.com.

[8] http://www.yummly.com.

[9] https://www.csiro.au/en/Research/Health/CSIRO-diets/CSIRO-Total-Wellbeing-Diet

[10] http://im2recipe.csail.mit.edu.

for food classification tasks, such as the *image2recipe* task, where the aim is to predict the recipe and the ingredients of a recipe given a picture alone as input to a system. The dataset has not been used by many recommender systems researchers as no interaction data is included.

## 6.2   Grocery Datasets

In the context of grocery shopping only a handful of datasets exist that have been officially published to be used for research purposes.

The first to mention is the so-called instacard dataset made available via Kaggle[11] in 2017. The dataset contains information about three million purchases via instacard in several grocery stores in the US. The dataset contains session information about purchases, as well as item meta-data. The dataset has been used to predict future purchases mostly employing machine learning [20].

Another interesting data source for potential future research in the context of grocery food recommenders is the Tesco Grocery 1.0 dataset,[12] published in February 2020 in Nature's Scientific Data repository. The dataset consists of a record of 420 million food items purchased by 1.6 million fidelity card owners that shopped in 411 Tesco shops in London. While the dataset is very interesting, a downside is that it is currently only available in an aggregated format. Session or per user data is not available. The dataset only allows predictions to be made on an area level. A further downside of the dataset is that it does not contain detailed information about the actual grocery items. Only categorical information is available.

A dataset that does contain meta-data information is the grocery dataset published by Klasson et al. in 2019 [93]. Included in the dataset are grocery images and explanations.[13] Moreover the OpenFoodsFacts[14] dataset may be the largest open access database of groceries all around the globe with extensive meta-date available. Again, the dataset provides no user interaction data that would it make possible to test personalised algorithms.

With the exception of the mentioned resources, research relies on proprietary Web crawls from Amazon[15] and other large online retailers that allow the crawling of their data or provide a dedicated API [77].

---

[11] https://www.kaggle.com/c/instacart-market-basket-analysis.

[12] https://www.nature.com/articles/s41597-020-0397-7.

[13] https://github.com/marcusklasson/GroceryStoreDataset.

[14] https://world.openfoodfacts.org/.

[15] https://amazon.com.

## 6.3 Meals, Menus and Restaurant Datasets

To implement meal plan recommender systems, research typically relies on single item representations as, for example, available in the datasets as discussed above, most commonly recipes. Dedicated meal plans as a ground truth data can, however, also be obtained from online sources such as Eatingwell,[16] FitBit[17] and many other popular health platforms. These sources provide the possibility to obtain general meal plan templates for a day or a week.

In the context of meals, i.e., a combination of, for example, a starter, main dish and dessert, Allrecipes has been used as a resource in research [28, 46, 180, 181]. Others researchers have used Foods, TudoGostoso[18] as well as giallozafferano as Italian food website.[19] In [126] the authors also released a detailed behavioral dataset, including user interactions with the items.[20]

Finally, Yelp[21] is prominently used as a data source in the context of restaurant recommendations for eating out [186]. Other researchers have used Twitter [163],[22] Foursquare [163],[23] Tripadvisor [208],[24] as well as Dianping [207][25] and Baidu [205].[26]

All of these data sources are preparatory and no standardised sources exist.

## 6.4 Flavour Resources

Another important aspect of food one is often confronted with when building food recommendation services is flavour. To compute flavour of a given recipe or ingredient the current approach is to rely on food chemicals. To extract these chemicals, Ahn et al. [4] rely on Fenaroli's handbook of flavor ingredients [21]. The dataset can be downloaded directly from their article in Scientific Reports. A newer approach to the same problem is the online resource FLAVORDB.[27] FlavorDB is a database that comprises over 25 thousand flavor molecules representing an array

---

[16] http://www.eatingwell.com/.

[17] https://fitbit.com.

[18] https://wnw.tudogostoso.com.br/.

[19] https://www.giallozafferano.it/ricette-cat/.

[20] http://www.di.uniba.it/~swap/datasets/recipes.csv.

[21] https://yelp.com.

[22] https://twitter.com.

[23] https://foursquare.com.

[24] https://tripadvisor.com.

[25] http://dianping.com.

[26] http://Baidu.com.

[27] https://cosylab.iiitd.edu.in/flavordb/.

of tastes and odor associated with over 900 natural ingredients. The repository has been made available through an online app and is accessible through a Creative Commons licence.

Another useful resources available in the food recommender systems context, though less used yet for that purpose are the online service FOODPAIRING,[28] which allows to estimate which types of food go well together with based on taste and smell. While the platform offers promising service, the science behind the service is rather in-transparent and not free of charge.

## 6.5   Software Frameworks

To date, no dedicated software package exists to build a food recommender system. All of the existing research in the food recommender context either has implemented their own system or has been building their prototypes upon existing recommendation frameworks. This of course creates issues such as, for example, algorithms being developed in the movie domain and used in the food domain, but not suitable for also recommending healthy food items [180]. Another issue is obviously that the results obtained in one work cannot be compared to another study, even on the same dataset as protocols and evaluation metrics often differ. To work towards resolving this issue the well-know LibRec library has been extended especially for the purpose of food recommendations [181], e.g., by integrating recipe content and collaborative features. A downside of the framework, however, is that the framework yet only is available in the Java programming language. Similar to the well-known LensKit[29] framework it is planned to transition from Java to Python.

A selection of other recommender systems frameworks in other programming languages can be found on Graham Jenson's Github page.[30] These frameworks typically provide standard recommender systems algorithms, such as user- and item-based collaborative filtering with nearest neighborhood search as well as BPR, SVD and many others. In general, these existing algorithms work more or less well in predicting the users preferences. Further research, however, is needed to make these more accurate as current experiments show that standard algorithms perform significantly less well for the purpose of recommending foods (e.g., recipes) than movies, their original application domain [180].

---

[28] https://foodparing.com.

[29] https://arxiv.org/abs/1809.03125.

[30] https://github.com/grahamjenson/list_of_recommender_systems.

## 6.6  Nutrition Resources

In the context of food recommender systems, it is often essential to know more about a certain food item, such as their energy value or other nutritional properties. These are often essential to develop health-aware food recommender systems or systems that aim for certain food constraints or goals.

To measure and compute the nutrition of a given food item the usual way is to map ingredients to standard databases. Examples of such standard databases used in food computing are the ones provided by the USDA[31] (US) and the BLS[32] (Germany). The USDA database is also used in Google's knowledge graph.

A typical issue associated with these direct mapping principles is that food items can only be mapped correctly and computed if they exactly match the entries in the database. That often causes issues and calls for NLP techniques that are aimed to normalise words [96]. A more practical solution to the problem is to relay on existing frameworks or Web services. Examples of such systems are SPOONACULAR,[33] a Web service that is able to extract ingredients, nutrition and correct amounts from noise text inputs as well as EDAMAM.[34] A downside of both are that they are commercial services. In [149], however, a method to predict nutrition of recipes based noisy data is discussed. Software and data can be obtained for free from the authors.

## 6.7  Health Resources

Health resources are needed when implementing health-aware food recommender systems. A typical approach to make a food recommender system "health-aware" is to filter food items in terms of their healthiness. To measure healthiness one can rely on a variety of resources. In the following we discuss the most commonly used these days in research.

A common approach in the real world to inform someone of to what extent a particular food item is healthy via food labels and standards as set by food safety authorities. The latter is typically country dependent. Successfully used so far in the food recommendation context have been the standards as set by the Food Standard Agency (FSA)[35] in the UK as well as standards as set by the World Healthcare Organisation (WHO)[36] [180]. Both standards are based on a 2000kcal diet and

---

[31] https://ndb.nal.usda.gov/ndb.

[32] https://www.blsdb.de.

[33] https://market.mashape.com/spoonacular.

[34] https://www.edamam.com.

[35] https://www.food.gov.uk/sites/default/files/media/document/fop-guidance_0.pdf.

[36] https://www.who.int/publications/guidelines/nutrition/en/.

**Table 6** FSA front of package guidelines for healthy eating

| Text | Low | Medium | High |
|------|-----|--------|------|
| Color code | Green | Amber | Red |
| Fat | ≤ 3.0 g/100 g | > 3.0 g to ≤ 17.5 g/100 g | > 17.5 g/100 g or > 21 g/portion |
| Saturates | ≤ 1.5 g/100 g | > 1.5 g to ≤ 5.0 g/100 g | > 5.0 g/100 g or > 6.0 g/portion |
| Sugars | ≤ 5.0 g/100 g | > 5.0 g to ≤ 22.5 g/100 g | > 22.5 g/100 g or > 27 g/portion |
| Salt | ≤ 0.3 g/100 g | > 0.3 g to ≤ 1.5 g/100 g | > 1.5 g/100 g or > 1.8 g/portion |

**Table 7** WHO dietary guidelines

| Dietary factor | Range (percentage of kcal per meal/recipe) |
|----------------|--------------------------------------------|
| Protein | 10–15 |
| Carbohydrates | 55–75 |
| Sugar | < 10 |
| Fat | 15–30 |
| Saturated Fat | < 10 |
| Fiber density (g/MJ) | > 3.0[a] |
| Sodium density (g/MJ) | < 0.2[b] |

[a] Based on 8.4 MJ/day (2000 kcal/day) diet and recommended daily fiber intake of >25 g
[b] Based on 8.4 MJ/day (2000 kcal/day) diet and recommended daily sodium intake of <2 g

account for different nutritional properties of a food, see Tables 6 and 7. While the FSA accounts for Fat, Saturated Fat, Sugars and Salt, the WHO guidelines also take into consideration Fibre, Proteins and Carbs. Both metrics have been used in previous research in the context of food recommender systems. Other metrics used to date to measure or compute healthiness is the 'Healthy Eating Index'[37] proposed by the USDA to target the US population. More and more such standards are being developed now all over the world, per region and country, and it is expected that they find their ways in food recommender systems of the future.

## 6.8   Food Sustainability Resources

A currently emerging hot topic in the food recommender systems context is sustainability [13]. Only a handful of tools exist to measure to what extent a food item is sustainable or not. One of the most recently released resources is the NAHGAST online[38] tool that considers the following 4 dimensions of sustainability [167]:

- *Environment:* Material Footprint (<2670 g/<4000 g), Carbon Footprint (<800 g/<1200 g), Water use (<640 L/<975 L), Land use (<1.25 m2/<1.875 m2).

---

[37] https://www.fns.usda.gov/resource/healthy-eating-index-hei.
[38] https://www.nahgast.de/rechner/.

- *Social:* Share of fair ingredients (>90%/>85%)l Share of animal-based food that foster animal welfare (>60%/>55%).
- *Health:* Energy (<670 kcal/<830 kcal), Fat (<24 g/<30 g), Carbohydrates (<90 g/<95 g), Sugar (<17 g/<19 g), Fibers (<8 g/> 6 g).
- *Economic:* Popularity (without quantified target value), Cost recovery (without quantified target value).

Other related resources in that context can be found at GREENHOUSE GAS AND DIETARY CHOICES OPEN SOURCE TOOLKIT[39] and TAKE A BITE OUT OF CLIMATE CHANGE.[40]

## 6.9  Other Resources

Another useful resource one would like to use when building a food recommender system is provided by FOODSUBS.[41] The service includes a food thesaurus that can suggest food substitutes. This is a particularly useful resource to develop future similar item food recommender systems [183] that are able to display recommendations with ingredient alternatives [182].

Furthermore useful resources to develop food recommender systems are food word lists such as provided by ENCHANTEDLEARNING[42] and WIKIPEDIA.[43] They are typically used in content-based recommender approaches to normalise ingredient list from noisy data sources such as the Web.

Last but not least, one may also want to employ regional data in a food recommender system such, for example, provided by the Centers for Disease Control and Prevention (CDC) in the US to implement, for example, regional and health-aware food recommender systems [153, 185].

Finally, one would like to employ knowledge-graphs for the purpose of recommending foods to people. Sources in this context are still at an early stage but can be more and more found online see, for example, FOODKG[44] and in the literature [76]. These may be useful in the future to provide semantic recommendations.

Plenty of further resources can be found on the Web to construct food recommender systems. It is though recommended that one validates the quality of these sources with care as they often lacking scientific foundation.

---

[39] https://www.ggdot.org/.

[40] https://www.takeabitecc.org/.

[41] http://www.foodsubs.com.

[42] http://www.enchantedlearning.com/wordlist/food.shtml.

[43] https://en.wikipedia.org/wiki/Lists_of_foods.

[44] https://foodkg.github.io/.

# 7  Conclusion

**Lessons Learned** In the following, we attempt to extract key takeaways from the literature. In particular we use our experience to provide guidance for practitioners unfamiliar with the research and who wish to develop a food recommender system. In our view the biggest mistake one could make would be to reduce the problem to a single algorithmic problem. While there is no evidence of one particular algorithm being best suited to different food recommendation problems, differences do occur in terms of what should be shown, how items should be shown and to whom. More concretely, the research results so far encourage practitioners to:

- **Think about the task:** What people want to be recommended differs depending on the application (i.e. the type of item the system recommends. See Sect. 4)
- **Consider who the users are and what they may need:** The evidence shows that different groups of users have different preferences and priorities when looking for food items (see Sect. 3.3)
- **Consider context:** What people want to eat changes depending on a wide range of context variables and it is important to consider how these may affect your system (see Sect. 3.2).
- **Exploit the visual nature of food choice:** Food choices in digital environments are highly visual (see Sect. 3.4) and this is one of the few constant findings across different settings. This is even true across cultures (see Sect. 3.2).

**Summary and Future Work** In the following, we shed light on the food recommendation problem. We have provided a review of the current state-of-the-art in the field from an algorithm, interface, evaluation, and implementation resources perspective. We try to summarize the most important findings and points of our work:

- At the moment, no theory about food recommender systems exists. There is research that tries to combine theory about food choice but no work that has tried to integrate systems into a theory.
- Most research on food recommender systems is in the context of health, but less has been done in sub-fields such as cooking, grocery shopping, or restaurant visits, including menu items.
- Algorithms are mostly based on standard recommender systems approaches, and not many specialized algorithmic developments exist.
- Most recommender system approaches in food typically integrate a collaborative component and use algorithms such as collaborative filtering.
- Algorithms in food recommender systems are mostly focused on single users. The group context exists but is not the main focus of research at the moment.
- When it comes to the evaluation of recommender system algorithms, there is furthermore a lack of standardized datasets. Still, there is no initiative that tries to collect these preparatory datasets. The reasons for this are GDPR or copyright limitations for behavioral data or recipe data.

**Table 8** Challenges and suggestions for future research listed in previous food recommender surveys.

| Year | 2017 | 2017 | 2018 | 2018 | 2018 | 2019 | 2019 | 2019 | |
|---|---|---|---|---|---|---|---|---|---|
| Ref | [158] | [179] | [12] | [178] | [119] | [124] | [125] | [176] | $\sum$ |
| Health personalization | | ✓ | | | ✓ | ✓ | ✓ | ✓ | 5 |
| Erroneous user input and eating behavior data | ✓ | | | ✓ | ✓ | | ✓ | ✓ | 5 |
| Deep learning on food images | | ✓ | ✓ | | | ✓ | ✓ | ✓ | 5 |
| Sensors and physical interfaces | | ✓ | ✓ | | | ✓ | | ✓ | 4 |
| Preference prediction and algorithmic performance | | ✓ | | ✓ | | ✓ | | ✓ | 4 |
| Contextualization of algorithms | ✓ | ✓ | ✓ | | | | | ✓ | 4 |
| Sparse/biased data | | | ✓ | ✓ | | | ✓ | | 3 |
| Social context and groups | | ✓ | | ✓ | | | ✓ | | 3 |
| Persuasion and behaviour change | ✓ | | | ✓ | ✓ | | | | 3 |
| Multiple contexts mixed | | | ✓ | | ✓ | | ✓ | | 3 |
| Multi-objective and multi modality | | | | | | ✓ | ✓ | ✓ | 3 |
| Explanation of algorithm | ✓ | | | ✓ | | ✓ | | | 3 |
| Integrating knowledge graphs | | | | | | ✓ | ✓ | | 2 |
| Evaluation methods and benchmark data | | ✓ | | | | ✓ | | | 2 |

- Furthermore, when it comes to evaluation, offline assessments are still dominating, but online evaluations have been catching up, which is commendable, as online evaluation protocols capture the real world better than offline simulations.
- There has been little research on how different interfaces change the effectiveness and acceptance rate of food recommender systems. Only a few studies test interfaces to their algorithms at all, and none compare different variants of interfaces for the same recommender system.
- Finally, when it comes to the availability of implementation resources, many tools exist, but no standardized food recommender systems framework, which would allow the community to build upon and advance the research field.

Besides these main findings, our review of existing work reconfirms the gaps and challenges identified by previous more specialized surveys of food recommender systems, as shown in Table 8. For example, personalization of health-focused food recommender systems by the user's genome, microbiome, blood values, or changing behavioral patterns is a major gap identified by five surveys. In behavioral personalization, this challenge is further complicated by the accurate and effortless extraction of user data regarding past eating behavior. One solution to this issue and significant challenge itself is the extraction of information from food images and the accurate integration of food-related sensor data, e.g., volatile organic compounds sensors. Besides health-related issues, both the prediction of preference or other success criteria and standard algorithms' performance is still below that in other fields of recommender systems, especially when regarding multiple criteria at once. We further confirm missing contextualization as one large research gap across different food recommender systems that might lead to much higher acceptance rates.

In summary, there are many challenges ahead when it comes to solving the food recommendation problem. While this may sound limiting for practitioners, this is, on the other hand, exciting from a scientific point of view, as there are many open issues and questions that need to be resolved. Therefore, it may be some more time before we see such systems all over in our everyday lives, as is the case with other commercial recommender systems, such as media content recommender systems.

# References

1. J. Aberg, Dealing with malnutrition: a meal planning system for elderly, in *AAAI Spring Symposium: Argumentation for Consumers of Healthcare* (2006), pp. 1–7
2. P. Achananuparp, I. Weber, Extracting food substitutes from food diary via distributional similarity (2016). arXiv:1607.08807
3. G. Agapito, B. Calabrese, P.H. Guzzi, M. Cannataro, M. Simeoni, Ilaria Caré, T. Lamprinoudi, G. Fuiano, and A. Pujia, Dietos: A recommender system for adaptive diet monitoring and personalized food suggestion, in *2016 IEEE 12th International Conference on Wireless and Mobile Computing, Networking and Communications (WiMob)* (IEEE, Piscataway, 2016), pp. 1–8
4. Y.-Y. Ahn, S.E. Ahnert, J.P. Bagrow, A.-L. Barabási, Flavor network and the principles of food pairing. Sci. Rep. **1**, 196 (2011)
5. J. Ahn, J. Williamson, M. Gartrell, R. Han, Q. Lv, S. Mishra, Supporting healthy grocery shopping via mobile augmented reality. ACM Trans. Multimed. Comput. Commun. Appl. **12**(1s), 1–24 (2015)
6. S. Akkoyunlu, C. Manfredotti, A. Cornuéjols, N. Darcel, F. Delaere, Investigating substitutability of food items in consumption data. In *Second International Workshop on Health Recommender Systems Co-located with ACM RecSys*, vol. 5 (2017)
7. S.I. Ali, M.B. Amin, S. Kim, S. Lee, A hybrid framework for a comprehensive physical activity and diet recommendation system, in *International Conference on Smart Homes and Health Telematics* (Springer, Berlin, 2018), pp. 101–109
8. S. Alian, J. Li, V. Pandey, A personalized recommendation system to support diabetes self-management for american indians. IEEE Access **6**, 73041–73051 (2018)
9. B. Aljbawi, Health-aware food planner: a personalized recipe generation approach based on GPT-2. Theses and Dissertations (2020)
10. G.M. Almerico, Food and identity: food studies, cultural, and personal identity. J. Int. Bus. Cultural Stud. **8**, 1 (2014)

11. E.N. Anderson, *Everyone Eats: Understanding Food and Culture* (NYU Press, New York, 2014)
12. C. Anderson, A survey of food recommenders (2018). arXiv:1809.02862
13. Y.M. Asano, G. Biermann, Rising adoption and retention of meat-free diets in online recipe data. Nat. Sustain. **2**(7), 621–627 (2019)
14. N. Baghaei, S. Kimani, J. Freyne, E. Brindal, S. Berkovsky, G. Smith, Engaging families in lifestyle changes through social networking. Int. J. Hum.-Comput. Interact. **27**(10), 971–990 (2011)
15. L. Baltrunas, M. Kaminskas, B. Ludwig, O. Moling, F. Ricci, A. Aydin, K.-H. Lüke, R. Schwaiger, Incarmusic: context-aware music recommendations in a car. In *International Conference on Electronic Commerce and Web Technologies* (Springer, Berlin, 2011), pp. 89–100
16. S. Barko-Sherif, D. Elsweiler, M. Harvey, Conversational agents for recipe recommendation, in *Proceedings of the 2020 Conference on Human Information Interaction and Retrieval* (2020), pp. 73–82
17. S. Berkovsky, J. Freyne, Group-based recipe recommendations: analysis of data aggregation strategies, in *Proceedings of the Fourth ACM Conference on Recommender Systems* (2010), pp. 111–118
18. A. Bharadwaj, A.N. Rao, A. Kulhalli, K.S. Mehta, N. Bhattacharya, P. Ramkumar, N. Nag, R. Jain, D. Sitaram, Flavour based food recommendation (2019). arXiv:1904.05331
19. D. Bianchini, V. De Antonellis, N. De Franceschi, M. Melchiori, Prefer: a prescription-based food recommender system. Comput. Standards Interfaces **54**, 64–75 (2017)
20. Y. Bodike, D. Heu, B. Kadari, B. Kiser, M. Pirouz, A novel recommender system for healthy grocery shopping, in *Future of Information and Communication Conference* (Springer, Berlin, 2020), pp. 133–146
21. G.A. Burdock, *Fenaroli's Handbook of Flavor Ingredients*, vol. 2 (CRC Press, Boca Raton, 2019)
22. J. Caldeira, R.S. Oliveira, L. Marinho, C. Trattner, Healthy menus recommendation: optimizing the use of the pantry, in *Proceedings of the 3rd International Workshop on Health Recommender Systems (HealthRecSys' 18) Co-located with the 12th ACM Conference on Recommender Systems (ACM RecSys 2018)(CEUR Workshop Proceedings)* (2018)
23. C. Celis-Morales, K.M. Livingstone, C.F.M. Marsaux, H. Forster, C.B. O'Donovan, C. Woolhead, A.L. Macready, R. Fallaize, S. Navas-Carretero, R. San-Cristobal, et al., Design and baseline characteristics of the food4me study: a web-based randomised controlled trial of personalised nutrition in seven european countries. Genes Nutr. **10**(1), 450 (2015)
24. C. Celis-Morales, K.M. Livingstone, C.F.M. Marsaux, A.L. Macready, R. Fallaize, C.B. O'Donovan, C. Woolhead, H. Forster, M.C. Walsh, S. Navas-Carretero, et al., Effect of personalized nutrition on health-related behaviour change: evidence from the food4me european randomized controlled trial. Int. J. Epidemiol. **46**(2), 578–588 (2017)
25. W.-Y. Chao, Z. Hass, Choice-based user interface design of a smart healthy food recommender system for nudging eating behavior of older adult patients with newly diagnosed type ii diabetes, in *International Conference on Human-Computer Interaction* (Springer, Berlin, 2020), pp. 221–234
26. P. Chavan, B. Thoms, J. Isaacs, A recommender system for healthy food choices: building a hybrid model for recipe recommendations using big data sets, in *Proceedings of the 54th Hawaii International Conference on System Sciences* (2021), p. 3774
27. L. Chen, H.K. Tsoi, Users' decision behavior in recommender interfaces: impact of layout design, in *RecSys' 11 Workshop on Human Decision Making in Recommender Systems* (2011)
28. M. Chen, X. Jia, E. Gorbonos, C.T. Hong, X. Yu, Y. Liu, Eating healthier: exploring nutrition information for healthier recipe recommendation. Inf. Process. Manag. **57**(6) 102051 (2019)
29. H. Cheng, M. Rokicki, E. Herder, The influence of city size on dietary choices and food recommendation, in *Proceedings of the 25th Conference on User Modeling, Adaptation and Personalization* (2017), pp. 359–360

30. N.A. Christakis, J.H. Fowler, The spread of obesity in a large social network over 32 years. New England J. Med. **357**(4), 370–379 (2007)
31. K. Christakopoulou, F. Radlinski, K. Hofmann, Towards conversational recommender systems, in *Proceedings of the 22nd ACM SIGKDD International Conference on Knowledge Discovery and Data Mining* (2016), pp. 815–824
32. P. Christodoulou, K. Christodoulou, A.S. Andreou, A Real-time Targeted Recommender System for Supermarkets, in *ICEIS 2017 - Proceedings of the 19th International Conference on Enterprise Information Systems*, vol. 2, (Porto, Portugal, Apr. 26-29, 2017, 2017), pp. 703–712. https://doi.org/10.5220/0006309907030712
33. M. Clark, D. Tilman, Comparative analysis of environmental impacts of agricultural production systems, agricultural input efficiency, and food choice. Environ. Res. Lett. **12**(6), 064016 (2017)
34. A.K. Clear, A. Friday, M. Rouncefield, A. Chamberlain, Supporting sustainable food shopping. IEEE Pervasive Comput. **14**(4), 28–36 (2015)
35. F.M. Clydesdale, Color as a factor in food choice. Critical Rev. Food Sci. Nutr. **33**(1), 83–101 (1993)
36. M. Connors, C.A. Bisogni, J. Sobal, C.M. Devine, Managing values in personal food systems. Appetite **36**(3), 189–200 (2001)
37. F. Cordeiro, E. Bales, E. Cherry, J. Fogarty, Rethinking the mobile food journal: exploring opportunities for lightweight photo-based capture, in *Proceedings of the 33rd Annual ACM Conference on Human Factors in Computing Systems* (2015), pp. 3207–3216
38. P.F. Cueto, M. Roet, A. Słowik, Completing partial recipes using item-based collaborative filtering to recommend ingredients (2019). arXiv:1907.12380
39. M. De Choudhury, S. Sharma, E. Kiciman, Characterizing dietary choices, nutrition, and language in food deserts via social media, in *Proceedings of the 19th ACM Conference on Computer-Supported Cooperative Work & Social Computing* (2016), pp. 1157–1170
40. T. De Pessemier, S. Dooms, L. Martens, A food recommender for patients in a care facility, in *Proceedings of the 7th ACM Conference on Recommender Systems* (2013), pp. 209–212
41. M. Deudon. On food, bias and seasons: A recipe for sustainability. 2020. (hal-02532348) https://hal.archives-ouvertes.fr/hal-02532348, last accessed on 03.01.2022
42. Effect of an internet-based, personalized nutrition randomized trial on dietary changes associated with the mediterranean diet: the food4me study. Am. J. Clin. Nutrition **104**(2), 288–297 (2016)
43. M. Elahi, M. Ge, F. Ricci, D. Massimo, S. Berkovsky, Interactive food recommendation for groups, in *Recsys Posters* (Citeseer, 2014)
44. D. Elsweiler, M. Harvey, Towards automatic meal plan recommendations for balanced nutrition, in *Proceedings of the 9th ACM Conference on Recommender Systems* (2015), pp. 313–316
45. D. Elsweiler, M. Harvey, B. Ludwig, A. Said, Bringing the "healthy" into food recommenders, in *2nd International Workshop on Decision Making and Recommender Systems (DMRS)* (2015), pp. 33–36
46. D. Elsweiler, C. Trattner, M. Harvey, Exploiting food choice biases for healthier recipe recommendation, in *Proceedings of the 40th International ACM SIGIR Conference on Research and Development in Information Retrieval* (2017), pp. 575–584
47. V. Espín, M.V. Hurtado, M. Noguera, Nutrition for elder care: a nutritional semantic recommender system for the elderly. Exp. Syst. **33**(2), 201–210 (2016)
48. R. Fallaize, R.Z. Franco, F. Hwang, J.A. Lovegrove, Evaluation of the enutri automated personalised nutrition advice by users and nutrition professionals in the UK. PloS One **14**(4), e0214931 (2019)
49. D. Fister, I. Fister, S. Rauter, Generating eating plans for athletes using the particle swarm optimization, in *2016 IEEE 17th International Symposium on Computational Intelligence and Informatics (CINTI)*, (IEEE, Piscataway, 2016), pp. 000193–000198
50. C. Fjellström, Mealtime and meal patterns from a cultural perspective. Scand. J. Nutrition **48**(4), 161–164 (2004)

51. J.S. Foer, *We Are the Weather: Saving the Planet Begins at Breakfast* (Penguin, New York, 2019)
52. J.M. Fontana, M. Farooq, E. Sazonov, Automatic ingestion monitor: a novel wearable device for monitoring of ingestive behavior. IEEE Trans. Bio-med. Eng. **61**(6), 1772 (2014)
53. P. Forbes, M. Zhu, Content-boosted matrix factorization for recommender systems: experiments with recipe recommendation, in *Proceedings of the Fifth ACM Conference on Recommender Systems* (2011), pp. 261–264
54. J.H. Freeland-Graves, S. Nitzke, Position of the academy of nutrition and dietetics: total diet approach to healthy eating. J. Acad. Nutr. Diet. **113**(2), 307–317 (2013)
55. J. Freyne, S. Berkovsky, Intelligent food planning: personalized recipe recommendation, in *Proceedings of the 15th International Conference on Intelligent User Interfaces* (2010), pp. 321–324
56. J. Freyne, S. Berkovsky, Recommending food: reasoning on recipes and ingredients, in *International Conference on User Modeling, Adaptation, and Personalization* (Springer, Berlin, 2010), pp. 381–386
57. J. Freyne, S. Berkovsky, N. Baghaei, S. Kimani, G. Smith, Personalized techniques for lifestyle change, in *Conference on Artificial Intelligence in Medicine in Europe* (Springer, Berlin, 2011), pp. 139–148
58. J. Freyne, S. Berkovsky, G. Smith, Recipe recommendation: accuracy and reasoning, in *International Conference on User Modeling, Adaptation, and Personalization* (Springer, Berlin, 2011)
59. A. Frummet, D. Elsweiler, B. Ludwig, Detecting domain-specific information needs in 1214 conversational search dialogues, in *Proceedings of the 3rd Workshop on Natural Language for Artificial Intelligence. Ceur Workshop Proceedings*, vol. 2521, (2019) http://ceur-ws.org/Vol-2521/paper-02.pdf
60. K. Fuchs, M. Haldimann, T. Grundmann, E. Fleisch, Supporting food choices in the internet of people: automatic detection of diet-related activities and display of real-time interventions via mixed reality headsets. Futur. Gener. Comput. Syst. **113**, 343–362 (2020)
61. T. Furst, M. Connors, C.A. Bisogni, J. Sobal, L.W. Falk, Food choice: a conceptual model of the process. Appetite **26**(3), 247–266 (1996)
62. E. Gaillard, J. Lieber, E. Nauer, Improving ingredient substitution using formal concept analysis and adaptation of ingredient quantities with mixed linear optimization, in *Computer Cooking Contest Workshop*, (2015., Ceur Workshop proceedings), vol. 1520. http://ceur-ws.org/Vol-1520/paper22.pdf
63. M.B. Garcia, Plan-cook-eat: a meal planner app with optimal macronutrient distribution of calories based on personal total daily energy expenditure, in *2019 IEEE 11th International Conference on Humanoid, Nanotechnology, Information Technology, Communication and Control, Environment, and Management (HNICEM)* (IEEE, Piscataway, 2019), pp. 1–5
64. M. Ge, M. Elahi, I. Fernaández-Tobías, F. Ricci, D. Massimo, Using tags and latent factors in a food recommender system, in *Proceedings of the 5th International Conference on Digital Health 2015* (2015), pp. 105–112
65. M. Ge, F. Ricci, D. Massimo, Health-aware food recommender system, in *Proceedings of the 9th ACM Conference on Recommender Systems* (2015), pp. 333–334
66. G. Geleijnse, P. Nachtigall, P. van Kaam, L. Wijgergangs, A personalized recipe advice system to promote healthful choices, in *Proceedings of the 16th International Conference on Intelligent User Interfaces* (2011), pp. 437–438
67. S.A. Golder, M.W. Macy, Diurnal and seasonal mood vary with work, sleep, and daylength across diverse cultures. Science **333**(6051), 1878–1881 (2011)
68. E. Gorbonos, Y. Liu, C.T. Hoàng, NutRec: nutrition oriented online recipe recommender, in *2018 IEEE/WIC/ACM International Conference on Web Intelligence (WI)* (IEEE, Piscataway, 2018), pp. 25–32
69. F. Gutiérrez, K. Verbert, N.N. Htun, PHARA: an augmented reality grocery store assistant, in *Proceedings of the 20th International Conference on Human-Computer Interaction with Mobile Devices and Services Adjunct* (2018), pp. 339–345

70. F. Gutiérrez, N.N. Htun, S. Charleer, R. De Croon, K. Verbert, Designing augmented reality applications for personal health decision-making, in *Proceedings of the 52nd Hawaii International Conference on System Sciences* (2019)
71. D. Halpern, *Inside the Nudge Unit: How Small Changes Can Make a Big Difference* (Random House, New York, 2015)
72. K.J. Hammond, CHEF: a model of case-based planning, in *Fifth National Conference on Artificial Intelligence (AAAI)* (1986), pp. 267–271
73. M. Harvey, D. Elsweiler, Automated recommendation of healthy, personalised meal plans, in *Proceedings of the 9th ACM Conference on Recommender Systems* (2015), pp. 327–328
74. M. Harvey, B. Ludwig, D. Elsweiler, Learning user tastes: a first step to generating healthy meal plans. Proc. LIFESTYLE **12**, 18 (2012)
75. M. Harvey, B. Ludwig, D. Elsweiler, You are what you eat: learning user tastes for rating prediction, in *International Symposium on String Processing and Information Retrieval* (Springer, Berlin, 2013), pp. 153–164
76. S. Haussmann, O. Seneviratne, Y. Chen, Y. Ne'eman, J. Codella, C.-H. Chen, D.L. McGuinness, M.J. Zaki, FoodKG: a semantics-driven knowledge graph for food recommendation, in *International Semantic Web Conference* (Springer, Berlin, 2019), pp. 146–162
77. Y. Heng, Z. Gao, Y. Jiang, X. Chen, Exploring hidden factors behind online food shopping from amazon reviews: a topic mining approach. J. Retail. Consum. Serv. **42**, 161–168 (2018)
78. N. Henze, T. Olsson, S. Schneegass, A.S. Shirazi, K. Väänänen-Vainio-Mattila, Augmenting food with information, in *Proceedings of the 14th International Conference on Mobile and Ubiquitous Multimedia* (2015), pp. 258–266
79. J. Herrera, Sustainable recipes. A food recipe sourcing and recommendation system to minimize food miles (2020). arXiv:2004.07454
80. R. Hertwig, T. Grüne-Yanoff, Nudging and boosting: steering or empowering good decisions. Perspect. Psychol. Sci. **12**(6), 973–986 (2017)
81. N. Rastogi, O. Seneviratne, D. Gruen, C.-H. Chen, C. Yu, J. Harris, D. Li, et al., Applying learning and semantics for personalized food recommendations, in *ISWC (Demos/Industry)*, (2020), pp. 305–310
82. T.R. Hinrichs. J.L. Kolodner, The roles of adaptation in case-based design, in *AAAI Proceedingss*, vol. 91 (1991), pp. 28–33
83. T. Ivaşcu, A. Diniş, K. Cincar, A disease-driven nutrition recommender system based on a multi-agent architecture, in *Proceedings of the 8th International Conference on Web Intelligence, Mining and Semantics* (2018), pp. 1–5
84. J. Jermsurawong, N. Habash, Predicting the structure of cooking recipes, in *Proceedings of the 2015 Conference on Empirical Methods in Natural Language Processing* (2015), pp. 781–786
85. H. Jiang, W. Wang, M. Liu, L. Nie, L.-Y. Duan, C. Xu, Market2dish: a health-aware food recommendation system, in *Proceedings of the 27th ACM International Conference on Multimedia* (2019), pp. 2188–2190
86. A.K. Kant, Indexes of overall diet quality: a review. J. Am. Dietetic Assoc. **96**(8), 785–791 (1996)
87. D. Karpati, A. Najjar, D.A. Ambrossio, Ethics of food recommender applications, in *Proceedings of the AAAI/ACM Conference on AI, Ethics, and Society* (2020), pp. 313–319
88. T. Kashima, S. Matsumoto, H. Ishii, Recommendation method with rough sets in restaurant point of sales system, in *Proceedings of the International MultiConference of Engineers and Computer Scientists*, vol. 3 (2010)
89. P.D. Kaur, et al., A context-aware recommender engine for smart kitchen, in *Smart Innovations in Communication and Computational Sciences* (Springer, Berlin, 2019), pp. 161–170
90. A.S. Khan, A. Hoffmann, Building a case-based diet recommendation system without a knowledge engineer. Artif. Intell. Med. **27**(2), 155–179 (2003)
91. M.A. Khan, E. Rushe, B. Smyth, D. Coyle, Personalized, health-aware recipe recommendation: an ensemble topic modeling based approach (2019). arXiv:1908.00148

92. J. Kim, D. Lee, K.-Y. Chung, Item recommendation based on context-aware model for personalized u-healthcare service. Multimed. Tools Appl. **71**(2), 855–872 (2014)

93. M. Klasson, C. Zhang, H. Kjellström, A hierarchical grocery store image dataset with visual and semantic labels, in *IEEE Winter Conference on Applications of Computer Vision (WACV)* (2019)

94. F.-F. Kuo, C.-T. Li, M.-K. Shan, S.-Y. Lee, Intelligent menu planning: recommending set of recipes by ingredients, in *Proceedings of the ACM Multimedia 2012 Workshop on Multimedia for Cooking and Eating Activities* (2012), pp. 1–6

95. T. Kusmierczyk, K. Nørvåg, Online food recipe title semantics: combining nutrient facts and topics, in *Proceedings of the 25th ACM International on Conference on Information and Knowledge Management* (2016), pp. 2013–2016

96. T. Kusmierczyk, C. Trattner, K. Nørvåg, Temporal patterns in online food innovation, in *Proceedings of the 24th International Conference on World Wide Web* (2015), pp. 1345–1350

97. T. Kusmierczyk, C. Trattner, K. Nørvåg, Temporality in online food recipe consumption and production, in *Proceedings of the 24th International Conference on World Wide Web* (2015), pp. 55–56

98. T. Kusmierczyk, C. Trattner, K. Nørvåg, Understanding and predicting online food recipe production patterns, in *Proceedings of the 27th ACM Conference on Hypertext and Social Media* (2016), pp. 243–248

99. J. Kwon, Y.M. Lee, H. Wen, Knowledge, attitudes, behaviors about dining out with food allergies: a cross-sectional survey of restaurant customers in the united states. Food Control **107**, 106776 (2020)

100. R.D. Lawrence, G.S. Almasi, V. Kotlyar, M. Viveros, S.S. Duri, Personalization of supermarket product recommendations, in *Applications of Data Mining to Electronic Commerce* (Springer, Berlin, 2001), pp. 11–32

101. C.-S. Lee, M.-H. Wang, H. Hagras, A type-2 fuzzy ontology and its application to personal diabetic-diet recommendation. IEEE Trans. Fuzzy Syst. **18**(2), 374–395 (2010)

102. H.I. Lee, I.Y. Choi, H.S. Moon, J.K. Kim, A multi-period product recommender system in online food market based on recurrent neural networks. Sustainability **12**(3), 969 (2020)

103. N. Leipold, M. Madenach, H. Schäfer, M. Lurz, N. Terzimehic, G. Groh, M. Böhm, K. Gedrich, H. Krcmar, Nutrilize a personalized nutrition recommender system: an enable study. Health Recommend. Syst. **2216**, 24–29 (2018)

104. A. Levi, O. Mokryn, C. Diot, N. Taft, Finding a needle in a haystack of reviews: cold start context-based hotel recommender system, in *Proceedings of the Sixth ACM Conference on Recommender Systems* (2012), pp. 115–122

105. M. Li, B.M. Dias, I. Jarman, W. El-Deredy, P.J.G. Lisboa, Grocery shopping recommendations based on basket-sensitive random walk, in *Proceedings of the 15th ACM SIGKDD International Conference on Knowledge Discovery and Data Mining* (2009), pp. 1215–1224

106. X. Li, W. Jia, Z. Yang, Y. Li, D. Yuan, H. Zhang, M. Sun, Application of intelligent recommendation techniques for consumers' food choices in restaurants. Front. Psychiatry **9**, 415 (2018)

107. D.G. Liem, L. Zandstra, A. Thomas, Prediction of children's flavour preferences. Effect of age and stability in reported preferences. Appetite **55**(1), 69–75 (2010)

108. C.-J. Lin, T.-T. Kuo, S.-D. Lin, A content-based matrix factorization model for recipe recommendation, in *Pacific-Asia Conference on Knowledge Discovery and Data Mining* (Springer, Berlin, 2014), pp. 560–571

109. K. Lin, N. Sonboli, B. Mobasher, R. Burke, Calibration in collaborative filtering recommender systems: a user-centered analysis, in *Proceedings of the 31st ACM Conference on Hypertext and Social Media* (2020), pp. 197–206

110. Y. Linné, B. Barkeling, S. Rössner, P. Rooth, Vision and eating behavior. Obes. Res. **10**(2), 92–95 (2002)

111. Y. Liu, H. Lee, P. Achananuparp, E.-P. Lim, T.-L. Cheng, S.-D. Lin, Characterizing and predicting repeat food consumption behavior for just-in-time interventions, in *Proceedings of the 9th International Conference on Digital Public Health* (2019), pp. 11–20

112. Y. Luo, L. Tang, E. Kim, X. Wang, Finding the reviews on yelp that actually matter to me: innovative approach of improving recommender systems. Int. J. Hospital. Manag. **91**, 102697 (2020)
113. B.P. Majumder, S. Li, J. Ni, J. McAuley, Generating personalized recipes from historical user preferences (2019). arXiv:1909.00105
114. J. Mankoff, G. Hsieh, H.C. Hung, S. Lee, E. Nitao, Using low-cost sensing to support nutritional awareness, in *International Conference on Ubiquitous Computing* (Springer, Berlin, 2002), pp. 371–378
115. K. Maruyama, M. Spranger, "Interpretable Relational Representations for Food Ingredient Recommendation Systems." submitted for publication ICLR 2021, available via openreview https://openreview.net/pdf?id=48goXfYCVFX last accessed 03.01.2022
116. W. Mason, S. Suri, Conducting behavioral research on amazon's mechanical turk. Behav. Res. Methods **44**(1), 1–23 (2012)
117. D. Massimo, M. Elahi, M. Ge, F. Ricci, Item contents good, user tags better: empirical evaluation of a food recommender system, in *Proceedings of the 25th Conference on User Modeling, Adaptation and Personalization* (2017), pp. 373–374
118. A. Mathur, S.K. Juguru, M. Eirinaki, A graph-based recommender system for food products, in *2019 First International Conference on Graph Computing (GC)* (IEEE, Piscataway, 2019), pp. 83–87
119. C.E. Mauch, T.P. Wycherley, R.A. Laws, B.J. Johnson, L.K. Bell, R.K. Golley, Mobile apps to support healthy family food provision: systematic assessment of popular, commercially available apps. JMIR Mhealth Uhealth **6**(12), e11867 (2018)
120. R. McCarney, J. Warner, S. Iliffe, R. Van Haselen, M. Griffin, P. Fisher, The hawthorne effect: a randomised, controlled trial. BMC Med. Res. Methodol. **7**(1), 30 (2007)
121. B. McFerran, D.W. Dahl, G.J. Fitzsimons, A.C. Morales, I'll have what she's having: effects of social influence and body type on the food choices of others. J. Consum. Res. **36**(6), 915–929 (2010)
122. L. Meng, F. Feng, X. He, X. Gao, T.-S. Chua, Heterogeneous fusion of semantic and collaborative information for visually-aware food recommendation, in *Proceedings of the 28th ACM International Conference on Multimedia* (2020), pp. 3460–3468
123. W. Min, S. Jiang, J. Sang, H. Wang, X. Liu, L. Herranz, Being a supercook: joint food attributes and multimodal content modeling for recipe retrieval and exploration. IEEE Trans. Multimed. **19**(5), 1100–1113 (2016)
124. W. Min, S. Jiang, R.C. Jain, Food recommendation: framework, existing solutions and challenges. IEEE Trans. Multimed. **22**, 2659–2671 (2019)
125. W. Min, S. Jiang, L. Liu, Y. Rui, R. Jain, A survey on food computing. ACM Comput. Surv. **52**(5), 1–36 (2019)
126. C. Musto, C. Trattner, A. Starke, G. Semeraro, Towards a knowledge-aware food recommender system exploiting holistic user models, in *Proceedings of the 28th ACM Conference on User Modeling, Adaptation and Personalization* (2020), pp. 333–337
127. N. Nag, A.N. Rao, A. Kulhalli, K.S. Mehta, N. Bhattacharya, P. Ramkumar, A. Bharadwaj, D. Sitaram, R. Jain, Flavour enhanced food recommendation, in *Proceedings of the 5th International Workshop on Multimedia Assisted Dietary Management* (2019), pp. 60–66
128. Y. Nam, Y. Kim, Individualized exercise and diet recommendations: an expert system for monitoring physical activity and lifestyle interventions in obesity. J. Electr. Eng. Technol. **10**(6), 2434–2441 (2015)
129. K. Namgung, T.-H. Kim, Y.-S. Hong, Menu recommendation system using smart plates for 1394 well-balanced diet habits of young children. Wireless Commun. Mobile Comput. **2019** (2019) https://doi.org/10.1155/2019/7971381
130. A. Naresh, M.S.S. Shaastry, B.P. Yadav, K. Bhaskar, Understanding user taste preferences for food recommendation. Int. J. Eng. Res. Technol. **9**(6) (2020). ISSN: 2278-0181

131. Y.-K. Ng, M. Jin, Personalized recipe recommendations for toddlers based on nutrient intake and food preferences, in *Proceedings of the 9th International Conference on Management of Digital EcoSystems* (2017), pp. 243–250

132. K.H. Ng, V. Shipp, R. Mortier, S. Benford, M. Flintham, T. Rodden, Understanding food consumption lifecycles using wearable cameras. Pers. Ubiquit. Comput. **19**(7), 1183–1195 (2015)

133. B.T. Nguyen, D.T. Dang Nguyen, T.X. Dang, P. Thai, C. Gurrin, A deep learning based food recognition system for lifelog images, in *Proceedings of the 7th International Conference on Pattern Recognition Applications and Methods (ICPRAM 2018)*, (2018), pp. 657–664. ISBN: 978-989-758-276-9. https://doi.org/10.5220/0006749006570664

134. M. Kümpel Nørgaard, K. Bruns, P. Haudrup Christensen, M. Romero Mikkelsen, Children's influence on and participation in the family decision process during food buying. Young Consum., **8**(3), 197–216 (2007). https://doi.org/10.1108/17473610710780945

135. A. Ooi, T. Iiba, K. Takano, Ingredient substitute recommendation for allergy-safe cooking based on food context, in *2015 IEEE Pacific Rim Conference on Communications, Computers and Signal Processing (PACRIM)* (2015), pp. 444–449

136. T. Osadchiy, I. Poliakov, P. Olivier, M. Rowland, E. Foster, Recommender system based on pairwise association rules. Exp. Syst. Appl. **115**, 535–542 (2019)

137. T. Osadchiy, I. Poliakov, P. Olivier, M. Rowland, E. Foster, Validation of a recommender system for prompting omitted foods in online dietary assessment surveys, in *Proceedings of the 13th EAI International Conference on Pervasive Computing Technologies for Healthcare* (2019), pp. 208–215

138. M.-H. Park, H.-S. Park, S.-B. Cho, Restaurant recommendation for group of people in mobile environments using probabilistic multi-criteria decision making, in *Asia-Pacific Conference on Computer Human Interaction* (Springer, Berlin, 2008), pp. 114–122

139. F. Pecune, L. Callebert, S. Marsella, A socially-aware conversational recommender system for personalized recipe recommendations, in *Proceedings of the 8th International Conference on Human-Agent Interaction* (2020), pp. 78–86

140. M.P. Poelman, I.H.M. Steenhuis, Food choices in context, in *Context* (Elsevier, Amsterdam, 2019), pp. 143–168

141. N. Rajabpour, A. Naserasadi, M. Estilayee, TFR: a tourist food recommender system based on collaborative filtering. Int. J. Comput. Appl. **975**, 8887 (2018)

142. J. Ranganathan, D. Vennard, R. Waite, T. Searchinger, P. Dumas, B. Lipinski, Shifting diets: Toward a sustainable food future, in *2016 Global Food Policy Report. IFPRI*, (IFPRI, Washington, 2016), pp. 66–79. (Global Food Policy Report) ISBN 978-0-89629-582-7. https://doi.org/10.2499/9780896295827_08

143. S. Rendle, C. Freudenthaler, Z. Gantner, L. Schmidt-Thieme, BPR: Bayesian personalized ranking from implicit feedback (2012). arXiv:1205.2618

144. D. Ribeiro, J. Machado, J. Ribeiro, M.J.M. Vasconcelos, E.F. Vieira, A.C. de Barros, Souschef: mobile meal recommender system for older adults, in *3rd International Conference on Information and Communication Technologies for Ageing Well and e-Healt (ICT4AgeingWell)* (2017), pp. 36–45

145. D. Ribeiro, J. Ribeiro, M.J.M. Vasconcelos, E.F. Vieira, A.C. de Barros, Souschef: improved meal recommender system for portuguese older adults, in *International Conference on Information and Communication Technologies for Ageing Well and e-Health* (Springer, Berlin, 2017), pp. 107–126

146. J. Ribeiro, D. Ribeiro, A. Schwarz, M.J.M. Vasconcelos, F. Gerardo, C. Van Harten, R. Succu, R. Davison, T. Oliveira, T. Silva, et al., Cordon gris: integrated solution for meal recommendations, in *2018 IEEE International Conference on Pervasive Computing and Communications Workshops (PerCom Workshops)* (IEEE, Piscataway, 2018), pp. 46–51

147. A. Rivolli, L.C. Parker, A.C.P.L.F. de Carvalho, Food truck recommendation using multilabel classification, in *EPIA Conference on Artificial Intelligence* (Springer, Berlin, 2017), pp. 585–596

148. M. Rokicki, E. Herder, T. Kuśmierczyk, C. Trattner, Plate and prejudice: gender differences in online cooking, in *Proceedings of the 2016 Conference on User Modeling Adaptation and Personalization* (2016), pp. 207–215
149. M. Rokicki, C. Trattner, E. Herder, The impact of recipe features, social cues and demographics on estimating the healthiness of online recipes, in *Twelfth International AAAI Conference on Web and Social Media (ICWSM)* (2018), pp. 310–319
150. A. Rostami, V. Pandey, N. Nag, V. Wang, R. Jain, Personal food model, in *Proceedings of the 28th ACM International Conference on Multimedia* (2020), pp. 4416–4424
151. F. Ruis, Spilling the beans: food recipe popularity prediction using ingredient networks. B.S. Thesis, University of Twente (2019)
152. A. Rusu, M. Randriambelonoro, C. Perrin, C. Valk, B. Álvarez, A.-K. Schwarze, Aspects influencing food intake and approaches towards personalising nutrition in the elderly. J. Popul. Ageing **13**, 239–256 (2020)
153. A. Said, A. Bellogín, You are what you eat! Tracking health through recipe interactions, in *Proceedings of the 6th Workshop on Recommender Systems and the Social Web (Rsweb@recsys)* (2014)
154. A. Said, S. Berkovsky, E.W. De Luca, Putting things in context: challenge on context-aware movie recommendation, in *Proceedings of the Workshop on Context-Aware Movie Recommendation* (2010), pp. 2–6
155. S. Sajadmanesh, S. Jafarzadeh, S.A. Ossia, H.R. Rabiee, H. Haddadi, Y. Mejova, M. Musolesi, E.D. Cristofaro, G. Stringhini, Kissing cuisines: exploring worldwide culinary habits on the web, in *Proceedings of the 26th International Conference on World Wide Web Companion* (2017), pp. 1013–1021
156. Á.M.F.M. Samagaio, (2020). Chatbot for Food Preferences Modelling and Recipe Recommendation (Thesis) (2020). Retrieved January 4, 2022, from https://repositorio-aberto.up.pt/bitstream/10216/128328/2/411632.1.pdf.
157. H. Schäfer, M.C. Willemsen, Rasch-based tailored goals for nutrition assistance systems, in *Proceedings of the 24th International Conference on Intelligent User Interfaces* (2019), pp. 18–29
158. H. Schäfer, M. Elahi, D. Elsweiler, G. Groh, M. Harvey, B. Ludwig, F. Ricci, A. Said, User nutrition modelling and recommendation: balancing simplicity and complexity, in *Adjunct Publication of the 25th Conference on User Modeling, Adaptation and Personalization* (2017), pp. 93–96
159. H. Schösler, J. de Boer, J.J. Boersema, Can we cut out the meat of the dish? Constructing consumer-oriented pathways towards meat substitution. Appetite **58**, 39–47 (2012)
160. S.K. Shabanabegum, P. Anusha, E. Seethalakshmi, M. Shunmugam, K. Vadivukkarasi, P. Vijayakumar, IoT enabled food recommender with NIR system, in *Materials Today: Proceedings* (2020)
161. A.-L. Sibony, A. Alemanno, The emergence of behavioural policy-making: a european perspective. *Nudge and the Law: A European Perspective* (Hart Publishing, Oregon, 2015)
162. E. Simons, C.C. Weiss, T.J. Furlong, S.H. Sicherer, Impact of ingredient labeling practices on food allergic consumers. Ann. Allergy Asthma Immunol. **95**(5), 426–428 (2005)
163. P. Siriaraya, Y. Nakaoka, Y. Wang, Y. Kawai, A food venue recommender system based on multilingual geo-tagged tweet analysis, in *2018 IEEE/ACM International Conference on Advances in Social Networks Analysis and Mining (ASONAM)* (IEEE, Piscataway, 2018), pp. 686–689
164. J. Sobal, C.A. Bisogni, Constructing food choice decisions. Ann. Behav. Med. **38**(suppl. 1), s37–s46 (2009)
165. R. Sookrah, J.D. Dhowtal, S.D. Nagowah, A dash diet recommendation system for hypertensive patients using machine learning, in *2019 7th International Conference on Information and Communication Technology (ICoICT)* (IEEE, Piscataway, 2019), pp. 1–6
166. R.A. Sowah, A.A. Bampoe-Addo, S.K. Armoo, F.K. Saalia, F. Gatsi, B. Sarkodie-Mensah, Design and development of diabetes management system using machine learning. Int. J. Telemed. Appl. **5**, 1–17 (2020)

167. M. Speck, K. Bienge, L. Wagner, T. Engelmann, S. Schuster, P. Teitscheid, N. Langen, Creating sustainable meals supported by the NAHGAST online tool approach and effects on GHG emissions and use of natural resources. Sustainability **12**(3), 1136 (2020)
168. J. Starychfojtu, L. Peska, Smartrecepies: towards cooking and food shopping integration via mobile recipes recommender system, in *Proceedings of the 22nd International Conference on Information Integration and Web-based Applications & Services* (2020), pp. 144–148
169. R.I. Steim, C.J. Nemeroff, Moral overtones of food: judgments of others based on what they eat. Personal. Soc. Psychol. Bull. **21**(5), 480–490 (1995)
170. S. Stoll-Kleemann, U.J. Schmidt, Reducing meat consumption in developed and transition countries to counter climate change and biodiversity loss: a review of influence factors. Reg. Environ. Chang. **17**(5), 1261–1277 (2017)
171. L. Sun, J. Guo, Y. Zhu, Applying uncertainty theory into the restaurant recommender system based on sentiment analysis of online chinese reviews. World Wide Web **22**(1), 83–100 (2019)
172. C.R. Sunstein, *The Ethics of Influence: Government in the Age of Behavioral Science* (Cambridge University Press, Cambridge, 2016)
173. M. Svensson, K. Höök, R. Cöster, Designing and evaluating kalas: a social navigation system for food recipes. ACM Trans. Comput.-Hum. Interact. **12**(3), 374–400 (2005)
174. C.-Y. Teng, Y.-R. Lin, L.A. Adamic, Recipe recommendation using ingredient networks, in *Proceedings of the 4th Annual ACM Web Science Conference* (2012), pp. 298–307
175. R.H. Thaler, C.R. Sunstein, *Nudge: Improving Decisions About Health, Wealth, and Happiness* (Penguin, New York, 2009)
176. T. Theodoridis, V. Solachidis, K. Dimitropoulos, L. Gymnopoulos, P. Daras, A survey on AI nutrition recommender systems, in *Proceedings of the 12th ACM International Conference on Pervasive Technologies Related to Assistive Environments* (2019), pp. 540–546
177. R.Y. Toledo, A.A. Alzahrani, L. Martínez, A food recommender system considering nutritional information and user preferences. IEEE Access **7**, 96695–96711 (2019)
178. T.N.T. Tran, M. Atas, A. Felfernig, M. Stettinger, An overview of recommender systems in the healthy food domain. J. Intell. Inf. Syst. **50**(3), 501–526 (2018)
179. C. Trattner, D. Elsweiler, Food recommender systems: important contributions, challenges and future research directions (2017). arXiv:1711.02760
180. C. Trattner, D. Elsweiler, Investigating the healthiness of internet-sourced recipes: implications for meal planning and recommender systems, in *Proceedings of the 26th International Conference on World Wide Web* (2017), pp. 489–498
181. C. Trattner, D. Elsweiler, An evaluation of recommendation algorithms for online recipe portals, in *Health Recommender Systems (HealthRecSys@ RecSys)* (2019), pp. 24–28
182. C. Trattner, D. Elsweiler, What online data say about eating habits. Nat. Sustain. **2**(7), 545–546 (2019)
183. C. Trattner, D. Jannach, Learning to recommend similar items from human judgments. User Model. User Adapt. Interact. **30**(1), 1–49 (2020)
184. C. Trattner, D. Moesslang, D. Elsweiler, On the predictability of the popularity of online recipes. EPJ Data Sci. **7**(1), 20 (2018)
185. C. Trattner, D. Parra, D. Elsweiler, Monitoring obesity prevalence in the united states through bookmarking activities in online food portals. PloS One **12**(6), e0179144 (2017)
186. M. Trevisiol, L. Chiarandini, R. Baeza-Yates, Buon appetito: recommending personalized menus, in *Proceedings of the 25th ACM Conference on Hypertext and Social Media* (2014), pp. 327–329
187. M. Ueda, S. Asanuma, Y. Miyawaki, S. Nakajima, Recipe recommendation method by considering the users preference and ingredient quantity of target recipe, in *Proceedings of the International Multi Conference of Engineers and Computer Scientists*, vol. 1 (2014), pp. 12–14
188. T. Ueta, M. Iwakami, T. Ito, A recipe recommendation system based on automatic nutrition information extraction, in *International Conference on Knowledge Science, Engineering and Management* (Springer, Berlin, 2011), pp. 79–90

189. Y. van Pinxteren, G. Geleijnse, P. Kamsteeg, Deriving a recipe similarity measure for recommending healthful meals, in *Proceedings of the 16th International Conference on Intelligent User Interfaces* (2011), pp. 105–114
190. D.G. Vico, W. Woerndl, R. Bader, A study on proactive delivery of restaurant recommendations for android smartphones, in *ACM RecSys Workshop on Personalization in Mobile Applications, Chicago, USA* (2011)
191. C. Wagner, P. Singer, M. Strohmaier, The nature and evolution of online food preferences. EPJ Data Sci. **3**, 1–22 (2014)
192. G. Waltner, M. Schwarz, S. Ladstätter, A. Weber, P. Luley, H. Bischof, M. Lindschinger, I. Schmid, L. Paletta, Mango-mobile augmented reality with functional eating guidance and food awareness, in *International Conference on Image Analysis and Processing* (Springer, Berlin, 2015), pp. 425–432
193. G. Waltner, M. Schwarz, S. Ladstätter, A. Weber, P. Luley, M. Lindschinger, I. Schmid, W. Scheitz, H. Bischof, L. Paletta, Personalized dietary self-management using mobile vision-based assistance, in *International Conference on Image Analysis and Processing* (Springer, Berlin, 2017), pp. 385–393
194. Z. Wang, C. Meng, S. Ji, T. Li, Y. Zheng, Food package suggestion system based on multi-objective optimization: A case study on a real-world restaurant. Appl. Soft Comput. (2020), p. 106369
195. B. Wansink, J. Sobal, Mindless eating: the 200 daily food decisions we overlook. Environ. Behav. **39**(1), 106–123 (2007)
196. E. Wayman, S. Madhvanath, Nudging grocery shoppers to make healthier choices, in *Proceedings of the 9th ACM Conference on Recommender Systems* (2015), pp. 289–292
197. R. West, R.W. White, E. Horvitz, From cookies to cooks: insights on dietary patterns via analysis of web usage logs, in *Proceedings of the 22nd International Conference on World Wide Web* (2013)
198. H. Westhoek, J.P. Lesschen, T. Rood, S. Wagner, A. De Marco, D. Murphy-Bokern, A. Leip, H. van Grinsven, M.A. Sutton, O. Oenema, Food choices, health and environment: effects of cutting europe's meat and dairy intake. Global Environ. Change **26**, 196–205 (2014)
199. R. Xu, I.P. Cvijikj, T. Kowatsch, F. Michahelles, D. Büchter, B. Brogle, A. Dintheer, D. I'Allemand, W. Maass, Tell me what to eat–design and evaluation of a mobile companion helping children and their parents to plan nutrition intake, in *European Conference on Ambient Intelligence* (Springer, Berlin, 2014), pp. 100–113
200. L. Yang, Y. Cui, F. Zhang, J.P. Pollak, S. Belongie, D. Estrin, Plateclick: bootstrapping food preferences through an adaptive visual interface, in *Proceedings of the 24th ACM International on Conference on Information and Knowledge Management* (2015), pp. 183–192
201. L. Yang, C.-K. Hsieh, H. Yang, J.P. Pollak, N. Dell, S. Belongie, C. Cole, D. Estrin, Yum-me: a personalized nutrient-based meal recommender system. ACM Trans. Inf. Syst. **36**(1), 1–31 (2017)
202. M.R. Yeomans, A. Jackson, M.D. Lee, B. Steer, E. Tinley, P. Durlach, P.J. Rogers, Acquisition and extinction of flavour preferences conditioned by caffeine in humans. Appetite **35**(2), 131–141 (2000)
203. M. Zanker, D. Ninaus, Knowledgeable explanations for recommender systems, In *2010 IEEE/WIC/ACM International Conference on Web Intelligence and Intelligent Agent Technology*, vol. 1 (IEEE, Piscataway, 2010), pp. 657–660
204. D. Zeevi, T. Korem, N. Zmora, D. Israeli, D. Rothschild, A. Weinberger, O. Ben-Yacov, D. Lador, T. Avnit-Sagi, M. Lotan-Pompan, et al., Personalized nutrition by prediction of glycemic responses. Cell **163**(5), 1079–1094 (2015)
205. J. Zeng, F. Li, H. Liu, J. Wen, S. Hirokawa, A restaurant recommender system based on user preference and location in mobile environment, in *2016 5th IIAI International Congress on Advanced Applied Informatics (IIAI-AAI)* (IEEE, Piscataway, 2016), pp. 55–60
206. R. Zenun Franco, Online recommender system for personalized nutrition advice, in *Proceedings of the Eleventh ACM Conference on Recommender Systems* (2017), pp. 411–415

207. F. Zhang, N.J. Yuan, K. Zheng, D. Lian, X. Xie, Y. Rui, Exploiting dining preference for restaurant recommendation, in *Proceedings of the 25th International Conference on World Wide Web* (2016), pp. 725–735

208. C. Zhang, H. Zhang, J. Wang, Personalized restaurant recommendation method combining group correlations and customer preferences. Inf. Sci. **454**, 128–143 (2018)

209. Q. Zhang, C. Trattner, B. Ludwig, D. Elsweiler, Understanding cross-cultural visual food tastes with online recipe platforms, in *Proceedings of the International AAAI Conference on Web and Social Media*, vol. 13 (2019), pp. 671–674

210. Q. Zhang, D. Elsweiler, C. Trattner, Visual cultural biases in food classification. Foods **9**(6), 823 (2020)

211. B. Zhou, J. Cheng, P. Lukowicz, A. Reiss, O. Amft, Monitoring dietary behavior with a smart dining tray. IEEE Pervasive Comput. **14**(4), 46–56 (2015)

212. Y.-X. Zhu, J. Huang, Z.-K. Zhang, Q.-M. Zhang, T. Zhou, Y.-Y. Ahn, Geography and similarity of regional cuisines in China. PloS One **8**(11), e79161 (2013)

# Music Recommendation Systems: Techniques, Use Cases, and Challenges

Markus Schedl, Peter Knees, Brian McFee, and Dmitry Bogdanov

## 1 Introduction

In the past decade, we have experienced a drastic change in the way how people search for and consume music. The rise of digital music distribution, followed by the spiraling success of music streaming services such as those offered by Spotify,[1] Pandora,[2] Apple,[3] Amazon,[4] YouTube,[5] and Deezer,[6] has led to the ubiquitous

---

[1] https://www.spotify.com.

[2] https://www.pandora.com.

[3] https://www.apple.com/apple-music.

[4] https://music.amazon.com.

[5] https://www.youtube.com.

[6] https://www.deezer.com.

---

M. Schedl (✉)
Johannes Kepler University Linz, Institute of Computational Perception, Multimedia Mining and Search Group, Linz, Austria

LIT AI Lab, Human centered AI Group, Linz, Austria
e-mail: markus.schedl@jku.at

P. Knees
TU Wien, Faculty of Informatics, Institute of Information Systems Engineering, Vienna, Austria
e-mail: peter.knees@tuwien.ac.at

B. McFee
New York University, Center for Data Science and Music and Audio Research Lab, New York, NY, USA
e-mail: brian.mcfee@nyu.edu

D. Bogdanov
Universitat Pompeu Fabra, Music Technology Group, Barcelona, Spain
e-mail: dmitry.bogdanov@upf.edu

© Springer Science+Business Media, LLC, part of Springer Nature 2022
F. Ricci et al. (eds.), *Recommender Systems Handbook*,
https://doi.org/10.1007/978-1-0716-2197-4_24

availability of music. While the catalogs of these big players are mostly geared towards Western music, in recent years, worldwide, many platforms focusing on domestic markets and music emerged, including Taiwanese KKBOX,[7] Korean Melon,[8] Nigerian Boomplay Music,[9] and Brazilian Superplayer.[10]

As a result, music listeners are suddenly faced with an unprecedented scale of readily available musical content, which can easily become burdensome. Addressing this issue, music recommender systems (MRS) provide support to users accessing large collections of music items and additional music-related content. Music items that are most commonly recommended include artists, albums, tracks, and playlists. Moreover, integrating additional music-related content into their catalogs has become more and more important for streaming providers, to offer their users a unique selling proposition. Such additional content include lyrics, music video clips and animated video backgrounds, album cover images, and information about concert venues.

This chapter gives an introduction to music recommender systems research. In the remainder of this section, we next discuss the unique characteristics of the music recommendation domain (Sect. 1.1), as compared to other content domains, such as videos or books. Then, we define the scope and structure of the subsequent sections (Sect. 1.2).

## 1.1  Characteristics of the Music Recommendation Domain

There exist several distinguishing characteristics of the music domain that differentiates MRS from other kinds of recommender systems. We summarize the major ones in the following. For a more detailed treatment, we refer the reader to [158, 161].

*Duration of consumption:* The amount of time required for a user to consume a single media item strongly differs between different categories of items: an image (typically a few seconds), a song (typically a few minutes), a movie (typically one to a few hours), a book (typically days or weeks). Since music ranges at the lower end of the duration scale, the time it takes for a user to form opinions on a music item can be much shorter than in most other domains. As a result, music items may be considered more disposable.

*Catalog size:* Typical commercial music catalogs contain tens of millions of songs or other musical pieces while catalogs of movies and TV series are several magnitudes smaller. The scalability of commercially used MRS algorithms is, therefore, a more important requirement in the music domain than in other domains.

---

[7] https://www.kkbox.com.

[8] https://www.melon.com.

[9] https://www.boomplay.com.

[10] https://www.superplayer.fm.

*Different representations and abstraction levels:* Another distinguishing property is that music recommendations can be made at different item abstraction levels and modalities. While movie recommender systems typically suggest individual items of one specific category (e.g., movies or series) to the user, MRS may recommend music items of various representations and modalities (most commonly, the audio of a song, but also music videos or even digital score sheets offered by providers such as OKTAV[11] or Chordify[12]). Music recommendations can also be effected at different levels of granularity (e.g., at the level of artist, album, or song). Furthermore, non-standard recommendation tasks exist in the music domain, such as recommending radio stations or concert venues.

*Repeated consumption:* A single music item is often consumed repeatedly by a user, even multiple times in a row. In contrast, other media items are commonly consumed at most a few times. This implies that a user might not only tolerate, but actually appreciate recommendations of already known items.

*Sequential consumption:* Unlike movies or books, songs are frequently consumed in sequence, e.g., in a listening session or playlist. As a result, sequence-aware recommendation problems [143] such as automatic playlist continuation or next-track recommendation play a crucial role in MRS research. Because of the unique constraints and modeling assumptions of serial consumption, also the evaluation criteria substantially differ from the more standard techniques found in the recommender systems literature [71].

*Passive consumption:* Unlike most other media content, music is often consumed passively, in the background, which can affect the quality of preference indications. Especially when relying on implicit feedback to infer music preferences of users, the situation where a listener is not paying attention to the music (and therefore does not skip a disliked song) might be misinterpreted as a positive feedback.

*Importance of content:* In traditional recommendation domains such as movie recommendation, collaborative filtering (CF) techniques have been predominantly used and refined over the years, not least thanks to initiatives such as the *Netflix Prize*.[13] In contrast, research on music recommendation has emerged to a large extent from the fields of audio signal processing and music information retrieval (MIR), and is still strongly connected to these areas. This is one of the reasons why content-based recommendation approaches, such as content-based filtering (CBF), are more important in the music domain than in other domains. Such approaches aim at extracting semantic information from or about music at different representation levels (e.g., the audio signal, artist or song name, album cover, lyrics, album reviews, or score sheet), and subsequently leverage similarities computed on these semantic music descriptors, between items and user profiles, to effect recommendations.[14]

---

[11] https://www.oktav.com.

[12] https://www.chordify.net.

[13] https://www.netflixprize.com.

[14] To avoid confusion, we note that *content* has different connotations within the MIR and recommender systems communities. MIR makes an explicit distinction between (content-based)

Another reason for the importance of content-based approaches in MRS is the fact that explicit rating data is relatively rare in this domain, and even when available, tends to be sparser than in other domains [44]. Therefore, research in music recommendation techniques tend to rely more upon content descriptions of items than techniques in other domains.

## 1.2   Scope and Structure of the Chapter

In this chapter, we first categorize in Sect. 2 music recommendation tasks into three major types of use cases. Section 3 subsequently explains the major categories of MRS from a technical perspective, including content-based filtering, sequential recommendation, and recent psychology-inspired approaches. Section 4 is devoted to a discussion of challenges that are faced in MRS research and practice, and of approaches that address these challenges. Finally, in Sect. 5 we conclude by summarizing the main recent trends and open challenges in MRS.

Please note that this chapter substantially differs from the previous version that was published in the second edition of the Recommender Systems Handbook [159]. While the previous version was generally structured according to different techniques and types of music recommender systems, in the version at hand, we take a more user-centric perspective, by organizing our discussion with respect to current use cases and challenges.

## 2   Types of Use Cases

Research on and development of MRS has evolved significantly over the last decade, owed to changes in the typical use cases of MRS. We can categorize these use cases broadly into basic music recommendation (Sect. 2.1), lean-in (Sect. 2.3), and lean-back experiences (Sect. 2.2).

We refer to the most traditional use case as *basic music recommendation*, which aims at providing recommendations to support users in browsing a music collection. Technically, corresponding tasks are common to other domains, and include predicting a user's explicit rating (rating prediction task) or predicting whether a given user will listen to a particular song (predicting item consumption behavior). Requiring a higher degree of attention and engagement, use cases pertaining to *lean-in* exploration refer to supporting users in searching particular music based on a semantic query that expresses a user intent, e.g., finding music

---

approaches that operate directly on audio signals and (metadata) approaches that derive item descriptors from external sources, e.g., web documents [90]. In recommender systems research, as in the remainder of this chapter, both types of approaches are described as "content-based".

that fits a certain activity or affective state. In contrast, by providing a *lean-back* experience to the user no specific user task is addressed rather than indulging, for instance, an endless music listening session.

## 2.1 Basic Music Recommendation

A typical function of a recommender system is to assist in actively browsing the catalog of items through item-to-item recommendations. For an MRS, this implies providing *lists of relevant artists, albums, and tracks*, when a user browses item pages of a music shop or a streaming service. Commonly, such recommendations rely on similarity inferred from the consumption patterns of the users, and they are presented to a user in the form of a list of "people who played that also played this" items.

Another basic functionality is to generate *personalized recommendation lists* on the platform's landing page to engage a user in a session even without their active navigation of the content in the first place. Such recommendations are generated based on the user's previous behavior on the platform, which is a core research task in the recommender systems community. At the same time, it is the topic of lots of user interface (UI) and user experience (UX) design decisions in the industry, often out of the scope of academic research. For example, the system interface may provide contextual "shelves", grouping recommendations by a particular reason, or time span of user activity (e.g., recommendations based on global user profile versus recent user activity).

Figures 1 and 2 demonstrate both types of basic approaches, on the example of Soundcloud[15] and Last.fm.[16] In both cases, such *basic music recommendation* systems deal with artist, album, or track recommendations using the information about previous user interactions and their feedback for the items in the music catalog.

### 2.1.1 Interaction and Feedback Data

Music services can gather explicit user feedback, including rating provided by a user for artist, album, or track items (e.g., using 1–10 or 1–5 rating scales), or binary Likes, Loves, or Favorites reactions, as well as the information about items purchased or saved to the user's library. Therefore, a common task is predicting those explicit user ratings and reactions for items in the system's music catalog, which is useful to estimate relevance and generate ranked recommendation lists. Historically, user ratings have been associated with online music shops (e.g.,

---

[15] https://www.soundcloud.com.

[16] https://www.last.fm.

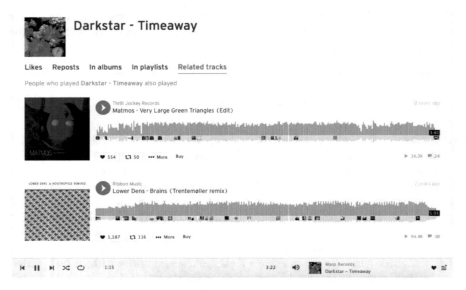

**Fig. 1** Soundcloud's "Related tracks" web pages provide playlists of top 50 track recommendations for the tracks in their catalog. The list is pre-computed according to the seed in advance. The user is able to scroll and listen to the entire list, start and stop the playback (▶/❚❚), as well as navigate by clicking on a particular track or the next or previous track buttons (▶▶/◀◀)

iTunes[17] or Amazon[18]) and music metadata websites that allow users to organize their music collections (e.g., Discogs[19] or RateYourMusic[20]). They became much less common nowadays due to the advance of streaming platforms. Even in the shop scenario, explicit ratings may be too difficult to gather for the majority of the user base, and instead, systems rely on purchase history.

In the case of music streaming services, it is common to gather implicit user feedback. These systems often strive to minimize the required interaction effort while asking for explicit ratings can be tedious. Instead, they register each track played by a user (user listening events), compute play counts or total time listened for different items (tracks, albums, and artists), and track skips within the music player UI.

Such implicit feedback represents music preferences only indirectly. It can be dependent on user activity, context, and engagement, and there may be other reasons for user behavior unknown to the system. A played track in the user history does not necessarily mean the user actively liked it, and a skipped track does not necessarily

---

[17] https://www.apple.com/itunes.

[18] https://www.amazon.com.

[19] https://www.discogs.com.

[20] https://www.rateyourmusic.com.

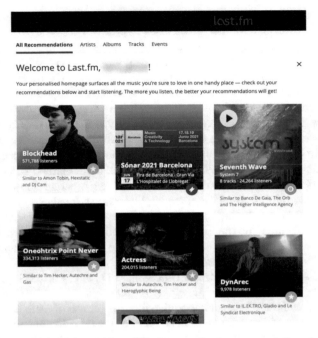

**Fig. 2** Last.fm's landing page provides a list of personalized recommendations (artists, albums, tracks, and events) based on gathered listening statistics in a user profile. To give some context, the justification for the recommendations is provided by referencing similar music items listened by the user. The user can navigate the entire list as well as listen to a playlist with recommendations

imply negative preference. Still, this feedback is often the only information available to the system, and therefore it is used as a proxy for music preference.[21] Different criteria can be applied to consider a track as played and relevant for a user. For example, one can rely on the fraction of the total track duration that is reproduced within the system's UI and define a threshold to identify fully or almost fully played tracks. Also, the raw play count values can be normalized and thresholded to define relevant items (for example, in the simplest case, consider all items with at least one play as relevant).

### 2.1.2 Evaluation Metrics and Competitions

In essence, the basic recommendation task is the prediction of relevant items for a user and generation of recommendation lists with items ranked by relevance. The evaluation is commonly done in the offline setting, retaining part of the user

---

[21] Note that explicit ratings can be estimated from implicit feedback such as play counts, as investigated by Parra and Amatriain [137].

behavior data (e.g., user-item relevance ratings or play counts) as a ground truth and measuring the error in relevance predictions (typically via RMSE), or assessing the quality of generated ranked lists in terms of the position of relevant items therein, e.g. via precision at $k$, recall at $k$, mean average precision (MAP) at $k$, average percentile rank, or normalized discounted cumulative gain (NDCG). A few official research challenges described below have addressed these aspects.

Formulated as a purely collaborative filtering task, the problem has been addressed by the recommender systems community in the *KDD Cup 2011* challenge [44].[22] It featured a large-scale dataset of user-item ratings provided by *Yahoo!*[23] with different levels of granularity of the ratings (tracks, albums, artists, and genres) and a very high sparsity (99.96%) making the task particularly challenging. There were two objectives in the challenge, addressed on separate sub-tracks: predict unknown music ratings based on given explicit ratings (evaluated by RMSE) and distinguish highly rated songs from songs never rated by a user (evaluated by error rate). Unfortunately, the dataset for the challenge is anonymized, including all descriptive metadata, which made it impossible to try any approaches based on content analysis and music domain knowledge.

The *Million Song Dataset (MSD) Challenge*[24] [115] organized in 2012 opened the possibility to work with a wide variety of data sources (for instance, including web crawling, audio analysis, collaborative filtering, or use of metadata). Given full listening histories of one million users and half of the listening histories for another 110,000 test users, the task was to predict the missing hidden listening events for the test users. Mean average precision computed on the top 500 recommendations for each listener (MAP@500) was used as main performance measure.

## 2.2   Lean-in Exploration

Other music consumption settings emphasize more active and engaged user interaction. In these *lean-in* scenarios, the user is often exploring a collection and the found tracks in-depth to select candidates for listening immediately or at a later point. This can be used e.g. to find music that fits a certain affective state, activity, or setting such as a workout or a road trip. In many cases these scenarios are also tied to building and maintaining "personal" music collections within online platforms for individual or shared use, cf. [36, 37]. A recommender system can support such a process by presenting candidate tracks based on the user's behavior and adapting to the selection of tracks, cf. [80]. An example of a lean-in interface is Spotify's playlist creation interface, as shown in Fig. 3, in which recommendations are made to complement the tracks already added to a playlist.

---

[22] https://www.kdd.org/kdd2011/kddcup.shtml.

[23] http://music.yahoo.com.

[24] http://labrosa.ee.columbia.edu/millionsong/challenge.

**Fig. 3** Spotify's playlist creation application as an example of a *lean-in* experience. The user can search the catalog using the textual search box on top. Based on the songs added to the list, further songs are recommended at the bottom for consideration. Hovering over a track allows to play/pause the audio (▶/❚❚). Further recommendations can be requested by clicking the "Refresh" button. Before any songs are added, the user is prompted for a title and description for the playlist, which provide the basis for making initial recommendations. (Note that further playback control, navigation, advertisement, and social network panels in the interface are omitted in this screenshot.)

Lean-in-oriented interfaces provide a higher degree of control and are therefore richer in terms of user interface components, demanding more attention and a higher cognitive load from the user [126]. Controls often include search functionalities that add possibilities to retrieve specific tracks based on their metadata or via a "semantic" query that expresses the user intent, cf. Fig. 3. To index music pieces for semantic textual search, several sources can be tapped, such as knowledge graphs [134], various forms of community metadata such as tags or websites [90], or playlist titles given by other users [116]. This not only enriches the descriptions of individual tracks (e.g., to allow for queries like "90's band with female singer"), but also introduces information on context and usage purposes (e.g., "sleep" or "party").

Beyond playlist creation, lean-in interfaces have been proposed for a variety of tasks, e.g., exploration of musical structure, beat structure, melody line, and chords of a track [65], or exploration of tracks based on similar lyrics content [130]. These "active music-listening interfaces" [64], however, have not seen much adoption in commercial streaming platforms as they are often targeted at specific music consumers and non-traditional tasks, cf. [93].

### 2.2.1  Evaluation Metrics and Competitions

Evaluation metrics for lean-in scenarios are measuring similar aspects as in a less targeted browsing scenario, i.e. mostly retrieval-oriented metrics, see Sect. 2.1.2. In a focused task like the above mentioned playlist creation, however, feedback is available more explicitly, as selections are made from a known pool of tracks and a selection more strongly indicates a positive feedback as much as an omission a negative than is the case in general browsing. This also impacts evaluation, as information retrieval metrics like precision and recall can be meaningfully calculated.

This is reflected in the *2018 RecSys Challenge* [33],[25] which was centered on the task of recommending tracks for playlist creation, as performed in the Spotify application shown in Fig. 3. More specifically, provided with a playlist of specific length $k$ (and optionally a playlist title indicating a context), the task was to recommend up to 500 tracks that fit the target characteristics of the given playlist. Different scenarios were addressed in terms of playlist types: by varying $k$, whether or not a playlist title was provided, and whether or not the playlist was shuffled. As part of the challenge, Spotify released a collection of 1 million user-generated playlists to be used for model development and evaluation.[26] Evaluation metrics used were R-precision, NDCG, and a Spotify-specific metric called "recommended songs clicks" (defined as the number of times a user has to request 10 more songs before the first relevant track is encountered). A detailed description and analysis of the top approaches can be found in [188].

### 2.2.2  Discussion

Lean-in experiences relate most closely to traditional directed information retrieval tasks like search and extend to all activities where a user is willing to devote time and attention to a system to enhance the personal experience. The role of the recommender system is to support the user in this specific scenario, e.g. by suggesting complementary items, without interfering, distracting, or persuading the user.

In terms of designing the recommendation algorithm, a lean-in scenario provides a good opportunity to favor *exploration* over *exploitation*. That is, the recommender system might not optimize for positive feedback only, but "probe" the user with potentially negatively perceived items. As such, these closer interactions between user and system provide an opportunity to develop the user profile for future recommendations, resulting in a longer-term reward than just the exploitation of items known to please the user (however, likely in a different context). Immediate

---

[25] https://www.recsyschallenge.com/2018.

[26] The *Million Playlist Dataset* is available from https://www.aicrowd.com/challenges/spotify-million-playlist-dataset-challenge.

user satisfaction does not necessarily suffer from this strategy, as more diverse recommendations are acceptable in a setting in which the user is open to reflect upon the suggestions made.

## 2.3 Lean-Back Listening

In contrast to the recommendation use cases outlined in the previous sections, the so-called *lean-back* formulation is designed to address use cases in which user interaction is minimized, such as automatic playlist generation or streaming radio. In lean-back settings, users typically are presented a single song at a time, which is selected automatically by the MRS. Often, it is expected that users do not have the interface directly in view, but rather are consuming recommendations on a smartphone application with the device out of view (e.g., in a pocket or bag, or while driving).

Typical lean-back user interfaces, such as the one depicted in Fig. 4 (left), tend to be minimal, and severely limit how the user can control the system. Although some

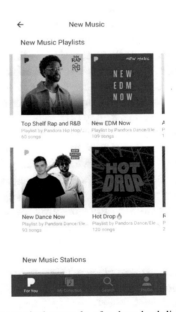

**Fig. 4** Pandora's mobile application provides a prototypical example of a *lean-back* listening interface. Left: An image representing the current song (or an advertisement) is displayed, along with a progress bar and controls to switch "stations". The user can play/pause the audio (▶/❚❚), provide thumbs-up/down feedback (👍/👎), skip to the next track (▶▶), or replay a track (↻). No information is provided for alternative track selections: the next track is selected automatically. Right: recommendations are organized by *stations* or *playlists*, allowing users to select a stream with minimal interactions

interfaces afford either positive or negative explicit feedback (likes, hearts, thumbs-up/thumbs-down), the fact that users are typically disengaged during consumption implies that explicit feedback is relatively rare. As a result, methods and evaluation criteria in lean-back settings tend to rely more upon implicit feedback, such as song completion, skipping, or session termination.

### 2.3.1 Lean-Back Data and Evaluation

Compared to the basic recommender system setup, the treatment of implicit feedback in lean-back settings requires a bit more nuance. In the basic setting, users are typically presented with a collection of items simultaneously, from which positive, negative, and relative interactions can be inferred efficiently. For example, if a user is recommended ten songs, and purchases only one, it is relatively safe to infer that the remaining nine were less relevant than the purchased song—if not outright *irrelevant*—which facilitates rapid and large-scale data collection [77]. Because lean-back interfaces do not present alternatives to the user, implicit feedback can only be inferred for a single item at a time, which significantly reduces the efficiency of data collection.

Lean-back music recommenders are often used in certain *contexts*: for example, while a user is exercising or working. Users are therefore expected to be inattentive, at least some portion of the time, and this can make it difficult to properly interpret even the weak signals that come from play, stop, and skip interactions. Accurately making sense of feedback gained in a lean-back setting is therefore highly challenging, and such data may not be suited to exhaustively model user preferences. For example, listening to a full song may be construed as a *positive* interaction, but it may also happen because a user was completely disengaged and forgot to turn off the stream. Alternatively, skipping a song may suggest a *negative* interaction, but a user may also skip a song that they otherwise like because they recently heard it elsewhere. Finally, a user may abandon a session because they are dissatisfied with the song selections (a negative interaction), or because they are finished with whatever outside activity they were performing. While these issues can be impossible to fully work around (barring invasive user surveillance), it is common to assume that play/skip/stop behavior, on average, provide weak positive and negative signals that can be used for model development and evaluation. However, these issues do highlight an important characteristic of lean-back recommendation: interactions take place within a particular *context*, and interpreting the resulting data outside its context can easily become problematic.

Lean-back interfaces are often designed around concepts of *playlists*, *radio stations*, or other similar abstractions which allow a user to express preference for groups of songs, e.g., by selecting a pre-generated playlist or station by description, or by choosing a genre, artist, or song to *seed* the session. This is depicted in Fig. 4 (right). As a result, much of the research on lean-back systems has focused on modeling playlists, which can be defined as either ordered or shuffled selections of songs that are meant to be heard together in a session. Playlists can be composed by expert curators (like traditional disc jockeys), amateur users, or collaboratively by groups of users. Playlist authors often select songs deliberately to reflect some specific intended use or context [38]. Playlist data therefore provides an attractive source of high-quality positive interactions: co-occurrence of songs in a playlist can be a strong positive indicator of similarity. An algorithm which can accurately predict which songs should go into a playlist—perhaps conditioned on a context, user preference, or pre-selected songs—should therefore be useful for generating recommendations [59].

Note that modeling playlist *composition* is not equivalent to modeling playlist *consumption*. In the generation case, it is (perhaps arguably) justified to infer negative interactions: the playlist author presumably decided which songs to include. This in turn justifies the use of information retrieval metrics (precision, recall, etc.), which rely upon having a well-defined notion of what constitutes a *relevant* (included) or *irrelevant* (excluded) example. Playlist consumption, however, does not inherit relevance and irrelevance from playlist composition. The user was not exposed to alternative selections, so it is not justified to infer a negative interaction from tracks which were not included in the playlist. Evaluating a playlist generation method therefore relies on comparisons to playlist *authors*, not playlist *consumers*. In many commercial settings, authors can be employees of the service, while consumers are the customers, and it is important to bear this distinction in mind when developing a recommender system.

As an alternative to playlist data, *listening log* data can provide a more direct measurement of actual user behavior, as it captures how users behave in response to specific song selections. This makes listening log data attractive from a modeling perspective, but care must be taken to ensure that the data is interpreted correctly, and does not unduly propagate inductive bias from the system which selected the songs in the first place. That said, an algorithm which could accurately predict which songs are likely or unlikely to be skipped by a user (in a particular session or context) could be used to power a recommendation system. Developing and evaluating such a method requires large volumes of log data to capture the diversity of listening preferences and contexts. Fortunately, log data can be collected passively: unlike playlist generation, which requires deliberate intent from the playlist creator, log data is generated automatically by users interacting naturally with the system.

While user-generated playlist data has been relatively abundant and freely available [117, 139, 188], high-quality listening log data has been scarce outside of private, commercial environments. Until recently, the main source of openly available log data has been Last.fm scrobbles [28, 151, 181], which provide a large volume of data, but little in the way of transparency or provenance with respect to how tracks are selected and interacted with by users. This has recently changed with the MRS-related challenge in the *2019 WSDM Cup*:[27] participating teams had to predict whether a user will listen to or skip the music recommended next by a MRS. Participants were provided with a set of 130 million listening sessions by Spotify

---

[27] https://www.crowdai.org/challenges/spotify-sequential-skip-prediction-challenge.

users [23]. For each listening session, the first half of the session was observed, and the objective was to predict whether or not each track in the second half of the session was actually consumed or skipped. The adopted evaluation metric was mean average accuracy, where average accuracy was computed over all items in the unseen part of each session. The winning approaches are described in [32, 69, 195].

### 2.3.2 Discussion

In contrast to the basic recommendation formulation, lean-back settings require careful attention to several unique factors arising from the characteristics of music consumption.

First, as previously mentioned, users in lean-back scenarios are assumed to want minimal interaction with the recommender system. Concretely, lean-back music recommenders are often designed to be more conservative with recommendations, prioritizing *exploitation* over *exploration* to minimize negative feedback (i.e., skips) [29]. While the exploration–exploitation trade-off is a well-known concept in recommender systems generally [163], one must bear in mind that the right trade-off fundamentally depends on the mode of delivery and interaction with the user.

Second, the sequential nature of lean-back recommendation scenarios presents a substantial methodological challenge to evaluation. For example, skip prediction methods are trained and evaluated on historical log data, which is almost always biased by whichever recommendation algorithm was used at the time of the interactions. Left unchecked, this can propagate inductive bias from previous algorithms, and skew the evaluation results. While not inherently unique to the lean-back setting, the rapid sequential consumption of recommendations in this context renders typical simplifying assumptions (e.g., independence between interactions) suspicious at best. Compensating for this source of bias is generally challenging, though in some situations, counterfactual risk minimization [170] can be employed during training [120].

Finally, playlist data may carry biases beyond what are commonly found in standard collaborative filter data. In particular, several streaming platforms employ content curators to create playlists which can be shared to users, or allow (and promote) users to share playlists with each other. While some amount of curation is undoubtedly desirable in many situations, it is also important to understand the sources of bias in artist and track selection within the data when developing and evaluating a playlist generation algorithm.

## 2.4   Other Applications

There also exist various applications beyond the main lines of research on music recommendation. In particular, *music event recommendation* is addressed in [176], where the authors consider recommending events/venues with local long-tail artists, from which metadata available to the system may be lacking. A related task is the recommendation of music for particular venues, addressed in [35]. Last.fm and Spotify are examples of industrial systems that provide event recommendations.

*Playlist discovery* and *playlist recommendation* are other recent directions [136]. They have not yet received as much attention as playlist continuation (Sect. 2.3): even though there is research on track recommendation for playlists and listening sessions, a lack of studies on recommending entire playlists to a user is evident.

We can also highlight the future role of recommender systems in music and video production. An exemplary use case here is *recommending background music for video*. The goal of this task is to assist the user in finding music that fits video scenes in their video production in terms of semantics, rhythm, and motion. Such systems rely on multi-modal analysis of audio and video [100, 110]. In turn, in music production, recommender systems can enrich the users' workflow, helping navigate extensive audio collections. In particular, *sound recommendation* has been considered in [134, 165]. Based on audio analysis and domain knowledge in music composition, recommender systems will open promising possibilities for building more intelligent digital audio workstations with sound, loop, and audio effect recommendation functionalities.

Another task related to MRS is to *recommend digital score sheets*, such as implemented in the system offered by OKTAV[28] for piano players; or to *recommend chords* for guitar players, offered by Chordify.[29]

# 3 Types of Music Recommender Systems

In the following, we briefly characterize the major types of MRS and summarize the input data and techniques adopted in each type. Please note that the research works we point to by no means represent an exhaustive list. We intentionally kept this general part rather short and point the reader, for instance, to our chapter in the second edition of the Recommender Systems Handbook [159] for a more detailed description of the different types of MRS.

## 3.1 Collaborative Filtering

Similar to other recommendation domains, CF-based approaches are often used for music recommendation. They operate solely on data about user–item interactions, which are either given as explicit ratings (e.g., on a rating scale) or as implicit feedback (e.g., statistics on play counts or skipped songs). In the most common

---

[28] https://www.oktav.com.

[29] https://www.chordify.net.

variant, CF approaches create a model that predicts whether a given user will interact or not with a (previously unseen) item. Since approaches solely based on CF are domain-agnostic, we invite the reader to consider the surveys provided in [34, 73, 86, 96] for a more detailed general treatment of the topic; and the surveys provided in [152, 161] that review CF approaches in the music domain, among others. Also consider chapter "Advances in Collaborative Filtering" of this book.

Note that CF-based approaches are particularly prone to several kinds of biases, such as *data bias* (e.g., community bias and popularity bias) and *algorithmic bias*. *community bias* refers to a distortion of the data (user–item interactions) caused by the fact that the users of a certain MRS platform do not form a representative sample of the population at large. An example is provided by a study of classical music on Last.fm and Twitter, carried out in [160]. The over- or under-representation of a user group, and in turn resulting user preferences, influences the quality of CF-based algorithms. *Popularity bias* occurs when certain items receive many more user interactions than others. This relates to the long-tail property of consumption behavior and can favor such highly popular items [5, 98]. Other data biases and artifacts in music usage data stem from additional factors such as record label associations [89].

In terms of *algorithmic bias*, Ekstrand et al. [48] find an effect of age and gender on recommendation performance of CF algorithms, even when equalizing the amount of data considered in each user group. Similarly, Melchiorre et al. [124] identify considerable personality bias in MRS algorithms, i.e., users with different personalities receive recommendations of different quality levels which depend on the adopted recommendation algorithm. To alleviate these undesirable effects, devising methods for debiasing is one of the current big challenges in MRS research (see Sect. 4.3).

## 3.2   Content-Based Filtering

While CF-based approaches operate solely on user–item interaction data, the main ingredient to CBF algorithms is content information about items, which is used to create the user profiles. To compute recommendations for a target user, no other users' interaction data is needed.

Recommendations are commonly made based on the similarity between the user profile and the item representations in a top-$k$ fashion, where the former is created from individual statistics of item content interacted with by the user. In other words, given the content-based user profile of the target user, the items with best matching content representations are recommended. Alternatively, a machine learning model can be trained to directly predict the preference of a user for an item.

A crucial task for every CBF approach is the representation of content information, which is commonly provided in form of a feature vector that can be composed of (1) handcrafted features or (2) latent embeddings from deep learning tasks such as auto-tagging. In the former case, commonly leveraged features include computational audio descriptors of rhythm, melody, harmony, or timbre, but also metadata such as genre, style, or epoch (either extracted from user-generated tags or provided as editorial information), e.g. [20, 21, 42, 66, 118]. We refer to [128] for an overview of music audio descriptors typically used in MIR. As for the latter (2), latent item embeddings are often computed from low-level audio signal representations such as spectrograms, e.g. [178], or from textual metadata such as words in artist biographies, e.g. [133]. Higher-level semantic descriptors (such as genres, moods, and instrumentation) retrieved from audio features by machine learning can also be used and have been shown to correlate with preference models in music psychology [60].

One of the earliest and most cited deep-learning-based approached to CBF in the music domain is van den Oord et al.'s work [178]. The authors train a convolutional neural network (CNN) to predict user–item latent factors from matrix factorization of collaborative filtering data. The resulting CNN is then able to infer these latent representations from audio only.

More recent deep learning approaches based on content representations include [74, 109, 133, 148, 177, 177]. To give one example, Vall et al. [177] propose a content-based approach that predicts whether a track fits a given user's listening profile (or playlist). To this end, all tracks are first represented as a feature vector (constituting of, e.g., text embeddings of Last.fm tags or audio features such as i-vectors based on MFCCs [47]). These track vectors are subsequently fed into a CNN to transform both tracks and profiles into a latent factor space; profiles by averaging the network's output of their constituting tracks. As common in other approaches too, tracks and user profiles are eventually represented in a single vector space, which allows to compute standard distance metrics in order to identify the best fitting tracks given a user profile.

## 3.3 Hybrid Approaches

There exist several perspectives as to what makes a recommendation approach a hybrid one: either (1) the consideration of several, complementary data sources or (2) the combination of two or more recommendation techniques.[30] In the former case, complementary data sources can also be leveraged when only a single recommendation technique is used. For instance, a CBF-based MRS can exploit both textual information (e.g., tags) and acoustic clues (e.g., MFCC features), cf. [41].

As for the latter perspective, i.e., combining two or more recommendation techniques, a common strategy is to integrate a CBF with a CF component. Traditionally, this has often been achieved in a late fusion manner, i.e., the recommendations

---

[30] Note that perspective (2) most commonly also entails (1) since different recommendation techniques require different data to operate on.

made by two separate recommendation models are merged by an aggregation function to create the final recommendation list. Examples in the area of music recommendation include [84, 112, 114, 171]. Tiemann and Pauws [171] propose an item-based memory-based CF and an audio-based CBF that make independent rating predictions, which are aggregated based on rating vector similarities. Lu and Tseng [112] fuse the output of three nearest-neighbor (top-$k$) recommenders: two CBF approaches based on similarity of musical scores and of emotion tags, and one CF approach; the final recommendation list is then created by reranking items based on personalized weighting of the three components. Mcfee et al. [114] optimize a content-based similarity metric (based on MFCC audio features) by learning from a sample of collaborative data. Kaminskas et al. [84] use Borda rank aggregation to combine into a single recommendation list the items recommended by an auto-tagging-based CBF recommender that performs matching via emotion labels and a knowledge-based approach that exploits DBpedia.[31] For a comprehensive treatment of hybridization techniques, we refer to [25, 75].

In contrast, most current deep learning approaches integrate into a deep neural network architecture audio content information and collaborative information such as co-listening or co-rating of songs or artists, e.g. [74, 133]. Furthermore, these approaches can incorporate other types of (textual) metadata. For instance, Oramas et al. [133] first use weighted matrix factorization to obtain, from users' implicit feedback (track play counts), artist and track latent factors. These latent factors are used as a prediction target to train two neural networks: one to create track embeddings, the other to obtain artist embeddings, exploiting spectrograms and biographies, respectively. Based on the spectrograms (constant-Q transformed), a CNN is used to create track embeddings. Biographies are represented as TF-IDF vectors and a multilayer perceptron (MLP) is trained to obtain the latent artist embeddings. Eventually, the resulting track and artist embeddings are concatenated and fed into another MLP, trained to predict final track latent factors. To create ranked recommendations, the dot product between a given user latent factor and the final track factors is computed.

## 3.4   Context-Aware Approaches

Definitions of what constitutes the "context" of an item, a user, or an interaction between the two are manifold. So are recommendation approaches that are named "context-based" or "context-aware". For a meta-study on different taxonomies and categories of context, see for instance [16]. Here, in the context of MRS, we adopt a pragmatic perspective and distinguish between item-related context, user-related

---

[31] https://wiki.dbpedia.org.

context, and interaction-related context.[32] Item-related context may constitute, for instance, of the position of a track in a playlist or listening session. User-related context includes demographics, cultural background, activity, or mood of the music listener. Interaction-related or situational context refers to the characteristics of the very listening event, and include aspects of time and location, among others.

There exist various strategies to integrate context information into a MRS, which vary dependent on the type of context considered. Simple variants include *contextual prefiltering* and *contextual postfiltering*, cf. [7]. In the former case, only the portion of the data that fits the user context is chosen to create a recommender model and effect recommendations, e.g. [17, 157]; or users or items are duplicated and considered in different recommendation models if their ratings differ for different contexts, e.g. [15, 194]. In contrast, when adopting contextual postfiltering, a recommendation model that disregards context information is first created; subsequently, the predictions made by this model are adjusted, conditioned on the context, to make the final recommendations, e.g. [193].

An alternative to contextual filtering approaches is to extend latent factor models by contextual dimensions. If user–item interactions are provided "in context" and differ between contexts, a common approach is to extend matrix factorization to *tensor factorization*, i.e., instead of a matrix of user–item ratings, a tensor of user–item–context ratings is factorized, so that each item, user, and context can be represented. More details and examples can be found, among others, in [1, 14, 62, 85].

Recently, deep neural network approaches to context-aware MRS have emerged. They often simply concatenate the content- or interaction-based input vector to the network with a contextual feature vector, e.g. [182]. Another approach is to integrate context through a gating mechanism, e.g., by computing the element-wise product between context embeddings and the neural network's hidden states [19]. An example in the music domain is [153], where the authors propose a variational autoencoder architecture extended by a gating mechanism that is fed with different models created from users' country information.

## 3.5   Sequential Recommendation

Sequence-aware recommender systems play a crucial role in music recommendation, in particular for tasks such as *next-track recommendation* or *automatic playlist continuation* that aims at creating a coherent sequence of music items. Corresponding approaches consider sequential patterns of songs, e.g., based on playlists or listening sessions, and create a model thereof. Note that such approaches can also be considered a variant of context-aware recommendation in which item

---

[32] Note that we use the term "interaction data" in Sect. 4 to refer to data belonging to the latter kind of context.

context is leveraged in form of preceding and subsequent tracks in the sequence. To create such a system, most state-of-the-art algorithms employ variants of recurrent or convolutional neural networks, or autoencoders [152, 188]. For a more detailed treatment of the subject matter, we refer the reader to recent survey articles on sequence-aware recommendation, which also review approaches to sequential music recommendation, e.g. [143, 183].

## 3.6 Psychology-Inspired Approaches

Recently, research on recommender systems emerged that aims at enhancing the traditional data-driven techniques (based on user–item interactions like in CF or item content information like in CBF) with psychological constructs. Examples include the use of models of human memory, personality traits, or affective states (mood or emotion) of the user in the recommendation algorithm. There exist several psychological models to formalize human memory, including the adaptive control of thought-rational (ACT-R) [10] and the inverted-U model [131], which have been studied in the context of music preferences. These also relate to familiarity and novelty aspects discussed in Sect. 4.1.2.

For instance, Kowald et al. [97] propose an approach to MRS that integrates a *psychological model of human memory*, i.e., ACT-R [10]. They identify two factors that are important for remembering music: (1) frequency of exposure and (2) recentness of exposure. Their ACT-R-based approach outperforms a popularity baseline, several CF variants, and models that only consider one of the two factors mentioned above.

MRS approaches that consider the users' *personality traits* rely on insights gained from studies that relate personality traits to music preferences, e.g. [45, 56, 123, 144, 145, 155]. Building upon such work, Lu and Tintarev in [113] propose a MRS that adapts the level of diversity in the recommendation list according to the personality traits of the user, by reranking the results of a CF system. The proposed MRS builds upon their finding that users with different personalities prefer different levels of diversity in terms of music key, genres, and artists. Another personality-aware approach to music recommendation (and recommendation in other media domains) is presented by Fernández-Tobías et al. in [50]. The authors integrate into a traditional matrix factorization-based CF approach a user latent factor that describes the user's personality traits.

Research on MRS that considers the user's *affective state*, such as mood or emotion, when computing a recommendation list rely on results of studies in music psychology and cognition that identified correlations between perceived or induced emotion on the one hand, and musical properties of the music listened to on the other hand, e.g. [46, 72, 79, 174, 190]. Such insights are exploited in MRS, for instance in [12, 43]. Deng et al. [43] acquire the users' emotional state and music listening information by applying natural language processing techniques on a corpus of microblogs. Leveraging the temporal vicinity of extracted emotions and listening events, the authors consider emotion as a contextual factor of the user–song interaction. They integrate this contextual information into a user-based CF model, an item-based CF model, a hybrid of the two, and a random walk approach. Ayata et al. [12] propose a MRS architecture that uses the affective response of a user to the previously recommended songs in order to adapt future recommendations. Their system leverages data from wearable sensors as physiological signals, which are used to infer the user's emotional state.

For a comprehensive survey on psychology-informed recommender systems, we refer the reader to [108].

# 4    Challenges

In the following, we provide a discussion of major challenges that are faced in MRS research and practice, and present approaches to address these challenges.

## 4.1    How to Ensure and Measure Multi-Faceted Qualities of Recommendation Lists?

The music items that constitute the recommendation list of a MRS should fulfill a variety of quality criteria. Obviously, they should match the user's preferences or needs. Additional criteria, depending on the situational context or state of the listener (cf. lean-in and lean-back tasks in Sect. 2), are equally important, though. In the following, we identify and discuss several of these characteristics that contribute to a good recommendation list.

### 4.1.1    Similarity Versus Diversity

Items in the recommendation list should be similar to the user's preferred tracks, and also show a certain extent of similarity between them. Determining similarity of music items is a multi-faceted and non-trivial, if not actually elusive, task [90, 91]. However, operational models of acoustic similarity have been used in the past.[33] Notwithstanding all optimization for similarity and consistency, it has been shown that music recommendation lists containing only highly similar items are often perceived as boring [104]. At the same time, diversity in recommendation lists has a positive impact on conversion and retention [9]. Therefore, the right balance between similarity and diversity of items is key to optimize user satisfaction. Note that the level of diversity (and accordingly similarity) itself can be personalized,

---

[33] https://www.music-ir.org/mirex/wiki.

e.g., using a linear weighting on similarity and diversity metrics. In fact, it has been shown that different users prefer different levels of diversity [172, 186].

An overview of diversification strategies in recommender systems is given in [27]. A common approach to measure diversity is to compute the inverse of average pairwise similarities between all items in the recommendation list [166, 196]. In particular in the music domain, similarity and diversity are highly multi-faceted constructs; user preferences for them should therefore also be investigated and modeled in a multi-faceted manner. To give an example, a user may want to receive recommendations of songs with the same rhythm (similarity) but from different artists or genres (diversity).

### 4.1.2  Novelty Versus Familiarity

On the one hand, a recommendation list should contain content the user is familiar with, e.g., known songs or songs by a known artist, not least because familiarity seems vital to engage listeners emotionally with music [138]. On the other hand, a recommendation list should typically also contain a certain amount of items that are novel to the target user. This can be easily measured as the fraction of artists, albums, or tracks in the user's recommendation list that the user has not been interacted with or is not aware of [27, 156]. However, whether or not a user already knows a track is not always easy to determine. A recommended item—novel according to the data— may, in contrast, be already known to the user. For instance, a user may have listened to the item on another platform. In this case, the MRS does not know that the item is, in fact, not novel to the user. On the other hand, a user might have forgotten a track that he or she has not listened to for a while, i.e., the user is (no longer) familiar with the content.[34] In this case, the track may be a novel and interesting recommendation for the user, even though the user has already interacted with it.

An interesting insight into users' preferences for novelty has been gained in [184], where Ward et al. found that even for users who indicate that they prefer novel music, familiarity is a positive preference predictor for songs, playlists, and radio stations.

### 4.1.3  Popularity, Hotness, and Trendiness

In contrast to novelty and familiarity, which are defined for individual users (is a particular user already familiar with an item?), popularity, hotness, or trendiness refer to aspects that are global measures, commonly computed on system level. These terms are often used interchangeably; though, sometimes hotness and trendiness refer to a more recent or current scope, related to music charts, while popularity is considered

---

[34] This scenario is addressed in MRS that leverage cognitive models of frequency and recentness of exposure, discussed in Sect. 3.6.

time-independent. Offering track lists created by a popularity-based recommender is a common feature of many commercial MRS, to mitigate cold start or keep the user up-to-date about trending music. Even though such lists are not personalized, they allow to engage users, serving as an entry point to the system and as basic discovery tool. For this reason, studies in MRS often consider recommendations based on popularity as a baseline.

A concept related to popularity is "mainstreaminess" [17], also known as "mainstreamness" [180], which is sometimes leveraged in MRS. Assuming that users prefer to different extents music that is considered mainstream, such systems tailor the amount of highly popular and of long-tail items in the recommendation list to the user's preference for mainstream music. What constitutes the music mainstream can further be contextualized, for instance, by defining the mainstreaminess of an item or user at the scope of a country or an age group [17, 180].

### 4.1.4 Serendipity

The topic of serendipity has attracted some attention a few years ago, but less nowadays. Serendipity is often defined rather vaguely as items being relevant and useful but at the same time surprising to or unexpected by the user [161]. The notion of unexpectedness, which is central to serendipity, is often interpreted as being unknown to the user or far away from the user's regular music taste [82]. For instance, in their proposal for a serendipitous MRS, Zhang et al. use the average similarity between the user's known music and the candidates for song recommendation to define a measure of "unserendipity" [191].

### 4.1.5 Sequential Coherence

The coherence of music items in the recommendation list is another qualitative aspect of a MRS. What is considered a coherent sequence of songs in a listening session or playlist is highly subjective and influenced by individual preferences, though [104]. Some findings on aspects of coherence that recur in different studies include a common theme, story, or intention (e.g., soundtracks of a movie, music for doing sports), a common era (e.g., songs of the 1990s), and the same (or similar) artist, genre, style, or orchestration [38, 40, 104].

Focusing on music playlists, recent studies conducted by Kamehkhosh et al. investigated the criteria that are applied when music lovers create playlists, either with or without support by a recommender engine [80, 81]. In a study involving 270 participants, the following characteristics of playlists were judged by the participants according to their importance: homogeneity of musical features such as tempo, energy, or loudness (indicated as important by 64% of participants), diversity of artists (57%), lyrics' fit to the playlist's topic (37%), track order (25%), the transition between tracks (24%), tracks popularity (21%), and freshness (20%).

## 4.2   How to Consider Intrinsic User Characteristics?

Modeling the user is central to providing personalized recommendations. Traditionally, a user model for recommender systems consists of the history of recorded interactions or expressions of preference (cf. Sect. 2.1), a latent space embedding derived from these, or meta-features aggregating usage patterns based on domain knowledge, such as "exploratoryness", "genderedness", "fringeness", or the above mentioned "mainstreamness" [17, 181]. Describing and grouping users based on such dimensions then allows to tailor recommendations accordingly, e.g. by balancing similarity and diversity, cf. Sect. 4.1.1. However, in a cold-start setting, this information can not be resorted to. In addition, one might argue that none of these descriptions model "the user" in a task- or domain-independent manner, i.e. incorporate intrinsic user characteristics.

In order to avoid user cold-start and include individual user information into models, external data can be used. Such individual aspects cover *demographic information*, such as age or sex, which have been shown to have an impact on music preference. For instance, there is evidence that younger users explore a larger number of genres, while, with increasing age, music preferences become more stable [55, 144]. A further direction in this line of research is to make use of psychological models of *personality* to inform the recommender system, also see chapter "Personality and Recommender Systems". Personality is a stable, general model relating to the behavior, preferences, and needs of people [78] and is commonly expressed via the five factor model, based on the dimensions openness to experience, conscientiousness, extraversion, agreeableness, and neuroticism. For interaction with online music systems, personality has been related to music browsing preferences [57] and diversity needs [53], among others. Another stable attribute to describe users is to identify them from a *cultural* point of view, i.e. by associating them with their country or culture, and building upon aggregated usage patterns [17].

More dynamic aspects of user characteristics concern *affect and emotional state* and *listening intent* of the user. Short-time music preference, i.e. what people want to listen to in a specific moment, depends strongly on the affective state of the user, e.g. [83]. This, again, can partly be traced to questions of personality, e.g. there is indication that when being sad, open and agreeable persons prefer happy music, whereas introverts prefer sad music [54]. Vice versa, music influences the user's affective state. Affect regulation is therefore not surprisingly one of the main listening intents people have [95, 111, 132, 149]. Other important motivations for listening to music are, obviously, discovery and serendipitous encounters [39, 101] and social interaction [87]. Differences in goals and intended outcomes do not only impact the type of music recommender to be built, but also the type of evaluation strategy and success criteria to be chosen (see Sects. 2 and 4.5). To facilitate this line of research further, a deeper understanding of how people seek and discover new music in their everyday life, mediated through online platforms is needed. Pointers can be found in existing work utilizing user-centric evaluations, in-situ interviews, and ethnographic studies, cf. [39, 102, 103, 105, 106, 168]. A more thorough discussion on user aspects in music recommendation can be found in [92].

To include models of user characteristics into a recommendation systems, the more static dimensions providing demographic and personality information can be treated as side information or even as fixed-target user dimensions in collaborative filtering matrix factorization approaches. Other, feature-driven methods like factorization machines or deep neural networks can benefit by incorporating this information as user dimensions. Section 3.6 highlights existing work that integrates these dimensions into MRS. The more dynamic dimensions of user affect and intent can also be incorporated as contextual information, cf. Sect. 3.4.

## 4.3  How to Make a Music Recommender Fair?

The increasing adoption of machine learning methods and data-driven recommender systems in real-world applications has exhibited outcomes discriminating against protected groups and enforced stereotypes and biases (cf. Sect. 3.1). These outcomes are considered unfair, as they treat individuals and groups of people differently based on sensitive characteristic (or even violate anti-discrimination laws), and irrespective of personal preference, cf. [26]. To address this, existing work builds upon operational definitions of fairness emphasizing parity at different machine learning stages [179] and aims at improving fairness by overcoming inadequacy of metrics and bias in datasets, optimization objectives, or evaluation procedures [187].

To address the question of fairness in MRS, the bigger picture of the music industry ecosystem needs to be taken into account, cf. [4, 158]. As are other recommendation domains, music recommendation is a multi-stakeholder setting, in which different stakeholders have different, contradictory goals [3, 24]. Stakeholders in the music domain include the listeners, the artists, the composers, the right owners, the publishers, the record labels, the distributors, and the music streaming services, to mention but the most important. The question of fairness therefore first needs to be phrased as "fairness for whom"?

Although a large number of stakeholders is involved in the recommendation process, fairness in music recommendation is often reduced to being a two-sided setting where fairness towards artists (as providers of the items) and the users (as consumers) needs to be maintained [122]. Even in this reduced setting, the opportunity of the artists as item providers, i.e. fairness with respect to exposure to the customer, is prioritized over fairness for users. This can be attributed to a fairness definition for users which is simply equated with high user satisfaction and prediction accuracy, e.g. by calibrating for diversity in user histories [122, 169]. A common restraint for such a definition of user fairness can be found in item popularity [6, 98]. This two-sided view, however, neglects the very important role of the recommender system platform as an intermediary between provider and consumer, cf. [2]. Non-neutrality of the platform, due to utility associated with recommending certain items, presents another potential source of unfairness to both artists and listeners, cf. [88].

It is obvious that in order to optimize MRS towards fairness, several perspectives and notions can be adopted and need to be taken into consideration. Investigating perceptions and definitions of fairness for different tasks and different stakeholders is therefore a highly relevant question for future research.

## 4.4 How to Explain Recommendations?

Providing justifications of recommendations offer the user a means to understand why the system recommended a certain item. It has been shown that this can increase user trust and engagement [11, 61, 146, 173]. Depending on the recommendation task and technique, such justifications can include similar users' preferences, predicted rating values, common properties between liked items and recommendations (e.g., "We are recommending this song because you seem to like Viking Metal.") [125, 129], or natural language explanations based on item and/or contextual features (e.g., "We are recommending this song because it is energetic and will stimulate your current sports activity.") [31].

Early works on the topic of explainability in the music domain commonly adopt a content-based approach to explain artist or track similarities. For instance, Green et al. [67] use social tags and Wikipedia descriptions to create a tag cloud explaining artist similarities. Likewise, Pampalk and Goto [135] integrate user-generated artist tags into a music recommendation and discovery system, thereby enabling users to steer the artist recommendation process. Turnbull et al. use an auto-tagger to predict semantic labels from audio signal features in order to generate verbal descriptions of tracks [175].

More recent examples of explainability in MRS include *Moodplay* [11], a recommendation and exploration interface which builds upon a visualized latent mood space created from artist mood tags. Andjelkovic et al. integrate a CBF recommender based on acoustic artist similarity into the audiovisual interface. This interface can be used both to interact with and to explain the recommended artists. Another method is BAndits for Recsplanations as Treatments (*Bart*) by Spotify [120]. Adopting a reinforcement learning strategy, *Bart* learns interactions between items and explanations conditioned on the user or item context. User engagement as result of an explanation is used as a reward function. Explanations include, for instance, time ("Because it's Monday"), novelty ("Because it's a new release"), genre ("Because you like Jazz"), or popularity ("Because it's popular"). McInerney et al. also find that personalizing not only the recommendations, but also the explanations substantially increase user satisfaction.

While research on explainability in MRS is still scarce, though recently seeing a strong increase, a vital aspect to keep in mind is that users differ in their demand for and acceptance of explanations. In fact, Millecamp et al. [125] find that personal characteristics such as musical expertise, tech savviness, or need for cognition influence how users interact with MRS and perceive explanations. Therefore, a vital challenge to address is not only to devise methods for explaining recommendation results, but also tailor these methods to the needs of the MRS' users.

## 4.5 How to Evaluate a Music Recommender System?

In the recommender systems literature, evaluation strategies are commonly divided into *offline testing*, *online testing*, and *user studies*, cf. [68]. In the following, these are briefly discussed in the context of MRS. On a critical note, most if not all of the methods discussed below do not consider bias and fairness, cf. Sect. 4.3.

### 4.5.1 Offline Evaluation

Offline evaluation relies on existing datasets of user–item interactions, and is carried out without involvement of users during evaluation. In this way, it enables gaining quantitative and objective insights into the performance of preference prediction algorithms, and is similar to quantitative evaluation strategies found in the machine learning and (music) information retrieval literature [13, 161].

Like in offline evaluation of other types of recommender systems, *performance metrics* commonly used when evaluating MRS include *error measures* such as root-mean-squared error (RMSE) computed between predicted and true ratings,[35] *item relevance measures* such as recall and precision, and *rank-based metrics* such as mean average precision (MAP), normalized discounted cumulative gain (NDCG), and mean percentile rank (MPR). More details can be found in [161], for instance. In addition to these measures of accuracy, beyond-accuracy measures that gauge some of the characteristics described in Sect. 4.1 are tailored to the music domain and include metrics for diversity, familiarity, popularity, and serendipity.

Public *datasets* available for academic research are summarized in Tables 1 and 2. Table 1 contains basic statistics of the datasets, such as number of songs, albums, artists, users, and user–item interactions. Table 2 provides more details on the composition of each dataset, including release year, origin of the data, and the kinds of item-, user-, and interaction-related data that is included: interaction data (e.g., ratings or timestamps), item data (e.g., tags), and user data (e.g., demographics). Note that the available datasets are dominated by industrial data released for research. There is no publicly gathered data except for the ListenBrainz initiative that strives for building an open alternative to gathering listener behavior

---

[35] Note that these ratings can also be binary (1 if the user interacted with the item; 0 otherwise).

**Table 1** Statistics of public data sets for music recommendation research

| Dataset | Songs | Albums | Artists | Users | Interactions |
|---|---|---|---|---|---|
| Yahoo! Music [44] | – | 625K in total | – | 1M | 262.81M |
| MSD [18] | 1M | | | 1.02M | 48.37M |
| Last.fm–360K [28] | | | 187K | 359K | |
| Last.fm–1K [28] | | | 108K | 1K | 19.15M |
| MusicMicro [150] | 71K | | 20K | 137K | 594K |
| MMTD [70] | 134K | | 25K | 215K | 1.09M |
| AotM-2011 [117] | 98K | | 17K | 16K | 859K |
| LFM-1b [151, 154] | 32M | 16M | 3M | 120K | 1B |
| MSSD [23] | 3.7M | | | | 150M |
| MLHD [181] | 7M | 900K | 555K | 583K | 27B |
| ListenBrainz[a] | >7.58M | >1.32M | >776K | 15K | 507.84M |
| MPD [188] | 2.26M | 735K | 296K | | |
| Spotify Playlists [139] | 1.88M | | 277K | 15K | 144K |
| #nowplaying [189] | 1.21M | | | 4.15M | 46.05M |
| #nowplaying-RS [140] | 346K | | | 139K | 11.64M |
| Melon Playlist Dataset [52] | 649K | 269K | 108K | | |

[a] https://listenbrainz.org, statistics as of January 3, 2022. Only tracks mapped to MBIDs with direct string matching are reported

data. As a result, all these dataset have a clear bias towards Western music as they primarily originate from Western companies.[36]

Following offline strategies is still the predominant way of evaluating MRS in academia, not least due to the lack of contacts to (large numbers of) real users. However, despite their obvious advantages, offline evaluations do not provide sufficient clues on the perceived quality of recommendations and their actual usefulness for the listener [162], and there is research evidence that high recommender accuracy does not always correlate with user satisfaction [121]. They also do not account for biases on neither the consumer nor the artist side [98]. Furthermore, if recommender systems are targeted towards discovery, it is fundamental to assess the listener's familiarity with the recommended items apart from their relevance, which is problematic using existing datasets. Another critical point of offline studies is the overly high confidence in results, often seen in publications. This probably arises due to the computational and seemingly "objective" nature of the metrics. With huge amounts of evaluation data on user–item interactions available, it has become common to train the recommendation models and compute evaluation metrics only on a randomly drawn sample of the data, in particular to select negative/irrelevant items. However, results obtained by this kind of evaluation often show low variance (when metrics are computed across different sets of random samples), but high

---

[36] The Melon Playlist Dataset is a notable exception, containing data from a South Korean music streaming service.

**Table 2** Features of public data sets for music recommendation research. The following symbols are used to denote the featured categories of data: ♪ single listening events, ♫ playlists or listening sessions, 🖒 ratings, 🏷 tags, ♫🏷 playlist titles or annotations, ◀◔ audio features, ☉ temporal information/timestamps, ♀ location, ♂ gender, 🧑+ age, ▦ MusicBrainz or Twitter identifiers. Please note that we always denote the entity for which the data is provided, e.g., user-generated tags can be provided as *item data* (ignoring the information about which users provides them), but also as *interaction data* (indicating which user assigned which tag to which item)

| Dataset | Year | Source | Interaction data | Item data | User data |
|---|---|---|---|---|---|
| Yahoo! Music [44] | 2011 | *Yahoo* | ♪🖒 | ✗ | ✗ |
| MSD [18] | 2011 | *Echo Nest* | ♪ | ▦ 🏷 ◀◔ | ✗ |
| Last.fm – 360K [28] | 2010 | *Last.fm* | ♪ | ▦ | ♀ ♂ 🧑+ |
| Last.fm – 1K [28] | 2010 | *Last.fm* | ♪ | ▦ | ♀ ♂ 🧑+ |
| MusicMicro [150] | 2013 | *Twitter* | ♪ ☉ ♀ | ▦ | ▦ |
| MMTD [70] | 2013 | *Twitter* | ♪ ☉ | ▦ | ▦ |
| AotM-2011 [117] | 2011 | *Art of the Mix* | ♫🏷 ☉ | ▦ 🏷 ◀◔ | ☉ |
| LFM-1b [151, 154] | 2016 | *Last.fm* | ♪ ☉ | 🏷 | ♀ ♂ 🧑+ |
| MSSD [23] | 2019 | *Spotify* | ♫ ☉ | ▦ ◀◔ | ✗ |
| MLHD [181] | 2017 | *Last.fm* | ♪ | ▦ | ♀ ♂ 🧑+ |
| ListenBrainz | 2015- | *ListenBrainz* | ♪ | ▦ | ✗ |
| MPD [188] | 2018 | *Spotify* | ♫🏷 | ▦ | ✗ |
| Spotify Playlists [139] | 2015 | *Spotify* | ♫🏷 | ✗ | ✗ |
| #nowplaying [189] | 2014 | *Twitter* | ♪ ☉ | ▦ | ✗ |
| #nowplaying-RS [140] | 2018 | *Twitter* | ♪ ☉ 🏷 | ◀◔ | ▦ |
| Melon Playlist Dataset [52] | 2021 | *Melon* | ♫🏷 | ◀◔ 🏷 | ✗ |

bias, the former being particularly dangerous as it is likely to lead to an unjustified confidence in evaluation results [99].

### 4.5.2 Online Evaluation

While offline evaluation is still the predominantly adopted evaluation methodology in academia, evaluation of MRS in industrial settings is nowadays dominated by online studies, involving real users. This, of course, does not come as a surprise since MRS providers such as Spotify, Deezer, or Amazon Music have millions of customers and can involve them into the evaluation, even without the need to let them know. In contrast, most academic research lacks such possibility, focusing instead on offline experiments.

The most common variant of online evaluation is *A/B testing*, i.e., a comparative evaluation of two (or more) recommendation algorithms in a productive system [68].[37] A/B testing is the most efficient way to evaluate MRS as it allows to measure the system's performance or impact according to the final goals of the system directly in the experiment, using measures such as user retention, click through rate, amount of music streamed, etc. A recent example of A/B testing in an MRS is provided by Spotify in [120], where a multi-armed bandit approach to balance exploration and exploitation, which also provides explanations for recommendations, is proposed and evaluated both offline and online.

### 4.5.3 User Studies

Evaluation via user studies allows to investigate the user experience of a MRS, including aspects of user engagement [107] and user satisfaction [105]. Pu et al. [142] as well as Knijnenburg et al. [94] propose evaluation frameworks for user-centric evaluation of recommender systems via user studies, which are partly adopted in the MRS domain too. Pu et al.'s framework [142], called *ResQue*, includes aspects of perceived recommendation quality (e.g., attractiveness, novelty, diversity, and perceived accuracy), interface adequacy (e.g., information sufficiency and layout clarity), interaction adequacy (e.g., preference elicitation and revision), as well as perceived usefulness, ease of use, user control, transparency, explicability, and trust. While the authors do not explicitly showcase their framework on a music streaming platform, some results obtained for Youtube[38] and Douban[39]— both platforms heavily used for music consumption—are likely to generalize to dedicated MRS. Knijnenburg et al. [94] propose a different instrument to investigate user experience of recommender systems, which includes aspects such as perceived recommendation quality, perceived system effectiveness, perceived recommendation variety, choice satisfaction, intention to provide feedback, general trust in technology, and system-specific privacy concern, among others.

Conducting user studies to evaluate MRS presents obvious advantages over offline and online experiments because the respective questionnaires are capable of uncovering intrinsic characteristics of user experience to a much deeper degree than the other mentioned strategies. However, such evaluations are rare in the MRS literature as it is difficult to gather a number of participants large enough to draw significant and usable conclusions, due to the required effort on the user side. Existing user studies are typically restricted to a small number of subjects (tens to a few hundreds) and tested approaches or systems. Although the number of user studies has increased [185], conducting such studies on real-world MRS remains time-consuming, expensive, and impractical, particularly for academic researchers.

Consequently, relatively few studies measuring aspects related to user satisfaction have been published, even though their number has been increasing in the

---

[37] Notwithstanding, there also exist offline variants of A/B testing strategies, e.g. [63].

[38] https://www.youtube.com.

[39] https://www.douban.com.

past years. The study by Celma and Herrera [30] serves as an early example of a subjective evaluation experiment, carried out on a larger scale. Each of the 288 participants provided *liking* (enjoyment of the recommended music) and *familiarity* ratings for 19 tracks recommended by three recommendation approaches. Bogdanov et al. [20] use four subjective measures addressing different aspects of user preference and satisfaction: *liking*, *familiarity* with the recommended tracks, *listening intention* (readiness to listen to the same track again in the future), and "*give-me-more*" (indicating a request for or rejection of more music that is similar to the recommended track). These subjective ratings are analyzed for consistency (a user may like a recommended track, but will not want to listen to it again) and are additionally re-coded into recommendation outcomes: *trusts* (relevant recommendations already known to a user), *hits* (relevant novel tracks), and *fails* (disliked tracks).

Other examples include the recent study by Robinson et al. [147] on perceived diversity in music recommendation lists created by a CF system. The authors investigate the extent to which applying an algorithmic diversification strategy, adopting an intra-list diversity metric, transfers to actual user-perceived diversity. They find a clear difference between diversity preferences in recommendation lists within the user's bounds of music preferences and outside of these bounds.

Another recent study by Jin et al. [76] investigates the extent to which user control over contextual factors that are considered by the recommendation algorithm influences perceived quality, diversity, effectiveness, and cognitive load. In their study, 114 participants are either given no control over the algorithm (realized via Spotify's API) or they could chose a specific context and recommendations are reranked accordingly. The authors find that the users' ability to control whether recommendations are contextualized by mood, weather, and location influences their perception of the MRS. For instance, the ability to consider mood positively affects perceived recommendation quality and diversity.

## 4.6    How to Deal with Missing and Negative Feedback in Evaluation?

As mentioned throughout this chapter, music recommendation poses several challenges for evaluation. In particular, MRS's are often faced with both implicit feedback (e.g., from the *lean-back* setting) and extreme sparsity of observation. These both contribute to a lack of strong *negative* feedback, which makes standard ranking metrics (derived from precision and recall) difficult to estimate.

The most common approach to cope with sparsity in evaluation is to exploit structure in the content provided by *meta-data*. For example, rather than evaluate a recommender according to its ability to predict interactions between users and songs, the evaluation can be abstracted to measure interactions between users and *artists*. In this view, an item is considered *relevant* if the user interacted with other items by the same artist. This evaluation obviously over-simplifies the task from the user's perspective, but it can be considerably more stable in practice than a pure item-based evaluation. It is also possible to combine song- and artist-level relevance, as was done in the 2018 RecSys Challenge [33]. Of course, there are other relational structures present in music meta-data that can be leveraged in similar ways. Slaney et al. [164] measured artist, album, and blog co-occurrence as a proxy for (content-based) item similarity, which each gave varying degrees of specificity to the evaluation. Zheleva et al. [192] evaluated playlist generation models by their ability to select songs of appropriate *genres*. While these approaches do not measure utility in the traditional sense, but in the absence of sufficient interaction data, they can at least act as stable proxy metrics.

## 4.7 How to Design User Interfaces That Match the Use Case and Increase User Experience?

Like most recommender systems domains, music recommendation can benefit from having recommendations presented in a meaningfully organized manner. Many commercial music services present recommendations grouped together by genre, similarity to a popular artist, year of release, etc. Similar experiences are provided by movie recommenders, book recommenders, or general online shopping sites where products may be grouped by "department" (e.g., *kitchen*, *apparel*, *toys*, etc.). The MRS experience differs principally by the high variability of expertise and familiarity with terminology possessed by users. While movies, books, and department stores generally have fairly consistent and familiar "sections" (or shelves), the taxonomies used to categorize and organize music can be both deep and obscure [49, 167]. For example, a casual jazz listener may not understand (or care about) the distinctions between sub-genres like *bebop* or *cool jazz*, but these differences would be obvious to listeners with a bit more familiarity with the genre. This has ramifications for interface design in music recommendation: the groupings used to organize a set of recommendations should adapt to the user's experience and prior knowledge.

Prior knowledge and listening histories are not the only user characteristics that inform MRS (and MRS interface) design. User studies have demonstrated that personality traits (e.g., the Big Five taxonomy [78]) correlate with different preferences for organizational principles in music collections, such as genre, mood, or intended context/activity [58]. This observation motivates the use of hybrid methods which adaptively personalize the combination of recommender algorithms based on each user's listening patterns [51].

Finally, content curation and simple rule-based systems can play a substantial role in music recommendation. Some familiar examples include: removing Christmas music from streams outside the month of December, avoiding the mixing of religious and secular music, filtering music with potentially offensive lyrics, moderating user-generated content (e.g., podcasts), restricting access to content based on licensing terms and availability, and so on. Streaming radio services additionally may need to comply with regional broadcast regulations, such as avoiding playing multiple songs from the same album within a given window of time. Providing a satisfying user experience under these constraints typically requires a mixture of high-quality meta-data, careful curation and moderation of content, and logical constraints in the recommendation algorithm.

## 4.8 Which Open Tools and Data Sources can be Used to Build a Music Recommender System?

Many generic cross-domain software tools for recommender systems are available to build MRS. LightFM,[40] Implicit,[41] Surprise,[42] and LibRec[43] provide Python and Java implementations of some popular recommendation algorithms for both implicit and explicit feedback data. Annoy,[44] NMSLIB,[45] and faiss[46] can be used for efficient nearest neighbor search in feature spaces of item and user representations, and are implemented in C++ with Python wrappers.

Here we focus on content annotation tools and data sources particular to music. Specifically, we discuss open-source tools and publicly accessible data that allow bootstrapping MRS, even though some commercial services provide API endpoints to gather music content data and even create recommendations out of the box.

Researchers in MIR have developed *tools* for music audio content analysis and feature extraction, that can be used for music recommendation. Essentia[47] [22] and Librosa[48] [119] provide signal processing algorithms for computation of MIR features and audio representations suitable as inputs for CBF and hybrid approaches (both traditional and deep learning-based). Both libraries provide flexibility of use for fast prototyping in Python. Essentia provides feature extractors implemented in C++ for fast analysis on the large scale. It also includes pre-trained TensorFlow models for auto-tagging and music annotation tasks [8], outputs and latent feature embeddings of which can be useful features for MRS tasks. More audio analysis libraries available to researchers are reviewed in [127].

---

[40] https://making.lyst.com/lightfm/docs.

[41] https://implicit.readthedocs.io.

[42] http://surpriselib.com.

[43] https://guoguibing.github.io/librec.

[44] https://github.com/spotify/annoy.

[45] https://github.com/nmslib/nmslib.

[46] https://github.com/facebookresearch/faiss.

[47] https://essentia.upf.edu.

[48] https://librosa.org.

The *datasets* presented in Tables 1 and 2 can be used for prototyping MRS. However, their characteristics (provider, origin, release year, composition of user base, type and quality of data, etc.) will bias any recommendation model created from these datasets, likely resulting in problems when translating to the newly built system. Furthermore, it is advisable to check the licenses of the available datasets, as there may be potential legal uncertainty of their usage outside of academic research.

Given the limitations of access to commercial data for researchers and practitioners, there is an initiative to build an open data and open source ecosystem suitable for building recommender systems. MusicBrainz[49] is one of the largest databases of editorial music metadata, including information and relations between millions of artists, recording labels, releases, and particular tracks. It is coupled with ListenBrainz,[50] a database of user listening events, and AcousticBrainz[51] [141], which contains results of automatic music audio analysis and annotation, including high-level music concepts (e.g., genre, mood, instrumentation, key, rhythm). Similar to MusicBrainz, Discogs[52] [21] also provides rich editorial metadata relations and genre/style annotations at the very large scale. All of these data sources are available under open licenses, and they can be used to enrich the data about a particular music collection at hand.

# 5   Conclusions

To summarize, research and development in music recommender systems has seen a paradigm shift in recent years: away from traditional recommendation tasks such as predicting ratings or impressions, towards more specific *use cases* to satisfy a lean-in or lean-back demand of the user and continuously provide a listening experience. As such, typical applications relevant in music recommendation are, for instance, automatic playlist continuation, cross-modal tasks such as creating a session or playlist based on a textual query, and contextual recommendations based on user intent recognition.

As for *techniques*, we focused on methods that leverage characteristics of the music domain, which distinguishes the music recommendation task from recommendation tasks in other domains. Owed to recent developments, we put a focus on hybrid methods that integrate both co-listening data and content-based information. Furthermore, we reviewed recent research on context-aware, sequential, and psychology-inspired approaches.

Current *challenges* we eventually discuss include considering multi-faceted qualities in recommendation lists, adapting recommendations based on intrinsic

---

[49] https://www.musicbrainz.org.

[50] https://www.listenbrainz.org.

[51] https://www.acousticbrainz.org.

[52] https://www.discogs.com.

user characteristics, considering fairness and bias, explaining recommendations, evaluating music recommender systems, publicly available datasets, dealing with missing and negative feedback, and designing user interfaces that increase user experience. We are sure that addressing those challenges will yield exciting research results and products in the near future.

**Acknowledgments** We would like to thank Marius Kaminskas for contributing to the previous version of this chapter, in the second edition of this book.

# References

1. M.H. Abdi, G.O. Okeyo, R.W. Mwangi, Matrix factorization techniques for context-aware collaborative filtering recommender systems: a survey. Comput. Inf. Sci. **11**(2), 1–10 (2018)
2. H. Abdollahpouri, G. Adomavicius, R. Burke, I. Guy, D. Jannach, T. Kamishima, J. Krasnodebski, L. Pizzato, Beyond personalization: Research directions in multistakeholder recommendation (2019). arXiv:1905.01986
3. H. Abdollahpouri, R. Burke, B. Mobasher, Recommender systems as multistakeholder environments. in *Proceedings of the 25th Conference on User Modeling, Adaptation and Personalization, UMAP '17*, New York, NY, 2017 (Association for Computing Machinery, New York, 2017), pp. 347–348
4. H. Abdollahpouri, S. Essinger, Multiple stakeholders in music recommender systems (2017). arXiv:1708.00120
5. H. Abdollahpouri, M. Mansoury, R. Burke, B. Mobasher, The unfairness of popularity bias in recommendation, in *Proceedings of the Workshop on Recommendation in Multi-stakeholder Environments co-located with the 13th ACM Conference on Recommender Systems (RecSys 2019)*, Copenhagen, Denmark, September 20, 2019, ed. by R. Burke, H. Abdollahpouri, E.C. Malthouse, K.P. Thai, Y. Zhang. CEUR Workshop Proceedings, vol. 2440 (CEUR-WS.org, Amsterdam, 2019)
6. H. Abdollahpouri, M. Mansoury, R. Burke, B. Mobasher, The connection between popularity bias, calibration, and fairness in recommendation, in *Fourteenth ACM Conference on Recommender Systems*, RecSys '20, New York, NY, 2020 (Association for Computing Machinery, New York, 2020), pp. 726–731
7. G. Adomavicius, A. Tuzhilin, Context-aware recommender systems, in *Recommender Systems Handbook*, ed. by F. Ricci, L. Rokach, B. Shapira (Springer, New York, 2015), pp. 191–226
8. P. Alonso-Jiménez, D. Bogdanov, J. Pons, X. Serra, Tensorflow audio models in essentia, in *ICASSP 2020-2020 IEEE International Conference on Acoustics, Speech and Signal Processing (ICASSP)* (IEEE, New York, 2020), pp. 266–270
9. A. Anderson, L. Maystre, I. Anderson, R. Mehrotra, M. Lalmas, Algorithmic effects on the diversity of consumption on spotify, in *WWW '20: The Web Conference 2020, Taipei, Taiwan, April 20–24, 2020*, ed. by Y. Huang, I. King, T. Liu, M. van Steen (ACM/IW3C2, New York, 2020), pp. 2155–2165
10. J.R. Anderson, M. Matessa, C. Lebiere, Act-r: a theory of higher level cognition and its relation to visual attention. Human-Computer Interact. **12**(4), 439–462 (1997)
11. I. Andjelkovic, D. Parra, J. O'Donovan, Moodplay: Interactive mood-based music discovery and recommendation, in *Proceedings of the 2016 Conference on User Modeling Adaptation and Personalization*, UMAP '16, New York, NY (ACM, New York, 2016), pp. 275–279
12. D. Ayata, Y. Yaslan, M.E. Kamasak, Emotion based music recommendation system using wearable physiological sensors. IEEE Trans. Consum. Electron. **64**(2), 196–203 (2018)

13. R. Baeza-Yates, B.A. Ribeiro-Neto, *Modern Information Retrieval - The Concepts and Technology Behind Search*, 2nd edn. (Pearson Education Ltd., Harlow, 2011)
14. L. Baltrunas, B. Ludwig, F. Ricci, Matrix factorization techniques for context aware recommendation, in *Proceedings of the 2011 ACM Conference on Recommender Systems, RecSys 2011, Chicago, IL, October 23–27, 2011*, ed. by B. Mobasher, R.D. Burke, D. Jannach, G. Adomavicius, pp. 301–304 (ACM, New York, 2011)
15. L. Baltrunas, F. Ricci, Context-based splitting of item ratings in collaborative filtering, in *Proceedings of the 2009 ACM Conference on Recommender Systems, RecSys 2009, New York, NY, October 23–25, 2009*, ed. by L.D. Bergman, A. Tuzhilin, R.D. Burke, A. Felfernig, L. Schmidt-Thieme (ACM, New York, 2009), pp. 245–248
16. C. Bauer, A. Novotny, A consolidated view of context for intelligent systems. J. Ambient Intell. Smart Environ. **9**(4), 377–393 (2017)
17. C. Bauer, M. Schedl, Global and country-specific mainstreaminess measures: definitions, analysis, and usage for improving personalized music recommendation systems. PLoS One **14**(6), 1–36 (2019)
18. T. Bertin-Mahieux, D.P. Ellis, B. Whitman, P. Lamere, The million song dataset, in *Proceedings of the 12th International Society for Music Information Retrieval Conference*, Miami, October 24–28 2011, pp. 591–596
19. A. Beutel, P. Covington, S. Jain, C. Xu, J. Li, V. Gatto, E.H. Chi, Latent cross: Making use of context in recurrent recommender systems. In ed. by Y. Chang, C. Zhai, Y. Liu, Y. Maarek, *Proceedings of the Eleventh ACM International Conference on Web Search and Data Mining, WSDM 2018, Marina Del Rey, CA, USA, February 5–9, 2018* (ACM, New York, 2018), pp. 46–54
20. D. Bogdanov, M. Haro, F. Fuhrmann, A. Xambó, E. Gómez, P. Herrera, Semantic audio content-based music recommendation and visualization based on user preference examples. Inf. Process. Manag. **49**(1), 13–33 (2013)
21. D. Bogdanov, P. Herrera, Taking advantage of editorial metadata to recommend music, in *Int. Symp. on Computer Music Modeling and Retrieval (CMMR'12)*, 2012
22. D. Bogdanov, N. Wack, E. Gómez Gutiérrez, S. Gulati, H. Boyer, O. Mayor, G. Roma Trepat, J. Salamon, J. R. Zapata González, X. Serra, et al., Essentia: an audio analysis library for music information retrieval, in *Britto A, Gouyon F, Dixon S, editors. 14th Conference of the International Society for Music Information Retrieval (ISMIR); 2013 Nov 4–8; Curitiba, Brazil.[place unknown]: ISMIR; 2013. p. 493–498.* International Society for Music Information Retrieval (ISMIR), 2013.
23. B. Brost, R. Mehrotra, T. Jehan, The music streaming sessions dataset, in L. Liu, R.W. White, A. Mantrach, F. Silvestri, J.J. McAuley, R. Baeza-Yates, L. Zia, editors, *The World Wide Web Conference, WWW 2019, San Francisco, CA, May 13–17, 2019* (ACM, New York, 2019), pp. 2594–2600
24. Burke, R., Multisided fairness for recommendation (2017). CoRR abs/1707.00093. arXiv
25. R.D. Burke, Hybrid recommender systems: Survey and experiments. User Model. User Adapt. Interact. **12**(4), 331–370 (2002)
26. R.D. Burke, M. Mansoury, N. Sonboli, Experimentation with fairness-aware recommendation using librec-auto: Hands-on tutorial, in *Proceedings of the 2020 Conference on Fairness, Accountability, and Transparency*, FAT* '20, p. 700, New York, NY 2020. Association for Computing Machinery.
27. P. Castells, N.J. Hurley, S. Vargas, Novelty and diversity in recommender systems, in *Recommender Systems Handbook* (Springer, Boston, MA, 2015), pp. 881–918
28. Ò. Celma, *Music Recommendation and Discovery – The Long Tail, Long Fail, and Long Play in the Digital Music Space* (Springer, Berlin, 2010)
29. O. Celma, The exploit-explore dilemma in music recommendation. In *Proceedings of the 10th ACM Conference on Recommender Systems* (2016), pp. 377–377
30. O. Celma, P. Herrera, A new approach to evaluating novel recommendations, in *ACM Conference on Recommender Systems (RecSys'08)* (2008), pp. 179–186

31. S. Chang, F.M. Harper, L.G. Terveen, Crowd-based personalized natural language explanations for recommendations, in *Proc. ACM Conf. on Recommender Systems, RecSys '16*, pp. 175–182 (ACM, New York, 2016)
32. S. Chang, S. Lee, K. Lee, Sequential skip prediction with few-shot in streamed music contents. CoRR abs/1901.08203, 2019.
33. C.-W. Chen, P. Lamere, M. Schedl, and H. Zamani. Recsys challenge 2018: Automatic music playlist continuation. In *Proceedings of the 12th ACM Conference on Recommender Systems*, RecSys '18, page 527–528, New York, NY, USA, 2018. Association for Computing Machinery.
34. R. Chen, Q. Hua, Y. Chang, B. Wang, L. Zhang, X. Kong, A survey of collaborative filtering-based recommender systems: from traditional methods to hybrid methods based on social networks. IEEE Access **6**, 64301–64320 (2018)
35. Z. Cheng, J. Shen, On effective location-aware music recommendation. ACM Trans. Inf. Syst. (TOIS) **34**(2), 1–32 (2016)
36. S.J. Cunningham, Interacting with personal music collections. in *Information in Contemporary Society*, 2019 (Springer International Publishing, Cham, 2019), pp. 526–536
37. S.J. Cunningham, D. Bainbridge, A. Bainbridge, Exploring personal music collection behavior, in ed. by S. Choemprayong, F. Crestani, S.J. Cunningham, *Digital Libraries: Data, Information, and Knowledge for Digital Lives* (Springer International Publishing, Cham, 2017), pp. 295–306
38. S.J. Cunningham, D. Bainbridge, A. Falconer, 'More of an art than a science': supporting the creation of playlists and mixes, in *ISMIR 2006, 7th International Conference on Music Information Retrieval, Victoria, 8–12 October 2006, Proceedings* (2006), pp 240–245.
39. S.J. Cunningham, D. Bainbridge, D. Mckay, Finding new music: a diary study of everyday encounters with novel songs, in *Proceedings of the 8th International Conference on Music Information Retrieval*, pp. 83–88, Vienna, September 23–27 (2007)
40. S.J. Cunningham, J.S. Downie, D. Bainbridge, The pain, the pain: modelling music information behavior and the songs we hate, in *ISMIR 2005, 6th International Conference on Music Information Retrieval, London, 11–15 September 2005, Proceedings* (2005), pp. 474–477
41. Y. Deldjoo, M. Schedl, P. Cremonesi, G. Pasi, Recommender systems leveraging multimedia content. ACM Computing Surv. **53**(5) (2020)
42. Y. Deldjoo, M. Schedl, P. Knees, Content-driven music recommendation: evolution, state of the art, and challenges (2021). Preprint. arXiv
43. S. Deng, D. Wang, X. Li, G. Xu, Exploring user emotion in microblogs for music recommendation. Expert Syst. Appl. **42**(23), 9284–9293 (2015)
44. G. Dror, N. Koenigstein, Y. Koren, M. Weimer, The Yahoo! Music Dataset and KDD-Cup'11. J. Mach. Learn. Res. Proc. KDD-Cup 2011 Compet. **18**, 3–18 (2012)
45. P.G. Dunn, B. de Ruyter, D.G. Bouwhuis, Toward a better understanding of the relation between music preference, listening behavior, and personality. Psychol. Music **40**(4), 411–428 (2012)
46. T. Eerola, J. Vuoskoski, A comparison of the discrete and dimensional models of emotion in music. Psychol. Music **39**(1), 18–49 (2011)
47. H. Eghbal-zadeh, B. Lehner, M. Schedl, G. Widmer, I-vectors for timbre-based music similarity and music artist classification, in *Proceedings of the 16th International Society for Music Information Retrieval Conference, ISMIR 2015, Málaga, October 26–30, 2015*, ed. by M. Müller, F. Wiering (2015), pp. 554–560
48. M.D. Ekstrand, M. Tian, I.M. Azpiazu, J.D. Ekstrand, O. Anuyah, D. McNeill, M.S. Pera, All the cool kids, how do they fit in?: Popularity and demographic biases in recommender evaluation and effectiveness, in *Conference on Fairness, Accountability and Transparency, FAT 2018, 23–24 February 2018, New York, NY*, ed. by S.A. Friedler, C. Wilson. Proceedings of Machine Learning Research, vol. 81 (PMLR, 2018), pp. 172–186
49. F. Fabbri, A theory of musical genres: two applications. Popul. Mus. Perspect. **1**, 52–81 (1982)

50. I. Fernández-Tobías, M. Braunhofer, M. Elahi, F. Ricci, I. Cantador, Alleviating the new user problem in collaborative filtering by exploiting personality information. User Model. User-Adapt. Interact. **26**(2–3), 221–255 (2016)
51. A. Ferraro, D. Bogdanov, K. Choi, X. Serra, Using offline metrics and user behavior analysis to combine multiple systems for music recommendation. in *Proceedings of the RecSys 2018 Workshop on Offline Evaluation of Recommender Systems (REVEAL)* (2018), pp. 6
52. A. Ferraro, Y. Kim, S. Lee, B. Kim, N. Jo, S. Lim, S. Lim, J. Jang, S. Kim, X. Serra, et al., Melon playlist dataset: a public dataset for audio-based playlist generation and music tagging. in *ICASSP 2021-2021 IEEE International Conference on Acoustics, Speech and Signal Processing (ICASSP)* (IEEE, New York, 2021), pp. 536–540
53. B. Ferwerda, M. Graus, A. Vall, M. Tkalčič, M. Schedl, The influence of users' personality traits on satisfaction and attractiveness of diversified recommendation lists. in *4th Workshop on Emotions and Personality in Personalized Systems (EMPIRE) 2016* (2016), p. 43
54. B. Ferwerda, M. Schedl, M. Tkalčič, Personality & emotional states: understanding users' music listening needs, in *Extended Proceedings of the 23rd International Conference on User Modeling, Adaptation and Personalization (UMAP)*, Dublin, June–July 2015
55. B. Ferwerda, M. Tkalčič, M. Schedl, Personality traits and music genre preferences: How music taste varies over age groups, in *Proceedings of the 1st Workshop on Temporal Reasoning in Recommender Systems (RecTemp) at the 11th ACM Conference on Recommender Systems, Como, August 31, 2017*, 2017
56. B. Ferwerda, M. Tkalcic, M. Schedl, Personality traits and music genres: What do people prefer to listen to? in *Proceedings of the 25th Conference on User Modeling, Adaptation and Personalization*, UMAP '17, New York, NY, (ACM, New York, 2017), pp. 285–288
57. B. Ferwerda, E. Yang, M. Schedl, M. Tkalčič, Personality traits predict music taxonomy preferences, in *Proceedings of the 33rd Annual ACM Conference Extended Abstracts on Human Factors in Computing Systems* (ACM, New York, 2015), pp. 2241–2246
58. B. Ferwerda, E. Yang, M. Schedl, M. Tkalcic, Personality and taxonomy preferences, and the influence of category choice on the user experience for music streaming services. Multim. Tools Appl. **78**(14), 20157–20190 (2019)
59. B. Fields, Contextualize your listening: the playlist as recommendation engine. PhD thesis, Department of Computing Goldsmiths, University of London, 2011
60. K.R. Fricke, D.M. Greenberg, P.J. Rentfrow, P.Y. Herzberg, Computer-based music feature analysis mirrors human perception and can be used to measure individual music preference. J. Res. Personal. **75**, 94–102 (2018)
61. G. Friedrich, M. Zanker, A taxonomy for generating explanations in recommender systems. AI Mag. **32**(3), 90–98 (2011)
62. A. Gautam, P. Chaudhary, K. Sindhwani, P. Bedi, CBCARS: content boosted context-aware recommendations using tensor factorization, in *2016 International Conference on Advances in Computing, Communications and Informatics, ICACCI 2016, Jaipur, September 21–24, 2016* (IEEE, New York, 2016), pp. 75–81
63. A. Gilotte, C. Calauzènes, T. Nedelec, A. Abraham, S. Dollé, Offline a/b testing for recommender systems, in *Proceedings of the Eleventh ACM International Conference on Web Search and Data Mining*, WSDM '18, New York, NY (Association for Computing Machinery, New York, 2018), pp. 198–206
64. M. Goto, R.B. Dannenberg, Music interfaces based on automatic music signal analysis: new ways to create and listen to music. IEEE Signal Process. Mag. **36**(1), 74–81 (2019)
65. M. Goto, K. Yoshii, H. Fujihara, M. Mauch, T. Nakano, Songle: a web service for active music listening improved by user contributions, in *Proceedings of the 12th International Society for Music Information Retrieval Conference*, pp. 311–316, Miami, October. 2011. ISMIR.
66. S.J. Green, P. Lamere, J. Alexander, F. Maillet, S. Kirk, J. Holt, J. Bourque, X. Mak, Generating transparent, steerable recommendations from textual descriptions of items, in *Proceedings of the 2009 ACM Conference on Recommender Systems, RecSys 2009*, New York, NY, October 23–25, 2009, ed. by L.D. Bergman, A. Tuzhilin, R.D. Burke, A. Felfernig, L. Schmidt-Thieme (ACM, New York, 2009), pp. 281–284

67. S.J. Green, P. Lamere, J. Alexander, F. Maillet, S. Kirk, J. Holt, J. Bourque, X.-W. Mak, Generating transparent, steerable recommendations from textual descriptions of items, in *Proc. ACM Conf. on Recommender Systems, RecSys '09* (ACM, New York, 2009), pp. 281–284

68. A. Gunawardana, G. Shani, Evaluating recommender systems, in *Recommender Systems Handbook*, ed. by F. Ricci, L. Rokach, B. Shapira (Springer, New York, 2015), pp. 265–308

69. C. Hansen, C. Hansen, S. Alstrup, J.G. Simonsen, C. Lioma, Modelling sequential music track skips using a multi-rnn approach. CoRR abs/1903.08408, 2019

70. D. Hauger, M. Schedl, A. Košir, M. Tkalčič, The million musical tweets dataset: what can we learn from microblogs, in *Proceedings of the 14th International Society for Music Information Retrieval Conference (ISMIR 2013)*, Curitiba, November 2013

71. J.L. Herlocker, J.A. Konstan, L.G. Terveen, J.T. Riedl, Evaluating collaborative filtering recommender systems. ACM Trans. Inf. Syst. **22**(1), 5–53 (2004)

72. K. Hevner, Expression in music: a discussion of experimental studies and theories. Psychol. Rev. **42**, 186–204 (1935)

73. Y. Hu, Y. Koren, C. Volinsky, Collaborative filtering for implicit feedback datasets, in *Proceedings of the 8th IEEE International Conference on Data Mining (ICDM 2008)*, December 15–19, 2008, Pisa (IEEE Computer Society, Washington, 2008), pp. 263–272

74. Q. Huang, A. Jansen, L. Zhang, D.P.W. Ellis, R.A. Saurous, J.R. Anderson, Large-scale weakly-supervised content embeddings for music recommendation and tagging, in *2020 IEEE International Conference on Acoustics, Speech and Signal Processing, ICASSP 2020*, Barcelona, May 4–8, 2020 (IEEE, New York, 2020), pp. 8364–8368

75. D. Jannach, M. Zanker, A. Felfernig, G. Friedrich, *Recommender Systems - An Introduction* (Cambridge University Press, Cambridge, 2010)

76. Y. Jin, N.N. Htun, N. Tintarev, K. Verbert, Contextplay: Evaluating user control for context-aware music recommendation, in *Proceedings of the 27th ACM Conference on User Modeling, Adaptation and Personalization, UMAP 2019*, Larnaca, Cyprus, June 9–12, 2019, ed. by G.A. Papadopoulos, G. Samaras, S. Weibelzahl, D. Jannach, O.C. Santos (ACM, New York, 2019)

77. T. Joachims, Optimizing search engines using clickthrough data, in *Proceedings of the Eighth ACM SIGKDD International Conference on Knowledge Discovery and Data Mining* (2002), pp. 133–142

78. O.P. John, E.M. Donahue, R.L. Kentle, The big five inventory—versions 4a and 54 (1991)

79. P. Juslin, P. Laukka, Expression, perception, and induction of musical emotions: a review and a questionnaire study of everyday listening. J. New Music Res. **33**(2), 217–238 (2004)

80. I. Kamehkhosh, G. Bonnin, D. Jannach, Effects of recommendations on the playlist creation behavior of users, in *User Modeling and User-Adapted Interaction*, 2019

81. I. Kamehkhosh, D. Jannach, G. Bonnin, How automated recommendations affect the playlist creation behavior of users, in *Joint Proceedings of the ACM IUI 2018 Workshops co-located with the 23rd ACM Conference on Intelligent User Interfaces (ACM IUI 2018)*, Tokyo, March 11, 2018, ed. by A. Said, T. Komatsu. CEUR Workshop Proceedings, vol. 2068 (CEUR-WS.org, Amsterdam, 2018)

82. M. Kaminskas, D. Bridge, Diversity, serendipity, novelty, and coverage: a survey and empirical analysis of beyond-accuracy objectives in recommender systems. ACM Trans. Interact. Intell. Syst. **7**(1), 2:1–2:42 (2017)

83. M. Kaminskas, F. Ricci, Contextual music information retrieval and recommendation: state of the art and challenges. Comput. Sci. Rev. **6**, 89–119 (2012)

84. M. Kaminskas, F. Ricci, M. Schedl, Location-aware music recommendation using auto-tagging and hybrid matching, in *Proceedings of the 7th ACM Conference on Recommender Systems (RecSys 2013)*, Hong Kong, October 2013

85. A. Karatzoglou, X. Amatriain, L. Baltrunas, N. Oliver, Multiverse recommendation: n-dimensional tensor factorization for context-aware collaborative filtering, in *Proceedings of the 2010 ACM Conference on Recommender Systems, RecSys 2010*, Barcelona, September 26–30, 2010, ed. by X. Amatriain, M. Torrens, P. Resnick, M. Zanker (eds.) (ACM, New York, 2010), pp. 79–86

86. E. Karydi, K.G. Margaritis, Parallel and distributed collaborative filtering: a survey. ACM Comput. Surv. **49**(2), 37:1–37:41 (2016)
87. Y. Kjus, Musical exploration via streaming services: The norwegian experience. Popul. Commun. **14**(3), 127–136 (2016)
88. P. Knees, A proposal for a neutral music recommender system, in , *Proceedings of the 1st Workshop on Designing Human-Centric Music Information Research Systems*, ed. by M. Miron (2019), pp. 4–7
89. P. Knees, M. Hübler, Towards uncovering dataset biases: investigating record label diversity in music playlists, in *Proceedings of the 1st Workshop on Designing Human-Centric Music Information Research Systems*, ed. by M. Miron (2019), pp. 19–22
90. P. Knees, M. Schedl, A survey of music similarity and recommendation from music context data. ACM Trans. Multimedia Comput. Commun. Appl. **10**(1), 2:1–2:21 (2013)
91. P. Knees, M. Schedl, *Music Similarity and Retrieval - An Introduction to Audio- and Web-based Strategies*, vol. 36. The Information Retrieval Series (Springer, New York, 2016)
92. P. Knees, M. Schedl, B. Ferwerda, A. Laplante, User awareness in music recommender systems, in *Personalized Human-Computer Interaction*, ed. by M. Augstein, E. Herder, W. Wörndl (DeGruyter, Berlin, Boston, 2019), pp. 223–252
93. P. Knees, M. Schedl, M. Goto, Intelligent user interfaces for music discovery. Trans. Int. Soc. Music Inf. Retriev. **3**, 165—179 (2020)
94. B.P. Knijnenburg, M.C. Willemsen, Z. Gantner, H. Soncu, C. Newell, Explaining the user experience of recommender systems. User Model. User Adapt. Interact. **22**(4–5), 441–504 (2012)
95. V.J. Konecni, Social interaction and musical preference, in *The Psychology of Music* (Academic, New York, 1982), pp. 497–516
96. Y. Koren, R.M. Bell, Advances in collaborative filtering, in *Recommender Systems Handbook*, ed. by F. Ricci, L. Rokach, B. Shapira (Springer, New York, 2015), pp. 77–118
97. D. Kowald, E. Lex, M. Schedl, Utilizing human memory processes to model genre preferences for personalized music recommendations (2020). CoRR abs/2003.10699
98. D. Kowald, M. Schedl, E. Lex, The unfairness of popularity bias in music recommendation: a reproducibility study, in *Advances in Information Retrieval - 42nd European Conference on IR Research, ECIR 2020*, Lisbon, Portugal, April 14–17, 2020, Proceedings, Part II, ed. by J.M. Jose, E. Yilmaz, J. Magalhães, P. Castells, N. Ferro, M. J. Silva, F. Martins. Lecture Notes in Computer Science, vol. 12036 (Springer, New York, 2020), pp. 35–42
99. W. Krichene, S. Rendle, On sampled metrics for item recommendation. in *KDD '20: The 26th ACM SIGKDD Conference on Knowledge Discovery and Data Mining, Virtual Event*, CA, August 23–27, 2020, ed. by R. Gupta, Y. Liu, J. Tang, B.A. Prakash (ACM, New York, 2020), pp. 1748–1757
100. F.-F. Kuo, M.-K. Shan, S.-Y. Lee, Background music recommendation for video based on multimodal latent semantic analysis, in *2013 IEEE International Conference on Multimedia and Expo (ICME)* (IEEE, New York, 2013), pp. 1–6
101. A. Laplante, Everyday life music information-seeking behaviour of young adults: An exploratory study. Doctoral dissertation, 2008
102. A. Laplante, Improving music recommender systems: What we can learn from research on music tastes? in *15th International Society for Music Information Retrieval Conference*, Taipei, Taiwan, October 2014
103. A. Laplante, J.S. Downie, Everyday life music information-seeking behaviour of young adults, in *Proceedings of the 7th International Conference on Music Information Retrieval*, Victoria (BC), October 8–12, 2006
104. J.H. Lee, How similar is too similar?: Exploring users' perceptions of similarity in playlist evaluation, in *Proceedings of the 12th International Society for Music Information Retrieval Conference, ISMIR 2011*, Miami, FL, October 24–28, 2011, ed. by A. Klapuri, C. Leider (University of Miami, Miami, 2011), pp. 109–114

105. J.H. Lee, H. Cho, Y.-S. Kim, Users' music information needs and behaviors: Design implications for music information retrieval systems. J. Assoc. Inf. Sci. Technol. **67**(6), 1301–1330 (2016)
106. J.H. Lee, R. Wishkoski, L. Aase, P. Meas, C. Hubbles, Understanding users of cloud music services: selection factors, management and access behavior, and perceptions. J. Assoc. Inf. Sci. Technol. **68**(5), 1186–1200 (2017)
107. J. Lehmann, M. Lalmas, E. Yom-Tov, G. Dupret, Models of user engagement, in *User Modeling, Adaptation, and Personalization - 20th International Conference, UMAP 2012*, Montreal, July 16–20, 2012. Proceedings, ed. by J. Masthoff, B. Mobasher, M.C. Desmarais, R. Nkambou. Lecture Notes in Computer Science, , vol. 7379, pp. 164–175 (Springer, New York, 2012)
108. E. Lex, D. Kowald, P. Seitlinger, T.N.T. Tran, A. Felfernig, M. Schedl, Psychology-informed recommender systems, in *Foundations and Trends in Information Retrieval*, 2021
109. Q. Lin, Y. Niu, Y. Zhu, H. Lu, K.Z. Mushonga, Z. Niu, Heterogeneous knowledge-based attentive neural networks for short-term music recommendations. IEEE Access **6**, 58990–59000 (2018)
110. Y.-T. Lin, T.-H. Tsai, M.-C. Hu, W.-H. Cheng, J.-L. Wu, Semantic based background music recommendation for home videos, in *International Conference on Multimedia Modeling* (Springer, New York, 2014), pp. 283–290
111. A.J. Lonsdale, A.C. North, Why do we listen to music? A uses and gratifications analysis. Br. J. Psychol. **102**(1), 108–134 (2011)
112. C.-C. Lu, V.S. Tseng, A novel method for personalized music recommendation. Expert Syst. Appl. **36**(6), 10035–10044 (2009)
113. F. Lu, N. Tintarev, A diversity adjusting strategy with personality for music recommendation, in *Proceedings of the 5th Joint Workshop on Interfaces and Human Decision Making for Recommender Systems, co-located with ACM Conference on Recommender Systems (RecSys 2018)*, October 2018, pp. 7–14
114. B. McFee, L. Barrington, G. Lanckriet, Learning content similarity for music recommendation. IEEE Trans. Audio Speech Lang. Process. **20**(8), 2207–2218 (2012)
115. B. McFee, T. Bertin-Mahieux, D. Ellis, and G. Lanckriet. The million song dataset challenge. In *Proc. of the 4th International Workshop on Advances in Music Information Research (AdMIRe)*, April 2012.
116. B. McFee, G. Lanckriet, The natural language of playlists, in *Proceedings of the 12th International Society for Music Information Retrieval Conference (ISMIR)*, Miami, FL, 2011
117. B. McFee, G. Lanckriet, Hypergraph models of playlist dialects, in *Proceedings of the 13th International Society for Music Information Retrieval Conference (ISMIR 2012)*, Porto, October 2012
118. B. McFee, G.R.G. Lanckriet, Learning multi-modal similarity. J. Mach. Learn. Res. **12**, 491–523 (2011)
119. B. McFee, C. Raffel, D. Liang, D.P. Ellis, M. McVicar, E. Battenberg, O. Nieto, librosa: audio and music signal analysis in python, in *Proceedings of the 14th Python in Science Conference*, vol. 8 (2015), pp. 18–25
120. J. McInerney, B. Lacker, S. Hansen, K. Higley, H. Bouchard, A. Gruson, R. Mehrotra, Explore, exploit, and explain: Personalizing explainable recommendations with bandits, in *Proceedings of the 12th ACM Conference on Recommender Systems*, RecSys '18, New York, NY, (Association for Computing Machinery, New York, 2018), pp. 31–39
121. S. McNee, J. Riedl, J. Konstan, Being accurate is not enough: how accuracy metrics have hurt recommender systems, in *CHI'06 Extended Abstracts on Human Factors in Computing Systems* (2006), p. 1101
122. R. Mehrotra, J. McInerney, H. Bouchard, M. Lalmas, F. Diaz, Towards a fair marketplace: Counterfactual evaluation of the trade-off between relevance, fairness & satisfaction in recommendation systems, in *Proceedings of the 27th ACM International Conference on Information and Knowledge Management*, CIKM '18 (Association for Computing Machinery, New York, NY, 2018), pp. 2243–2251

123. A.B. Melchiorre, M. Schedl, Personality correlates of music audio preferences for modelling music listeners, in *Proceedings of the 28th ACM Conference on User Modeling, Adaptation and Personalization*, UMAP '20 (Association for Computing Machinery, New York, NY, 2020), pp. 313–317

124. A.B. Melchiorre, E. Zangerle, M. Schedl, Personality bias of music recommendation algorithms, in *Fourteenth ACM Conference on Recommender Systems, RecSys '20* (Association for Computing Machinery, New York, NY, 2020), pp. 533–538

125. M. Millecamp, N.N. Htun, C. Conati, K. Verbert, To explain or not to explain: the effects of personal characteristics when explaining music recommendations, in *Proceedings of the 24th International Conference on Intelligent User Interfaces, IUI 2019*, Marina del Ray, CA, March 17–20, 2019, ed. by W. Fu, S. Pan, O. Brdiczka, P. Chau, G. Calvary (ACM, New York, 2019), pp. 397–407

126. M. Millecamp, N.N. Htun, Y. Jin, K. Verbert, Controlling spotify recommendations: Effects of personal characteristics on music recommender user interfaces, in *Proceedings of the 26th Conference on User Modeling, Adaptation and Personalization, UMAP '18* (Association for Computing Machinery, New York, NY, 2018), pp. 101–109

127. D. Moffat, D. Ronan, J.D. Reiss, An evaluation of audio feature extraction toolboxes, in *18th International Conference on Digital Audio Effects (DAFx-15)* (2015), p. 7

128. M. Müller, *Fundamentals of Music Processing: Audio, Analysis, Algorithms, Applications* (Springer, New York, 2015)

129. C. Musto, F. Narducci, P. Lops, M. De Gemmis, G. Semeraro, ExpLOD: a framework for explaining recommendations based on the LOD cloud, in *Proc. ACM Conf. on Recommender Systems, RecSys '16* (ACM, New York, 2016), pp. 151–154

130. T. Nakano, M. Goto, LyricListPlayer: a consecutive-query-by-playback interface for retrieving similar word sequences from different song lyrics, in *Proceedings of the 13th Sound and Music Computing Conference (SMC2016)*, Hamburg, August 2016, Zenodo

131. A.C. North, D.J. Hargreaves, Subjective complexity, familiarity, and liking for popular music. Psychomusicol. Music Mind Brain **14**(1–2), 77–93 (1995)

132. A.C. North, D.J. Hargreaves, Situational influences on reported musical preference. Psychomusicol. J. Res. Music Cogn. **15**(1–2), 30 (1996)

133. S. Oramas, O. Nieto, M. Sordo, X. Serra, A deep multimodal approach for cold-start music recommendation, in *Proceedings of the 2nd Workshop on Deep Learning for Recommender Systems, DLRS@RecSys 2017, Como, August 27, 2017*, ed. by B. Hidasi, A. Karatzoglou, O.S. Shalom, S. Dieleman, B. Shapira, D. Tikk (ACM, New York, 2017), pp. 32–37

134. S. Oramas, V.C. Ostuni, T.D. Noia, X. Serra, E.D. Sciascio, Sound and music recommendation with knowledge graphs. ACM Trans. Intell. Syst. Technol. **8**(2), 1–2 (2016)

135. E. Pampalk, M. Goto, Musicsun: a new approach to artist recommendation, in *Proceedings of the 8th International Conference on Music Information Retrieval, ISMIR 2007*, Vienna, September 23–27, 2007, ed. by S. Dixon, D. Bainbridge, R. Typke (Austrian Computer Society, Vienna, 2007), pp. 101–104

136. P. Papreja, H. Venkateswara, S. Panchanathan, Representation, exploration and recommendation of music playlists (2019). Preprint. arXiv:1907.01098

137. D. Parra, X. Amatriain, Walk the talk, in *International Conference on User Modeling, Adaptation, and Personalization* (Springer, New York, 2011), pp. 255–268

138. C.S. Pereira, J. Teixeira, P. Figueiredo, J. Xavier, S.L. Castro, E. Brattico, Music and emotions in the brain: familiarity matters. PLOS One **6**(11), 1–9 (2011)

139. M. Pichl, E. Zangerle, G. Specht, Towards a Context-Aware Music Recommendation Approach: What is Hidden in the Playlist Name? in *2015 IEEE International Conference on Data Mining Workshop (ICDMW)*, November 2015, Atlantic City, NJ (IEEE, New York, 2015), pp. 1360–1365

140. A. Poddar, E. Zangerle, Y.-H. Yang, #nowplaying-rs: A new benchmark dataset for building context-aware music recommender systems, in *Proceedings of the 15th Sound & Music Computing Conference*, Limassol, Cyprus, 2018. Code at https://github.com/asmitapoddar/nowplaying-RS-Music-Reco-FM

141. A. Porter, D. Bogdanov, R. Kaye, R. Tsukanov, X. Serra, Acousticbrainz: a community platform for gathering music information obtained from audio, in *International Society for Music Information Retrieval Conference (ISMIR'15)*, 2015
142. P. Pu, L. Chen, R. Hu, A user-centric evaluation framework for recommender systems, in *Proceedings of the 2011 ACM Conference on Recommender Systems, RecSys 2011*, Chicago, IL, October 23–27, 2011, ed. by B. Mobasher, R.D. Burke, D. Jannach, G. Adomavicius (ACM, New York, 2011), pp. 157–164
143. M. Quadrana, P. Cremonesi, D. Jannach, Sequence-aware recommender systems. ACM Comput. Surv. **51**(4), 66:1–66:36 (2018)
144. P.J. Rentfrow, S.D. Gosling, The do re mi's of everyday life: The structure and personality correlates of music preferences. J. Personal. Soc. Psychol. **84**(6), 1236–1256 (2003)
145. P.J. Rentfrow, S.D. Gosling, The content and validity of music-genre stereotypes among college students. Psychol. Music **35**(2), 306–326 (2007)
146. M. T. Ribeiro, S. Singh, and C. Guestrin. "Why Should I Trust You?". In *Proc. Intl. Conf. on Knowledge Discovery and Data Mining* (ACM, New York, 2016), pp. 1135–1144
147. K. Robinson, D. Brown, M. Schedl, User insights on diversity in music recommendation lists, in *Proceedings of the 21st International Society for Music Information Retrieval Conference (ISMIR 2020)*, Virtual, October 2020
148. N. Sachdeva, K. Gupta, V. Pudi, Attentive neural architecture incorporating song features for music recommendation, in *Proceedings of the 12th ACM Conference on Recommender Systems, RecSys 2018*, Vancouver, BC, Canada, October 2–7, 2018, ed. by S. Pera, M.D. Ekstrand, X. Amatriain, J. O'Donovan (ACM, New York, 2018), pp. 417–421
149. T. Schäfer, P. Sedlmeier, C. Städtler, D. Huron, The psychological functions of music listening. Front. Psychol. **4**(511), 1–34 (2013)
150. M. Schedl, Leveraging microblogs for spatiotemporal music information retrieval, in *Proceedings of the 35th European Conference on Information Retrieval (ECIR 2013)*, Moscow, March 24–27 (2013)
151. M. Schedl, The lfm-1b dataset for music retrieval and recommendation, in *Proceedings of the 2016 ACM on International Conference on Multimedia Retrieval, ICMR 2016*, New York, New York, June 6–9, 2016, ed. by J.R. Kender, J.R. Smith, J. Luo, S. Boll, W.H. Hsu (ACM, New York, 2016), pp. 103–110
152. M. Schedl, Deep learning in music recommendation systems. Front. Appl. Math. Stat. **5**, 44 (2019)
153. M. Schedl, C. Bauer, W. Reisinger, D. Kowald, E. Lex, Listener modeling and context-aware music recommendation based on country archetypes. Front. Artif. Intell. **3**, 508725 (2020)
154. M. Schedl, B. Ferwerda, Large-scale analysis of group-specific music genre taste from collaborative tags, in *19th IEEE International Symposium on Multimedia, ISM 2017*, Taichung, December 11–13, 2017 (IEEE Computer Society, New York, 2017), pp. 479–482
155. M. Schedl, E. Gómez, E.S. Trent, M. Tkalcic, H. Eghbal-Zadeh, A. Martorell, On the interrelation between listener characteristics and the perception of emotions in classical orchestra music. IEEE Trans. Affect. Comput. **9**(4), 507–525 (2018)
156. M. Schedl, D. Hauger, Tailoring music recommendations to users by considering diversity, mainstreaminess, and novelty, in *Proceedings of the 38th International ACM SIGIR Conference on Research and Development in Information Retrieval*, Santiago, August 9–13, 2015, ed. by R. Baeza-Yates, A. Lalmas, A. Moffat, B.A. Ribeiro-Neto (ACM, New York, 2015), pp. 947–950
157. M. Schedl, D. Hauger, K. Farrahi, M. Tkalcic, On the influence of user characteristics on music recommendation algorithms, in *Advances in Information Retrieval - 37th European Conference on IR Research, ECIR 2015*, , Vienna, Austria, March 29 - April 2, 2015. Proceedings, ed. by A. Hanbury, G. Kazai, A. Rauber, N. Fuhr. Lecture Notes in Computer Science, vol. 9022 (2015), pp. 339–345
158. M. Schedl, P. Knees, F. Gouyon, New paths in music recommender systems research, in *Proceedings of the Eleventh ACM Conference on Recommender Systems, RecSys 2017*, Como, August 27–31, 2017, ed. by P. Cremonesi, F. Ricci, S. Berkovsky, A. Tuzhilin (ACM, New York, 2017), pp. 392–393

159. M. Schedl, P. Knees, B. McFee, D. Bogdanov, M. Kaminskas, Music recommender systems, in *Recommender Systems Handbook*, 2nd edn., ed. by F. Ricci, L. Rokach, B. Shapira. (Springer, New York, 2015), pp. 453–492

160. M. Schedl, M. Tkalcic,  Genre-based analysis of social media data on music listening behavior: are fans of classical music really averse to social media? in *Proceedings of the First International Workshop on Internet-Scale Multimedia Management, WISMM '14*, , Orlando, FL, November 7, 2014, ed. by R. Zimmermann, Y. Yu (ACM, New York, 2014), pp. 9–13

161. M. Schedl, H. Zamani, C. Chen, Y. Deldjoo, M. Elahi,  Current challenges and visions in music recommender systems research. Int. J. Multim. Inf. Retr. **7**(2), 95–116 (2018)

162. G. Shani, A. Gunawardana,  Evaluating recommender systems, in *Recommender Systems Handbook* (Springer, New York, 2009), pp. 257–298

163. G. Shani, D. Heckerman, R.I. Brafman,  An MDP-based recommender system. J. Mach. Learn. Res. **6**, 1265–1295 (2005)

164. M. Slaney, K. Weinberger, W. White, Learning a metric for music similarity, in *Int. Symp. on Music Information Retrieval (ISMIR'08)* (2008), pp. 313–318

165. J. Smith, D. Weeks, M. Jacob, J. Freeman, B. Magerko,  Towards a hybrid recommendation system for a sound library, in *IUI Workshops* (2019)

166. B. Smyth, P. McClave, Similarity vs. diversity, in *Case-Based Reasoning Research and Development, 4th International Conference on Case-Based Reasoning, ICCBR 2001*, Vancouver, BC, Canada, July 30 - August 2, 2001, Proceedings, ed. by D.W. Aha, I.D. Watson. Lecture Notes in Computer Science, vol. 2080 (Springer, New York, 2001), pp. 347–361

167. M. Sordo, O. Celma, M. Blech, E. Guaus, The quest for musical genres: Do the experts and the wisdom of crowds agree? in *Int. Conf. of Music Information Retrieval (ISMIR'08)* (2008), pp. 255–260

168. L. Spinelli, J. Lau, L. Pritchard, J.H. Lee,  Influences on the social practices surrounding commercial music services: a model for rich interactions, in *Proceedings of the 19th International Society for Music Information Retrieval Conference (ISMIR)*, Paris, 2018

169. H. Steck,  Calibrated recommendations,  in *Proceedings of the 12th ACM Conference on Recommender Systems, RecSys '18* (Association for Computing Machinery, New York, NY, 2018), pp. 154–162

170. A. Swaminathan, T. Joachims, Counterfactual risk minimization: learning from logged bandit feedback, in *International Conference on Machine Learning* (2015), pp. 814–823

171. M. Tiemann, S. Pauws,  Towards ensemble learning for hybrid music recommendation,  in *ACM Conf. on Recommender Systems (RecSys'07)* (2007), pp. 177–178

172. N. Tintarev, M. Dennis, J. Masthoff,  Adapting recommendation diversity to openness to experience: a study of human behaviour, in *User Modeling, Adaptation, and Personalization*, ed. by S. Carberry, S. Weibelzahl, A. Micarelli, G. Semeraro (Springer, Berlin, Heidelberg, 2012), pp. 190–202

173. N. Tintarev, J. Masthoff,  Explaining recommendations: design and evaluation,  in *Recommender Systems Handbook* (Springer, New York, 2015), pp. 353–382

174. W. Trost, T. Ethofer, M. Zentner, P. Vuilleumier, Mapping aesthetic musical emotions in the brain. Cerebral Cortex **22**(12), 2769–2783 (2012)

175. D. Turnbull, L. Barrington, D. Torres, G. Lanckriet,  Semantic annotation and retrieval of music and sound effects. Trans. Audio Speech Lang. Process. **16**(2), 467–476 (2008)

176. D. Turnbull, L. Waldner,  Local music event recommendation with long tail artists (2018). Preprint. arXiv:1809.02277

177. A. Vall, M. Dorfer, H. Eghbal-zadeh, M. Schedl, K. Burjorjee, G. Widmer,  Feature-combination hybrid recommender systems for automated music playlist continuation. User Model. User Adapt. Interact. **29**(2), 527–572 (2019)

178. A. van den Oord, S. Dieleman, B. Schrauwen, Deep content-based music recommendation, in *Advances in Neural Information Processing Systems 26: 27th Annual Conference on Neural Information Processing Systems 2013. Proceedings of a meeting held December 5–8, 2013*, Lake Tahoe, Nevada, United States, ed. by C.J.C. Burges, L. Bottou, Z. Ghahramani, K.Q. Weinberger (2013), pp. 2643–2651

179. S. Verma, J. Rubin, Fairness definitions explained, in *Proceedings of the International Workshop on Software Fairness, FairWare '18* (Association for Computing Machinery, New York, NY, 2018), pp. 1–7
180. G. Vigliensoni, I. Fujinaga, Automatic music recommendation systems: Do demographic, profiling, and contextual features improve their performance? in *Proceedings of the 17th International Society for Music Information Retrieval Conference, ISMIR 2016*, New York City, August 7–11, 2016, ed. by M.I. Mandel, J. Devaney, D. Turnbull, G. Tzanetakis (2016), pp. 94–100
181. G. Vigliensoni, I. Fujinaga, The music listening histories dataset, in *Proceedings of the 18th International Society for Music Information Retrieval Conference*, Suzhou, People's Republic of China, 2017, pp. 96–102
182. D. Wang, S. Deng, X. Zhang, G. Xu, Learning to embed music and metadata for context-aware music recommendation. World Wide Web **21**(5), 1399–1423 (2018)
183. S. Wang, L. Hu, Y. Wang, L. Cao, Q.Z. Sheng, M.A. Orgun, Sequential recommender systems: Challenges, progress and prospects. CoRR abs/2001.04830 (2020)
184. M. Ward, J. Goodman, J. Irwin, The same old song: the power of familiarity in music choice. Market. Lett. **25**, 1–11 (2013)
185. D. Weigl, C. Guastavino, User Studies in the Music Information Retrieval Literature, in *Proceedings of the 12th International Society for Music Information Retrieval Conference (ISMIR 2011)*, Miami, FL, USA, October 2011
186. W. Wu, L. Chen, Y. Zhao, Personalizing recommendation diversity based on user personality. User Model. User-Adapt. Interact. **28**(3), 237–276 (2018)
187. S. Yao, B. Huang, Beyond parity: fairness objectives for collaborative filtering, in *Advances in Neural Information Processing Systems 30*, ed. by I. Guyon, U.V. Luxburg, S. Bengio, H. Wallach, R. Fergus, S. Vishwanathan, R. Garnett (Curran Associates, Inc., Red Hook, 2017), pp. 2921–2930
188. H. Zamani, M. Schedl, P. Lamere, C. Chen, An analysis of approaches taken in the ACM recsys challenge 2018 for automatic music playlist continuation. ACM Trans. Intell. Syst. Technol. **10**(5), 57:1–57:021 (2019)
189. E. Zangerle, M. Pichl, W. Gassler, G. Specht, #nowplaying music dataset: extracting listening behavior from twitter, in *Proceedings of the First International Workshop on Internet-Scale Multimedia Management, WISMM '14* (Association for Computing Machinery, New York, NY, 2014), pp. 21–26
190. M. Zenter, D. Grandjean, K. Scherer, Emotions evoked by the sound of music: characterization, classification, and measurement. Emotion **8**, 494 (2008)
191. Y.C. Zhang, D.O. Séaghdha, D. Quercia, T. Jambor, Auralist: Introducing serendipity into music recommendation, in *Proceedings of the Fifth ACM International Conference on Web Search and Data Mining, WSDM '12*. (ACM,New York, NY, 2012), pp. 13–22
192. E. Zheleva, J. Guiver, E. Mendes Rodrigues, N. Milić-Frayling, Statistical models of music-listening sessions in social media. in *Int. Conf. on World Wide Web (WWW'10)* (2010), pp. 1019–1028
193. Y. Zheng, Context-aware mobile recommendation by A novel post-filtering approach, in *Proceedings of the Thirty-First International Florida Artificial Intelligence Research Society Conference, FLAIRS*, 2018, Melbourne, FL, May 21–23 2018, ed. by K. Brawner, V. Rus (AAAI Press, New york, 2018), pp. 482–485
194. Y. Zheng, R.D. Burke, B. Mobasher, Splitting approaches for context-aware recommendation: an empirical study, in *Symposium on Applied Computing, SAC 2014, Gyeongju, Republic of Korea - March 24–28, 2014*, ed. by Y. Cho, S.Y. Shin, S. Kim, C. Hung, J. Hong (ACM, New York, 2014), pp. 274–279
195. L. Zhu, Y. Chen, Session-based sequential skip prediction via recurrent neural networks. CoRR abs/1902.04743 (2019)
196. C. Ziegler, S.M. McNee, J.A. Konstan, G. Lausen, Improving recommendation lists through topic diversification, in *Proceedings of the 14th international conference on World Wide Web, WWW 2005*, Chiba, May 10–14, 2005, ed. by A. Ellis, T. Hagino (ACM, New York, 2005), pp. 22–32

# Multimedia Recommender Systems: Algorithms and Challenges

Yashar Deldjoo, Markus Schedl, Balázs Hidasi, Yinwei Wei, and Xiangnan He

## 1 Introduction

With the increasing popularity of online services that provide access to multimedia collections (e.g., music catalogs or user-generated videos offered by major streaming providers), both creation and consumption of expert- and user-generated multimedia content have skyrocketed. The number of users and the amount of available multimedia content, e.g., audio, visual, and textual data, have reached an unprecedented level. This calls for personalized recommender systems (RS) that help users find desirable media items (e.g., newspaper articles, music pieces, or movies) without the burden of active search.

In this chapter, we provide an introduction to multimedia recommender systems (MMRSs) [37, 41]. Such systems leverage data of multiple modalities (e.g., text,

Y. Deldjoo (✉)
Polytechnic University of Bari, Bari, Italy
e-mail: deldjooy@acm.org; yashar.deldjoo@poliba.it

M. Schedl
Johannes Kepler University Linz, Institute of Computational Perception,
Multimedia Mining and Search Group, Linz, Austria

LIT AI Lab, Human centered AI Group, Linz, Austria
e-mail: markus.schedl@jku.at

B. Hidasi
Gravity R&D, Budapest, Hungary
e-mail: balazs.hidasi@gravityrd.com

Y. Wei
National University of Singapore, Singapore, Singapore

X. He
University of Science and Technology, Hefei, China

© Springer Science+Business Media, LLC, part of Springer Nature 2022
F. Ricci et al. (eds.), *Recommender Systems Handbook*,
https://doi.org/10.1007/978-1-0716-2197-4_25

**Table 1** Various application domains and example research works where multimedia content—in particular audio and visual—is used for a recommendation task

| Target item | Feature extraction content | Recommended items | Domain | Research examples |
|---|---|---|---|---|
| Image | Image | Photos | Social media, search | [61, 109, 154] |
|  | Image | Paintings | Art, culture | [3, 8] |
| Video | Video | Movies | Entertainment | [19, 34, 38, 51] |
|  | Video | Short/micro-videos | User-generated content | [27, 98, 164] |
|  | Video | News | Information | [97] |
| Audio | Audio | Sounds | Music creation | [112, 133] |
|  | Audio | Music items, playlists | Music consumption | [75, 111, 113, 141] |
|  | Audio | Podcasts | Information | [156] |
| Non-media | Image | Clothes, shoes, furniture | Fashion, interior design | [43, 72, 99, 139] |
|  | Image | Food | Health | [54, 60, 104, 158] |
|  | Image | Points-of-interest | Tourism | [56, 146] |
|  | Image | Tags, news | Information | [90] |
|  | Image | Friends, partners | Social media, dating | [153] |

audio, image) to extract computational features and create the item representation. This representation is subsequently used in a content-based filtering (CBF) or hybrid system, combining collaborative and content information, to recommend either (1) media items from which the features have been extracted, or (2) non-media items using the features that have been extracted from a proxy multimedia representation of the item (e.g., images of clothes). As a third variant, cross-media recommender systems leverage multimedia content of one item type to recommend another item type, often considering a user's context such as his or her emotion as a catalyst. Examples of this type of research include, e.g., recommending background music for a slide show and recommending music based on the emotions predicted from the article the user reads or writes [16, 81].

MMRSs are used in various application domains. Depending on the domain, item features are either extracted directly from the target item or from a proxy representation thereof. Table 1 provides an overview of the most important application domains and lists a few exemplary research works addressing them. The top three rows (image, video, and audio) refer to the first category, application domains where content features are extracted from the target items. For the target item *image*, common examples include recommending photos in social media platforms and recommending paintings to visit in a (virtual) museum. As for *video* as target item, the most common application domain is movie recommendation in

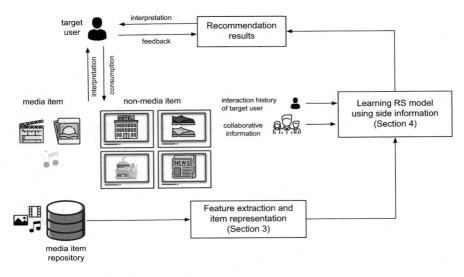

**Fig. 1** Typical pipeline of a multimedia recommender system

streaming platforms. Other domains include recommending news videos and short user-generated videos. Typical domains of *audio* recommendation include music creation (e.g., recommending sounds such as drum loops to a creator of electronic music), music consumption (e.g., automatic playlist generation), and information (e.g., podcast recommendation).

The bottom row in Table 1 relates to the research that uses features extracted from a proxy multimedia representation of items to recommend *non-media items*. Corresponding approaches or systems predominantly leverage the visual modality, for tasks such as recommending fashion, food, places (points-of-interest), news, or even persons (friends or partners in online social networks).

**Generic Pipeline of a Multimedia Recommender System**

To give the reader an idea of the typical components and pipeline of an MMRS, its standard workflow is shown in Fig. 1.

The box at the bottom ("Feature extraction and item representation") is the core part that discriminates MMRSs from systems that purely use collaborative filtering and from knowledge-based systems. This component is responsible for the computation of features for each item (or the proxy representation thereof), and the resulting item descriptors are used in the recommendation model. The individual steps of the multimedia content processing part are detailed in Sect. 3. The (hybrid) recommender system then leverages these item descriptors and user–item interactions to learn one or several model(s),[1] and uses the final model to generate recommendations. From the user perspective (upper left part of the figure),

---

[1] In the latter case, the individual models are combined to form an ensemble model.

either media items or generic (non-media) items are recommended; examples are provided in Table 1. Furthermore, the user can typically provide feedback to the system, either explicitly or implicitly (e.g., via ratings or clicking behavior, respectively).

In Sect. 2, we identify and summarize the major aspects and challenges that have to be considered when creating an MMRS. Since feature extraction and item representation are key ingredients to build such a system, in Sect. 3, we present the most common multimedia content processing techniques to obtain item representations that can be used as side information in an MMRS. In Sect. 4, we discuss how the challenges introduced in Sect. 2 are addressed in the literature by reviewing traditional hybrid approaches, neural approaches, and graph-based approaches to MMRS.

Please also note that this book contains other chapters that are dedicated to certain sub-tasks of multimedia recommendation, in particular music recommendation Chapter "Music Recommendation Systems: Techniques, Use Cases, and Challenges", fashion recommender systems Chapter "Fashion Recommender Systems", and food recommendation Chapter "Food Recommender Systems".

## 2  Key Challenges

MMRSs have to face several challenges. Their key challenges can be studied from complementary perspectives (e.g., in a user-centric, algorithmic, or even sociological fashion). Here, we adopt a technical point of view and correspondingly discuss some of the most prominent and pressing challenges in MMRS from a technological perspective. We distinguish the challenges that relate to warm-start situations and those that concern cold-start situations (new-item or new-user problem), which we briefly summarize below.

**Warm-Start Challenges**
- C1: How to model the interaction between user and multimedia content, in terms of data and data structures (e.g., graph, feature vector, matrix) based on collaborative and content features?
- C2: How to fuse multimedia representations with collaborative representations?
- C3: How to integrate additional information such as contextual information into the recommendation model?
- C4: How to learn multimedia representations effective for recommender systems (or user behavior modeling), which are different from the representations learned for computer vision.

**Cold-Start Challenges**
- C5: How to handle the training–testing discrepancy [50]? (the training stage does not have cold-start items, whereas the testing stage has) How to ensure the model components used in training are still suited for the testing data?

- C6: How to perform continuous learning (quick model update) with new—and few—interactions with cold-start items?

  In the main part of our technical review (Sect. 4) we discuss solutions that address these challenges.

# 3 Fundamentals of Feature Extraction and Item Representation

A prerequisite to building an MMRS is to create item representations that encode or reflect characteristics of the multimedia material in a way that they can be leveraged by the recommendation algorithm. To this end, techniques from multimedia content processing (i.e., signal processing and machine learning) are used. In the following, we discuss for the three media types image, video, and audio, state-of-the-art approaches that address the computational steps mentioned in the box in the lower part of Fig. 1.

The goal of item representation, as illustrated in Fig. 2 for the image domain and in Fig. 4 for the audio domain, is to transform the input media item—or its proxy—into a feature vector that describes some of its properties. This feature vector can then be used directly to compute recommendations, or leveraged to solve an intermediate classification task (e.g., tag prediction), whose results can be finally used.

To connect these item representation techniques initially introduced for generic multimedia processing to their actual use in MMRSs, Tables 2, 3, 4, and 5 introduce common methods for the various steps involved in item representation, which have been used over the last years in RSs research. The tables also highlights research papers and domains in which these techniques have been adopted in the RSs literature. For a detailed description of the approaches and papers, interested readers are invited to read our recent survey on RS that leverage multimedia content [41].

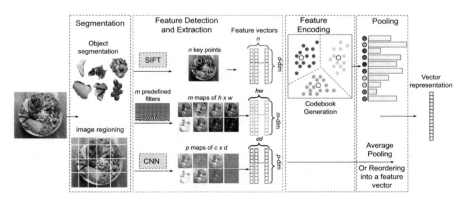

**Fig. 2** Generic computational pipeline for image feature extraction and item representation

**Table 2** Common approaches for item *segmentation* in recommender systems involving image and video modalities, and references to publications covering various target items

| Approach | Reference research for domains and target items |
| --- | --- |
| Regioning (image) | fashion [22, 64] |
| Shot-level segmentation (video) | Movie [31, 35, 38, 53] |
| Clip-level segmentation (video) | Movie [39] |

While the exact algorithmic steps involved in creating item representations of multimedia objects in MMRSs varies between domains, types of media, and recommendation methods, we nevertheless provide a brief introduction to the typical steps involved when creating such item representations, for the image, video, and audio modalities. Note that depending on the recommendation domain and algorithm, some steps can be omitted or additional steps can be necessary. For instance, in the domain of fashion recommendation, understanding the content in (garment) images is a vital requirement for good recommendations while precisely understanding the content of all frames in a music video clip is less important when recommending music videos. For the latter, in contrast, accurate descriptors of musical properties (e.g., timbre, key, or melody) or predictions made by pretrained music genre classifiers commonly require additional steps. Notwithstanding limited generalizability to arbitrary domains, we identify the most common subtasks involved in item representation for the most important modalities used in multimedia items, i.e., image, video, and audio (see [41]); and we discuss common methods to approach these subtasks of item segmentation, feature detection and extraction, feature encoding, and data fusion.

Table 2 provides an overview of the studied approaches for **item segmentation**; Table 3 for **detecting and extracting features**; Table 4 for **feature encoding**, i.e. creating a fixed-length representation for each item; and Table 5 for **fusing the resulting features** (e.g., features of different modalities) into a joint descriptor. Note that in Table 5, we include this additional data fusion step (that does not exist in Fig. 1) for completeness.

More precisely, segmentation aims to segment the MM signal into fine-grained units, which facilitates the description of the MM item and can improve computational efficiency (Table 2). We divide the feature extraction approaches (Table 3) into signal-level methods (operating on the raw pixel information in case of images), local feature extractors (describing small coherent regions of an image), and CNN-based approaches (typically using pre-trained models) [41]. Approaches to feature encoding and pooling, whose goals are, respectively, representing the extracted features using a specific representation schema (e.g., a codebook) and producing fixed-length descriptors, are summarized in Table 4. Finally, for data fusion, i.e., fusing the pooled item representation, we differentiate between early fusion, late fusion, and other (specific) methods that cannot be categorized into any of the former two (Table 5). In early fusion, different features are combined and jointly used to create a model (e.g., for rating prediction). In contrast, late fusion refers to creating different prediction models and combining their outputs. Note that

**Table 3** Common approaches for *feature detection and extraction* in the recommendation domains involving image and video modalities, and references to publications covering various target items

| Feature category | Type/approach | Reference research for domains and target items |
|---|---|---|
| Signal-level features | Color | Fashion [22, 161], eye-glasses [64], make-up [2, 24], food [54], restaurant [23], tags [88], social media [26], painting [3, 8], photography [10, 121], movie [13], user-generated video [101] |
| | Texture | Fashion [22], make-up [2, 24], food [54], points-of-interest [146], social media [26], search query [154], photography [121], generic images [11, 65], painting [3, 8], tags [88], movie [13], user-generated video [101] |
| | Motion | Movie [13, 30, 31, 51], user-generated video [102] |
| Local features | SIFT/SURF [9, 96] | Fashion [22], points-of-interest [146], tags [90], social media [135], search query [154], generic image [11], movie [47, 166], user-generated video [27, 51] |
| | LBP [110] | Eye-glasses [64], make-up [2] |
| | HOG [29] | Eye-glasses [64], make-up [2], tourism [146], photography [121] |
| | DWT/DFT | Generic images [65], tags [88], eye-glasses [64] |
| Pretrained or fine-tuned CNN | Caffe reference model | Fashion [68, 69, 99] |
| | AlexNet [80] | Social curation network [61], dance background [152], movie [34, 38] |
| | ResNet [67] | Fashion [84] |
| | VGG [132] | Food [158], tourism [56, 146], movie [19, 51, 98] |
| | GoogleNet [126] | Movie [33] |
| | VGG-face [115] | Police photo lineup [117] |
| | DenseNet [73] | Fashion clothes [89] |
| | Places-CNN [165] | Geolocated images [109], movies [125] |
| | Aesthetic network | Fashion clothes [161] |
| | Siamese-GoogleNet [126] | Fashion clothes [136] |
| | Siamese-AlexNet [80] | Fashion clothes [72] |

**Table 4** Common approaches for *feature encoding* in the recommendation domains involving image and video modalities, and references to publications covering various target items

| Approach | Reference research for domains and target items |
|---|---|
| $k$-means | Fashion [22], generic images [11, 26], photography [121], user-generated video [47] |
| GMM | Photography [121], movie [34, 35, 38] |

**Table 5** Common approaches for *data fusion* in the recommendation domains involving image and video modalities, and references to publications covering various target items

| Fusion type | Approach | Reference research for domains and target items |
|---|---|---|
| Early | Feature vector concatenation | Fashion [22, 161], generic image [109], food [54] |
| | Canonical correlation analysis | Movie [38] |
| Late | Borda count | Movie [34] |
| | Priority-aware aggregation | movie [34] |
| Specific | Similarity-level | Movie [13], paintings [8] |
| | Latent space model | Tags [90] |
| | Neural network-based | User-generated video [98] |
| | Kernel-based | User-generated video [47] |
| | Optimization-based | User-generated video [27] |
| | Graph-based | Venue [57], generic image [26] |

in practice, the position of data fusion in the pipeline can differ based on the fusion type.

In the following, we present the fundamentals necessary to understand the concepts in item representation, categorized according to media type, i.e., in Sect. 3.1 for image, in Sect. 3.2 for video, and in Sect. 3.3 for audio.

## 3.1   Image Representation

The common image representation pipeline is sketched in Table 2 and consists of the following basic steps:

**1. Segmentation** For a given image, $N$ smaller segments are extracted, which are homogeneous in some feature space.[2] The extracted segments—called patches—can be a set of points of interest, a fixed grid, or a set of densely sampled points at each pixel location [94]. Segmentation can also be performed at a semantic level aimed toward object detection. The region proposal network (RPN) [140], for example, has been proposed to suggest, for a given arbitrary-sized image, a set of rectangular object candidates, each indexed with an "objectness" score. Also,

---

[2] For images, segmentation is carried out in the image/pixel space, a process known as *spatial segmentation*.

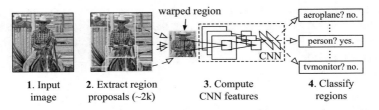

**Fig. 3** Object detection using the R-CNN approach. Courtesy of [63]

convolutional neural networks (CNNs) have been successfully applied for semantic segmentation tasks. For example, the region-based CNN (R-CNN) [63] extracts several candidate regions—also known as region of interest (RoI)—and then extracts CNN-based features for each RoI and evaluates them for object classification, as illustrated in Fig. 3. Several variations of R-CNN have been proposed, namely, Fast R-CNN [62], Faster R-CNN [122], single shot multibox detector (SSD) [92], and R-FCN [28], which improve R-CNN's speed and predictive accuracy. Segmentation can also be used to infer the global context of the image [91], pooled at the top-most part in the hierarchy of the image segments, where this information can be leveraged to build context-aware RS.

Please note that segmentation of images is an optional step for MMRSs tasks. Currently, the most popular approach for feature extraction is to apply CNN features on the entire image and extract the last layer outputs. Alternatively, CNN-based object detectors can be used to generate candidate boxes for RoIs and, subsequently, features are extracted from these RoIs.

**2. Feature Detection and Extraction** Obtaining high quality image descriptors plays a critical role in the image representation pipeline. Given the segmented or non-segmented image, the goal is to extract a set of features that describe the content of the media item. For $N$ extracted patches and feature dimensionality of $D$, we denote the local features of $N$ patches as $\{x_i\}_{i=1}^{N}$, $x_i \in \mathbf{R}^D$, where $N = 1$ if no segmentation is carried out. For instance, raw image pixels may be fed into deep CNN models, such as AlexNet [80], ResNet [67], or VGG [132], pre-trained on millions of images for generalized classification tasks and to obtain a more discriminating feature representation.

**3. Feature Encoding** The output of the last step is a series of feature vectors $\{x_i\}$. This step now represents each feature $x_i$ using a codebook, typically by mapping $x_i$ into one or several feature coding vector(s) $\{v_i\}$ ($v_i \in \mathbf{R}^K$), where $K$ is the number of codewords determined based on training data. The codewords may be predefined or learned (e.g., by clustering the training data with $k$-means). The size and characteristics of the codebook impact the subsequent item description and thereby its discrimination power [94]. In the case of images, common feature encoding methods include vector quantization (VQ), $k$-means, sparse coding, vector of locally aggregated descriptors (VLAD), and Fisher vectors (FV).

**4. Pooling** The goal of this step is to produce a fixed-length global feature representation that statistically summarizes the properties of the item's segments by aggregating the feature coding vectors $\{v_i\}$. The most common pooling methods include average pooling and max pooling.

## 3.2   Video Representation

Video representation is a more challenging task compared to image representation due to the additional time dimension and computational constraints imposed by the common need to process a large number of video frames.

**1. Segmentation** For videos, the segmentation can be performed in space (spatial segmentation), in time (temporal segmentation), or in both space and time (spatio-temporal segmentation). The motivation is efficiency (e.g., a video can contain thousands of similar frames) and improving discriminating power of the final descriptor. Spatial segmentation of video uses the same techniques already discussed in the previous section for images. As for temporal segmentation, the standard approach is to cut the full video into shorter semantic segments that are homogeneous in some semantic/feature space according to:

- **Frame**: Video frames are the smallest units of a video; typically videos are made with 25 or 30 frames-per-second (fps).
- **Scene**: Each shot represents a sequence of frames that represent that camera shot; scenes can therefore be extracted by detecting shot changes [53].
- **Clip**: Clips are natural cuts of a movie with certain semantics, e.g., a dance performance. A clip can contain many scenes, e.g., a conversation can show different faces in each turn of the conversation [39].

Please note that the above temporal segmentation in the context of item representation is nowadays more and more often circumvented by the use of deep neural networks (DNNs). In this case, the preferred option is to down-sample videos to lower fps rate (e.g., 1 fps) and use all resulting frames as input to a DNN. This approach relies on the assumption that down-sampling does not result in losing a significant amount of information because of the high similarity between consecutive video frames. Another consideration related to movie segmentation is that often, instead of the full movie, only a representative trailer [30, 34, 38] or a set of movie clips [39] is used for feature extraction. The reasons are reduced computational demand, reduced time for the user to watch, and circumventing possible copyright issues. For a more detailed discussion, please see Section 2.3.1 of the recent survey on MMRSs [41].

**2. Feature Detection and Extraction** Extracting features from the visual modality of a video is the most common approach to represent its content, though in recent years, a few works have leveraged the audio modality in addition, to build *multimodal* video recommender systems [13, 34, 38, 51]. Visual features extracted from video are essentially quite similar to those used for images, as shown in Table 3. In addition, motion features can be leveraged for the video items. Such motion features include descriptors of camera movements and object movements.

**3. Feature Encoding and Pooling** This step in the video domain predominantly adopts the same approaches as in the image domain, but adds a temporal dimension. Encoding and pooling can therefore be performed on segment-level features within video frames and temporally, i.e., aggregating image features over time. Generally, temporal aggregation in video is a more important task since it can dramatically increase the computational efficiency of video processing. Corresponding methods can be classified into classical approaches (e.g., trajectory-based) [143] and methods relying on neural architectures (e.g., two-stream architectures or 3D-CNNs) [58, 77, 131].

As a final remark on video representation, we would like to point out that the recent technical solutions in MM systems have not been adopted in RSs. The research in RSs has been traditionally and predominantly focused on understanding the *user side* of the system [32, 36, 55, 130] and using approaches based on metadata or CF to build video RS. Conversely, the MM community has made a tremendous effort on the *computational* aspect of the system by focusing on building automated video understanding algorithms based on multimedia (visual and audio) content toward tasks such as human action recognition.[3] Research that focuses on extracting audio visual content from videos to build video RS has recently become popularized and is reviewed in [41].

## 3.3 Audio Representation

We will not provide here a detailed account of the topic of item representation in the audio domain, rather give an overview. We refer the interested reader to the book [167] for a comprehensive treatment of audio signal processing, to [45, 79, 106] for a discussion within the context of the music domain, and in particular to Chapter "Music Recommendation Systems: Techniques, Use Cases, and Challenges" of this book on the use on music recommender systems. The general pipeline of the common audio processing steps involved in audio item representation is provided in Fig. 4.

**1. Segmentation** Segmenting an audio stream is performed temporally instead of spatially as in the case of images. There exist two variants for such a temporal segmentation. Structural segmentation aims at cutting an audio stream at semantically meaningful boundaries, such as when the chorus in music, or the speaker, in

---

[3] Consider an overview of these advances on the multimedia recommender systems tutorial presented by He et al. at the ACM ICMR conference 2018: (http://staff.ustc.edu.cn/~hexn/icmr18-recsys.pdf).

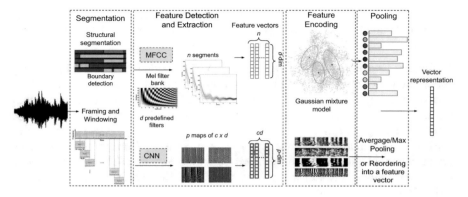

**Fig. 4** General pipeline of stages involved in audio feature extraction and item representation. Some images by courtesy of [79]

spoken dialog, changes. Alternatively, in the absence of such semantic information about segments (or the absence of techniques to detect them), the audio stream can also be cut into equally sized blocks whose duration typically ranges from tens of milliseconds to a few seconds. This process is called *framing* since it cuts the stream at a predefined number of audio frames. Larger segments, such as those used in the block-level feature (BLF) framework [129] have been shown to be powerful in capturing temporal aspect in audio and music retrieval and similarity tasks.

**2. Feature Detection and Extraction** There exist two main strategies for feature extraction: (1) using handcrafted features and (2) automatic feature learning. As for the former, a very common approach in the literature is to extract a given number $d$ of Mel frequency cepstral coefficients (MFCCs) for each of the $n$ segments of the audio stream. MFCCs have their origin in the speech processing domain [119], but were later also used frequently in music information retrieval [95]. Automatic feature learning is usually achieved by a DNN architecture, by adopting approaches from the image domain. Instead of an arbitrary image, in the case of audio, the image of a spectrogram is fed into the neural network of choice, e.g., a CNN. This last step is today often done via pre-trained CNNs, such as Musicnn [118] for music auto-tagging.

**3. Feature Encoding** The feature extraction process results in a constant-length-vector representation of each segment. In the subsequent encoding step, an intermediate representation of these vectors is created. In case of MFCCs, a common choice is to model all MFCC vectors via a Gaussian mixture model (GMM). In case of automatically learned features via deep architectures, this step is not necessary.

**4. Pooling** The final step to obtain a single-vector-representation of constant dimensionality is to statistically summarize or pool the segments based on the intermediate representation. For a Gaussian (or a GMM), this can be done by concatenating the mean vector and the flattened covariance matrix. In case of CNNs, or other variants of DNNs, typically the penultimate or the output layer of the network is used as feature vector.

From an algorithmic point-of-view such item representations for audio recommendation have been used in various types of approaches, e.g., memory-based [4, 134], model-based [160], graph-based [12, 100], and deep neural algorithms [144]. Also, they have been adopted for recommending different types of audio content, such as music [128], sound [134], and speech [21] (e.g., podcast recommendation).

# 4 Recommendation by Incorporating Multimedia Side Information

Recommendations based on information on user–item interactions, e.g., via collaborative filtering methods, has dominated the research in the RS community for years. However, these methods do not leverage descriptive item attributes to compute recommendations, which can be regarded as a waste of information. After all, if a fictional user Jane likes the movie Fast and Furious, which features very fast paced car chases, then it is likely she will also like other movies that show similar scenes. Similarly, if she likes a dress that follows a particular, visually appealing style, there are high chances that she will like to keep the same style in her future shopping behavior. And if her music playlist predominantly contains upbeat pop music, it is expected that similar-sounding music will please her in the future too.

As "*recommendation is not a one-size-fits-all problem*" [52], over the last years, we have witnessed the emergence of recommendation models that go beyond exploiting the user–item interaction matrix [130] and incorporate more domain features into CF models. In the context of multimedia recommendation, most commonly, additional information on the items (or users) is integrated into the recommendation algorithm as side information. Traditionally, such side information has been some form of textual description or attribute of the item (e.g., genre or tags associated with a movie), neglecting the rich audiovisual signals multimedia items are composed of. In recent years, computational power has increased, and so we have the possibilities to process large amounts of multimedia data, which is required to build item representations that fuel CBF and hybrid MMRS. In particular, in the video domain and with state-of-the-art deep learning methods, current systems still face computational limitations, though. As a solution, a common strategy to build a hybrid system is depicted in Fig. 5. In this two-stage system, the first (left) stage it uses a CF model that creates a list of candidate recommendations based on collaborative information, reducing the search space from millions to hundreds of items. Leveraging multimedia side information, a CBF component (on the right side) subsequently reranks these candidates and selects at most tens of items to finally recommend to the user [71].

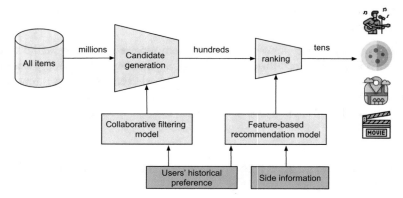

**Fig. 5** Illustration of a hybrid MMRS, containing a CF component and a CBF reranking component fueled with side information

In the following, we survey algorithms for multimedia recommender systems, structured according to the used technique into: traditional hybrids (Sect. 4.1), approaches based on neural networks (Sect. 4.2), and graph-based approaches (Sect. 4.3).

## 4.1   Traditional Hybrid Approaches

Collaborative filtering-based approaches that are able to incorporate side information of items (and users) can be classified into the following two broad categories, according to [130]: *extended memory-based CF* and *extended model-based CF approaches*. Note that, traditionally, these methods have leveraged textual side information such as reviews, tags, or metadata, which comes as a single modality. In contrast, multimedia data constitutes of multi-channel signals and therefore calls for efficient algorithms that can scale up.

### 4.1.1   Extending Memory-Based Collaborative Filtering

Traditional memory-based CF strategies compute recommendations based on a matrix containing pairwise similarities between users or between items, referred to as user-based or item-based CF, respectively. This similarity matrix is computed from user–item interaction data. To build hybrid MMRS, item side information (content features) can be integrated into the item similarity matrix of item-based CF systems (see "Item-based CF with side information" below). Alternatively, multimedia content features can be used to augment the sparse user rating vector (see "CBCF").

- **Item-based collaborative filtering with side information:** Item-based CF with side information (see baseline in [108] as an example) is a simple extension of the traditional item-based CF for top-N recommendation. Here, item-item similarities are not purely computed from the interaction data, instead, they are calculated as a linear combination between interaction similarities and content-based item similarities.
- **Content-boosted collaborative filtering (CBCF):** Melville et al. [103] tightly integrate textual side information into a memory-based CF architecture for movie recommendation. Based on metadata (e.g., title and cast) and each user's existing ratings, their CBCF system trains a classifier to predict the user's missing ratings. These predictions, named "pseudo-ratings", are used to densify the user–item rating matrix, which is eventually used in a user-based memory-based CF algorithm, which the authors extend by several confidence weights (depending on the number of items the target user rated, the co-rated items between target user and other users, and the confidence into the content-based rating prediction).

### 4.1.2 Extending Model-Based Collaborative Filtering

In addition to memory-based CF, a large body of research has been dedicated to extending model-based CF approaches to incorporate rich side information. These methods train a predictive model based on all available information, including the URM interaction data and side information of items (and users). Here, we list and describe the most prominent approaches.

- **Factorization Machine (FM):** FM [123] models encode user–item collaborative data and corresponding side information into real-valued feature vectors and numeric target variables, thereby modeling all interactions between features as $n$-way interactions, where $n = 2$ (i.e., second-order model) is the most common. Unlike a polynomial regression that computes interaction individually, FM uses factorized interaction parameters. FMs are powerful methods, in particular in sparse scenarios. However, they are limited in scaling up to high-dimensional multimedia content features.
- **Extending BPR with visual side information:** The visual Bayesian personalized ranking (VBPR) method [69] is built upon BPR, a state-of-the-art pairwise ranking optimization framework for implicit feedback (e.g., clicked vs. non-clicked). VBPR extends BPR by incorporating a latent content-based preference factor, where the item content features are typically extracted from a pre-trained CNN (e.g., AlexNet or ResNet 50). As an example, He et al. [69] show that VBPR improves the performance of BPR-MF on Amazon fashion and Amazon phones datasets. Interestingly, the amount of improvement is higher in the fashion domain, where the visual factor is likely a more dominant key in users' decision making. The advent of VBPR has led to the design of a wide range of VBPR-based extensions, typically known as *visually-aware recommender systems* in the community. These include the following methods: DeepStyle (2017) [93], which

extends VBPR by a style factor learned from user–item interaction data; attentive collaborative filtering (ACF) (2017) [17], which uses an attention network to learn an item-level and component-level attention (such as the regions in an image and frames of a video) in the core VBPR predictor; DVBPR (2017) [76], which replaces the pretrained CNNs used in VBPR with an end-to-end model, whose visual feature extractor is trained together with the recommendation model in a pair-wise manner; and TVBPR (2016) [68], which uses a time-aware predictor to account for seasonality and evolving fashion trends.

- **Set of sparse linear methods with side information (SSLIM):** Ning and Karypis [108] propose several extensions to the popular SLIM model [107], which is an efficient CF model that learns sparse aggregation coefficients, i.e., the users' scores are modeled as a sparse aggregation of their item interactions. These extensions include collective SLIM (cSLIM), relaxed collective SLIM (rcSLIM), side information induced SLIM (fSLIM), and side information induced double SLIM (f2SLIM). They have in common that they are linear models created from collaborative interaction data, which are constrained by multimedia side information. This can be achieved in different ways: aggregation coefficients can either be learned in a way so that user–item interactions and item side information should be reproducible by the same aggregation coefficients (collective SLIM, cSLIM), the aggregation coefficient matrix can be represented as linear combinations of the item feature vectors and user preferences (fSLIM), or a combination of both strategies (f2SLIM).

- **Collaborative-filtering enriched content-based filtering (CFeCBF):** Deldjoo et al. [38] propose a two-stage recommendation approach to learn the weights of a feature weighting CBF approach based on interaction data [14]. The key insight of their approach is the semantic enrichment of the cold items (i.e., items with no/few interactions), leveraged to address the cold-start problem. In particular, a two-stage model is proposed that trains CF on warm items (i.e., items with sufficient interactions in the training phase) and leverages the learned model to recommend cold movie items in the test phase. The authors test their approach on a wide range of visual and audio content features extracted from movie trailers together with metadata (tags and genre) named *Movie Genome*. The evaluation results show the merits of their approach for recommending cold items.

## 4.2  Neural Approaches

Multimedia recommenders can significantly benefit from using tools from deep learning, not just for feature extraction (see Sect. 3), but also for the recommendation model itself. The modular nature of deep learning algorithms allows for the easy integration of different data sources, such as the content features, the event history (past behavior) of the users, and (bandit) feedback collected from interactions with the recommender [163]. It is easy to connect different neural networks, use one or more as the input of others. Therefore, adding a new aspect of the items (such as the content features) to an existing deep learning-based recommender system can be easier than adding the same information to traditional recommendation algorithms. Combining the aforementioned data sources is crucial since purely content-based recommenders often underperform compared to collaborative filtering ones. However, on the other hand, pure CF cannot leverage the rich information derived from the multimedia content and often ignores essential connections between the items. Collaborative filtering is also susceptible to the cold-start problem, meaning that it cannot recommend any item with no (or only a few) events associated with it.

The distinction between "recommending [an item]" and "recommending to [a user]" is an important question: the former means that the item in question is shown as a recommendation for some requests, while the latter means that when the item in question is considered as part of a specific request (i.e., recommendations are tailored to each users' request), it can be used to influence recommendations. The relative importance of these two aspects depends on the domain, but for most of the domains being able to recommend a new item is less important than recommending a new user. The reason for this is that it is usually possible for the users to quickly discover new items through other means, for example through a specific recommender widget focusing on new items or using search or landing on the page of the new items from external search engines. However, if the new item cannot influence recommendations due to cold-start, item-to-item recommendations cannot be provided; the quality of session-based recommendations will be hurt severely since sessions are usually short; user history-based recommendations are also affected but to a lesser degree. Therefore when we discuss cold-start in this section, we will mostly focus on the "recommending to [a user]" aspect. We will explicitly mention when we talk about the "recommending [an item]" aspect.

This section focuses on techniques that help utilize content features together with behavior data in recommender systems.

**How to Train a Complex Model (End-to-End vs. Multi-Phase Learning)?** One of the first questions during the design of a hybrid multimedia recommender is whether the content features are learned at the same time as the user preferences or separately because it affects the architecture of the deep neural network. Most of the algorithms fall into two categories. They either utilize end-to-end learning, meaning that content features, preferences, and other representations are trained in the same network simultaneously. The other choice is to pretrain a separate neural network and use it to create the content features before preference learning.

*Multi-Phase Learning* Different parts of the model are trained separately. The most common setup is to create designated neural networks for feature extraction and pretrain those. Subsequently, features can be generated using these pretrained networks, and these features can be used as (part of) the training set for a separate network that learns the preferences. Since the training of the feature extracting network is disjoint from preference learning, these models need some form of objective other than preferences to train on.

One standard solution is to utilize neural networks developed for other supervised tasks, such as image classification. The idea is that the activations of the hidden layers of these kinds of networks are good generic representations of the input rather than task-specific ones, and only the final layer that predicts the label is specific for the classification task. Therefore, usually, the activations of the final hidden layer (i.e., the penultimate layer of the entire network) are used as the representation. Another way to do pretraining is to use unsupervised methods, e.g., Word2Vec [105].[4] These methods learn the representation of their inputs so that related inputs will have similar representations. Whether two inputs are related is defined by rules set by the algorithm. For example, Word2Vec considers two words to be related if they frequently appear close to one another in the document(s). Multi-phase learning is often the solution in practical systems due to resource constraints and scalability. The feature extractor part of the model does not require the same retraining frequency as the preference estimator, but it is often much more expensive to train the feature extractor part. With multi-phase learning, one can train the expensive feature extractor a few times every few months and use this to generate features from the content; meanwhile, the cheaper preference estimator can be trained several times a day on the cached feature vectors that we precomputed using the fully trained feature extractor. Another advantage of this setup is that it is easy to cache the feature vectors, which is often necessary for fast inference. The main disadvantage of multi-phase learning is that the learned representations are not entirely in line with the task; therefore, systems using this technique are usually inferior to end-to-end learning-based methods in terms of recommendation accuracy. In Table 3 we have pointed readers to the most common pretrained or fine-tuned networks adopted in the RS research.

*End-to-End Learning* Training a recommender model end-to-end means that the feature extractor and the preference estimator parts are merged into a single model and every parameter of this joint model is trained at the same time. If the feature extractor part has a lot of parameters or a complex structure, it is often a good idea to warm up these parts, so that the joint model will estimate the preference on somewhat sensible representations instead of random ones. One way to do this is called fine-tuning: pretrained parameters of the feature extractor part can be inputted as its initial parameters. These parameters can be acquired by pretraining and then they will be changed (i.e. fine-tuned) by the end-to-end learning. Another way is performing some form of warm-up at the beginning of the training. End-to-end learning is usually very accurate since the learned representations are task-specific. However, using this approach on the scale can be challenging. Training is expensive and has to be done frequently (due to changing preferences). Also, direct inference through the complex architecture of the network might be slow. Therefore caching the activations of the feature extractors is often vital. Readers are referred to research on MMRSs employing end-to-end learning here, e.g., [86] for music recommendation, [26, 109] for generic photos recommendation, [89] for outfit recommendation, and [19] for video recommendation.

---

[4] https://code.google.com/archive/p/word2vec/.

**How to Adapt Features Using Attention?** Content features are usually extracted using the appropriate architecture (e.g., convolutional networks for images) with a simple learning task, such as classification. The features come from internal layers before the final layer and they are generally a good representation of the content, but they are still somewhat tuned towards the original task that was used to learn them. Besides, the length of pretrained features rarely matches with the dimensionality of the recommender model. In many cases, it is a good idea to adapt these features in the recommender part of the model. We already discussed one way of doing this, which is fine-tuning. However, fine-tuning changes the original pretrained features, which might not be ideal for the specific target application. For example, if we want to use content features of cold-start items in the model during inference time, it is crucial that the model trains on pretrained features that are consistent with the ones given to it during inference. Since fine-tuning cannot be applied during inference the same way as during the training, it should not be used in this scenario. Besides this shortcoming, it might be also a good idea to focus on different parts of the content features, depending on the actual user and context, which requires a different approach.

A different way to adapt the features is using an attention mechanism [145]. Attention is a generic term in deep learning describing many different techniques. The common part of these techniques is that the input is transformed using learnable weights or weight functions. Different attention techniques use different weighting functions, some reduce the number of inputs, some keep them the same. Nowadays, the most common technique formulates attention using a query (Q), keys (K), and values (V); and a portion of the literature refers to this technique when mentioning attention.

Attention can be used to address several tasks/challenges in MMRSs, from reducing dimensionality to aggregating items to personalized reweighting of features. It is important to choose a technique that matches what is aimed to achieve during the algorithm design. In the following, we provide a brief guide to some common techniques that are useful during recommendations, and we describe some of the best practices. However, the in-depth discussion of various attention mechanisms falls outside of the scope of the chapter.

**Alignment and Dimensionality Reduction** The length of multimedia content features usually does not match the dimensionality used in the recommendation model as user and context- feature vectors tend to have much smaller dimensionality. An easy way to reduce the dimensionality from $d$ to $k$ is to multiply the content feature vector with a weight matrix of size $d \times k$. Even if the dimensionality of the content feature matches the model ($d = k$), adding this extra parameter matrix to the model can serve as an alternative for the fine-tuning process. Instead of modifying the features during training, this parameter matrix is modified to adapt the content features to the recommender model. This technique—while dealing with learnable weights—is usually not considered to be an attention mechanism due to its simplicity.

**Aggregating Items** Having multiple items associated with a user or session and aggregating them into a single vector (i.e., user or session history) is common in recommender systems. It is possible to use a global weight matrix to do this, but personalized or context-aware aggregation [127] requires a more sophisticated solution and this is where attention comes into play. The techniques can be classified into *location-based* and *content-based*. Lets assume that there are $N$ content feature vectors $\{x_i\}_{i=1}^N$ ($x_i \in \mathscr{R}^d$) that should be aggregated into a single vector $f \in \mathscr{R}^d$ and let $X \in \mathscr{R}^{N \times d}$ so the $i$th row of $X$ is $x_i^T$.

*Location-based attention* techniques commonly use some form of sequential processing to compute the neural network weights, such as recurrent neural networks or masked CNNs where the weight depends on the input processed and its location in the sequence. For example, an RNN can be used to compute the weights as $c_i, h_i = \mathrm{RNN}(x_i, h_{i-1})$, then these weights are normalized using softmax $a_i = \frac{e^{c_i}}{\sum_{j=1}^N e^{c_j}}$ and $f = \sum_{i=1}^N a_i x_i$. Additional information, such as context or user embedding can also be provided to the RNN, so those can also influence the weight computation.

*Content-based attention* is usually formulated using a query (Q), keys (K), and values (V). The core of the technique is that the normalized similarities between the query and the keys serve as weights, which are then used to weight the values. Similarities can be computed in many ways, the most common technique nowadays uses scaled dot product, thus $A = \mathrm{softmax}\left(\frac{QK^T}{\sqrt{d_k}}\right)$, where softmax is applied in a row-wise manner and $d_k$ is the dimensionality of the key. Much like in natural language processing, recommender systems also derive both $K$ and $V$ from the same source using linear transformation. In the context of our example both the keys and values are derived from the content feature vectors ($K = XW_K$ and $V = XW_V$, $W_K \in \mathscr{R}^{d \times d_k}$, $W_V \in \mathscr{R}^{d \times d_v}$). The user (or session or context) feature vector can serve as the query ($Q = u^T W_Q$, $u \in \mathscr{R}^{d_u}$, $W_Q \in \mathscr{R}^{d_k \times d_u}$). The aggregated content feature is computed as $f^T = AV$. Note that we can also reduce the dimensionality of the content features in the same step by setting $d_v < d$. The idea behind using content-based attention for aggregation is that the more similar the user feature to the item representations, the higher weight should be assigned to that specific item because the similarity in the latent space is often associated with preference. Note that what we described above is a single attention head for content-based attention, but multiple of such attentions can be used with different $W_K, W_Q, W_V$ matrices to get different aggregations of the same set of content features. These can be aggregated again by simple methods, such as averaging or used at a different point in the model.

**Reweighting Specific Features** It is possible that some of the content features are more important for the recommender than others. The content feature vector can be reweighted by multiplying it element-wise with a global weight vector ($x' = x \circ w$). A more sophisticated vector would be to have this weight vector depend on the user (or session or context) feature vector and compute it via a function, such as $w_u = \tanh(Wu + b)$ and $x_u = x \circ w_u$. It is also possible to use content-based attention for this task by computing interaction between the user and content features as $A_u = \text{softmax}\left(\frac{QK^T}{\sqrt{d_k}}\right)$, $Q = u^T W_Q$, $K = X' W_K$, where each column of $X'$ is $x$ and $x_u = x \circ A_u^T$.

*How to Assemble Deep Learning Modules Fusing Data Sources?* As already mentioned, one advantage of neural approaches is that they can easily leverage heterogeneous data sources, such as different aspects of the content (e.g. text, image) and behavior data in the same model. Different representations for the same item can be built based on these data sources. Fusing different content representations is fairly trivial. We can use operators such as concatenation, element-wise sum or product (with optional dimension reduction when needed). However, fusing content-based and collaborative filtering representations is not always straightforward due to the latter not always being available due to cold start. Not every model architecture fuses content and behavior (event) data on the representation level. For example, the model could use behavior data for user–item interaction only and use content data to create item representations. However, fusing on the representation level is worth discussing since it is a common problem to solve for many models, and different techniques influence the cold-start compliance of the model. Common techniques include [78, 120]:

- **Initializing:** The features extracted from the content are used to initialize item features. In this case, there is no separate CF-based representation in the model, and thus the behavior data might be used to fine-tune the representations. Representations can be either fixed during training or fine-tuned using behavior data. With fixed representations, the model can tackle the cold-start problem, but its accuracy on non-cold-start items may be suboptimal. If the initial representations are modified during training, the model loses its ability to handle the cold-start problem but becomes better on non-cold-start items.
- **Aggregation:** The content and CF-based item representation are generated separately and then aggregated into a single representation. Common methods include concatenation, element-wise product, or (weighted) sum. The model has limited capacity to handle the cold-start problem. On the one hand, the feature vector can be computed even if the CF part is missing. However, these vectors are not present during training, therefore, other model parameters are not fully optimized for the cold start. Still, the model is able to compute scores for cold-start items, even if they are suboptimal.
- **Joint training:** A single item representation is trained for a joint multi-task learning optimization problem, where one of the tasks is related to the behavior data (e.g., preference prediction) and the other is to content data (e.g., reconstruction). The model will not be able to handle the cold-start problem, since behavior data at training time is required to compute the item representation.
- **Regularization:** Content-based representations are computed before CF-based ones either by pretraining, or they are updated in the model before CF-based representations. A term is added to the loss function that is based on the difference between the content and the CF representations. This way the CF representations

are regularized to be close to their content-based counterparts, thus the model can work fairly well with new items since content-based representations are somewhat compatible with the CF representations used by the model. However, performance on cold-start items might be sub-optimal. This approach can be viewed as the relaxed version of joint training because the content of behavior-based item features does not have to be exactly identical, but should be close in terms of similarity.

• **Mapping content:** After learning the CF-based representations in the preference estimator, a separate network is used to map the content (or content features) of the training items to their CF representation. This separate mapping network can be used to create representation for cold-start items during inference time that is compatible with CF-based representations used by the model (see [111]).

### 4.2.1 Answers to Challenges

(C1) Representations learned for multimedia content by neural approaches are latent feature vectors in the vast majority of models. Interaction with users can be as simple as computing the dot product between the content representation and a learned user feature vector, or the user and content features can be used in a classifier that estimates the preference or more complex approaches can be utilized, such as personalized attention on the content feature vector, etc.

(C2) If the model utilizes CF-based and content-based representations for the same items, those can be fused in many different ways from aggregation and concatenation to mapping (see Sect. 4.2).

(C3) Context can be added by modifying the structure of the non-context-aware model. There are various methods, but this is not always a straightforward approach to perform and often results in designing a new model architecture.

(C4) Representation can be learned by either multi-phase or end-to-end learning. Both approaches have their advantages and disadvantages: multi-phase learning requires fewer resources and it is easier to apply on scale, but end-to-end learning learns task-specific representations and is, therefore, more accurate.

(C5) If the model does not rely on CF-based item representations and the feature extractor part of the model managed to learn sensible representations, the discrepancy between training and testing becomes irrelevant. The model architecture should always be tailored to the domain and the problem we would like to solve. If the cold-start problem is prominent with the specific task, the final model should not build representations based on behavior data, but rather use this data somewhere else in the model, such as to model user–item interactions, or to train/fine-tune the feature extractor subnetwork.

(C6) Doing quick model updates is not straightforward, but with the appropriate model architecture, it is not even needed.

## 4.3   Graph-Based Approaches

Besides neural approaches, recent works propose to model the user–item interactions with a graph structure and incorporate the multimedia information with the collaborative signal. Conducting some carefully designed operations over the graph enables us to capture direct and/or indirect collaborative signals and inject them into the representations of users and items. We group state-of-the-art graph-based methods into three categories—random walk-based [7, 15, 70, 147, 157], graph neural network-based [18, 85, 150, 159], and graph autoencoder [142, 155, 162] approaches, which are summarized in Table 6.

### 4.3.1   Random Walk-Based Method

Both CF and CBF recommendation methods are built upon the assumption that a user's preference vector should be close to the representation vector of items he or she adopted before. For example, Personal PageRank [66] extends the standard random walk algorithm in PageRank [114] to implement personalized recommendation. Specifically, given a graph $\mathcal{G} = \{\mathcal{V}, \mathcal{E}\}$, $\mathcal{V}$ is the set of vertices and $\mathcal{E}$ is the set of edges. For instance, $\mathcal{V}$ may represent the set of songs in a music catalog and edges $\mathcal{E}$ between nodes represent co-listening behavior of users, i.e., there is an edge between two songs' vertices if at least one user listened to both songs. Each random walk starts from a root node $v \in \mathcal{V}$ and continually moves to the next node with probability $\alpha$. Iteratively performing that operation, each node $u$ achieves a convergent value denoted as $\mathbf{PR}(u)$, which reflects the affinity between the root node $v$ and end node $u$. It is formally defined as,

$$\mathbf{PR}(u) = (1 - \alpha) \, r_u + \alpha \sum_{i \in \mathcal{N}(u)} \frac{\mathbf{PR}(i)}{\|\mathcal{N}(i)\|}, \quad r_u = \begin{cases} 0, & u = v \\ 1, & u \neq v \end{cases} \quad (1)$$

where $\mathcal{N}(u)$ denotes the set of neighbors of node $u$, which indicates the nodes immediately connect to $u$.

For instance, in personalized video recommendation, Baluja et al. [7] argue that the high-degree nodes (e.g., popular items) in the user–video graph negatively affect the user–item similarity computing, since, for one path, it has much larger probability to end with the high-degree nodes than other nodes. For example, there is the shortest path between two users due to a popular video co-watched by them. Even though they both have no interest in the video, the method tends to recommend the video watched by one of them to the other. To alleviate the disadvantage of the model based on the shortest path detection, Baluja et al.'s method suggests a video to a user if:

1. there is a short path between them;
2. there are several paths between them;
3. there are paths avoiding high-degree nodes.

**Table 6** Summary of graph-based multimedia recommendation approaches. In the column "Input", we summarize the external information fed into models except for the historical interaction records

| Category | Approach | Graph | Input | Output | Cold start |
|---|---|---|---|---|---|
| Random walk | Adsorption algorithm [7] | Co-view/user-video graph | Item content | Ratings | × |
| | Tripartite propagation [15] | User-query-video graph | Query content | Ratings | × |
| | Co-ranking algorithm [157] | User-user and tweet-tweet and user-tweet graphs | Tweet content | Ratings | × |
| | BiRank [70] | User-aspects-item graph | User reviews | Ratings | × |
| | BGE & GES [147] | Direct item graph | Session-based user behaviors and side-information | Ranking | ✓ |
| Graph convolutional network | ACF [18] | User-item graph | Item content | Ranking | × |
| | PinSage [159] | Pin-board graph | Item content | Ranking | × |
| | MMGCN [150] | Modal-specific user-video graph | Micro-video content | Ranking | × |
| | HFGN [85] | Hierarchical user-outfit-item graph | Item content | Ranking | × |
| Graph Autoencoder | GCMC [142] | User-item graph | Item Content | Ratings | × |
| | STAR-GCN [162] | User-item graph | Item content | Ratings | ✓ |
| | GCM [155] | User-context-item graph | Context information and side-information | Ratings | ✓ |

Therefore, the authors propose a novel absorption algorithm to implement the video recommendation on a co-view video graph, where edges between two nodes indicate whether the corresponding videos are co-watched by some users. In the co-view graph, the weight of the edge, denoted as $w(u, v)$, is the number of users who interacted with both videos $u$ and $v$. In addition, one shadow node is created for each user/item node and connects them via an edge with weight one. The algorithm takes a starting vertex $v$ for the random walk, moves to the next node with

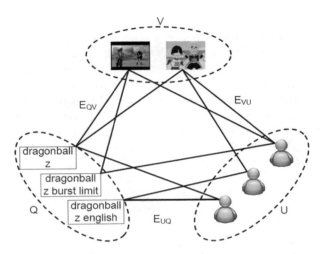

**Fig. 6** The user–query–video tripartite graph. Courtesy of [15]

$w(u, v)/\sum_{i\in\mathcal{N}(v)} w(i, v)$, and outputs a video recommendation distribution $L_v$ for vertex $v$ when it reaches a shadow node, which is treated as the absorbing state.

Rather than only exploring the user–video graph, Chen et al. [15] investigate the user–query and query–video connections in the video click-through prediction. As a result, a tripartite graph over (user, video, query) is introduced, as shown in Fig. 6, which provides the potential bridge to connect users and videos. As such, the Tripartite Propagation algorithm is proposed to exploit the tripartite graph. In this approach, video nodes in the graph act as the message sources and continuously send their ids as the message along the edges to their neighbors. During the iterative propagation procedure, a trade-off parameter $\alpha$ is used to control the behavior of video nodes on how they treat the received messages. It is formulated as $\mathbf{V} = (1 - \alpha)\mathbf{V} + \alpha\mathbf{I}$, which simulates the random walk algorithm to update the representation of the video node. After a few iterations, the messages of user nodes are learned, which contain the recommended videos' identifiers for the users.

In the same period, Yan et al. [157] also construct a heterogeneous graph that connects the users and their posts for personalized tweet recommendation. Beyond the links between tweets and their authors, the user–user and tweet–tweet links present the behaviors of users following or followed by others and content similarity of tweets, respectively. Following the idea that there is a mutually reinforcing relationship between authors and tweets, a co-ranking algorithm is presented to reflect the rankings of tweets. Technically, the model couples two random walks, one on the tweet graph and the other on the author graph, into a unified one, which extends the random walk method to individual users and produces personalized recommendations. Different from Personal PageRank, the co-ranking algorithm represents users with their topic preference and encodes the representations into a simple random walk to implement the personalized recommendation.

Analyzing the disadvantages of existing works, He et al. [70] develop a generic method, termed BiRank, to specifically address the ranking problem on bipartite graphs. It accounts for both the graph's structural information and the proper incorporation of any prior information for vertices, where such vertex priors can be used to encode any features of vertices. Further, the method is generalized to n-partite graphs and deployed on a personalized ranking scenario. Compared with the seminal-based algorithms, like PageRank and HITS, the proposed model uses symmetric normalization to score the vertices, which follows the smoothness convention. In particular, it smooths the edge weight by the degree of its two connected vertices simultaneously:

$$\mathbf{V}_v = \sum_{u \in \mathcal{N}(v)} \frac{\omega_{vu}}{\sqrt{d_v}\sqrt{d_u}} \mathbf{V}_u, \tag{2}$$

where $\sqrt{d_v}$ and $\sqrt{d_u}$ are the degrees of vertices $v$ and $u$, respectively. In addition, $\mathbf{V}_v$ and $\mathbf{V}_u$ denote the representations of vertices $v$ and $u$, respectively. Inspired by word2vec, Perozzi et al. [116] propose a DeepWalk framework, which incorporates random walk and skip-gram algorithms to learn the embedding of each node in the graph. Wang et al. [147] transfer this idea into the e-commerce recommendation domain. To address the technical challenges (i.e., scalability, sparsity, and cold start) in Taobao,[5] they propose two novel graph embedding models, which are dubbed base graph embedding (BGE) and graph embedding with side information (GES). Their method constructs the item graph based upon user behaviors in sessions and adopts the DeepWalk method to learn the embedding of each node in the item graph. Moreover, the embedding vector of each node is aggregated with its side information to address the cold start problem of new items.

### 4.3.2 Graph Convolutional Network-Based Method

Inspired by the success of convolutional neural networks in the computer vision domain, researchers extended the convolutional operations to design the architecture of graph convolutional networks (GCNs) for graph-structured data representation learning. For each node in the graph, GCNs iteratively aggregate the information from its local graph neighborhoods and encode the information into its embedding vector to gain the informative representation [148]. This can be formulated as

$$\mathbf{h}_v^{(l)} = \phi(\mathbf{W}^{(l)} \cdot aggregate(\mathbf{h}_v^{(l-1)}, \{\mathbf{h}_u^{(t-1)}, \forall u \in \mathcal{N}(v)\}), \tag{3}$$

---

[5] https://world.taobao.com/.

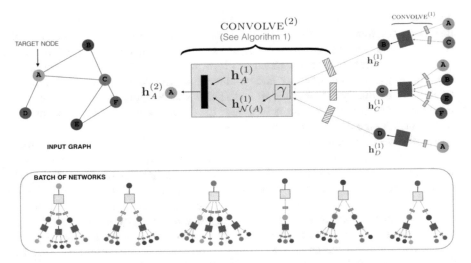

**Fig. 7** Overview of the PinSage architecture. Courtesy of [159]

where $\mathbf{h}_v^{(l)}$ ($\mathbf{h}_u^{(l)}$) represents the representation of vertex $v$ ($u$) at $l$-th layer and *aggregate* denotes the function aggregating the neighborhood information.

Benefiting from the powerful representation ability, GCN-based models have been widely used in RS applications. Chen et al. [18] develop a hierarchical attention neural network to address the challenge of leveraging item- and component-level implicit feedback for multimedia recommendation. Their model is dubbed attentive collaborative filtering (ACF). In this model, each user collects the item- and component-level features from his or her historical items to enhance the representation. Although only the direct neighboring items are aggregated to enrich the representation, ACF also achieves a significant improvement with respect to the hit ratio, which demonstrates the effectiveness of GCN-based models in multimedia recommendation.

To alleviate the neighbor explosion problem, Ying et al. [159] propose a data-efficient graph convolutional network algorithm (PinSage) to implement a large-scale deep recommendation engine, as shown in Fig. 7. The method combines random walk and graph convolutional operations to generate embeddings of nodes, which contain both the content features and local structure information of items. Besides, they design a novel training strategy that relies on harder-and-harder training examples to improve the robustness and convergence of the model. By deploying the proposed method on the Pinterest dataset with billions of nodes, the authors demonstrate that the method can effectively represent the nodes for high-quality recommendations.

More recently, micro-video, as an emerging social multimedia form, has attracted increasing attention in academia and industry [74, 83, 138, 149–151]. In the micro-video domain, the multi-modal graph convolutional network (MMGCN) [150] has been proposed to provide a quality micro-video recommendation service. In this work, the multi-modal information is incorporated into historical interactions between users and micro-videos to establish an in-depth understanding of user preference. Specifically, as illustrated in Fig. 8, MMGCN constructs the user–item bipartite graph in each modality and conducts the graph convolutional operations on them individually, which aims to distinguish and consider model-specific user preference. Moreover, the authors introduce the shared user and item embeddings, which interchange the information in different modalities.

Another research direction where GCNs are used is personalized fashion recommendation. Li et al. [85] propose a hierarchical fashion graph neural network (HFGN) upon a hierarchical user–outfit–item graph for personalized outfit recommendation. Distinct from traditional item recommendation, the personalized outfit recommendation should satisfy two requirements: compatibility of fashion and consistency with personal taste. As such, the graph consists of three levels (i.e., user, outfit, and item levels), where each level contains a corresponding type of nodes, and connects the cross-level nodes according to the historical records. To refine the representation of the user, outfit, and item, HFGN performs embedding propagation operations from the bottom level to the top one, which models the compatibility of the outfit, aggregates item semantics into outfit embeddings, and integrates historical outfits as user representations.

### 4.3.3 Graph Autoencoder-Based Method

Integrating GCNs into the autoencoder framework, graph autoencoders represent (encode) each node by injecting its local topological information and reconstruct (decode) the interactions of node pairs by computing the distance of their representation. The encoder could be defined as

$$\mathbf{Z} = GCN(\mathbf{X}, \mathbf{A}) \tag{4}$$

and the decoder is defined as

$$\hat{\mathbf{A}} = \theta(\mathbf{Z}\mathbf{Z}^{\mathsf{T}}), \tag{5}$$

where $\mathbf{X}$, $\mathbf{A}$, and $\mathbf{Z}$ denote the input feature vectors, adjacent matrix, and encoded node representation, respectively. Treating the recommendation problem as link prediction task, Berg et al. [142] propose a new graph-based autoencoder framework, dubbed graph convolutional matrix completion, as shown in Fig. 9. It learns the latent features of user and item nodes through a form of message passing on the bipartite interaction graph. Moreover, side information can be combined with the learned user and item embeddings to reconstruct the rating links through a bilinear decoder. Based on this idea, Zhang et al. propose the stacked and reconstructed graph convolutional network (STAR-GCN) [162] to learn node representations for boosting performance. Compared with GCMC, STAR-GCN represents users and

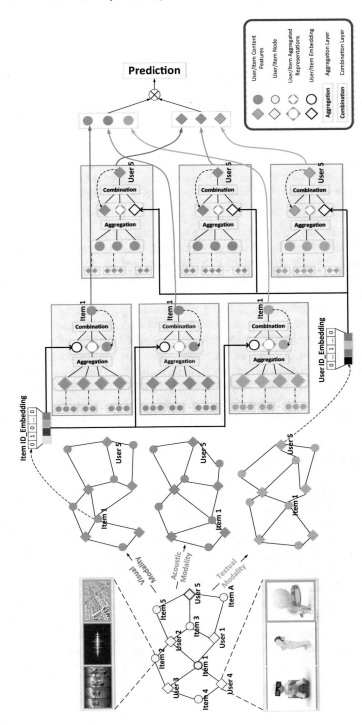

**Fig. 8** Schematic illustration of MMGCN. Courtesy of [150]

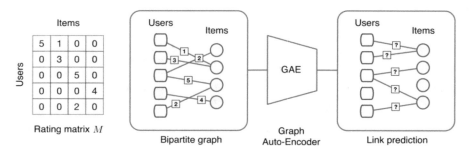

**Fig. 9** Illustration of the GCMC framework. Personalized recommendation can be cast as a link prediction problem and implemented using a graph autoencoder. Courtesy of [142]

items not only in the transductive setting but in the inductive one to tackle the cold start problem. As such, the approach randomly masks some percentage of the input nodes and then reconstructs the clean node embedding. Moreover, the authors propose a training strategy to solve the problem of label leakage, which leads to substantial performance improvement.

Despite the remarkable performance of standard graph convolutions, Wu et al. [155] argue that edges between the user and item nodes could carry context information. Therefore, they propose an end-to-end framework named graph convolutional machine (GCM). In particular, their framework consists of three components: (1) an encoder which projects user, item, and context features into embedding vectors; (2) graph convolutional layers which refine the representation of the user and item with context-aware graph convolutions on the user–item graph; (3) a decoder that predicts the interactions of user–item pairs under certain contexts.

### 4.3.4   Answers to Challenges

In summary, the graphs defined in the aforementioned approaches consist of the user and item nodes, where the presence of an edge between the user and item nodes signals interaction between them. As such, constructing the graph upon consumed records of users to items could be used to model their interactions. The item nodes contain the multi-modal features and the user nodes represent the users' preferences to the content (C1). By conducting the information aggregation operations on the graph, each node not only collects the feature distribution of its neighbors but captures the local structure around it. Moreover, the multi-modal features and collaborative signals are fused and encoded into the centric nodes during graph constructions (C2). Beyond users and items, some other types of information also could be organized into the graph. For instance, context information can be used as features of edges, or side information can be integrated as a new type of node into the graph (C3).

Despite the significant progress of graph-based recommendation methods, there are still some challenges. For instance, in multimodal feature extraction, the item features currently tend to be captured from the content information by some pre-trained deep learning model, like ResNet [67] or BERT [48]. However, such models are designed and trained for other tasks (e.g., image classification and natural language processing), different from personalized recommendation. Therefore, how to leverage the graph-based model to guide the effective feature extraction is still challenging (C4).

**Discussion**  *How to choose the most suitable class of MMRS algorithm?*

Graphs are data structures that can be ingested by various algorithms, notably deep neural networks, learning to uncover complex relationships between users, items, or user and items. Let us assume a fashion recommendation scenario, where the goal is for a given skirt liked by the user before, to recommend a complete fashion set (e.g., shoes, top, and bag). The first idea can be to obtain a latent representation of fashion items by using a bidirectional LSTM. However, such models do not necessarily lead to a good representation since selection of fashion items in an outfit may not have a fixed order.

For RS and in particular MMRSs, the key to success is to find a proper representation learning strategy for users and items, typically performed in some latent space. Classical recommendation models treat each (user, item, preference) instance as an independent instance. Although implicitly capturing the collaborative signal hidden in the user–item interactions, classical models ignore the important high-order collaborative signal aggregated from the multi-hop neighbors of each node, which should be taken into account in the representation learning for users and items.

As a remedy, different types of graph learning methods can be applied to explore and exploit relationships among instances.

One of the main issues with graph-based models is their size. For instance, a social network with millions of users would result in a graph of millions of nodes and perhaps billions of edges. This causes high computational costs for the companies if remained unaddressed.

# 5   Conclusions and Open Challenges

In this chapter, we surveyed the state-of-the-art research related to multimedia recommender systems (MMRS), focusing on methods that integrate item or user side information into a hybrid recommender. We presented methods to extract features and create item representations based on image, video, and audio modalities. An MMRS then leverages these features (together with user–item interaction data) to recommend either (1) media items from which the features have been extracted, or (2) non-media items using the features that have been extracted from a proxy multimedia representation of the item (e.g., images of clothes). In terms of recommendation approaches, we discussed traditional hybrid methods, neural approaches, and graph-based approaches; and we detailed how each of these addresses a list of crucial challenges we defined related to warm-start and cold-start scenarios.

In addition to the technical challenges we discussed in Sect. 2, we identify the following general grand challenges in MMRS:

- **Exploiting additional kinds of user–item interaction:** Current recommender systems are limited in the type of interaction data they leverage, i.e., only explicit or implicit preference indications, such as ratings, clicks, or skips, are used. In practice, however, there exist many more forms of interaction, which are addressed in HCI research, for instance. These include looking at a particular part of a movie scene (revealed by eye movements), zooming into an RoI of an image (using the mouse wheel), or increasing the volume when listening to a certain motif in a classical music piece (via keyboard shortcuts). Whether and how such additional forms of interaction can and should be leveraged in MMRS is subject to future research.
- **Considering multi-modality:** Most MMRSs leverage a single modality, even in cases where the item is composed of multi-channel data. For instance, the vast majority of video recommendation systems leverage only the visual channel. While this is often due to computational limitations, it still neglects the wealth of information present in other related modalities, e.g., for music, these include text (genre, biographies, tags), audio (the recording itself), image (band photographs, album covers), and video (music video clip). Investigating the usefulness of such multi-modal representations and side information and elaborating highly multi-modal MMRS remains an open challenge [20, 138].
- **Multi-modal conversational information seeking:** Conversational information seeking (CIS) research has typically concentrated on uni-modal interactions (e.g., textual) and information items. Human conversations, on the other hand, are well-known to be multi-modal. These characteristics necessitate the development of CIS systems that use multi-modal products and interact with users through channels with multiple modalities [46, 87].
- **Cold-start recommendation** The ability to suggest cold items (items with insufficient interactions) is the main advantage of MMRSs compared to behavior-based RSs such as CF. A key challenge is to learn effective item representation in the presence of a discrepancy between training and test items, as cold items exist in the test phase. Therefore, finding representation for cold and non-cold items in the testing stage remains an open research area [38, 50].
- **Security, privacy and biases:** Recent research revealed remarkable successes of adversarial attacks to against different MMRSs, e.g. [5, 25, 49, 137]. How to devise countermeasures to such attacks and research ways to improve the security of MMRSs, thereby alleviating justified privacy concerns of users, remains a significant challenge. More information on this topic is provided in the recent survey by Deldjoo et al. [44], and the chapter on Adversarial RSs "Adversarial Recommender Systems: Attack, Defense, and Advances". Finally, detecting

biases and mitigating fairness in MMRSs and generally content-driven RS is a topic that has received much less attention compared to CF models, see e.g., [1, 40, 42, 82].

- **Improving transparency and explainability:** Fueled by the current trend towards using deep neural networks in MMRS, systems are perceived increasingly as "black boxes", and decisions made by them are no longer understandable by the end users. However, being able to interpret the results made by an MMRS increases trust in and engagement with the system, as reported in several works [6, 59, 124]. Therefore, comprehensive explanation models tailored to MMRSs and their users' requirements are urgently needed. Readers are referred to the chapter on Explanation in RSs "Beyond Explaining Single Item Recommendations" for an overview of recent advances in the field.

Bridging the areas of multimedia and recommender systems will surely lead to further new and exciting research endeavors.

**Acknowledgments** We would like to thank Jan Schlüter for providing image examples.

# References

1. H. Abdollahpouri, R. Burke, B. Mobasher, Managing popularity bias in recommender systems with personalized re-ranking, in *The Thirty-Second International Flairs Conference* (2019)
2. T. Alashkar, S. Jiang, S. Wang, Y. Fu, Examples-rules guided deep neural network for makeup recommendation, in *AAAI* (2017), pp. 941–947
3. M. Albanese, A. d'Acierno, V. Moscato, F. Persia, A. Picariello, A multimedia recommender system. ACM Trans. Int. Technol. **13**(1), 1–32 (2013)
4. I. Andjelkovic, D. Parra, J. O'Donovan, Moodplay: interactive music recommendation based on artists' mood similarity. Int. J. Human-Comput. Stud. **121**, 142–159 (2019)
5. V.W. Anelli, Y. Deldjoo, T.D. Noia, D. Malitesta, F.A. Merra, A study of defensive methods to protect visual recommendation against adversarial manipulation of images, in *The 44th International ACM SIGIR Conference on Research and Development in Information Retrieval, SIGIR*, vol. 21 (2021), pp. 11–15
6. K. Balog, F. Radlinski, Measuring recommendation explanation quality: The conflicting goals of explanations, in *Proceedings of the 43rd International ACM SIGIR Conference on Research and Development in Information Retrieval SIGIR*, Virtual, (ACM, New York, 2020), pp. 329–338
7. S. Baluja, R. Seth, D. Sivakumar, Y. Jing, J. Yagnik, S. Kumar, D. Ravichandran, M. Aly, Video suggestion and discovery for youtube: taking random walks through the view graph, in *Proceedings of the 17th International Conference on World Wide Web* (2008), pp. 895–904
8. I. Bartolini, V. Moscato, R.G. Pensa, A. Penta, A. Picariello, C. Sansone, M.L. Sapino, Recommending multimedia objects in cultural heritage applications, in *International Conference on Image Analysis and Processing* (Springer, Berlin, 2013), pp. 257–267
9. H. Bay, T. Tuytelaars, L.V. Gool, Surf: Speeded up robust features, in *European Conference on Computer Vision* (Springer, Berlin, 2006), pp. 404–417
10. S. Bourke, K. McCarthy, B. Smyth, The social camera: A case-study in contextual image recommendation, in *Proceedings of the 16th International Conference on Intelligent User Interfaces* (ACM, New York, 2011), pp. 13–22
11. S. Boutemedjet, D. Ziou, A graphical model for context-aware visual content recommendation. IEEE Trans. Multimedia **10**(1), 52–62 (2008)

12. J. Bu, S. Tan, C. Chen, C. Wang, H. Wu, L. Zhang, X. He, Music recommendation by unified hypergraph: combining social media information and music content, in *Proceedings of the 18th International Conference on Multimedia 2010, Firenze, Italy, October 25–29, 2010.* (ACM, New York, 2010), pp. 391–400
13. L. Canini, S. Benini, R. Leonardi, Affective recommendation of movies based on selected connotative features. IEEE Trans. Circ. Syst. Video Technol. **23**(4), 636–647 (2013)
14. L. Cella, S. Cereda, M. Quadrana, P. Cremonesi, Deriving item features relevance from past user interactions, in *Proceedings of the 25th Conference on User Modeling, Adaptation and Personalization, UMAP 2017, Bratislava, Slovakia, July 09–12, 2017* (ACM, New York, 2017), pp. 275–279
15. B. Chen, J. Wang, Q. Huang, T. Mei, Personalized video recommendation through tripartite graph propagation, in *Proceedings of the 20th ACM International Conference on Multimedia* (2012), pp. 1133–1136
16. C.-M. Chen, M.-F. Tsai, J.-Y. Liu, Y.-H. Yang, Using emotional context from article for contextual music recommendation, in *Proceedings of the 21st ACM International Conference on Multimedia, MM '13, New York, NY* (ACM, New York, 2013), pages 649–652
17. J. Chen, H. Zhang, X. He, L. Nie, W. Liu, T.-S. Chua, Attentive collaborative filtering: Multimedia recommendation with item- and component-level attention, in *SIGIR* (ACM, New York, 2017), pp. 335–344
18. J. Chen, H. Zhang, X. He, L. Nie, W. Liu, T.-S. Chua, Attentive collaborative filtering: Multimedia recommendation with item-and component-level attention, in *Proceedings of the 40th International ACM SIGIR Conference on Research and Development in Information Retrieval*, (2017), pp. 335–344
19. X. Chen, P. Zhao, J. Xu, Z. Li, L. Zhao, Y. Liu, V.S. Sheng, Z. Cui, Exploiting visual contents in posters and still frames for movie recommendation. IEEE Access **6**, 68874–68881 (2018)
20. X. Chen, H. Chen, H. Xu, Y. Zhang, Y. Cao, Z. Qin, H. Zha, Personalized fashion recommendation with visual explanations based on multimodal attention network: Towards visually explainable recommendation, in *Proceedings of the 42nd International ACM SIGIR Conference on Research and Development in Information Retrieval* (2019), pp. 765–774
21. C.-W. Chen, L. Yang, H. Wen, R. Jones, V. Radosavljevic, H. Bouchard, Podrecs: Workshop on podcast recommendations, in *Fourteenth ACM Conference on Recommender Systems, RecSys '20, New York, NY* (Association for Computing Machinery, New York, 2020), pp. 621–622
22. H.-Y. Chi, C.-C. Chen, W.-H. Cheng, M.-S. Chen, Ubishop: commercial item recommendation using visual part-based object representation. Multimedia Tools Appl. **75**(23), 16093–16115 (2016)
23. W.-T. Chu, Y.-L. Tsai, A hybrid recommendation system considering visual information for predicting favorite restaurants. World Wide Web **20**(6), 1313–1331 (2017)
24. K.-Y. Chung, Effect of facial makeup style recommendation on visual sensibility. Multimedia Tools Appl. **71**(2), 843–853 (2014)
25. R. Cohen, O. Sar Shalom, D. Jannach, A. Amir, A black-box attack model for visually-aware recommender systems, in *Proceedings of the 14th ACM International Conference on Web Search and Data Mining* (2021), pp. 94–102
26. B. Cui, A.K.H. Tung, C. Zhang, Z. Zhao, Multiple feature fusion for social media applications, in *Proceedings of the 2010 ACM SIGMOD International Conference on Management of Data* (ACM, New York, 2010), pp. 435–446
27. P. Cui, Z. Wang, Z. Su, What videos are similar with you?: Learning a common attributed representation for video recommendation, in *Proceedings of the 22nd ACM International Conference on Multimedia* (ACM, New York, 2014), pp. 597–606

28. J. Dai, Y. Li, K. He, J. Sun, R-FCN: Object detection via region-based fully convolutional networks, in *Advances in Neural Information Processing Systems 29: Annual Conference on Neural Information Processing Systems 2016, December 5–10, 2016, Barcelona* (2016), pp. 379–387
29. N. Dalal, B. Triggs, Histograms of oriented gradients for human detection, in *IEEE Computer Society Conference on Computer Vision and Pattern Recognition, 2005. CVPR 2005*, vol. 1 (IEEE, Piscataway, 2005), pp. 886–893
30. Y. Deldjoo, M. Elahi, P. Cremonesi, Using visual features and latent factors for movie recommendation, in *CBRecSys@ RecSys* CEUR-WS (2016)
31. Y. Deldjoo, M. Elahi, P. Cremonesi, F. Garzotto, P. Piazzolla, M. Quadrana, Content-based video recommendation system based on stylistic visual features. J. Data Semantics **5**(2), 99–113 (2016)
32. Y. Deldjoo, C. Frà, M. Valla, A. Paladini, D. Anghileri, M. Anil Tuncil, F. Garzotta, P. Cremonesi, et al., Enhancing children's experience with recommendation systems, in *Workshop on Children and Recommender Systems (KidRec'17)-11th ACM Conference of Recommender Systems* (2017), pp. N–A
33. Y. Deldjoo, M. Elahi, M. Quadrana, P. Cremonesi, Using visual features based on MPEG-7 and deep learning for movie recommendation. Int. J. Multim. Inf. Retr. **7**(4), 207–219 (2018)
34. Y. Deldjoo, M. Gabriel Constantin, H. Eghbal-Zadeh, B. Ionescu, M. Schedl, P. Cremonesi, Audio-visual encoding of multimedia content for enhancing movie recommendations, in *Proceedings of the 12th ACM Conference on Recommender Systems, RecSys 2018, Vancouver, BC, October 2–7, 2018* (ACM, New York, 2018), pp. 455–459
35. Y. Deldjoo, M. Gabriel Constantin, B. Ionescu, M. Schedl, P. Cremonesi, MMTF-14k: A multifaceted movie trailer feature dataset for recommendation and retrieval, in *Proceedings of the 9th ACM Multimedia Systems Conference* (ACM, New York, 2018), pp. 450–455
36. Y. Deldjoo, M. Schedl, P. Cremonesi, G. Pasi, Content-based multimedia recommendation systems: Definition and application domains, in *Proceedings of the 9th Italian Information Retrieval Workshop, Rome, May, 28–30, 2018*. CEUR Workshop Proceedings, vol. 2140. CEUR-WS.org (2018)
37. Y. Deldjoo, M. Schedl, B. Hidasi, P. Knees, Multimedia recommender systems, in ed. by S. Pera, M.D. Ekstrand, X. Amatriain, J. O'Donovan, *Proceedings of the 12th ACM Conference on Recommender Systems, RecSys 2018, Vancouver, BC, October 2–7, 2018* (ACM, New York, 2018), pp. 537–538
38. Y. Deldjoo, M. Ferrari Dacrema, M. Gabriel Constantin, H. Eghbal-zadeh, S. Cereda, M. Schedl, B. Ionescu, P. Cremonesi, Movie genome: alleviating new item cold start in movie recommendation. User Model. User Adapt. Interact. **29**(2), 291–343 (2019)
39. Y. Deldjoo, M. Schedl, Retrieving relevant and diverse movie clips using the MFVCD-7K multifaceted video clip dataset, in *2019 International Conference on Content-Based Multimedia Indexing, CBMI 2019, Dublin, September 4–6, 2019* (IEEE, Piscataway, 2019), pp. 1–4
40. Y. Deldjoo, V.W. Anelli, H. Zamani, A. Bellogin, T. Di Noia, A flexible framework for evaluating user and item fairness in recommender systems. User Model. User-Adap. Interac., 1–55 (2021)
41. Y. Deldjoo, M. Schedl, P. Cremonesi, G. Pasi, Recommender systems leveraging multimedia content. ACM Comput. Surv. **53**(5), 38 (2020)
42. Y. Deldjoo, A. Bellogin, T. Di Noia, Explaining recommender systems fairness and accuracy through the lens of data characteristics. Inf. Proc. Manag. **58**, 102662 (2021)
43. Y. Deldjoo, T. Di Noia, D. Malitesta, F. Antonio Merra, A study on the relative importance of convolutional neural networks in visually-aware recommender systems, in *CVPRW-CVFAD 2021 :The 4th CVPR Workshop on Computer Vision for Fashion, Art, and Design*. CVPR Proceedings (2021)
44. Y. Deldjoo, T. Di Noia, F. Antonio Merra, A survey on adversarial recommender systems: from attack/defense strategies to generative adversarial networks. ACM Comput. Surveys **54**(2), 1–38 (2021)

45. Y. Deldjoo, M. Schedl, P. Knees, Content-Driven Music Recommendation: Evolution, State of the Art, and Challenges. Preprint arXiv: 2107.11803 (2021)
46. Y. Deldjoo, J.R. Trippas, H. Zamani, Towards multi-modal conversational information seeking, in *Proceedings of the 44th International ACM SIGIR Conference on Research and Development in Information Retrieval* (2021)
47. Z. Deng, J. Sang, C. Xu, Personalized video recommendation based on cross-platform user modeling, in *2013 IEEE International Conference on Multimedia and Expo (ICME)* (IEEE, Piscataway, 2013), pp. 1–6
48. J. Devlin, M.-W. Chang, K. Lee, K. Toutanova, Bert: Pre-training of deep bidirectional transformers for language understanding. Preprint arXiv:1810.04805 (2018)
49. T. Di Noia, D. Malitesta, F. Antonio Merra, Taamr: Targeted adversarial attack against multimedia recommender systems, in *2020 50th Annual IEEE/IFIP International Conference on Dependable Systems and Networks Workshops (DSN-W)* (IEEE, Piscataway, 2020), pp. 1–8
50. X. Du, X. Wang, X. He, Z. Li, J. Tang, T.-S. Chua, How to learn item representation for cold-start multimedia recommendation? in *Proceedings of the 28th ACM International Conference on Multimedia* (2020), pp. 3469–3477
51. X. Du, H. Yin, L. Chen, Y. Wang, Y. Yang, X. Zhou, Personalized video recommendation using rich contents from videos. IEEE Trans. Knowl. Data Eng. **32**(3), 492–505 (2020)
52. M.D. Ekstrand, J.T. Riedl, J.A. Konstan, et al., Collaborative filtering recommender systems. Foundations Trends® Human–Comput. Int. **4**(2), 81–173 (2011)
53. M. Elahi, Y. Deldjoo, F. Bakhshandegan Moghaddam, L. Cella, S. Cereda, P. Cremonesi, Exploring the semantic gap for movie recommendations, in *Proceedings of the Eleventh ACM Conference on Recommender Systems* (ACM, New York, 2017), pp. 326–330
54. D. Elsweiler, C. Trattner, M. Harvey, Exploiting food choice biases for healthier recipe recommendation, in *Proceedings of the 40th International ACM SIGIR Conference on Research and Development in Information Retrieval* (ACM, New York, 2017), pp. 575–584
55. J.A. Fails, M.S. Pera, N. Kucirkova, F. Garzotto, International and interdisciplinary perspectives on children & recommender systems (kidrec), in *Proceedings of the 17th ACM Conference on Interaction Design and Children* (2018), pp. 705–712
56. A. Farseev, L. Nie, M. Akbari, T.-S. Chua, Harvesting multiple sources for user profile learning: A big data study, in *Proceedings of the 5th ACM on International Conference on Multimedia Retrieval* (ACM, New York, 2015), pp. 235–242
57. A. Farseev, I. Samborskii, A. Filchenkov, T.-S. Chua, Cross-domain recommendation via clustering on multi-layer graphs, in *Proceedings of the 40th International ACM SIGIR Conference on Research and Development in Information* (ACM, New York, 2017), pp. 195–204
58. C. Feichtenhofer, A. Pinz, A. Zisserman, Convolutional two-stream network fusion for video action recognition, in *2016 IEEE Conference on Computer Vision and Pattern Recognition, CVPR 2016, Las Vegas, NV, June 27–30, 2016* (IEEE Computer Society, Washington, DC, 2016), pp. 1933–1941
59. G. Friedrich, M. Zanker, A taxonomy for generating explanations in recommender systems. AI Mag. **32**(3), 90–98 (2011)
60. X. Gao, F. Feng, X. He, H. Huang, X. Guan, C. Feng, Z. Ming, T.-S. Chua, Hierarchical attention network for visually-aware food recommendation. IEEE Trans. Multim. **22**(6), 1647–1659 (2020)
61. X. Geng, H. Zhang, J. Bian, T.-S. Chua, Learning image and user features for recommendation in social networks, in *2015 IEEE International Conference on Computer Vision, ICCV 2015, Santiago, December 7–13, 2015* (2015), pp. 4274–4282
62. R.B. Girshick, Fast R-CNN, in *2015 IEEE International Conference on Computer Vision, ICCV 2015, Santiago, December 7–13, 2015* (IEEE Computer Society, Washington, DC, 2015), pp. 1440–1448
63. R.B. Girshick, J. Donahue, T. Darrell, J. Malik, Rich feature hierarchies for accurate object detection and semantic segmentation, in *2014 IEEE Conference on Computer Vision and*

*Pattern Recognition, CVPR 2014, Columbus, OH, June 23–28, 2014* (IEEE Computer Society, Washington, DC, 2014), pp. 580–587

64. X. Gu, L. Shou, P. Peng, K. Chen, S. Wu, G. Chen, iGlasses: A novel recommendation system for best-fit glasses, in *Proceedings of the 39th International ACM SIGIR Conference on Research and Development in Information Retrieval* (ACM, New York, 2016), pp. 1109–1112

65. S.C. Guntuku, S. Roy, L. Weisi, Personality modeling based image recommendation, in *International Conference on Multimedia Modeling* (Springer, Berlin, 2015), pp. 171–182

66. T.H. Haveliwala, Topic-sensitive pagerank: a context-sensitive ranking algorithm for web search. IEEE Trans. Knowl. Data Eng. **15**(4), 784–796 (2003)

67. K. He, X. Zhang, S. Ren, J. Sun, Deep residual learning for image recognition, in *2016 IEEE Conference on Computer Vision and Pattern Recognition, CVPR 2016, Las Vegas, NV, June 27–30, 2016* (IEEE Computer Society, Washington, DC, 2016), pp. 770–778

68. R. He, J. McAuley, Ups and downs: Modeling the visual evolution of fashion trends with one-class collaborative filtering, in *Proceedings of the 25th International Conference on World Wide Web* (2016), pp. 507–517

69. R. He, J. McAuley, VBPR: visual bayesian personalized ranking from implicit feedback, in *Proceedings of the Thirtieth AAAI Conference on Artificial Intelligence, February 12-17, 2016, Phoenix, Arizona* (2016), pp. 144–150

70. X. He, M. Gao, M.-Y. Kan, D. Wang, Birank: towards ranking on bipartite graphs. IEEE Trans. Knowl. Data Eng. **29**(1), 57–71 (2016)

71. X. He, H. Zhang, T.-S. Chua, Recommendation technologies for multimedia content, in *ICMR* (2018), p. 8

72. M. Hou, L. Wu, E. Chen, Z. Li, V.W. Zheng, Q. Liu, Explainable fashion recommendation: A semantic attribute region guided approach, in *Proceedings of the Twenty-Eighth International Joint Conference on Artificial Intelligence, IJCAI 2019, Macao, August 10–16, 2019*. ijcai.org (2019), pp. 4681–4688

73. G. Huang, Z. Liu, L. van der Maaten, K.Q. Weinberger, Densely connected convolutional networks, in *2017 IEEE Conference on Computer Vision and Pattern Recognition, CVPR 2017, Honolulu, HI, July 21–26, 2017* (IEEE Computer Society, Washington, DC, 2017), pp. 2261–2269

74. H. Jiang, W. Wang, Y. Wei, Z. Gao, Y. Wang, L. Nie, What aspect do you like: Multi-scale time-aware user interest modeling for micro-video recommendation, in *Proceedings of the 28th ACM International Conference on Multimedia* (2020), pp. 3487–3495

75. M. Kaminskas, F. Ricci, M. Schedl, Location-aware music recommendation using auto-tagging and hybrid matching, in *Proceedings of the 7th ACM Conference on Recommender Systems* (ACM, New York, 2013), pp. 17–24

76. W.-C. Kang, C. Fang, Z. Wang, J.J. McAuley, Visually-aware fashion recommendation and design with generative image models, in *ICDM* (IEEE Computer Society, Washington, DC, 2017), pp. 207–216

77. A. Karpathy, G. Toderici, S. Shetty, T. Leung, R. Sukthankar, F.-F. Li, Large-scale video classification with convolutional neural networks, in *2014 IEEE Conference on Computer Vision and Pattern Recognition, CVPR 2014, Columbus, OH, June 23–28, 2014* (IEEE Computer Society, Washington, DC, 2014), pp. 1725–1732

78. R. Kaur, S. Kautish, Multimodal sentiment analysis: A survey and comparison. Int. J. Serv. Sci. Manag. Eng. Technol. **10**(2), 38–58 (2019)

79. P. Knees, M. Schedl, *Music Similarity and Retrieval: An Introduction to Audio- and Web-based Strategies*. The Information Retrieval Series, vol. 36 (Springer, Berlin, 2016)

80. A. Krizhevsky, I. Sutskever, G.E. Hinton, Imagenet classification with deep convolutional neural networks, in *Advances in Neural Information Processing Systems* (2012), pp. 1097–1105

81. C.-T. Li, M.-K. Shan, Emotion-based impressionism slideshow with automatic music accompaniment, in *Proceedings of the 15th ACM International Conference on Multimedia, MM '07, New York, NY*, (ACM, New York, 2007), pp. 839–842

82. J. Li, K. Lu, Z. Huang, H.T. Shen, Two birds one stone: On both cold-start and long-tail recommendation, in *Proceedings of the 25th ACM International Conference on Multimedia* (2017), pp. 898–906
83. Y. Li, M. Liu, J. Yin, C. Cui, X.-S. Xu, L. Nie, Routing micro-videos via a temporal graph-guided recommendation system, in *Proceedings of the 27th ACM International Conference on Multimedia* (2019), pp. 1464–1472
84. X. Li, X. Wang, X. He, L. Chen, J. Xiao, T.-S. Chua, Hierarchical fashion graph network for personalized outfit recommendation. CoRR abs/2005.12566 (2020)
85. X. Li, X. Wang, X. He, L. Chen, J. Xiao, T.-S. Chua, Hierarchical fashion graph network for personalized outfit recommendation, in *Proceedings of the 43rd International ACM SIGIR Conference on Research and Development in Information Retrieval* (2020)
86. D. Liang, M. Zhan, D.P.W. Ellis, Content-aware collaborative music recommendation using pre-trained neural networks, in ed. by M. Müller, F. Wiering, *Proceedings of the 16th International Society for Music Information Retrieval Conference, ISMIR 2015, Málaga, October 26–30, 2015* (2015), pp. 295–301
87. L. Liao, L. Le Hong, Z. Zhang, M. Huang, T.S. Chua, MMConv: an environment for multimodal conversational search across multiple domains, in *Proceedings of the 44th International ACM SIGIR Conference on Research and Development in Information Retrieval* (2021)
88. Z. Lin, G. Ding, J. Wang, Image annotation based on recommendation model, in *Proceedings of the 34th international ACM SIGIR conference on Research and development in Information Retrieval* (ACM, New York, 2011), pp. 1097–1098
89. Y. Lin, M. Moosaei, H. Yang, Outfitnet: Fashion outfit recommendation with attention-based multiple instance learning, in *WWW '20: The Web Conference 2020, Taipei, April 20–24, 2020* (ACM/IW3C2, New York/Geneva, 2020), pp. 77–87
90. J. Liu, Z. Li, J. Tang, Y. Jiang, H. Lu, Personalized geo-specific tag recommendation for photos on social websites. IEEE Trans. Multim. **16**(3), 588–600 (2014)
91. W. Liu, A. Rabinovich, A.C. Berg, Parsenet: Looking wider to see better. CoRR abs/1506.04579 (2015)
92. W. Liu, D. Anguelov, D. Erhan, C. Szegedy, S.E. Reed, C.-Y. Fu, A.C. Berg, SSD: Single shot multibox detector, in *Computer Vision - ECCV 2016 - 14th European Conference, Amsterdam, October 11–14, 2016, Proceedings, Part I*. Lecture Notes in Computer Science, vol. 9905 (Springer, Berlin, 2016), pp. 21–37
93. Q. Liu, S. Wu, L. Wang, Deepstyle: Learning user preferences for visual recommendation, in *SIGIR* (ACM, New York, 2017), pp. 841–844
94. L. Liu, J. Chen, P. Fieguth, G. Zhao, R. Chellappa, M. Pietikäinen, From bow to CNN: Two decades of texture representation for texture classification. Int. J. Comput. Vision **127**(1), 74–109 (2019)
95. B. Logan, Mel frequency cepstral coefficients for music modeling, in *ISMIR 2000, 1st International Symposium on Music Information Retrieval, Plymouth, Massachusetts, October 23–25, 2000, Proceedings* (2000)
96. D.G. Lowe, Distinctive image features from scale-invariant keypoints. Int. J. Comput. Vision **60**(2), 91–110 (2004)
97. H. Luo, J. Fan, D.A. Keim, S. Satoh, Personalized news video recommendation, in *International Conference on Multimedia Modeling* (Springer, Berlin, 2009), pp. 459–471
98. J. Ma, G. Li, M. Zhong, X. Zhao, L. Zhu, X. Li, LGA: latent genre aware micro-video recommendation on social media. Multim. Tools Appl. **77**(3), 2991–3008 (2018)
99. J. McAuley, C. Targett, Q. Shi, A. Van Den Hengel, Image-based recommendations on styles and substitutes, in *Proceedings of the 38th International ACM SIGIR Conference on Research and Development in Information Retrieval* (ACM, New York, 2015), pp. 43–52
100. B. McFee, G.R.G. Lanckriet, The natural language of playlists, in ed. by A. Klapuri, C. Leider, *Proceedings of the 12th International Society for Music Information Retrieval Conference, ISMIR 2011, Miami, Florida, October 24–28, 2011* (University of Miami, Coral Gables, 2011), pp. 537–542

101. T. Mei, B. Yang, X.-S. Hua, L. Yang, S.-Q. Yang, S. Li, Videoreach: An online video recommendation system, in *Proceedings of the 30th Annual International ACM SIGIR Conference on Research and Development in Information Retrieval* (ACM, New York, 2007), pp. 767–768
102. T. Mei, B. Yang, X.-S. Hua, S. Li, Contextual video recommendation by multimodal relevance and user feedback. ACM Trans. Inf. Syst. **29**(2), 10 (2011)
103. P. Melville, R.J. Mooney, R. Nagarajan, Content-boosted collaborative filtering for improved recommendations, in *Eighteenth National Conference on Artificial Intelligence*, (American Association for Artificial Intelligence, Menlo Park, 2002), pp. 187–192
104. L. Meng, F. Feng, X. He, X. Gao, T.-S. Chua, Heterogeneous fusion of semantic and collaborative information for visually-aware food recommendation, in *MM '20: The 28th ACM International Conference on Multimedia, Virtual Event/Seattle, WA, October 12–16, 2020* (ACM, New York, 2020), pp. 3460–3468
105. T. Mikolov, K. Chen, G. Corrado, J. Dean, Efficient estimation of word representations in vector space, in *1st International Conference on Learning Representations, ICLR 2013, Scottsdale, Arizona, May 2–4, 2013* (2013)
106. M. Müller, *Fundamentals of Music Processing: Audio, Analysis, Algorithms, Applications* (Springer, Berlin, 2015)
107. X. Ning, G. Karypis, SLIM: sparse linear methods for top-n recommender systems, in D.J. Cook, J. Pei, W. Wang, O.R. Zaïane, X. Wu, *11th IEEE International Conference on Data Mining, ICDM 2011, Vancouver, BC, December 11–14, 2011* (IEEE Computer Society, Washington, DC, 2011), pp. 497–506
108. X. Ning, G. Karypis, Sparse linear methods with side information for top-n recommendations, in *Proceedings of the Sixth ACM Conference on Recommender Systems*, RecSys '12 New York, NY, (Association for Computing Machinery, New York, 2012), pp. 155–162
109. W. Niu, J. Caverlee, H. Lu, Neural personalized ranking for image recommendation, in *Proceedings of the Eleventh ACM International Conference on Web Search and Data Mining* (ACM, New York, 2018), pp. 423–431
110. T. Ojala, M. Pietikainen, T. Maenpaa, Multiresolution gray-scale and rotation invariant texture classification with local binary patterns. IEEE Trans. Pattern Analy. Mach. Intell. **24**(7), 971–987 (2002)
111. A.v.d. Oord, S. Dieleman, B. Schrauwen, Deep content-based music recommendation, in ed. by C. Burges, L. Bottou, M. Welling, Z. Ghahramani, K. Weinberger, *Advances in Neural Information Processing Systems 26 (NIPS)* (Curran Associates, Lake Tahoe, NV, 2013), pp. 2643–2651
112. S. Oramas, V.C. Ostuni, T. Di Noia, X. Serra, E. Di Sciascio, Sound and music recommendation with knowledge graphs. ACM Trans. Intell. Syst. Technol. **8**(2), 21:1–21:21 (2016)
113. S. Oramas, O. Nieto, M. Sordo, X. Serra, A deep multimodal approach for cold-start music recommendation, in *Proceedings of the 2Nd Workshop on Deep Learning for Recommender Systems*, DLRS 2017, New York, NY (ACM, New York, 2017), pp. 32–37
114. L. Page, S. Brin, R. Motwani, T. Winograd, The pagerank citation ranking: Bringing order to the web. Technical report, Stanford InfoLab (1999)
115. O.M. Parkhi, A. Vedaldi, A. Zisserman, Deep face recognition, in *Proceedings of the British Machine Vision Conference 2015, BMVC 2015, Swansea, September 7–10, 2015* (BMVA Press, Swansea, 2015), pp. 41.1–41.12
116. B. Perozzi, R. Al-Rfou, S. Skiena, Deepwalk: Online learning of social representations, in *Proceedings of the 20th ACM SIGKDD International Conference on Knowledge Discovery and Data Mining* (2014), pp. 701–710
117. L. Peska, H. Trojanova, Towards recommender systems for police photo lineup. Preprint arXiv:1707.01389 (2017)
118. J. Pons, X. Serra, musicnn: Pre-trained convolutional neural networks for music audio tagging. Preprint arXiv:1909.06654 (2019)
119. L.R. Rabiner, B.-H. Juang, *Fundamentals of Speech Recognition*. Prentice Hall Signal Processing Series (Prentice Hall, Hoboken, 1993)

120. D. Ramachandram, G.W. Taylor, Deep multimodal learning: a survey on recent advances and trends. IEEE Signal Proc. Mag. **34**(6), 96–108 (2017)
121. Y.S. Rawat, M.S. Kankanhalli, Clicksmart: a context-aware viewpoint recommendation system for mobile photography. IEEE Trans. Circuits Syst. Video Techn. **27**(1), 149–158 (2017)
122. S. Ren, K. He, R.B. Girshick, J. Sun, Faster R-CNN: towards real-time object detection with region proposal networks, in *Advances in Neural Information Processing Systems 28: Annual Conference on Neural Information Processing Systems 2015, December 7–12, 2015, Montreal, Quebec* (2015), pp. 91–99
123. S. Rendle, Factorization machines with libFM. ACM Trans. Intell. Syst. Technol. **3**(3), 57 (2012)
124. M.T. Ribeiro, S. Singh, C. Guestrin, Why should i trust you?, in *Proceeding of the International Conference on Knowledge Discovery and Data Mining (KDD)* (ACM, New York, 2016), pp. 1135–1144
125. S. Roy, S.C. Guntuku, Latent factor representations for cold-start video recommendation, in *Proceedings of the 10th ACM Conference on Recommender Systems* (ACM, New York, 2016), pp. 99–106
126. O. Russakovsky, J. Deng, H. Su, J. Krause, S. Satheesh, S. Ma, Z. Huang, A. Karpathy, A. Khosla, M.S. Bernstein, A.C. Berg, F.-F. Li, Imagenet large scale visual recognition challenge. Int. J. Comput. Vis. **115**(3), 211–252 (2015)
127. P. Sánchez, A. Bellogín, On the effects of aggregation strategies for different groups of users in venue recommendation. Inf. Proc. Manag. **58**(5), 102609 (2021)
128. M. Schedl, H. Zamani, C.-W. Chen, Y. Deldjoo, M. Elahi, Current challenges and visions in music recommender systems research. Int. J. Multim. Inf. Retr. **7**(2), 95–116 (2018)
129. K. Seyerlehner, G. Widmer, T. Pohle, Fusing block-level features for music similarity estimation, in *Proceedings of the 13th International Conference on Digital Audio Effects (DAFx-10), Graz, September 6–10* (2010)
130. Y. Shi, M. Larson, A. Hanjalic, Collaborative filtering beyond the user-item matrix: A survey of the state of the art and future challenges. ACM Comput. Surv. **47**(1), 3 (2014)s
131. K. Simonyan, A. Zisserman, Two-stream convolutional networks for action recognition in videos, in *Advances in Neural Information Processing Systems 27: Annual Conference on Neural Information Processing Systems 2014, December 8–13 2014, Montreal, Quebec* (2014), pp. 568–576
132. K. Simonyan, A. Zisserman, Very deep convolutional networks for large-scale image recognition, in ed. by Y. Bengio, Y. LeCun, *3rd International Conference on Learning Representations, ICLR 2015, San Diego, CA,May 7–9, 2015, Conference Track Proceedings* (2015)
133. J. Smith, D. Weeks, M. Jacob, J. Freeman, B. Magerko, Towards a hybrid recommendation system for a sound library, in *Joint Proceedings of the ACM IUI 2019 Workshops co-located with the 24th ACM Conference on Intelligent User Interfaces (ACM IUI 2019), Los Angeles, March 20, 2019* (2019)
134. J. Smith, D. Weeks, M. Jacob, J. Freeman, B. Magerko, Towards a hybrid recommendation system for a sound library, in *IUI Workshops* (2019)
135. J. Song, Y. Yang, Z. Huang, H.T. Shen, J. Luo, Effective multiple feature hashing for large-scale near-duplicate video retrieval. IEEE Trans. Multim. **15**(8), 1997–2008 (2013)
136. G.-L. Sun, Z.-Q. Cheng, X. Wu, Q. Peng, Personalized clothing recommendation combining user social circle and fashion style consistency. Multim. Tools Appl. **77**(14), 17731–17754 (2018)
137. J. Tang, X. Du, X. He, F. Yuan, Q. Tian, T.-S. Chua, Adversarial training towards robust multimedia recommender system. IEEE Trans. Knowl. Data Eng. **32**, 855–867 (2019)
138. Z. Tao, Y. Wei, X. Wang, X. He, X. Huang, T.-S. Chua, MGAT: Multimodal graph attention network for recommendation. Inf. Proc. Manag. **57**(5), 102277 (2020)
139. I. Tautkute, A. Możejko, W. Stokowiec, T. Trzciński, Ł. Brocki, K. Marasek, What looks good with my sofa: Multimodal search engine for interior design, in ed. by M. Ganzha,

L. Maciaszek, M. Paprzycki, *Proceedings of the 2017 Federated Conference on Computer Science and Information Systems*. Annals of Computer Science and Information Systems, vol. 11 (IEEE, Piscataway, 2017), pp. 1275–1282

140. J.R.R. Uijlings, K.E.A. van de Sande, T. Gevers, A.W.M. Smeulders, Selective search for object recognition. Int. J. Comput. Vis. **104**(2), 154–171 (2013)

141. A. Vall, M. Dorfer, H. Eghbal-zadeh, M. Schedl, K. Burjorjee, G. Widmer, Feature-combination hybrid recommender systems for automated music playlist continuation. User Model. User-Adap. Interac. J. Personaliz. Res. **29**, 527–572 (2019)

142. R. van den Berg, T.N. Kipf, M. Welling, Graph convolutional matrix completion, in *Proceedings of the 24th ACM SIGKDD International Conference on Knowledge Discovery & Data Mining* (2018)

143. H. Wang, C. Schmid, Action recognition with improved trajectories, in *IEEE International Conference on Computer Vision, ICCV 2013, Sydney, December 1–8, 2013* (IEEE Computer Society, Washington, DC, 2013), pp. 3551–3558

144. X. Wang, Y. Wang, Improving content-based and hybrid music recommendation using deep learning, in *Proceedings of the ACM International Conference on Multimedia, MM '14, Orlando, FL, November 03–07, 2014* (ACM, New York, 2014), pp. 627–636

145. F. Wang, M. Jiang, C. Qian, S. Yang, C. Li, H. Zhang, X. Wang, X. Tang, Residual attention network for image classification, in *Proceedings of the IEEE Conference on Computer Vision and Pattern Recognition* (2017), pp. 3156–3164

146. S. Wang, Y. Wang, J. Tang, K. Shu, S. Ranganath, H. Liu, What your images reveal: Exploiting visual contents for point-of-interest recommendation, in *Proceedings of the 26th International Conference on World Wide Web* (2017), pp. 391–400

147. J. Wang, P. Huang, H. Zhao, Z. Zhang, B. Zhao, D.L. Lee, Billion-scale commodity embedding for e-commerce recommendation in alibaba, in *Proceedings of the 24th ACM SIGKDD International Conference on Knowledge Discovery & Data Mining* (2018), pp 839–848

148. X. Wang, X. He, M. Wang, F. Feng, T.-S. Chua, Neural graph collaborative filtering, in *Proceedings of the 42nd International ACM SIGIR Conference on Research and Development in Information Retrieval* (2019), pp. 165–174

149. Y. Wei, Z. Cheng, X. Yu, Z. Zhao, L. Zhu, L. Nie, Personalized hashtag recommendation for micro-videos, in *Proceedings of the 27th ACM International Conference on Multimedia* (2019), pp. 1446–1454

150. Y. Wei, X. Wang, L. Nie, X. He, R. Hong, T.-S. Chua, MMGCN: Multi-modal graph convolution network for personalized recommendation of micro-video, in *Proceedings of the 27th ACM International Conference on Multimedia* (2019), pp. 1437–1445

151. Y. Wei, X. Wang, L. Nie, X. He, T.-S. Chua, Graph-refined convolutional network for multimedia recommendation with implicit feedback, in *Proceedings of the 28th ACM International Conference on Multimedia* (2020), pp. 3541–3549

152. J. Wen, J. She, X. Li, H. Mao, Visual background recommendation for dance performances using deep matrix factorization. TOMCCAP **14**(1), 11:1–11:19 (2018)

153. Z. Wu, S. Jiang, Q. Huang, Friend recommendation according to appearances on photos, in ed. by W. Gao, Y. Rui, A. Hanjalic, C. Xu, E.G. Steinbach, A. El-Saddik, M.X. Zhou, *Proceedings of the 17th International Conference on Multimedia 2009, Vancouver, British Columbia, October 19–24, 2009* (ACM, New York, 2009), pp. 987–988

154. C.-C. Wu, T. Mei, W.H. Hsu, Y. Rui, Learning to personalize trending image search suggestion, in *Proceedings of the 37th International ACM SIGIR Conference on Research & Development in Information Retrieval* (ACM, New York, 2014), pp. 727–736

155. J. Wu, X. He, X. Wang, Q. Wang, W. Chen, J. Lian, X. Xie, Y. Zhang, Graph convolution machine for context-aware recommender system. Preprint arXiv:2001.11402 (2020)

156. Z. Xing, M. Parandehgheibi, F. Xiao, N. Kulkarni, C. Pouliot, Content-based recommendation for podcast audio-items using natural language processing techniques, in *2016 IEEE International Conference on Big Data (Big Data)* (IEEE, Piscataway, 2016), pp. 2378–2383

157. R. Yan, M. Lapata, X. Li, Tweet recommendation with graph co-ranking, in *Proceedings of the 50th Annual Meeting of the Association for Computational Linguistics (Volume 1: Long Papers)* (2012), pp. 516–525

158. L. Yang, C.-K. Hsieh, H. Yang, J.P. Pollak, N. Dell, S. Belongie, C. Cole, D. Estrin, Yumme: a personalized nutrient-based meal recommender system. ACM Trans. Inf. Syst. **36**(1), 7 (2017)

159. R. Ying, R. He, K. Chen, P. Eksombatchai, W.L. Hamilton, J. Leskovec, Graph convolutional neural networks for web-scale recommender systems, in *Proceedings of the 24th ACM SIGKDD International Conference on Knowledge Discovery & Data Mining* (2018), pp. 974–983

160. K. Yoshii, M. Goto, K. Komatani, T. Ogata, H.G. Okuno, IEEE Trans. Audio Speech Language Proc. **16**(2), 435–447 (2008)

161. W. Yu, H. Zhang, X. He, X. Chen, L. Xiong, Z. Qin, Aesthetic-based clothing recommendation, in *Proceedings of the 2018 World Wide Web Conference* (2018), pp. 649–658

162. J. Zhang, X. Shi, S. Zhao, I. King, Star-gcn: Stacked and reconstructed graph convolutional networks for recommender systems, in *The 28th International Joint Conference on Artificial Intelligence* (2019), pp. 4264–4270

163. S. Zhang, L. Yao, A. Sun, Y. Tay, Deep learning based recommender system: a survey and new perspectives. ACM Comput. Surv. **52**(1), 1–38 (2019)

164. X. Zhao, H. Luan, J. Cai, J. Yuan, X. Chen, Z. Li, Personalized video recommendation based on viewing history with the study on youtube, in *Proceedings of the 4th International Conference on Internet Multimedia Computing and Service* (ACM, New York, 2012), pp. 161–165

165. B. Zhou, À. Lapedriza, J. Xiao, A. Torralba, A. Oliva, Learning deep features for scene recognition using places database, in *Advances in Neural Information Processing Systems 27: Annual Conference on Neural Information Processing Systems 2014, December 8–13 2014, Montreal, Quebec* (2014), pp. 487–495

166. Q. Zhu, M.-L. Shyu, H. Wang, Videotopic: Content-based video recommendation using a topic model, in *2013 IEEE International Symposium on Multimedia (ISM)* (IEEE, Piscataway, 2013), pp. 219–222

167. U. Zoelzer, *Digital Audio Signal Processing* (Wiley, Hoboken, 2008)

# Fashion Recommender Systems

**Shatha Jaradat, Nima Dokoohaki, Humberto Jesús Corona Pampín, and Reza Shirvany**

## 1 Introduction

Fashion is a broad domain where the expectations and needs of customers are very diverse, from buying mass-produced pieces such as blue jeans, to experiencing *haute couture* pieces that have been elevated to art, to the must-haves of every season that become ubiquitous in social networks. More importantly, through the history, and still nowadays, fashion is a tool of self expression and identity. Wearing a t-shirt with a message is still as important now as the color purple, white and green was for the suffragettes movement in the early twentieth century. Fashion is also ever-changing, guided by the seasons, and trends. What is fashionable today will most certainly not be fashionable tomorrow. Finally, fashion still plays a very functional need in the lives of billions of people, who choose what to wear every day based on many different factors such as the weather, comfort or occasion. This makes fashion one of the most interesting and challenging applications of recommender systems, specially thanks to the explosion of online e-commerce retailers, online personal styling services, and fashion-focused social networks.

Within this context, recommender systems can generally be applied to solve several core user problems. The most well-known problem within the recommender systems literature is that of a single item recommendation. In fashion, single item

S. Jaradat · N. Dokoohaki (✉)
KTH - Royal Institute of Technology, Stockholm, Sweden
e-mail: shatha@kth.se; nimad@kth.se

H. J. Corona Pampín
Spotify, Amsterdam, The Netherlands
e-mail: humbertoc@spotify.com

R. Shirvany
Zalando SE, Berlin, Germany
e-mail: reza.shirvany@zalando.de

© Springer Science+Business Media, LLC, part of Springer Nature 2022      1015
F. Ricci et al. (eds.), *Recommender Systems Handbook*,
https://doi.org/10.1007/978-1-0716-2197-4_26

recommendations have two main applications; Firstly, users often need to find an alternative to a given item that has the same function, while looking and feeling similarly. For example, recommender systems can help a user finding a suitable alternative to an evening dress the user likes, or finding the same model of running shoes in a different colors. Secondly, users rarely think about what to wear in isolation, so the complementary item recommendations enable users find fashion items that complement a given one. For example, hiking boots that go with a given hiking jacket, or the perfect tie for a given suit. We provide deeper insights on these problems and the proposed solutions in Sect. 2.

In the last decade, the way fashion is consumed and marketed has had a huge shift towards social networks, both in importance and volume. *Fashion influencers* now have millions of followers and have even become brand themselves. With the affluence of content in social networks, social recommendations play a big role on how people discover and consume fashion. We discuss the main trends in Sect. 3.

Recent advances in the field of size and fit, discussed in Sect. 4 enable customers to find fashion not only matching their style but also in their size, in an otherwise complex, confusing and non-standardized sizing environment. Indeed, size and fit is considered as one of the biggest customer problems in online fashion. It plays a critical role in tackling the challenge of article returns due to wrong sizing, and can thus strongly improve the customers' shopping experience, the environmental footprint, and the profitability of online fashion retailers.

Finally, outfit recommendations, discussed in Sect. 5 enable users to find the complete set of clothes they can wear for a given occasion, all matching the same style, or finding the perfect fashion piece to complement a partially built outfit. Outfit recommendations combine the complexities of all aforementioned types of fashion recommendations as it can be described as the task of deciding and recommending compatible single clothing items that can be personalised with respect to size and fit individual measures and that can be reflecting preferences of individuals from their favorite fashion influencers in social media.

Research in the fashion domain is built on various in-house and public datasets trying to represent the diversity of the fashion domain. Section 6 presents an overview of such datasets that can be used for the research of the most fashion recommender systems domains. To conclude, Sect. 7 presents an outlook of the future of the field and discusses challenges and opportunities ahead based on the studies presented in this work.

Most of the current fashion recommendations state-of-the-art that is explored in this chapter is based on deep learning approaches. The recent advances in deep learning have been changing in a significant way the recommendation systems that were based on matrix/tensor factorization and traditional computer vision methods. This is explained by the ability of deep learning in overcoming the obstacles of the conventional recommendation methods and outperforming their performance. The power of using deep learning techniques in recommendation systems stems from the fact that they allow for enhanced feature extraction from items (such as image, video, text, audio). This consequently allows for a more accurate modeling of items and can strongly affect the content-based and hybrid recommendation methods performance when compared to the traditional techniques.

# 2 Single Fashion Item Recommendations

Single item recommendations are widely used by users of e-commerce platforms to find items they want to buy and wear; it helps them narrow down their selection or easily compare items in a visual way. For example, it might help users find items that are very similar visually, but in a different material, items that are cheaper than a given one, from a known brand, or items that are labeled as sustainable. In this section, we discuss recent advances in solving the problem of single item recommendations that are specific to the fashion domain. Particularly, similar fashion item recommendations in Sect. 2.1, and complementary fashion item recommendations in Sect. 2.2

## *2.1 Similar Fashion Item Recommendations*

The problem of similar item recommendations in fashion can be defined as follows: Given a user $u$ at a given moment of time $t$, and an item defined as anchor item $i \in I$, the set of all items, the goal of similar recommendations is to find a subset of items $\overline{I} \subset I$ that achieve the same functionality for user $u$ as the original item $i$. From an intuitive perspective, we can describe items in $\overline{I}$ as being part of the same clothing category (e.g, t-shirts, jeans, etc), being from the same color (e.g, light-pink, black), being from the same brand, or belonging to the same clothing style as the anchor item $i$. Moreover, in cases in which the order in which recommendations are displayed, instead of a set, $\overline{I}$ can be defined as an ordered list.

A visual example is presented in Fig. 1, which is taken from Zalando [1], a popular European online fashion e-commerce platform. Here, the anchor item $i$, which is a t-shirt belonging to the *Pride* collection, is shown in the top-left corner. In this figure, the recommended similar items $\overline{I}$, are presented in a carousel under the heading *"How about these?, Similar Items"*. All the items displayed as similar are also t-shirts, of different colors, and all of them belong to the *Pride* theme.

There are several approaches that researchers and practitioners have taken to solve the problem of Similar Item recommendations in Fashion; these can be classified depending on the information used to generate recommendations. In Session-based Recommender systems, introduced in Sect. 2.1.1 information from the user's interaction history with the items is used when calculating the recommendation. In visual recommender systems, discussed in Sect. 2.1.2, characteristics of the items, including visual and stylistic properties are exploited to find similarities across them.

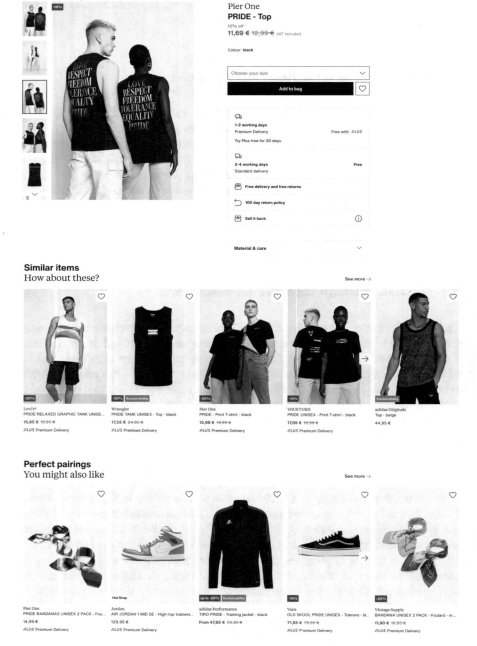

**Fig. 1** Example of a Product Display Page showing different recommendations. The similar item recommendations are shown under the heading *"How about these?: Similar Items"* The complementary item recommendations are shown under the heading *"Perfect pairings: You might also like"*. These type of recommendations allow customers to continue the fashion discovery journey by finding items that can be worn together. Source: [1]

### 2.1.1 Collaborative Filtering Similar Fashion Items Recommendations

In the area of collaborative filtering recommendations, session-based approaches that exploit a Recurrent Neural Networks architecture (RNNs), such as the introduced by Devooght et al. [2] showed success in the field of sequence modeling. The idea of using session-based approaches to solve the problem of recommendations is based on the intuition that the actions performed by the user within a session can model the inherent intent to be achieved within that session. These approaches are gaining popularity; specially in domains such as online retailers, where very often, only a limited amount of information (that of a web session) is known about a given user.

One of the first attempts of using the session-based user-item interaction sequences as input data is proposed in [2]. The authors define the recommendation problem as a prediction problem for the next action the user will perform. The proposed approach is based on a Long Short Term Memory (LSTM) Recurrent Neural Network, which allows the reformulation of the problem of recommendation into a sequence prediction problem. While this approach outperforms other state-of-the art approaches, the authors use a video and movies datasets for evaluation, which differs in nature from the fashion applications of recommender systems.

The Gated Recurrent Unit (GRU) architecture family described by Chung et al. [3] has proven to be very effective when using longer sequences to model recommendations. One of the main advances in this family of approaches is introduced by the NARM architecture (Neural Attentive session-based Recommendation Model) [4], which further increased the prediction power of GRU-based session recommender by adding an attention layer on top of the output. Despite of the high recall and Mean Reciprocal Rank (MRR), the training process of NARM takes longer time than pure GRU approaches. The Short-term attention/memory priority model for session-based recommendation introduced by Liu et al. [5] tackles the shortfalls of previous approaches by exploiting the user's current actions to generate future recommendations within the same session, on real-time or near real-time. It does so by employing an attention model that can model both the long-term session properties, while also modeling the actions from the user's last clicks to capture any short-term attention drift. The authors claim state-of-the-art performance in several e-commerce datasets that contain no metadata on the user or item level. This now popular idea approach has inspired several authors to propose variations on the base model.

More recently, Rodriguez Sanchez et al. [6] proposed improvements over the STAMP enhancing the rank of the few most relevant items from the original similar item recommendations produced by a simple item similarity [7], by employing the information of the session of the user encoded by an attention network. Together with a novel Candidate Rank Embedding that encodes the global ranking information of each candidate in the re-ranking process, this approach showed improvement in the specific task of similar items fashion recommendations, evaluated in several datasets, including an increase of performance in an online experiment in a proprietary dataset [1].

### 2.1.2   Content-Based Similar Fashion Items Recommendations

Content-Based recommender systems have gained new attention in the field of fashion recommendations thanks to the recent advances of deep learning techniques, particularly in image processing, understanding and segmentation, as discussed by Guo et al. [8]. This effect can be amplified in fashion, where the visual appearance of the items is of utmost importance, and in online fashion retailers, where metadata about the fashion items, such as color, brand, category or season is often readily available. Moreover, there is recent promising early work that goes beyond, and focuses on computer-generated design of fashion garments [9] thanks, in part to the applications of Generative Adversarial Networks [10–12].

The meta-representation of items is an important source of information for content-based recommender systems. Within this area, *FashionDNA* [13] proposes the combination of both textual information and images to create a representation of fashion items in a latent space which is designed to predict the non-personalized likelihood of purchase for each fashion item. Thanks to the addition of purchase information in the space, this approach performs well in the case of cold-start recommendations, and has proven successful in the task of searching images in the wild (Lasserre et al. [14]), from a source of professionally shot images (such as those in a fashion retailer catalog), or those taken by any smartphone user taking pictures of an outfit on the street.

In *Fashionista: A Fashion-aware Graphical System for Exploring Visually Similar Items*, He et al. [15] present a search system for fashion that uses visual features to find items that are both similar to a given query and also considered fashionable. The fashion similarity is calculated as a K-nearest-neighbours problem within low-dimension visual representation space where the fashionability is learned from user-item interactions derived from a fashion dataset. The proposed system enables users to explore visually and stylistically similar fashion items in a graphical interface that visually represents how visually similar the two fashion items are, while also representing the fashionability score in a heatmap.

More recently, Goncalves et al. [16], show that similar item recommendation can be aided by exploiting embeddings that describe fashion-specific attributes such as texture prints or materials, as these are a proxy for item style. The authors propose to use feature reduction methods by sampling feature maps to use these embeddings. When representing fashion items in a low-dimensional space, they achieve a reduction in memory use of the item representation in 99.91% in a real-world proprietary dataset from the online luxury fashion e-commerce platform Farfetch [17]. Moreover, using this low-dimensional representation of items as side-information into the similar-item recommendation tasks can potentially improve the style-related facet of similar recommendations, while not hurting the accuracy of the results returned.

In a study by Ji et al. [18], Pinterest's [19] approach to image-based search at scale is presented. This approach exploits modern cloud-based infrastructures and open-source tools to build the system, and has proven to increase the engagement with users in a live A/B testing [20]. Similarly, the e-commerce platform Alibaba

[21] describes the main challenges and solutions while building visual search at scale in [22] handling a wide range of images; from in-the-wild looks to professionally shot catalog images for a wide range of product categories, including fashion. They detail how to build large-scale indexing for seamless data updating, or how to train models with as little human annotation as possible.

New work is already trying to solve not only recommendation problems within the fashion domain, but also to autonomously design fashion garments that can be tailored to a person's taste. For example, in in *Visually-Aware Fashion Recommendation and Design with Generative Image Models*, Kang et al. [12] propose a fashion aware visual representation that, in contrast with previous work, exploits Siamese convolutions neural networks (CNN) to create a fashion-aware visual representation of the items. Moreover, the system can create new fashion items by understanding the user's past interactions to understand taste and a set of constrains. More recently, Kato et al. [11] propose the use of Progressive Growing of GANs (P-GANs) in the task of fashion design to serve as base to dress-making or pattern-making. However, while this approach shows some interesting and promising results, it was found through a user study that the technology does not yield the same results as an experienced pattern maker in the same tasks, highlighting the importance of the craft in the the design of fashion garments.

## 2.2 Complementary Clothing Item Recommendations

Thanks to the nature of fashion, where users generally desire to wear fashion items that stylistically or visually match when they are worn together [23], its importance in today's culture and its never-ending change [24], the recommendation of complementary clothing items has become a core problem within the fashion recommendations domain, which has led to an increase of attention in this topic of research. Moreover, due to the business model of modern fashion e-commerce platforms, it is also generally desirable that customers purchase more than a single fashion item at once, making these type of recommendations very interesting.

The problem of complementary item recommendations can be defined as a relaxation of the Outfit recommendations problem described in Sect. 5. Here, it is not necessary to match a set of clothes together, but rather to find a match for a single item. Formally, we can define it as: given a user $u$ at a given moment of time $t$, and an item defined as anchor item $i$ in $I$, the set of all available items, the goal is to recommend a list of items $I_c \subset I$ for user $u$, where item $i$ in can be worn together with any of the items in $I_c$.

The example in Fig. 1 depicts a carousel of fashion items complementary to the anchor white t-shirt displayed in the top-left corner. Complementary recommendations in this case are shown under the heading *"Perfect pairings: You might also like"*. This carousel contains several fashion items that can be worn together with the anchor item. Moreover, they share a similar style, and most of the items are part of the *Pride* theme.

The approaches proposed to solve this problem are often similar to the ones we have introduced in Sect. 2, as this problem can also be modelled adding an additional constrain in the similar item recommendation problem. However, when it comes to complementary fashion item recommendations, most approaches rely on hybrid models that use both user-item interactions and content-based features to produce recommendations.

The style relationship for complementary recommendations is exploited in [25], where Zhao et al. infer this relationship between fashion items based on the title description, taking the assumption that the title carries the most relevant information of the item. The proposed Siamese Convolutional Neural Network, which performs the task of finding compatible pairs of items in a words space that is later mapped into an embedded style space. The proposed approach takes only words as the input, which makes it very light in terms of computation, and it requires only a few preprocessing steps and no feature engineering.

*ENCORE (Neural Complementary Recommendation)*, by Zhang et al. [26], is a hybrid approach to generate complementary item recommendations, which is non-fashion specific. The authors propose a model that can learn the complementary relation between items exploiting both image and text of the items, while also learning the inherent quality of an item using product ratings. The two are then jointly learned using a neural network transformation, allowing it to combine both stylistic and functional facets of complementary items across categories. The results show that ENCORE outperforms state-of-the art approaches in terms of accuracy.

More recently, [27], Wu et al. propose a session based approach that includes modifications of the *STAMP model* [5] to produce complementary personalized item recommendations which outperforms the existing recommender system in production at the time [7]. Moreover, the authors propose a method to sample data from existing production systems from an existing dataset, in order to create training data that can be used to learn a training set for the complementary fashion items problem. This can be particularly interesting when due to the quality or existence of the complementary recommendations user-item interaction data can not be used to create a training set to learn upon.

Kunh et al. [28] present a real-world recommender for complementary fashion items used by stylists in Outfittery [29]. The proposed approach can recommend complementary items taking into account the style relationships between both individual fashion items. A *word2vec* [30] model is used to find similarities in a latent feature space, while a style fit score is applied to each item pair based on their similarity in this latent feature space. In order to ensure the complementary of recommendations, candidates are selected based on their functional characteristics, similarly to using categories.

While most of the approaches introduced above rely on content-extraction approaches to understand the relationship of complement, the approach introduced by McAuley et al. [31] tries to mimic the way humans visually understand similarity or complementary relationships between fashion items. The problem is formalized as a network inference problem defined on graphs of images that have a relationship with each others. The authors also propose a fashion item encoder which can understand the complementary suitability based on visual and textual features. This approach is shown to improve against other state-of-the-art methods in several datasets.

## 2.3 Evaluating Single Item Fashion Recommendations

The definition of single item fashion recommendation problem described in this Section is well known in the recommender systems domain. Particularly, ranking metrics such as Normalized Discounted Cumulative Gain (NDCG) or Mean Reciprocal Rank (MRR), as well as traditional accuracy metrics such as Precision (PRC) or Recall (RCL) are widely used in offline evaluation settings, with time-aware training, validating and testing splits. However, there are several particularities of evaluating single item fashion recommendations that should be considered.

### 2.3.1 Offline Evaluation of Single Item Fashion Recommendations

Fashion is time-aware and time-dependent. At the macro-level, new fashion is introduced at several times in the year, every year more frequently [32]; new items go into fashion and old items go out of fashion. At a micro-level, sequence modeling, and in particularly, attention-based models have been gaining popularity thanks to their accuracy performance [5, 27], as the task of finding what to wear, or what to buy are often carried out in a sequence of steps. Thus, when performing offline evaluation, it is very important to preserve and mimic the macro and micro level time dependencies, and always perform a time-aware evaluation, as described by Campos et al. [33].

As shown in Fig. 1 recommendations are often displayed in carousels, in which an implicit order is defined: left-most items are the most relevant and right-most items are least relevant. Thus, if the task is to enable users finding their desired items with the least effort possible, that item should be placed in the left-most positions of the carousel. Other type of commonly used representations are a grid of items, usually seen on e-mails or in search results pages. In this representation, recommendations also follow an order of left-most results being perceived as the top results. Thus, position bias [34] plays a huge role on how users interact with fashion recommender systems online, and therefore should be considered during evaluation.

Sherman et al. [35] highlight the importance of segmenting the offline evaluation results with respect to the user's available interaction history. They define three categories: *new users* who have not bought or seen any item. *view users* who have seen items but not made a purchase, and *sale users* who have made a purchase. All the metrics reported in their analysis are aggregated at group level in a proprietary dataset from True Fit [36], showing how content-based solutions, like the ones introduced in Sect. 2.1.2 outperform collaborative approaches for *new users*.

Finally, the availability of labels for success might arrive at different times in the evaluation process. In a traditional e-commerce scenario, the most widely available signal for success is a *click* on an item. Next, a stronger signal comes when the user makes a *purchase*. As the user's final goal is to wear an item, and not to return it, return information could be potentially considered as the ultimate signal of success for a recommender systems. However, such labels are often delayed days or weeks, as the users have a long period in which they can decide to return a given item. This can potentially slow-down the evaluation process, depending on the metrics of success that the practitioner is interested on evaluating.

### 2.3.2    Online Evaluation of Single Item Fashion Recommendations

The importance of online evaluation of Single Item Recommendation systems in controlled A/B experiments [20] is mentioned in the literature, where authors try to ensure the offline and online evaluation approaches are as similar as possible. Moreover, there are several advantages that should be taken into account when performing controlled A/B experiments.

In these type of experiments, the interaction between the recommendations being evaluated and the rest of the elements or content in the same page are easier to monitor than in offline evaluations. For example, while the introduction of a new recommendation carousel in a given page might lead to more accurate recommendations, it might also hurt the performance of other carousels in the same page, or in other subsequent pages the user might visit, and ultimately don't lead to a global conversion.

Moreover, the interaction of single item recommendations with other type of recommendations, particularly the size and fit recommendations described in Sect. 4 can also be studied in more detail in controlled A/B experiments. While in offline evaluation often assumes infinite available stock for a person's preferred size, one can only observe the interactions between both systems in online experiments.

### 2.3.3    Beyond Accuracy Evaluation of Single Item Fashion Recommendations

Finally, while the topic of evaluation beyond accuracy is well known in the field of recommender systems, it is however often overlooked in Single Item fashion recommendations literature. Topics such as evaluating recommendations based on their diversity, novelty or popularity biases, as explained by Corona Pampin et al. [37] should be carefully considered. Moreover, fairness and ethical aspect of recommendation should also be considered during the evaluation and design of the recommender system. For example, recommendations of beauty products can show racial bias, either because of the data available contains bias, or because the products available are not suited for a wide range of skin tones [38].

# 3    Social Fashion Recommendations

## 3.1    Leveraging Social Networking for Fashion Recommendation

Social networks have had a visible impact on many consumer industries. Amongst them fashion brands, retailers and influencers have been heavily impacted by the emergence of social networks. Social networking sites have established the standard mean for consumers to directly interact with fashion designers, browse catalogs of high-end clothing items, shoes and accessory businesses. This has been achieved by turning social data into a source for detecting trends, adapting user recommendations, and for marketing purposes according to Berthon et al. [39]. Major social networking platforms playing crucial roles in fashion domain include Facebook [40], Instagram [41], Pinterest [19], Polyvore [42], Chictopia [43], and Lookbook [44].

Amongst them, Instagram is seen as one of the most powerful means in directing the way consumers discern and recognize brands. Instagram allows a variety of media to be uploaded onto its platform including photos which become the subject of interaction between followers and content producers. Mohr [45], highlights that Instagram specifically has become a key player for establishing near real-time, direct engagement between brands and consumers. This has been a niche way of showcasing fashion as there is little proxy to the content shared online. Instagram has became the favorite new tool for both designers and models who gain more popularity every day through sharing content and connecting with their followers.

As an example, Fig. 2 depicts a sample fashion post on Instagram that clearly shows how followers engage with brands and get influenced by the content shared.

**Fig. 2** A fashion related Instagram post, depicting the visual content of image as well as multi-lingual comments as interaction means between fashion brand influencer and followers. Source: [46]

This anatomy serves as the skeleton for majority of social networking style platforms used by fashion brands, retailers, and influencers.

## 3.2   Methods for Social Fashion Recommendations

With increasing attention to fashion across the breadth of social networking platforms, a number of papers focus on understanding how to leverage such rich content as well as context, to improve personalized recommendations on behalf online shoppers. Existing techniques can be grouped into two major categories:

1. *Single and Multi-Modal Social Fashion Embeddings*: This category of techniques tend to address the problem of how to detect and extract visual features pertaining to garment parts, helping us to better understand fashion clues from the multi-modal fashion images and their accompanied textual data (such as captions, hashtags and comments).
2. *Visual and Dynamic Social Influence*: This category of techniques try to capture influencers dynamics as well as visual information in order to extract fashion features and provide users with visual influence based recommendations.

Within following subsections we go deeper into each categories.

### 3.2.1   Embeddings for Fashion Recommendations

Traditional recommender systems paradigms seen on fashion e-commerce platforms, as those discussed in Sect. 2, rely on user's online shopping history and metadata from professionally-shot product images to make recommendations. By using only this data, such systems tend to ignore the hidden taste of users that gets reflected within their own uploaded photos as well as their interactions with fashion producers online (Fig. 3).

As such, developing recommendation systems that can combine information from both set of sources can introduce possibilities for generating more personalized

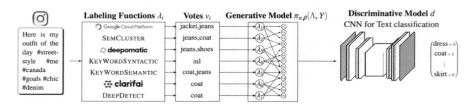

**Fig. 3**   A weakly supervised architecture for classification of Instagram posts. Source: [46]

recommendations. In addition to visual aspects of the posts, a large volume of content accompanying the posts are that of the user generated text.

On many social networking platforms, a post can be annotated with a caption written by the author of the post, by comments written by other users, and by "tags" on the image which can refer to users as well as products on online shops of brands and retailers. This has been under looked until recently as most of the existing work have been focused solely on visual analysis of social media content [47–49]. However, understanding and classification of fashion attributes on social platforms is important for modern applications focusing on tasks such as detection of fashion trends and user recommendation. To this end Hammar et al. [50], direct their attention on fashion domain-specific embeddings which were trained on an Instagram text corpus. They present a novel architecture utilizing weak supervision leveraging a deep classifier to recognize fashion clothing based only on text accompanying Instagram posts. The weak supervision signals were then combined with the data programming paradigm. Moreover, the original model for binary classification was extended to the multi-label setting by learning a separate generative model for each class. At the last step, weak supervision signals were combined with generative models that outperformed a baseline using a majority voting.

Zheng et al. [51], argue that the weak supervised framework can be replaced by a self-supervised framework capable of learning to measure the similarity between historical fashion posts of a user to matching potential outfits. Central to their framework is a joint embedding module that performs multi-modality feature extraction by combining cues from both image and tags as shown in Fig. 4. Similar to [50] this in turn assists with the extraction of visual features for garment parts to better understand fashion clues.

While learned embeddings by themselves are useful to understand fashion context such as material or styles, they need to be fused with recommendation framework properly to help generating more accurate recommendations of individual clothing items or outfit sets. For instance, Zheng et al. [51], leverage embedding of all posts of a user along with embeddings from potential outfits using a similarity metric generating distance scores for potentially matching outfits. Yang et al. [52], argue that collaborative filtering algorithms can benefit from additional data accompanying social fashion posts which can be advantageous for recommendations of sets of fashion items to users. In addition to matching user preferences, the fashion items in an outfit should also match with each other and thus it is important to integrate auxiliary information about the fashion items to capture the relationships amongst items. To this end they propose a functional tensor factorization method to model the interactions between user and fashion items. To effectively utilize the multi-modal features of the fashion items, they use a gradient boosting based method to learn nonlinear functions to map the feature vectors from the feature space into some low dimensional latent space.

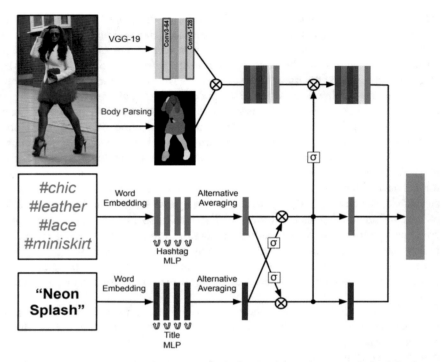

**Fig. 4** Embedding module takes into account both visual and textual aspects of a LookBook post and in turn generates a joint embedding. Source: [51]

### 3.2.2 Social Influence for Fashion Recommendations

Another important aspect of social networks is the inherent network interactions along the social ties in between users and their fashion brands. Social influence has been a driving factor behind the decisions customers make towards their fashion purchases online. Domingos and Richardson [53] emphasize that ignoring network effects constituting interactions of customers when deciding which segment of customers to market towards can lead to low-quality decisions. In addition to inherent value from the purchases made, a customer effectively has a network value that originates from his or her influence on other customers. As a result a number of work has focused on analyzing behaviors of potential customers of fashion brands on social networks to improve marketing strategies. For instance, Chinchilla et al. [54] leverage clustering and association rules in order to discern patterns of common behaviors of customers of a fashion brand on Instagram. They show that the learned models can provide useful information for devising personalized marketing strategies.

Jiménez-Castillo and Sánchez-Fernández [55] focus on understanding how effective influencers are in recommending brands and their products via the word-of-mouth by revisiting whether the potential influence they have on their followers may affect brand engagement and intention to purchase recommended brands. Their study concludes that the perceived influential power of influencers helps in increasing expected value and behavioral intention around the recommended brands.

As such measuring the social influence and leveraging it for improving recommendations have been a focus of a body of research over the years. Often the generic metric of audience size is taken as a measure for estimating influence on social networks. Segev et al. [56], challenge this concept under the context of Instagram, and demonstrate that features extracted from public information allows to use various regression models to create an influence ranking algorithm based on an intuitive score derived from network-oblivious user and posts statistics.

Li et al. [57], propose a social recommendation framework that given query terms describing a user's preference, discovers a set of targeted influencers who have the maximum activation probability on those nodes related to the query in the social network. Based on the discovered set of influencers as key experts, the framework recommends items via an endorsement network. Li et al. [58] use multiple features in user interactions, such as the number of comments and the number of followers and introduces the mechanisms of macroscopic and microscopic influences. They introduce the microscopic and macroscopic social influence mechanisms as matrix factorization to improve recommendations.

Zhang et al. [59], argue that a user's purchase behavior is not always determined by their directly connected peers and may be significantly influenced by the high reputation of people they do not know in the network, or others who have expertise in specific domains. They leverage this idea for social recommendation taking into account the global and local influence using regularization terms, and incorporating them into a matrix factorization-based recommendation model. Eskandanian et al. [60], argue that identifying influential users and studying their impact on other users helps in understanding how small groups can inadvertently or intentionally affect the behavior of the system as a whole. They formalize the notion of Influence Discrimination Model and empirically identify and characterize influential users and analyze their impact on the system under different underlying recommendation algorithms across various recommendation domains. Pipergias Analytis et al. [61], study various mechanisms of social influence behind both offline and online recommender systems. Using a weighted k-nearest neighbors algorithm (KNN) to represent an array of social learning strategies, they show through network science methods how the KNN algorithm discerns networks of social influence in various real-world domains. They argue that their results help improve the understanding of notion of social influence.

A number of recent work focuses explicitly on fashion influencers active on social networking platforms, and study their respective impact on recommendations. Bertini et al. [62] propose a user recommendation framework to use a user similarity metric that is computed through analysing the images shared by users on Instagram. They consider that influencers and users posting similar content share the same taste. Similarity between influencers' photo collections is estimated through deep embeddings, using a neural network trained to classify photo collections in

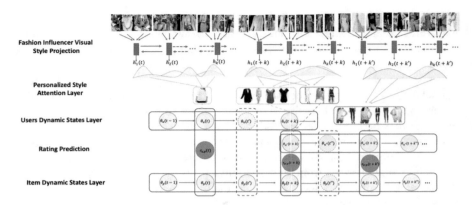

**Fig. 5** Architecture of the recurrent neural fashion recommender which leverages influencers dynamic and visual information to extract fashion features and provide users with personalized visual and influence based recommendations. Source: [63]

categories of interest. They propose a hybrid recommendation which combines collaborative filtering and results from this compact representation of visual content of photo collections. They demonstrate that their approach enhances the collaborative filtering baseline.

Zhang and Caverlee [63] propose a visual influence-aware recommender system that leverages on fashion influencers and their dynamic visual posts on Instagram. To do so, they extract dynamic fashion features highlighted by bloggers via a BiLSTM model, integrating a large corpus of visual posts and community influence. Subsequently this helps with learning the implicit visual influence funnel from bloggers to individual users via a personalized attention layer, as depicted in Fig. 5. Eventually, they incorporate perceived personal style of user and respective preferred fashion features across time in a recurrent recommendation network for dynamic fashion-updated clothing recommendation. They state that the framework is very effective especially for users who are most impacted by fashion influencers, and utilizing fashion influencers can bring significant improvement in recommendations generated, compared to other potential sources of visual information.

## 3.3 Evaluating Social Fashion Recommendations

As discussed in Sect. 3.2, existing body of research on social fashion recommendation can be categorized into two main stream views. In both categories quantitative as well as qualitative metrics have been used to validate the performance of solutions proposed.

In case of former category, embeddings play crucial role on understanding and relating semantic concepts underlying the textual cues of fashion under the social posts. Hammar et al. [50], argues that embeddings are particularly handy in the social networking context given that content can be multilingual as well as the syntactic variety of the spoken words. There are two sets of challenges that commonly undermine the impact of such techniques. First, high syntactic variety makes it difficult to use syntactic word similarity for information extraction. Second, mining text in supervised manner relies on having annotated training data, which in practice is difficult and costly to produce. This is due to several reasons including sheer size of social posts, lack of dedicated tools for annotating fashion items and most importantly finding human annotators with proper domain knowledge of design and fashion to tag clothing items as argued by Jaradat et al. [64]. The main focus of the embedding techniques is extraction of fashion related attributes, and as such learning to rank metrics such as Normalized Discounted Cumulative Gain (NDGC), Precision (P), Mean Average Precision (MAP) could be used to validate recommendations task performance, as demonstrated in Hammar et al. [50]. Given that such mode of retrieval is often combined with visual features to enable multi-modal recommendations, the challenge shifts onto assessing the benefit of producing and leveraging multi-modal features. To this end Zheng et al. [51] evaluate their approach using different combinations of modalities as well as fusion methods, and report their recommendation performance using recall scores in turn.

With respect to the latter category, a number of challenges arise from understanding how influence of bloggers are modelled, and in turn how that can be leveraged to affect the recommendations generated for the users. For instance, Zhang and Caverlee [63] argue that it is difficult to benchmark if the blogger's implicit visual influence really helps the recommendation. This is mainly due to the fact that explicit or direct signals of how customers have been affected by the bloggers' recommendation are often not available. As such, more work is needed to see how such measure actually improves the performance of fashion recommendation, compared to using other sources of visual information when available. Another challenge is in understanding how different types of fashion users (i.e. users with high interest in fashion versus casual shoppers) perceive the quality of recommendation generated. This turns out to be difficult, given that for such task one needs to know personal information of users before hand.

In both categories, the role of qualitative evaluation is well-emphasized. Zhang et al. [51], emphasize this by measuring performance of Top-K recommendations in recall scores against different size of input social posts. Zhang and Caverlee [63] take a varying size of selected fashion savvy users and compare the performance of recommendations in Root of the Mean Square Error (RMSE) scores against various baselines such as visual [65] and blogger influence [66] baselines. In both cases authors showcase visually how recommended clothing items relate to their input user data to validate their approaches.

# 4 Size and Fit in Fashion

Article returns are a crucial part of any retail industry where customers return those items deemed unsatisfactory for a variety of reasons. In turn, customers receive either a full or partial refund for their returns according to the retailer's specific return policy. Over the past years, customer returns have been growing at a significant rate reaching up to 50% increase year over year for certain categories [67, 68]. Article return rates vary greatly between retail industries, categories, brands, and distribution channels [69–71]. However, not surprisingly, the highest return rates are in online fashion apparel with $25 - 40\%$ overall return rates that can reach up to 75% for specific categories and brands [71, 72]. The underlying reason is that fashion apparel involves complex factors such as size, fit, color, style, taste and unquantifiable factors such as "it's not me" [71].

Indeed, fashion is a way to express identity, moods and opinions. Customers also tend to use fashion to either emphasize certain parts of their body or hide others. In this context, size and fit has been shown to be among factors influencing the overall satisfaction of the customers [73] the most, and in turn poor size and fit is cited as the number one factor causing online fashion returns [74]. Physical examination of fashion articles to assess their size, fit, fabric, design and pairing with other fashion articles is crucial for customers in their order decisions [75]. On the other hand, in online fashion, customers need to order garments and shoes without the possibility of trying them on and this crucial sensory and visual feedback is delayed to the unboxing experience. The absence of "feel and touch" experience therefore leads to major uncertainties in the buying process and to the hurdle of returning articles. As such, many customers either hesitate to place an order, or opt for various strategies for reducing the uncertainty, such as ordering multiple sizes or colors of the same article and then returning those that did not match their criteria—even more so for fashion categories and brands they are less familiar with.

Most fashion retailers create garments for consumers they have never met, not having the opportunity to interact with them directly, by using body measurement data aggregated and sold by third parties intended to represent size and fit requirements of entire populations in different countries and regions. Fashion articles suffer important sizing variations due to multiple factors starting with limited and coarse definition of size systems (e.g Small, Medium, Large for garments). To make matters worse, each brand has their own idea and definition of their target customer base and the correlating aggregated sizes that represent that customer base. A brand might specifically target their customer base to be on the younger side, and thus, represented by say a size 8, so the products are fitted on a size 8 fit model and then "graded", i.e. sized up and down in increments to generate other sizes of the garment. This leads to distortion of the larger sizes because they have now been increased, e.g. in width, without ever being fitted on a larger model [76–78].

Furthermore there is essentially no consistency with regards to sizing systems between brands (or in fact often even within brands) resulting in different physical specifications for the same nominal size according to the article and brand. Adding to this the complexity of radically different sizing systems (Alpha versus Numeric), different country conventions (EU, FR, IT, UK), different ways of converting a local size system to another (where brands don't always use the same conversion logic), and "vanity sizing" [79]. i.e. uncommunicated increase in physical measurements of a nominal size to boost customer's self-esteem, exacerbating the problem of providing correct fashion size advice. From another perspective, customers' perception of size and fit for their body also remains highly personal and subjective which influences what the right size is for each customer. The combination of the aforementioned factors leaves the customers alone to face a highly challenging problem of determining the right size and fit during their order journey. This in turn results in having size and fit to be the number one factor for apparel returns in fashion e-commerce.

The problem of size and fit has a major impact on the customer's shopping experience, the environment, and the profitability of online fashion retailers. A crucial issue in sustainability is the substantial carbon footprint incurred in the logistic process of returning articles [80]. Research indicates that online fashion retailers are strongly inclined to offer lenient return policies to lower customer perceptions of uncertainty [81] which inherently leads to an increase in return rates. As customer expectations around fit and sustainability evolve, online fashion retailers have to quickly adapt and find innovative techniques to conclusively reduce the high size and fit related return rates.

Over the past few years there has been an emerging body of studies on customer returns. Diggins et al. and Cullinane et al. [82, 83] address issues related to customer returns in online fashion retailing and discuss their implications while Walsh et al. [84] studies customer-based preventive article return management. Among instruments to reduce returns, size tables show no clear effectiveness and thus Zhang et al. [85] suggests a crowd sourcing approach where customers get suggestions on garments by matching the existing articles in their wardrobes to those of other customers in the community, with a strong limitation arising from customers' willingness to share their wardrobes with others. The research work on determining the right size and fit for customers is still in its infancy and remains very challenging due to the following factors:

1. Data sparsity is severe for both customers and articles- a customer only buys a tiny fraction of all the possible articles and sizes that exist and articles on the other hand have a limited stock and may not receive any return data for days (if not weeks) after their activation on the platform.
2. Data is also very noisy—a customer can buy various articles for multiple friends and family members in close and neighbouring sizes to their own.
3. The right size for a customer is very subjective- two customers with the exact same order history or body measurements might still buy two different sizes for the same article based on their perception of the right size and fit [86].
4. Customers may have a high degree of emotional engagement with the size and fit topic—even an accurate size advice can come with a high emotional cost for

the customer when the advised size differs from customers own expectation in an emotionally charged way.

**Supporting Customers with Size and Fit**  Being able to accurately find the right size and fit can therefore significantly contribute to increasing customer satisfaction and business profitability through reducing returns, which also reduces the carbon footprint of fashion e-commerce platforms. As an increasing number of people use online fashion stores to shop for articles, online platforms try to both support their customers and steer customers' behavior by providing data driven size advice and recommendations in various passive or active forms as following.

1. Size tables, size charts, and aggregated article measurements [87], provided per brand and article category, which map physical body measurements to the article size system. This approach is one of the simplest methods used and requires customers to find what fits them best themselves by measuring different parts of their body, such as "bust", "waist", "hip", to determine the right size themselves by cross referencing their measurements with the size tables for each article. Interestingly, even if the customer gets accurate measurements with the aid of tailor-like tutorials and expert explanations, the size tables themselves almost always suffer from high variance, even within a single brand. These differences stem from either different datasets used for size tables (e.g. German vs. UK population) or are due to vanity sizing, which in turn represent major influences on the body measurements [88–90]. These problems are especially intensified for fast fashion brands that represent the largest part of sales volume. In a fast moving fashion environment, designers strive to beat competition by continuously serving consumers with the latest trends at competitive prices. To meet time, cost, and design constraints, same articles with varying attributes (e.g., color, material, etc.) are often sourced from different production channels, causing inconsistencies in size and fit characteristics. Therefore, in practice size tables rarely help customers to find the right size [91] and fashion e-commerce platforms strive to develop data-driven systems for providing data driven size and fit advice to their customers.
2. Article based size advice, leveraging article attributes and return data from existing customers to provide article based size advice [92–95]. Here, customers receive aggregated non-personalized size advice on the article level (e.g. this article usually runs small/big, or $x$ percent of customers find this article to run small/big) and customers are required to make an informed decision themselves based on the given article based advice.
3. Personalized order history based size recommendations where customers' order history is leveraged to generate a personalized size advice [76–78, 96–101]. In this category of approaches, for eligible customers with an order history a size tailored to the customer is automatically pre-selected for each article based on the size recommendation predictions, without any requirement for customers to provide explicit personal data or images.
4. Customer in the loop personalized size recommendations [102–104]. Here, customers are asked to provide various explicit personal data through ques-

tionnaires and dialogue-like mechanisms. The required data may include age, height, weight, tummy shape, hips form, body type, favorite brand, usual sizes, fit preference, etc. and is necessary for receiving a personalized size recommendation.

5. Computer vision and 3D approaches [103, 105–107], providing virtual try-on solutions based on the recent progress in 3D human body estimations [108, 109]. To do so, customers are required to provide detailed personal information (as in approach 4) and/or to take one or multiple pictures of their bodies in tight fitting clothes so that a simulated avatar of their body can be built enabling the algorithm to predict customer's 3D body or body measurements. In turn, customers receive either a visual display of how the garment may fit on their body or a personalized size recommendation for each article. Although bringing this category of approaches to customers is currently of high challenge—in part due to their lack of accuracy at scale combined with their high requirements on private customer data (such as customer images or videos in tight clothes)—the future of these approaches remains very promising in particular in the context of high fidelity tailor-like recommendations and made-to-measure applications.

Each of these radically different approaches rely on disparate assumptions and require different levels of engagement from the customers, making each approach more or less appropriate for a particular customer segment and experience. In this chapter we dive deeper to provide details on the second, third and fourth category of aforementioned mechanisms where customers are either not required to actively engage in the solution or they do so through light weight questionnaires and dialog style mechanisms.

## 4.1 Evaluating Size and Fit Recommendations

Evaluating size and fit recommendations remains highly challenging due to multiple underlying factors specific to this problem space, in particular:

1. Starting with the customer, the "true" size of a customer is often unknown, remains subjective for each customer, may depend on the context and the occasion for which a customer is shopping, and can vary greatly by external factors such as fashion trends, seasonality, and life changing events impacting a customer's physical body, and/or mindset around what fits best.

2. Looking at the problem from the article side, the right size for a customer is not a unique quantity and varies greatly both within and across brands, in different sizing systems, in different countries, and for different fashion categories.

3. In online fashion, there is a significant delay between the time that a size and fit recommendation is provided to a customer, and the feedback signal coming from the customer once they have actually tried the recommended size on. Through the lens of size and fit related returns, in order to evaluate the quality and effect of a

recommendation, one needs to wait for many days if not weeks for the customer's return to reach the platform.

4. From the customer's experience point of view, customer satisfaction from a size and fit recommendation is not solely based on the recommendation's quality itself but is often intertwined with the physical characteristics and the experienced "feel" of the shoe or the garment once it is worn. This satisfaction is time-dependent and may change radically after wearing an article for a few weeks, or after a washing experience.

Considering the above challenges, the size and fit recommenders literature proposes some recent directions for evaluating current approaches at scale. In here we mention two of the major directions: (1) Evaluating the quality of the recommendations on a combination of the customer based metrics and business related indicators. Within this body of work A/B tests [20] or continuous evaluation frameworks are set up to measure a change in the customers' conversion rate, in the number of products added to the cart, in the revenues per visit, in selection orders (where a customer orders the same article in multiple sizes), and in reorders (where a customer returns an article and reorders it in a different size). In the same spirit, the literature also considers the customer's acceptance of the recommendations (i.e. how often they order in a recommended size), and accuracy of such recommendations (i.e. how often they keep or return a recommended size when they have accepted or not that recommendation). For more details on this direction, readers are invited to see [76–78, 97–101] and references therein. (2) Evaluating the quality of the recommendations on their impact in reducing the size and fit returns. Within this body of work evaluation frameworks are set up following A/B testing [20] or more recently using causal inference methods [110] to assess how and to what extent a recommendation is reducing size and fit related returns. Often deemed counter-intuitive, it is important to mention that while methods successful on the metrics mentioned in (1) bring a great deal of value and support to the customers and the business, they do not necessarily perform well in reducing returns, due to the inherit tension in the challenges mentioned above. For more details on this direction, readers are invited to see [93–95] and references therein.

# 5  Outfit Recommendations

Choosing a complete outfit can be challenging considering the time and efforts it usually requires. Deciding whether the clothing items in the outfit are compatible, the colors are suitable to the person's skin tone, and whether the outfit is suitable to the occasion, budget, and the weather conditions are some of the factors that are usually evaluated before choosing the perfect look. The automation of this process can reduce the burden of outfit selection and save the customer's purchasing time during online shopping. However, most of the previous research in fashion recommendations has focused on individual items' recommendation and a limited

(a)                                                    (b)

**Fig. 6** Examples of whole outfits' recommendation for a summer outfit (**a**) and a winter outfit (**b**) The example showcases the variations of the categories and the numbers of clothing items in an outfit that change depending on the season and other factors. Images are licensed—Adobe Stock

amount of work has been conduced on whole outfits' recommendations. This can be explained by the following reasons:

- An outfit consists of multiple clothing items (see Fig. 6), and customers might have different views on the individual items while considering whether they like the whole outfit or not. Colors, fabrics, style, fit and other factors that are difficult to be inferred by the recommendation system, yet are often considered by the user when purchasing an outfit.
- The compatibility between the clothing items that make an outfit can be decided using styling experts advice and latest trends. However, customers might have their own policies or preferences when combining items that need to be learnt by the system. Different people might have different perspectives of styles, which is a reflection of their occupations, ages, culture and place of living.
- Generating valuable personalized outfit recommendations to customers require learning from their past purchases and possibly explicit feedback on which regions of the outfits they like, or their outfits' pictures that can give an idea about their general style and body shape. However, there are significant reasons behind the difficulty of processing customers' provided pictures as this process requires more efforts from customers to take pictures in neutral backgrounds and with acceptable quality. Moreover, many customers might not provide such input due to the lack of time, or for privacy reasons.

Most of the approaches that have been proposed for personalized outfit recommendation in research are based on the idea of suggesting a whole outfit using a seed clothing item that the user is browsing, or by suggesting a complementary item that completes the look. Moreover, there has been another category that is based on selections from the personal wardrobe (smart digital closet). In this section, we

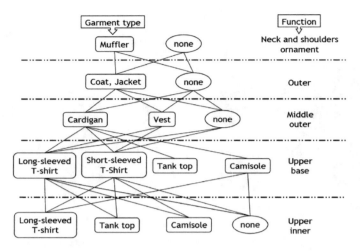

**Fig. 7** Example of a defined policy on garments' combinations in an outfit. Source: [112]

discuss the state of the art in outfits recommendation covering the aforementioned categories. Generally, the outfit recommendation problem can be formulated as follows: Given a set of fashion clothing items $I = \{i_1, \ldots, i_N\}$, a set of fashion outfits $O = \{o_1, \ldots, o_M\}$ can be generated. An outfit $o_m$ with $t$ fashion items can be presented as $o_m = \{i_1, i_2, \ldots i_t\}$ where the outfit can be presented as *a set, a sequence or as pairs of clothing items*. Each clothing item belongs to a certain clothing category that is denoted as $c_i$ and possibly has a set of metadata attributes that we denote as $A_i = \{a_{i1}, a_{i2}, .., a_{iK}\}$. Examples of attributes are: material, pattern, length, and brand. Additionally, style information at the outfit level can be provided or inferred from the existing attributes and denoted as $S_o$. Assume there are U users where u = { $u_1, \ldots, u_U$ }. The user's u preference towards the outfit $o_m$ is represented as $p_{u,m}$. The objective is to learn a model so that the preferences scores of any pair of user $u$ and outfit $o$ can be predicted. The pairs of outfits that have the highest predicted preferences will be recommended to the user $u$.

Following the problem definition, outfit recommendation can be compromised of the following stages:

- **Learning Outfit Representation**: Learning the visual representations of the clothing items and/or their textual metadata attributes. Each clothing item $i$ in an outfit $o_m$ will be transformed into an embedding $T_i$. The transformation might be a joint representation of the image and its metadata. Additionally, the style representation can be inferred and provided as input in the next steps.
- **Learning Compatibility**: Learning the compatibility between several clothing items in a single outfit which can be based on: (a) defined policies recommended by stylists and designers (e.g. [111]) (an example is shown in Fig. 7) or defined by the customers to show their preferences on combinations of garments (e.g. [112]) or (b) can be learnt by the system from existing examples of positive/negative

compatible items or by measuring similarity between latent representations between items.

- **Personalization**: Learning personal preferences based on direct input from customers or through inference from their input and behaviour and integrating the representation of the users' preferences in the learning model.

Early work on whole outfits' recommendation (e.g. *The Complete Fashion Coordinator* [113], *What Am I Gonna Wear* [114], and *Magic Closet* [115]) was based on input provided by users on the clothing items they have in their wardrobes and the occasions of wearing these items, and then suggesting outfits based on historical data, inferred personal style and matching with the occasion. For example, in the *Complete Fashion Coordinator* proposed by Tsujita et al. [113], the user registers the pictures of their wardrobe clothes with some input such as the occasion for which each item was worn, and they can share them with their friends in a social network to get advice and evaluation of the outfits. Liu et al. [115] proposed *Magic Closet*, a system that enables the retrieval of matching clothes from online shops that pair with an item from the user's personal wardrobe. Other similar examples are: [112, 116, 117].

While the early work provided the foundation and definition of the problem, the research direction has shifted afterwards to providing inspirations of outfits to users from online shops and fashion blogging websites. Thus, we can categorize research in outfits recommendations based on the mechanisms of generating and modeling representations of clothing items and outfits as follows:

## 5.1 Outfits' Compatibility Scoring

A large amount of research [31, 111, 116, 118–126] in outfits recommendation is based on outfit scoring using uni- or multi-modal neural architectures for extracting and learning the feature representations of outfits and then applying a classifier network to predict a score that describes the outfit's style compatibility and adherence to the user's personal style. An example of such an approach is shown in Fig. 8, which reflects the general steps in this direction. The outfit representation in these approaches is usually learnt by concatenation or pooling to aggregate the clothing item embedding representations. For example, He et al. [118], project pairs of the clothing items from which an outfit is composed to the embedding space and then they feed the compound features to the CNN model following a pairwise compatibility matching approach. Li et al. [119] jointly fuse learnt embeddings from images and their metadata to further improve the fashion outfit representation. Another interesting approach is proposed by Chen et al. [120], where they learn a multimodal attention model over pre-segmented images to emphasize the parts of the outfit in which the user has shown explicit interest in their reviews of the outfit. For example, given the user review: *"Great Material, loose fit around the waist. Nice wide neck opening, very stylish look"*, additional attention will be on the

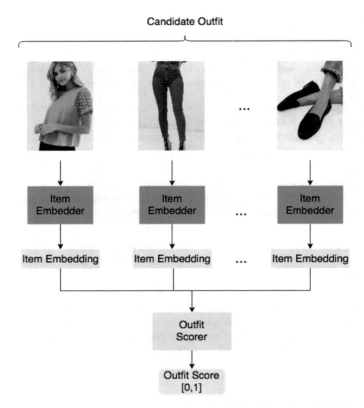

**Fig. 8** Example of the general steps in outfit scoring systems where the latent representations of the clothing items are learnt and fed as input to a scoring step that evaluates the compatibility of the whole outfit. Source: [125]

neck part of the outfit. However, such approaches can be time-consuming due to the segmentation process that is needed to obtain the annotations at different levels of the outfit.

The compatibility model that Song et al. [111] propose is based on the teacher-student network scheme where they encode the clothing matching policies that are learnt from domain knowledge in the teacher network using the attention mechanism to regularize the learning of compatibility between clothing items in an outfit. However, the construction of the compatibility matching rules can be a challenging process due to the large amount of fine-grained details in clothing items that might affect the rules.To address this issue, Bettaney et al. [125] combine embeddings of multi-modal features for all items in an outfit and outputs a single score that reflects the compatibility between the representations. To summarise, most of the research has agreed that the visual and textual information of clothing items are great supplements to the outfit score prediction.

Some researchers have proposed approaches for calculating pairwise compatibility between clothing items in an outfit and then learning the overall compatibility score from enumerating pairwise similarities. However, these approaches are usually criticized by the high computational cost of enumerating all the pairs for a large candidate item and their inefficiency in capturing the global compatibility between the sets of clothing items. Examples of research that follow this direction are [121–124] where the authors use a Siamese convolutional neural network approach to learn feature transformation for measuring clothing pair compatibility. [127–130] have the objective of providing interpretability to the outfit scores by analysing the clothing pairwise compatibility. For example, Wang et al. [129] follow a backpropagation gradient analysis from the outfit score output to the input pairwise similarities for the purpose of outfit diagnosis in order to point out the incompatible items in the outfit and provide explanation for outfit scores. Song et al. [131] employ the Bayesian personalized ranking framework to model the pairwise preferences among the fashion items of the outfit. Following the idea of outfit grading, in [132], the *Fashion++* approach is proposed to suggest edits for both fit and aesthetic details such as color, patterns and material of an outfit. In the *Fashion++* model, the clothing items of an original outfit are preprocessed and then fed to a discriminative fashionability module to gradually update the outfit composition in the direction that maximizes the outfit's score, and consequently improves its style. Similar to the idea of providing outfit's improvement tips to the user, Simo-Serra et al. [133] propose a Conditional Random Field model to reason about the outfit's score and the garments composing the outfit. An interesting factor has been effectively integrated in *BOXREC* [134] where Banerjee et al. consider the users' preference of price range for individual clothing types and their overall shopping budget for a set of items while generating outfits' recommendations (Fig. 9). In *BOXREC*, a set of preferred outfits is generated by retrieving all types of preferred items from the

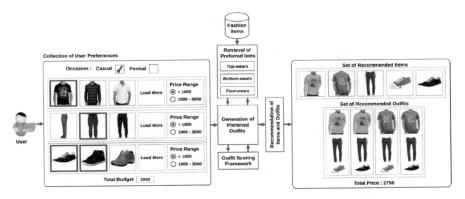

**Fig. 9** An overview of the BOXREC framework where users decide their preferred price ranges of individual clothing items and their total budget to receive recommendations customised based on their preferences. Source: [134]

clothing database customised according to users' price ranges, and then all possible combinations of clothing items are created and verified using an outfit scoring framework.

## 5.2 Sequential Outfits Representations and Predictors

Most existing research [135–139] follows a sequential approach to modeling the fashion outfit (usually from top to bottom to accessories). Here, each clothing item is a time step, and a fixed order in terms of clothing categories is applied to ensure that items in a single outfit are neither redundant or missing. Then, using the property of Bi-directional Long short-term memory (LSTM) architectures [140] that allows modeling the connection among current items, past items, and future items through forward and backward LSTMs, the global dependencies between the clothing items can be detected. An example of such approach is shown in Fig. 10, where the images of clothing items are transformed to feature representations in 1D vectors and fed to Bi-LSTM. Wang et al. [137] provide the visual-semantic embedding (VSE) and the aesthetic embedding (AE) to supervise the Bi-LSTM learning by incorporating meaningful category and personal aesthetic information that is learnt from outfit features. The visual-semantic embedding is learnt by projecting the image features and their semantic textual representations into a joint space. Consequently, the method can generate an outfit that is visually compatible to the given query clothing item and also semantically relevant to the text query. Nakamura et al. [139] incorporate the outfit's style embedding in the Bi-LSTM model by first reducing the item features that are included in the outfit then

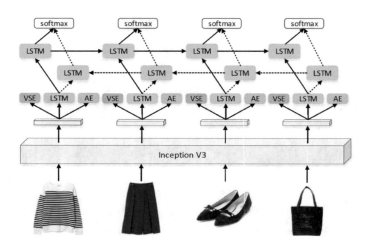

**Fig. 10** Example of the general steps used in handling fashion outfit recommendations as sequences. Source: [137]

encoding and decoding the obtained vector using an autoencoder to extract the style representation that should conform with each of the outfit's clothing items' style. Paragraph Vector neural embeddings model has been used by Jaradat et al. [141] in which they leverage the complex relationship between user's fashion metadata while generating outfits recommendations. Generally, the aforementioned sequential approaches require fixing the order of items in terms of their categories in an outfit. However, modeling outfits as an ordered sequence might be practically different than reality as items in an outfit do not have any specific order. To allow for a flexible representation of outfits, more recently, researchers modelled outfits as sets or graphs rather than sequences as we explain in the next section.

## 5.3 Outfits' Flexible Representations

Lin et al. [142] formulate *OutfitNet*; a fashion outfit recommendation model as a multiple instance learning problem, in which the outfits are handled as sets with the assumption that the labels that indicate the users' preferences at the outfit level are known, while the labels that indicate their preferences for the individual clothing items in the specific outfit are unknown. The compatibility among the fashion items is learnt through the semantic similarity of their embedded representations. Specifically, given a subset of fashion items, a positive item that occurred in the outfit, and a negative item that never occurred in the outfit before, a relevancy embedding of those items will be generated. Then, the learnt embeddings are fused together with users' outfits' preferences where an attention layer is applied to capture the emphasis of users on certain parts of the outfit that is learnt by pooling from the multiple inputs of the model, and hence it achieves personalized results.

Collaborative filtering approaches are proposed in [112, 143] to model the interactions between users and fashion outfits. Lin et al. [144] propose a fashion recommendation framework that is based on an image generation model by incorporating a layer-to-layer matching to fuse the generation features from different visual layers to improve the recommendation performance of the candidate item. Chen et al. [145] use user clicks on clothing items to capture users' interests which is integrated with their history of purchase to generate personalized fashion outfits. In their model they use a transformer encoder-decoder architecture to model users' preferences and compatibility among clothing items.

Aforementioned research represent the outfit as a sequence, a set or as pairs of clothing items. However, Cui et al. [146] chose a different representation by modeling the outfit as a sub-graph of a fashion graph (example shown in Fig. 11) where each node represents a category and each edge represents interactions between the categories. Then, the task of outfit compatibility can be converted into a graph inference problem. A topic modeling approach was proposed by Hsiao et al. [147] to learn style representations from fine-grained fashion attributes where an outfit is represented as a *document*, a visual attribute (e.g. polka dotted) is a *word* and the style is the inferred *topic*.

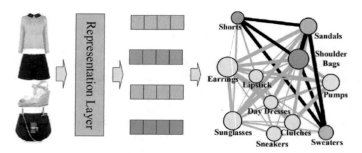

**Fig. 11** Representation of an outfit as subgraph of Fashion Graph where each node represents a category and each edge represents the interaction between two nodes. Source: [146]

## 5.4 Evaluating Outfits Recommendations

Generally, the accuracy of the outfit recommendation task can be evaluated with different quantitative methods such as, fill in the blanks (FITB) or compatibility prediction (CP). When the order of the outfits shown to a user is important, the outfit recommendation task can be evaluated using ranking accuracy metrics.

**Fill in the Blanks** is a widely standard test used in fashion compatibility research. It is usually formulated as follows: Given an outfit that is composed of $x_1, \ldots x_{i-1}, x_i, x_{i+1} \ldots, x_n$ clothing items, one of the items (e.g. $x_i$) is masked out randomly. Then, the task is to predict the item that is most compatible with the remaining items in the partial outfit. The performance is measured by the accuracy of the prediction compared to a ground truth dataset.

**Compatibility Prediction** is an evaluation technique for the model's capability at distinguishing compatible from non-compatible outfits. Usually, for each compatible outfit, a non-compatible example can be generated by replacing one item at a randomly selected position by another item from the clothing items vocabulary. This generated example is then labelled as non-compatible. Then, the task becomes a binary classification problem where the model is trained with the compatible and non-compatible outfits. Area under the curve (AUC) of the receiver operating characteristic (ROC) is a commonly used evaluation metric in this technique.

**Outfits ranking accuracy** can be evaluated using metrics as Normalized Discounted Cumulative Gain (NDCG). Applying ranking measures in outfits' ranking tasks can be difficult, as it requires defining a way of considering an outfit which is composed of several clothing items to be a relevant recommendation to the user. This implicitly requires defining rules for positive or neutral choices for the user. Some researchers considered an outfit that is created or evaluated by the user (like/view) to be a positive outfit, while the other outfits to be neutral. Then using NDCG which is a widely used criteria for comparing ranked lists, the relevance of top N recommended outfits is evaluated.

In order to easily compare the performance of outfit recommendation algorithms across different tasks, several datasets have been released. These originate from fashion online websites where users can create their outfits by composing them from different clothing items and receiving feedback from other users such as. Several examples of widely used datasets are: Polyvore [42], ShopLook [148], iFashion and (Alibaba) [21]. More details about datasets used in fashion recommendation domain are provided in Sect. 6.

## 6  Fashion Recommendation Datasets

In this Section we introduce a selection of well known datasets in the field of fashion recommendations, together with their main characteristics, possible uses as well as the direct links to download them. We hope that this collection enables and facilitates the repricability and comparability of the work done by practitioners in several of the tasks discussed in this chapter; size and fit recommendations, single item recommendations and outfit recommendations.

- **Amazon Reviews dataset** introduced by McAuley et al. [15, 31], contains product reviews and metadata from Amazon, including products in the *Clothing, Shoes and Jewelry* category. It includes 142.8 million reviews, ratings, product metadata as well as bought and viewed actions at product level. This dataset can be used in the tasks of single item recommendation or similarity calculation.
- **Large-scale Fashion (DeepFashion)** is a fashion dataset built by Lui et al. [149], which contains a set of 800,000 diverse fashion images that can be used for different tasks such as fashion similarity, or fashion item recognition.
- **DeepFashion2 dataset** is a dataset collected by Ge et al, [150] which contains 491,000 images, mapping to 801,000 clothing items and more detailed metadata associated with them. It has useful metadata information that can be used for tasks such as fashion similarity, fashion item recognition and single item recommendation.
- **Clothing Fit Datasets for Size Recommendation** are datasets collected by by Misra et al. [98] from sever fashion e-commerce sites ModCloth [151] and RentTheRunway [152]. It contains size and self-reported fit information, as well as reviews and ratings for 1,738 items and 47,958 users. The dataset can be used for research in size and fit related recommendation problems.
- **ViBE (Dressing for Diverse Body Shapes)** is a Dataset collected by [153] Hsiao et al. from the fashion site Birdsnest [154] which contains photography of garments worn by a variety of models of different body shapes. The dataset also contains medatada for both the item and the model. Thus, this dataset is very interesting for solving size and fit related recommendation problems
- **Street2Shop** is a dataset collected by M. Hadi Kiapour et al. [155], which contains 20,357 labeled images of clothing worn by people in the real world, and 404,683 images of clothing from shopping websites. The dataset contains 39,479 pairs of exactly matching items worn in street photos and shown in shop

images, which can be used for the task of similar item detection from images, complementary item recommendations, or outfit recommendations.

- **Pinterest's Shop The Look Dataset**, collected by Kang et al. [126]. The dataset contains 47,739 scenes of people wearing fashion, which are labeled and linked to the corresponding 38,111 items. The dataset also has style labels, which makes it useful for the task of outfit recommendations or complementary item recommendations.
- **Alibaba iFashion** is a dataset collected by Xu et al. [145]. The dataset contains 1.01 million outfits that are created by fashion experts. It also contains the clicks behaviors from 3.57 million users over a period of three months. This dataset can be useful for the task of outfits' personalised recommendations.
- **PolyVore Outfits** is a dataset collected by Vasileva et al. [123]. It contains 68,306 outfits and 365,054 clothing items with their descriptions and types. This dataset is suitable for outfit recommendations or complementary item recommendations.

## 7    Conclusion and Challenges Ahead

In the growing world of fashion e-commerce it is becoming increasingly important for online fashion retailers to provide more personalized clothing recommendations that match the expectations of customers. Moreover, modern strategies in merchandising and marketing aim to ensure that the unique needs and preferences of consumers are addressed. Yet, with the huge information overload and the wide variety in products and services, personalized recommendations become more challenging to achieve. As business, brands and online retailers are investing in social media, influencer-based marketing can be effectively employed in generating personalised recommendations for the followers by adapting their recommended content based on their interactions with influencers which reflect some of their fashion preferences. Due to the novelty of the digital influencers domain and its constantly evolving nature, the topic of enhancing consumers' personalization through their relation with influencers has not been discussed enough by scholarly researchers which leaves a space for more research contributions on this type of cross-domain recommendations.

Future work on the size and fit recommendations will open the doors to deliver substantially superior customer experience with targeted improvements on the size and fit related returns, the recommendations' accuracy, interpretability, and explainability. Experimentation with deep leaning models in this domain have the great potential to unlock rich size and fit recommendations across categories, customers, and brands. From a different perspective such accurate recommendations may still come with an emotional cost for the customers, for example where a customer is faced with a size recommendation that largely differs from their own expectation. Such emotional costs are of particular importance for recommendation algorithms where customers have a high degree of emotional engagement with the topic, as is the case in the size and fit domain. Future experiments will put greater focus on the interpretability of the recommendations to describe the machine's reasoning behind the predictions, critical in decreasing the emotional cost for customers, and playing a major role in increasing customer's understanding and trust and acceptance of those recommendations. Computer vision and 3D methods, when successfully implemented at scale, can radically change the game in the size and fit domain by creating highly personalized and accurate solutions tailored to each customer's unique body, taste, and preferences. They can also prove essential for creating made-to-measure solutions at scale where articles are created and personalized for each customer on demand.

We believe that the major challenges and future directions in outfits recommendations research would be on providing more scalable and automated solutions that minimize the individual efforts by stylists and at the same time minimize the required input from customers which can be replaced by inferring their preferences from previous images, purchase history or possibly their social networks posts. Detecting users' preferences from their images require efficient localisation or segmentation of clothing items, and matching with similar items from fashion shops databases

Finally, the more traditional task of item recommendations is likely to further gain from the interactions between this area and the outfits, social media and size and fit recommendations areas. We imagine that while ranking optimization models will continue to drive the state-of-the art when it comes to accuracy metrics, in the future it will be more important to built a digital experience where customers can see fashion products in context (either by automatically generated outfits or influencer outfits) and they always fit perfectly when ordering them.

Taking a moment to zoom out, we see that Fashion plays different roles in people's life. Some are just looking for a comfortable piece of apparel to wear, while others are deeply interested in the latest fashion, and as we dress up every morning, there is no doubt that fashion very personal to each of us. Because of this, the effect that recommendations of size and fit, influencer content or outfits can have on someone's well being needs to be more widely studied. Moreover, the production and distribution of clothes generates substantial waste and in its current form can negatively impact the climate and our society. While recent advances on tackling the size and fit problem has shown great gains in saving unnecessary returns (and their associated environmental and economic costs), many of ethical concerns that affect our society also directly affect the fashion industry. In this context, what is recommended can play an important role in amplify or mitigating the positive or the negative impact. A big area of research that will grow within the next years is thus ethical and fair AI applied to the problem of physical good recommendations, particularly intimate to each of us and greatly present in fashion.

# References

1. Zalando. https://en.zalando.de/, 2020. Accessed: 2020
2. R. Devooght, H. Bersini, Long and short-term recommendations with recurrent neural networks, in *Proceedings of the 25th Conference on User Modeling, Adaptation and Personalization*, UMAP '17 (ACM, New York, 2017), pp. 13–21. Code available in: https://github.com/rdevooght/sequence-based-recommendations. Accessed: 2020.
3. J. Chung, C. Gulcehre, K. Cho, Y. Bengio, Empirical evaluation of gated recurrent neural networks on sequence modeling, in *NIPS 2014 Workshop on Deep Learning, December 2014*, 2014
4. J. Li, P. Ren, Z. Chen, Z. Ren, T. Lian, J. Ma, Neural attentive session-based recommendation, in *Proceedings of the 2017 ACM on Conference on Information and Knowledge Management*, 2017, pp. 1419–1428
5. Q. Liu, R. Mokhosi, Y. Zeng, H. Zhang, STAMP: Short-term attention/memory priority model for session-based recommendation, in *Proceedings of the 24th ACM SIGKDD International Conference on Knowledge Discovery and Data Mining*, 2018
6. J.A.S. Rodríguez, J.-C. Wu, M. Khandwawala, Two-stage session-based recommendations with candidate rank embeddings, in *Fashion Recommender Systems* (Springer, New York, 2020)
7. F. Aiolli, A preliminary study on a recommender system for the million songs dataset challenge, in *IIR*, 2013, pp. 73–83
8. Y. Guo, Y. Liu, A. Oerlemans, S. Lao, S. Wu, M.S. Lew, Deep learning for visual understanding: A review. *Neurocomputing, 187*, 27–48 (2016)
9. Data-Driven Fashion Design. https://multithreaded.stitchfix.com/blog/2016/07/14/data-driven-fashion-design, 2016. Accessed: 2020
10. S. Mahdizadehaghdam, A. Panahi, H. Krim, Sparse generative adversarial network. *Proceedings of the 2019 IEEE/CVF International Conference on Computer Vision Workshop (ICCVW)*, 2019, pp. 3063–3071
11. N. Kato, H. Osone, K. Oomori, C.W. Ooi, Y. Ochiai, GANs-based clothes design: pattern maker is all you need to design clothing, in *Proceedings of the 10th Augmented Human International Conference*, 2019, pp. 1–7.
12. W.C. Kang, C. Fang, Z. Wang, J. McAuley, Visually-aware fashion recommendation and design with generative image models. *2017 IEEE International Conference on Data Mining (ICDM)*, 2017 November, pp. 207–216
13. C. Bracher, S. Heinz, R. Vollgraf, Fashion DNA: merging content and sales data for recommendation and article mapping. *KDD 2016 Fashion Workshop, August 14, 2016*, San Francisco, CA, 2016.
14. J. Lasserre, K. Rasch, R. Vollgraf, Studio2Shop: from studio photo shoots to fashion articles, in *ICPRAM 2018 - Proc. 7th Int. Conf. Pattern Recognit. Appl. Methods*, 2018
15. K. He, X. Zhang, S. Ren, J. Sun, Deep residual learning for image recognition, in *2016 IEEE Conference on Computer Vision and Pattern Recognition (CVPR)*, 2016, pp. 770–778
16. D. Goncalves, L. Liu, A. Magalhães, How big can style be? Addressing high dimensionality for recommending with style, in *First Workshop on Recommender Systems in Fashion*, 2019
17. Farfetch, an online luxury fashion retail platform. https://www.farfetch.com/, 2020. Accessed: 2020
18. Y. Jing, D. Liu, D. Kislyuk, A. Zhai, J. Xu, J. Donahue, S. Tavel, Visual search at Pinterest, in *Proceedings of the ACM SIGKDD International Conference on Knowledge Discovery and Data Mining*, 2015 August, pp. 1889–1898
19. Pinterest. https://www.pinterest.com, 2020. Pinterest. Accessed: 2020
20. S. Young, Improving library user experience with A/B testing: principles and process, in *Weave: Journal of Library User Experience*, 2014
21. Farfetch, a Chinese multinational technology company specializing in e-commerce, retail, Internet, and technology. https://www.alibaba.com/, 2020. Accessed: 2020

22. Y. Zhang, P. Pan, Y. Zheng, K. Zhao, Y. Zhang, X. Ren, R. Jin, Visual search at Alibaba, in *Proceedings of the 24th ACM SIGKDD International Conference on Knowledge Discovery and Data Mining*, 2018, pp. 993–1001
23. J. Craik, *Fashion: The Key Concepts* (Bloomsbury Academic, New Delhi, 2009)
24. K.T. Hansen, The world in dress: anthropological perspectives on clothing, fashion, and culture. Annu. Rev. Anthropol. **33**, 369–392 (2004)
25. K. Zhao, X. Hu, J. Bu, C. Wang, Deep style match for complementary recommendation. *AAAI Workshop*, WS-17-01 - WS-17-15, 464–470 (2017)
26. Y. Zhang, H. Lu, W. Niu, J. Caverlee, in *Quality-Aware Neural Complementary Item Recommendation*, 2018, pp. 77–85
27. J.-C. Wu, J.A.S. Rodríguez, H.J.C. Pampín, in *Session-Based Complementary Fashion Recommendations*, 2019, pp. 2–6
28. T. Kuhn, S. Bourke, L. Brinkmann, T. Buchwald, C. Digan, H. Hache, S. Jaeger, P. Lehmann, O. Maier, S. Matting, Y. Okulovsky, Supporting stylists by recommending fashion style, (2019)
29. Outfittery, an online personal shopping service for men. https://www.outfittery.de/, 2020. Accessed: 2020
30. K.W. Church, Word2vec. Nat. Lang. Eng. **23**(1), 155–162 (2017)
31. J. McAuley, C. Targett, Q. Shi, A. Van Den Hengel, Image-based recommendations on styles and substitutes, in *Proceedings of the 38th International ACM SIGIR Conference on Research and Development in Information Retrieval*, 2015, pp. 43–52. Dataset available in: http://jmcauley.ucsd.edu/data/amazon/links.html. Accessed: 2020
32. V. Bhardwaj, A. Fairhurst, Fast fashion: response to changes in the fashion industry, in *International Review of Retail, Distribution and Consumer Research*, 2010
33. P.G. Campos, F. Díez, I. Cantador, Time-aware recommender systems: a comprehensive survey and analysis of existing evaluation protocols. User Model. User-Adapt. Interact. **24**(1–2), 67–119 (2014)
34. K. Hofmann, A. Schuth, A. Bellogin, M. De Rijke, Effects of position bias on click-based recommender evaluation, in *European Conference on Information Retrieval* (Springer, New York, 2014), pp. 624–630
35. J. Sherman, C. Shukla, R. Textor, S. Zhang, A.A. Winecoff, Assessing fashion recommendations: A multifaceted offline evaluation approach, 2019. arXiv:1909.04496v1 [cs.IR]
36. TrueFit, a personalization platform for footwear and apparel retailers. https://www.truefit.com/, 2020. Accessed: 2020
37. H.J.C. Pampín, H. Jerbi, M.P. O'Mahony, Evaluating the relative performance of collaborative filtering recommender systems. J. Univ. Comput. Sci. **21**(13), 1849–1868 (2015)
38. Pinterest's new skin tone beauty search capability. https://newsroom.pinterest.com/en/skintoneranges2020, 2020. Accessed: 2020
39. P.R. Berthon, L.F. Pitt, K. Plangger, D. Shapiro, Marketing meets web 2.0, social media, and creative consumers: implications for international marketing strategy. Bus. Horiz. **55**(3), 261–271 (2012)
40. Facebook. https://www.facebook.com/, 2020. Facebook. Accessed: 2020
41. Instagram. https://www.instagram.com, 2020. Instagram. Accessed: 2020
42. Polyvore. https://www.polyvore.com, 2020. Polyvore. Accessed: 2020
43. Chictopia. http://www.chictopia.com/, 2020. Chictopia. Accessed: 2020
44. LookBook. https://www.lookbook.nu/, 2020. LookBook. Accessed: 2020
45. I. Mohr, The impact of social media on the fashion industry. J. Appl. Bus. Econ. **15**(2), 17–22 (2013)
46. K. Hammar, S. Jaradat, N. Dokoohaki, M. Matskin, Deep text mining of instagram data without strong supervision, in *2018 IEEE/WIC/ACM International Conference on Web Intelligence (WI)* (IEEE, New York, 2018), pp. 158–165.

47. S. Bakhshi, D.A. Shamma, E. Gilbert, Faces engage us: photos with faces attract more likes and comments on instagram, in *Proceedings of the SIGCHI Conference on Human Factors in Computing Systems*, CHI '14, New York, NY, 2014 (Association for Computing Machinery, New York, 2014), pp. 965–974

48. B. Ferwerda, M. Schedl, M. Tkalcic, Predicting personality traits with instagram pictures, in *Proceedings of the 3rd Workshop on Emotions and Personality in Personalized Systems 2015*, EMPIRE '15, New York, NY, 2015 (Association for Computing Machinery, New York, 2015), pp. 7–10

49. L. Wang, R. Liu, S. Vosoughi, Salienteye: maximizing engagement while maintaining artistic style on instagram using deep neural networks, in *Proceedings of the 2020 International Conference on Multimedia Retrieval*, ICMR '20, New York, NY, 2020 (Association for Computing Machinery, New York, 2020), pp. 331–335

50. K. Hammar, S. Jaradat, N. Dokoohaki, M. Matskin, Deep text classification of instagram data using word embeddings and weak supervision, in *Web Intelligence*. Preprint (IOS Press, New York, 2020), pp. 1–15

51. H. Zheng, K. Wu, J.-H. Park, W. Zhu, J. Luo, Personalized fashion recommendation from personal social media data: An item-to-set metric learning approach (2020). Preprint. arXiv:2005.12439

52. Y. Hu, X. Yi, L.S. Davis, Collaborative fashion recommendation: A functional tensor factorization approach, in *Proceedings of the 23rd ACM International Conference on Multimedia*, MM '15, New York, NY, 2015 (Association for Computing Machinery, New York), pp. 129–138

53. P. Domingos, M. Richardson, Mining the network value of customers, in *Proceedings of the Seventh ACM SIGKDD International Conference on Knowledge Discovery and Data Mining*, KDD '01, New York, NY, 2001 (Association for Computing Machinery, New York, 2001), pp. 57–66

54. L. del Carmen Contreras Chinchilla, K.A.R. Ferreira, Analysis of the behavior of customers in the social networks using data mining techniques, in *2016 IEEE/ACM International Conference on Advances in Social Networks Analysis and Mining (ASONAM)*, 2016, pp. 623–625.

55. D. Jiménez-Castillo, R. Sánchez-Fernández, The role of digital influencers in brand recommendation: Examining their impact on engagement, expected value and purchase intention. Int. J. Inf. Manage. **49**, 366–376 (2019)

56. N. Segev, N. Avigdor, E. Avigdor, Measuring influence on instagram: A network-oblivious approach, in *The 41st International ACM SIGIR Conference on Research & Development in Information Retrieval*, SIGIR '18, New York, NY, 2018 (Association for Computing Machinery, New York, 2018), pp. 1009–1012

57. C.-T. Li, S.-D. Lin, M.-K. Shan, Exploiting endorsement information and social influence for item recommendation, in *Proceedings of the 34th International ACM SIGIR Conference on Research and Development in Information Retrieval*, SIGIR '11, New York, 2011 (Association for Computing Machinery, New York, 2011), pp. 1131–1132

58. C. Li, F. Xiong, Social recommendation with multiple influence from direct user interactions. *IEEE Access, 5*, 16288–16296 (2017)

59. Q. Zhang, J. Wu, Q. Zhang, P. Zhang, G. Long, C. Zhang, Dual influence embedded social recommendation. World Wide Web **21**(4), 849–874 (2018)

60. F. Eskandanian, N. Sonboli, B. Mobasher, Power of the few: analyzing the impact of influential users in collaborative recommender systems, in *Proceedings of the 27th ACM Conference on User Modeling, Adaptation and Personalization*, UMAP '19, New York, NY, 2019 (Association for Computing Machinery, New York, 2019), pp. 225–233

61. P.P. Analytis, D. Barkoczi, P. Lorenz-Spreen, S. Herzog, The structure of social influence in recommender networks, in *Proceedings of the Web Conference 2020*, WWW '20, New York, NY, 2020 (Association for Computing Machinery, New York, 2020), pp. 2655–2661

62. M. Bertini, A. Ferracani, R. Papucci, A.D. Bimbo, Keeping up with the influencers: Improving user recommendation in instagram using visual content, in *Adjunct Publication*

of the 28th ACM Conference on User Modeling, Adaptation and Personalization, UMAP '20 Adjunct, pp. 29–34, New York, NY, 2020 (Association for Computing Machinery, New York, 2020)

63. Y. Zhang, J. Caverlee, Instagrammers, fashionistas, and me: Recurrent fashion recommendation with implicit visual influence, in *Proceedings of the 28th ACM International Conference on Information and Knowledge Management*, CIKM '19, New York, NY, 2019 (Association for Computing Machinery, New York, 2019), pp. 1583–1592

64. S. Jaradat, N. Dokoohaki, U. Wara, M. Goswami, K. Hammar, M. Matskin, TALs: a framework for text analysis, fine-grained annotation, localisation and semantic segmentation, in *2019 IEEE 43rd Annual Computer Software and Applications Conference (COMPSAC)*, vol. 2 (2019), pp. 201–206

65. R. He, J. McAuley, Ups and downs: modeling the visual evolution of fashion trends with one-class collaborative filtering, in *Proceedings of the 25th International Conference on World Wide Web*, WWW '16, page 507–517, Republic and Canton of Geneva, CHE, 2016. International World Wide Web Conferences Steering Committee.

66. X. Wang, X. He, L. Nie, T.-S. Chua, Item silk road: recommending items from information domains to social users, in *Proceedings of the 40th International ACM SIGIR Conference on Research and Development in Information Retrieval*, SIGIR '17, New York, NY, 2017 (Association for Computing Machinery, New York, 2017), pp. 185–194

67. G. Stalk, Customer returns top $10 billion in 2005: most canadian retailers fail to capitalize on this key customer relationship. *Canada NewsWire*, 2006

68. T.-M. Choi, *Analytical Modeling Research in Fashion Business* (Springer, New York, 2016)

69. C. John Langley Jr., J.J. Coyle, B.J. Gibson, R.A. Novack, E.J. Bardi, *Supply Chain Management: A Logistics Perspective*, 8th edn. (South Western College, Kentucky, 2008)

70. E. Ofek, Z. Katona, M. Sarvary, "Bricks and Clicks": The impact of product returns on the strategies of multichannel retailers. Market. Sci. **30**(1), 42–60 (2011)

71. C. Barry, Happy returns: how to reduce customer returns-and their costs. *Catalog Age*, 2000. arXiv:2106.03532v1 [cs.LG]

72. J. Mostard, R. Teunter, The newsboy problem with resalable returns: A single period model and case study. *European Journal of Operational Research, 169*(1), 81–96 (2006)

73. G. Pisut, L.J. Connell, Fit preferences of female consumers in the usa. J. Fash. Market. Manage. Int. J. **11**(3), 366–379 (2007)

74. C. Ratcliff, How fashion ecommerce retailers can reduce online returns. *Blog text, Econsultancy. Saatavissa*, 2014

75. Y. Ha, L. Stoel, Internet apparel shopping behaviors: the influence of general innovativeness. Int. J. Retail. Distrib. Manage. **32**(8), 377–385 (2004)

76. R. Guigourès, Y.K. Ho, E. Koriagin, A.-S. Sheikh, U. Bergmann, R. Shirvany, A hierarchical Bayesian model for size recommendation in fashion, in *Proceedings of the 12th ACM Conference on Recommender Systems* (ACM, New York, 2018), pp. 392–396

77. A.-S. Sheikh, R. Guigourès, E. Koriagin, Y.K. Ho, R. Shirvany, U. Bergmann, A deep learning system for predicting size and fit in fashion e-commerce, in *Proceedings of the 13th ACM Conference on Recommender Systems* (ACM, New York, 2019)

78. G. Mohammed Abdulla, S. Borar, Size recommendation system for fashion e-commerce, in *KDD Workshop on Machine Learning Meets Fashion*, 2017

79. N Weidner, Vanity sizing, body image, and purchase behavior: A closer look at the effects of inaccurate garment labeling (2010)

80. R. Velazquez, S.M. Chankov, Environmental impact of last mile deliveries and returns in fashion e-commerce: a cross-case analysis of six retailers, in *2019 IEEE International Conference on Industrial Engineering and Engineering Management (IEEM)*, 2019, pp. 1099–1103

81. Y. Yu, H.-S. Kim, Online retailers' return policy and prefactual thinking: an exploratory study of USA and China e-commerce markets. J. Fash. Market. Manage. **23**(4), 504–518 (2019)

82. M.A. Diggins, C. Chen, J. Chen, A review: customer returns in fashion retailing, in *Analytical Modeling Research in Fashion Business* (Springer, New York, 2016), pp. 31–48

83. S. Cullinane, M. Browne, E. Karlsson, Y. Wang, Retail clothing returns: a review of key issues, in *Contemporary Operations and Logistics* (Springer, New York, 2019), pp. 301–322

84. G. Walsh, M. Möhring, C. Koot, M. Schaarschmidt, Preventive product returns management systems: a review and a model, in *ECIS* (2014)

85. Y. Zhang, O. Juhlin, Using crowd sourcing to solve the fitting problem in online fashion sales. Global Fashion Manage. Conf. **1**, 62–66 (2015)

86. A. Vecchi, F. Peng, M. Al-Sayegh, et al., Looking for the perfect fit? Online fashion retail-opportunities and challenges, in *Conference Proceedings: The Business & Management Review*, vol. 6 (The Academy of Business & Retail Management, London, 2015), pp. 134–146

87. S. Charts, https://www.adidas.com.sg/help-topics-size_charts.html, 2020. Accessed: 2020 Jan 13

88. D. Ujević, L. Szirovicza, I. Karabegović, Anthropometry and the comparison of garment size systems in some european countries. Coll. Antropol. **29**(1), 71–78 (2005)

89. S.-J.H. Shin, C.L. Istook, The importance of understanding the shape of diverse ethnic female consumers for developing jeans sizing systems. Int. J. Consum. Stud. **31**(2), 135–143 (2007)

90. M.-E. Faust, S. Carrier, *Designing Apparel for Consumers: The Impact of Body Shape and Size* (Woodhead Publishing, Sawston, 2014)

91. One size fits none. https://time.com/how-to-fix-vanity-sizing, 2020. Accessed: 2020 Jan 28

92. S. Baier, Analyzing customer feedback for product fit prediction (2019). Preprint. arXiv:1908.10896

93. N. Karessli, R. Guigourès, R. Shirvany, Sizenet: weakly supervised learning of visual size and fit in fashion images, in *IEEE Conference on Computer Vision and Pattern Recognition (CVPR) Workshop on FFSS-USAD*, 2019

94. N. Karessli, R. Guigourès, R. Shirvany, Learning size and fit from fashion images, in *Springer's Special Issue on Fashion Recommender Systems*, 2020

95. A. Nestler, N. Karessli, K. Hajjar, R. Weffer, R. Shirvany, Sizeflags: reducing size and fit related returns in fashion e-commerce, in *Proceedings of the 27th ACM SIGKDD Conference on Knowledge Discovery and Data Mining*.

96. V. Sembium, R. Rastogi, A. Saroop, S. Merugu, Recommending product sizes to customers, in *Proceedings of the Eleventh ACM Conference on Recommender Systems*, pp. 243–250 (ACM, New York, 2017)

97. V. Sembium, R. Rastogi, L. Tekumalla, A. Saroop, Bayesian models for product size recommendations, in *Proceedings of the 2018 World Wide Web Conference*, WWW '18, 2018, pp. 679–687

98. R. Misra, M. Wan, J. McAuley, Decomposing fit semantics for product size recommendation in metric spaces, in *Proceedings of the 12th ACM Conference on Recommender Systems*, 2018, pp. 422–426. Dataset available in: https://cseweb.ucsd.edu/~jmcauley/datasets.html#clothing_fit. Accessed: 2020

99. K. Dogani, M. Tomassetti, S. De Cnudde, S. Vargas, B. Chamberlain, Learning embeddings for product size recommendations, in *SIGIR eCom, Paris*, July 2019

100. J. Lasserre, A.-S. Sheikh, E. Koriagin, U. Bergmann, R. Vollgraf, R. Shirvany, Meta-learning for size and fit recommendation in fashion, in *SIAM International Conference on Data Mining (SDM20)*, 2020

101. K. Hajjar, J. Lasserre, A. Zhao, R. Shirvany, Attention gets you the right size and fit in fashion, in *14th ACM Conference on Recommender Systems Workshops, Recsys'20 - fashionXrecsys'20* (ACM, New York, 2020)

102. Y. Yuan, J.-H. Huh, *Cloth Size Coding and Size Recommendation System Applicable for Personal Size Automatic Extraction and Cloth Shopping Mall: MUE/FutureTech 2018* (2019), pp. 725–731

103. M. Januszkiewicz, C.J. Parker, S.G. Hayes, S. Gill, Online virtual fit is not yet fit for purpose: an analysis of fashion e-commerce interfaces (2017)

104. L. Lefakis, E. Koriagin, J. Lasserre, R. Shirvany, Personalized size recommendations with human in the loop, in *Second ICML Workshop on Human in the Loop Learning (HILL)* (2020)

105. N. Thalmann, B. Kevelham, P. Volino, M. Kasap, E. Lyard, 3d web-based virtual try on of physically simulated clothes. Comput.-Aided Des. Appl. **8**, 163–174 (2011)
106. J. Surville, T. Moncoutie, 3d virtual try-on: The avatar at center stage. 2013
107. F. Peng, A.-S. Mouhannad, Personalised size recommendation for online fashion, in *6th International conference on mass customization and personalization in Central Europe*, 2014, pp. 1–6
108. F. Bogo, A. Kanazawa, C. Lassner, P.V. Gehler, J. Romero, M.J. Black, Keep it SMPL: automatic estimation of 3d human pose and shape from a single image (2016). CoRR abs/1607.08128
109. G. Pavlakos, L. Zhu, X. Zhou, K. Daniilidis, Learning to estimate 3d human pose and shape from a single color image (2018). CoRR abs/1805.04092
110. J.J. Heckman, H. Ichimura, P.E. Todd, Matching as an econometric evaluation estimator: Evidence from evaluating a job training programme. Rev. Econ. Stud. **64**(4), 605–654 (1997)
111. X. Song, F. Feng, X. Han, X. Yang, W. Liu, L. Nie, Neural compatibility modeling with attentive knowledge distillation, in *The 41st International ACM SIGIR Conference on Research & Development in Information Retrieval*, 2018, pp. 5–14
112. F. Harada, H. Shimakawa, Outfit recommendation with consideration of user policy and preference on layered combination of garments. Int. J. Adv. Comput. Sci. **2**, 49–55 (2012)
113. H. Tsujita, K. Tsukada, K. Kambara, I. Siio, Complete fashion coordinator: a support system for capturing and selecting daily clothes with social networks, in *Proceedings of the International Conference on Advanced Visual Interfaces*, 2010, pp. 127–132
114. E. Shen, H. Lieberman, F. Lam, What am i gonna wear? scenario-oriented recommendation, in *Proceedings of the 12th International Conference on Intelligent User Interfaces*, 2007, pp. 365–368
115. S. Liu, J. Feng, Z. Song, T. Zhang, H. Lu, C. Xu, S. Yan, Hi, magic closet, tell me what to wear! in *Proceedings of the 20th ACM international conference on Multimedia*, 2012, pp. 619–628
116. P. Tangseng, K. Yamaguchi, T. Okatani, Recommending outfits from personal closet, in *Proceedings of the IEEE International Conference on Computer Vision Workshops* (2017), pp. 2275–2279
117. C. Zagel, Product experience wall: a context-adaptive outfit recommender system, in *Mensch & Computer*, 2014, pp. 367–370
118. T. He, Y. Hu, Fashionnet: Personalized outfit recommendation with deep neural network (2018). Preprint. arXiv:1810.02443
119. Y. Li, L. Cao, J. Zhu, J. Luo, Mining fashion outfit composition using an end-to-end deep learning approach on set data. IEEE Trans. Multimedia **19**(8), 1946–1955 (2017)
120. X. Chen, H. Chen, H. Xu, Y. Zhang, Y. Cao, Z. Qin, H. Zha, Personalized fashion recommendation with visual explanations based on multimodal attention network: Towards visually explainable recommendation, in *Proceedings of the 42nd International ACM SIGIR Conference on Research and Development in Information Retrieval*, 2019, pp. 765–774
121. Y.-G. Shin, Y.-J. Yeo, M.-C. Sagong, S.-W. Ji, S.-J. Ko, Deep fashion recommendation system with style feature decomposition, in *2019 IEEE 9th International Conference on Consumer Electronics (ICCE-Berlin)* (IEEE, New York, 2019), pp. 301–305
122. A. Veit, B. Kovacs, S. Bell, J. McAuley, K. Bala, S. Belongie, Learning visual clothing style with heterogeneous dyadic co-occurrences, in *Proceedings of the IEEE International Conference on Computer Vision*, 2015, pp. 4642–4650
123. M.I. Vasileva, B.A. Plummer, S. Dusad, S. Rajpal, R. Kumar, D. Forsyth, Learning type-aware embeddings for fashion compatibility, in *Proceedings of the European Conference on Computer Vision (ECCV)*, 2018, pp. 390–405. Dataset available in: https://github.com/xthan/polyvore. Accessed: 2020
124. L.F. Polanía, S. Gupte, Learning fashion compatibility across apparel categories for outfit recommendation, in *2019 IEEE International Conference on Image Processing (ICIP)* (IEEE, New York, 2019), pp. 4489–4493

125. E.M. Bettaney, S.R. Hardwick, O. Zisimopoulos, B.P. Chamberlain, Fashion outfit generation for e-commerce (2019). Preprint arXiv:1904.00741
126. W.-C. Kang, E. Kim, J. Leskovec, C. Rosenberg, J. McAuley, Complete the look: scene-based complementary product recommendation, in *Proceedings of the IEEE Conference on Computer Vision and Pattern Recognition*, 2019, pp. 10532–10541. Dataset available in: https://github.com/kang205/STL-Dataset. Accessed: 2020
127. P. Tangseng, T. Okatani, Toward explainable fashion recommendation, in *The IEEE Winter Conference on Applications of Computer Vision*, 2020, pp. 2153–2162
128. Z. Feng, Z. Yu, Y. Yang, Y. Jing, J. Jiang, M. Song, Interpretable partitioned embedding for customized multi-item fashion outfit composition, in *Proceedings of the 2018 ACM on International Conference on Multimedia Retrieval*, 2018, pp. 143–151
129. X. Wang, B. Wu, Y. Zhong, Outfit compatibility prediction and diagnosis with multi-layered comparison network, in *Proceedings of the 27th ACM International Conference on Multimedia*, 2019, pp. 329–337
130. Y. Lin, P. Ren, Z. Chen, Z. Ren, J. Ma, M. de Rijke, et al., Explainable fashion recommendation with joint outfit matching and comment generation (2018). Preprint. arXiv:1806.08977
131. X. Song, F. Feng, J. Liu, Z. Li, L. Nie, J. Ma, Neurostylist: Neural compatibility modeling for clothing matching, in *Proceedings of the 25th ACM International Conference on Multimedia*, 2017, pp. 753–761
132. W.-L. Hsiao, I. Katsman, C.-Y. Wu, D. Parikh, K. Grauman, Fashion++: minimal edits for outfit improvement, in *Proceedings of the IEEE International Conference on Computer Vision*, 2019, pp. 5047–5056
133. E. Simo-Serra, S. Fidler, F. Moreno-Noguer, R. Urtasun, Neuroaesthetics in fashion: modeling the perception of fashionability, in *Proceedings of the IEEE Conference on Computer Vision and Pattern Recognition*, 2015, pp. 869–877
134. D. Banerjee, K.S. Rao, S. Sural, N. Ganguly, Boxrec: recommending a box of preferred outfits in online shopping. ACM Trans. Intell. Syst. Technol. **11**(6), 1–28 (2020)
135. Y. Jiang, X. Qianqian, X. Cao, Outfit recommendation with deep sequence learning. in *2018 IEEE Fourth International Conference on Multimedia Big Data (BigMM)* (IEEE, New York, 2018), pp. 1–5
136. Y. Jiang, Q. Xu, X. Cao, Q. Huang, Who to ask: An intelligent fashion consultant, in *Proceedings of the 2018 ACM on International Conference on Multimedia Retrieval*, 2018, pp. 525–528
137. Z. Wang, H. Quan, Fashion outfit composition combining sequential learning and deep aesthetic network, in *2019 International Joint Conference on Neural Networks (IJCNN)* (IEEE, New York, 2019), pp. 1–7
138. X. Han, Z. Wu, Y.-G. Jiang, L.S. Davis, Learning fashion compatibility with bidirectional LSTMs, in *Proceedings of the 25th ACM International Conference on Multimedia*, 2017, pp. 1078–1086
139. T. Nakamura, R. Goto, Outfit generation and style extraction via bidirectional LSTM and autoencoder (2018). Preprint. arXiv:1807.03133
140. A. Graves, N. Jaitly, A.-R. Mohamed, Hybrid speech recognition with deep bidirectional LSTM, in *2013 IEEE workshop on automatic speech recognition and understanding* (IEEE, New York, 2013), pp. 273–278
141. S. Jaradat, N. Dokoohaki, M. Matskin, Outfit2vec: Incorporating clothing hierarchical metadata into outfits' recommendation, in *Fashion Recommender Systems* (Springer, New York, 2020), pp. 87–107
142. Y. Lin, M. Moosaei, H. Yang, Outfitnet: fashion outfit recommendation with attention-based multiple instance learning, in *Proceedings of the Web Conference 2020*, 2020, pp. 77–87
143. Y. Hu, X. Yi, L.S. Davis, Collaborative fashion recommendation: a functional tensor factorization approach, in *Proceedings of the 23rd ACM International Conference on Multimedia*, 2015, pp. 129–138
144. Y. Lin, P. Ren, Z. Chen, Z. Ren, J. Ma, M. de Rijke, Improving outfit recommendation with co-supervision of fashion generation, in *The World Wide Web Conference*, 2019, pp. 1095–1105

145. W. Chen, P. Huang, J. Xu, X. Guo, C. Guo, F. Sun, C. Li, A. Pfadler, H. Zhao, B. Zhao, Pog: Personalized outfit generation for fashion recommendation at alibaba ifashion, in *Proceedings of the 25th ACM SIGKDD International Conference on Knowledge Discovery & Data Mining*, 2019, pp. 2662–2670. Dataset available in: https://github.com/wenyuer/POG

146. Z. Cui, Z. Li, S. Wu, X.-Y. Zhang, L. Wang, Dressing as a whole: Outfit compatibility learning based on node-wise graph neural networks, in *The World Wide Web Conference*, 2019, pp. 307–317

147. W.-L. Hsiao, K. Grauman, Learning the latent "look": unsupervised discovery of a style-coherent embedding from fashion images, in *2017 IEEE International Conference on Computer Vision (ICCV)* (IEEE, New York, 2017), pp. 4213–4222

148. ShopLook: Outfits Inspirations. https://shoplook.io/home, 2020. Accessed: 2020

149. Z. Liu, P. Luo, S. Qiu, X. Wang, X. Tang, DeepFashion: powering robust clothes recognition and retrieval with rich annotations, in *Proceedings of the IEEE Computer Society Conference on Computer Vision and Pattern Recognition*, 2016. Dataset available in: http://mmlab.ie.cuhk.edu.hk/projects/DeepFashion.html. Accessed: 2020

150. Y. Ge, R. Zhang, X. Wang, X. Tang, P. Luo, Deepfashion2: A versatile benchmark for detection, pose estimation, segmentation and re-identification of clothing images, in *Proceedings of the IEEE Computer Society Conference on Computer Vision and Pattern Recognition*, 2019. Dataset available in: https://github.com/switchablenorms/DeepFashion2. Accessed: 2020

151. ModCloth, an American online retailer of indie and vintage-inspired women's clothing. https://www.modcloth.com, 2020. Accessed: 2020

152. RentTheRunway, an online service that provides designer dress and accessory rentals. https://www.renttherunway.com/, 2020. Accessed: 2020

153. W.-L. Hsiao, K. Grauman, Vibe: dressing for diverse body shapes, in *Computer Vision and Pattern Recognition*, 2020. Dataset available in: https://github.com/facebookresearch/VIBE. Accessed: 2020

154. Birdsnest, an Australian women's online fashion retailer. https://www.birdsnest.com.au/, 2020. Accessed: 2020

155. M. Hadi Kiapour, X. Han, S. Lazebnik, A.C. Berg, T.L. Berg, Where to buy it: matching street clothing photos in online shops. in *International Conference on Computer Vision*, 2015. Dataset available in http://www.tamaraberg.com/street2shop. Accessed: 2020

# Index

**A**

A/B testing, 955
Accuracy, 571
Active user, 2
Adaptivity, 594
Adversarial machine learning, 336, 337, 365
Affect, 946
Affective state, 381, 401
Aggregating information, 381
Aggregation, 388, 404
Agreeableness, 757, 759
Algorithmic bias, 942
Alternating least squares, 104
Ambient intelligence, 384
ARCADES model, 792, 803, 821
Aspect-based diversity, 633
ASPECT model, 792, 794, 815
Attacks, 336
Audio, 977
Automatic playlist continuation, 945
Average precision, 579
Average Reciprocal Hit Rank (ARHR), 582


**B**

Bartle model, 760
Baseline predictors, 96
Beyond-accuracy, 953
Biases, 96, 635
Big-Five model, 759
Blog recommendation, 837
Bundle recommendation, 486

**C**

Choice support, 790, 792, 803, 821
Click-Through-Rate, 530
Coherence, 949
Cold Start, 410, 486, 489, 584
Collaborative filtering, 40, 91, 143, 424, 929,
      941, 1019
Community-based recommender systems, 15
Community bias, 942
Complementary clothing, 1021
Concept drift, 99
Confidence, 585
Conscientiousness, 757, 759
Consumers, 690
Content-based recommender systems, 12, 251,
      424, 929, 942, 1020
Content-collaborative, 433
Context aware ranking, 143
Context-aware recommendation, 944
Context-aware recommender systems, 16, 211,
      217, 227
Context-dependent food, 891
Contextual factors, 217
Contextual information, 214, 223
Contextual modeling, 232
Contextual post-filtering, 231, 945
Contextual pre-filtering, 229, 945
Conversational systems, 253, 381, 463
Cooking recommender, 875
Correlation coefficient, 109
Co-training, 504
Coverage, 583

Critiquing, 464
Cross domain, 862
Cross-domain recommendation, 486, 493, 511,
    779
Cross-domain recommender systems, 486
Cross-selling, 486
Cross-selling recommendations, 490
Cross-system personalization, 495
Crowdsourcing, 859
Customer ROC, 575

**D**
Data bias, 942
Data features, 92
Decision making, 790, 793, 815
Decision support, 790, 792, 803, 821
Deep learning, 173
Deep neural networks, 173
Defense, 337
Demographic recommender systems, 15
Distributional Semantics, 262
Diversity, 590, 603, 774, 947
    aggregate, 612
    enhancement, 618
    evaluation, 609
    in information retrieval, 616
    intra-list, 610
    temporal, 614

**E**
Embedding, 253
Emotion, 946
Enterprise, 841
Entropy, 613
Eudaimonic quality, 763
Evaluation, 859, 862
Evaluation metric, 548
Evaluation protocols, 365
Expected utility, 591
Expertise, 406
Expertise location, 856
Explainability, 952
Explanation, 260, 857
Explanation styles, 720
Explicit feedback, 91
Explicit user preferences, 428, 438
Extraversion, 757, 759

**F**
Fairness, 637, 679, 951
Fairness-aware recommendation, 672
Familiarity, 948

Fashion, 973
Feature-Level Explanations, 475
Five Factor Model, 757, 759
F-measure, 574
Food, 973
Food choices, 874
Food recommendation, 871
Food recommender, 871
Framing, 984

**G**
Gender, 685
Generative adversarial network, 349
Gini index, 584, 613
Global ROC, 575
Gradient descent, 104
Graph learning, 1003
Grocery recommender, 875
Group and individual attributes, 381
Group decision making, 808
Group dynamics, 381
Group fairness, 681
Group model, 387, 412
Group recommendation, 381, 840
Group recommender systems, 381, 383, 413,
    792, 808

**H**
Health-aware food, 894
Health recommender, 875
Hedonic quality, 763
Hotness, 949
Human-computer interaction, 21
Human in the Loop, 539
Human understandable, 711
Hybrid recommender systems, 16, 943

**I**
Image, 977
Implicit feedback, 92, 95, 143
Implicit preferences, 440
Implicit user preferences, 428
Individual fairness, 681
Information overload, 2
Intelligible, 711
Interactive recommendations, 719
Interactive television, 383
International Personality Item Pool, 758
Interpretability, 711
Inverse propensity scoring (IPS), 636
Item-based recommendation, 48

Item-item, 108
Item recommendation, 40, 143

**J**
Justification, 712, 952

**K**
Kendall's $\tau$, 580
Knowledge-based recommender systems, 15
Knowledge graphs, 261
Knowledge transfer, 499

**L**
Latent factor models, 102
Lean-back experience, 931
Lean-in exploration, 930
Link prediction, 856
Long tail, 610, 942

**M**
Mainstreaminess, 949
Markov models, 308
Matrix factorization, 102, 360
Mean absolute error (MAE), 572
Memory based methods, 127
Model based methods, 127
Mood, 946
Multi-agent negotiation, 381
Multimedia recommender systems, 25, 986
Multiple criteria, 408
Multi-sided platforms, 647, 650, 653–655
Multistakeholder evaluation, 648, 650, 651,
        657, 658, 660, 661, 669, 670, 672
Multistakeholder recommendation, 647, 648,
        651, 652, 654, 655, 657, 659, 660,
        672
Music information retrieval, 929
Music recommendation, 927

**N**
Natural language, 252
Natural language processing (NLP), 256, 447,
        581
NBTree, 438
Nearest neighbors, 307
Neighborhood-based recommendation, 45
Neighborhood methods, 108
Netflix data, 94
Netflix prize, 92, 94

Neuroticism, 757, 759
News recommendation, 839
New-user problem, 772
Normalized discounted cumulative gain
        (NDCG), 580
Normalized distance based performance
        measure (NDPM), 578
Novelty, 587, 603, 948
        enhancement, 618–625, 631, 639, 694
        evaluation, 603, 604, 608–617, 635–637
        Long tail, 530, 606, 607, 610–611, 615,
        616, 621, 635, 637, 639, 655

**O**
Offline evaluation, 315
Offline experiments, 550
Online dating, 421
Openness, 757, 759
Organization, 841
Outfit recommendations, 1036

**P**
Pearson coefficient, 109
People recommendation, 846
People-to-people, 421
People-to-people reciprocal, 441
Personality, 385, 402, 405, 406, 757, 946
        datasets, 769
Personality acquisition
        explicit, 763
        implicit, 763
Personality Markup Language, 779
Persuasion, 736
Popularity, 949
Popularity bias, 659, 661, 663, 665, 666, 670,
        942
Precision at N, 574
Precision-recall, 573, 574
Prediction accuracy, 571
Prediction quality, 21
Preference elicitation, 462
Privacy, 414, 593, 858
Providers, 693
Psychology-inspired recommendation, 946
Psychology of choice, 790, 793, 815

**R**
Rating
        binary, 10
        numerical, 10
        ordinal, 10
        unary, 10

Realistic, investigative, artistic, social,
    enterprising, and conventional
    (RIASEC), 760
Receiver operating characteristic (ROC), 574
Recipe recommender, 889
Reciprocal recommenders, 421, 422
Recommendation, 174
    non-personalized, 1
    personalized, 1
Recommendation techniques, 11
Recommender system, 1, 40, 173, 335, 789
    data, 8
    function, 4
Recurrent Neural Networks, 309
Reference Ranking, 577
Regularization, 94
Reinforcement Learning, 311
Relationship, 385, 402, 407, 408
Representation learning, 174
Reputation, 859
Re-ranking, 700
Restaurant recommender, 875
Review-Level Explanations, 472
Risk, 592
Robustness, 593
Root mean squared error (RMSE), 95, 572
R-score, 581

S
Satisfaction, 381, 395, 399, 401, 402, 404, 740
Scalability, 595
Scrutability, 734
Semantic-based recommender systems, 13
Semantics, 252
Sequence-aware recommendation, 304
Sequence learning, 307
Sequence order, 399
Sequential recommendation, 304, 945
Serendipity, 589, 615, 622, 774, 949
Session, 301, 303
Shannon entropy, 595
Shrinkage, 109
Similar item, 1017
Similarity, 947
Similarity measure, 109
Singular value decomposition (SVD), 103
Social fashion, 1025
Social influence, 1028
Social media, 835
Social network, 858
Social network sites (SNS), 858
Social overload, 835
Social-recommender-systems, 24

Social relationships, 841, 858
Social streams, 861
Social web, 835
Sparsity problem, 489
Spearman's $\rho$, 580
Stakeholders, 521, 683
Stochastic gradient descent, 104
Subtopic recall, 616
SVD++, 104

T
Tags, 843, 858
Temporal dynamics, 97, 123
Temporal effects, 97, 123
Thomas-Kilmann model, 760
Time decay function, 124
TimeSvd++, 106
Top-n recommendations, 143
Transfer learning, 485
Transparency, 414, 733, 952
Trendiness, 949
Trust, 385, 408, 587, 735, 859

U
UGC extraction, 863
Unexpectedness, 611
Unfairness, 684
User-centric evaluation, 321
User engagement, 534
User experience, 931, 956, 958
User generated reviews, 448
User interactions, 428
User interface, 931, 958
User modeling, 553
User profile, 422
User roles, 405
Users, 9, 690
User study, 554, 956
User-user, 108
Utility, 591

V
Video, 977
Visual and audio content, 988

W
Web 2.0, 835
Word embedding, 257
Workplace, 841

Printed in the United States
by Baker & Taylor Publisher Services